SYMBOL	DESCRIPTION
A'	Complement of event A 112
A, B	Events 99
$A \cup B$	Union of events A and B 109
$A \cap B$	Intersection of events A and B 109
α (alpha)	Probability of rejecting H_0 when H_0 is true (Type I error) 327
b	Number of blocks in a randomized block design 480
B_i	Total of observations for block i in an analysis of variance 907
β (beta)	Probability of accepting H_0 when H_0 is false (Type II error) 329
β_0	y-intercept in regression models 640
$\hat{\beta}_0$	Least squares estimator of β_0 644
β_1	Slope of straight-line regression model 640
$\hat{\beta}_1$	Least squares estimator of β_1 644
β_i	Coefficient of independent variable x_i in a multiple regression model 708
$\hat{\beta}_i$	Least squares estimator of β_i 709
c_j	Total count in column j of a contingency table 614
CM	Correction for the mean in an analysis of variance 906
χ^2 (chi-square)	Probability distribution of various test statistics 367
df	Degrees of freedom for the t, χ^2, and F distributions 303
D_0	Hypothesized difference between population means or binomial probabilities 382
$E(n_i)$	Expected number of outcomes of type i under the null hypothesis in a multinomial test 605
$E(n_{ij})$	Expected count in cell (i, j) of a contingency table 615
$\hat{E}(n_{ij})$	Estimated expected count in cell (i, j) of a contingency table 615
$E(x)$	Expected value of the random variable x 176
ε (epsilon)	Random error component in regression models 640
$f(x)$	Probability density or frequency function for a continuous random variable x 212
F	Test statistic used to compare two variances, to compare p population means, and to test several terms in a multiple regression 428
F_r	Test statistic for the Friedman nonparametric analysis of variance 567
F_α	Value of F distribution with area α to its right 430
H	Test statistic for the Kruskal–Wallis nonparametric analysis of variance 559
H_a	Alternative (or research) hypothesis 326
H_0	Null hypothesis 326
IQR	Interquartile range 70

SYMBOL	DESCRIPTION
λ (lambda)	Mean of the Poisson distribution 197
MSB	Mean square for blocks in a randomized block design 488
MSE	Estimator of variance, σ^2, in a multiple regression or analysis of variance model (SSE divided by error degrees of freedom) 460
MST	Mean square for treatments in an analysis of variance 460
μ (mu)	Population mean 40
μ_D	Population mean of differences in a paired difference experiment 402
μ_0	Hypothesized value of a population mean 337
$\mu_{\bar{x}}$	Mean of the sampling distribution of $\bar{x}$ 268
n	(1) Sample size, total number of measurements in a sample 40 (2) Number of trials in a binomial experiment 183
n_D	Number of differences (or pairs) in a paired difference experiment 403
n_i	(1) Number of observations for treatment i in an analysis of variance 459 (2) Count in ith cell in a multinomial sample 604
n_{ij}	Observed count in cell (i, j) in a contingency table 614
$n!$	n factorial; product of first n integers $(0! = 1)$ 141
$\binom{N}{n}$	Number of possible combinations when n elements are drawn without replacement from N elements 141
ν_1 (nu)	Numerator degrees of freedom for an F statistic 428
ν_2	Denominator degrees of freedom for an F statistic 428
p	(1) Observed significance level for a hypothesis test 341 (2) Probability of Success for the binomial distribution 183 (3) Number of treatments being compared in a completely randomized or randomized block design 458
$\hat{p}$	(1) Sample estimator of binomial probability p; proportion of successes in the sample 311 (2) Pooled estimator of $p_1 = p_2 = p$ in test of equality of two population probabilities 417
$p(x)$	Probability distribution for a discrete random variable x 171
p_0	Hypothesized value for binomial probability 352
p_i	Probability of outcome i in a multinomial probability distribution 604
p_{ij}	Probability associated with the (i, j) cell in a contingency table 614
$p_1 - p_2$	Difference between true probabilities of success in two independent binomial experiments 416
$\hat{p}_1 - \hat{p}_2$	Difference between sample probabilities of success in two independent binomial samples; sample estimator of $(p_1 - p_2)$ 416

Continued inside back cover

FIFTH EDITION

Statistics

JAMES T. MCCLAVE

University of Florida

FRANK H. DIETRICH, II

Northern Kentucky University

DELLEN PUBLISHING COMPANY

San Francisco

a division of

MACMILLAN PUBLISHING
COMPANY

New York

COLLIER MACMILLAN CANADA

Toronto

MAXWELL MACMILLAN INTERNATIONAL

New York Oxford Singapore Sydney

On the cover: Painted wood sculptures by San Francisco artist and designer Robert Hutchinson. Photography by John Jensen.

Copyright 1991 by Dellen Publishing Company.

Printed in the United States of America

Permissions: Dellen Publishing Company
400 Pacific Avenue
San Francisco, California 94133

Orders: Dellen Publishing Company
c/o Macmillan Publishing Company
Front and Brown Streets
Riverside, New Jersey 08075

Macmillan Publishing Company
866 Third Avenue, New York, New York 10022

Collier Macmillan Canada, Inc.
1200 Eglinton Avenue East
Suite 200
Don Mills, Ontario M3C 3N1

Library of Congress Cataloging in Publication Data
McClave, James T.
    Statistics / James T. McClave, Frank H. Dietrich, II.—5th ed.
    Includes index.
    ISBN 0-02-379185-3
    1. Statistics.   I. Dietrich, Frank H.   II. Title.
QA276.12.M4   1991
519.5—dc20   91-9614
                                                             CIP

Printing: 1 2 3 4 5 6 7 8 9   Year: 1 2 3 4 5

ISBN 0-02-379185-3

Contents

CHS. 1, 2, 6, 8
PARTS OF 9

CHAPTER 13 **Simple Linear Regression**   637

CHAPTER 14 **Multiple Regression**   707

CHAPTER 15 **Model Building**   791

APPENDIX A **Tables**

APPENDIX B **Calculation Formulas for Analysis of Variance**

APPENDIX C **Demographic Data Set**

Preface

A number of important changes have been made in the fifth edition of *Statistics*:

1. We introduce the concept of *variables* in Chapter 1. We do this by redefining the population to consist of "units" rather than "data," and then defining a variable to be the characteristic of these units. Data are then obtained by measuring the variable of interest. This will focus more attention on measurement and will make an easier transition to the concept of random variables and their probability distributions.

2. In the same vein, in Chapter 2 we have added more on the concept of measuring different types of variables, introducing and defining the four types of data: nominal, ordinal, interval, and ratio.

3. The introduction to the classical inferential techniques of estimation and tests of hypotheses had made for a very long Chapter 7 in the fourth edition, especially with the (optional) material on Type II errors. In the fifth edition we have separated single-sample inference into two chapters: one on estimation (Chapter 7) and the second on tests of hypotheses (Chapter 8).

4. In previous editions, no nonparametric methods for making single-sample inferences were presented. With this edition we have added a new section on the use of the sign test for single-sample inferences to the Nonparametric Statistics chapter (Chapter 11).

5. New exercises have been added throughout, this time concentrating on new and better mechanical exercises in the Learning the Mechanics sections at the beginning of most exercise sets. Some new applied exercises have been added as well, especially accompanying the new material.

6. We have also been made aware by our users that the text was getting too bulky. This is a problem that "grows" with each new edition, when new material is added but old material is rarely removed. We have tried to deal with the problem in the fifth edition by adopting a new design that enables more material to be presented on each page. We have also eliminated some optional material and reduced the art when clarity was not sacrificed.

Although the fifth edition represents a minor revision relative to the fourth edition, we believe that the new design and increased emphasis on inferential statistics are important changes in our continuing commitment to the theme of statistical inference.

The flexibility of past editions is maintained in this edition. Sections that are not prerequisite to succeeding sections and chapters are marked "(Optional)." For example, an instructor who wishes to devote significant time to exploratory data analysis might cover all topics in Chapter 2. In contrast, an instructor who wishes to move rapidly into inferential procedures might omit the optional section on box plots and devote only several lectures to this chapter.

We have maintained the features of this text that we believe make it unique among introductory statistics texts. These features, which assist the student in achieving an overview of statistics and an understanding of its relevance in the sciences, business, and everyday life, are as follows:

1. **Case Studies.** (See the list of case studies on page xv.) Many important concepts are emphasized by the inclusion of case studies, which consist of brief summaries of actual applications of statistical concepts and are often drawn directly from the research literature. These case studies allow the student to see applications of important statistical concepts immediately after their introduction. The case studies also help to answer by example the often-asked questions, "Why should I study statistics? Of what relevance is statistics to my program?" Finally, the case studies constantly remind the student that each concept is related to the dominant theme—statistical inference.

2. **The Use of Examples as a Teaching Device.** We have introduced and illustrated almost all new ideas by examples. Our belief is that most students will better understand definitions, generalizations, and abstractions *after* seeing an application. In most sections, an introductory example is followed by a general discussion of the procedures and techniques, and then a second example is presented to solidify the understanding of the concepts.

3. **A Simple, Clear Style.** We have tried to achieve a simple and clear writing style. Subjects that are tangential to our objective have been avoided, even though some may be of academic interest to those well versed in statistics. We have not taken an encyclopedic approach in the presentation of material.

4. **Many Exercises—Labeled by Type.** The text has a large number (more than 1,200) of exercises illustrating applications in almost all areas of research. However, we believe that many students have trouble learning the mechanics of statistical techniques when problems are all couched in terms of realistic applications—the concept becomes lost in the words. Thus, the exercises at the ends of all sections are divided into two parts:

   a. **Learning the Mechanics.** These exercises are intended to be straightforward applications of the new concepts. They are introduced in a few words and are unhampered by a barrage of background information designed to make them "practical," but which often detracts from instructional objectives. Thus, with a minimum of labor, the student can recheck his or her ability to comprehend a concept or a definition.

   b. **Applying the Concepts.** The mechanical exercises described above are followed by realistic exercises that allow the student to see applications of statistics across a broad spectrum. Once the mechanics are mastered, these exercises develop the student's skills at comprehending realistic problems that describe situations to which the techniques may be applied.

5. **On Your Own . . .** Each chapter ends with an exercise entitled **On Your Own . . . .** The intent of this exercise is to give the student some hands-on experience with an application of the statistical concepts introduced in the chapter. In most cases, the student is required to collect, analyze, and interpret data relating to some real application.

6. **Using the Computer.** A new feature has been added at the end of most chapters to encourage the use of computers in the analysis of real data. A demographic database, consisting of 1,000 observations on 15 variables, has been described in Appendix C and is available on diskette from the publisher.

Most chapters include a suggested computer application, **Using the Computer** . . . , which provides one or more computer exercises that utilize the data in Appendix C and enhance the new material covered in the chapter.

7. **A Choice in Level of Coverage of Probability.** One of the most troublesome aspects of an introductory statistics course is the study of probability. Probability is troublesome for instructors because they must decide on the level of presentation, and it is troublesome for students because they (often) find it difficult at any level. We believe that one cause for these problems is the mixture of probability and counting rules that occurs in most introductory texts. We have included the counting rules in a separate and optional section at the end of the chapter on probability. In addition, all exercises that require the use of counting rules are marked with an asterisk (*) to indicate this. Thus, the instructor can control the level of coverage of probability.

A word should be added about the length of the probability chapter. Although more space is devoted to probability than in many introductory texts, there are three simple explanations for this: more examples, more exercises, and the optional counting rule section mentioned above. We have included the usual die and coin examples to introduce concepts, but we also work many practical examples so that the connection between probability and statistics is clearly made. This same pattern is followed in the exercise sections. Thus, the 34 examples and 121 exercises account for the length of the chapter. We trust that these will make this troublesome subject easier to learn and to teach.

8. **Where We've Been . . . Where We're Going** . . . The first page of each chapter is a "unification" page. Our purpose is to allow the student to see how the chapter fits into the scheme of statistical inference. First, we briefly show how the material presented in previous chapters helps us to achieve our goal (Where We've Been). Then, we indicate what the next chapter (or chapters) contributes to the overall objective (Where We're Going). This feature allows us to point out that we are constructing the foundation block by block, with each chapter an important component in the structure of statistical inference. Furthermore, this feature provides a series of brief résumés of the material covered as well as glimpses of future topics.

9. **An Extensive Coverage of Multiple Regression Analysis and Model Building.** This topic represents one of the most useful statistical tools for the solution of applied problems. Although an entire text could be devoted to regression modeling, we feel that we have presented a coverage that is understandable, usable, and much more comprehensive than the presentations in other introductory statistics texts. We devote three chapters to discussing the major types of inferences that can be derived from a regression analysis, showing how these results appear in computer printouts and, most important, selecting multiple regression models to be used in an analysis. Thus, the instructor has the choice of a one-chapter coverage of simple regression, a two-chapter treatment of simple and multiple regression, or a complete three-chapter coverage of simple regression, multiple regression, and model building. This extensive coverage of such useful statistical tools will provide added evidence to the student of the relevance of statistics to the solution of applied problems.

10. **Footnotes.** Although the text is designed for students with a noncalculus background, footnotes explain the role of calculus in various derivations.

Footnotes are also used to inform the student about some of the theory underlying certain results. The footnotes allow additional flexibility in the mathematical and theoretical level at which the material is presented.

11. **Supplementary Material.** A solutions manual, a study guide, a Minitab supplement, an integrated companion software system, a computer-generated test system, and a 1,000-observation demographic database are available.

   a. **Solutions Manual** (by Nancy Shafer Boudreau). The solutions manual presents detailed solutions to most odd-numbered exercises in the text. Many points are clarified and expanded to provide maximum insight into and benefits from each exercise.

   b. **Study Guide** (by Susan L. Reiland). For each chapter, the study guide includes (1) a brief summary that highlights the concepts and terms introduced in the textbook; (2) section-by-section examples with detailed solutions; and (3) exercises (with answers provided at the end of the study guide) that allow the student to check mastery of the material in each section.

   c. **Minitab Supplement** (by Ruth K. Meyer and David D. Krueger). The Minitab computer supplement was developed to be used with Minitab Release 7.0, a general-purpose statistical computing system. The supplement, which was written especially for the student with no previous experience with computers, provides step-by-step descriptions of how to use Minitab effectively as an aid in data analysis. Each chapter begins with a list of new commands introduced in the chapter. Brief examples are then given to explain new commands, followed by examples from the text illustrating the new and previously learned commands. Where appropriate, simulation examples are included. Exercises, many of which are drawn from the text, conclude each chapter.

   A special feature of the supplement is a chapter describing a survey sampling project. The objectives of the project are to illustrate the evaluation of a questionnaire, provide a review of statistical techniques, and illustrate the use of Minitab for questionnaire evaluation.

   d. **DellenStat** (by Michael Conlon). DellenStat is an integrated statistics package consisting of a workbook and an IBM PC floppy diskette with software and example sets of data. The system contains a file creation and management facility, a statistics facility, and a presentation facility. The software is menu-driven and has an extensive help facility. It is completely compatible with the text.

   The DellenStat workbook describes the operation of the software and uses examples from the text. After an introductory chapter for new computer users, the remaining chapters follow the outline of the text. Additional chapters show how to create new sets of data. Technical appendices cover material for advanced users and programmers.

   DellenStat runs on any IBM PC or close compatible with at least 256K of memory and at least one floppy disk drive.

   e. **DellenTest.** This unique computer-generated random test system is available to instructors without cost. Utilizing an IBM PC computer and a number of commonly used dot-matrix printers, the system will generate an almost unlimited number of quizzes, chapter tests, final examinations, and drill exercises. At the same time, the system produces an answer key and student worksheet with an answer column that exactly matches the column on the answer key.

f. **Database.**   A demographic data set was assembled based on a systematic random sample of 1,000 U.S. zip codes. Demographic data for each zip code area selected were supplied by CACI, an international demographic and market information firm. Fifteen demographic measurements (including population, number of households, median age, median household income, variables related to the cost of housing, educational levels, the work force, and purchasing potential indexes based on the Bureau of the Census Consumer Expenditure Surveys) are presented for each zip code area.

Some of the data are referenced in the **Using the Computer** sections. The objectives are to enable the student to analyze real data in a relatively large sample using the computer, and to gain experience using the statistical techniques and concepts on real data.

We have organized the text in what we believe is a logical sequence. For example, rather than place analysis of variance and nonparametric statistics at the end of the book, we have placed them immediately following the chapters on making inferences about one and two populations. The analysis of variance extends these methods to p populations, and nonparametric statistics gives procedures for making the inferences when the assumptions necessary for parametric methods are in doubt. The three regression and model building chapters, which we consider an essential and unique component, are the final three chapters, because regression represents a greater change in direction than the other inference-making techniques.

Because many introductory courses in statistics are of one-term duration, we show several possible sequences of coverage in the diagram below:

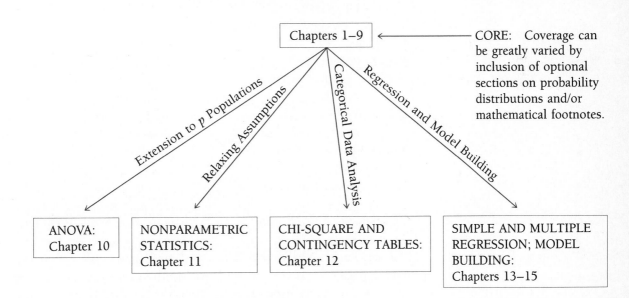

Acknowledgments

Thanks are due to many individuals who helped in the preparation of this text. Among them are the reviewers, whose names are listed below. Special thanks to John Dirkse, California State University at Bakersfield. Susan Reiland has our appreciation and admiration for managing the production of this book. Her work defies explanation; you have to see to believe the care and professionalism with which she works. We thank Brenda Dobson for converting our scribbling to a polished manuscript. Finally, we thank the thousands of students who have helped us to form our ideas about teaching statistics. Their most common complaint seems to be that texts are written for the instructor rather than the student. We hope that this book is an exception.

David Atkinson
Olivet Nazarene University

William H. Beyer
University of Akron

Patricia M. Buchanan
Pennsylvania State University

Kathryn Chaloner
University of Minnesota

N. B. Ebrahimi
Northern Illinois University

Rudy Gideon
University of Montana

David Groggel
Miami University at Oxford

John E. Groves
California Polytechnic State
University at San Luis Obispo

Jean L. Holton
Virginia Commonwealth
University

John H. Kellermeier
Northern Illinois University

Timothy J. Killeen
University of Connecticut

William G. Koellner
Montclair State University

James R. Lackritz
San Diego State University

Diane Lambert
Carnegie–Mellon University

James Lang
Valencia Junior College

John J. Lefante, Jr.
University of South Alabama

Pi-Erh Lin
Florida State University

R. Bruce Lind
University of Puget Sound

Linda C. Malone
University of Central Florida

Allen E. Martin
California State University at Los
Angeles

Leslie Matekaitis
Northern Illinois University

A. Mukherjea
University of South Florida

Bernard Ostle
University of Central Florida

William B. Owen
Central Washington University

Won J. Park
Wright State University

John J. Peterson
Smith Kline & French Laboratories

Rita Schillaber
University of Alberta

James R. Schott
University of Central Florida

Susan C. Schott
University of Central Florida

George Schultz
St. Petersburg Junior College

Carl James Schwarz
University of Manitoba

Charles W. Sinclair
Portland State University

Robert K. Smidt
California Polytechnic State
University at San Luis Obispo

Vasanth B. Solomon
Drake University

W. Robert Stephenson
Iowa State University

Barbara Treadwell
Western Michigan University

Dan Voss
Wright State University

Augustin Vukov
University of Toronto

Case Studies

What Is Statistics?

CONTENTS

WHERE WE'RE GOING...

Statistics? Is it a field of study, a group of numbers that summarizes the state of our national economy, the performance of a football team, the social conditions in a particular locale, or, as the title of a popular book (Tanur et al., 1989) suggests, "a guide to the unknown"? We will see in Chapter 1 that each of these descriptions has some applicability in understanding statistics. We will see that **descriptive statistics** focuses on developing numerical summaries that describe some phenomenon, whereas **inferential statistics** uses these numerical summaries to assist in making decisions. The primary theme of this text is inferential statistics. Thus, we concentrate on showing you how statistics can be used to interpret and use data to make decisions. Since many jobs in industry, government, medicine, and other fields require this facility, we show you how statistics can be beneficial to you.

1.1 Statistics: What Is It?

What does statistics mean to you? Does it bring to mind batting averages, Gallup polls, unemployment figures, numerical distortions of facts (lying with statistics!), or simply a college requirement you have to complete? We hope to convince you that statistics is a meaningful, useful science with a broad, almost limitless scope of application to business, government, and the sciences. We also want to show that statistics lie only when they are misapplied. Finally, our objective is to paint a unified picture of statistics to leave you with the impression that your time was well spent studying a subject that will prove useful to you in many ways.

Statistics means "numerical descriptions" to most people. Monthly unemployment figures, the failure rate of a particular type of steel-belted automobile tire, and the proportion of women who favor the Equal Rights Amendment all represent statistical descriptions of large sets of data collected on some phenomenon. Often the purpose of calculating these numbers goes beyond the description of the set of data. Frequently, the data are regarded as a sample selected from some larger set of data whose characteristics we may wish to estimate. For example, a sampling of unpaid accounts for a large merchandiser would allow you to calculate an estimate of the average value of unpaid accounts. This estimate could be used as an audit check on the total value of all unpaid accounts held by the merchandiser. So, the applications of statistics can be divided into two broad areas: (1) describing large masses of data and (2) drawing conclusions (making estimates, decisions, predictions, etc.) about some set of data based on sampling. So, the applications of statistics can be divided into two broad areas: *descriptive* and *inferential* statistics.

Descriptive statistics utilizes numerical and graphical methods to look for patterns, to summarize, and to present the information in a set of data.

Inferential statistics utilizes sample data to make estimates, decisions, predictions, or other generalizations about a larger set of data.

Although both descriptive and inferential statistics are discussed in the following chapters, the primary theme of the text is *inference*. Let us examine some case studies that illustrate applications of statistics.

CASE STUDY 1.1

A SURVEY: WHERE "WOMEN'S WORK" IS DONE BY MEN

The 1980 February/March issue of *Public Opinion* describes the results of a survey of several hundred married men from each of nine countries who responded to the following question:

> In the following list, which household jobs would you say it would be reasonable that the man would often take over from his wife: washing up (doing dishes), changing baby's napkin (diaper), cleaning house, ironing, organizing meal, staying at home with sick child, shopping, none of these?

The graphs in Figure 1.1 provide an effective summary of the thousands of opinions obtained and allow for an easy comparison of attitudes across countries.

F I G U R E 1.1

"Women's Work" Is Rarely Done by
Men in Italy, Germany . . .
Note: The sample size for each country
exceeded 900, except in Luxembourg,
where the sample size was 334.
Source: Survey by the European
Economic Community Commission.
"Women and Men of Europe in 1978,"
October–November 1977, as shown in
Public Opinion, February–March 1980,
p. 37. © Copyright American Enterprise
Institute.

Question: In the following list, which household jobs would you say it would be reasonable that the man would often take over from his wife: washing up (doing dishes), changing baby's napkin (diaper), cleaning house, ironing, organizing meal, staying at home with sick child, shopping, none of these?

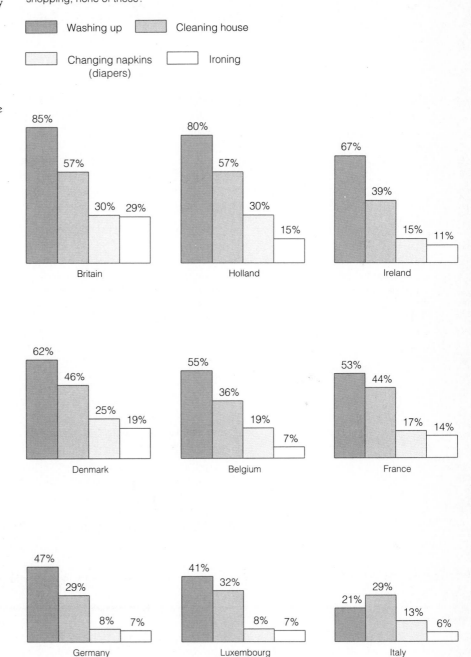

The graphs, which summarize and describe the data, are examples of *descriptive statistics*. The area of statistics concerned with the summarization and description of data is called **descriptive statistics**.

AN EXPERIMENT: INVESTIGATING AN EFFECT OF SMOKING DURING PREGNANCY

In an article in the *Journal of the American Medical Association*, M. Sexton and J. R. Hebel* reported on their research into the effects of maternal smoking on the birth weight of babies. Their experiment randomly assigned 935 pregnant women smokers into two groups; one continued smoking throughout pregnancy and the other, the **control group**, received smoking intervention (i.e., assistance to reduce or eliminate smoking). From the measured baby weights of these two groups of women, Sexton and Hebel inferred that "some fetal growth retardation can be overcome by the provision of antismoking assistance to pregnant women."

In making this statement, Sexton and Hebel made an *inference* about the impact of a smoking intervention program on all pregnant women who smoke, an inference based on the comparison of the samples of baby weights for the smoking and the control groups of women. Thus, these authors applied *inferential statistics* in the analysis of their data.

DOES JUDICIAL ACTION AFFECT THE PROBABILITY OF CONVICTION?

Defense attorneys often remind juries that accused criminals are innocent until proven guilty. But if the judge waits to deliver his or her own version of the reminder until after the testimony is in, jurors may be more likely to render a guilty verdict than if they had heard the reminder before the trial.

This quote is from an article in the June 1980 issue of *Psychology Today*† that discusses a study conducted by two social psychologists at the University of Kansas.

Involved in the study were 107 student jurors, some of whom heard the judge's reminder at the start of a taped trial, some at the end, and some not at all. An analysis of the 107 student verdicts prompted the quote just given.

Note that a **sample** of 107 student jurors was observed and that their verdicts were used to *infer* that the jurors in any trial may be more likely to render a guilty verdict if they hear a reminder from the judge at the end of the trial. This case study, like Case Study 1.2, is an example of the use of *inferential statistics*.

TASTE PREFERENCE FOR BEER: BRAND IMAGE OR PHYSICAL CHARACTERISTICS OF THE BEER?

Two sets of data of interest to the marketing department of a food products firm are (1) the set of taste-preference scores given by consumers to their product

---

*Reprinted from Sexton, M. and Hebel, J. R. "A clinical trial of change in maternal smoking and its effect on birth weight." *Journal of the American Medical Association*, Feb. 17, 1984, Vol. 251, No. 7, pp. 911–915. Copyright 1984, American Medical Association.

†Reprinted from Rubenstein, C. "The presumption of innocence needs prompting." *Psychology Today*, June 1980, p. 30. Copyright © 1980 by Ziff Davis Publishing Co.

and their competitors' products when all brands are clearly labeled and (2) the taste-preference scores given by the same set of consumers when all brand labels have been removed and the consumer's only means of product identification is taste. With such information the marketing department should be able to determine whether taste preference arose because of perceived physical differences in the products or as a result of the consumer's image of the brand. (Brand image is, of course, largely a result of a firm's marketing efforts.) Such a determination should help the firm develop marketing strategies for their product.

A study using these two sets of data was conducted by Ralph Allison and Kenneth Uhl* in an effort to determine whether beer drinkers could distinguish among major brands of unlabeled beer. A sample of 326 beer drinkers was randomly selected from the set of beer drinkers identified as males who drank beer at least three times a week. During the first week of the study each of the 326 participants was given a six-pack of unlabeled beer containing three major brands and was asked to taste-rate each beer on a scale from 1 (poor) to 10 (excellent). During the second week the same set of drinkers was given a six-pack containing six major brands. This time, however, each bottle carried its usual label. Again, the drinkers were asked to taste-rate each beer from 1 to 10. From a statistical analysis of the two sets of data yielded by the study, Allison and Uhl concluded that the 326 beer drinkers studied could not distinguish among brands by taste on an overall basis. This result enabled them to infer statistically that such was also the case for beer drinkers in general. Their results also indicated that brand labels and their associations did significantly influence the tasters' evaluations. These findings suggest that physical differences in the products have less to do with their success or failure in the marketplace than the image of the brand in the consumers' mind. As to the benefits of such a study, Allison and Uhl note, "to the extent that product images, and their changes, are believed to be a result of advertising . . . the ability of firms' advertising programs to influence product images can be more thoroughly examined."

This case study, like Case Studies 1.2 and 1.3, is an example of the use of *inferential statistics*. Using data collected from a sample of 326 beer drinkers, Allison and Uhl make inferences about the ability of all beer drinkers to distinguish among major brands of unlabeled beer.

These case studies provide four examples of the uses of statistics. Note that each involves an analysis of data, either for the purpose of describing the data set (Case Study 1.1) or for making inferences about a much larger data set based on sampling (Case Studies 1.2, 1.3, and 1.4). They thus provide realistic examples of the two broad areas of statistical applications.

1.2 The Elements of Statistics

Statistical methods are particularly useful for studying, analyzing, and learning about *populations*.

---

*Reprinted from Allison, R. I. and Uhl, K. P. "Influence of beer brand identification on taste perception." *Journal of Marketing Research*, August 1965, 36–39.

> **Definition 1.1**
>
> A **population** is a set of units (usually, people, objects, transactions, or events).

Examples of populations include (1) all employed workers in the United States, (2) all registered voters in California, (3) everyone who is afflicted with AIDS, (4) all the cars produced last year by a particular assembly line, (5) the current stock of spare parts at United Airlines' maintenance facility, (6) all sales made at the drive-in window of a fast-food restaurant during a given year, and (7) the set of all accidents occurring on a particular stretch of interstate highway during a holiday period. Notice that the first three population examples (1–3) are sets (groups) of people; the next two (4–5) are sets of objects; the next (6) is a set of transactions; and the last (7) is a set of events.

In studying a population, we focus on one or more characteristics or properties of the units in the population. For example, we may be interested in the age, income, and/or the number of years of education of the people currently unemployed in the United States. We call such characteristics *variables.*

> **Definition 1.2**
>
> A **variable** is a characteristic or property of an individual population unit. The name *variable* is derived from the fact that any particular characteristic may "vary" among the units in a population.

In studying a particular variable it is helpful—as we will see in forthcoming chapters—to be able to obtain a numerical representation for the variable. Thus, when numerical representations are not readily available, the process of measurement plays an important supporting role in statistical studies. **Measurement** is the process by which numbers are assigned to variables of individual population units. Measurement may entail asking a registered voter to rate the performance of the president on a scale from 1 to 10 or simply asking a worker how old she is. Frequently, however, it involves the use of instruments such as stop watches, scales, and calipers. We discuss measurement in more detail in the next chapter.

If the population you wish to study is small in size, then it is feasible to measure a variable for every unit in the population. For example, if you are measuring the GPA for all incoming freshmen at your university, it is at least feasible to obtain every GPA. When we measure a variable for every unit of a population, it is called a **census** of the population. However, the populations of interest in most applications are typically much larger, involving perhaps many thousands or even an infinite number of units. Examples of large populations are those given under Definition 1.1, as well as all graduates of your university or college, all potential buyers of a new facsimile machine, and all pieces of first-class mail handled by the U.S. Post Office. In studying such populations it would typically be too time-consuming and/or too costly to conduct a census. A more reasonable alternative would be to select and study a subset (a portion) of the units in the population.

> **Definition 1.3**
>
> A **sample** is a subset of the units of a population.

Thus, for example, instead of polling all 120,000,000 registered voters in the United States during a presidential election year, a pollster may select and examine a sample of only 1,500 voters. If he is interested in the variable "presidential preference," then he would record (measure) the preference of each sampled vote.

The method of selecting the sample is called the sampling procedure, or sampling plan. One very important sampling procedure is **random sampling**, one that assures every unit in the population has the same chance of being included in the sample. Thus, if the pollster samples 1,500 of the 120,000,000 voters in the population so that every voter has an equal chance of being included in the sample, he has devised a random sample. Random sampling is discussed in Chapter 3.

After selecting the sample and measuring the variable(s) of interest for every sampled unit, the information contained in the sample is used to make *inferences* about the population.

> **Definition 1.4**
>
> A **statistical inference** is an estimate, prediction, or other generalization about a population based on information contained in a sample.

That is, *we use the information contained in the smaller sample to learn about the larger population.** Thus, from the sample of 1,500 voters, the pollster may estimate the percentage of all the voters who would vote for each presidential candidate if the election were held on the day the poll was conducted or predict the outcome on election day.

The preceding definitions identify four of the five elements of an inferential statistical problem: a population, one or more variables of interest, a sample, and an inference. The fifth, and perhaps most important, is a measure of reliability for the inference. This is the topic of Section 1.3.

EXAMPLE 1.1

A sociologist hypothesizes that the average annual income of households in a particular large city is less than $25,000 per year. To test her hypothesis, she samples 500 households in the city and determines the income of each.

a. Describe the population.
b. Describe the variable of interest.
c. Describe the sample.
d. Describe the inference.

---

*The terms population and sample are often used to refer to the sets of measurements themselves, in addition to the units on which the measurements are made. For applications in which a single variable of interest is being measured, this will cause little confusion. When the terminology is potentially ambiguous, the measurements are referred to as *population data sets* and *sample data sets*, respectively.

Solution

a. The population is the set of units of interest to the sociologist, which is the set of all households in the city.

b. The total annual income of each household is the variable of interest to the sociologist.

c. The sample must be a subset of the population. In this case, it is the 500 households selected by the sociologist. If the 500 households represent a random sample, then the sampling procedure used must be such that each of the households in the city had an equal chance of being included in the sample.

d. The inference of interest involves the *generalization* of the information contained in the sample of 500 households to the population of all households in the city. In particular, the sociologist wants to estimate the average income of the households in the city in order to determine whether it is less than $25,000. This might be accomplished by calculating the average income in the sample and using the sample average to estimate the population average. Of course, this sample estimate is not likely to be exact, and therefore some measure of its reliability is needed. Measuring the reliability of an inference is the subject of Section 1.3.

EXAMPLE 1.2

Cola wars is the popular media term for the intense competition between the marketing campaigns of Coca-Cola and Pepsi. The campaigns have featured movie and television stars, rock videos, athletic endorsements, and claims of consumer preference based on taste tests. Suppose, as part of a Pepsi marketing campaign, 1,000 cola consumers are given a "blind" taste test (i.e., a taste test in which the two brand names are disguised). Each consumer is asked to state a preference for brand A or brand B.

a. Describe the population.
b. Describe the variable of interest.
c. Describe the sample.
d. Describe the inference.

Solution

a. The population of interest is the collection or set of all cola consumers.

b. The characteristic of each cola consumer that Pepsi wants to measure is the consumer's cola preference as revealed under the conditions of a blind taste test. Thus, cola preference is the variable of interest.

c. The sample is the 1,000 cola consumers selected from the population of all cola consumers.

d. The inference of interest is the *generalization* of the cola preferences of the 1,000 sampled consumers to the population of all cola consumers. In particular, the preferences of the consumers in the sample can be used to *estimate* the percentage of all cola consumers who prefer each brand.

1.3   Statistics: Witchcraft or Science?

The *primary objective of statistics is inference*. In the previous section we described inference as making generalizations about populations based on information contained in a sample. But making the inference is only part of the story. We also need to know how good the inference is. The only way we can be reasonably certain that an inference about a population is correct is to include the entire population in our sample. But, due to resource constraints (i.e., insufficient time and/or money), this is generally not an option. In basing inferences on only a portion of the population (a sample), we introduce an element of uncertainty into our inferences. In general, the smaller the sample size, the less certain we are about the inference. Thus, an inference based on a sample of size 5 is (usually) less reliable than an inference based on a sample of size 100. Consequently, whenever possible, it is important to determine and report the **reliability** of each inference made; this is the fifth element of inferential statistical problems.

The measure of reliability that accompanies an inference separates the science of statistics from the art of fortune-telling. A palm reader, like a statistician, may examine a sample (your hand) and make inferences about the population (your life). However, unlike statistical inferences, no measure of reliability can be attached to the palm-reader's inferences.

Suppose as in Example 1.1 we are interested in estimating the average income of a population of households from the average weight of a sample of households. Using statistical methods we can determine a *bound on the estimation error*. This bound is simply a number that our estimation error (the difference between the average income of the sample and the average income of the population of households) is not likely to exceed. We will see in later chapters that this bound is a measure of the uncertainty of our inference. The reliability of statistical inferences is discussed throughout this text. For now, we simply want you to realize that an inference is incomplete without a measure of its reliability.

We conclude this section with a summary of the elements of inferential statistical problems and an example to illustrate a measure of reliability.

> ### Five Elements of Inferential Statistical Problems
>
> 1. The population of interest.
> 2. One or more variables (characteristics of the population units) that are to be investigated.
> 3. The sample of population units.
> 4. The inference about the population based on information contained in the sample.
> 5. A measure of reliability for the inference.

EXAMPLE 1.3

Refer to Example 1.2, in which 1,000 consumers indicated their cola preferences in a taste test. Describe how the reliability of an inference concerning the preferences of all cola consumers in the Pepsi bottler's marketing region could be measured.

Solution

When the preferences of 1,000 consumers are used to estimate the preferences of all consumers in the region, the estimate will not exactly mirror the preferences of the population. For example, if the taste test shows that 56% of the 1,000 consumers preferred Pepsi, it does not follow (nor is it likely) that exactly 56% of all cola drinkers in the region prefer Pepsi. Nevertheless, we may be able to use sound statistical reasoning (which is presented later in the text) to ensure that the sampling procedure used will generate estimates that are almost certainly within a specified limit of the true percentage of all consumers who prefer Pepsi. For example, such reasoning might assure us that the estimate of the preference for Pepsi is almost certainly within 5% of the actual population preference. The implication is that the actual preference for Pepsi is between 51% [i.e., (56 − 5)%] and 61% [i.e., (56 + 5)%]. This interval represents a measure of reliability for the inference.

EXAMPLE 1.4

Refer to Case Study 1.2, in which research on the effects of maternal smoking on birth weight is discussed. Assume that 500 of the pregnant women receive smoking intervention therapy during their pregnancies, whereas 435 receive no therapy or treatment and continue to smoke during their pregnancies.

a. Describe the population for this research.
b. Describe the variable of interest.
c. Describe the sample.
d. Describe the type of inference of interest to the researchers.
e. Discuss how the reliability of the inference could be measured.

Solution

a. Because the objective of this research is to compare these data between two groups of babies, those born to smoking mothers and those born to mothers receiving smoking intervention therapy, two populations are pertinent: The first consists of all babies born to smoking mothers, and the second consists of all babies born to mothers receiving smoking intervention therapy.
b. The characteristic of the babies being measured in this research is their birth weight. Thus, the variable of interest is *birth weight*.
c. Two samples were selected, one from each population. The first sample consisted of 435 babies born to smoking mothers; the second sample consisted of the 500 babies born to mothers receiving intervention therapy.
d. The inference of interest might be to compare the average weight of babies born to smoking mothers to the average weight of babies born to mothers receiving smoking intervention therapy. Specifically, the research is aimed at determining whether the average weight of babies born to smoking mothers is less than that of those born to mothers receiving therapy.
e. The reliability of the inference must address the issue of the precision with which the difference between the average weights of the two populations is estimated. For example, suppose the two samples indicate that babies born to smoking mothers average .8 pound less at birth than those born to mothers receiving smoking intervention therapy. It is premature to assert that this implies that the average weight of babies born to *all* mothers receiving therapy will exceed that of babies born to all smoking mothers; we must first know the precision associated with the difference in sample averages. For example, suppose that sound statistical reasoning (which we develop in this text) is used to show that the sample estimate of the difference is almost certainly within .5 pound of the true difference between the two population averages. This implies that the true difference is (almost certainly) between (.8 − .5) = .3 and (.8 + .5) = 1.3 pounds, which supports the inference that babies born to smoking mothers will, on average, weigh between .3 and 1.3 pounds less than those born to mothers receiving therapy. This interval

is therefore a measure of reliability. Note that there remains uncertainty not only about the exact value of the difference but also about the cause of the difference. The cause might be attributed to smoking, but the reliability of that aspect of the inference depends on more than statistical reasoning. Might another factor, such as diet, be different for smoking mothers than for mothers receiving therapy, and might diet rather than smoking be the *cause* of the babies' average weight difference? The issue of causation requires the expertise of medical scientists as well as statistical inference.

1.4 Why Study Statistics?

Why study statistics? The growth in data collection associated with scientific phenomena as well as the operations of business and government (quality control, statistical auditing, forecasting, etc.) has been truly remarkable over the past several decades. Published results of political, economic, and social surveys as well as increasing government emphasis on drug and product testing provide vivid evidence of the need to be able to evaluate data sets intelligently. Consequently, you will want to develop a discerning sense of rational thought that will enable you to evaluate numerical data. You may be called upon to use this ability to make intelligent decisions, inferences, and generalizations. For this reason, the study of statistics is an essential preparation for a role in modern society.

EXERCISES 1.1–1.18

LEARNING THE MECHANICS

1.1 Explain the difference between descriptive and inferential statistics.

1.2 List and define the five elements of an inferential statistical analysis.

1.3 Explain how populations and variables differ.

1.4 Explain how populations and samples differ.

1.5 Why would a statistician consider an inference incomplete without an accompanying measure of its reliability?

1.6 Consider the set of all students enrolled in your statistics course this term. Suppose you are interested in learning about the current grade point averages (GPAs) of this group.
a. Define the population and variable of interest.
b. Suppose you determine the GPA of every member of the class. Would this represent a census or a sample?
c. Suppose you determine the GPA of ten members of the class. Would this represent a census or a sample?
d. If you determine the GPA of every member of the class and then calculate the average, how much reliability does this have as an "estimate" of the class average GPA?
e. If you determine the GPA of ten members of the class and then calculate the average, will the number you get necessarily be the same as the average GPA for the whole class? On what factors would you expect the reliability of the estimate to depend?

1.7 Refer to Exercise 1.6. What must be true in order for the sample of ten students you select from your class to be considered a random sample?

1.8   Pollsters regularly conduct opinion polls to determine the popularity rating of the current president. Suppose a poll is to be conducted tomorrow in which 2,000 individuals will be asked whether the president is doing a good or bad job.
    a. What is the relevant population?
    b. What is the variable of interest? Is it numerical or nonnumerical?
    c. What is the sample?
    d. What is the inference of interest to the pollster?

1.9   Refer to Exercise 1.8. Suppose the poll is conducted as described therein, except that each of the 2,000 individuals polled is asked to rate the job performance of the president on a scale from 0 to 100. How do your answers to parts **a–d** change, if at all?

1.10   To evaluate the current status of the dental health of schoolchildren, the American Dental Association conducted a survey to estimate the average number of cavities per child in grade school in the United States. One thousand schoolchildren from across the country were selected, and the number of cavities for each was recorded.
    a. Describe the population of interest to the American Dental Association.
    b. What is the variable of interest?
    c. Describe the sample.

1.11   A first-year chemistry student conducts an experiment to determine the amount of hydrochloric acid necessary to neutralize 2 ounces of a basic solution. The student prepares five 2-ounce portions of the solution and adds a known concentration of hydrochloric acid to each. The amount of acid necessary to achieve neutrality of the solution is recorded for each of the five portions.
    a. Describe the population of interest to the student.
    b. What is the variable of interest?
    c. Describe the sample.

1.12   A manufacturer of vacuum cleaners has decided that an assembly line is operating satisfactorily if less than 2% of the cleaners produced per day are defective. If 2% or more of the cleaners are defective, the line must be shut down and proper adjustments made. To check every cleaner as it comes off the line would be costly and time-consuming. The manufacturer decides to choose 30 cleaners at random from a specific day's production and test for defects.
    a. Describe the population of interest to the manufacturer.
    b. Identify the variable of interest.
    c. Describe the sample.
    d. Give an example of an inference the manufacturer might make.

1.13   An insurance company would like to determine the proportion of all medical doctors who have been involved in one or more malpractice suits. The company selects 500 doctors at random from a professional directory and determines the number in the sample who have ever been involved in a malpractice suit.
    a. Describe the population of interest to the insurance company.
    b. Identify the variable of interest.
    c. Describe the sample.
    d. Give an example of an inference the insurance company might make.

1.14   A Gallup Youth Poll was conducted to determine the topics that teenagers most want to discuss with their parents. The findings show that 46% would like more discussion about the family's financial situation, 37% would like to talk about school, and 30% would like to talk about religion. The survey was based on a national sampling of 505 teenagers.

a. Describe the sample.
b. Describe the population from which the sample was selected.
c. What is the variable of interest?
d. How is the inference expressed?
e. Newspaper accounts of most polls usually give a margin of error (a percentage) for the survey result. What is the purpose of the "margin of error" and what is its interpretation?

1.15 Myron Gable and Martin T. Topol (1988) sampled 218 department store executives in order to study the relationship between job satisfaction and the degree of *Machiavellian* orientation. Briefly, the Machiavellian orientation is one in which the executive exerts very strong control, even to the point of deception and cruelty, over the employees he or she supervises. The authors administered a questionnaire to each of the sampled executives and obtained both a job-satisfaction score and a Machiavellian rating. They concluded that those with higher job satisfaction scores are likely to have a lower "Mach" rating.
a. What is the population from which the sample was selected?
b. What variables were measured by the authors?
c. Identify the sample.
d. What inference was made by the authors?

1.16 Dr. A. Lewis Rhodes conducted research on the relationship between religion and environmental concern, and presented a report of the results at the 1985 meeting of the Society for the Study of Social Problems. Surveys were conducted to determine both religion and environmental concern, where environmental concern was measured by the willingness to have government spend more money for environmental protection.
a. Describe the population of interest for this research.
b. Describe the sample for the study.
c. What is the variable of interest?
d. What inference do you think is of interest in this research?

1.17 A *U.S. News & World Report* (June 21, 1982) article describes a new method of treating a major form of blindness in elderly people. The process, using laser beams to seal abnormal blood vessels in the eye, was tried on 224 patients. Of these, only 14% went blind in 1 year. In a control group of similar untreated patients, 42% went blind in 1 year. Therefore, to determine whether the laser beam treatment was effective, research physicians wished to compare the proportions of patients going blind in 1 year for *two* different populations.
a. Describe the two populations that the research physicians want to compare.
b. Identify the samples.
c. What is the variable of interest?
d. Describe the inference that the researchers will make.

1.18 In order to monitor the quality of care provided to Medicare recipients, 5% of all Medicare surgical cases performed in hospital outpatient departments and ambulatory surgical centers are to be sampled each year on an ongoing basis (Cronin, 1988). Peer-review organizations will evaluate the sampled surgeries and will assign a quality rating to each.
a. Describe the population being studied.
b. Describe the variable of interest.
c. Describe the sample in terms of process output.
d. Describe the inference of interest.
e. Before any sound statistical conclusions about the quality of Medicare can be drawn, what should accompany the inference?

ON YOUR OWN...

Scan a recent issue of a daily newspaper and look for articles that contain numerical data. The data might be a summary of the results of a public opinion poll, the results of a vote by the United States Senate, crime rates, birth or death rates, an election result, etc. For each article containing data that you find, answer the following questions:

a. Do the data constitute a sample or an entire population? If a sample has been taken, clearly identify both the sample and the population; otherwise, identify the population.

b. If a sample has been observed, does the article present an explicit (or implied) inference about the population of interest? If so, state the inference made in the article.

c. If an inference has been made, has a measure of reliability been included? What is it?

References

Careers in Statistics. American Statistical Association and the Institute of Mathematical Statistics, 1974.

Cronin, Carol. "PRO's focus on quality sharpened with new federal contracts." *Business and Health*, March 1988, 47.

Gable, Myron, and Topol, Martin T. "Machiavellianism and the Department Store Executive." *Journal of Retailing*, Spring 1988, 68–84.

Tanur, J. M., Mosteller, F. Kruskal, W. H., Link, R. F., Pieters, R. S., and Rising, G. R. *Statistics: A Guide to the Unknown*. (E. L. Lehmann, special editor.) San Francisco: Holden-Day, 1989.

Methods for Describing Sets of Data

CONTENTS

WHERE WE'VE BEEN...

In Chapter 1 we examined some typical examples of the use of statistics. We discussed the role that statistics plays in supporting decision-making. We introduced you to descriptive and inferential statistics and to the five elements of inferential statistics: a population, one or more variables, a sample, an inference, and a measure of reliability for the inference. We described the primary goal of inferential statistics as using sample data to make inferences (estimates, predictions, or other generalizations) about the population from which the sample was drawn.

WHERE WE'RE GOING...

Before we make an inference, we must be able to describe a data set. Both graphic and numerical methods for describing sets of data are discussed in this chapter. As you will learn in Chapter 7, we will use some sample numerical descriptive measures to estimate the values of corresponding population descriptive measures. Therefore, our efforts in this chapter will ultimately lead to statistical inference.

Suppose we wish to evaluate the mathematical capabilities of a class of 1,000 college freshmen based on their quantitative Scholastic Aptitude Test (SAT) scores. How would you describe these 1,000 measurements? You can see that this is not an easy question to answer. The 1,000 scores provide too many bits of information for our minds to comprehend. It is clear that we need some method for summarizing the information in a data set. Methods for describing data sets are also essential for statistical inference. Most populations are large data sets. Consequently, if we are going to make descriptive statements (inferences) about a population based on information contained in a sample, we will once again need methods for describing a data set.

Two methods for describing data are presented in this chapter, one **graphic** and the other **numerical**. As you will subsequently see, both play an important role in statistics.

In Section 2.1 we define four different types of data. Then we present graphical methods for describing data in Sections 2.2 and 2.3. Numerical descriptive methods are presented in Sections 2.4–2.8. We conclude this chapter with a section on the *misuse* of descriptive techniques.

2.1 Types of Data

In Chapter 1 you learned that statistics, both descriptive and inferential, is concerned with the measurements of one or more variables of a sample of units drawn from a population. These measurements are referred to as **data**. We generally classify data as one of four types: *nominal*, *ordinal*, *interval*, or *ratio*.

> **Definition 2.1**
>
> **Nominal data** are measurements that simply classify the units of the sample (or population) into categories.

Nominal data (also referred to as **categorical data**) are labels or names that identify the category to which each unit belongs. The following are examples of nominal data:

1. The political party affiliation of each individual in a sample of 50 registered voters.
2. The gender of each individual in a sample of seven applicants for a computer programming job.
3. The brand of toothpaste preferred by each individual in a sample of 100 consumers.

Note that in each case—political party, gender, and brand—the measurement is no more than a categorization of each sample unit. Nominal data are often reported as nonnumerical labels, such as Democrat, women, and Crest. Even if the labels are converted to numbers, as they often are for ease of computer entry and analysis, the numerical values are simply codes. They cannot be meaningfully added, subtracted, multiplied, or divided. For example, we might code Democrat = 1, Republican = 2, and other = 3. These are simply numerical codes for each of the categories into which units may fall and have no utility beyond that.

Definition 2.2

Ordinal data are measurements that enable the units of the sample (or population) to be ordered with respect to the variable of interest.

Ordinal data are rankings that indicate when one unit has more of a property than another unit. That is, the measurements indicate the relative amount of a property possessed by the units. The following are examples of ordinal data:

1. The size of car rented by each individual in a sample of 30 business travelers: compact, subcompact, midsize, or full-size
2. A taste-tester's ranking of four brands of barbecue sauce for a panel of ten tasters
3. A supervisor's annual performance rating on a scale of 1 (lowest) to 10 (highest) for each of 20 employees

Note that in each case—size, flavor preference, and performance rating—more than a categorization of units is involved. In addition to a categorization, the measurements actually rank the units. For example, we know that a midsize car is larger than a subcompact, and that an employee with a performance ranking of 9 performed better, in the opinion of the supervisor, than one with a ranking of 7. We also know that a taster preferred brand C barbecue sauce to brand A if he gives brand C a higher flavor preference ranking than brand A.

Ordinal data are said to represent a "higher" level of measurement than nominal data because ordinal data contain all the information of nominal data (i.e., category labels that differentiate units) *plus* an ordering of the units. As with nominal measurements, the distance between ordinal measurements is not meaningful. For example, we do not know whether the difference in size between a full-size and midsize car is the same as the difference between a midsize and subcompact. Nor do we know whether the flavor preference between barbecue sauce brands C and A is the same as that between brands A and B if a taster reports his flavor preference as $C > A > B > D$, where ">" means "more flavorful than."

As with nominal data, ordinal data can be reported with or without numbers. For example, the automobile sizes are nonnumerically labeled, whereas the supervisor's performance ratings are numerical. Even if numbers are used, we must again be careful; they simply provide an ordering or ranking of the units in the sample or population. The arithmetic operations of addition, subtraction, multiplication, and addition are not meaningful for ordinal data.

Definition 2.3

Interval data are measurements that enable the determination of how much more or less of the characteristic being measured is possessed by one unit of the sample (or population) than another.

Interval data are always numerical, and the numbers assigned to two units can be subtracted to determine the difference between the units with respect to the variable being measured. The following are examples of interval data:

1. The temperature (in degrees Fahrenheit) at which each of a sample of 20 pieces of heat-resistant plastic begins to melt
2. The scores of a sample of 150 law school applicants on the LSAT, a standardized law school entrance exam administered nationwide
3. The time at which the 5 P.M. Washington to New York air shuttle arrives at LaGuardia on each of a sample of 30 weekdays

Note that in each case—temperature, score, and arrival time—more than a ranking is involved. The difference between the numerical values assigned to the units is meaningful. For example, the difference between scores of 600 and 580 on the LSAT is the same as that between scores of 520 and 500. Also, the morning shuttle due at 9 A.M. but arriving at 9:20 A.M. is just as late as the afternoon shuttle due at 5:30 P.M and arriving at 5:50 P.M. Note in each case that the difference is the key, not the numerical measurement itself.

Interval data represent a higher level of measurement than ordinal data, because in addition to ranking the units, interval data reflect the difference between the units with respect to the variable being measured. Although adding and subtracting interval data are valid, multiplying and dividing them are not. This is because the zero point (the origin, or 0) is not meaningful for such data. For example, the origin on the temperature scale differs for the Fahrenheit and Celsius scales and is not meaningful (in an absolute sense) for either. Zero degrees does not mean "no heat." Temperatures lower than $0°$ (e.g., $-10°C$ and $-10°F$) indicate that less heat is present, so $0°$ cannot mean no heat. The result is that we cannot say that a temperature of $100°F$ indicates twice the heat of $50°F$. Similarly, since LSAT scores range from 200 to 800, a zero score is not meaningful. The result is that a score of 600 cannot be interpreted as 50% higher than a score of 400.

Most numerical business data are measured on scales for which the origin is meaningful. Thus, most numerical measurements encountered in business are *ratio data*.

Definition 2.4

Ratio data are measurements that enable the determination of how many times as much of the characteristic being measured is possessed by one unit of the sample (or population) than another.

Ratio data are always numerical, and the ratio between the numbers assigned to two units can be interpreted as the multiple by which the units differ. The following are examples of ratio data:

1. The annual income for each member of a sample of 500 households
2. The unemployment rate (reported as a percentage) in the United States for each of the past 60 months

3. The number of female executives employed in each of a sample of 50 industrial companies

Note that in each case—dollars of income, unemployed percentage, and count of female executives—the scale measures the absolute amount of the characteristic possessed by the unit. The result is that the ratio of measurements between units is meaningful. That is, a household with annual income of $60,000 has twice that of a household with $30,000 annual income. Similarly, an unemployment rate of 8% means twice as many unemployed as a rate of 4%. And a company with 30 female executives has 1.5 times as many as one with 20 female executives.

Ratio data represent the highest level of measurement. The numbers can be used to categorize, rank, differentiate, and measure multiples of one unit with respect to another. All arithmetic operations performed on ratio data are meaningful. *The key to differentiating interval and ratio data is that the zero point, or origin, is meaningful for ratio data.* For example, zero income, zero unemployment, and zero female executives all have readily discerned meaning. Most measurement scales yield ratio data: measures of monetary value, distance, weight, height, percentages, and numerical counts all usually generate ratio data.

The four types of data are often combined into two classes that are sufficient for most statistical applications. Nominal and ordinal data are often referred to as **qualitative data**, whereas interval and ratio data are called **quantitative data**.

The properties of the four types of data are summarized in the box. As you would expect, the methods for describing and reporting data depend on the type of data being analyzed. Describing qualitative data generally consists of calculating the percentage of the sample measurements falling in each category. However, most data are quantitative, and a number of methods exist to summarize and describe these data. We devote the remainder of this chapter to graphical and numerical methods for describing quantitative data.

Type of Data	Description
Nominal	Classification of sample (or population) units into categories
	Often labels rather than numbers
Ordinal	Rank orders the sample (or population) units
	May be verbal labels or numbers
Interval	Enables comparison of sample (or population) units according to differences between values
	Always numerical, but the zero point on the scale is not meaningful
Ratio	Enables comparison of sample (or population) units according to multiples of the values
	Always numerical, and the zero point on the scale is meaningful
Qualitative	Includes nominal and ordinal data types.
Quantitative	Includes interval and ratio data types.

EXERCISES 2.1–2.8

LEARNING THE MECHANICS

2.1 **a.** Explain the difference between nominal and ordinal data.
   b. Explain the difference between interval and ratio data.
   c. Explain the difference between qualitative and quantitative data.

2.2 Each of the descriptions of data defines one of the following types: nominal, ordinal, interval, ratio. Match the correct type to each description.
   a. Data that enable the units of the sample to be compared by the differences between their numerical values
   b. Data that enable the units of the sample to be classified into categories
   c. Data that enable the units of the sample to be rank-ordered
   d. Data that enable the units of the sample to be compared by computing the ratios of the numerical values

2.3 Suppose you are provided a data set that classifies each sample unit into one of four categories: A, B, C, or D. You plan to create a computer database consisting of these data, and you decide to code the data as A = 1, B = 2, C = 3, and D = 4 for inputting them into the computer. Are the data consisting of the classifications A, B, C, and D qualitative or quantitative? After the data are input as 1, 2, 3, or 4, are they qualitative or quantitative? Explain your answers.

APPLYING THE CONCEPTS

2.4 A food-products company is considering marketing a new snack food. To see how consumers react to the product, the company conducted a taste test using a sample of 100 randomly selected shoppers at a suburban shopping mall. The shoppers were asked to taste the snack food and then fill out a short questionnaire that requested the following information:
   a. What is your age?
   b. Are you the person who typically does the food shopping for your household?
   c. How many people are in your family?
   d. How would you rate the taste of the snack food on a scale of 1 to 10, where 1 is least tasty?
   e. Would you purchase this snack food if it were available on the market?
   f. If you answered yes to part **e**, how often would you purchase it?
   Each of these questions defines a variable of interest to the company. Classify the data generated for each variable as nominal, ordinal, interval, or ratio. Justify your classifications.

2.5 Classify the following samples of data as nominal, ordinal, interval, or ratio. Justify your classifications.
   a. Ten college freshmen were asked to indicate the brand of jeans they prefer.
   b. Fifteen television cable companies were asked how many hours of sports programming they carry in a typical week.
   c. Fifty students were asked what percentage of their time they spend studying each week.
   d. The number of long-distance phone calls made from each of 100 public telephone booths on a particular day was recorded.
   e. The SAT scores of 250 incoming freshmen to a small college were recorded.

2.6 Classify the following examples of data as either qualitative or quantitative:
   a. The bacteria count in the water at each of 30 city swimming pools
   b. The occupation of each of 200 shoppers at a supermarket
   c. The marital status of each person living on a city block
   d. The number of months between auto maintenance for each of 100 sales representatives

2.7 Classify the following examples of data as nominal, ordinal, interval, or ratio.
  a. The brand of stereo speaker for which each of 25 college students indicated a preference
  b. The loss (in dollars) incurred in each of the last 5 years by a department store as a result of shoplifting
  c. The color of interior house paint (other than white) that each of the five largest manufacturers of paint says generates the most sales revenue for the firm
  d. The final ranking of the ten football teams in the Southeastern Conference at the end of the season

2.8 State whether the following types of data are qualitative or quantitative:
  a. The height of each student in your class
  b. The length of time each of 30 patients must stay in a hospital
  c. The political party of each United States senator
  d. The religious affiliation of each patient of a psychologist

2.2 Graphic Methods for Describing Quantitative Data: Stem and Leaf Displays

Recall from Section 2.1 that quantitative data sets consist of either interval or ratio data. Most data are quantitative, so that methods for summarizing quantitative data are especially important.

Before we can use the information in a sample to make inferences about a population, we need methods to summarize, or describe, a set of data. For example, the Environmental Protection Agency (EPA) performs extensive tests on all new car models to determine their mileage rating. Suppose that the 100 measurements in Table 2.1 represent the results of such tests on a certain new car model. How can we summarize the information in this rather large sample?

TABLE 2.1 **EPA Mileage Ratings on 100 Cars**

36.3	41.0	36.9	37.1	44.9	36.8	30.0	37.2	42.1	36.7
32.7	37.3	41.2	36.6	32.9	36.5	33.2	37.4	37.5	33.6
40.5	36.5	37.6	33.9	40.2	36.4	37.7	37.7	40.0	34.2
36.2	37.9	36.0	37.9	35.9	38.2	38.3	35.7	35.6	35.1
38.5	39.0	35.5	34.8	38.6	39.4	35.3	34.4	38.8	39.7
36.3	36.8	32.5	36.4	40.5	36.6	36.1	38.2	38.4	39.3
41.0	31.8	37.3	33.1	37.0	37.6	37.0	38.7	39.0	35.8
37.0	37.2	40.7	37.4	37.1	37.8	35.9	35.6	36.7	34.5
37.1	40.3	36.7	37.0	33.9	40.1	38.0	35.2	34.8	39.5
39.9	36.9	32.9	33.8	39.8	34.0	36.8	35.0	38.1	36.9

A visual inspection of the data indicates some obvious facts. Most of the mileages are in the 30's, for example, with a smaller fraction in the 40's. But it is difficult to provide much additional information on the 100 mileage ratings without resorting to some method of summarizing the data. A graphic method of summarizing is provided by a **stem and leaf display**.

Figure 2.1 on page 22 shows a stem and leaf display for the data. To construct this display, we first partition a typical observation into a **stem** and a **leaf**. For our display we chose the stem portion of an observation as all digits at or to the

FIGURE 2.1

A stem and leaf display for the EPA
mileage ratings on 100 cars

STEM	LEAF
30	0
31	8
32	7 5 9 9
33	9 1 8 9 2 6
34	8 0 4 8 2 5
35	5 9 3 9 7 6 2 0 6 1 8
36	3 2 3 5 8 9 9 0 7 6 4 8 5 4 6 1 8 7 7 9
37	0 1 3 9 2 6 3 1 9 4 0 0 1 6 8 7 0 2 4 7 5
38	5 6 2 3 0 2 7 8 4 1
39	9 0 8 4 0 7 3 5
40	5 3 7 2 5 1 0
41	0 0 2
42	1
43	
44	9

left of the 1's digit place. The remaining portion of the observation to the right
of the stem is the leaf. The stems and leaves for the mileage readings 36.3, 32.7,
and 40.5 are shown here:

STEM	LEAF		STEM	LEAF		STEM	LEAF
36	3		32	7		40	5

After choosing the stem and leaf for an observation, we list the set of possible
stems for the data set in a column from the smallest (30) to the largest (44).
Then the leaf for each observation is recorded in the row of the display corresponding to the observation's stem. For example, the leaf (3) of the first observation in Table 2.1 is written in the row corresponding to the stem (36). Similarly,
the leaf (7) for the second observation in Table 2.1 is recorded in the row of
Figure 2.1 corresponding to the stem (32).

The stem and leaf display (Figure 2.1) presents a compact picture of the data
set. You can see at a glance that the 100 mileage readings were distributed between
30.0 and 44.9, with most of them falling in stem rows 35 to 39. The 6 leaves in
stem row 34 indicate that 6 of the 100 readings were at least 34.0 but less than
35.0. Similarly, the 11 leaves in stem row 35 indicate that 11 of the 100 readings
were at least 35.0 but less than 36. Only five cars had readings equal to 41 or
larger, and only one was as low as 30.

The definitions of the stem and leaf for a data set can be modified to alter the
graphic description. For example, suppose we had defined the stem as the tens
digit for the gas mileage data, rather than the ones and tens digits. With this
definition, the stems and leaves corresponding to the measurements 36.3 and
32.7 would be as follows:

STEM	LEAF		STEM	LEAF
3	6		3	2

Note that the decimal portion of the numbers has been dropped. Generally, only
one digit is displayed in the leaf.

If you look at the data, you will see why we did not define the stem this way. All the mileage measurements fall in the 30's and 40's, so all the leaves would fall into only two stem rows in this display. The resulting picture would not be nearly as informative as Figure 2.1. Generally we try to construct the stem and leaf display so that there are between 5 and 20 stems.

Guidelines for constructing a stem and leaf display are given in the box.

How to Construct a Stem and Leaf Display

1. Define the stem and leaf you wish to use. Choose the stem so that the number of stems in the display is between 5 and 20.

2. Write the stems in a column arranged from the smallest stem at the top to the largest stem at the bottom. Include all stems in the range of the data, even if there are some stems with no corresponding leaves.

3. If the leaves consist of more than one digit, drop the digits after the first. You may round the numbers to be more precise, but this is not necessary for the graphic description to be useful.

4. Record the leaf for each measurement in the row corresponding to its stem. You may wish to order the leaves corresponding to each stem to obtain a more informative display.

EXAMPLE 2.1

The data in Table 2.2 give the percentages of the total number of college or university student loans by state that are in default. Construct a stem and leaf display for the data.

TABLE 2.2 **Percentage of Student Loans (per State) in Default**

STATE	%	STATE	%	STATE	%	STATE	%
Ala.	12.0	Ill.	9.3	Mont.	6.4	R.I.	8.8
Alaska	19.7	Ind.	6.7	Nebr.	4.9	S.C.	14.1
Ariz.	12.1	Iowa	6.2	Nev.	10.1	S.Dak.	5.5
Ark.	12.9	Kans.	5.7	N.H.	7.9	Tenn.	12.3
Calif.	11.4	Ky.	10.3	N.J.	12.0	Tex.	15.2
Colo.	9.5	La.	13.5	N.Mex.	7.5	Utah	6.0
Conn.	8.8	Maine	9.7	N.Y.	11.3	Vt.	8.3
Del.	10.9	Md.	16.6	N.C.	15.5	Va.	14.4
D.C.	14.7	Mass.	8.3	N.Dak.	4.8	Wash.	8.4
Fla.	11.8	Mich.	11.4	Ohio	10.4	W.Va.	9.5
Ga.	14.8	Minn.	6.6	Okla.	11.2	Wis.	9.0
Hawaii	12.8	Miss.	15.6	Oreg.	7.9	Wyo.	2.7
Idaho	7.1	Mo.	8.8	Pa.	8.7		

Source: National Direct Student Loan Program.

Solution

The first step is to define the stem for an observation. Since the smallest observation in the data set is 2.7 and the largest is 19.7, a good choice for the stem would be all digits to the left of the decimal point. Thus, the stems and leaves for 2.7 and 19.7 would be

STEM	LEAF		STEM	LEAF
2	7		19	7

This choice generates a total of 18 stems, but only 15 are actually used to represent one or more observations. The complete stem and leaf display is shown in Figure 2.2.

FIGURE 2.2

Stem and leaf display for the percentage of student loans (per state) in default

STEM	LEAF
2	7
3	
4	9 8
5	7 5
6	7 2 6 4 0
7	1 9 5 9
8	8 3 8 7 8 3 4
9	5 3 7 5 0
10	9 3 1 4
11	4 8 4 3 2
12	0 1 9 8 0 3
13	5
14	7 8 1 4
15	6 5 2
16	6
17	
18	
19	7

You can see that Figure 2.2 provides a good graphic description of the percentage of student loans in default by state. Most of the observations fall in stem rows 6 through 12, indicating percentages of loans in default from 6.0% to 12.9%. All but two fall in stem rows 4 through 16, i.e., from 4.0% to 16.9%. Interestingly, one observation clearly stands out from the main body of data. The highest rate is 19.7% in the "independent" state of Alaska.

EXAMPLE 2.2

Refer to Example 2.1 and the student loan default data in Table 2.2. Use a statistical computer software package to create a stem and leaf display for the student loan default rates.

Solution

Most statistical software packages now have stem and leaf options available. We used Minitab on a microcomputer to generate the stem and leaf display in Figure

FIGURE 2.3

Computer-generated stem and leaf display
for defaulted student loan data

```
STEM-AND-LEAF DISPLAY OF DEFAULT
LEAF DIGIT UNIT =      .1000
1 2 REPRESENTS 1.2


         1      2   7
         1      3
         3      4   89
         5      5   57
        10      6   02467
        14      7   1599
        21      8   3347888
       (5)      9   03557
        25     10   1349
        21     11   23448
        16     12   001389
        10     13   5
         9     14   1478
         5     15   256
         2     16   6

               HI   197,
```

2.3. Note that the program selected the same stem (second column in printout) and leaf (third column in printout) that we did in Example 2.1. Also, the Minitab program labels the extreme observation, 19.7, using "HI" to indicate that the measurement is well away from the main body of data.* Finally, Minitab indicates the cumulative number of measurements (first column of printout) from the nearest tail of the distribution to each stem. For the stem that is in the middle of the data set the number of leaves is indicated in parentheses. The program also prints a key, giving the units of the leaf and an example showing where the decimal goes.

Stem and leaf displays possess a number of advantages over other forms of graphic description. First and foremost, each of the original measurements is "visible" in the display. The stem and leaf display arranges the data in ascending order, which enables easy location of individual measurements. Finally, for data sets that are not too large, the construction of stem and leaf displays is relatively simple.

A disadvantage of the stem and leaf display is that for very large data sets the display can become unwieldy, with leaves that extend off the edge of the page. As an alternative, analysts typically use a *relative frequency histogram*. This method is discussed in Section 2.3.

---

*The designation of a "HI" or "LO" measurement by Minitab is related to definitions of "outliers" in box plots of the data. Box plots (and outliers) are discussed in optional Section 2.8. You should exercise caution in the interpretation of these designations until we have covered their definitions.

EXERCISES 2.9–2.14

[Note: Starred (*) exercises require the use of a computer.]

2.9 Construct a stem and leaf display for the following measurements:

| 2.6 | 3.3 | 2.4 | 1.1 | .8 | 3.5 | 3.9 | 1.6 | 2.8 | 2.6 |
| 3.4 | 4.1 | 2.0 | 1.7 | 2.9 | 1.9 | 2.9 | 2.5 | 4.5 | 5.0 |

2.10 Consider the following data:

18.3	27.3	25.6	19.0	27.4
22.7	21.9	22.4	23.7	20.9
17.5	38.7	30.1	24.6	28.2
33.6	26.4	42.5	33.5	16.4
35.1	40.1	34.1	31.2	21.9

a. Construct a stem and leaf display with the stem defined to be all digits at or to the left of the 10's digit.

b. Construct a stem and leaf display with the stem defined to be all digits to the left of the decimal point.

c. Which stem and leaf display seems to give a better description of the data?

2.11 Construct a stem and leaf display for the following data. Define the stem to be all digits at or to the left of the 10's digit.

8.4	32.5	17.6	33.9	20.6
16.2	12.0	9.4	42.2	19.5
29.7	28.2	24.3	20.7	29.1
10.3	23.9	22.1	21.5	52.8

Explain how this stem and leaf display describes the data set.

2.12 The SAS System was used to generate the stem and leaf display shown in here. Note that SAS arranges the stems in descending order.

STEM	LEAF
5	1
4	4 5 7
3	0 0 0 3 6
2	1 1 3 4 5 9 9
1	2 2 4 8
0	0 1 2

a. How many observations were in the original data set?

b. In the bottom row of the stem and leaf display, identify the stem, the leaves, and the numbers in the original data set represented by this stem and its leaves.

2.13 According to the U.S. Department of Education, the national dropout rate for high school students fell by more than 1%, from 30.3% to 29.1%, between 1982 and 1984. The accompanying table shows the dropout rate, defined as the percentage of ninth graders that do not graduate, for each state (and the District of Columbia) in 1982 and 1984.

a. Construct two stem and leaf displays, side by side, to describe the 1982 and 1984 dropout rates, respectively.

b. Use the displays you constructed in part **a** to find the highest and lowest dropout rates in each of the two years. With which states are they associated? Do the states having the extreme rates change from 1982 to 1984?

STATE	1982	1984	STATE	1982	1984	STATE	1982	1984
Ala.	36.6	37.9	Ky.	34.1	31.6	N.D.	16.1	13.7
Alaska	35.7	25.3	La.	38.5	43.3	Ohio	22.5	20.0
Ariz.	36.6	35.4	Maine	27.9	22.8	Okla.	29.2	26.9
Ark.	26.6	24.8	Md.	25.2	22.2	Ore.	27.6	26.1
Calif.	39.9	36.8	Mass.	23.6	25.7	Pa.	24.0	22.8
Colo.	29.1	24.6	Mich.	28.4	27.8	R.I.	27.3	31.3
Conn.	29.4	20.9	Minn.	11.8	10.7	S.C.	37.4	35.5
Del.	25.3	28.9	Miss.	38.7	37.6	S.D.	17.3	14.5
D.C.	43.1	44.8	Mo.	25.8	23.8	Tenn.	32.2	29.5
Fla.	39.8	37.8	Mont.	21.3	17.9	Texas	36.4	35.4
Ga.	35.0	36.9	Neb.	18.1	13.7	Utah	25.0	21.3
Hawaii	25.1	26.8	Nev.	35.2	33.5	Vt.	20.4	16.9
Idaho	25.6	24.2	N.H.	23.0	24.8	Va.	26.2	25.3
Ill.	23.9	25.5	N.J.	23.5	22.3	Wash.	23.9	24.9
Ind.	28.3	23.0	N.M.	30.6	29.0	W.Va.	33.7	26.9
Iowa	15.9	14.0	N.Y.	36.6	37.8	Wis.	16.9	15.5
Kansas	19.3	18.3	N.C.	32.9	30.7	Wyo.	27.6	24.0

c. Use the stem and leaf display to locate the "middle" dropout rate in each of the two years. Note that this is the 26th measurement when the data are arranged in ascending order. How much did the middle dropout rate move between 1982 and 1984? [*Note:*   The middle measurement is called the *median*. We discuss the median in Section 2.4.]

*d. Use a statistical software package to generate stem and leaf displays for the two years' dropout rates.

2.14 While producing many economic benefits to the state of Florida, gypsum and phosphate mines also produce a harmful byproduct: radiation. It has been known for a number of years that the mine tailings (waste) contain radioactive radon 222. In fact, new housing complexes built over the leveled piles of residue have shown disturbing radiation levels within the houses. The radiation levels in waste gypsum and phosphate mounds in Polk County, Florida, are regularly monitored by the Eastern Environmental Radiation Facility (EERF), and by the Polk County Health Department (PCHD), Winter Haven, Florida. Shown in the table are measurements of the exhalation rate (a measure of radiation) of soil samples taken on waste piles in Polk County, Florida. They represent part of the data contained in a report by Thomas R. Horton of EERF.

EXHALATION RATE OF SOIL SAMPLES							
1,709.79	4,132.28	2,996.49	2,796.42	3,750.83	961.40	1,096.43	1,774.77
357.17	1,489.86	2,367.40	11,968.23	178.99	5,402.35	2,315.52	2,617.57
1,150.94	3,017.48	599.84	2,758.84	3,764.96	1,888.22	2,055.20	205.84
1,572.69	393.55	538.37	1,830.78	878.56	6,815.69	752.89	1,977.97
558.33	880.84	2,770.23	1,426.57	1,322.76	1,480.04	9,139.21	1,698.39

Source: Horton, T. R. "A preliminary radiological assessment of radon exhalation from phosphate gypsum piles and inactive uranium mill tailings piles." EPA–520/5–79–004. Washington, D.C.: Environmental Protection Agency, 1979.

SPSS/PC+ was used to generate the following stem and leaf display for these data:

```
Stem-and-leaf display for variable .. EXHLRATE

   0 . 22445668990123455677889
   2 . 013468880088
   4 . 14
   6 . 8
   8 . 1
  10 .
  12 . 0
```

a. Interpret the display. Which digit(s) was used for the stem and which for the leaf? Find the largest measurement in the data set, and interpret its representation in the display.

b. Note that the presence of a measurement well removed from the main body of data somewhat distorts the display, since most of the data set is compressed into a small portion of it. The largest measurement was removed from the data set, and a new SPSS/PC+ stem and leaf display was generated, as shown here. Interpret this display, identifying the stem and leaf used. Using both displays, give a verbal description of the data.

```
Stem-and-leaf display for variable .. EXHLRATE

   0 . 2244566899
   1 . 0123455677889
   2 . 01346888
   3 . 0088
   4 . 1
   5 . 4
   6 . 8
   7 .
   8 .
   9 . 1
```

2.3 Graphic Methods for Describing Quantitative Data: Histograms

A **relative frequency histogram** for the 100 EPA mileage readings is shown in Figure 2.4. The horizontal axis of Figure 2.4, which gives the miles per gallon for a given automobile, is divided into intervals commencing with the interval from 29.95 to 31.45 and proceeding in intervals of equal size to 43.45 to 44.95 miles per gallon. The vertical axis gives the proportion (or **relative frequency**) of the 100 readings that fall in each interval. Thus, you can see that .33, or 33%, of the owners obtained a mileage between 35.95 and 37.45. This interval contains the highest relative frequency, and the intervals tend to contain a smaller fraction of the measurements as the mileages get smaller or larger.

By summing the relative frequencies in the intervals 34.45–35.95, 35.95–37.45, and 37.45–38.95, you can see that 65% of the mileages are between 34.45 and 38.95. Similarly, only 2% of the cars obtained a mileage rating over 41.95. Many other summary statements can be made by further study of the histogram. When constructing a histogram, the general rules listed in the accompanying box should be followed.

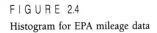

FIGURE 2.4

Histogram for EPA mileage data

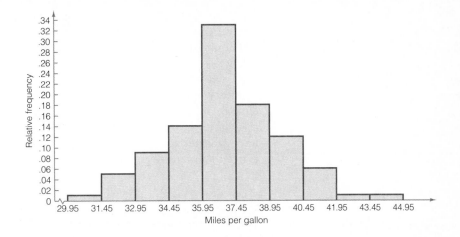

How to Construct a Histogram

1. Identify the smallest and the largest measurements in the set of data.

2. Divide the interval between the smallest and the largest measurement into between 5 and 20 equal subintervals called **classes**. These classes should satisfy the following requirements:

   a. Each measurement falls into one and only one subinterval.

   b. No measurement falls on a boundary of a subinterval.

   Although the choice of the number of classes is arbitrary, you will obtain a better description of the data if you use a small number of subintervals when you have a small amount of data and use a large number of subintervals for a large amount of data.

3. Compute the proportion (relative frequency) of measurements in each subinterval.*

4. Using a vertical axis of about three-fourths the length of the horizontal axis, plot each relative frequency as a rectangle over the corresponding subinterval.

To construct the relative frequency histogram in Figure 2.4 we must first define the measurement classes. Since the number of measurements, $n = 100$, is moderately large, we arbitrarily choose to construct ten measurement classes. The ten classes must span the distance between the smallest measurement, 30.0, and the largest measurement, 44.9. Thus, each class should have a width of

$$\text{Approximate class width} = \frac{\text{Largest measurement} - \text{Smallest measurement}}{\text{Number of classes}}$$

$$= \frac{44.9 - 30.0}{10} \approx 1.49$$

---

*Note that *frequencies* rather than relative frequencies may be used in constructing a histogram. The **frequency** is the actual number of measurements in each interval.

or, rounding upward to be certain of including all the observations, the class interval width is approximately equal to 1.50.

Locating the lower class boundary of the first class interval at 29.95 (slightly below the smallest measurement) and adding the class width of 1.5, we find the upper class boundary to be 31.45. Adding 1.5 again, we find the upper class boundary of the second class to be 32.95. Continuing this process, we obtain the ten class intervals shown in Table 2.3. Note that the points which locate the class boundaries are written with a 5 in the second decimal place. This makes it impossible for any of the miles per gallon observations to fall on a class boundary (since they were given only to the nearest tenth).

T A B L E 2.3 **Measurement Classes, Frequencies, and Relative Frequencies for the Car Mileage Data**

MEASUREMENT CLASS	FREQUENCY	RELATIVE FREQUENCY
29.95–31.45	1	.01
31.45–32.95	5	.05
32.95–34.45	9	.09
34.45–35.95	14	.14
35.95–37.45	33	.33
37.45–38.95	18	.18
38.95–40.45	12	.12
40.45–41.95	6	.06
41.95–43.45	1	.01
43.45–44.95	1	.01
	100	1.00

The next step is to determine the class frequencies and calculate the class relative frequencies. These quantities are defined as follows.

Definition 2.5

The **class frequency** for a given class, say class i, is equal to the total number of measurements that fall in that class. The class frequency for class i is denoted by the symbol f_i.

Definition 2.6

The **class relative frequency** for a given class, say class i, is equal to the class frequency divided by the total number n of measurements, i.e.,

$$\text{Relative frequency for class } i = \frac{f_i}{n}$$

Scan the data and count the number of measurements in each class interval. This number, the *class frequency*, is entered in the second column of Table 2.3. Finally, calculate the *class relative frequency*, the proportion of the total number of measurements that fall in each interval. The class relative frequencies for the mileage data are obtained by dividing each of the class frequencies by the total

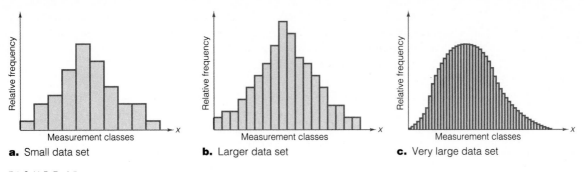

a. Small data set   **b.** Larger data set   **c.** Very large data set

FIGURE 2.5

The effect of the size of a data set on the outline of a histogram

number of measurements (100); these are listed in the third column of Table 2.3. As noted at the outset of this discussion, these relative frequencies were used to construct the relative frequency histogram of Figure 2.4.

By looking at a histogram (say, the relative frequency histogram in Figure 2.4), you can see two important facts. First, note the total area under the histogram and then note the proportion of the total area that falls over a particular interval of the x-axis. You will see that the proportion of the total area that falls above an interval is equal to the relative frequency of measurements falling in the interval. For example, the relative frequency for the class interval 35.95–37.45 is .33. Consequently, the rectangle above the interval contains .33 of the total area under the histogram.

Second, you can imagine the appearance of the relative frequency histogram for a very large set of data (say, a population). As the number of measurements in a data set is increased, you can obtain a better description of the data by decreasing the width of the class intervals. When the class intervals become small enough, a relative frequency histogram will (for all practical purposes) appear as a smooth curve (see Figure 2.5).

EXAMPLE 2.3

The data consisting of the percentages of college or university loans in default for each state and the District of Columbia are given in Table 2.4. Construct a relative frequency histogram for these data.

TABLE 2.4 **Defaulted Student Loans (in Millions of Dollars)**

STATE	%	STATE	%	STATE	%	STATE	%
Ala.	12.0	Ill.	9.3	Mont.	6.4	R.I.	8.8
Alaska	19.7	Ind.	6.7	Nebr.	4.9	S.C.	14.1
Ariz.	12.1	Iowa	6.2	Nev.	10.1	S.Dak.	5.5
Ark.	12.9	Kans.	5.7	N.H.	7.9	Tenn.	12.3
Calif.	11.4	Ky.	10.3	N.J.	12.0	Tex.	15.2
Colo.	9.5	La.	13.5	N.Mex.	7.5	Utah	6.0
Conn.	8.8	Maine	9.7	N.Y.	11.3	Vt.	8.3
Del.	10.9	Md.	16.6	N.C.	15.5	Va.	14.4
D.C.	14.7	Mass.	8.3	N.Dak.	4.8	Wash.	8.4
Fla.	11.8	Mich.	11.4	Ohio	10.4	W.Va.	9.5
Ga.	14.8	Minn.	6.6	Okla.	11.2	Wis.	9.0
Hawaii	12.8	Miss.	15.6	Oreg.	7.9	Wyo.	2.7
Idaho	7.1	Mo.	8.8	Pa.	8.7		

Source: National Direct Student Loan Program.

Solution

There are 51 observations, so we (arbitrarily) choose to use eight classes. Often, the final selection of this number is based on trial and error, and the exact number of classes selected is, to some extent, judgmental.

Using eight classes, we calculate

$$\text{Class width} = \frac{\text{Largest measurement} - \text{Smallest measurement}}{8}$$

$$= \frac{19.7 - 2.7}{8} = \frac{17}{8} = 2.125 \approx 2.2$$

(round up to assure the entire range is covered by eight intervals).

We start at .05 below the smallest measurement, using the second decimal place to assure that no measurement falls on a class boundary. The resulting frequencies and relative frequencies are shown in Table 2.5.

TABLE 2.5 **Measurement Classes, Frequencies, and Relative Frequencies for the Defaulted Student Loan Data**

MEASUREMENT CLASS	FREQUENCY	RELATIVE FREQUENCY
2.65– 4.85	2	.04
4.85– 7.05	8	.16
7.05– 9.25	12	.24
9.25–11.45	12	.24
11.45–13.65	8	.16
13.65–15.85	7	.14
15.85–18.05	1	.02
18.05–20.25	1	.02
	51	1.02*

*The column of relative frequencies does not add to 1.00 due to rounding.

The relative frequency histogram appears in Figure 2.6. Compare this with the stem and leaf display for the same data (Figure 2.2) Which do you think provides a more descriptive graphic display?

FIGURE 2.6

Relative frequency histogram for student loan default data

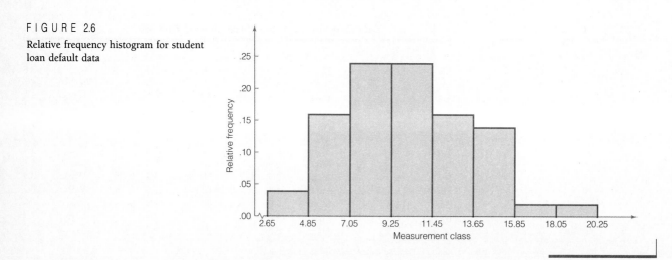

EXAMPLE 2.4

Solution

Refer to Example 2.3. Use a statistical computer software package to create a relative frequency histogram for the student loan default data.

We used SAS on a microcomputer to generate the relative frequency histogram in Figure 2.7. Note that six classes were found by the SAS program, illustrating the judgmental aspect in the selection of the number of classes. Note, too, that the SAS program labels the vertical axis "Percentage" rather than "Relative frequency." Perhaps the most significant difference between Figure 2.6 and Figure 2.7 is that the measurement classes formed by the SAS program are identified by their midpoints rather than their endpoints. Thus, the first interval has a midpoint of 3, the second of 6, etc. The corresponding class intervals are therefore 1.5 to 4.5, 4.5 to 7.5, etc.

FIGURE 2.7

SAS printout of relative frequency histogram for student loan default data

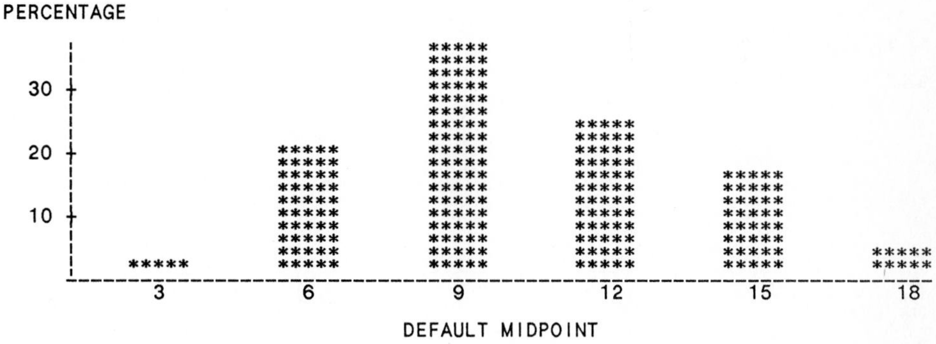

Compare the computer stem and leaf display (Figure 2.3) to the computer-generated histogram (Figure 2.7). Which do you prefer as a graphic description of these data?

CASE STUDY 2.1

MERCURY POISONING AND THE DENTAL PROFESSION

The hazards to health traced to environmental pollution constitute an area of major national concern. Mercury has been identified as one source of environmental contamination. Recognition of mercury as a hazard can be traced to Theophrastus, a Greek scientist living about 400 B.C. The physiologic effects of mercury poisoning are a matter of record with death as the ultimate possibility.

This is the introductory paragraph to an article that appeared in the November 1974 issue of the *Journal of the American Dental Association*.* The article discusses mercury vapor contamination levels in the air of dental offices and the

*Miller, S. L., Domey, R. G., Elston, S. F. A., and Milligan, G. "Mercury vapor levels in the dental office: A survey," *Journal of the American Dental Association*, November 1974, 89, 1084–1091. Copyright by the American Dental Association. Reprinted by permission.

resulting threat to the health of dentists and dental assistants. Such contamination might result from spills in handling mercury, the unprotected storage of scrap amalgam, or aerosols created by the use of high-speed rotary cutting instruments in removing old amalgam fillings.

The cumulative absorption of small quantities of mercury can result in serious medical problems. The constant daily exposure of dentists and their auxiliary personnel to possible mercury contamination is therefore an important concern. A level of .05 milligram of mercury per cubic meter of air is the largest amount considered safe for those working a 40-hour week.

A determination of mercury vapor levels was made in 60 dental offices in San Antonio, Texas. A relative frequency histogram summarizing the information provided by these measurements is given in Figure 2.8. The histogram clearly shows that an alarming fraction $\left(\frac{6}{60}, \text{ or } \frac{1}{10}\right)$ were above the danger level of .05 milligram. You can see that these data have been effectively summarized and clearly indicate a need for strict policing of mercury vapor levels in dental offices.

FIGURE 2.8
Relative frequency histogram

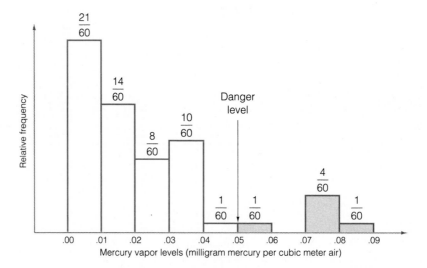

Case Study 2.1 illustrates a shortcoming of relative frequency histograms and graphic data displays in general. Suppose you wish to use the relative frequency histogram to infer the nature of the population relative frequency distribution— that is, the distribution of the mercury vapor levels in the offices of all dentists practicing in the United States. It is true that the sample and population relative frequency distributions will be similar. But how similar? How can you explain how similar the two figures will be? Or, equivalently, how can you measure the reliability of the inference? In Section 2.4 we will explain how you can use one or more numbers (numerical descriptive measures) to describe a distribution of measurements. Further, you will see in Chapter 7 that we can use sample numerical descriptive measures to make inferences about their population counterparts and that we can measure the reliability of these inferences. In short, you will see that numerical descriptive measures are superior to graphic descriptive measures when you want to use the sample data to make inferences about the population from which the sample was selected.

EXERCISES 2.15–2.26

[*Note: Starred (*) exercises require the use of a computer.*]

LEARNING THE MECHANICS

2.15   Construct a relative frequency histogram for the data summarized in the accompanying relative frequency table.

MEASUREMENT CLASS	RELATIVE FREQUENCY
.5– 2.5	.10
2.5– 4.5	.15
4.5– 6.5	.25
6.5– 8.5	.20
8.5–10.5	.05
10.5–12.5	.10
12.5–14.5	.10
14.5–16.5	.05

2.16   Use the following data to construct a relative frequency histogram with eight measurement classes:

5.9	5.3	1.6	7.4	9.8	1.7
8.6	1.2	2.1	4.0	6.5	7.2
7.3	8.4	8.9	6.7	9.2	2.8
4.5	6.3	7.6	9.7	9.4	8.8
3.5	1.1	4.3	3.3	3.1	1.3
8.4	1.6	8.2	6.5	4.1	3.1
1.1	5.0	9.4	6.4	7.7	2.7

2.17   Construct a relative frequency histogram for the following data:

23	12	82	12	67
52	24	17	15	60
32	49	37	55	81
39	99	88	24	30
12	19	53	50	18
16	40	51	61	35

2.18   Construct a relative frequency histogram for the data of Exercise 2.10. Compare your histogram with the stem and leaf display of Exercise 2.10. Which seems to describe the data set better?

2.19   Construct a relative frequency histogram for the data of Exercise 2.11. Compare your histogram with the stem and leaf display of Exercise 2.11. Which seems to describe the data set better?

APPLYING THE CONCEPTS

2.20   The graph summarizes the scores obtained by 100 students on a questionnaire designed to measure aggressiveness. (Scores are integer values that range from 0 to 20. A high score indicates a high level of aggression.)
a. Which measurement class contains the highest proportion of test scores?
b. What proportion of the scores lie between 3.5 and 5.5?

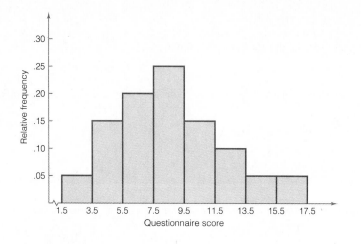

c. What proportion of the scores are higher than 11.5?

d. How many students scored less than 5.5?

2.21   The numbers of years of experience (to the nearest half year) for the faculty of a certain statistics department are recorded as follows:

5.0	4.5	19.5	1.0	6.5
.5	4.0	1.0	2.5	9.0
5.5	5.0	3.5	4.5	1.5
8.0	21.0	7.0	13.5	3.5

a. Construct a relative frequency histogram for these data. Use six measurement classes, commencing with .45 and ending with 21.45.

b. In order to be eligible for certain benefits, a faculty member must have more than 13 years of experience. What proportion of the statistics faculty are eligible for these benefits?

2.22   In the late 1970s and early 1980s gasoline prices rose at an incredible rate. The following is a sample of the prices charged (in cents per gallon) for regular grade gasoline at 25 gas stations in the greater Cincinnati area in January 1981:

128.9	121.9	133.9	119.9	115.9
115.9	127.9	119.8	116.8	122.8
135.9	118.8	115.9	121.9	126.9
117.9	115.8	121.9	124.9	129.9
122.8	131.9	132.8	120.8	124.9

a. Construct a relative frequency histogram for these data.

b. Suppose that you were to sample 25 gas stations from another area of the country, say New York City or rural Kansas. Do you think a histogram for the 25 measurements would look similar to the one you constructed in part a? Explain.

c. Explain the correspondence between areas under the relative frequency histogram and relative frequencies.

2.23   Considering the climate, is it economically feasible to start an orange grove in northern Florida? If the temperature falls below 32°F, oil-burning smudge pots must be lit to keep the orange trees from freezing. Suppose a prospective grower decides that a grove would be economically feasible if the pots have to be lit an average of 15 days or less each year. The grower selects 20 years since 1900 at random and obtains the total number of days per year that the temperature fell below 32°F.

20	16	13	12
9	25	16	6
15	10	18	11
14	12	17	13
13	28	14	15

a. Construct a relative frequency histogram for these data.

b. Based on these sample data, estimate the proportion of years in which the pots have to be lit 15 days or less. [*Note:* We show you how to evaluate the reliability of this estimate in Chapter 7.]

2.24 Refer to the data (Exercise 2.13, page 26) on the high school dropout rates in 1982 and 1984. These data for each state and the District of Columbia are reproduced here.

STATE	1982	1984	STATE	1982	1984	STATE	1982	1984
Ala.	36.6	37.9	Ky.	34.1	31.6	N.D.	16.1	13.7
Alaska	35.7	25.3	La.	38.5	43.3	Ohio	22.5	20.0
Ariz.	36.6	35.4	Maine	27.9	22.8	Okla.	29.2	26.9
Ark.	26.6	24.8	Md.	25.2	22.2	Ore.	27.6	26.1
Calif.	39.9	36.8	Mass.	23.6	25.7	Pa.	24.0	22.8
Colo.	29.1	24.6	Mich.	28.4	27.8	R.I.	27.3	31.3
Conn.	29.4	20.9	Minn.	11.8	10.7	S.C.	37.4	35.5
Del.	25.3	28.9	Miss.	38.7	37.6	S.D.	17.3	14.5
D.C.	43.1	44.8	Mo.	25.8	23.8	Tenn.	32.2	29.5
Fla.	39.8	37.8	Mont.	21.3	17.9	Texas	36.4	35.4
Ga.	35.0	36.9	Neb.	18.1	13.7	Utah	25.0	21.3
Hawaii	25.1	26.8	Nev.	35.2	33.5	Vt.	20.4	16.9
Idaho	25.6	24.2	N.H.	23.0	24.8	Va.	26.2	25.3
Ill.	23.9	25.5	N.J.	23.5	22.3	Wash.	23.9	24.9
Ind.	28.3	23.0	N.M.	30.6	29.0	W.Va.	33.7	26.9
Iowa	15.9	14.0	N.Y.	36.6	37.8	Wis.	16.9	15.5
Kansas	19.3	18.3	N.C.	32.9	30.7	Wyo.	27.6	24.0

a. Construct a relative frequency histogram for the dropout rates in each of the two years. Use the same measurement classes for the histograms and show the histograms on the same paper, using different colors to distinguish between 1982 and 1984.

b. Can you perceive a shift in the distribution of dropout rates from 1982 to 1984? If so, in which direction is the shift?

c. Compare the relative frequency histograms to the stem and leaf displays in Exercise 2.13. Which do you prefer as a descriptive display of these data?

2.25 The table contains the populations (in thousands) of 30 sunbelt cities in 1960 and 1980:

a. Construct two relative frequency histograms, one for the 1960 data set and one for the 1980 data set. Use the same width for the measurement classes of the two histograms.

b. Compare your histograms and describe what they reveal about the change in the populations of sunbelt cities between 1960 and 1980.

CITY	1960	1980	CITY	1960	1980
Albuquerque, NM	201	332	Las Vegas, NV	64	165
Anaheim, CA	104	219	Lubbock, TX	129	174
Arlington, TX	45	160	Miami, FL	292	347
Atlanta, GA	487	425	New Orleans, LA	628	558
Austin, TX	187	345	Oklahoma City, OK	324	403
Charlotte, NC	202	314	Orlando, FL	88	128
Dallas, TX	680	904	Phoenix, AZ	439	790
Fort Lauderdale, FL	84	153	St. Petersburg, FL	181	239
Fort Worth, TX	356	385	San Antonio, TX	588	786
Fresno, CA	134	218	San Diego, CA	573	876
Honolulu, HI	294	365	San Francisco, CA	740	679
Houston, TX	938	1,595	San Jose, CA	204	629
Huntington Beach, CA	11	171	Tampa, FL	275	272
Huntsville, AL	72	143	Tucson, AZ	213	331
Jacksonville, FL	201	541	Tulsa, OK	262	361

Source: U.S. Bureau of the Census, *Statistical Abstract of the United States: 1981* (pp. 21–23), *1985* (pp. 23–25).

2.26   Refer to Exercise 2.14 (page 27), where we constructed a stem and leaf display for the 40 exhalation rate measurements on waste gypsum and phosphate mounds in Florida. The accompanying figure is a SAS-generated relative frequency histogram for these data.

PERCENTAGE

```
 80 +       *****
            *****
            *****
            *****
 60 +       *****
            *****
            *****
 40 +       *****
            *****
            *****
            *****
 20 +       *****
            *****
            *****
            *****       *****       *****                    *****       *****
            *****       *****       *****                    *****       *****
       -----+-----------+-----------+-----------+-----------+-----------+-----
          2000        4000        6000        8000       10000       12000
                       EXHLRATE MIDPOINT
```

a. Interpret the histogram. What fraction of the measurements fall in the first measurement class? What are the boundaries for that class? [*Note:*   Many computer programs, including SAS, do not require that the first and last class have well-defined outer boundaries.]

b. Compare the histogram to the stem and leaf display for these data in Exercise 2.14a.

2.4 Numerical Measures of Central Tendency

Now that we have presented some graphic techniques for summarizing and describing data sets, we turn to numerical methods for accomplishing this objective. When we speak of a data set, we refer to either a sample or a population. If statistical inference is our goal, we will wish ultimately to use sample numerical descriptive measures to make inferences about the corresponding measures for a population.

As you will see, there are a large number of numerical methods available to describe data sets. Most of these methods measure one of two data characteristics:

1. The **central tendency** of the set of measurements, i.e., the tendency of the data to cluster or to center about certain numerical values.
2. The **variability** of the set of measurements, i.e., the spread of the data.

In this section we concentrate on measures of central tendency. In the next section, we discuss measures of variability.

The most popular and best understood measure of central tendency for a quantitative data set is the *arithmetic mean* (or simply the *mean*) of a data set.

Definition 2.7

The **mean** of a set of quantitative data is equal to the sum of the measurements divided by the number of measurements contained in the data set.

In everyday terms, the mean is the average value of the data set.

| EXAMPLE 2.5

Calculate the mean of the following five sample measurements: 5, 3, 8, 5, 6.

Solution

The mean is the average of the five measurements—i.e.,

$$\frac{5 + 3 + 8 + 5 + 6}{5} = 5.4$$

Thus, the mean of this sample is 5.4.*

At this point, it is advantageous to present some shorthand notation that will simplify calculation instructions for the mean as well as other more complicated numerical descriptive measures we will subsequently encounter. This notation is summarized in the box on page 40. Remember that such notation is used for one reason only—to avoid having to repeat the same verbal descriptions over and over again. If you mentally substitute the verbal definition of a symbol each time you read it, you will soon become accustomed to its use.

---

*In the examples given here, $\bar{x}$ is sometimes rounded to the nearest tenth, sometimes the nearest hundredth, sometimes the nearest thousandth. There is no specific rule for rounding when calculating $\bar{x}$ because $\bar{x}$ is specifically defined to be the sum of all measurements divided by n; i.e., it is a specific fraction. When $\bar{x}$ is used for descriptive purposes, it is often convenient to round the calculated value of $\bar{x}$ to the number of significant figures used for the original measurements. When $\bar{x}$ is to be used in other calculations, however, it may be necessary to retain more significant figures.

> x: The letter x is used to represent an arbitrary measurement of a sample.
>
> n: A lowercase n is used to represent the number of measurements in a sample.
>
> $\sum x$:   The Greek capital letter sigma followed by the letter x is used to mean "add all the measurements of a sample."
>
> $\bar{x}$:   The letter x with a bar over it (read "x bar") is used to denote the sample mean.
>
> Thus, using this notation, the sample mean is given by the formula*
>
> $$\bar{x} = \frac{\sum x}{n} = \frac{\text{Sum of all the sample measurements}}{\text{Number of measurements in the sample}}$$
>
> μ: The Greek letter mu is used to denote the mean of all the measurements in the population.

EXAMPLE 2.6

Calculate the sample mean for the 100 EPA mileages given in Table 2.1.

Solution

The mean gas mileage for the 100 cars is

$$\bar{x} = \frac{36.3 + 41.0 + \cdots + 38.1 + 36.9}{100} = \frac{3{,}699.4}{100} = 36.99$$

Given this information, you would be able to visualize a distribution of gas mileage readings centered in the vicinity of $\bar{x} = 37.0$. An examination of the relative frequency histogram (Figure 2.4) confirms that $\bar{x}$ does in fact fall near the center of the distribution.

In a practical situation, we rarely know the population mean μ, but we can use the sample mean $\bar{x}$ to estimate its value. Thus, we would infer that the mean gas mileage for all cars (of the model included in the sample) is near 37.0 miles per gallon. We will show you how to evaluate the reliability of this estimate in Chapter 7.

CASE STUDY 2.2

HOTELS: A RATIONAL METHOD FOR OVERBOOKING

The most outstanding characteristic of the general hotel reservation system is the option of the prospective guest, without penalty, to change or cancel his reservation or even to "no-show" (fail to arrive without notice). Overbooking (taking reservations in excess of the hotel capacity) is practiced widely throughout the industry as a compensating economic measure. This has motivated our research into the problem of determining policies for over-booking which are based on some set of rational criteria.

---

*We omit the formula for calculating the mean for grouped data (data presented in a relative frequency table). The reader interested in this special topic should consult the references at the end of this chapter.

So says Marvin Rothstein* in an article that appeared in the journal for the American Institute for Decision Sciences. In this paper Rothstein introduces a method for scientifically determining hotel booking policies and applies it to the booking problems of the 133-room Sheraton Pocono Inn at Stroudsburg, Pennsylvania.

From the Sheraton Pocono Inn's records the number of reservations, walk-ins (people without reservations who expect to be accommodated), cancellations, and no-shows were tabulated for each day during the period August 1–28, 1971. The inn's records for this period included approximately 3,100 guest histories. From the tabulated data the mean or average number of room reservations per day for each of the 7 days of the week were computed. These appear in Table 2.6.

TABLE 2.6 **Mean Number of Room Reservations, August 1–28, 1971, 133 Rooms**

SUNDAY	MONDAY	TUESDAY	WEDNESDAY	THURSDAY	FRIDAY	SATURDAY
138	126	149	160	150	150	169

In applying his booking policy decision method to the Sheraton's data, Rothstein used the means listed in Table 2.6 to help portray the inn's demand for rooms.

The mean number of Saturday reservations during August 1–28, 1971, is 169. This may be interpreted as an estimate of μ, the mean number of rooms demanded via reservations (walk-ins also contribute to the demand for rooms) on a Saturday during 1971. If this reservation data for all Saturdays during 1971 had been tabulated, μ could have been computed. But since only August data are available, they were used to estimate μ. Can you think of some problems associated with using August's data to estimate the mean for the entire year?

> **Definition 2.8**
>
> The **median** is another important measure of central tendency. In general terms, the median is the middle number when the measurements in a data set are arranged in ascending (or descending) order.

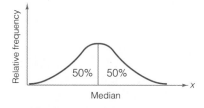

FIGURE 2.9
Location of the median

The median is of most value in describing large data sets. If the data set is characterized by a relative frequency histogram (Figure 2.9), the median is the point on the x-axis such that half the area under the histogram lies above the median and half lies below. [*Note:* In Section 2.3 we observed that the relative frequency associated with a particular interval on the x-axis is proportional to the amount of area under the histogram that lies above the interval.]

*Reprinted from Rothstein, M. "Hotel overbooking as a Markovian sequential process," *Decision Sciences*, July 1974, 5, 389–405.

> **Calculating a Median**
>
> Arrange the n measurements from the smallest to the largest.
>
> **1.** If n is odd, the median is the middle number.
> **2.** If n is even, the median is the mean of the middle two numbers.

EXAMPLE 2.7

Consider the following sample of $n = 7$ measurements: 5, 7, 4, 5, 20, 6, 2

a. Calculate the median of this sample.
b. Eliminate the last measurement (the 2) and calculate the median of the remaining $n = 6$ measurements.

Solution

a. The seven measurements in the sample are ranked in ascending order:

   2,   4,   5,   5,   6,   7,   20

Because the number of measurements is odd, the median is the middle measurement. Thus, the median of this sample is 5.

b. After removing the 2 from the set of measurements, we rank the sample measurements in ascending order as follows:

   4,   5,   5,   6,   7,   20

Now the number of measurements is even, so we average the middle two measurements. The median is $(5 + 6)/2 = 5.5$.

In certain situations, the median may be a better measure of central tendency than the mean. In particular, the median is less sensitive than the mean to extremely large or small measurements. To illustrate, note that all but one of the measurements in part **a** of Example 2.7 center about $x = 5$. The single relatively large measurement, $x = 20$, does not affect the value of the median, 5, but it causes the mean, $\bar{x} = 7$, to lie to the right of most of the measurements. As another example, if you were interested in computing a measure of central tendency of the incomes of a company's employees, the mean might be misleading. If all blue-collar and white-collar employees' incomes are included in the data set, the high incomes of a few executives will influence the mean more than the median. Thus, the median will provide a more accurate picture of the typical income for employees of the company. Similarly, the median yearly sales for a sample of companies would locate the middle of the sales data. However, a few companies with very large yearly sales would greatly influence the mean, making it deceptively large. That is, the mean could exceed a vast majority of the sample measurements, making it a misleading measure of central tendency.

If you were to calculate the median for the 100 gas mileage readings in Table 2.1, you would find that the median, 37.0, and the mean, 36.99, are almost equal. This fact indicates that the data form an approximately **symmetric** distribution. (Compare the center figure in the box on page 43 with Figure 2.4 on page 29.) As indicated in the box, a comparison of the mean and median gives an indication of the **skewness** (nonsymmetry) of a data set.

Comparing the Mean and the Median

If the median is less than the mean, the data set is skewed to the right:

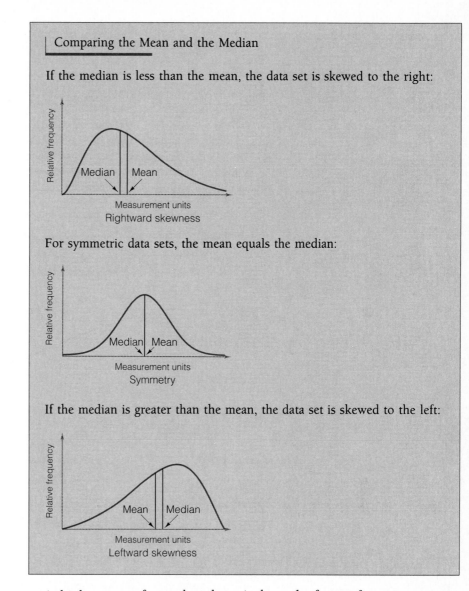

For symmetric data sets, the mean equals the median:

If the median is greater than the mean, the data set is skewed to the left:

A third measure of central tendency is the *mode* of a set of measurements.

Definition 2.9

The **mode** is the measurement that occurs most frequently in the data set.

Therefore, the mode shows where the data tend to concentrate.

EXAMPLE 2.8

Calculate the mode for the following ten quiz grades:

    8,   7,   9,   6,   8,   10,   9,   9,   5,   7

Solution

Since 9 occurs most often, the mode is 9.

The mode is of primary value in describing large data sets. When a large data set has been described using a relative frequency histogram, the mode will be located in the class containing the largest relative frequency. This is called the **modal class**. Several definitions exist for locating the position of the mode within a modal class, but the simplest is to define the mode as the midpoint of the modal class. For example, examine the relative frequency histogram for the EPA mileage readings (Figure 2.4). You can see that the modal class is the interval 35.95–37.45. The mode (the midpoint) is 36.7. Note that this measure of central tendency is very close to the mean, 36.99, and the median, 37.0.

Because it emphasizes data concentration, the mode is often used with large data sets to locate the region in which much of the data is concentrated. A retailer of men's clothing would be interested in the modal neck size and sleeve length of potential customers. A supermarket manager is interested in the cereal brand with the largest share of the market, i.e., the modal brand. The modal income class of the American worker is of interest to the Labor Department. Thus, the mode provides a useful measure of central tendency for various applications.

EXERCISES 2.27–2.43

LEARNING THE MECHANICS

2.27   Calculate the mode, mean, and median of the following data:

   18   10   15   13   17   15   12   15   18   16   11

2.28   Calculate the mean and median of the following grade-point averages:

   3.2   2.5   2.1   3.7   2.8   2.0

2.29   Explain the difference between the calculation of the median for an odd and an even number of measurements. Construct one data set consisting of five measurements and another consisting of six measurements for which the medians are equal.

2.30   Explain how the relationship between the mean and median provides information about the symmetry or skewness of the data's distribution.

2.31   Calculate the mean for samples where:
        a. $n = 10$, $\Sigma x = 85$ b. $n = 16$, $\Sigma x = 400$ c. $n = 45$, $\Sigma x = 35$
        d. $n = 18$, $\Sigma x = 242$

2.32   Calculate the mean, median, and mode for each of the following samples:
        a. 7, −2, 3, 3, 0, 4
        b. 2, 3, 5, 3, 2, 3, 4, 3, 5, 1, 2, 3, 4
        c. 51, 50, 47, 50, 48, 41, 59, 68, 45, 37

2.33   Describe how the mean compares to the median for a distribution as follows:
        a. Skewed to the left   b. Skewed to the right   c. Symmetric

APPLYING THE CONCEPTS

2.34   According to *Consumers' Digest* (Nov.–Dec. 1980), at that time sugar was the leading food additive in the U.S. food supply. Sugar may be listed more than once on a product's ingredient list since it goes by different names depending on its source (e.g., sucrose, corn sweetener, fructose, and dextrose). Thus, when you read a product's label you may have to total up the sugar in the product to see how much sweetener it contains. The table gives a list of candy bars and the percentage of sugar they contain relative to their weight.

BRAND	PERCENTAGE OF SUGAR BY WEIGHT	BRAND	PERCENTAGE OF SUGAR BY WEIGHT
Baby Ruth	23.7	Power House	30.6
Butterfinger	29.5	Bit-O-Honey	23.5
Mr. Goodbar	34.2	Chunky	38.4
Milk Duds	36.0	Milk Chocolate Covered	
Mello Mint	79.6	Raisinettes	24.7
M & M Plain Chocolate		Oh Henry!	31.2
Candies	52.2	Borden Cracker Jack	14.7
Mars Chocolate Almond	36.4	Good & Plenty Licorice	28.2
Milky Way	26.8	Nestle's Crunch	43.5
Marathon	36.7	Planter Jumbo Block	
Snickers	28.0	Peanut Candy	21.5
3 Musketeers	36.1	Switzer Licorice	8.4
Junior Mints	45.3	Switzer Red Licorice	2.8
Pom Poms	29.5	Tootsie Pop Drops	54.1
Sugar Babies	41.0	Tootsie Roll	21.1
Sugar Daddy	22.0	Fancy Fruit Lifesavers	77.6
Almond Joy	20.0	Spear-O-Mint Lifesavers	67.6

Source: *National Confectioners Association Brand Name Guide to Sugar* (Nelson Hall Paperback).
Secondary source: *Consumers' Digest*, Nov.–Dec. 1980, p. 11.

a. Calculate the mean percentage of sugar per bar for the candy bars listed.
b. Find the median for the data set.
c. What do the mean and median indicate about the skewness of the data set?
d. Construct a relative frequency histogram for the data set. Indicate the location of the mean, median, and modal class of the data set on your histogram.

2.35 A psychologist has developed a new technique intended to improve rote memory. To test the method against other standard methods, 20 high school students are selected at random, and each is taught the new technique. The students are then asked to memorize a list of 100 word phrases using the technique. The following are the number of word phrases memorized correctly by the students:

91 64 98 66 83 87 83 86 80 93
83 75 72 79 90 80 90 71 84 68

a. Define the terms *mean*, *median*, and *mode* in the context of this problem.
b. Construct a relative frequency histogram for the data.
c. Compute the mean, median, and mode for the data set and locate them on the histogram. Do these measures of central tendency appear to locate the center of the distribution of data?

2.36 Would you expect the data sets described below to possess relative frequency distributions that are symmetric, skewed to the right, or skewed to the left? Explain.
a. The salaries of all persons employed by a large university
b. The grades on an easy test
c. The grades on a difficult test
d. The amounts of time students in your class studied last week
e. The ages of automobiles on a used car lot
f. The amounts of time spent by students on a difficult examination (maximum time is 50 minutes)

2.37   During the 1980s the pharmaceutical industry placed an increased emphasis on producing revolutionary new products. As a result, research and development (R&D) costs have increased, and companies are taking a greater interest in R&D management. The table lists the 1984 R&D expenditures (in millions of dollars) of the world's largest pharmaceutical manufacturers.

COMPANY	R&D EXPENDITURES	COMPANY	R&D EXPENDITURES
Merck	$290	Upjohn	$200
American Home	90	Takeda	125
Bristol–Myers	162	Hoffman–LaRoche	363
Pfizer	159	Sandoz	181
Ciba–Geigy	230	Johnson & Johnson	187
Hoechst	274	Boehringer Ingelheim	176
Warner–Lambert	162	Squibb	114
Abbott	110	Schering–Plough	129
Smith Kline–Beckman	158	Rhone–Poulenc	110
Bayer	200		

Source: Business Quarterly, Fall 1985, p. 81.

a. Calculate the mean and median for this data set.
b. What do the mean and median indicate about the skewness of this data set?
c. Will the median of a data set always be equal to an actual value in the data set, as was the case in part a?

2.38   The scores for a statistics test are as follows:

87   76   96   77   94   92   88   85   66   89
79   95   50   91   83   88   82   58   18   69

a. Compute the mean, median, and mode for these data.
b. Which of the three measures of central tendency do you think would best represent the achievement of the class?
c. Eliminate the two lowest scores, and again compute the mean, median, and mode. Which measure of central tendency is most affected by extremely low scores?

2.39   Ten presumably trained rats were released in a maze. Their times to escape (in seconds) are recorded below. The N's represent two rats that had still not escaped by the end of the experiment.

100   38   N   122   95   116   56   135   104   N

a. Can you calculate the mean for these data? Explain.
b. Is the median a meaningful measure of central tendency for these data? Explain. Calculate the median.

2.40   The salaries of superstar professional athletes receive much attention in the media. The million dollar annual contract is becoming more commonplace among this elite group with each passing year. Nevertheless, rarely does a year pass without one or more of the players' associations negotiating with team owners for additional salary and fringe benefit considerations for *all* players in their particular sports.
a. If a players' association wanted to support its argument for higher "average" salaries, which measure of central tendency do you think it should use? Why?
b. To refute the argument, which measure of central tendency should the owners apply to the players' salaries? Why?

2.41   In 1985, U.S. consumers redeemed 6.49 billion manufacturers' coupons and saved themselves $2.24 billion. Find the mean amount saved per coupon.

2.42 The table contains the price per acre of farmland for a sample of states that includes nine eastern states and eleven western states (i.e., west of the Mississippi River).

STATE	PRICE PER ACRE (1984)	STATE	PRICE PER ACRE (1984)
Arizona	$ 265	Nebraska	$ 444
California	1,726	Nevada	229
Colorado	435	New Hampshire	1,419
Connecticut	3,208	New Jersey	3,525
Delaware	1,642	New Mexico	163
Florida	1,527	North Dakota	360
Kansas	466	Pennsylvania	1,510
Massachusetts	2,372	Rhode Island	3,335
Maryland	2,097	South Dakota	250
Montana	222	Wyoming	177

Data: U.S. Department of Agriculture.

a. Find the mean and median price per acre for the sample of 20 states.

b. Find and compare the mean price per acre for the eastern states and the western states. Also, compare these means to the mean you found in part a. What do your comparisons reveal about the value of farmland in the United States?

c. As a measure of central tendency, the mean of a data set is frequently used to characterize the data set or to represent a typical measurement in the data set. Examine the data set and determine whether you would use the mean of these data to characterize a typical measurement. Explain.

2.43 In Exercises 2.13 (page 26) and 2.24 (page 37) we constructed stem and leaf displays and relative frequency histograms for the high school dropout rates in 1982 and 1984 for all the states and the District of Columbia. We used Minitab to compute the mean and median for the dropout rates in 1982 and 1984. Part of the Minitab output is shown:

```
        DROP1982 DROP1984
N            51       51
MEAN      28.12    26.50
MEDIAN    27.60    25.30
```

a. Interpret these measures of central tendency.

b. What, if anything, do the relationships between the mean and median tell you about the skewness in these two data sets? Is your answer supported by your stem and leaf displays (Exercise 2.13) and histograms (Exercise 2.24)?

c. In which direction does the distribution appear to have shifted from 1982 to 1984? Do you think the shift is indicative of some trend due to changes in policies and/or attitudes, or is the shift possibly nothing more than a result of random variation in the rates over time? [*Note:* We learn in Chapter 9 how to distinguish between sample differences caused by true population differences and those caused by random sampling variation.]

2.5 Numerical Measures of Variability

Measures of central tendency provide only a partial description of a quantitative data set. The description is incomplete without a measure of the variability, or spread, of the data set.

If you examine the two histograms in Figure 2.10, you will notice that both hypothetical data sets are symmetric with equal modes, medians, and means.

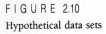

Hypothetical data sets

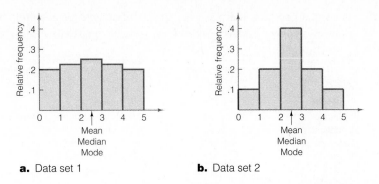

a. Data set 1   **b.** Data set 2

However, data set 1 in Figure 2.10(a) has measurements spread with almost equal relative frequency over the measurement classes, while data set 2 in Figure 2.10(b) has most of its measurements clustered about its center. Thus, data set 2 is *less variable* than data set 1. Consequently, you can see that we need a measure of variability as well as a measure of central tendency to describe a data set.

Perhaps the simplest measure of the variability of a quantitative data set is its **range**.

Definition 2.10

The **range** of a data set is equal to the largest measurement minus the smallest measurement.

The range is easy to compute and easy to understand, but it is a rather insensitive measure of data variation when the data sets are large. This is because two data sets can have the same range and be vastly different with respect to data variation. This phenomenon is demonstrated in Figure 2.10. Both distributions of data shown in the figure have the same range, but most of the measurements in data set 2 tend to concentrate near the center of the distribution. Consequently, the data are much less variable than the data in set 1. Thus, you can see that the range does not always detect differences in data variation for large data sets.

Let us see if we can find a measure of data variation that is more sensitive than the range. Consider the two samples in Table 2.7; each has five measurements. (We have ordered the numbers for convenience.) Note that both samples have a mean of 3 and that we have also calculated the distance between each measurement and the mean. What information do these distances contain? If they tend to be large in magnitude, as in sample 1, the data are spread out, or highly variable. If the distances are mostly small, as in sample 2, the data are clustered around the mean, $\bar{x}$, and therefore do not exhibit much variability. You can see that these distances, displayed graphically in Figure 2.11, provide information about the variability of the sample measurements.

The next step is to condense the information in these distances into a single numerical measure of variability. Averaging the distances from $\bar{x}$ will not help

TABLE 2.7   **Two Hypothetical Data Sets**

	SAMPLE 1	SAMPLE 2
Measurements	1, 2, 3, 4, 5	2, 3, 3, 3, 4
Mean	$\bar{x} = \dfrac{1 + 2 + 3 + 4 + 5}{5} = \dfrac{15}{5} = 3$	$\bar{x} = \dfrac{2 + 3 + 3 + 3 + 4}{5} = \dfrac{15}{5} = 3$
Distances of measurement values from $\bar{x}$	$(1 - 3), (2 - 3), (3 - 3), (4 - 3), (5 - 3)$ or $-2, -1, 0, 1, 2$	$(2 - 3), (3 - 3), (3 - 3), (3 - 3), (4 - 3)$ or $-1, 0, 0, 0, 1$

FIGURE 2.11

Dot diagrams for two data sets

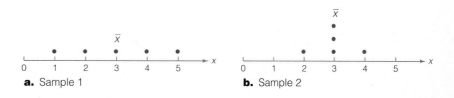

a. Sample 1   b. Sample 2

because the negative and positive distances cancel; i.e., the sum of the deviations (and thus the average deviation) is always equal to zero.

Two methods come to mind for dealing with the fact that positive and negative distances from the mean cancel. The first is to treat all the distances as though they were positive, ignoring the sign of the negative distances. We do not pursue this line of thought because the resulting measure of variability (the mean of the absolute values of the distances) presents analytical difficulties beyond the scope of this text. A second method of eliminating the minus signs associated with the distances is to square them. The quantity we can calculate from the squared distances will provide a meaningful description of the variability of a data set and presents fewer analytical difficulties in inference-making.

To use the squared distances calculated from a data set, we first calculate the **sample variance**.

> ### Definition 2.11
>
> The **sample variance** for a sample of n measurements is equal to the sum of the squared distances from the mean divided by $(n - 1)$. In symbols, using s^2 to represent the sample variance,
>
> $$s^2 = \frac{\sum (x - \bar{x})^2}{n - 1}$$

Referring to the two samples in Table 2.7, you can calculate the variance for sample 1 as follows:

$$s^2 = \frac{(1 - 3)^2 + (2 - 3)^2 + (3 - 3)^2 + (4 - 3)^2 + (5 - 3)^2}{5 - 1}$$

$$= \frac{4 + 1 + 0 + 1 + 4}{4} = 2.5$$

The second step in finding a meaningful measure of data variability is to calculate the **standard deviation** of the data set:

Definition 2.12

The **sample standard deviation**, s, is defined as the positive square root of the sample variance, s^2. Thus,

$$s = \sqrt{s^2} = \sqrt{\frac{\sum (x - \bar{x})^2}{n - 1}}$$

The population variance, denoted by the symbol σ^2 (sigma squared), is the average of the squared distances of the measurements on *all* units in the population from the mean, μ, and σ (sigma) is the square root of this quantity. Since we never really compute σ^2 or σ from the population (the object of sampling is to avoid this procedure), we simply denote these two quantities by their respective symbols.

---

s^2 = Sample variance s = Sample standard deviation

σ^2 = Population variance σ = Population standard deviation

Notice that in contrast to the variance, the standard deviation is expressed in the original units of measurement. For example, if the original measurements are in dollars, the variance is expressed in the peculiar units "dollars squared," but the standard deviation is expressed in dollars.

You may wonder why we use the divisor $(n - 1)$ instead of n when calculating the sample variance. This is because by using the divisor $(n - 1)$ we obtain a better estimate* of σ^2 than we do by dividing the sum of the squared distances by n. Since we will ultimately want to use sample statistics to make inferences about the corresponding population parameters, $(n - 1)$ is preferred to n when defining the sample variance.

EXAMPLE 2.9

Calculate the standard deviation of the following sample: 2, 3, 3, 3, 4

Solution

For this data set, $\bar{x} = 3$. Then

$$s = \sqrt{\frac{(2 - 3)^2 + (3 - 3)^2 + (3 - 3)^2 + (3 - 3)^2 + (4 - 3)^2}{5 - 1}}$$

$$= \sqrt{\frac{2}{4}} = \sqrt{.5} = .71$$

Better here means that s^2 with a divisor of $(n - 1)$ is an **unbiased estimator** of σ^2. We define and discuss unbiasedness of estimators in Chapter 6.

As the number of measurements increases, calculating s^2 and s becomes very tedious. Fortunately, as we show in Example 2.12, we can get around the difficulty of calculating s^2 and s by using a statistical software package (or most calculators). However, if you must calculate it by hand, there is a shortcut formula for computing the sample variance.

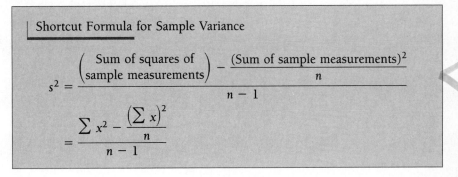

Shortcut Formula for Sample Variance

$$s^2 = \frac{\left(\begin{array}{c}\text{Sum of squares of} \\ \text{sample measurements}\end{array}\right) - \dfrac{(\text{Sum of sample measurements})^2}{n}}{n - 1}$$

$$= \frac{\sum x^2 - \dfrac{\left(\sum x\right)^2}{n}}{n - 1}$$

EXAMPLE 2.10

Use the shortcut formula to calculate the standard deviation for the sample used in Example 2.9:   2, 3, 3, 3, 4.

Solution

We need two summations to use the shortcut formula: $\sum x$ and $\sum x^2$. These can easily be obtained from the following type of tabulation:

x	x^2
2	4
3	9
3	9
3	9
4	16
$\sum x = 15$	$\sum x^2 = 47$

Then we use

$$s^2 = \frac{\sum x^2 - \dfrac{\left(\sum x\right)^2}{n}}{n - 1} = \frac{47 - \dfrac{(15)^2}{5}}{5 - 1} = \frac{2}{4} = .5$$

$$s = \sqrt{.5} = .71$$

Note that the result is identical to that obtained in Example 2.9.

EXAMPLE 2.11

Calculate the sample variance s^2 and the sample standard deviation s for the 100 gas mileage readings given in Table 2.1.

Solution

Recall that $\bar{x} = 36.99$. Clearly it would be a formidable task to calculate $(x - \bar{x})^2$ for all 100 readings. Likewise, a tabulation such as that used in Example 2.10 would be tedious. However, we can use the shortcut formula to follow a step-by-step procedure using a calculator.

STEP 1 Calculate the sum of the squared measurements. With the aid of a calculator, we obtain

$$\sum x^2 = 137,434.38$$

STEP 2 Calculate the square of the sum of the measurements.

$$\left(\sum x\right)^2 = (3,699.4)^2 = 13,685,560.36$$

STEP 3 Find $\sum (x - \bar{x})^2 = \sum x^2 - [(\sum x)^2/n]$, the numerator of s^2 in the shortcut formula.

$$\sum (x - \bar{x})^2 = \sum x^2 - \frac{\left(\sum x\right)^2}{n}$$

$$= 137{,}434.38 - \frac{13{,}685{,}560.36}{100} = 578.7764$$

We can now calculate s^2 by dividing this quantity by $(n - 1)$:

$$s^2 = \frac{578.7764}{99} = 5.85$$

$$s = +\sqrt{5.85} = 2.42$$

Note that the shortcut formula requires only the sum of the sample measurements, $\sum x$, and the sum of the squares of the sample measurements, $\sum x^2$. Be careful when you calculate these two sums. Rounding the values of x^2 that appear in $\sum x^2$ or rounding the quantity $(\sum x)^2/n$ can lead to substantial errors in the calculation of s^2.

EXAMPLE 2.12

Use a statistical software package to compute the mean, median, variance, and standard deviation of the gas mileage data in Table 2.1.

Solution

The SPSS/PC+™ printout for the median, mean, variance, and standard deviation of the gas mileage data is shown in Figure 2.12.

FIGURE 2.12

SPSS/PC+ printout of descriptive statistics

Mean	36.994	Std Err	.242	Median	37.000
Mode	37.000	Std Dev	2.418	Variance	5.846
Kurtosis	.770	S E Kurt	.478	Skewness	.051
S E Skew	.241	Range	14.900	Minimum	30.000
Maximum	44.900	Sum	3699.400		

Valid Cases	100	Missing Cases	0

Note that the answers are identical to those obtained previously, with the exception that different numbers of decimal places are carried in some instances.

One question occurs over and over again about the calculation of s^2. How many decimal places should you carry? The answer has *nothing* to do with the number of decimal places retained in the sample measurements. If the sample measurements are rounded to the nearest hundredth, tenth, or whatever, rounding simply adds to the variability of the measurements. This added variation is reflected in the value of s^2, a number calculated precisely according to Definition 2.11.

Theoretically, s^2 is a number regardless of how many decimal places it takes to specify its value. In practice, we round the calculated value. There are no rules for the rounding procedure but, keeping in mind that we wish to calculate the standard deviation s, it is reasonable to retain twice as many decimal places in s^2 as you wish to have in s. If you wish to calculate s to the nearest hundredth, for example, you should calculate s^2 to the nearest ten-thousandth.

Generally, in our examples and exercises we round final answers so that one or two more significant digits are displayed in the answers than in the original measurements. For example, we displayed the gas mileage answers in Examples 2.6 and 2.11 to two decimals, even though the data were given to the nearest decimal.

You now know that the standard deviation measures the variability of a set of data and how to calculate it. But how can we interpret and use the standard deviation? This is the topic of Section 2.6.

EXERCISES 2.44–2.57

LEARNING THE MECHANICS

2.44 What is the primary disadvantage of using the range to compare the variability of data sets?

2.45 Describe the sample variance using words rather than a formula. Do the same with the population variance.

2.46 Can the variance of a data set ever be negative? Explain. Can the variance ever be smaller than the standard deviation? Explain.

2.47 Calculate the range, variance, and standard deviation for the following samples:
a. 4, 2, 1, 0, 1
b. 1, 6, 2, 2, 3, 0, 3
c. 8, −2, 1, 3, 5, 4, 4, 1, 3, 3
d. 0, 2, 0, 0, −1, 1, −2, 1, 0, −1, 1, −1, 0, −3, −2, −1, 0, 1

2.48 Calculate the variance and standard deviation for samples where:
a. $n = 10$, $\Sigma x^2 = 84$, $\Sigma x = 20$ b. $n = 40$, $\Sigma x^2 = 380$, $\Sigma x = 100$
c. $n = 20$, $\Sigma x^2 = 18$, $\Sigma x = 17$

2.49 Calculate the range, variance, and standard deviation for the following samples:
a. 39, 42, 40, 37, 41 b. 100, 4, 7, 96, 80, 3, 1, 10, 2 c. 100, 4, 7, 30, 80, 30, 42, 2

2.50 Given the following information about two data sets, compute $\bar{x}$, the sample variance, and the standard deviation for each:

a. $n = 25$, $\displaystyle\sum_{i=1}^{n} x_i^2 = 1{,}000$, $\displaystyle\sum_{i=1}^{n} x_i = 50$

b. $n = 80$, $\displaystyle\sum_{i=1}^{n} x_i^2 = 270$, $\displaystyle\sum_{i=1}^{n} x_i = 100$

2.51 Compute $\bar{x}$, s^2, and s for each of the following data sets. If appropriate, specify the units in which your answer is expressed.
a. 3, 1, 10, 10, 4
b. 8 feet, 10 feet, 32 feet, 5 feet
c. −1, −4, −3, 1, −4, −4
d. $\frac{1}{5}$ ounce, $\frac{1}{5}$ ounce, $\frac{1}{5}$ ounce, $\frac{2}{5}$ ounce, $\frac{1}{5}$ ounce, $\frac{4}{5}$ ounce

2.52 Using only integers between 0 and 10, construct two data sets with at least ten observations each that have the same mean but different variances. Construct dot diagrams for each of your data sets (see Figure 2.11), and mark the mean of each data set on its dot diagram.

2.53   Using only integers between 0 and 10, construct two data sets with at least ten observations each that have the same range but different means. Construct a dot diagram for each of your data sets (see Figure 2.11), and mark the mean of each data set on its dot diagram.

APPLYING THE CONCEPTS

2.54   Consider the following sample of five measurements: 2, 1, 1, 0, 3
  a. Calculate the range, s^2, and s.
  b. Add 3 to each measurement and repeat part **a.**
  c. Subtract 4 from each measurement and repeat part **a.**
  d. Considering your answers to parts **a**, **b**, and **c**, what seems to be the effect on the variability of a data set by adding the same number to or subtracting the same number from each measurement?

2.55   The final grades given by two professors in introductory statistics courses have been carefully examined. The students in the first professor's class had a grade-point average of 3.0 and a standard deviation of .2. Those in the second professor's class had grade points with an average of 3.0 and a standard deviation of 1.0. If you had a choice, which professor would you take for this course? Explain.

2.56   Consider the following two samples:

  Sample 1:   10, 0, 1, 9, 10, 0, 8, 1, 1, 9
  Sample 2:   0, 5, 10, 5, 5, 5, 6, 5, 6, 5

  a. Examine both samples and identify the one that you believe has the greater variability.
  b. Calculate the range for each sample. Does the result agree with your answer to part **a**? Explain.
  c. Calculate the standard deviation for each sample. Does the result agree with your answer to part **a**? Explain.
  d. Which of the two, the range or the standard deviation, provides a better measure of variability? Why?

2.57   The accompanying table lists the yearly steel production (in millions of metric tons) for 1975, 1980, and 1985 for the five leading steel-producing countries in Europe.

	1975	1980	1985
Germany	6.5	7.3	7.9
Italy	21.8	26.5	23.9
France	21.5	23.2	19.1
United Kingdom	20.1	11.3	15.7
Belgium	11.6	12.3	10.8

Source: *Statistical Abstract of the United States, 1989*, p. 832.

  a. Measure the variation of each country's yearly steel production using the range.
  b. Repeat part **a** using the standard deviation as the measure of variation.
  c. Rank the countries in terms of the variablity of their production using the ranges from part **a**, and then rank them according to the standard deviations from part **b**.
  d. Do your two sets of rankings agree? Will the range and standard deviation always yield rankings that agree? Explain.

2.6 Interpreting the Standard Deviation

As we have seen, if we are comparing the variability of two samples selected from a population, the sample with the larger standard deviation is the more variable of the two. Thus, we know how to interpret the standard deviation on a relative or comparative basis, but we have not explained how it provides a measure of variability for a single sample.

To understand how the standard deviation provides a measure of variability of a data set, consider a specific data set and answer the following questions: How many measurements are within 1 standard deviation of the mean? How many measurements are within 2 standard deviations? For example, consider the 100 mileage per gallon readings given in Table 2.1.

Recall that $\bar{x} = 36.99$ and $s = 2.42$. Then

$$\bar{x} - s = 34.57 \qquad \bar{x} - 2s = 32.15$$
$$\bar{x} + s = 39.41 \qquad \bar{x} + 2s = 41.83$$

If we examine the data, we find that 68 of the 100 measurements, or 68% of the measurements, are in the interval

$$\bar{x} - s \quad \text{to} \quad \bar{x} + s$$

Similarly, we find that 96, or 96%, of the 100 measurements are in the interval

$$\bar{x} - 2s \quad \text{to} \quad \bar{x} + 2s$$

These intervals are usually written $(\bar{x} - s, \bar{x} + s)$ and $(\bar{x} - 2s, \bar{x} + 2s)$.

These observations identify criteria for interpreting a standard deviation that apply to any set of data, whether a population or a sample. The criteria, expressed as a mathematical theorem and as a rule of thumb, are presented in Tables 2.8 and 2.9. In these tables we give two sets of answers to the questions of how many measurements fall within 1, 2, and 3 standard deviations of the mean. The first, which applies to *any* set of data, is derived from a theorem proved by the Russian mathematician Chebyshev. The second, which applies only to mound-shaped distributions of data, is based upon empirical evidence that has accumulated over the years. The frequency histogram of a mound-shaped sample is approximately symmetric, with a clustering of measurements about the midpoint of the distribution (the mean, median, and mode should all be about the same), tailing off rapidly as we move away from the center of the histogram. Thus, the histogram

TABLE 2.8 **An Aid to Interpretation of a Standard Deviation: Chebyshev's Rule**

| Chebyshev's Rule

Chebyshev's rule applies to any sample of measurements, regardless of the shape of the frequency distribution:

a. It is possible that very few of the measurements will fall within 1 standard deviation of the mean $(\bar{x} - s, \bar{x} + s)$.

b. At least $\frac{3}{4}$ of the measurements will fall within 2 standard deviations of the mean $(\bar{x} - 2s, \bar{x} + 2s)$.

c. At least $\frac{8}{9}$ of the measurements will fall within 3 standard deviations of the mean $(\bar{x} - 3s, \bar{x} + 3s)$.

d. Generally, at least $1 - 1/k^2$ of the measurements will fall within k standard deviations of the mean $(\bar{x} - ks, \bar{x} + ks)$ for any number k greater than 1.

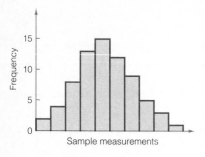

FIGURE 2.13

Histogram of a mound-shaped sample

TABLE 2.9 **An Aid to Interpretation of a Standard Deviation: The Empirical Rule**

The Empirical Rule
The Empirical Rule is a rule of thumb that applies to samples with frequency distributions that are mound-shaped:
a. Approximately 68% of the measurements will fall within 1 standard deviation of the mean $(\bar{x} - s, \bar{x} + s)$.
b. Approximately 95% of the measurements will fall within 2 standard deviations of the mean $(\bar{x} - 2s, \bar{x} + 2s)$.
c. Essentially all the measurements will fall within 3 standard deviations of the mean $(\bar{x} - 3s, \bar{x} + 3s)$.

will have the appearance of a mound or bell, as shown in Figure 2.13. The percentages given for the intervals in Table 2.9 provide remarkably good approximations even when the distribution of the data is slightly skewed or asymmetric.

EXAMPLE 2.13

Here is a sample of earnings per share data for 30 *Fortune* 500 companies:

1.97	.60	4.02	3.20	1.15	6.06
4.44	2.02	3.37	3.65	1.74	2.75
3.81	9.70	8.29	5.63	5.21	4.55
7.60	3.16	3.77	5.36	1.06	1.71
2.47	4.25	1.93	5.15	2.06	1.65

The mean and standard deviation of these data are 3.74 and 2.20, respectively. Calculate the fraction of the 30 measurements in the intervals $\bar{x} \pm s$, $\bar{x} \pm 2s$, and $\bar{x} \pm 3s$, and compare the results with those in Tables 2.8 and 2.9.

Solution

We first form the interval

$$(\bar{x} - s, \bar{x} + s) = (3.74 - 2.20, 3.74 + 2.20) = (1.54, 5.94)$$

A check of the measurements shows that 23 measurements are within this 1 standard deviation interval around the mean. This number represents $\frac{23}{30} \approx 77\%$ of the sample measurements.

The next interval of interest is

$$(\bar{x} - 2s, \bar{x} + 2s) = (3.74 - 4.40, 3.74 + 4.40) = (-.66, 8.14)$$

All but two measurements are within this interval, so $\frac{28}{30}$, or approximately 93%, are within 2 standard deviations of $\bar{x}$.

Finally, the 3 standard deviation interval around $\bar{x}$ is

$$(\bar{x} - 3s, \bar{x} + 3s) = (3.74 - 6.60, 3.74 + 6.60) = (-2.86, 10.34)$$

All the measurements fall within 3 standard deviations of the mean.

These 1, 2, and 3 standard deviation percentages (77, 93, and 100) agree fairly well with the approximations of 68%, 95%, and 100% given by the Empirical Rule (Table 2.9) for mound-shaped distributions. If you look at the frequency histogram for this data set in Figure 2.14, you will note that the distribution is

FIGURE 2.14

Histogram for earnings per share data

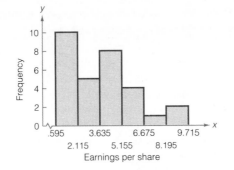

not really mound-shaped, nor is it extremely skewed. Thus, we get reasonably good results from the mound-shaped approximations. Of course, we know from Chebyshev's rule (Table 2.8) that no matter what the shape of the distribution, we would expect at least 75% and 89% of the measurements to lie within 2 and 3 standard deviations of $\bar{x}$, respectively.

EXAMPLE 2.14

Chebyshev's rule and the Empirical Rule (Tables 2.8 and 2.9) are useful as a check on the calculation of the standard deviation. For example, suppose we calculated the standard deviation for the gas mileage data (Table 2.1) to be 5.85. Are there any "clues" in the data that enable us to judge whether this number is reasonable?

Solution

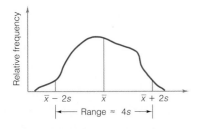

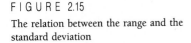

FIGURE 2.15

The relation between the range and the standard deviation

The range of the mileage data is $44.9 - 30.0 = 14.9$. From Chebyshev's rule and the Empirical Rule we know that most of the measurements (approximately 95% if the distribution is mound-shaped) will be within 2 standard deviations of the mean. And, regardless of the shape of the distribution and the number of measurements, almost all of them will fall within 3 standard deviations of the mean. Consequently, we would expect the range of the measurements to be between 4 (i.e., $\pm 2s$) and 6 (i.e., $\pm 3s$) standard deviations in length (see Figure 2.15). For the car mileage data, this means that s should fall between

$$\frac{\text{Range}}{6} = \frac{14.9}{6} = 2.48$$

and

$$\frac{\text{Range}}{4} = \frac{14.9}{4} = 3.73$$

In particular, the standard deviation should not be much larger than one-fourth of the range, particularly for a data set with 100 measurements. Thus, we have reason to believe that the calculation of 5.85 is too large. A check of our work reveals that 5.85 is the variance s^2, not the standard deviation s (see Example 2.11). We "forgot" to take the square root (a common error); the correct value is $s = 2.42$. Note that this value is slightly smaller than the range divided by 6 (2.48). The larger the data set, the greater the tendency for very large or very small measurements (extreme values) to appear, and when they do, the range may exceed 6 standard deviations.

In examples and exercises we will sometimes use $s \approx$ range/4 to obtain a crude, and usually conservatively large, approximation for s. However, we stress that this is no substitute for calculating the exact value of s when possible.

Finally, and most importantly, we will use the concepts in Chebyshev's rule and the Empirical Rule to build the foundation for statistical inference-making. The method is illustrated in Example 2.15.

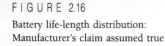

EXAMPLE 2.15

A manufacturer of automobile batteries claims that the average length of life for its grade A battery is 60 months. However, the guarantee on this brand is for only 36 months. Suppose the standard deviation of the life length is known to be 10 months, and the frequency distribution of the life-length data is known to be mound-shaped.

a. Approximately what percentage of the manufacturer's grade A batteries will last more than 50 months, assuming the manufacturer's claim is true?

b. Approximately what percentage of the manufacturer's batteries will last less than 40 months, assuming the manufacturer's claim is true?

c. Suppose your battery lasts 37 months. What could you infer about the manufacturer's claim?

Solution

If the distribution of life length is assumed to be mound-shaped with a mean of 60 months and a standard deviation of 10 months, it would appear as shown in Figure 2.16. Note that we can take advantage of the fact that mound-shaped distributions are (approximately) symmetric about the mean, so that the percentages given by the Empirical Rule can be split equally between the halves of the distribution on each side of the mean. The approximations given in Figure 2.15 are dependent on the mound-shaped assumption because the approximations in Figure 2.16 depend on the (approximate) symmetry of the mound-shaped distribution. We saw in Example 2.13 that the Empirical Rule can yield good approximations even for skewed distributions. This will *not* be true of the approximations in Figure 2.16; the distribution *must* be mound-shaped and (approximately) symmetric.

For example, since approximately 68% of the measurements will fall within 1 standard deviation of the mean, the distribution's symmetry implies that approximately $\frac{1}{2}(68\%) = 34\%$ of the measurements will fall between the mean and 1 standard deviation on each side. This concept is illustrated in Figure 2.16. The figure also shows that 2.5% of the measurements lie beyond 2 standard deviations in each direction from the mean. This result follows from the fact that if approximately 95% of the measurements fall within 2 standard deviations of the mean,

FIGURE 2.16

Battery life-length distribution: Manufacturer's claim assumed true

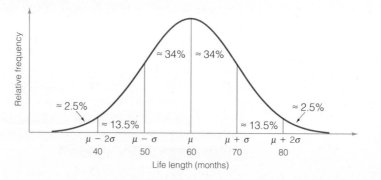

then about 5% fall outside 2 standard deviations; if the distribution is approximately symmetric, then about 2.5% of the measurements fall beyond 2 standard deviations on each side of the mean.

a. It is easy to see in Figure 2.16 that the percentage of batteries lasting more than 50 months is approximately 34% (between 50 and 60 months) plus 50% (greater than 60 months). Thus, approximately 84% of the batteries should have life length exceeding 50 months.

b. The percentage of batteries that last less than 40 months can also be easily determined from Figure 2.16. Approximately 2.5% of the batteries should fail prior to 40 months, assuming the manufacturer's claim is true.

c. If you are so unfortunate that your grade A battery fails at 37 months, one of two inferences can be made. Either your battery was one of the approximately 2.5% that fail prior to 40 months, or something about the manufacturer's claim is not true. Because the chances are so small that a battery fails before 40 months, you would have good reason to have serious doubts about the manufacturer's claim. A mean smaller than 60 months and/or a standard deviation longer than 10 months would both increase the likelihood of failure prior to 40 months.*

Example 2.15 is our initial demonstration of the statistical inference-making process. At this point you should realize that we will use sample information (in Example 2.15, your battery's failure at 37 months) to make inferences about the population (in Example 2.15, the manufacturer's claims about the life length for the population of all batteries). We will build on this foundation as we proceed.

EXERCISES 2.58–2.73

LEARNING THE MECHANICS

2.58   To what kind of data sets can Chebyshev's rule be applied? The Empirical Rule?

2.59   The output from a statistical computer program indicates that the mean and standard deviation of a data set consisting of 200 measurements are $1,500 and $300, respectively.
a. What are the units of measurement of the variable of interest? Based on the units, what type of data is this: nominal, ordinal, interval, or ratio?
b. What can be said about the number of measurements between $900 and $2,100? Between $600 and $2,400? Between $1,200 and $1,800? Between $1,500 and $2,100?

2.60   For any set of data, what can be said about the percentage of the measurements contained in each of the following intervals?
a. $\bar{x} - s$ to $\bar{x} + s$ b. $\bar{x} - 2s$ to $\bar{x} + 2s$ c. $\bar{x} - 3s$ to $\bar{x} + 3s$

2.61   For a set of data with a mound-shaped relative frequency distribution, what can be said about the percentage of the measurements contained in each of the intervals specified in Exercise 2.60?

---

*The assumption that the distribution is mound-shaped and symmetric may also be incorrect. However, if the distribution were skewed to the right, as life-length distributions often tend to be, the percentage of measurements more than 2 standard deviations *below* the mean would be even less than 2.5%.

2.62   The following is a sample of 25 measurements:

7	6	6	11	8	9	11	9	10	8	7	7	5
9	10	7	7	7	7	9	12	10	10	8	6	

a. Compute $\bar{x}$, s^2, and s for this sample.

b. Count the number of measurements in the intervals $\bar{x} \pm s$, $\bar{x} \pm 2s$, $\bar{x} \pm 3s$. Express each count as a percentage of the total number of measurements.

c. Compare the percentages found in part **b** to the percentages given by the Empirical Rule and Chebyshev's rule.

d. Calculate the range, and use it to obtain a rough approximation for s. Does the result compare favorably with the actual value for s found in part **a**?

2.63   Given a data set with a largest value of 760 and a smallest value of 135, what would you estimate the standard deviation to be? Explain the logic behind the procedure you used to estimate the standard deviation. Suppose the standard deviation is reported to be 25. Is this feasible? Explain.

APPLYING THE CONCEPTS

2.64   In Exercise 2.13 we gave the high school dropout rates in 1982 and 1984 for each state and the District of Columbia. The data are reproduced here.

STATE	1982	1984	STATE	1982	1984	STATE	1982	1984	STATE	1982	1984
Ala.	36.6	37.9	Ill.	23.9	25.5	Mont.	21.3	17.9	R.I.	27.3	31.3
Alaska	35.7	25.3	Ind.	28.3	23.0	Neb.	18.1	13.7	S.C.	37.4	35.5
Ariz.	36.6	35.4	Iowa	15.9	14.0	Nev.	35.2	33.5	S.D.	17.3	14.5
Ark.	26.6	24.8	Kansas	19.3	18.3	N.H.	23.0	24.8	Tenn.	32.2	29.5
Calif.	39.9	36.8	Ky.	34.1	31.6	N.J.	23.5	22.3	Texas	36.4	35.4
Colo.	29.1	24.6	La.	38.5	43.3	N.M.	30.6	29.0	Utah	25.0	21.3
Conn.	29.4	20.9	Maine	27.9	22.8	N.Y.	36.6	37.8	Vt.	20.4	16.9
Del.	25.3	28.9	Md.	25.2	22.2	N.C.	32.9	30.7	Va.	26.2	25.3
D.C.	43.1	44.8	Mass.	23.6	25.7	N.D.	16.1	13.7	Wash.	23.9	24.9
Fla.	39.8	37.8	Mich.	28.4	27.8	Ohio	22.5	20.0	W.Va.	33.7	26.9
Ga.	35.0	36.9	Minn.	11.8	10.7	Okla.	29.2	26.9	Wis.	16.9	15.5
Hawaii	25.1	26.8	Miss.	38.7	37.6	Ore.	27.6	26.1	Wyo.	27.6	24.0
Idaho	25.6	24.2	Mo.	25.8	23.8	Pa.	24.0	22.8			

a. Verify that the mean and standard deviation for the 1982 dropout rates are 28.12 and 7.29, respectively, and that the mean and standard deviation for the 1984 dropout rates are 26.50 and 7.95, respectively.

b. Locate the means on the stem and leaf displays (Exercise 2.13) and on the relative frequency histograms (Exercise 2.24). Does the mean appear to be located near the center of the distribution of each data set?

c. Calculate the percentage of measurements in the intervals $\bar{x} \pm s$, $\bar{x} \pm 2s$, and $\bar{x} \pm 3s$ for each data set (1982 and 1984). Check the agreement of these percentages with both Chebyshev's rule and the Empirical Rule.

2.65   Better Homes and Gardens® Real Estate Service conducted a nationwide survey of real estate agents to determine the cost of a 1,600-square-foot home with $1\frac{1}{2}$ or 2 baths. The table contains a portion of the resulting sample data.

CITY	SALE PRICE	CITY	SALE PRICE	CITY	SALE PRICE
Albany, N.Y.	$ 72,500	Des Moines, Iowa	$ 93,900	Owensboro, Ky.	$ 72,500
Allentown, Pa.	82,500	Fargo, N.D.	100,000	Pittsburgh, Pa.	72,500
Amarillo, Tex.	75,000	Greenville, S.C.	76,800	Pleasanton, Calif.	185,000
Baltimore, Md.	100,000	Honolulu, Hawaii	165,000	Raleigh, N.C.	94,000
Baton Rouge, La.	100,000	Jacksonville, Fla.	77,000	Reno, Nev.	93,950
Biloxi, Miss.	78,500	Joplin, Mo.	65,000	Rockford, Ill.	76,000
Bowling Green, Ky.	67,400	Lake Tahoe, Nev.	190,000	San Diego, Calif.	140,000
Cheyenne, Wyo.	82,000	Little Rock, Ark.	85,000	Sheboygan, Wis.	72,000
Cincinnati, Ohio	85,000	Manhattan, Kans.	85,000	South Bend, Ind.	71,300
Columbus, Ga.	69,000	Montgomery, Ala.	65,000	Springfield, Mass.	75,000
Cranston, R.I.	98,500	Nashville, Tenn.	84,000	Trenton, N.J.	100,000
Darien, Conn.	290,000	Olympia, Wash.	75,000	Vancouver, Wash.	68,000

Source: *Consumers Research*, February 1985, p. 19. Courtesy Meredith Corporation, Des Moines, Iowa.

a. In describing this data set, is it more appropriate to apply Chebyshev's rule or the Empirical Rule? Justify your answer.

b. The mean and standard deviation of this data set are $96,732 and $45,322, respectively. Interpret each in the context of the problem.

c. What proportion of the sale prices in the sample would you expect to find within 2 standard deviations of the sample mean? More than 3 standard deviations from the mean? How many sale prices actually fall in these intervals?

d. In terms of numbers of standard deviations, how far does the price of a home in Darien, Connecticut, lie from the mean price of the sample?

2.66 In Exercises 2.14 and 2.26 we used a computer-generated stem and leaf display and relative frequency histogram to describe a sample of 40 radiation measurements (exhalation rates) from gypsum and phosphate waste mounds. These data are reproduced here:

1,709.79	4,132.28	2,996.49	2,796.42	3,750.83	961.40	1,096.43	1,774.77
357.17	1,489.86	2,367.40	11,968.23	178.99	5,402.35	2,315.52	2,617.57
1,150.94	3,017.48	599.84	2,758.84	3,764.96	1,888.22	2,055.20	205.84
1,572.69	393.55	538.37	1,830.78	878.56	6,815.69	752.89	1,977.97
558.33	880.84	2,770.23	1,426.57	1,322.76	1,480.04	9,139.21	1,698.39

The mean and standard deviation of these data are 2,384.842 and 2,379.366, respectively.

a. Locate the approximate positions of $\bar{x} - 2s$, $\bar{x} - s$, $\bar{x}$, $\bar{x} + s$, and $\bar{x} + 2s$ on the computer-generated stem and leaf display in Exercise 2.14a and the computer-generated relative frequency histogram in Exercise 2.26. Use these displays to approximate the percentages of measurements in $\bar{x} \pm s$ and $\bar{x} \pm 2s$.

b. Calculate the exact percentages in the intervals $\bar{x} \pm s$ and $\bar{x} \pm 2s$, and compare them with your approximations in part a.

c. Based on the displays in Exercises 2.14a and 2.26, is the distribution mound-shaped? How do the percentages you obtained in part b agree with the approximations given by the Empirical Rule?

2.67 In Case Study 10.2, we discuss a study designed to investigate the effects of two variables—(1) a student's level of mathematical anxiety and (2) teaching method—on a student's achievement in a course in mathematics. Students who had a low level of mathematical anxiety and were taught using the traditional expository method obtained a mean score of 183.43 and a standard deviation of 52.27. Use the Empirical Rule, along with this information, to give a verbal description of the data set.

2.68   For each day of last year, the number of vehicles passing through a certain intersection was recorded by a city engineer. One objective of this study was to determine the percentage of days that more than 425 vehicles used the intersection. If the mean for the data was 375 vehicles per day and the standard deviation was 25 vehicles:

a. What can be said about the percentage of days that more than 425 vehicles used the intersection? Assume that nothing is known about the shape of the relative frequency distribution for the data.

b. What is your answer to part a if you know that the relative frequency distribution for the data is mound-shaped?

2.69   A buyer for a lumber company must decide whether to buy a piece of land containing 5,000 pine trees. If 1,000 of the trees are at least 40 feet tall, the buyer will purchase the land; otherwise, he will not. The owner of the land reports that the height of the trees has a mean of 30 feet and a standard deviation of 3 feet. Based on this information, what is the buyer's decision?

2.70   A chemical company produces a substance composed of 98% cracked corn particles and 2% zinc phosphide for use in controlling rat populations in sugarcane fields. Production must be carefully controlled to maintain the 2% zinc phosphide because too much zinc phosphide will cause damage to the sugarcane and too little will be ineffective in controlling the rat population. Records from past production indicate that the distribution of the actual percentage of zinc phosphide present in the substance is approximately mound-shaped, with a mean of 2.0% and a standard deviation of .08%. If the production line is operating correctly, approximately what proportion of batches from a day's production will contain less than 1.84% zinc phosphide? Suppose one batch chosen randomly actually contains 1.80% zinc phosphide. Does this indicate that there is too little zinc phosphide in today's production? Explain your reasoning.

2.71   Solar energy is considered by many to be the energy of the future. A recent survey was taken to compare the cost of solar energy to the cost of gas electric energy. Results of the survey revealed that the distribution of the amount of the monthly utility bill of a 3-bedroom house using gas or electric energy had a mean of $125 and a standard deviation of $10.

a. If nothing is known about the distribution of the amounts of monthly utility bills, what can you say about the fraction of all 3-bedroom homes using gas or electric energy having bills between $95 and $155?

b. If it is reasonable to assume that the distribution of the amounts of monthly utility bills is mound-shaped, approximately what proportion of 3-bedroom homes would have monthly bills less than $135?

c. Suppose that three houses with solar energy units had the following monthly utility bills: $101, $98, $104. Does this suggest that solar energy units might result in lower utility bills? Explain. [*Note:* We present a statistical method in Chapter 8 for testing this conjecture.]

2.72   The following data sets have been invented to demonstrate that the lower bounds given by Chebyshev's rule are appropriate. Notice that the data are contrived and would not be encountered in a real-life problem.

a. Consider a data set that contains ten 0's, two 1's, and ten 2's. Calculate $\bar{x}$, s^2, and s. What percentage of the measurements are in the interval $\bar{x} \pm s$? Compare this result to Chebyshev's rule.

b. Consider a data set that contains five 0's, thirty-two 1's, and five 2's. Calculate $\bar{x}$, s^2, and s. What percentage of the measurements are in the interval $\bar{x} \pm 2s$? Compare this result to Chebyshev's rule.

c. Consider a data set that contains three 0's, fifty 1's, and three 2's. Calculate $\bar{x}$, s^2, and s. What percentage of the measurements are in the interval $\bar{x} \pm 3s$? Compare this result to Chebyshev's rule.

d. Draw a histogram for each of the data sets in parts a, b, and c. What do you conclude from these graphs and the answers to parts a, b, and c?

2.73 In Exercise 10.41 (Chapter 10) we present the results of a study to compare the effects of fructose and glucose on the high endurance of women athletes. Six women athletes received 300 milliliters each of a certain drink (say, water and glucose) and then ran until exhausted. Various measurements were then taken on each athlete. The mean and standard deviation of the performance times (in minutes) for the six athletes after receiving the glucose drink were 61.9 and 20.3, respectively. What do the mean and standard deviation tell you about this small data set?

2.7 Measures of Relative Standing

As we have seen, numerical measures of central tendency and variability describe the general nature of a data set (either a sample or a population). We may also be interested in describing the relative quantitative location of a particular measurement within a data set. Descriptive measures of the relationship of a measurement to the rest of the data are called **measures of relative standing**.

One measure of the relative standing of a measurement is its **percentile ranking**. For example, suppose you scored an 80 on an examination and you want to know how you fared in comparison with others in your class. If the instructor tells you that you scored in the 90th percentile, it means that 90% of the examination grades were less than yours and 10% were greater. Thus, if the examination scores were described by the relative frequency histogram in Figure 2.17, the 90th percentile would be located at a point such that 90% of the total area under the relative frequency histogram lies below the 90th percentile and 10% lies above. If the instructor tells you that you scored in the 50th percentile (the median of the data set), 50% of the examination grades would be less than yours and 50% would be greater.

Percentile rankings are of practical value only for large data sets. Finding them involves a process similar to the one used in finding a median. The measurements are ranked in order and a rule is selected, similar to that used in locating a median, to define the location of each percentile. Since we are primarily interested in interpreting the percentile rankings of measurements (rather than finding particular percentiles for a data set), we will define the pth percentile of a data set as shown in the box on page 64.

FIGURE 2.17

Location of 90th percentile for examination grades

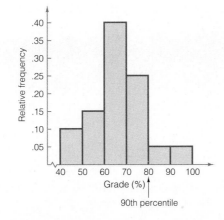

> **Definition 2.13**
>
> For any set of n measurements (arranged in ascending or descending order), the **pth percentile** is a number such that $p\%$ of the measurements fall below the pth percentile and $(100 - p)\%$ fall above it.

Another measure of relative standing in popular use is the **z-score**. As you can see in Definition 2.14, the z-score makes use of the mean and standard deviation of the data set in order to specify the relative location of a measurement:

> **Definition 2.14**
>
> The **sample z-score** for a measurement x is
>
> $$z = \frac{x - \bar{x}}{s}$$
>
> The **population z-score** for a measurement x is
>
> $$z = \frac{x - \mu}{\sigma}$$

Note that the z-score is calculated by subtracting $\bar{x}$ (or μ) from the measurement x and then dividing the result by s (or σ). The final result, the z-score, represents the distance between a given measurement x and the mean, expressed in standard deviations.

EXAMPLE 2.16

Suppose a sample of 2,000 high school seniors' verbal SAT (Scholastic Aptitude Test) scores is selected. The mean and standard deviation are

$$\bar{x} = 550 \qquad s = 75$$

Suppose Joe Smith's score is 475. What is his sample z-score?

Solution

You can see that Joe Smith's score lies below the mean score of the 2,000 seniors:

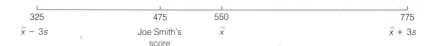

We compute

$$z = \frac{x - \bar{x}}{s} = \frac{475 - 550}{75} = -1.0$$

which tells us that Joe Smith's score is 1.0 standard deviation *below* the sample mean, or in short, his sample z-score is -1.0.

The numerical value of the z-score reflects the relative standing of the measurement. A large positive z-score implies that the measurement is larger than almost all other measurements, whereas a large negative z-score indicates that the measurement is smaller than almost every other measurement. If a z-score is 0 or near 0, the measurement is located at or near the mean of the sample or population.

We can be more specific if we know that the frequency distribution of the measurements is mound-shaped. In this case, the following interpretation of the z-score can be given:

Interpretation of z-Scores for Mound-Shaped Distributions of Data

1. Approximately 68% of the measurements will have a z-score between -1 and 1.
2. Approximately 95% of the measurements will have a z-score between -2 and 2.
3. All or almost all the measurements will have a z-score between -3 and 3.

Note that this interpretation of z-scores is identical to that given by the Empirical Rule for samples from mound-shaped distributions. The statement that a measurement falls in the interval $(\mu - \sigma)$ to $(\mu + \sigma)$ is equivalent to the statement that a measurement has a population z-score between -1 and 1, since all measurements between $(\mu - \sigma)$ and $(\mu + \sigma)$ are within 1 standard deviation of μ.

We end this section with an example and a case study that indicate how z-scores may be used to accomplish our primary objective—the use of sample information to make inferences about the population.

EXAMPLE 2.17

Suppose a female bank employee believes that her salary is low as a result of sex discrimination. To substantiate her belief, she collects information on the salaries of her male counterparts in the banking business. She finds that their salaries have a mean of $34,000 and a standard deviation of $2,000. Her salary is $27,000. Does this information support her claim of sex discrimination?

Solution

The analysis might proceed as follows: First, we calculate the z-score for the woman's salary with respect to those of her male counterparts. Thus,

$$z = \frac{\$27,000 - \$34,000}{\$2,000} = -3.5$$

The implication is that the woman's salary is 3.5 standard deviations *below* the mean of the male salary distribution. Furthermore, if a check of the male salary data shows that the frequency distribution is mound-shaped, we can infer that very few salaries in this distribution should have a z-score less than -3, as shown in Figure 2.18 on page 66. Therefore, a z-score of -3.5 represents either a measurement from a distribution different from the male salary distribution or a very unusual (highly improbable) measurement for the male salary distribution.

FIGURE 2.18

Male salary distribution

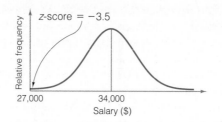

Which of the two situations do you think prevails? Do you think the woman's salary is simply unusually low in the distribution of salaries, or do you think her claim of sex discrimination is justified? Most people would probably conclude that her salary does not come from the male salary distribution. However, the careful investigator should require more information before inferring sex discrimination as the cause. We would want to know more about the data collection technique the woman used and more about her competence at her job. Also, perhaps other factors such as length of employment should be considered in the analysis.

Examples 2.15 and 2.17 exemplify an approach to statistical inference that might be called the **rare event approach.** An experimenter hypothesizes a specific frequency distribution to describe a population of measurements. Then a sample of measurements is drawn from the population. If the experimenter finds it unlikely that the sample came from the hypothesized distribution, the hypothesis is concluded to be false. Thus, in Example 2.17 the woman believes her salary reflects sex discrimination. She hypothesizes that her salary should be just another measurement in the distribution of her male counterparts' salaries if no discrimination exists. However, it is so unlikely that the sample (in this case, her salary) came from the male frequency distribution that she rejects that hypothesis, concluding that the distribution from which her salary was drawn is different from the distribution for the men.

This rare event approach to inference-making is discussed further in later chapters. Proper application of the approach requires a knowledge of probability, the subject of our next chapter.

CASE STUDY 2.3

STATISTICS AND AIR QUALITY STANDARDS

H. E. Neustadter and S. M. Sidik* discuss calculation procedures for obtaining an air quality standard (AQS) to use in judging whether a company is exceeding air pollution standards. A simplification of a method they discuss is outlined here:

1. Obtain a large set of daily measurements from a company known to be complying with established air pollution standards.

*From Neustadter, H. E., and Sidik, S. M. "On evaluating compliance with air pollution levels 'not to be exceeded more than once a year'," *Journal of Air Pollution Control Association*, 1974, Vol. 24, No. 6, 559–563.

2. If necessary, transform the data so that they form a mound-shaped distribution to which the aids in Table 2.9 may be applied. For example, if the logarithm of each measurement is calculated for typical air quality data, the new set of data often forms a mound-shaped distribution.

3. Calculate the mean $\bar{x}$ and standard deviation s of the mound-shaped set of data and use this information to obtain the AQS. For example, if it is desired to obtain a standard that would be exceeded on approximately 2.5% of the industry's operating days, calculate

$$AQS = \bar{x} + 2s$$

The number of standard deviations to be added to $\bar{x}$ can be adjusted to reflect the percentage of days a company is to be permitted to exceed the standard. As Neustadter and Sidik point out, "a major objective of air quality monitoring is often to determine compliance with air quality standards which may in part consist of a 24-hour level not to be exceeded more than once a year." They show that one should calculate the mean plus approximately 2.7 standard deviations to meet this standard.

EXERCISES 2.74–2.85

LEARNING THE MECHANICS

2.74   Compute the z-score corresponding to each of the following values of x:
   a. $x = 40$, $s = 5$, $\bar{x} = 30$ b. $x = 90$, $\mu = 89$, $\sigma = 2$
   c. $\mu = 50$, $\sigma = 5$, $x = 50$ d. $s = 4$, $x = 20$, $\bar{x} = 30$
   e. In parts a–d, state whether the z-score locates x within a sample or a population.
   f. In parts a–d, state whether each value of x lies above or below the mean and by how many standard deviations.

2.75   Give the percentage of measurements in a data set that are above and below each of the following percentiles:
   a. 75th percentile   b. 50th percentile   c. 20th percentile   d. 84th percentile

2.76   What is the 50th percentile of a quantitative data set called? What is another name for the 50th percentile?

2.77   Compare the z-scores to decide which of the following x values lie the greatest distance above the mean and the greatest distance below the mean.
   a. $x = 100$, $\mu = 50$, $\sigma = 25$ b. $x = 1$, $\mu = 4$, $\sigma = 1$
   c. $x = 0$, $\mu = 200$, $\sigma = 100$ d. $x = 10$, $\mu = 5$, $\sigma = 3$

2.78   Consider the following data set:

0	5	6	3	1
2	3	2	1	8
1	7	1	5	4
4	3	1	6	0
5	0	20	5	2

   a. Calculate $\bar{x}$ and s for the data set.
   b. Suppose you were to draw an additional observation, and its value was 100. Calculate the z-score for this observation, and explain why you would or would not classify it as an extreme observation.

2.79   At one university, the students are given z-scores at the end of each semester rather than the traditional GPAs. The mean and standard deviation of all students' cumulative GPAs, on which the z-scores are based, are 2.7 and .5, respectively.

    a. Translate each of the following z-scores to a corresponding GPA: $z = 2.0$, $z = -1.0$, $z = .5$, $z = -2.5$.

    b. Students with z-scores below -1.6 are put on probation. What is the corresponding probationary GPA?

    c. The president of the university wishes to graduate the top 16% of the students with *cum laude* honors and the top 2.5% with *summa cum laude* honors. Where (approximately) should the limits be set in terms of z-scores? In terms of GPAs? What assumption, if any, did you make about the distribution of the GPAs at the university?

2.80   Suppose that 40 and 90 are two elements of a population data set and that their z-scores are -2 and 3, respectively. Using only this information, is it possible to determine the population's mean and standard deviation? If so, find them. If not, explain why it is not possible.

APPLYING THE CONCEPTS

2.81   The distribution of scores on a nationally administered college achievement test has a median of 520 and a mean of 540.

    a. Explain why it is possible for the mean to exceed the median for this distribution of measurements.

    b. Suppose you are told that the 90th percentile is 660. What does this mean?

    c. Suppose you are told that you scored at the 94th percentile. Interpret this statement.

2.82   Many firms use on-the-job training to teach their employees computer programming. Suppose you work in the personnel department of a firm that just finished training a group of its employees to program, and you have been requested to review the performance of one of the trainees on the final test that was given to all trainees. The mean and standard deviation of the test scores are 80 and 5, respectively, and the distribution of scores is mound-shaped.

    a. The employee in question scored 65 on the final test. Compute the employee's z-score.

    b. Approximately what percentage of the trainees will have z-scores equal to or less than the employee of part **a**?

    c. If a trainee were arbitrarily selected from those who had taken the final test, is it more likely that he or she would score 90 or above, or 65 or below?

2.83   The table lists the unemployment rates in 1983 for a sample of nine countries.

COUNTRY	PERCENT UNEMPLOYED
Australia	10.0
Canada	11.9
France	8.8
Germany	7.3
Great Britain	13.4
Italy	5.2
Japan	2.7
Sweden	3.5
United States	9.6

Source: *Statistical Abstract of the United States: 1985*, p. 852.

    a. Calculate the mean and standard deviation of this data set.

    b. Calculate the z-scores of the unemployment rates of the United States, Australia, and Japan.

    c. Describe the information conveyed by the sign (positive or negative) of the z-scores you calculated in part **b**.

2.84 A city librarian claims that books have been checked out an average of 7 (or more) times in the last year. You suspect he has exaggerated the checkout rate (book usage) and that the mean number of checkouts per book per year is, in fact, less than 7. Using the card catalog, you randomly select one book and find that it has been checked out 4 times in the last year. Assume that you know from previous records that the standard deviation of the number of checkouts per book per year is 1.

    a. If the mean number of checkouts per book per year really is 7, what is the z-score corresponding to 4?

    b. Considering your answer to part **a**, is there evidence to indicate that the librarian's claim is incorrect?

    c. If you knew that the distribution of the number of checkouts were mound-shaped, would your answer to part **b** change? Explain.

    d. If the standard deviation of the number of checkouts per book per year were 2 (instead of 1), would your answers to parts **b** and **c** change? Explain.

2.85 Polychlorinated biphenyls (PCBs), considered to be extremely hazardous to humans, are often used in the insulation of large electrical transformers. On March 24, 1984, the *Gainesville Sun* reported on the discovery of a particularly high PCB count at a salvage company in Clay County, Florida. The company, which salvaged the copper in electrical transformers, allowed oil contaminated with PCBs to seep into the soil in and around the salvage site. One soil sample in the vicinity registered 200 parts per million (ppm) of PCBs, four times the safe limit established by the Florida Department of Environmental Regulation.

    Suppose that the PCB count in samples of soil in the vicinity of the salvage operation has a distribution with mean equal to 25 ppm and standard deviation equal to 5 ppm of PCBs. Would a soil sample showing 200 ppm be classified as an extreme observation? Explain.

2.8 Box Plots: Graphic Descriptions Based on Quartiles (Optional)

The **box plot**, a relatively recent introduction to the methodology of descriptive measures, is based on the **quartiles** of a data set. Quartiles are values that partition the data set into four groups, each containing 25% of the measurements. The lower quartile Q_L is the 25th percentile, the middle quartile is the median M (the 50th percentile), and the upper quartile Q_U is the 75th percentile (see Figure 2.19).

> **Definition 2.15**
>
> The **lower quartile Q_L** is the 25th percentile of a data set. The **middle quartile M** is the median. The **upper quartile Q_U** is the 75th percentile.

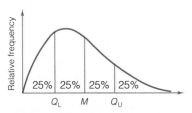

FIGURE 2.19

The quartiles for a data set

A box plot is based on the *interquartile range* (IQR), the distance between the lower and upper quartiles:

$$IQR = Q_U - Q_L$$

> **Definition 2.16**
>
> The **interquartile range (IQR)** is the distance between the lower and upper quartiles:
>
> $$IQR = Q_U - Q_L$$

The box plot for the gas mileage data (Table 2.1) is given in Figure 2.20. It was generated by the Minitab statistical software package for personal computers.* Note that a rectangle (the **box**) is drawn, with the ends of the rectangle (the **hinges**, represented by the "I's" at the ends of the box drawn by the Minitab program) drawn at the quartiles Q_L and Q_U. By definition, then, the "middle" 50% of the observations—those between Q_L and Q_U—fall inside the box. For the gas mileage data, these quartiles appear to be at (approximately) 35.5 and 38.5. Thus,

$$IQR = 38.5 - 35.5 = 3.0 \quad \text{(approximately)}$$

Note that the median is shown at about 37 by a + sign within the box.

FIGURE 2.20

Minitab box plot for gas mileage data

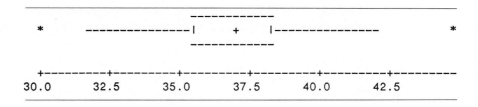

To guide the construction of the "tails" of the box plot, two sets of limits, called **inner fences and outer fences**, are used. Neither set of fences actually appears on the box plot. Inner fences are located at a distance of 1.5(IQR) from the hinges. Emanating from the hinges of the box are dashed lines called the **whiskers**. The two whiskers extend to the most extreme observation inside the inner fences. For example, the inner fence on the lower side of the gas mileage box plot is (approximately):

$$\begin{aligned} \text{Lower inner fence} &= \text{Lower hinge} - 1.5(IQR) \\ &\approx 35.5 - 1.5(3.0) \\ &= 35.5 - 4.5 = 31.0 \end{aligned}$$

The smallest measurement *inside* this fence is the second smallest measurement, 31.8. Thus, the lower whisker extends to 31.8. Similarly, the upper whisker extends to 42.1, the largest measurement inside the upper inner fence at about $38.5 + 4.5 = 43.0$.

---

*Although box plots can be generated by hand, the amount of detail required makes them particularly well suited for computer-generation. We use computer software to generate the box plots in this section.

Values that are beyond the inner fences receive special attention because they are extreme values that represent relatively rare occurrences. In fact, for mound-shaped distributions, fewer than 1% of the observations are expected to fall outside the inner fences. Two of the 100 gasoline mileage measurements, 30.0 and 44.9, fall beyond the inner fences, one on each end of the distribution. These measurements are represented by asterisks (*).

The other pair of imaginary fences, the outer fences, are defined at a distance 3(IQR) from each end of the box. Measurements that fall beyond the outer fences are represented by 0's and are very extreme measurements that require special analysis. Less than one-hundredth of 1% (.01%, or .0001) of the measurements from mound-shaped distributions are expected to fall beyond the outer fences. Since no measurement in the gas mileage box plot (Figure 2.20) is represented by a 0, we know that none of the measurements fall outside the outer fences.

Generally, any measurements that fall beyond the inner fences—and certainly any that fall beyond the outer fences—are considered potential **outliers**. Outliers are extreme measurements that stand out from the rest of the sample and may be faulty—incorrectly recorded observations, members of a different population than the rest of the sample or, at the least, very unusual measurements from the same population. For example, the two gasoline mileage measurements beyond the inner fences may be considered outliers. When we analyze these measurements, we find that they are correctly recorded. Perhaps they represent mileages that correspond to exceptional models of the automobile being tested or to unusual gasoline mixtures. Outlier analysis often reveals useful information of this kind and therefore plays an important role in the statistical inference-making process.

The elements (and nomenclature) of box plots are summarized in the box. Some aids to the interpretation of box plots are also given on page 72.

Elements of a Box Plot

1. A rectangle (the **box**) is drawn with the ends (the **hinges**) drawn at the lower and upper quartiles (Q_L and Q_U). The median of the data is shown in the box, usually by a "+".

2. The points at distances 1.5(IQR) from each hinge mark the **inner fences** of the data set. Horizontal lines (the **whiskers**) are drawn from each hinge to the most extreme measurement inside the inner fence.

3. A second pair of fences, the **outer fences**, exist at a distance of 3 interquartile ranges, 3(IQR), from the hinges. One symbol (usually "*") is used to represent measurements falling between the inner and outer fences, and another (usually "0") is used to represent measurements beyond the outer fences. Thus, outer fences are not shown unless one or more measurements lie beyond them.

4. The symbols used to represent the median and the extreme data points (those beyond the fences) will vary depending on the software you use to construct the box plot. (You may use your own symbols if you are constructing a box plot by hand.) You should consult the program's documentation to determine exactly which symbols are used.

Aids to the Interpretation of Box Plots

1. Examine the length of the box. The IQR is a measure of the sample's variability and is especially useful for the comparison of two samples (see Example 2.19).

2. Visually compare the lengths of the whiskers. If one is clearly longer, the distribution of the data is probably skewed in the direction of the longer whisker.

3. Analyze any measurements that lie beyond the fences. Fewer than 5% should fall beyond the inner fences, even for very skewed distributions. Measurements beyond the outer fences are probably *outliers*, with one of the following explanations:
   a. The measurement is incorrect. It may have been observed, recorded, or entered into the computer incorrectly.
   b. The measurement belongs to a population different from that from which the rest of the sample was drawn (see Example 2.19).
   c. The measurement may be correct and from the same population as the rest but represents a rare event. Generally, we accept this explanation only after carefully ruling out all others.

EXAMPLE 2.18

Solution

Use a statistical software package to draw a box plot for the student loan default data, Table 2.2.

The Minitab box plot for the student loan default rates is shown in Figure 2.21. Note that the median appears to be about 9.5, and, with the exception of a single extreme observation, the distribution appears to be symmetrically distributed between approximately 3% and 17%. The single outlier is beyond the inner fence but inside the outer fence. Examination of the data reveals that this observation corresponds to Alaska's default rate of 19.7%.

FIGURE 2.21

Minitab box plot for student loan default rates

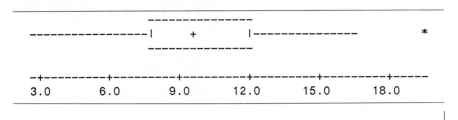

EXAMPLE 2.19

A Ph.D. student in psychology conducted a stimulus reaction experiment as a part of her dissertation research. She subjected 50 subjects to a threatening stimulus and 50 to a nonthreatening stimulus. The reaction times of all 100 students were recorded electronically to the nearest tenth of a second. Box plots of the two resulting samples of reaction times are shown in Figure 2.22. Interpret the box plots.

FIGURE 2.22

SAS box plots for reaction time data

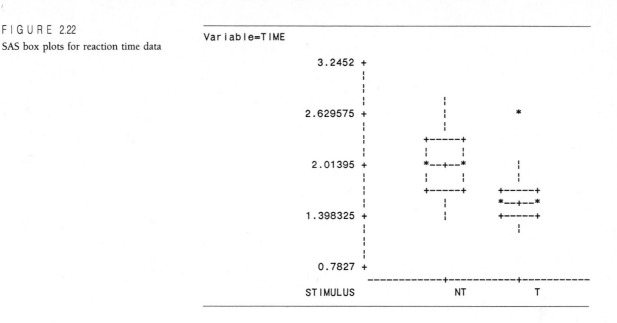

Solution

Perhaps the first thing you notice about the two box plots is that they are arranged vertically rather than horizontally. Some statistical software packages, including the SAS System used here, use this arrangement. Also, note that the median is represented by a dashed line through the box. The plus (+) symbol represents the mean in the SAS box plot. Analysis of the box plots on the same numerical scale reveals that the distribution of times corresponding to the threatening stimulus lies below that of the nonthreatening stimulus. The implication is that the reaction times tend to be faster to the threatening stimulus. Note, too, that the upper whiskers of both samples are longer than the lower whiskers, indicating that the reaction times are skewed to the right. The box length corresponding to the threatening stimulus is smaller than that for the nonthreatening stimulus, indicating less variability in the reaction times to the threatening stimulus.

No observations in the two samples fall between the inner and outer fences (denoted by 0 in SAS). However, note that one of the observations corresponding to the threatening stimulus is beyond the outer fence (denoted by *). When the researcher carefully examined her notes for the experiments, she found that the subject whose time was beyond the outer fence had mistakenly been given the nonthreatening stimulus. You can see in Figure 2.22 that his time would have been within the upper whisker if moved to the box plot corresponding to the nonthreatening stimulus. Of course, the box plots should be reconstructed since they will both change slightly when the misclassified reaction time is moved from one sample to the other.

The researcher concluded that the reactions to the threatening stimulus were faster and more predictable (less variable) than those to the nonthreatening stimulus. However, she was asked by her Ph.D. committee whether the results were *statistically significant*. Their question addresses the issue of whether the observed difference between the samples might be attributable to chance or sampling variation rather than to real differences between the populations. To answer their question, the researcher must use inferential statistics rather than graphic descriptions. We discuss how to compare two samples using inferential statistics in Chapter 9.

EXERCISES 2.86–2.99

[*Note: Starred (*) exercises require the use of a computer.*]

LEARNING THE MECHANICS

2.86   Define the 25th, 50th, and 75th percentiles of a data set. Explain how they provide a description of the data.

2.87   Suppose a data set consisting of exam scores has a lower quartile $Q_L = 60$, a median $M = 75$, and an upper quartile $Q_U = 85$. The scores on the exam ranged from 18 to 100. Without having the actual scores available to you, construct as much of the box plot as possible.

2.88   Minitab was used to generate the following box plot:

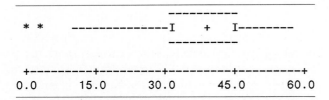

a. What is the median of the data set (approximately)?
b. What are the upper and lower quartiles of the data set (approximately)?
c. What is the interquartile range of the data set (approximately)?
d. Is the data set skewed to the left, skewed to the right, or symmetric?
e. What percentage of the measurements in the data set lie to the right of the median? To the left of the upper quartile?

2.89   Minitab was used to generate the accompanying box plots. Compare and contrast the frequency distributions of the two data sets. Your answer should include comparisons of the following characteristics: central tendency, variation, skewness, and outliers.

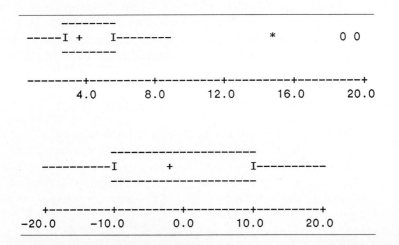

*2.90   Use a statistical software package to construct a box plot for the following set of sample measurements:

1.11	1.24	1.35	1.31	1.55	1.49
1.72	1.66	1.40	2.00	1.24	1.14
1.55	1.26	1.30	1.86	1.25	1.28
1.39	1.65	1.46	1.41	1.31	2.10
1.36	1.33	1.50	1.12	1.38	1.82

*2.91   Consider the following data set:

21	42	30	65	51	35	28	10	34	55
44	49	47	99	33	34	32	33	33	72

a. Use a statistical software package to construct a box plot for the data set.
b. Identify any outliers that may exist in the data set.

*2.92   Consider the following two sample data sets:

SAMPLE A			SAMPLE B		
121	171	158	171	152	170
173	184	163	168	169	171
157	85	145	190	183	185
165	172	196	140	173	206
170	159	172	172	174	169
161	187	100	199	151	180
142	166	171	167	170	188

a. Use a statistical software package to construct a box plot for each data set.
b. Using the information reflected in your box plots, describe the similarities and differences in the two data sets.
c. Identify any outliers that may exist in the two data sets.

APPLYING THE CONCEPTS

2.93   In Exercises 2.14 and 2.26 we constructed a stem and leaf display and a relative frequency histogram for 40 radiation (exhalation rate) measurements on waste gypsum and phosphate mounds in Florida. The accompanying figure, constructed using SAS, is a box plot for the same data.

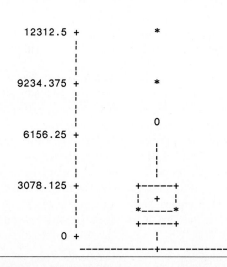

a. Use the box plot to estimate the lower quartile, median, and upper quartile of these data.
b. Does the distribution appear to be skewed? Explain.
c. Are there any outliers among these data? Explain.
d. Examine the graphic displays for these data in Exercise 2.14 (stem and leaf display), Exercise 2.26 (relative frequency histogram), and the box plot here. Which do you prefer as a description of these radiation measurements?

2.94   In 1990 *Fortune* magazine ranked the nation's 500 largest companies by 1989 sales. Owens-Illinois fell at the upper quartile with sales of $3,692 million. Describe Owens-Illinois' location within the sales distribution.

*2.95 A manufacturer of minicomputer systems is interested in improving its customer support services. As a first step, its marketing department has been charged with the responsibility of summarizing the extent of customer problems in terms of system down time. The 40 most recent customers were surveyed to determine the amount of down time (in hours) they had experienced during the previous month. These data are listed in the table.

CUSTOMER NUMBER	DOWN TIME	CUSTOMER NUMBER	DOWN TIME	CUSTOMER NUMBER	DOWN TIME
230	12	244	2	257	18
231	16	245	11	258	28
232	5	246	22	259	19
233	16	247	17	260	34
234	21	248	31	261	26
235	29	249	10	262	17
236	38	250	4	263	11
237	14	251	10	264	64
238	47	252	15	265	19
239	0	253	7	266	18
240	24	254	20	267	24
241	15	255	9	268	49
242	13	256	22	269	50
243	8				

a. Use a statistical software package to construct a box plot for these data. Use the information reflected in the box plot to describe the frequency distribution of the data set. Your description should address central tendency, variation, and skewness.
b. Use your box plot to determine which customers are having unusually lengthy down times.
c. Find and interpret the z-scores associated with customers you identified in part **b**.

2.96   In Exercises 2.13 and 2.24 we used stem and leaf displays and relative frequency histograms to compare the high school dropout rates for the 50 states and the District of Columbia in 1982 and 1984. At the top of page 77 we show box plots, constructed using SAS, for these rates.
a. How do the median dropout rates compare for the two years? [*Hint:* Recall that the SAS program used a dashed line through the box to represent the median.]
b. How do the variabilities of the rates compare for the two years?
c. Refer to Exercise 2.64 (page 60), where the standard deviations of the rates are given as 7.29 and 7.95 for 1982 and 1984, respectively. Do the standard deviations agree with the interquartile ranges (part **b**) with regard to the comparison of the variabilities of the rates?
d. Is there evidence of outliers in either of the distributions?

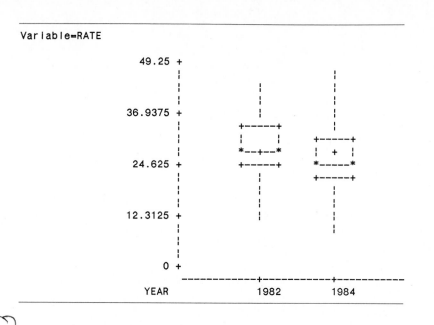

*2.97 In Exercise 2.25 you constructed relative frequency histograms to compare the distributions of population sizes of 30 sunbelt cities in 1960 and 1980. The data are repeated in the table.

CITY	1960	1980	CITY	1960	1980
Albuquerque, NM	201	332	Las Vegas, NV	64	165
Anaheim, CA	104	219	Lubbock, TX	129	174
Arlington, TX	45	160	Miami, FL	292	347
Atlanta, GA	487	425	New Orleans, LA	628	558
Austin, TX	187	345	Oklahoma City, OK	324	403
Charlotte, NC	202	314	Orlando, FL	88	128
Dallas, TX	680	904	Phoenix, AZ	439	790
Fort Lauderdale, FL	84	153	St. Petersburg, FL	181	239
Fort Worth, TX	356	385	San Antonio, TX	588	786
Fresno, CA	134	218	San Diego, CA	573	876
Honolulu, HI	294	365	San Francisco, CA	740	679
Houston, TX	938	1,595	San Jose, CA	204	629
Huntington Beach, CA	11	171	Tampa, FL	275	272
Huntsville, AL	72	143	Tucson, AZ	213	331
Jacksonville, FL	201	541	Tulsa, OK	262	361

Source: U.S. Bureau of the Census, *Statistical Abstract of the United States: 1981* (pp. 21–23), *1985* (pp. 23–25).

a. Use a statistical software package to construct box plots for these two data sets.
b. How do the distributions compare? Your comparison should address the central tendency, the variability, and the skewness of the distributions.
c. Based on these box plots alone, can you comment on the growth of these cities compared to cities not in the sunbelt? Explain.

2.98 The accompanying Minitab-generated box plots describe the U.S. Environmental Protection Agency's 1986 automobile mileage estimates for all models manufactured by Ford and Honda.
a. Which manufacturer has the higher median mileage estimate?
b. Which manufacturer's mileage estimates have the greater range?

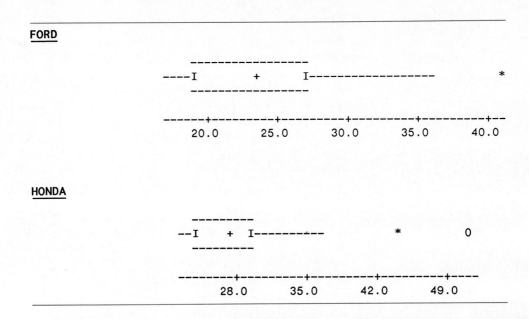

FORD

HONDA

c. Which manufacturer's mileage estimates have the greater interquartile range?
d. Which manufacturer has the model with the highest mileage estimate? Approximately what is that mileage?

*2.99 A hydraulic metal-stamping machine is set to produce disks that are between 15.75 and 16.25 centimeters in diameter. At the start of each work shift, 30 disks are sampled from the production process and measured by an inspector to be sure the machine is operating properly. The measurements shown in the table were obtained at the start of the current work shift.

| DISK DIAMETERS | | | | |
(Centimeters)				
16.22	15.95	15.92	16.05	16.10
15.82	16.15	16.05	15.96	16.02
15.74	16.07	16.13	16.00	16.09
16.01	16.13	16.19	15.84	15.92
16.13	16.25	15.94	16.04	16.07
16.00	15.99	16.24	16.09	16.02

a. Use a statistical software package to construct a box plot for these data. What does your box plot reveal about the current operation of the stamping machine?
b. Does your box plot indicate that the set of 30 measurements contains any outliers? Explain.

2.9 Distorting the Truth With Descriptive Techniques

While it may be true in telling a story that "a picture is worth a thousand words," it is also true that pictures can be used to convey a colored and distorted message to the viewer. So the old adage applies: "Let the buyer (reader) beware." Examine relative frequency histograms and, in general, all graphic descriptions with care.

We will mention a few of the pitfalls to watch for when interpreting a chart or graph. But first we should mention the **time series graph**, which is often the object of distortion. This type of graph records the behavior of some variable of interest recorded over time. Examples of variables commonly graphed as time

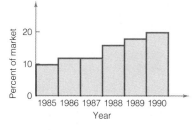

F I G U R E 2.23

Firm A's market share from 1985 to 1990—packed vertical axis

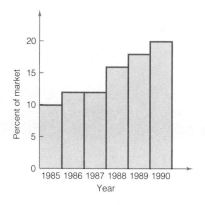

F I G U R E 2.24

Firm A's market share from 1985 to 1990—stretched vertical axis

series abound: economic indices, the United States food surplus, defense spending, presidential popularity index, etc. Since time series graphs often appear in newspapers or magazines, we will use some of them to demonstrate several ways in which pictures are commonly distorted.

One common way to change the impression conveyed by a graph is to change the scale on the vertical axis, the horizontal axis, or both. For example, Figure 2.23 is a **bar graph** that shows the market share of sales for a company for each of the years 1985 to 1990. If you want to show that the change in firm A's market share over time is moderate, you should pack in a large number of units per inch on the vertical axis. That is, make the distance between successive units on the vertical scale small, as shown in Figure 2.23. You can see that a change in the firm's market share over time is barely apparent.

If you want to use the same data to make the changes in firm A's market share appear large, you should increase the distance between successive units on the vertical axis. That is, stretch the vertical axis by graphing only a few units per inch as in Figure 2.24. A telltale sign of stretching is a long vertical axis, but this is often hidden by starting the vertical axis at some point above 0, as shown in Figure 2.25(a). The same effect can be achieved by using a broken line—called a *scale break*—for the vertical axis, as shown in Figure 2.25(b).

Stretching the horizontal axis (increasing the distance between successive units) may also lead you to incorrect conclusions. For example, at the top of page 80, Figure 2.26(a) depicts the change in the Gross National Product (GNP) from the first quarter of 1979 to the last quarter of 1980. If you increase the size of the horizontal axis, as in Figure 2.26(b), the change in the GNP over time seems less pronounced.

The changes in categories indicated by a bar graph can also be emphasized or deemphasized by stretching or shrinking the vertical axis. Another method of achieving visual distortion with bar graphs is by making the width of the bars proportional to their height. For example, look at the bar chart in Figure 2.27(a), page 80, which depicts the percentage of a year's total automobile sales attributable to each of the four major manufacturers. Now suppose we make the width as well as the height grow as the market share grows. This change is shown in Figure 2.27(b). The reader may tend to equate the *area* of the bars with the relative market share of each manufacturer. In fact, the true relative market share is proportional only to the *height* of the bars.

We have presented only a few of the ways that graphs can be used to convey misleading pictures of phenomena. However, the lesson is clear. Examine all

F I G U R E 2.25

Changes in money supply from January to June

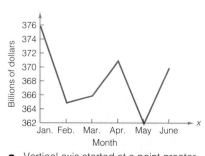

a. Vertical axis started at a point greater than zero

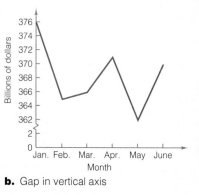

b. Gap in vertical axis

FIGURE 2.26

Gross national product
from 1979 to 1980

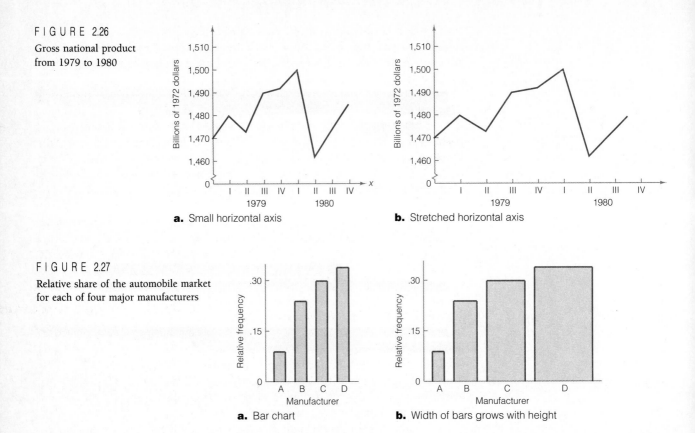

a. Small horizontal axis   **b.** Stretched horizontal axis

FIGURE 2.27

Relative share of the automobile market
for each of four major manufacturers

a. Bar chart   **b.** Width of bars grows with height

graphic descriptions of data with care. Particularly, check the axes and the size of the units on each axis. Ignore the visual changes and concentrate on the actual numerical changes indicated by the graph or chart. The information in a data set can also be distorted by using numerical descriptive measures, as Example 2.20 indicates.

EXAMPLE 2.20

Suppose you are considering working for a small law firm that presently has a senior member and three junior members. You inquire about the salary you could expect to earn if you join the firm. Unfortunately, you receive two answers:

Answer A: The senior member tells you that an "average employee" earns $57,500.

Answer B: One of the junior members later tells you that an "average employee" earns $45,000.

Which answer can you believe? The confusion exists because the phrase "average employee" has not been clearly defined. Suppose the four salaries paid are $45,000 for each of the three junior members and $95,000 for the senior member. Thus,

$$\bar{x} = \frac{3(\$45,000) + \$95,000}{4} = \frac{\$230,000}{4} = \$57,500$$

Median = $45,000

You can now see how the two answers were obtained. The senior member reported the mean of the four salaries, and the junior member reported the median. The information you received was distorted because neither person stated which measure of central tendency was being used.

Another distortion of information in a sample occurs when *only* a measure of central tendency is reported. Both a measure of central tendency and a measure of variability are needed to obtain an accurate mental image of a data set.

Suppose you want to buy a new car and are trying to decide which of two models to purchase. Since energy and economy are both important issues, you decide to purchase model A because its EPA mileage rating is 32 miles per gallon in the city, whereas the mileage rating for model B is only 30 miles per gallon in the city.

However, you may have acted too quickly. How much variability is associated with the ratings? As an extreme example, suppose that further investigation reveals that the standard deviation for model A mileages is 5 miles per gallon, whereas that for model B is only 1 mile per gallon. If the mileages form a mound-shaped distribution, they might appear as shown in Figure 2.28. Note that the larger amount of variability associated with model A implies that more risk is involved in purchasing model A. That is, the particular car you purchase is more likely to have a mileage rating that will greatly differ from the EPA rating of 32 miles per gallon if you purchase model A, while a model B car is not likely to vary from the 30 miles per gallon rating by more than 2 miles per gallon.

FIGURE 2.28

Mileage distributions for two car models

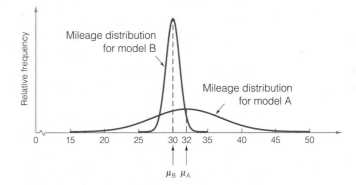

CASE STUDY 2.4

CHILDREN OUT OF SCHOOL IN AMERICA:
MAKING AN UGLY PICTURE LOOK WORSE

David L. Martin* points out another method of distorting the truth with descriptive techniques in his article, "Firsthand report: How flawed statistics can make an ugly picture look even worse." In his critique of the Children Defense Fund's (CDF) 1973 report, *Children Out of School in America*, Martin quotes (italicized comments are Martin's):

*Reprinted from Martin, D. L. "Children out of school: Firsthand report: how flawed statistics can make an ugly picture look even worse," *American School Board Journal*, 1975, 162, 57–59.

25 percent of the 16 and 17 year olds in the Portland, Me., Bayside East Housing Project were out of school. *Only eight children were surveyed; two were found to be out of school.*

Of all the secondary school students who had been suspended more than once in census tract 22 in Columbia, S.C., 33 percent had been suspended two times and 67 percent had been suspended three or more times. *CDF found only three children in that entire census tract who had been suspended; one child was suspended twice and the other two children, three or more times.*

In the Portland Bayside East Housing Project, CDF says that 50 percent of all the secondary school children who had been suspended more than once had been suspended three or more times. *The survey found two secondary school children had been suspended in that area; one of them had been suspended three or more times.*

In each of these examples the reporting of percentages instead of the numbers themselves is misleading. Any inference one might draw from the cited examples would not be reliable. (We will see how to measure the reliability of estimated percentages in Chapter 7.) In short, either the numbers alone should be reported instead of percentages, or, better yet, the report should state that the numbers were too small to report by region. If several regions were combined, the numbers (and percentages) would be more meaningful.

Summary

Data may be classified as one of four types: **nominal**, **ordinal**, **interval**, and **ratio**. Nominal data simply classify the sample or population units, whereas ordinal data enable the units to be ranked. Nominal and ordinal data are often referred to collectively as **qualitative**. Interval data are numbers that enable the comparison of sample or population units by calculation of their numerical differences, whereas ratio data enable the comparison using multiples, or ratios, of the numerical values. Interval and ratio data are often referred to collectively as **quantitative**. Since we want to use sample data to make inferences about the population from which it is drawn, it is important for us to be able to describe the data. **Graphic methods** are important and useful tools for describing data sets. Our ultimate goal, however, is to use the sample to make inferences about the population. We are wary of using graphic techniques to accomplish this goal, since they do not lend themselves to a measure of the reliability for an inference. We therefore developed **numerical measures** to describe a data set.

These numerical methods for describing **quantitative** data sets can be grouped as follows:

1. Measures of central tendency
2. Measures of variability

The measures of central tendency we presented were the **mean**, **median**, and **mode**. The relationship between the mean and median provides information about the **skewness** of the frequency distribution. For making inferences about the population, the sample mean will usually be preferred to the other measures of central tendency. The **range**, **variance**, and **standard deviation** all represent numerical measures of variability. Of these, the variance and standard deviation are in most common use, especially when the ultimate objective is to make inferences about a population.

The mean and standard deviation may be used to make statements about the fraction of measurements in a given interval. For example, we know that at least 75% of the measurements in a data set lie within 2 standard deviations of the mean. If the frequency distribution of the data set is mound-shaped, approximately 95% of the measurements will lie within 2 standard deviations of the mean.

Measures of relative standing provide still another dimension on which to describe a data set. The objective of these measures is to describe the location of a specific measurement relative to the rest of the data set. By doing so, you can construct a mental image of the relative frequency distribution. **Percentiles** and **z-scores** are important examples of measures of relative standing.

The **rare event** concept of statistical inference means that if the chance that a particular sample came from a hypothesized population is very small, we can conclude either that the sample is extremely rare or that the hypothesized population is not the one from which the sample was drawn. The more unlikely it is that the sample came from the hypothesized population, the more strongly we favor the conclusion that the hypothesized population is not the true one. We need to be able to assess accurately the rarity of a sample, and this requires knowledge of probability, the subject of our next chapter.

Finally, we gave some examples that demonstrated how descriptive statistics may be used to distort the truth. You should be very critical when interpreting graphic or numerical descriptions of data sets.

SUPPLEMENTARY EXERCISES 2.100–2.128

[*Note: Starred (*) exercises require the use of a computer. Exercises marked with a dagger (†) refer to an optional section.*]

LEARNING THE MECHANICS

2.100   Classify the following data as one of four types: nominal, ordinal, interval, or ratio.
a. The length of time it takes each of 15 telephone installers to hook up a wall phone
b. The style of music preferred by each of 30 randomly selected radio listeners
c. The arrival time of the 5 P.M. train from New York to Newark
d. A sample of 100 customers in a fast-food restaurant is asked to rate their hamburger on the following scale: poor, fair, good, excellent.
e. Classify each of the data sets in parts **a–d** as qualitative or quantitative.

2.101   Discuss the conditions under which the median is preferred to the mean as a measure of central tendency.

2.102   Consider the following data set:

50	38	56	57	48	64
41	51	52	46	51	50
60	46	41	53	65	59
42	37	50	45	52	42
56	47	63	58	49	55

a. Construct a stem and leaf display for the data. Use it to give a verbal description of the data set.

b. Construct a relative frequency histogram for this set of data.

c. Calculate $\bar{x}$, s^2, and s.

d. Calculate the intervals $\bar{x} \pm s$, $\bar{x} \pm 2s$, and $\bar{x} \pm 3s$. What percentage of the measurements would you expect to fall in each interval?

e. Calculate the actual percentage of the number of measurements falling in each interval. How does the result compare with your answers to part **c**?

*f. Construct a box plot for the data and use it to identify any outliers.

2.103   Construct a relative frequency histogram for the data summarized in the accompanying table.

MEASUREMENT CLASS	RELATIVE FREQUENCY	MEASUREMENT CLASS	RELATIVE FREQUENCY
.00– .75	.02	5.25–6.00	.15
.75–1.50	.01	6.00–6.75	.12
1.50–2.25	.03	6.75–7.50	.09
2.25–3.00	.05	7.50–8.25	.05
3.00–3.75	.10	8.25–9.00	.04
3.75–4.50	.14	9.00–9.75	.01
4.50–5.25	.19		

2.104   If it is not examined carefully, the graphical description of U.S. peanut production shown here can be misleading.

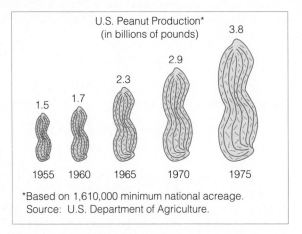

U.S. Peanut Production*
(in billions of pounds)

1.5   1.7   2.3   2.9   3.8

1955   1960   1965   1970   1975

*Based on 1,610,000 minimum national acreage.
Source: U.S. Department of Agriculture.

a. Explain why the graph may mislead some readers.

b. Construct an undistorted graph of U.S. peanut production for the given years.

2.105   Consider the following three measurements: 50, 70, 80. Find the z-score for each measurement if they are from a population with a mean and standard deviation equal to:

a. $\mu = 60$, $\sigma = 10$ **b.** $\mu = 60$, $\sigma = 5$ **c.** $\mu = 40$, $\sigma = 10$ **d.** $\mu = 40$, $\sigma = 100$

2.106   If the range of a set of data is 20, find a rough approximation to the standard deviation of the data set.

2.107   Compute s^2 for data sets with the following characteristics:

a. $\sum_{i=1}^{n} x_i^2 = 246$, $\sum_{i=1}^{n} x_i = 63$, $n = 22$

b. $\sum_{i=1}^{n} x_i^2 = 666, \quad \sum_{i=1}^{n} x_i = 106, \quad n = 25$

c. $\sum_{i=1}^{n} x_i^2 = 76, \quad \sum_{i=1}^{n} x_i = 11, \quad n = 7$

2.108 For each of the following data sets, compute $\bar{x}$, s^2, and s:
   a. 13, 1, 10, 3, 3   **b.** 13, 6, 6, 0   **c.** 1, 0, 1, 10, 11, 11, 15   **d.** 3, 3, 3, 3
   e. For each data set in parts **a–d**, form the interval $\bar{x} \pm 2s$, and calculate the percentage of the measurements that fall in the interval.

2.109 For each of the following data sets, compute $\bar{x}$, s^2, and s. If appropriate, specify the units in which your answers are expressed.
   a. 4, 6, 6, 5, 6, 7
   b. −$1, $4, −$3, $0, −$3, −$6
   c. $\frac{3}{5}$%, $\frac{4}{5}$%, $\frac{2}{5}$%, $\frac{1}{5}$%, $\frac{1}{16}$%
   d. Calculate the range of each data set in parts **a–c**.

2.110 Identify each of the following data sets as nominal, ordinal, interval, or ratio:
   a. The types of metal that are used to make automobiles
   b. The amount of an anesthesia required for each of three surgical patients
   c. The planting distances for certain types of garden vegetables
   d. The nationality of each of 20 immigrants to the United States
   e. Which of the preceding data sets are qualitative, and which are quantitative?

2.111 Explain why we generally prefer the standard deviation to the range as a measure of variability for quantitative data.

APPLYING THE CONCEPTS

2.112 The owner of a service station decided to conduct a survey of service records to determine the length of time (in months) between customer oil changes. A random sample of the station records produced the following times between oil changes:

   6   6   24   8   6   6   6   16   6   12   18   8   4   12   12

   Compute the sample mean, variance, and standard deviation.

2.113 In Exercise 2.112, if the individual who changed the oil in his or her car every 24 months had instead changed it every 18 months, would the sample variance increase or decrease? Why? If instead of every 24 or 18 months, the person had changed the oil every 6 months, how would the resulting sample variance compare to the sample variance when the oil was changed every 24 months? Every 18 months?

2.114 As a result of government pressure, automobile manufacturers in the United States are deeply involved in research to improve their products' gas mileage. One manufacturer, hoping to achieve 30 miles per gallon on one of its full-size models, measured the mileage obtained by 30 test versions of the model with the following results (rounded to the nearest mile for convenience):

30	30	32	31	30	27	28	31	33	28
30	31	29	31	27	33	31	30	35	29
32	27	31	34	30	27	26	29	28	30

   a. Construct a stem and leaf display for the data. Use it to give a verbal description of the data set.
   b. Construct a relative frequency histogram to describe the data set.

c. Compute $\bar{x}$, s^2, and s for the data set.

d. If the manufacturer would be satisfied with a (population) mean of 30 miles per gallon, how would it react to the preceding test data?

e. What percentage of the measurements would you expect to find in the intervals $\bar{x} \pm s$, $\bar{x} \pm 2s$, and $\bar{x} \pm 3s$?

f. Count the number of measurements that actually fall within the intervals of part **e** and express each interval count as a percentage of the total number of measurements. Compare these results with the results of part **e**.

2.115   A radio station claims that the amount of advertising per hour of broadcast time has an average of 3 minutes and a standard deviation equal to 2.1 minutes. You listen to the radio station for 1 hour, at a randomly selected time, and carefully observe that the amount of advertising time is equal to 7 minutes. Does this observation appear to disagree with the radio station's claim? Explain.

2.116   Various state and national automobile associations regularly survey gasoline stations to determine the current retail price of gasoline. Suppose one such national association decides to survey 200 stations in the United States and intends to determine the price of regular unleaded gasoline at each station.

a. Identify the population of interest.

b. Identify the sample.

c. Identify the variable of interest.

d. In the context of this problem, define the following numerical descriptive measures: μ, σ, $\bar{x}$, s.

e. Suppose the sample of 200 stations is selected, and the mean and standard deviation of their regular unleaded prices (per gallon) are $1.39 and $.12, respectively. Interpret these descriptive statistics and describe the probable distribution of the 200 prices at the time of the survey.

f. One station in the southeast priced unleaded gasoline at $1.09 per gallon at the time of the survey. Describe the relative standing of this price in the national price distribution as indicated by the sample.

2.117   A severe drought affected several western states for 3 years. A Christmas tree farmer is worried about the drought's effect on the size of his trees. To decide whether the growth of the trees has been retarded, the farmer decides to take a sample of the heights of 25 trees and obtains the following results (recorded in inches):

60	57	62	69	46
54	64	60	59	58
75	51	49	67	65
44	58	55	48	62
63	73	52	55	50

a. Construct a stem and leaf display for the data. Use it to give a verbal description of the data set. Do any of the measurements appear to be outliers?

b. Construct a relative frequency histogram for these data.

c. Compute $\bar{x}$, s^2, and s for these data.

d. What percentage of the tree heights would you expect to find in the intervals $\bar{x} \pm s$, $\bar{x} \pm 2s$, and $\bar{x} \pm 3s$?

e. Count the number of measurements that actually fall in each interval of part **d** and express each interval count as a percentage of the total number of measurements. Compare these results to your estimates from part **d**.

2.118   The following table shows the average Scholastic Aptitude Test scores for 21 states and the District of Columbia in 1982 and 1985:

STATE	1982	1985	STATE	1982	1985
California	889	904	New Hampshire	925	939
Connecticut	896	915	New Jersey	869	889
Delaware	897	918	New York	896	900
D.C.	821	844	North Carolina	827	833
Florida	889	884	Oregon	908	928
Georgia	823	837	Pennsylvania	885	893
Hawaii	857	877	Rhode Island	877	895
Indiana	860	875	South Carolina	790	815
Maine	890	898	Texas	868	878
Maryland	889	910	Vermont	904	919
Massachusetts	888	906	Virginia	888	908

Source: U.S. Department of Education.

Stem and leaf displays for these two sets of scores are given here, using SPSS/PC+:

```
Stem-and-leaf display .. SAT_1982        Stem-and-leaf display .. SAT_1985

    7 . 9
    8 . 223                              8 . 2344
    8 . 667789999999                     8 . 888899
    9 . 000013                           9 . 000011122234
```

a. Identify the stem and leaf used for each display.
b. Locate the largest and smallest score in each year, and identify the state to which each belongs.
c. Use the displays to calculate the median for each year.
d. Generally, how do the distributions of scores compare?

2.119 Refer to Exercise 2.118. Minitab was used to create the accompanying stem and leaf displays for the same 1982 and 1985 SAT scores. Recall that Minitab shows the cumulative number of observations from the nearest end of the distribution in the left column, and the stem and leaf appear in the next two columns, respectively. The number in the left column in parentheses is the number of measurements corresponding to the stem that contains the median. Minitab also indicates potential outliers by separating them from the display, and labeling them either LO or HI.

```
STEM-AND-LEAF DISPLAY OF SAT_1982        STEM-AND-LEAF DISPLAY OF SAT_1985
  LEAF DIGIT UNIT =   1.0000               LEAF DIGIT UNIT =   1.0000
    1 2 REPRESENTS 12.                       1 2 REPRESENTS 12.

        LO   790,                                LO   815,

    4    82   137
    4    83                              3    83   37
    4    84                              4    84   4
    5    85   7                          4    85
    8    86   089                        4    86
    9    87   7                          7    87   578
  (6)    88   588999                     9    88   49
    7    89   0667                     (3)    89   358
    3    90   48                        10    90   0468
    1    91                              6    91   0589
    1    92   5                          2    92   8
                                         1    93   9
```

a. Identify the stem and leaf used for each display.
b. Locate the largest and smallest score in each year, and identify the state to which each belongs.

c. Use the displays to calculate the median for each year.

d. Generally, how do the distributions of scores compare?

e. The subjectivity involved with the choice of the number of stems to display is evident in Exercise 2.118. The SPSS displays use 4 and 3 stems, respectively, whereas the Minitab displays use 11 stems. Which do you prefer as a description of these data sets?

†2.120 Refer to Exercises 2.118 and 2.119. Box plots, generated by Minitab, for the two years' SAT scores are given here.

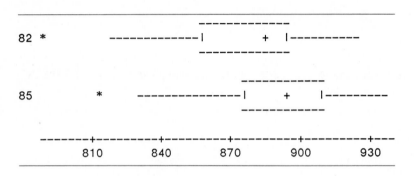

a. Compare the two distributions. Comment on the central tendency, variability, and skewness of the two years' scores.

b. Is there evidence of outliers in either distribution?

c. Compare the box plots to the stem and leaf displays in Exercises 2.118 and 2.119. Which do you prefer?

2.121 A small computing center has found that the number of jobs submitted per day to its computers has a distribution that is approximately mound-shaped, with a mean of 83 jobs and a standard deviation of 10.

a. On approximately what percentage of days will the number of jobs submitted be between 73 and 93?

b. On approximately what percentage of days will the number of jobs submitted be between 63 and 83?

c. On approximately what percentage of days will the number of jobs submitted be greater than 93?

2.122 The Community Attitude Assessment Scale (CAAS) measures citizens' attitudes toward 15 life areas (e.g., education, employment, and health) on four dimensions—importance, influence, equality of opportunity, and satisfaction. In order to develop the CAAS, a number of households in each of 25 communities were randomly selected and sent questionnaires. Because relatively low response rates suggest that there could be a substantial but unknown opinion bias in the reported data, the percentage of the sample responding to the survey was determined in each community. The results are given here (in percent):

21	14	18	20	14
16	6	22	28	16
26	14	13	15	25
21	14	7	12	8
15	14	21	22	10

    a. Construct a stem and leaf dislay for the data. Use it to give a verbal description of the data set.

    b. Construct a relative frequency histogram for the data given, locating the mean, median, and mode.

    c. Find the range for the data and use it to calculate an approximate value for s. Use this value to check your answer to part **d**.

    d. Calculate the variance and standard deviation for the data.

    e. Find the proportion of the measurements that fall in the interval $\bar{x} \pm 2s$.

*2.123   Refer to Exercise 2.112. Use a statistical software package to construct a box plot for the percent responses. Use the box plot to describe the distribution of responses.

2.124   A professor believes that if a class is allowed to work on an examination as long as desired, the times spent by the students would be approximately mound-shaped with mean 40 minutes and standard deviation 6 minutes. Approximately how long should be allotted for the examination if the professor wants almost all (say, 97.5%) of the class to finish?

2.125   By law a box of cereal labeled as containing 16 ounces must contain at least 16 ounces of cereal. It is known that the machine filling the boxes produces a distribution of fill weights that is mound-shaped, with mean equal to the setting on the machine and with a standard deviation equal to .03 ounce. To ensure that most of the boxes contain at least 16 ounces, the machine is set so that the mean fill per box is 16.09 ounces.

    a. What percentage of the boxes will contain less than 16 ounces if the machine is set so that $\mu = 16.09$?

    b. If the machine is set so that $\mu = 16.09$, is it likely that a randomly selected box would contain less than 16 ounces?

    c. If the machine is set so that $\mu = 16.09$, is it likely that a randomly selected box of cereal would contain as little as 16.05 ounces? Explain.

2.126   Most people living in metropolitan areas receive impressions of what is happening in their area primarily through their major newspapers. A study was conducted to determine whether the *Uniform Crime Report*, compiled by the Federal Bureau of Investigation, and the daily newspaper gave consistent information about the trend and distribution of crime in a metropolitan area. An attention score, based on the amount of space devoted to a story, was calculated for each paper's coverage of murders, assaults, robberies, etc. Suppose μ, the average murder attention score of metropolitan newspapers across the country in 1990, was 60, with $\sigma = 4.5$. One metropolitan newspaper in the midwest had a 1990 murder attention score of 69.

    a. Approximately what percentage of the newspapers had a murder attention score higher than 69 in 1990? (Make no assumptions about the nature of the distribution of scores.)

    b. Repeat part **a**, assuming attention scores were mound-shaped.

2.127   The advertising expenditures (in thousands of dollars) by media are described for the 16 top-selling brandies and cordials by the Minitab-generated box plots on page 90.

    a. For which medium are expenditures the highest?

    b. For which medium is the range of expenditures the largest? The smallest?

    c. For which medium is the interquartile range the largest? The smallest?

    d. Describe the shape of each data set's frequency distribution.

    e. Do any of the media receive advertising expenditures from all 16 brands of brandies and cordials? Explain.

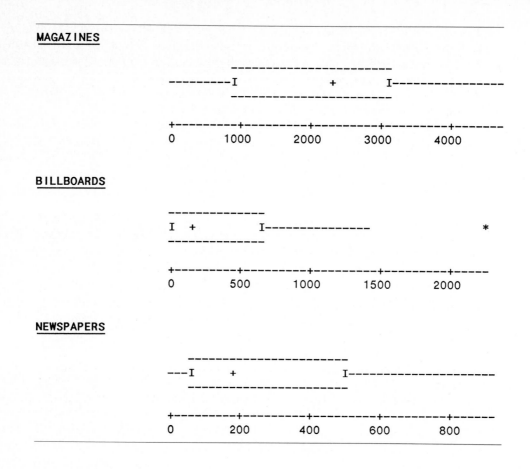

†*2.128 "In most industries, goods are shipped, the product works, and the customer pays. But in high-technology [industries], confusion abounds over when a sale is really a sale. Sometimes, the product is shipped, and a sale is recorded—but the product has bugs, and customers balk at paying" ("High-tech sales: Now you see them, now you don't," *Business Week*, Nov. 18, 1985, p. 106). The table lists a sample of high-technology companies and the number of days (on average) it takes each to collect payment for sales made.

COMPANY	AVERAGE NO. OF DAYS TO COLLECT ON SALE	COMPANY	AVERAGE NO. OF DAYS TO COLLECT ON SALE
CPT	155	Ungermann-Bass	131
Data Switch	143	Radiation Systems	184
C3	231	Porta Systems	141
Scan-Optics	144	Silicon General	140
Aydin	178	Network Systems	181
Computer Entry Systems	137	Masstor Systems	163
UTL	242	Intecom	131
Seagate Technology	143	Microdyne	154
DSC Communications	217	T-Bar	144
Computervision	132	Applied Data Research	126

Data: Standard & Poor's Compustat Services, Inc.

a. Use a statistical software package to construct a box plot for these data.
b. Based on the box plot, what can be said about the shape of the frequency distribution of the data set?
c. Use the box plot to identify outliers in the data set. Which firm(s) have outlying values?
d. The standard deviation of the data set is 34.3. Find the z-score for the outlier(s) you identified in part c.
e. Considering the z-score you obtained in part d, is it likely that the average number of days to collect on a sale for a high-technology firm would be larger than 242? Explain.

ON YOUR OWN...

We list here several sources of real-life data sets that have been obtained from Wasserman and Bernero's *Statistics Sources*. This index of data sources is very complete and is a useful reference for anyone interested in finding almost any type of data. First we list some almanacs:

CBS News Almanac
Information Please Almanac
World Almanac and Book of Facts

United States Government publications are also rich sources of data:

Agricultural Statistics
Digest of Educational Statistics
Handbook of Labor Statistics
Housing and Urban Development Yearbook
Social Indicators
Uniform Crime Reports for the United States
Vital Statistics of the United States
Business Conditions Digest
Economic Indicators
Monthly Labor Review
Survey of Current Business
Bureau of the Census Catalog

Many data sources are published on an annual basis:

Commodity Yearbook
Facts and Figures on Government Finance
Municipal Yearbook
*Standard and Poor's Corporation, Trade and
    Securities: Statistics*

Some sources contain data that are international in scope:

Compendium of Social Statistics
Demographic Yearbook
United Nations Statistical Yearbook
World Handbook of Political and Social Indicators

Utilizing the data sources listed, sources suggested by your instructor, or your own resourcefulness, find one real-life quantitative data set that stems from an area of particular interest to you.

a. Describe the data set by using a relative frequency histogram.
b. Find the mean, median, variance, standard deviation, and range of the data set.
c. Use Tables 2.8 and 2.9 to describe the distribution of this data set. Count the actual number of observations that fall within 1, 2, and 3 standard deviations of the mean of the data set and compare these counts with the description of the data set you developed in part b.

USING THE COMPUTER...

We have described a set of demographic data in Appendix C (available on magnetic tape or diskette from the publisher) that will be used as a source for the "Using the Computer" exercises at the end of most chapters. Briefly, the data consist of 15 demographic variables measured for 1,000 U.S. zip code areas. Included are such variables as population size, number of households, average household size, average income, percentage of college graduates, percentage of women in the work force, and purchasing potential indexes for groceries, sporting goods, and home improvements.

a. Consider the percentage of women in the work force. Use a statistical software package to generate a stem and leaf display, a relative frequency histogram, and a box plot for these 1,000 percentages. Compare the graphical descriptions you obtain, and discuss what each reveals about the distribution of the percentage of women in the work force in the sample of 1,000 zip codes.

b. Repeat part **a** for one of the census regions, perhaps the one in which you currently reside. Compare the regional distribution to the national distribution you obtained in part **a**.

c. Finally, repeat part **a** for the zip codes corresponding to one state, perhaps the one in which you currently reside. Compare the state distribution to the national and regional distributions of parts **a** and **b**.

d. Compute the mean and standard deviation of the percentage data over the 1,000 zip codes and over the zip codes in the region you selected in part **b**. Use the computer to count the number of zip codes' percentages in the intervals $\bar{x} \pm s$, $\bar{x} \pm 2s$, and $\bar{x} \pm 3s$. Compare the results with those given by Chebyshev's theorem and the Empirical Rule (Tables 2.8 and 2.9).

References

Huff, D. *How to Lie with Statistics*. New York: Norton, 1954.

Koopmans, L. H. *An Introduction to Contemporary Statistics*. North Scituate, Mass.: Duxbury, 1981, Chapters 1 and 2.

Mendenhall, W. *Introduction to Probability and Statistics*, 7th ed. North Scituate, Mass.: Duxbury, 1987, Chapter 3.

Wasserman, P., and Bernero, J. *Statistics Sources*, 5th ed. Detroit: Gale Research Company, 1978.

Probability

CONTENTS

WHERE WE'VE BEEN...

We have identified inference, from a sample to a population, as the goal of statistics. To reach this goal, we must be able to describe a set of measurements. The use of graphic and numerical methods for describing data sets and for phrasing inferences was the topic of Chapter 2.

WHERE WE'RE GOING...

Now that we know how to phrase an inference about a population, we turn to the problem of making the inference. What is it that permits us to make the inferential jump from sample to population and then to give a measure of reliability for the inference? As you will subsequently see, the answer is *probability*. This chapter is devoted to a study of probability—what it is and some of the basic concepts of the theory behind it.

You will recall that statistics is concerned with decisions about a population based on sample information. Understanding how this is accomplished will be easier if you understand the relationship between population and sample. This understanding is enhanced by reversing the statistical procedure of making inferences from sample to population. In this chapter we assume the population *known* and calculate the chances of obtaining various samples from the population. Thus, probability is the reverse of statistics: In probability we use the population information to infer the probable nature of the sample.

Probability plays an important role in inference making. To illustrate, suppose you have an opportunity to invest in an oil exploration company. Past records show that out of ten previous oil drillings (a sample of the company's experiences), all ten resulted in dry wells. What do you conclude? Do you think the chances are better than 50–50 that the company will hit a producing well? Should you invest in this company? We think your answer to these questions will be an emphatic no. If the company's exploratory prowess is sufficient to hit a producing well 50% of the time, a record of ten dry wells out of ten drilled is an event that is just too improbable. Do you agree?

As another illustration, suppose you are playing poker with what your opponents assure you is a well-shuffled deck of cards. In three consecutive 5-card hands, the person on your right is dealt 4 aces. Based on this sample of three deals, do you think the cards are being adequately shuffled? Again, we think your answer will be no and that you will reach this conclusion because dealing three hands of 4 aces is just too improbable, assuming that the cards were properly shuffled.

Note that the decisions concerning the potential success of the oil drilling company and the decision concerning the card shuffling were both based on probabilities, namely, the probabilities of certain sample results. Both situations were contrived so that you could easily conclude that the probabilities of the sample results were small. Unfortunately, the probabilities of many observed sample results are not so easy to evaluate intuitively. For these cases we will need the assistance of a theory of probability.

3.1 Events, Sample Spaces, and Probability

We begin our treatment of probability with simple examples that are easily described, thus eliminating any discussion that could be distracting. With the aid of simple examples, important definitions are introduced and the notion of probability is more easily developed.

Suppose a coin is tossed once and the up face is recorded. This is an **observation**, or **measurement**. Any process of making an observation is called an **experiment**. Our definition of experiment is broader than that used in the physical sciences, where you would picture test tubes, microscopes, etc. Other practical examples of statistical experiments are recording whether a customer prefers one of two brands of electronic calculators, recording a voter's opinion on an important political issue, measuring the amount of dissolved oxygen in a polluted river, observing the closing price of a stock, counting the number of errors in an inventory, and observing the fraction of insects killed by a new insecticide. This list of statistical experiments could be continued, but the point is that our definition of experiment is very broad.

> **Definition 3.1**
>
> An **experiment** is an act or process that leads to a single outcome that cannot be predicted with certainty.

Consider another simple experiment consisting of tossing a die and observing the number on the up face. The six basic possible outcomes to this experiment are:

1. Observe a 1   2. Observe a 2
3. Observe a 3   4. Observe a 4
5. Observe a 5   6. Observe a 6

Note that if this experiment is conducted once, *you can observe one and only one of these six basic outcomes, and the outcome cannot be predicted with certainty.* Also, these possibilities cannot be decomposed into more basic outcomes. The basic possible outcomes to an experiment are called **simple events**.

> **Definition 3.2**
>
> A **simple event** is the most basic outcome of an experiment.

EXAMPLE 3.1

Two coins are tossed, and their up faces are recorded. List all the simple events for this experiment.

Solution

Even for a seemingly trivial experiment, we must be careful when listing the simple events. At first glance the basic outcomes seem to be: Observe two heads; Observe two tails; or Observe one head and one tail. However, further reflection reveals that the last of these, Observe one head and one tail, can be decomposed into Head on coin 1, Tail on coin 2 and Tail on coin 1, Head on coin 2.* Thus, the simple events are as follows:

1. Observe *HH*   2. Observe *HT*
3. Observe *TH*   4. Observe *TT*

(where *H* in the first position means "Head on coin 1," *H* in the second position means "Head on coin 2," etc.).

We often wish to refer to the collection of all the simple events of an experiment. This collection is called the *sample space* of the experiment. For example, there are six simple events in the sample space associated with the die-toss experiment. The sample spaces for the experiments discussed thus far are shown in Table 3.1 on page 96.

---

*Even if the coins are identical in appearance, there are, in fact, two distinct coins. Thus, the designation of one coin as coin 1 and the other as coin 2 is legitimate in any case.

T A B L E 3.1   **Experiments and Their Sample Spaces**

Experiment: Observe the up face on a coin.

Sample space: 1. Observe a head
                  2. Observe a tail

This sample space can be represented in set notation as a set containing two simple events:

   S: {*H*, *T*}

where *H* represents the simple event Observe a head and *T* represents the simple event Observe a tail.

Experiment: Observe the up face on a die.

Sample space: 1. Observe a 1
                  2. Observe a 2
                  3. Observe a 3
                  4. Observe a 4
                  5. Observe a 5
                  6. Observe a 6

This sample space can be represented in set notation as a set of six simple events:

   S: {1, 2, 3, 4, 5, 6}

Experiment: Observe the up faces on two coins.

Sample space: 1. Observe *HH*
                  2. Observe *HT*
                  3. Observe *TH*
                  4. Observe *TT*

This sample space can be represented in set notation as a set of four simple events:

   S: {*HH*, *HT*, *TH*, *TT*}

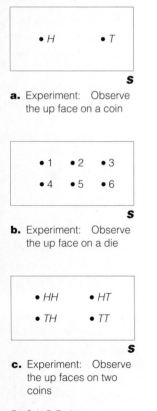

a. Experiment: Observe the up face on a coin

b. Experiment: Observe the up face on a die

c. Experiment: Observe the up faces on two coins

FIGURE 3.1

Venn diagrams for the three experiments from table 3.1

| Definition 3.3 |

The **sample space** of an experiment is the collection of all its simple events.

Just as graphs are useful in describing sets of data, a pictorial method for presenting the sample space and its simple events will often be useful. Figure 3.1 shows such a representation for each of the experiments in Table 3.1. In each case, the sample space is shown as a closed figure, labeled *S*, containing a set of points, called **sample points**, with each point representing one simple event. Note that the number of sample points in a sample space *S* is equal to the number of simple events associated with the respective experiment: two for the coin toss, six for the die toss, and four for the two-coin toss. These graphic representations are called **Venn diagrams**.

Now that we have defined simple events as the basic outcomes of the experiment and the sample space as the collection of all the simple events, we are prepared to discuss the probabilities of simple events. You have undoubtedly used the term *probability* and have some intuitive idea about its meaning. Probability is generally used synonymously with "chance," "odds," and similar concepts. We will begin our treatment of probability using these informal concepts and then solidify what we mean later. For example, if a fair coin is tossed, we might reason that both the simple events, Observe a head and Observe a tail, have the same chance of occurring. Thus, we might state that "the probability of observing a head is 50%" or "the odds of seeing a head are 50–50." Both these statements are based on an informal knowledge of probability.

The probability of a simple event is a number between 0 and 1 that measures the likelihood that the event will occur when the experiment is performed. This number is usually taken to be the relative frequency of the occurrence of a simple event in a very long series of repetitions of an experiment. When this information is not available, we select the number based on experience. For example, if we are assigning probabilities to the two simple events in the coin-toss experiment (Observe a head and Observe a tail), we might reason that if we toss a balanced coin a very large number of times, the simple events Observe a head and Observe a tail will occur with the same relative frequency of .5. Thus, the probability of each simple event is .5.

For some experiments we may assign probabilities to the simple events based on general information about the experiment. For example, if the experiment is to invest in a business venture and to observe whether it succeeds or fails, the sample space would appear as in Figure 3.2. We are unlikely to be able to assign probabilities to the simple events of this experiment based on a long series of repetitions since unique factors govern each performance of this kind of experiment. Instead, we may consider factors such as the personnel managing the venture, the general state of the economy at the time, the rate of success of similar ventures, and any other information deemed pertinent. If we finally decide that the venture has an 80% chance of succeeding, we assign a probability of .8 to the simple event Success. This probability can be interpreted as a measure of our degree of belief in the outcome of the business venture. Such subjective probabilities should be based on expert information and must be carefully assessed. If they are not, we may be misled on any decisions based on these probabilities or based on any calculations in which they appear.*

FIGURE 3.2

Experiment: Invest in a business venture and observe whether it succeeds (S) or fails (F)

CASE STUDY 3.1

BLOOM COUNTY PROBABILITIES

The issue of whether probability should be defined as the relative frequency in a long series of repetitions of an experiment or as a subjective measure of belief is one that has been debated for many years by probabilists, statisticians, and even philosophers. A considerably lighter side of this debate was illustrated in a Bloom County comic strip (see page 98).

No matter how you assign the probabilities to simple events, the probabilities assigned must obey two rules:

1. All simple event probabilities *must* lie between 0 and 1.
2. The probabilities of all the simple events within a sample space *must* sum to 1.

*For a text that deals in detail with the subjective evaluation of probabilities, see Winkler (1972) or Lindley (1985).

Assigning probabilities to simple events is easy for some experiments. For example, if the experiment is to toss a fair coin and observe the up face, we would probably all agree to assign a probability of $\frac{1}{2}$ to the two simple events, Observe a head and Observe a tail. However, many experiments have simple events whose probabilities are more difficult to assign.

E X A M P L E 3.2

A retail computer store owner sells two basic types of microcomputers: IBM personal computers (IBM PCs) and IBM compatibles (PCs that run all or most of the same software as an IBM PC but that are not manufactured by IBM). One problem facing the owner is deciding how many of each type of PC to stock. An important factor affecting the solution is the proportion of customers who purchase each type of PC. Show how this problem might be formulated in the framework of an experiment with simple events and a sample space. Indicate how probabilities might be assigned to the simple events.

Solution

If we use the term *customer* to refer to a person who purchases one of the two types of PCs, the experiment can be defined as the entrance of a customer and the observation of which type of PC is purchased. There are two simple events in the sample space corresponding to this experiment:

1. I: {The customer purchases an IBM PC}
2. C: {The customer purchases a compatible}

The difference between this and the coin-toss experiment becomes apparent when we attempt to assign probabilities to the two simple events. What probability should we assign to the simple event I? If you answer .5, you are assuming that the events I and C should occur with equal likelihood, just as the simple events Heads and Tails in the coin-toss experiment. The assignment of simple event probabilities for the PC purchase experiment is not so easy. Suppose a check of the store's records indicates that 80% of its customers purchase IBM PCs. Then it might be reasonable to approximate the probability of the simple event I as .8 and that of the simple event C as .2. The important points are that simple events are not always equally likely and that the probabilities of simple events are not always easy to assign, particularly for experiments that represent real applications (as opposed to coin- and die-toss experiments).

Although the probabilities of simple events are often of interest in their own right, it is usually probabilities of collections of simple events that are important. Example 3.3 demonstrates this point.

EXAMPLE 3.3

A fair die is tossed, and the up face is observed. If the face is even, you win $1. Otherwise, you lose $1. What is the probability that you win?

Solution

Recall that the sample space for this experiment contains six simple events:

S: {1, 2, 3, 4, 5, 6}

Since the die is balanced, we assign a probability of $\frac{1}{6}$ to each of the simple events in this sample space. An even number will occur if one of the simple events, Observe a 2, Observe a 4, or Observe a 6, occurs. A collection of simple events such as this is called an *event*, and we denote this event by the letter A. Since the event A contains three simple events—all with probability $\frac{1}{6}$—and since no simple events can occur simultaneously, we reason that the probability of A is the sum of the probabilities of the simple events in A. Thus, the probability of A is $\frac{1}{6} + \frac{1}{6} + \frac{1}{6} = \frac{1}{2}$. This implies that *in the long run* you will win $1 half the time and lose $1 half the time.

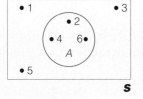

FIGURE 3.3

Die-toss experiment with event A: Observe an even number

Figure 3.3 is a Venn diagram depicting the sample space associated with a die-toss experiment and the event A, Observe an even number. The event A is represented by the closed figure inside the sample space S. This closed figure A contains all the simple events that comprise it.

How do you decide which simple events belong to the set associated with an event A? Test each simple event in the sample space S. If event A occurs when a particular simple event occurs, then that simple event is in the event A. For example, the event A, Observe an even number, in the die-toss experiment will occur if the simple event Observe a 2 occurs. By the same reasoning, the simple events Observe a 4 and Observe a 6 are also in event A.

To summarize, we have demonstrated that an event can be defined in words or it can be defined as a specific set of simple events. This leads us to the following general definition of an event:

> **Definition 3.4**
>
> An **event** is a specific collection of simple events.

EXAMPLE 3.4

SIMPLE EVENT	PROBABILITY
HH	$\frac{4}{9}$
HT	$\frac{2}{9}$
TH	$\frac{2}{9}$
TT	$\frac{1}{9}$

Consider the experiment of tossing two coins. Suppose the coins are *not* balanced and the correct probabilities associated with the simple events are given in the table. [*Note:* The necessary properties for assigning probabilities to simple events are satisfied.]

Consider the events

A: {Observe exactly one head}

B: {Observe at least one head}

Calculate the probability of A and the probability of B.

Event A contains the simple events HT and TH. Since two or more simple events cannot occur at the same time, we can easily calculate the probability of event A by summing the probabilities of the two simple events. Thus, the probability of observing exactly one head (event A), denoted by the symbol $P(A)$, is

$$P(A) = P(\text{Observe } HT) + P(\text{Observe } TH)$$

$$= \frac{2}{9} + \frac{2}{9} = \frac{4}{9}$$

Similarly, since B contains the simple events HH, HT, and TH,

$$P(B) = \frac{4}{9} + \frac{2}{9} + \frac{2}{9} = \frac{8}{9}$$

The preceding example leads us to a general procedure for finding the probability of an event A:

> The probability of an event A is calculated by summing the probabilities of the simple events in A.

Thus, we can summarize the steps for calculating the probability of any event, as indicated in the next box.

> **Steps For Calculating Probabilities of Events**
>
> 1. Define the experiment, i.e., describe the process used to make an observation and the type of observation that will be recorded.
> 2. List the simple events.
> 3. Assign probabilities to the simple events.
> 4. Determine the collection of simple events contained in the event of interest.
> 5. Sum the simple event probabilities to get the event probability.

EXAMPLE 3.5

In a poll of "computer-familiar" adults who do not own a home computer, each was asked to identify which of ten electronic appliances was his or her highest priority purchase, if any. The results are summarized in Table 3.2.

a. Define the experiment that generated the data in Table 3.2, and list the simple events.
b. Assign probabilities to the simple events.
c. What is the probability that a telephonic appliance is of highest priority?
d. What is the probability that a video appliance is of highest priority?

TABLE 3.2 **Electronic Appliances of Highest Priority to Purchase by Computer-Familiar Adults**

ELECTRONIC APPLIANCE	PERCENT RESPONSE[a]
Home computer (HC)	24
Microwave oven (MO)	13
Compact disc player (CDP)	4
Phone-answering machine (PAM)	6
Car telephone (CT)	5
Programmable phone (PP)	3
Video cassette recorder (VCR)	17
Video camera (VC)	9
Big screen TV (BSTV)	7
Movie camera (MC)	3
None (N)	9

[a]Response percentages in the *USA Today* article did not add to 100% due to rounding. We have added 1% to the two smallest responses to facilitate the solution to this example.
Source: "Buying a computer is in our budget," *USA Today*, Sept. 25, 1985. Copyright 1985, USA Today. Excerpted with permission.

Solution

a. The experiment is the act of polling a computer-familiar adult. The simple events, the simplest outcomes of the experiment, are the 11 response categories listed in Table 3.2. They are shown in the Venn diagram in Figure 3.4.

FIGURE 3.4

Venn diagram for the electronic appliance poll

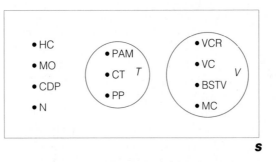

S

b. In Example 3.1 the simple events were assigned equal probabilities. If we were to assign equal probabilities in this case, each of the response categories would be assigned a probability of one-eleventh $\left(\frac{1}{11}\right)$, or .09. However, you can see by examining Table 3.2 that equal probabilities are not reasonable in this case because the response percentages were not the same in the 11 classifications. Instead, we assign a probability equal to the actual response percentage in each class. These are shown in Table 3.3.

c. The event T that a telephonic appliance is the highest priority is not a simple event because it consists of more than one of the response classifications (the simple events). In fact, as shown in Figure 3.4, T consists of three simple events. The probability of T is defined to be the sum of the probabilities of the simple events in T:

$$P(T) = P(PAM) + P(CT) + P(PP)$$
$$= .06 + .05 + .03 = .14$$

TABLE 3.3 **Simple Event Probabilities for Electronic Appliance Poll**

SIMPLE EVENT	PROBABILITY
HC	.24
MO	.13
CDP	.04
PAM	.06
CT	.05
PP	.03
VCR	.17
VC	.09
BSTV	.07
MC	.03
N	.09

d. The event *V* that a video appliance is identified as the highest-priority purchase consists of four simple events, and the probability is the sum of the corresponding simple event probabilities:

$$P(V) = P(VCR) + P(VC) + P(BSTV) + P(MC)$$
$$= .17 + .09 + .07 + .03 = .36$$

For the experiments discussed thus far, listing the simple events has been easy. For more complex experiments, the number of simple events may be so large that listing them is impractical. In solving probability problems for experiments with many simple events we employ the same principles as for experiments with few simple events. The only difference is that we need **counting rules** for determining the number of simple events without actually enumerating all of them. In Section 3.8 (an optional section), we present several of the more useful counting rules.

CASE STUDY 3.2

COMPARING SUBJECTIVE PROBABILITY ASSESSMENTS WITH RELATIVE FREQUENCIES

Preston and Baratta* performed an experiment with the objective of comparing how an individual's subjective assessment of the probability of an event compares with the known probability (relative frequency of occurrence) of the event. The individuals selected for the experiment ranged from undergraduates with no training in probability theory to professors of mathematics and statistics with a "substantial acquaintance with probability theory." Each individual participated in a game in which he or she bet part of an initial stake on one of seven different outcomes of a combination card–dice game. The probabilities of these seven different events, known only to the experimenters, were .01, .05, .25, .50, .75, .95, and .99. From the amount the subject is willing to bet, the individual's subjective probabilities can be determined and then compared with the actual probabilities of the events.

In Figure 3.5 we reproduce the author's figure depicting the average subjective probability assessed by the experimental subjects compared to the true probabilities. Some of the conclusions reached were:

1. Events with probabilities less than .25 were subjectively overestimated.
2. Events with probabilities more than .25 were subjectively underestimated.
3. These conclusions were the same for both the probabilistically naive and sophisticated subjects.

For events with probabilities that are not clearly defined (such as the probability of rain tomorrow), it is important to have information on general

*Reprinted from Preston, M. G. and Baratta, P. "An experimental study of the auction-value of an uncertain outcome," *American Journal of Psychology*, 1948, 61, 183–193.

FIGURE 3.5

FIGURE 3.5

Observed relationship between true and subjective probabilities

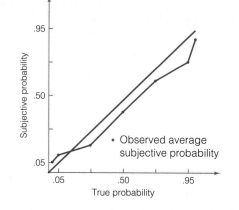

tendencies in subjectively evaluating the probabilities. Of course, much more evidence has been collected on the subjective evaluation of probabilities since the Preston and Baratta article (not all of which corroborate their conclusions), but it remains an interesting evaluation of a person's ability to evaluate probabilities subjectively.

EXERCISES 3.1–3.23

LEARNING THE MECHANICS

3.1 An experiment results in one of the following simple events: E_1, E_2, E_3, E_4, or E_5.
  a. Find $P(E_3)$ if $P(E_1) = .1$, $P(E_2) = .3$, $P(E_4) = .1$, and $P(E_5) = .1$.
  b. Find $P(E_3)$ if $P(E_1) = P(E_3)$, $P(E_2) = .1$, $P(E_4) = .2$, and $P(E_5) = .2$.
  c. Find $P(E_3)$ if $P(E_1) = P(E_2) = P(E_4) = P(E_5) = .1$.

3.2 The diagram describes the sample space of a particular experiment and events A and B.

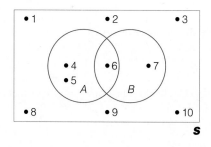

  a. What is this type of diagram called?
  b. Suppose the simple events are equally likely. Find $P(A)$ and $P(B)$.
  c. Suppose $P(1) = P(2) = P(3) = P(4) = P(5) = \frac{1}{20}$ and $P(6) = P(7) = P(8) = P(9) = P(10) = \frac{3}{20}$. Find $P(A)$ and $P(B)$.

3.3   The sample space for an experiment contains five simple events with probabilities as shown in the table. Find the probability of each of the following events:

SIMPLE EVENTS	PROBABILITIES
1	.05
2	.25
3	.30
4	.25
5	.15

A: {Either 1, 2, or 3 occurs}

B: {Either 1, 3, or 5 occurs}

C: {4 does not occur}

3.4   A nickel, dime, and quarter are tossed, and the up faces are noted after each toss.
a. List the simple events in the sample space for this experiment.
b. Assign reasonable probabilities to the simple events.
c. Find the probability of each of the following events:

A: {At least one head appears}

B: {Exactly one head appears}

C: {The first toss is a head}

3.5   Repeat the experiment in Exercise 3.4 a total of 200 times and record the numbers of times events A, B, and C occur. Compare these observed relative frequencies with the probabilities calculated in Exercise 3.4. If you were to repeat the experiment many millions of times, how would the observed relative frequencies differ from those calculated from the 200 repetitions?

3.6   Consider the experiment of tossing a die and observing the up face.
a. Draw a Venn diagram for the experiment. On your diagram indicate the event Observe a number greater than 4. Call this event A. Also indicate the event Observe an even number. Call this event B.
b. We would all agree that the probability of observing a 3 on the toss of a "fair" die is $\frac{1}{6}$. Explain what it means for a die to be fair (or balanced) and explain how knowing that a die is fair leads us to $P(3) = \frac{1}{6}$.
c. For your Venn diagram of part a, assume the die is fair and find $P(A)$ and $P(B)$.
d. If you knew that a particular die were unfair (i.e., "loaded"), how would you determine the probability of observing a 3?

3.7   The Venn diagram depicts an experiment with six simple events. The events A and B are also shown.

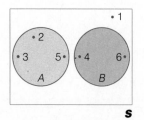

S

The probabilities of the simple events are as follows:

$$P(1) = P(2) = P(4) = \frac{2}{9} \qquad P(3) = P(5) = P(6) = \frac{1}{9}$$

a. Find $P(A)$. **b.** Find $P(B)$.
c. Find the probability that the events A and B occur *simultaneously*.

3.8 Two fair dice are tossed, and the up face on each die is recorded.
a. List the 36 simple events contained in the sample space.
b. Find the probability of observing each of the following events:

A: {A 3 appears on each of the two dice}

B: {The sum of the numbers is even}

C: {The sum of the numbers is equal to 7}

D: {A 5 appears on at least one of the dice}

E: {The sum of the numbers is 10 or more}

3.9 Consider the experiment composed of one roll of a fair die followed by one toss of a fair coin. List the simple events. Assign a logical probability to each simple event. Determine the probability of observing each of the following events:

A: {6 on the die; H on the coin}

B: {Even number on the die; T on the coin}

C: {Even number on the die}

D: {T on the coin}

3.10 Two marbles are randomly drawn from a box containing two blue marbles and three red marbles. Determine the probability of observing each of the following events:

A: {Two blue marbles are drawn}

B: {A red and a blue marble are drawn}

C: {Two red marbles are drawn}

3.11 Simulate the experiment described in Exercise 3.10 using any five identically shaped objects, two of which are one color and three are another. Mix the objects, randomly draw two, record the results, and then replace the objects. Repeat the experiment a large number of times (at least 100). Calculate the proportion of time events A, B, and C occur. How do these proportions compare with the probabilities you calculated in Exercise 3.10? Should these proportions equal the probabilities? Explain.

APPLYING THE CONCEPTS

3.12 A hospital reports that two patients have been admitted who have contracted Legionnaire's disease. Suppose our experiment consists of observing whether the patients survive or die as a result of the disease. The simple events and probabilities of their occurrence are shown in the table (where S in

SIMPLE EVENTS	PROBABILITIES
SS	.81
SD	.09
DS	.09
DD	.01

the first position means that patient 1 survives, *D* in the first position means that patient 1 dies, etc.). Find the probabilities of each of the following events:

A: {Both patients survive the disease}

B: {At least one patient dies}

C: {Exactly one patient survives the disease}

3.13 In its "Intelligence Report," *Parade Magazine* (October 23, 1983) comments on a rare occurrence: the birth of twins, one white and one black, to an English couple. *Parade Magazine* notes that the odds that a pregnant woman will give birth to twins are 1 in 80. The probability that one of the twins is white and the other black is much smaller, even if you are given the fact that the children are the product of a mixed marriage.

Suppose a child born of a mixed marriage has a 50–50 chance of being white or black and the outcome for one of a pair of twins is independent of the outcome for the other. What is the probability that a woman in a mixed marriage who is pregnant with twins will:

a. Have twins, both black?   **b.** Have twins, one white and one black?

c. Have twins, both white?

3.14 The corporations in the highly competitive razor blade industry do a tremendous amount of advertising each year. Corporation G gave a supply of the three top name brands, G, S, and W, to a consumer and asked him to use them and rank them in order of preference. The corporation was, of course, hoping the consumer would prefer its brand and rank it first, thereby giving them some material for a consumer interview advertising campaign. If the consumer did not prefer one blade over any other, but was still required to rank the blades, what is the probability that:

a. The consumer ranked brand G first?

b. The consumer ranked brand G last?

c. The consumer ranked brand G last and brand W second?

d. The consumer ranked brand W first, brand G second, and brand S third?

3.15 Approximately 77 million Visa and 60 million Mastercard credit cards had been issued in the United States by the end of 1984. Both are issued by thousands of banks, including Citicorp. Citicorp also issued competing cards of its own. Diners Club, Carte Blanche, and Choice are wholly owned by Citicorp. The table describes the population of credit cards issued by Citicorp.

CREDIT CARD	NUMBER ISSUED (in millions)
Visa and Mastercard	6.0
Diners Club	2.2
Carte Blanche	.3
Choice	1.0

Source: *Fortune*, February 4, 1985, p. 21. © 1985 Time, Inc. All rights reserved.

One Citicorp credit card customer is to be selected at random and the type of credit card recorded. (Assume each customer has only one Citicorp-issued card.)

a. List the simple events in this experiment.

b. Find the probability of each simple event.

c. What is the probability that the customer selected uses one of Citicorp's own credit cards?

3.16 An individual's genetic makeup is determined by the genes obtained from each parent. For every genetic trait, each parent possesses a gene pair; and each contributes one-half of this gene pair, with equal probability, to their offspring, forming a new gene pair. The offspring's traits (eye color, baldness, etc.) come from this new gene pair, where each gene in this pair possesses some characteristic.

For the gene pair that determines eye color, each gene trait may be one of two types: dominant brown (*B*) or recessive blue (*b*). A person possessing the gene pair *BB* or *Bb* has brown eyes, whereas the gene pair *bb* produces blue eyes.

  a. Suppose both parents of an individual are brown-eyed, each with a gene pair of the type *Bb*. What is the probability that a randomly selected child of this couple will have blue eyes? [*Hint:* Construct the sample space for the experiment.]

  b. If one parent has brown eyes, type *Bb*, and the other has blue eyes, what is the probability that a randomly selected child of this couple will have blue eyes?

  c. Suppose one parent is brown-eyed, type *BB*. What is the probability that a child has blue eyes?

3.17 Highway Traffic Safety Administration officials estimate that used car buyers are losing $2 billion a year because of falsified odometer readings (*Orlando Sentinel*, April 13, 1984)—a cost that, on the average, adds $750 to the price of a used car. Although it is against the law to falsify these readings, they estimate that nine out of ten heavily used cars, those owned by leasing companies, etc., have had the odometer mileage reduced. Suppose you are looking for two used cars and the dealer has four cars in stock that were formerly owned by an auto leasing company. If two of the four cars have had their odometers set back and you purchase two cars from among these four, list the simple events that can occur. Assign reasonable probabilities to the simple events and then find the probability that:

  a. You select the two leasing cars that have had their odometer mileage set back.

  b. You select the two with untampered odometers.

  c. Exactly one of the two cars has had its odometer mileage set back.

3.18 In 1987, unemployment rates in the United States ranged from 2.5% in Vermont to 12% in Louisiana. The nationwide unemployment rate was 6.2%. The table lists the 1987 and 1986 unemployment rates for the 15 Atlantic coast states. Suppose one of these 15 states is to be selected and the direction and amount of change in its unemployment rate from 1986 to 1987 is to be observed. Assume each state has an equal probability of being selected.

STATE	1987 (%)	1986 (%)	STATE	1987 (%)	1986 (%)
Connecticut	3.3	3.8	New Jersey	5.0	4.0
Delaware	3.2	4.3	New York	6.3	4.9
Florida	5.3	5.7	North Carolina	5.3	4.5
Georgia	5.5	5.9	Pennsylvania	6.8	5.7
Maine	5.3	4.4	Rhode Island	4.0	3.8
Maryland	4.2	4.5	South Carolina	6.2	5.6
Massachusetts	3.8	3.2	Virginia	5.0	4.2
New Hampshire	2.8	2.5			

Source: *Statistical Abstract of the United States: 1989*, p. 396.

  a. What is the probability that Pennsylvania will be selected? Florida? Virginia?

  b. What is the probability of selecting a state that had no change in its unemployment rate?

  c. What is the probability of selecting a state whose unemployment rate increased? Decreased?

  d. What is the probability of selecting a state whose unemployment rate increased 1% or more? Decreased 1% or more?

3.19 Three people play a game called "Odd Man Out." In this game, each player flips a fair coin until the outcome (heads or tails) for one of the players is not the same as the other two players'. This player is then the odd man out and loses the game. Find the probability that the game ends (i.e., either exactly one of the coins will fall heads or exactly one of the coins will fall tails) after only one toss by each player. Suppose one of the players, hoping to reduce his chances of being the odd man, uses a two-headed coin. Will this ploy be successful? Solve by listing the simple events in the sample space.

3.20   The breakdown of workers in a particular state according to their political affiliation and type of job held is shown here. Suppose a worker is selected at random within the state and the worker's political affiliation and type of job are noted.

| | | POLITICAL AFFILIATION | | |
		Republican	Democrat	Independent
TYPE OF JOB	White-collar	12%	12%	6%
	Blue-collar	23%	43%	4%

   a. List all simple events for this experiment.
   b. What is the set of all simple events called?
   c. Let A be the event that the worker is a white-collar worker. Find $P(A)$.
   d. Let B be the event that the worker is a Republican. Find $P(B)$.
   e. Let C be the event that the worker is a Democrat. Find $P(C)$.
   f. Let D be the event that the worker is a white-collar worker and a Democrat. Find $P(D)$.

3.21   According to David Dreman (*Forbes*, October 27, 1980, pp. 202–203) investment in new issues (the stock of newly formed companies) can be both suicidal and rewarding. Dreman based his comments on a Securities and Exchange Commission (SEC) study of 500 new issues that went public during the 1961–1962 stock boom. The SEC found that of the 500 companies, 43% went bankrupt, 25% were operating at losses, and only 20% showed a profit. Only 12 companies of the 500 appeared to have outstanding prospects. Suppose back in 1961 you had randomly selected two of five new issues for investment and that, unknown to you, only two of the five would eventually show a profit. What is the probability that:
   a. Both of the new issues in which you invested will eventually show a profit?
   b. Neither of the two issues will eventually show a profit?
   c. At least one of the two issues you selected will eventually show a profit?

3.22   Before placing a person in a highly skilled position, a company gives the applicants a series of three examinations. The first is a physical examination, and each applicant is classified as satisfactory or unsatisfactory. The other two are verbal and quantitative examinations, and the scores are used to classify each applicant as high, medium, or low in each area. Thus, each individual will receive a health score, a verbal score, and a quantitative score.
   a. List the different sets of classifications that can result from this battery of examinations.
   b. If all applicants who take the examinations are equally qualified, and all the variation in test scores is random, what is the probability that an applicant receives the lowest classification on all three examinations?
   c. If an applicant scores in the highest category on at least two of the three examinations, the applicant will get a position. What is the probability that a randomly selected applicant will get a position? (Make the same assumption as in part **b**.)

3.23   Often probabilities are expressed in terms of **odds**, especially in gambling settings. For example, handicappers for horse races express their belief about the probabilities of each horse winning a race in terms of odds. If the probability of event E is $P(E)$, then the **odds in favor of** E are $P(E)$ to $1 - P(E)$. Thus, if a handicapper assesses a probability of .25 that Snow Chief will win the Belmont Stakes, the odds in favor of Snow Chief are $\frac{25}{100}$ to $\frac{75}{100}$, or 1 to 3. It follows that the **odds against** E are $1 - P(E)$ to $P(E)$, or 3 to 1 against a win by Snow Chief. In general, if the odds in favor of event E are a to b, then $P(E) = a/(a + b)$.
   a. A second handicapper assesses the probability of a win by Snow Chief to be $\frac{1}{3}$. According to the second handicapper, what are the odds in favor of a Snow Chief win?

b. A third handicapper assesses the odds in favor of Snow Chief to be 1 to 1. According to the third handicapper, what is the probability of a Snow Chief win?

c. A fourth handicapper assesses the odds against Snow Chief winning to be 3 to 2. Find this handicapper's assessment of the probability that Snow Chief will win.

3.2 Compound Events

An event can often be viewed as a composition of two or more other events. Such events are called **compound events**; they can be formed (composed) in two ways, as defined in the boxes and illustrated here.

Definition 3.5

The **union** of two events A and B is the event that occurs if either A or B or both occur on a single performance of the experiment. We denote the union of events A and B by the symbol $A \cup B$.

Definition 3.6

The **intersection** of two events A and B is the event that occurs if both A and B occur on a single performance of the experiment. We write $A \cap B$ for the intersection of events A and B.

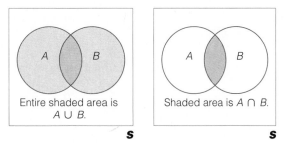

Entire shaded area is $A \cup B$.

Shaded area is $A \cap B$.

EXAMPLE 3.6

Consider the die-toss experiment. Define the following events:

A: {Toss an even number}

B: {Toss a number less than or equal to 3}

a. Describe $A \cup B$ for this experiment.

b. Describe $A \cap B$ for this experiment.

c. Calculate $P(A \cup B)$ and $P(A \cap B)$ assuming the die is fair.

Solution

Draw the Venn diagram as shown in the margin on page 110.

a. The union of A and B is the event that occurs if we observe either an even number, a number less than or equal to 3, or both on a single throw of the

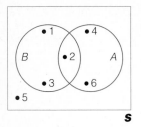

die. Consequently, the simple events in the event $A \cup B$ are those for which A occurs, B occurs, or both A and B occur. Testing the simple events in the entire sample space, we find that the collection of simple events in the union of A and B is

$$A \cup B = \{1, 2, 3, 4, 6\}$$

b. The intersection of A and B is the event that occurs if we observe *both* an even number and a number less than or equal to 3 on a single throw of the die. Testing the simple events to see which imply the occurrence of *both* events A and B, we see that the intersection contains only one simple event:

$$A \cap B = \{2\}$$

In other words, the intersection of A and B is the simple event Observe a 2.

c. Recalling that the probability of an event is the sum of the probabilities of the simple events of which the event is composed, we have

$$P(A \cup B) = P(1) + P(2) + P(3) + P(4) + P(6)$$
$$= \frac{1}{6} + \frac{1}{6} + \frac{1}{6} + \frac{1}{6} + \frac{1}{6} = \frac{5}{6}$$

and

$$P(A \cap B) = P(2) = \frac{1}{6}$$

Unions and intersections can be defined for more than two events. For example, the event $A \cup B \cup C$ represents the union of three events, A, B, and C. This event, which includes the set of simple events in A, B, or C, will occur if any one or more of the events A, B, or C occurs. Similarly, the intersection $A \cap B \cap C$ is the event that all three of the events A, B, and C occur. Therefore, $A \cap B \cap C$ is the set of simple events that are in all three of the events, A, B, and C.

EXAMPLE 3.7

Refer to Example 3.6 and define the event

C: {Toss a number greater than 1}

Find the simple events in:

a. $A \cup B \cup C$ b. $A \cap B \cap C$

where

A: {Toss an even number}

B: {Toss a number less than or equal to 3}

Solution

a. Event C contains the simple events corresponding to tossing a 2, 3, 4, 5, or 6, and event B contains the simple events 1, 2, and 3. Therefore, the event that either A, B, or C occurs contains all six simple events in S—that is, those corresponding to tossing a 1, 2, 3, 4, 5, or 6.

b. You can see that you will observe all of the events A, B, and C only if you observe a 2. Therefore, the intersection $A \cap B \cap C$ contains the single simple event Toss a 2.

EXAMPLE 3.8

Many firms have undertaken direct marketing campaigns to promote their products. The campaigns typically involve mailing information to millions of households. The response rates are carefully monitored to determine the demographic characteristics of respondents. By studying tendencies to respond, the firms can better target future mailings to those segments of the population most likely to purchase the products.

Suppose a distributor of mail-order tools is analyzing the results of a recent mailing. The probability of response is believed to be related to income and age. The percentages of the total number of respondents to the mailing are given by income and age classification in Table 3.4.

TABLE 3.4 **Percentages of Respondents in Age–Income Classes**

| | | INCOME | | |
		< $25,000	$25,000–$50,000	> $50,000
	< 30 y	5%	12%	10%
AGE	30–50 y	14%	22%	16%
	> 50 y	8%	10%	3%

Define the following events:

    A: {A respondent's income is more than $50,000}

    B: {A respondent's age is 30 or more}

a. Find $P(A)$ and $P(B)$.
b. Find $P(A \cup B)$.
c. Find $P(A \cap B)$.

Solution

Following the steps for calculating probabilities of events, we first note that the objective is to characterize the income and age distribution of respondents to the mailing. To accomplish this, we define the experiment to consist of selecting a respondent from the collection of all respondents and observing which income and age class he or she occupies. The simple events are the nine different age–income classifications:

    E_1: {< 30 y, < $25,000}
    E_2: {30–50 y, < $25,000}
      $\vdots$ $\vdots$
    E_9: {> 50 y, > $50,000}

Next, we assign probabilities to the simple events. If we blindly select one of the respondents, the probability that he or she will occupy a particular age–income classification is just the proportion, or relative frequency, of respondents in the classification. These proportions are given (as percentages) in Table 3.4. Thus,

$$P(E_1) = \text{Relative frequency of respondents in}$$
$$\text{age–income class } \{< 30 \text{ y}, < \$25,000\}$$
$$= .05$$
$$P(E_2) = .14$$

and so forth. You may verify that the simple event probabilities add to 1.

a. To find $P(A)$, we first determine the collection of simple events contained in event A. Since A is defined as $\{> \$50,000\}$, we see from Table 3.4 that A contains the three simple events represented by the last column of the table. In words, the event A consists of the income classification $\{> \$50,000\}$ in all three age classifications. The probability of A is the sum of the probabilities of the simple events in A:

$$P(A) = .10 + .16 + .03 = .29$$

Similarly, B consists of the six simple events in the second and third rows of Table 3.4:

$$P(B) = .14 + .22 + .16 + .08 + .10 + .03 = .73$$

b. The union of events A and B, $A \cup B$, consists of all simple events in *either A or B or both A and B*. That is, the union of A and B consists of all respondents whose income exceeds \$50,000 *or* whose age is 30 or more. In Table 3.4 this is any simple event found in the third column *or* the last two rows. Thus,

$$P(A \cup B) = .10 + .14 + .22 + .16 + .08 + .10 + .03$$
$$= .83$$

c. The intersection of events A and B, $A \cap B$, consists of all simple events in *both A and B*. That is, the intersection of A and B consists of all respondents whose income exceeds \$50,000 *and* whose age is 30 or more. In Table 3.4 this is any simple event found in the third column *and* the last two rows. Thus,

$$P(A \cap B) = .16 + .03 = .19$$

3.3 Complementary Events

A very useful concept in the calculation of event probabilities is the notion of *complementary events*:

Definition 3.7

The **complement** of an event A is the event that A does not occur—i.e., the event consisting of all simple events that are not in event A. We denote the complement of A by A'.

An event A is a collection of simple events, and the simple events included in A' are those that are not in A. Figure 3.6 demonstrates this idea. You will note from the figure that all simple events in S are included in *either A or A'* and that *no* simple event is in both A and A'. This leads us to conclude that the probabilities of an event and its complement *must sum to* 1:

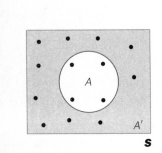

FIGURE 3.6

Venn diagram of complementary events

The sum of the probabilities of complementary events equals 1; i.e.,

$$P(A) + P(A') = 1$$

In many probability problems it is easier to calculate the probability of the complement of the event of interest rather than the event itself. Then, since

$$P(A) + P(A') = 1$$

we can calculate $P(A)$ by using the relationship

$$P(A) = 1 - P(A')$$

EXAMPLE 3.9

Consider the experiment of tossing two fair coins. Calculate the probability of event A: {Observing at least one head} by using the complementary relationship.

Solution

We know that the event A: {Observing at least one head} consists of the simple events

$$A: \quad \{HH, HT, TH\}$$

The complement of A is defined as the event that occurs when A does not occur. Therefore,

$$A': \quad \{\text{Observe no heads}\} = \{TT\}$$

This complementary relationship is shown in Figure 3.7. Assuming the coins are balanced,

$$P(A') = P(TT) = \frac{1}{4}$$

and

$$P(A) = 1 - P(A') = 1 - \frac{1}{4} = \frac{3}{4}$$

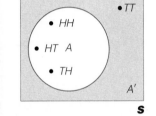

FIGURE 3.7

Complementary events in the toss of two coins

EXAMPLE 3.10

A fair coin is tossed ten times, and the up face is recorded after each toss. What is the probability of event A: {Observe at least one head}?

Solution

We solve this problem by following the five steps for calculating probabilities of events (see Section 3.1).

STEP 1 Define the experiment. The experiment is to record the results of the ten tosses of the coin.

STEP 2 List the simple events. A simple event consists of a particular sequence of ten heads and tails. Thus, one simple event is

$$HHTTTHTHTT$$

which denotes head on first toss, head on second toss, tail on third toss, etc. Others are $HTHHHTTTTT$ and $THHHTHTHTH$. There is obviously a very large number of simple events—too many to list. It can be shown (see Section 3.8) that there are $2^{10} = 1,024$ simple events for this experiment.

STEP 3 Assign probabilities. Since the coin is fair, each sequence of heads and tails has the same chance of occurring, and therefore all the simple events are equally likely. Then

$$P(\text{Each simple event}) = \frac{1}{1,024}$$

STEP 4 Determine the simple events in event A. A simple event is in A if at least one H appears in the sequence of ten tosses. However, if we consider the complement of A, we find that

A': {No heads are observed in ten tosses}

Thus, A' contains only one simple event:

$$A': \{TTTTTTTTTT\} \quad \text{and} \quad P(A') = \frac{1}{1,024}$$

STEP 5 Since we know the probability of the complement of A, we use the relationship for complementary events:

$$P(A) = 1 - P(A') = 1 - \frac{1}{1,024} = \frac{1,023}{1,024} = .999$$

That is, we are virtually certain of observing at least one head in ten tosses of the coin.

EXERCISES 3.24–3.36

LEARNING THE MECHANICS

3.24 A fair coin is tossed three times and the events A and B are defined as follows:

   A: {At least one head is observed}

   B: {The number of heads observed is odd}

   a. Identify the simple events in the events A, B, $A \cup B$, A', and $A \cap B$.
   b. Find $P(A)$, $P(B)$, $P(A \cup B)$, $P(A')$, and $P(A \cap B)$ by summing the probabilities of the appropriate simple events.

3.25 A pair of fair dice is tossed. Define the following events:

   A: {You will roll a 7}   (i.e., the sum of the numbers of dots on the upper faces of the two dice is equal to 7)

   B: {At least one of the two dice is showing a 4}

   a. Identify the simple events in the events A, B, $A \cap B$, $A \cup B$, and A'.
   b. Find $P(A)$, $P(B)$, $P(A \cap B)$, $P(A \cup B)$, and $P(A')$ by summing the probabilities of the appropriate simple events.

3.26 Consider the Venn diagram, where $P(E_1) = P(E_2) = P(E_3) = \frac{1}{5}$, $P(E_4) = P(E_5) = \frac{1}{20}$, $P(E_6) = \frac{1}{10}$, and $P(E_7) = \frac{1}{5}$. Find each of the following probabilities.

Venn diagram for Exercise 3.26

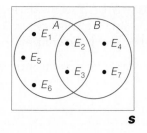

S

a. $P(A)$ **b.** $P(B)$ **c.** $P(A \cup B)$ **d.** $P(A \cap B)$
e. $P(A')$ **f.** $P(B')$ **g.** $P(A \cup A')$ **h.** $P(A' \cap B)$

3.27 Consider the Venn diagram, where $P(E_1) = .13$, $P(E_2) = .05$, $P(E_3) = P(E_4) = .2$, $P(E_5) = .06$, $P(E_6) = .3$, and $P(E_7) = .06$. Find each of the following probabilities:
a. $P(A')$ **b.** $P(B')$ **c.** $P(A' \cap B)$
d. $P(A \cup B)$ **e.** $P(A \cap B)$ **f.** $P(A' \cup B')$

Venn diagram for Exercise 3.27

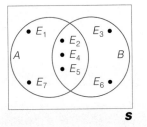

3.28 The table describes the adult population of a small suburb of a large southern city.

		INCOME		
		Under $20,000	$20,000–$50,000	Over $50,000
	Under 25	950	1,000	50
AGE	25–45	450	2,050	1,500
	Over 45	50	950	1,000

A marketing research firm plans to randomly select one adult from the suburb to evaluate a new food product. For this experiment the nine age–income categories are the simple events. Consider the following events:

A: {Person is under 25}

B: {Person is between 25 and 45}

C: {Person is over 45}

D: {Person has income under $20,000}

E: {Person has income of $20,000–$50,000}

F: {Person has income over $50,000}

Convert the frequencies in the table to relative frequencies and use them to calculate the following probabilities:
a. $P(B)$ **b.** $P(F)$ **c.** $P(C \cap F)$ **d.** $P(B \cup C)$ **e.** $P(A')$ **f.** $P(A' \cap F)$

3.29 Refer to Exercise 3.28. Use the same event definitions to solve the following.
a. Write the event that the person selected is under 25 with an income over $50,000 as an intersection of two events.
b. Write the event that the person selected is age 25 or more as the union of two events. As the complement of an event.

APPLYING THE CONCEPTS

3.30 A buyer for a large metropolitan department store must choose two firms from the four available to supply the store's fall line of men's slacks. The buyer has not dealt with any of the four firms before and considers their products equally attractive. Unknown to the buyer, two of the four firms are having serious financial problems that may result in their not being able to deliver the fall line of slacks as soon as promised. The four firms are identified as G_1 and G_2 (firms in good financial condition) and P_1 and P_2 (firms in poor financial condition). Simple events identify the pairs of firms selected. If the probability of the buyer selecting a particular firm from among the four is the same for each firm, the simple events and their probabilities for this buying experiment are those listed in the table. We will define the following events:

A: {At least one of the selected firms is in good financial condition}

B: {Firm P_1 is selected}

SIMPLE EVENTS	PROBABILITY
G_1G_2	$\frac{1}{6}$
G_1P_1	$\frac{1}{6}$
G_1P_2	$\frac{1}{6}$
G_2P_1	$\frac{1}{6}$
G_2P_2	$\frac{1}{6}$
P_1P_2	$\frac{1}{6}$

a. Define the event $A \cap B$ as a specific collection of simple events.
b. Define the event $A \cup B$ as a specific collection of simple events.
c. Define the event A' as a specific collection of simple events.
d. Find $P(A)$, $P(B)$, $P(A \cap B)$, $P(A \cup B)$, and $P(A')$ by summing the probabilities of the appropriate simple events.

3.31 A state energy agency mailed questionnaires on energy conservation to 1,000 homeowners in the state capital. Five hundred questionnaires were returned. Suppose an experiment consists of randomly selecting one of the returned questionnaires. Consider the events:

A: {The home is constructed of brick}

B: {The home is more than 30 years old}

C: {The home is heated with oil}

Describe each of the following events in terms of unions, intersections, and complements ($A \cup B$, $A \cap B$, A', etc.):
a. The home is more than 30 years old and is heated with oil.
b. The home is not constructed of brick.
c. The home is heated with oil or is more than 30 years old.
d. The home is constructed of brick and is not heated with oil.

3.32 The types of occupations of the 112,440,000 employed workers (age 16 years and older) in the United States in 1987 are described in the table, and their relative frequencies are listed. A worker is to be selected at random from this population and his or her occupation is to be determined. (Assume that each worker in the population has only one occupation.)
a. What is the probability that the worker will be a male service worker?
b. What is the probability that the worker will be a manager or a professional?
c. What is the probability that the worker will be a female professional or a female operator/fabricator?
d. What is the probability that the worker will not be in a technical/sales/administrative occupation?

OCCUPATION	RELATIVE FREQUENCY	
Male workers	.552	
Managerial/professional		.137
Technical/sales/administrative		.110
Service		.053
Precision production, craft, and repair		.110
Operators/fabricators		.115
Farming, forestry, and fishing		.027
Female workers	.448	
Managerial/professional		.109
Technical/sales/administrative		.202
Service		.081
Precision production, craft, and repair		.010
Operators/fabricators		.040
Farming, forestry, and fishing		.006

Source: *Statistical Abstract of the United States, 1989*, p. 388.

3.33 One game that is very popular in many American casinos is roulette. Roulette is played by spinning a ball on a circular wheel that has been divided into 38 arcs of equal length, bearing the numbers 00, 0, 1, 2, . . . , 35, 36. The number of the arc on which the ball comes to rest is the outcome of one play of the game. The numbers are also colored in the following manner:

Red: 1, 3, 5, 7, 9, 12, 14, 16, 18, 19, 21, 23, 25, 27, 30, 32, 34, 36

Black: 2, 4, 6, 8, 10, 11, 13, 15, 17, 20, 22, 24, 26, 28, 29, 31, 33, 35

Green: 00, 0

Players may place bets on the table in a variety of ways, including bets on odd, even, red, black, high, low, etc. Define the following events:

A: {Outcome is an odd number}   (00 and 0 are not considered odd or even)

B: {Outcome is a black number}

C: {Outcome is a low number (1–18)}

a. Define the event $A \cap B$ as a specific set of simple events.
b. Define the event $A \cup B$ as a specific set of simple events.
c. Find $P(A)$, $P(B)$, $P(A \cap B)$, $P(A \cup B)$, and $P(C)$ by summing the probabilities of the appropriate simple events.
d. Define the event $A \cap B \cap C$ as a specific set of simple events.
e. Find $P(A \cap B \cap C)$ by summing the probabilities of the simple events given in part d.
f. Define the event $(A \cup B \cup C)$ as a specific set of simple events.
g. Find $P(A \cup B \cup C)$ by summing the probabilities of the simple events given in part f.

3.34 After completing an inventory of three warehouses, a golf club shaft manufacturer described its stock of 12,246 shafts with the percentages given in the table. Suppose a shaft is selected at random from the 12,246 currently in stock, and the warehouse number and type of shaft are observed.

		TYPE OF SHAFT		
		Regular	Stiff	Extra stiff
WAREHOUSE	1	19%	8%	3%
	2	14%	8%	2%
	3	28%	18%	0%

a. List all the simple events for this experiment.

b. What is the set of all simple events called?

c. Let C be the event that the shaft selected is from warehouse 3. Find $P(C)$ by summing the probabilities of the simple events in C.

d. Let F be the event that the shaft chosen is an extra stiff type. Find $P(F)$.

e. Let A be the event that the shaft selected is from warehouse 1. Find $P(A)$.

f. Let D be the event that the shaft selected is a regular type. Find $P(D)$.

g. Let E be the event that the shaft selected is a stiff type. Find $P(E)$.

3.35 Refer to Exercise 3.34. Define the characteristics of a golf club shaft portrayed by the following events, and then find the probability of each.

a. $A \cap F$ b. $C \cup E$ c. $C \cap D$ d. $A \cup F$ e. $A \cup D$

3.36 Identifying managerial prospects who are both talented and motivated is difficult. A personnel manager constructed the following two-way table to define nine combinations of talent–motivation levels. The number in a cell is the manager's estimate of the probability that a managerial prospect will fall in that category. Suppose the personnel manager has decided to hire a new manager. Define the following events:

A: {Prospect places in high motivation category}

B: {Prospect places in high talent category}

C: {Prospect is average or better in both categories}

D: {Prospect rates poor in at least one category}

E: {Prospect places highest in both categories}

		TALENT		
		High	Medium	Low
	High	.05	.16	.05
MOTIVATION	Medium	.19	.32	.05
	Low	.11	.05	.02

a. Does the sum of the cell probabilities equal 1?

b. List the simple events in each of the events described above and find their probabilities.

c. Find $P(A \cup B)$, $P(A \cap B)$, and $P(A \cup C)$.

d. Find $P(A')$ and explain what this means from a practical point of view.

3.4 Conditional Probability

The event probabilities we have been discussing give the relative frequencies of the occurrences of the events when the experiment is repeated a very large number of times. They are called **unconditional probabilities** because no special conditions are assumed, other than those that define the experiment.

Sometimes we may wish to alter the probability of an event when we have additional knowledge that might affect its outcome. This probability is called the **conditional probability** of the event. For example, we have shown that the probability of observing an even number (event A) on a toss of a fair die is $\frac{1}{2}$. However, suppose you are given the information that on a particular throw of the die the result was a number less than or equal to 3 (event B). Would you still believe that the probability of observing an even number on that throw of the die is equal to $\frac{1}{2}$? If you reason that making the assumption that B has occurred

FIGURE 3.8

Reduced sample space for the die toss experiment—given that event B has occurred

reduces the sample space from six simple events to three simple events (namely, those contained in event B), the reduced sample space is as shown in Figure 3.8. Because the simple events for the die-toss experiment are equally likely, each of the three simple events in the reduced sample space is assigned an equal *conditional probability* of $\frac{1}{3}$. Since the only even number of the three in the reduced sample space B is the number 2 and the die is fair, we conclude that the probability that A occurs *given that B occurs* is $\frac{1}{3}$. We use the symbol $P(A \mid B)$ to represent the probability of event A given that event B occurs. For the die-toss example

$$P(A \mid B) = \frac{1}{3}$$

To get the probability of event A given that event B occurs, we proceed as follows. We divide the probability of the part of A that falls within the reduced sample space B, namely, $P(A \cap B)$, by the total probability of the reduced sample space, namely, $P(B)$. Thus, for the die-toss example with event A: {Observe an even number} and event B: {Observe a number less than or equal to 3}, we find

$$P(A \mid B) = \frac{P(A \cap B)}{P(B)} = \frac{P(2)}{P(1) + P(2) + P(3)} = \frac{\frac{1}{6}}{\frac{3}{6}} = \frac{1}{3}$$

The formula for $P(A \mid B)$ is true in general:

> To find the *conditional probability that event A occurs given that event B occurs*, divide the probability that *both* A and B occur by the probability that B occurs, that is,
>
> $$P(A \mid B) = \frac{P(A \cap B)}{P(B)} \qquad \text{[We assume that } P(B) \neq 0.\text{]}$$

The formula adjusts the probability of $A \cap B$ from its orginal value in the complete sample space S to a conditional probability in the reduced sample space B. If the simple events in the complete sample space are equally likely, then the formula will assign equal probabilities to the simple events in the reduced sample space, as in the die-toss experiment. If, on the other hand, the simple events have unequal probabilities, the formula will assign conditional probabilities proportional to the probabilities in the complete sample space. This is illustrated by the following examples.

EXAMPLE 3.11

SIMPLE EVENTS	PROBABILITIES
$A \cap C$	.15
$A \cap C'$	.25
$A' \cap C$	.10
$A' \cap C'$	.50

Many medical researchers have conducted experiments to examine the relationship between cigarette smoking and cancer. Let A represent the event that an individual smokes, and let C represent the event that an individual develops cancer. Thus, $A \cap C$ is the simple event that an individual smokes and develops cancer; $A \cap C'$ is the simple event that an individual smokes and does not develop cancer, etc. Assume that the probabilities associated with the four simple events are as shown in the table for a certain section of the United States. How can these simple event probabilities be used to examine the relationship between smoking and cancer?

Solution

One method of determining whether these probabilities indicate that smoking and cancer are related is to compare the conditional probability that an individual acquires cancer given that he or she smokes with the conditional probability that an individual acquires cancer given that he or she does not smoke.

First, we consider the reduced sample space A corresponding to smokers. The two simple events $A \cap C$ and $A \cap C'$ are contained in this reduced sample space, and the adjusted probabilities of these two simple events are the two conditional probabilities:

$$P(C \mid A) = \frac{P(A \cap C)}{P(A)} \quad \text{and} \quad P(C' \mid A) = \frac{P(A \cap C')}{P(A)}$$

The probability of event A is the sum of the probabilities of the simple events in A:

$$P(A) = P(A \cap C) + P(A \cap C') = .15 + .25 = .40$$

Then the values of the two conditional probabilities in the reduced sample space A are

$$P(C \mid A) = \frac{.15}{.40} = .375 \quad \text{and} \quad P(C' \mid A) = \frac{.25}{.40} = .625$$

These two numbers represent the probabilities that a smoker develops cancer and does not develop cancer, respectively. Notice that the conditional probabilities .625 and .375 are in the same 5 to 3 proportion as the original (unconditional) probabilities, .25 and .15. The conditional probability formula simply adjusts the unconditional probabilities so that they add to 1 in the reduced sample space, A, of smokers.

In a like manner, the conditional probabilities of a nonsmoker developing cancer and not developing cancer are:

$$P(C \mid A') = \frac{P(A' \cap C)}{P(A')} = \frac{.10}{.60} = .167$$

$$P(C' \mid A') = \frac{P(A' \cap C')}{P(A')} = \frac{.50}{.60} = .833$$

Notice that the conditional probabilities .833 and .167 are in the same 5 to 1 ratio as the unconditional probabilities .5 and .1.

Two of the conditional probabilities give some insight into the relationship between cancer and smoking: the probability of developing cancer given that the individual is a smoker, and the probability of developing cancer given that the individual is not a smoker. The conditional probability that a smoker develops cancer (.375) is more than twice the probability that a nonsmoker develops cancer (.167). This does not imply that smoking *causes* cancer, but it does suggest a pronounced link between smoking and cancer.

EXAMPLE 3.12

The investigation of consumer product complaints by the Federal Trade Commission (FTC) has generated much interest by manufacturers in the quality of their products. A manufacturer of an electromechanical kitchen aid conducted an analysis of a large number of consumer complaints and found that they fell into the six categories shown in Table 3.5. If a consumer complaint is received, what is the probability that the cause of the complaint was product appearance given that the complaint originated during the guarantee period?

TABLE 3.5 **Distribution of Product Complaints**

	REASON FOR COMPLAINT			TOTALS
	Electrical	Mechanical	Appearance	
During Guarantee Period	18%	13%	32%	63%
After Guarantee Period	12%	22%	3%	37%
TOTALS	30%	35%	35%	100%

Solution

Let A represent the event that the cause of a particular complaint is product appearance, and let B represent the event that the complaint occurred during the guarantee period. Checking Table 3.5, you can see that $(18 + 13 + 32)\% = 63\%$ of the complaints occur during the guarantee period. Hence, $P(B) = .63$. The percentage of complaints that were caused by appearance and occurred during the guarantee period (the event $A \cap B$) is 32%. Therefore, $P(A \cap B) = .32$.

Using these probability values, we can calculate the conditional probability $P(A \mid B)$ that the cause of a complaint is appearance given that the complaint occurred during the guarantee time:

$$P(A \mid B) = \frac{P(A \cap B)}{P(B)} = \frac{.32}{.63} = .51$$

Consequently, you can see that slightly more than half the complaints that occurred during the guarantee period were due to scratches, dents, or other imperfections in the surface of the kitchen devices.

CASE STUDY 3.3

PURCHASE PATTERNS AND THE CONDITIONAL PROBABILITY OF PURCHASING

In his doctoral dissertation, Alfred A. Kuehn* examined sequential purchase data to gain some insight into consumer brand switching. He analyzed the frozen orange juice purchases of approximately 600 Chicago families during 1950–1952. The data were collected by the *Chicago Tribune* Consumer Panel. Kuehn was interested in determining the influence of a consumer's last four orange juice purchases on the next purchase. Thus, sequences of five purchases were analyzed.

Table 3.6 (page 122) summarizes the data collected for Snow Crop brand orange juice and part of Kuehn's analysis of the data. In the column labeled "Previous Purchase Pattern" an S stands for the purchase of Snow Crop by a consumer and an O stands for the purchase of a brand other than Snow Crop. Thus,

*Reprinted from Kuehn, A. A. "An analysis of the dynamics of consumer behavior and its implications for marketing management," Unpublished doctoral dissertation, Graduate School of Industrial Administration, Carnegie Institute of Technology, 1958.

for example, SSSO is used to represent the purchase of Snow Crop three times in a row followed by the purchase of some other brand of frozen orange juice. The column labeled "Sample Size" lists the number of occurrences of the purchase sequences in the first column. The column labeled "Frequency" lists the number of times the associated purchase sequence in the first column led to the next purchase (i.e., the fifth purchase in the sequence) being Snow Crop.

T A B L E 3.6 **Observed Approximate Conditional Probability of Purchasing Snow Crop Given the Four Previous Brand Purchases**

PREVIOUS PURCHASE PATTERN S = Snow Crop O = Other brand	SAMPLE SIZE	FREQUENCY	OBSERVED APPROXIMATE CONDITIONAL PROBABILITY OF PURCHASE $P\{$Purchase | Previous purchase pattern$\}$
SSSS	1,047	844	.806
OSSS	277	191	.690
SOSS	206	137	.665
SSOS	222	132	.595
SSSO	296	144	.486
OOSS	248	137	.552
SOOS	138	78	.565
OSOS	149	74	.497
SOSO	163	66	.405
OSSO	181	75	.414
SSOO	256	78	.305
OOOS	500	165	.330
OOSO	404	77	.191
OSOO	433	56	.129
SOOO	557	86	.154
OOOO	8,442	405	.048

The column labeled "Observed Approximate Conditional Probability of Purchase" contains the relative frequency with which each sequence of the first column led to the next purchase being Snow Crop. These relative frequencies, which give approximate conditional probabilities, are computed for each sequence of the first column by dividing the frequency of the sequence by the sample size of the sequence. For example, .806 is the approximate conditional probability that the next purchase will be Snow Crop given that the previous four purchases were also Snow Crop.

An examination of the approximate conditional probabilities in the fourth column indicates that both the most recent brand purchased and the number of times a brand is purchased have an effect on the next brand purchased. It appears that the influence on the next brand of orange juice purchased by the second most recent purchase is not as strong as the most recent purchase, but it is stronger than the third most recent purchase. In general, it appears that the probability of a particular consumer purchasing Snow Crop the next time he or she buys orange juice is inversely related to the number of consecutive purchases of another brand he or she made since last purchasing Snow Crop and is directly proportional to the number of Snow Crop purchases among the four purchases.

Kuehn conducts a more formal statistical analysis of these data, which we do not pursue here. We simply want you to see that probability is a basic tool for making inferences about populations using sample data.

EXERCISES 3.37–3.47

LEARNING THE MECHANICS

3.37 Consider the experiment defined by the accompanying Venn diagram, with the sample space S containing five simple events. The simple events are assigned the following probabilities:

$$P(E_1) = .1 \qquad P(E_2) = .1 \qquad P(E_3) = .3 \qquad P(E_4) = .4 \qquad P(E_5) = .1$$

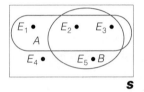

a. Calculate $P(A)$, $P(B)$, and $P(A \cap B)$.
b. Suppose we know event A has occurred, so the reduced sample space consists of the three simple events in A: E_1, E_2, and E_3. Use the formula for conditional probability to determine the probabilities of these three simple events given that A has occurred. Verify that the conditional probabilities are in the same proportion to one another as the original simple event probabilities.
c. Calculate the conditional probability $P(B \mid A)$ in two ways. First, add the adjusted (conditional) probabilities of the simple events in the intersection $A \cap B$, since these represent the event that B occurs given that A has occurred. Second, use the formula for conditional probability:

$$P(B \mid A) = \frac{P(A \cap B)}{P(A)}$$

Verify that the two methods yield the same result.

3.38 Given that $P(A) = .3$, $P(B) = .7$, and $P(A \cap B) = .15$, find $P(A \mid B)$ and $P(B \mid A)$.

3.39 A sample space contains six simple events and events A, B, and C as shown in the Venn diagram. The probabilities of the simple events are $P(1) = .20$, $P(2) = .05$, $P(3) = .30$, $P(4) = .10$, $P(5) = .10$, $P(6) = .25$. Use the Venn diagram and the probabilities of the simple events to find:

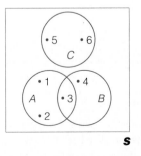

a. $P(A)$, $P(B)$, and $P(C)$
b. $P(A \cap B)$, $P(A \cap C)$, and $P(B \cap C)$

c. Suppose you know that event *A* has occurred. Assign conditional probabilities to the three simple events contained in *A*. Verify that they add to 1 and are in the same proportion as the original (unconditional) probabilities.

d. Use the conditional probabilities from part **c** to calculate $P(B \mid A)$. Use the formula for $P(B \mid A)$ to verify your answer.

e. Use the formula for conditional probability to calculate $P(C \mid A)$ and $P(C \mid A')$. Verify the results by inspection of the Venn diagram, remembering that the "given event" is a reduced sample space for a conditional probability.

3.40 Two fair coins are tossed and the events *A* and *B* are defined as follows:

A: {At least one head appears}

B: {Exactly one head appears}

a. Draw a Venn diagram for the experiment, labeling each simple event and showing events *A* and *B*. Assign probabilities to the simple events.

b. Find $P(A)$, $P(B)$, and $P(A \cap B)$.

c. Use the formula for conditional probability to find $P(A \mid B)$ and $P(B \mid A)$. Verify your answers by inspecting the Venn diagram and using the concept of reduced sample spaces.

3.41 A box contains two white, two red, and two blue poker chips. Two chips are randomly chosen without replacement and their colors are noted. Define the following events:

A: {Both chips are of the same color}

B: {Both chips are red}

C: {At least one chip is red or white}

Find $P(B \mid A)$, $P(B \mid A')$, $P(B \mid C)$, $P(A \mid C)$, and $P(C \mid A')$.

APPLYING THE CONCEPTS

3.42 A soap manufacturer has decided to market two new brands. An analysis of current market conditions and a review of the firm's past successes and failures with new brands have led the manufacturer to believe that the simple events and the probabilities of their occurrence in this marketing experiment are as shown in the table (where *S* means the brand succeeds and *F* means the brand fails in the first year). Define the following events:

A: {Both new brands are successful in the first year}

B: {At least one new brand is successful in the first year}

SIMPLE EVENTS	PROBABILITIES
SS	.09
SF	.21
FS	.21
FF	.49

a. Find $P(A)$, $P(B)$, and $P(A \cap B)$.   b. Find $P(A \mid B)$ and $P(B \mid A)$.

3.43 The table describes the 101.5 million U.S. federal tax returns filed with the Internal Revenue Service (IRS) in 1987 and the percentage of those returns that were audited by the IRS.

a. If a tax filer were to be randomly selected from the population of tax filers, what is the probability that the tax filer would have been audited?

b. If a tax filer were to be randomly selected from the population of tax filers described in the table, what is the probability that the tax filer had an income of $10,000–$24,999 in 1987 and was

INCOME	NUMBER OF TAX FILERS (millions)	PERCENTAGE AUDITED
Under $10,000	29.8	.50
$10,000–$24,999	33.1	.85
$25,000–$49,999	27.2	1.40
$50,000 or more	11.4	2.24

Source: *Statistical Abstract of the United States, 1989*, p. 388.

audited? What is the probability that the tax filer had an income of $50,000 or more in 1987 or was not audited?

c. Refer to part **b**. If it were known that the randomly selected tax filer has been audited, what is the probability that the person had an income of $50,000 or more? Under $10,000?

d. What is the probability that a tax filer with an income of $50,000 or more in 1987 would have been audited?

3.44 Six people apply for two identical positions in a company. Four are minority applicants and the remainder are nonminority. Define the following events:

A: {Both persons selected are nonminority candidates}

B: {Both persons selected are minority candidates}

C: {At least one of the persons selected is a minority candidate}

If all the applicants are equally qualified and the choice is essentially a random selection of two applicants from the six available, find:

a. $P(A)$ **b.** $P(B)$ **c.** $P(C)$ **d.** $P(B \mid C)$

e. Assume that the minority candidates are numbered 1, 2, 3, 4 for purposes of identification. Define the event

D: {Minority candidate 1 is selected}

Find $P(D \mid C)$.

3.45 An article in *Business Week* (September 12, 1983) reports on the problems that evolve from the failure to inform patients adequately of both the proper application of prescription drugs and the precautions to take in order to avoid potential side effects. This failure results in numerous cases of serious illness and, in some cases, even death. One study revealed that 300,000 U.S. hospital admissions each year are caused by adverse reactions to prescription drugs. Another study concluded that 7% of all hospital admissions are related to drug-induced problems resulting from imprudent prescriptions. One method of increasing patients' awareness of the problem is for physicians to provide Patient Medication Instruction (PMI) sheets. The American Medical Association, however, has found that only 20% of the doctors who prescribe drugs frequently distribute PMI sheets to their patients. Assume that 20% of all patients receive the PMI sheet with their prescriptions and that 12% receive the PMI sheet and are hospitalized because of a drug-related problem. What is the probability that a person will be hospitalized for a drug-related problem given that the person has received the PMI sheet?

3.46 A fast-food restaurant chain with 700 outlets in the United States describes the geographic location of its restaurants with the accompanying table of percentages. A restaurant is to be chosen at random from the 700 to test market a new style of chicken.

		REGION			
		NE	SE	SW	NW
POPULATION OF CITY	Under 10,000	5%	6%	3%	0%
	10,000–100,000	15%	15%	12%	5%
	Over 100,000	20%	4%	10%	5%

a. Given that the restaurant chosen is in a city with population over 100,000, what is the probability that it is located in the northeast?

b. Given that the restaurant chosen is in the southeast, what is the probability that it is located in a city with population under 10,000?

c. If the restaurant selected is located in the southwest, what is the probability that the city it is in has a population of 100,000 or less?

d. If the restaurant selected is located in the northwest, what is the probability that the city it is in has a population of 10,000 or more?

3.47   There are several methods of typing, or classifying, human blood. The most common procedure types blood into the general classifications of A, B, O, or AB. A method that is not as well known examines phosphoglucomutase (PGM) and classifies the blood into one of three main categories, 1-1, 2-1, or 2-2. Suppose a certain geographic region of the United States has the PGM percentages shown in the accompanying table. A person is to chosen at random from this region.

		1-1	2-1	2-2
RACE	White	46.3%	39.2%	4.0%
	Black	6.7%	3.4%	.4%

a. What is the probability that a black person is chosen?

b. Given that a black is chosen, what is the probability he or she is PGM type 1-1?

c. Given that a white is chosen, what is the probability he or she is PGM type 1-1?

3.5 Probabilities of Unions and Intersections

Since unions and intersections of events are themselves events, we can always calculate their probabilities by adding the probabilities of the simple events that compose them. However, when the probabilities of certain events are known, it is easier to use two rules—the additive rule and the multiplicative rule—to calculate the probability of unions and intersections. How and why these rules work will be illustrated by example.

EXAMPLE 3.13

A loaded (unbalanced) die is tossed and the up face is observed. The following two events are defined:

A: {Observe an even number}

B: {Observe a number less than 3}

Suppose $P(A) = .4$, $P(B) = .2$, and $P(A \cap B) = .1$. Find $P(A \cup B)$. [*Note:* Assuming that we would know these probabilities in a practical situation is not very realistic, but the example will illustrate a point.]

Solution

By studying the Venn diagram in Figure 3.9, we can obtain information that will help us find $P(A \cup B)$. We can see that

$$P(A \cup B) = P(1) + P(2) + P(4) + P(6)$$

Also, we know that

$$P(A) = P(2) + P(4) + P(6) = .4$$
$$P(B) = P(1) + P(2) = .2$$
$$P(A \cap B) = P(2) = .1$$

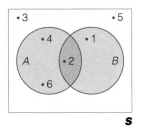

FIGURE 3.9
Venn diagram for die toss

If we add the probabilities of the simple events that comprise events A and B, we find

$$P(A) + P(B) = \overbrace{P(2) + P(4) + P(6)}^{P(A)} + \overbrace{P(1) + P(2)}^{P(B)}$$

$$= \overbrace{P(1) + P(2) + P(4) + P(6)}^{P(A \cup B)} + \overbrace{P(2)}^{P(A \cap B)}$$

Thus, by subtraction, we have

$$P(A \cup B) = P(A) + P(B) - P(A \cap B)$$
$$= .4 + .2 - .1 = .5$$

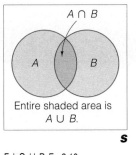

FIGURE 3.10
Venn diagram of union

By studying the Venn diagram in Figure 3.10, you can see that the method used in Example 3.13 may be generalized to find the union of two events for any experiment. The probability of the union of two events, A and B, can always be obtained by summing $P(A)$ and $P(B)$ and subtracting $P(A \cap B)$. Note that we must subtract $P(A \cap B)$ because the simple event probabilities in $A \cap B$ have been included twice—once in $P(A)$ and once in $P(B)$.

The formula for calculating the probability of the union of two events, often called the **additive rule of probability**, is given in the box.

Additive Rule of Probability

The probability of the union of events A and B is the sum of the probability of events A and B minus the probability of the intersection of events A and B; i.e.,

$$P(A \cup B) = P(A) + P(B) - P(A \cap B)$$

EXAMPLE 3.14

Hospital records show that 12% of all patients are admitted for surgical treatment, 16% are admitted for obstetrics, and 2% receive both obstetrics and surgical treatment. If a new patient is admitted to the hospital, what is the probability that the patient will be admitted either for surgery, obstetrics, or both?

Solution

Consider the following events:

   A: {A patient admitted to the hospital receives surgical treatment}
   B: {A patient admitted to the hospital receives obstetrics treatment}

Then, from the given information,

   $P(A) = .12$ $P(B) = .16$

and the probability of the event that a patient receives both obstetrics and surgical treatment is

   $P(A \cap B) = .02$

The event that a patient admitted to the hospital receives either surgical treatment, obstetrics treatment, or both is the union $A \cup B$. The probability of $A \cup B$ is given by the additive rule of probability:

$$P(A \cup B) = P(A) + P(B) - P(A \cap B)$$
$$= .12 + .16 - .02 = .26$$

Thus, 26% of all patients admitted to the hospital receive either surgical treatment, obstetrics treatment, or both.

A very special relationship exists between events A and B when $A \cap B$ contains no simple events. In this case, we call the events A and B *mutually exclusive* events.

> **Definition 3.8**
>
> Events A and B are **mutually exclusive** if $A \cap B$ contains no simple events.

Figure 3.11 shows a Venn diagram of two mutually exclusive events. The events A and B have no simple events in common, i.e., A and B cannot occur simultaneously, and $P(A \cap B) = 0$. Thus, we have the important relationship given in the box.

> *If two events A and B are mutually exclusive*, the probability of the union of A and B equals the sum of the probabilities of A and B; that is,
>
> $$P(A \cup B) = P(A) + P(B)$$

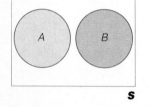

FIGURE 3.11
Venn diagram of mutually exclusive events

EXAMPLE 3.15

Consider the experiment of tossing two balanced coins. Find the probability of observing *at least* one head.

Solution

Define the events

    A: {Observe at least one head}

    B: {Observe exactly one head}

    C: {Observe exactly two heads}

Note that

    $A = B \cup C$

and that $B \cap C$ contains no simple events (see Figure 3.12). Thus, B and C are mutually exclusive, so that

$$P(A) = P(B \cup C) = P(B) + P(C)$$
$$= \frac{1}{2} + \frac{1}{4} = \frac{3}{4}$$

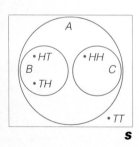

FIGURE 3.12
Venn diagram for coin toss experiment

Although Example 3.15 is very simple, the concept of writing events with verbal descriptions that include the phrases "at least" or "at most" as unions of mutually exclusive events is a very useful one. This enables us to find the probability of the event by adding the probabilities of the mutually exclusive events.

The second rule of probability, which will help us find the probability of the intersection of two events, is illustrated by Example 3.16.

EXAMPLE 3.16

An agriculturist, who is interested in planting wheat next year, is concerned with the following events:

A: {The production of wheat will be profitable}

B: {A serious drought will occur}

Based on available information, the agriculturist believes that the probability is .01 that production of wheat will be profitable *assuming* a serious drought will occur in the same year and that the probability is .05 that a serious drought will occur. That is,

$$P(A \mid B) = .01$$

and

$$P(B) = .05$$

Based on the information provided, what is the probability that a serious drought will occur *and* that a profit will be made? That is, find $P(A \cap B)$, the probability of the intersection of events A and B.

Solution

As you will see, we have already developed a formula for finding the probability of an intersection of two events. Recall that the conditional probability of A given B is

$$P(A \mid B) = \frac{P(A \cap B)}{P(B)}$$

Multiplying both sides of this equation by $P(B)$, we obtain a formula for the probability of the intersection of events A and B. This is often called the **multiplicative rule of probability** and is given by

$$P(A \cap B) = P(A \mid B)P(B)$$

Thus,

$$P(A \cap B) = (.01)(.05)$$
$$= .0005$$

The probability that a serious drought occurs *and* the production of wheat is profitable is only .0005. As we might expect, this intersection is a very rare event.

Multiplicative Rule of Probability

$$P(A \cap B) = P(A \mid B)P(B) = P(B \mid A)P(A)$$

Intersections often contain only a few simple events. In this case, the probability of an intersection is easy to calculate by summing the appropriate simple event probabilities. However, the formula for calculating intersection probabilities plays a very important role, particularly in an area of statistics known as **Bayesian statistics**. (More detailed discussions of Bayesian statistics are contained in the references at the end of the chapter.)

EXAMPLE 3.17

Consider the experiment of tossing a fair coin twice and recording the up face on each toss. The following events are defined:

A: {First toss is a head}

B: {Second toss is a head}

Does *knowing* that event A has occurred affect the probability that B will occur?

Solution

Intuitively the answer should be no, since what occurs on the first toss should in no way affect what occurs on the second toss. Let us check our intuition. Recall the sample space for this experiment:

1. Observe *HH*   2. Observe *HT*
3. Observe *TH*   4. Observe *TT*

Each of these simple events has a probability of $\frac{1}{4}$. Thus,

$$P(B) = P(HH) + P(TH) \quad \text{and} \quad P(A) = P(HH) + P(HT)$$

$$= \frac{1}{4} + \frac{1}{4} = \frac{1}{2} \qquad\qquad\qquad = \frac{1}{4} + \frac{1}{4} = \frac{1}{2}$$

Now, what is $P(B \mid A)$?

$$P(B \mid A) = \frac{P(A \cap B)}{P(A)} = \frac{P(HH)}{P(A)}$$

$$= \frac{\frac{1}{4}}{\frac{1}{2}} = \frac{1}{2}$$

We can now see that $P(B) = \frac{1}{2}$ and $P(B \mid A) = \frac{1}{2}$. Knowing that the first toss resulted in a head does not affect the probability that the second toss will be a head. The probability is $\frac{1}{2}$ whether or not we know the result of the first toss. When this occurs, we say that the two events A and B are *independent*.

Definition 3.9

Events A and B are **independent** if the occurrence of B does not alter the probability that A has occurred; i.e., events A and B are independent if

$$P(A \mid B) = P(A)$$

When events A and B are independent, it is also true that

$$P(B \mid A) = P(B)$$

Events that are not independent are said to be **dependent**.

EXAMPLE 3.18

Consider the experiment of tossing a fair die and let

A: {Observe an even number}

B: {Observe a number less than or equal to 4}

Are events A and B independent?

Solution

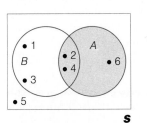

FIGURE 3.13
Venn diagram for die-toss experiment

The Venn diagram for this experiment is shown in Figure 3.13. We first calculate

$$P(A) = P(2) + P(4) + P(6) = \frac{1}{2}$$

$$P(B) = P(1) + P(2) + P(3) + P(4) = \frac{4}{6} = \frac{2}{3}$$

$$P(A \cap B) = P(2) + P(4) = \frac{2}{6} = \frac{1}{3}$$

Now assuming B has occurred, the conditional probability of A given B is

$$P(A \mid B) = \frac{P(A \cap B)}{P(B)} = \frac{\frac{1}{3}}{\frac{2}{3}} = \frac{1}{2} = P(A)$$

Thus, assuming that event B occurs does not alter the probability of observing an even number—it remains $\frac{1}{2}$. Therefore, the events A and B are independent. Note that if we calculate the conditional probability of B given A, our conclusion is the same:

$$P(B \mid A) = \frac{P(A \cap B)}{P(A)} = \frac{\frac{1}{3}}{\frac{1}{2}} = \frac{2}{3} = P(B)$$

EXAMPLE 3.19

Refer to the consumer product complaint study in Example 3.12. The percentages of complaints of various types during and after the guarantee period are shown in Table 3.5. Define the following events:

A: {Cause of complaint is product appearance}

B: {Complaint occurred during the guarantee term}

Are A and B independent events?

Solution

Events A and B are independent if $P(A \mid B) = P(A)$. We calculated $P(A \mid B)$ in Example 3.12 to be .51, and from Table 3.5 we see that

$$P(A) = .32 + .03 = .35$$

Therefore, $P(A \mid B)$ is not equal to $P(A)$, and A and B are not independent events.

To gain an intuitive understanding of independence, think of situations in which the occurrence of one event does not alter the probability that a second event will occur. For example, new medical procedures are often tested on laboratory animals. The scientists conducting the tests generally try to perform the procedures on the animals so the results for one animal do not affect the results for the others. That is, the event that the procedure is successful on one animal is *independent* of the result for another. In this way, the scientists can get a more accurate idea of the efficacy of the procedure than if the results were dependent, with the success or failure for one animal affecting the results for other animals.

As a second example, consider an election poll in which 1,000 registered voters are asked their preference between two candidates. Pollsters try to use procedures for selecting a sample of voters so that the responses are independent. That is, the objective of the pollster is to select the sample so the event that one polled voter prefers candidate *A* does not alter the probability that a second polled voter prefers candidate *A*.

In the world of sports, do you think the results of a batter's successive trips to the plate in baseball or of a basketball player's successive shots at the basket, are independent? If a basketball player makes two successive shots, is the probability of making the next shot altered from its value if the result of the first shot were not known? If a player makes two shots in a row, we are likely to assign a probability to his making the third shot different from what we would assign if we knew nothing about the first two shots. Research has shown that many such results in sports tend to be *dependent* because players (and even teams) tend to get on "hot" and "cold" streaks, during which their probabilities of success may increase or decrease significantly.

We will make three final points about independence. The first is that the property of independence, unlike the mutually exclusive property, cannot be shown on or gleaned from a Venn diagram, and you cannot trust your intuition. In general, the only way to check for independence is by performing the calculations of the probabilities in the definition.

The second point concerns the relationship between the mutually exclusive and independence properties. Suppose that events *A* and *B* are mutually exclusive, as shown in Figure 3.11. Are these events independent or dependent? That is, does the assumption that *B* occurs alter the probability of the occurrence of *A*? It certainly does, because if we assume that *B* has occurred, it is impossible for *A* to have occurred simultaneously. *Thus, mutually exclusive events are dependent events*.

The third point is that the probability of the intersection of independent events is very easy to calculate. Referring to the formula for calculating the probability of an intersection, we find

$$P(A \cap B) = P(B)P(A \mid B)$$

Thus, since $P(A \mid B) = P(A)$ when *A* and *B* are independent, we have the following useful rule:

If events A and B are independent, the probability of the intersection of *A* and *B* equals the product of the probabilities of *A* and *B*; that is,

$$P(A \cap B) = P(A)P(B)$$

In the die-toss experiment, we showed in Example 3.18 that the events A: {Observe an even number} and B: {Observe a number less than or equal to 4} are independent if the die is fair. Thus,

$$P(A \cap B) = P(A)P(B) = \left(\frac{1}{2}\right)\left(\frac{2}{3}\right) = \frac{1}{3}$$

This agrees with the result

$$P(A \cap B) = P(2) + P(4) = \frac{2}{6} = \frac{1}{3}$$

that we obtained in the example.

EXAMPLE 3.20

Almost every retail business has the problem of determining how much inventory to purchase. Insufficient inventory may result in lost business, and excess inventory may have a detrimental effect on profits. Suppose a retail computer store owner is planning to place an order for personal computers (PCs). She is trying to decide how many IBM PCs and how many IBM compatibles (personal computers that run all or most of the same software as the IBM PC but are not manufactured by IBM) to order.

The owner's records indicate that 80% of the previous PC customers purchased IBM PCs and 20% purchased compatibles.

a. What is the probability that the next two customers will purchase compatibles?
b. What is the probability that the next ten customers will purchase compatibles?

Solution

a. Let C_1 represent the event that customer 1 will purchase a compatible and C_2 represent the event that customer 2 will purchase a compatible. The event that *both* customers purchase compatibles is the intersection of the two events, $C_1 \cap C_2$. From the records the store owner could reasonably conclude that $P(C_1) = .2$ (based on the fact that 20% of past customers have purchased compatibles), and the same reasoning would apply to C_2. However, in order to compute the probability of $C_1 \cap C_2$, we need more information. Either the records must be examined for the occurrence of consecutive purchases of compatibles, or some assumption must be made to enable the calculation of $P(C_1 \cap C_2)$ from the multiplicative rule. It seems reasonable to make the assumption that the two events are independent, since the decision of the first customer is not likely to affect the decision of the second customer. Assuming independence, we have

$$P(C_1 \cap C_2) = P(C_1)P(C_2) = (.2)(.2) = .04$$

b. To see how to compute the probability that ten consecutive purchases will be compatibles, first consider the event that three consecutive customers purchase compatibles. If C_3 represents the event that the third customer purchases a compatible, then we want to compute the probability of the intersection of $C_1 \cap C_2$ with C_3. Again assuming independence of the purchasing decisions, we have

$$P(C_1 \cap C_2 \cap C_3) = P(C_1 \cap C_2)P(C_3)$$
$$= (.2)^2(.2) = .008$$

Similar reasoning leads to the conclusion that the intersection of ten such events can be calculated as follows:

$$P(C_1 \cap C_2 \cap \cdots \cap C_{10}) = P(C_1)P(C_2) \cdots P(C_{10})$$
$$= (.2)^{10}$$
$$= .0000001024$$

Thus, the probability that ten consecutive customers purchase IBM compatibles is about 1 in 10 million, assuming the probability of each customer's purchase of a compatible is .2 and the purchase decisions are independent.

EXERCISES 3.48–3.66

LEARNING THE MECHANICS

3.48 An experiment results in one of three mutually exclusive events, A, B, or C. It is known that $P(A) = .40$, $P(B) = .25$, and $P(C) = .35$.
a. Find $P(A \cup B)$. b. Find $P(A \cap C)$. c. Find $P(A \mid B)$. d. Find $P(B \cup C)$.
e. Are B and C independent events? Explain.

3.49 An experiment results in one of five simple events with the following probabilities: $P(E_1) = .22$, $P(E_2) = .31$, $P(E_3) = .15$, $P(E_4) = .22$, and $P(E_5) = .1$. The following events have been defined:

A: $\{E_1, E_3\}$

B: $\{E_2, E_3, E_4\}$

C: $\{E_1, E_5\}$

Find each of the following probabilities:
a. $P(A)$ b. $P(B)$ c. $P(A \cap B)$ d. $P(A \mid B)$ e. $P(B \cap C)$ f. $P(C \mid B)$

3.50 Three fair coins are tossed and the following events are defined:

A: $\{$Observe at least one head$\}$

B: $\{$Observe exactly two heads$\}$

C: $\{$Observe exactly two tails$\}$

D: $\{$Observe at most one head$\}$

a. Sum the probabilities of the appropriate simple events to find: $P(A)$, $P(B)$, $P(C)$, $P(D)$, $P(A \cap B)$, $P(A \cap D)$, $P(B \cap C)$, and $P(B \cap D)$.
b. Use the formulas of this section and your answers to part **a** to calculate: $P(A \cup B)$, $P(D \cup A)$, $P(B \cup C)$, $P(B \mid A)$, $P(A \mid D)$, and $P(C \mid B)$.
c. Are events A and B independent? Mutually exclusive?
d. Are events A and D independent? Mutually exclusive?
e. Are events B and C independent? Mutually exclusive?

3.51 Two hundred shoppers at a large suburban mall were asked two questions: (1) Did you see a television ad for the sale at department store X during the past 2 weeks? (2) Did you shop at department store X during the past 2 weeks? The responses to the questions are summarized in the table.

	SHOPPED AT X	DID NOT SHOP AT X
Saw ad	100	25
Did not see ad	25	50

One of the 200 shoppers questioned is to be chosen at random.

a. What is the probability that the person selected saw the ad?

b. What is the probability that the person selected saw the ad and shopped at store X?

c. Find the conditional probability that the person shopped at store X given that the person saw the ad.

d. What is the probability that the person selected shopped at store X?

e. Use your answers to parts **a, b,** and **d** to check the independence of the events Saw ad and Shopped at X.

f. Are the two events Did not see ad and Did not shop at X mutually exclusive? Explain.

3.52 Two fair dice are tossed, and the following events are defined:

> A: {Sum of the numbers showing is odd}
>
> B: {Sum of the numbers showing is 9, 11, or 12}

a. Are events A and B independent? Why?

b. Are events A and B mutually exclusive? Why?

3.53 A sample space contains six simple events and events A, B, and C as shown in the Venn diagram. The probabilities of the simple events are $P(1) = .20$, $P(2) = .05$, $P(3) = .30$, $P(4) = .10$, $P(5) = .10$, $P(6) = .25$.

Venn diagram for Exercise 3.53

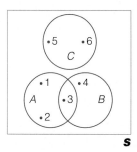

a. Which pairs of events, if any, are mutually exclusive? Why?

b. Which pairs of events, if any, are independent? Why?

c. Find $P(A \cup B)$ by adding the probabilities of the simple events and then by using the additive rule. Verify that the answers agree. Repeat for $P(A \cup C)$.

3.54 For two events, A and B, $P(A) = .4$ and $P(B) = .2$.

a. If A and B are independent, find $P(A \cap B)$, $P(A \mid B)$, and $P(A \cup B)$.

b. If A and B are dependent, with $P(A \mid B) = .6$, find $P(A \cap B)$ and $P(B \mid A)$.

APPLYING THE CONCEPTS

3.55 Each of 50 people in a random sample was asked to name his or her favorite soft drink. The responses are shown below:

Pepsi-Cola	18
Coca-Cola	16
Seven-Up	6
Mr. Pibb	4
Sprite	4
Nehi Orange	1
Dr Pepper	1

Suppose a person is selected at random from the survey. Let A be the event that the person preferred a soft drink bottled by the Coca-Cola Company (Coca-Cola, Mr. Pibb, or Sprite). Let B be the event that the person did *not* choose a cola (either Pepsi-Cola or Coca-Cola).

a. Find $P(A)$.   b. Find $P(B)$.

c. Describe the event $A \cap B$, and find its probability.

d. Describe the event $A \cup B$ and use the additive rule to find its probability.

e. Use the multiplicative rule to find $P(A \mid B)$.

f. Use the multiplicative rule to find $P(A \mid B')$.

3.56   The percentages of all teenagers aged 14 to 19 in two small communities fell in the following six community–delinquency categories:

		COMMUNITY	
		A	B
DELINQUENCY	Nondelinquents	28%	42%
	First offenders	5%	15%
	Repeat offenders	7%	3%

Suppose a teenager was chosen at random from one of the two communities and the following events are defined:

A: {Teenager chosen is from community A}

B: {Teenager chosen is from community B}

C: {Teenager chosen is not delinquent}

D: {Teenager chosen has committed at least one crime}

Find the following probabilities:

a. $P(C')$ b. $P(A \cup D)$ c. $P(C \mid B)$ d. $P(D \mid A)$

3.57   The probability that a certain microchip for a computer fails when first used is .10. If it does not fail immediately, the probability that it lasts 1 year is .99. Define the experiment as observing whether a microchip fails or not during its first year, and define the events:

A: {Microchip lasts through first use}

B: {Microchip lasts from first use through end of first year}

a. List the simple events of the experiment in terms of the events A and B. [*Hint:* The event that the chip fails during the first year is *not* a simple event because it can be decomposed into two more basic events. There are three simple events.]

b. Assign probabilities to the simple events. [*Hint:* You will need to use the multiplicative rule for some of the simple events.]

c. Use the simple event probabilities to find the probability that the microchip does not fail during the first year.

3.58   Some strings of holiday lights are wired in series; thus, if one bulb fails the entire string goes out. Suppose the probability of an individual bulb failing during a certain period of time is .05. What is the probability that a string of ten lights goes out during that period of time? What assumption did you make concerning the light bulbs? [*Hint:* The complement of the event, At least one of the ten lights fails, is the event, None of the ten lights fails.]

3.59   Refer to Exercise 3.17 and the problem of purchasing used automobiles that have had their odometer mileage set back. Suppose you are shopping for a used car, and the dealer has three good-looking cars in stock that were formerly owned by an auto leasing company. If, as stated by the Highway

Traffic Safety Administration, nine out of ten of these cars have had their odometer mileage set back, what is the probability that:

a. All three cars have had their odometer mileage set back?

b. None have had their odometer mileage set back?

c. Exactly one has had its odometer mileage set back?

d. The car you select (from the three used cars available) has had its odometer mileage set back?

3.60 On April 25, 1980, the *Bakersfield Californian* reported on a record payout in the Pennsylvania lottery. The winning number in the state lottery that day, with options from 000 to 999, was 666. Harried state officials were denying the possibility of tampering with the lottery, but many bookies in Pittsburgh were refusing bets on any numbers involving 4's or 6's. Each digit of the winning number was determined by allowing a blast of air to propel one of ten Ping-Pong balls, numbered 0, 1, 2, . . . , 9, into a basket. On this particular day, the TV host (eventually indicted and convicted) and some friends injected liquid into all the balls except those numbered 4 and 6. Thus, only numbers involving the 4 and the 6 digits could be winners.

a. If the balls had not been fixed, what is the probability that a number involving only 4's and 6's would win? What assumption about the determination of the three digits did you make to solve this problem?

b. Given the conditions of part **a**, what is the probability that the number 666 would win?

c. Since the TV announcer and collaborators knew that the winning number would involve only 4's and 6's, what is the probability that the winning number would be 666?

3.61 A popular dice game called craps is played in the following manner. A player starts by rolling two dice. If the result is a 7 or 11, the player wins. For any other sum appearing on the dice, the player continues to roll the dice until that outcome reoccurs (in which case the player wins) or until a 7 or 11 occurs (in which case the player loses). If on any roll the outcome is 2 (snake-eyes), the game is over, and the player loses.

a. What is the probability that a player wins the game on the first roll of the dice? (Assume the dice are balanced.)

b. What is the probability that a player loses the game on the first roll of the dice?

c. If the player throws a total of 3 on the first roll, what is the probability that the game ends on the next roll?

3.62 Despite penicillin and other antibiotics, bacterial pneumonia still kills thousands of Americans every year. Just recently, the United States Food and Drug Administration approved the use of a new antipneumonia vaccine called Pneumovax. It is designed especially for elderly or debilitated patients who are usually the most vulnerable to bacterial pneumonia. Field trials proved the new vaccine to be 90% effective in stimulating the production of antibodies to pneumonia-producing bacteria (i.e., 90% successful in preventing a person exposed to pneumonia-producing bacteria from acquiring the disease). Suppose the probability of an elderly or debilitated person being exposed to these bacteria is .40 (whether inoculated or not), and after being exposed, the probability of each person contracting bacterial pneumonia if not inoculated with the vaccine is .95. Find the probability that an elderly or debilitated person inoculated with this new vaccine acquires pneumonia. What is the probability if this person has not been inoculated?

3.63 According to an article in the *Bakersfield Californian* (May 10, 1982), as many as 1 in every 500 blacks in the United States has sickle-cell anemia. One in 10 is a carrier of the trait, and those who marry have a 1 in 4 chance of giving the disease to their children. Suppose a carrier has three children and the transmission of the disease from the carrier to any one child is independent of whether or not the carrier transmits to another. Find the probability that:

a. None of the children acquire the disease.

b. All three acquire the disease.

c. Exactly one acquires the disease.

3.64   One of the problems encountered in organ transplants is rejection by the body of the transplanted tissue, i.e., the tendency of the white blood cells to attack the foreign tissue. An article in *Newsweek* ("The New Era of Transplants," August 29, 1983) notes that the key to reception or rejection is the nature of the antigens attached to the tissue cells. If the antigens of the donor and receiver match, the body will accept the transplanted tissue. The article goes on to note that the antigens in twins always match. The probability of a match in siblings is .25, and of a match in two people from the population at large is .001. Suppose you need a kidney, and you have two brothers and a sister.
   a. If one of the three siblings offers a kidney, what is the probability that the antigens will match?
   b. If all three siblings offer a kidney, what is the probability that all three antigens will match?
   c. If all three siblings offer a kidney, what is the probability that none of the antigens will match?

3.65   Refer to Exercise 3.64. Answer parts **b** and **c** but this time assume that the three donors were obtained from the population at large.

3.66   Refer to the kidney transplant exercise (Exercise 3.64). The *Newsweek* article (August 29, 1983) reports that Sandoz, a pharmaceutical firm, has developed a new drug, cyclosporine, that appears to retard a body's immune system from rejecting transplanted organs. Sandoz reports that kidney transplant patients who receive the drug have an 80% chance of living through the first year, and it has also improved the survival rates for other types of transplants. Suppose a hospital performs four kidney transplants, and all patients receive the drug cyclosporine. What is the probability that:
   a. All four patients are alive at the end of 1 year?
   b. None of the four patients is alive at the end of 1 year?
   c. At least one of the patients is alive at the end of 1 year?

3.6 Probability and Statistics: An Example

We have introduced a number of new concepts in the preceding sections, and this makes the study of probability a particularly arduous task. It is therefore essential to establish clearly the connection between probability and statistics, which we will do in the remaining chapters. However, we present one brief example in this section so that you can begin to understand why some knowledge of probability is important in the study of statistics.

Suppose a psychologist is researching the hypothesis that rats which have been trained will pass on at least part of the training to their offspring. To test the hypothesis, three offspring (no two with the same parents) of trained rats are randomly selected and subjected to a training test. It is known from many previous experiments that the relative frequency distribution of the scores for untrained rats is mound-shaped with a mean of 60 and a standard deviation of 10. Suppose all three of the trained rats' offspring score more than 70 on the test. What can the research psychologist conclude?

The relative frequency distribution of the scores for untrained rats is shown in Figure 3.14. If the distribution is mound-shaped and approximately symmetric about the mean, we can conclude that approximately 16% of untrained rats will score more than 70 on the test (see Table 2.9). Now define the events

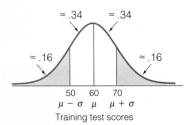

A_1:   {Offspring 1 scores more than 70}

A_2:   {Offspring 2 scores more than 70}

A_3:   {Offspring 3 scores more than 70}

FIGURE 3.14

Relative frequency distribution of training test scores

We want to find $P(A_1 \cap A_2 \cap A_3)$, the probability that all three offspring score more than 70 on the training test.

Since the offspring are selected so that they have different parents, it may be plausible to assume that the events A_1, A_2, and A_3 are independent. That is,

$$P(A_2 \mid A_1) = P(A_2)$$

In words, knowing that the first offspring scores more than 70 on the test does not affect the probability that the second offspring scores more than 70. With the assumption of independence, we can calculate the probability of the intersection by multiplying the individual probabilities:

$$P(A_1 \cap A_2 \cap A_3) = P(A_1)P(A_2)P(A_3)$$
$$\approx (.16)(.16)(.16) = .004096$$

Thus, the probability that the research psychologist will observe all three offspring scoring more than 70 is only about .004 *if the offspring are untrained.* If this event were to occur, the psychologist might conclude that it lends credence to the theory that the offspring inherit some of the parents' training *since it is so unlikely to occur if they are untrained.* Such a conclusion would be an application of the rare event approach to statistical inference. You can see that the basic principles of probability play an important role.

3.7 Random Sampling

How a sample is selected from a population is of vital importance in statistical inference because the probability of an observed sample will be used to infer the characteristics of the sampled population. To illustrate, suppose you deal yourself 4 cards from a deck of 52 cards and all 4 cards are aces. Do you conclude that your deck is an ordinary bridge deck, containing only 4 aces, or do you conclude that the deck is stacked with more than 4 aces? It depends on how the cards were drawn. If the 4 aces were always placed at the top of a standard bridge deck, drawing 4 aces is not unusual—it is certain. On the other hand, if the cards are thoroughly mixed, drawing 4 aces in a sample of 4 cards is highly improbable. The point, of course, is that in order to use the observed sample of 4 cards to draw inferences about the population (the deck of 52 cards), you need to know how the sample was selected from the deck.

One of the simplest and most frequently employed sampling procedures is implied in the previous examples and exercises. It produces what is known as a *random sample.*

Definition 3.10

If n elements are selected from a population in such a way that every set of n elements in the population has an equal probability of being selected, the n elements are said to be a **random sample.***

---

*Strictly speaking, this is a **simple random sample**. There are many different types of random samples. The simple random sample is the most common.

EXAMPLE 3.21

Suppose a lottery consists of ten tickets. (This number is small to simplify our example.) One ticket stub is to be chosen, and the corresponding ticket holder will receive a generous prize. How would you select this ticket stub so that the prize will be awarded fairly?

Solution

If the prize is to be awarded fairly, it seems reasonable to require that each ticket stub have the same probability of being drawn. That is, each stub should have a probability of $\frac{1}{10}$ of being selected. A method to achieve the objective of equal selection probabilities is to *mix* the ten stubs thoroughly and *blindly* pick one of the stubs. If this procedure were repeatedly used, each time replacing the selected stub, a particular stub should be chosen approximately $\frac{1}{10}$ of the time in a long series of draws. This method of sampling is known as *random sampling*.

If a population is not too large and the elements can be numbered on slips of paper, poker chips, etc., you can physically mix the slips of paper or chips and remove n elements from the total. The numbers that appear on the chips selected would indicate the population elements to be included in the sample. Such a procedure will not guarantee a random sample because it is often difficult to achieve a thorough mix (see Case Study 3.4), but it usually provides a reasonably good approximation to random sampling.

Sampling can be thought of as an experiment consisting of drawing n elements from a population of N elements, with each different sample representing a simple event of the experiment. Thus, in Example 3.21, in which a lottery consisted of drawing one of ten tickets, there are ten different samples, or simple events. If the drawing is held so that each simple event is equally likely with probability $\frac{1}{10}$, then the result of the experiment is a *random sample*.

Of course, for most applications the population will consist of more than $N = 10$ elements, and the sample will consist of more than $n = 1$ element. Often, the total number of possible samples will not be easy to visualize, so a method for counting the number of samples is needed. For example, suppose the lottery consists of drawing two tickets from ten. We can list the possible samples as in Table 3.7, where T_1 represents ticket 1, T_2 ticket 2, . . . , and T_0 ticket 10. The systematic listing in Table 3.7 shows the 45 possible samples, the simple events of the experiment of sampling two elements from ten. However, the listing is tedious and only gets more so as the values of the population size N and the sample size n are increased.

TABLE 3.7 **Listing of All Possible Samples of Two Tickets Drawn from Ten Tickets**

T_1, T_2	T_2, T_3	T_3, T_4	T_4, T_5	T_5, T_6	T_6, T_7	T_7, T_8	T_8, T_9	T_9, T_0
T_1, T_3	T_2, T_4	T_3, T_5	T_4, T_6	T_5, T_7	T_6, T_8	T_7, T_9	T_8, T_0	
T_1, T_4	T_2, T_5	T_3, T_6	T_4, T_7	T_5, T_8	T_6, T_9	T_7, T_0		
T_1, T_5	T_2, T_6	T_3, T_7	T_4, T_8	T_5, T_9	T_6, T_0			
T_1, T_6	T_2, T_7	T_3, T_8	T_4, T_9	T_5, T_0				
T_1, T_7	T_2, T_8	T_3, T_9	T_4, T_0					
T_1, T_8	T_2, T_9	T_3, T_0						
T_1, T_9	T_2, T_0							
T_1, T_0								

A second method of determining the number of samples is to use **combinatorial mathematics**. The combinatorial symbol for the number of different ways of selecting n elements from N elements is $\binom{N}{n}$, which is read "the number of combinations of N elements taken n at a time." The formula* for calculating the number is

$$\binom{N}{n} = \frac{N!}{n!(N-n)!}$$

where "!" is the factorial symbol and is a shorthand for the following multiplication:

$$n! = n(n-1)(n-2) \cdot \cdots \cdot (3)(2)(1)$$

Thus, for example, $5! = 5 \cdot 4 \cdot 3 \cdot 2 \cdot 1 = 120$. (The quantity $0!$ is defined to be equal to 1.)

EXAMPLE 3.22

a. Use the combinatorial formula to count the number of different ways of drawing one lottery ticket from a total of ten tickets.
b. Use the combinatorial formula to count the number of ways of drawing two lottery tickets from a total of ten tickets.

Solution

a. Substituting $n = 1$ and $N = 10$ into the formula, we find

$$\binom{N}{n} = \binom{10}{1} = \frac{10!}{1!(10-1)!} = \frac{10!}{1!9!}$$
$$= \frac{10 \cdot 9 \cdot 8 \cdot 7 \cdot 6 \cdot 5 \cdot 4 \cdot 3 \cdot 2 \cdot 1}{(1)(9 \cdot 8 \cdot 7 \cdot 6 \cdot 5 \cdot 4 \cdot 3 \cdot 2 \cdot 1)}$$
$$= 10$$

Thus, as we know intuitively, there are ten different samples that can be selected when drawing one ticket from ten.
b. Substituting $n = 2$ and $N = 10$ into the formula, we find

$$\binom{N}{n} = \binom{10}{2} = \frac{10!}{2!(10-2)!} = \frac{10!}{2!8!}$$
$$= \frac{10 \cdot 9 \cdot 8 \cdot 7 \cdot 6 \cdot 5 \cdot 4 \cdot 3 \cdot 2 \cdot 1}{(2 \cdot 1)(8 \cdot 7 \cdot 6 \cdot 5 \cdot 4 \cdot 3 \cdot 2 \cdot 1)} = \frac{10 \cdot 9}{2 \cdot 1}$$
$$= 45$$

This agrees with the number obtained by listing all possible samples in Table 3.7 but requires much less effort. And, as the next example illustrates, the combinatorial formula works long after listing has ceased to be a viable option.

EXAMPLE 3.23

Suppose you wish to randomly sample 5 households from a population of 100,000 households.

a. How many different samples can be selected?
b. Give a procedure for selecting a random sample.

---

*For a more thorough discussion of the reasoning behind this and other counting rules, see (optional) Section 3.8.

Solution

a. Using the combinatorial rule, we find

$$\binom{100,000}{5} = \frac{100,000!}{5!\,99,995!}$$

$$= \frac{100,000 \cdot 99,999 \cdot 99,998 \cdot 99,997 \cdot 99,996}{5 \cdot 4 \cdot 3 \cdot 2 \cdot 1}$$

$$= 8.33 \times 10^{22}$$

Thus, there are 83.3 billion trillion different samples of 5 households that can be selected from 100,000.

b. How can we ensure that each of the possible samples has an equal chance of being selected, as required for random sampling, when there are so many? Generally, we will use a **random number table**, such as the one in Table I of Appendix A.

First, we number the households in the population from 1 to 100,000. Then, we turn to a page of Table I, say the first page. (A partial reproduction of the first page of Table I is shown in Figure 3.15.) Now, randomly select a starting number, say the random number appearing in the third row, second column. This number is 48360. Proceed down the second column to obtain the remaining four random numbers. The five selected random numbers are shaded in Figure 3.15. Using the first five digits to represent the households from 1 to 99,999 and the number 00000 to represent household 100,000, you can see that the households numbered

48,360   93,093   39,975   6,907   72,905

should be included in your sample.

FIGURE 3.15

Partial reproduction of Table I in Appendix A

ROW \ COLUMN	1	2	3	4	5	6
1	10480	15011	01536	02011	81647	91646
2	22368	46573	25595	85393	30995	89198
3	24130	48360	22527	97265	76393	64809
4	42167	93093	06243	61680	07856	16376
5	37570	39975	81837	16656	06121	91782
6	77921	06907	11008	42751	27756	53498
7	99562	72905	56420	69994	98872	31016
8	96301	91977	05463	07972	18876	20922
9	89579	14342	63661	10281	17453	18103
10	85475	36857	53342	53988	53060	59533
11	28918	69578	88231	33276	70997	79936
12	63553	40961	48235	03427	49626	69445
13	09429	93969	52636	92737	88974	33488

Can we be sure that all 83.3 billion trillion samples have an equal chance of being selected? We cannot, but to the extent that the random number table contains truly random sequences of digits, the sample should be very close to random.

Table I in Appendix A is just one example of a *table of random numbers*. Most samplers use such a table to obtain random samples. Random number tables are constructed in such a way that every number occurs with (approximately) equal probability. Further, the occurrence of any one number in a position is independent of any of the other numbers that appear in the table. To use a table of random numbers, number the *N* elements in the population from 1 to *N*. Then turn to Table I and select a starting number in the table. Proceeding from this number either across the row or down the column, remove and record *n* numbers from the table. Use only the necessary number of digits in each random number to identify the element to be included in the sample. If, in the course of recording the *n* numbers from the table, you select a number that has already been selected, simply discard the duplicate and select a replacement at the end of the sequence. Thus, you may have to record more than *n* numbers from the table to obtain a sample of *n* unique numbers.

CASE STUDY 3.4

THE 1970 DRAFT LOTTERY

From 1948 through the early years of the Vietnam War, the Selective Service System drafted men into military service by age—oldest first, starting with 25-year-olds. A network of local draft boards was used to implement the selection process. Then, on the evening of December 1, 1969, the Selective Service System conducted a lottery to determine the order of selection for 1970 in an attempt to overcome what many believed were inequities in the system. (Such lotteries had been used during World Wars I and II, but it had been 27 years since the last one.)

The objective of the lottery was to randomly order the induction sequence of men between the ages of 19 and 26. To do this, the 366 possible days in a year were written on slips of paper and placed in egg-shaped capsules that were stored in monthly lots. The monthly lots were placed one by one into a wooden box that was ". . . turned end over end several times to mix the numbers" ("Random or Not? Judge Studies Lottery Protest," *The National Observer*, January 12, 1970, p. 2). The capsules were then dumped into a large glass bowl and drawn one by one to obtain the order of induction. All men born on the first day drawn would be inducted first; those born on the second day drawn would be inducted next, etc. Thus, the lottery assigned a rank to each of the 366 birthdays. The results of the lottery are shown in Table 3.8 on pages 144 and 145.

In order to generate a random sequence of numbers with this procedure, it is necessary for each (remaining) capsule in the bowl to have an equal probability of being selected on each draw. That is, by means of thorough mixing, each capsule must have an equal opportunity to come to rest precisely where the sampler's hand closes within the bowl. Although a mixing procedure with this property is almost impossible to achieve, the ideal can be closely approximated. Unfortunately, this was apparently not the case in the 1970 lottery. Even though the sequence of dates in Table 3.8 may appear to be random, there is ample statistical evidence to indicate a nonrandom selection of induction dates.*

To obtain an understanding of the problem with the 1970 lottery, we calculated the median rank for each month and plotted them in Figure 3.16(a), page 145. If the sequence of ranks were randomly generated, there should be no relationship

---

*Formal statistical tests to detect nonrandomness are beyond the scope of this text.

T A B L E 3.8   **1970 Draft Lottery Results**

1	Sept. 14	57	Feb. 23	113	Sept. 15	169	Mar. 15	225	Sept. 1	281	Nov. 28
2	April 24	58	Jan. 19	114	Aug. 6	170	Mar. 26	226	May 29	282	Nov. 10
3	Dec. 30	59	Jan. 24	115	July 3	171	Oct. 15	227	July 19	283	Oct. 8
4	Feb. 14	60	June 21	116	Aug. 23	172	July 23	228	June 2	284	July 10
5	Oct. 18	61	Aug. 29	117	Oct. 22	173	Dec. 26	229	Oct. 29	285	Feb. 29
6	Sept. 6	62	April 21	118	Jan. 23	174	Nov. 30	230	Nov. 24	286	Aug. 25
7	Oct. 26	63	Sept. 20	119	Sept. 23	175	Sept. 13	231	April 14	287	July 30
8	Sept. 7	64	June 27	120	July 16	176	Oct. 25	232	Sept. 4	288	Oct. 17
9	Nov. 22	65	May 10	121	Jan. 16	177	Sept. 19	233	Sept. 27	289	July 27
10	Dec. 6	66	Nov. 12	122	Mar. 7	178	May 14	234	Oct. 7	290	Feb. 22
11	Aug. 31	67	July 25	123	Dec. 28	179	Feb. 25	235	Jan. 17	291	Aug. 21
12	Dec. 7	68	Feb. 12	124	April 13	180	June 15	236	Feb. 24	292	Feb. 18
13	July 8	69	June 13	125	Oct. 2	181	Feb. 8	237	Oct. 11	293	Mar. 5
14	April 11	70	Dec. 21	126	Nov. 13	182	Nov. 23	238	Jan. 14	294	Oct. 14
15	July 12	71	Sept. 10	127	Nov. 14	183	May 20	239	Mar. 20	295	May 13
16	Dec. 29	72	Oct. 12	128	Dec. 18	184	Sept. 8	240	Dec. 19	296	May 27
17	Jan. 15	73	June 17	129	Dec. 1	185	Nov. 20	241	Oct. 19	297	Feb. 3
18	Sept. 26	74	April 27	130	May 15	186	Jan. 21	242	Sept. 12	298	May 2
19	Nov. 1	75	May 19	131	Nov. 15	187	July 20	243	Oct. 21	299	Feb. 28
20	June 4	76	Nov. 6	132	Nov. 25	188	July 5	244	Oct. 3	300	Mar. 12
21	Aug. 10	77	Jan. 28	133	May 12	189	Feb. 17	245	Aug. 26	301	June 3
22	June 26	78	Dec. 27	134	June 11	190	July 18	246	Sept. 18	302	Feb. 20
23	July 24	79	Oct. 31	135	Dec. 20	191	April 29	247	June 22	303	July 26
24	Oct. 5	80	Nov. 9	136	Mar. 11	192	Oct. 20	248	July 11	304	Dec. 17
25	Feb. 19	81	April 4	137	June 25	193	July 31	249	June 1	305	Jan. 1
26	Dec. 14	82	Sept. 5	138	Oct. 13	194	Jan. 9	250	May 21	306	Jan. 7
27	July 21	83	April 3	139	Mar. 6	195	Sept. 24	251	Jan. 3	307	Aug. 13
28	June 5	84	Dec. 25	140	Jan. 18	196	Oct. 24	252	April 23	308	May 28
29	Mar. 2	85	June 7	141	Aug. 18	197	May 9	253	April 6	309	Nov. 26
30	Mar. 31	86	Feb. 1	142	Aug. 12	198	Aug. 14	254	Oct. 16	310	Nov. 5
31	May 24	87	Oct. 6	143	Nov. 17	199	Jan. 8	255	Sept. 17	311	Aug. 19
32	April 1	88	July 28	144	Feb. 2	200	Mar. 19	256	Mar. 23	312	April 8
33	Mar. 17	89	Feb. 15	145	Aug. 4	201	Oct. 23	257	Sept. 28	313	May 31
34	Nov. 2	90	April 18	146	Nov. 18	202	Oct. 4	258	Mar. 24	314	Dec. 12
35	May 7	91	Feb. 7	147	April 7	203	Nov. 19	259	Mar. 13	315	Sept. 30
36	Aug. 24	92	Jan. 26	148	April 16	204	Sept. 21	260	April 17	316	April 22
37	May 11	93	July 1	149	Sept. 25	205	Feb. 27	261	Aug. 3	317	Mar. 9
38	Oct. 30	94	Oct. 28	150	Feb. 11	206	June 10	262	April 28	318	Jan. 13
39	Dec. 11	95	Dec. 24	151	Sept. 29	207	Sept. 16	263	Sept. 9	319	May 23
40	May 3	96	Dec. 16	152	Feb. 13	208	April 30	264	Oct. 27	320	Dec. 15
41	Dec. 10	97	Nov. 8	153	July 22	209	June 30	265	Mar. 22	321	May 8
42	July 13	98	July 17	154	Aug. 17	210	Feb. 4	266	Nov. 4	322	July 15
43	Dec. 9	99	Nov. 29	155	May 6	211	Jan. 31	267	Mar. 3	323	Mar. 10
44	Aug. 16	100	Dec. 31	156	Nov. 21	212	Feb. 16	268	Mar. 27	324	Aug. 11
45	Aug. 2	101	Jan. 5	157	Dec. 3	213	Mar. 8	269	April 5	325	Jan. 10
46	Nov. 11	102	Aug. 15	158	Sept. 11	214	Feb. 5	270	July 29	326	May 22
47	Nov. 27	103	May 30	159	Jan. 2	215	Jan. 4	271	April 2	327	July 6
48	Aug. 8	104	June 19	160	Sept. 22	216	Feb. 10	272	June 12	328	Dec. 2
49	Sept. 3	105	Dec. 8	161	Sept. 2	217	Mar. 30	273	April 15	329	Jan. 11
50	July 7	106	Aug. 9	162	Dec. 23	218	April 10	274	June 16	330	May 1
51	Nov. 7	107	Nov. 16	163	Dec. 13	219	April 9	275	Mar. 4	331	July 14
52	Jan. 25	108	Mar. 1	164	Jan. 30	220	Oct. 10	276	May 4	332	Mar. 18
53	Dec. 22	109	June 23	165	Dec. 4	221	Jan. 12	277	July 9	333	Aug. 30
54	Aug. 5	110	June 6	166	Mar. 16	222	Jan. 28	278	May 18	334	Mar. 21
55	May 16	111	Aug. 1	167	Aug. 28	223	Mar. 28	279	July 4	335	June 9
56	Dec. 5	112	May 17	168	Aug. 7	224	Jan. 6	280	Jan. 20	336	April 19

(continued)

TABLE 3.8 (continued)

337 Jan. 22	342 Oct. 9	347 Feb. 6	352 Aug. 27	357 May 26	362 Mar. 29
338 Feb. 9	343 Mar. 25	348 Nov. 3	353 June 29	358 June 24	363 Feb. 21
339 Aug. 22	344 Aug. 20	349 Jan. 29	354 Mar. 14	359 Oct. 1	364 May 5
340 April 26	345 April 20	350 July 2	355 Jan. 27	360 June 20	365 Feb. 26
341 June 18	346 April 12	351 April 25	356 June 14	361 May 25	366 June 8

FIGURE 3.16

Median plots for lottery results: 1970 and 1971

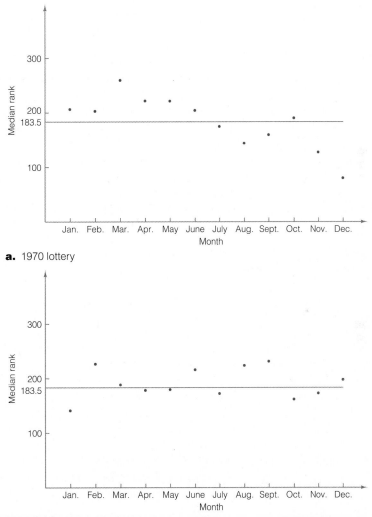

a. 1970 lottery

b. 1971 lottery

between the size of the median ranks and the months of the year. The medians should vary randomly above and below a horizontal line with intercept 183.5 (the median of the integers 1 through 366). However, Figure 3.16(a) reveals a general downward trend in the medians. Men born later in the year were more likely to be drafted before men born early in the year. Furthermore, since not all men between 19 and 25 years old would be drafted in 1970, men born earlier in the year were more likely not to be drafted at all. Although it is possible to observe such a sequence of medians when random sampling is employed, it is highly unlikely. It is more likely that the capsules were not mixed thoroughly enough to give every capsule an equal chance of selection on each draw. The graph indicates that capsules tended to be drawn in monthly groups, and thus suggests that monthly lots of capsules in the wooden boxes were not mixed thoroughly enough after they were dumped into the large glass bowl (Williams, 1978).

The following year, the Selective Service System used more sophisticated mixing techniques to guard against the monthly clustering of selections. The median plot for the 1971 lottery is shown in Figure 3.16(b). [*Note:* The 1971 lottery involved only men born in 1951. Thus, only 365 birthdays were ranked, and the sequence of monthly ranked medians should be compared to 183 instead of 183.5.] Notice that no apparent trend remains; the new mixing technique appears to have been successful.

This case study emphasizes the value of using a random-number table or a random numer generator of a statistical software package in the selection of a random sample. Most important, it points out the problems that may be encountered when attempting to acquire a random sample by a mechanical selection process.

You can now better understand what we mean when we use the phrases "assume the selection is made at random," or simply "at random" in probability examples and exercises. In most cases we are indicating that each simple event, or sample, has an equal probability of occurring.

EXERCISES 3.67–3.71

LEARNING THE MECHANICS

3.67 Suppose you wish to sample $n = 3$ elements from a total of $N = 6$ elements.
   a. Count the number of different samples that can be drawn, first by listing them, and then by using combinatorial mathematics.
   b. If random sampling is to be employed, what is the probability that any particular sample will be selected?
   c. Show how to use the random-number table, Table I in Appendix A, to select a random sample of 3 elements from a population of 6 elements. Perform the sampling procedure 20 times. Do any two of the samples contain the same three elements? Given your answer to part **b**, did you expect repeated samples?

3.68 Suppose you wish to sample $n = 3$ elements from a total of $N = 600$ elements.
   a. Count the number of different samples by using combinatorial mathematics.
   b. If random sampling is to be employed, what is the probability that any particular sample will be selected?
   c. Show how to use the random-number table, Table I in Appendix A, to select a random sample of 3 elements from a population of 600 elements. Perform the sampling procedure 20 times. Do any two of the samples contain the same three elements? Given your answer to part **b**, did you expect repeated samples?

APPLYING THE CONCEPTS

3.69 Archaeologists plan to perform test digs at a location believed to have been inhabited several thousand years ago. The location is approximately 10,000 meters long by 5,000 meters wide. They first draw rectangular grids over the area, consisting of lines every 100 meters, creating a total of $100 \cdot 50 = 5{,}000$ intersections (not counting one of the outer boundaries). The plan is to randomly sample 50 intersection points and dig at the sampled intersections. Explain how you could use the random-number table to obtain a random sample of 50 intersections. Develop at least two plans: one that numbers the intersections from 1 to 5,000 prior to selection and another that selects the row and column of each sampled intersection (from the total of 100 rows and 50 columns).

3.70 To ascertain the effectiveness of their advertising campaigns, firms frequently conduct telephone interviews with consumers. Random samples of telephone numbers may be randomly or systematically selected from telephone directories, or a recent innovation called *random-digit dialing* may be employed. This approach involves using a random-number generator to mechanically create the sample of phone numbers to be called. An advantage of random-digit dialing is that it makes it possible to obtain a representative sample from the population of all households with telephones, whereas with telephone directory sampling, it is only possible to obtain a sample from the population of households that have *listed* telephone numbers.

a. Explain how the random-number table (Table I of Appendix A) could be used to generate a sample of 7-digit telephone numbers.

b. Use the procedure you described in part **a** to generate a sample of ten 7-digit telephone numbers.

c. Use the procedure you described in part **a** to generate five 7-digit telephone numbers whose first three digits are 373.

3.71 In addition to its decennial enumeration of the population, the U.S. Bureau of the Census regularly samples the population to estimate level of and changes in a number of other attributes, such as income, family size, employment, and marital status. Suppose the Bureau plans to sample 1,000 households in a city that has a total of 534,322 households. Although it would use computer-generated random numbers to obtain such a sample, show how the random-number table in Appendix A could be used to generate the sample. Select the first ten households to be included in the sample.

3.8 Some Counting Rules (Optional)

In Section 3.1 we pointed out that experiments sometimes have so many simple events that it is impractical to list them all. However, many of these experiments possess simple events with identical characteristics. If you can develop a **counting rule** to count the number of simple events for such an experiment, it can be used to aid in the solution of the problems.

EXAMPLE 3.24

A product can be shipped by four airlines and each airline can ship via three different routes. How many distinct ways exist to ship the product?

Solution

A pictorial representation of the different ways to ship the product will aid in counting them. The representation, called a **decision tree**, is shown in Figure 3.17. At the starting point (stage 1), there are four choices—the different airlines—to begin the journey. Once we have chosen an airline (stage 2), there are three choices—the different routes—to complete the shipment and reach the final destination. Thus, the decision tree clearly shows that there are $(4)(3) = 12$ distinct ways to ship the product.

Decision tree for shipping problem

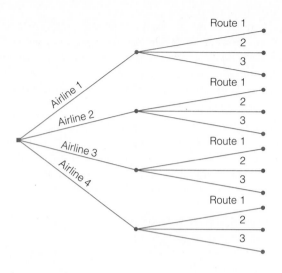

The method of solving this example can be generalized to any number of stages with sets of different elements. This method is given by the **multiplicative rule**.

> ### The Multiplicative Rule
>
> You have k sets of elements, n_1 in the first set, n_2 in the second set, . . . , and n_k in the kth set. Suppose you wish to form a sample of k elements *by taking one element from each* of the k sets. The number of different samples that can be formed is the product
>
> $$n_1 n_2 n_3 \cdots \cdot n_k$$

E X A M P L E 3.25

There are 20 candidates for three different executive positions, E_1, E_2, and E_3. How many different ways could you fill the positions?

Solution

For this example, there are $k = 3$ sets of elements corresponding to:

Set 1: Candidates available to fill position E_1

Set 2: Candidates remaining (after filling E_1) that are available to fill E_2

Set 3: Candidates remaining (after filling E_1 and E_2) that are available to fill E_3

The numbers of elements in the sets are $n_1 = 20$, $n_2 = 19$, $n_3 = 18$. Therefore, the number of different ways of filling the three positions is given by the multiplicative rule as $n_1 n_2 n_3 = (20)(19)(18) = 6,840$.

E X A M P L E 3.26

Consider the experiment discussed in Example 3.10 of tossing a coin ten times. Show how we found that there were $2^{10} = 1,024$ simple events for this experiment.

Solution

There are $k = 10$ sets of elements for this experiment. Each set contains two elements: a head and a tail. Thus, there are

$$(2)(2)(2)(2)(2)(2)(2)(2)(2)(2) = 2^{10} = 1,024$$

different outcomes (simple events) of this experiment.

EXAMPLE 3.27

Suppose there are five dangerous military missions, each requiring one soldier. In how many different ways can five soldiers from a squadron of 100 be assigned to these five missions?

Solution

We can solve this problem by using the multiplicative rule. The entire set of 100 soldiers is available for the first mission, and after the selection of one soldier for that mission, 99 are available for the second mission, etc. Thus, the total number of different ways of choosing five soldiers for the five missions is

$$n_1 n_2 n_3 n_4 n_5 = (100)(99)(98)(97)(96) = 9,034,502,400$$

The arrangement of elements in a distinct order is called a *permutation*. Thus, in Example 3.27 there are more than 9 billion different *permutations* of 5 elements (soldiers) drawn from a set of 100 elements!

Permutations Rule

Given a *single set* of N distinctly different elements, you wish to select n elements from the N and *arrange* them within n positions. The number of different **permutations** of the N elements taken n at a time is denoted by P_n^N and is equal to

$$P_n^N = N(N-1)(N-2) \cdots \cdots (N-n+1) = \frac{N!}{(N-n)!}$$

where $n! = n(n-1)(n-2) \cdots \cdots (3)(2)(1)$ and is called n factorial. (Thus, for example, $5! = 5 \cdot 4 \cdot 3 \cdot 2 \cdot 1 = 120$.) The quantity $0!$ is defined to be equal to 1.

EXAMPLE 3.28

Consider the following transportation problem. You wish to drive, in sequence, from a starting point to each of five cities, and you wish to compare the distances—and ultimately the costs—of the different routes. How many different routes would have to be compared?

Solution

Denote the cities as $C_1, C_2, \ldots, C_5$. Then a route moving from the starting point to C_2 to C_1 to C_3 to C_4 to C_5 would be represented as $C_2 C_1 C_3 C_4 C_5$. The total number of routes would equal the number of ways you could rearrange the $N = 5$ cities in $n = 5$ positions. This number is

$$P_n^N = P_5^5 = \frac{5!}{(5-5)!} = \frac{5!}{0!} = \frac{5 \cdot 4 \cdot 3 \cdot 2 \cdot 1}{1} = 120$$

(Recall that $0! = 1$.)

EXAMPLE 3.29

Solution

There are four construction workers. You must assign three to job 1 and one to job 2. In how many different ways can you make this assignment?

To begin, suppose each worker is to be assigned to a distinct job. Then, using the multiplicative rule, we obtain $(4)(3)(2)(1) = 24$ ways of assigning the workers to four distinct jobs. The 24 ways are listed in four groups in Table 3.9 (where ABCD signifies that worker A was assigned the first distinct job, worker B the second, etc.).

TABLE 3.9

GROUP 1	GROUP 2	GROUP 3	GROUP 4
ABCD	ABDC	ACDB	BCDA
ACBD	ADBC	ADCB	BDCA
BACD	BADC	CADB	CBDA
BCAD	BDAC	CDAB	CDBA
CABD	DABC	DACB	DBCA
CBAD	DBAC	DCAB	DCBA

Now suppose the first three positions represent job 1 and the last position represents job 2. We can now see that all the listings in group 1 represent the same outcome of the experiment of interest. That is, workers A, B, and C are assigned to job 1 and worker D to job 2. Similarly, group 2 listings are equivalent, as are group 3 and group 4 listings. Thus, there are only four different assignments of four workers to the two jobs. These are shown in Table 3.10.

To generalize this result, we point out that the final result can be found by

$$\frac{(4)(3)(2)(1)}{(3)(2)(1)(1)} = 4$$

TABLE 3.10

JOB 1	JOB 2
ABC	D
ABD	C
ACD	B
BCD	A

The $(4)(3)(2)(1)$ is the number of different ways (*permutations*) the workers could be assigned four distinct jobs. The division by $(3)(2)(1)$ is to remove the duplicated permutations resulting from the fact that three workers are assigned the same jobs. And the division by 1 is associated with the worker assigned to job 2.

| The Partitions Rule

There exists a *single* set of N distinctly different elements. You wish to partition them into k sets, with the first set containing n_1 elements, the second containing n_2 elements, . . . , and the kth set containing n_k elements. The number of different partitions is

$$\frac{N!}{n_1! n_2! \cdots \cdot n_k!} \qquad \text{where } n_1 + n_2 + n_3 + \cdots + n_k = N$$

EXAMPLE 3.30

You have 12 construction workers. You wish to assign three to job site 1, four to job site 2, and five to job site 3. In how many different ways can you make this assignment?

Solution

For this example, $k = 3$ (corresponding to the $k = 3$ different job sites), $N = 12$, and $n_1 = 3$, $n_2 = 4$, and $n_3 = 5$. Then the number of different ways to assign the workers to the job sites is

$$\frac{N!}{n_1! n_2! n_3!} = \frac{12!}{3! 4! 5!} = \frac{12 \cdot 11 \cdot 10 \cdot \cdots \cdot 3 \cdot 2 \cdot 1}{(3 \cdot 2 \cdot 1)(4 \cdot 3 \cdot 2 \cdot 1)(5 \cdot 4 \cdot 3 \cdot 2 \cdot 1)} = 27{,}720$$

EXAMPLE 3.31

How many samples of four courses can be selected from a list of 25 courses in a university catalog?

Solution

For this example, $k = 2$ (corresponding to the $n_1 = 4$ classes you *do* choose and the $n_2 = 21$ classes you *do not* choose) and $N = 25$. Then, the number of different ways to choose the four classes from 25 is

$$\frac{N!}{n_1! n_2!} = \frac{25!}{(4!)(21!)} = \frac{25 \cdot 24 \cdot 23 \cdot \cdots \cdot 3 \cdot 2 \cdot 1}{(4 \cdot 3 \cdot 2 \cdot 1)(21 \cdot 20 \cdot \cdots \cdot 2 \cdot 1)} = 12{,}650$$

(Perhaps you now understand better the confusion that surrounds registration time at many schools.)

This special application of the partitions rule—partitioning a set of N elements into $k = 2$ groups (the elements that appear in a sample and those that do not)—is very common. Therefore, we give a different name to the rule for counting the number of different ways of partitioning a set of elements into two parts: the **combinations rule**.

The Combinations Rule

A sample of n elements is to be chosen from a set of N elements. Then the number of different samples of n elements that can be selected from N is denoted by $\binom{N}{n}$ and is equal to

$$\binom{N}{n} = \frac{N!}{n!(N - n)!}$$

Note that the order in which the n elements are drawn is not important.

EXAMPLE 3.32

Five soldiers from a squadron of 100 are to be chosen for a dangerous mission. In how many ways can groups of five be formed?

This problem is equivalent to sampling $n = 5$ elements from a set of $N = 100$ elements. Thus, the number of ways is the number of possible combinations of five soldiers selected from 100, or

$$\binom{100}{5} = \frac{100!}{(5!)(95!)} = \frac{100 \cdot 99 \cdot 98 \cdot 97 \cdot 96 \cdot 95 \cdot 94 \cdots \cdots 2 \cdot 1}{(5 \cdot 4 \cdot 3 \cdot 2 \cdot 1)(95 \cdot 94 \cdots \cdots 2 \cdot 1)}$$

$$= \frac{100 \cdot 99 \cdot 98 \cdot 97 \cdot 96}{5 \cdot 4 \cdot 3 \cdot 2 \cdot 1}$$

$$= 75,287,520$$

Compare this result with that of Example 3.7, where we found that the number of permutations of five elements drawn from 100 was more than 9 billion. Because the order of the elements does not affect combinations, there are *fewer* combinations than permutations.

Summary of Counting Rules

1. *Multiplicative rule.*   If you are drawing *one element from each of k sets of elements*, with the sizes of the sets $n_1, n_2, \ldots, n_k$, the number of different results is

$$n_1 n_2 n_3 \cdots \cdots n_k$$

2. *Permutations rule.*   If you are drawing *n elements from a set of N elements and arranging the n elements in a distinct order*, the number of different results is

$$P_n^N = \frac{N!}{(N - n)!}$$

3. *Partitions rule.*   If you are partitioning the *elements of a set of N elements into k groups consisting of* $n_1, n_2, \ldots, n_k$ elements ($n_1 + \cdots + n_k = N$), the number of different results is

$$\frac{N!}{n_1! n_2! \cdots \cdots n_k!}$$

4. *Combinations rule.*   If you are drawing *n elements from a set of N elements without regard to the order of the n elements*, the number of different results is

$$\binom{N}{n} = \frac{N!}{n!(N - n)!}$$

[*Note:*   The combinations rule is a special case of the partitions rule when $k = 2$.]

When working a probability problem, you should carefully examine the experiment to determine whether you can use one or more of the rules we have discussed in this section. We will illustrate how these rules can help solve a probability problem.

EXAMPLE 3.33

A consumer testing service is commissioned to rank the top three brands of laundry detergent. A total of ten brands are to be included in the study.

a. In how many different ways can the consumer testing service arrive at the final ranking?
b. If the testing service can distinguish no difference among the brands and therefore arrives at the final ranking by random selection, what is the probability that company Z's brand is ranked first? In the top three?

Solution

a. Since the testing service is drawing three elements (brands) from a set of ten elements and arranging the three elements in a distinct order, we use the permutations rule to find the number of different results:

$$P^{10}_3 = \frac{10!}{(10-3)!} = 10 \cdot 9 \cdot 8 = 720$$

b. The steps for calculating the probability of interest are as follows:

STEP 1 The experiment is to select and rank three brands of detergent from ten brands.

STEP 2 There are too many simple events to list. However, we know from part **a** that there are 720 different outcomes (i.e., simple events) of this experiment.

STEP 3 If we assume the testing service determines the rankings at random, each of the 720 simple events should have an equal probability of occurrence. Thus,

$$P(\text{Each simple event}) = \frac{1}{720}$$

STEP 4 One event of interest to company Z is that their brand receives top ranking. We will call this event A. The list of simple events that result in the occurrence of event A is long, but the *number* of simple events contained in event A is determined by breaking event A into two parts:

Company Z brand   9 other brands
↓   ↙   ↘
Rank 1   Rank 2   Rank 3

Only one possibility $P^9_2 = \frac{9!}{(9-2)!} = 72$ possibilities

→ Event A ←
$1 \cdot 72 = 72$ simple events

Thus, event A can occur in 72 different ways.

Now define B as the event that company Z's brand is ranked in the top three. Since event B specifies only that brand Z appear in the top three, we repeat the calculations above, fixing brand Z in position 2 and then in position 3. We conclude that the number of simple events contained in event B is $3(72) = 216$.

STEP 5 The final step is to calculate the probabilities of events A and B. Since the 720 simple events are equally likely to occur, we find

$$P(A) = \frac{\text{Number of simple events in } A}{\text{Total number of simple events}} = \frac{72}{720} = \frac{1}{10}$$

Similarly,

$$P(B) = \frac{216}{720} = \frac{3}{10}$$

EXAMPLE 3.34

Refer to Example 3.33. Suppose the consumer testing service is to choose the top three laundry detergents from the group of ten *but is not to rank the three.*

a. In how many different ways can the testing service choose the three to be designated as top detergents?
b. Assuming that the testing service makes its choice by a random selection and that company X has two brands in the group of ten, what is the probability that exactly one of the company X brands is selected in the top three? At least one?

Solution

a. The testing service is selecting three elements (brands) from a set of ten elements *without regard to order*, so we can apply the combinations rule to determine the number of different results:

$$\binom{10}{3} = \frac{10!}{3!(10-3)!} = \frac{10 \cdot 9 \cdot 8}{3 \cdot 2 \cdot 1} = 120$$

b. The steps for calculating the probability of interest are as follows:

STEP 1 The experiment is to select (*but not rank*) three brands from ten.

STEP 2 There are 120 simple events for this experiment.

STEP 3 Since the selection is made at random,

$$P(\text{Each simple event}) = \frac{1}{120}$$

STEP 4 Define events A and B as follows:

A: {Exactly one company X brand is selected}
B: {At least one company X brand is selected}

Since each of the simple events is equally likely to occur, we need only know the number of simple events in A and B to determine their probabilities.

In order for event A to occur, exactly one company X brand must be selected, along with two of the remaining eight brands. We thus break A into two parts:

2 company X brands   8 other brands
↓   ↓
Select 1   Select 2

$$\binom{2}{1} = 2 \text{ different possibilities} \qquad \binom{8}{2} = \frac{8!}{2!(8-2)!} = 28 \text{ different possibilities}$$

Event A
$2 \cdot 28 = 56$ simple events

Note that the one company X brand can be selected in two ways, whereas the two other brands can be selected in 28 ways. (We use the combinations rule because the order of selection is not important.) Then, we use the multiplicative rule to combine one of the two ways to select a company X brand with one of the 28 ways to select two other brands, yielding a total of 56 simple events for event A.

To count the simple events contained in B, we first note that

$$B = A \cup C$$

where A is defined as before and

    C: {Both company X brands are selected}

We have determined that A contains 56 simple events. Using the same method for event C, we find:

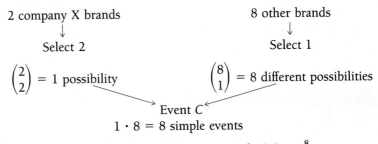

Thus, event C contains eight simple events and $P(C) = \frac{8}{120}$.

STEP 5 Since all the simple events are equally likely,

$$P(A) = \frac{\text{Number of simple events in } A}{\text{Total number of simple events}} = \frac{56}{120} = \frac{7}{15}$$

$$P(B) = P(A \cup C) = P(A) + P(C) - P(A \cap C)$$

where $P(A \cap C) = 0$ because the testing service cannot select exactly one *and* exactly two of the company X brands; that is, A and C are mutually exclusive events. Thus,

$$P(B) = \frac{56}{120} + \frac{8}{120} = \frac{64}{120} = \frac{8}{15}$$

Learning how to decide whether a particular counting rule applies to an experiment takes patience and practice. If you wish to develop this skill, attempt to use the rules to solve the following exercises and especially the optional exercises (marked with an asterisk) at the end of this chapter. Proofs of the counting rules can be found in the text by W. Feller (1968, Chapter 3).

EXERCISES 3.72–3.90

LEARNING THE MECHANICS

3.72 Use the multiplicative rule to determine the number of simple events in the sample space corresponding to the experiment of tossing a coin the following number of times:

    a. 2 times   **b.** 3 times   **c.** 5 times   **d.** n times

3.73   Find the numerical values of:

    a. $\binom{6}{3}$ **b.** P_2^5 **c.** P_2^4 **d.** $\binom{100}{98}$ **e.** $\binom{50}{50}$ **f.** $\binom{50}{0}$ **g.** P_3^5 **h.** P_0^{10}

3.74   An experiment consists of choosing objects without regard to order. How many simple events are there if you choose the following?

    a. 3 objects from 7   **b.** 2 objects from 6   **c.** 2 objects from 30
    d. 8 objects from 10   **e.** r objects from q

3.75   Determine the number of simple events contained in the sample space when you toss the following:
    a. 1 die   **b.** 2 dice   **c.** 4 dice   **d.** n dice

3.76   Suppose you are managing ten employees, and you need to form three teams to work on different projects. Assume that each employee may serve on any team. In how many different ways can the teams be formed if the number of members on each project team are as follows:
    a. 3, 3, 4   **b.** 2, 3, 5   **c.** 1, 4, 5   **d.** 2, 4, 4

APPLYING THE CONCEPTS

3.77   Flying into New York from a certain city, you can choose one of three airlines and can travel either first class or economy. How many travel options do you have?

3.78   A computer retail store has nine personal computers (PCs) in stock. A university buyer plans to purchase three of them. Unknown to either buyer or seller, two of the PCs in stock have defective disk drives. The three computers are selected at random from the nine available.
    a. In how many different ways can the three computers be selected?
    b. What is the probability that none of the computers selected will have a defective disk drive?
    c. What is the probability that at least one of the computers selected will have a defective disk drive?

3.79   How many different 5-card hands can be dealt from a 52-card bridge deck?

3.80   Suppose an automobile license plate is designed to show a number from 1 to 26, corresponding to the 26 counties in a state, and then is followed by a 5-digit number. How many different license plates can the state issue?

3.81   The Republican governor of a state with no income tax is appointing a committee of 5 to consider changes in the income tax law. There are 15 state representatives available for appointment to the committee, 7 Democrats and 8 Republicans. Assume that the governor selects the committee of 5 members randomly from the 15 representatives.
    a. In how many different ways can the committee members be selected?
    b. What is the probability that no Democrat is appointed to the committee? If this were to occur, what would you conclude about the assumption that the appointments were made at random? Why?
    c. What is the probability that the majority of the committee members are Republican? If this were to occur, would you have reason to doubt that the governor made the selections randomly? Why?

3.82   A company sells five different styles of intermediate model automobile. The buyer can get an automobile in one of eight colors with either standard or automatic transmission. Would it be reasonable to expect a dealer to stock at least one automobile in every style, color, transmission combination? At a minimum, how many automobiles would the dealer have to stock?

3.83   A salesperson living in city A wishes to visit four cities, B, C, D, and E.
    a. If the cities are all interconnected by airlines, how many different travel plans could be constructed to visit each city exactly once and then return home?
    b. Suppose all cities are connected except that B and C are not directly connected. How many different flight plans would be available to the salesperson?

3.84　In an article titled "State's License Plates Will Soon Look Different," the *Gainesville Sun* (April 11, 1984) notes that the state of Florida is running out of combinations of letters and numbers for its license plates. The current license plate contains three letters of the alphabet followed by three digits selected from the ten digits, 0, 1, 2, . . . , 9. This system, according to the article, allows a maximum of 17,576,000 combinations of letters and numbers or, equivalently, 17,576,000 license plates. Is this number, 17,576,000, correct?

3.85　Refer to Exercise 3.84. According to the *Gainesville Sun* article, new Florida tags will reverse the existing procedure, starting with three digits followed by three letters.
a. How many new tags will the new system provide?
b. Since the new tag numbers will be added to the old numbers, what is the total number of licenses available for registration in Florida?

3.86　The 5-digit zip code has become an integral part of the United States postal system. How many different zip codes are available for use by the postal service? (Assume any 5-digit number can be used as a zip code.) The first three digits of a Chicago zip code are 606. If no other city in the United States has these first three digits as part of its zip, how many different zip codes may possibly exist in Chicago?

3.87　Suppose you are to choose a basketball team (five players) from eight available athletes.
a. How many ways can you choose a team (ignoring positions)?
b. How many ways can you choose a team composed of two guards, two forwards, and a center?
c. How many ways can you choose a team composed of tall guard, short guard, power forward, shooting forward, and center?

3.88　Intercollegiate volleyball rules require that after the opposing team has lost its serve, each of the six members of the serving team must rotate into new positions on the court. Hence, each player must be able to play all six different positions. How many different team combinations by player and position are possible during a volleyball game? If players are initially assigned to the positions in a random manner, find the probability that the best server on the team is in the serving position.

3.89　A college professor hands out a list of ten questions, five of which will appear on the final examination for the course. One of the students taking the course is pressed for time and can prepare for only seven of the ten questions. If the professor chooses the five questions at random from the ten, then:
a. What is the probability that the student will be prepared for all five questions that appear on the final examination?
b. What is the probability that the student will be prepared for less than three questions?
c. What is the probability that the student will be prepared for exactly four questions?

3.90　Consider 5-card poker hands dealt from a standard 52-card bridge deck. Two important events are:

　　A:　{You draw a flush}

　　B:　{You draw a straight}

a. Find $P(A)$.　　b. Find $P(B)$.
c. The event that both A and B occur, that is, $A \cap B$, is called a straight flush. Find $P(A \cap B)$.

[*Note:*　A *flush* consists of any 5 cards of the same suit. A *straight* consists of any 5 cards with values in sequence. In a straight, the cards may be of any suit and an ace may be considered as having a value of 1 or a value higher than a king.]

Summary

We have developed some of the basic tools of probability to enable us to assess the probability of various sample outcomes given a specific population structure. Although many of the examples we presented were of no practical importance, they accomplished their purpose if you now understand the basic concepts and definitions of probability.

In the next several chapters we will present probability models that can be used to solve practical problems. You will see that for most applications, we will need to make inferences about unknown aspects of these probability models, i.e., we will need to apply inferential statistics to the problem.

SUPPLEMENTARY EXERCISES 3.91–3.121

[*Note: Starred (*) exercises refer to the optional section in this chapter.*]

LEARNING THE MECHANICS

3.91   A fair die is tossed and the up face is noted. If the number is even, the die is tossed again; if the number is odd, a fair coin is tossed. Define the events:

   A: {A head appears on the coin}

   B: {The die is tossed only one time}

   a.   List the simple events in the sample space.
   b.   Give the probability for each of the simple events.
   c.   Find $P(A)$ and $P(B)$.
   d.   Identify the simple events in A', B', $A \cap B$, and $A \cup B$.
   e.   Find $P(A')$, $P(B')$, $P(A \cap B)$, $P(A \cup B)$, $P(A \mid B)$, and $P(B \mid A)$.
   f.   Are A and B mutually exclusive events? Independent events? Why?

3.92   The Venn diagram shown below illustrates a sample space containing six simple events and three events, A, B, and C. The probabilities of the simple events are

$$P(1) = .4 \qquad P(2) = .1 \qquad P(3) = .1$$
$$P(4) = .2 \qquad P(5) = .1 \qquad P(6) = .1$$

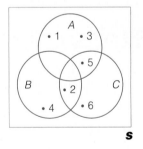

   a.   Find $P(A \cap B)$, $P(B \cap C)$, $P(A \cup C)$, $P(A \cup B \cup C)$, $P(B')$, $P(A' \cap B)$, $P(B \mid C)$, and $P(B \mid A)$.
   b.   Are A and B independent? Mutually exclusive? Why?
   c.   Are B and C independent? Mutually exclusive? Why?

3.93   Which of the following pairs of events are mutually exclusive? Justify your response.
   a.   The Dow Jones Industrial Average increases on Monday
        A large New York bank decreases its prime interest rate on Monday
   b.   An IBM microcomputer is purchased
        An Apple microcomputer is purchased
   c.   Subject A in a psychology experiment responds to a stimulus within 5 seconds
        Subject A in a psychology experiment records the fastest response to a stimulus (2.3 seconds)

3.94   Psychologists tend to believe that there is a relationship between aggressiveness and order of birth.
        To test this belief, a psychologist chose 500 elementary school students at random and administered
        each a test designed to measure the student's aggressiveness. Each student was classified according
        to one of four categories. The percentages of students falling in the four categories are shown here.

	Firstborn	Not firstborn
Aggressive	15%	15%
Not agressive	25%	45%

     a. If one student is chosen at random from the 500, what is the probability that the student is
        firstborn?
     b. What is the probability that the student is aggressive?
     c. What is the probability that the student is aggressive, given the student was firstborn?
     d. If

         A: {Student chosen is aggressive}

         B: {Student chosen is firstborn}

     are A and B independent events? Explain.

3.95   Two events, A and B, are independent, with $P(A) = .3$ and $P(B) = .1$.
     a. Are A and B mutually exclusive? Why?
     b. Find $P(A \mid B)$ and $P(B \mid A)$.
     c. Find $P(A \cup B)$.

3.96   A balanced die is thrown once. If a 4 appears, a ball is drawn from urn 1; otherwise a ball is drawn
        from urn 2. Urn 1 contains four red, three white, and three black balls. Urn 2 contains six red and
        four white balls.
     a. Find the probability that a red ball is drawn.
     b. Find the probability that urn 1 was used given that a red ball was drawn.

*3.97   Find the numerical value of:
     a. $6!$ **b.** $\binom{10}{9}$ **c.** $\binom{10}{1}$ **d.** P_2^6 **e.** $\binom{6}{3}$ **f.** $0!$ **g.** P_4^{10} **h.** P_2^{50}

3.98   A random sample of five students is to be selected from 50 sociology majors for participation in a
        special program.
     a. In how many different ways can the sample be drawn?
     b. Show how the random-number table, Table I of Appendix A, can be used to select the sample of
         students.

APPLYING THE CONCEPTS

3.99   In college basketball games a player may be afforded the opportunity to shoot two consecutive foul
        shots (free throws).
     a. Suppose a player who scores on 80% of his foul shots has been awarded two free throws. If the
        two throws are considered independent, what is the probability that the player scores on both
        shots? Exactly one? Neither shot?
     b. Suppose a player who scores on 80% of his first attempted foul shots has been awarded two free
        throws, and the outcome on the second shot is dependent on the outcome of the first shot. In
        fact, if this player makes the first shot he makes 90% of the second shots, and if he misses the
        first shot, he makes 70% of the second shots. In this case, what is the probability that the player
        scores on both shots? Exactly one? Neither shot?

c. In parts **a** and **b**, we considered two ways of *modeling* the probability a basketball player scores on two consecutive foul shots. Which model do you think is a more realistic attempt to explain the outcome of shooting foul shots, i.e., do you think two consecutive foul shots are independent or dependent? Explain.

3.100 Two companies, A and B, package and market a chemical substance and claim .15 of the total weight of the substance is sodium. However, a careful survey of 4,000 packages (half from each company) indicates the proportion varies around .15, with the results shown here.

		PROPORTION OF SODIUM			
		Less than .100	.100–.149	.150–.199	Over .200
CHEMICAL BRAND	A	25%	10%	10%	5%
	B	5%	5%	10%	30%

Suppose a package is chosen at random from the 4,000 packages. If

    A: {Package chosen is brand A}

    B: {Package chosen is brand B}

    C: {Package chosen contains less than .100 sodium}

    D: {Package chosen contains between .100 and .149 sodium}

    E: {Package chosen contains between .150 and .199 sodium}

    F: {Package chosen contains over .200 sodium}

then describe the characteristics of a package portrayed by the following events:
a. $A \cup B$ **b.** $B \cup F$ **c.** $A \cap D$ **d.** $E \cap B$ **e.** $(A \cap C) \cup (A \cap D)$

3.101 Use your intuitive understanding of independence to form an opinion about whether each of the following scenarios represents independent events.
a. The results of consecutive tosses of a coin
b. The opinions of randomly selected individuals in a pre-election poll
c. A major league baseball player's results in two consecutive at-bats
d. The amount of gain or loss associated with investments in different stocks that are bought on the same day and sold on the same day 1 month later
e. The amount of gain or loss associated with investments in different stocks that are bought and sold in different time periods, 5 years apart
f. The responses of two different subjects to the same stimulus in a psychology experiment

3.102 A survey of the faculty of a large university yielded the breakdown according to sex and marital status shown here.

		MARITAL STATUS			
		Married	Single	Divorced	Widowed
SEX	Male	60%	12%	8%	5%
	Female	10%	3%	2%	0%

Suppose a faculty member is chosen at random. If

    A: {Faculty member chosen is female}

    B: {Faculty member chosen is male}

    C: {Marital status of chosen faculty member is married}

D: {Marital status of chosen faculty member is single}

E: {Marital status of chosen faculty member is widowed}

F: {Marital status of chosen faculty member is divorced}

then describe the characteristics of a faculty member portrayed by the following events:

a. $A \cup B$ **b.** $B \cup F$ **c.** $C \cup D$ **d.** $E \cap F$ **e.** $F \cup B$ **f.** $E \cup F$

3.103 A local country club has a membership of 600 and operates facilities that include an 18-hole championship golf course and 12 tennis courts. Before deciding whether to accept new members, the club president would like to know how many members regularly use each facility. A survey of the membership indicates that 70% regularly use the golf course, 50% regularly use the tennis courts, and 5% use neither of these facilities regularly.

a. Construct a Venn diagram to describe the results of the survey.

b. If one member is chosen at random, what is the probability that the member uses the golf course or the tennis courts or both?

c. If one member is chosen at random, what is the probability that the member uses both the golf and the tennis facilities?

d. A member is chosen at random from among those known to use the tennis courts regularly. What is the probability that the member also uses the golf facilities regularly?

3.104 In the game of Parcheesi each player rolls a pair of dice on each turn. In order to begin the game you must throw a 5 on at least one of the dice, or a total of 5 on the two dice. What is the probability that you can begin the game on your first turn? The second turn? The third turn? The nth turn?

3.105 Entomologists are often interested in studying the effect of chemical attractants (pheromones) on insects. One common technique is to release several insects equidistant from the pheromone being studied and from a control substance. If the pheromone has an effect, more insects will travel toward it rather than toward the control. Otherwise, the insects are equally likely to travel in either direction. Suppose the pheromone under study has no effect so that it is equally likely that an insect will move toward either the pheromone or the control.

a. If five insects are released, what is the probability that all five travel toward the pheromone?

b. Exactly four?

3.106 The probability of the union of three events, A, B, and C, is given by this expression (proof omitted):

$$P(A \cup B \cup C) = P(A) + P(B) + P(C) - P(A \cap B) - P(A \cap C) - P(B \cap C) + P(A \cap B \cap C)$$

a. Sketch a Venn diagram showing the events A, B, and C and their intersections. Let the areas in the diagram corresponding to events A, B, C, $A \cap B$, $A \cap C$, $B \cap C$, and $A \cap B \cap C$ represent their probabilities. Then use the diagram to justify the formula for $P(A \cup B \cup C)$.

b. A national poll of biostatisticians was conducted to ascertain their professional responsibilities. An analysis of the responses gave the following distribution of professional responsibilities:

A: {Research}   40%

B: {Professional consultation}   64%

C: {Data collection and analysis}   36%

Suppose 10% are involved in all three activities; 15% are involved in both A and C; and 17% are involved in both A and B. Using this information, find the percentage of all biostatisticians that are involved in both B and C. Assume that

$$P(A \cup B \cup C) = 1$$

3.107 The performance of quality inspectors affects both the quality of outgoing products and the cost of the products. A product that passes inspection is assumed to meet quality standards; a product that fails inspection may be reworked, scrapped, or reinspected. Quality engineers at Westinghouse Electric

Corporation evaluated performances of inspectors in judging the quality of solder joints by comparing each inspector's classifications of a set of 153 joints with the consensus evaluation of a panel of experts. The results for a particular inspector are shown in the table.

| | | INSPECTOR'S JUDGMENT | |
		Joint acceptable	Joint rejectable
COMMITTEE'S JUDGMENT	Joint acceptable	101	10
	Joint rejectable	23	19

Source: Meagher, J. J. and Scazzero, J. A. "Measuring inspector variability," *39th Annual Quality Congress Transactions*, May 1985, 75–81. ©1985 American Society for Quality Control. Reprinted with permission.

One of the 153 solder joints is to be selected at random.
  a. What is the probability that the inspector judges the joint to be acceptable? That the committee judges the joint to be acceptable?
  b. What is the probability that both the inspector and the committee judge the joint to be acceptable? That neither judge the joint to be acceptable?
  c. What is the probability that the inspector and the committee disagree? Agree?

3.108 A manufacturer of 35-mm cameras knows that a shipment of 30 cameras sent to a large discount store contains six defective cameras. The manufacturer also knows that the store will choose two of the cameras at random, test them, and accept the shipment if neither is defective.
  a. What is the probability that the first camera chosen by the store will be defective?
  b. Given that the first camera chosen passed inspection, what is the probability that the second camera chosen will fail inspection?
  c. What is the probability that the shipment will be accepted?

3.109 The probability that a microcomputer salesperson sells a computer to a prospective customer on the first visit to the customer is .4. If the salesperson fails to make the sale on the first visit, the probability that the sale will be made on the second visit is .65. The salesperson never visits a prospective customer more than twice. What is the probability that the salesperson will make a sale to a particular customer?

3.110 Seventy-five percent of all women who submit to pregnancy tests are really pregnant. A certain pregnancy test gives a false positive result with probability .02 and a valid positive result with probability .99. If a particular woman's test is positive, what is the probability that she really is pregnant? [*Hint:* If A is the event that a woman is pregnant and B is the event that the pregnancy test is positive, then B is the union of the two mutually exclusive events $A \cap B$ and $A' \cap B$. Also, the probability of a "false positive result" may be written as $P(B \mid A') = .02$.]

3.111 Blackjack, a favorite game of gamblers, is played by a dealer and at least one opponent and uses a 52-card bridge deck. At the outset of the game, two cards are dealt to the player and two cards to the dealer. Drawing an ace and a face card is called "blackjack." If the dealer draws it, he or she automatically wins. If the dealer does not draw a blackjack and the player does, the player wins.
  a. What is the probability that the dealer will draw a blackjack?
  b. What is the probability that the player wins with a blackjack?

3.112 The figure shown here is a schematic representation of a system comprised of three components. The system operates properly only if all three components operate properly. The three components are said to operate *in series*. The components could be mechanical or electrical; they could be work stations in an assembly process; or they could represent the functions of three different departments in an organization. The probability of failure for each component is listed in the table. Assume the components operate independently of each other.

A System Comprised of Three Components in Series

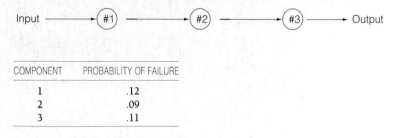

COMPONENT	PROBABILITY OF FAILURE
1	.12
2	.09
3	.11

a. Find the probability that the system operates properly.
b. What is the probability that at least one of the components will fail and therefore that the system will fail?

3.113   The accompanying figure is a representation of a system comprised of two subsystems that are said to operate *in parallel*. Each subsystem has two components that operate in series (refer to Exercise 3.112). The system will operate properly as long as at least one of the subsystems functions properly. The probability of failure for each component in the system is .1. Assume the components operate independently of each other.

A System Comprised of Two Parallel Subsystems

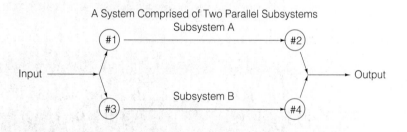

a. Find the probability that the system operates properly.
b. Find the probability that exactly one subsystem fails.
c. Find the probability that the system fails to operate properly.
d. How many parallel subsystems like the two shown here would be required to guarantee that the system would operate properly at least 99% of the time?

3.114   A small brewery has two bottling machines. Machine A produces 75% of the bottles and machine B produces 25%. One out of every 20 bottles filled by A is rejected for some reason, while one out of every 30 bottles from B is rejected. What proportion of bottles is rejected? What is the probability that a bottle comes from machine A, given that it is accepted?

3.115   A county welfare agency employs 30 welfare workers who interview prospective food stamp recipients. Periodically, the supervisor selects, at random, the forms completed by two workers to audit for illegal deductions. Unknown to the supervisor, six of the workers have regularly been giving illegal deductions to applicants.
a. What is the probability that the first worker chosen has been giving illegal deductions?
b. Given that the first worker chosen has been giving illegal deductions, what is the probability that the second worker chosen has also been giving illegal deductions?
c. What is the probability that neither of the two workers chosen has been giving illegal deductions?

3.116   A clinical psychologist is asked to view tapes in which each of six experimental subjects is discussing his or her recent dreams. Three of the six subjects have previously been classified as "high-anxiety" individuals and the other three as "low-anxiety." The psychologist is told only that there are three of each type and is asked to select the three high-anxiety subjects.
   a. List all possible outcomes (simple events) for this experiment.
   b. Assuming that the psychologist guesses at the classifications of the subjects, assign probabilities to the simple events.
   c. Find the probability that the psychologist guesses all classifications correctly.
   d. Find the probability that the psychologist guesses at least two of the three high-anxiety subjects correctly.

*3.117 Refer to Exercise 3.116. Suppose the clinical psychologist is given tapes of 20 subjects and is told only that ten are high-anxiety individuals and ten are low-anxiety individuals. The psychologist is asked to select the ten high-anxiety subjects.
   a. Using the appropriate counting rule from Section 3.8, count the number of simple events for this experiment.
   b. Assuming that the psychologist is guessing, assign probabilities to each of the simple events.
   c. Find the probability that the psychologist guesses all classifications correctly.
   d. Find the probability that the psychologist guesses at least nine of the ten high-anxiety subjects correctly.

*3.118 Suppose a manufacturer of television sets is prepared to send 20 new sets to three retail dealers. Dealer A is to get ten of the sets, dealer B six of the sets, and dealer C four of the sets.
   a. In how many different ways can the 20 sets be divided among the dealers in the prescribed numbers?
   b. Assuming the sets are randomly divided among the dealers and there are three defective sets among the 20, what is the probability that dealer A gets all three defective sets?
   c. Making the same assumptions as in part b, what is the probability that dealer A gets two defective sets and dealer B the other defective set?

*3.119 Suppose there are 500 applicants for five equivalent positions at a factory and the company is able to narrow the field to 30 equally qualified applicants. Seven of the finalists are minority candidates. Assume that the five who are chosen are selected at random from this final group of 30.
   a. In how many different ways can the selection be made?
   b. What is the probability that none of the minority candidates is hired?
   c. What is the probability that no more than one minority candidate is hired?

*3.120 A poll is taken in an attempt to determine the top three prime-time television shows. Each person interviewed is asked to rank his or her first, second, and third favorite show from a list of 42 shows (14 from each of the three major networks, ABC, CBS, NBC).
   a. In how many different ways can the ranking be made by an interviewee?
   b. If an interviewee ranks the top three shows by making a random selection, what is the probability that the first two programs selected are ABC programs while the third is an NBC program?
   c. If the ranking is random, what is the probability that all three program selections are from the same network?

*3.121 According to a morning news program, a very rare event recently occurred in Dubuque, Iowa. Each of four women playing bridge was astounded to note that she had been dealt a perfect bridge hand. That is, one woman was dealt all 13 spades, another all 13 hearts, another all diamonds, and another all the clubs. What is the probability of this rare event?

ON YOUR OWN...

Obtain a standard deck of 52 playing cards (the kind commonly used for bridge, poker, or solitaire). An experiment will consist of drawing 1 card at random from the deck of cards and recording which card was observed. This will be simulated by shuffling the deck thoroughly and observing the top card. Consider the following two events:

A: {Card observed is a heart}

B: {Card observed is an ace, king, queen, or jack}

a. Find $P(A)$, $P(B)$, $P(A \cap B)$, and $P(A \cup B)$.

b. Conduct the experiment ten times and record the observed card each time. Be sure to return the observed card each time and thoroughly shuffle the deck before making the draw. After ten cards have been observed, calculate the proportion of observations that satisfy event A, event B, event $A \cap B$, and event $A \cup B$. Compare the observed proportions with the true probabilities calculated in part **a**.

c. Conduct the experiment 40 more times to obtain a total of 50 observed cards. Now calculate the proportion of observations that satisfy event A, event B, event $A \cap B$, and event $A \cup B$. Compare these proportions with those found in part **b** and the true probabilities found in part **a**.

d. Conduct the experiment 50 more times to obtain a total of 100 observations. Compare the observed proportions for the 100 trials with those found previously. What comments do you have concerning the different proportions found in parts **b**, **c**, and **d** as compared to the true probabilities found in part **a**? How do you think the observed proportions and true probabilities would compare if the experiment were conducted 1,000 times? 1 million times?

USING THE COMPUTER...

Suppose a sociologist is studying income patterns across the United States. She will use the demographic data set described in Appendix C to determine factors that affect the median income for zip code areas.

a. If one of the 1,000 zip codes were to be randomly selected, what is the probability that the selected zone is one for which the median income exceeds $35,000?

b. Suppose the zip code were to be selected from the Northeast census region (from those in the sample). What is the probability that the selected zone is one for which the median income exceeds $35,000?

c. Are the events described in parts **a** and **b** independent? Why or why not? What are the practical implications of the events' independence, or lack thereof?

References

Feller, W. *An Introduction to Probability Theory and Its Applications*, 3d ed., Vol. I. New York: Wiley, 1968, Chapters 1, 3, 4, and 5.

Lindley, D. V. *Making Decisions*, 2d ed. London: Wiley, 1985.

Parzen, E. *Modern Probability Theory and Its Applications*. New York: Wiley, 1960, Chapters 1 and 2.

Rosenblatt, J. R., and Filliben, J. J. "Randomization and the draft lottery," *Science*, 171, pp. 306–308.

Scheaffer, R. L., and Mendenhall, W. *Introduction to Probability: Theory and Applications*. North Scituate, Mass.: Duxbury, 1975, Chapters 1 and 2.

Williams, B. *A Sampler on Sampling*. New York: Wiley, 1978, pp. 5–8.

Winkler, R. L. *An Introduction to Bayesian Inference and Decision*. New York: Holt, Rinehart and Winston, 1972, Chapter 2.

Discrete Random Variables

CONTENTS

WHERE WE'VE BEEN...

By illustration we indicated in Chapter 3 how probability would be used to make an inference about a population from data contained in an observed sample. We also noted that probability would be used to measure the reliability of the inference.

WHERE WE'RE GOING...

Most experimental events in Chapter 3 were events described in words and denoted by capital letters. In real life, most sample observations are numerical—in other words, numerical data. In this chapter, we learn that data are observed values of random variables. We study two important random variables and learn how to find the probabilities of specific numerical outcomes.

You may have noticed that many of the examples of experiments in Chapter 3 generated quantitative (numerical) observations. The unemployment rate, the percentage of voters favoring a particular candidate, the cost of textbooks for a school term, and the amount of pesticide in the discharge waters of a chemical plant are all examples of numerical measurements of some phenomenon. Thus, most experiments have simple events that correspond to values of some numerical variable.

> Definition 4.1
>
> A **random variable** is a rule that assigns one (and only one) numerical value to each simple event of an experiment.*

The term *random variable* is more meaningful than just the term *variable* because the adjective *random* indicates that the experiment may result in one of the several possible values of the variable, according to the *random* outcome of the experiment. For example, if the experiment is to count the number of customers who use the drive-up window of a bank each day, the random variable (the number of customers) will vary from day to day, partly because of the random phenomena that influence whether customers use the drive-up window. Thus, the possible values of this random variable range from 0 to the maximum number of customers the window could possibly serve in a day.

We define two different types of random variables, *discrete* and *continuous*, in Section 4.1. Then we spend the remainder of this chapter discussing specific types of discrete random variables and the aspects that make them important to the statistician. We discuss continuous random variables in Chapter 5.

4.1 Two Types of Random Variables

Recall that the simple event probabilities corresponding to an experiment must sum to 1. Dividing one unit of probability among the simple events in a sample space and consequently assigning probabilities to the values of a random variable is not always as easy as the examples in Chapter 3 might lead you to believe. If the number of simple events is finite, that is, if they can be completely listed, the job is relatively easy. However, some experiments result in an infinite number of sample points, in which case assignment of probabilities is more difficult. In fact, we have to use different probability models depending on the number of values that a random variable can assume.

EXAMPLE 4.1

A panel of ten wine experts is asked to taste a new white wine and assign a rating of 0, 1, 2, or 3. A score is then obtained by adding together the ratings of the ten experts. How many values can this random variable assume?

Solution

A simple event is a sequence of ten numbers associated with the rating of each expert. The random variable assigns a score to each one of these simple events

---

*By *experiment*, we mean an experiment that yields random outcomes (as defined in Chapter 3).

by adding the ten numbers together. Thus, the smallest score is 0 (if all ten ratings are 0) and the largest score is 30 (if all ten ratings are 3). Since every integer between 0 and 30 is a possible score, the random variable x can assume 31 values.

This is an example of a *discrete random variable*, since there is a finite number of distinct possible values. Whenever all the possible values a random variable can assume can be listed (or *counted*), the random variable is *discrete*.

EXAMPLE 4.2

Suppose the Environmental Protection Agency (EPA) takes readings once a month on the amount of pesticide in the discharge water of a chemical company. If the amount of pesticide exceeds the maximum level set by the EPA, the company is forced to take corrective action and may be subject to penalty. Consider the following random variable:

Number, x, of months before the company's discharge exceeds the EPA's maximum level

What values can x assume?

Solution

The company's discharge of pesticide may exceed the maximum allowable level on the first month of testing, the second month of testing, etc. It is possible that the company's discharge will *never* exceed the maximum level. Thus, the set of possible values for the number of months until the level is first exceeded is the set of all positive integers:

1, 2, 3, 4 . . .

If we can list the values of a random variable x, even though the list is never-ending, we call the list **countable** and the corresponding random variable *discrete*. Thus, the number of months until the company's discharge first exceeds the limit is a *discrete random variable*.

EXAMPLE 4.3

Refer to Example 4.2. A second random variable of interest is the amount x of pesticide (in milligrams per liter) found in the monthly sample of discharge waters from the chemical company. What values can this random variable assume?

Solution

Unlike the *number* of months before the company's discharge exceeds the EPA's maximum level, the set of all possible values for the *amount* of discharge *cannot be listed*—i.e., is not countable. The possible values for the amount x of pesticide would correspond to the points on the interval between 0 and the largest possible value the amount of the discharge could attain, the maximum number of milligrams that could occupy 1 liter of volume. (Practically, the interval would be much smaller, say, between 0 and 500 milligrams per liter.) When the values of a random variable are not countable but instead correspond to the points on some interval, we call it a *continuous random variable*. Thus, the *amount* of pesticide in the chemical plant's discharge waters is a *continuous random variable*.

Definition 4.2

Random variables that can assume a *countable* number of values are called **discrete**.

Definition 4.3

Random variables that can assume values coresponding to any of the points contained in one or more intervals are called **continuous**.

The following are examples of discrete random variables:

1. The number of sales made by a salesperson in a given week: $x = 0, 1, 2, \ldots$
2. The number of students in a sample of 500 who favor an increase in student activities and, correspondingly, an increase in student activity fees: $x = 0, 1, 2, \ldots, 500$
3. The number of students applying to medical schools this year: $x = 0, 1, 2, \ldots$
4. The number of errors on a page of an accountant's ledger: $x = 0, 1, 2, \ldots$
5. The number of customers waiting to be served in a restaurant at a particular time: $x = 0, 1, 2, \ldots$

Note that each of the examples of discrete random variables begins with the words "The number of . . . ." This wording is very common, since the discrete random variables most frequently observed are counts. The following are examples of continuous random variables:

1. The length of time between arrivals at a hospital clinic: $0 \leq x < \infty$ (infinity)
2. For a new apartment complex, the length of time from completion until a specified number of apartments are rented: $0 \leq x < \infty$
3. The amount of carbonated beverage loaded into a 12-ounce can in a can-filling operation: $0 \leq x \leq 12$
4. The depth at which a successful oil drilling venture first strikes oil: $0 \leq x \leq c$, where c is the maximum depth obtainable
5. The weight of a food item bought in a supermarket: $0 \leq x \leq 500$
   [*Note:* Theoretically, there is no upper limit on x, but it is unlikely that it would exceed 500 pounds.]

Discrete random variables and their probability distributions are discussed in this chapter. Continuous random variables and their probability distributions are the topic of Chapter 5.

EXERCISES 4.1–4.5

APPLYING THE CONCEPTS

4.1   What is a random variable?

4.2   How do discrete and continuous random variables differ?

4.3   Classify the following random variables according to whether they are discrete or continuous:
   a. The number of words spelled correctly by a student on a spelling test
   b. The amount of liquid waste a plant purifies daily
   c. The length of time an employee is late for work
   d. The number of bacteria per cubic centimeter of drinking water
   e. The amount of carbon monoxide produced per gallon of unleaded gas
   f. Your weight

4.4   Identify the following random variables as discrete or continuous:
   a. The number of patients entering a doctor's office during any given hour
   b. The heart rate (number of beats per minute) of an American male
   c. The time it takes a student to complete an examination
   d. The barometric pressure at a given location
   e. The number of persons in a given town who are disabled
   f. Your heart rate

4.5   Identify the following variables as discrete or continuous:
   a. The length of time until recovery from a tonsillectomy
   b. The number of violent crimes committed per month in your community
   c. The number of commercial aircraft near-misses per month
   d. The number of winners each week in a state lottery
   e. Your blood pressure
   f. Your height

4.2   Probability Distributions for Discrete Random Variables

A complete description of a discrete random variable requires that we *specify the possible values the random variable can assume* and *the probability associated with each value*. To illustrate, consider Example 4.4.

EXAMPLE 4.4

Recall the experiment of tossing two coins (Chapter 3), and let x be the number of heads observed. Find the probability associated with each value of the random variable x, assuming the two coins are fair.

Solution

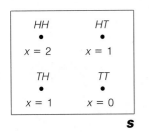

FIGURE 4.1

Venn diagram for the two-coin-toss experiment

Recall from Chapter 3 that the sample space and simple events for this experiment are as shown in Figure 4.1, and the probability associated with each of the four simple events is $\frac{1}{4}$. The random variable x can assume values 0, 1, 2. Then, identifying the probabilities of the simple events associated with each of these values of x, we have

$$P(x = 0) = P(TT) = \frac{1}{4}$$

$$P(x = 1) = P(TH) + P(HT) = \frac{1}{4} + \frac{1}{4} = \frac{1}{2}$$

$$P(x = 2) = P(HH) = \frac{1}{4}$$

Thus, we now know the values the random variable can assume (0, 1, 2) and how the probability is *distributed over* these values $\left(\frac{1}{4}, \frac{1}{2}, \frac{1}{4}\right)$. This completely describes the random variable and is referred to as the *probability distribution*, denoted by the symbol $p(x)$. The probability distribution for the coin-toss example is shown in tabular form in Table 4.1 and in graphic form in Figure 4.2. Since

TABLE 4.1 **Probability Distribution for Coin-Toss Experiment: Tabular Form**

x	$p(x)$
0	$\frac{1}{4}$
1	$\frac{1}{2}$
2	$\frac{1}{4}$

the probability distribution for a discrete random variable is concentrated at specific points (values of x), the graph in Figure 4.2(a) represents the probabilities as the heights of vertical lines over the corresponding values of x. Although the representation of the probability distribution as a histogram, as in Figure 4.2(b), is less precise (since the probability is spread over a unit interval), the histogram representation will prove useful when we approximate probabilities of certain discrete random variables in Section 4.4.

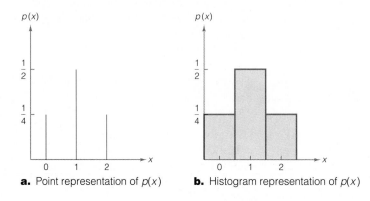

a. Point representation of $p(x)$ **b.** Histogram representation of $p(x)$

FIGURE 4.2

Probability distribution for coin-toss experiment: Graphic form

We could also present the probability distribution for x as a formula, but this would unnecessarily complicate a very simple example. We give the formulas for the probability distributions of some common discrete random variables later in this chapter.

Definition 4.4

The **probability distribution** of a discrete random variable is a graph, table, or formula that specifies the probability associated with each possible value the random variable can assume.

Two requirements must be satisfied by all probability distributions for discrete random variables.

Requirements for the Probability Distribution of a Discrete Random Variable x

$$p(x) \geq 0 \quad \text{for all values of } x$$
$$\sum p(x) = 1$$

where the summation* of $p(x)$ is over all possible values of x.

---

*Unless otherwise indicated, summations will always be over all possible values of x.

Example 4.4 illustrates how the probability distribution for a discrete random variable can be derived, but for many practical situations the task is much more difficult. Fortunately, many experiments and associated discrete random variables observed in nature possess identical characteristics. Thus, you might observe a random variable in a psychology experiment that would possess the same probability distribution as a random variable observed in an engineering experiment or a social sample survey. We classify random variables according to type of experiment, derive the probability distribution for each of the different types, and then use the appropriate probability distribution when a particular type of random variable is observed in a practical situation. The probability distributions for most commonly occurring discrete random variables have already been derived. This fact simplifies the problem of finding the probability distributions for random variables.

In Sections 4.4 and 4.5, we describe two important types of discrete random variables, give their probability distributions, and explain where and how they can be applied in business. (Mathematical derivations of the probability distributions will be omitted, but these details can be found in the references at the end of the chapter.)

But first, in Section 4.3, we discuss some descriptive measures of these sometimes complex probability distributions. Since probability distributions are analogous to the relative frequency distributions of Chapter 2, it should be no surprise that the mean and standard deviation are useful descriptive measures.

EXERCISES 4.6–4.18

LEARNING THE MECHANICS

4.6   Consider the following probability distribution:

x	−4	0	1	3
$p(x)$	.1	.3	.4	.2

a. List the values that x may assume.
b. What value of x is most probable?
c. What is the probability that x is greater than 0?
d. What is the probability that $x = -2$?

4.7   A discrete random variable x can assume five possible values: 2, 3, 5, 8, and 10. Its probability distribution is shown here:

x	2	3	5	8	10
$p(x)$	.20	.10		.25	.15

a. What is $p(5)$?
b. What is the probability that x equals 2 or 10?
c. What is $P(x \leq 8)$?

4.8   Explain why each of the following is or is not a valid probability distribution for a discrete random variable x:

a.
x	0	1	2	3
$p(x)$	.1	.3	.3	.2

b.
x	−2	−1	0
$p(x)$	.25	.50	.25

c.
x	4	9	20
$p(x)$	−.3	.4	.3

d.
x	2	3	5	6
$p(x)$	.15	.15	.45	.35

4.9   The random variable x has the following discrete probability distribution:

x	10	11	12	13	14
$p(x)$	.1	.2	.5	.1	.1

Since the values that x can assume are mutually exclusive events, the event $\{x \le 12\}$ is the union of three mutually exclusive events:

$$\{x = 10\} \cup \{x = 11\} \cup \{x = 12\}$$

a. Find $P(x \le 12)$.   b. Find $P(x > 12)$.
c. Find $P(x \le 14)$.   d. Find $P(x = 14)$.
e. Find $P(x \le 11 \text{ or } x > 12)$.

4.10   The random variable x has the discrete probability distribution shown here.

x	−2	−1	0	1	2
$p(x)$	.05	.20	.40	.20	.15

a. Find $P(x \le 0)$.   b. Find $P(x > -1)$.   c. Find $P(-1 \le x \le 1)$.
d. Find $P(x < 2)$.   e. Find $P(-1 < x < 2)$.   f. Find $P(x < 1)$.

4.11   Toss three fair coins and let x equal the number of heads observed.
a. Identify the simple events associated with this experiment and assign a value of x to each simple event.
b. Calculate $p(x)$ for each value of x.
c. Construct a probability histogram for $p(x)$.
d. What is $P(x = 2 \text{ or } x = 3)$?

4.12   A fair die is tossed twice and x, the sum of the up faces, is recorded.
a. Give the probability distribution for x in tabular form.
b. Find $P(x \ge 8)$.
c. Find $P(x < 8)$.
d. What is the probability that x is odd? Even?
e. What is $P(x = 7)$?

APPLYING THE CONCEPTS

4.13   In Exercise 3.63 we noted that 1 in every 500 blacks in the United States has sickle-cell anemia. One in 10 is a carrier of the trait, and those who marry have a 1 in 4 chance of giving the disease to their children. Suppose a carrier has three children and let x equal the number who acquire the disease.
a. Find $p(x)$ for $x = 0, 1, 2, 3$.   b. Graph $p(x)$.   c. Find $P(x \ge 1)$.

4.14 Every human possesses two sex chromosomes. A copy of one or the other (equally likely) is contributed to an offspring. Males have one X chromosome and one Y chromosome. Females have two X chromosomes. If a couple has three children, what is the probability that they have at least one boy? [*Hint:* Define the random variable z as the number of male offspring, and find the probability distribution of z.]

4.15 Coach "Bear" Bryant of the University of Alabama, a legendary figure in college football, was known for his winning seasons. He consistently won nine or more games per season. Suppose x equals the number of games won up to the halfway mark (six games) in a 12-game season. If Coach Bryant and his team had a probability $p = .70$ of winning any one game (and the winning or losing of one game was independent of another), then the probability distribution of the number x of winning games in a series of six games (we show how to calculate these probabilities in Section 4.4) is

x	0	1	2	3	4	5	6
$p(x)$	.001	.010	.060	.185	.324	.302	.118

Find the probability that the number of games won by Coach Bryant in the first half of a randomly selected season is:
a. 6   **b.** 5   **c.** Less than or equal to 4

4.16 A May 1986 Florida Statewide Household Survey conducted by the University of Florida's Bureau of Economic and Business Research revealed that (approximately) 10% of Florida households intended to purchase a house within the next 6 months. Let x be the number of households contacted by a realtor until one that intends to buy a house in the next 6 months is found. Assume that the probability any household contacted has this intention is exactly .10 and that the intentions of different households are independent.
a. What is the range of possible values for x?
b. Find $P(x < 2)$ and $P(x < 5)$.
c. Find the probability distribution for values of x from 1 to 10 and graph it over that domain. Can x exceed 10?

4.17 In Exercise 3.66 we noted that a new drug, cyclosporine, appears to retard a body's immune system from rejecting transplanted organs. The drug's developer, Sandoz, reports that kidney transplant patients who receive the drug have an 80% chance of living through the first year (*Newsweek*, August 29, 1983). Suppose four kidney transplant patients are given cyclosporine, and let x equal the number living through the first year.
a. Calculate $p(x)$ for $x = 0, 1, 2, 3, 4$.   **b.** Graph $p(x)$.   **c.** Find $P(x < 2)$.

4.18 To entice potential real estate buyers to attend one of its sales presentations, Recreation Resorts of America mails advertising brochures that guarantee that all attendees will win one prize from the groups of prizes listed in the table. The brochure contains a coded "winning number" that can be turned in at the sales presentation to determine which prize has been won. The fine print of the brochure explains that for each 100,000 brochures mailed, there is one $5,000 winner, one microwave oven winner, one television set winner, one home computer winner, one video recorder winner, one

	PRIZE	VALUE OF PRIZE
Group 1:	$5,000	$5,000.00
Group 2:	Microwave oven	769.00
Group 3:	2-person inflatable raft	49.95
Group 4:	19-inch color television	699.00
Group 5:	Home computer	699.00
Group 6:	Home video recorder	795.00
Group 7:	Moped	500.00

moped winner, and 99,994 inflatable raft winners. It also explains that the winning numbers have been "randomly assigned" to mailing labels.

a. Suppose 300,000 brochures were mailed and you received "winning number" 2,823. Construct a probability distribution in graphic form that describes the potential value of your winning number.

b. What is the probability that you will not win an inflatable raft?

c. Does your answer to part **b** change if 1,000,000 brochures were mailed? Explain.

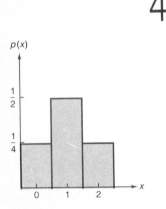

FIGURE 4.3

Probability distribution for a two-coin toss

4.3 Expected Values of Discrete Random Variables

If a discrete random variable x were observed a very large number of times and the data generated were arranged in a relative frequency distribution, the relative frequency distribution would be indistinguishable from the probability distribution for the random variable. Thus, the probability distribution for a random variable is a theoretical model for the relative frequency distribution of a population. To the extent that the two distributions are equivalent (and we will assume they are), the probability distribution for x possesses a mean μ and a variance σ^2 that are identical to the corresponding descriptive measures for the population. This section explains how you can find the mean value for a random variable. We illustrate the procedure with an example.

Examine the probability distribution for x (the number of heads observed in the toss of two fair coins) in Figure 4.3. Try to locate the mean of the distribution intuitively. We may reason that the mean μ of this distribution is equal to 1 as follows: In a large number of experiments, $\frac{1}{4}$ should result in $x = 0$, $\frac{1}{2}$ in $x = 1$, and $\frac{1}{4}$ in $x = 2$ heads. Therefore, the average number of heads is

$$\mu = 0\left(\tfrac{1}{4}\right) + 1\left(\tfrac{1}{2}\right) + 2\left(\tfrac{1}{4}\right)$$
$$= 0 + \tfrac{1}{2} + \tfrac{1}{2} = 1$$

Note that to get the population mean of the random variable x, we multiply each possible value of x by its probability $p(x)$, and then we sum this product over all possible values of x. The *mean of x* is also referred to as the *expected value of x*, denoted $E(x)$.

Definition 4.5

The **mean**, or **expected value**, of a discrete random variable x is

$$\mu = E(x) = \sum xp(x)$$

The term "expected" is a mathematical term and should not be interpreted as it is typically used. Specifically, a random variable might never be equal to its "expected value." Rather, the expected value is the mean of the probability distribution, or a measure of its central tendency.

EXAMPLE 4.5

Suppose you work for an insurance company, and you sell a $10,000 whole-life insurance policy at an annual premium of $290. Actuarial tables show that the probability of death during the next year for a person of your customer's age, sex, health, etc., is .001. What is the expected gain (amount of money made by the company) for a policy of this type?

Solution

The experiment is to observe whether the customer survives the upcoming year. The probabilities associated with the two simple events, Live and Die, are .999 and .001, respectively. The random variable you are interested in is the gain x, which can assume the values shown in the table.

GAIN x	SIMPLE EVENT	PROBABILITY
$290	Customer lives	.999
$290 − $10,000	Customer dies	.001

If the customer lives, the company gains the $290 premium as profit. If the customer dies, the gain is negative because the company must pay $10,000, for a net "gain" of $(290 − 10,000). The expected gain is therefore

$$\mu = E(x) = \sum xp(x)$$
$$= (290)(.999) + (290 - 10,000)(.001)$$
$$= 290(.999 + .001) - 10,000(.001)$$
$$= 290 - 10 = \$280$$

In other words, if the company were to sell a very large number of 1-year $10,000 policies to customers possessing the characteristics described above, it would (on the average) net $280 per sale in the next year.

Example 4.5 illustrates that the expected value of a random variable x need not equal a possible value of x. That is, the expected value is $280, but x will equal either $290 or −$9,710 each time the experiment is performed (a policy is sold and a year elapses). The expected value is a measure of central tendency, and in this case represents the average over a very large number of 1-year policies—but is not a possible value of x.

We learned in Chapter 2 that the mean and other measures of central tendency tell only part of the story about a set of data. The same is true about probability distributions. We need to measure variability as well. Since a probability distribution can be viewed as a representation of a population, we will use the population variance to measure its variability.

The *population variance* σ^2 is defined as the average of the squared distance of x from the population mean μ. Since x is a random variable, the squared distance, $(x - \mu)^2$, is also a random variable. Using the same logic used to find the mean value of x, we find the mean value of $(x - \mu)^2$ by multiplying all possible values of $(x - \mu)^2$ by $p(x)$ and then summing over all possible x values.* This quantity,

$$E[(x - \mu)^2] = \sum_{\text{all } x} (x - \mu)^2 p(x)$$

is also called the **expected value of the squared distance from the mean**; that is, $\sigma^2 = E[(x - \mu)^2]$. The standard deviation of x is defined as the square root of the variance σ^2.

---

*It can be shown that $E[(x - \mu)^2] = E(x^2) - \mu^2$, where $E(x^2) = \sum x^2 p(x)$. Note the similarity between this expression and the shortcut formula $\sum (x - \bar{x})^2 = \sum x^2 - (\sum x)^2/n$ given in Chapter 2.

> **Definition 4.6**
>
> The **variance** of a random variable x is
>
> $$\sigma^2 = E[(x - \mu)^2] = \sum (x - \mu)^2 p(x)$$

> **Definition 4.7**
>
> The **standard deviation** of a discrete random variable is equal to the square root of the variance, i.e., to $\sigma = \sqrt{\sigma^2}$.

EXAMPLE 4.6

Medical research has shown that a certain type of chemotherapy is successful 70% of the time when used to treat skin cancer. Suppose five skin cancer patients are treated with this type of chemotherapy and let x equal the number of successful cures out of the five. The probability distribution for the number x of successful cures out of five is given in the table:

x	0	1	2	3	4	5
$p(x)$	.002	.029	.132	.309	.360	.168

a. Find $\mu = E(x)$.   b. Find $\sigma = \sqrt{E[(x - \mu)^2]}$.

c. Graph $p(x)$. Locate μ and the interval $\mu \pm 2\sigma$ on the graph. Explain how μ and σ can be used to describe $p(x)$.

Solution

a. Applying the formula,

$$\mu = E(x) = \sum xp(x)$$
$$= 0(.002) + 1(.029) + 2(.132) + 3(.309) + 4(.360) + 5(.168)$$
$$= 3.50$$

b. Now we calculate the variance of x:

$$\sigma^2 = E[(x - \mu)^2] = \sum (x - \mu)^2 p(x)$$
$$= (0 - 3.5)^2(.002) + (1 - 3.5)^2(.029) + (2 - 3.5)^2(.132)$$
$$+ (3 - 3.5)^2(.309) + (4 - 3.5)^2(.360) + (5 - 3.5)^2(.168)$$
$$= 1.05$$

Thus, the standard deviation is

$$\sigma = \sqrt{\sigma^2} = \sqrt{1.05} = 1.02$$

c. The graph of $p(x)$ is shown in Figure 4.4. Note that the mean μ and the interval $\mu \pm 2\sigma$ are shown on the graph. We can use μ and σ to describe the probability distribution $p(x)$ in the same way that we used $\bar{x}$ and s to describe a relative frequency distribution in Chapter 2. Note particularly that $\mu = 3.5$ locates the center of the probability distribution. If five skin cancer patients receive the chemotherapy treatment, we expect the number x that

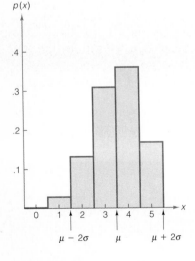

FIGURE 4.4

Graph of $p(x)$ for Example 4.6

are cured to be near 3.5. Similarly, $\sigma = 1.02$ measures the spread of the probability distribution $p(x)$. Since this distribution is a theoretical relative frequency distribution that is moderately mound-shaped (see Figure 4.4), we expect (see Tables 2.8 and 2.9) at least 75% and, more likely, near 95% of observed x values to fall in the interval $\mu \pm 2\sigma$—that is, between 1.46 and 5.54. Compare this result with the actual probability that x falls in the interval $\mu \pm 2\sigma$. From Figure 4.4 you can see that this probability includes the sum of $p(x)$ for all values of x except $p(0) = .002$ and $p(1) = .029$. Therefore, 96.9% of the probability distribution lies within 2 standard deviations of the mean. This percentage is consistent with Tables 2.8 and 2.9.

CASE STUDY 4.1

A RESTAURANT CHAIN FIGHTS SALES TAX CLAIM

The June 1, 1977, business section of the Orlando, Florida, *Sentinel Star* featured the following headline: "Red Lobster to Fight Tax Claim." According to the *Sentinel Star*, the Red Lobster Inns of America, a national seafood chain, had decided to take the state of Florida to court. The dispute concerned the 4% sales tax levied on most purchases in the state and focused mainly on the state's "bracket collection system." According to the bracket system, a merchant must collect 1¢ for sales between 10¢ and 25¢, 2¢ for sales from 26¢ to 50¢, 3¢ for sales between 51¢ and 75¢, and 4¢ for sales between 76¢ and 99¢. Red Lobster contended that if this system is followed, merchants will always collect more than 4%. That is, if a sale were made for $10.41, 4% will be collected on the $10, but more than 4% will be collected on the 41¢. This, they contend, would amount to more than 4% on the total sale and therefore is not consistent with the 4% tax required by law.

Concrete evidence supplied by the state of Florida tax records does indeed support the contention that the amount of tax collected using the bracket system exceeds the 4% specified by law. It appears that the sales tax receipts exceeded expected revenue (based on 4%) by $9.5 million in a single year.

What percent sales tax should the state expect to receive using the bracket system for computing the tax? (As noted, the tax on the whole dollar portion of the sale will be 4%.) Using the formula for calculating expected values, you can show (see Exercise 4.95) that the expected percent tax paid on the cents portion of a sale is 4.6%.

EXERCISES 4.19–4.33

LEARNING THE MECHANICS

4.19 Consider the probability distribution for the random variable x shown here.

x	10	20	30	40	50	60
$p(x)$	.10	.25	.30	.20	.10	.05

a. Find μ, σ^2, and σ.   **b.** Graph $p(x)$.

c. Locate μ and the interval $\mu \pm 2\sigma$ on your graph. What is the probability that x will fall within the interval $\mu \pm 2\sigma$?

4.20   Consider the probability distribution for the random variable x shown here.

x	1	2	3	4	5
$p(x)$	.05	.30	.35	.20	.10

a. Find μ, σ^2, and σ.   **b.** Graph $p(x)$.
c. Locate μ and the interval $\mu \pm \sigma$ on your graph. What is the probability that x will fall within the interval $\mu \pm \sigma$?
d. Locate the interval $\mu \pm 3\sigma$ on your graph. What is the probability that x falls within this interval?

4.21   Consider the probability distribution shown here.

x	1	2	4	10
$p(x)$	.3	.5	.1	.1

a. Find $\mu = E(x)$.   **b.** Find $\sigma^2 = E[(x - \mu)^2]$.   **c.** Find σ.
d. Interpret the value you obtained for μ.
e. In this case, can the random variable x ever assume the value μ? Explain.
f. In general, can a random variable ever assume a value equal to its expected value? Explain.

4.22   Consider the probability distribution shown here.

x	0	1	2	3	4
$p(x)$	.1	.4	.3	.1	.1

a. Find the mean of the distribution.   **b.** Find the variance.
c. Find the standard deviation.

4.23   Consider the probability distribution shown here.

x	−4	−3	−2	−1	0	1	2	3	4
$p(x)$	.02	.07	.10	.15	.30	.18	.10	.06	.02

a. Calculate μ, σ^2, and σ.
b. Graph $p(x)$. Locate μ, $\mu - 2\sigma$, and $\mu + 2\sigma$ on the graph.
c. What is the probability that x is in the interval $\mu \pm 2\sigma$?

4.24   Consider the probability distributions shown here.

x	0	1	2	y	0	1	2
$p(x)$	.3	.4	.3	$p(y)$	.1	.8	.1

a. Use your intuition to find the mean for each distribution. How did you arrive at your choice?
b. Which distribution appears to be more variable? Why?
c. Calculate μ and σ^2 for each distribution. Compare these answers to your answers in parts **a** and **b**.

4.25   A rehabilitation officer at a county jail questioned each inmate to determine how many previous convictions, x, each had prior to the one for which he or she was now serving. The relative frequencies corresponding to x are given in the probability distribution shown here.

x	0	1	2	3	4
$p(x)$	.16	.53	.20	.08	.03

If we can regard the relative frequencies as the approximate values for $p(x)$, find the expected number of previous convictions for an inmate.

4.26   A hospital research laboratory purchases rats from a distributor for use in experiments. The distributor sells four different strains of rats, each at different prices, and fills requests with one of the four strains depending on availability. The laboratory would like to estimate how much it will have to spend on rats in the next year. The table lists the four strains, price per 50 rats for each strain, and the probability of the purchase of each strain.

STRAIN	PRICE PER 50 RATS	PROBABILITY
A	$50.00	$\frac{1}{10}$
B	$75.00	$\frac{2}{5}$
C	$87.50	$\frac{3}{10}$
D	$100.00	$\frac{1}{5}$

a. If x is the price of a shipment of 50 rats, what is the expected price?
b. What is the variance of the price?
c. Graph $p(x)$. Locate μ and $\mu \pm 2\sigma$ on the graph. What proportion of the time does x fall in this interval?

4.27   In 1965 the U.S. Weather Bureau (now called the National Weather Service) initiated a nationwide program in which precipitation probabilities were included in all public weather forecasts. This was the first time in any field of application that probabilities were issued on such a large scale. The program continues today. Probabilistic precipitation forecasts indicate the likelihood of measurable precipitation ($\geq$.01 inch) at a specific point (the official rain gauge) during a given time period (Murphy and Winkler, 1984). Suppose that if a measurable amount of rain falls during the next 24 hours, a river will reach flood stage and a business will incur damages of $300,000. The National Weather Service has indicated that there is a 30% chance of a measurable amount of rain during the next 24 hours.
a. Construct the probability distribution that describes the potential flood damages faced by the firm.
b. Find the firm's expected loss due to flood damage.

4.28   Exercise 4.15 gives a hypothetical probability distribution for the number x of football games that Coach Bear Bryant of Alabama would win in the first half of his season. The probability distribution is reproduced here:

x	0	1	2	3	4	5	6
$p(x)$	.001	.010	.060	.185	.324	.302	.118

a. Find the expected number of games that Alabama would win in the first half of the season.

b. Find σ.

c. Find the probability that x is in the interval $\mu \pm 2\sigma$.

4.29   A company's marketing and accounting departments have determined that if the company markets its newly developed line of party favors, the probability distribution in the table will describe the contribution of the new line to the firm's profit during the next 6 months. The company has decided it should market the new line of party favors if the expected contribution to profit for the next 6 months is over $10,000. Based on the probability distribution, will the company market the new line?

PROFIT CONTRIBUTION	p(PROFIT CONTRIBUTION)
$-$5,000^a	.3
$10,000	.4
$30,000	.3

^aA negative contribution is a loss.

4.30   On one busy holiday weekend, a national airline has many requests for standby flights at half of the usual one-way air fare. However, past experience has shown that these passengers have only about a 1 in 5 chance of getting on the standby flight. When they fail to get on a flight as a standby, their only other choice is to fly first class on the next flight out. Suppose that the usual one-way air fare to a certain city is $70 and the cost of flying first class is $90. Should a passenger who wishes to fly to this city opt to fly as a standby? [*Hint:*   Find the expected cost of the trip for a person flying standby.]

4.31   Odds makers try to predict which football teams will win and by how much (the *spread*). If the odds makers do this accurately, adding the spread to the underdog's score should make the final score a tie. Suppose a bookie will give you $6 for every $1 you risk if you pick the winners in three ballgames (adjusted by the spread). Thus, for every $1 bet you will either lose $1 or gain $5. What is the bookie's expected earnings per dollar wagered?

4.32   A rock concert producer has scheduled an outdoor concert for Saturday, May 24. If it does not rain, the producer expects to make $20,000 profit from the concert. If it does rain, the producer will be forced to cancel the concert and will lose $12,000 (rock star's fee, advertising costs, stadium rental, administrative costs, etc.). The producer has learned from the National Weather Service that the probability of rain on May 24 is .4.

a. Find the producer's expected profit from the concert.

b. For a fee of $1,000, an insurance company has offered to insure the producer against all losses resulting from a rained-out concert. If the producer buys the insurance, what is her expected profit from the concert?

c. Assuming the National Weather Service's forecast is accurate, do you believe the insurance company has charged too much or too little for the policy? Explain.

4.33   An automobile insurance company estimates the following loss probabilities for the next year on a $25,000 sports car:

Total loss: .001
50% loss: .01
25% loss: .05
10% loss: .10

Assuming the company will sell only a $500 deductible policy for this model (i.e., the owner covers the first $500 damage), how much annual premium should the company charge in order to average $250 profit per policy sold?

4.4 The Binomial Random Variable

Many experiments result in dichotomous responses—i.e., responses for which there exist two possible alternatives, such as Yes–No, Pass–Fail, Defective–Non-defective, or Male–Female. A simple example of such an experiment is the coin-toss experiment. A coin is tossed a number of times, say ten. Each toss results in one of two outcomes, Head or Tail, and the probability of observing each of these two outcomes remains the same for each of the ten tosses. Ultimately, we are interested in the probability distribution of x, the number of heads observed. Many other experiments are equivalent to tossing a coin (either balanced or unbalanced) a fixed number n of times and observing the number x of times that one of the two possible outcomes occurs. Random variables that possess these characteristics are called **binomial random variables**.

Public opinion and consumer preference polls (e.g., the Gallup and Harris polls) frequently yield observations on binomial random variables. For example, suppose a sample of 100 students is selected from a large student body and each person is asked whether he or she favors (a Head) or opposes (a Tail) a certain campus issue. Ultimately, we are interested in x, the number of people in the sample who favor the issue. If each student is randomly selected from the student body and if the (unknown) proportion of students favoring the issue is p, then observing whether a student favors or is opposed to the issue is analogous to tossing an unbalanced coin. The chance that any randomly selected student favors the issue is p; the probability that he or she opposes the issue is $(1 - p)$. Sampling 100 students is analogous to tossing the coin 100 times. Thus, you can see that opinion polls which record the number of people who favor a certain issue are real-life equivalents of coin-toss experiments.

The experiment we have been describing is called a **binomial experiment** and is identified by the following characteristics:

Characteristics of a Binomial Random Variable

1. The experiment consists of n identical trials.
2. There are only two possible outcomes on each trial. We will denote one outcome by S (for Success) and the other by F (for Failure).
3. The probability of S remains the same from trial to trial. This probability is denoted by p, and the probability of F is denoted by q. Note that $q = 1 - p$.
4. The trials are independent.
5. The binomial random variable x is the number of S's in n trials.

EXAMPLE 4.7

For each of the following examples, decide whether x is a binomial random variable.

a. Suppose a university scholarship committee must select two students to receive a scholarship for the next academic year. The committee receives ten applications for the scholarships—six from male students and four from female students. Suppose the applicants are all equally qualified, so that the selections are randomly made. Let x be the number of female students who receive a scholarship.

b. Before marketing a new product on a large scale, many companies will conduct a consumer-preference survey to determine whether the product is likely to be successful. Suppose a company develops a new diet soda and then conducts a taste-preference survey with 100 randomly chosen consumers stating their preference among the new soda and the two leading sellers. Let x be the number of the 100 who choose the new brand over the two others.

c. Some surveys are conducted by using a method of sampling other than simple random sampling (defined in Chapter 3). For example, suppose a television cable company plans to conduct a survey to determine the fraction of households in the city that would use the cable television service. The sampling method is to choose a city block at random and then survey every household on that block. This sampling technique is called **cluster sampling**. Suppose ten blocks are so sampled, producing a total of 124 household responses. Let x be the number of the 124 households that would use the television cable service.

Solution

a. In checking the binomial characteristics, a problem arises with independence (characteristic 4 in the preceding box). Given that the first student selected is female, the probability that the second chosen is female is $\frac{3}{9}$. On the other hand, given that the first selection is a male student, the probability that the second is female is $\frac{4}{9}$. Thus, the conditional probability of a Success (choosing a female student to receive a scholarship) on the second trial (selection) depends on the outcome of the first trial, and the trials are therefore dependent. Since the trials are *not independent*, this is not a binomial random variable.

b. Surveys that produce dichotomous responses and use random sampling techniques are classic examples of binomial experiments. In our example, each randomly selected consumer either states a preference for the new diet soda or does not. The sample of 100 consumers is a very small proportion of the totality of potential consumers, so the response of one would be, for all practical purposes, independent of another. Thus, x is a binomial random variable.

c. This example is a survey with dichotomous responses (Yes or No to the cable service), but the sampling method is not simple random sampling. Again, the binomial characteristic of independent trials would probably not be satisfied. The responses of households within a particular block almost surely would be dependent, since households within a block tend to be similar with respect to income, level of education, and general interests. Thus, the binomial model would not be satisfactory for x if the cluster sampling technique were employed.

EXAMPLE 4.8

The Heart Association claims that only 10% of adults over 30 years of age in the United States can pass the minimum requirements established by the president's Physical Fitness Commission. Suppose four adults are randomly selected, and each is given the fitness test. Let x be the number of the four who pass the minimum requirements. Find the probability distribution for x, assuming that the Heart Association's claim is true.

Solution

Recall that a probability distribution describes a discrete random variable by assigning probabilities to each of its values. In this example, the possible values of x, the number of four adults who pass the minimum requirements, are 0, 1, 2, 3, 4. Furthermore, there are four identical trials (the sampling and observation of the four adults), each with two possible outcomes (pass or fail minimum requirements). We are assuming that the probability that each adult passes is .1, and the result for each adult will be (at least approximately) independent of that for the others. Thus, the number, x, of adults who pass the minimum requirements is a binomial random variable.

Let us first consider the event $x = 0$, that is, the event that none of the four tested adults passes the test. You can see that the event $x = 0$ is equivalent to the simple event

$FFFF$

where F in the first position implies that adult 1 fails, F in the second position implies that adult 2 fails, etc. Since the trials are independent in this binomial experiment (knowing whether adult 1 passes should not affect the probability that adult 2 passes), we can find the probability of an intersection by multiplying the probabilities of the events. Thus,

$$P(x = 0) = P(FFFF) = P(F)P(F)P(F)P(F)$$
$$= (.9)(.9)(.9)(.9) = (.9)^4 = .6561$$

The event $x = 1$ implies that one of the four adults passes the physical fitness test and three fail it. The following list of simple events contains all simple events that imply $x = 1$:

$SFFF \qquad FSFF \qquad FFSF \qquad FFFS$

where S in the first position corresponds to adult 1 passing the test, S in the second position corresponds to adult 2 passing the test, etc. Note that each of these simple events will have the same probability, $(.1)(.9)^3$, where .1 corresponds to the one adult who passes the test and $(.9)^3$ corresponds to the three who fail it. Remembering from Chapter 3 that we obtain the probability of an event by summing the probabilities of the simple events of which it is composed, we get

$$P(x = 1) = 4[(.1)(.9)^3] = .2916$$

The event $x = 2$ implies that two adults pass the test and two fail it; this event consists of the following six simple events:

$SSFF \qquad SFSF \qquad SFFS \qquad FSSF \qquad FSFS \qquad FFSS$

Each of these simple events has probability $(.1)^2(.9)^2$, so that

$$P(x = 2) = 6[(.1)^2(.9)^2] = .0486$$

Similarly,

$$P(x = 3) = 4[(.1)^3(.9)] = .0036$$
$$P(x = 4) = (.1)^4 = .0001$$

The complete probability distribution is given in Table 4.2 and shown in Figure 4.5.

TABLE 4.2 **Probability Distribution for Physical Fitness Example: Tabular Form**

x	$p(x)$
0	.6561
1	.2916
2	.0486
3	.0036
4	.0001

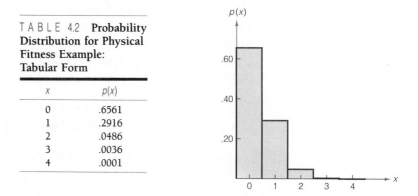

FIGURE 4.5

Probability distribution for physical fitness example: Graphic form

Before we give a formula for $p(x)$, we will refresh your memory on factorial notation. Particularly, the symbol $n!$ is to be read "n factorial" and is calculated by

$$n! = n(n - 1)(n - 2) \cdots 3 \cdot 2 \cdot 1$$

We define $0! = 1$. Thus, for example, $4! = 4 \cdot 3 \cdot 2 \cdot 1 = 24$.

Using factorial notation, we write the formula for a binomial probability distribution with $n = 4$ and $p = .2$:

$$p(x) = \frac{4!}{x!(4 - x)!}(.1)^x(.9)^{4-x}$$

Then, for $x = 2$, we have

$$p(2) = \frac{4!}{2!(4 - 2)!}(.1)^2(.9)^{4-2} = \frac{4 \cdot 3 \cdot 2 \cdot 1}{(2 \cdot 1)(2 \cdot 1)}(.1)^2(.9)^2$$
$$= 6(.1)^2(.9)^2 = .0486$$

which agrees with our simple event calculation. Note that the first part of the formula, $4!/[x!(4 - x)!]$, counts the number of simple events that result in x adults passing the physical fitness test. The second part of the formula, $(.1)^x(.9)^{4-x}$, is the probability assigned to each simple event that has x adults passing and $(4 - x)$ failing. When we multiply the *number* of simple events by the *probability* assigned to each simple event, we get the probability that x adults pass the test. We will see that this formula can be generalized to give the probability distribution of any binomial random variable.

Note that $\binom{n}{x}$, shorthand for $n!/[x!(n - x)!]$, is the number of simple events that have x successes and $(n - x)$ failures (see Section 3.8 for a discussion of the combinations rule for counting simple events), and $p^x q^{n-x}$ is the probability assigned to each simple event that has x successes and $(n - x)$ failures. The product of these two quantities, $\binom{n}{x}p^x q^{n-x}$, is the probability that x successes and $(n - x)$ failures are observed.

The Binomial Probability Distribution

$$p(x) = \binom{n}{x}p^x q^{n-x} \qquad (x = 0, 1, 2, \ldots, n)$$

where

$p = $ Probability of a success on a single trial
$q = 1 - p$
$n = $ Number of trials
$x = $ Number of successes in n trials
$$\binom{n}{x} = \frac{n!}{x!(n - x)!}$$

The binomial probability distribution is so named because the probabilities, $p(x)$, $x = 0, 1, \ldots, n$, are terms of the binomial expansion, $(q + p)^n$.

CASE STUDY 4.2

THE SPACE SHUTTLE *CHALLENGER*: CATASTROPHE IN SPACE

On January 28, 1986, at 11:39.13 A.M., while traveling at Mach 1.92 at an altitude of 46,000 feet, the space shuttle *Challenger* was totally enveloped in an explosive burn that destroyed the shuttle and resulted in the deaths of all seven astronauts aboard. What happened? What was the cause of this catastrophe? This was the 25th shuttle mission. The preceding 24 missions had all been successful.

The report of the Presidential Commission assigned to investigate the accident concluded that the explosion was caused by the failure of the O-ring seal in the joint between the two lower segments of the right solid rocket booster. The seal is supposed to prevent superhot gases from leaking through the joint during the propellant burn of the booster rocket. The failure of the seal permitted a jet of white-hot gases to escape and to ignite the liquid fuel of the external fuel tank. The fuel tank fireburst destroyed the *Challenger*.

What were the chances of this event occurring? In a 1985 report, the National Aeronautics and Space Administration (NASA) claimed that the probability of such a failure was about $\frac{1}{60,000}$, or about once in every 60,000 flights. But a 1983 risk assessment study conducted for the Air Force assessed the probability of a shuttle catastrophe due to booster rocket "burn-through" to be $\frac{1}{35}$, or about once in every 35 missions.

If it is assumed that (1) p, the probability of shuttle catastrophe due to booster failure, remains the same from mission to mission, and (2) the performance of the booster rockets on one mission is independent of the performance of the boosters on other missions, then the number, x, of shuttle catastrophes due to booster failure in n missions can be treated as a binomial random variable. Accordingly, the probability that no disasters would have occurred during 25 missions is

$$P(x = 0) = \binom{25}{0}p^0(1 - p)^{25-0} = \frac{25!}{0!25!}p^0(1 - p)^{25}$$

$$= (1 - p)^{25}$$

If we use NASA's probability of shuttle catastrophe ($p = \frac{1}{60,000} = .0000167$), the probability of no catastrophes in 25 missions is approximately .9996. If we use the probability of catastrophe from the study prepared for the Air Force ($p = \frac{1}{35} = .02857$), the probability of no catastrophes in 25 missions is .4845. Or, if we consider the complementary event that at least one catastrophe occurs in 25 missions, the chances are .0004, or about 4 in 10,000, given NASA's assumptions. On the other hand, the probability of at least one catastrophe under the Air Force's assumptions is .5155, or slightly more than 50–50. Given the events of January 28, 1986, which risk assessment—NASA's or the Air Force's—appears to be more appropriate? The probability of one or more disasters in 25 missions is so remote using NASA's assessment that it casts serious doubt on the risk assessment practices used by NASA prior to the *Challenger*'s fatal mission (McKean, 1986; Biddle, 1986; Robinson, 1986; *Minneapolis Star and Tribune*, February 11, 1986).

EXAMPLE 4.9

Refer to Example 4.8. Calculate μ and σ, the mean and standard deviation, respectively, of the number of the four adults who pass the test.

Solution

From Section 4.3 we know that the mean of a discrete probability distribution is

$$\mu = \sum xp(x)$$

Referring to Table 4.2, the probability distribution for the number x who pass the fitness test, we find

$$\mu = 0(.6561) + 1(.2916) + 2(.0486) + 3(.0036) + 4(.0001)$$
$$= .4 = 4(.1) = np$$

The relationship $\mu = np$ holds in general for a binomial random variable.
   The variance is

$$\sigma^2 = \sum (x - \mu)^2p(x) = \sum (x - .4)^2p(x)$$
$$= (0 - .4)^2(.6561) + (1 - .4)^2(.2916) + (2 - .4)^2(.0486)$$
$$+ (3 - .4)^2(.0036) + (4 - .4)^2(.0001)$$
$$= .104976 + .104976 + .124416 + .024336 + .001296$$
$$= .36 = 4(.1)(.9) = npq$$

The relationship $\sigma^2 = npq$ holds in general for a binomial random variable.
   Finally, the standard deviation of the number who pass the fitness test is

$$\sigma = \sqrt{\sigma^2} = \sqrt{.36} = .6$$

We emphasize that you need not use the expectation summation rules to calculate μ and σ^2 for a binomial random variable. You can find them easily using the formulas $\mu = np$ and $\sigma^2 = npq$.

Mean, Variance, and Standard Deviation for a Binomial Random Variable

$$Mean: \quad \mu = np$$
$$Variance: \quad \sigma^2 = npq$$
$$Standard\ deviation: \quad \sigma = \sqrt{npq}$$

As we demonstrated in Chapter 2, the mean and standard deviation provide measures of the central tendency and variability, respectively, of a distribution. Thus, we can use μ and σ to obtain a rough visualization of the probability distribution for x when the calculation of the probabilities is too tedious. To illustrate the use of the binomial probability distribution, consider Example 4.10.

EXAMPLE 4.10

A poll of 20 voters is taken in a large city. The purpose is to determine x, the number in favor of a certain candidate for mayor. Suppose that (unknown to us) 60% of all the city's voters favor this candidate.

a. Find the mean and standard deviation of x.
b. Find the probability that x is less than or equal to 10 ($x \leq 10$).
c. Find the probability that x exceeds 12 ($x > 12$).
d. Find the probability that x equals 11 ($x = 11$).
e. Graph the probability distribution of x and locate the interval $\mu - 2\sigma$ to $\mu + 2\sigma$ on the graph.

Solution

a. Given that the sample of 20 was randomly selected from a large number of voters, x, the number of the 20 who favor the candidate, is (approximately) a binomial random variable. The value of p is the fraction of the total voters who favor the candidate; i.e., $p = .6$. Therefore, we calculate the mean and variance:

$$\mu = np = 20(.6) = 12 \qquad \sigma^2 = npq = 20(.6)(.4) = 4.8$$

The standard deviation is then

$$\sigma = \sqrt{4.8} = 2.2$$

b. Calculating binomial probabilities when n is large is a formidable task. For example, to find the probability that $x \leq 10$, we would calculate*

$$P(x \leq 10) = p(0) + p(1) + p(2) + \cdots + p(10)$$
$$= \sum_{x=0}^{10} p(x) = \sum_{x=0}^{10} \binom{20}{x}(.6)^x(.4)^{20-x}$$

---

*The value of x below the Σ symbol, $x = 0$, is the **first member**, or **lower limit**, of the summation. The value of x above the Σ symbol, $x = 10$, is the **last member**, or **upper limit**, of the summation. Thus, $\sum_{x=0}^{10} p(x) = p(0) + p(1) + \cdots + p(9) + p(10)$. We include these limits when the summation extends over only some of the possible values of x.

We can avoid these tedious calculations by making use of cumulative binomial probability tables (Table II in Appendix A). Part of Table II is shown in Figure 4.6. The columns correspond to values of p, and the rows correspond to values of the random variable x.

The entries in Table II are the cumulative sums

$$P(x \leq k) = p(0) + p(1) + p(2) + \cdots + p(k)$$

for values of $k = 0, 1, 2, \ldots, (n - 1)$. Observe that the bottom row of the table, the one corresponding to $k = n$, is omitted. This is because the sum of $p(x)$ from $x = 0$ to $x = n$ is always equal to 1; i.e., $P(x \leq n) = 1$ for any binomial random variable.

FIGURE 4.6

(a) Graph of binomial probability distribution for $n = 20$ and $p = .6$

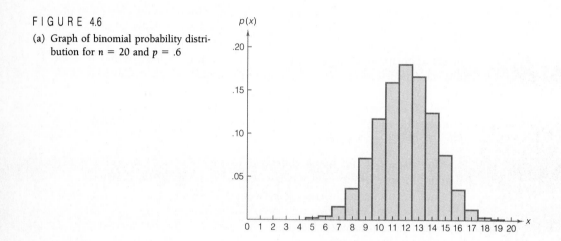

(b) Partial reproduction of Table II in Appendix A

k \ p	.01	.05	.10	.20	.30	.40	.50	.60	.70	.80	.90	.95	.99
0	.818	.358	.122	.012	.001	.000	.000	.000	.000	.000	.000	.000	.000
1	.983	.736	.392	.069	.008	.001	.000	.000	.000	.000	.000	.000	.000
2	.999	.925	.677	.206	.035	.004	.000	.000	.000	.000	.000	.000	.000
3	1.000	.984	.867	.411	.107	.016	.001	.000	.000	.000	.000	.000	.000
4	1.000	.997	.957	.630	.238	.051	.006	.000	.000	.000	.000	.000	.000
5	1.000	1.000	.989	.804	.416	.126	.021	.002	.000	.000	.000	.000	.000
6	1.000	1.000	.998	.913	.608	.250	.058	.006	.000	.000	.000	.000	.000
7	1.000	1.000	1.000	.968	.772	.416	.132	.021	.001	.000	.000	.000	.000
8	1.000	1.000	1.000	.990	.887	.596	.252	.057	.005	.000	.000	.000	.000
9	1.000	1.000	1.000	.997	.952	.755	.412	.128	.017	.001	.000	.000	.000
10	1.000	1.000	1.000	.999	.983	.872	.588	.245	.048	.003	.000	.000	.000
11	1.000	1.000	1.000	1.000	.995	.943	.748	.404	.113	.010	.000	.000	.000
12	1.000	1.000	1.000	1.000	.999	.979	.868	.584	.228	.032	.000	.000	.000
13	1.000	1.000	1.000	1.000	1.000	.994	.942	.750	.392	.087	.002	.000	.000
14	1.000	1.000	1.000	1.000	1.000	.998	.979	.874	.584	.196	.011	.000	.000
15	1.000	1.000	1.000	1.000	1.000	1.000	.994	.949	.762	.370	.043	.003	.000
16	1.000	1.000	1.000	1.000	1.000	1.000	.999	.984	.893	.589	.133	.016	.000
17	1.000	1.000	1.000	1.000	1.000	1.000	1.000	.996	.965	.794	.323	.075	.001
18	1.000	1.000	1.000	1.000	1.000	1.000	1.000	.999	.992	.931	.608	.264	.017
19	1.000	1.000	1.000	1.000	1.000	1.000	1.000	1.000	.999	.988	.878	.642	.182

To find $P(x \leq 10)$ for $n = 20$ and $p = .6$, we first find the column corresponding to $p = .6$ and then the row corresponding to $k = 10$. The recorded value, shaded in both the graph and table in Figure 4.6, is

$$P(x \leq 10) = .245$$

c. To find the probability

$$P(x > 12) = p(13) + p(14) + \cdots + p(19) + p(20) = \sum_{x=13}^{20} p(x)$$

we use the fact that for all probability distributions, $\Sigma p(x) = 1$. Therefore, using the complementary event, we have

$$P(x > 12) = 1 - [p(0) + p(1) + \cdots + p(12)]$$
$$= 1 - P(x \leq 12) = 1 - \sum_{x=0}^{12} p(x)$$

Consulting Table II, we find the entry in row $k = 12$, column $p = .6$ to be .584. Thus,

$$P(x > 12) = 1 - .584 = .416$$

d. To find the probability that exactly 11 voters favor the candidate, recall that the entries in Table II are cumulative probabilities and use the relationship

$$P(x = 11) = [p(0) + p(1) + \cdots + p(10) + p(11)]$$
$$- [p(0) + p(1) + \cdots + p(9) + p(10)]$$
$$= P(x \leq 11) - P(x \leq 10)$$

Then

$$P(x = 11) = .404 - .245 = .159$$

e. The probability distribution for x is shown in Figure 4.7. Note that

$$\mu - 2\sigma = 12 - 2(2.2) = 7.6$$
$$\mu + 2\sigma = 12 + 2(2.2) = 16.4$$

F I G U R E 4.7

The binomial probability distribution for x in Example 4.11

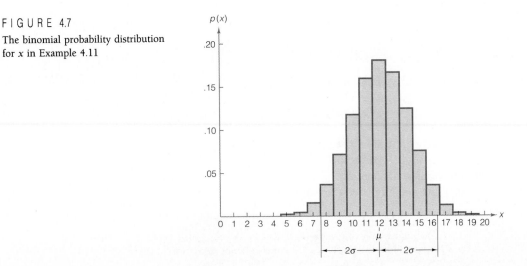

The interval $\mu - 2\sigma$ to $\mu + 2\sigma$ is shown in Figure 4.7. The probability that x falls in the interval, $\mu \pm 2\sigma$, that is, $P(x = 8, 9, 10, \ldots, 16) = P(x \leq 16) - P(x \leq 7) = .984 - .021 = .963$. Note that this probability is very close to the .95 given by the Empirical Rule.

<hr>

CASE STUDY 4.3

A SURVEY OF CHILDREN'S POLITICAL KNOWLEDGE

Children's images of political leaders in Britain, France, and the United States were studied by Fred I. Greenstein.* Data were collected by means of interviews with small samples of children in the three countries "in order to examine various standard assumptions about political culture and socialization among the three nations, as well as black–white differences in the United States." During one phase of the study, 25 black children from the United States were asked to name the president of their country. This phase represents a binomial experiment with $n = 25$ trials and p equal to the proportion of all black children who could correctly name the president at that time (1969–1970). One objective of the experiment was to obtain an estimate of the value of p.

Of the sample of 25 black children, 24 correctly identified Richard Nixon as president. The implication of this result is that the proportion of all black children who could have made a correct identification must have been quite high. In fact, if the true proportion were equal to .8, the probability that at least 24 out of 25 would correctly identify the president is only .027. (You can verify this result by using Table II in Appendix A.) Thus, unless the observed outcome represents a rare event, the proportion of all black children who could have correctly identified the president was probably in excess of .8 at that time. In Chapter 7 we will develop a systematic approach for making inferences about proportions.

<hr>

EXERCISES 4.34–4.57

LEARNING THE MECHANICS

4.34 Compute the following:

a. $\dfrac{5!}{3!(5-3)!}$ b. $\dbinom{6}{3}$ c. $\dbinom{8}{0}$ d. $\dbinom{5}{5}$ e. $\dbinom{6}{1}$

4.35 Consider the following probability distribution:

$$p(x) = \binom{6}{x}(.4)^x(.6)^{6-x} \qquad (x = 0, 1, 2, \ldots, 6)$$

a. Is x a discrete or a continuous random variable? Explain.
b. What is the name of this probability distribution?
c. Graph the probability distribution.
d. Find the mean and standard deviation of x.

<hr>

*Reprinted from Greenstein, F. I. "The benevolent leader revisited: Children's images of political leaders in three democracies." *American Political Science Review*, December 1975, 69, 1371–1398.

      e. Show the mean and the 2-standard-deviation interval on each side of the mean on the graph you drew in part **c**.

4.36   If x is a binomial random variable, compute $p(x)$ for each of the following cases:
      a. $n = 5$, $x = 1$, $p = .3$ **b.** $n = 4$, $x = 2$, $q = .5$
      c. $n = 3$, $x = 0$, $p = .2$ **d.** $n = 5$, $x = 3$, $p = .4$
      e. $n = 4$, $x = 2$, $q = .6$ **f.** $n = 3$, $x = 1$, $p = .3$

4.37   Suppose x is a binomial random variable with $n = 4$ and $p = .2$.
      a. Calculate the value of $p(x)$, $x = 0, 1, 2, 3, 4, 5$, using the formula for a binomial probability distribution.
      b. Using your answers to part **a**, give the probability distribution for x in tabular form.

4.38   If x is a binomial random variable, calculate μ, σ^2, and σ for each of the following:
      a. $n = 30$, $p = .4$ **b.** $n = 80$, $p = .3$
      c. $n = 100$, $p = .3$ **d.** $n = 70$, $p = .9$
      e. $n = 60$, $p = .5$ **f.** $n = 1,000$, $p = .03$

4.39   If x is a binomial random variable, use Table II in Appendix A to find the following probabilities:
      a. $P(x = 2)$ for $n = 10$, $p = .3$
      b. $P(x \le 5)$ for $n = 15$, $p = .5$
      c. $P(x > 1)$ for $n = 5$, $p = .2$
      d. $P(x < 10)$ for $n = 25$, $p = .8$
      e. $P(x \ge 10)$ for $n = 15$, $p = .7$
      f. $P(x = 2)$ for $n = 20$, $p = .1$

4.40   Suppose x is a binomial random variable with $n = 5$ and $p = .5$. Compute $p(x)$ for $x = 0, 1, 2, 3, 4$, and 5 using the following three methods:
      a. List the simple events (using S for a Success and F for a Failure on each trial) corresponding to each value of x, assign probabilities to each simple event, and obtain $p(x)$ by adding simple event probabilities.
      b. Use the formula for the binomial probability distribution to obtain $p(x)$.
      c. Use Table II to obtain $p(x)$.

4.41   Given that x is a binomial random variable, $n = 15$, and $p = .4$, use Table II to find the following probabilities:
      a. $P(x \le 1)$ **b.** $P(x \ge 3)$ **c.** $P(x \le 5)$
      d. $P(x < 10)$ **e.** $P(x > 10)$ **f.** $P(x = 6)$

4.42   The binomial probability distribution is a family of probability distributions with each single distribution depending on the values of n and p. Assume that x is a binomial random variable with $n = 4$.
      a. Determine a value of p such that the probability distribution of x is symmetric.
      b. Determine a value of p such that the probability distribution of x is skewed to the right.
      c. Determine a value of p such that the probability distribution of x is skewed to the left.
      d. Graph each of the binomial distributions you obtained in parts **a**, **b**, and **c**. Locate the mean for each distribution on its graph.
      e. In general, for what values of p will a binomial distribution be symmetric? Skewed to the right? Skewed to the left?

APPLYING THE CONCEPTS

4.43   A large southern university has determined from past records that the probability a student who registers for fall classes will have his or her schedule rejected (due to overfilled classrooms, clerical error, etc.) is .2.

a. Suppose 25,000 students register for fall classes, and x is the number of students who have their schedules rejected. Is x a binomial random variable? Explain.

b. Suppose a random sample of 20 students is selected from the total of 25,000, and x is the number of these students who have their schedules rejected. Is x a binomial random variable? Explain.

c. Suppose you sample the results of the first 1,000 students who register next fall and record x, the number of rejected registrations. Is x a binomial random variable? Explain.

d. For the random variables in parts **a**, **b**, and **c** that you identified as being binomial, find μ, σ^2, and σ.

4.44   Suppose 60% of all people who are eligible for jury duty in a large Florida city are in favor of capital punishment. How does this finding affect the composition of a jury in a murder trial? Suppose a jury of 12 is to be randomly selected from among all the prospective jurors in this city.

a. What is the expected number of jurors who favor capital punishment?

b. What is the probability that none of the 12 jurors selected favors capital punishment?

c. If the jury were really selected at random, would you be surprised if none of the jurors favored capital punishment? Explain.

4.45   An automobile manufacturer has determined that 30% of all gas tanks that were installed on its 1988 compact model are defective.

a. If 15 of the cars are recalled by a particular dealer, what is the probability that more than 10 of the 15 will need new gas tanks?

b. If 10,000 of the cars are recalled by the manufacturer, what is the probability that fewer than 3,000 will need new gas tanks? Set up the solution but do not perform the calculations. [*Note:* In Section 5.4, we discuss a procedure that can be used to obtain an approximate answer to this question without having to perform the tedious calculations required by the binomial distribution.] Calculate the mean and standard deviation of the distribution, and give an "intuitive" approximation to this probability.

4.46   A problem of considerable impact on the economy is the burgeoning cost of Medicare and other public-funded medical services. One aspect of this problem, reported in the "Behavior" section of *Time* (April 18, 1983), concerns the high percentage of people seeking medical treatment who, in fact, have no physical basis of their ailment. One conservative estimate is that the percentage of people who seek medical assistance but have no real physical ailment is 10%, and some doctors believe that it may be as high as 40%. Suppose we randomly sample the records of a doctor and find that five of 15 patients seeking medical assistance are physically healthy.

a. What is the probability of observing five or more physically healthy patients in a sample of 15 if the proportion p that the doctor normally sees is 10%?

b. What is the probability of observing five or more physically healthy patients in a sample of 15 if the proportion p that the doctor normally sees is 40%?

c. Why might your answer to part **a** make you believe that p is larger than .1?

4.47   The *Orlando Sentinel* (April 12, 1984) reports that "after a celebrity passes out of the limelight, the public quickly assumes that he has died." To substantiate this statement, they give the results of an experiment conducted by *Psychology Today*. The magazine asked its readers to guess whether 20 celebrities of the past were alive. On the average, the respondents guessed correctly on 9 and did not even recognize 6 of the 20. Suppose a respondent possessed no knowledge about the state of health of any of the celebrities and simply guessed whether they were alive or dead. What is the probability that a respondent would guess correctly on 9 or more of the 20 celebrities?

4.48   The *Wall Street Journal* (March 8, 1984) notes that a "blood clot dissolver that researchers hope can stop heart attacks passed its first test in humans but researchers said the substance must still undergo extensive trials before any life-saving potential can be determined." One aspect of the research by Frans Van de Werf, M.D., and colleagues (1984), reported in the *New England Journal of Medicine*, involved actual tests on seven humans aged 50 or more who suffered heart attacks and were treated with the new drug, t-PA. After the treatment, the blood clots in six of the seven patients were dissolved. The blood clot did not dissolve in the seventh patient.

Assume that the drug is ineffective in dissolving blood clots and that without the drug the probability p that a heart attack patient's blood clot would dissolve of its own accord, in a short time, is very small, say less than .1. Let x equal the number of heart attack patients in the sample of seven whose blood clots dissolved in a short time.

a. If the drug is ineffective and $p = .1$, what is the probability that blood clots would have dissolved of their own accord in as many as six of seven heart attack patients?

b. If p really is equal to .1 and if the drug t-PA is ineffective in treating heart attacks, would you conclude that $x \geq 6$ is a rare event? Using the logic of Section 2.7, what do you think about the utility of t-PA in dissolving blood clots in heart attack patients?

4.49 An accountant believes that 90% of the company's invoices are free of errors. To check this theory the accountant randomly samples 25 invoices and finds that seven contain errors.

a. If the accountant's theory is correct, what is the probability that of the 25 invoices written, seven or more contain errors?

b. What assumptions do you have to make to solve this problem using the methodology of this section?

c. If these assumptions are satisfied and if the sample of 25 invoices produces seven that contain errors, do you think that more than 10% of the company's invoices contain errors? Explain.

4.50 A particular system in a space vehicle must work properly in order for it to gain reentry into the earth's atmosphere. One component of the system operates successfully only 85% of the time. To increase the reliability of the system, four of the components are installed in such a way that the system will operate successfully if at least one component is working. What is the probability that the system will fail? Assume the components operate independently.

4.51 A problem of great concern to a manufacturer is the cost of repair and replacement required under a product's guarantee agreement. Assume it is known that 10% of all electronic pocket calculators purchased are returned for repair while their guarantee is still in effect. If a firm purchased 25 pocket calculators for its salespeople, what is the probability that five or more of these calculators will need repair while their guarantees are still in effect?

4.52 Suppose you are a purchasing officer for a large company. You have purchased 5 million electrical switches and have been guaranteed by the supplier that the shipment will contain no more than .1% defectives. To check the shipment, you randomly sample 500 switches, test them, and find that four are defective. If the switches are as represented, calculate μ and σ for this sample of 500. Based on this evidence, do you think the supplier has complied with the guarantee? Explain. [*Hint:* Calculate μ and σ for this binomial random variable with $p = .001$ to see if a value of x as large as 4 is probable.]

4.53 According to the "January" theory, if the stock market is up in January it will be up for the whole year (and vice versa). Believe it or not, this indicator of stock market behavior has been correct for 29 of the last 34 years and 100% correct in odd-numbered years.* Suppose that there is no truth whatever in this theory and that stock prices are just as likely to move up or down in any given year, regardless of the direction of movement in January.

a. Find the probability of perfect agreement between the January and annual movements in stock prices over the 15 odd-numbered years.

b. What is the probability of perfect agreement between the January and annual movements in stock prices in at least 10 of the 15 years?

---

*From the "Heard on the Street" section of the *Wall Street Journal* (February 1, 1984).

4.54   An experiment is to be conducted to see whether an acclaimed psychic has extrasensory perception (ESP). Five different cards are shuffled, and one is chosen at random. The psychic will then try to identify which card was drawn without seeing it. The experiment is to be repeated 20 times and x, the number of correct decisions, is recorded. (Assume that the 20 trials are independent.)

   a. If the psychic is guessing—i.e., if the psychic does not possess ESP—what is the value of p, the probability of a correct decision on each trial?

   b. If the psychic is guessing, what is the expected number of correct decisions in 20 trials?

   c. If the psychic is guessing, what is the probability of six or more correct decisions in 20 trials?

   d. Suppose that the psychic makes six correct decisions in 20 trials. Is there evidence to indicate that the psychic is *not* guessing and actually has ESP? Explain.

4.55   A literature professor decides to give a 20-question true–false quiz to determine who has read an assigned novel. She wants to choose the passing grade such that the probability of passing a student who guesses on every question is less than .05. What score should she set as the lowest passing grade?

4.56   A new drug has been synthesized that is designed to reduce a person's blood pressure. Twenty randomly selected hypertensive patients receive the new drug. Suppose 18 or more of the patients' blood pressures drop.

   a. Suppose the probability that a hypertensive patient's blood pressure drops if he or she is *untreated* is .5. Then what is the probability of observing 18 or more blood pressure drops in a random sample of 20 treated patients if the new drug is in fact ineffective in reducing blood pressure?

   b. Considering this probability (part **a**), do you think you have observed a rare event, or do you conclude that the drug is effective in reducing hypertension?

4.57   Most firms use sampling plans to control the quality of manufactured items ready for shipment or the quality of items that have been purchased. To illustrate the use of a sampling plan, suppose you are shipping electrical fuses in lots, each containing 10,000 fuses. The plan specifies that you will randomly sample 25 fuses from each lot and accept (or ship) the lot if x, the number of defective fuses in the sample, is less than 3. If $x \geq 3$, you will reject the lot and hold it for a complete reinspection.

   a. What is the probability of accepting a lot ($x = 0, 1,$ or 2) if the actual fraction defective in the lot is:

   (i) 1   **(ii)** .8   **(iii)** .5   **(iv)** .2   **(v)** .05   **(vi)** 0

   b. Construct a graph showing $P(A)$, the probability of lot acceptance, as a function of lot fraction defective, p. This graph is called the **operating characteristic curve** for the sampling plan.

   c. Suppose the sampling plan calls for sampling $n = 25$ fuses and accepting a lot if $x \leq 3$. Calculate the quantities specified in part **a** and construct the operating characteristic curve for this sampling plan. Compare this curve with the curve obtained in part **b**. (Note how the curve characterizes the ability of the plan to screen bad lots from shipment.)

 ## The Poisson Random Variable (Optional)

A type of probability distribution that is often useful in describing the number of events that will occur in a specific period of time or in a specific area or volume is the **Poisson distribution** (named after the 18th-century physicist and mathematician, Siméon Poisson). Typical examples of random variables for which the Poisson probability distribution provides a good model are

1. The number of traffic accidents per month at a busy intersection
2. The number of noticeable surface defects (scratches, dents, etc.) found by quality inspectors on a new automobile

3. The parts per million of some toxicant found in the water or air emission from a manufacturing plant
4. The number of diseased trees per acre of a certain woodland
5. The number of death claims received per day by an insurance company
6. The number of unscheduled admissions per day to a hospital

---

| Characteristics of a Poisson Random Variable

1. The experiment consists of counting the number of times a certain event occurs during a given unit of time or in a given area or volume (or weight, distance, or any other unit of measurement).
2. The probability that an event occurs in a given unit of time, area, or volume is the same for all the units.
3. The number of events that occur in one unit of time, area, or volume is independent of the number that occur in other units.
4. The mean (or expected) number of events in each unit is denoted by the Greek letter lambda, λ.

The characteristics of the Poisson random variable are usually difficult to verify for practical examples. The examples given satisfy them well enough that the Poisson distribution provides a good model in many instances. As with all probability models, the real test of the adequacy of the Poisson model is in whether it provides a reasonable approximation to reality—that is, whether empirical data support it.

The Poisson probability distribution also provides a good approximation to a binomial probability distribution with mean

$$\lambda = np$$

when n is large, p is small, and $np \leq 7$. To illustrate with a relatively small value of n, if $n = 25$ and $p = .05$, then the exact value of the binomial probability $p(2)$ is .231. The Poisson approximation is .224. The approximations are better for $n \geq 100$.

The probability distribution, mean, and variance for a Poisson random variable are shown in the next box.

| Probability Distribution, Mean, and Variance for a
| Poisson Random Variable

$$p(x) = \frac{\lambda^x e^{-\lambda}}{x!} \qquad (x = 0, 1, 2, \ldots)$$

$$\mu = \lambda \qquad \sigma^2 = \lambda$$

where

λ = Mean number of events during given unit of time, area, volume, etc.

$e = 2.71828\ldots$

The calculation of Poisson probabilities is made easier by the use of Table III in Appendix A, which gives the cumulative probabilities $P(x \leq k)$ for various values of λ. The use of Table III is illustrated in Example 4.11.

EXAMPLE 4.11

Ecologists often use the number of reported sightings of a rare species of animal to estimate the remaining population size. For example, suppose the number, x, of reported sightings per week of blue whales is recorded. Assume that x has (approximately) a Poisson probability distribution. Furthermore, assume that the average number of weekly sightings is 2.6.

a. Find the mean and standard deviation of x, the number of blue whale sightings per week.
b. Use Table III to find the probability that fewer than two sightings are made during a given week.
c. Use Table III to find the probability that more than five sightings are made during a given week.
d. Use Table III to find the probability that exactly five sightings are made during a given week.

Solution

a. The mean and variance of a Poisson random variable are both equal to λ. Thus, for this example,

$$\mu = \lambda = 2.6 \qquad \sigma^2 = \lambda = 2.6$$

Then the standard deviation of x is

$$\sigma = \sqrt{2.6} = 1.61$$

Remember that the mean measures the central tendency of the distribution and does not necessarily equal a possible value of x. In this example, the mean is 2.6 sightings, and although there cannot be 2.6 sightings during a given week, the average number of weekly sightings is 2.6. Similarly, the standard deviation of 1.61 measures the variability of the number of sightings per week. Perhaps a more helpful measure is the interval $\mu \pm 2\sigma$, which in this case stretches from $-.62$ to 5.82. We expect the number of sightings to fall in this interval most of the time—with at least 75% relative frequency and probably with more than 90% relative frequency. The mean and the 2-standard-deviation interval around it are shown in Figure 4.8(a).

b. A partial reproduction of Table III is shown in Figure 4.8(b). The rows of the table correspond to different values of λ, and the columns correspond to different values of the Poisson random variable x. The entries in the table are cumulative probabilities (much like the binomial probabilities in Table II). To find the probability that fewer than two sightings are made during a given week, we first note that

$$P(x < 2) = P(x \leq 1)$$

This probability is a cumulative probability and therefore is the entry in Table III in the row corresponding to $\lambda = 2.6$ and the column corresponding to $x = 1$. The entry is .267, shown shaded in Figure 4.8(b). This probability corresponds to the shaded area in Figure 4.8(a) and may be interpreted as meaning that there is a 26.7% chance that fewer than two sightings will be made during a given week.

FIGURE 4.8

(a) Probability distribution for number of blue whale sightings

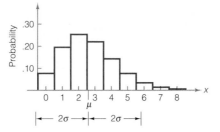

(b) Reproduction of part of Table III of Appendix A

λ \ x	0	1	2	3	4	5	6	7	8	9
2.2	.111	.355	.623	.819	.928	.975	.993	.998	1.000	1.000
2.4	.091	.308	.570	.779	.904	.964	.988	.997	.999	1.000
2.6	.074	.267	.518	.736	.877	.951	.983	.995	.999	1.000
2.8	.061	.231	.469	.692	.848	.935	.976	.992	.998	.999
3.0	.050	.199	.423	.647	.815	.916	.966	.988	.996	.999
3.2	.041	.171	.380	.603	.781	.895	.955	.983	.994	.998
3.4	.033	.147	.340	.558	.744	.871	.942	.977	.992	.997
3.6	.027	.126	.303	.515	.706	.844	.927	.969	.988	.996
3.8	.022	.107	.269	.473	.668	.816	.909	.960	.984	.994
4.0	.018	.092	.238	.433	.629	.785	.889	.949	.979	.992
4.2	.015	.078	.210	.395	.590	.753	.867	.936	.972	.989
4.4	.012	.066	.185	.359	.551	.720	.844	.921	.964	.985
4.6	.010	.056	.163	.326	.513	.686	.818	.905	.955	.980
4.8	.008	.048	.143	.294	.476	.651	.791	.887	.944	.975
5.0	.007	.040	.125	.265	.440	.616	.762	.867	.932	.968
5.2	.006	.034	.109	.238	.406	.581	.732	.845	.918	.960
5.4	.005	.029	.095	.213	.373	.546	.702	.822	.903	.951
5.6	.004	.024	.082	.191	.342	.512	.670	.797	.886	.941
5.8	.003	.021	.072	.170	.313	.478	.638	.771	.867	.929
6.0	.002	.017	.062	.151	.285	.446	.606	.744	.847	.916

c. To find the probability that more than five sightings are made during a given week, we consider the complementary event:

$$P(x > 5) = 1 - P(x \le 5) = 1 - .951 = .049$$

where .951 is the entry in Table III corresponding to λ = 2.6 and x = 5 [see Figure 4.8(b)]. Note from Figure 4.8(a) that this is the area in the interval $\mu \pm 2\sigma$, or −.62 to 5.82. Then the number of sightings should exceed 5—or, equivalently, should be more than 2 standard deviations from the mean—during only about 4.9% of all weeks. Note that this percentage agrees remarkably well with that given by the Empirical Rule for mound-shaped distributions, which tells us to expect approximately 5% of the measurements (values of the random variable) to lie further than 2 standard deviations from the mean.

d. To use Table III to find the probability that *exactly* five sightings are made during a given week, we must write the probability as the difference between two cumulative probabilities:

$$P(x = 5) = P(x \le 5) - P(x \le 4) = .951 - .877$$
$$= .074$$

Note that the probabilities in Table III are all rounded to three decimal places. Thus, although in theory a Poisson random variable can assume infinitely large values, the values of x in Table III are extended only until the cumulative probability is 1.000. This does not mean that x *cannot* assume larger values but only that the likelihood is less than .001 (in fact, less than .0005) that it will do so.

Finally, you may need to calculate Poisson probabilities for values of λ not found in Table III. You may be able to obtain an adequate approximation by interpolation, but if not, consult more extensive tables for the Poisson distribution.

EXERCISES 4.58–4.72

LEARNING THE MECHANICS

4.58   Consider the probability distribution shown here:

$$p(x) = \frac{2^x e^{-2}}{x!} \qquad (x = 0, 1, 2, \ldots)$$

a. Is x a discrete or continuous random variable? Explain.
b. What is the name of this probability distribution?
c. Graph the probability distribution.
d. Find the mean and standard deviation of x.
e. Find the mean and standard deviation of the probability distribution.

4.59   Given that x is a random variable for which a Poisson probability distribution provides a good approximation, use Table III to compute the following:
a. $P(x \leq 2)$ when $\lambda = 1$ **b.** $P(x \leq 2)$ when $\lambda = 2$
c. $P(x \leq 2)$ when $\lambda = 3$
d. What happens to the probability of the event $\{x \leq 2\}$ as λ increases from 1 to 3? Is this intuitively reasonable?

4.60   Assume that x is a random variable having a Poisson probability distribution with a mean of 1.5. Use Table III to find the following probabilities:
a. $P(x \leq 2)$ **b.** $P(x \geq 2)$ **c.** $P(x = 2)$
d. $P(x = 0)$ **e.** $P(x > 0)$ **f.** $P(x > 5)$

4.61   Suppose x is a random variable for which a Poisson probability distribution with $\lambda = 2$ provides a good characterization.
a. Graph $p(x)$ for $x = 0, 1, 2, \ldots, 9$.
b. Find μ and σ for x, and locate μ and the interval $\mu \pm 2\sigma$ on the graph.
c. What is the probability that x will fall within the interval $\mu \pm 2\sigma$?

4.62   Suppose x is a random variable for which a Poisson probability distribution with $\lambda = 4$ provides a good characterization.
a. Graph $p(x)$ for $x = 0, 1, 2, \ldots, 9$.
b. Find μ and σ for x, and locate μ and the interval $\mu \pm 2\sigma$ on the graph.
c. What is the probability that x will fall within the interval $\mu \pm 2\sigma$?

4.63   As mentioned in Section 4.5, when n is large, p is small, and $np \leq 7$, the Poisson probability distribution provides a good approximation to the binomial probability distribution. Since we provide exact binomial probabilities (Table II in Appendix A) for relatively small values of n, you can investigate the adequacy of the approximation for $n = 25$. Use Table II to find $p(0)$, $p(1)$, and $p(2)$ for

$n = 25$ and $p = .05$. Calculate the corresponding Poisson approximations using $\lambda = \mu = np$. [*Note:* These approximations are reasonably good for *n* as small as 25, but to use the approximation in a practical situation we would prefer to have $n \geq 100$.]

APPLYING THE CONCEPTS

4.64 The mean number of patients admitted per day to the emergency room of a small hospital is 2.5. If, on a given day, there are only four beds available for new patients, what is the probability the hospital will not have enough beds to accommodate its newly admitted patients?

4.65 The Environmental Protection Agency (EPA) was established in 1970 as part of the executive branch of the federal government. Its mission is to abate and control pollution in the areas of air, water, solid waste, pesticides, radiation, and toxic substances. The EPA issues pollution standards that vitally affect the safety of consumers and the operations of industry (*The United States Government Manual 1985–1986*). For example, the EPA states that manufacturers of vinyl chloride and similar compounds must limit the amount of these chemicals in plant air emissions to no more than 10 parts per million. Suppose the mean emission of vinyl chloride for a particular plant is 4 parts per million. Assume that the number of parts per million of vinyl chloride in air samples, *x*, follows a Poisson probability distribution.
   a. What is the standard deviation of *x* for the plant?
   b. Is it likely that a sample of air from the plant would yield a value of *x* that would exceed the EPA limit? Explain.
   c. Discuss conditions that would make the Poisson assumption plausible.

4.66 A can company reports that the number of breakdowns per 8-hour shift on its machine-operated assembly line follows a Poisson distribution with a mean of 1.5.
   a. What is the probability of exactly two breakdowns on the midnight shift?
   b. What is the probability of fewer than two breakdowns on the afternoon shift?
   c. What is the probability that more than two breakdowns occur on the midnight shift?
   d. What is the probability of no breakdowns during three consecutive 8-hour shifts? (Assume that the machine operates independently across shifts.)

4.67 A certain automatic car wash takes exactly 5 minutes to wash a car. On the average, 10 cars per hour arrive at the car wash. Suppose that, 30 minutes before closing time, five cars are in line. If the car wash is in continuous use until closing time, what is the probability that no one will be in line at closing time?

4.68 The safety supervisor at a large manufacturing plant believes the expected number of industrial accidents per month is 3.4.
   a. What is the probability of exactly two accidents occurring next month?
   b. What is the probability of three or more accidents occurring next month?
   c. What assumptions do you need to make to solve this problem using the methodology of this chapter?

4.69 The National Transportation Safety Board is responsible for investigating aviation accidents. In 1986, 302.1 billion passenger-miles were flown by commercial airlines in the United States. During this period there were no fatalities. In 1987, 324.5 billion passenger-miles were flown and there were 231 fatalities. Based on data from 1983 to 1987, it is known that the average number of fatalities per 100 million passenger-miles flown is approximately .033 (*Statistical Abstract of the United States: 1989*, p. 611). Assuming that airlines fly approximately 25 billion passenger-miles per month, the mean number of fatalities for a 1-month period is 8.25 (250 times the mean per 100 million miles). Suppose the probability distribution for *x*, the number of fatalities per month, can be approximated by a Poisson probability distribution.

a. What is the probability that no fatalities will occur during any given month? [*Hint:*   Either use Table III and interpolate to approximate the probability, or use a calculator or computer to calculate the probability exactly.]

b. Find $E(x)$ and the standard deviation of x.

c. Use your answers to part **b** to describe the probability that as many as 20 fatalities will occur in any given month.

d. Discuss conditions that would make the Poisson assumption plausible.

4.70   As a check on the quality of the wooden doors produced by a company, its owner requested that each door undergo inspection for defects before leaving the plant. The plant's quality control inspector found that on the average 1 square foot of door surface contains .5 minor flaw. Subsequently, 1 square foot of each door's surface was examined for flaws. The owner decided to have all doors reworked that were found to have two or more minor flaws in that square foot of surface.

a. What is the probability that a door will fail inspection and be sent back for reworking?

b. What is the probability that a door will pass inpection?

4.71   The number x of people who arrive at a cashier's counter in a bank during a specified period of time often possesses (approximately) a Poisson probability distribution. If we know the mean arrival rate λ, the Poisson probability distribution can be used to aid in the design of the customer service facility. Suppose you estimate that the mean number of arrivals per minute for cashier service at a bank is one person per minute.

a. What is the probability that in a given minute the number of arrivals will equal three or more?

b. Can you tell the bank manager that the number of arrivals will rarely exceed two per minute?

4.72   The Federal Deposit Insurance Corporation (FDIC), established in 1933, insures deposits of up to $100,000 in banks that are members of the Federal Reserve System (and others that voluntarily join the insurance fund) against losses due to bank failure or theft. From 1978 through 1987, the average number of bank failures per year among insured banks was (approximately) 67.6 (*Statistical Abstract of the United States: 1989*, p. 493). Assume that the number of bank failures per year, x, among insured banks can be adequately characterized by a Poisson probability distribution with mean 67.6.

a. Find the expected value and standard deviation of x.

b. In 1985, 120 insured banks failed. Use the Poisson probability distribution to show how you could compute $P(x = 120)$. Do not actually perform the calculation.

c. How far (in standard deviations) does $x = 120$ lie above the mean of the Poisson distribution? That is, find the z-score for $x = 120$.

d. In 1983, 48 insured banks failed. Indicate how to calculate $P(x \leq 48)$. Do not actually perform the calculation.

Summary

Random variables are rules that assign numerical values to the outcomes of an experiment. **Discrete random variables** have countable numbers of possible values, and **continuous random variables** can assume an uncountable number of values corresponding to the points in one or more intervals. For purposes of distinguishing the two, the values of a discrete random variable can be listed, whereas those of a continuous random variable cannot.

The **probability distribution** of a discrete random variable specifies the probability associated with each possible value the random variable can assume. The mean and standard deviation of the probability distribution provide numerical descriptive measures of the distribution. Many applications of statistics involve the estimation of the mean and standard deviation of a probability distribution based on sample data.

The **binomial** and **Poisson** probability distributions describe two specific types of discrete random variables that have many applications in business. The characteristics of these important random variables were presented in this chapter, along with many examples of business data for which each would be an appropriate model. The formulas for the probability distributions, means, and variances were also presented, along with tables to assist you in calculating probabilities associated with each of these distributions.

SUPPLEMENTARY EXERCISES 4.73–4.99

[*Note: Starred (*) exercises refer to optional sections in this chapter.*]

LEARNING THE MECHANICS

*4.73 Identify the type of random variable—binomial or Poisson—described by each of the following probability distributions:

a. $p(x) = \dfrac{.5^x e^{-.5}}{x!}$ $(x = 0, 1, 2, \ldots)$

b. $p(x) = \dbinom{6}{x}(.2)^x(.8)^{6-x}$ $(x = 0, 1, 2, \ldots, 6)$

c. $p(x) = \dfrac{10!}{x!(10 - x)!}(.9)^x(.1)^{10-x}$ $(x = 0, 1, 2, \ldots, 10)$

4.74 Which of the following describe discrete random variables, and which describe continuous random variables?
a. The number of damaged inventory items
b. The average monthly sales revenue generated by a salesperson over the past year
c. The number of square feet of warehouse space a company rents
d. The length of time a firm must wait before its copying machine is fixed

4.75 Suppose x is a binomial random variable. Find $p(x)$ for each of the following combinations of x, n, and p.
a. $x = 1$, $n = 3$, $p = .1$ b. $x = 4$, $n = 20$, $p = .3$
c. $x = 0$, $n = 2$, $p = .4$ d. $x = 4$, $n = 5$, $p = .5$
e. $n = 15$, $x = 12$, $p = .9$ f. $n = 10$, $x = 8$, $p = .6$

4.76 For each of the following examples, decide whether x is a binomial random variable and explain your decision:
a. A manufacturer of computer chips randomly selects 100 chips from each hour's production in order to estimate the proportion defective. Let x represent the number of defectives in the 100 sampled chips.
b. Of five applicants for a job, two will be selected. Although all applicants appear to be equally qualified, only three have the ability to fulfill the expectations of the company. Suppose that the two selections are made at random from the five applicants, and let x be the number of qualified applicants selected.
c. A software developer establishes a support hotline for customers to call in with questions regarding use of the software. Let x represent the number of calls received on the support hotline during a specified workday.
d. Florida is one of a minority of states with no state income tax. A poll of 1,000 registered voters is conducted to determine how many would favor a state income tax in light of the state's current fiscal condition. Let x be the number in the sample who would favor the tax.

4.77   Consider the discrete probability distribution shown here.

x	10	12	18	20
$p(x)$	.2	.2	.1	.5

    a. Calculate μ, σ^2, and σ.
    b. What is $P(x < 15)$?
    c. Calculate $\mu \pm 2\sigma$.
    d. What is the probability that x is in the interval $\mu \pm 2\sigma$?

4.78   Suppose x is a binomial random variable with $n = 20$ and $p = .6$.
    a. Find $P(x = 14)$.   b. Find $P(x \le 10)$.   c. Find $P(x > 10)$.
    d. Find $P(8 \le x \le 17)$.   e. Find $P(8 < x < 17)$.   f. Find μ, σ^2, and σ.
    g. What is the probability that x is in the interval $\mu \pm 2\sigma$?

*4.79   Suppose x is a Poisson random variable. Compute $p(x)$ for each of the following cases:
    a. $\lambda = 2$, $x = 3$ b. $\lambda = 1$, $x = 4$ c. $\lambda = .5$, $x = 2$

*4.80   Suppose x is a random variable for which a Poisson probability distribution with $\lambda = 3$ provides a good characterization. Compute the following:
    a. $P(x = 1)$ b. $P(x = 3)$ c. $P(x = 0)$
    d. $P(x = 5)$ e. $P(x \le 2)$ f. $P(x \ge 2)$

APPLYING THE CONCEPTS

4.81   The U.S. Census Bureau reported that one wife in five earned more than her husband in 1983, among couples where both husband and wife had earnings. Suppose a random sample of 20 working couples in an industrial community is selected, and the number, x, of wives who earn more than their husbands is recorded.
    a. Check the assumptions that must be satisfied for x to be a binomial random variable.
    b. Assuming that the probability distribution of x is binomial, what are the expected value and standard deviation of x?
    c. What is the probability that x falls within 2 standard deviations of its expected value?
    d. Suppose none of the 20 couples sampled has a wife who earns more than her husband. Does this event cast doubt on the applicability of the Census Bureau's nationwide estimate to this industrial community? Explain.

4.82   An important function in any business is long-range planning. Additions to a firm's physical plant, for example, cannot be achieved overnight; their construction must be planned years in advance. Anticipating a substantial growth in sales over the next 5 years, a printing company is planning today for the warehouse space it will need 5 years hence. It obviously cannot be certain exactly how many square feet of storage space, x, it will need in 5 years, but the company can project its needs by using a probability distribution such as the following:

x	10,000	15,000	20,000	25,000	30,000	35,000
$p(x)$	.05	.15	.35	.25	.15	.05

What is the expected number of square feet of storage space the printing company will need in 5 years?

4.83   Many minor operations at a hospital can be performed the same day the patient is admitted. A hospital serving a large metropolitan area has found that in the past, 20% of newly admitted patients needing an operation are scheduled for same-day surgery. Suppose that ten patients are randomly selected from those admitted to the hospital for surgery over the past year. If x, the number in the sample of ten who receive same-day surgery, possesses a binomial probability distribution:
   a. What is the probability that exactly five of these patients have same-day surgery?
   b. What is the probability that at most one has same-day surgery?
   c. If the ten patients are selected from the admissions on a single given day, is it reasonable to expect x to possess the characteristics of a binomial random variable? That is, is this a binomial experiment? Explain.

4.84   An employee of a firm has an option to invest $1,000 in the company's bonds. At the end of 1 year, the company will buy back the bonds at a price determined by its profits for the year. From past years, the company predicts it will buy the bonds back at the following prices with the associated probabilities (x = price paid for bonds):

x	$0	$500	$1,000	$1,500	$2,000
$p(x)$	.01	.22	.30	.22	.25

   a. What is the probability the employees will receive $1,000 or less for the investment?
   b. What is the expected price paid for the bonds?
   c. What is the employee's expected profit?
   d. Find σ^2 and σ for this probability distribution.

4.85   In a 1986 article in the *Journal of Applied Psychology*, J. Near and M. Miceli report the results of an extensive survey conducted to determine the extent of "whistle blowing" among federal employees and to study the factors that are associated with retaliation against whistle blowers. Whistle blowing refers to an employee's reporting of wrongdoing by coworkers. Among other things, the survey found that about 5% of employees contacted had reported wrongdoing during the past 12 months. Assume that a sample of 25 employees in one agency are contacted, and let x be the number who have observed and reported wrongdoing in the last 12 months. Assume that the probability of whistle blowing is .05 for any federal employee over the past 12 months.
   a. Find the mean and standard deviation of x. Can x be equal to its expected value? Explain.
   b. Write the event that at least five of the employees are whistle blowers in terms of x. Find the probability of the event.
   c. If five of the 25 contacted have been whistle blowers over the past 12 months, what would you conclude about the applicability of the 5% assumption to this agency? Use your answer to part b to justify your conclusion.

4.86   The probability that a person responds to a mailed questionnaire is .4.
   a. What is the probability that of 20 questionnaires, more than 12 will be returned?
   b. How many questionnaires should be mailed if you want to be reasonably certain that at least 100 will be returned?

4.87   The state highway patrol has determined that one out of every six calls for help originating from roadside call boxes is a hoax. Five calls for help have been received and five tow trucks dispatched.
   a. What is the probability that none of the calls was a hoax?
   b. What is the probability that only three of the callers really needed assistance?
   c. What assumptions do you have to make in order to solve this problem?
   d. If the highway patrol answers 10,000 calls for help next year and each call costs the patrol about $30 (labor, gas, etc.), approximately how much money will be wasted answering false alarms?

4.88   A sales manager has determined that a salesperson makes a sale to 70% of the retailers visited.
   a. If the salesperson visits 5 retailers today and 20 tomorrow, what is the probability that she makes exactly four sales today and more than ten tomorrow?
   b. If the salesperson visits 4 retailers today and 5 tomorrow, what is the probability that in these two days she will make exactly two sales?

4.89   The owner of construction company A makes bids on jobs so that if awarded the job, company A will make a $10,000 profit. The owner of construction company B makes bids on jobs so that if awarded the job, company B will make a $15,000 profit. Each company describes the probability distribution of the number of jobs the company is awarded per year as shown in the table.

COMPANY A		COMPANY B	
2	.05	2	.15
3	.15	3	.30
4	.20	4	.30
5	.35	5	.20
6	.25	6	.05

   a. Find the expected number of jobs each will be awarded in a year.
   b. What is the expected profit for each company?
   c. Find the variance and standard deviation of the distribution of number of jobs awarded per year for each company.
   d. Graph $p(x)$ for both companies A and B. For each company, what proportion of the time will x fall in the interval $\mu \pm 2\sigma$?

4.90   A large cigarette manufacturer has determined that the probability of a new brand of cigarettes obtaining a large enough market share to make production profitable is .3. Over the next 3 years, this manufacturer will introduce one new brand a year.
   a. What is the probability that at least one new brand will obtain sufficient market share to make its production profitable?
   b. What is the probability that all three new brands will obtain sufficient market share?
   c. What assumptions do you have to make in order to solve this problem?

4.91   In recent years, the use of the telephone as a data collection instrument for public opinion polls has been steadily increasing. However, one of the major factors bearing on the extent to which the telephone will become an acceptable data collection tool in the future is the refusal rate, i.e., the percentage of the eligible subjects actually contacted who refuse to take part in the poll. Suppose that past records indicate a refusal rate of 20% in a large city. A poll of 25 residents is to be taken and x is the number of residents contacted by telephone who refuse to take part in the poll.
   a. Find the mean and variance of x.
   b. Find $P(x \le 5)$.
   c. Find $P(x > 10)$.

4.92   As part of a study of how pricing decisions are made by fabricare firms (the dry cleaning and laundry industry), Joe F. Goetz, Jr. contacted and questioned 103 fabricare firms. The accompanying table describes the sales volumes of these firms. Assume that the sales volume, x, for each of the six intervals can be adequately approximated by the midpoint of the interval. These midpoints are shown in the second column of the table.
        One of the 103 firms is to be randomly selected for more intensive questioning. In this situation the relative frequencies in the table can be interpreted as probabilities that describe the approximate likelihood of the sampled firm having sales volume x.
   a. Find $E(x)$.   b. Find $E[(x - \mu)^2]$.
   c. Interpret the values you obtained in parts a and b in the context of the problem.

SALES VOLUME	x	RELATIVE FREQUENCY
Over $0 to $75,000	$37,500	.14
Over $75,000 to $150,000	$112,500	.30
Over $150,000 to $300,000	$225,000	.20
Over $300,000 to $500,000	$400,000	.17
Over $500,000 to $1,000,000	$750,000	.10
Over $1,000,000 to $5,000,000	$3,000,000	.09

Source: Goetz, J. F. Jr., "The pricing decision: A service industry's experience," *Journal of Small Business Management*, April 1985, 23, 63.

4.93 The efficacy of insecticides is often measured by the dose necessary to kill a certain percentage of insects. Suppose a certain dose of a new insecticide is supposed to kill 80% of the exposed insects. To test the claim, 25 insects are put in contact with the insecticide.
   a. If the insecticide really kills 80% of the exposed insects, what is the probability that fewer than 15 die?
   b. If you observed such a result, what would you conclude about the new insecticide? Explain your logic.

4.94 A physical fitness specialist claims that the probability is greater than .5 that an average adult male can improve his physical condition by spending 5 minutes per day on a certain exercise program. To test the claim, 15 randomly selected adult males follow the program for a specified amount of time. Maximal oxygen uptake is measured before and after the program for each male and serves as the criterion for assessing physical condition. If the program is not really beneficial (i.e., the probability of improvement is only .5), what is the probability that 11 or more of the 15 men have improved maximal oxygen uptake?

   Suppose 11 or more showed increased maximal oxygen uptake. Assuming that the specialist's exercise program is ineffective, would you regard $x \geq 11$ as a rare event, or would you conclude that, in actuality, the probability of improvement exceeds $p = \frac{1}{2}$ and that the program is effective?

4.95 [*Warning:* This exercise is realistic, but the computations involved are tedious.] Refer to Case Study 4.1—the Red Lobster sales tax problem. Let $x = 0, 1, 2, \ldots, 99$ be the number of cents (exceeding whole dollars) involved in a sale, and assume that the sales tax is assessed using the bracket system listed in the table. Suppose that x has a probability distribution $p(x) = .01, x = 0, 1, 2, \ldots, 99$ (an assumption that might be fairly accurate for restaurant sales). Find the expected value of the percentage of tax paid on the cents portion of a sale. [*Hint:* You can write the percent tax—call it y—for each value of x. You also know the probabilities associated with each value of x and, consequently, each value of y. Then the expected percentage of tax paid is $E(y) = \Sigma y p(y)$.]

Values of x	Tax (¢)
0, 1, . . . , 9	0
10, 11, . . . , 25	1
26, 27, . . . , 50	2
51, 52, . . . , 75	3
76, 77, . . . , 99	4

*4.96 A small life insurance company has determined that on the average it receives five death claims per day.
   a. What is the probability that the company will receive three claims or less on a particular day?
   b. What is the probability that the company will receive exactly five claims on a particular day?
   c. What assumptions must you make to find these probabilities?

*4.97 An emergency rescue vehicle is used an average of 1.3 times daily.
 a. What is the probability that the vehicle will be used exactly twice tomorrow?
 b. What is the probability that it will be used more than twice?
 c. Exactly three times?

*4.98 A wholesale office equipment outlet claims that on an average it sells 2.5 typewriters per day. If it has only 5 typewriters in stock at the close of business today and does not expect to receive a shipment of new typewriters until some time after the close of business tomorrow, what is the probability that the outlet's current supply of typewriters will not be sufficient to meet tomorrow's demand?

*4.99 Large bakeries typically have fleets of delivery trucks. It was determined by one such bakery that the expected number of delivery truck breakdowns per day is 1.5. Assume that the number of breakdowns is independent from day to day.
 a. What is the probability that there will be exactly 2 breakdowns today and exactly 3 tomorrow?
 b. Fewer than 2 today and more than 2 tomorrow?

ON YOUR OWN...

Consider the following random variables:

1. The number x of people who recover from a certain disease out of a sample of five patients
2. The number x of voters in a sample of five who favor a method of tax reform
3. The number x of hits a baseball player gets in five official times at bat

In each case, x is a binomial random variable (or approximately so) with $n = 5$ trials. Assume that in each case the probability of Success is .3. (What is a Success in each of the examples?) If this were true, the probability distribution for x would be the same for each of the three examples. To obtain a relative frequency histogram for x, conduct the following experiment: Place ten poker chips (pennies, marbles, or any ten *identical* items) in a bowl and mark three of the ten Success—the remaining seven will represent Failure. Randomly select a chip from the ten, observing whether it was a Suc-

cess or Failure. Then return the chip and randomly select a second chip from the ten available chips, and record this outcome. Repeat this process until a total of five trials has been conducted. Count the number x of Successes observed in the five trials. Repeat the entire process 100 times to obtain 100 observed values of x.

a. Use the 100 values of x obtained from the simulation to construct a relative frequency histogram for x. Note that this histogram is an approximation to $p(x)$.
b. Calculate the exact values of $p(x)$ for $n = 5$ and $p = .3$ and compare these values with the approximations found in part a.
c. If you were to repeat the simulation an extremely large number of times (say 100,000), how do you think the relative frequency histogram and true probability distribution would compare?

USING THE COMPUTER...

Calculate the mean percentage of college graduates over all the zip codes for one of the census regions listed in Appendix C. Round the mean to the nearest integer, and use the result as an estimate of the percentage of college graduates among all people at least 25 years old in the region.

a. Suppose 20 individuals from the region respond to an advertisement by an employment agency. If these people represent a random sample of all individuals age 25 or older in the region, what is the probability that more than half of them have college degrees?

b. If 2,000 individuals respond, find the mean μ and standard deviation σ of the number of them who have college degrees. Calculate $\mu \pm 2\sigma$ and $\mu \pm 3\sigma$, and use Chebyshev's theorem and the Empirical Rule to estimate the probability that the number of college degree responses falls in each of the intervals. Based on your answers, assess the likelihood that more than half the applicants will have college degrees.

c. What are the potential problems with using the mean percentage of college graduates as an estimate of the percentage for the region? How could the estimate be improved?

References

Alexander, G. J., and Francis, J. C. *Portfolio Analysis*. Englewood Cliffs, N.J.: Prentice Hall, 1966.

Biddle, W. "What destroyed *Challenger*?" *Discover*, April 1986, pp. 40–47.

"'83 report put booster accident as most likely." *Minneapolis Star and Tribune*, Feb. 11, 1986, p. 1.

Hogg, R. V., and Craig, A. T. *Introduction to Mathematical Statistics*, 4th ed. New York: Macmillan, 1978, Chapters 1 and 3.

McKean, K. "They fly in the face of danger." *Discover*, April 1986, pp. 48–58.

Mendenhall, W., Scheaffer, R. L., and Wackerly, D. *Mathematical Statistics with Applications*, 2d ed. North Scituate, Mass.: Duxbury, 1980, Chapter 3.

Mood, A. M., Graybill, F. A., and Boes, D. C. *Introduction to the Theory of Statistics*, 3d ed. New York: McGraw-Hill, 1963, Chapter 3.

Murphy, A. H., and Winkler, R. L. "Probability, forecasting in meteorology," *Journal of the American Statistical Association*, September 1984, 79, pp. 489–500.

Robinson, W. V. "NASA blamed for shuttle disaster," *Boston Globe*, June 10, 1986, p. 1.

The United States Government Manual 1985–1986. Office of the Federal Register, revised July 1, 1985, pp. 485–487.

Van de Werf, F., et al. "Coronary thrombosis with tissue-type plasminogen activator in patients with evolving myocardial infarction," *New England Journal of Medicine*, 1984, 310.

Continuous Random Variables

WHERE WE'VE BEEN...

Because numerical data represent observed values of random variables, we needed to find the probabilities associated with specific sample observations. The probability theory of Chapter 3 provided the mechanism for finding the probabilities associated with discrete random variables. Finding and describing this set of probabilities—the probability distribution for a discrete random variable—was the subject of Chapter 4.

WHERE WE'RE GOING...

Since data may be derived from observations on continuous as well as discrete random variables, we need to know about probability distributions associated with continuous random variables and also how to use the mean and standard deviation to describe these distributions. Chapter 5 addresses the problem and, in particular, introduces the *normal probability distribution*. As you will subsequently see, the normal probability distribution is one of the most useful distributions in statistics.

In this chapter we will consider some continuous random variables that are commonly encountered. Recall that a continuous random variable is one that can assume any value within some interval or intervals. For example, the length of time between a person's visits to a doctor, the thickness of sheets of steel produced in a rolling mill, and the yield of wheat per acre of farmland are all continuous random variables. The methodology we employ to describe continuous random variables will necessarily be somewhat different from that used to describe discrete random variables. We first discuss the general form of **continuous probability distributions**, and then we present three specific types that are used in making statistical decisions. The **normal probability distribution**, which plays a basic and important role in both the theory and application of statistics, is essential to the study of most of the subsequent chapters in this book. The other types have practical applications, but a study of these topics is optional.

5.1 Continuous Probability Distributions

The graphic form of the probability distribution for a continuous random variable x is a smooth curve that might appear as shown in Figure 5.1. This curve, a function of x, is denoted by the symbol $f(x)$ and is variously called a **probability density function**, a **frequency function**, or a **probability distribution**.

FIGURE 5.1

A probability distribution $f(x)$ for a continuous random variable x

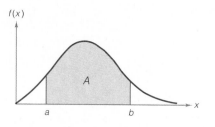

The areas under a probability distribution correspond to probabilities for x. For example, the area A beneath the curve between the two points a and b, as shown in Figure 5.1, is the probability that x assumes a value between a and b ($a < x < b$). Because there is no area over a point, say $x = a$, it follows that (according to our model) the probability associated with a particular value of x is equal to 0; that is, $P(x = a) = 0$ and hence $P(a < x < b) = P(a \leq x \leq b)$. In other words, the probability is the same regardless of whether you include the endpoints of the interval. Also, because areas over intervals represent probabilities, it follows that the total area under a probability distribution, the probability assigned to all values of x, should equal 1. Note that probability distributions for continuous random variables possess different shapes depending on the relative frequency distributions of real data that the probability distributions are supposed to model.

The areas under most probability distributions are obtained by use of the calculus or numerical methods.* Because this is often a difficult procedure, we

---

*Students with a knowledge of calculus should note that the probability that x assumes a value in the interval $a < x < b$ is $P(a < x < b) = \int_a^b f(x)\, dx$, assuming the integral exists. Similar to the requirements for a discrete probability distribution, we require $f(x) \geq 0$ and $\int_{-\infty}^{\infty} f(x)\, dx = 1$.

will give the areas for some of the most common probability distributions in tabular form in Appendix A. Then to find the area between two values of x, say $x = a$ and $x = b$, you simply have to consult the appropriate table.

For each of the continuous random variables presented in this chapter, we will give the formula for the probability distribution along with its mean and standard deviation. These two numbers, μ and σ, will enable you to make some approximate probability statements about a random variable even when you do not have access to a table of areas under the probability distribution.

5.2 The Uniform Distribution

All the probability problems discussed in Chapter 3 had sample spaces that contained a finite number of simple events. In many of these problems, the simple events were assigned equal probabilities—for example, the die toss or the coin toss. For continuous random variables there is an infinite number of values in the sample space, but in some cases the values may appear to be equally likely. For example, if a short exists in a 5-meter stretch of electrical wire, it may have an equal probability of being in any particular 1-centimeter segment along the line. Or if a safety inspector plans to choose a time at random during the four afternoon work-hours to pay a surprise visit to a certain area of a plant, then each 1-minute time interval in this 4-work-hour period will have an equally likely chance of being selected for the visit.

Continuous random variables that appear to have equally likely outcomes over their range of possible values possess a **uniform probability distribution**, perhaps the simplest of all continuous probability distributions. Suppose the random variable x can assume values only in an interval $c \leq x \leq d$. Then the uniform frequency function has a rectangular shape, as shown in Figure 5.2. Note that the possible values of x consist of all points in the interval between point c and point d. The height of $f(x)$ is constant in that interval and equals $1/(d - c)$. Therefore, the total area under $f(x)$ is given by

Total area of rectangle = (Base)(Height)

$$= (d - c)\left(\frac{1}{d - c}\right) = 1$$

The uniform probability distribution provides a model for continuous random variables that are *evenly distributed* over a certain interval. That is, a uniform random variable is one that is just as likely to assume a value in one interval as it is to assume a value in any other interval of equal size. There is no clustering of values around any value; instead, there is an even spread over the entire region of possible values.

The uniform distribution is sometimes referred to as the **randomness distribution**, since one way of generating a uniform random variable is to perform an experiment in which a point is *randomly selected* on the horizontal axis between the points c and d. If we were to repeat this experiment infinitely often, we would create a uniform probability distribution like that shown in Figure 5.2. The random selection of points in an interval can also be used to generate random numbers such as those in Table I of Appendix A. Recall that random numbers are selected in such a way that every number would have an equal probability of selection. Therefore, random numbers are realizations of a uniform random variable. (Random numbers were used to draw random samples in Section 3.7.)

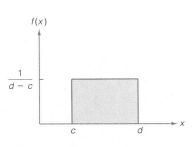

FIGURE 5.2

The uniform probability distribution

The formulas for the uniform probability distribution, its mean, and standard deviation are shown in the box.

Probability Distribution, Mean, and Standard Deviation of a Uniform Random Variable x

$$f(x) = \frac{1}{d - c} \quad (c \leq x \leq d)$$

$$\mu = \frac{c + d}{2} \qquad \sigma = \frac{d - c}{\sqrt{12}}$$

Suppose the interval $a < x < b$ lies within the domain of x; that is, it falls within the larger interval $c \leq x \leq d$. Then the probability that x assumes a value within the interval $a < x < b$ is the area of the rectangle over the interval, namely, $(b - a)/(d - c)$.*

EXAMPLE 5.1

An unprincipled used car dealer sells a car to an unsuspecting buyer, even though the dealer knows that the car will have a major breakdown within the next 6 months. The dealer provides a warranty of 45 days on all cars sold. Let x represent the length of time until the breakdown occurs. Assume that x is a uniform random variable with values between 0 and 6 months.

a. Calculate the mean and standard deviation of x. Graph the probability distribution of x, and show the mean on the horizontal axis. Also show 1- and 2-standard-deviation intervals around the mean.
b. Calculate the probability that the breakdown occurs while the car is still under warranty.

Solution

a. To calculate the mean and standard deviation for x, we substitute 0 and 6 months for c and d, respectively, in the formulas for uniform random variables. Thus,

$$\mu = \frac{c + d}{2} = \frac{0 + 6}{2} = 3 \text{ months}$$

and

$$\sigma = \frac{d - c}{\sqrt{12}} = \frac{6 - 0}{\sqrt{12}} = \frac{6}{3.464} = 1.73 \text{ months}$$

The uniform probability distribution is

$$f(x) = \frac{1}{d - c} = \frac{1}{6 - 0} = \frac{1}{6} \quad (0 \leq x \leq 6)$$

---

*The student with a knowledge of calculus should note that

$$P(a < x < b) = \int_a^b f(x) \, dx = \int_a^b 1/(d - c) \, dx = (b - a)/(d - c)$$

The graph of this function is shown in Figure 5.3. The mean and 1- and 2-standard-deviation intervals around the mean are shown on the horizontal axis.

FIGURE 5.3

Distribution for x in Example 5.1

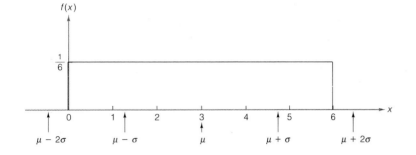

b. To find the probability that the car is still under warranty when it breaks down, we must find the probability that x, the time until the breakdown occurs, is less than 45 days or (about) 1.5 months. As indicated in Figure 5.4, we need to calculate the area under the frequency function $f(x)$ between the points $x = 0$ and $x = 1.5$. This is the area of a rectangle with base $1.5 - 0 = 1.5$ and height $\frac{1}{6}$. The probability that the unsuspecting buyer will be able to have the car repaired at the dealer's expense is then

$$P(0 < x < 1.5) = (\text{Base})(\text{Height}) = (1.5)\left(\frac{1}{6}\right) = .25$$

FIGURE 5.4

Probability that car breaks down within 1.5 months of purchase

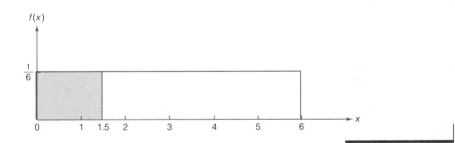

EXERCISES 5.1–5.14

LEARNING THE MECHANICS

5.1 Suppose x is a random variable best described by a uniform probability distribution with $c = 20$ and $d = 45$.
   a. Find $f(x)$.
   b. Find the mean and standard deviation of x.
   c. Graph $f(x)$ and locate μ and the interval $\mu \pm 2\sigma$ on the graph. Note that the probability that x assumes a value within the interval $\mu \pm 2\sigma$ is equal to 1.

5.2 Refer to Exercise 5.1. Find the following probabilities:
   a. $P(20 \leq x \leq 35)$ b. $P(20 < x < 35)$
   c. $P(x \geq 35)$ d. $P(x \leq 20)$
   e. $P(x \leq 25)$ f. $P(10 \leq x \leq 40)$

g. $P(x \geq 36)$ **h.** $P(x \geq 35.5)$

i. $P(20.5 \leq x \leq 35.5)$ **j.** $P(x < 20.5)$

5.3 Suppose x is a random variable best described by a uniform probability distribution with $c = 2$ and $d = 4$.

a. Find $f(x)$.

b. Find the mean and standard deviation of x.

c. Find $P(\mu - \sigma \leq x \leq \mu + \sigma)$.   **d.** Find $P(x > 2.78)$.

e. Find $P(2.4 \leq x \leq 3.7)$.   **f.** Find $P(x < 2)$.

5.4 Refer to Exercise 5.3. Find the value of a that makes each of the following probability statements true.

a. $P(x \geq a) = .5$ **b.** $P(x \leq a) = .2$

c. $P(x \leq a) = 0$ **d.** $P(2.5 \leq x \leq a) = .5$

5.5 The random variable x is best described by a uniform probability distribution with $c = 100$ and $d = 200$. Find the probability that x assumes a value:

a. More than 2 standard deviations from μ.

b. Less than 3 standard deviations from μ.

c. Within 2 standard deviations of μ.

5.6 The random variable x is best described by a uniform probability distribution with mean 10 and standard deviation 1. Find c, d, and $f(x)$. Graph the probability distribution.

APPLYING THE CONCEPTS

5.7 The manager of a large department store with three floors reports that the time a customer on the second floor must wait for an elevator has a uniform distribution ranging from 0 to 4 minutes. Find the mean and standard deviation of x, the time a customer on the second floor waits for an elevator. If it takes the elevator 15 seconds to go from floor to floor, find the probability that a hurried customer can reach the first floor in less than 1.5 minutes after pushing the second-floor elevator button.

5.8 A bus is scheduled to stop at a certain bus stop every half hour on the hour and the half hour. At the end of the day, buses still stop about every 30 minutes, but due to delays that often occur earlier in the day, the bus is likely to be late. The director of the bus line claims that the length of time a bus is late is uniformly distributed and the maximum time that a bus is late is 20 minutes.

a. If the director's claim is true, what is the expected number of minutes a bus will be late?

b. If the director's claim is true, what is the probability that the last bus on a given day will be more than 19 minutes late?

c. If you arrive at the bus stop at the end of a day at exactly half-past the hour and must wait more than 19 minutes for the bus, what would you conclude about the director's claim? Why?

5.9 Probability distributions and event probabilities can be used to express uncertainty about future events. For example, security analysts use probability distributions to forecast stock prices and quarterly earnings; weather forecasters use event probabilities to forecast precipitation, temperature ranges, and the location of a hurricane's landfall. Such forecasts are known as *probability forecasts* (Murphy and Winkler, 1984). Probability forecasts can be derived from sample data, expert judgment, or a combination of both. When a forecaster possesses relatively little information about the future value of a random variable, the uniform distribution is frequently used to make a probability forecast. Suppose a security analyst believes that the closing price of a particular stock 3 months from today will be between $30 and $40, but has no idea where within the interval the price will be.

a. Graph the uniform distribution the security analyst should use as a probability forecast.

b. According to the analyst's forecast, what is the probability that the stock will close higher than $35? Higher than $38? Between $34 and $36 inclusive? Higher than $50?

5.10 The manager of a local soft-drink bottling company believes that when a new beverage-dispensing machine is set to dispense 7 ounces, it in fact dispenses an amount x at random anywhere between 6.5 and 7.5 ounces inclusive. Suppose x has a uniform probability distribution.
  a. Is the amount dispensed by the beverage machine a discrete or a continuous random variable? Explain.
  b. Graph the frequency function for x, the amount of beverage the manager believes is dispensed by the new machine when it is set to dispense 7 ounces.
  c. Find the mean and standard deviation for the distribution graphed in part b, and locate the mean and the interval $\mu \pm 2\sigma$ on the graph.

5.11 Refer to Exercise 5.10. Find the following probabilities.
  a. $P(x \geq 7)$ b. $P(6.5 \leq x \leq 7.5)$ c. $P(x \leq 6.75)$
  d. $P(x > 7.25)$ e. $P(x < 6)$ f. $P(6.5 \leq x \leq 7.25)$

5.12 Refer to Exercises 5.10 and 5.11. What is the probability that each of the next six bottles filled by the new machine will contain more than 7.25 ounces of beverage? Assume that the amount of beverage dispensed in one bottle is independent of the amount dispensed in another bottle.

5.13 The weather on a tropical island in January is fairly constant. Records indicate that the high temperatures for each day of the month tend to have a uniform distribution over the interval from 75°F to 90°F. A tourist arrives on the island on a randomly selected day in January.
  a. What is the probability that the temperature will be above 80°F?
  b. What is the probability that the temperature will be between 80°F and 85°F?
  c. What is the expected temperature?

5.14 Rapid advances in technology in recent years have led to the development and manufacture of extremely complex equipment and, consequently, to the need to evaluate the equipment's reliability. The **reliability** of a piece of equipment is frequently defined to be the probability p that the equipment performs its intended function successfully for a given period of time under specific conditions (Martz and Waller, 1982). Because p varies from one point in time to another, some reliability analysts treat p as if it were a random variable with a uniform probability distribution over the interval from 0 to 1. This assumption implies, for example, that p is equally likely to be below .2 as it is to be above .8. Suppose an analyst characterizes the total uncertainty about the reliability of a particular robotic device used in an automobile assembly line using the following distribution:

$$f(p) = \begin{cases} 1 & 0 \leq p \leq 1 \\ 0 & \text{otherwise} \end{cases}$$

  a. Graph the analyst's probability distribution for p.
  b. Find the mean and variance of p.
  c. According to the analyst's probability distribution for p, what is the probability that p is greater than .95? Less than .95?
  d. Suppose the analyst receives the additional information that p is definitely between .90 and .95, but there is complete uncertainty about where it lies between these values. Describe the probability distribution the analyst should use to describe the available information about p.

5.3 The Normal Distribution

One of the most commonly observed continuous random variables has a **bell-shaped** probability distribution as shown in Figure 5.5 (page 218). It is known as a **normal random variable** and its probability distribution is called a **normal distribution**.

You will see during the remainder of this text that the normal distribution plays a very important role in the science of statistical inference. Moreover, many

FIGURE 5.5

A normal probability distribution

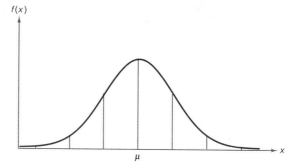

phenomena generate random variables with probability distributions that are very well approximated by a normal distribution. For example, the error made in measuring a person's blood pressure may be a normal random variable, and the probability distribution for the yearly rainfall in a certain region might be approximated by a normal probability distribution. The normal distribution might also provide an accurate model for the distribution of the scores on an aptitude test. You can determine the adequacy of the normal approximation to an existing population of data by comparing the relative frequency distribution of a large sample of the data to the normal probability distribution. Tests to detect disagreement between a set of data and the assumption of normality are available, but they are beyond the scope of this book.

The normal distribution is perfectly symmetric about its mean μ, as can be seen in the examples in Figure 5.6. Its spread is determined by the value of its standard deviation σ.

FIGURE 5.6

Several normal distributions with different means and standard deviations

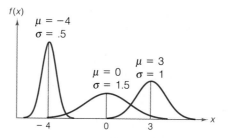

The formula for the normal probability distribution is shown in the box. When plotted, this formula yields a curve like that shown in Figure 5.5.

Probability Distribution for a Normal Random Variable x

$$f(x) = \frac{1}{\sigma\sqrt{2\pi}} e^{-(1/2)[(x - \mu)/\sigma]^2}$$

where

μ = Mean of the normal random variable x

σ = Standard deviation

π = 3.1416. . .

e = 2.71828. . .

Note that the mean μ and standard deviation σ appear in this formula, so that no separate formulas for μ and σ are necessary. To graph the normal curve we have to know the numerical values of μ and σ.

Computing the area over intervals under the normal probability distribution is a difficult task.* Consequently, we will use the computed areas listed in Table IV of Appendix A (and inside the front cover). Although there are an infinitely large number of normal curves—one for each pair of values for μ and σ—we have formed a single table that will apply to any normal curve. This is done by constructing the table of areas as a function of the z-score (presented in Section 2.7). The population z-score for a measurement was defined as the *distance* between the measurement and the population mean, divided by the population standard deviation. Thus, the z-score gives the distance between a measurement and the mean in units equal to the standard deviation. In symbolic form, the z-score for the measurement x is

$$z = \frac{x - \mu}{\sigma}$$

Note that when $x = \mu$, we obtain $z = 0$.

To illustrate the use of Table IV, suppose we know that the length of time x between charges of a pocket calculator has a normal distribution with a mean of 50 hours and a standard deviation of 15 hours. If we were to observe the length of time that elapses before the need for the next charge, what is the probability that this measurement will assume a value between 50 and 70 hours? This probability is the area under the normal probability distribution between 50 and 70, as shown in the shaded area A of Figure 5.7.

FIGURE 5.7

Normal distribution:
$\mu = 50$, $\sigma = 15$

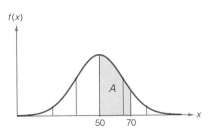

The first step in finding the area A is to calculate the z-score corresponding to the measurement 70. We calculate

$$z = \frac{x - \mu}{\sigma} = \frac{70 - 50}{15} = \frac{20}{15} = 1.33$$

Thus, the measurement 70 is 1.33 standard deviations above the mean, $\mu = 50$. The second step is to refer to Table IV (a partial reproduction of this table is shown in Figure 5.8). Note that z-scores are listed in the left-hand column of the table. To find the area corresponding to a z-score of 1.33, we first locate the value 1.3 in the left-hand column. Since this column lists z-values to one decimal place only, we refer to the top row of the table to get the second decimal place,

---

*The student with a knowledge of calculus should note that there is not a closed-form expression for $P(a < x < b) = \int_a^b f(x)\, dx$ for the normal probability distribution. The value of this definite integral can be obtained to any desired degree of accuracy by numerical approximation procedures. For this reason, it is tabulated for the user.

FIGURE 5.8

Reproduction of part of Table IV in Appendix A

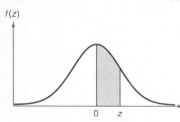

z	.00	.01	.02	.03	.04	.05	.06	.07	.08	.09
.0	.0000	.0040	.0080	.0120	.0160	.0199	.0239	.0279	.0319	.0359
.1	.0398	.0438	.0478	.0517	.0557	.0596	.0636	.0675	.0714	.0753
.2	.0793	.0832	.0871	.0910	.0948	.0987	.1026	.1064	.1103	.1141
.3	.1179	.1217	.1255	.1293	.1331	.1368	.1406	.1443	.1480	.1517
.4	.1554	.1591	.1628	.1664	.1700	.1736	.1772	.1808	.1844	.1879
.5	.1915	.1950	.1985	.2019	.2054	.2088	.2123	.2157	.2190	.2224
.6	.2257	.2291	.2324	.2357	.2389	.2422	.2454	.2486	.2517	.2549
.7	.2580	.2611	.2642	.2673	.2704	.2734	.2764	.2794	.2823	.2852
.8	.2881	.2910	.2939	.2967	.2995	.3023	.3051	.3078	.3106	.3133
.9	.3159	.3186	.3212	.3238	.3264	.3289	.3315	.3340	.3365	.3389
1.0	.3413	.3438	.3461	.3485	.3508	.3531	.3554	.3577	.3599	.3621
1.1	.3643	.3665	.3686	.3708	.3729	.3749	.3770	.3790	.3810	.3830
1.2	.3849	.3869	.3888	.3907	.3925	.3944	.3962	.3980	.3997	.4015
1.3	.4032	.4049	.4066	.4082	.4099	.4115	.4131	.4147	.4162	.4177
1.4	.4192	.4207	.4222	.4236	.4251	.4265	.4279	.4292	.4306	.4319
1.5	.4332	.4345	.4357	.4370	.4382	.4394	.4406	.4418	.4429	.4441

decimal place, .03. Finally, we locate the number where the row labeled $z = 1.3$ and the column labeled .03 meet. This number represents the area between the mean μ and the measurement that has a z-score of 1.33:

$$A = .4082$$

Hence, the probability that the calculator operates between 50 and 70 hours before needing a charge is .4082.

The use of the z-score simplifies the calculation of normal probabilities because if x is normally distributed with any mean and standard deviation, z is *always* a normal random variable with a mean of 0 and a standard deviation of 1. For this reason z is often referred to as a *standard normal random variable*.

Definition 5.1

The **standard normal random variable** z is defined by the formula

$$z = \frac{x - \mu}{\sigma}$$

where x is a normal random variable with mean μ and standard deviation σ. The standard normal random variable z is normally distributed with mean 0 and standard deviation 1 and can be described as the number of standard deviations between x and μ.

Since we will convert all normal random variables to standard normal in order to use Table IV to find probabilities, it is important that you learn to use Table IV well. The following examples illustrate the use of Table IV.

EXAMPLE 5.2

Find the probability that the standard normal random variable z falls between -1.33 and $+1.33$.

Solution

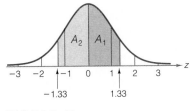

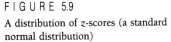

FIGURE 5.9

A distribution of z-scores (a standard normal distribution)

The standard normal distribution is shown in Figure 5.9. Since all probabilities associated with standard normal random variables can be depicted as areas under the standard normal curve, you should always draw the curve and then equate the desired probability to an area.

In this example we want to find the probability that z falls between -1.33 and $+1.33$, which is equivalent to the area between -1.33 and $+1.33$, shown shaded in Figure 5.9. Table IV provides the area between $z = 0$ and any value of z looked up, so that if we look up $z = 1.33$, we find that the area between $z = 0$ and $z = 1.33$ is .4082. This is the area labeled A_1 in Figure 5.9. To find the area A_2 between $z = 0$ and $z = -1.33$, we note that the symmetry of the normal distribution implies that the area between $z = 0$ and any point to the left is equal to the area between $z = 0$ and the point equidistant to the right. Thus, in this example the area between $z = 0$ and $z = -1.33$ is equal to the area between $z = 0$ and $z = +1.33$. That is,

$$A_1 = A_2 = .4082$$

The probability that z falls between -1.33 and $+1.33$ is the sum of the areas A_1 and A_2. We summarize in probabilistic notation:

$$P(-1.33 < z < +1.33) = P(-1.33 < z < 0) + P(0 \leq z < 1.33)$$
$$= A_1 + A_2 = .4082 + .4082$$
$$= .8164$$

Remember that "<" and "≤" are equivalent in events involving z, because the inclusion (or exclusion) of a single point does not alter the probability of an event involving a continuous random variable.

EXAMPLE 5.3

Find the probability that a standard normal random variable exceeds 1.64; i.e., find $P(z > 1.64)$.

Solution

The area under the standard normal distribution to the right of 1.64 is the shaded area labeled A_1 in Figure 5.10. This area represents the desired probability that z exceeds 1.64. However, when we look up $z = 1.64$ in Table IV, we must remember that the probability given in the table corresponds to the area between $z = 0$ and $z = 1.64$ (the area labeled A_2 in Figure 5.10). From Table IV we find

FIGURE 5.10

Standard normal distribution: $\mu = 0$, $\sigma = 1$

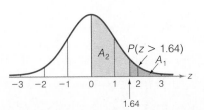

that $A_2 = .4495$. To find the area A_1 to the right of 1.64, we make use of two facts:

1. The standard normal distribution is symmetric about its mean, $z = 0$.
2. The total area under the standard normal probability distribution equals 1.

Taken together, these two facts imply that the areas on either side of the mean $z = 0$ equal .5; thus, the area to the right of $z = 0$ in Figure 5.10 is $A_1 + A_2 = .5$. Then

$$P(z > 1.64) = A_1 = .5 - A_2 = .5 - .4495 = .0505$$

To attach some practical significance to this probability, note that the implication is that the chance of a standard normal random variable exceeding 1.64 is approximately .05. Or, since z represents the number of standard deviations between *any* normal random variable and its mean, a normal random variable will exceed its mean by more than 1.64 standard deviations only about 5% of the time.

EXAMPLE 5.4

Find the probability that a normal random variable lies to the right of a point $-.74$ standard deviation from its mean.

Solution

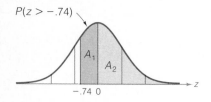

FIGURE 5.11
Standard normal distribution:
$\mu = 0, \sigma = 1$

We first must interpret the event in terms of a standard normal random variable, so that we can use Table IV to find the event's probability. Since the standard normal random variable z is simply the number of standard deviations between an observed value of a normal random variable and its mean, the event that a normal random variable lies to the right of a point $-.74$ standard deviation from the mean is equivalent to the event that the standard normal random variable z exceeds $-.74$. The event is shown as the shaded area in Figure 5.11, and we want to find $P(z > -.74)$.

We divide the shaded area into two parts: the area A_1 between $z = -.74$ and $z = 0$, and the area A_2 to the right of $z = 0$. We must always make such a division when the desired area lies on both sides of the mean ($z = 0$) because Table IV contains areas between $z = 0$ and the point you look up. To find A_1, we remember that the sign of z is unimportant when determining the area, because the standard normal distribution is symmetric about its mean. We look up $z = .74$ in Table IV to find that $A_1 = .2704$. The symmetry also implies that half the distribution lies on each side of the mean, so the area A_2 to the right of $z = 0$ is .5. Then,

$$P(z > -.74) = A_1 + A_2 = .2704 + .5 = .7704$$

EXAMPLE 5.5

Find the probability that a normal random variable lies more than 1.96 standard deviations from its mean *in either direction*.

Solution

The event that a normal random variable exceeds 1.96 standard deviations in either direction is equivalent to the event that the standard normal random variable z exceeds 1.96 in absolute value. That is, we want to find

$$P(z > |1.96|) = P(z < -1.96 \text{ or } z > 1.96)$$

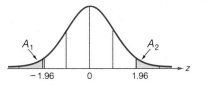

F I G U R E 5.12
Standard normal distribution:
$\mu = 0, \sigma = 1$

This probability is the shaded area in Figure 5.12. Note that the total shaded area is the sum of two areas, A_1 and A_2—areas that are equal because of the symmetry of the normal distribution.

We look up $z = 1.96$ and find the area between $z = 0$ and $z = 1.96$ to be .4750. Then the area to the right of 1.96, A_2, is $.5 - .4750 = .0250$, so that

$$P(z > |1.96|) = A_1 + A_2$$
$$= .0250 + .0250 = .05$$

The implication is that any normal random variable lies more than 1.96 standard deviations from its mean 5% of the time. Recall from Chapter 2 that the Empirical Rule tells us that about 5% of the measurements in mound-shaped distributions will lie beyond 2 standard deviations from the mean; the normal distribution, which is certainly mound-shaped, has 5% of its area beyond 1.96 standard deviations. In fact, the normal distribution provides the model on which the Empirical Rule is based, along with much "empirical" experience with real data that often approximately obey the rule, whether drawn from a normal distribution or not.

EXAMPLE 5.6

Assume that the length of time, x, between charges of a pocket calculator is normally distributed with a mean of 50 hours and a standard deviation of 15 hours. Find the probability that the calculator will last between 30 and 70 hours between charges.

Solution

The normal distribution with mean $\mu = 50$ and $\sigma = 15$ is shown in Figure 5.13. The desired probability that the calculator lasts between 30 and 70 hours is shaded. In order to find the probability, we must first convert the distribution to standard normal, which we do by calculating the z-score:

$$z = \frac{x - \mu}{\sigma}$$

F I G U R E 5.13
Normal probability distribution:
$\mu = 50, \sigma = 15$

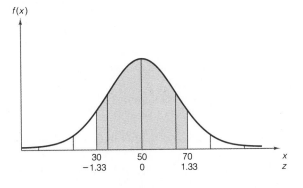

The z-scores corresponding to the important values of x are shown beneath the x values on the horizontal axis in Figure 5.13. Note that $z = 0$ corresponds to the mean of $\mu = 50$ hours, whereas the x values 30 and 70 yield z-scores of -1.33 and $+1.33$, respectively. Thus, the event that the calculator lasts between 30 and 70 hours is equivalent to the event that a standard normal random variable lies between -1.33 and $+1.33$. We found this probability in Example 5.2 (see Figure 5.9) by doubling the area corresponding to $z = 1.33$ in Table IV. That is,

$$P(30 \le x \le 70) = P(-1.33 \le z \le 1.33) = 2(.4082)$$
$$= .8164$$

> **Steps for Finding a Probability Corresponding to a Normal Random Variable**
>
> 1. Sketch the normal distribution and indicate the mean of the random variable x. Then shade the area corresponding to the probability you want to find.
> 2. Convert the boundaries of the shaded area from x values to standard normal random variable z values using the formula
>
> $$z = \frac{x - \mu}{\sigma}$$
>
> Show the z values under the corresponding x values on your sketch.
> 3. Use Table IV in Appendix A (and inside the front cover) to find the areas corresponding to the z values. If necessary, use the symmetry of the normal distribution to find areas corresponding to negative z values and the fact that the total area on each side of the mean equals .5 to convert the areas from Table IV to the probabilities of the event you have shaded.

The steps to follow when calculating a probability corresponding to a normal random variable are shown in the box.

EXAMPLE 5.7

Suppose an automobile manufacturer introduces a new model that has an advertised mean in-city mileage of 27 miles per gallon. Although such advertisements seldom report any measure of variability, suppose you write the manufacturer for the details of the tests, and you find that the standard deviation is 3 miles per gallon. This information leads you to formulate a probability model for the random variable x, the in-city mileage for this car model. You believe that the probability distribution of x can be approximated by a normal distribution with a mean of 27 and a standard deviation of 3.

a. If you were to buy this model of automobile, what is the probability that you would purchase one that averages less than 20 miles per gallon for in-city driving? In other words, find $P(x < 20)$.
b. Suppose you purchase one of these new models and it does get less than 20 miles per gallon for in-city driving. Should you conclude that your probability model is incorrect?

Solution

a. The probability model proposed for x, the in-city mileage, is shown in Figure 5.14. We are interested in finding the area A to the left of 20 since this area

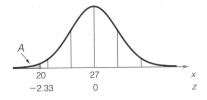

F I G U R E 5.14

Normal probability distribution for x in Example 5.7: $\mu = 27$ miles per gallon, $\sigma = 3$ miles per gallon

corresponds to the probability that a measurement chosen from this distribution falls below 20. In other words, if this model is correct, the area A represents the fraction of cars that can be expected to get less than 20 miles per gallon for in-city driving. To find A, we first calculate the z value corresponding to $x = 20$. That is,

$$z = \frac{x - \mu}{\sigma} = \frac{20 - 27}{3} = -\frac{7}{3} = -2.33$$

Then

$$P(x < 20) = P(z < -2.33)$$

as indicated by the shaded area in Figure 5.14. Since Table IV gives only areas to the right of the mean (and because the normal distribution is symmetric about its mean), we look up 2.33 in Table IV and find that the corresponding area is .4901. This is equal to the area between $z = 0$ and $z = -2.33$, so we find

$$P(x < 20) = A = .5 - .4901 = .0099 \approx .01$$

According to this probability model, you should have only about a 1% chance of purchasing a car of this make with an in-city mileage under 20 miles per gallon.

b. Now you are asked to make an inference based on a sample—the car you purchased. You are getting less than 20 miles per gallon for in-city driving. What do you infer? We think you will agree that one of two possibilities is true:

The probability model is correct. You simply were unfortunate to have purchased one of the cars in the 1% that get less than 20 miles per gallon in the city.

The probability model is incorrect. Perhaps the assumption of a normal distribution is unwarranted, or the mean of 27 is an overestimate, or the standard deviation of 3 is an underestimate, or some combination of these errors was made. At any rate, the form of the actual probability model certainly merits further investigation.

You have no way of knowing with certainty which possibility is correct, but the evidence points to the second one. We are again relying on the rare event approach to statistical inference that we introduced earlier. The sample (one measurement in this case) was so unlikely to have been drawn from the proposed probability model that it casts serious doubt on the model. We would be inclined to believe that the model is somehow in error.

Occasionally you will be given a probability and will want to find the values of the normal random variable that correspond to the probability. For example, suppose the scores on a college entrance examination are known to be normally distributed, and a certain prestigious university will consider for admission only those applicants whose scores exceed the 90th percentile of the test score distribution. To determine the minimum score for admission consideration, you will need to be able to use Table IV in reverse, as demonstrated in the following example.

EXAMPLE 5.8

Solution

Find the value of z, call it z_0, in the standard normal distribution that will be exceeded only 10% of the time. That is, find z_0 such that $P(z \geq z_0) = .10$.

In this case we are given a probability, or an area, and asked to find the value of the standard normal random variable that corresponds to the area. Specifically, we want to find the value z_0 such that only 10% of the standard normal distribution exceeds z_0 (see Figure 5.15).

FIGURE 5.15
Standard normal distribution for Example 5.8

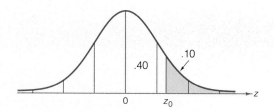

We know that the total area to the right of the mean $z = 0$ is .5, which implies that z_0 must lie to the right of (above) 0. To pinpoint the value, we use the fact that the area to the right of z_0 is .10, which implies that the area between $z = 0$ and z_0 is $.5 - .1 = .4$. But areas between $z = 0$ and some other z value are exactly the types given in Table IV. Therefore, we look up the area .4000 in the body of Table IV and find that the corresponding z value is (to the closest approximation) $z_0 = 1.28$. The implication is that the point 1.28 standard deviations above the mean is the 90th percentile of a normal distribution.

EXAMPLE 5.9

Solution

Find the value of z_0 such that 95% of the standard normal z values lie between $-z_0$ and $+z_0$—i.e., $P(-z_0 \leq z \leq z_0) = .95$.

Here we wish to move an equal distance z_0 in the positive and negative direction from the mean $z = 0$ until 95% of the standard normal distribution is enclosed. This means that the area on each side of the mean will be equal to $\frac{1}{2}(.95) = .475$, as shown in Figure 5.16. Since the area between $z = 0$ and z_0 is .475, we look up .475 in the body of Table IV to find the value $z_0 = 1.96$. Thus, as we found in the reverse order in Example 5.5, 95% of a normal distribution lies between plus and minus 1.96 standard deviations of the mean.

FIGURE 5.16
Standard normal distribution: $\mu = 0, \sigma = 1$

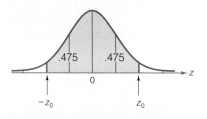

Now that you have learned to use Table IV to find a standard normal z value that corresponds to a specified probability, we demonstrate a practical application in Example 5.10.

EXAMPLE 5.10

Suppose the scores, x, on a college entrance examination are normally distributed with a mean of 550 and a standard deviation of 100. A certain prestigious university will consider for admission only those applicants whose scores exceed the 90th percentile of the distribution. Find the minimum score an applicant must achieve in order to receive consideration for admission to the university.

Solution

First, we find the 90th percentile in the standard normal distribution. We did this in Example 5.8 (see Figure 5.15) and found that $z = 1.28$ is the standard normal value that will be exceeded only 10% of the time.

Next, we want to convert the standard normal value to an x value, a test score in this example. We know that the minimum test score will be $z = 1.28$ standard deviations above the mean score. To determine the minimum score, remember that

$$z = \frac{x - \mu}{\sigma}$$

If we solve this equation for x, we find

$$x = \mu + z\sigma$$

In this example, $\mu = 550$, $z = 1.28$, and $\sigma = 100$, so

$$x = 550 + 1.28(100) = 550 + 128$$
$$= 678$$

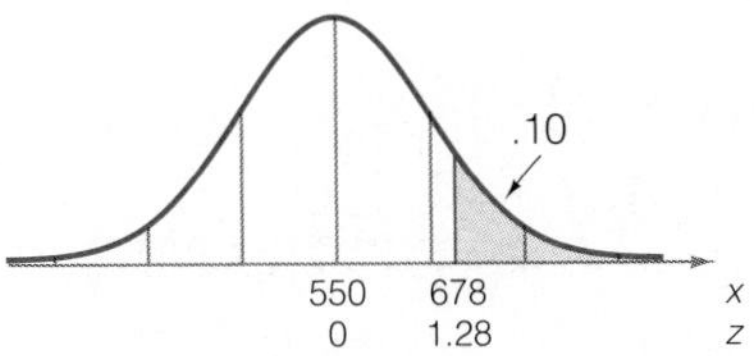

FIGURE 5.17
Normal distribution for Example 5.10

This x value is shown in Figure 5.17 corresponding to the z value of 1.28. Thus, the 90th percentile of the test score distribution is 678. An applicant must score at least 678 on the entrance exam to receive consideration for admission by the university.

CASE STUDY 5.1

GRADING ON THE CURVE

How did teachers ever suppose that the statistician's bell-shaped curve ought somehow to be imposed on the results of their work? The famous curve was developed to describe the distribution of natural phenomena. If we weigh 10,000 grains of corn, or measure the heights of men and women, or their ability to learn nonsense syllables, the findings will cluster around some central value and taper at both ends. The bell-shaped curve is descriptive of raw or unselected phenomena, and then only if vast numbers of cases are used. When the teacher receives his pupils, their distribution with respect to some characteristics may follow the "normal curve." But, having received his charges, the skillful teacher sets out as fast as he can to destroy the "natural" state of affairs.

This statement is the introduction to Clyde W. Bresee's article* about "grading on the curve." His main point is that many teachers who consistently grade on the curve are assuming that the measures of learning (test scores and the like) will always form a normal, or bell-shaped, distribution. These teachers may even use

*Reprinted from Bresee, C. W. "On 'grading on the curve,'" *Clearing House*, November 1976, 50, 108–110.

z-scores to determine grades and the corresponding areas under the normal curve (those in Table IV) to obtain the percentages of students who will receive each grade. The implication is that a student's performance will be measured only in relation to the other students in the class. Bresee relates the following anecdote:

> After the completion of a particularly successful unit, a teacher was heard to say, "I don't know how I'll grade this thing because they all did so well." This is a sad and dangerous statement because this teacher is on the verge of undoing a month's or even a year's work. Here is a teacher so indoctrinated by an erroneous concept that he questions his own accomplishment even in the face of clear evidence. But question it he must, if he has been trained to grade "on the curve." In a class of 20, for example, the only possible explanation for 15 A's is that he is a weak teacher or a "soft grader," or both.

The caveat presented by Bresee can be more generally applied. The normal distribution is widely used, and in many situations it provides an adequate approximation to reality. However, before each new application of the normal distribution, the situation should be carefully studied, and the user must be confident that all attendant assumptions are satisfied. When using the normal distribution to "grade on a curve," Bresee worries that the day may come when a supervisor admonishes a teacher, "Last year we sent you twenty children to teach. We spent a year's time and upwards of $20,000 on them, and all we have to show for it is the same old bell-shaped curve!"

EXERCISES 5.15–5.50

LEARNING THE MECHANICS

5.15 Find the area under the standard normal probability distribution between the following pairs of z-scores.
 a. $z = 0$ and $z = 2.00$
 b. $z = 0$ and $z = 1.00$
 c. $z = 0$ and $z = 3.00$
 d. $z = 0$ and $z = .58$
 e. $z = -2.00$ and $z = 0$
 f. $z = -1.00$ and $z = 0$
 g. $z = -1.69$ and $z = 0$
 h. $z = -.58$ and $z = 0$

5.16 Find each of the following.
 a. $P(0 \leq z \leq 2.00)$
 b. $P(-1.33 < z < 0)$
 c. $P(0 \leq z < 3.06)$
 d. $P(-.75 < z < 0)$

5.17 Define the standard normal random variable z. Why is the standard normal random variable useful?

5.18 Find the following probabilities for the standard normal random variable z.
 a. $P(-1 \leq z \leq 1)$
 b. $P(-2 \leq z \leq 2)$
 c. $P(-1.73 < z \leq .64)$
 d. $P(-1.45 < z < .33)$
 e. $P(z \geq -1.95)$
 f. $P(z < 2.57)$

5.19 Find the following probabilities for the standard normal random variable z.
 a. $P(z > 1.20)$
 b. $P(z < -1.40)$
 c. $P(.75 \leq z \leq 2.11)$
 d. $P(-1.96 \leq z < -.61)$
 e. $P(z \geq 0)$
 f. $P(-2.33 < z < 1.10)$

5.20 Find each of the following probabilities for a standard normal random variable z.
 a. $P(z = 1)$
 b. $P(z \leq 1)$
 c. $P(z < 1)$
 d. $P(z > 1)$

5.21 Find each of the following probabilities for the standard normal random variable z.
 a. $P(-1 \leq z \leq 1)$
 b. $P(-1.96 \leq z \leq 1.96)$
 c. $P(-1.645 \leq z \leq 1.645)$
 d. $P(-2 \leq z \leq 2)$

5.22 Find each of the following probabilities for the standard normal random variable z, and compare your answers to those of Exercise 5.21.
    a. $P(-1 \leq z < 1)$ **b.** $P(-1.96 < z < 1.96)$
    c. $P(-1.645 < z \leq 1.645)$ **d.** $P(-2 < z < 2)$

5.23 Find a value of the standard normal random variable z, call it z_0, such that
    a. $P(z \geq z_0) = .05$ **b.** $P(z \geq z_0) = .025$
    c. $P(z \leq z_0) = .025$ **d.** $P(z \geq z_0) = .10$
    e. $P(z > z_0) = .10$

5.24 Find a value of the standard normal random variable z, call it z_0, such that
    a. $P(z \leq z_0) = .0301$ **b.** $P(-z_0 \leq z \leq z_0) = .95$
    c. $P(-z_0 \leq z < z_0) = .90$ **d.** $P(-z_0 \leq z \leq z_0) = .6826$
    e. $P(z_0 \leq z \leq 0) = .1628$ **f.** $P(-.75 < z < z_0) = .7026$

5.25 Find a z-score, call it z_0, such that
    a. $P(z \geq z_0) = .5$ **b.** $P(z \geq z_0) = .0228$
    c. $P(0 \leq z \leq z_0) = .4803$ **d.** $P(z < z_0) = .0401$

5.26 Find the following probabilities for the standard normal random variable z.
    a. $P(z \leq 1.86)$ **b.** $P(-1.68 \leq z \leq .85)$
    c. $P(z \geq 2.77)$ **d.** $P(z \geq -1.62)$
    e. $P(-1.96 \leq z \leq -1.12)$ **f.** $P(z \leq 3.00)$
    g. $P(z \leq -3.05)$ **h.** $P(-1.45 \leq z \leq 1.45)$

5.27 Find a value of the standard normal random variable z, call it z_0, such that
    a. $P(z \leq z_0) = .0212$ **b.** $P(z \geq z_0) = .0885$
    c. $P(z \leq z_0) = .7704$ **d.** $P(-z_0 \leq z \leq z_0) = .8414$
    e. $P(-2 \leq z \leq z_0) = .6722$ **f.** $P(z_0 \leq z \leq 2.6) = .0312$

5.28 Give the z-score for a measurement from a normal distribution for the following.
    a. 1 standard deviation above the mean
    b. 1 standard deviation below the mean
    c. Equal to the mean
    d. 2.5 standard deviations below the mean
    e. 3 standard deviations above the mean

5.29 Suppose the random variable x is best described by a normal distribution with $\mu = 25$ and $\sigma = 5$. Find the z-score that corresponds to each of the following x values:
    a. $x = 25$ **b.** $x = 30$ **c.** $x = 37.5$
    d. $x = 10$ **e.** $x = 50$ **f.** $x = 32$

5.30 A random variable x is normally distributed with $\mu = 22$ and $\sigma = 5$. Determine the distance, in units of standard deviations, between each of the following values of x and the mean, $\mu = 22$.
    a. $x = 10$ **b.** $x = 20$ **c.** $x = 25$
    d. $x = 15$ **e.** $x = 0$ **f.** $x = 60$

5.31 The random variable x has a normal distribution with $\mu = 1,000$ and $\sigma = 10$.
    a. Find the probability that x assumes a value more than 2 standard deviations from its mean. More than 3 standard deviations from μ.
    b. Find the probability that x assumes a value within 1 standard deviation of its mean. Within 2 standard deviations of μ.
    c. Find the value of x that represents the 80th percentile of this distribution. The 10th percentile.

5.32 Suppose x is a normally distributed random variable with $\mu = 11$ and $\sigma = 2$. Find each of the following:
    a. $P(10 \leq x \leq 12)$ **b.** $P(6 \leq x \leq 10)$ **c.** $P(13 \leq x \leq 16)$
    d. $P(7.8 \leq x \leq 12.6)$ **e.** $P(x \geq 13.24)$ **f.** $P(x \geq 7.62)$

5.33   Suppose x is a normally distributed random variable with $\mu = 45$ and $\sigma = 10$. Find each of the following:

   a. $P(x \leq 50)$ **b.** $P(x \leq 35.6)$ **c.** $P(40.7 \leq x \leq 65.8)$
   d. $P(22.9 \leq x \leq 33.2)$ **e.** $P(x \geq 25.3)$ **f.** $P(x \leq 25.3)$

5.34   Suppose x is a normally distributed random variable with $\mu = 30$ and $\sigma = 8$. Find a value of the random variable, call it x_0, such that

   a. $P(x \geq x_0) = .5$ **b.** $P(x < x_0) = .025$
   c. $P(x > x_0) = .10$ **d.** $P(x > x_0) = .95$
   e. 10% of the values of x are less than x_0
   f. 80% of the values of x are less than x_0
   g. 1% of the values of x are greater than x_0

5.35   Suppose x is a normally distributed random variable with mean 100 and standard deviation 8. Draw a rough graph of the distribution of x. Locate μ and the interval $\mu \pm 2\sigma$ on the graph. Find the following probabilities:

   a. $P(\mu - 2\sigma \leq x \leq \mu + 2\sigma)$ **b.** $P(x \geq \mu + 2\sigma)$ **c.** $P(x \leq 92)$
   d. $P(92 \leq x \leq 116)$ **e.** $P(92 \leq x \leq 96)$ **f.** $P(76 \leq x \leq 124)$

5.36   The random variable x has a normal distribution with standard deviation 25. It is known that the probability that x exceeds 150 is .90. Find the mean μ of the probability distribution.

APPLYING THE CONCEPTS

5.37   Personnel tests are designed to test a job applicant's cognitive and/or physical abilities. An IQ test is an example of the former; a speed test involving the arrangement of pegs on a peg board is an example of the latter. During the 1970s, the proportion of employers using personnel tests dropped from more than 90% to less than 50%, in part due to concerns with equal-rights laws. As a result of improved testing procedures and the realization that such tests could hold down the costs associated with hiring the wrong person, the use of personnel tests has increased dramatically during the 1980s (Dessler, 1986).

   A particular dexterity test is administered nationwide by a private testing service. It is known that for all tests administered last year the distribution of scores was approximately normal with mean 75 and standard deviation 7.5.

   a. A particular employer requires job candidates to score at least 80 on the dexterity test. Approximately what percentage of the test scores during the past year exceeded 80?
   b. The testing service reported to a particular employer that one of its job candidate's scores fell at the 98th percentile of the distribution (i.e., approximately 98% of the scores were lower than the candidate's, and only 2% were higher). What was the candidate's score?

5.38   In a survey of 2,000 long-distance telephone calls reported in the *Orlando Sentinel* (March 12, 1984), it was found that seven of eight long-distance phone companies were overcharging (charging for additional time) for a given call and that six were charging for unconnected calls. Suppose the additional time being charged to a long-distance phone call has a normal distribution with a mean of 25 seconds and a standard deviation of 8 seconds.

   a. Find the probability that a given long-distance call will be overcharged by at least 40 seconds.
   b. By no more than 10 seconds.
   c. *Fill in the blank in the following statement:*   Eighty percent of long-distance calls are overcharged by _____ seconds or more.

5.39   The average salary for a major league baseball player has risen steadily from $19,000 per year in 1967 to $600,000 in 1990.

   a. If the 1990 distribution of salaries is normally distributed with a standard deviation equal to $200,000, what percentage of major league baseball players are making $1,000,000 per year or more?

b. Can you give a reason why it is unlikely that the distribution of major league baseball salaries is normally distributed?

5.40 When economic times are difficult, people keep their automobiles. During the recession of 1982–1983, the R. L. Polk statistics showed that the nation's automobile population was the oldest it had ever been, with a median age of 6.2 years in 1983—up from 4.9 years in 1970 (*Bakersfield Californian*, August 3, 1983). Suppose the age of automobiles in 1983 was (approximately) normally distributed with standard deviation equal to 2.1 years.

a. Give (approximately) the mean age of the automobiles on the road.

b. Give the approximate percentage of automobiles 10 years old or older.

c. Give the approximate percentage of automobiles less than 3 years old.

5.41 The amount of oxygen dissolved in rivers and streams depends on the water temperature and on the amounts of decaying organic matter from natural processes or human disturbances that are present in the water. The Council on Environmental Quality (CEQ) considers a dissolved oxygen content of less than 5 milligrams per liter of water to be undesirable because it is unlikely to support aquatic life. Suppose an industrial plant discharges its waste into a river and the downstream daily oxygen content measurements are normally distributed with a mean equal to 6.3 milligrams per liter and a standard deviation of .6 milligram per liter.

a. What percentage of the days would the dissolved oxygen content in the river be considered undesirable by the CEQ?

b. Within what limits would we expect the dissolved oxygen content to fall?

5.42 The pulse rate per minute of the adult male population between 18 and 25 years of age in the United States is known to have a normal distribution with a mean of 72 beats per minute and a standard deviation of 9.7. If the requirements for military service state that anyone with a pulse rate over 100 is medically unsuitable for service, what proportion of the males between 18 and 25 years of age would be declared unfit because their pulse rates are too high?

5.43 Ideally, a worker seeking a new job should acquire information about available wage rates in the industry in order to be able to compare wage rates offered by different firms. However, such a search could be time-consuming and costly. In particular, the longer an unemployed worker searches for a higher wage, the greater will be the loss in income. Therefore, workers may not find it worthwhile to search until they find the highest available wage rate, and managers may not have to pay top dollar to attract workers. These factors help explain the existing dispersion in wage rates. Suppose the distribution of wage rates nationwide that would be offered to a certain skilled worker can be approximated by a normal distribution with $\mu = \$10.50$ per hour and $\sigma = \$1.25$ per hour. In addition, assume that the worker is offered $12.00 per hour by the first firm contacted.

a. Suppose the worker were to undertake a nationwide job search. What proportion of the wage rates that would be offered to the worker would be greater than $12.00 per hour?

b. If the worker were to complete a nationwide job search and then randomly select one of the many job offers received, what is the probability that the wage rate would be more than $10.00 per hour?

c. The *median*, call it x_m, of a continuous random variable x is the value such that $P(x \geq x_m) = P(x \leq x_m) = .5$. That is, the median is the value x_m such that half the area under the probability distribution lies above x_m and half lies below it. Find the median of the random variable corresponding to the wage rate and compare it to the mean wage rate.

5.44 Do security analysts do a good job of forecasting corporate earnings growth and advising their clientele? David Dreman, a *Forbes* columnist, addresses this question in an article titled "Astrology Might Be Better" (*Forbes*, March 26, 1984). The basis of Dreman's article is a study by Professors Michael Sandretto of Harvard and Sudhir Milkrishnamurthi of MIT. The study surveys security analysts' forecasts of annual earnings for the (then) current year for more than 769 companies with five or more forecasts per company per year. The average forecast error for this large number of forecasts was plus or minus 31.3%. To apply this information to a practical situation, suppose the population of analysts' forecast errors is normally distributed with a mean of 31.3% and a standard deviation of 10%.

a. If you obtain a security analyst's forecast for a certain company, what is the probability that it will be in error by more than 50%?

b. If three analysts make the forecast, what is the probability that at least one of the analysts will err by more than 50%?

5.45 A machine used to regulate the amount of dye dispensed for mixing shades of paint can be set so that it discharges an average of μ milliliters of dye per can of paint. The amount of dye discharged is known to have a normal distribution with a standard deviation of .4 milliliter. If more than 6 milliliters of dye are discharged when making a certain shade of blue paint, the shade is unacceptable. Determine the setting for μ so that only 1% of the cans of paint will be unacceptable.

5.46 A physical fitness association is including the mile run in their secondary school fitness test for boys. The time for this event for boys in secondary school is approximately normally distributed with a mean of 450 seconds and a standard deviation of 40 seconds. If the association wants to designate the fastest 10% as "excellent," what time should the association set for this criterion?

5.47 The board of examiners that administers the real estate brokers' examination in a certain state found that the mean score on the test was 435 and the standard deviation was 72. If the board wants to set the passing score so that only the best 30% of all applicants pass, what is the passing score? Assume that the scores are normally distributed.

5.48 An important quality characteristic for soft-drink bottlers is the amount of soft drink injected into each bottle. This volume is determined (approximately) by measuring the height of the soft drink in the neck of the bottle and comparing it to a scale that converts the height measurement to a volume measurement (Montgomery, 1985). In a particular filling process, the number of ounces injected into 8-ounce bottles is approximately normally distributed with mean 8.00 ounces and standard deviation .05 ounce. Bottles that contain less than 7.9 ounces do not meet the bottler's quality standard and are sold at a substantial discount.

a. If 20,000 bottles are filled, approximately how many will fail to meet the quality standard?

b. Suppose that, due to the failure of one of the filling system's components, the mean of the filling process shifts to 7.95 ounces. (Assume that the standard deviation remains .05 ounce.) If 20,000 bottles are filled, approximately how many will fail to meet the quality standard?

c. Suppose a different component fails and, although the mean of the filling process remains 8.00 ounces, the standard deviation increases to .1 ounce. If 20,000 bottles are filled, approximately how many will fail to meet the quality standard?

5.49 The distribution of the demand (in number of units per unit time) for a product can often be approximated by a normal probability distribution. For example, a bakery has determined that the number of loaves of its white bread demanded daily has a normal distribution with mean 7,200 loaves and standard deviation 300 loaves. Based on cost considerations, the company has decided that its best strategy is to produce a sufficient number of loaves so that it will fully supply demand on 94% of all days.

a. How many loaves of bread should the company produce?

b. Based on the production in part **a**, on what percentage of days will the company be left with more than 500 loaves of unsold bread?

5.50 What relationship exists between the standard normal distribution and the box-plot methodology (Section 2.8) for describing distributions of data using quartiles? The answer depends on the true underlying probability distribution of the data. Assume for the remainder of this exercise that the distribution is normal.

a. Calculate the values of the standard normal random variable z, call them z_L and z_U, that correspond to the hinges of the box plot—i.e., the lower and upper quartiles, Q_L and Q_U—of the probability distribution.

b. Calculate the z values that correspond to the inner fences of the box plot for a normal probability distribution.

c. Calculate the z values that correspond to the outer fences of the box plot for a normal probability distribution.

d. What is the probability that an observation lies beyond the inner fences of a normal probability distribution? The outer fences?

e. Can you better understand why the inner and outer fences of a box plot are used to detect outliers in a distribution? Explain.

5.4 Approximating a Binomial Distribution with a Normal Distribution

When a binomial random variable can assume a large number of values, the calculation of its probabilities may become very tedious. To contend with this problem, we provide tables in Appendix A to give the probabilities for some values of n and p, but these tables are by necessity incomplete. In particular, the binomial probability table (Table II) can be used only for $n = 5$, 6, 7, 8, 9, 10, 15, 20, or 25. To deal with this limitation, we seek approximation procedures for calculating the probabilities associated with a binomial probability distribution.

When n is large, a normal probability distribution may be used to provide a good approximation to the probability distribution of a binomial random variable. To show how this approximation works, we refer to Example 4.10, in which we used the binomial distribution to model the number x of 20 voters who favor a candidate. We assumed that 60% of all the eligible voters favored the candidate. The mean and standard deviation of x were found to be $\mu = 12$ and $\sigma = 2.2$. The binomial distribution for $n = 20$ and $p = .6$ is shown in Figure 5.18, and the approximating normal distribution with mean $\mu = 12$ and standard deviation $\sigma = 2.2$ is superimposed.

FIGURE 5.18

Binomial distribution for $n = 20, p = .6$ and normal distribution with $\mu = 12$, $\sigma = 2.2$

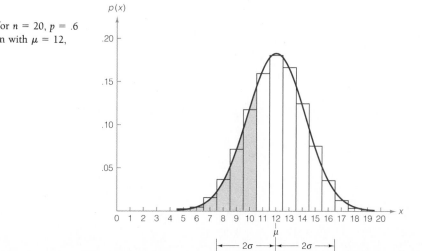

As part of Example 4.10, we used Table II to find the probability that $x \le 10$. This probability, which is equal to the sum of the areas contained in the rectangles (shown in Figure 5.18) that correspond to $p(0)$, $p(1)$, $p(2)$, . . . , $p(10)$, was found to equal .245. The portion of the approximating normal curve that would be used to approximate the area $p(0) + p(1) + \cdots + p(10)$ is shaded in Figure 5.18. Note that this shaded area lies to the left of 10.5 (not 10), so we may include all of the probability in the rectangle corresponding to $p(10)$. Because we are approximating a discrete distribution (the binomial) with a continuous distribution (the normal), we call the use of 10.5 (instead of 10 or 11) a **correction for continuity**. That is, we are correcting the discrete distribution so that it can be approximated by the continuous one. The use of the correction for continuity leads to the calculation of the following standard normal z value:

$$z = \frac{x - \mu}{\sigma} = \frac{10.5 - 12}{2.2} = -.68$$

Using Table IV, we find the area between $z = 0$ and $z = .68$ to be .2517. Then the probability that x is less than or equal to 10 is approximated by the area under the normal distribution to the left of 10.5, shown shaded in Figure 5.18. That is,

$$P(x \le 10) \approx P(z \le -.68) = .5 - P(-.68 < z \le 0)$$
$$= .5 - .2517 = .2438$$

FIGURE 5.19

Rule of thumb for normal approximation to binomial probabilities

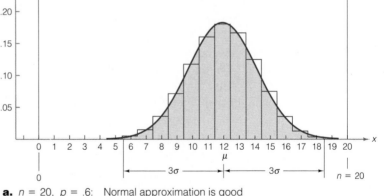

a. $n = 20$, $p = .6$:   Normal approximation is good

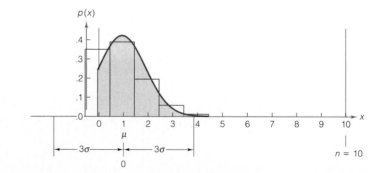

b. $n = 10$, $p = .1$:   Normal approximation is poor

The approximation differs only slightly from the exact binomial probability, .245. Of course, when tables of exact binomial probabilities are available, we will use the exact value rather than a normal approximation.

Use of the normal distribution will not always provide a good approximation for binomial probabilities. The following is a useful rule of thumb to determine when n is large enough for the approximation to be effective: the interval $\mu \pm 3\sigma$ should lie within the range of the binomial random variable x (i.e., 0 to n) in order for the normal approximation to be adequate. The rule works well because almost all of the normal distribution falls within 3 standard deviations of the mean, so if this interval is contained within the range of x values, there is "room" for the normal approximation to work.

As shown in Figure 5.19(a) for the preceding example with $n = 20$ and $p = .6$, the interval $\mu \pm 3\sigma = 12 \pm 3(2.19) = (5.43, 18.57)$ lies within the range 0 to 20. However, if we were to try to use the normal approximation with $n = 10$ and $p = .1$, the interval $\mu \pm 3\sigma$ is $1 \pm 3(.95)$, or $(-1.85, 3.85)$. As shown in Figure 5.19(b), this interval is not contained within the range of x since $x = 0$ is the lower bound for a binomial random variable. Note in Figure 5.19(b) that the normal distribution will not "fit" in the range of x, and therefore it will not provide a good approximation to the binomial probabilities.

The steps for approximating a binomial probability by a normal probability are given in the accompanying box.

Using a Normal Distribution to Approximate Binomial Probabilities

1. After you have determined n and p for the binomial distribution, calculate the interval

$$\mu \pm 3\sigma = np \pm 3\sqrt{npq}$$

If the interval lies in the range 0 to n, the normal distribution will provide a reasonable approximation to the probabilities of most binomial events.

2. If the binomial probability to be approximated is of the form $P(x \le a)$ or $P(x > a)$, the correction for continuity is $(a + .5)$, and the approximating standard normal z value is

$$z = \frac{(a + .5) - \mu}{\sigma}$$

See Figure 5.20(a).

FIGURE 5.20

Approximating binomial probabilities by normal probabilities

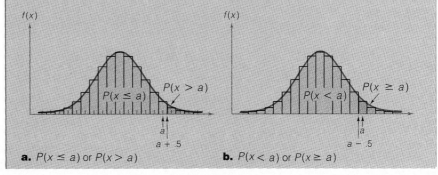

a. $P(x \le a)$ or $P(x > a)$ **b.** $P(x < a)$ or $P(x \ge a)$

3. If the binomial probability to be approximated is of the form $P(x \geq a)$ or $P(x < a)$, the correction for continuity is $(a - .5)$, and the approximating standard normal z value is

$$z = \frac{(a - .5) - \mu}{\sigma}$$

See Figure 5.20(b).

4. If the binomial probability to be approximated is an interval, such as $P(a \leq x < b)$, treat the ends of the interval separately, calculating two distinct z values according to either step 3 or step 4, whichever is appropriate.

5. Sketch the approximating normal distribution and shade the area corresponding to the probability of the event of interest, as in Figure 5.20. Verify that the rectangles you have included in the shaded area correspond to the event probability you wish to approximate. Using Table IV and the z value(s) you calculated in steps 2–4, find the shaded area. This is the approximate probability of the binomial event.

EXAMPLE 5.11

The pocket calculator has become relatively inexpensive because its solid-state circuitry is stamped by machine, thus making mass production feasible. One problem with anything that is mass-produced is quality control. The process must somehow be monitored or audited to be sure the output of the process conforms to requirements.

One method of dealing with this problem is **lot acceptance sampling**, in which items being produced are sampled at various stages of the production process and are carefully inspected. The lot of items from which the sample is drawn is then accepted or rejected, based on the number of defectives in the sample. Lots that are accepted may be sent forward for further processing or may be shipped to customers; lots that are rejected may be reworked or scrapped. For example, suppose a manufacturer of calculators chooses 200 stamped circuits from the day's production and determines x, the number of defective circuits in the sample. Suppose that up to a 6% rate of defectives is considered acceptable for the process.

a. Find the mean and standard deviation of x, assuming the defective rate is 6%.
b. Use the normal approximation to determine the probability that 20 or more defectives are observed in the sample of 200 circuits (i.e., find the approximate probability that $x \geq 20$).

Solution

a. The random variable x is binomial with $n = 200$ and the fraction defective $p = .06$. Thus,

$$\mu = np = 200(.06) = 12$$
$$\sigma = \sqrt{npq} = \sqrt{200(.06)(.94)} = \sqrt{11.28} = 3.36$$

We first note that

$$\mu \pm 3\sigma = 12 \pm 3(3.36) = 12 \pm 10.08 = (1.92, 22.08)$$

lies completely within the range from 0 to 200. Therefore, a normal probability distribution should provide an adequate approximation to this binomial distribution.

b. To find the approximating area corresponding to $x \geq 20$, refer to Figure 5.21. Note that we want to include all the binomial probability histogram from 20 to 200, inclusive. But in order to include the entire rectangle corresponding to $x = 20$, we must begin the approximating area at $20 - .5 = 19.5$. In other words, since the event is of the form $x \geq a$, with $a = 20$, the correction for continuity is $a - .5 = 20 - .5 = 19.5$. Thus, the z value is

$$z = \frac{(a - .5) - \mu}{\sigma} = \frac{19.5 - 12}{3.36} = \frac{7.5}{3.36} = 2.23$$

FIGURE 5.21

Normal approximation to the binomial distribution with $n = 200$, $p = .06$

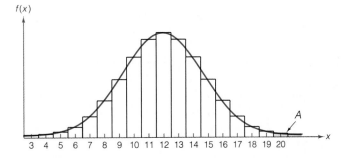

Referring to Table IV in Appendix A, we find that the area to the right of the mean corresponding to $z = 2.23$ (see Figure 5.22) is .4871. So the area A is

$$A = .5 - .4871 = .0129$$

FIGURE 5.22

Standard normal distribution

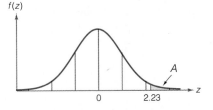

Thus, the normal approximation to the binomial probability is

$$P(x \geq 20) \approx .0129$$

In other words, the probability is extremely small that 20 or more defectives will be observed in a sample of 200 circuits—*if in fact the true fraction of defectives is .06.* If the manufacturer observes $x \geq 20$, the likely reason is that the process is producing more than the acceptable 6% defectives. The lot acceptance sampling procedure is another example of using the rare event approach to make inferences.

EXERCISES 5.51–5.70

LEARNING THE MECHANICS

5.51   Why might you want to use a normal distribution to approximate a binomial distribution?

5.52   What conditions must be satisfied in order for the normal distribution to provide a good approximation to the binomial probability distribution?

5.53 Assume that x is a binomial random variable with n and p as specified in parts **a–f**. For which cases would it be appropriate to use a normal distribution to approximate the binomial distribution?
   a. $n = 50$, $p = .01$ **b.** $n = 20$, $p = .45$
   c. $n = 10$, $p = .4$ **d.** $n = 1,000$, $p = .1$
   e. $n = 200$, $p = .8$ **f.** $n = 35$, $p = .7$

5.54 Suppose x is a binomial random variable with $p = .4$ and $n = 25$.
   a. Would it be appropriate to approximate the probability distribution of x with a normal distribution? Explain.
   b. Assuming that a normal distribution provides an adequate approximation to the distribution of x, what are the mean and variance of the approximating normal distribution?
   c. Use Table II of Appendix A to find the exact value of $P(x \geq 9)$.
   d. Use the normal approximation to find $P(x \geq 9)$.

5.55 Assume that x is a binomial random variable with $n = 25$ and $p = .5$. Use Table II of Appendix A and the normal approximation to find the exact and approximate values, respectively, for the following probabilities:
   a. $P(x \leq 12)$ **b.** $P(x \geq 15)$ **c.** $P(9 \leq x \leq 15)$

5.56 Assume that x is a binomial random variable with $n = 100$ and $p = .45$. Use a normal approximation to find the following:
   a. $P(x \leq 45)$ **b.** $P(40 \leq x \leq 50)$ **c.** $P(x \geq 38)$

5.57 Assume that x is a binomial random variable with $n = 1,000$ and $p = .50$. Find each of the following probabilities.
   a. $P(x > 500)$ **b.** $P(490 \leq x < 500)$ **c.** $P(x > 1,000)$

APPLYING THE CONCEPTS

5.58 The *Statistical Abstract of the United States: 1989* reports that 24% of the country's 91,061,000 households are inhabited by one person. If 1,000 randomly selected homes are to participate in a Nielsen survey to determine television ratings, find the approximate probability that no more than 250 of these homes are inhabited by one person.

5.59 In 1982, General Motors Corporation's auto-assembly plant in Fremont, California, was shut down and turned over to Toyota Motor Corporation as part of a joint venture with the Japanese firm. Before the shutdown, 5,000 workers produced 240,000 cars a year and had an absentee rate of 20%. After the Japanese took over and introduced their distinctive management style, the same number of cars were produced by only 2,500 workers and the absentee rate dropped to 2% ("The difference Japanese management makes," *Business Week*, July 14, 1986, p. 47).
   a. With an absentee rate of 20%, what is the probability that at least 90% of a random sample of 50 workers will be on the job on a particular day?
   b. What assumption must we make in order to use the normal distribution to calculate the probability required in part **a**?
   c. If the absentee rate were 2%, should the probability of part **a** be approximated using a normal distribution? Explain.

5.60 In Case Study 4.2, the number of shuttle catastrophes due to booster failure in n missions was treated as a binomial random variable. Using the binomial distribution and the probability of catastrophe determined by the Air Force's risk assessment study $\left(\frac{1}{35}\right)$, we determined the probability of at least 1 shuttle catastrophe in 25 missions to be .5155.
   a. Based on the guidelines presented in this section, would it have been advisable to approximate this probability using the normal approximation to the binomial distribution? Explain.
   b. Regardless of your answer to part **a**, use the normal distribution to approximate the binomial probability. Comment on the difference between the exact and approximate probabilities.

c. Refer to part **a**. Would the normal approximation be advisable if $n = 100$? If $n = 500$? If $n = 1,000$?

d. Approximate the probability that more than 25 catastrophes occur in 1,000 flights, assuming that the probability of a catastrophe in any given flight remains $\frac{1}{35}$.

5.61 If you are attracted to lotteries, one of the best is run by the U.S. government. Furthermore, with some research you can improve your chances of winning. We refer to the Interior Department's lottery for oil and gas leases. The *Wall Street Journal* (March 29, 1984) reports on abuses in the lottery—particularly the tendency of the Interior Department to include valuable oil leases, some worth millions of dollars, among the many included in the lottery. The Interior Department claims that only 5% to 10% of the leases included in the $75 per ticket lottery should be salable to an oil company. Yet 184 of 328 winners in the July 1980 lottery of Wyoming lands were able to sell their leases to oil companies.

Suppose as many as 10% of all leases awarded by the Interior Department are salable to oil companies to answer the following questions:

a. What is the probability that as many as 50 leases (i.e., 50 or more) would be salable to oil companies?

b. If 184 of the 328 leases awarded in the July 1980 Wyoming lottery were in fact salable to oil companies, would you regard this as a rare event? Explain.

c. Given the outcome of the Wyoming lottery and considering your answer to part **b**, do you believe the Interior Department's claim? Explain.

5.62 According to the "January" theory for investing in corporate stocks, the Dow Jones Industrial (stock) Average will show an increase (or decrease) for the full year if it rises (or falls) in January. This barometer of stock market behavior has been correct for 29 of the past 34 years (*Wall Street Journal*, "Heard on the Street," February 1, 1984). The coincidence in the January and the full year movements in the Dow Jones Average may not be pure chance. If the Dow Jones Average is up (or down) in January, it certainly increases the probability that it will be up (or down) for the full year. Suppose the probability of coincidence in any single year is .6. Use the normal curve approximation to the binomial probability distribution to find the probability that the January and the full year movements in the Dow Jones Average will coincide 29 or more times in 34 years.

5.63 Melanoma, a malignant form of skin cancer, strikes more than 15,000 Americans per year and kills 45% of this number (*Time*, May 30, 1983).

a. What are the expected value and variance of x, the number of the 15,000 annual melanoma patients who die of the affliction?

b. Find the probability that x will exceed 6,900 patients per year.

c. Would you expect the number x dying of melanoma to exceed 7,000 in any single year? Explain.

5.64 It is against the law to discriminate against job applicants because of race, religion, sex, or age. Of the individuals who apply for an accountant's position in a large corporation, 40% are over 45 years of age. If the company decides to choose 50 of a very large number of applicants for closer credential screening, claiming that the selection will be random and not age-biased, what is the approximate probability that fewer than 15 of those chosen are over 45 years of age? (Assume that the applicant pool is large enough so that x, the number in the sample over 45 years of age, has a binomial probability distribution.)

5.65 Recent market research reveals that an estimated 60% of U.S. households own one or more personal computers (PCs). Suppose that in a sample of 500 households in a large high-tech community, 325 own PCs.

a. Approximate the probability that 325 or more households in the sample would own PCs if in fact 60% of all households in the community own PCs.

b. Would you conclude on the basis of your answer to part **a** that more than 12% of this community's households own PCs? Explain.

5.66   An advertising agency was hired to introduce a new product. It claimed that after its campaign, 30% of all consumers were familiar with the product. To check the claim, the manufacturer of the product surveyed 2,000 consumers. Of this number, 527 consumers had learned about the product through sources attributable to the campaign. What is the approximate probability that as few as 527 (i.e., 527 or less) would have learned about the product if the campaign was really 30% effective?

5.67   A recent study involving attrition rates at a major university has shown that 43% of all incoming freshmen do not graduate within 4 years of entrance.
    a. If 200 freshmen are randomly sampled this year and their progress through college is followed, what is the approximate probability that no less than half will graduate within the next 4 years?
    b. What is the approximate probability that the number of sampled freshmen graduating within 4 years will be between 40 and 80?

5.68   To check on the effectiveness of a new production process, 700 photoflash devices were randomly selected from a large number that had been produced. If the process actually produces 6% defectives, what is the approximate probability that
    a. More than 50 defectives appear in the sample of 700?
    b. The number of defectives in the sample of 700 is 45 or less?

5.69   The median time a patient waits to see a doctor in a large clinic is 20 minutes. On a day when 150 patients visit the clinic, what is the approximate probability that
    a. More than half will have to wait more than 20 minutes?
    b. More than 85 will have to wait more than 20 minutes?
    c. More than 60 but less than 90 will have to wait more than 20 minutes?

5.70   The percentage of fat in the bodies of American men is an approximate normal random variable with mean equal to 15% and standard deviation equal to 2%.
    a. If these values were used to describe the body fat of men in the United States Army and if 20% or more body fat is characterized as obese, what is the approximate probability that a random sample of 10,000 soldiers will contain fewer than 50 who would be characterized as obese?
    b. If the army actually were to check the percentage of body fat for a random sample of 10,000 men and if only 30 contained 20% (or higher) body fat, would you conclude that the army was successful in reducing the percentage of obese men below the percentage in the general population? Explain your reasoning.

5.5 The Exponential Distribution (Optional)

The length of time between emergency arrivals at a hospital, the length of time between breakdowns of manufacturing equipment, the length of time between catastrophic events (floods, earthquakes, etc.), and the distance traveled by a wildlife ecologist between sightings of an endangered species are all random phenomena that we might want to describe probabilistically. The amount of time or distance between occurrences of random events like these can often be described by the **exponential probability distribution**. For this reason, the exponential distribution is sometimes called the **waiting time distribution**. The formula for the exponential probability distribution is shown in the box along with its mean and standard deviation.

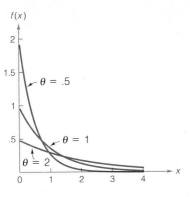

FIGURE 5.23

Exponential distributions

> **Probability Distribution, Mean, and Standard Deviation for an Exponential Random Variable x**
>
> $$f(x) = \frac{1}{\theta}\, e^{-x/\theta} \quad (x > 0)$$
>
> $$\mu = \theta \qquad \sigma = \theta$$

Unlike the normal distribution, which has a shape and location determined by the values of the two quantities μ and σ, the shape of the exponential distribution is governed by a single quantity, θ. Further, it is a probability distribution with the property that its mean equals its standard deviation. Exponential distributions corresponding to $\theta = .5$, 1, and 2 are shown in Figure 5.23.

To calculate probabilities for exponential random variables, we need to be able to find areas under the exponential probability distribution. Suppose we want to find the area A to the right of some number a, as shown in Figure 5.24. This area can be calculated by using the following formula:

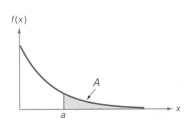

FIGURE 5.24

The area A to the right of a number a for an exponential distribution

> **Finding the Area A to the Right of a Number a for an Exponential Distribution***
>
> $$A = P(x \ge a) = e^{-a/\theta}$$

Use Table V in Appendix A to find the value of $e^{-a/\theta}$ after substituting the appropriate numerical values for θ and a.

EXAMPLE 5.12

Suppose the length of time (in hours) between emergency arrivals at a certain hospital is modeled as an exponential distribution with $\theta = 2$. What is the probability that more than 5 hours pass without an emergency arrival?

Solution

The probability we want is the area A to the right of $a = 5$ in Figure 5.25. To find this probability, use the formula given for area:

$$A = e^{-a/\theta} = e^{-(5/2)} = e^{-2.5}$$

Referring to Table V, we find

$$A = e^{-2.5} = .082085$$

Our exponential model indicates that the probability that more than 5 hours pass between emergency arrivals is about .08 for this hospital.

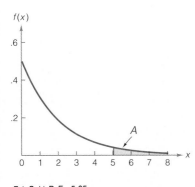

FIGURE 5.25

Exponential distribution for Example 5.12: $\theta = 2$

---

*For students with a knowledge of calculus, the shaded area in Figure 5.24 corresponds to the integral

$$\int_a^\infty \frac{1}{\theta} e^{-x/\theta}\, dx = -e^{-x/\theta}\,\Big|_a^\infty = e^{-a/\theta}$$

EXAMPLE 5.13

A microwave oven manufacturer is trying to determine the length of warranty period it should attach to its magnetron tube, the most critical component in the oven. Preliminary testing has shown that the length of life (in years), x, of a magnetron tube has an exponential probability distribution with $\theta = 6.25$.

a. Find the mean and standard deviation of x.
b. Suppose a warranty period of 5 years is attached to the magnetron tube. What fraction of tubes must the manufacturer plan to replace, assuming that the exponential model with $\theta = 6.25$ is correct?
c. Find the probability that the length of life of a magnetron tube will fall within the interval $\mu - 2\sigma$ to $\mu + 2\sigma$.

Solution

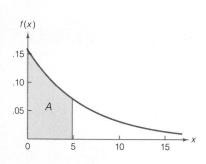

FIGURE 5.26

Exponential distribution for Example 5.13: $\theta = 6.25$

a. Since $\theta = \mu = \sigma$, both μ and σ equal 6.25.
b. To find the fraction of tubes that will have to be replaced before the 5-year warranty period expires, we need to find the area between 0 and 5 under the distribution. This area, A, is shown in Figure 5.26. To find the required probability, we recall the formula

$$P(x > a) = e^{-a/\theta}$$

Using this formula, we can find

$$P(x > 5) = e^{-a/\theta} = e^{-5/6.25} = e^{-.80} = .449329$$

(see Table V). To find the area A, we use the complementary relationship:

$$P(x \le 5) = 1 - P(x > 5) = 1 - .449329 = .550671$$

So approximately 55% of the magnetron tubes will have to be replaced during the 5-year warranty period.
c. We would expect the probability that the life of a magnetron tube, x, falls within the interval $\mu - 2\sigma$ to $\mu + 2\sigma$ to be quite large. A graph of the exponential distribution showing the interval $\mu - 2\sigma$ to $\mu + 2\sigma$ is given in Figure 5.27. Since the point $\mu - 2\sigma$ lies below $x = 0$, we need to find only the area between $x = 0$ and $x = \mu + 2\sigma = 6.25 + 2(6.25) = 18.75$. This area, P, which is shaded in Figure 5.27, is

$$P = 1 - P(x > 18.75)$$
$$= 1 - e^{-18.75/\theta} = 1 - e^{-18.75/6.25}$$
$$= 1 - e^{-3}$$

FIGURE 5.27

Exponential distribution for Example 5.13: $\theta = 6.25$

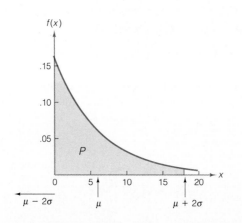

Checking Table V for the value of e^{-3}, we find $e^{-3} = .049787$. Therefore, the probability that the life x of a magnetron tube will fall within the interval $\mu - 2\sigma$ to $\mu + 2\sigma$ is

$$P = 1 - e^{-3}$$
$$= 1 - .049787 = .950213$$

You can see that this probability agrees very well with the interpretation of a standard deviation given by the Empirical Rule in Table 2.9 even though this probability distribution is not mound-shaped. (It is strongly skewed to the right.)

CASE STUDY 5.2

ASSESSING THE RELIABILITY OF COMPUTER SOFTWARE

In a discussion about the reliability of computer software, G. J. Schick* says the following:

> Custom software . . . is expensive to develop and requires extensive testing—the goal being to certify that the software is in error-free condition, ready to support the mission for which it was designed. Similar economies should also be expected from an integrated statistical software test program. Traditionally, there are never enough time or resources to test all possible branches and data combinations in a computer program of reasonable size.
>
> Current practice is to design and develop a software system and then to test it to detect errors, until the amount of time and expense required to discover remaining errors is too great to justify further testing. . . . In principle, few large real-time computer programs ever have been tested completely and unequivocally in the sense that every logical data path has been successfully executed under every logical combination for the data at hand for all possible options. One management objective would be to test every logical path in the computer program at least once with some kind of numerical check. At the present state of the art, such a degree of testing is neither feasible nor realistic. In practice, the contractor must be willing to release and the customer willing to accept a level of risk associated with a program that has been less than completely checked.

In finding and correcting errors in a computer program (debugging) and determining the program's reliability, Schick and others have noted the importance of the distribution of the time until the next program error is found. If this distribution is assumed to be exponential, with

$$f(x) = \frac{1}{\theta} e^{-x/\theta} \qquad (x > 0, \theta > 0)$$

then, as Schick points out, its mean, θ, would be the average time required to find the next error.

In his article, Schick describes a method relative to software reliability for estimating the parameter θ of the exponential distribution. Using computer

---

*Reprinted from Schick, G. J. "The search for a software reliability model," *Design Sciences*, October 1974, 5, 529.

debugging data supplied by the United States Navy, Schick demonstrates how this estimation procedure and the exponential distribution can be used to estimate the reliability of a computer program. [*Note:* The model used by Schick to represent the distribution for the time until the next error is based on the exponential distribution, but it is slightly more complicated because he assumes that θ varies. For our purposes, however, nothing is lost by assuming the distribution he uses to be exponential.]

After 26 program errors were found, Schick estimated θ to be 23.8 days. This means that the average time it would take to find the next (27th) error is approximately 24 days. Thus, using the estimated 23.8 days, the probability of it taking, say, 60 or more days to find the next error is approximately

$$P(x \geq 60) = e^{-60/23.8} = .08046$$

Over the next 290 days, five more errors were detected. Since this is a rate of about one error every 60 days and since $P(x \geq 60) \approx .08$, it seems unlikely that an exponential distribution with $\theta = 23.8$ is an appropriate representation of the distribution for the time until the next error. Based on the number of new errors found and the length of time it took to find them, Schick reestimated θ to be 278 days, meaning that on an average the next error (32nd) would not occur for 278 days. At this point, the length of time and, therefore, the cost required to find any remaining program errors may be prohibitive. Debugging should probably be discontinued.

EXERCISES 5.71–5.85

LEARNING THE MECHANICS

5.71 The random variables x and y have exponential distributions with $\theta = 3$ and $\theta = .75$, respectively. Using Table V in Appendix A, carefully plot both distributions on the same set of axes.

5.72 Use Table V in Appendix A to determine the value of $e^{-a/\theta}$ for each of the following cases.
a. $\theta = 1$, $a = 1$ b. $\theta = 1$, $a = 2.5$
c. $\theta = .4$, $a = 3$ d. $\theta = .2$, $a = .3$

5.73 Suppose x has an exponential distribution with $\theta = .5$. Find the following probabilities.
a. $P(x > .5)$ b. $P(x \leq 3)$ c. $P(x > 1.5)$ d. $P(x \leq 5)$

5.74 Suppose x has an exponential distribution with $\theta = 2.0$. Find the following probabilities.
a. $P(x \leq 4)$ b. $P(x > .5)$ c. $P(x \leq 2)$ d. $P(x > 3)$

5.75 Suppose the random variable x has an exponential probability distribution with $\theta = .5$. Find the mean and standard deviation of x. Find the probability that x will assume a value within the interval $\mu \pm 2\sigma$.

5.76 The random variable x can be adequately approximated by an exponential probability distribution with $\theta = 1$. Find the probability that x assumes a value:
a. More than 3 standard deviations from μ
b. Less than 2 standard deviations from μ
c. Within .5 standard deviation of μ

APPLYING THE CONCEPTS

5.77   A taxi service based at an airport can be characterized as a transportation system with one source terminal and a fleet of vehicles that take passengers from the terminal to different destinations. Each vehicle returns to the terminal after some random trip time and makes another trip. In order to improve the vehicle-dispatching decisions involved in such a system (e.g., How many passengers should be allocated to a waiting taxi?), Sims and Templeton (1985) assumed travel times of successive trips are independent exponential random variables to model the system. Assume $\theta = 20$ minutes.
    a.   What is the mean trip time for the taxi service?
    b.   What is the probability that a particular trip will take more than 30 minutes?
    c.   Two taxis have just been dispatched. What is the probability that both will be gone for more than 30 minutes? That at least one of the taxis will return within 30 minutes?

5.78   After bacteria are subjected to a certain drug, the length of time until the bacteria die follows an exponential distribution with a mean of $\frac{1}{2}$ hour.
    a.   How long after the drug is administered will half the bacteria be dead?
    b.   What proportion of bacteria will die between .2 and 1 hour?

5.79   The length of time x it takes to produce a human reaction to tear gas has an exponential distribution with a mean of 3 minutes.
    a.   If tear gas is fired into a house to subdue a terrorist, how long should the police wait so that the probability is .99 that the tear gas has taken effect?
    b.   What proportion of people will be affected within 5 minutes?

5.80   An article in the Jacksonville, Florida, *Times Union* (March 11, 1984) reports on the unexplained crash of a small plane on takeoff and the resulting injury to its 24-year-old student pilot. The article notes that Shields Aviation, the renter of the plane, inspects (and presumably performs maintenance) on the aircraft every 100 flight hours. The plane had flown 30 hours since its last inspection. Suppose x, the time between malfunctions for this particular plane, has an exponential distribution with mean equal to 300 hours. What is the probability that a plane of this type will malfunction within 30 hours after the last inspection?

5.81   The probability distribution of the length of service time is important in the design of service facilities, and it has a definite effect on sales. Suppose the time an individual has to wait in line to be served at a fast-food franchise is assumed to have an exponential distribution with mean equal to 1 minute.
    a.   What is the probability that an individual would have to wait more than 2 minutes before being served?
    b.   What is the probability that an individual will be served within 30 seconds after arriving in line?

5.82   An outbreak of a certain species of caterpillar, the spruce budworm, can cause extensive damage to the timberlands of the northern United States. It is known that an outbreak of this type of caterpillar occurs, on the average, every 30 years. Assuming that this phenomenon obeys an exponential probability law, what is the probability that catastrophic outbreaks of spruce budworm will occur within 6 years of each other?

5.83   The shelf life of a perishable product is a random variable that is related to consumer acceptance and, ultimately, to sales and profit. Suppose the shelf life of bread is best approximated by an exponential distribution with mean equal to 2 days. What fraction of the loaves stocked today would you expect to still be salable (i.e., not stale) 3 days from now?

5.84   The length of time between arrivals at a hospital clinic and the length of clinical service time are two random variables that play important roles in designing a clinic and deciding how many physicians and nurses are needed for its operation. The probability distributions of both the length of time between arrivals and the length of service time are often approximately exponential. Suppose the mean time between arrivals for patients at a clinic is 4 minutes.

a. What is the probability that a particular interarrival time (the time between the arrival of two patients) is less than 1 minute?

b. What is the probability that the next four interarrival times are all less than 1 minute?

c. What is the probability that an interarrival time will exceed 10 minutes?

5.85 The following is a sample of the lengths of time between arrivals (rounded to the nearest minute) at the emergency room of a hospital:

1	10	3	3	9	7	1	4	3	14
6	4	5	7	9	5	1	9	13	4
2	6	12	5	1	10	5	8	3	23
11	14	18	15	12	1	4	22	26	37
3	8	19	5	8	18	16	28	31	21

a. Draw a relative frequency histogram for these data. Does it appear that an exponential distribution could be used to characterize the length of time between arrivals? Explain.

b. If you were asked to model the length of time between arrivals for this emergency room, discuss how you would estimate θ.

Summary

Many **random variables** encountered in the real world have probability distributions that are well approximated by the **normal, uniform,** or **exponential probability distributions**. In this chapter we showed the graphic shape of these probability distributions, gave their means and standard deviations, and pointed out some practical applications for each of these probability models. Moreover, we showed that the normal probability distribution provides a good approximation for the binomial distribution when the number of trials, n, is sufficiently large.

SUPPLEMENTARY EXERCISES 5.86–5.117

[*Note: Starred (*) exercises refer to the optional section in this chapter.*]

LEARNING THE MECHANICS

5.86 Assume that x is a random variable best described by a uniform distribution with $c = 30$ and $d = 70$.

a. Find $f(x)$.

b. Find the mean and standard deviation of x.

c. Graph the probability distribution for x and locate its mean and the interval $\mu \pm 2\sigma$ on the graph.

d. Find $P(x \leq 45)$.   e. Find $P(x \geq 58)$.   f. Find $P(x \leq 100)$.

g. Find $P(\mu - \sigma \leq x \leq \mu + \sigma)$.   h. Find $P(x > 60)$.

5.87 Calculate the area under the standard normal probability distribution between the following pairs of z-scores.

a. 0 and 1.96   b. −1.96 and 1.96   c. −1.30 and 1.30

d. −2.14 and −1.07   e. .74 and 2.01   f. −1.43 and 1.95

5.88 Find the following probabilities for the standard normal random variable z.

a. $P(z \leq 1.1)$ b. $P(z \geq 1.1)$ c. $P(z \geq -1.86)$

d. $P(-2.75 \leq z \leq -.55)$ e. $P(-2.45 \leq z \leq .38)$ f. $P(z \leq -1.91)$

5.89 Find a z-score, call it z_0, such that
 a. $P(z \leq z_0) = .8023$ b. $P(z \geq z_0) = .0985$ c. $P(z \leq z_0) = .5$
 d. $P(-z_0 \leq z \leq z_0) = .6212$ e. $P(z \geq z_0) = .8508$ f. $P(z \geq z_0) = .0055$

5.90 The random variable x has a normal distribution with $\mu = 70$ and $\sigma = 10$. Find the following probabilities.
 a. $P(x \leq 75)$ b. $P(x \geq 90)$ c. $P(60 \leq x \leq 75)$
 d. $P(x > 75)$ e. $P(x = 75)$ f. $P(x \leq 95)$

5.91 The random variable x has a normal distribution with $\mu = 60$ and $\sigma^2 = 64$. Find a value of x, call it x_0, such that
 a. $P(x \geq x_0) = .5$ b. $P(x \leq x_0) = .9911$ c. $P(x \leq x_0) = .0028$
 d. $P(x \geq x_0) = .0228$ e. $P(x \leq x_0) = .1003$ f. $P(x \geq x_0) = .7995$

5.92 Assume that x is a binomial random variable with $n = 100$ and $p = .6$. Use the normal probability distribution to approximate the following probabilities.
 a. $P(x \leq 48)$ b. $P(50 \leq x \leq 65)$ c. $P(x \geq 70)$
 d. $P(55 \leq x \leq 58)$ e. $P(x = 62)$ f. $P(x \leq 49$ or $x \geq 72)$

*5.93 Assume that x has an exponential distribution with $\theta = 2.0$. Find
 a. $P(x \leq 1)$ b. $P(x > 1)$ c. $P(x = 1)$
 d. $P(x \leq 6)$ e. $P(2 \leq x \leq 10)$

APPLYING THE CONCEPTS

5.94 For a student to graduate, a high school requires that each student demonstrate competence in mathematics by scoring 70% or above on a mathematics achievement test. The scores of those students taking the test for the first time are normally distributed with a mean of 77% and a standard deviation of 7.3%. What percentage of students who take the test for the first time will pass the test?

5.95 Farmers often sell fruits and vegetables at roadside stands during the summer. One such roadside stand has a daily demand for tomatoes that is approximately normally distributed with a mean equal to 125 tomatoes per day and a standard deviation equal to 30 tomatoes per day.
 a. If there are 90 tomatoes available to be sold at the roadside stand at the beginning of a day, what is the probability that they will all be sold?
 b. If there are 200 tomatoes available to be sold, what is the probability that 50 or more will not be sold that day?
 c. How many tomatoes must be available on any given day so that there will be only a 10% chance that all tomatoes will be sold?

5.96 The *tolerance limits* for a particular quality characteristic (i.e., length, weight, strength) of a product are the minimum and/or maximum values at which the product will operate properly. Tolerance limits are set by the engineering design function of the manufacturing operation (Juran and Gryna, 1980). The tensile strength of a particular metal part can be characterized as being normally distributed with a mean of 25 pounds and a standard deviation of 2 pounds. The upper and lower tolerance limits for the part are 30 pounds and 21 pounds, respectively. A part that falls within the tolerance limits results in a profit of $10. A part that falls below the lower tolerance limit costs the company $2; a part that falls above the upper tolerance limit costs the company $1. Find the company's expected profit per metal part produced.

5.97 A firm believes the internal rate of return for its proposed investment can best be described by a normal distribution with mean 20% and standard deviation 3%. What is the probability that the internal rate of return for the investment will be the following?
 a. Greater than 26% or less than 14%   b. At least 16%   c. More than 28.5%

5.98 The probability distribution of the number of people per month who open a savings account in a large banking system is approximately normally distributed with $\mu = 1,280$ and $\sigma = 265$.

a. If the bank gives a $10 gift to each new account holder, what portion of the time will the bank's payout for gifts exceed $15,000 per month?

b. What is the bank's mean monthly payout for gifts?

5.99 On the average, the main chute fails in one of every 1,000 parachutes. Suppose during a lifetime a professional parachutist makes 4,000 jumps, and let x equal the number of times the main chute fails. What is the approximate probability that the parachutist's main chute fails on at least one jump? [*Note:*   Because n is large and p is so small, the Poisson probability distribution will also provide a good approximation to this probability. (See Exercise 4.63.) If you covered Section 4.5, find the Poisson approximation to $P(x > 0)$.]

5.100 The length of time required to assemble a photoelectric cell is normally distributed with $\mu = 18.1$ minutes and $\sigma = 1.3$ minutes. What is the probability that it will require more than 20 minutes to assemble a cell?

5.101 A local track club has decided to sponsor a 10,000-meter road race. From past results of races across the state, it is known that the length of time to complete the race has an approximately normal distribution with a mean of 49 minutes and a standard deviation of 8 minutes. The club has decided that everyone who completes the race will receive a T-shirt. Those who run between 34 and 60 minutes will also receive a medal, and those finishing under 34 minutes will receive a plaque.

a. What proportion of racers would you expect to receive a medal and a T-shirt?

b. What proportion of racers would you expect to receive a plaque and a T-shirt?

5.102 The net weight per package of a certain brand of corn chips is listed as 10 ounces. The weight actually delivered to each package by an automated machine is a normal random variable with mean 10.5 ounces and standard deviation .2 ounce. Suppose 100 packages are chosen at random and the net weights are ascertained. Let x be the number of the 100 selected packages that contain at least 10 ounces of corn chips. Then x is a binomial random variable with $n = 100$ and $p =$ probability that a randomly selected package contains at least 10 ounces. What is the probability that they all contain at least 10 ounces of corn chips? What is the probability that at least 90% of the packages contain 10 ounces or more?

5.103 After extensive testing of a new low-tar cigarette, a tobacco company concluded that the number of milligrams of tar yielded by each cigarette has a probability distribution that is approximately normal with $\mu = 8$ and $\sigma = 1.9$ milligrams.

a. What is the probability that one of the new low-tar cigarettes will yield more than 10 milligrams of tar?

b. If two of the new low-tar cigarettes are chosen at random and tested, what is the probability that both will yield less than 6 milligrams of tar?

5.104 Eighty 10-pound bags of sugar are randomly sampled from the stock of a local sugar wholesaler and weighed. The following are the results rounded to the nearest tenth of a pound:

10.2	10.2	9.6	9.9	10.0	10.0	10.0	10.1
10.4	9.9	9.6	10.0	9.9	10.1	9.9	9.9
9.8	10.0	10.4	10.0	9.7	9.9	10.2	9.8
9.9	10.3	10.0	10.0	9.8	9.8	10.1	9.9
10.1	9.9	10.2	9.8	10.0	10.1	9.5	10.0
10.0	10.1	10.1	10.1	10.2	10.5	9.9	10.1
9.5	9.9	9.7	9.9	9.9	10.1	9.9	10.3
10.1	10.1	9.8	10.0	10.0	10.0	9.7	10.0
10.0	10.2	10.0	10.1	10.3	9.8	10.0	10.2
9.6	9.9	9.8	10.0	9.8	10.2	10.0	9.9

a. Construct a relative frequency histogram for these data and suggest a continuous probability distribution that could be used to approximate the weight distribution of the wholesaler's stock of 10-pound bags of sugar.

b. Using the histogram, estimate μ and σ for the weight distribution of the wholesaler's stock of 10-pound bags of sugar.

c. Calculate $\bar{x}$ and s for the sample of 80 weights. How do these estimates of μ and σ compare to your answer in part b?

5.105 Sixteen percent of the American black population is known to suffer from sickle-cell anemia. If 1,000 American black people are sampled at random, what is the approximate probability that:

a. More than 175 have the disease?

b. Fewer than 140 have the disease?

c. The number of people in the sample with the disease is between 130 and 180 inclusive?

5.106 To help highway planners anticipate the need for road repairs and design future construction projects, data are collected on the volume and weight of truck traffic on specific roadways. Recently, equipment was developed that can be built into road surfaces to measure traffic volumes and to weigh trucks without requiring them to stop at roadside weigh stations. As with any measuring device, however, the "weigh-in-motion" equipment does not always record truck weights accurately. In an experiment performed by the Minnesota Department of Transportation involving repeated weighing of a 27,907-pound truck, it was found that the weights recorded by the weigh-in-motion equipment were approximately normally distributed with mean 27,315 and standard deviation 628 pounds (Dahlin, 1982; Wright, Owen, and Pena, 1983). It follows that the difference between the actual and recorded weight, the error of measurement, is normally distributed with mean 592 pounds and standard deviation 628 pounds.

a. What is the probability that the weigh-in-motion equipment understates the actual weight of the truck?

b. If a 27,907-pound truck were driven over the weigh-in-motion equipment 100 times, approximately how many times would the equipment overstate the truck's weight?

c. What is the probability that the error in the weight recorded by the weigh-in-motion equipment for a 27,907-pound truck exceeds 400 pounds?

d. It is possible to adjust (or *calibrate*) the weigh-in-motion equipment to control the mean error of measurement. At what level should the mean error be set so the equipment will understate the weight of a 27,907-pound truck 50% of the time? Only 40% of the time?

5.107 A company has a lump-sum incentive plan for salespeople that is dependent on their level of sales. If they sell less than $100,000 per year, they receive a $1,000 bonus; from $100,000 to $200,000, they receive $5,000; and above $200,000, they receive $10,000. Suppose the annual sales per salesperson has approximately a normal distribution with $\mu = \$180,000$ and $\sigma = \$50,000$.

a. Find p_1, the proportion of salespeople who receive a $1,000 bonus.

b. Find p_2, the proportion of salespeople who receive a $5,000 bonus.

c. Find p_3, the proportion of salespeople who receive a $10,000 bonus.

d. What is the mean value of the bonus payout for the company? [*Hint:* See the definition for the expected value of a random variable in Chapter 4.]

5.108 It is quite common for the standard deviation of a random variable to increase proportionally as the mean increases. When this occurs, the **coefficient of variation**,

$$CV = \frac{\sigma}{\mu}$$

which is the ratio of σ to μ, is the **proportionality constant**. To illustrate, the error (in dollars) in assessing the value of a house increases as the house increases in value. Suppose that long experience with assessors in a given region has shown that the coefficient of variation is .08 and that the probability distribution of assessed valuations on the same house by many different assessors is approximately normal with a mean we will call the "true value" of the house. Suppose the true value of your house is $50,000 and it is being assessed for taxation purposes. What is the probability the assessor will assess your house in excess of $55,000?

5.109 An admissions officer for a law school indicates that 35% of the applicants meet all ten requirements and 95% meet at least eight of the ten requirements.
  a. If a random sample of 300 applicants is taken, what is the approximate probability that fewer than 250 will fail to meet all ten requirements?
  b. What is the approximate probability that more than 280 will meet at least eight of the ten requirements?

*5.110 Based on sample data collected in the Denver area, Nicholas Kiefer (1985) found that in some cases the exponential distribution is an adequate approximation for the distribution of the time (in weeks) an individual is unemployed. In particular, he found the exponential distribution to be appropriate for white and black workers but not for Hispanics. Use $\theta = 13$ to answer the following questions.
  a. What is the mean time workers are unemployed according to the exponential distribution?
  b. Find the probability that a white worker who just lost her job will be unemployed for at least 2 weeks. For more than 6 weeks.
  c. What is the probability that an unemployed worker will find a new job within 12 weeks?

*5.111 Blending feeders are machines used to break up tobacco that has been aged in tightly packed hogsheads. One cigarette manufacturer determines that the time between breakdowns for each of its blending feeders is best represented by an exponential distribution with mean equal to 100 hours of operation. A particular feeder has just been repaired and put back into service.
  a. What is the probability that it will not break down for at least 50 more hours?
  b. What is the probability that it will break down within the next 100 hours?

5.112 A loan officer in a large bank has been assigned to screen 60 loan applications during the next week.
  a. If her past record indicates that she turns down 20% of the applicants, what is the approximate probability that 41 or more of the 60 applications will be approved?
  b. What is the approximate probability that between 45 and 50 of the applications will be approved?

5.113 Contrary to our intuition, very reliable decisions concerning the proportion of a large group of consumers favoring a certain product or a certain social issue can be based on relatively small samples. For example, suppose the target population of consumers contains 50 million people and we wish to decide whether the proportion of consumers, p, in the population that favor some product (or issue) is as large as some value, say .2. Suppose you randomly select a sample as small as 1,600 from the 50 million and you observe the number, x, of consumers in the sample who favor the new product. Assuming that $p = .2$, find the mean and standard deviation of x. Suppose that 400 (or 25%) of the sample of 1,600 consumers favor the new product. Why might this sample result lead you to conclude that p (the proportion of consumers favoring the product in the population of 50 million) is at least as large as .2? [Hint: Find the values of μ and σ for $p = .2$, and use them to decide whether the observed value of x is unusually large.]

*5.114 The lifetime (in hours) of a certain electronics component has an exponential distribution with $\theta = 2,000$.
  a. What is the probability that a randomly chosen component lasts more than 2,500 hours?
  b. Given that a component has lasted 1,000 hours, what is the probability it will last at least 2,500 hours longer?
  [Note: Compare the answers to parts a and b. This is the result of a property of the exponential distribution. Items that possess an exponential lifetime distribution are never subject to fatigue. The probability that they will live for 100 hours is the same for an item that is 1,000 hours old as for a new item. Certainly, not too many objects in nature possess this lifetime distribution, but machines or equipment that are subject to periodic maintenance often do.]

5.115 The median income of residents in a certain community is $20,000. If 1,000 people are randomly sampled from the population of residents, let x equal the number whose incomes are less than $20,000. Compute approximate values for the following:
   a. $P(x > 500)$ b. $P(x \leq 480)$ c. $P(475 \leq x \leq 525)$

*5.116 The number of serious accidents in a manufacturing plant has (approximately) a Poisson probability distribution with a mean of two serious accidents per month. If x, the number of events per unit time, has a Poisson distribution with mean λ, then it can be shown that the time between two successive events has an exponential probability distribution with mean $\theta = 1/\lambda$.
   a. If an accident occurs today, what is the probability that the next serious accident will not occur within the next month?
   b. What is the probability that more than one serious accident will occur within the next month?

*5.117 The Poisson probability distribution, like the binomial, can be approximated by a normal probability distribution. This approximation, using $\mu = \lambda$ and $\sigma = \sqrt{\lambda}$, will be good when λ is large (large enough so that the distance between $x = 0$ and λ is at least $3\sigma = 3\sqrt{\lambda}$, or $\lambda \geq 9$). The number of union complaints per month at a certain manufacturing plant has a Poisson probability distribution with $\lambda = 40$ complaints per month. Use the normal approximation to the Poisson probability distribution.
   a. Approximate the probability that the number of complaints in a given month will be less than 35.
   b. Approximate the probability that the number of complaints in a given month exceeds 40.
   c. If the mean number of complaints per month remains constant and if the number of complaints in 1 month is independent of the number in any other, what is the probability that in each of 3 successive months, the number of complaints exceeds 40?

ON YOUR OWN...

For large values of n the computational effort involved in working with the binomial probability distribution is considerable. Fortunately, in many instances the normal distribution provides a good approximation to the binomial distribution. This exercise was designed to enable you to demonstrate to yourself how well the normal distribution approximates the binomial distribution.

a. Let the random variable x have a binomial probability with $n = 10$ and $p = .5$. Using the binomial distribution, find the probability that x takes on a value in each of the following intervals: $\mu \pm \sigma$, $\mu \pm 2\sigma$, and $\mu \pm 3\sigma$.

b. Find the probabilities requested in part a by using a normal approximation to the given binomial distribution.

c. Determine the magnitude of the difference between each of the three probabilities as determined by the binomial distribution and by the normal approximation.

d. Letting x have a binomial distribution with $n = 20$ and $p = .5$, repeat parts a, b, and c. Notice that the probability estimates provided by the normal distribution are more accurate for $n = 20$ than for $n = 10$.

e. Letting x have a binomial distribution with $n = 20$ and $p = .01$, repeat parts a, b, and c. Notice that the probability estimates provided by the normal distribution are very poor in this case. Explain why this occurs.

USING THE COMPUTER...

Refer to **Using the Computer** in Chapter 4. Again use the mean percentage of college graduates as an estimate of the percentage among all people at least 25 years old in the region you selected.

a. Use the normal approximation to the binomial to estimate the probability that fewer than 10% of 20 applicants (still assuming they represent a random sample) have a college education. Compare your answer with the exact binomial probability of the same event.

b. Use the normal approximation to the binomial to estimate the probability that fewer than 10% of 200 applicants have a college education. If your statistical software package has a function that calculates exact binomial probabilities, use it to calculate the same probability you approximated, and compare the results.

References

Dahlin, C. *Minnesota's Experience with Weighing Trucks in Motion*. St. Paul: Minnesota Department of Transportation, 1982.

Dessler, G. "Personnel tests gain in popularity," *St. Paul Pioneer Press and Dispatch*, March 17, 1986, p. 12.

Hogg, R. V., and Craig, A. T. *Introduction to Mathematical Statistics*, 4th ed. New York: Macmillan, 1978, Chapter 1.

Juran, J. M., and Gryna, F. M., Jr. *Quality Planning and Analysis*, 2d ed. New York: McGraw-Hill, 1980.

Kiefer, N. M. "Specification diagnostics based on Laguerre alternatives for econometric models of duration," *Journal of Econometrics*, 28, 1985, pp. 135–154.

Martz, H. F., and Waller, R. A. *Bayesian Reliability Analysis*. New York: Wiley, 1982, pp. 1 and 256.

Mendenhall, W., Scheaffer, R. L., and Wackerly, D. *Mathematical Statistics with Applications*, 2d ed. North Scituate, Mass.: Duxbury, 1980, Chapter 4.

Montgomery, D. C. *Introduction to Statistical Quality Control*. New York: Wiley, 1985.

Murphy, A. H., and Winkler, R. L. "Probability forecasting in meteorology," *Journal of the American Statistical Association*, Vol. 79, September 1984, pp. 489–500.

Sims, S. H., and Templeton, J. G. C. "Steady state results for the $M/M(a, b)/c$ batch-service system," *European Journal of Operational Research*, Vol. 21, 1985, pp. 260–267.

Wright, J. L., Owen, F., and Pena, D. *Status of MN/DOT's Weigh-in-motion Program*. St. Paul: Minnesota Department of Transportation, January 1983.

Sampling Distributions

WHERE WE'RE GOING...

We have learned in earlier chapters that the objective of most statistical investigations is inference—that is, making decisions or predictions about a population based on information in a sample. To actually make the decision, we use the sample data to compute sample statistics, such as the sample mean or variance. The knowledge of random variables and their probability distributions enables us to construct theoretical models of populations.

WHERE WE'VE BEEN...

Because sample measurements are observed values of random variables, the value for a sample statistic will vary in a random manner from sample to sample. In other words, since sample statistics are random variables, they therefore possess probability distributions that are either discrete or continuous. These probability distributions, called *sampling distributions* because they characterize the distribution of values of the various statistics over a very large number of samples, are the topic of this chapter. In particular, you will learn why many sampling distributions tend to be approximately normal, and you will see how sampling distributions can be used to evaluate the reliability of inferences made using the statistics. Then in subsequent chapters we will show how to use sampling distributions to make inferences about populations.

In Chapters 4 and 5 we assumed that we knew the probability distribution of a random variable, and using this knowledge we were able to compute the mean, variance, and probabilities associated with the random variable. However, in most practical applications, this information is not available. To illustrate, in Example 4.10 we calculated the probability that the binomial random variable x, the number of 20 polled voters who favor a certain mayoral candidate, assumed specific values. To do this, it was necessary to assume some value for p, the proportion of all voters who favor the candidate. Thus, for the purposes of illustration, we assumed $p = .6$ when, in all likelihood, the exact value of p would be unknown. In fact, the probable purpose of taking the poll is to estimate p. Similarly, when we modeled the in-city gas mileage of a certain automobile model, we used the normal probability distribution with an *assumed* mean and standard deviation of 27 and 3 miles per gallon, respectively. In most situations, the true mean and standard deviation are unknown quantities that would have to be estimated. Numerical quantities that describe probability distributions are called *parameters*. Thus, p, the probability of a success in a binomial experiment, and μ and σ, the mean and standard deviation of a normal distribution, are examples of parameters.

Definition 6.1

A **parameter** is a numerical descriptive measure of a population. It is calculated from the observations in the population.

We have also discussed the sample mean, $\bar{x}$, sample variance s^2, sample standard deviation s, etc., which are numerical descriptive measures calculated from the sample. We will often use the information contained in these *sample statistics* to make inferences about the parameters of a population.

Definition 6.2

A **sample statistic** is a numerical descriptive measure of a sample. It is calculated from the observations in the sample.

Note that the term *statistic* refers to a *sample* quantity and the term *parameter* refers to a *population* quantity.

Before we can show you how to use sample statistics to make inferences about population parameters, we need to be able to evaluate their properties. Does one sample statistic contain more information than another about a population parameter? On what basis should we choose the "best" statistic for making inferences about a parameter? The purpose of this chapter is to answer these questions.

6.1 What Is a Sampling Distribution?

If we want to estimate a parameter of a population—say, the population mean μ—there are a number of sample statistics that could be used for the estimate. Two possibilities are the sample mean $\bar{x}$ and the sample median m. Which of these do you think will provide a better estimate of μ?

Before answering this question, consider the following example: Toss a fair die, and let x equal the number of dots showing on the up face. Suppose the die is tossed three times, producing the sample measurements 2, 2, 6. The sample mean is $\bar{x} = 3.33$ and the sample median is $m = 2$. Since the population mean of x is $\mu = 3.5$, you can see that for this sample of three measurements, the sample mean $\bar{x}$ provides an estimate that falls closer to μ than does the sample median [see Figure 6.1(a)]. Now suppose we toss the die three more times and obtain the sample measurements 3, 4, 6. The mean and median of this sample are $\bar{x} = 4.33$ and $m = 4$, respectively. This time m is closer to μ [see Figure 6.1(b)].

FIGURE 6.1

Comparing the sample mean ($\bar{x}$) and sample median (m) as estimators of the population mean (μ)

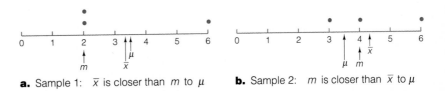

a. Sample 1: $\bar{x}$ is closer than m to μ **b.** Sample 2: m is closer than $\bar{x}$ to μ

This simple example illustrates an important point: Neither the sample mean nor the sample median will *always* fall closer to the population mean. Consequently, we cannot compare these two sample statistics, or, in general, any two sample statistics, on the basis of their performance for a single sample. Instead, we need to recognize that sample statistics are themselves random variables, because different samples can lead to different values for the sample statistics. As random variables, sample statistics must be judged and compared on the basis of their probability distributions, i.e., the *collection* of values and associated probabilities of each statistic that would be obtained if the sampling experiment were repeated a *very large number of times*. We will illustrate this concept with an example.

Suppose it is known that in a certain part of Canada the daily high temperature recorded for all past months of January has a mean of $\mu = 10°F$ and a standard deviation of $\sigma = 5°F$. Consider an experiment consisting of randomly selecting 25 daily high temperatures from the records of past months of January and calculating the sample mean $\bar{x}$. If this experiment were repeated a very large number of times, the value of $\bar{x}$ would vary from sample to sample. For example, the first sample of 25 temperature measurements might have a mean $\bar{x} = 9.8$, the second sample a mean $\bar{x} = 11.4$, the third sample a mean $\bar{x} = 10.5$, etc. If the sampling experiment were repeated a very large number of times, the resulting histogram of sample means would be approximately the probability distribution of $\bar{x}$. If $\bar{x}$ is a good estimator of μ, we would expect the values of $\bar{x}$ to cluster around μ as shown in Figure 6.2. This probability distribution is called a *sampling distribution* because it is generated by repeating a sampling experiment a very large number of times.

FIGURE 6.2

Sampling distribution for $\bar{x}$ based on a sample of $n = 25$ measurements

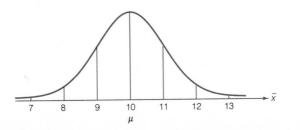

256   6 SAMPLING DISTRIBUTIONS

Definition 6.3

The **sampling distribution** of a sample statistic calculated from a sample of n measurements is the probability distribution of the statistic.

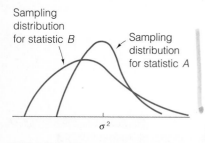

FIGURE 6.3

Two sampling distributions for estimating the population variance, σ^2

In actual practice, the sampling distribution of a statistic is obtained mathematically or (at least approximately) by simulating the sample on a computer using a procedure similar to that just described.

If $\bar{x}$ has been calculated from a sample of $n = 25$ measurements selected from a population with mean $\mu = 10$ and standard deviation $\sigma = 5$, the sampling distribution (Figure 6.2) provides information about the behavior of $\bar{x}$ in repeated sampling. For example, the probability that you will draw a sample of 25 measurements and obtain a value of $\bar{x}$ in the interval $9 \le \bar{x} \le 10$ will be the area under the sampling distribution over that interval.

Since the properties of a statistic are typified by its sampling distribution, it follows that to compare two sample statistics you compare their sampling distributions. For example, if you have two statistics, A and B, for estimating the same parameter (for purposes of illustration, suppose the parameter is the population variance σ^2) and if their sampling distributions are as shown in Figure 6.3, you would choose statistic A in preference to statistic B. You would make this choice because the sampling distribution for statistic A centers over σ^2 and has less spread (variation) than the sampling distribution for statistic B. When you draw a single sample in a practical sampling situation, the probability is higher that statistic A will fall nearer σ^2.

Remember that in practice we will not know the numerical value of the unknown parameter σ^2, so we will not know whether statistic A or statistic B is closer to σ^2 for a sample. We have to rely on our knowledge of the theoretical sampling distributions to choose the best sample statistic and then use it sample after sample. The procedure for finding the sampling distribution for a statistic is demonstrated in Example 6.1.

EXAMPLE 6.1

Consider a population consisting of the measurements 0, 3, and 12 and described by the probability distribution shown here. A random sample of $n = 3$ measurements is selected from the population.

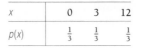

x	0	3	12
$p(x)$	$\frac{1}{3}$	$\frac{1}{3}$	$\frac{1}{3}$

a. Find the sampling distribution of the sample mean $\bar{x}$.
b. Find the sampling distribution of the sample median m.

Solution

Every possible sample of $n = 3$ measurements is listed in Table 6.1 along with the sample mean and median. Also, because any one sample is as likely to be selected as any other (random sampling), the probability of observing any particular sample is $\frac{1}{27}$. The probability is also listed in Table 6.1.

TABLE 6.1

POSSIBLE SAMPLES	$\bar{x}$	m	PROBABILITY	POSSIBLE SAMPLES	$\bar{x}$	m	PROBABILITY
0, 0, 0	0	0	$\frac{1}{27}$	3, 3, 12	6	3	$\frac{1}{27}$
0, 0, 3	1	0	$\frac{1}{27}$	3, 12, 0	5	3	$\frac{1}{27}$
0, 0, 12	4	0	$\frac{1}{27}$	3, 12, 3	6	3	$\frac{1}{27}$
0, 3, 0	1	0	$\frac{1}{27}$	3, 12, 12	9	12	$\frac{1}{27}$
0, 3, 3	2	3	$\frac{1}{27}$	12, 0, 0	4	0	$\frac{1}{27}$
0, 3, 12	5	3	$\frac{1}{27}$	12, 0, 3	5	3	$\frac{1}{27}$
0, 12, 0	4	0	$\frac{1}{27}$	12, 0, 12	8	12	$\frac{1}{27}$
0, 12, 3	5	3	$\frac{1}{27}$	12, 3, 0	5	3	$\frac{1}{27}$
0, 12, 12	8	12	$\frac{1}{27}$	12, 3, 3	6	3	$\frac{1}{27}$
3, 0, 0	1	0	$\frac{1}{27}$	12, 3, 12	9	12	$\frac{1}{27}$
3, 0, 3	2	3	$\frac{1}{27}$	12, 12, 0	8	12	$\frac{1}{27}$
3, 0, 12	5	3	$\frac{1}{27}$	12, 12, 3	9	12	$\frac{1}{27}$
3, 3, 0	2	3	$\frac{1}{27}$	12, 12, 12	12	12	$\frac{1}{27}$
3, 3, 3	3	3	$\frac{1}{27}$				

a. From Table 6.1 you can see that $\bar{x}$ can assume the values 0, 1, 2, 3, 4, 5, 6, 8, 9, and 12. Because $\bar{x} = 0$ occurs in only one sample, $P(\bar{x} = 0) = \frac{1}{27}$. Similarly, $\bar{x} = 1$ occurs in three samples: (0, 0, 3), (0, 3, 0), and (3, 0, 0). Therefore, $P(\bar{x} = 1) = \frac{3}{27} = \frac{1}{9}$. Calculating the probabilities of the remaining values of $\bar{x}$ and arranging them in a table, we obtain the probability distribution shown here.

$\bar{x}$	0	1	2	3	4	5	6	8	9	12
$p(\bar{x})$	$\frac{1}{27}$	$\frac{3}{27}$	$\frac{3}{27}$	$\frac{1}{27}$	$\frac{3}{27}$	$\frac{6}{27}$	$\frac{3}{27}$	$\frac{3}{27}$	$\frac{3}{27}$	$\frac{1}{27}$

This is the sampling distribution for $\bar{x}$ because it specifies the probability associated with each possible value of $\bar{x}$.

b. In Table 6.1 you can see that the median m can assume one of the three values 0, 3, or 12. The value $m = 0$ occurs in seven different samples. Therefore, $P(m = 0) = \frac{7}{27}$. Similarly, $m = 3$ occurs in 13 samples and $m = 12$ occurs in seven samples. Therefore, the probability distribution (i.e., the sampling distribution) for the median m is as shown in the margin.

m	0	3	12
$p(m)$	$\frac{7}{27}$	$\frac{13}{27}$	$\frac{7}{27}$

Example 6.1 demonstrates the procedure for finding the exact sampling distribution of a statistic when the number of different samples that could be selected from the population is relatively small. In the real world, populations often consist of a large number of different values, making samples difficult (or impossible) to enumerate. When this situation occurs, we may choose to obtain the approximate sampling distribution for a statistic by simulating the sampling over and

over again and recording the proportion of times different values of the statistic occur. Example 6.2 illustrates this procedure.

| EXAMPLE 6.2

Suppose we perform the following experiment over and over again: Take a sample of 11 measurements from the uniform distribution shown in Figure 6.4. Calculate the two sample statistics

$$\bar{x} = \text{Sample mean} = \frac{\sum x}{11}$$

$m = \text{Median} = \text{Sixth sample measurement when the 11 measurements are arranged in ascending order}$

Obtain approximations to the sampling distributions of $\bar{x}$ and m.

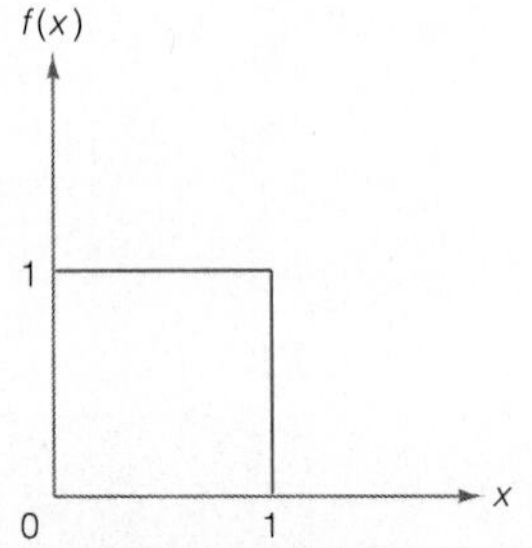

FIGURE 6.4

Uniform distribution from 0 to 1

Solution

We use a computer to generate 1,000 samples, each with $n = 11$ observations. Then we compute $\bar{x}$ and m for each sample. Our goal is to obtain approximations to the sampling distributions of $\bar{x}$ and m and to find out which sample statistic ($\bar{x}$ or m) contains more information about μ. (Note that, in this particular example, we *know* the population mean is $\mu = .5$.) The first 10 of the 1,000 samples generated are presented in Table 6.2. For example, the first computer-generated sample from the uniform distribution (arranged in ascending order) contained the following measurements: .125, .138, .139, .217, .419, .506, .516, .757, .771, .786, and .919. The sample mean $\bar{x}$ and median m computed for this sample are:

$$\bar{x} = \frac{.125 + .138 + \cdots + .919}{11} = .481$$

$m = \text{Sixth ordered measurement} = .506$

TABLE 6.2 **First Ten Samples of $n = 11$ Measurements from a Uniform Distribution**

SAMPLE	MEASUREMENTS										
1	.217	.786	.757	.125	.139	.919	.506	.771	.138	.516	.419
2	.303	.703	.812	.650	.848	.392	.988	.469	.632	.012	.065
3	.383	.547	.383	.584	.098	.676	.091	.535	.256	.163	.390
4	.218	.376	.248	.606	.610	.055	.095	.311	.086	.165	.665
5	.144	.069	.485	.739	.491	.054	.953	.179	.865	.429	.648
6	.426	.563	.186	.896	.628	.075	.283	.549	.295	.522	.674
7	.643	.828	.465	.672	.074	.300	.319	.254	.708	.384	.534
8	.616	.049	.324	.700	.803	.399	.557	.975	.569	.023	.072
9	.093	.835	.534	.212	.201	.041	.889	.728	.466	.142	.574
10	.957	.253	.983	.904	.696	.766	.880	.485	.035	.881	.732

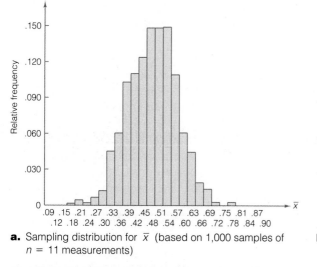

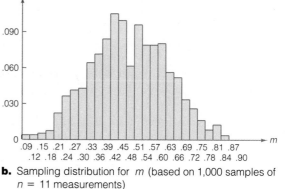

a. Sampling distribution for $\bar{x}$ (based on 1,000 samples of $n = 11$ measurements)

b. Sampling distribution for m (based on 1,000 samples of $n = 11$ measurements)

FIGURE 6.5

Relative frequency histograms for $\bar{x}$ and m, Example 6.2

The relative frequency histograms for $\bar{x}$ and m for the 1,000 samples of size $n = 11$ are shown in Figure 6.5.

You can see that the values of $\bar{x}$ tend to cluster around μ to a greater extent than do the values of m. Thus, on the basis of the observed sampling distributions, we conclude that $\bar{x}$ contains more information about μ than m does—at least for samples of $n = 11$ measurements from the uniform distribution.

As noted earlier, many sampling distributions can be derived mathematically, but the theory necessary to do this is beyond the scope of this text. Consequently, when we need to know the properties of a statistic, we will present its sampling distribution and simply describe its properties. Several of the important properties we look for in sampling distributions are discussed in the next section.

EXERCISES 6.1–6.11

[*Note: Starred (*) exercises require the use of a computer.*]

6.1 The probability distribution shown here describes a population of measurements that can assume values of 0, 2, 4, and 6, each of which occurs with the same relative frequency:

x	0	2	4	6
$p(x)$	$\frac{1}{4}$	$\frac{1}{4}$	$\frac{1}{4}$	$\frac{1}{4}$

a. List all the different samples of $n = 2$ measurements that can be selected from this population.
b. Calculate the mean of each different sample listed in part **a**.
c. If a sample of $n = 2$ measurements is randomly selected from the population, what is the probability that a specific sample will be selected?

    d. Assume that a random sample of $n = 2$ measurements is selected from the population. List the different values of $\bar{x}$ found in part **b**, and find the probability of each. Then give the sampling distribution of the sample mean $\bar{x}$ in tabular form.

    e. Construct a probability histogram for the sampling distribution of $\bar{x}$.

6.2   Simulate sampling from the population (Exercise 6.1) by marking the values of x, one on each of four identical coins (or poker chips, etc.). Place the coins (marked 0, 2, 4, and 6) into a bag, randomly select one, and observe its value. Replace this coin, draw a second coin, and observe its value. Finally, calculate the mean $\bar{x}$ for this sample of $n = 2$ observations randomly selected from the population (Exercise 6.1). Replace the coins, mix, and using the same procedure, select a sample of $n = 2$ observations from the population. Record the numbers and calculate $\bar{x}$ for this sample. Repeat this sampling process until you acquire 100 values of $\bar{x}$. Construct a relative frequency distribution for these 100 sample means. This distribution will be an approximation to the exact sampling distribution of $\bar{x}$ found in part **e** of Exercise 6.1. Compare the two distributions. The distribution obtained in this exercise will not be exactly the same as the exact sampling distribution (Exercise 6.1, part **e**).

    If you were to repeat the sampling procedure, drawing two coins not 100 times but 10,000 times, the relative frequency distribution for the 10,000 sample means would be almost identical to the sampling distribution of $\bar{x}$ found in Exercise 6.1, part **e**.

6.3   Consider the population described by the probability distribution shown here.

x	1	2	3	4	5
$p(x)$	.2	.3	.2	.2	.1

The random variable x is observed twice. If these observations are independent, verify that the different samples of size 2 and their probabilities are as shown here.

SAMPLE	PROBABILITY	SAMPLE	PROBABILITY	SAMPLE	PROBABILITY
1, 1	.04	3, 1	.04	5, 1	.02
1, 2	.06	3, 2	.06	5, 2	.03
1, 3	.04	3, 3	.04	5, 3	.02
1, 4	.04	3, 4	.04	5, 4	.02
1, 5	.02	3, 5	.02	5, 5	.01
2, 1	.06	4, 1	.04		
2, 2	.09	4, 2	.06		
2, 3	.06	4, 3	.04		
2, 4	.06	4, 4	.04		
2, 5	.03	4, 5	.02		

    a. Find the sampling distribution of the sample mean $\bar{x}$.

    b. Construct a probability histogram for the sampling distribution of $\bar{x}$.

    c. What is the probability that $\bar{x}$ is 4.5 or larger?

    d. Would you expect to observe a value of $\bar{x}$ equal to 4.5 or larger? Explain.

6.4   Refer to Exercise 6.3 and find $E(x) = \mu$. Then use the sampling distribution of $\bar{x}$ found in Exercise 6.3 to find the expected value of $\bar{x}$. Note that $E(\bar{x}) = \mu$.

6.5   Refer to Exercise 6.3. Assume that a random sample of $n = 2$ measurements is randomly selected from the population.

    a. List the different values that the sample median m may assume and find the probability of each. Then give the sampling distribution of the sample median.

    b. Construct a probability histogram for the sampling distribution of the sample median and compare it with the probability histogram for the sample mean (Exercise 6.3, part **b**).

6.6 Consider the population described by the probability distribution shown here. The random variable x is observed three times. Suppose the observations are independent.

x	1	2	3
$p(x)$	.6	.3	.1

a. List all possible values of $n = 3$ observations and give their probabilities.
b. List all possible values of the sample mean $\bar{x}$ and give their probabilities. Present the sampling distribution of $\bar{x}$ in tabular form.
c. Construct a probability histogram of the sampling distribution of $\bar{x}$.

6.7 Refer to Exercise 6.6.
a. Find the sampling distribution of the sample median m for samples of $n = 3$ observations.
b. Construct a probability histogram for the sampling distribution of m, and compare it with the probability histogram for the sampling distribution of the sample mean (Exercise 6.6, part c).

6.8 Refer to the probability distribution of Exercise 6.6.
a. List all possible samples of $n = 4$ observations and give their probabilities.
b. List all possible values of the sample mean $\bar{x}$ and give their probabilities. Present the sampling distribution of $\bar{x}$ in tabular form.
c. Construct a probability histogram for the sampling distribution of $\bar{x}$, and compare it with the corresponding sampling distribution based on samples of $n = 3$ observations (Exercise 6.6, part c). What is the effect of increasing the sample size on the sampling distribution of the sample mean $\bar{x}$?

6.9 Repeat the instructions of Exercise 6.8 for the sample median m. Compare the probability histogram of the sampling distribution of m based on $n = 4$ observations with the one based on $n = 3$ observations (Exercise 6.7, part b). What is the effect of increasing the sample size on the sampling distribution of the sample median m?

*6.10 In Example 6.2 we used the computer to generate 1,000 samples, each containing $n = 11$ observations, from a uniform distribution over the interval from 0 to 1. For this exercise, generate 500 samples, each containing $n = 15$ observations, from this population.
a. Calculate the sample mean for each sample. To approximate the sampling distribution of $\bar{x}$, construct a relative frequency histogram for the 500 values of $\bar{x}$.
b. Repeat part a for the sample median. Compare this approximate sampling distribution with the approximate sampling distribution of $\bar{x}$ found in part a.

*6.11 Consider a population that contains values of x equal to 00, 01, 02, 03, . . . , 96, 97, 98, 99. Assume that these values of x occur with equal probability. Generate 500 samples, each containing $n = 25$ measurements, from this population. Calculate the sample mean $\bar{x}$ and sample variance s^2 for each of the 500 samples.
a. To approximate the sampling distribution of $\bar{x}$, construct a relative frequency histogram for the 500 values of $\bar{x}$.
b. Repeat part a for the 500 values of s^2.

6.2 Properties of Sampling Distributions: Unbiasedness and Minimum Variance

The simplest type of statistic used to make inferences about a population parameter is a *point estimator*. A point estimator is a rule or formula that tells us how to use the sample data to calculate a single number that is intended to estimate the value of some population parameter. For example, the sample mean $\bar{x}$ is a point estimator of the population mean μ. Similarly, the sample variance s^2 is a point estimator of the population variance σ^2.

> ### Definition 6.4
>
> A **point estimator** of a population parameter is a rule or formula that tells us how to use the sample data to calculate a single number that can be used as an *estimate* of the population parameter.

Often, many different point estimators can be found to estimate the same parameter. Each will have a sampling distribution that provides information about the point estimator. By examining the sampling distribution, we can determine how large the difference between an estimate and the true value of the parameter (called the **error of estimation**) is likely to be. We can also tell whether an estimator is more likely to overestimate or to underestimate a parameter.

EXAMPLE 6.3

Suppose two statistics, A and B, exist to estimate the same population parameter, θ (theta). (Note that θ could be any parameter, μ, σ^2, σ, etc.) Suppose the two statistics have sampling distributions as shown in Figure 6.6. Based on these sampling distributions, which statistic is more attractive as an estimator of θ?

FIGURE 6.6

Sampling distributions of unbiased and biased estimators

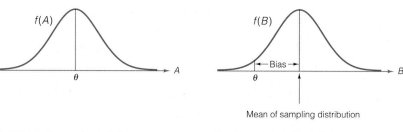

a. Unbiased sample statistic for the parameter θ

b. Biased sample statistic for the parameter θ

Solution

As a first consideration, we would like the sampling distribution to center over the value of the parameter we wish to estimate. One way to characterize this property is in terms of the mean of the sampling distribution. Consequently, we say that a statistic is *unbiased* if the mean of the sampling distribution is equal to the parameter it is intended to estimate. This situation is shown in Figure 6.6(a), where the mean μ_A of statistic A is equal to θ. If the mean of a sampling distribution is not equal to the parameter it is intended to estimate, the statistic is said to be *biased*. The sampling distribution for a biased statistic is shown in Figure 6.6(b). The mean μ_B of the sampling distribution for statistic B is not equal to θ; in fact, it is shifted to the right of θ.

You can see that biased statistics tend either to overestimate or to underestimate a parameter. Consequently, when other properties of statistics tend to be equivalent, we will choose an unbiased statistic to estimate a parameter of interest.*

Definition 6.5

If the sampling distribution of a sample statistic has a mean equal to the population parameter the statistic is intended to estimate, the statistic is said to be an **unbiased** estimate of the parameter.

If the mean of the sampling distribution is not equal to the parameter, the statistic is said to be a **biased** estimate of the parameter.

The standard deviation of a sampling distribution measures another important property of statistics—the spread of these estimates generated by repeated sampling. Suppose two statistics, A and B, are both unbiased estimators of the population parameter. Since the means of the two sampling distributions are the same, we turn to their standard deviations in order to decide which will provide estimates that fall closer to the unknown population parameter we are estimating. Naturally, we will choose the sample statistic that has the smaller standard deviation. Figure 6.7 depicts sampling distributions for A and B. Note that the standard deviation of the distribution of A is smaller than the standard deviation for B, indicating that over a large number of samples, the values of A cluster more closely around the unknown population parameter than do the values of B.

In summary, to make an inference about a population parameter, we use the sample statistic with a sampling distribution that is unbiased and has a small standard deviation (usually smaller than the standard deviation of other unbiased sample statistics). The derivation of this sample statistic will not concern us, because the "best" statistic for estimating specific parameters is a matter of record. We will simply present an unbiased estimator with its standard deviation for each population parameter we consider. [*Note:* The standard deviation of the sampling distribution of a statistic is also called the **standard error of the statistic**.]

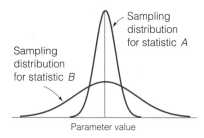

FIGURE 6.7

Sampling distributions for two unbiased estimators

EXAMPLE 6.4

In Example 6.1, we found the sampling distributions of the sample mean $\bar{x}$ and the sample median m for random samples of $n = 3$ measurements from a population defined by the probability distribution shown here.

x	0	3	12
$p(x)$	$\frac{1}{3}$	$\frac{1}{3}$	$\frac{1}{3}$

---

*Unbiased statistics do not exist for all parameters of interest, but they do exist for the parameters considered in this text.

The sampling distributions of $\bar{x}$ and m were found to be as shown here.

$\bar{x}$	0	1	2	3	4	5	6	8	9	12
$p(\bar{x})$	$\frac{1}{27}$	$\frac{3}{27}$	$\frac{3}{27}$	$\frac{1}{27}$	$\frac{3}{27}$	$\frac{6}{27}$	$\frac{3}{27}$	$\frac{3}{27}$	$\frac{3}{27}$	$\frac{1}{27}$

m	0	3	12
$p(m)$	$\frac{7}{27}$	$\frac{13}{27}$	$\frac{7}{27}$

a. Show that $\bar{x}$ is an unbiased estimator of μ in this situation.
b. Show that m is a biased estimator of μ in this situation.

Solution

a. The expected value of a discrete random variable x (see Section 4.3) is defined to be $E(x) = \Sigma\, xp(x)$, where the summation is over all values of x. Then

$$E(x) = \mu = \sum xp(x) = (0)(\tfrac{1}{3}) + (3)(\tfrac{1}{3}) + (12)(\tfrac{1}{3}) = 5$$

The expected value of the discrete random variable $\bar{x}$ is

$$E(\bar{x}) = \sum (\bar{x})p(\bar{x})$$

summed over all values of $\bar{x}$. Or

$$E(\bar{x}) = (0)(\tfrac{1}{27}) + (1)(\tfrac{3}{27}) + 2(\tfrac{3}{27}) + \cdots + (12)(\tfrac{1}{27})$$
$$= 5$$

Since $E(\bar{x}) = \mu$, we see that $\bar{x}$ is an unbiased estimator of μ.

b. The expected value of the sample median m is

$$E(m) = \sum mp(m) = (0)(\tfrac{7}{27}) + (3)(\tfrac{13}{27}) + (12)(\tfrac{7}{27}) = 4.56$$

Since the expected value of m is not equal to μ ($\mu = 5$), the sample median m is a biased estimator of μ.

EXAMPLE 6.5

Refer to Example 6.4 and find the standard deviations of the sampling distributions of $\bar{x}$ and m. Which statistic would appear to be a better estimator for μ?

Solution

The variance of the sampling distribution of $\bar{x}$ (we denote it by the symbol $\sigma_{\bar{x}}^2$) is found to be

$$\sigma_{\bar{x}}^2 = E\{[\bar{x} - E(\bar{x})]^2\} = \sum (\bar{x} - \mu)^2 p(\bar{x})$$

where, from Example 6.4,

$$E(\bar{x}) = \mu = 5$$

Then

$$\sigma_{\bar{x}}^2 = (0 - 5)^2(\tfrac{1}{27}) + (1 - 5)^2(\tfrac{3}{27}) + (2 - 5)^2(\tfrac{3}{27}) + \cdots + (12 - 5)^2(\tfrac{1}{27})$$
$$= 8.6667$$

and

$$\sigma_{\bar{x}} = \sqrt{8.6667} = 2.94$$

Similarly, the variance of the sampling distribution of m (we denote it by σ_m^2) is

$$\sigma_m^2 = E\{[m - E(m)]^2\}$$

where, from Example 6.4, the expected value of m is $E(m) = 4.56$. Then

$$\sigma_m^2 = E\{[m - E(m)]^2\} = \sum [m - E(m)]^2 p(m)$$

$$= (0 - 4.56)^2\left(\tfrac{7}{27}\right) + (3 - 4.56)^2\left(\tfrac{13}{27}\right) + (12 - 4.56)^2\left(\tfrac{7}{27}\right)$$

$$= 20.9136$$

and

$$\sigma_m = \sqrt{20.9136} = 4.57$$

Which statistic appears to be the better estimator for the population mean μ: the sample mean $\bar{x}$ or the median m? To answer this question, we compare the sampling distributions of the two statistics. The sampling distribution of the sample median m is biased (i.e., it is located to the left of the mean μ) and its standard deviation $\sigma_m = 4.57$ is much larger than the standard deviation of the sampling distribution of $\bar{x}$, $\sigma_{\bar{x}} = 2.94$. Consequently, the sample mean $\bar{x}$ would be a better estimator of the population mean μ, for the population in question, than would be the sample median m.

EXERCISES 6.12–6.18

[Note: Starred (*) exercises require the use of a computer.]

6.12 Consider the probability distribution:

x	0	1	5
$p(x)$	$\tfrac{1}{3}$	$\tfrac{1}{3}$	$\tfrac{1}{3}$

a. Find μ and σ^2.
b. Find the sampling distribution of the sample mean $\bar{x}$ for a random sample of $n = 2$ measurements from this distribution.
c. Show that $\bar{x}$ is an unbiased estimator for μ. [Hint: Show that $E(\bar{x}) = \sum \bar{x}p(\bar{x}) = \mu$.]
d. Find the sampling distribution of the sample variance s^2 for a random sample of $n = 2$ measurements from this distribution.
e. Show that s^2 is an unbiased estimator for σ^2.

6.13 Consider the probability distribution shown here.

x	2	4	9
$p(x)$	$\tfrac{1}{3}$	$\tfrac{1}{3}$	$\tfrac{1}{3}$

a. Calculate μ for this distribution.

b. Find the sampling distribution of the sample mean $\bar{x}$ for a random sample of $n = 3$ measurements from this distribution, and show that $\bar{x}$ is an unbiased estimator of μ.

c. Find the sampling distribution of the sample median m for a random sample of $n = 3$ measurements from this distribution, and show that the median is a biased estimator of μ.

d. If you wanted to estimate μ using a sample of three measurements from this population, which estimator would you use? Why?

6.14   Consider the probability distribution shown here.

x	0	1	2
$p(x)$	$\frac{1}{3}$	$\frac{1}{3}$	$\frac{1}{3}$

a. Find μ.

b. For a random sample of $n = 3$ observations from this distribution, find the sampling distribution of the sample mean.

c. Find the sampling distribution of the median of a sample of $n = 3$ observations from this population.

d. Refer to parts b and c and show that both the mean and median are unbiased estimators of μ for this population.

e. Find the variances of the sampling distributions of the sample mean and the sample median.

f. Which estimator would you use to estimate μ? Why?

*6.15   Generate 500 samples, each containing $n = 25$ measurements, from a population that contains values of x equal to 01, 02, . . . , 48, 49, 50. Assume that these values of x are equally likely. Calculate the sample mean $\bar{x}$ and median m for each sample. Construct relative frequency histograms for the 500 values of $\bar{x}$ and the 500 values of m. Use these approximations to the sampling distributions of $\bar{x}$ and m to answer the following questions:

a. Does it appear that $\bar{x}$ and m are unbiased estimators of the population mean? [*Note:* $\mu = 25.5$.]

b. Which sampling distribution displays greater variation?

6.16   Refer to Exercise 6.6.

a. Show that $\bar{x}$ is an unbiased estimator of μ.

b. Find $\sigma_{\bar{x}}^2$.

c. Find the probability that $\bar{x}$ will fall within $2\sigma_{\bar{x}}$ of μ.

6.17   Refer to Exercise 6.6.

a. Find the sampling distribution of s^2.

b. Find the population variance σ^2.

c. Show that s^2 is an unbiased estimator of σ^2.

d. Find the sampling distribution of the sample standard deviation s.

e. Show that s is a biased estimator of σ.

6.18   Refer to Exercise 6.7, where we found the sampling distribution of the sample median. Show that the median is or is not an unbiased estimator of the population mean μ.

6.3   The Central Limit Theorem

Estimating the mean useful life of automobiles, the mean number of crimes per month in a large city, and the mean yield per acre of a new soybean hybrid are practical problems with something in common. In each case we are interested in making an inference about the mean μ of some population. As we mentioned in Chapter 2, the sample mean $\bar{x}$ is, in general, a good estimator of μ. We now develop pertinent information about the sampling distribution for this useful statistic.

EXAMPLE 6.6

Suppose a population has the uniform probability distribution given in Figure 6.8. The mean and standard deviation of this probability distribution are $\mu = .5$ and $\sigma = .29$. (See Section 5.2 for the formulas for μ and σ.) Now suppose a sample of 11 measurements is selected from this population. Describe the sampling distribution of the sample mean $\bar{x}$ based on the 1,000 sampling experiments discussed in Example 6.2.

FIGURE 6.8

Sampled uniform population

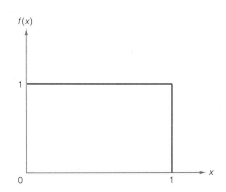

Solution

You will recall that in Example 6.2 we generated 1,000 samples of $n = 11$ measurements each. The relative frequency histogram for the 1,000 sample means is shown in Figure 6.9 with a normal probability distribution superimposed. You

FIGURE 6.9

Relative frequency histogram for $\bar{x}$ in 1,000 samples of $n = 11$ measurements with normal distribution superimposed

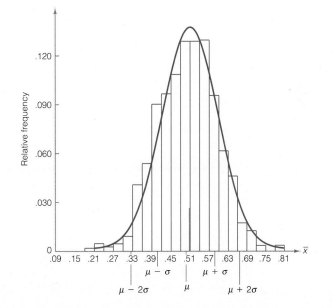

can see that this normal probability distribution approximates the computer-generated sampling distribution very well.

To describe fully a normal probability distribution, it is necessary to know its mean and standard deviation. Inspection of Figure 6.9 indicates that the mean of the distribution of $\bar{x}$, $\mu_{\bar{x}}$, appears to be very close to .5, the mean of the sampled uniform population. Furthermore, for a mound-shaped distribution such as that shown in Figure 6.9, almost all the measurements should fall within 3 standard deviations of the mean. Since the number of values of $\bar{x}$ is very large (1,000), the range of the observed $\bar{x}$'s divided by 6 (rather than 4) should give a reasonable approximation to the standard deviation of the sample means, $\sigma_{\bar{x}}$. The values of $\bar{x}$ range from about .2 to .8, so we calculate

$$\sigma_{\bar{x}} \approx \frac{\text{Range of } \bar{x}\text{'s}}{6} = \frac{.8 - .2}{6} = .1$$

To summarize our findings based on 1,000 samples, each consisting of 11 measurements from a uniform population, the sampling distribution of $\bar{x}$ appears to be approximately normal with a mean of about .5 and a standard deviation of about .1.

It can be shown in general that the sampling distribution of $\bar{x}$ has the properties given in the box.

> ### Properties of the Sampling Distribution of $\bar{x}$
>
> 1. Mean of sampling distribution = Mean of sampled population
>   That is, $\mu_{\bar{x}} = E(\bar{x}) = \mu$.
> 2. Standard deviation of sampling distribution equals
>
> $$\frac{\left(\begin{array}{c} \text{Standard deviation of} \\ \text{sampled population} \end{array} \right)}{\text{Square root of sample size}}$$
>
>   That is, $\sigma_{\bar{x}} = \sigma/\sqrt{n}$.
>
>   The standard deviation $\sigma_{\bar{x}}$ is often referred to as the **standard error of the mean**.
> 3. The sampling distribution of $\bar{x}$ is approximately normal for large sample sizes.

You can see that our approximation to $\mu_{\bar{x}}$ in Example 6.6 was precise, since property 1 assures us that the mean is the same as that of the sampled population: .5. Property 2 tells us how to calculate the standard deviation of the sampling distribution of $\bar{x}$. Substituting $\sigma = .29$, the standard deviation of the sampled uniform distribution, and the sample size $n = 11$ into the formula for $\sigma_{\bar{x}}$, we find

$$\sigma_{\bar{x}} = \frac{\sigma}{\sqrt{n}} = \frac{.29}{\sqrt{11}} = .09$$

Thus, the approximation we obtained in Example 6.6, $\sigma_{\bar{x}} \approx .1$, is very close to the exact value, $\sigma_{\bar{x}} = .09$.

The justification for property 3 is contained in one of the most important theoretical results in statistics, the **Central Limit Theorem**.

Central Limit Theorem

If a random variable of n observations is selected from a population (*any* population), then when n is sufficiently large, the sampling distribution of $\bar{x}$ will be approximately a normal distribution. The larger the sample size, the better will be the normal approximation to the sampling distribution of $\bar{x}$.*

Thus, for sufficiently large samples the sampling distribution of $\bar{x}$ is approximately normal. How large must the sample size n be so that the normal distribution provides a good approximation for the sampling distribution of $\bar{x}$? The answer depends on the shape of the distribution of the sampled population, as shown by Figure 6.10 on page 270. Generally speaking, the greater the skewness of the sampled population distribution, the larger the sample size must be before the normal distribution is an adequate approximation for the sampling distribution of $\bar{x}$. For most sampled populations, sample sizes of $n \geq 30$ will suffice for the normal approximation to be reasonable. We will use the normal approximation for the sampling distribution of $\bar{x}$ when the sample size is at least 30.

EXAMPLE 6.7

Suppose we have selected a random sample of $n = 25$ observations from a population with mean equal to 80 and standard deviation equal to 5. It is known that the population is not extremely skewed.

a. Sketch the relative frequency distributions for the population and for the sampling distribution of the sample mean, $\bar{x}$.

b. Find the probability that $\bar{x}$ will be larger than 82.

---

*Moreover, because of the Central Limit Theorem the sum of a random sample of n observations, Σx, will possess a sampling distribution that is approximately normal for large samples. This distribution will have a mean equal to $n\mu$ and a variance equal to $n\sigma^2$. Proof of the Central Limit Theorem is beyond the scope of this book, but it can be found in many mathematical statistics texts.

FIGURE 6.10

Sampling distributions of $\bar{x}$ for different populations and different sample sizes

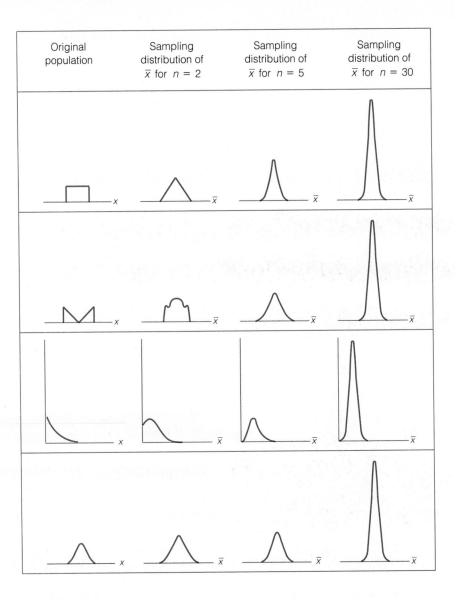

Solution

a. We do not know the exact shape of the population relative frequency distribution, but we do know that it should be centered about $\mu = 80$, its spread should be measured by $\sigma = 5$, and it is not highly skewed. One possibility is shown in Figure 6.11(a). From the Central Limit Theorem, we know that the sampling distribution of $\bar{x}$ will be approximately normal since the sampled population distribution is not extremely skewed. We also know that the sampling distribution will have mean and standard deviation

$$\mu_{\bar{x}} = \mu = 80 \quad \text{and} \quad \sigma_{\bar{x}} = \frac{\sigma}{\sqrt{n}} = \frac{5}{\sqrt{25}} = 1$$

The sampling distribution of $\bar{x}$ is shown in Figure 6.11(b).

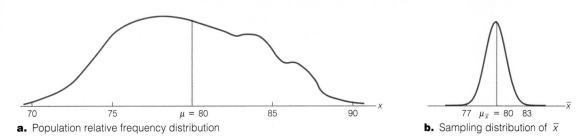

a. Population relative frequency distribution   **b.** Sampling distribution of $\bar{x}$

FIGURE 6.11

A population relative frequency distribution and the sampling distribution for $\bar{x}$

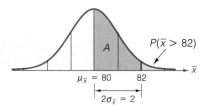

FIGURE 6.12

The sampling distribution of $\bar{x}$

b. The probability that $\bar{x}$ will exceed 82 is equal to the darker shaded area in Figure 6.12. To find this area, we need to find the z value corresponding to $\bar{x} = 82$. Recall that the standard normal random variable z is the difference between any normally distributed random variable and its mean, expressed in units of its standard deviation. Since $\bar{x}$ is a normally distributed random variable with mean $\mu_{\bar{x}} = \mu$ and standard deviation $\sigma_{\bar{x}} = \sigma/\sqrt{n}$, it follows that the standard normal z value corresponding to the sample mean, $\bar{x}$, is

$$z = \frac{(\text{Normal random variable}) - (\text{Mean})}{\text{Standard deviation}} = \frac{\bar{x} - \mu_{\bar{x}}}{\sigma_{\bar{x}}}$$

Therefore, for $\bar{x} = 82$, we have

$$z = \frac{\bar{x} - \mu_{\bar{x}}}{\sigma_{\bar{x}}} = \frac{82 - 80}{1} = 2$$

The area A in Figure 6.12 corresponding to $z = 2$ is given in the table of areas under the normal curve (see Table IV of Appendix A) as .4772. Therefore, the tail area corresponding to the probability that $\bar{x}$ exceeds 82 is

$$P(\bar{x} > 82) = P(z > 2) = .5 - .4772 = .0228$$

EXAMPLE 6.8

A manufacturer of automobile batteries claims that the distribution of the lengths of life of its best battery has a mean of 54 months and a standard deviation of 6 months. Suppose a consumer group decides to check the claim by purchasing a sample of 50 of these batteries and subjecting them to tests that determine their lives.

a. Assuming that the manufacturer's claim is true, describe the sampling distribution of the mean lifetime of a sample of 50 batteries.

b. Assuming that the manufacturer's claim is true, what is the probability the consumer group's sample has a mean life of 52 or fewer months?

Solution

a. Even though we have no information about the shape of the probability distribution of the lives of the batteries, we can use the Central Limit Theorem to deduce that the sampling distribution for a sample mean lifetime of 50 batteries is approximately normally distributed. Furthermore, the mean of this sampling distribution is the same as the mean of the sampled population,

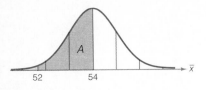

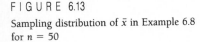

FIGURE 6.13

Sampling distribution of $\bar{x}$ in Example 6.8 for $n = 50$

which is $\mu = 54$ months according to the manufacturer's claim. Finally, the standard deviation of the sampling distribution is given by

$$\sigma_{\bar{x}} = \frac{\sigma}{\sqrt{n}} = \frac{6}{\sqrt{50}} = .85 \text{ month}$$

Note that we used the claimed standard deviation of the sampled population, $\sigma = 6$ months. Thus, if we assume that the claim is true, the sampling distribution for the mean life of the 50 batteries sampled is as shown in Figure 6.13.

b. If the manufacturer's claim is true, the probability that the consumer group observes a mean battery life of 52 or fewer months for their sample of 50 batteries, $P(\bar{x} \le 52)$, is equivalent to the darker shaded area in Figure 6.13. Since the sampling distribution is approximately normal, we can find this area by computing the standard normal z value:

$$z = \frac{\bar{x} - \mu_{\bar{x}}}{\sigma_{\bar{x}}} = \frac{\bar{x} - \mu}{\sigma_{\bar{x}}} = \frac{52 - 54}{.85} = -2.35$$

where $\mu_{\bar{x}}$, the mean of the sampling distribution of $\bar{x}$, is equal to μ, the mean of the lives of the sampled population, and $\sigma_{\bar{x}}$ is the standard deviation of the sampling distribution of $\bar{x}$. Note that z is the familiar standardized distance (z-score) of Section 2.7 and, since $\bar{x}$ is approximately normally distributed, it will possess the standard normal distribution of Section 5.3.

The area A shown in Figure 6.13 between $\bar{x} = 52$ and $\bar{x} = 54$ (corresponding to $z = -2.35$) is found in Table IV of Appendix A to be .4906. Therefore, the area to the left of $\bar{x} = 52$ is

$$P(\bar{x} \le 52) = .5 - A = .5 - .4906 = .0094$$

Thus, the probability the consumer group will observe a sample mean of 52 or less is only .0094 if the manufacturer's claim is true. If the 50 tested batteries do result in a mean of 52 or fewer months, the consumer group will have strong evidence that the manufacturer's claim is untrue, because such an event is very unlikely to occur if the claim is true. (This is still another application of the *rare event approach* to statistical inference.)

In addition to providing a very useful approximation for the sampling distribution of a sample mean, the Central Limit Theorem offers an explanation for the fact that many relative frequency distributions of data possess mound-shaped distributions. Many of the macroscopic measurements we take in various areas of research are really means or sums of many microscopic phenomena. For example, a year's growth of a pine seedling is the total of the many individual components that affect the plant's growth. Similarly, the length of time a construction company takes to complete a house might be viewed as the total of the times taken to complete each of the distinct jobs necessary to build the house. The monthly demand for blood at a hospital may be viewed as the total of the individual patients' needs. Whether the observations entering into these sums satisfy the assumptions basic to the Central Limit Theorem is questionable, but it is a fact that many distributions of data are mound-shaped and possess the appearance of normal distributions. Thus, the Central Limit Theorem offers one explanation for the frequent occurrence of mound-shaped distributions in nature.

[Note: Starred (*) exercises require the use of a computer.]

LEARNING THE MECHANICS

6.19 Suppose a random sample of n measurements is selected from a population with mean $\mu = 50$ and variance $\sigma^2 = 50$. For each of the following values of n, give the mean and standard deviation of the sampling distribution of the sample mean $\bar{x}$.
a. $n = 4$ b. $n = 25$ c. $n = 100$ d. $n = 50$ e. $n = 500$ f. $n = 1,000$

6.20 Suppose a random sample of $n = 25$ measurements is selected from a population with mean μ and standard deviation σ. For each of the following values of μ and σ, give the values of $\mu_{\bar{x}}$ and $\sigma_{\bar{x}}$.
a. $\mu = 5$, $\sigma = 3$ b. $\mu = 100$, $\sigma = 2$ c. $\mu = 10$, $\sigma = 40$ d. $\mu = 10$, $\sigma = 200$

6.21 Consider the probability distribution shown here.

x	1	2	3	10
$p(x)$	.1	.4	.4	.1

a. Find μ, σ^2, and σ.
b. Find the sampling distribution of $\bar{x}$ for random samples of $n = 2$ measurements from this distribution by listing all possible values of $\bar{x}$, and find the probability associated with each.
c. Use the results of part b to calculate $\mu_{\bar{x}}$ and $\sigma_{\bar{x}}$. Confirm that $\mu_{\bar{x}} = \mu$ and that $\sigma_{\bar{x}} = \sigma/\sqrt{n} = \sigma/\sqrt{2}$.

6.22 Will the sampling distribution of $\bar{x}$ always be approximately normally distributed? Explain.

6.23 A random sample of $n = 64$ observations is drawn from a population with a mean equal to 15 and standard deviation equal to 4.
a. Give the mean and standard deviation of the (repeated) sampling distribution of $\bar{x}$.
b. Describe the shape of the sampling distribution of $\bar{x}$. Does your answer depend on the sample size?
c. Calculate the standard normal z-score corresponding to a value of $\bar{x} = 15.5$.
d. Calculate the standard normal z-score corresponding to $\bar{x} = 14$.

6.24 Refer to Exercise 6.23. Find the probability that
a. $\bar{x}$ is greater than 16   b. $\bar{x}$ is less than 16
c. $\bar{x}$ is greater than 14.7   d. $\bar{x}$ falls between 14 and 16
e. $\bar{x}$ is less than 14

6.25 A random sample of $n = 100$ observations is selected from a population with $\mu = 30$ and $\sigma = 16$. Approximate the following probabilities:
a. $P(\bar{x} \geq 28)$ b. $P(22.1 \leq \bar{x} \leq 26.8)$ c. $P(\bar{x} \leq 28.2)$ d. $P(\bar{x} \geq 27.0)$

6.26 A random sample of $n = 900$ observations is selected from a population with $\mu = 100$ and $\sigma = 10$.
a. What are the largest and smallest values of $\bar{x}$ that you would expect to see?
b. How far, at the most, would you expect $\bar{x}$ to deviate from μ?
c. Did you have to know μ to answer part b? Explain.

*6.27 Consider a population that contains values of x equal to 00, 01, 02, ..., 97, 98, 99. Assume that the values of x are equally likely. For each of the following values of n, generate 500 random samples and calculate $\bar{x}$ for each sample. For each sample size, construct a relative frequency histogram of the 500 values of $\bar{x}$. What changes occur in the histograms as the value of n increases? What similarities exist? Use $n = 2$, $n = 5$, $n = 10$, $n = 30$, and $n = 50$.

APPLYING THE CONCEPTS

6.28 Suppose a sample of $n = 50$ items is drawn from a population of manufactured products and the weight x of each item is recorded. Prior experience has shown that the weight has a probability distribution with $\mu = 6.0$ ounces and $\sigma = 2.5$ ounces. Then $\bar{x}$, the sample mean, will be approximately normally distributed (because of the Central Limit Theorem).

   a. Calculate $\mu_{\bar{x}}$ and $\sigma_{\bar{x}}$.

   b. What is the approximate probability that the manufacturer's sample has a mean weight of between 5.75 and 6.25 ounces?

   c. What is the approximate probability that the manufacturer's sample has a mean weight of less than 5.5 ounces?

   d. How would the sampling distribution of $\bar{x}$ change if the sample size n were increased from 50 to, say, 100?

6.29 The relationship among socioeconomic status, IQ, and juvenile delinquency has long been the subject of psychological research. In one recent study (Moffitt, Gabrielli, Mednick, and Schulsinger, 1981), researchers examined the relationship between delinquent behavior and poor verbal abilities. Assume that scores on the verbal IQ test administered during this research have a population mean $\mu = 107$ and a population standard deviation $\sigma = 15$.

   a. What shape would you expect the (repeated) sampling distribution of $\bar{x}$ for $n = 84$ juveniles to have? Does your answer depend on the shape of the distribution of verbal IQ scores for all juveniles?

   b. Assuming that the population mean and standard deviation for juveniles with no record of delinquency are the same as that for all juveniles, approximate the probability that the sample mean verbal IQ for $n = 84$ juveniles will be 110 or more. State any assumptions you make.

   c. As part of the cited study, the researchers found that a sample of $n = 84$ juveniles with no record of delinquency had a mean verbal IQ of $\bar{x} = 110$. Considering your answer to part **b**, do you think that the population mean and standard deviation for nondelinquent juveniles are the same as those for all juveniles? Explain.

6.30 Refer to Exercise 6.29. As part of the study cited there, the researchers reported the sample mean verbal IQ for $n = 22$ juveniles with records of two or more offenses.

   a. What shape would you expect the (repeated) sampling distribution of $\bar{x}$ for $n = 22$ juveniles to have? Does your answer depend on the shape of the distribution of verbal IQ scores for all juveniles?

   b. Assuming that the population mean and standard deviation for juveniles with no record of delinquency are the same as those for all juveniles, approximate the probability that the sample mean verbal IQ for $n = 22$ juveniles will be 98 or less. State any assumptions you make.

   c. As part of the cited study, the researchers found that a sample of $n = 22$ juveniles with no record of delinquency had a mean verbal IQ of $\bar{x} = 98$. Considering your answer to part **b**, do you think that the population mean and standard deviation for nondelinquent juveniles are the same as for all juveniles? Explain.

6.31 The number of violent crimes per day in a certain city possesses a mean equal to 1.3 and a standard deviation equal to 1.7. A random sample of 50 days is observed, and the daily mean number of crimes for this sample, $\bar{x}$, is calculated.

   a. Give the mean and standard deviation of the sampling distribution of $\bar{x}$.

   b. Will the sampling distribution of $\bar{x}$ be approximately normal? Explain.

   c. Find an approximate value of $P(\bar{x} < 1)$.

   d. Find an approximate value of $P(\bar{x} > 1.9)$.

6.32 Last year a company initiated a program to compensate its employees for unused sick days, paying each employee a bonus of one-half the usual wage earned for each unused sick day. The question that naturally arises is: "Did this policy motivate employees to use fewer allotted sick days?" *Before* last year, the number of sick days used by employees had a distribution with a mean of 7 days and a standard deviation of 2 days.

  a. Assuming that these parameters did not change last year, find the approximate probability that the sample mean number of sick days used by 100 employees chosen at random was less than or equal to 6.4 last year.

  b. Suppose the sample mean for the 100 employees was, in fact, 6.4. How would you interpret this result?

6.33 Purchased materials, parts, and services account for over 50% of the manufacturing costs of many companies. As a result, it is important for a company's overall quality program to extend to the vendors (i.e., suppliers) from whom the purchases are made. In recent years, many firms have begun including vendor surveillance as part of their quality programs (Juran and Gryna, 1980).

  For example, suppose a soft drink bottler requires bottles with an internal pressure strength of at least 150 pounds per square inch (psi). A prospective bottle vendor claims that its production process yields bottles with a mean internal pressure strength of 157 psi and a standard deviation of 3 psi. As part of its vendor surveillance program, the bottler strikes an agreement with the vendor that permits the bottler to sample from the vendor's production process to verify the vendor's claim. The bottler randomly selects 40 bottles from the last 10,000 produced, measures the internal pressure strength of each, and finds the mean strength for the sample to be 1.3 psi below the process mean cited by the vendor.

  a. Assuming the vendor's claim to be true, what is the probability of obtaining a sample mean this far or farther below the process mean? What does your answer suggest about the validity of the vendor's claim?

  b. If the process standard deviation were 3 psi as claimed by the vendor, but the mean were 156 psi, would the observed sample result be more or less likely than in part **a**? What if the mean were 158 psi?

  c. If the process mean were 157 psi as claimed, but the process standard deviation were 2 psi, would the sample result be more or less likely than in part **a**? What if instead the standard deviation were 6 psi?

6.4 The Relation Between Sample Size and a Sampling Distribution

Suppose you draw two random samples, one containing $n = 5$ and the second $n = 10$ observations, from a population with mean μ and standard deviation σ. If you compute the mean $\bar{x}$ for each sample and obtain the results shown in the table, which estimate do you think contains more information about μ?

Intuitively, it would seem that the sample mean based on ten measurements contains more information than the mean based on five, but to answer the question correctly we need to compare the sampling distributions of these two statistics.

SAMPLE 1	SAMPLE 2
$n = 5$	$n = 10$
$\bar{x} = 12.6$	$\bar{x} = 13.1$

From Section 6.3 we know that the expected value of the sample means in repeated sampling is μ, regardless of the sample size. That is, both sample means are unbiased estimators of μ. The main difference in the sampling distributions lies in their standard deviations: The standard deviation of the mean based on $n = 5$ is $\sigma/\sqrt{5}$, whereas that based on $n = 10$ is $\sigma/\sqrt{10}$. Since the second standard deviation is smaller, we expect the sample means based on ten measurements to cluster more closely around μ in repeated sampling than those based on five measurements. Thus, our intuitive feeling that $\bar{x}$ for $n = 10$ contains more information about μ is justified.

For most of the statistics you will encounter in this text, the variance of a statistic's sampling distribution is inversely proportional to the sample size. Or, since the standard deviation of $\bar{x}$ is equal to $\sigma/\sqrt{n}$, you can say that the standard deviation of the sampling distribution is proportional to $1/\sqrt{n}$. To reduce the standard deviation of the sampling distribution of a statistic by one-half, you will therefore need four times as many observations in your sample. To reduce the standard deviation to one-third its original value, you will need nine times as many observations.

The sampling distributions for the sample mean $\bar{x}$, based on random samplings from a normally distributed population, are shown in Figure 6.14 for $n = 1, 4$, and 16 observations. The curve for $n = 1$ represents the probability distribution for the population. Those for $n = 4$ and $n = 16$ are sampling distributions for $\bar{x}$. Note how the distributions contract (variation decreases) as n increases from 1 to 4 to 16. The standard deviation for $\bar{x}$ based on $n = 16$ measurements is one-half the corresponding standard deviation for the distribution based on $n = 4$ measurements.

FIGURE 6.14

Three sampling distributions for $\bar{x}$

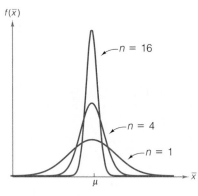

EXAMPLE 6.9

Consider the **Bernoulli random variable** x that can assume the values 1 or 0 with probabilities p and $q = 1 - p$, respectively. The distribution is summarized in Table 6.3, where we have associated the term "Success" with the outcome $x = 1$ and "Failure" with the outcome $x = 0$. The Bernoulli random variable is just a binomial random variable with $n = 1$ trial.

Suppose a random sample of n measurements is drawn from this Bernoulli distribution, and the sample mean $\bar{x}$ is calculated. Note that

TABLE 6.3   Bernoulli Distribution

OUTCOME	x	$p(x)$
Failure	0	q
Success	1	p

$$\bar{x} = \frac{\sum\limits_{i=1}^{n} x_i}{n}$$

and that $\sum\limits_{i=1}^{n} x_i$ is the total number of Successes (the number of 1's in the sample) in the random sample of n measurements (or "trials"). This sum is therefore a binomial random variable, with n trials and probability of Success p. For example, the Bernoulli random variable might be the *status* of a single computer microchip (nondefective or defective), x the *number* of n such chips that are nondefective, and $\bar{x}$ the *fraction* of nondefective chips in a set of n.

a. Find the mean and standard deviation of the sampling distribution of $\bar{x}$.

b. Simulate the distribution of $\bar{x}$ using $p = .8$, and $n = 1, 10, 25$, and 100 by generating 1,000 samples for each sample size and creating a histogram of the 1,000 sample means.

Solution

a. The mean and variance of the Bernoulli random variable are:

$$\mu = E(x) = 0(q) + 1(p) = p$$
$$\sigma^2 = E[(x - \mu)^2] = (0 - p)^2(q) + (1 - p)^2(p)$$
$$= p^2 q + q^2 p = pq(p + q) = pq(1) = pq$$

Note that these are the mean and variance of a binomial random variable with $n = 1$, which is another description of a Bernoulli random variable.

We know that the mean $\bar{x}$ of a random sample is unbiased, so that

$$E(\bar{x}) = \mu = p$$

and the standard error is

$$\sigma_{\bar{x}} = \frac{\sigma}{\sqrt{n}} = \frac{\sqrt{pq}}{\sqrt{n}} = \sqrt{\frac{pq}{n}}$$

Because $\bar{x}$ provides an unbiased estimate of the probability of Success p and has a standard error that decreases as the sample size increases, we will use it to estimate p in subsequent chapters, where we will refer to it as the sample fraction of Successes, $\hat{p}$.

b. The mean and standard deviation of $\bar{x}$ when $p = .8$ are

$$\mu_{\bar{x}} = p = .8$$
$$\sigma_{\bar{x}} = \sqrt{\frac{pq}{n}} = \sqrt{\frac{(.8)(.2)}{n}} = \frac{.4}{\sqrt{n}}$$

You can see how the standard error of $\bar{x}$ decreases as n increases in Table 6.4.

TABLE 6.4   **Mean and Standard Error of $\bar{x}$**

n	$\mu_{\bar{x}}$	$\sigma_{\bar{x}}$
1	.8	$.4/\sqrt{1} = .4000$
10	.8	$.4/\sqrt{10} = .1265$
25	.8	$.4/\sqrt{25} = .0800$
100	.8	$.4/\sqrt{100} = .0400$

The simulation of 1,000 sample means with each of the sample sizes given in Table 6.4 resulted in the sampling distributions shown in Figure 6.15 (page 278), where a relative frequency histogram is used to display each sampling distribution. Note that each sampling distribution more closely resembles a normal distribution than the previous one. As the Central Limit Theorem promises, the distribution of $\bar{x}$ becomes approximately normal for large n, and the approxima-

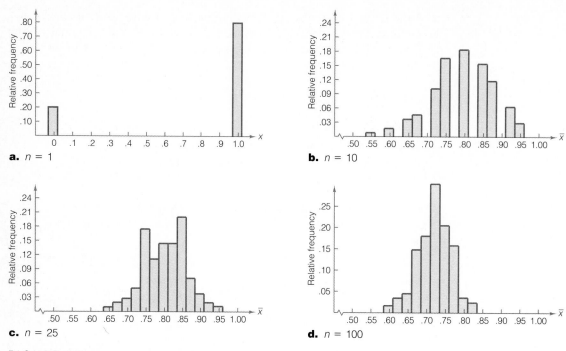

FIGURE 6.15
Sampling distributions for Bernoulli sample means, Example 6.9

tion improves as n increases. Note, too, that the values of $\bar{x}$ cluster more closely around the mean (.8) as n is increased. We make use of these properties to estimate the probability of success p for binomial random variables in Chapter 7.

EXAMPLE 6.10

Refer to Example 6.9. Suppose $n = 400$ microchips are to be sampled from a very large batch, of which a proportion p are good. The proportion of nondefective chips in the sample will be determined by assigning a 1 to each nondefective chip, a 0 to each defective chip, and calculating the sample mean of the 400 Bernoulli observations. Assuming that the approximate value of p is .8, what is the probability that $\bar{x}$ will fall within .03 of the exact value of p?

Solution

We first note (see Example 6.9) that the mean and standard deviation of $\bar{x}$ are

$$E(\bar{x}) = p$$

$$\sigma_{\bar{x}} = \sqrt{\frac{pq}{n}} = \sqrt{\frac{pq}{400}}$$

Using the approximate value of p to obtain an approximation for the standard deviation of the sampling distribution of $\bar{x}$, we find

$$\sigma_{\bar{x}} \approx \sqrt{\frac{(.8)(.2)}{400}} = .02$$

Next, the Central Limit Theorem implies that the distribution of $\bar{x}$ based on a sample of size 400 is approximately normal. The properties are summarized in

Figure 6.16. Note that the mean of the sampling distribution is p, so that the standard normal z value is

FIGURE 6.16

Sampling distribution of $\bar{x}$

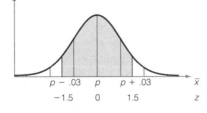

given by the formula

$$z = \frac{\bar{x} - \mu_{\bar{x}}}{\sigma_{\bar{x}}} = \frac{\bar{x} - p}{.02}$$

Even though the value of p is unknown, we can find the probability that $\bar{x}$ is within .03 of p by first calculating the z value corresponding to the $\bar{x}$ value that is .03 greater than p:

$$z = \frac{(p + .03) - p}{.02} = \frac{.03}{.02} = 1.5$$

Similarly, the value .03 less than p is 1.5 standard deviations *below* the mean, and has a z value of -1.5. Thus, the event that $\bar{x}$ falls within .03 of p is equivalent to the event that the standard normal random variable z is between -1.5 and 1.5. Using Table IV in Appendix A, we find the probability that $\bar{x}$ falls within 1.5 standard deviations of p is $2(.4332) = .8664$, the area shown shaded in Figure 6.16. The interpretation of this probability is that there is about an 87% chance that the proportion of nondefective chips in the sample of 400 will fall within .03 of the exact proportion of nondefective chips in the entire batch.

For most sampling distributions, the standard deviation of the distribution decreases as the sample size increases. We will use this result in Chapter 7 to help us determine the sample size needed to obtain a specified accuracy of estimation.

Summary

Many practical problems require that an inference be made about some population **parameter** (for example, the mean or the standard deviation). If we want to make this inference on the basis of sample information, we need to compute a **sample statistic** that contains information about the population parameter. The amount of such information contained in a sample statistic is reflected in its **sampling distribution**, the probability distribution of the sample statistic. The sampling distribution describes the behavior of the statistic in repeated sampling. In particular, we want a sample statistic that is an **unbiased** estimator of the population parameter and has a smaller variance than any other unbiased sample statistic.

When the population parameter of interest is the mean μ, the sample mean provides an unbiased estimator with a standard deviation of $\sigma/\sqrt{n}$. Moreover, the **Central Limit Theorem** assures us that the sampling distribution for the mean of a large sample is approximately normally distributed, no matter what the shape of the relative frequency distribution of the sampled population.

The amount of information in a sample that is relevant to some population parameter is related to the sample size. Another way of saying this is that as n gets larger, the standard deviation of most sample statistics gets smaller.

The sampling distributions for all the many statistics that can be computed from sample data could be discussed in detail, but this would delay discussion of the practical objective of this course—the role of statistical inference in decision-making. Consequently, we will comment further on the sampling distributions of statistics when we use them as estimators or decision-makers in the following chapters.

SUPPLEMENTARY EXERCISES 6.34–6.65

[*Note: Starred* (*) *exercises require the use of a computer.*]

LEARNING THE MECHANICS

6.34 Consider a sample statistic A. As with all sample statistics, A is computed by utilizing a specified function (formula) of the sample measurements. (For example, if A were the sample mean, the specified formula would be to sum the measurements and divide by the number of measurements.)
   a. Describe what we mean by the phrase "the sampling distribution of the sample statistic A."
   b. Suppose A is to be used to estimate a population parameter α. What is meant by the assertion that A is an unbiased estimator of α?
   c. Consider another sample statistic, B. Assume that B is also an unbiased estimator of the population parameter α. How can we use the sampling distributions of A and B to decide which is the better estimator of α?
   d. If the sample sizes on which A and B are based are large, can we apply the Central Limit Theorem and assert that the sampling distributions of A and B are approximately normal? Why or why not?

6.35 The standard deviation (or, as it is usually called, the standard error) of the sampling distribution for the sample mean, $\bar{x}$, is equal to the standard deviation of the population from which the sample was selected divided by the square root of the sample size. That is,

$$\sigma_{\bar{x}} = \frac{\sigma}{\sqrt{n}}$$

   a. As the sample size is increased, what happens to the standard error of $\bar{x}$? Why is this property considered important?
   b. Suppose that a sample statistic has a standard error that is not a function of the sample size. In other words, the standard error remains constant as n changes. What would this imply about the statistic as an estimator of a population parameter?

c. Suppose another unbiased estimator (call it A) of the population mean is a sample statistic with a standard error equal to

$$\sigma_A = \frac{\sigma}{\sqrt[3]{n}}$$

Which of the sample statistics, $\bar{x}$ or A, is preferable as an estimator of the population mean? Why?

d. Suppose that the population standard deviation σ is equal to 10 and that the sample size is 64. Calculate the standard errors of $\bar{x}$ and A. Assuming that the sampling distribution of A is approximately normal, interpret the standard errors. Why is the assumption of (approximate) normality unnecessary for the sampling distribution of $\bar{x}$?

6.36 A random sample of $n = 3$ observations is selected from a population that is described by the probability distribution given here.

x	0	2	3
$p(x)$	$\frac{1}{3}$	$\frac{1}{3}$	$\frac{1}{3}$

a. Calculate μ for this distribution.

b. List all the possible samples of $n = 3$ measurements from this population, and calculate the sample mean and median for each sample.

c. Find the sampling distributions of the sample mean and median.

d. Show that the mean is an unbiased estimator of μ and that the median is not.

6.37 A random sample of 40 observations is to be drawn from a large population of measurements. It is known that 30% of the measurements in the population are 1's, 20% are 2's, 20% are 3's, and 30% are 4's.

a. Give the mean and standard deviation of the (repeated) sampling distribution of $\bar{x}$, the sample mean of the 40 observations.

b. Describe the shape of the sampling distribution of $\bar{x}$. Does your answer depend on the sample size?

*6.38 Using a statistical software package, generate 100 random samples of size $n = 40$ from the population described in Exercise 6.37.

a. Compute $\bar{x}$ for each sample, and plot a frequency distribution for the 100 values of $\bar{x}$.

b. Compute the mean and standard deviation for the 100 values of $\bar{x}$.

c. How does the approximation for the sampling distribution of $\bar{x}$ developed in parts **a** and **b** compare with the approximation you obtained in Exercise 6.37? Explain why differences may exist.

6.39 A random sample of $n = 75$ observations is selected from a population with $\mu = 120$ and $\sigma^2 = 410$.

a. Find $\mu_{\bar{x}}$ and $\sigma_{\bar{x}}$.

b. What is the shape of the sampling distribution of $\bar{x}$? Does your answer depend on the shape of the population distribution?

c. Find the approximate value of $P(\bar{x} \le 118)$.

d. Find the approximate value of $P(115 \le \bar{x} \le 123)$.

e. Find the approximate value of $P(\bar{x} \le 124.6)$.

f. Find the approximate value of $P(\bar{x} \ge 122.7)$.

6.40 A random sample of $n = 52$ observations is selected from a population with $\mu = 19.6$ and $\sigma = 2.5$. Approximate each of the following probabilities.

a. $P(\bar{x} \le 19.6)$ b. $P(\bar{x} \le 19)$ c. $P(\bar{x} \ge 20.1)$ d. $P(19.2 \le \bar{x} \le 20.6)$

6.41   Suppose two fair dice are tossed. Let $\bar{x}$ be the mean of the two numbers that appear on the up faces of the dice.
    a. Give the sampling distribution of $\bar{x}$.
    b. Find the approximate value of $P(\bar{x} \leq 5)$.

6.42   Refer to Exercise 6.41. Simulate the sampling distribution of $\bar{x}$ by tossing a pair of balanced dice 100 times. Calculate $\bar{x}$ for each toss of the dice, and construct a relative frequency histogram for these 100 values of $\bar{x}$. Compare this graph with the exact sampling distribution for $\bar{x}$ of Exercise 6.41.

6.43   Suppose x equals the number of heads observed when a single coin is tossed; that is, $x = 0$ or $x = 1$. The population corresponding to x is the set of 0's and 1's generated when the coin is tossed repeatedly a large number of times. Suppose we select $n = 2$ observations from this population. (That is, we toss the coin twice and observe two values of x.)
    a. List the three different samples (combinations of 0's and 1's) that could be obtained.
    b. Calculate the value of $\bar{x}$ for each of the samples.
    c. List the values that $\bar{x}$ can assume, and find the probabilities of observing these values.
    d. Construct a graph of the sampling distribution of $\bar{x}$.

*6.44   Use a statistical software package to generate 100 random samples of size $n = 2$ from a population characterized by a uniform probability distribution (Section 5.2) with $c = 0$ and $d = 10$. Compute $\bar{x}$ for each sample, and plot a frequency distribution for the 100 $\bar{x}$ values. Repeat this process for $n = 5, 10, 30$, and 50. Explain how your plots illustrate the Central Limit Theorem.

*6.45   Use a statistical software package to generate 100 random samples of size $n = 2$ from a population characterized by a normal probability distribution with mean 100 and standard deviation 10. Compute $\bar{x}$ for each sample and plot a frequency distribution for the 100 values of $\bar{x}$. Repeat this process for $n = 5, 10, 30$, and 50. How does the fact that the sampled population is normal affect the sampling distribution of $\bar{x}$?

6.46   Suppose x equals the number of heads observed when a single coin is tossed; that is, $x = 0$ or $x = 1$. The population corresponding to x is the set of 0's and 1's generated when the coin is tossed repeatedly a large number of times. Suppose we select $n = 3$ observations from this population. (That is, we toss the coin three times and observe three values of x.)
    a. List the four different samples (combinations of 0's and 1's) that could be obtained.
    b. Calculate the value of $\bar{x}$ for each of the samples.
    c. List the values that $\bar{x}$ can assume, and find the probabilities of observing these values.
    d. Construct a graph of the sampling distribution of $\bar{x}$.

6.47   A random sample of size n is to be drawn from a large population with mean 100 and standard deviation 10, and the sample mean $\bar{x}$ is to be calculated. To see the effect of different sample sizes on the standard deviation of the sampling distribution of $\bar{x}$, plot $\sigma/\sqrt{n}$ against n for $n = 1, 5, 10, 20, 30, 40$, and 50.

6.48   A random sample of size $n = 30$ is to be drawn from a population with $\mu = 500$ and $\sigma = 200$.
    a. What is the standard deviation of the sampling distribution of $\bar{x}$?
    b. In order to reduce the standard deviation of $\bar{x}$ to 50% of the value in part **a**, how much larger would n need to be?
    c. In order to reduce $\sigma_{\bar{x}}$ to 75% of the value in part **a**, how much larger would n need to be?

APPLYING THE CONCEPTS

6.49   Over the last month a large supermarket chain has received many consumer complaints about the quantity of chips in 9-ounce bags of a particular brand of potato chips. Suspecting that the complaints are merely the result of the potato chips settling to the bottom of the bags during shipping but wanting to be able to assure its customers they are getting their money's worth, the chain decides to examine the next shipment of chips received by their largest store. Thirty-five 9-ounce bags are randomly selected from the shipment, their contents weighed, and the sample mean weight computed. The chain's management decides that if the sample mean is less than 8.95 ounces, the shipment will be refused and a complaint registered with the potato chip company. Assume that the distribution of weights of the contents of all the potato chip bags in question has a mean of 8.9 ounces and a standard deviation of .13 ounce.

   a. What is the approximate probability that the supermarket chain's investigation will lead to refusal of the shipment?

   b. What assumptions did you have to make in order to answer part **a**? Justify the assumptions.

6.50   The distribution of the number of barrels of oil produced by a certain oil well each day for the past 3 years has a mean of 400 and a standard deviation of 75.

   a. Describe the sampling distribution of the mean number of barrels produced per day for samples of 40 production days drawn from the past 3 years.

   b. What is the approximate probability that the sample mean will be greater than 425?

   c. What is the approximate probability that the sample mean will be less than 400?

6.51   Water availability is of prime importance in the life cycle of most reptiles. To determine the rate of evaporative water loss of a certain species of lizard at a particular desert site, 34 such lizards were randomly collected, weighed, and placed under the appropriate experimental conditions. After 24 hours, each lizard was removed, reweighed, and its total water loss was calculated as the difference between initial body weight and body weight after treatment. Previous studies have shown that the distribution of water loss for the lizards has a mean of 3.1 grams and a standard deviation of .8 gram.

   a. Find the approximate probability that the 34 lizards have a mean water loss of less than 3.0 grams.

   b. Between 3.15 and 3.25 grams.

6.52   Electric power plants that use water for cooling their condensers sometimes discharge heated water into rivers, lakes, or oceans. It is known that water heated above certain temperatures has a detrimental effect on the plant and animal life in the water. Suppose it is known that the increased temperature of the heated water discharged by a certain power plant on any given day has a distribution with a mean of 5°C and a standard deviation of .5°C.

   a. For 50 randomly selected days, what is the approximate probability that the average increase in temperature of the discharged water is greater than 5.0°C?

   b. Less than 4.8°C?

   c. What assumptions must be made for you to answer the questions?

6.53   The primary responsibility of government tax assessors is to estimate the market values of all properties in their respective jurisdictions. These estimates form the basis upon which property tax bills are determined. In order to estimate the mean market value for single-family homes in a particular county in 1987, a county tax assessor uses the mean sale price for all single-family homes that sold during 1987. Such a sample is typically treated by tax assessors as if it were a random sample from the population of properties in question (*Standard on Assessment-Ratio Studies*, 1980). Suppose 400 properties sold during 1987 and that the standard deviation of the market values for the population of properties is about $50,000.

   a. What is the population in question?

   b. What is the probability that the assessor's sample mean will overestimate the actual mean market value of the population?

   c. What is the probability that the assessor's sample mean will fall within $4,000 of the actual mean market value for the population?

6.54   **Random-number generators** have many uses in statistics.* One type is designed to produce a sequence of numbers between 0 and 1. A number x can assume any value in the interval from 0 to 1 with

---

*Random-number generators are used to produce random numbers such as those that appear in Table 1 of Appendix A.

equal probability, and any value of x is independent of the values of previous numbers that appear in the sequence. Furthermore, the probability distribution of x has a mean $\mu = .5$ and a standard deviation $\sigma = .29$. Let y be the average of n such random numbers.

a. Graph the probability distribution for x.

b. Give the mean and standard deviation of the sampling distribution of y.

c. What is the approximate form of the sampling distribution of y when n is large?

d. Sketch the sampling distribution of y and compare it with your graph from part **a**.

6.55   To determine whether a metal lathe that produces machine bearings is properly adjusted, a random sample of 25 bearings is collected and the diameter of each is measured.

a. If the standard deviation of the diameters of the bearings measured over a long period of time is .001 inch, what is the approximate probability that the mean diameter $\bar{x}$ of the sample of 25 bearings will lie within .0001 inch of the population mean diameter of the bearings?

b. If the population of diameters has an extremely skewed distribution, how will your approximation in part **a** be affected?

6.56   Refer to Exercise 6.55. The mean diameter of the bearings produced by the machine is supposed to be .5 inch. The company decides to use the sample mean (from Exercise 6.55) to decide whether the process is in control, i.e., whether it is producing bearings with a mean diameter of .5 inch. The machine will be considered out of control if the mean of the sample of $n = 25$ diameters is less than .4994 inch or larger than .5006 inch. If the true mean diameter of the bearings produced by the machine is .501 inch, what is the approximate probability that the test will imply that the process is out of control?

6.57   Suppose you wish to purchase a case of expensive wine. You plan to open two bottles for immediate use and you will keep the remaining bottles if the two are acceptable. Suppose there are ten bottles in the case, and, unknown to you, the condition of the wine in the bottles is as shown ($1 =$ good, $0 =$ bad):

Bottle	1	2	3	4	5	6	7	8	9	10
Condition	1	0	0	1	1	1	1	1	0	1

Since you are interested only in the ten bottles in the case, the collection of ten 0 or 1 responses is the population of interest to you.

a. If you randomly sample two bottles from the case, how many different samples (different pairs of bottles) could you select? List them.

b. Suppose you are going to accept the case only if both bottles in the sample are good. Identify all samples containing two good bottles. What is the probability that you will accept the case? [*Hint:*   See the definition of a random sample in Section 3.7.]

c. Let x equal the number of good bottles in the sample of $n = 2$. Construct the sampling distribution of x.

6.58   Much emphasis has recently been placed on preventative health behavior. In one study at a health fair (Price, O'Connell, and Kukulka, 1985), 100 attendees were administered a questionnaire on preventative health behaviors. Assume that the population mean and standard deviation of the questionnaire scores in the area of preventing heart disease are 38 and 5, respectively, when administered to the general public.

a. Describe the sampling distribution of the sample mean questionnaire score of 100 health fair attendees, assuming that they are a random sample of the general public.

    b. What is the probability that the sample mean score of the 100 health fair attendees will exceed 39.1, if they represent a random sample of the general public?

    c. The scores of the 100 health fair attendees on the preventative behavior questionnaire had a mean of 39.1. Considering your answer to part **b**, do you think the attendees represent a random sample of the general public with regard to the heart disease questionnaire?

6.59 Refer to Exercise 6.58. The 100 health fair attendees were also administered a questionnaire on malignancies. Assume that the mean and standard deviation of scores for the general public are 42 and 6.5, respectively.

    a. Describe the sampling distribution of the sample mean score of 100 attendees, assuming that they are a random sample of the general public.

    b. What is the probability that the sample mean questionnaire score of the 100 attendees will exceed 43.6, if they represent a random sample of the general public?

    c. The scores of the 100 health fair attendees on the preventative behavior questionnaire related to malignancies had a mean of 43.6. Considering your answer to part **b**, do you think the attendees represent a random sample of the general public with regard to the malignancy questionnaire?

6.60 The distribution of the number of characters printed per second by a particular kind of line printer at a computer terminal has the following parameters: $\mu = 45$ characters per second, $\sigma = 2$ characters per second.

    a. Describe the sampling distribution of the mean number of characters printed per second for random samples of 1-minute intervals.

    b. Find the approximate probability that the sample mean for a random sample of 60 seconds will be between 44.5 and 45.3 characters per second.

    c. Find the approximate probability that the sample mean will be less than 44 characters per second.

6.61 This past year, an elementary school began using a new method to teach arithmetic to first graders. A standardized test, administered at the end of the year, was used to measure the effectiveness of the new method. The distribution of past scores on the standardized test produced a mean of 75 and a standard deviation of 10.

    a. If the new method is no different from the old method, what is the approximate probability that the mean score $\bar{x}$ of a random sample of 36 students will be greater than 79?

    b. What assumptions must be satisfied to make your answer valid?

6.62 In any production process, some variation in the quality of the product is unavoidable. Variation in product quality can be divided into two categories: variation due to *special causes* and variation due to *common causes*. The former includes variation due to a specific worker or group of workers, a specific machine, or a specific local condition. The latter includes variation due to faults of the overall production system, such as poor working conditions for workers, use of raw materials of inferior quality, poor design of the product, etc. The discovery and removal of special causes is generally the responsibility of a person directly involved with the production operation in question, and common causes are the responsibility of management.

    A production process in which all special causes of variation have been eliminated is said to be *stable* or *in statistical control*. The variation that remains is simply random variation. If the magnitude of the random variation is unacceptable to management, it can be reduced through the elimination of common causes (Deming, 1986).

    It is common practice to monitor the variation in the quality characteristic of a product over time by plotting the quality characteristic on a *control chart*. For example, the amount of alkali in soap might be monitored each hour by randomly selecting from the production process and measuring the quantity of alkali in $n = 5$ test specimens of soap. The mean, $\bar{x}$, of the sample alkaline measurements would be plotted against time, as shown in the figure on page 286. If the process is in statistical control, $\bar{x}$ should assume a distribution with a mean equal to the process mean, μ, with standard deviation equal to the process standard deviation divided by the square root of the sample size, $\sigma_{\bar{x}} = \sigma/\sqrt{n}$. The control chart includes a horizontal line to locate the process mean and two lines, called *control limits*, located $3\sigma_{\bar{x}}$ above and below μ. If $\bar{x}$ falls within the control limits, the process is deemed to be in control. If $\bar{x}$ is outside the limits, there is strong evidence that special causes of variation are present, and the process is deemed to be out of control.

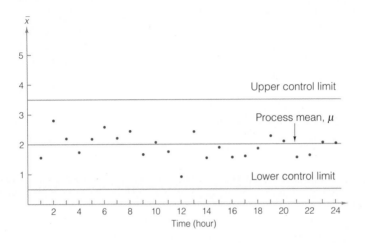

Suppose experience has shown that the percentage of alkali in a test specimen of soap follows approximately a normal distribution with $\mu = 2\%$ and $\sigma = 1\%$.

a. If $n = 5$, how far from μ should the upper and lower control limits be located?

b. If the process is in control, what is the probability that $\bar{x}$ will fall outside the control limits?

c. If the process mean is $\mu = 3\%$, and therefore out of control, what is the probability that the sample will lead to the (correct) conclusion that the process is out of control?

*6.63   [*Note:*   This exercise refers to an optional section in Chapter 4.] A building contractor has decided to purchase a load of factory-reject aluminum siding as long as the average number of flaws per piece of siding in a sample of size 35 from the factory's reject pile is 2.1 or less. If it is known that the number of flaws per piece of siding in the factory's reject pile has a Poisson probability distribution with a mean of 2.5, find the approximate probability that the contractor will not purchase a load of siding. [*Hint:*   If x is a Poisson random variable with mean λ, then σ_x^2 also equals λ.]

6.64   The distribution of the number of loaves of bread sold per day by a large grocery store over the past 5 years has a mean of 250 and a standard deviation of 45.

a. Describe the sampling distribution of the total number of loaves of bread sold per 30 randomly selected shopping days. [*Hint:*   See the footnote in Section 6.3 that gives the application of the Central Limit Theorem to the sum of the measurements in a sample.]

b. Give the approximate probability that the total number of loaves sold per 30 shopping days is between 7,000 and 8,000.

c. Give the approximate probability that the total is greater than 8,100 loaves.

*6.65   [*Note:*   This exercise refers to an optional section in Chapter 5.] The exponential random variable (described in Section 5.5) is a continuous random variable with probability distribution $f(x) = (1/\theta)e^{-x/\theta}$. The mean and standard deviation of the exponential distribution are both equal to θ. Suppose n observations are randomly selected from such a distribution.

a. If n is large, what is the shape of the sampling distribution of the sample mean?

b. What are the mean and standard deviation of this sampling distribution?

ON YOUR OWN...

To understand the Central Limit Theorem and sampling distribution, consider the following experiment: Toss four identical coins, and record the number of heads observed. Then repeat this experiment four more times, so that you end up with a total of five observations for the random variable x, the number of heads when four coins are tossed.

Now derive and graph the probability distribution for x, assuming the coins are balanced. Note that the mean of this distribution is $\mu = 2$ and the standard deviation is $\sigma = 1$. This probability distribution represents the one from which you are drawing a random sample of five measurements.

Next, calculate the mean $\bar{x}$ of the five measurements—i.e., calculate the mean number of heads you observed in five repetitions of the experiment. Although you have repeated the basic experiment five times, you have only one observed value of $\bar{x}$.

To derive the probability distribution or sampling distribution of $\bar{x}$ empirically, you have to repeat the entire process (of tossing four coins five times) many times. Do it 100 times.

The approximate sampling distribution of $\bar{x}$ can be derived theoretically by making use of the Central Limit Theorem. We expect at least an approximate normal probability distribution with a mean $\mu = 2$ and a standard deviation

$$\sigma_{\bar{x}} = \frac{\sigma}{\sqrt{n}} = \frac{1}{\sqrt{5}} = .45$$

Count the number of your 100 $\bar{x}$'s that fall in each of the intervals in the figure below. Use the normal probability distribution with $\mu = 2$ and $\sigma_{\bar{x}} = .45$ to calculate the expected number of the 100 $\bar{x}$'s in each of the intervals. How closely does the theory describe your experimental results?

Interval	Interval	Interval	Interval	Interval	Interval
←—1—→	←—2—→	←—3—→	←—4—→	←—5—→	←—6—→

1.10	1.55	2	2.45	2.90
$\mu - 2\sigma_{\bar{x}}$	$\mu - \sigma_{\bar{x}}$	μ	$\mu + \sigma_{\bar{x}}$	$\mu + 2\sigma_{\bar{x}}$

USING THE COMPUTER...

Calculate the mean and standard deviation for the 1,000 zip codes' median household incomes. We will treat these quantities as the population mean μ and the population standard deviation σ for this exercise.

a. Draw 100 random samples of $n = 20$ observations from the 1,000 zip codes' median incomes. Select the samples with replacement—i.e., replace each measurement before selecting the next.* Calculate the 100 sample means. Gener-

ate a stem and leaf display or a relative frequency histogram for the 100 means. Then count the number of the 100 sample means that fall in the intervals $\mu \pm \sigma/\sqrt{n}$, $\mu \pm 2\sigma/\sqrt{n}$, and $\mu \pm 3\sigma/\sqrt{n}$. How do the graphic description and the percentage of means falling in the intervals agree with a normal distribution having mean μ and standard deviation $\sigma/\sqrt{n}$?

b. Repeat part a using a sample size of $n = 50$. Is the sampling distribution of the sample means closer to normal for the larger sample size?

*In this and future exercises we will specify sampling with replacement to simulate the sampling from very large or infinite populations.

References

Deming, W. E. *Out of the Crisis*. Cambridge, Mass.: M.I.T. Center for Advanced Study of Engineering, 1986.

Hogg, R. V., and Craig, A. T. *Introduction to Mathematical Statistics*, 4th ed., New York: Macmillan, 1978, Chapter 4.

Juran, J. M., and Gryna, F. M., Jr. *Quality Planning and Analysis*. New York: McGraw-Hill Book Company, 1980, Chapter 9.

Lindgren, B. W. *Statistical Theory*, 3d ed. New York: Macmillan, 1976, Chapter 2.

Mendenhall, W., Scheaffer, R. L., and Wackerly, D. *Mathematical Statistics with Applications*, 2d ed. North Scituate, Mass.: Duxbury, 1980, Chapter 7.

Moffitt, T. E., Gabrielli, W. F., Mednick, S. A., and Schulsinger, F. "Socioeconomic status, IQ, and delinquency," *Journal of Abnormal Psychology*, 90, 1981, 152–156.

Price, J. H., O'Connell, J., and Kukulka, G. "Preventative health behaviors related to the ten leading causes of mortality of health-fair attenders and nonattenders," *Psychological Reports*, 56, 1985, 131–135.

Standard on Assessment-Ratio Studies. Chicago: International Association of Assessing Officers, 1980.

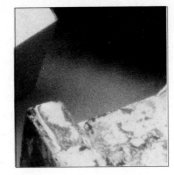

Inferences Based on a Single Sample: Estimation

WHERE WE'VE BEEN...

In the preceding chapters we learned that populations are characterized by numerical descriptive measures (called *parameters*) and that decisions about their values are based on sample statistics computed from sample data. Since statistics vary in a random manner from sample to sample, inferences based on them will be subject to uncertainty. This property is reflected in the sampling (probability) distribution of a statistic.

WHERE WE'RE GOING...

This chapter puts all the preceding material into practice; that is, we estimate population means and proportions based on a single sample selected from the population of interest. Most importantly, we use the sampling distribution of a sample statistic to assess the reliability of an estimate.

The estimation of the mean gas mileage for a new car model, the estimation of the expected life of a computer monitor, and the estimation of the mean yearly sales for companies in the steel industry are problems with a common element. In each case, we are interested in estimating the mean of a population of measurements. This important problem constitutes the primary topic of this chapter.

You will see that different techniques are used for estimating a mean, depending on whether a sample contains a large or small number of measurements. Regardless, our objectives remain the same: We want to use the sample information to estimate the mean and to assess the reliability of the estimate.

In Sections 7.1 and 7.2 we consider a method of estimating a population mean using a large sample and develop a formula to determine just how large the sample must be to achieve a specified degree of reliability. In Section 7.3 we show how a mean can be estimated when only a small sample is available. Finally, estimation of binomial probabilities and the determination of sample sizes necessary to make reliable estimates are covered in Sections 7.4 and 7.5, respectively.

7.1   Large-Sample Estimation of a Population Mean

We illustrate the **large-sample method** of estimating a population mean with an example. Suppose a large hospital wants to estimate the average length of time patients remain in the hospital. To accomplish this objective, the hospital administrators plan to sample 100 of all previous patients' records and to use the sample mean, $\bar{x}$, of the lengths of stay to estimate the mean stay, μ, of *all* patients' visits. The sample mean $\bar{x}$ represents a *point estimator* of the population mean μ (Definition 6.4). How can we assess the accuracy of this point estimator?

According to the Central Limit Theorem, the sampling distribution of the sample mean is approximately normal for large samples, as shown in Figure 7.1. Let us calculate the interval

$$\bar{x} \pm 2\sigma_{\bar{x}} = \bar{x} \pm \frac{2\sigma}{\sqrt{n}}$$

That is, we will form an interval 4 standard deviations wide—from 2 standard deviations below the sample mean to 2 standard deviations above the mean. What are the chances (answer before we have drawn a sample) that this interval will enclose μ, the population mean?

To answer this question, refer to Figure 7.1. If the 100 measurements yield a value of $\bar{x}$ that falls between the two lines on either side of μ—i.e., within 2 standard deviations of μ—then the interval $\bar{x} \pm 2\sigma_{\bar{x}}$ will contain μ; if $\bar{x}$ falls outside these boundaries, the interval $\bar{x} \pm 2\sigma_{\bar{x}}$ will not contain μ. Since the area under the normal curve (the sampling distribution of $\bar{x}$) between these boundaries is about .95 (more precisely, from Table IV in Appendix A the area is .9544), we

FIGURE 7.1

Sampling distribution of $\bar{x}$

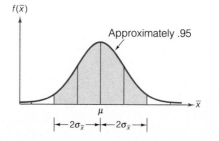

know that the interval $\bar{x} \pm 2\sigma_{\bar{x}}$ will contain μ with a probability approximately equal to .95.

To illustrate, suppose the sum and the sum of squared deviations for the sample of 100 lengths of time spent in the hospital are

$$\sum x = 465 \text{ days} \quad \text{and} \quad \sum (x - \bar{x})^2 = 2,387$$

Then

$$\bar{x} = \frac{\sum x}{n} = \frac{465}{100} = 4.65$$

$$s^2 = \frac{\sum (x - \bar{x})^2}{n - 1} = \frac{2,387}{99} = 24.11 \quad \text{and} \quad s = 4.9$$

Then, we consider the interval

$$\bar{x} \pm 2\sigma_{\bar{x}} = 4.65 \pm 2\frac{\sigma}{\sqrt{100}}$$

But now we face a problem. You can see that without knowing the standard deviation σ of the original population—i.e., the standard deviation of the lengths of stay of *all* patients—we cannot calculate this interval. However, since we have a large sample ($n = 100$ measurements), we can approximate the interval by using the sample standard deviation s to approximate σ. Thus,

$$\bar{x} \pm 2\frac{\sigma}{\sqrt{100}} \approx \bar{x} \pm 2\frac{s}{\sqrt{100}} = 4.65 \pm 2\left(\frac{4.9}{10}\right) = 4.56 \pm .98$$

That is, we estimate the mean length of stay in the hospital for all patients to fall in the interval 3.67 to 5.63 days.

Can we be sure that μ, the true mean, is in the interval 3.67 to 5.63? We cannot be certain, but we can be reasonably confident that it is. This confidence is derived from the knowledge that if we were to draw repeated random samples of 100 measurements from this population and form the interval $\bar{x} \pm 2\sigma_{\bar{x}}$ each time, approximately 95% of the intervals would contain μ. We have no way of knowing (without looking at all the patients' records) whether our sample interval is one of the 95% that contain μ or one of the 5% that do not, but the odds certainly favor its containing μ. Consequently, the interval 3.67 to 5.63 provides an estimate of the mean length of patient stay in the hospital. The formula that tells us how to calculate an interval estimate based on sample data is called an *interval estimator*. The probability, .95, that measures the confidence we can place in the interval estimate is called a *confidence coefficient*. The percentage, 95%, is called the *confidence level* for the interval estimate. It is not usually possible to assess precisely the reliability of point estimators because they are single points rather than intervals. Since we prefer to use estimators for which a measure of reliability can be calculated, interval estimators will generally be used.

Definition 7.1

An **interval estimator** is a formula that tells us how to use sample data to calculate an interval that estimates a population parameter.

> **Definition 7.2**
>
> The **confidence coefficient** is the probability that an interval estimator encloses the population parameter—i.e., the relative frequency with which the interval estimator encloses the population parameter when the estimator is used repeatedly a very large number of times.
>
> The **confidence level** is the confidence coefficient expressed as a percentage.

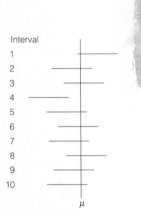

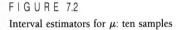

FIGURE 7.2

Interval estimators for μ: ten samples

Now we have seen how an interval can be used to estimate a population mean. When we use an interval estimator, we can usually calculate the probability that the estimation *process* will result in an interval that contains the true value of the population mean. That is, the probability that the interval contains the parameter in repeated usage is usually known. Figure 7.2 shows what happens when ten different samples are drawn from a population, and a confidence interval for μ is calculated from each. The location of μ is indicated by the vertical line in the figure. Ten confidence intervals, each based on one of ten samples, are shown as horizontal line segments. Note that the confidence intervals move from sample to sample—sometimes containing μ and other times missing μ. If our confidence level is 95%, then in the long run, 95% of our sample confidence intervals will contain μ.

Suppose you wish to choose a confidence coefficient other than .95. Notice in Figure 7.1 that the confidence coefficient .95 is equal to the total area under the sampling distribution, less .05 of the area, which is divided equally between the two tails. Using this idea, we can construct a confidence interval with any desired confidence coefficient by increasing or decreasing the area (call it α) assigned to the tails of the sampling distribution (see Figure 7.3). For example,

FIGURE 7.3

Locating $z_{\alpha/2}$ on the standard normal curve

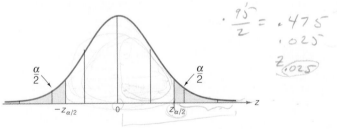

if we place area $\alpha/2$ in each tail and if $z_{\alpha/2}$ is the z value such that the area $\alpha/2$ will lie to its right, then the confidence interval with confidence coefficient $(1 - \alpha)$ is

$$\bar{x} \pm z_{\alpha/2}\sigma_{\bar{x}}$$

To illustrate, for a confidence coefficient of .90 we have $(1 - \alpha) = .90$, $\alpha = .10$, and $\alpha/2 = .05$; $z_{.05}$ is the z value that locates area .05 in the upper tail of the sampling distribution. Recall that Table IV in Appendix A (and inside the front cover of the book) gives the areas between the mean and a specified z value. Since the total area to the right of the mean is .5, we find that $z_{.05}$ will be the z value corresponding to an area of $.5 - .05 = .45$ to the right of the mean (see Figure 7.4). This z value is $z_{.05} = 1.645$. Confidence coefficients used in practice (in published articles) range from .90 to .99. The most commonly used confidence coefficients with corresponding values of α and $z_{\alpha/2}$ are shown in Table 7.1.

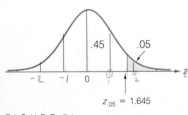

FIGURE 7.4

The z value ($z_{.05}$) corresponding to an area equal to .05 in the upper tail of the z-distribution

TABLE 7.1   **Commonly Used Values of $z_{\alpha/2}$**

CONFIDENCE LEVEL $100(1 - \alpha)$	α	$\alpha/2$	$z_{\alpha/2}$
90%	.10	.05	1.645
95%	.05	.025	1.96
99%	.01	.005	2.58

Large-Sample $100(1 - \alpha)$% Confidence Interval for μ

$$\bar{x} \pm z_{\alpha/2}\sigma_{\bar{x}} = \bar{x} \pm z_{\alpha/2}\frac{\sigma}{\sqrt{n}}$$

where $z_{\alpha/2}$ is the z value with an area $\alpha/2$ to its right (see Figure 7.3) and $\sigma_{\bar{x}} = \sigma/\sqrt{n}$. The parameter σ is the standard deviation of the sampled population and n is the sample size.

When n is equal to 30 or more, the confidence interval is approximately equal to

$$\bar{x} \pm z_{\alpha/2}\left(\frac{s}{\sqrt{n}}\right)$$

where s is the sample standard deviation.

EXAMPLE 7.1

Unoccupied seats on flights cause airlines to lose revenue. Suppose a large airline wants to estimate its average number of unoccupied seats per flight over the past year. To accomplish this, the records of 225 flights are randomly selected, and the number of unoccupied seats is noted for each of the sampled flights. The sample mean and standard deviation are

$$\bar{x} = 11.6 \text{ seats} \qquad s = 4.1 \text{ seats}$$

Estimate μ, the mean number of unoccupied seats per flight during the past year, using a 90% confidence interval.

Solution

The general form of the 90% confidence interval for a population mean is

$$\bar{x} \pm z_{\alpha/2}\sigma_{\bar{x}} = \bar{x} \pm z_{.05}\sigma_{\bar{x}} = \bar{x} \pm 1.645\left(\frac{\sigma}{\sqrt{n}}\right)$$

For the 225 records sampled, we have

$$11.6 \pm 1.645\left(\frac{\sigma}{\sqrt{225}}\right)$$

Since we do not know the value of σ (the standard deviation of the number of unoccupied seats per flight for all flights of the year), we use our best approximation—the sample standard deviation s. Then the 90% confidence interval is, approximately,

$$11.6 \pm 1.645\left(\frac{4.1}{\sqrt{225}}\right) = 11.6 \pm .45$$

or from 11.15 to 12.05. That is, at the 90% confidence level, we estimate the mean number of unoccupied seats per flight to be between 11.15 and 12.05 during the sampled year. We stress that the confidence level refers to the procedure used. If we were to apply this procedure repeatedly to different samples, approximately 90% of the intervals would contain μ. We do not know whether this particular interval (11.15, 12.05) is one of the 90% that contain μ or one of the 10% that do not.

The interpretation of confidence intervals for a population mean is summarized in the accompanying box.

Interpretation of a Confidence Interval for a Population Mean

When we form a $100(1 - \alpha)\%$ confidence interval for μ, we usually express our confidence in the interval with a statement such as, "We can be $100(1 - \alpha)\%$ confident that μ lies between the lower and upper bounds of the confidence interval," where for a particular application, we substitute the appropriate numerical values for the confidence, and the lower and upper bounds. *The statement reflects our confidence in the estimation process rather than in the particular interval that is calculated from the sample data.* We know that repeated application of the same procedure will result in different lower and upper bounds on the interval. Furthermore, we know that $100(1 - \alpha)\%$ of the resulting intervals will contain μ. There is (usually) no way to determine whether any particular interval is one of those that contain μ, or one that does not. However, unlike point estimators, confidence intervals have some measure of reliability, the confidence coefficient, associated with them. For that reason they are generally preferred to point estimators.

CASE STUDY 7.1

DANCING TO THE CUSTOMER'S TUNE: THE NEED TO ASSESS CUSTOMER PREFERENCES

The following quotations have been extracted from the December 13, 1976, issue of *Business Week*:*

"We're dancing to the tune of the customer as never before," says J. Janvier Wetzel, vice-president for sales promotion at Los Angeles-based Broadway Department Stores. "With population growth down to a trickle compared with its previous level, we're no longer spoiled with instant success every time we open a new store. Traditional department stores are locked in the biggest competitive battle in their history."

The nation's retailers are becoming uncomfortably aware that today's operating environment is vastly different from that of the 1960s. Population growth is slowing, a growing singles market is emerging, family formations are coming at later ages, and more women are embarking on careers. Of the 71

---

*Reprinted by special permission. All rights reserved.

million households in the U.S. today, the dominant consumer buying segment is families headed by persons over 45. But by 1980 this group will have lost its majority status to the 25 to 40-year-old group. Merchants must now reposition their stores to attract these new customers.

To do so retailers are using market research to ferret out new purchasing attitudes and lifestyles and then translating this into customer buying segments. . . . Department stores are taking a hard look at some of the basics of their business by . . . spending heavily for far more elaborate market research. Data on demographics, psychographics (measurement of attitudes), and lifestyle are being fed into retailers' computers so they can make marketing decisions based on actual spending patterns and estimate their inventory needs with less risk.

In order to stock its various departments with the type and style of goods that appeal to its potential group of customers, a downtown department store should be interested in estimating the average age of downtown shoppers, not shoppers in general. Suppose a downtown department store questions 49 downtown shoppers concerning their age (the offer of a small gift certificate may help persuade shoppers to respond to such questions). The sample mean and standard deviation are found to be 40.1 years and 8.6 years, respectively. The store could then estimate the mean age μ of all downtown shoppers with a 95% confidence interval as follows:

$$\bar{x} \pm 1.96 \left(\frac{s}{\sqrt{n}}\right) = 40.1 \pm 1.96 \left(\frac{8.6}{\sqrt{49}}\right) = 40.1 \pm 2.4$$

Thus, the department store should gear its sales to consumers with average age between 37.7 and 42.5.

EXERCISES 7.1–7.19

LEARNING THE MECHANICS

7.1 Find $z_{\alpha/2}$ for each of the following:
a. $\alpha = .10$ b. $\alpha = .01$ c. $\alpha = .05$ d. $\alpha = .20$

What is the confidence level of each of the following confidence intervals for μ?

7.2 a. $\bar{x} \pm 1.96\left(\frac{\sigma}{\sqrt{n}}\right)$ b. $\bar{x} \pm 1.645\left(\frac{\sigma}{\sqrt{n}}\right)$ c. $\bar{x} \pm 2.575\left(\frac{\sigma}{\sqrt{n}}\right)$

d. $\bar{x} \pm 1.282\left(\frac{\sigma}{\sqrt{n}}\right)$ e. $\bar{x} \pm .99\left(\frac{\sigma}{\sqrt{n}}\right)$

7.3 A random sample of 64 observations from a population produced the following summary statistics:

$$\sum x = 500 \qquad \sum (x_i - \bar{x})^2 = 3{,}566$$

a. Find a 95% confidence interval for μ.
b. Interpret the confidence interval you found in part a.

7.4 A random sample of 60 observations produced a mean $\bar{x} = 30.4$ and a standard deviation $s = 1.6$.
a. Find a 95% confidence interval for the population mean μ.
b. Find a 90% confidence interval for μ.
c. Find a 99% confidence interval for μ.

7.5   A random sample of 100 observations from a normally distributed population possesses a mean equal to 76.9 and a standard deviation equal to 4.8.
   a. Find a 95% confidence interval for μ.
   b. What is meant when you say that a confidence coefficient is .95?
   c. Find a 99% confidence interval for μ.
   d. What happens to the width of a confidence interval as the value of the confidence coefficient is increased while the sample size is held fixed?
   e. Would your confidence intervals of parts **a** and **c** be valid if the distribution of the original population was not normal? Explain.

7.6   A random sample of n measurements was selected from a population with unknown mean μ and standard deviation σ. Calculate a 95% confidence interval for μ for each of the following situations.
   a. $n = 50, \quad \bar{x} = 28, \quad s^2 = 12$ b. $n = 200, \quad \bar{x} = 102, \quad s^2 = 12$
   c. $n = 100, \quad \bar{x} = 17, \quad s = .1$ d. $n = 100, \quad \bar{x} = 1.02, \quad s = 1.10$
   e. Is the assumption that the underlying population of measurements is normally distributed necessary to assure the validity of the confidence intervals in parts **a**–**d**? Explain.

7.7   Explain the difference between an interval estimator and a point estimator for μ.

7.8   Explain what is meant by the statement, "We are 95% confident that an interval estimate contains μ."

7.9   Will a large-sample confidence interval be valid if the population from which the sample is taken is not normally distributed? Explain.

7.10   The mean and standard deviation of a random sample of n measurements are equal to 27.4 and 2.5, respectively.
   a. Find a 95% confidence interval for μ if $n = 100$.
   b. Find a 95% confidence interval for μ if $n = 400$.
   c. Find the widths of the confidence intervals found in parts **a** and **b**. What is the effect on the width of a confidence interval of quadrupling the sample size while holding the confidence coefficient fixed?

APPLYING THE CONCEPTS

7.11   Moffitt, Gabrielli, Mednick, and Schulsinger (1981) studied the relationship among socioeconomic status, IQ, and delinquency in Denmark youths. A random sample of 84 Danish youths with no offenses was selected, and the mean IQ was 113. Suppose the standard deviation is 21. Use a 90% confidence interval to estimate the true mean IQ of Danish youths with no history of offenses.

7.12   A fact long known but little understood is that twins, in their early years, tend to have lower IQ's and pick up language more slowly than nontwins. Recently, psychologists have found that the slower intellectual growth of most twins may be caused by benign parental neglect. Suppose it is desired to estimate the mean attention time given to twins per week by their parents. A sample of 46 sets of $2\frac{1}{2}$-year-old twin boys is taken, and at the end of 1 week the attention time given to each pair is recorded. The results are as follows:

$$\bar{x} = 22 \text{ hours} \qquad s = 16 \text{ hours}$$

Using the data, find a 90% confidence interval for the mean attention time given to all twin boys by their parents. Interpret the confidence interval.

7.13   Suppose a large labor union wishes to estimate the mean number of hours per month a union member is absent from work. The union decides to sample 320 of its members at random and monitor their working time for 1 month. At the end of the month, the total number of hours absent from work is recorded for each employee. If the mean and standard deviation of the sample are $\bar{x} = 9.6$ hours and $s = 6.4$ hours, find a 95% confidence interval for the true mean number of hours absent per month per employee. Interpret the confidence interval.

7.14 An article in the *Wall Street Journal* (March 1, 1984) states that of 48 sources surveyed by blue chip economic indicators, Prudential Bache Securities gave the lowest estimate (3%) of the rise in the Consumer Price Index for 1984, and Prudential Insurance (Prudential Bache's owner) gave the highest (5.6%).

    a. Find an approximate value for the sample standard deviation of the sample of 48 estimates. [*Hint:* Assume that the range is approximately equal to $4s$.]

    b. Assume that the 48 estimates represent a random sample of estimates from a large number of estimate sources. If the mean of the sample was 4.3%, find a 90% confidence interval for the mean estimated increase in the Consumer Price Index for this population of estimates associated with the estimate sources.

7.15 Automotive engineers are continually improving their products. Suppose a new type of brake light has been developed by General Motors. As part of a product safety evaluation program, General Motors' engineers wish to estimate the mean driver response time to the new brake light. (Response time is the length of time from the point that the brake is applied until the driver in the following car takes some corrective action.) Fifty drivers are selected at random and the response time (in seconds) for each driver is recorded, yielding the following results: $\bar{x} = .72$, $s^2 = .022$. Estimate the mean driver response time to the new brake light using a 99% confidence interval. Interpret the confidence interval.

7.16 To assess the magnitude of recent rent increases in metropolitan Minneapolis, an apartment referral service randomly sampled and interviewed 32 apartment building owners. They obtained the following data (adapted from *Minneapolis Star and Tribune*, July 26, 1986) concerning the change in rent for two-bedroom apartments between May 1985 and May 1986:

1.13%	3.32%	.62%	2.03%	1.82%
2.61	1.79	3.59	2.27	1.80
−1.22	2.06	−.45	.05	.34
7.33	5.20	1.25	1.42	.68
2.03	.68	.00	1.41	1.39
2.42	1.00	.90	.61	.00
1.00	−1.07			

    a. Carefully describe the population from which the sample was drawn.

    b. Use a 95% confidence interval to estimate the mean percentage change in rent for two-bedroom apartments.

    c. In constructing the confidence interval of part **b**, was it necessary to assume that the population was normally distributed? Explain.

7.17 As an aid in the establishment of personnel requirements, the director of a hospital wishes to estimate the mean number of people who are admitted to the emergency room during a 24-hour period. The director randomly selects 64 different 24-hour periods and determines the number of admissions for each. For this sample, $\bar{x} = 19.8$ and $s^2 = 25$. Estimate the mean number of admissions per 24-hour period with a 95% confidence interval. Interpret the result.

7.18 An article titled "Scientists Seek Upper Hand in Insect Wars" (*Wall Street Journal*, March 6, 1984) notes that the cockroach has had 300 million years to develop a resistance to destruction. That they are "reproductively brilliant" is evidenced by a study conducted by researchers for S. C. Johnson & Son, Inc. (manufacturers of Raid and Off). Five thousand roaches, the expected number in a roach-infested house, were released in the Raid test kitchen. One week later the kitchen was fumigated and 16,298 dead roaches were counted, a gain of 11,298 roaches for the 1-week period. Assume that none of the original roaches died during the 1-week period and that the standard deviation of x, the number of roaches produced per roach in a 1-week period, is 1.5. Use the number of roaches produced by the sample of 5,000 roaches to find a 95% confidence interval for the mean number of roaches produced per week for each roach in a typical roach-infested house.

7.19 Nasser Arshadi and Edward Lawrence (1984) investigated the profiles (i.e., career patterns, social backgrounds, and so forth) of the top executives in the U.S. banking industry. They sampled 96 executives and found, among other things, that 80% studied business or economics and that 45% had a graduate degree. With respect to the number of years of service, x, at the same bank, the group had a mean of 23.43 years and a standard deviation of 10.82 years.

a. Construct a 95% confidence interval for $E(x) = \mu$.

b. Interpret the interval in the context of the problem.

c. What assumption(s) was it necessary to make in order to construct the confidence interval of part a?

7.2 Determining the Sample Size Necessary to Estimate a Population Mean

In many practical applications of statistics, the analyst observes data that have been collected by an outside agency not under his or her direction. Examples are data collected by federal agencies such as the Bureau of the Census or the Bureau of Labor Statistics, survey data collected by a national pollster, and data published by another researcher. Such data are called **observational**, since they are observed rather than collected by the analyst.

Sometimes the analyst must plan the sampling experiment that generates the data used to make inferences about the population. Such data are generated by **designed experiments**, and perhaps the most important design decision faced by the analyst is to determine the size of the sample. We will show in this section that the appropriate sample size for making an inference about a population mean depends on the desired reliability.

To see this, consider the example from Section 7.1 in which we estimated the mean length of stay for patients in a large hospital. A sample of 100 patients' records produced an estimate $\bar{x}$ that was within .98 day of the true mean length of stay, μ, for all the hospital's patients at the 95% confidence level. That is, the 95% confidence interval for μ was $2(.98) = 1.96$ days wide when 100 accounts were sampled. This is illustrated in Figure 7.5(a).

Now suppose we want to estimate μ to within .25 day with 95% confidence. That is, we want to narrow the width of the confidence interval from 1.96 days to .50 day, as shown in Figure 7.5(b). How much will the sample size have to be increased to accomplish this? If we want the estimator $\bar{x}$ to be within .25 day of μ, we must have

$$2\sigma_{\bar{x}} = .25$$

or, equivalently,

$$2\left(\frac{\sigma}{\sqrt{n}}\right) = .25$$

The necessary sample size is obtained by solving this equation for n. To do this we need an approximation for σ. We have an approximation from the initial

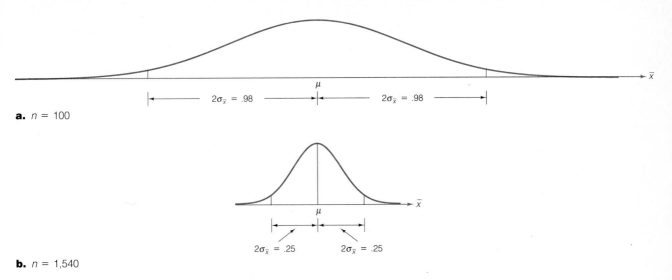

a. $n = 100$

b. $n = 1,540$

FIGURE 7.5

Relationship between sample size and width of confidence interval: Hospital stay example

sample of 100 patients' records—namely, the sample standard deviation, $s = 4.9$. Thus,

$$2\left(\frac{\sigma}{\sqrt{n}}\right) \approx 2\left(\frac{s}{\sqrt{n}}\right) = 2\left(\frac{4.9}{\sqrt{n}}\right) = .25$$

$$\sqrt{n} = \frac{2(4.9)}{.25} = 39.2$$

$$n = (39.2)^2 = 1,536.64$$

Approximately 1,540 patients' records will have to be sampled to estimate the mean length of stay μ to within .25 day with (approximately) 95% confidence. The confidence interval resulting from a sample of this size will be approximately .50 day wide [see Figure 7.5(b)].

In general, we can express the reliability associated with a confidence interval for the population mean μ in one of two equivalent ways. We can specify the bound, B, within which we want to estimate μ with $100(1 - \alpha)\%$ confidence. The bound B then is equal to the half-width of the confidence interval, as shown in Figure 7.6. Equivalently, we can specify the total width, W, of the confidence interval for μ, also shown in Figure 7.6. Note that $W = 2B$.

FIGURE 7.6

Specifying the bound B or total width W for a confidence interval

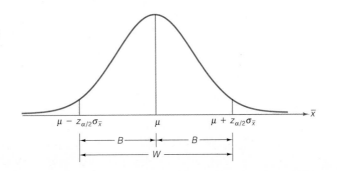

The procedure for finding the sample size necessary to estimate μ to within a given bound B or with a total interval width W is given in the accompanying box.

Sample Size Determination For $100(1 - \alpha)\%$
Confidence Intervals for μ

In order to estimate μ to within a bound B or, equivalently, with a confidence interval of total width W with $100(1 - \alpha)\%$ confidence, the required sample size is found as follows:

$$z_{\alpha/2}\left(\frac{\sigma}{\sqrt{n}}\right) = B \quad \text{or} \quad z_{\alpha/2}\left(\frac{\sigma}{\sqrt{n}}\right) = \frac{W}{2}$$

The solution can be written in terms of either B or W as follows:

$$n = \frac{(z_{\alpha/2})^2\sigma^2}{B^2} \quad \text{or} \quad n = \frac{4(z_{\alpha/2})^2\sigma^2}{W^2}$$

The value of σ is usually unknown. It can be estimated by the standard deviation, s, from a prior sample. Alternatively, we may approximate the range R of observations in the population, and (conservatively) estimate $\sigma \approx R/4$. In any case, you should round the value of n obtained *upward* to ensure that the sample size will be sufficient to achieve the specified reliability.

EXAMPLE 7.2

Suppose the manufacturer of official NFL footballs uses a machine to inflate the new balls to a pressure of 13.5 pounds. When the machine is properly calibrated, the mean inflation pressure is 13.5 pounds, but uncontrollable factors cause the pressures of individual footballs to vary randomly from about 13.3 and 13.7 pounds. For quality control purposes, the manufacturer wishes to estimate the true mean inflation pressure with a 99% confidence interval that is only .05 pounds wide. What sample size should be specified for the experiment?

Solution

For a 99% confidence interval, we have $z_{\alpha/2} = z_{.005} = 2.575$. To estimate σ, we note that the range of observations is $R = 13.7 - 13.3 = .4$ and use $\sigma \approx R/4 = .1$. Thus, in order to have a confidence interval of width $W = .05$, we use the formula derived in the box to find the sample size n:

$$n = \frac{4(z_{\alpha/2})^2\sigma^2}{W^2} \approx \frac{4(2.575)^2(.1)^2}{(.05)^2} = 106.09$$

We round this up to $n = 107$. Realizing that σ was approximated by $R/4$, we might even advise that the sample size be specified as $n = 110$ to be more certain of attaining the objective of a 99% confidence interval with width $W = .05$ pound or less.

Sometimes the formulas will lead to a solution that indicates a small sample size is sufficient to achieve the confidence interval goal. As we will see in Section 7.3, the procedures and assumptions for small samples differ from those for large samples. Therefore, if the formulas yield a small sample size ($n < 30$), one simple strategy is to select a sample size $n \geq 30$. Of course, the cost of sampling must also be considered when the sample size is being determined. Although more complex formulas can be derived to take sampling costs into account, these are beyond the scope of this text. For our purposes, it is sufficient to realize that a sampling budget may prove to be a restriction on the sample size, and therefore on the reliability of the confidence interval.

EXERCISES 7.20–7.30

LEARNING THE MECHANICS

7.20 If you wish to estimate a population mean correct to within a bound $B = .2$ with probability .95 and you know from prior sampling that σ^2 is approximately equal to 6.1, how many observations would have to be included in your sample?

7.21 If you wish to estimate a population mean with a 95% confidence interval of width $W = .2$ and you know from prior sampling that σ^2 is approximately equal to 6.1, how many observations would have to be included in your sample? Compare your answer to that for Exercise 7.20. Explain the difference.

7.22 Suppose you wish to estimate the mean of a normal population using a 95% confidence interval, and you know from prior information that $\sigma^2 \approx 1$.
a. To see the effect of the sample size on the width of the confidence interval, calculate the width W of the confidence interval for $n = 16, 25, 49, 100,$ and 400.
b. Plot the width as a function of sample size n on graph paper. Connect the points by a smooth curve and note how the width decreases as n increases.

7.23 Suppose you wish to estimate a population mean correct to within a bound $B = .15$ with probability equal to .90. You do not know σ^2, but you know that the observations will range in value between 24 and 27.
a. Find the approximate sample size that will produce the desired accuracy of the estimate. You wish to be conservative to ensure that the sample size will be ample to achieve the desired accuracy of the estimate. [Hint: Using your knowledge of data variation from Section 2.6, assume that the range of the observations will equal 4σ.]
b. Calculate the approximate sample size making the less conservative assumption that the range of the observations is equal to 6σ.

7.24 It costs you $10 to draw a sample of size $n = 1$ and measure the attribute of interest. You have a budget of $1,200.
a. Do you have sufficient funds to estimate the population mean for the attribute of interest with a 95% confidence interval 4 units in width? Assume $\sigma = 12$.
b. If a 90% confidence level were used, would your answer to part a change? Explain.

APPLYING THE CONCEPTS

7.25 It costs more to produce defective items—since they must be scrapped or reworked—than it does to produce nondefective items. This simple fact suggests that manufacturers should ensure the quality of their products by perfecting their production processes rather than through inspection of finished products (Deming, 1986). In order to better understand a particular metal stamping process, a manufacturer wishes to estimate the mean length of items produced by the process during the past 24 hours.

a. How many parts should be sampled in order to estimate the population mean to within .1 mm with 90% confidence? Previous studies of this machine have indicated that the standard deviation of lengths produced by the stamping operation is about 2 mm.

b. Time permits the use of a sample size no larger than 100. If a 90% confidence interval for μ is constructed using $n = 100$, will it be wider or narrower than would have been obtained using the sample size determined in part **a**? Explain.

c. If management requires that μ be estimated to within .1 mm and that a sample size of no more than 100 be used, what is (approximately) the maximum confidence level that could be attained for a confidence interval that meets management's specifications?

7.26   The EPA standard on the amount of suspended solids that can be discharged into rivers and streams is a maximum of 60 milligrams per liter daily, with a maximum monthly average of 30 milligrams per liter. Suppose you want to test a randomly selected sample of n water specimens and estimate the mean daily rate of pollution produced by a mining operation. If you want a 95% confidence interval estimate of width 2 milligrams, how many water specimens would you have to include in your sample? Assume prior knowledge indicates that pollution readings in water samples taken during a day are approximately normally distributed with a standard deviation equal to 5 milligrams.

7.27   Refer to Exercise 7.11, in which the mean IQ of Danish youths with no history of juvenile delinquency offenses was estimated. How many youths would have to be sampled in order to estimate the mean IQ to within 1.5 points with 90% confidence? Use the standard deviation based on the sample of $n = 84$ youths, $s = 21$, to approximate the true standard deviation σ.

7.28   Suppose a department store wants to estimate μ, the average age of the customers in its contemporary apparel department, correct to within 2 years with probability equal to .95. Approximately how large a sample would be required? [*Note:* Management does not know the standard deviation σ but guesses that the ages of its customers range from 15 to 45. Use a conservative approximation for σ to calculate n.]

7.29   According to a Food and Drug Administration (FDA) study, a cup of coffee contains an average of 115 milligrams of caffeine, with the amount per cup ranging from 60 to 180 milligrams. In contrast, sugar-free Mr. Pibb tested at 58.8 milligrams of caffeine per 12-ounce serving, Coca-Cola and Diet Coke at 45.6 milligrams, and Pepsi at 38.4 milligrams. Suppose you want to repeat the FDA experiment to obtain an estimate of the mean caffeine content in a cup of coffee correct to within 5 milligrams with 95% confidence. How many cups of coffee would have to be included in your sample?

7.30   The United States Golf Association (USGA) tests all new brands of golf balls to assure that they meet USGA specifications. One test conducted is intended to measure the average distance traveled when the ball is hit by a machine called "Iron Byron," a name inspired by the swing of the famous golfer Byron Nelson. Suppose the USGA wishes to estimate the mean distance for a new brand with a 90% confidence interval of width 2 yards. Assume that past tests have indicated that the standard deviation of the distances Iron Byron hits golf balls is approximately 10 yards. How many golf balls should be hit by Iron Byron to achieve the desired accuracy in estimating the mean?

7.3   Small-Sample Estimation of a Population Mean

Federal legislation requires pharmaceutical companies to perform extensive tests on new drugs before they can be marketed. Initially, a new drug is tested on animals. If the drug is deemed safe after this first phase of testing, the pharmaceutical company is then permitted to begin human testing on a limited basis. During this second phase, inferences must be made about the safety of the drug based on information in very small samples.

Suppose a pharmaceutical company must estimate the average increase in blood pressure of patients who take a certain new drug. Assume that only six patients can be used in the initial phase of human testing. The use of a *small sample* in making an inference about μ presents two immediate problems when we attempt to use the standard normal z as a test statistic.

PROBLEM 1

The shape of the sampling distribution of the sample mean $\bar{x}$ (and the z statistic) now depends on the shape of the population that is sampled. We can no longer assume that the sampling distribution of $\bar{x}$ is approximately normal, because the Central Limit Theorem assures normality only for samples that are sufficiently large.

PROBLEM 2

The population standard deviation σ is almost always unknown. Although it is still true that $\sigma_{\bar{x}} = \sigma/\sqrt{n}$, the sample standard deviation s may provide a poor approximation for σ when the sample size is small.

Solution to Problem 1

The sampling distribution of $\bar{x}$ (and z) is exactly normal even for relatively small samples if the sampled population is normal. It is approximately normal if the sampled population is approximately normal.

Solution to Problem 2

Instead of using the standard normal statistic

$$z = \frac{\bar{x} - \mu}{\sigma_{\bar{x}}} = \frac{\bar{x} - \mu}{\sigma/\sqrt{n}}$$

which requires knowledge of or a good approximation to σ, we define and use the statistic

$$t = \frac{\bar{x} - \mu}{s/\sqrt{n}}$$

in which the sample standard deviation, s, replaces the population standard deviation, σ.

The distribution of the *t* **statistic** in repeated sampling was discovered by W. S. Gosset, a scientist in the Guinness brewery in Ireland, who published his discovery in 1908 under the pen name of Student. The main result of Gosset's work is that if we are sampling from a normal distribution, the t statistic has a sampling distribution very much like that of the z statistic: mound-shaped, symmetric, with mean 0. The primary difference between the sampling distributions of t and z is that the t statistic is more variable than the z, which follows intuitively when you realize that t contains two random quantities ($\bar{x}$ and s), whereas z contains only one ($\bar{x}$).

The actual amount of variability in the sampling distribution of t depends on the sample size n. A convenient way of expressing this dependence is to say that the t statistic has $(n - 1)$ **degrees of freedom**. Recall that the quantity $(n - 1)$ is the divisor that appears in the formula for s^2. This number plays a key role in the sampling distribution of s^2 and appears in discussions of other statistics in later chapters. Particularly, the smaller the number of degrees of freedom associated with the t statistic, the more variable will be its sampling distribution.

FIGURE 7.7

Standard normal (z) distribution and
t-distribution with 4 df

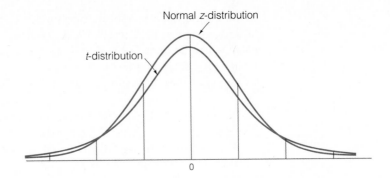

In Figure 7.7 we show both the sampling distribution of z and the sampling distribution of a t statistic with 4 degrees of freedom (df). You can see that the increased variability of the t statistic means that the t value, t_α, that locates an area α in the upper tail of the t-distribution is larger than the corresponding value z_α. For any given value of α, the t value t_α increases as the number of degrees of freedom (df) decreases. Values of t that will be used in forming small-sample confidence intervals for μ are given in Table VI of Appendix A and inside the back cover of the text. A partial reproduction of this table is shown in Figure 7.8.

Note that t_α values are listed for degrees of freedom from 1 to 29, where α refers to the tail area under the t-distribution to the right of t_α. For example, if we want the t value with an area of .025 to its right and 4 df, we look in the table under the column $t_{.025}$ for the entry in the row corresponding to 4 df. This

FIGURE 7.8

Reproduction of part of Table VI in
Appendix A

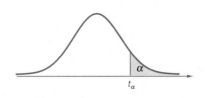

DEGREES OF FREEDOM	$t_{.100}$	$t_{.050}$	$t_{.025}$	$t_{.010}$	$t_{.005}$	$t_{.001}$	$t_{.0005}$
1	3.078	6.314	12.706	31.821	63.657	318.31	636.62
2	1.886	2.920	4.303	6.965	9.925	22.326	31.598
3	1.638	2.353	3.182	4.541	5.841	10.213	12.924
4	1.533	2.132	2.776	3.747	4.604	7.173	8.610
5	1.476	2.015	2.571	3.365	4.032	5.893	6.869
6	1.440	1.943	2.447	3.143	3.707	5.208	5.959
7	1.415	1.895	2.365	2.998	3.499	4.785	5.408
8	1.397	1.860	2.306	2.896	3.355	4.501	5.041
9	1.383	1.833	2.262	2.821	3.250	4.297	4.781
10	1.372	1.812	2.228	2.764	3.169	4.144	4.587
11	1.363	1.796	2.201	2.718	3.106	4.025	4.437
12	1.356	1.782	2.179	2.681	3.055	3.930	4.318
13	1.350	1.771	2.160	2.650	3.012	3.852	4.221
14	1.345	1.761	2.145	2.624	2.977	3.787	4.140
15	1.341	1.753	2.131	2.602	2.947	3.733	4.073

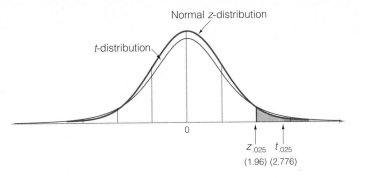

entry is $t_{.025} = 2.776$, as shown in Figure 7.9. The corresponding standard normal z-score is $z_{.025} = 1.96$.

Note that the last row of Table VI, where df = infinity, contains the standard normal z values. This follows from the fact that as the sample size n grows very large, s becomes closer to σ and thus t becomes closer in distribution to z. In fact, when df = 29, there is little difference between corresponding tabulated values of z and t. Thus, we choose the arbitrary cutoff of $n = 30$ (df = 29) to distinguish between the large-sample and small-sample inferential techniques.

Returning to the example of testing a new drug, suppose that the six test patients have blood pressure increases of 1.7, 3.0, .8, 3.4, 2.7, and 2.1 points. We calculate

$$\bar{x} = \frac{\sum x}{n} = \frac{13.7}{6} = 2.28$$

$$s^2 = \frac{\sum (x - \bar{x})^2}{n - 1} = \frac{\sum x^2 - \dfrac{\left(\sum x\right)^2}{n}}{n - 1} = \frac{35.79 - \dfrac{(13.7)^2}{6}}{5} = .9017$$

$$s = \sqrt{s^2} = .950$$

How can we use this information to construct a 95% confidence interval for μ, the mean increase in blood pressure associated with the new drug for all patients in the population?

First, we know that we are dealing with a sample too small to assume that the sample mean $\bar{x}$ is approximately normally distributed by the Central Limit Theorem. That is, we do not get the normal distribution of $\bar{x}$ "automatically" from the Central Limit Theorem when the sample size is small. Instead, we must assume that the measured variable, in this case the increase in blood pressure, is normally distributed in order for the distribution of $\bar{x}$ to be normal.

Second, unless we are fortunate enough to know the population standard deviation σ, which in this case represents the standard deviation of *all* the patients' increase in blood pressure when they take the new drug, we cannot use the standard normal z statistic to form our confidence interval for μ. Instead, we must use the t-distribution, with $(n - 1)$ degrees of freedom.

In this case, $n - 1 = 5$ df, and the t value is found in Figure 7.8 (or inside back cover) to be

$$t_{.025} = 2.571 \qquad \text{with 5 df}$$

Recall that the large-sample confidence interval would have been of the form

$$\bar{x} \pm z_{\alpha/2}\sigma_{\bar{x}} = \bar{x} \pm z_{\alpha/2}\frac{\sigma}{\sqrt{n}} = \bar{x} \pm z_{.025}\frac{\sigma}{\sqrt{n}}$$

where 95% is the desired confidence level. To form the interval for a small sample *from a normal distribution, we simply substitute t for z and s for σ in the preceding formula:*

$$\bar{x} \pm t_{\alpha/2}\frac{s}{\sqrt{n}}$$

Substituting the numerical values, we get

$$2.28 \pm (2.571)\left(\frac{.95}{\sqrt{6}}\right) = 2.28 \pm 1.00$$

or 1.28 to 3.28 points. That is, we can be 95% confident that the mean increase in blood pressure associated with taking this new drug is between 1.28 and 3.28 points. As with our large-sample interval estimates, our confidence is in the process, not in this particular interval. We know that the procedure we used will produce an interval that contains the true mean μ 95% of the time we utilize it, *assuming that the probability distribution of changes in blood pressure from which our sample was selected is normal.* The latter assumption is necessary for the small-sample interval to be valid.

What price did we pay for having to utilize a small sample to make the inference? First, we had to assume the underlying population is normally distributed, and if the assumption is invalid, our interval might also be invalid.* Second, we had to form the interval using a t value of 2.571 rather than a z value of 1.96, resulting in a wider interval to achieve the same 95% level of confidence. If the interval from 1.28 to 3.28 is too wide to be of much use, then we know how to remedy the situation: increase the number of patients sampled in order to decrease the interval width (on average).

The procedure for forming a small-sample confidence interval is summarized in the accompanying box.

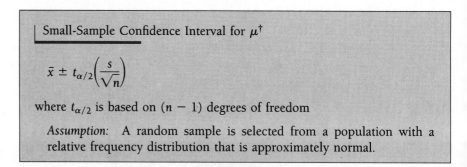

Small-Sample Confidence Interval for $\mu^{\dagger}$

$$\bar{x} \pm t_{\alpha/2}\left(\frac{s}{\sqrt{n}}\right)$$

where $t_{\alpha/2}$ is based on $(n - 1)$ degrees of freedom

Assumption: A random sample is selected from a population with a relative frequency distribution that is approximately normal.

---

*By *invalid*, we mean that the probability that the procedure will yield an interval that contains μ is not equal to $(1 - \alpha)$. Generally, if the underlying population is approximately normal, then the confidence coefficient will approximate the probability that the interval contains μ.

†The procedure given in the box assumes that the population standard deviation σ is unknown, which is almost always the case. If σ is known, we can form the small-sample confidence interval just as we would a large-sample confidence interval using a standard normal z value instead of t. However, we must still assume that the underlying population is approximately normal.

EXAMPLE 7.3

Some quality control experiments require *destructive sampling* (i.e., the test to determine whether the item is defective destroys the product) in order to measure some particular characteristic of the product. For example, suppose a manufacturer of printers for personal computers wishes to estimate the mean number of characters printed before the printhead fails. The cost of destructive sampling often dictates small samples. Suppose the printer manufacturer tests $n = 15$ printheads and calculates the following statistics:

$$\bar{x} = 1.23 \text{ million characters} \qquad s = .27 \text{ million characters}$$

Form a 99% confidence interval for the mean number of characters printed before the printhead fails.

Solution

If we assume that the number of characters printed before printhead failure is normally distributed, we can use the t statistic to form the confidence interval. We use a confidence coefficient of .99 and $n - 1 = 14$ degrees of freedom to find in Table VI:

$$t_{\alpha/2} = t_{.005} = 2.977$$

Thus, the small sample forces us to assume normality and extend the interval almost 3 standard deviations (of $\bar{x}$) on each side of the sample mean in order to form the 99% confidence interval. For these data, the interval is

$$\bar{x} \pm t_{.005}\left(\frac{s}{\sqrt{n}}\right) = 1.23 \pm 2.977\left(\frac{.27}{\sqrt{15}}\right)$$

$$= 1.23 \pm .21 \quad \text{or} \quad (1.02, 1.44)$$

Thus, the manufacturer can be 99% confident that the printhead has a mean life between 1.02 and 1.44 million characters. If the manufacturer were to advertise that the mean life of its printheads is (at least) 1 million characters, the interval would support such a claim. Our confidence is derived from the fact that 99% of the intervals formed in repeated applications of this procedure will contain μ.

We have emphasized throughout this section that an assumption that the population is approximately normally distributed is necessary for making small-sample inferences about μ when using the t statistic. Although many phenomena do have approximately normal distributions, it is also true that many random phenomena have distributions that are not normal or even mound-shaped. Empirical evidence acquired over the years has shown that the t-distribution is rather insensitive to moderate departures from normality. That is, use of the t statistic when sampling from mound-shaped populations generally produces credible results; however, for cases in which the distribution is distinctly nonnormal, either take a large sample or use a *nonparametric method* (the topic of Chapter 11).

> What Do You Do When the Population Relative Frequency Distribution Departs Greatly From Normality?
>
> *Answer:* Use the nonparametric statistical methods of Chapter 11.

EXERCISES 7.31–7.44

7.31   Explain the differences in the sampling distributions of $\bar{x}$ for large and small samples under the following assumptions.
   a. The variable of interest, x, is normally distributed.
   b. Nothing is known about the distribution of the variable x.

7.32   Suppose you have selected a random sample of $n = 5$ measurements from a normal distribution. Compare the standard normal z values with the corresponding t values if you were forming the following confidence intervals.
   a. 80% confidence interval
   b. 90% confidence interval
   c. 95% confidence interval
   d. 98% confidence interval
   e. 99% confidence interval
   f. Use the table values you obtained in parts a–e to sketch the z- and t-distributions. What are the similarities and differences?

7.33   Let t_0 be a specific value of t. Use Table VI in Appendix A to find t_0 values such that the following statements are true.
   a. $P(t \geq t_0) = .025$ where df = 8   b. $P(t \geq t_0) = .01$ where df = 10
   c. $P(t \leq t_0) = .005$ where df = 17   d. $P(t \leq t_0) = .05$ where df = 14

7.34   Let t_0 be a particular value of t. Use Table VI of Appendix A to find t_0 values such that the following statements are true.
   a. $P(-t_0 < t < t_0) = .95$ where df = 11
   b. $P(t \leq -t_0 \text{ or } t \geq t_0) = .05$ where df = 11
   c. $P(t \leq t_0) = .05$ where df = 11
   d. $P(t \leq -t_0 \text{ or } t \geq t_0) = .10$ where df = 20
   e. $P(t \leq -t_0 \text{ or } t \geq t_0) = .01$ where df = 6

7.35   The following random sample was selected from a normal distribution: 4, 7, 3, 4, 8, 3.
   a. Construct a 90% confidence interval for the population mean μ.
   b. Construct a 95% confidence interval for the population mean μ.
   c. Construct a 99% confidence interval for the population mean μ.
   d. Assume that the sample mean $\bar{x}$ and sample standard deviation s remain exactly the same as those you just calculated but that they are based on a sample of $n = 25$ observations rather than $n = 6$ observations. Repeat parts a–c. What is the effect of increasing the sample size on the width of the confidence intervals?

7.36   The following sample of 24 measurements was selected from a population that is approximately normally distributed:

91	80	99	110	95	106	78	121
106	100	97	82	100	83	115	104
114	118	96	101	79	130	94	101

   a. Construct an 80% confidence interval for the population mean.
   b. Construct a 95% confidence interval for the population mean and compare the width of this interval with that of part a.
   c. Carefully interpret each of the confidence intervals, and explain why the 80% confidence interval is narrower.

APPLYING THE CONCEPTS

7.37   Pulse rate is an important measure of the fitness of a person's cardiovascular system. The mean pulse rate for American adult males is approximately 72 heart beats per minute. A random sample of 21 American adult males who jog at least 15 miles per week had a mean pulse rate of 52.6 beats per minute and a standard deviation of 3.22 beats per minute.
   a. Find a 95% confidence interval for the mean pulse rate of all American adult males who jog at least 15 miles per week.
   b. Interpret the interval found in part a.
   c. What assumptions are required for the validity of the confidence interval?

7.38   Health insurers and the federal government are both putting pressure on hospitals to shorten the average length of stay (LOS) of their patients. A random sample of 20 hospitals in one state had a mean LOS in 1990 of 3.8 days and a standard deviation of 1.2 days.
   a. Use a 90% confidence interval to estimate the population mean of the LOS for the state's hospitals in 1990.
   b. Interpret the interval in terms of this application.
   c. What is meant by the phrase "90% confidence interval"?

7.39   United States firms are increasingly looking to the international market for expansion and growth. A random sample of 15 of the *Fortune* 1,000 firms is selected, and the percentage of their 1990 revenues from foreign sales is recorded for each. The mean is 23.8% and the standard deviation is 9.4%.
   a. Use a 99% confidence interval to estimate the mean percentage of 1990 foreign sales for all large U.S. firms.
   b. Interpret the interval in terms of this application.
   c. What assumption is necessary to ensure the validity of this confidence interval?

7.40   A company purchases large quantities of naphtha in 50-gallon drums. Because the purchases are ongoing, small shortages in the drums can represent a sizable loss to the company. The weights of the drums vary slightly from drum to drum, so the weight of the naphtha is determined by removing it from the drums and measuring it. Suppose the company samples the contents of 20 drums, measures the naphtha in each, and calculates $\bar{x} = 49.70$ gallons and $s = .32$ gallon. Find a 95% confidence interval for the mean number of gallons of naphtha per drum. What assumptions are necessary to assure the validity of the confidence interval?

7.41   In Exercise 4.48 we reported on research by Frans Van de Werf, M.D., and colleagues (*New England Journal of Medicine*, March 8, 1984) on a new drug, *t*-PA, which may prove to be effective in dissolving blood clots in heart attack patients. One aspect of their research involved measuring the length x of time for a heart attack patient's blood clot to be dissolved after treatment with *t*-PA. These times, recorded for $n = 7$ patients, were 50, 75, 0, 33, 57, 35, and 19 minutes.
   a. Assume that the length x of time until a blood clot dissolves is normally distributed. Find a 90% confidence interval for μ, the mean length of time for a blood clot to be dissolved after treatment with *t*-PA.
   b. Explain why the distribution of x might not be normally distributed.
   c. If the distribution of x is not normally distributed, what effect would this information have on your confidence interval in part a?

7.42   A study indicated that the cost of hiring an employee (excluding salary) ranges from about $1,500 for a secretary to more than $40,000 for a manager (Dessler, 1986). In order to estimate its mean cost of hiring an entry-level secretary, a large corporation randomly selected eight of the entry-level secretaries it had hired during the last 2 years and determined the costs (in dollars) involved in hiring each. The following data were obtained:

2,100	1,650	1,315	2,035
2,245	1,980	1,700	2,190

Assume that the population from which these data were sampled is approximately normally distributed.

a. Describe the population from which the corporation collected the sample data.

b. Use a 90% confidence interval to estimate the mean of interest to the corporation.

c. How wide is the confidence interval you constructed in part b? Would a 95% confidence interval be wider or narrower? Explain.

7.43 A *mortgage* is a type of loan that is secured by a designated piece of property. If the borrower defaults on the loan, the lender can sell the property to recover the outstanding debt. The home mortgage is the most important type of personal loan in the United States. In a home mortgage, the borrower pledges the home in question as security for the loan. A federal bank examiner is interested in estimating the mean outstanding principal balance of all home mortgages foreclosed by the bank due to default by the borrower during the last three years. A random sample of 12 foreclosed mortgages yielded the following data (in dollars):

| 95,982 | 81,422 | 39,888 | 46,836 | 66,899 | 69,110 |
| 59,200 | 62,331 | 105,812 | 55,545 | 56,635 | 72,123 |

a. Describe the population from which the bank examiner collected the sample data. What characteristic must this population possess to enable us to construct a confidence interval for the mean outstanding principal balance using the method described in this section?

b. Construct a 90% confidence interval for the mean of interest.

c. Carefully interpret your confidence interval in the context of the problem.

7.44 As we enter the 1990s, the U.S. Congress annually considers legislation requiring companies to provide more benefits for their employees, while U.S. industry bemoans the increasing cost of health insurance for its workers. A random sample of 23 small companies (companies with less than $10 million in annual revenues) that offer paid health insurance as a benefit was selected. The mean health insurance cost per worker per month was $135, and the standard deviation was $32.

a. Use a 95% confidence interval to estimate the mean cost per worker per month for all small companies.

b. What assumption is necessary to ensure the validity of the confidence interval?

c. What is meant by the phrase "95% confidence interval"?

7.4 Large-Sample Estimation of a Binomial Probability

In recent years the number of public opinion polls has grown at an astounding rate. Almost daily, the news media report the results of some poll. Pollsters regularly determine the percentage of people in favor of the president's energy program, the fraction of voters in favor of a certain candidate, the fraction of customers who favor a particular brand of wine, and the proportion of people who smoke cigarettes. In each case, we are interested in estimating the percentage (or proportion) of some group with a certain characteristic. In this section we will consider methods for making inferences about population proportions.

EXAMPLE 7.4

The mid-1970s may well be remembered for political unrest across the country. Since the days of Watergate, public opinion polls have been conducted to estimate the fraction of Americans who trust the president. Suppose 1,000 people are randomly chosen and 637 answer that they trust the president. How would you estimate the true fraction of *all* American people who trust the president?

Solution

What we have really asked is how would you estimate the probability p of success in a binomial experiment, where p is the probability that a person chosen trusts the president. One logical method of estimating p for the population is to use the proportion of successes in the sample. That is, we can estimate p by calculating

$$\hat{p} = \frac{\text{Number of people sampled who trust the president}}{\text{Number of people sampled}}$$

where $\hat{p}$ is read "p hat." Thus, in this case,

$$\hat{p} = \frac{637}{1{,}000} = .637$$

To determine the reliability of the estimator $\hat{p}$, we need to know its sampling distribution. That is, if we were to draw samples of 1,000 people over and over again, each time calculating a new estimate $\hat{p}$, what would be the frequency distribution of all the $\hat{p}$ values? The answer lies in viewing $\hat{p}$ as the average, or mean, number of successes per trial over the n trials. If each success is assigned a value equal to 1 and a failure is assigned a value of 0, then the sum of all n sample observations is x, the total number of successes, and $\hat{p} = x/n$ is the average, or mean, number of successes per trial in the n trials. The Central Limit Theorem tells us that the relative frequency distribution of the sample mean for any population is approximately normal for sufficiently large samples.

We demonstrated the properties of $\hat{p}$ in Example 6.9. The repeated sampling distribution of $\hat{p}$ has the characteristics listed in the accompanying box and shown in Figure 7.10.

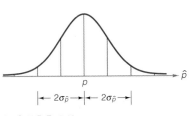

FIGURE 7.10

Sampling distribution of $\hat{p}$

Sampling Distribution of $\hat{p}$

1. The mean of the sampling distribution of $\hat{p}$ is p; that is, $\hat{p}$ is an unbiased estimator of p.

2. The standard deviation of the sampling distribution of $\hat{p}$ is $\sqrt{pq/n}$; that is, $\sigma_{\hat{p}} = \sqrt{pq/n}$, where $q = 1 - p$.

3. For large samples, the sampling distribution of $\hat{p}$ is approximately normal. A sample size is considered large if the interval $\hat{p} \pm 3\sigma_{\hat{p}}$ does not include 0 or 1. [*Note:* This requirement is almost equivalent to that given in Section 5.4 for approximating a binomial distribution with a normal one. The difference is that we assumed p to be known in Section 5.4; now we are trying to make inferences about an unknown p, so we use $\hat{p}$ to estimate p in checking the adequacy of the normal approximation.]

The fact that $\hat{p}$ is a "sample mean fraction of successes" allows us to form confidence intervals about p in a manner that is completely analogous to that used for large-sample estimation of μ.

> Large-Sample Confidence Interval for p
>
> $$\hat{p} \pm z_{\alpha/2}\sigma_{\hat{p}} = \hat{p} \pm z_{\alpha/2}\sqrt{\frac{pq}{n}} \approx \hat{p} \pm z_{\alpha/2}\sqrt{\frac{\hat{p}\hat{q}}{n}}$$
>
> where $\hat{p} = \dfrac{x}{n}$ and $\hat{q} = 1 - \hat{p}$
>
> [*Note:*   When n is large, we can use $\hat{p}$ to approximate the value of p in the formula for $\sigma_{\hat{p}}$.]

Thus, if 637 of 1,000 Americans say they trust the president, a 95% confidence interval for the proportion of *all* Americans who trust the president is

$$\hat{p} \pm z_{\alpha/2}\sigma_{\hat{p}} = .637 \pm 1.96\sqrt{\frac{pq}{1,000}}$$

TABLE 7.2   **Values of pq for Several Different Values of p**

p	pq	$\sqrt{pq}$
.5	.25	.50
.6 or .4	.24	.49
.7 or .3	.21	.46
.8 or .2	.16	.40
.9 or .1	.09	.30

where $q = 1 - p$. Just as we needed an approximation for σ in calculating a large-sample confidence interval for μ, we now need an approximation for p. As Table 7.2 shows, the approximation for p does not have to be especially accurate, because the value of $\sqrt{pq}$ needed for the confidence interval is relatively insensitive to changes in p. Therefore, we can use $\hat{p}$ to approximate p. Keeping in mind that $\hat{q} = 1 - \hat{p}$, we substitute these values into the formula for the confidence interval:

$$\hat{p} \pm 1.96\sqrt{pq/1,000} \approx \hat{p} \pm 1.96\sqrt{\hat{p}\hat{q}/1,000}$$
$$= .637 \pm 1.96\sqrt{(.637)(.363)/1,000} = .637 \pm .030$$
$$= (.607, .667)$$

Then we can be 95% confident that the interval from 60.7% to 66.7% contains the true percentage of *all* Americans who trust the president. That is, in repeated construction of confidence intervals, approximately 95% of all samples would produce confidence intervals that enclose p. Note that the guidelines for interpreting a confidence interval about μ also apply to interpreting a confidence interval for p because p is the "population mean fraction of successes" in a binomial experiment.

EXAMPLE 7.5

United States law requires companies to treat minorities fairly in their hiring, promotion, and pay practices. The Equal Employment Opportunity Commission (EEOC) is the agency charged with the responsibility of monitoring the treatment of minorities in the workplace. Suppose the EEOC samples 135 recent hires of one large company and determines that 12 are minorities. Use a 90% confidence interval to estimate the proportion of all new hires of the company that are minorities.

Solution

The number, x, of the 135 sampled hires who are minorities is a binomial random variable if we can assume that the sample was randomly selected from all the company's hires and that the process by which the company hires is relatively stable during the period of interest (so that the probability of a minority hire remains constant over the period).

Then the point estimate of the proportion of minority hires for this company is

$$\hat{p} = \frac{x}{n} = \frac{12}{135} = .089$$

We first check to be sure that the sample size is sufficiently large that the normal distribution provides a reasonable approximation for this binomial proportion. We check the 3-standard-deviation interval around $\hat{p}$:

$$\hat{p} \pm 3\sigma_{\hat{p}} \approx \hat{p} \pm 3\sqrt{\frac{\hat{p}\hat{q}}{n}}$$

$$= .089 \pm 3\sqrt{\frac{(.089)(.911)}{135}}$$

$$= .089 \pm .074$$

$$= (.015, .163)$$

Since this interval is wholly contained in the interval (0, 1), we may conclude that the normal approximation is reasonable.

We now form the 90% confidence interval for p, the true proportion of minority hires for the company:

$$\hat{p} \pm z_{\alpha/2}\sigma_{\hat{p}} = \hat{p} \pm z_{\alpha/2}\sqrt{\frac{pq}{n}}$$

$$\approx \hat{p} \pm z_{\alpha/2}\sqrt{\frac{\hat{p}\hat{q}}{n}}$$

$$= .089 \pm 1.645\sqrt{\frac{(.089)(.911)}{135}}$$

$$= .089 \pm .040$$

$$= (.049, .129)$$

Thus, we can be 90% confident that the proportion of all the companies' hires (under the current policy) who are minorities is between .049 and .129. As always, our confidence stems from the fact that 90% of all similarly formed intervals will contain the true proportion p and not form any knowledge about whether this particular interval does.

Suppose the EEOC knows that the minority proportion of the qualified work force for jobs in the industry represented by this company is 15%. Can we conclude that the company is hiring minorities at a rate below the 15% qualification rate? Since the 90% confidence interval lies completely below .15 (the upper confidence limit is only .129), we would conclude based on this interval that the company's minority hire rate is below the 15% qualification rate. In fact, we would conclude, with 90% confidence, that the minority hire rate is between 4.9% and 12.9%.

In Example 7.5 we used the confidence interval to make an inference about whether the true value of p is less than .15. That is, we used the sample to **test** whether p is less than .15. When we want to use sample information to test the value of a population parameter, we usually conduct a *test of hypothesis*, the subject of Chapter 8.

We conclude Chapter 7 by showing how to determine the sample size necessary to estimate a binomial proportion with a specified precision.

EXERCISES 7.45–7.57

7.45 Describe the sampling distribution of $\hat{p}$ based on large samples of size n. That is, give the mean, the standard deviation, and the (approximate) shape of the distribution of $\hat{p}$ when large samples of size n are (repeatedly) selected from a binomial distribution with probability of success p.

7.46 Explain the meaning of the phrase "$\hat{p}$ is an unbiased estimator of p."

7.47 A random sample of size $n = 400$ yielded $\hat{p} = .42$.
   a. Is the sample size large enough to use the methods of this section to construct a confidence interval for p? Explain.
   b. Construct a 95% confidence interval for p.
   c. Interpret the 95% confidence interval.
   d. Explain what is meant by the phrase "95% confidence interval."

7.48 A random sample of size $n = 144$ yielded $\hat{p} = .76$.
   a. Is the sample size large enough to use the methods of this section to construct a confidence interval for p? Explain.
   b. Construct a 90% confidence interval for p.
   c. What assumption is necessary to ensure the validity of this confidence interval?

7.49 For the binomial sample information summarized in each part, indicate whether the sample size is large enough to use the methods of this chapter to construct a confidence interval for p.
   a. $n = 500$, $\hat{p} = .05$ b. $n = 100$, $\hat{p} = .05$
   c. $n = 10$, $\hat{p} = .5$ d. $n = 10$, $\hat{p} = .3$

7.50 A random sample of 50 consumers taste-tested a new snack food. Their responses were coded (0: do not like; 1: like; 2: indifferent) and recorded as follows:

1	0	0	1	2	0	1	1	0	0
0	1	0	2	0	2	2	0	0	1
1	0	0	0	0	1	0	2	0	0
0	1	0	0	1	0	0	1	0	1
0	2	0	0	1	1	0	0	0	1

   a. Use an 80% confidence interval to estimate the proportion of consumers who like the snack food.
   b. Provide a statistical interpretation for the confidence interval you constructed in part **a**.

7.51 A random sample of 122 Illinois law firms was selected to determine their degree of computer usage (Wentling, 1988). She found that 76 of the firms used microcomputers (PCs).
   a. Use a 95% confidence interval to estimate the proportion of all Illinois law firms that used microcomputers at the time of the survey.
   b. Interpret the interval in the terms of this application.
   c. What is meant by the phrase "95% confidence interval"?

    d. Do you think this interval provides an estimate for the proportion of all U.S. law firms that were using microcomputers at the time of the survey? Why or why not?

7.52 Refer to Exercise 7.51. In a similar sample survey of 71 Illinois banks, 55 used microcomputers (Wentling and Wentling, 1988).

    a. Use a 95% confidence interval to estimate the proportion of all Illinois banks that used microcomputers at the time of the survey.

    b. Interpret the interval in terms of this application.

    c. How does this interval compare to the one you constructed for the proportion of Illinois law firms that used microcomputers at the time of the survey (Exercise 7.51)?

7.53 In February and March of 1985, the Gallup Organization conducted telephone interviews with a random sample of 258 owners and managers of small U.S. businesses (i.e., firms with 20 or more employees but with less than $50 million in sales). Forty percent of those interviewed were under 45 years of age, and 30% had worked at four or more companies (Graham, 1985).

    a. Describe the population of interest to the Gallup Organization.

    b. Is the sample size large enough to construct a confidence interval for the proportion of owners and managers who are under 45 years of age? Explain.

    c. Construct a 95% confidence interval for the proportion of interest in part **b.**

    d. If a 95% confidence interval were used to estimate the proportion of owners and managers who have worked at four or more companies, would the interval be wider, narrower, or the same width as the interval estimate of part **c**? Explain.

7.54 The U.S. Commission on Crime wishes to estimate the fraction of crimes related to firearms in an area with one of the highest crime rates in the country. The commission randomly selects 600 files of recently committed crimes in the area and finds 380 in which a firearm was reportedly used. Find a 99% confidence interval for p, the true fraction of crimes in the area in which some type of firearm was reportedly used.

7.55 An article titled "Searching for a forever home" (*Time*, May 2, 1983) reports on how television programs aid in the adoption of orphans who have physical or mental handicaps. One method of stimulating the adoption program is to present television profiles of the children, one or more per program. The article documents the success of these programs and notes that Oklahoma City's television station KOCO helped to place 92 of the 119 it profiled, New York's WCBS placed 21 of 35, and Atlanta's WXIA placed 79 of 177. How effective were these three television stations in promoting the adoption of children?

    a. Check to see whether each of the sample sizes is large enough to use the normal approximation for the sampling distributions of the sample proportions.

    b. Construct a 95% confidence interval for each television station's placement success rate.

    c. Suppose the total of 331 children profiled by the three stations could be regarded as a random sample from the population of all similar profiles that might be presented by television stations throughout the country. Construct a 90% confidence interval for the national placement success rate.

7.56 According to *U.S. News and World Report* (May 10, 1982), major infections are the fifth leading cause of death in the United States and approximately 2 million of these cases occur in the nation's hospitals. A powerful new antibiotic, piperacillin, was developed by Lederle Laboratories and approved by the Food and Drug Administration in 1982. Tests on 600 patients showed the drug to be 92% effective in curing serious bacterial infections.

    a. Find a 99% confidence interval for the probability p that a patient with a serious bacterial infection will be cured after treatment with piperacillin.

    b. If all 2 million hospital patients with serious bacterial infections were treated with piperacillin, what is the expected value of the number x that would be cured?

c. Find the standard deviation of x.

d. Give an upper limit on the number x that would be cured.

7.57   An article in the *Gainesville Sun* (March 14, 1984) reports on the topics that teenagers most want to discuss with their parents. The findings, the results of a Gallup Youth Poll, showed that 46% would like more discussion about the family's financial situation, 37% would like to talk about school, and 30% would like to talk about religion. These and other percentages were based on a national sampling of 505 teenagers. What "margin of error" (using the language of the media) would you attach to these findings? Explain.

7.5 Determining the Sample Size Necessary to Estimate a Binomial Probability

We showed in Section 7.2 that sampling experiments can be designed to estimate a population mean μ with a specified degree of reliability. An analogous situation exists for estimating a binomial probability p. For example, in Section 7.4 a pollster used a sample of 1,000 Americans to calculate a 95% confidence interval for the proportion who trust the president, obtaining the interval .607 to .667. Note that the total width of the interval is $.667 - .607 = .06$. Suppose the pollster wishes to estimate more precisely the proportion who trust the president, say with a 95% confidence interval having a width of .03.

The pollster wants a confidence interval width W of .03. This corresponds to a half-width, or bound B on the estimate of p, of $B = W/2 = .015$. The sample size n to generate such an interval is found by solving the following equation for n:

$$z_{\alpha/2}\sigma_{\hat{p}} = B \quad \text{or} \quad z_{\alpha/2}\sqrt{\frac{pq}{n}} = .015 \qquad \text{(see Figure 7.11)}$$

FIGURE 7.11

Specifying the total width W (or bound B) of a confidence interval for a binomial probability p

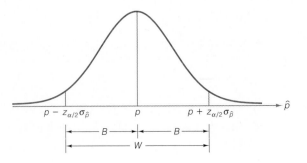

Since a 95% confidence interval is desired, the appropriate z value is $z_{\alpha/2} = z_{.025} = 1.96$. We must approximate the value of the product pq before we can solve the equation for n. As shown in Table 7.2, the closer the values of p and q to .5, the larger the product pq. Thus, to find a conservatively large sample size that will generate a confidence interval with the specified reliability, we generally choose an approximation of p close to .5. In the case of the proportion of Americans who trust the president, however, we have an initial sample estimate of $\hat{p} = .637$. A conservatively large estimate of pq can therefore be obtained by using $p = .60$. We now substitute into the equation and solve for n:

$$1.96 \sqrt{\frac{(.60)(.40)}{n}} = .015$$

$$n = \frac{(1.96)^2(.60)(.40)}{(.015)^2}$$

$$n = 4,097.7 \approx 4,098$$

The pollster must sample about 4,098 Americans to estimate the percentage who trust the president with a confidence interval of width .03.

The procedure for finding the sample size necessary to estimate a binomial probability p to within a given bound B or with a total interval width W is given in the box.

Sample Size Determination for $100(1 - \alpha)\%$ Confidence Interval for p

In order to estimate a binomial probability p to within a bound B or, equivalently, with a confidence interval of total width W with $100(1 - \alpha)\%$ confidence, the required sample size is found by solving one of the following equations for n:

$$z_{\alpha/2} \sqrt{\frac{pq}{n}} = B \quad \text{or} \quad z_{\alpha/2} \sqrt{\frac{pq}{n}} = \frac{W}{2}$$

The solution can be written in terms of either B or W:

$$n = \frac{(z_{\alpha/2})^2(pq)}{B^2} \quad \text{or} \quad n = \frac{4(z_{\alpha/2})^2(pq)}{W^2}$$

The value of the product pq is usually unknown. It can be estimated by using the sample fraction of successes, $\hat{p}$, from a prior sample. Remember (Table 7.2) that the value of pq is at its maximum when p equals .5, so that you can obtain conservatively large values of n by approximating p by .5 or values close to .5. In any case, you should round the value of n obtained *upward* to ensure that the sample size will be sufficient to achieve the specified reliability.

EXAMPLE 7.6

Since the deregulation of the telephonic communications industry, many firms have begun manufacturing telephones. Suppose one large manufacturer that entered the market quickly has an initial problem with excessive customer complaints and consequent returns of the phones for repair or replacement. The manufacturer wants to estimate the magnitude of the problem in order to design a quality control program. How many telephones should be sampled and checked in order to estimate the fraction defective p to within .01 with 90% confidence?

Solution

In order to estimate p to within a bound of .01, we set the half-width of the confidence interval equal to $B = .01$, as shown in Figure 7.12 on page 318.

The equation for the sample size n requires an estimate of the product pq. We could most conservatively estimate $pq = .25$ (i.e., use $p = .5$), but this may be

FIGURE 7.12

Specified reliability for estimate of
fraction defective in Example 7.6

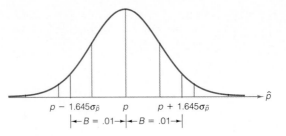

overly conservative when estimating a fraction defective. A value of .1, corresponding to 10% defective, will probably be conservatively large for this application. The solution is therefore

$$n = \frac{(z_{\alpha/2})^2(pq)}{B^2} = \frac{(1.645)^2(.1)(.9)}{(.01)^2} = 2,435.4$$

$$\approx 2,436$$

Thus, the manufacturer should sample 2,436 telephones in order to estimate the fraction defective p to within .01 with 90% confidence. Remember that this answer depends on our approximation for pq, where we used .09. If the fraction defective is closer to .05 than .10, we can use a sample of 1,286 telephones (check this) to estimate p to within .01 with 90% confidence.

The cost of sampling will play an important role in the final determination of the sample size to be selected to estimate a binomial probability p. Although more complex formulas can be derived to balance the reliability and cost considerations, we will solve for the necessary sample size and note that the sampling budget may be a limiting factor. Consult the references for a more complete treatment of this problem.

EXERCISES 7.58–7.68

LEARNING THE MECHANICS

7.58 If nothing is known about p, .5 can be substituted for p in the sample-size formula. But when this is done, the resulting sample-size may be larger than needed. Under what circumstances will using $p = .5$ in the sample-size formula yield a sample size larger than needed to construct a confidence interval for p with a specified bound and a specified confidence level?

7.59 In each case, find the approximate sample size necessary to estimate a binomial proportion p correct to within .02 with probability equal to .90.
a. Assume you know p is near .8.
b. Assume you have no knowledge of the value of p, but you wish to be certain that your sample is large enough to achieve the specified accuracy for the estimate.

7.60 In each case, find the approximate sample size required to construct a 95% confidence interval for p that has width .12.
a. Assume p is near .3.
b. Assume you have no prior knowledge about p.

7.61 The following is a 90% confidence interval for p: (.26, .54). How large was the sample used to construct this interval?

7.62 According to estimates made by the General Accounting Office, the Internal Revenue Service (IRS) answered 18.3 million telephone inquiries during a recent tax season, and 17% of the IRS offices provided answers that were wrong. These estimates were based on data collected from sample calls to numerous IRS offices. How many IRS offices should be randomly selected and contacted in order to estimate the proportion of IRS offices that fail to correctly answer questions about gift taxes with a 90% confidence interval of width .06?

7.63 Although corporate executives are probably not as highly stressed as air-traffic controllers or inner-city police, research has indicated that they are among the more highly pressured work groups. In order to estimate p, the proportion of managers who perceive themselves to be frequently under stress, Hall and Savery (1986) sampled 532 managers in western Australian corporations. One hundred ninety of these managers fell into the high-stress group. Assume that random sampling was used in this study. Was the sample size large enough to estimate p to within .03 with 95% confidence? Explain.

7.64 If you want to estimate the proportion of operating automobiles that are equipped with air-pollution devices, approximately how large a sample would be required to estimate p to within .02 with probability equal to .95?

7.65 A marketing research organization wishes to estimate the proportion of television viewers who watch a particular prime-time situation comedy on May 24. The proportion is expected to be approximately .30. At a minimum, how many viewers should be randomly selected to ensure that a 95% confidence interval for the true proportion of viewers will have a width of .01 or less?

7.66 Before a bill to increase federal price supports for farmers comes before the U.S. Congress, a congressperson would like to know how nonfarmers feel about the issue. Approximately how many nonfarmers should the congressperson survey in order to estimate the true proportion favoring this bill to within .05 with 90% confidence?

7.67 Suppose you are a retailer and you want to estimate the proportion of your customers who are shoplifters. You decide to select a random sample of shoppers and check closely to determine whether they steal any merchandise while in the store. Suppose experience suggests that the percentage of customers who are shoplifters is near 5%. How many customers should you include in your sample if you want to estimate the proportion of shoplifters in your store correct to within .02 with 90% confidence?

7.68 Some quality control testing involves destructive sampling—i.e., the test to determine whether the item is defective destroys the product. This type of sampling is generally expensive, and high costs often prohibit large sample sizes. For example, suppose the National Highway Safety Administration (NHSA) wishes to determine the proportion of new tires that will fail when subjected to hard braking at a speed of 60 miles per hour. NHSA can obtain the tires for $25 (wholesale price) each. Suppose the budget for the experiment is $10,000, and NHSA wishes to estimate the percentage that will fail to within .02 with 95% confidence. If the entire $10,000 can be spent on tires (ignoring other costs) and assuming that the true fraction that will fail is approximately .05, can NHSA attain its goal while staying within the budget? Explain.

Summary

From the beginning we have stressed that the theme of this text is the use of sample information to make inferences about the parameters of a population. For the first seven chapters we developed the tools necessary to perform inference-making procedures. In Chapter 7 we discussed the inferential procedure called **estimation**.

Estimation consists of both **point estimates** and **interval estimates**. Point estimates are single points calculated from a sample to estimate a population parameter. Because we cannot assess the reliability of a point estimate, we prefer to use intervals to estimate the unknown population parameters. Interval estimates utilize **confidence coefficients** to express the reliability of the estimate. The confidence coefficient is the probability that the estimation procedure will generate an interval that contains the parameter being estimated. Thus, confidence is expressed in terms of the long-run performance of the estimation procedure.

We used the standard normal z-distribution to develop large-sample interval estimates for the population mean μ and the binomial probability p. The Student t-distribution was used to form confidence intervals for the mean of normal distributions when n is small.

SUPPLEMENTARY EXERCISES 7.69–7.89

[*Note: List the assumptions necessary for the valid implementation of the statistical procedures you use in solving all these exercises. Starred (*) exercises refer to the optional sections.*]

LEARNING THE MECHANICS

7.69   In each of the following instances, determine whether you would use a z or t statistic (or neither) to form a 95% confidence interval, and then look up the appropriate z or t value.
    a. Random sample of size $n = 23$ from a normal distribution with unknown mean μ and standard deviation σ
    b. Random sample of size $n = 135$ from a normal distribution with unknown mean μ and standard deviation σ
    c. Random sample of size $n = 10$ from a normal distribution with unknown mean μ and standard deviation $\sigma = 5$
    d. Random sample of size $n = 73$ from a distribution about which nothing is known
    e. Random sample of size $n = 12$ from a distribution about which nothing is known

7.70   Let t_0 represent a particular value of t from Table VI of Appendix A. Find the tabled values such that the following statements are true.
    a. $P(t \leq t_0) = .05$ where df $= 20$ **b.** $P(t \geq t_0) = .005$ where df $= 9$
    c. $P(t \leq -t_0 \text{ or } t \geq t_0) = .10$ where df $= 8$ **d.** $P(t \leq -t_0 \text{ or } t \geq t_0) = .01$ where df $= 17$

7.71   A large New York City bank is interested in estimating (1) the proportion of weeks in which it processes more than 100,000 checks and (2) the average number of checks it processes per week. The bank maintains records of x, the number of checks processed each week. Suppose the bank records the number of checks, x, processed per week for 50 weeks randomly sampled from among the past 6 years. Define each of the following *in the context of the problem.*
    a. $\bar{x}$ **b.** $\hat{p}$ **c.** σ_x **d.** μ_x **e.** n **f.** $\sigma_{\bar{x}}$ **g.** p **h.** s_x

APPLYING THE CONCEPTS

7.72   In 1978, only 110 U.S. companies offered any form of child-care assistance to their employees. By 1985, 2,500 companies offered day-care advice, day-care programs, or financial assistance (*Minneapolis Star and Tribune*, November 30, 1986, p. 11A). A large corporation randomly sampled 60 of its employees with children under 5 years of age and asked them whether they would prefer to have the company pay for day-care at a program of the parents' choosing or to have the company supply free on-site day-care. The responses were coded (1: prefer choice; 2: prefer on-site program):

1	1	2	1	2	1	1	2	2	2
2	1	2	2	2	1	2	1	1	2
1	2	1	2	2	2	2	2	1	2
1	1	2	2	2	2	2	2	2	1
1	2	1	1	1	2	2	2	2	1
2	2	1	2	2	2	1	2	2	2

a. Construct a 90% confidence interval for the true proportion of employees with children under 5 years of age who prefer an on-site program.

b. Interpret the interval in terms of this application.

c. Explain what is meant by the phrase "90% confidence interval."

7.73 A company is interested in estimating μ, the mean number of days of sick leave taken by all its employees. The firm's statistician selects at random 100 personnel files and notes the number of sick days taken by each employee. The following sample statistics are computed:

$\bar{x} = 12.2$ days $s = 10$ days

a. Estimate μ using a 90% confidence interval.

b. How many personnel files would the statistician have to select in order to estimate μ with a 99% confidence interval of width 4 days?

7.74 A university is considering a change in the way students pay for their education. Presently, the students pay $16 per credit hour. The university is contemplating charging each student a set fee of $240 per quarter, regardless of how many credit hours each takes. To see if this proposal would be economically feasible, the university would like to know how many credit hours, on the average, each student takes per quarter. A random sample of 250 students yields a mean of 14.1 credit hours per quarter and a standard deviation of 2.3 credit hours per quarter.

a. Estimate the mean credit hours per student per quarter using a 99% confidence interval.

b. Use this confidence interval to obtain an interval estimate for the mean tuition fee per student per quarter. Interpret the result.

7.75 In checking the reliability of a bank's records, auditing firms sometimes ask a sample of the bank's customers to confirm the accuracy of their savings account balances as reported by the bank. Suppose an auditing firm is interested in estimating p, the proportion of a bank's savings accounts on whose balances the bank and the customer disagree. Of 200 savings account customers questioned by the auditors, 15 said their balance disagreed with that reported by the bank.

a. Using a 95% confidence interval, estimate the actual proportion of the bank's savings accounts on whose balances the bank and customer disagree.

b. The bank claims that the fraction of accounts on which there is disagreement is no larger than .10. Does the confidence interval you calculated in part **a** support this claim?

7.76 Refer to Exercise 7.75. How many savings account customers should the auditors question if they want to estimate p to within .02 with probability equal to .95?

7.77 A meteorologist wishes to estimate the mean amount of snowfall per year in Spokane, Washington. A random sample of the recorded snowfalls for 20 years produces a sample mean equal to 54 inches and a standard deviation of 9.59 inches.

a. Estimate the true mean amount of snowfall in Spokane using a 99% confidence interval.

b. If you were purchasing snow-removal equipment for a city, what numerical descriptive measure of the distribution of depth of snowfall would be of most interest to you? Would it be the mean?

7.78 A university dean is interested in determining the proportion of students who receive some sort of financial aid. Rather than examine the records for all students, the dean randomly selects 200 students and finds that 118 of them are receiving financial aid. Use a 95% confidence interval to estimate the true proportion of students who receive aid.

7.79   Before approval is given for the use of a new insecticide, the United States Department of Agriculture (USDA) requires that several tests be performed to see how the substance will affect wildlife. In particular, the USDA would like to know the proportion of starlings that will die after being exposed to the insecticide. A random sample of 80 starlings were caught and fed their regular food, which had been treated with the substance. After 10 days, ten starlings had died. Use a 99% confidence interval to estimate the true proportion of starlings that will be killed by the substance.

7.80   Many people think that a national lobby's successful fight against gun control legislation is reflecting the will of a minority of Americans. A random sample of 4,000 citizens yielded 2,250 who are in favor of gun control legislation. Use a 99% confidence interval to estimate the true proportion of Americans who favor gun control legislation. Interpret the result.

7.81   To help consumers assess the risks they are taking, the Food and Drug Administration (FDA) publishes the amount of nicotine found in all commercial brands of cigarettes. A new cigarette has recently been marketed. The FDA tests on this cigarette gave a mean nicotine content of 26.4 milligrams and standard deviation of 2.0 milligrams for a sample of $n = 9$ cigarettes. Using a 95% confidence interval, estimate the true mean nicotine content per cigarette for the brand. Interpret the result.

7.82   A health researcher wishes to estimate the mean number of cavities per child for children under the age of 12 who live in a specified environment. The number of cavities per child for a random sample of 35 children under the age of 12 has a mean of 2 and a standard deviation of 1.7. Construct a 90% confidence interval for the mean number of cavities per child under the age of 12 who lives in the sampled environment.

7.83   Recycling is receiving increased emphasis among environmental concerns as a means of dealing with the significant growth in the trash and garbage of our "throw-away" society. In order to estimate the degree of awareness about recycling in one major city, a random sample of 346 households was selected, and 212 were found to use available recycling facilities. Use a 95% confidence interval to estimate the proportion of households in the city that are using available recycling facilities.

7.84   In 1988 tire sales in the United States reached a historic high (*Chemical Week*, 1989). Tires made of synthetic rubber accounted for much of this growth. Suppose one manufacturer was testing a new synthetic rubber design. Twenty tires of the new design were produced and subjected to wear tests. The results indicate that the mean wear for the test tires was 42,250 miles, and the standard deviation was 4,355 miles.
   a. What is the point estimate of the true mean wear for the new design?
   b. Construct a 90% confidence interval estimate for the mean wear associated with the new design.
   c. Which method of estimation is better, point estimation or interval estimation? Why?

7.85   Refer to Exercise 7.84. Suppose that 200 tires rather than 20 were tested, and assume that the sample mean and standard deviation for the 200 tires remain the same: $\bar{x} = 42{,}250$ miles, $s = 4{,}355$ miles. Repeat parts **a–c** of Exercise 7.84 and comment on the similarities and differences in your answers.

7.86   In 1987 a case of salmonella (bacterial) poisoning was traced to a particular brand of ice cream bar, and the manufacturer removed the bars from the market. Despite this response, many consumers refused to purchase any brand of ice cream bars for some period of time after the event (McClave, personal consulting). One manufacturer conducted a survey of consumers 6 months after the poisoning. A sample of 244 ice cream bar consumers were contacted, and 23 of them indicated that they would not purchase ice cream bars because of the potential for food poisoning.
   a. What is the point estimate of the true fraction of the entire market who refuse to purchase bars 6 months after the poisoning?
   b. Is the sample size large enough to use the normal approximation for the sampling distribution of the estimator of the binomial probability? Justify your response.
   c. Construct a 95% confidence interval for the true proportion of the market who still refuse to purchase ice cream bars 6 months after the event.

d. Interpret both the point estimate and confidence interval in terms of this application.

7.87   Refer to Exercise 7.86. Suppose it is now 1 year after the poisoning was traced to ice cream bars. The manufacturer wishes to estimate the proportion who still will not purchase bars using a 95% confidence interval of width .04. How many consumers should be sampled?

7.88   Medicaid health assistance programs are administered by the individual states, even though part of the funding is federal. The federal government requires that the states perform regular audits in order to assure that payments are accurate. One Florida hospital was audited by the Florida Department of Health and Rehabilitative Services (HRS) in 1989, and a random sample of 25 Medicaid claims was selected. The sample mean of the claims was $34.76 and the standard deviation was $11.34.
a. Use a 99% confidence interval to estimate the mean of all claims submitted by this hospital.
b. What assumptions are necessary to assure the validity of this confidence interval?

7.89   Refer to Exercise 7.88.
a. How many claims must be sampled if the HRS wants to estimate the mean size of the hospital's claims to within $1.00 using a 99% confidence interval?
b. If a sample of this size were to be selected and a 99% confidence interval constructed, what assumptions would be necessary to assure the validity of the interval?

ON YOUR OWN...

Choose a population pertinent to your major area of interest that has an unknown mean (or, if the population is binomial, that has an unknown probability of success). For example, a marketing major may be interested in the proportion of consumers who prefer a certain product. A sociology major may be interested in estimating the proportion of people in a certain socioeconomic group or the mean income of people living in a certain part of a city. A political scientist may wish to estimate the proportion of an electorate in favor of a certain candidate, a certain amendment, or a certain presidential policy. A person interested in medicine might want to find the average length of time patients stay in the hospital or the average number of people treated daily in the emergency room. We could continue with examples, but the point should be clear—choose something of interest to you.

Define the parameter you want to estimate and conduct a *pilot study* to obtain an initial estimate of the parameter of interest and, more importantly, an estimate of the variability associated with the estimator. A pilot study is a small experiment (perhaps 20 to 30 observations) used to gain some information about the population of interest. The purpose is to help plan more elaborate future experiments. Using the results of your pilot study, determine the sample size necessary to estimate the parameter to within a reasonable bound (of your choice) with a 95% confidence interval.

USING THE COMPUTER...

Refer to **Using the Computer** in Chapter 6. Recall the values of the "population" mean μ and standard deviation σ for the 1,000 zip code income measurements. Suppose our objective is to sample from this population and to estimate the mean μ using a 95% confidence interval.

a. Determine the sample size n_1 necessary to estimate μ to within $2,000 with 95% confidence. Then generate one hundred 95% confidence intervals by repeatedly drawing samples of size n_1 (with replacement) from the 1,000 measurements, and using the sample statistics to form a confidence interval. Treat σ as unknown when forming the confidence intervals. What percentage of confidence intervals contain μ?

b. Determine the sample size n_2 necessary to estimate μ to within \$500 with 95% confidence. Then generate one hundred 95% confidence intervals by repeatedly drawing samples of size n_2 (with replacement) from the 1,000 measurements, and using the sample statistics to form a confidence interval. Treat σ as unknown when forming the confidence intervals. What percentage of confidence intervals contain μ?

c. Repeat part **a**, but this time use an 80% confidence interval.

References

Arshadi, N., and Lawrence, E. C. "A characteristic appraisal of top bank executives," *Journal of Retail Banking*, Vol. 5, No. 4, Winter 1983–1984, pp. 19–25.

"Bounce for Synthetic Rubber: Tires Are Back in the Fast Lane." *Chemical Week*, April 19, 1989, pp. 40–44.

Deming, W. E. *Out of the Crisis.* Cambridge, Mass.: M.I.T. Center for Advanced Study of Engineering, 1986.

Dessler, G. "Personnel tests gain in popularity," *St. Paul Pioneer Press and Dispatch*, March 17, 1986, p. 12.

Freedman, D., Pisani, R., and Purves, R. *Statistics.* New York: W. W. Norton, 1978, Chapter 26.

Graham, E. "The entrepreneurial mystique," *Wall Street Journal*, May 20, 1985, p. 3C.

Hall, K., and Savery, L. K. "Tight reign, more stress," *Harvard Business Review*, January–February 1986, pp. 160–164.

Mendenhall, W. *Introduction to Probability and Statistics*, 7th ed. Boston: Duxbury, 1987, Chapters 8 and 9.

Moffitt, T. E., Gabrielli, W. F., Mednick, S. A., and Schulsinger, F. "Socioeconomic status, IQ, and delinquency," *Journal of Abnormal Psychology*, Vol. 90, 1981, pp. 152–156.

Wentling, R. M. "Master the Software Explosion," *Legal Assistant Today*, Vol 5, No. 4, March–April 1988, pp. 44–51.

Wentling, R. M., and Wentling, T. L. "Computer Technology in Banks: What Illinois Institutions are Using," *Bank Administration*, Vol. 64, No. 1, January 1988, pp. 42–45.

Inferences Based on a Single Sample: Tests of Hypotheses

WHERE WE'VE BEEN...

We showed how to use sample information to estimate population parameters in Chapter 7. The sampling distribution of a statistic was used to assess the reliability of an estimate, which was expressed in terms of a confidence interval.

WHERE WE'RE GOING...

We now show how to utilize sample information to test whether a population parameter is less than, equal to, or greater than a specified value. This type of inference is called a *test of hypothesis*. We show how to conduct a test of hypothesis about a population mean μ and a binomial probability *p*. But just as with estimation, we stress the measurement of the reliability of the inference. An inference without a measure of reliability is little more than a guess.

Suppose you wanted to determine whether the mean level of blood alcohol exceeds the legal limit after two drinks, or whether the mean breaking strength of sewer pipe exceeds 2,400 pounds per foot, or whether the majority of registered voters approve of the president's performance. In each case you are interested in making an inference about how the value of a parameter relates to a specific numerical value: Is it less than, equal to, or greater than the specified number? This type of inference is called a **test of hypothesis**, and testing hypotheses is the subject of this chapter.

We introduce the elements of a test of hypothesis in Section 8.1. We then show how to conduct a large-sample test of hypothesis about a population mean in Sections 8.2 and 8.3. In Section 8.4 we utilize small samples to conduct tests about means. Large-sample tests about binomial probabilities are the subject of Section 8.5, and some advanced methods for determining the reliability of a test are covered in optional Section 8.6. Finally, we show how to make inferences about a population variance in optional Section 8.7.

8.1   The Elements of a Test of Hypothesis

Suppose building specifications in a certain city require that the average breaking strength of residential sewer pipe be more than 2,400 pounds per foot of length (i.e., per lineal foot). Each manufacturer who wants to sell pipe in this city must demonstrate that its product meets the specification. Note that we are again interested in making an inference about the mean μ of a population. However, in this example we are less interested in estimating the value of μ than we are in testing a **hypothesis** about its value. That is, we want to decide whether the mean breaking strength of the pipe exceeds 2,400 pounds per lineal foot.

The method used to reach a decision is based on the rare event concept explained in earlier chapters. We define two hypotheses: (1) The **null hypothesis** is that which represents the status quo to the party performing the sampling experiment—the hypothesis that will be accepted unless the data provide convincing evidence that it is false. (2) The **research**, or **alternative**, **hypothesis** is that which will be accepted only if the data provide convincing evidence of its truth. From the point of view of the city conducting the tests, the null hypothesis is that the manufacturer's pipe does *not* meet specifications unless the tests provide convincing evidence otherwise. The null and alternative hypotheses are therefore

> *Null hypothesis* (H_0): $\mu \leq 2,400$ (i.e., the manufacturer's pipe does not meet specifications)
>
> *Alternative (research) hypothesis* (H_a): $\mu > 2,400$ (i.e., the manufacturer's pipe meets specifications)

How can the city decide when enough evidence exists to conclude that the manufacturer's pipe meets specifications? Since the hypotheses concern the value of the population mean μ, it is reasonable to use the sample mean $\bar{x}$ to make the inference, just as we did when forming confidence intervals for μ in Sections 7.1 and 7.2. The city will conclude that the pipe meets specifications only when the sample mean $\bar{x}$ convincingly indicates that the population mean exceeds 2,400 pounds per lineal foot.

"Convincing" evidence in favor of the alternative hypothesis will exist when the value of $\bar{x}$ exceeds 2,400 by an amount that cannot be readily attributed to sampling variability. To decide, we compute a **test statistic**, which is the z value that measures the distance between the value of $\bar{x}$ and the value of μ specified in the null hypothesis. When the null hypothesis contains more than one value of μ, as in this case (H_0: $\mu \leq 2{,}400$), we use the value of μ closest to the values specified in the alternative hypothesis. The idea is that if the hypothesis that μ equals 2,400 can be rejected in favor of $\mu > 2{,}400$, then μ *less than or equal to* 2,400 can certainly be rejected. Thus, the test statistic is

$$z = \frac{\bar{x} - 2{,}400}{\sigma_{\bar{x}}} = \frac{\bar{x} - 2{,}400}{\sigma/\sqrt{n}}$$

Note that a value of $z = 1$ means that $\bar{x}$ is 1 standard deviation above $\mu = 2{,}400$; a value of $z = 1.5$ means that $\bar{x}$ is 1.5 standard deviations above $\mu = 2{,}400$, etc. How large must z be before the city can be convinced that the null hypothesis can be rejected in favor of the alternative and conclude that the pipe meets specifications?

If you examine Figure 8.1, you will note that the chance of observing $\bar{x}$ more than 1.645 standard deviations above 2,400 is only .05—*if in fact the true mean μ is 2,400.* Thus, if the sample mean is more than 1.645 standard deviations above 2,400, either H_0 is true and a relatively rare event has occurred (.05 probability) or H_a is true and the population mean exceeds 2,400. Since we would most likely reject the notion that a rare event has occurred, we would reject the null hypothesis ($\mu \leq 2{,}400$) and conclude that the alternative hypothesis ($\mu > 2{,}400$) is true. What is the probability that this procedure will lead us to an incorrect decision?

Deciding that the null hypothesis is false when in fact it is true is called a **Type I decision error**. As indicated in Figure 8.1, the risk of making a Type I error—that is, deciding in favor of the research hypothesis when in fact the null hypothesis is true—is denoted by the symbol α. That is,

$$\alpha = P(\text{Type I error})$$
$$= P(\text{Rejecting the null hypothesis when in fact the null hypothesis is true})$$

In our example

$$\alpha = P(z > 1.645 \text{ when in fact } \mu = 2{,}400) = .05$$

FIGURE 8.1

The sampling distribution of $\bar{x}$, assuming $\mu = 2{,}400$

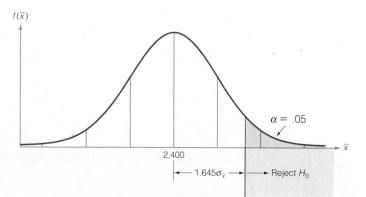

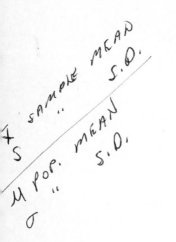

We now summarize the elements of the test:

$$H_0: \quad \mu = 2{,}400 \qquad\qquad H_a: \quad \mu > 2{,}400$$

Test statistic: $z = \dfrac{\bar{x} - 2{,}400}{\sigma_{\bar{x}}}$

Rejection region: $z > 1.645$, which corresponds to $\alpha = .05$

Note that the **rejection region** refers to the values of the test statistic for which we will **reject the null hypothesis**.

To illustrate the use of the test, suppose we test 50 sections of sewer pipe and find the mean and standard deviation for these 50 measurements to be

$$\bar{x} = 2{,}460 \text{ pounds per lineal foot} \qquad s = 200 \text{ pounds per lineal foot}$$

As in the case of estimation, we can use s to approximate σ when s is calculated from a large set of sample measurements.

The test statistic is

$$z = \frac{\bar{x} - 2{,}400}{\sigma_{\bar{x}}} = \frac{\bar{x} - 2{,}400}{\sigma/\sqrt{n}} \approx \frac{\bar{x} - 2{,}400}{s/\sqrt{n}}$$

Substituting $\bar{x} = 2{,}460$, $n = 50$, and $s = 200$, we have

$$z \approx \frac{2{,}460 - 2{,}400}{200/\sqrt{50}} = \frac{60}{28.28} = 2.12$$

Therefore, the sample mean lies $2.12\sigma_{\bar{x}}$ above the hypothesized value of μ, 2,400, as shown in Figure 8.2. Since this value of z exceeds 1.645, it falls in the rejection region. That is, we reject the null hypothesis that $\mu = 2{,}400$ and conclude that $\mu > 2{,}400$. Thus, it appears that the company's pipe has a mean strength that exceeds 2,400 pounds per lineal foot.

FIGURE 8.2

Location of the test statistic for a test of the hypothesis $H_0: \mu = 2{,}400$

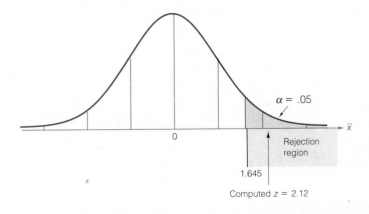

How much faith can be placed in this conclusion? What is the probability that our statistical test could lead us to reject the null hypothesis (and conclude that the company's pipe meets the city's specifications) when in fact the null hypothesis is true? The answer is $\alpha = .05$. That is, we selected the level of risk, α, of making a Type I error when we constructed the test. Thus, the chance is only 1 in 20 that our test would lead us to conclude the manufacturer's pipe satisfies the city's specifications when in fact the pipe does *not* meet specifications.

Now, suppose the sample mean breaking strength for the 50 sections of sewer pipe turned out to be $\bar{x} = 2{,}430$ pounds per lineal foot. Assuming that the sample standard deviation is still $s = 200$, the test statistic is

$$z = \frac{2{,}430 - 2{,}400}{200/\sqrt{50}} = \frac{30}{28.28} = 1.06$$

Therefore, the sample mean $\bar{x} = 2{,}430$ is only 1.06 standard deviations above the null hypothesized value of $\mu = 2{,}400$. As shown in Figure 8.3, this value does not fall into the rejection region ($z > 1.645$). Therefore, we know that we cannot reject H_0 using $\alpha = .05$. Even though the sample mean exceeds the city's specification of 2,400 by 30 pounds per lineal foot, it does not exceed the specification by enough to provide *convincing* evidence that the *population mean* exceeds 2,400.

FIGURE 8.3

Location of test statistic when $\bar{x} = 2{,}430$

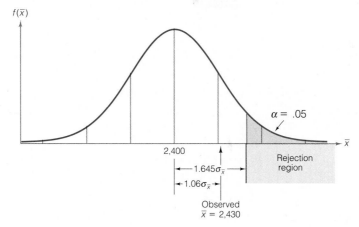

Should we accept the null hypothesis H_0: $\mu \le 2{,}400$ and conclude that the manufacturer's pipe does not meet specifications? To do so would be to risk a **Type II error**—that of concluding that the null hypothesis is true (the pipe does not meet specifications) when in fact it is false (the pipe does meet specifications). We denote the probability of committing a Type II error by β, and we show in optional Section 8.6 that β is often difficult to determine precisely. Rather than make a decision (accept H_0) for which the probability of error (β) is unknown, we avoid the potential Type II error by avoiding the conclusion that the null hypothesis is true. Instead, we will simply state that *the sample evidence is insufficient to reject H_0 at $\alpha = .05$.* Since the null hypothesis is the "status-quo" hypothesis, the effect of not rejecting H_0 is to maintain the status quo. In our pipe-testing example, the effect of having insufficient evidence to reject the null hypothesis that the pipe does not meet specifications is probably to prohibit the utilization of the manufacturer's pipe unless and until there is sufficient evidence that the pipe does meet specifications. That is, until the data indicate convincingly that the null hypothesis is false, we usually maintain the status quo implied by its truth.

Table 8.1 summarizes the four possible outcomes of a test of hypothesis. The "true state of nature" columns in Table 8.1 refer to the fact that either the null

TABLE 8.1   **Conclusions and Consequences for a Test of Hypothesis**

		TRUE STATE OF NATURE	
		H_0 true	H_a true
CONCLUSION	H_0 true	Correct decision	Type II error (probability β)
	H_a true	Type I error (probability α)	Correct decision

hypothesis H_0 is true or the alternative hypothesis H_a is true. Note that the true state of nature is unknown to the researcher conducting the test. The "decision" rows in Table 8.1 refer to the action of the researcher, assuming that he or she will either conclude that H_0 is true or that H_a is true, based on the results of the sampling experiment. Note that a Type I error can be made *only* when the alternative hypothesis is accepted (equivalently, when the null hypothesis is rejected), and a Type II error can be made *only* when the null hypothesis is accepted. Our policy will be to make a decision only when we know the probability of making the error that corresponds to that decision. Since α is usually specified by the analyst, we will generally be able to reject H_0 (accept H_a) when the sample evidence supports that decision. However, since β is usually not specified, we will generally avoid the decision to accept H_0, preferring instead to state that the sample evidence is insufficient to reject H_0 when the test statistic is not in the rejection region.

The elements of a test of hypothesis are summarized in the box. Note that the first four elements are all specified *before* the sampling experiment is performed. In no case will the results of the sample be used to determine the hypotheses—the data are collected to test the predetermined hypotheses, not to formulate them.

Elements of a Test of Hypothesis

1. *Null hypothesis* (H_0):   A theory about the values of one or more population parameters. The theory generally represents the status quo, which we accept until proven false.

2. *Alternative (research) hypothesis* (H_a):   A theory that contradicts the null hypothesis. The theory generally represents that which we will accept only when sufficient evidence exists to establish its truth.

3. *Test statistic:*   A sample statistic used to decide whether to reject the null hypothesis.

4. *Rejection region:*   The numerical values of the test statistic for which the null hypothesis will be rejected. The rejection region is chosen so that the probability is α that it will contain the test statistic when the null hypothesis is true, thereby leading to a Type I error. The value of α is usually chosen to be small (e.g., .01, .05, or .10), and is referred to as the **level of significance** of the test.

5. *Assumptions:*   Any assumptions made about the population(s) being sampled should be clearly stated.

6. *Experiment and calculation of test statistic:*   The sampling experiment is performed, and the numerical value of the test statistic is determined.

> 7. *Conclusion:*
>   a. If the numerical value of the test statistic falls in the rejection region, we reject the null hypothesis and conclude that the alternative hypothesis is true. We know that the hypothesis-testing process will lead to this conclusion incorrectly (Type I error) only $100\alpha\%$ of the time when H_0 is true.
>   b. If the test statistic does not fall in the rejection region, we reserve judgment about which hypothesis is true. We do not conclude that the null hypothesis is true because we do not (in general) know the probability β that our test procedure will lead to an incorrect acceptance of H_0 (Type II error).

| CASE STUDY 8.1 | STATISTICS IS MURDER! |

The jury trial of an accused murderer is analogous to the statistical hypothesis-testing process. Each of the elements of a test of hypothesis applies to the jury system of deciding the guilt or innocence of the accused:

1. H_0:   The null hypothesis in a jury trial is that the accused is innocent. The status-quo hypothesis in the American system of justice is innocence, which is assumed to be true until proven otherwise.
2. H_a: The alternative hypothesis is guilt, which is accepted only when sufficient evidence exists to establish its truth.
3. *Test statistic:*   The test statistic in a trial is the final vote of the jury—i.e., the number of the jury members who vote "guilty."
4. *Rejection region:*   In a murder trial the jury vote must be unanimous in favor of guilt before the null hypothesis of innocence is rejected in favor of the alternative hypothesis of guilt. Thus, for a 12-member jury trial, the rejection region is $x = 12$, where x is the number of "guilty" votes.
5. *Assumption:*   The primary assumption made in trials concerns the method of selecting the jury. The jury is assumed to represent a random sample of citizens who have no prejudice concerning the case.
6. *Experiment and calculation of the test statistic:*   The sampling experiment is analogous to the jury selection, the trial, and the jury deliberations. The final vote of the jury is analogous to the calculation of the test statistic.
7. *Conclusion:*
   a. If the vote of the jury is unanimous in favor of guilt, the null hypothesis of innocence is rejected and the court concludes that the accused murderer is guilty. Although the court does not, in general, know the probability α that the conclusion is in error, the system relies on the belief that the value is made very small by requiring a unanimous vote before guilt is concluded.
   b. Any vote other than a unanimous one for guilt results in the court reserving judgment about the hypotheses, either by declaring the accused "not guilty," or by declaring a mistrial and repeating the "test" with a new jury. (The latter is analogous to collecting more data and repeating a statistical test of hypothesis.) The court never accepts the null hypothesis by declaring the accused "innocent," perhaps recognizing both that innocence is the status-quo hypothesis and does not need to be proved and that the probability β of incorrectly concluding innocence may not be as small as α.

As in the case of tests of statistical hypotheses, we may never know whether the verdict in a murder trial is correct. Instead, we rely on the knowledge that the trial procedure will lead to incorrect conclusions (especially, guilt when the accused is in fact innocent) in only a very small percentage of trials.

EXERCISES 8.1–8.9

LEARNING THE MECHANICS

8.1   Which hypothesis, the null or the alternative, is the status-quo hypothesis? Which is the research hypothesis?

8.2   Which element of a test of hypothesis is used to decide whether to reject the null hypothesis in favor of the alternative hypothesis?

8.3   What is the level of significance of a test of hypothesis?

8.4   What is the difference between Type I and Type II errors in hypothesis testing? How do α and β relate to Type I and Type II errors?

8.5   List the four possible results of the combinations of decisions and true states of nature for a test of hypothesis.

8.6   Why do we (generally) reject the null hypothesis when the test statistic falls in the rejection region, but do not accept the null hypothesis when the test statistic does not fall in the rejection region?

8.7   If you test a hypothesis and reject the null hypothesis in favor of the alternative hypothesis, does your test prove that the alternative hypothesis is correct? Explain.

APPLYING THE CONCEPTS

8.8   In 1895 an Italian criminologist, Cesare Lombroso, proposed that blood pressure be used to test for truthfulness. In the 1930s, William Marston added the measurements of respiration and perspiration to the process and called his machine the *polygraph*—or lie detector. Today, the federal court system will not consider polygraph results as evidence, but nearly half of the state courts do permit polygraph tests under certain circumstances. In addition, its use in screening job applicants is on the rise (Dujack, 1986). Physicians Michael Phillips, Allan Brett, and John Beary subjected the polygraph to the same careful testing given to medical diagnostic tests. They found that if 1,000 people were subjected to the polygraph and 500 told the truth and 500 lied, the polygraph would indicate that approximately 185 of the truth-tellers were liars and that approximately 120 of the liars were truth-tellers ("Lie Detectors Can Make a Liar of You," *Discover*, June 1986, p. 7).
   a. In the application of a polygraph test, an individual is presumed to be a truth-teller (H_0) until "proven" a liar (H_a). In this context, what is a Type I error? A Type II error?
   b. According to Phillips, Brett, and Beary, what is the probability (approximately) that a polygraph test will result in a Type I error? A Type II error?

8.9   When a new drug is formulated, the pharmaceutical company must subject it to lengthy and involved testing before receiving the necessary permission from the Food and Drug Administration (FDA) to market the drug. The FDA's policy is that the pharmaceutical company must provide substantial evidence that a new drug is safe prior to receiving FDA approval, so that the FDA can confidently certify the safety of the drug to potential consumers.
   a. If the new drug testing were to be placed in a test of hypothesis framework, would the null hypothesis be that the drug is safe or unsafe? The alternative hypothesis?
   b. Given the choice of null and alternative hypotheses in part a, describe Type I and Type II errors in terms of this application. Define α and β in terms of this application.

c. If the FDA wants to be very confident that the drug is safe before permitting it to be marketed, is it more important that α or β be small? Explain.

8.2 Large-Sample Test of Hypothesis About a Population Mean

In Section 8.1 we learned that the null and alternative hypotheses form the basis for a test of hypothesis inference. The null and alternative hypotheses may take one of several forms. In the sewer pipe example we tested the null hypothesis that the population mean strength of the pipe is less than or equal to 2,400 pounds per lineal foot against the alternative hypothesis that the mean strength exceeds 2,400. That is, we tested

$$H_0: \quad \mu \leq 2,400$$
$$H_a: \quad \mu > 2,400$$

This is a **one-tailed** (or **one-sided**) statistical test because the alternative hypothesis specifies that the population parameter (the population mean μ, in this example) is strictly greater than a specified value (2,400, in this example). If the null hypothesis had been $H_0: \mu \geq 2,400$ and the alternative hypothesis had been $H_a: \mu < 2,400$, the test would still be one-sided, because the parameter is still specified to be on "one side" of the null hypothesis value. Some statistical investigations seek to show that the population parameter is *either larger or smaller* than some specified value. Such an alternative hypothesis is called a **two-tailed** (or **two-sided**) hypothesis.

While alternative hypotheses are always specified as strict inequalities, such as $\mu < 2,400$, $\mu > 2,400$, or $\mu \neq 2,400$, null hypotheses are usually specified as equalities, such as $\mu = 2,400$. Even when the null hypothesis is an inequality, such as $\mu \leq 2,400$, we specify $H_0: \mu = 2,400$, reasoning that if sufficient evidence exists to show that $H_a: \mu > 2,400$ is true when tested against $H_0: \mu = 2,400$, then surely sufficient evidence exists to reject $\mu < 2,400$ as well. Therefore, the null hypothesis is specified as the value of μ closest to a one-sided alternative hypothesis and as the only value *not* specified in a two-tailed alternative hypothesis. The steps for selecting the null and alternative hypotheses are summarized in the box.

| Steps for Selecting the Null and Alternative Hypotheses

1. Select the *alternative hypothesis* as that which the sampling experiment is intended to establish. The alternative hypothesis will assume one of three forms:
   a. One-tailed, upper-tailed   *Example:* H_a: $\mu > 2,400$
   b. One-tailed, lower-tailed   *Example:* H_a: $\mu < 2,400$
   c. Two-tailed   *Example:* H_a: $\mu \neq 2,400$

2. Select the *null hypothesis* as the status quo, that which will be presumed true unless the sampling experiment conclusively establishes the alternative hypothesis. The null hypothesis will be specified as that parameter value closest to the alternative in one-tailed tests, and as the complementary (or only unspecified) value in two-tailed tests.

   Example: H_0: $\mu = 2,400$

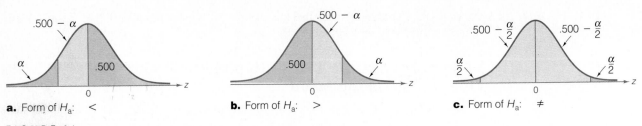

FIGURE 8.4

Rejection regions corresponding to one- and two-tailed tests

The rejection region for a two-tailed test differs from that for a one-tailed test. When we are trying to detect departure from the null hypothesis in *either* direction, we must establish a rejection region in both tails of the sampling distribution of the test statistic. Figures 8.4(a) and (b) show the one-tailed rejection regions for lower- and upper-tailed tests, respectively. The two-tailed rejection region is illustrated in Figure 8.4(c). Note that a rejection region is established in each tail of the sampling distribution for a two-tailed test.

The rejection regions corresponding to typical values selected for α are shown in Table 8.2 for one- and two-tailed tests. Note that the smaller α you select, the more evidence (the larger z) you will need before you can reject H_0.

TABLE 8.2   **Rejection Regions for Common Values of α**

	ALTERNATIVE HYPOTHESES		
	Lower-tailed	Upper-tailed	Two-tailed
$\alpha = .10$	$z < -1.28$	$z > 1.28$	$z < -1.645$ or $z > 1.645$
$\alpha = .05$	$z < -1.645$	$z > 1.645$	$z < -1.96$ or $z > 1.96$
$\alpha = .01$	$z < -2.33$	$z > 2.33$	$z < -2.575$ or $z > 2.575$

EXAMPLE 8.1

The effects of drug and alcohol on the nervous system have been the subject of considerable research recently. Suppose a research neurologist is testing the effects of a drug on response time by injecting 100 rats with a unit dose of the drug, subjecting each to a neurological stimulus, and recording its response time. The neurologist knows that the mean response time for rats not injected with the drug (the "control" mean) is 1.2 seconds. She wishes to test whether the mean response time for drug-injected rats differs from 1.2 seconds. Set up the test of hypothesis for this experiment, using $\alpha = .01$.

Solution

Since the neurologist wishes to detect whether the mean response time, μ, for drug-injected rats differs from the control mean of 1.2 seconds in *either* direction—i.e., $\mu < 1.2$ or $\mu > 1.2$—we conduct a two-tailed statistical test. Following the procedure for selecting the null and alternative hypotheses, we specify as the alternative hypothesis that the mean differs from 1.2 seconds, since determining whether the drug-injected mean differs from the control mean is the purpose of the experiment. The null hypothesis is the presumption that drug-injected rats have the same mean response time as control rats unless the research indicates otherwise. Thus,

$$H_0: \quad \mu = 1.2$$
$$H_a: \quad \mu \neq 1.2 \quad \text{(i.e., } \mu < 1.2 \text{ or } \mu > 1.2)$$

The test statistic measures the number of standard deviations between the observed value of $\bar{x}$ and the null hypothesized value $\mu = 1.2$:

$$\textit{Test statistic:} \quad z = \frac{\bar{x} - 1.2}{\sigma_{\bar{x}}}$$

The rejection region must be designated to detect a departure from $\mu = 1.2$ in *either* direction, so we will reject H_0 for values of z that are either too small (negative) or too large (positive). To determine the precise values of z that comprise the rejection region, we first select α, the probability that the test will lead to incorrect rejection of the null hypothesis. Then we divide α equally between the lower and upper tail of the distribution of z, as shown in Figure 8.5. In this example, $\alpha = .01$, so $\alpha/2 = .005$ is placed in each tail. The areas in the tails correspond to $z = -2.575$ and $z = 2.575$, respectively (from Table 8.2):

Rejection region: $\quad z < -2.575 \quad$ or $\quad z > 2.575 \quad$ (Figure 8.5)

Assumptions: Since the sample size of the experiment is large enough ($n > 30$), the Central Limit Theorem will apply, and no assumptions need be made about the population of response time measurements. The sampling distribution of the sample mean response of 100 rats will be approximately normal regardless of the distribution of the individual rats' response times.

FIGURE 8.5
Two-tailed rejection region: $\alpha = .01$

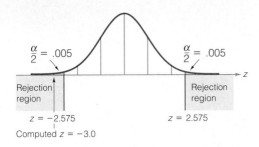

Note that the test in Example 8.1 is set up *before* the sampling experiment is conducted. The data are not used to develop the test. Evidently, the neurologist wants to conclude that the mean response time for the drug-injected rats differs from the control mean only when the evidence is very convincing, because the value of α has been set quite low at .01. If the experiment results in the rejection of H_0, she can be 99% confident that the mean response time of the drug-injected rats differs from the control mean.

Once the test is set up, she is ready to perform the sampling experiment and conduct the test. The test is performed in Example 8.2.

EXAMPLE 8.2

Refer to the neurological response-time test set up in Example 8.1. Suppose the sampling experiment is conducted with the following results:

$n = 100$ response times for drug-injected rats

$\bar{x} = 1.05$ seconds $s = .5$ second

Use the results of the sampling experiment to conduct the test of hypothesis.

Solution

Since the test is completely specified in Example 8.1, we simply substitute the sample statistics into the test statistic:

$$z = \frac{\bar{x} - 1.2}{\sigma_{\bar{x}}} = \frac{\bar{x} - 1.2}{\sigma/\sqrt{n}} = \frac{1.05 - 1.2}{\sigma/\sqrt{100}}$$

$$\approx \frac{1.05 - 1.2}{s/10} = \frac{-.15}{.5/10} = -3.0$$

The implication is that the sample mean, 1.05, is (approximately) 3 standard deviations below the null hypothesized value of 1.2 in the sampling distribution of $\bar{x}$. You can see in Figure 8.5 that this value of z is in the lower-tail rejection region, which consists of all values of $z < -2.575$. This sampling experiment provides sufficient evidence to reject H_0 and conclude, at the $\alpha = .01$ level of significance, that the mean response time for drug-injected rats differs from the control mean of 1.2 seconds. It appears that the rats receiving an injection of this drug have a mean response time that is less than 1.2 seconds.

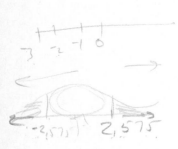

Two final points about the test of hypothesis in Example 8.2 apply to all statistical tests:

1. Since z is less than -2.575, it is tempting to state our conclusion at a significance level lower than $\alpha = .01$. We resist the temptation because the level of α is determined *before* the sampling experiment is performed. If we decide that we are willing to tolerate a 1% Type I error rate, the result of the sampling experiment should have no effect on that decision. *In general, the same data should not be used both to set up and to conduct the test.*
2. When we state our conclusion at the .01 level of significance, we are referring to the failure rate of the *procedure*, not the result of this particular test. We know that the test procedure will lead to the rejection of the null hypothesis only 1% of the time when in fact $\mu = 1.2$. *Therefore, when the test statistic falls in the rejection region, we infer that the alternative $\mu \neq 1.2$ is true and express our confidence in the procedure by quoting the α level of significance, or the $100(1 - \alpha)\%$ confidence level.*

The setup of a large-sample test of hypothesis about a population mean is summarized in the next box. Both the one- and two-tailed tests are shown.

> ### Large-Sample Test of Hypothesis About μ
>
> ONE-TAILED TEST
>
> H_0: $\mu = \mu_0$*
>
> H_a: $\mu < \mu_0$
> (or H_a: $\mu > \mu_0$)
>
> Test statistic: $z = \dfrac{\bar{x} - \mu_0}{\sigma_{\bar{x}}}$
>
> Rejection region: $z < -z_\alpha$
> (or $z > z_\alpha$ when H_a: $\mu > \mu_0$)
>
> where z_α is chosen so that
> $P(z > z_\alpha) = \alpha$
>
> TWO-TAILED TEST
>
> H_0: $\mu = \mu_0$*
>
> H_a: $\mu \neq \mu_0$
>
> Test statistic: $z = \dfrac{\bar{x} - \mu_0}{\sigma_{\bar{x}}}$
>
> Rejection region: $z < -z_{\alpha/2}$
> or $z > z_{\alpha/2}$
>
> where $z_{\alpha/2}$ is chosen so that
> $P(z > z_{\alpha/2}) = \alpha/2$
>
> *Assumptions:*   No assumptions need to be made about the probability distribution of the population because the Central Limit Theorem assures us that, for large samples, the test statistic will be approximately normally distributed regardless of the shape of the underlying probability distribution of the population.

Once the test has been set up, the sampling experiment is performed and the test statistic is calculated. The next box contains possible conclusions for a test of hypothesis, depending on the result of the sampling experiment.

> ### Possible Conclusions for a Test of Hypothesis
>
> 1. If the calculated test statistic falls in the rejection region, reject H_0 and conclude that the alternative hypothesis H_a is true. State that you are rejecting H_0 at the α level of significance. Remember that the confidence is in the testing *process*, not the particular result of a single test.
>
> 2. If the test statistic does not fall in the rejection region, conclude that the sampling experiment does not provide sufficient evidence to reject H_0 at the α level of significance. [Generally, we will not "accept" the null hypothesis unless the probability β of a Type II error has been calculated (see optional Section 8.6).]

| CASE STUDY 8.2

HYPOTHESIS TESTS IN COMPUTER SECURITY SYSTEMS

Hackers are able to crack password codes and wander the phone lines, trespassing through data banks. Industrial spies raid secret data files, snatching new product designs. Computer criminals enrich themselves,

---

*Note: μ_0 is the symbol for the numerical value assigned to μ under the null hypothesis.

plundering electronic fund transfer networks. The Defense Department is positive—almost—that its top-security computer systems are invulnerable to penetration by outsiders or, worse, insiders with evil intentions.

This paragraph introduces an article by Daniel Kagan* on one of the most pressing problems in high-technology industries today: computer security. The movie *War Games* is a fictional account of a young man's successful attempt to crack the security codes of Defense Department computers, gaining access to highly confidential, sensitive, and, as it turns out in the movie, potentially dangerous information. Perhaps more disturbing, however, are the many factual reports of computer thieves obtaining access to bank accounts, government data bases, and university research computers. Several proposed solutions to the growing computer security problem are presented in Kagan's article "Locking Up Data."

The objective of computer security is to allow only authorized personnel access to the computer's data files. This goal is typically achieved by use of a *password*—a collection of symbols (usually letters and numbers) that must be supplied by the user before the computer permits access to the account. The problem is that persistent computer thieves can program their computers to generate millions of combinations of letters and numbers and to enter them into the computer into which access is desired until the correct password is found. Some accounts are doubly protected with an unlisted phone number providing a first level of protection, followed by the password. This measure simply means that gaining illegal access may take longer, since the thief must now wait until his computer cracks both levels of security. The problem is augmented by the more common, less intriguing, but equally damaging illegal access by former employees and friends of current employees who steal the necessary passwords.

Omni reports on several innovative new proposals for solving the computer security problem. One school of thought consists of "a move away from I.D. verification based on what the operator *has* (like a magnetic stripe card) and what he *knows* (like a password)—toward using what the operator *is*. . . . Authorized users are identified by unique body characteristics, 'You can't leave your body at home like a card or key; and no one can steal it,' quips Tom Catto of Palmguard, Inc., in Beaverton, Oregon." Palmguard's security system consists of computer identification of the user's palm before access is permitted. The system tests the hypothesis

H_0:   The proposed user is authorized

versus

H_a: The proposed user is unauthorized

by checking characteristics of the proposed user's palm against those stored in the authorized users' data bank. Palmguard reports that the Type I error rate is less than 1%, whereas the Type II error rate is .00025%.

Since the Type I error refers to the rejection of H_0 when H_0 is true, the implication is that an authorized user will be denied access by the computer less

---

*Reprinted from Kagan, D. "Locking up data," *Omni*, March 1984. Reprinted by permission of Omni International Ltd.

than 1% of the time. This is primarily an inconvenience, since the authorized user can try to gain access again with reasonable assurance of success. Of more importance is the Type II error rate, which refers to the acceptance of H_0 when H_0 is false, or the probability that an unauthorized user will be granted access to the account. Palmguard's claim implies that a Type II error will occur only 25 times in 10 million.

Another new security system, the EyeDentifyer, "spots authorized computer users by reading the one-of-a-kind patterns formed by the network of minute blood vessels across the retina at the back of the eye." The Type I and II error rates are reported to be .01% (1 in 10,000) and .005% (5 in 100,000), respectively.

As new security systems are developed, the Type I and Type II error rates provide objective standards with which to compare them. Of course, the ultimate security system would have a Type II error rate of 0, but as with tests of statistical hypotheses, this is probably an impractical (and unachievable) goal.

EXERCISES 8.10–8.21

LEARNING THE MECHANICS

8.10 For each of the following rejection regions, sketch the sampling distribution for z and indicate the location of the rejection region.
a. $z > 1.96$ b. $z > 1.645$ c. $z > 2.58$ d. $z < -1.28$
e. $z < -1.645$ or $z > 1.645$ f. $z < -2.58$ or $z > 2.58$
g. For each of the rejection regions specified in parts a–f, what is the probability that a Type I error will be made?

8.11 Suppose you are interested in conducting the statistical test of H_0: $\mu = 200$ against H_a: $\mu > 200$, and you have decided to use the following decision rule: Reject H_0 if the sample mean of a random sample of 100 items is more than 212. Assume that the standard deviation of the population is 80.
a. Express the decision rule in terms of z.
b. Find α, the probability of making a Type I error, by using this decision rule.

8.12 A random sample of 100 observations from a population with standard deviation 60 yielded a sample mean of 110.
a. Test the null hypothesis that $\mu = 100$ against the alternative hypothesis that $\mu > 100$ using $\alpha = .05$. Interpret the results of the test.
b. Test the null hypothesis that $\mu = 100$ against the alternative hypothesis that $\mu \neq 100$ using $\alpha = .05$. Interpret the results of the test.
c. Compare the results of the two tests you conducted. Explain why the results differ.

8.13 A random sample of 49 observations produced the following sums:

$$\sum x = 20.7 \qquad \sum (x_i - \bar{x})^2 = 2.155$$

a. Test the null hypothesis that $\mu = .47$ against the alternative hypothesis that $\mu < .47$ using $\alpha = .10$. Interpret the result of the test.
b. Test the null hypothesis that $\mu = .47$ against the alternative hypothesis that $\mu \neq .47$ using $\alpha = .10$. Interpret the result.

8.14   The Environmental Protection Agency (EPA) estimated that the 1991 G-car obtains a mean of 35 miles per gallon on the highway, and the company that manufactures the car claims that it exceeds the EPA estimate in highway driving. To support its assertion, the company randomly selects thirty-six 1991 G-cars and records the mileage obtained for each car over a driving course similar to that used by the EPA. The following data resulted:

$$\bar{x} = 36.8 \text{ miles per gallon} \qquad s = 6.0 \text{ miles per gallon}$$

a. If the auto manufacturer wishes to show that the mean miles per gallon for 1991 G-cars is greater than 35 miles per gallon, what should it choose for the alternative hypothesis? The null hypothesis?
b. Do the data provide sufficient evidence to support the auto manufacturer's claim? Test using $\alpha = .05$.

8.15   Refer to Exercise 8.14. Repeat the test using $\alpha = .10$. How does your conclusion differ from that in Exercise 8.14? Why?

8.16   A pain reliever currently being used in a hospital is known to bring relief to patients in a mean time of 3.5 minutes. To compare a new pain reliever with the one currently being used, the new drug is administered to a random sample of 50 patients. The mean time to relief for the sample of patients is 2.8 minutes and the standard deviation is 1.14 minutes. Do the data provide sufficient evidence to conclude that the new drug was effective in reducing the mean time until a patient receives relief from pain? Test using $\alpha = .10$.

8.17   Small increases in the mean level of bills for monthly long-distance telephone calls produce substantial increases in the profits for telephone companies. A telephone company's records indicate that the amounts paid by private customers per month for long-distance telephone calls have a distribution with mean $17.10 and standard deviation $21.21.
a. If a random sample of 50 bills is taken, what is the approximate probability that the sample mean is greater than $20?
b. If a random sample of 100 bills is taken, what is the approximate probability that the sample mean is greater than $20?
c. Suppose a random sample of 100 customers' bills during a given month produced a sample mean of $22.10 expended for long-distance calls. Do these data suggest that the mean level of billing per private customer for long-distance calls is in excess of $17.10? Test using $\alpha = .10$.

8.18   In 1986, basic cable television service cost an average of $11.09 per month in the United States (*Statistical Abstract of the U.S.: 1990*, p. 554). In the spring of 1990, the Federal Communications Commission (FCC) noted that the cost of basic cable service had risen since 1986, but that it cost on average no more than $15.00 per month. A consumer advocacy group doubts the FCC's claim. In order to investigate the claim, the group randomly samples 33 of the more than 4,000 cable systems in the United States and asks each what its basic service charge was in early 1990. The mean and standard deviation of the sample of basic rates are

$$\bar{x} = \$15.33 \qquad s = \$1.18$$

a. Specify the null and alternative hypotheses that should be used by the consumer advocacy group in investigating the FCC's claim.
b. With respect to the hypotheses you specified in part **a**, explain the practical implications of making a Type I error and a Type II error.
c. In conducting this hypothesis test, what is the probability of committing a Type I error?
d. Conduct the hypothesis test you described in part **a** and interpret the test's result in the context of this exercise. Use $\alpha = .10$.

8.19   The University of Minnesota uses thousands of fluorescent light bulbs each year. The brand of bulb it currently uses has a mean life of 900 hours. A manufacturer claims that its new brand of bulbs, which cost the same as the brand the university currently uses, has a mean life of more than 900 hours. The university has decided to purchase the new brand if, when tested, the test evidence supports the manufacturer's claim at the .05 significance level. Suppose 64 bulbs were tested with the following results:

$\bar{x} = 920$ hours $s = 80$ hours

Will the University of Minnesota purchase the new brand of fluorescent bulbs?

8.20   The introduction of printed circuit boards (PCBs) in the 1950s revolutionized the electronics industry. However, solder-joint defects on PCBs have plagued electronics manufacturers since the introduction of the PCB. A single PCB may contain hundreds of solder joints. Until the 1980s, the only means of checking the quality of solder joints was by visual inspection. Because of the low reliability of visual inspection, some manufacturers required each of their joints to be inspected four times by four different people. Now both X-ray and laser technologies are available for use in inspection (Streeter, 1986). A particular manufacturer of laser-based inspection equipment claims that its product can inspect on average at least ten solder joints per second when the joints are spaced .1 inch apart. The equipment was tested by a potential buyer on 48 different PCBs. In each case, the equipment was operated for exactly 1 second. The number of solder joints inspected on each run are as follows:

10	9	10	10	11	9	12	8	8	9	6	10
7	10	11	9	9	13	9	10	11	10	12	8
9	9	9	7	12	6	9	10	10	8	7	9
11	12	10	0	10	11	12	9	7	9	9	10

a. The potential buyer wants to know whether the sample data refute the manufacturer's claim. Specify the null and alternative hypotheses that the buyer should test.

b. In the context of this exercise, what is a Type I error? A Type II error?

c. Conduct the hypothesis test you described in part **a**, and interpret the test's result in the context of this exercise. Use $\alpha = .05$.

8.21   A relatively high-priced chain of steakhouses has decided to sell new franchises only in metropolitan areas of at least 100,000 population with a mean household income of at least $30,000 per year. A group of investors interested in obtaining a franchise has identified an area with a population of 175,000. The investors select a random sample of 250 households from the community and calculate the following mean and standard deviation of the household incomes for this sample:

$\bar{x} = \$33,545$ $s = \$23,000$

Should the chain sell a franchise to these investors for this area?

8.3   Observed Significance Levels: *p*-Values

According to the statistical test procedure described in Section 8.2, the rejection region and, correspondingly, the value of α are selected prior to conducting the test, and the conclusions are stated in terms of rejecting or not rejecting the null hypothesis. A second method of presenting the results of a statistical test is one that reports the extent to which the test statistic disagrees with the null hypothesis and leaves to the reader the task of deciding whether to reject the null hypothesis. This measure of disagreement is called the *observed significance level* (or *p-value*) for the test.

> **Definition 8.1**
>
> The **observed significance level**, or ***p*-value**, for a specific statistical test is the probability (assuming H_0 is true) of observing a value of the test statistic that is at least as contradictory to the null hypothesis, and supportive of the alternative hypothesis, as the one computed from the sample data.

For example, the value of the test statistic computed for the sample of $n = 50$ sections of sewer pipe was $z = 2.12$. Since the test is one-tailed—i.e., the alternative (research) hypothesis of interest is H_a: $\mu > 2{,}400$—values of the test statistic even more contradictory to H_0 than the one observed would be values larger than $z = 2.12$. Therefore, the observed significance level (*p*-value) for this test is

$$p\text{-value} = P(z \geq 2.12)$$

or, equivalently, the area under the standard normal curve to the right of $z = 2.12$ (see Figure 8.6).

FIGURE 8.6

Finding the *p*-value for an upper-tailed test when $z = 2.12$

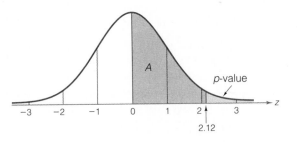

The area A in Figure 8.6 is given in Table IV in Appendix A as .4830. Therefore, the upper-tail area corresponding to $z = 2.12$ is

$$p\text{-value} = .5 - .4830 = .0170$$

Consequently, we say that these test results are "very significant"; i.e., they disagree rather strongly with the null hypothesis, H_0: $\mu = 2{,}400$, and favor H_a: $\mu > 2{,}400$. The probability of observing a z value as large as 2.12 is only .0170, if in fact the true value of μ is 2,400.

If you are inclined to select $\alpha = .05$ for this test, then you would reject the null hypothesis because the *p*-value for the test, .0170, is less than .05. In contrast, if you choose $\alpha = .01$ you would not reject the null hypothesis because the *p*-value for the test is larger than .01. Thus, the use of the observed significance level is identical to the test procedure described in the preceding sections except that the choice of α is left to you.

The steps for calculating the *p*-value corresponding to a test statistic for a population mean are given in the next box.

> | Steps for Calculating the *p*-Value for a Test of Hypothesis
>
> 1. Determine the value of the test statistic z corresponding to the result of the sampling experiment.
> 2. a. If the test is one-tailed, the *p*-value is equal to the tail area beyond z in the same direction as the alternative hypothesis. Thus, if the alternative hypothesis is of the form $>$, the *p*-value is the area to the right of, or above, the observed z value. Conversely, if the alternative is of the form $<$, the *p*-value is the area to the left of, or below, the observed z value.
>   b. If the test is two-tailed, the *p*-value is equal to twice the tail area beyond the observed z value in the direction of the sign of z. That is, if z is positive, the *p*-value is twice the area to the right of, or above, the observed z value. Conversely, if z is negative, the *p*-value is twice the area to the left of, or below, the observed z value.

EXAMPLE 8.3

Find the observed significance level for the test of the mean response time for drug-injected rats in Examples 8.1 and 8.2.

Solution

Example 8.1 presented a two-tailed test of the hypothesis

H_0: $\mu = 1.2$ seconds

against the alternative hypothesis

H_a: $\mu \neq 1.2$ seconds

The observed value of the test statistic in Example 8.2 was $z = -3.0$, and any value of z less than -3.0 or greater than $+3.0$ (because this is a two-tailed test) would be even more contradictory to H_0. Therefore, the observed significance level for the test is

p-value $= P(z < -3.0 \text{ or } z > +3.0)$

Thus, we calculate the area below the observed z value, $z = -3.0$, and double it. Consulting Table IV in Appendix A, we find that $P(z < -3.0) = .5 - .4987 = .0013$. Therefore, the *p*-value for this two-tailed test is

$2P(z < -3.0) = 2(.0013) = .0026$

We can interpret this *p*-value as a strong indication that the mean reaction time of drug-injected rats differs from the control mean ($\mu \neq 1.2$), since we would observe a test statistic this extreme or more extreme only 26 in 10,000 times if the drug-injected mean were equal to the control mean ($\mu = 1.2$). The extent to which the mean differs from 1.2 could be better determined by calculating a confidence interval for μ.

When publishing the results of a statistical test of hypothesis in journals, case studies, reports, etc., many researchers make use of *p*-values. Instead of selecting

α beforehand and then conducting a test, as outlined in this chapter, the researcher computes and reports the value of the appropriate test statistic and its associated p-value. It is left to the reader of the report to judge the significance of the result—i.e., the reader must determine whether to reject the null hypothesis in favor of the alternative hypothesis, based on the reported p-value. This p-value is often referred to as the **observed significance level** of the test. Usually, the null hypothesis is rejected if the observed significance level is *less* than the fixed significance level, α, chosen by the reader. The inherent advantages of reporting test results in this manner are twofold: (1) Readers are permitted to select the maximum value of α that they would be willing to tolerate if they actually carried out a standard test of hypothesis in the manner outlined in this chapter, and (2) a measure of the degree of significance of the result (i.e., the p-value) is provided.

Reporting Test Results as p-Values: How to Decide Whether to Reject H_0

1. Choose the maximum value of α that you are willing to tolerate.
2. If the observed significance level (p-value) of the test is less than the chosen value of α, reject the null hypothesis.

EXERCISES 8.22–8.37

LEARNING THE MECHANICS

8.22   If a hypothesis test were conducted using $\alpha = .05$, for which of the following p-values would the null hypothesis be rejected?
a. .06   b. .10   c. .01
d. .001   e. .251   f. .042

8.23   For each α and observed significance level (p-value) pair, indicate whether the null hypothesis would be rejected.
a. $\alpha = .05$, p-value $= .10$ b. $\alpha = .10$, p-value $= .05$
c. $\alpha = .01$, p-value $= .001$ d. $\alpha = .025$, p-value $= .05$
e. $\alpha = .10$, p-value $= .45$

8.24   Explain the difference between statistical significance and practical significance.

8.25   An analyst tested the null hypothesis $\mu \geq 20$ against the alternative hypothesis that $\mu < 20$. The analyst reported a p-value of .06. What is the smallest value of α for which the null hypothesis would be rejected?

8.26   In a test of H_0: $\mu \leq 100$ against H_a: $\mu > 100$, the sample data yielded the test statistic $z = 2.26$. Find the p-value for the test.

8.27   In a test of H_0: $\mu = 100$ against H_a: $\mu \neq 100$, the sample data yielded the test statistic $z = 2.08$. Find the p-value for the test.

8.28   In a test of H_0: $\mu \geq 100$ against H_a: $\mu < 100$, the sample data yielded the test statistic $z = -1.11$. Find the observed significance level of the test.

8.29 In a test of the hypothesis H_0: $\mu = 50$ versus H_a: $\mu > 50$, a sample of $n = 100$ observations possessed mean $\bar{x} = 50.5$ and standard deviation $s = 3.3$. Find and interpret the p-value for this test.

8.30 In a test of the hypothesis H_0: $\mu = 10$ versus H_a: $\mu \neq 10$, a sample of $n = 50$ observations possessed mean $\bar{x} = 9.5$ and standard deviation $s = 2.1$. Find and interpret the p-value for this test.

APPLYING THE CONCEPTS

8.31 The manufacturer of an over-the-counter pain reliever claims that its product brings pain relief to headache sufferers in less than 3.5 minutes, on average. In order to be able to make this claim in its television advertisements, the manufacturer was required by a particular television network to present statistical evidence in support of the claim. The manufacturer reported that for a random sample of 50 headache sufferers, the mean time to relief was 3.3 minutes and the standard deviation was 66 seconds.
a. Do these data support the manufacturer's claim? Test using $\alpha = .05$.
b. Report the p-value of the test.
c. In general, do large p-values or small p-values support the manufacturer's claim? Explain.

8.32 *The Chronicle of Higher Education Almanac* (Sept. 1990) reported that for the 1989–1990 academic year, 4-year private colleges charged students an average of $8,446 for tuition and fees, whereas at 4-year public colleges the average was $1,781. Suppose that for 1990–1991 a random sample of 30 colleges yields the following data on tuition and fees: $\bar{x} = \$9,392$ and $s = \$1,643$. Assume that $8,446 is the population mean for 1989–1990.
a. Specify the null and alternative hypotheses you would use to investigate whether the mean amount for tuition and fees in 1990–1991 was significantly larger (in the statistical sense) than it was in 1989–1990.
b. Calculate the p-value for the hypothesis test you described in part a, and explain what the p-value indicates about the statistical significance of the test results.
c. Explain the difference between statistical significance and practical significance in the context of this exercise.

8.33 Refer to the automobile mileage test in Exercise 8.14. Find the observed significance level for the test and interpret its value.

8.34 A new blood pressure drug is advertised to reduce, after 1 week of medication, a patient's blood pressure an average of 10 units. Blood pressure reductions were recorded for 37 patients after treatment with the drug. The mean and standard deviation for this sample were 8.7 and 6.8, respectively. Do the data appear to contradict the advertising claim? Explain. Find the observed significance level for the test.

8.35 Refer to the sampling of basic cable television services in Exercise 8.18. Find and interpret the p-value for the test.

8.36 According to advertisements, a strain of soybeans planted on soil prepared with a specified fertilizer treatment has a mean yield of 500 bushels per acre. Fifty farmers who belong to a cooperative plant the soybeans, each using a 40-acre plot, and each records the mean yield per acre. The mean and variance for the sample of 50 farms are $\bar{x} = 485$ and $s^2 = 10,045$.
a. Do the data provide sufficient evidence to indicate that the mean yield for the soybeans is different than advertised? Give the observed significance level for the test and interpret its value.
b. In reaching the conclusion in part a, would you have to qualify your conclusions because of the manner in which the sample was selected? Explain.

8.37 Refer to the test of the length of life of the fluorescent light bulbs in Exercise 8.19. Find the observed significance level for this test and interpret its value.

8.4 Small-Sample Test of Hypothesis About a Population Mean

Most water-treatment facilities monitor the quality of their drinking water on an hourly basis. One variable monitored is pH, which measures the degree of alkalinity or acidity in the water. A pH below 7.0 is acidic, one above 7.0 is alkaline, and a pH of 7.0 is neutral. One water-treatment plant has a target pH of 8.5 (most try to maintain a slightly alkaline level). The mean and standard deviation of one hour's test results, based on 17 water samples at this plant, are

$$\bar{x} = 8.42 \qquad s = .16$$

Does this sample provide sufficient evidence that the mean pH level in the water differs from 8.5?

This inference can be placed in a test of hypothesis framework. We establish the target pH as the null hypothesized value and then utilize a two-tailed alternative that the true mean pH differs from target:

H_0: $\mu = 8.5$

H_a: $\mu \neq 8.5$

Recall from Section 7.3 that when we are faced with making inferences about a population mean using the information in a small sample, two problems emerge:

1. The normality of the sampling distribution for $\bar{x}$ does not follow from the Central Limit Theorem when the sample size is small. We must assume that the distribution of measurements from which the sample was selected is approximately normally distributed in order to ensure the approximate normality of the sampling distribution of $\bar{x}$.
2. If the population standard deviation σ is unknown, as is usually the case, then we cannot assume that s will provide a good approximation for σ when the sample size is small. Instead, we must use the t-distribution rather than the standard normal z-distribution to make inferences about the population mean μ.

Therefore, the test statistic of a small-sample test of a population mean, we use the t statistic:

Test statistic: $\quad t = \dfrac{\bar{x} - \mu_0}{s/\sqrt{n}} = \dfrac{\bar{x} - 8.5}{s/\sqrt{n}}$

where μ_0 is the null hypothesized value of the population mean, μ. In our example, $\mu_0 = 8.5$.

To find the rejection region, we must specify the value of α, the probability that the test will lead to rejection of the null hypothesis when it is true, and then consult the t table (Table VI of Appendix A or inside the back cover of the text). Using $\alpha = .05$, the two-tailed rejection region is

Rejection region: $\quad t_{\alpha/2} = t_{.025} = 2.120$ with $n - 1 = 16$ degrees of freedom

Reject H_0 if $t < -2.120$ or $t > 2.120$

The rejection region is shown in Figure 8.7.

FIGURE 8.7

Two-tailed rejection region for
small-sample t-test

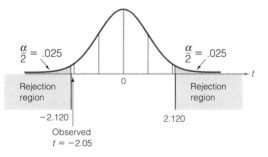

We are now prepared to calculate the test statistic and reach a conclusion:

$$t = \frac{\bar{x} - \mu_0}{s/\sqrt{n}} = \frac{8.42 - 8.50}{.16/\sqrt{17}} = \frac{-.08}{.039} = -2.05$$

Since the calculated value of t does not fall in the rejection region (Figure 8.7), we cannot reject H_0 at the $\alpha = .05$ level of significance. Thus, the water-treatment plant should not conclude that the mean pH differs from the 8.5 target based on the sample evidence.

It is interesting to note that the calculated t value, -2.05, is *less than* the .05 level z value, -1.96. The implication is that if we had *incorrectly* used a z statistic for this test, we would have rejected the null hypothesis at the .05 level, concluding that the mean pH level differs from 8.5. The important point is that the statistical procedure to be used must always be closely scrutinized and all the assumptions understood. Many statistical lies are the result of misapplications of otherwise valid procedures.

The technique for conducting a small-sample test of hypothesis about a population mean is summarized in the box.

Small-Sample Test of Hypothesis About μ

ONE-TAILED TEST

H_0: $\mu = \mu_0$

H_a: $\mu < \mu_0$
  (or H_a: $\mu > \mu_0$)

Test statistic: $t = \dfrac{\bar{x} - \mu_0}{s/\sqrt{n}}$

Rejection region: $t < -t_\alpha$
  (or $t > t_\alpha$ when
  H_a: $\mu > \mu_0$)

TWO-TAILED TEST

H_0: $\mu = \mu_0$

H_a: $\mu \neq \mu_0$

Test statistic: $t = \dfrac{\bar{x} - \mu_0}{s/\sqrt{n}}$

Rejection region: $t < -t_{\alpha/2}$
  or $t > t_{\alpha/2}$

where t_α and $t_{\alpha/2}$ are based on $(n - 1)$ degrees of freedom

Assumption: A random sample is selected from a population with a relative frequency distribution that is approximately normal.

EXAMPLE 8.4

A major car manufacturer wants to test a new engine to determine whether it meets new air-pollution standards. The mean emission μ of all engines of this type must be less than 20 parts per million of carbon. Ten engines are manufactured for testing purposes, and the mean and standard deviation of the emissions for this sample of engines are determined to be

$$\bar{x} = 17.1 \text{ parts per million} \qquad s = 3.0 \text{ parts per million}$$

Do the data supply sufficient evidence to allow the manufacturer to conclude that this type of engine meets the pollution standard? Assume that the manufacturer is willing to risk a Type I error with probability $\alpha = .01$.

Solution

The manufacturer wants to support the research hypothesis that the mean emission level μ for all engines of this type is less than 20 parts per million. The elements of this small-sample one-tailed test are

H_0: $\mu = 20$

H_a: $\mu < 20$

Test statistic: $t = \dfrac{\bar{x} - 20}{s/\sqrt{n}}$

Assumption: The relative frequency distribution of the population of emission levels for all engines of this type is approximately normal.

Rejection region: For $\alpha = .01$ and df $= n - 1 = 9$, the one-tailed rejection region (see Figure 8.8) is $t < -t_{.01} = -2.821$.

We now calculate the test statistic:

$$t = \frac{\bar{x} - 20}{s/\sqrt{n}} = \frac{17.1 - 20}{3.0/\sqrt{10}} = -3.06$$

FIGURE 8.8

A t-distribution with 9 df and the rejection region for Example 8.4

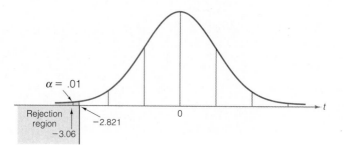

Since the calculated t falls in the rejection region (see Figure 8.8), the manufacturer concludes that $\mu < 20$ parts per million and the new engine type meets the pollution standard. Are you satisfied with the reliability associated with this inference? The probability is only $\alpha = .01$ that the test would support the research hypothesis if in fact it were false.

EXAMPLE 8.5

Find the observed significance level for the test described in Example 8.4.

Solution

The test of Example 8.4 was a lower-tail test: H_0: $\mu = 20$ versus H_a: $\mu < 20$. Since the value of t computed from the sample data was $t = -3.06$, the observed significance level (or p-value) for the test is equal to the probability that t would assume a value less than or equal to -3.06 if in fact H_0 were true. This is equal to the area in the lower tail of the t-distribution (shaded in Figure 8.9). To find this area—i.e., the p-value for the test—we consult the t-table (Table VI in Appendix A). Unlike the table of areas under the normal curve, Table VI gives only the t values corresponding to the areas .100, .050, .025, .010, .005, .001, and .0005. Therefore, we can only approximate the p-value for the test. Since the observed t value was based on 9 degrees of freedom, we use the df = 9 row in Table VI and move across the row until we reach the t values that are closest to the observed $t = -3.06$. [*Note:* We ignore the minus sign.] The t values corresponding to p-values of .010 and .005 are 2.821 and 3.250, respectively. Since the observed t value falls between $t_{.010}$ and $t_{.005}$, the p-value for the test lies between .005 and .010. We could interpolate to locate the p-value for the test more accurately, but it is easier and adequate for our purposes to choose the larger area as the p-value and report it as .010. Thus, we would reject the null hypothesis, H_0: $\mu = 20$ parts per million, for any value of α larger than .01.

FIGURE 8.9

The observed significance level for the test of Example 8.4

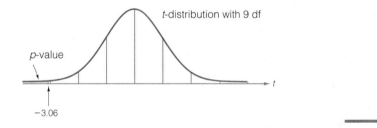

p-value

t-distribution with 9 df

-3.06

Small-sample inferences typically require more assumptions and provide less information about the population parameter than do large-sample inferences. Nevertheless, the t-test is a method of testing a hypothesis about a population mean of a normal distribution when only a small number of observations is available.

EXERCISES 8.38–8.49

LEARNING THE MECHANICS

8.38 In what ways are the distributions of the z-test statistic and t-test statistic alike? How do they differ?

8.39 Under what circumstances should you use the t-distribution in testing a hypothesis about a population mean?

8.40 A random sample of n observations is selected from a normal population to test the null hypothesis that $\mu = 10$. Specify the rejection region for each of the following combinations of H_a, α, and n:
  a. H_a: $\mu \neq 10$; $\alpha = .05$; $n = 15$ b. H_a: $\mu > 10$; $\alpha = .01$; $n = 20$
  c. H_a: $\mu > 10$; $\alpha = .10$; $n = 8$ d. H_a: $\mu < 10$; $\alpha = .01$; $n = 17$
  e. H_a: $\mu \neq 10$; $\alpha = .10$; $n = 22$ f. H_a: $\mu < 10$; $\alpha = .05$; $n = 6$

8.41   For each of the following rejection regions, sketch the sampling distribution of t, and indicate the location of the rejection region on your sketch:
a. $t > 1.440$ where df $= 6$ b. $t < -1.782$ where df $= 12$
c. $t < -2.060$ or $t > 2.060$ where df $= 25$

8.42   For each of the rejection regions defined in Exercise 8.41, what is the probability that a Type I error will be made?

8.43   The following sample of five measurements was randomly selected from a normally distributed population: 4, 7, 3, 4, 6
a. Test the null hypothesis that the mean of the population is 6 against the alternative hypothesis, $\mu < 6$. Use $\alpha = .05$.
b. Test the null hypothesis that the mean of the population is 6 against the alternative hypothesis, $\mu \neq 6$. Use $\alpha = .05$.
c. Find the observed significance level for each test.

8.44   The following sample of six measurements was randomly selected from a normally distributed population: 1, 3, -1, 5, 1, 2
a. Test the null hypothesis that the mean of the population is 3 against the alternative hypothesis, $\mu < 3$. Use $\alpha = .05$.
b. Test the null hypothesis that the mean of the population is 3 against the alternative hypothesis, $\mu \neq 3$. Use $\alpha = .05$.
c. Find the observed significance level for each test.

APPLYING THE CONCEPTS

8.45   A consumer protection group is concerned that a catsup manufacturer is filling its 20-ounce family-size containers with less than 20 ounces of catsup. The group purchases ten family-size bottles of this catsup, weighs the contents of each, and finds that the mean weight is equal to 19.86 ounces, and the standard deviation is equal to .22 ounce.
a. Do the data provide sufficient evidence for the consumer group to conclude that the mean fill per family-size bottle is less than 20 ounces? Test using $\alpha = .05$.
b. If the test in part a were conducted on a periodic basis by the company's quality control department, is the consumer group more concerned about making a Type I error or a Type II error? (The probability of making this type of error is called the *consumer's risk*.)
c. The catsup company is also interested in the mean amount of catsup per bottle. It does not wish to overfill them. For the test conducted in part a, which type of error is more serious from the company's point of view—a Type I error or a Type II error? (The probability of making this type of error is called the *producer's risk*.)
d. Find a 90% confidence interval for the mean number of ounces of catsup being dispensed.

8.46   A cigarette manufacturer advertises that its new low-tar cigarette "contains on average no more than 4 milligrams of tar." You have been asked to test the claim using the following sample information: $n = 25$, $\bar{x} = 4.16$ milligrams, $s = .30$ milligram. Does the sample information disagree with the manufacturer's claim? Test using $\alpha = .05$. List any assumptions you make.

8.47   One of the most feared predators in the ocean is the great white shark. Although it is known that the white shark grows to a mean length of 21 feet, a marine biologist believes that the great white sharks off the Bermuda coast grow much longer due to unusual feeding habits. To test this claim, a number of full-grown great white sharks are captured off the Bermuda coast, measured and then set free. However, because the capture of sharks is difficult, costly, and very dangerous, only three are sampled. Their lengths are 24, 20, and 22 feet.
a. Do the data provide sufficient evidence to support the marine biologist's claim? Use $\alpha = .10$.
b. Give the approximate observed significance level for the test in part a, and interpret its value.

c. What assumptions must be made in order to carry out the test?

d. Do you think these assumptions are likely to be satisfied in this sampling situation?

8.48 An important problem facing strawberry growers is the control of nematodes. These organisms compete with the plants for nutrients in the soil, thereby reducing yield. For this reason, fumigation is normally a part of field preparation. In the past, the fumigants used yielded an average of 8 pounds of marketable fruit for a certain standard sized plot. Recently, a new fumigant has been developed. It is applied to six standard plots of strawberries, and the yield of marketable fruit (in pounds) for each plot is 9, 9, 13, 9, 10, and 8.

a. Do the data indicate a significant increase in average yield at the .05 level of significance?

b. What assumptions are necessary for the procedure used to be valid?

8.49 A psychologist was interested in knowing whether male heroin addicts' assessments of self-worth differ from those of the general male population. On a test designed to measure assessment of self-worth, the mean score for males from the general population is 48.6. A random sample of 25 scores achieved by heroin addicts yielded a mean of 44.1 and a standard deviation of 6.2.

a. Do the data indicate a difference in assessment of self-worth between male heroin addicts and the general male population? Test using $\alpha = .01$.

b. Give the approximate observed significance level for the test and interpret its value.

8.5 Large-Sample Test of Hypothesis About a Binomial Probability

Inferences about proportions (or percentages) are often made in the context of the probability, p, of "success" for a binomial distribution. We showed how to use large samples from binomial distributions to form confidence intervals for p in Section 7.5. We now consider tests of hypotheses about a binomial probability p.

For example, consider the problem of *insider trading* in the stock market. Insider trading is the buying and selling of stock by an individual privy to inside information in a company, usually a high-level executive in the firm. The Securities and Exchange Commission (SEC) imposes strict guidelines about insider trading so that all investors can have equal access to information that may affect the stock's price. An investor wishing to test the effectiveness of the SEC guidelines monitors the market over a period of a year and records the number of times a stock price increases the day following a significant purchase of stock by an insider. For a total of 576 such transactions, the stock increased the following day 327 times. Does this sample provide evidence that the stock price may be affected by insider trading?

We first view this as a binomial experiment, with the 576 transactions as the trials and Success representing an increase in the stock's price the following day. Let p represent the probability that the stock price will increase following a large insider purchase. If the insider purchase has no effect on the stock price (that is, if the information available to the insider is identical to that available to the general market), then the investor expects the probability of a stock increase to be the same as that of a decrease, or $p = .5$. On the other hand, if insider trading affects the stock price (indicating that the market has not fully accounted for the information known to the insiders), then the investor expects the stock either to decrease or to increase more than half the time following significant insider transactions—i.e., $p \neq .5$.

We can now place the problem in the context of a test of hypothesis:

$$H_0: \quad p = .5$$
$$H_a: \quad p \neq .5$$

Recall that the sample proportion, $\hat{p}$, is really just the sample mean of the outcomes of the individual binomial trials and, as such, is approximately normally distributed (for large samples) according to the Central Limit Theorem. Thus, for large samples we can use the standard normal z as the test statistic:

$$Test\ statistic: \quad z = \frac{\text{Sample proportion} - \text{Null hypothesized proportion}}{\text{Standard deviation of sample proportion}}$$

$$= \frac{\hat{p} - p_0}{\sigma_{\hat{p}}}$$

where we use the symbol p_0 to represent the null hypothesized value of p.

Rejection region: We use the standard normal distribution to find the appropriate rejection region for the specified value of α. Using $\alpha = .05$, the two-tailed rejection region is

$$z < -z_{\alpha/2} = z_{.025} = -1.96 \quad \text{or} \quad z > z_{\alpha/2} = z_{.025} = 1.96$$

See Figure 8.10.

We are now prepared to calculate the value of the test statistic. Before doing so, we want to be sure that the sample size is large enough to ensure that the normal approximation for the sampling distribution of $\hat{p}$ is reasonable. To check this, we calculate a 3-standard-deviation interval around the null hypothesized value, p_0, which is assumed to be the true value of p until our test procedure proves otherwise. Recall that $\sigma_{\hat{p}} = \sqrt{pq/n}$ and that we need an estimate of the product pq in order to calculate a numerical value of the test statistic z. Since the null hypothesized value is generally the accepted-until-proven-otherwise value, we use the value of $p_0 q_0$ (where $q_0 = 1 - p_0$) to estimate pq in the calculation of z. Thus,

$$\sigma_{\hat{p}} = \sqrt{\frac{pq}{n}} \approx \sqrt{\frac{p_0 q_0}{n}} = \sqrt{\frac{(.5)(.5)}{576}}$$

$$= .021$$

and the 3-standard-deviation interval around p_0 is

$$p_0 \pm 3\sigma_{\hat{p}} \approx .5 \pm 3(.021) = (.437, .563)$$

As long as this interval does not contain 0 or 1 (i.e., is completely contained in the interval 0 to 1), as is the case here, then the normal distribution will provide a reasonable approximation for the sampling distribution of $\hat{p}$.

Returning to the hypothesis test at hand, the proportion of the sampled transactions that resulted in a stock increase is

$$\hat{p} = \frac{327}{576} = .568$$

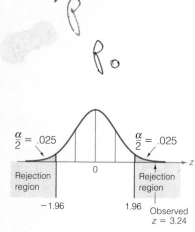

$\frac{\alpha}{2} = .025$ $\frac{\alpha}{2} = .025$

Rejection region   Rejection region

-1.96 1.96 Observed $z = 3.24$

FIGURE 8.10

Rejection region for insider trading example

Finally, we calculate the number of standard deviations (the z value) between the sampled and hypothesized value of the binomial probability:

$$z = \frac{\hat{p} - p_0}{\sigma_{\hat{p}}} \approx \frac{\hat{p} - p_0}{\sqrt{p_0 q_0/n}} = \frac{.568 - .5}{.021}$$

$$= \frac{.068}{.021} = 3.24$$

The implication is that the observed sample proportion is (approximately) 3.24 standard deviations above the null hypothesized probability, .5 (Figure 8.10). Therefore, we reject the null hypothesis, concluding at the .05 level of significance that the true probability of an increase or decrease in a stock's price differs from .5 the day following significant insider purchase of the stock. It appears that an insider purchase significantly *increases* the probability that the stock price will increase the following day.

The test of hypothesis about a binomial probability p is summarized in the box. Note that the procedure is entirely analogous to that used for conducting large-sample tests about a population mean.

Large-Sample Test of Hypothesis About p

ONE-TAILED TEST

H_0: $p = p_0$ (p_0 = hypothesized value of p)

H_a: $p < p_0$
(or H_a: $p > p_0$)

Test statistic: $z = \dfrac{\hat{p} - p_0}{\sigma_{\hat{p}}}$

TWO-TAILED TEST

H_0: $p = p_0$

H_a: $p \neq p_0$

Test statistic: $z = \dfrac{\hat{p} - p_0}{\sigma_{\hat{p}}}$

where, according to H_0, $\sigma_{\hat{p}} = \sqrt{p_0 q_0/n}$ and $q_0 = 1 - p_0$

Rejection region: $z < -z_\alpha$
(or $z > z_\alpha$
when H_a: $p > p_0$)

Rejection region: $z < -z_{\alpha/2}$
or $z > z_{\alpha/2}$

Assumption: The experiment is binomial, and the sample size is large enough that the interval $p_0 \pm 3\sigma_{\hat{p}}$ does not include 0 or 1.

EXAMPLE 8.6

The reputations (and hence sales) of many businesses can be severely damaged by shipments of manufactured items that contain a large percentage of defectives. For example, a manufacturer of alkaline batteries may want to be reasonably certain that fewer than 5% of its batteries are defective. Suppose 300 batteries are randomly selected from a very large shipment, each is tested, and ten defective batteries are found. Does this provide sufficient evidence for the manufacturer to conclude that the fraction defective in the entire shipment is less than .05? Use $\alpha = .01$.

Solution

Before conducting the test of hypothesis, we check to determine whether the sample size is large enough to use the normal approximation for the sampling distribution of $\hat{p}$. The criterion is tested by the interval

$$p_0 \pm 3\sigma_{\hat{p}} = p_0 \pm 3\sqrt{\frac{p_0 q_0}{n}} = .05 \pm 3\sqrt{\frac{(.05)(.95)}{300}}$$

$$= .05 \pm .04 \quad \text{or} \quad (.01, .09)$$

Since the interval lies within the interval $(0, 1)$, the normal approximation will be adequate.

The objective of the sampling is to determine whether there is sufficient evidence to indicate that the fraction defective, p, is less than .05. Consequently, we will test the null hypothesis that $p = .05$ against the alternative hypothesis that $p < .05$. The elements of the test are

H_0: $p = .05$

H_a: $p < .05$

Test statistic: $z = \dfrac{\hat{p} - p_0}{\sigma_{\hat{p}}}$

Rejection region: $z < -z_{.01} = -2.33$ (see Figure 8.11)

We now calculate the test statistic:

$$z = \frac{\hat{p} - .05}{\sigma_{\hat{p}}} = \frac{(10/300) - .05}{\sqrt{p_0 q_0 / n}} = \frac{.033 - .05}{\sqrt{p_0 q_0 / 300}}$$

Notice that we use p_0 to calculate $\sigma_{\hat{p}}$ because, in contrast to calculating $\sigma_{\hat{p}}$ for a confidence interval, the test statistic is computed on the assumption that the null hypothesis is true; that is, $p = p_0$. Therefore, substituting the values for $\hat{p}$ and p_0 into the z statistic, we obtain

$$z \approx \frac{-.017}{\sqrt{(.05)(.95)/300}} = \frac{-.017}{.0126} = -1.35$$

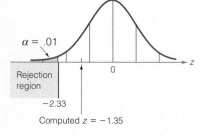

$\alpha = .01$

Rejection region

-2.33

Computed $z = -1.35$

FIGURE 8.11

Rejection region for Example 8.6

As shown in Figure 8.11, the calculated z value does not fall in the rejection region. Therefore, there is insufficient evidence at the .01 level of significance to indicate that the shipment contains fewer than 5% defective batteries.

EXAMPLE 8.7

In Example 8.6 we found that we did not have sufficient evidence, at the $\alpha = .01$ level of significance, to indicate that the fraction defective p of alkaline batteries was less than $p = .05$. How strong was the weight of evidence favoring the alternative hypothesis (H_a: $p < .05$)? Find the observed significance level for the test.

Solution

The computed value of the test statistic z was $z = -1.35$. Therefore, for this lower-tail test, the observed significance level is

Observed significance level $= P(z \leq -1.35)$

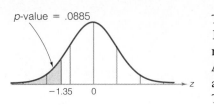

FIGURE 8.12

The observed significance level for Example 8.7

This lower-tail area is shown in Figure 8.12. The area between $z = 0$ and $z = 1.35$ is given in Table IV in Appendix A as .4115. Therefore, the observed significance level is $.5 - .4115 = .0885$. Note that this probability is quite small. Although we did not reject H_0: $p = .05$ at $\alpha = .01$, the probability of observing a z value as small as or smaller than -1.35 is only .0885 if in fact H_0 is true. Therefore, we would reject H_0 if we choose $\alpha = .10$ (since the observed significance level is less than .10), and we would not reject H_0 (the conclusion of Example 8.6) if we choose $\alpha = .05$ or $\alpha = .01$.

Small-sample test procedures are also available for p. These are omitted from our discussion because most surveys use samples that are large enough to employ the large-sample tests presented in this section.

EXERCISES 8.50–8.59

LEARNING THE MECHANICS

8.50   For the binomial sample sizes and null hypothesized values of p in each part, determine whether the sample size is large enough to use the normal approximation methodology presented in this section to conduct a test of the null hypothesis H_0: $p = p_0$.
  a. $n = 500, p_0 = .05$ **b.** $n = 100, p_0 = .99$ **c.** $n = 50, p_0 = .2$
  d. $n = 20, p_0 = .2$ **e.** $n = 10, p_0 = .4$

8.51   Suppose a random sample of 100 observations from a binomial population gives a value of $\hat{p} = .69$ and you wish to test the null hypothesis that the population parameter p is equal to .75 against the alternative hypothesis that $p < .75$.
  a. Noting that $\hat{p} = .69$, what does your intuition tell you? Does the value of $\hat{p}$ appear to contradict the null hypothesis?
  b. Use the large-sample z-test to test H_0: $p = .75$ against the alternative hypothesis, H_a: $p < .75$. Use $\alpha = .05$. How do the test results compare with your intuitive decision from part **a**?
  c. Find and interpret the observed significance level of the test you conducted in part **b**.

8.52   Suppose the sample in Exercise 8.51 has produced $\hat{p} = .84$ and we wish to test H_0: $p = .9$ against the alternative H_a: $p < .9$.
  a. Calculate the value of the z statistic for this test.
  b. Note that the numerator of the z statistic ($\hat{p} - p_0 = .84 - .90 = -.06$) is the same as for Exercise 8.51. Considering this, why is the absolute value of z for this exercise larger than that calculated in Exercise 8.51?
  c. Complete the test using $\alpha = .05$ and interpret the result.
  d. Find the observed significance level for the test and interpret its value.

8.53   A random sample of 100 observations is selected from a binomial population with unknown probability of success p. The computed value of $\hat{p}$ is equal to .74.
  a. Test H_0: $p = .65$ against H_a: $p > .65$. Use $\alpha = .01$.
  b. Test H_0: $p = .65$ against H_a: $p > .65$. Use $\alpha = .10$.
  c. Test H_0: $p = .90$ against H_a: $p \neq .90$. Use $\alpha = .05$.

    d. Form a 95% confidence interval for p.
    e. Form a 99% confidence interval for p.

8.54   Refer to Exercise 7.50.
    a. Test H_0: $p = .5$ against H_a: $p > .5$, where p is the proportion of customers who do not like the snack food. Use $\alpha = .10$.
    b. Report the observed significance level of your test.

APPLYING THE CONCEPTS

8.55   According to a spokesperson for General Mills, the company's "cents-off" coupon offers are designed to get people to buy its products, and its refund offers (money returned with proof of repeated purchases) are designed to encourage people to continue buying its products. In a national survey conducted by the Nielsen Clearing House in 1975, 65% of the respondents indicated that they used cents-off coupons when grocery shopping. In a 1980 survey, the Nielsen organization found that 76% of those surveyed used cents-off coupons (*Minneapolis Star*, November 29, 1981, p. 7F). Suppose the 1980 survey consisted of a random sample of 100 shoppers, of whom 76 indicated that they used cents-off coupons.
    a. Is the sample size large enough to use the inferential procedures presented in this section? Explain.
    b. Does the 1980 sample provide sufficient evidence that the percentage of shoppers using cents-off coupons exceeds 65%? Test using $\alpha = .05$.
    c. Find the observed significance level for the test you conducted in part **b**, and interpret its value.

8.56   A method currently used by doctors to screen women for possible breast cancer fails to detect cancer in 15% of the women who actually have the disease. A new method has been developed that researchers hope will be able to detect cancer more accurately. A random sample of 70 women known to have breast cancer were screened using the new method. Of these, the new method failed to detect cancer in six.
    a. Do the data provide sufficient evidence to indicate that the new screening method is better than the one currently in use? Test using $\alpha = .05$.
    b. Find the observed significance level for the test and interpret its value.

8.57   Major oil companies have been criticized for raising prices following major oil spills and Mideast crises. In the past, Standard Oil of California has used a sample survey to determine whether people's attitudes toward Standard's corporate image tended to be favorable or unfavorable. The sample results indicated that, for the first time in 30 years, more people had unfavorable than favorable attitudes. Standard Oil responded by initiating an institutional advertising campaign to help improve its image (*Marketing News*, 1976). In 1990, another large oil corporation conducted a similar survey, with the following results:

    Unfavorable opinions   3,465
    Favorable opinions   2,502
    No opinions   821

    a. Examine the data. Based on your intuition, would you say that more than 50% of the general public possess an unfavorable attitude toward the company?
    b. Do the sample data support the hypothesis that more than 50% of the general public hold unfavorable opinions about the company? Test at $\alpha = .05$.
    c. Construct a 90% confidence interval for the proportion of individuals with no opinion.
    d. List any assumptions that you made in answering parts **b** and **c**.

8.58   Refer to Exercise 8.57. Find the observed significance level for the test you conducted in part **b** and interpret its value.

8.59   Increasing numbers of businesses are offering child-care benefits for their workers. However, one union claims that at least 90% of firms in the manufacturing sector still do not offer any child-care benefits to their workers. A random sample of 350 manufacturing firms is selected, and only 28 of them offer child-care benefits.

    a. Does this sample result support the claim of the union? Test using $\alpha = .10$.

    b. Calculate the *p*-value associated with this test.

8.6 Calculating Type II Error Probabilities: More About β (Optional)

In our introduction to hypothesis testing in Section 8.1, we showed that the probability of committing a Type I error, α, can be controlled by the selection of the rejection region for the test. Thus, when the test statistic falls in the rejection region and we make the decision to reject the null hypothesis, we do so knowing the error rate for incorrect rejections of H_0. The situation corresponding to accepting the null hypothesis, and thereby risking a Type II error, is not generally as controllable. For that reason, we adopted a policy of nonrejection of H_0 when the test statistic does not fall in the rejection region, rather than risking an error of unknown magnitude.

To see how β, the probability of a Type II error, can be calculated for a test of hypothesis, recall the example in Section 8.1 in which a city tests a manufacturer's pipe to see whether it meets the requirement that the mean strength exceeds 2,400 pounds per lineal foot. The setup for the test is as follows:

H_0: $\mu = 2{,}400$

H_a: $\mu > 2{,}400$

Test statistic: $z = \dfrac{\bar{x} - 2{,}400}{\sigma/\sqrt{n}}$

Rejection region: $z > 1.645$ for $\alpha = .05$

Figure 8.13(a) on page 358 shows the rejection region for the **null distribution**—that is, the distribution of the test statistic assuming the null hypothesis is true. The area in the rejection region is .05, and this area represents α, the probability that the test statistic leads to rejection of H_0 when in fact H_0 is true.

The Type II error probability β is calculated assuming that the null hypothesis is false, because it is defined as the *probability of accepting H_0 when it is false*. Since H_0 is false for any value of μ exceeding 2,400, one value of β exists for each possible value of μ greater than 2,400 (an infinite number of possibilities). Figures 8.13(b), (c), and (d) show three of the possibilities, corresponding to alternative hypothesis values of μ equal to 2,425, 2,450, and 2,475, respectively. Note that β is the area in the *nonrejection* (or *acceptance*) *region* in each of these distributions and that β decreases as the true value of μ moves farther from the

FIGURE 8.13

Values of α and β for various values of μ

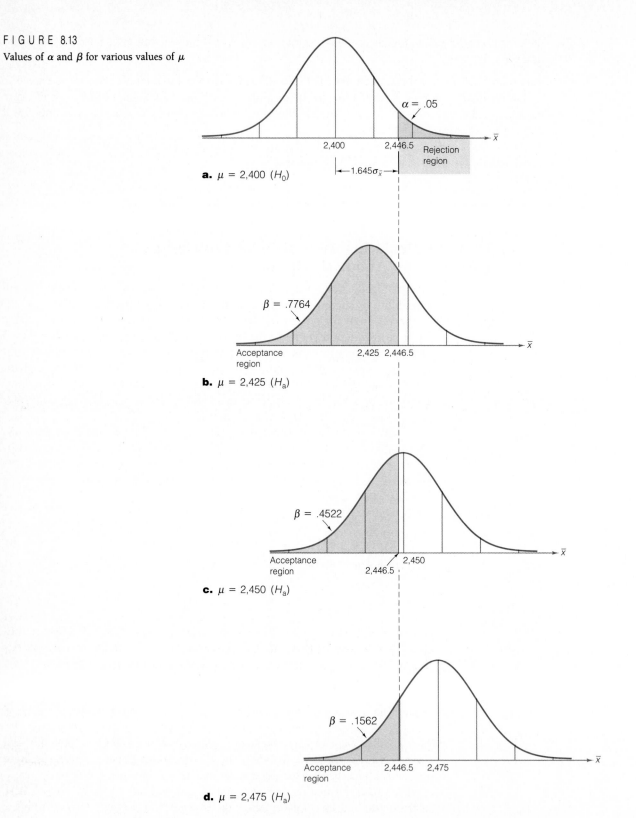

a. $\mu = 2,400$ (H_0)

b. $\mu = 2,425$ (H_a)

c. $\mu = 2,450$ (H_a)

d. $\mu = 2,475$ (H_a)

null hypothesized value of $\mu = 2,400$. This is sensible because the probability of incorrectly accepting the null hypothesis should decrease as the distance between the null and alternative values of μ increases.

In order to calculate the value of β for a specific value of μ in H_a, we proceed as follows:

1. Calculate the value of $\bar{x}$ that corresponds to the border between the acceptance and rejection regions. For the sewer pipe example, this is the value of $\bar{x}$ that lies 1.645 standard deviations above $\mu = 2,400$ in the sampling distribution of $\bar{x}$. Denoting this value by $\bar{x}_0$, corresponding to the largest value of $\bar{x}$ that supports the null hypothesis, we find (recalling that $s = 200$ and $n = 50$)

$$\bar{x}_0 = \mu_0 + 1.645\sigma_{\bar{x}} = 2,400 + 1.645\left(\frac{\sigma}{\sqrt{n}}\right)$$

$$\approx 2,400 + 1.645\left(\frac{s}{\sqrt{n}}\right) = 2,400 + 1.645\left(\frac{200}{\sqrt{50}}\right)$$

$$= 2,400 + 1.645(28.28) = 2,446.5$$

2. For a particular alternative distribution corresponding to a value of μ, denoted by μ_a, we calculate the z value corresponding to $\bar{x}_0$, the border between the rejection and acceptance regions. We then use this z value and Table IV of Appendix A to determine the area in the *acceptance* region under the alternative distribution. This area is the value of β corresponding to the particular alternative μ_a. For example, for the alternative $\mu_a = 2,425$, we calculate

$$z = \frac{\bar{x}_0 - 2,425}{\sigma_{\bar{x}}} = \frac{\bar{x}_0 - 2,425}{\sigma/\sqrt{n}}$$

$$\approx \frac{\bar{x}_0 - 2,425}{s/\sqrt{n}} = \frac{2,446.5 - 2,425}{28.28} = .76$$

Note in Figure 8.13(b) that the area in the acceptance region is the area to the left of $z = .76$. This area is

$$\beta = .5 + .2764 = .7764$$

Thus, the probability that the test procedure will lead to an incorrect acceptance of the null hypothesis $\mu = 2,400$ when in fact $\mu = 2,425$ is about .78. As the average strength of the pipe increases to 2,450, the value of β decreases to .4522 [Figure 8.13(c)]. If the mean strength is further increased to 2,475, the value of β is further decreased to .1562 [Figure 8.13(d)]. Thus, even if the true mean strength of the pipe exceeds the minimum specification by 75 pounds per lineal foot, the test procedure will lead to an incorrect acceptance of the null hypothesis (rejection of the pipe) approximately 16% of the time. The upshot is that the pipe must be manufactured so that the mean strength well exceeds the minimum requirement if the manufacturer wants the probability of its acceptance by the city to be large (i.e., β to be small).

The steps for calculating β for a large-sample test about a population mean are summarized in the box on page 360.

Steps for Calculating β for a Large-Sample Test About μ

1. Calculate the value(s) of $\bar{x}$ corresponding to the border(s) of the rejection region. There will be one border value for a one-tailed test, and two for a two-tailed test. The formula is one of the following, corresponding to a test with level of significance α:

   Upper-tailed test: $\bar{x}_0 = \mu_0 + z_\alpha \sigma_{\bar{x}} \approx \mu_0 + z_\alpha \left(\dfrac{s}{\sqrt{n}} \right)$

   Lower-tailed test: $\bar{x}_0 = \mu_0 - z_\alpha \sigma_{\bar{x}} \approx \mu_0 - z_\alpha \left(\dfrac{s}{\sqrt{n}} \right)$

   Two-tailed test: $\bar{x}_{0,L} = \mu_0 - z_{\alpha/2} \sigma_{\bar{x}} \approx \mu_0 - z_{\alpha/2} \left(\dfrac{s}{\sqrt{n}} \right)$

   $\bar{x}_{0,U} = \mu_0 + z_{\alpha/2} \sigma_{\bar{x}} \approx \mu_0 + z_{\alpha/2} \left(\dfrac{s}{\sqrt{n}} \right)$

2. Specify the value of μ_a in the alternative hypothesis for which the value of β is to be calculated. Then convert the border value(s) of $\bar{x}_0$ to z value(s) using the alternative distribution with mean μ_a. The general formula for the z value is

   $$z = \frac{\bar{x}_0 - \mu_a}{\sigma_{\bar{x}}}$$

   Sketch the alternative distribution (centered at μ_a), and shade the area in the acceptance (nonrejection) region. Use the z statistic(s) and Table IV of Appendix A to find the shaded area, which is β.

Following the calculation of β for a particular value of μ_a, you should interpret the value in the context of the hypothesis testing application. It is often useful to interpret the value of $1 - \beta$, which is known as the *power of the test* corresponding to a particular alternative, μ_a. Since β is the probability of accepting the null hypothesis when the alternative hypothesis is true with $\mu = \mu_a$, $1 - \beta$ is the probability of the complementary event, or the probability of rejecting the null hypothesis when the alternative $H_a: \mu = \mu_a$ is true. That is, the power $1 - \beta$ measures the likelihood that the test procedure will lead to the correct decision (reject H_0) for a particular value of the mean in the alternative hypothesis.

Definition 8.2

The **power** of a test is the probability that the test will correctly lead to the rejection of the null hypothesis for a particular value of μ in the alternative hypothesis. The power is equal to $1 - \beta$ for the particular alternative considered.

For example, in the sewer pipe example we found that $\beta = .7764$ when $\mu = 2{,}425$. This is the probability that the test leads to the (incorrect) acceptance of the null hypothesis when $\mu = 2{,}425$. Or, equivalently, the power of the test is $1 - .7764 = .2236$, which means that the test will lead to the (correct) rejection of the null hypothesis only 22% of the time when the pipe exceeds specifications by 25 pounds per lineal foot. When the manufacturer's pipe has a mean strength of 2,475 (that is, 75 pounds per lineal foot in excess of specifications), the power of the test increases to $1 - .1562 = .8438$. That is, the test will lead to the acceptance of the manufacturer's pipe 84% of the time if $\mu = 2{,}475$.

EXAMPLE 8.8

Recall the drug experiment in Examples 8.1 and 8.2, in which we tested to determine whether the mean response time for rats injected with a drug differs from the control mean response time of $\mu = 1.2$ seconds. The test setup is repeated here:

H_0: $\mu = 1.2$

H_a: $\mu \neq 1.2$ (i.e., $\mu < 1.2$ or $\mu > 1.2$)

Test statistic: $z = \dfrac{\bar{x} - 1.2}{\sigma_{\bar{x}}}$

Rejection region:

$z < -1.96$ or $z > 1.96$ for $\alpha = .05$

$z < -2.575$ or $z > 2.575$ for $\alpha = .01$

Note that two rejection regions have been specified corresponding to values of $\alpha = .05$ and $\alpha = .01$, respectively. Assume that $n = 100$ and $s = .5$.

a. Suppose drug-injected rats have a mean response time of 1.1 seconds, i.e., $\mu = 1.1$. Calculate the values of β corresponding to the two rejection regions. Discuss the relationship between the values of α and β.

b. Calculate the power of the test for each of the rejection regions when $\mu = 1.1$.

Solution

a. We first consider the rejection region corresponding to $\alpha = .05$. The first step is to calculate the border values of $\bar{x}$ corresponding to the two-tailed rejection region, $z < -1.96$ or $z > 1.96$:

$$\bar{x}_{0,L} = \mu_0 - 1.96\sigma_{\bar{x}} \approx \mu_0 - 1.96\left(\frac{s}{\sqrt{n}}\right)$$

$$= 1.2 - 1.96\left(\frac{.5}{10}\right) = 1.102$$

$$\bar{x}_{0,U} = \mu_0 + 1.96\sigma_{\bar{x}} \approx \mu_0 + 1.96\left(\frac{s}{\sqrt{n}}\right)$$

$$= 1.2 + 1.96\left(\frac{.5}{10}\right) = 1.298$$

These border values are shown in Figure 8.14(a) on page 362.

FIGURE 8.14

Calculation of β for drug-injected rats example

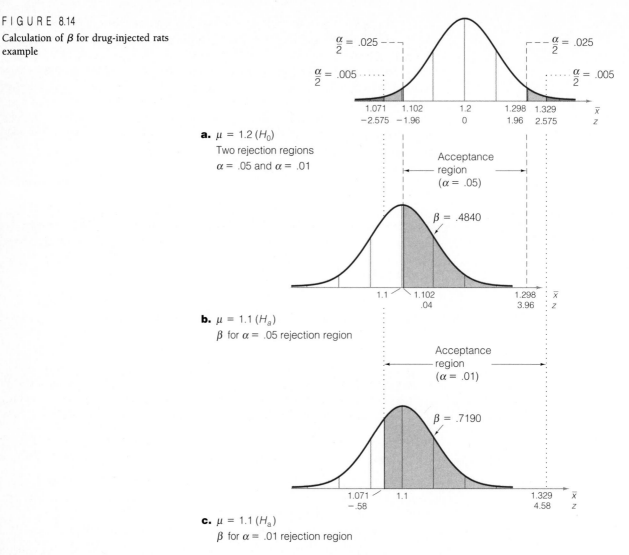

a. $\mu = 1.2 \, (H_0)$
Two rejection regions
$\alpha = .05$ and $\alpha = .01$

b. $\mu = 1.1 \, (H_a)$
β for $\alpha = .05$ rejection region

c. $\mu = 1.1 \, (H_a)$
β for $\alpha = .01$ rejection region

Next, we convert these values to z values in the alternative distribution with $\mu_a = 1.1$:

$$z_L = \frac{\bar{x}_{0,L} - \mu_a}{\sigma_{\bar{x}}} \approx \frac{1.102 - 1.1}{.05} = .04$$

$$z_U = \frac{\bar{x}_{0,U} - \mu_a}{\sigma_{\bar{x}}} \approx \frac{1.298 - 1.1}{.05} = 3.96$$

These z values are shown in Figure 8.14(b). You can see that the acceptance (or nonrejection) region is the area between them. Using Table IV of Appendix A, we find that the area between $z = 0$ and $z = .04$ is .0160, and the area between $z = 0$ and $z = 3.96$ is (approximately) .5 (since $z = 3.96$ is off the scale of Table IV). Then the area between $z = .04$ and $z = 3.96$ is, approximately,

$$\beta = .5 - .0160 = .4840$$

Thus, the test with $\alpha = .05$ will lead to a Type II error about 48% of the time when the mean reaction time for drug-injected rats is .1 second less than the control mean response time.

For the rejection region corresponding to $\alpha = .01$, $z < -2.575$ or $z > 2.575$, we find

$$\bar{x}_{0,L} = 1.2 - 2.575\left(\frac{.5}{10}\right)$$

$$= 1.0712$$

$$\bar{x}_{0,U} = 1.2 + 2.575\left(\frac{.5}{10}\right)$$

$$= 1.3288$$

These border values of the rejection region are shown in Figure 8.14(c).

Converting these to z values in the alternative distribution with $\mu_a = 1.1$, we find $z_L = -.58$ and $z_U = 4.58$. The area between these values is, approximately,

$$\beta = .2190 + .5 = .7190$$

Thus, the chance that the test procedure with $\alpha = .01$ will lead to an incorrect acceptance of H_0 is about 72%.

Note that the value of β increases from .4840 to .7190 when we decrease the value of α from .05 to .01. This is a general property of the relationship between α and β: *as α is decreased (increased), β is increased (decreased)*.

b. The power is defined to be the probability of (correctly) rejecting the null hypothesis when the alternative is true. When $\mu = 1.1$ and $\alpha = .05$, we find

$$\text{Power} = 1 - \beta$$

$$= 1 - .4840 = .5160$$

When $\mu = 1.1$ and $\alpha = .01$, we find

$$\text{Power} = 1 - \beta$$

$$= 1 - .7190 = .2810$$

You can see that the power of the test is decreased as the level of α is decreased. This means that as the probability of incorrectly rejecting the null hypothesis is decreased, the probability of correctly accepting the null hypothesis for a given alternative is also decreased. Thus, the value of α must be selected carefully, with the realization that a test is made less powerful to detect departures from the null hypothesis when the value of α is decreased.

We have shown that the probability of committing a Type II error, β, is inversely related to α (Example 8.8), and that the value of β decreases as the value of μ moves farther from the null hypothesis value (sewer pipe example). The sample size n also affects β. Remember that the standard deviation of the sampling distribution of $\bar{x}$ is inversely proportional to the square root of the sample size ($\sigma_{\bar{x}} = \sigma/\sqrt{n}$). Thus, as illustrated in Figure 8.15 on page 364, the variability of both the null and alternative sampling distributions is decreased as n is increased. If the value of α is specified and remains fixed, the value of β

FIGURE 8.15

Relationship between α, β, and n

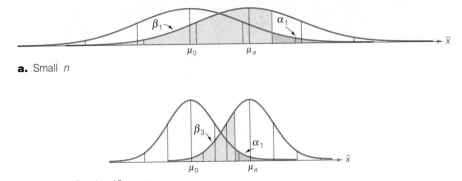

a. Small n

b. Large n, fixed α ($\beta_3 < \beta_1$)

decreases as n increases, as illustrated in Figure 8.15(b). Conversely, the power of the test for a given alternative hypothesis is increased as the sample size is increased.

The properties of β and power are summarized in the box.

Properties of β and Power

1. The value of β decreases and the power increases as the distance between the null and alternative values of μ increases (see Figure 8.13).

2. The value of β increases and the power decreases as the value of α is decreased (see Figure 8.14).

3. The value of β decreases and the power increases as the sample size is increased, assuming α remains fixed (see Figure 8.15).

EXERCISES 8.60–8.71

LEARNING THE MECHANICS

8.60   What is the relationship between β, the probability of committing a Type II error, and the power of a test?

8.61   List three factors that will increase the power of a test.

8.62   Suppose you want to test H_0: $\mu = 1,000$ against H_a: $\mu > 1,000$ using $\alpha = .05$. The population in question is normally distributed with standard deviation 120. A random sample of size $n = 36$ will be used.
   a. Sketch the sampling distribution of $\bar{x}$ assuming that H_0 is true.
   b. Find the value of $\bar{x}_0$, that value of $\bar{x}$ above which the null hypothesis will be rejected. Indicate the rejection region on your graph of part a. Shade the area above the rejection region and label it α.
   c. On your graph of part a, sketch the sampling distribution of $\bar{x}$ if $\mu = 1,020$. Shade the area under this distribution that corresponds to the probability that $\bar{x}$ falls in the nonrejection region when $\mu = 1,020$. Label this area β.

d. Find β.

e. Compute the power of this test for detecting the alternative H_a: $\mu = 1,020$.

8.63 Refer to Exercise 8.62.

    a. If $\mu = 1,040$ instead of $1,020$, what is the probability that the hypothesis test will incorrectly fail to reject H_0? That is, what is β?

    b. If $\mu = 1,040$, what is the probability that the test will correctly reject the null hypothesis? That is, what is the power of the test?

    c. Compare β and the power of the test when $\mu = 1,040$ to the values you obtained in Exercise 8.62 for $\mu = 1,020$. Explain the differences.

8.64 It is desired to test H_0: $\mu = 50$ against H_a: $\mu < 50$ using $\alpha = .10$. The population in question is uniformly distributed with standard deviation 20. A random sample of size 64 will be drawn from the population.

    a. Describe the (approximate) sampling distribution of $\bar{x}$ under the assumption that H_0 is true.

    b. Describe the (approximate) sampling distribution of $\bar{x}$ under the assumption that the population mean is 45.

    c. If μ were really equal to 45, what is the probability that the hypothesis test would lead the investigator to commit a Type II error?

    d. What is the power of this test for detecting the alternative H_a: $\mu = 45$?

8.65 Refer to Exercise 8.64. If the true value of the population mean is $\mu = 48$, what is the power of the test? How does it compare with the power when $\mu = 45$?

8.66 Suppose you want to conduct the two-tailed test of H_0: $\mu = 10$ against H_a: $\mu \neq 10$ using $\alpha = .05$. A random sample of size 100 will be drawn from the population in question. Assume the population has a standard deviation equal to 1.0.

    a. Describe the sampling distribution of $\bar{x}$ under the assumption that H_0 is true.

    b. Describe the sampling distribution of $\bar{x}$ under the assumption that $\mu = 9.9$.

    c. If μ were really equal to 9.9, find the value of β associated with the test.

    d. Find the value of β for the alternative H_a: $\mu = 10.1$.

8.67 Refer to Exercises 8.64 and 8.65.

    a. Find β for each of the following values of the population mean: 49, 47, 45, 43, and 41.

    b. Plot each value of β you obtained in part **a** against its associated population mean. Show β on the vertical axis and μ on the horizontal axis. Draw a curve through the five points on your graph.

    c. Use your graph of part **b** to find the approximate probability that the hypothesis test will lead to a Type II error when $\mu = 48$. Compare your answer to the result you obtained in Exercise 8.65.

    d. Convert each of the β values you calculated in part **a** to the power of the test at the specified value of μ. Plot the power on the vertical axis against μ on the horizontal axis. Compare the graph of part **b** to the *power curve* of this part.

    e. Examine the graphs of parts **b** and **d**. Explain what they reveal about the relationships among the distance between the true mean μ and the null hypothesized mean μ_0, the value of β, and the power.

APPLYING THE CONCEPTS

8.68 Refer to Exercise 8.20, in which the performance of a particular type of laser-based inspection equipment was investigated (Streeter, 1986). Assume that the standard deviation of the number of solder joints inspected on each run is 1.2. If $\alpha = .05$ is used in conducting the hypothesis test of interest using a sample of 48 circuit boards, and if the true mean number of solder joints that can be inspected is really equal to 9.5, what is the probability that the test will result in a Type II error?

8.69 Refer to Exercise 8.14, in which the alternative hypothesis that the mean miles per gallon achieved by 1991 G-cars exceeds 35 is tested against the null hypothesis that the mean is 35 (or less). A sample of 36 automobiles was tested. Assume that the resulting standard deviation of $s = 6$ is a good estimate of the true standard deviation.

   a. Calculate the power of the test for the mean values of 35.5, 36.0, 36.5, 37.0, and 37.5.

   b. Plot the power of the test on the vertical axis against the mean on the horizontal axis. Draw a curve through the points.

   c. Use the power curve of part b to estimate the power for the mean value $\mu = 36.75$. Calculate the power for this value of μ, and compare it to your approximation.

   d. Use the power curve to approximate the power of the test when $\mu = 40$. If the true value of the mean miles per gallon for this model is really 40, what (approximately) are the chances that the test will fail to reject the null hypothesis that the mean is 35?

8.70 Refer to Exercise 8.69. Show what happens to the power curve when the sample size is increased from $n = 36$ to $n = 100$. Assume that the standard deviation is $\sigma = 6$.

8.71 If a manufacturer (the vendee) buys all items of a particular type from a particular vendor, the manufacturer is practicing *sole sourcing*. Sole sourcing is a purchasing policy that is generally recognized as an important component of a firm's quality system. One of the major benefits of sole sourcing for the vendee is the improved communication that results from the closer vendee/vendor relationship (Treleven, 1986). As part of a sole sourcing arrangement, a vendor agrees to periodically supply its vendee with sample data from its production process. The vendee uses the data to investigate whether the mean length of rods produced by the vendor's production process is truly 5.0 mm or more, as claimed by the vendor and desired by the vendee.

   a. If the production process has a standard deviation of .01 mm, the vendor supplies $n = 100$ items to the vendee, and the vendee uses $\alpha = .05$ in testing H_0: $\mu = 5.0$ mm against H_a: $\mu < 5.0$ mm, what is the probability that the vendee's test will fail to reject the null hypothesis when in fact $\mu = 4.9975$ mm? What is the name given to this type of error?

   b. Refer to part a. What is the probability that the vendee's test will reject the null hypothesis when in fact $\mu = 5.0$? What is the name given to this type of error?

   c. What is the power of the test to detect a departure of .0025 mm below the specified mean rod length of 5.0 mm?

8.7 Inferences About a Population Variance (Optional)

Although many practical problems involve inferences about a population mean (or proportion), it is sometimes of interest to make an inference about a population variance, σ^2. To illustrate, a quality control supervisor in a cannery knows that the exact amount each can contains will vary since there are certain uncontrollable factors that affect the amount of fill. The mean fill per can is important, but equally important is the variation of fill. If σ^2, the variance of the fill, is large, some cans will contain too little and others too much. Suppose regulatory agencies specify that the standard deviation of the amount of fill should be less than .1 ounce. To determine whether the process is meeting this specification, the supervisor randomly selects ten cans, weighs the contents of each, and finds that the sample standard deviation of these measurements is .04. Do these data provide sufficient evidence to indicate that the variability is as small as desired? To answer this question, we need a procedure for testing a hypothesis about σ^2.

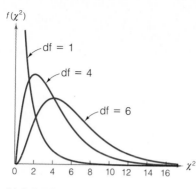

FIGURE 8.16

Several χ^2 probability distributions

Intuitively it seems that we should compare the sample variance s^2 to the hypothesized value of σ^2 (or s to σ) in order to make a decision about the population's variability. The quantity

$$\frac{(n-1)s^2}{\sigma^2}$$

has been shown to have a sampling distribution called a **chi-square (χ^2) distribution** when the population from which the sample is taken is *normally distributed*. Several chi-square distributions are shown in Figure 8.16.

The upper-tail areas for this distribution have been tabulated and are given in Table VII of Appendix A, a portion of which is reproduced in Figure 8.17. The table gives the values of χ^2, denoted as χ_α^2, that locate an area of α in the upper tail of the chi-square distribution; that is, $P(\chi^2 > \chi_\alpha^2) = \alpha$. In this case, as with the t statistic, the shape of the chi-square distribution depends on the degrees of freedom associated with s^2, namely $(n-1)$. Thus, for $n = 10$ and an upper-tail value $\alpha = .05$, you will have $n - 1 = 9$ df and $\chi_{.05}^2 = 16.9190$ (shaded area in Figure 8.17). To further illustrate the use of Table VII, we return to the can-filling example.

FIGURE 8.17

Reproduction of part of Table VII in Appendix A

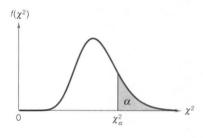

DEGREES OF FREEDOM	$\chi_{.100}^2$	$\chi_{.050}^2$	$\chi_{.025}^2$	$\chi_{.010}^2$	$\chi_{.005}^2$
1	2.70554	3.84146	5.02389	6.63490	7.87944
2	4.60517	5.99147	7.37776	9.21034	10.5966
3	6.25139	7.81473	9.34840	11.3449	12.8381
4	7.77944	9.48773	11.1433	13.2767	14.8602
5	9.23635	11.0705	12.8325	15.0863	16.7496
6	10.6446	12.5916	14.4494	16.8119	18.5476
7	12.0170	14.0671	16.0128	18.4753	20.2777
8	13.3616	15.5073	17.5346	20.0902	21.9550
9	14.6837	16.9190	19.0228	21.6660	23.5893
10	15.9871	18.3070	20.4831	23.2093	25.1882
11	17.2750	19.6751	21.9200	24.7250	26.7569
12	18.5494	21.0261	23.3367	26.2170	28.2995
13	19.8119	22.3621	24.7356	27.6883	29.8194
14	21.0642	23.6848	26.1190	29.1413	31.3193
15	22.3072	24.9958	27.4884	30.5779	32.8013
16	23.5418	26.2962	28.8454	31.9999	34.2672
17	24.7690	27.5871	30.1910	33.4087	35.7185
18	25.9894	28.8693	31.5264	34.8053	37.1564
19	27.2036	30.1435	32.8523	36.1908	38.5822

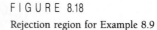

EXAMPLE 8.9

Solution

According to the previous discussion, the quality control supervisor sampled $n = 10$ cans and calculated $s = .04$. Does this value of s provide sufficient evidence to indicate that the standard deviation σ of the fill measurements is less than .1 ounce?

Since the null and alternative hypotheses must be stated in terms of σ^2 (rather than σ), we want to test the null hypothesis that $\sigma^2 = .01$ against the alternative that $\sigma^2 < .01$. Therefore, the elements of the test are

H_0: $\sigma^2 = .01$

H_a: $\sigma^2 < .01$

Test statistic: $\chi^2 = \dfrac{(n - 1)s^2}{\sigma^2}$

Assumption: The distribution of the amounts of fill is approximately normal.

Rejection region: The smaller the value of s^2 we observe, the stronger the evidence in favor of H_a. Thus, we reject H_0 for "small values" of the test statistic. With $\alpha = .05$ and 9 df, the χ^2 value for rejection is found in Table VII and pictured in Figure 8.18. We will reject H_0 if $\chi^2 < 3.32511$.

Remember that the area given in Table VII is the area to the *right* of the numerical value in the table. Thus, to determine the lower-tail value, which has $\alpha = .05$ to its *left*, we used the $\chi^2_{.95}$ column in Table VII.

FIGURE 8.18

Rejection region for Example 8.9

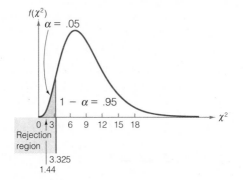

Since

$$\chi^2 = \frac{(n - 1)s^2}{\sigma^2} = \frac{9(.04)^2}{.01} = 1.44$$

is less than 3.32511, the supervisor can conclude that the variance σ^2 of the population of all amounts of fill is less than .01 ($\sigma < .1$) with probability of a Type I error equal to $\alpha = .05$. If this procedure is repeatedly used, it will incorrectly reject H_0 only 5% of the time. Thus, the quality control supervisor is confident in the decision that the cannery is operating within the desired limits of variability.

It is also possible to form a confidence interval for a population variance through the manipulation of the χ^2 distribution. This confidence interval, as well as the one-tailed and two-tailed tests of hypotheses for σ^2, are given in the accompanying boxes.

Test of a Hypothesis About σ^2

ONE-TAILED TEST

H_0: $\sigma^2 = \sigma_0^2$ ($\sigma_0^2 =$ hypothesized variance)

H_a: $\sigma^2 < \sigma_0^2$
(or H_a: $\sigma^2 > \sigma_0^2$)

Test statistic: $\chi^2 = \dfrac{(n-1)s^2}{\sigma_0^2}$

Rejection region: $\chi^2 < \chi^2_{(1-\alpha)}$
(or $\chi^2 > \chi^2_\alpha$ when H_a: $\sigma^2 > \sigma_0^2$)

TWO-TAILED TEST

H_0: $\sigma^2 = \sigma_0^2$

H_a: $\sigma^2 \neq \sigma_0^2$

Test statistic: $\chi^2 = \dfrac{(n-1)s^2}{\sigma_0^2}$

Rejection region: $\chi^2 < \chi^2_{(1-\alpha/2)}$
or $\chi^2 > \chi^2_{\alpha/2}$

where the distribution of χ^2 is based on $(n-1)$ degrees of freedom.

Assumption: The population from which the sample is drawn is approximately normal.

Confidence Interval for σ^2

$$\frac{(n-1)s^2}{\chi^2_{\alpha/2}} < \sigma^2 < \frac{(n-1)s^2}{\chi^2_{(1-\alpha/2)}}$$

where the distribution of χ^2 is based on $(n-1)$ degrees of freedom.

Assumption: The population from which the sample is drawn is approximately normal.

EXAMPLE 8.10

Test scores are often used to discriminate among individuals applying for the same job, students applying to graduate school, attorneys trying to become members of the bar, etc. Suppose an employment agency uses a 500-point examination to help in determining which job applicants are best qualified for certain positions. The variability in these test scores should be considered when evaluating the test results. For example, if all applicants should somehow score exactly the same score on the test (no variability among the scores), the test would be of no value in deciding which applicants should be employed. A large amount of variability among the test scores would be desirable in order to differentiate the relative merits of the applicants. To evaluate the variability of the test scores, the employment agency randomly selects 100 test scores and calculates $s^2 = 127$. Use this information to form a 95% confidence interval for σ^2, the variability for *all* test scores.

Solution

The general form of the confidence interval is

$$\frac{(n-1)s^2}{\chi^2_{\alpha/2}} < \sigma^2 < \frac{(n-1)s^2}{\chi^2_{(1-\alpha/2)}}$$

where we assume the test scores are approximately normally distributed. Using Table VII in Appendix A, we obtain the tabulated values of chi-square corresponding to $\alpha/2 = .025$ and $1 - \alpha/2 = .975$, based on $n - 1 = 100 - 1 = 99$ degrees of freedom:

$$\chi^2_{.025} \approx 129.56 \quad \text{and} \quad \chi^2_{.975} \approx 74.22$$

The interval of interest is thus

$$\frac{99(127)}{129.56} < \sigma^2 < \frac{99(127)}{74.22}$$

$$97.04 < \sigma^2 < 169.40$$

Thus, the employment agency can be 95% confident that the variance of all test scores is (approximately) between 97 and 170 or, equivalently, that the standard deviation is between 10 and 13, rounding to the nearest whole number.

This information can be used to determine whether the amount of variability is sufficient. If the standard deviation of 13 is deemed too small to allow the company to discriminate among the applicants, the structure of the test should be changed.

EXERCISES 8.72–8.84

LEARNING THE MECHANICS

8.72 Let χ_0^2 be a particular value of χ^2. Find the value of χ_0^2 such that:
  a. $P(\chi^2 > \chi_0^2) = .10$ for $n = 10$ **b.** $P(\chi^2 > \chi_0^2) = .05$ for $n = 13$
  c. $P(\chi^2 > \chi_0^2) = .025$ for $n = 6$

8.73 A random sample of n observations is selected from a normal population to test the null hypothesis that $\sigma^2 = 25$. Specify the rejection region for each of the following combinations of H_a, α, and n:
  a. $H_a: \sigma^2 \neq 25$; $\alpha = .05$; $n = 18$ **b.** $H_a: \sigma^2 > 25$; $\alpha = .01$; $n = 20$
  c. $H_a: \sigma^2 > 25$; $\alpha = .10$; $n = 15$ **d.** $H_a: \sigma^2 < 25$; $\alpha = .01$; $n = 12$
  e. $H_a: \sigma^2 \neq 25$; $\alpha = .10$; $n = 8$ **f.** $H_a: \sigma^2 < 25$; $\alpha = .05$; $n = 25$

8.74 A random sample of 15 observations is selected from a normal population with variance σ^2. Give the values of $\chi^2_{\alpha/2}$ and $\chi^2_{(1-\alpha/2)}$ that would be used to form a confidence interval for σ^2 for each of the following levels of confidence:
  a. 95%   **b.** 90%   **c.** 99%

8.75 A random sample of five measurements gave $\bar{x} = 9.4$ and $s^2 = 4.84$.
  a. What assumptions must you make concerning the population in order to test a hypothesis about (or estimate) σ^2?
  b. Suppose the assumptions in part **a** are satisfied. Test the null hypothesis, $\sigma^2 = 1$, against the alternative hypothesis, $\sigma^2 > 1$. Use $\alpha = .05$.
  c. Test the null hypothesis that $\sigma^2 = 1$ against the alternative hypothesis that $\sigma^2 \neq 1$. Use $\alpha = .05$.
  d. Find a 90% confidence interval for σ^2.

8.76 Refer to Exercise 8.75. Suppose we had $n = 100$, $\bar{x} = 9.4$, and $s^2 = 4.84$.
   a. Test the null hypothesis, H_0: $\sigma^2 = 1$, against the alternative hypothesis, H_a: $\sigma^2 > 1$.
   b. Compare your test result with those of Exercise 8.75.
   c. Find a 90% confidence interval for σ^2. Compare this confidence interval with the confidence interval obtained in Exercise 8.75 and note the effect of an increase in sample size on the width of the interval.

8.77 A random sample of $n = 5$ observations from a normal population produced the following measurements: 5, 7, 3, 8, 1
   a. Do the data provide sufficient evidence to indicate that $\sigma^2 > 20$? Test using $\alpha = .05$.
   b. Find a 90% confidence interval for σ^2.

8.78 A random sample of $n = 7$ observations from a normal population produced the following measurements: 4, 0, 6, 3, 3, 5, 9
   a. Do the data provide sufficient evidence to indicate that $\sigma^2 < 1$? Test using $\alpha = .05$.
   b. Find a 90% confidence interval for σ^2.

APPLYING THE CONCEPTS

8.79 An educational testing service designed an achievement test so that the range in student scores would be at least 300 points. To determine whether the objective was achieved, the testing service gave the test to a random sample of 30 students and found that the sample mean and variance were 759 and 1,943, respectively. Do the data provide sufficient evidence to indicate that the test does not achieve the desired dispersion in scores? Test using $\alpha = .05$. [*Hint:* Assume that range $= 6\sigma$.]

8.80 A marine biologist wishes to use male angelfish for experimental purposes due to the belief that their weight is fairly stable (i.e., the variability in weights among male angelfish is small). The biologist randomly samples 16 male angelfish and finds that their mean weight is 4.1 pounds and the standard deviation is 1.73 pounds. Find a 95% confidence interval for the variability in weights of all male angelfish. What assumptions must you make in order to form the interval?

8.81 Refer to Exercise 8.80. It is suggested that the marine biologist use parrotfish instead of male angelfish in the experiment. Since these are more difficult to obtain, the biologist decides to use parrotfish only if there is evidence that the variance of their weights is less than 4. A random sample of ten parrotfish produces a mean of 4.3 pounds and a variance of 2. Is there sufficient evidence for the biologist to claim that the variability in weights among parrotfish is small enough to justify their use in the experiment? Test at $\alpha = .05$, and state any assumptions that are needed.

8.82 A new gunlike apparatus has been devised to replace the needle in administering vaccines. The apparatus, which is connected to a large supply of vaccine, can be set to inject different amounts of the serum, but the variance in the amount of serum injected to a given person must not be greater than .06 to ensure proper inoculation. A random sample of 25 injections resulted in a variance of .135. Do the data provide sufficient evidence to indicate the gun is not working properly? Use $\alpha = .10$.

8.83 It is essential in the manufacture of machinery to utilize parts that conform to specifications. In the past, diameters of the ball-bearings produced by a certain manufacturer had a variance of .00156. To cut costs, the manufacturer instituted a less expensive production method. The variance of the diameters of 100 randomly sampled bearings produced by the new process was .00211. Do the data provide sufficient evidence to indicate that diameters of ball-bearings produced by the new process are more variable than those produced by the old process?

8.84 To perform an experiment, a chemist has to use a substance that contains 50% sodium nitrate. The chemist suspects that a particular batch of the substance has not been mixed thoroughly, thus causing the amount of sodium nitrate to vary from one portion of the batch to another. The results of 20 randomly selected 10-ounce samples yield a sample standard deviation equal to .05 ounce. Estimate the true variance of the amount of sodium nitrate in 10-ounce samples selected from the batch. Use a 95% confidence interval.

Summary

In this chapter we extended the concept of making inferences from using samples to estimate parameters of a distribution to performing tests of hypotheses about these parameters. The essential elements of a test of hypothesis are the **null** and **alternative hypotheses**, the **test statistic**, the **rejection region**, the **calculation of the test statistic**, and the **conclusion**.

The null hypothesis is the status-quo hypothesis, the hypothesis accepted until contradicted by sufficient sample evidence. The alternative is the research hypothesis, the hypothesis that will not be accepted until convincingly established by sample information. The test statistic is a value calculated from the sample information and utilized to decide whether to reject the null hypothesis. The rejection region is the collection of values of the test statistic that will cause us to reject the null hypothesis.

We design the test to have a specified **Type I error probability**, α, so that we know the probability of rejecting H_0 when H_0 is true. When the test statistic falls in the rejection region, we make the inference that the null hypothesis is false (that is, the alternative is true), at the α level of significance. If the test statistic does not fall in the rejection region, then we fail to reject the null hypothesis. We do not accept the null hypothesis in this case unless we know the **Type II error probability**, β.

We covered large-sample tests involving the population mean μ and the binomial probability p, both using the standard normal z as a test statistic. We also discussed the small-sample test involving the mean of a normal distribution using the Student t statistic.

SUPPLEMENTARY EXERCISES 8.85–8.119

[*Note:*   *List the assumptions necessary for the valid implementation of the statistical procedures you use in solving all these exercises. Starred (*) exercises refer to the optional sections.*]

LEARNING THE MECHANICS

8.85   Which of the elements of a test of hypothesis can and should be specified *prior* to analyzing the data that are to be utilized to conduct the test?

8.86   *Complete the following statement:*   The smaller the *p*-value associated with a test of hypothesis, the stronger the support for the _____ hypothesis. Explain your answer.

8.87   Specify the differences between a large-sample and small-sample test of hypothesis about a population mean μ. Focus on the assumptions and test statistics.

8.88   Medical tests have been developed to detect many serious diseases. A medical test is designed to minimize the probability that it will produce a "false positive" or a "false negative." A false positive refers to a positive test result when in fact the individual does not have the disease, whereas a false negative is a negative test result for an individual who does have the disease.

   a. If we treat a medical test for a disease as a statistical test of hypothesis, what are the null and alternative hypotheses for the medical test?

    b. What are the Type I and Type II errors for the test? Relate each to false positives and false negatives.

    c. Considering which of the errors has more grave consequences, is it more important to minimize α or β? Explain.

8.89   If the rejection of the null hypothesis of a particular test would cause your firm to go out of business, would you want α to be small or large? Explain.

8.90   *Complete the following statement:* The larger the *p*-value associated with a test of hypothesis, the stronger the support for the _____ hypothesis. Explain your answer.

8.91   A random sample of 20 observations selected from a normal population produced $\bar{x} = 72.6$ and $s^2 = 19.4$.

    a. Form a 90% confidence interval for the population mean.

    b. Test H_0: $\mu = 80$ against H_a: $\mu < 80$. Use $\alpha = .05$.

    c. Test H_0: $\mu = 80$ against H_a: $\mu \neq 80$. Use $\alpha = .01$.

    d. Form a 99% confidence interval for μ.

    e. How large a sample would be required to estimate μ to within 1 unit with 95% confidence?

8.92   A random sample of $n = 200$ observations from a binomial population yields $\hat{p} = .29$.

    a. Test H_0: $p = .35$ against H_a: $p < .35$. Use $\alpha = .05$.

    b. Test H_0: $p = .35$ against H_a: $p \neq .35$. Use $\alpha = .05$.

    c. Form a 95% confidence interval for *p*.

    d. Form a 99% confidence interval for *p*.

    e. How large a sample would be required to estimate *p* to within .05 with 99% confidence?

8.93   A random sample of 175 measurements possessed a mean $\bar{x} = 8.2$ and a standard deviation $s = .79$.

    a. Form a 95% confidence interval for μ.

    b. Test H_0: $\mu = 8.3$ against H_a: $\mu \neq 8.3$. Use $\alpha = .05$.

    c. Test H_0: $\mu = 8.4$ against H_a: $\mu \neq 8.4$. Use $\alpha = .05$.

*8.94   A random sample of 41 observations from a normal population possessed a mean $\bar{x} = 88$ and a standard deviation $s = 6.9$.

    a. Form a 90% confidence interval for σ^2.

    b. Form a 99% confidence interval for σ^2.

    c. Test H_0: $\sigma^2 = 30$ against H_a: $\sigma^2 > 30$. Use $\alpha = .05$.

    d. Test H_0: $\sigma^2 = 30$ against H_a: $\sigma^2 \neq 30$. Use $\alpha = .05$.

APPLYING THE CONCEPTS

8.95   Failure to meet payments on student loans guaranteed by the United States government has been a major problem for both banks and the government. Approximately 50% of all student loans guaranteed by the government are in default. A random sample of 350 loans to college students in one region of the United States indicates that 147 loans are in default.

    a. Do the data indicate that the proportion of student loans in default in this area of the country differs from the proportion of all student loans in the United States that are in default? Use $\alpha = .01$.

    b. Find the observed significance level for the test and interpret its value.

8.96   In order to be effective, the mean length of life of a certain mechanical component used in a spacecraft must be larger than 1,100 hours. Due to the prohibitive cost of the components, only three can be tested under simulated space conditions. The lifetimes (hours) of the components were recorded and the following statistics were computed: $\bar{x} = 1,173.6$ and $s = 36.3$. Do the data provide sufficient evidence to conclude that the component will be effective? Use $\alpha = .01$.

8.97   The mean score on a Peace Corps application test, based on many tests conducted over a long period of time, is 80. Ten prospective applicants have taken a course designed to improve their scores on the test. The scores of the ten applicants who completed the course had a mean equal to 86.1 and a standard deviation equal to 12.4.

    a. Do the data provide sufficient evidence to conclude that students taking the course will have a higher mean score than those who do not? Test using $\alpha = .10$.

    b. Find the approximate observed significance level for the test and interpret its value.

    c. What assumptions must be made in order for the procedure that you used in part **a** to be valid?

8.98 A sporting goods manufacturer who produces both white and yellow tennis balls claims that more than 75% of all tennis balls sold are yellow. A marketing study of the purchases of white and yellow tennis balls at a number of stores showed that of 470 cans sold, 410 were yellow and 60 were white.

    a. Is there sufficient evidence to support the manufacturer's claim? Test using $\alpha = .01$.

    b. Find the observed significance level for the test and interpret its value.

    *c. Calculate the probability, β, of a Type II error if in fact 80% of the tennis balls sold are yellow.

8.99 A discount store chain claims that its steel-belted radial tires last longer than those of a major tire company. The following experiment was performed to test this claim: On each of 40 cars, one discount tire and one major company tire were mounted on the rear axle. After each car was driven 8,000 miles, the tires were inspected for wear. Suppose the tires of the discount chain show less wear on 32 of the cars. Is there sufficient evidence to conclude that the discount chain's claim is correct? Test using $\alpha = .10$. [*Hint:* Let p equal the proportion of discount store tires that last longer than those of the major tire company. Then test $H_0: p = .5$ against $H_a: p > .5$.]

8.100 During past harvests, a farmer has averaged 68.2 bushels of corn per acre. A new fertilizer has been placed on the market, and after using the new fertilizer the farmer notes the yield of corn for four randomly selected fields of equal size. The mean yield is 72.4 bushels per acre and the standard deviation is 2.2 bushels.

    a. If these data truly represent a random sample of corn yields that the farmer might expect (now and in the future) when using the new fertilizer, do they suggest that the mean yield of corn per acre has changed from past years? Test using $\alpha = .05$.

    b. Note that the four yield measurements were selected from within the same year. Are these measurements a random sample selected from the population of interest to the farmer? If not, what information do the data provide the farmer?

8.101 The EPA sets a limit of 5 parts per million on PCB (a dangerous substance) in water. A major manufacturing firm producing PCB for electrical insulation discharges small amounts from the plant. The company management, attempting to control the PCB in its discharge, has given instructions to halt production if the mean amount of PCB in the effluent exceeds 3 parts per million. A random sample of 50 water specimens produced the following statistics: $\bar{x} = 3.1$ parts per million and $s = .5$ part per million.

    a. Do these statistics provide sufficient evidence to halt the production process? Use $\alpha = .01$.

    b. If you were the plant manager, would you want to use a large or a small value for α for the test in part **a**?

*8.102 Refer to Exercise 8.101.

    a. In the context of the problem, define a Type II error.

    b. Calculate β for the test described in part **a** of Exercise 8.101 assuming that the true mean is $\mu = 3.1$ parts per million.

    c. What is the power of the test to detect the effluent's departure from the standard of 3.0 parts per million when the mean is 3.1 parts per million?

    d. Repeat parts **b** and **c** assuming that the true mean is 3.2 parts per million. What happens to the power of the test as the plant's mean PCB departs further from the standard?

*8.103 Refer to Exercises 8.101 and 8.102.

    a. Suppose an α value of .05 is used to conduct the test. Does this change favor the manufacturer? Explain.

b. Determine the value of β and the power for the test when $\alpha = .05$ and $\mu = 3.1$.

c. What happens to the power of the test when α is increased?

8.104 One recent study (Sauer et al., 1988) of gambling newsletters that purport to improve a bettor's odds of winning bets on NFL football games indicates that the newsletters' betting schemes were not profitable. Suppose a random sample of 50 games is selected to test one gambling newsletter. Following the newsletter's recommendations, 30 of the 50 games produced winning wagers. Test whether the newsletter can be said to significantly increase the odds of winning over what one could expect by selecting the winner at random. Use $\alpha = .05$.

8.105 Refer to Exercise 8.104. Calculate and interpret the p-value for the test.

*8.106 Refer to Exercise 8.104.

a. Describe a Type II error in terms of this application.

b. Calculate the probability β of a Type II error for this test assuming that the newsletter really does increase the probability of winning a wager on an NFL game to $p = .55$.

c. Suppose the number of games sampled is increased from 50 to 100. How does this affect the probability of a Type II error for $p = .55$?

8.107 A large mail-order company has placed an order for 5,000 electric can openers with a supplier on condition that no more than 2% of the devices will be defective. To check the shipment, the company tests a random sample of 400 of the can openers and finds 11 are defective. Does this provide sufficient evidence to indicate that the proportion of defective can openers in the shipment exceeds 2%? Test using $\alpha = .05$.

8.108 Refer to Exercise 8.107. Find and interpret the observed significance level for the hypothesis test.

8.109 Refer to Exercise 8.107. Suppose the company wants to estimate the proportion, p, of defective can openers in the shipment using a 95% confidence interval of width .08. Approximately how large a sample would be required?

8.110 The mean grade-point average (GPA) at a certain university was 3.20 in 1980. To show that grade inflation has been reversed, a dean sets out to show that the mean GPA is now lower than 3.20. A random sample of 100 students yields a mean GPA of 3.05 and a standard deviation equal to .90. Do the data provide sufficient evidence to indicate that grade inflation has been reversed? (Test using $\alpha = .10$.)

8.111 A leading cigarette manufacturer claims its cigarettes contain an average of less than 16 milligrams of tar. To check this claim, a random sample of cigarettes will be chosen and the mean amount of tar per cigarette will be estimated. If previous information indicates that the amount of tar per cigarette has a standard deviation of 2.5 milligrams, how many cigarettes should be sampled to estimate the true mean to within .45 milligram with probability .95?

8.112 The "beta coefficient" of a stock is a measure of the stock's volatility (or risk) relative to the market as a whole. Stocks with beta coefficients greater than 1 generally bear greater risk (more volatility) than the market, whereas stocks with beta coefficients less than 1 are less risky (less volatile) than the overall market. A random sample of 15 high-technology stocks was selected at the end of 1990, and the mean and standard deviation of the beta coefficients were calculated:

$$\bar{x} = 1.23 \qquad s = .37$$

a. Set up the appropriate null and alternative hypothesis to test whether the average high-technology stock is riskier than the market as a whole.

b. Establish the appropriate test statistic and rejection region for the test. Use $\alpha = .10$.

c. What assumptions are necessary to assure the validity of the test?

d. Calculate the test statistic and state your conclusion.

e. What is the approximate p-value associated with this test? Interpret it.

8.113   Refer to Exercise 8.112. A number of statistical software packages can be used to conduct single-sample tests of hypotheses about a population mean. Most of them report both a value of the test statistic and a p-value associated with the test after the user inputs the data and the null hypothesis of interest. When the data from Exercise 8.112 were entered into one such program, the output was:

T = 2.408   P-VALUE = 0.0304

The P-VALUE corresponds to a two-tailed test of the null hypothesis $\mu = 1$.
a. Interpret the p-value on the computer output.
b. If the alternative hypothesis of interest is $\mu > 1$, what is the appropriate p-value of the test?

8.114   Officials from a high school claim that at least 85% of the students who have graduated from the school have received a college degree or are enrolled in a college degree program. A random sample of 60 former graduates indicates that 47 have received or are enrolled in a program to receive a college degree. Do the data contradict the school official's claim? (Use $\alpha = .05$.)

8.115   In the past, a chemical company produced 880 pounds of a certain type of plastic per day. Now, using a newly developed and less expensive process, the mean daily yield of plastic for the first 50 days of production is 871 pounds; the standard deviation is 21 pounds.
a. Do the data provide sufficient evidence to indicate that the mean daily yield for the new process is less than for the old procedure? (Test using $\alpha = .01$.)
b. What assumptions must you make in order to use the statistical test you employed?

*8.116   Refer to Exercise 8.115. Calculate the probability β that the test fails to reject the null hypothesis that the new process has a mean daily yield of 880 when in fact the true mean is 875. What is the power of the test to determine that $\mu < 880$ when $\mu = 875$?

*8.117   A machine used to fill beer cans must operate so that the amount of beer actually dispensed varies very little. If too much beer is released, the cans will overflow, causing waste. If too little beer is released, the cans will not contain enough beer, causing complaints from customers. A random sample of the fills for 20 cans yielded a standard deviation of .07 ounce. Using a 95% confidence interval, estimate the true variance of the fills.

*8.118   Ophthalmologists require an instrument that can rapidly measure interocular pressure for glaucoma patients. The device now in general use is known to yield readings of this pressure with a variance of 10.3. The variance of five pressure readings on the same eye by a newly developed instrument is equal to 9.8. Does this sample variance provide sufficient evidence to indicate that the new instrument is more reliable than the instrument currently in use? (Use $\alpha = .05$.)

*8.119   The standard deviation of the diameters of screw-top lids must not be larger than .6 millimeter to ensure that the lids will fit properly on glass jars. A random sample of 15 lids yields a sample standard deviation of .78 millimeter. Do the data provide sufficient evidence to conclude that the standard deviation of the lid diameters is larger than .6? (Use $\alpha = .01$.)

ON YOUR OWN...

The "efficient market" theory postulates that the best predictor of a stock's price at some point in the future is the current price of the stock (with some adjustments for inflation and transaction costs, which we shall assume to be negligible for the purpose of this exercise). To test this theory, select a random sample of 25 stocks on the New York Stock Exchange and record the closing prices on the last days of two recent consecutive months. Calculate the increase or decrease in the stock price over the 1-month period.

a. Define μ as the mean change in price of all stocks over a 1-month period. Set up the appropriate null and alternative hypotheses in terms of μ.

b. Use the sample of the 25 stock price differences to conduct the test of hypothesis established in part **a**. Use $\alpha = .05$.

c. Tabulate the number of rejections and nonrejections of the null hypothesis in your class. If the null hypothesis were true, how many rejections of the null hypothesis would you expect among those in your class? How does this expectation compare with the actual number of nonrejections? What does the result of this exercise indicate about the efficient market theory?

USING THE COMPUTER...

Refer to **Using the Computer** in Chapters 6 and 7. Let μ_0 be the mean of the population of 1,000 zip codes you selected—i.e., μ_0 is the "true" value of the population mean. In each of the following scenarios you will be testing the null hypothesis H_0: $\mu = \mu_0$ versus the two-tailed alternative hypothesis H_a: $\mu \neq \mu_0$. Thus, in this exercise you know that the null hypothesis is true.

a. Select 100 samples of size 50 (with replacement) from the 1,000 zip codes, and conduct the test for each sample using $\alpha = .05$. In how many of the tests did the sample lead to an incorrect rejection of the null hypothesis? How does this compare with what you expected to occur? Repeat the exercise using $\alpha = .10$.

b. Select 100 samples of size 20 (with replacement) from the 1,000 zip codes, and conduct the test for each sample using $\alpha = .05$. In how many of the tests did the sample lead to an incorrect rejection of the null hypothesis? How does this compare with what you expected to occur? Repeat the exercise using $\alpha = .10$.

References

Dujack, S. R. "Science leaves no doubt: The polygraph lies," *Minneapolis Star and Tribune*, July 31, 1986, p. 11A.

Environmental Protection Agency. *Environment Midwest*. September–October 1976, Region V.

Sauer, R. D., Brajer, V., Ferris, S. P., and Marr, M. W. "Hold Your Bets: Another Look at the Efficiency of the Gambling Market for National Football League Games," *Journal of Political Economy*, February 1988, Vol. 96, No. 1.

Snedecor, G. W., and Cochran, W. G. *Statistical Methods*, 7th ed. Ames: Iowa State University Press, 1980, Chapters 4 and 5.

Streeter, J. P. "Solder joint inspection using a laser inspector," *Quality Congress Transactions*. Milwaukee: American Society for Quality Control, 1986, pp. 507–515.

Treleven, M. "Sole sourcing from the vendor side," *Quality Congress Transactions*. Milwaukee: American Society for Quality Control, 1986, pp. 584–590.

Inferences Based on Two Samples: Estimation and Tests of Hypotheses

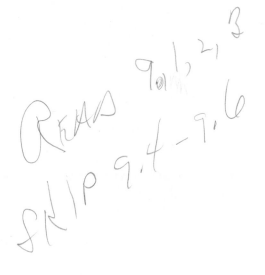

WHERE WE'VE BEEN...

Two methods for making statistical inferences, estimation and tests of hypotheses based on single samples, were presented in Chapters 7 and 8. In particular, we gave confidence intervals and tests of hypotheses concerning a population mean μ and a binomial proportion p, and we learned how to select the sample size necessary to obtain a specified amount of information concerning a parameter. (Inferences about a population variance σ^2 were included in an optional section.)

WHERE WE'RE GOING...

Now that we have learned to make inferences about a single population, we will learn how to compare two populations. Such problems often arise in practice. We may wish to compare the mean gas mileages for two models of automobiles, the mean retirement ages of workers in the public and private sectors, or the mean reaction times of men and women to a visual stimulus. We may also wish to compare two population proportions, say the proportions of subjects in a psychological experiment that respond to two different stimuli. How to decide whether differences exist and how to estimate the differences between population means and proportions are the subjects of this chapter.

Many experiments involve a comparison of two populations. For example, a sales manager for a steel company may want to estimate the difference in mean sales per customer between two different salespeople. A consumer group may want to test whether two major brands of food freezers differ in the mean amount of electricity they use. A political candidate may wish to estimate the difference in the proportions of voters in two districts who favor his or her candidacy. A professional golfer may wish to compare the variability in the distance that two competing brands of golf balls travel when struck with the same club. In this chapter we consider techniques for using two samples to compare the populations from which they were selected.

9.1 Large-Sample Inferences About the Difference Between Two Population Means: Independent Sampling

Many of the same procedures that are used to estimate and test hypotheses about a single parameter can be modified to make inferences about two parameters. Both the z and t statistics may be adapted to make inferences about the difference between two population means.

In this section we develop the large-sample z statistic for comparing two population means. The t statistic for making small-sample inferences about the difference between two population means is introduced in Section 9.2.

We use Example 9.1 to introduce the procedures for making large-sample inferences about the difference between two population means.

EXAMPLE 9.1

A dietitian has developed a diet that is low in fats, carbohydrates, and cholesterol. Although the diet was initially intended to be used by people with heart disease, the dietitian wishes to examine the effect this diet has on the weights of obese people. Two random samples of 100 obese people each are selected, and one group of 100 is placed on the low-fat diet. The other 100 are placed on a diet that contains approximately the same quantity of food but is not as low in fats, carbohydrates, and cholesterol. For each person, the amount of weight lost (or gained) in a 3-week period is recorded. Using the data given in the table, form a 95% confidence interval for the difference between the population mean weight losses for the two diets.

	LOW-FAT DIET	OTHER DIET
Sample size	100	100
Sample mean weight loss	9.3 pounds	3.7 pounds
Sample variance	22.4	16.3

Solution

Recall that the general form of a large-sample confidence interval for a single mean μ is $\bar{x} \pm z_{\alpha/2}\sigma_{\bar{x}}$. That is, we add and subtract $z_{\alpha/2}$ standard deviations of the sample estimate, $\bar{x}$, to the value of the estimate. We employ a similar procedure to form the confidence interval for the difference between two population means.

Let μ_1 represent the mean of the conceptual population of weight losses for all obese people who could be placed on the low-fat diet. Let μ_2 be similarly defined for the other diet. We wish to form a confidence interval for $(\mu_1 - \mu_2)$. An intuitively appealing estimator for $(\mu_1 - \mu_2)$ is the difference between the sample means $(\bar{x}_1 - \bar{x}_2)$. Thus, we will form the confidence interval of interest by

$$(\bar{x}_1 - \bar{x}_2) \pm z_{\alpha/2}\sigma_{(\bar{x}_1 - \bar{x}_2)}$$

Assuming the two samples are independent, the standard deviation of the difference between the sample means is

$$\sigma_{(\bar{x}_1 - \bar{x}_2)} = \sqrt{\frac{\sigma_1^2}{n_1} + \frac{\sigma_2^2}{n_2}} \approx \sqrt{\frac{s_1^2}{n_1} + \frac{s_2^2}{n_2}}$$

Using the sample data and noting that $\alpha = .05$ and $z_{.025} = 1.96$, we find that the 95% confidence interval is, approximately,

$$(9.3 - 3.7) \pm 1.96\sqrt{\frac{22.4}{100} + \frac{16.3}{100}} = 5.6 \pm (1.96)(.62)$$

$$= 5.6 \pm 1.22$$

or (4.38, 6.82). Using this estimation procedure over and over again for different samples, we know that approximately 95% of the confidence intervals formed in this manner will enclose the difference in population means $(\mu_1 - \mu_2)$. Therefore, we are reasonably confident that the mean weight loss for the low-fat diet is between 4.38 and 6.82 pounds more than the mean weight loss for the other diet. With this information, the dietitian better understands the potential of the low-fat diet as a weight-reducing diet.

The justification for the procedure used in Example 9.1 to estimate $(\mu_1 - \mu_2)$ relies on the properties of the sampling distribution of $(\bar{x}_1 - \bar{x}_2)$. These properties are summarized in the accompanying box, and the performance of the estimator in repeated sampling is pictured in Figure 9.1.

Properties of the Sampling Distribution of $(\bar{x}_1 - \bar{x}_2)$

1. The sampling distribution of $(\bar{x}_1 - \bar{x}_2)$ is approximately normal for *large samples* by the Central Limit Theorem.
2. The mean of the sampling distribution of $(\bar{x}_1 - \bar{x}_2)$ is $(\mu_1 - \mu_2)$.
3. If the two samples are independent, the standard deviation of the sampling distribution is

$$\sigma_{(\bar{x}_1 - \bar{x}_2)} = \sqrt{\frac{\sigma_1^2}{n_1} + \frac{\sigma_2^2}{n_2}}$$

where σ_1^2 and σ_2^2 are the variances of the two populations being sampled and n_1 and n_2 are the respective sample sizes.

FIGURE 9.1

Sampling distribution of $(\bar{x}_1 - \bar{x}_2)$

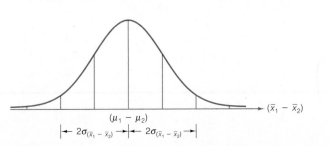

In Example 9.1, we noted the similarity in the procedures for forming a large-sample confidence interval for one population mean and a large-sample confidence interval for the difference between two population means. When we are testing hypotheses, the procedures are again very similar. The general large-sample procedures for forming confidence intervals and testing hypotheses about $(\mu_1 - \mu_2)$ are summarized in the accompanying boxes.

Large-Sample Confidence Interval for $(\mu_1 - \mu_2)$

$$(\bar{x}_1 - \bar{x}_2) \pm z_{\alpha/2}\sigma_{(\bar{x}_1 - \bar{x}_2)} = (\bar{x}_1 - \bar{x}_2) \pm z_{\alpha/2}\sqrt{\frac{\sigma_1^2}{n_1} + \frac{\sigma_2^2}{n_2}}$$

Assumptions: The two samples are randomly selected in an independent manner from the two populations. The sample sizes, n_1 and n_2, are large enough so that $\bar{x}_1$ and $\bar{x}_2$ each have approximately normal sampling distributions and so that s_1^2 and s_2^2 provide good approximations to σ_1^2 and σ_2^2. This will be true if $n_1 \geq 30$ and $n_2 \geq 30$.

Large-Sample Test of Hypothesis for $(\mu_1 - \mu_2)$

ONE-TAILED TEST

H_0: $(\mu_1 - \mu_2) = D_0$
H_a: $(\mu_1 - \mu_2) < D_0$
   [or H_a: $(\mu_1 - \mu_2) > D_0$]

TWO-TAILED TEST

H_0: $(\mu_1 - \mu_2) = D_0$
H_a: $(\mu_1 - \mu_2) \neq D_0$

where D_0 = Hypothesized difference between the means (this difference is
        often 0)

Test statistic: $z = \dfrac{(\bar{x}_1 - \bar{x}_2) - D_0}{\sigma_{(\bar{x}_1 - \bar{x}_2)}}$ *Test statistic:* $z = \dfrac{(\bar{x}_1 - \bar{x}_2) - D_0}{\sigma_{(\bar{x}_1 - \bar{x}_2)}}$

where $\sigma_{(\bar{x}_1 - \bar{x}_2)} = \sqrt{\dfrac{\sigma_1^2}{n_1} + \dfrac{\sigma_2^2}{n_2}}$

Rejection region: $z < -z_\alpha$ *Rejection region:* $z < -z_{\alpha/2}$
   [or $z > z_\alpha$ when   or $z > z_{\alpha/2}$
   H_a: $(\mu_1 - \mu_2) > D_0$]

Assumptions: Same as for the large-sample confidence interval.

EXAMPLE 9.2

In the late 1980s and early 1990s the United States and Japan were involved in very intense negotiations regarding restrictions on trade between the two countries. One of the claims made repeatedly by U.S. officials was that many Japanese manufacturers were pricing their goods higher in Japan than they were in the United States, in effect subsidizing low prices in the United States by extremely high prices in Japan. The basis of the U.S. argument was that Japan was able to do that only because they were keeping competitive U.S. goods from reaching the Japanese marketplace.

An economist decided to test the hypothesis that higher retail prices were being charged for Japanese automobiles in Japan than in the United States. She obtained random samples of 50 retail sales in the United States and 30 retail sales in Japan over the same time period and for the same model of automobile, converted the Japanese sales prices from yen to dollars using current conversion rates, and obtained the summary statistics shown:

U.S. SALES	JAPANESE SALES
$n_1 = 50$	$n_2 = 30$
$\bar{x}_1 = \$11{,}545$	$\bar{x}_2 = \$12{,}243$
$s_1 = \$1{,}989$	$s_2 = \$1{,}843$

Do these data provide sufficient evidence for the economist to conclude that the mean sales price for this model is higher in Japan than in the United States?

Solution

We can best answer this question by performing a test of hypothesis. Defining μ_1 as the mean sales price in the United States and μ_2 as the mean sales price in Japan during this period and for this model of automobile, we want to test whether the data support the alternative (research) hypothesis that $\mu_2 > \mu_1$ [i.e., that $(\mu_1 - \mu_2) < 0$]. Thus, we will test the null hypothesis, $(\mu_1 - \mu_2) = 0$. The evidence necessary to reject this hypothesis in favor of the alternative is a sufficiently large negative value of the difference between the sample means, $(\bar{x}_1 - \bar{x}_2)$.

The elements of the test are as follows:

H_0: $(\mu_1 - \mu_2) = 0$ (i.e., $\mu_1 = \mu_2$; note that $D_0 = 0$ for this hypothesis test)

H_a: $(\mu_1 - \mu_2) < 0$ (i.e., $\mu_1 < \mu_2$)

Test statistic: $z = \dfrac{(\bar{x}_1 - \bar{x}_2) - D_0}{\sigma_{(\bar{x}_1 - \bar{x}_2)}} = \dfrac{(\bar{x}_1 - \bar{x}_2) - 0}{\sigma_{(\bar{x}_1 - \bar{x}_2)}}$

Rejection region: $z < -z_\alpha = -1.645$ (see Figure 9.2)

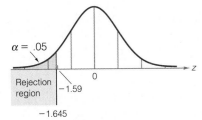

$\alpha = .05$

Rejection region

-1.59

-1.645

FIGURE 9.2
Rejection region for Example 9.2

Assuming the samples from the United States and Japan are independent, we now calculate

$$z = \frac{(\bar{x}_1 - \bar{x}_2) - 0}{\sigma_{(\bar{x}_1 - \bar{x}_2)}} = \frac{11{,}545 - 12{,}243}{\sqrt{\dfrac{\sigma_1^2}{n_1} + \dfrac{\sigma_2^2}{n_2}}}$$

$$\approx \frac{-698}{\sqrt{\dfrac{s_1^2}{n_1} + \dfrac{s_2^2}{n_2}}} = \frac{-698}{\sqrt{\dfrac{(1{,}989)^2}{50} + \dfrac{(1{,}843)^2}{30}}} = \frac{-698}{438.57} = -1.59$$

As you can see in Figure 9.2, the calculated z value does not fall in the rejection region. The samples do not provide sufficient evidence, at $\alpha = .05$ for the economist to conclude that the mean retail price in Japan exceeds that in the United States.

EXAMPLE 9.3

Solution

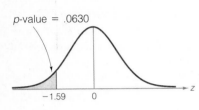

FIGURE 9.3

The observed significance level for
Example 9.2

Find the observed significance level for the test in Example 9.2.

The alternative hypothesis in Example 9.2, H_a: $\mu_1 - \mu_2 < 0$, required a lower one-tailed test using

$$z = \frac{\bar{x}_1 - \bar{x}_2}{\sigma_{(\bar{x}_1 - \bar{x}_2)}}$$

as a test statistic. Since the value z calculated from the sample data was -1.59, the observed significance level (p-value) for the test is the probability of observing a value of z at least as contradictory to the null hypothesis as $z = -1.59$; i.e.,

$$p\text{-value} = P(z \le -1.59)$$

This probability is computed assuming H_0 is true and is equal to the shaded area shown in Figure 9.3.

The tabulated area corresponding to $z = 1.59$ in Table IV of Appendix B is .4370. Therefore, the observed significance level for the test is

$$p\text{-value} = .5 - .4370 = .063$$

You will recall that in Example 9.2, we chose $\alpha = .05$ as the probability of a Type I error and consequently did not reject H_0; that is, we did not find sufficient evidence to indicate that $(\mu_1 - \mu_2) < 0$. If the results of the test had been presented in terms of an observed significance level and left for us to interpret, we might not have reached such an inflexible conclusion. Observing a value of z as small as $z = -1.59$ is an improbable event (rare event) if, in fact, $\mu_1 = \mu_2$. Since the probability is fairly small (.063)—in fact, quite close to .05— we would conclude that there is some evidence to suggest that $\mu_1 < \mu_2$. Naturally, we would be more certain of this conclusion if the observed significance level were smaller, say .05, .01, or, better yet, .001. However, the practical question to be answered is not whether the test results are statistically significant but whether the difference between μ_1 and μ_2 is large enough to have a practical business significance. To shed light on this question, we wish to estimate the difference, $(\mu_1 - \mu_2)$.

EXAMPLE 9.4

Solution

Refer to Example 9.2. Find a 95% confidence interval for the difference in the mean retail prices in the United States and Japan and discuss the implication of the confidence interval.

The 95% confidence interval for $(\mu_1 - \mu_2)$ is

$$(\bar{x}_1 - \bar{x}_2) \pm z_{\alpha/2} \sqrt{\frac{\sigma_1^2}{n_1} + \frac{\sigma_2^2}{n_2}}$$

Once again, we substitute s_1^2 and s_2^2 for σ_1^2 and σ_2^2, because these quantities provide good approximations to σ_1^2 and σ_2^2 for samples as large as $n_1 = 50$ and $n_2 = 30$. Then, the 95% confidence interval for $(\mu_1 - \mu_2)$ is

$$(11{,}545 - 12{,}243) \pm 1.96 \sqrt{\frac{(1{,}989)^2}{50} + \frac{(1{,}843)^2}{30}} = -698 \pm 859.60$$

Thus, we estimate the difference in retail prices to fall in the interval $-\$1,557.60$ to $\$161.60$. In other words, we estimate that μ_2, the mean retail price in Japan, could be larger than μ_1, the mean retail price in the United States, by as much as $\$1,557.60$, or it could be less than μ_1 by as much as $\$161.60$.

What is the practical interpretation of all this? By all appearances, the Japanese retail price for this model car was indeed higher than the U.S. retail price during the same time period. However, the data collected were insufficient to provide statistical support for this conclusion at the specified level of significance. What can be done to further refine and sharpen the analysis? The first and most obvious improvement would be to increase the sample sizes: Sample sizes of 50 and 30 yield a rather wide confidence interval (total width of $\$1,719.20$) for estimating the difference between the means. Increasing the sample sizes will decrease the standard error $\sigma_{(\bar{x}_1 - \bar{x}_2)}$, and so the width of the confidence interval will also be decreased.

A second, less obvious, refinement to the analysis would be to decrease somehow the rather large standard deviations associated with the samples, s_1 and s_2. Both exceed $\$1,800$, indicating that the retail prices contained within each sample vary widely. Perhaps the options available on this model result in significant price discrepancies, or perhaps the dealerships in urban settings have significantly different pricing strategies than those in rural settings. **Regression analysis** is a useful statistical methodology for comparing population means while controlling for other factors affecting the variability of the measurements, such as options and geographic location in this example. Regression analysis is the subject of Chapters 13–15.

CASE STUDY 9.1

PRODUCTIVITY AND MOBILITY

One theory regarding the mobility of college and university faculty members is that those who are most productive in the publishing of scholarly articles are also the most mobile. The logic behind this theory is that good researchers and publishers receive more job offers and are therefore more likely to move from one university to another. Since most people in the academic world are aware that a similar strong relationship does not exist for academics who are known to be fine teachers, we might wonder whether it holds for persons employed in industry. George F. Dreher* considered this question by examining the personnel records of a large national oil company. Dreher obtained early career performance records for 529 of the company's employees. Of these, 174 were classified as *stayers*, those who stayed with the company; the other 355, who left the company at varying points during a 15-year period, were classified as *leavers*. Dreher made a number of comparisons between these two groups. Among these were a comparison of measures of the mean initial performances of the stayers and leavers, their rates of career advancement (number of promotions per year), and their final performance appraisals. The means and standard deviations for these sample comparisons are shown in Table 9.1 on page 386.

---

*Reprinted from Dreher, G. F. "The role of performance in the turnover process," *Academy of Management Journal*, 1982, 25.

TABLE 9.1   **Difference Between Leavers and Continuing Employees**

| VARIABLE | STAYERS ($n_1 = 174$) | | LEAVERS ($n_2 = 355$) | | |
	$\bar{x}_1$	s_1	$\bar{x}_2$	s_2	t
Initial performance	3.51	.51	3.24	.52	5.23
Rate of career advancement	.43	.20	.31	.31	4.63
Final performance appraisal	3.78	.62	3.15	.68	9.76

Although a large-sample z-test would be appropriate, Dreher tests for differences between the means of stayers and leavers using a small-sample t-test. This test, to be discussed in the following section, is based on assumptions that are not necessary for the large-sample z-test. When the sample sizes are very large, as in Dreher's sampling, the numerical values of the t and z statistics are almost the same and lead to essentially the same results. Dreher's computed t values are shown in Table 9.1.

We have not computed the values of the large-sample z statistics for comparing the three pairs of population means, but it is clear (from examining Dreher's large t values) that there is ample evidence to indicate a difference in means for all three comparisons. (The p-values for all three tests are extremely small.) In fact, it appears that stayers had higher mean measures of initial and final performance and a higher mean rate of career advancement than the leavers.

EXERCISES 9.1–9.20

LEARNING THE MECHANICS

9.1   The purpose of this exercise is to compare the variability of $\bar{x}_1$ and $\bar{x}_2$ with the variability of $(\bar{x}_1 - \bar{x}_2)$.

a. Suppose the first sample is selected from a population with mean $\mu_1 = 150$ and variance $\sigma_1^2 = 900$. Within what range should the sample mean vary about 95% of the time in repeated samples of 100 measurements from this distribution? That is, construct an interval extending 2 standard deviations of $\bar{x}_1$ on each side of μ_1.

b. Suppose the second sample is selected independently of the first from a second population with mean $\mu_2 = 150$ and variance $\sigma_2^2 = 1,600$. Within what range should the sample mean vary about 95% of the time in repeated samples of 100 measurements from this distribution? That is, construct an interval extending 2 standard deviations of $\bar{x}_2$ on each side of μ_2.

c. Now consider the difference between the two sample means, $(\bar{x}_1 - \bar{x}_2)$. What are the mean and standard deviation of the sampling distribution of $(\bar{x}_1 - \bar{x}_2)$?

d. Within what range should the difference in sample means vary about 95% of the time in repeated independent samples of 100 measurements each from the two populations?

e. What, in general, can be said about the variability of the difference between independent sample means relative to the variability of the individual sample means?

9.2   Independent random samples of 64 observations each are chosen from two normal populations with the following means and standard deviations:

POPULATION 1	POPULATION 2
$\mu_1 = 12$	$\mu_2 = 10$
$\sigma_1 = 4$	$\sigma_2 = 3$

Let $\bar{x}_1$ and $\bar{x}_2$ denote the two sample means.
  a. Give the mean and standard deviation of the sampling distribution of $\bar{x}_1$.
  b. Give the mean and standard deviation of the sampling distribution of $\bar{x}_2$.
  c. Suppose you were to calculate the difference $(\bar{x}_1 - \bar{x}_2)$ between the sample means. Find the mean and standard deviation of the sampling distribution of $(\bar{x}_1 - \bar{x}_2)$.
  d. Will the statistic $(\bar{x}_1 - \bar{x}_2)$ be normally distributed? Explain.

9.3   Refer to Exercise 9.2.
  a. Give the z-score corresponding to $(\bar{x}_1 - \bar{x}_2) = 1.5$.
  b. Find the probability that $(\bar{x}_1 - \bar{x}_2)$ is greater than 1.5.
  c. Find the probability that $(\bar{x}_1 - \bar{x}_2)$ is less than 1.5.
  d. Find the probability that $(\bar{x}_1 - \bar{x}_2)$ is greater than 1 or less than -1.

9.4   Two independent random samples have been selected, 100 observations from population 1 and 100 from population 2. Sample means $\bar{x}_1 = 70$ and $\bar{x}_2 = 50$ were obtained. From previous experience with these populations, it is known that the variances are $\sigma_1^2 = 100$ and $\sigma_2^2 = 64$.
  a. Find $\sigma_{(\bar{x}_1-\bar{x}_2)}$.
  b. Sketch the approximate sampling distribution for $(\bar{x}_1 - \bar{x}_2)$ assuming $(\mu_1 - \mu_2) = 5$.
  c. Locate the observed value of $(\bar{x}_1 - \bar{x}_2)$ on the graph you drew in part b. Does it appear that this value contradicts the null hypothesis H_0: $(\mu_1 - \mu_2) = 5$?
  d. Use the z table on the inside of the front cover to determine the rejection region for the test of H_0: $(\mu_1 - \mu_2) = 5$ against H_a: $(\mu_1 - \mu_2) \neq 5$. Use $\alpha = .05$.
  e. Conduct the hypothesis test of part d and interpret your result.

9.5   Refer to Exercise 9.4. Construct a 95% confidence interval for $(\mu_1 - \mu_2)$. Interpret the interval. Which inference provides more information about the value of $(\mu_1 - \mu_2)$— the test of hypothesis in Exercise 9.4 or the confidence interval in this exercise?

9.6   Two independent random samples have been selected, 100 from population 1 and 150 from population 2. Sample means $\bar{x}_1 = 1,025$ and $\bar{x}_2 = 1,039$ were obtained. The sample standard deviations are $s_1 = 10$ and $s_2 = 12$.
  a. Describe the sampling distribution of $(\bar{x}_1 - \bar{x}_2)$. Assume that $(\mu_1 - \mu_2) = -5$.
  b. Test H_0: $(\mu_1 - \mu_2) = -5$ against H_a: $(\mu_1 - \mu_2) < -5$, using $\alpha = .01$. Interpret the result of your test.
  c. Report the p-value of your test.

9.7   Are the hypothesis test and confidence interval procedures given in this section valid if the sampled populations are not normally distributed? Explain.

APPLYING THE CONCEPTS

9.8   Lesley E. Tan investigated the relationship between handedness and motor competence in preschool children. Random samples of 41 right-handers and 41 left-handers were administered several tests of motor skills, yielding the means and standard deviations shown in the accompanying table.

	LEFT-HANDED	RIGHT-HANDED
n	41	41
$\bar{x}$	97.5	98.1
s	17.5	19.2

Source: Tan, L. E. "Laterality and motor skills in four-year-olds," *Child Development*, 56.

  a. Is there evidence of a difference between the motor skills of right- and left-handed preschool children based on this experiment? Use $\alpha = .10$.

b. Use a 90% confidence interval to estimate the true difference in mean motor skill scores between left- and right-handed preschoolers. Does the confidence interval support the result of the test you conducted in part **a**?

c. What assumptions about the distributions of the populations of test scores are necessary to assure the validity of the inferences you made in parts **a** and **b**?

d. What is the observed significance level of the test you conducted in part **a**?

e. Tan concluded, in part, that ". . . tests indicated no difference between left-handers and right-handers . . ." Does your analysis support this conclusion?

9.9   An experiment has been conducted at a university to compare the mean number of study hours expended per week by student athletes with the mean number of hours expended by nonathletes. A random sample of 55 athletes produced a mean equal to 20.6 hours studied per week and a standard deviation equal to 5.3 hours. A second random sample of 200 nonathletes produced a mean equal to 23.5 hours per week and a standard deviation equal to 4.1 hours.

a. Describe the two populations involved in the comparison.

b. Do the samples provide sufficient evidence to conclude that there is a difference in the mean number of hours of study per week between athletes and nonathletes? Test using $\alpha = .01$.

c. Construct a 99% confidence interval for $(\mu_1 - \mu_2)$.

d. Would a 95% confidence interval for $(\mu_1 - \mu_2)$ be narrower or wider than the one you found in part **c**? Why?

9.10 A new type of band has been developed by a dental laboratory for children who have to wear braces. The new brands are designed to be more comfortable, look better, and provide more rapid progress in realigning teeth. An experiment was conducted to compare the mean wearing time necessary to correct a specific type of misalignment between the old braces and the new bands. One hundred children were randomly assigned, 50 to each group. A summary of the data is shown in the table.

	OLD BRACES	NEW BANDS
$\bar{x}$	410 days	380 days
s	45 days	60 days

a. Is there sufficient evidence to conclude that the new bands do not have to be worn as long as the old braces? Use $\alpha = .01$.

b. Find a 95% confidence interval for the difference in wearing times for the two types of braces. Interpret the interval.

9.11 Suppose it is desired to compare two physical education training programs for preadolescent girls. A total of 80 girls are randomly selected, with 40 assigned to each program. After three 6-week periods on the program, each girl is given a fitness test that yields a score between 0 and 100. The means and variances of the scores for the two groups are shown in the table. Calculate a 99% confidence interval for the true difference in mean fitness scores for girls trained using these two programs.

	n	$\bar{x}$	s^2
Program 1	40	78.7	201.6
Program 2	40	75.3	259.2

9.12 It is often said that economic status is related to the commission of crimes. To test this theory, a sociologist selected a random sample of 70 people (who had no record of criminal conviction) from the census records of a certain city and recorded their annual incomes. Similarly, a random sample of 60 people, each of whom had committed their first crime, was selected from court records and the annual income (prior to arrest) was recorded for each. The means and variances of the annual incomes (in thousands of dollars) for the people in the two groups are shown in the table. Do the data provide sufficient evidence to indicate that the mean income of criminals, prior to committing their first offense, is lower than that for the noncriminal public? Test using $\alpha = .05$.

	$\bar{x}$	s^2
Criminals	13.3	24.2
Noncriminals	15.4	42.6

9.13 A large supermarket chain is interested in determining whether a difference exists between the mean shelf life (in days) of brand S bread and brand H bread. Random samples of 50 freshly baked loaves of each brand were tested, with the results shown in the table.

BRAND S	BRAND H
$\bar{x}_1 = 4.1$	$\bar{x}_2 = 5.2$
$s_1 = 1.2$	$s_2 = 1.4$

a. Is there sufficient evidence to conclude that a difference does exist between the mean shelf lives of brand S and brand H bread? Test at the $\alpha = .05$ level.

b. Find the observed significance level for the test and interpret its value.

c. Let μ_1 and μ_2 represent the mean shelf lives for brands S and H, respectively. Construct a 90% confidence interval for $(\mu_1 - \mu_2)$. Give an interpretation of your confidence interval.

9.14 Wansley, Roenfeldt, and Cooley compared the profiles of a sample of 44 firms that merged during 1975–1976 with those of a sample of 44 firms that did not merge. The table displays information obtained on the firms' price-earnings ratios.

	MERGED FIRMS	NONMERGED FIRMS
Sample mean	7.295	14.666
Sample standard deviation	7.374	16.089

Source: Wansley, J. W., Roenfeldt, R. L., and Cooley, P. L. "Abnormal returns from merger profiles," *Journal of Financial and Quantitative Analysis*, Vol. 18, No. 2, June 1983, pp. 149–162.

a. The analysis performed by Wansley, Roenfeldt, and Cooley indicated that "merged firms generally have smaller price-earnings ratios." Do you agree? Test using $\alpha = .05$.

b. Report the p-value of the test you conducted in part a.

c. What assumption(s) was it necessary to make in order to perform the test in part a?

d. Do you think that the distributions of the price-earnings ratios for the populations from which these samples were drawn are normally distributed? Why or why not? [*Hint:* Note the relative values of the sample means and standard deviations.]

9.15 Two manufacturers of corrugated fiberboard both claim that the strength of their product tests on the average at more than 360 pounds per square inch. As a result of consumer complaints, a consumer products testing firm suspects that firm A's product is at least 5 pounds per square inch stronger than firm B's. To test its suspicion, 100 fiberboards were chosen randomly from firm A's inventory and 100 were chosen from firm B's inventory. The results of tests run on the samples are shown in the table on the next page.

A	B
$\bar{x}_1 = 365$	$\bar{x}_2 = 352$
$s_1 = 23$	$s_2 = 41$

a. Set up the appropriate null and alternative hypotheses to test the testing firm's belief. How do they differ from the typical hypotheses used to compare two population means?

b. Does the sample information support the consumer products testing firm's suspicion? Test at the .05 significance level.

c. What assumptions did you make in conducting the test in part **a**? Do you think such assumptions could comfortably be made in practice? Why or why not?

d. Find the observed significance level for the test and interpret its value.

9.16 The *Orlando Sentinel* (April 12, 1984) presented an article titled "Books Hold Their Own Against Television." The article reports on an hour-long survey of 1,961 persons selected from among the American public. The survey was conducted for the Book Industry Study Group, a nonprofit organization representing book publishers, manufacturers, wholesalers, etc. Among the statistics computed from the survey data, they found that the amount of time each person claimed to read per week averaged 11.7 hours versus 16.3 hours spent watching television. However, the article notes that the A. C. Nielsen Co., which measures television viewing for sponsors, claims that the average American watched $28\frac{1}{2}$ hours of television per week. Suppose the standard deviations of the distributions of lengths of time spent per week reading and the lengths of time watching television are both equal to 5 hours. Assume also that the Nielsen estimate of mean television time is based on an independent random sample of size 1,961.

a. Use the survey results to find a 95% confidence interval for the difference between the mean length of time per week reading and the mean length of time watching television. Interpret your result.

b. If the survey and A. C. Nielsen are both sampling the same population, what (approximately) is the probability that their sample estimates of mean television viewing time would differ by as much as $(28.5 - 16.3) = 12.2$ hours?

c. Can you explain why the "rare event" in part **b** might have occurred?

9.17 Refer to Case Study 9.1 and the comparison of mean performance characteristics between corporate stayers and leavers.

a. Use Dreher's statistics, shown in Table 9.1, to test the null hypothesis that there is no difference in mean initial performance measure between stayers and leavers. Test using $\alpha = .01$. Interpret your conclusion.

b. Give the approximate *p*-value for the test and interpret it.

9.18 Refer to Case Study 9.1 and the comparision of mean performance characteristics between corporate stayers and leavers.

a. Use Dreher's statistics, shown in Table 9.1, to test the null hypothesis that there is no difference in the rate of career advancement—i.e., the mean number of promotions per year—between stayers and leavers. Test using $\alpha = .01$. Interpret your conclusion.

b. Give the approximate *p*-value for the test and interpret it.

9.19 Refer to Case Study 9.1 and the comparison of mean performance characteristics between corporate stayers and leavers.

a. Use Dreher's statistics, shown in Table 9.1, to test the null hypothesis that there is no difference in mean final performance measure between stayers and leavers. Test using $\alpha = .01$. Interpret your conclusion.

b. Give the approximate *p*-value for the test and interpret it.

9.20 As part of a study of participative management, George H. Hines sampled workers from two types of New Zealand sociocultural backgrounds: those who believed in the existence of a class system and those who believed that they lived and worked in a classless society. Each worker in the sampling was selected from a work environment with participatory management. Do workers who consider themselves to be social equals with their management superiors possess different levels of job satisfaction than those workers who see themselves as socially different from management? Each worker in the independent random samples was asked to answer this question by rating his or her job satisfaction on a scale of 1 (poor) to 7 (excellent). Using the results of this study (shown in the table), what can you say about the differences in job satisfaction for the two different sociocultural types of workers?

| | BELIEF IN EXISTENCE OF A CLASS SYSTEM | |
	Yes	No
Sample size	175	277
Mean	5.42	5.19
Standard deviation	1.24	1.17

Source: Hines, G. H. "Influences on employee expectancy and participative management," *Academy of Management Journal*, 1974, 17.

9.2 Small-Sample Inferences About the Difference Between Two Population Means: Independent Sampling

Suppose a television network wants to determine whether major sports events or first-run movies attract more viewers in the prime-time hours. It selects 28 prime-time evenings; of these, 13 have programs devoted to major sports events and the remaining 15 have first-run movies. The number of viewers (estimated by a television viewer rating firm) is recorded for each program. If μ_1 is the mean number of sports viewers per evening of sports programming and μ_2 is the mean number of movie viewers per evening of movie programming, we want to detect a difference between μ_1 and μ_2—if such a difference exists. Therefore, we want to test the null hypothesis

$$H_0: \quad (\mu_1 - \mu_2) = 0$$

against the alternative hypothesis

$$H_a: \quad (\mu_1 - \mu_2) \neq 0 \quad \text{(i.e., either } \mu_1 > \mu_2 \text{ or } \mu_2 > \mu_1)$$

Since the sample sizes are small, estimates of σ_1^2 and σ_2^2 are unreliable and the z-test statistic is inappropriate for the test. But as in the case of a single mean (Section 8.4), we can construct a Student t statistic. This statistic (formula to be given subsequently) has the familiar t-distribution described in Chapter 7. *To use the t statistic, both sampled populations must be approximately normally distributed with equal population variances, and the random samples must be selected independently of each other.* The normality and equal variances assumptions imply relative frequency distributions for the populations that would appear as shown in Figure 9.4.

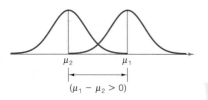

$(\mu_1 - \mu_2 > 0)$

FIGURE 9.4

Assumptions for the two-sample t:
(1) normal populations,
(2) equal variances

Will these assumptions be satisfied for the television-viewing problem? We think that both assumptions will be adequately satisfied in this sampling situation. Since we assume the two populations have equal variances ($\sigma_1^2 = \sigma_2^2 = \sigma^2$), it is reasonable to use the information contained in both samples to construct a **pooled sample estimator** of σ^2 for use in the t statistic. Thus, if s_1^2 and s_2^2 are the two sample variances (both estimating the variance σ^2 common to both populations), the pooled estimator of σ^2, denoted as s_p^2, is

$$s_p^2 = \frac{(n_1 - 1)s_1^2 + (n_2 - 1)s_2^2}{(n_1 - 1) + (n_2 - 1)} = \frac{(n_1 - 1)s_1^2 + (n_2 - 1)s_2^2}{n_1 + n_2 - 2}$$

or

$$s_p^2 = \frac{\overbrace{\sum (x_1 - \bar{x}_1)^2}^{\text{From sample 1}} + \overbrace{\sum (x_2 - \bar{x}_2)^2}^{\text{From sample 2}}}{n_1 + n_2 - 2}$$

where x_1 represents a measurement from sample 1 and x_2 represents a measurement from sample 2. Recall that the term *degrees of freedom* was defined in Section 8.4 as 1 less than the sample size. Thus, in this case, we have $(n_1 - 1)$ degrees of freedom for sample 1 and $(n_2 - 1)$ degrees of freedom for sample 2. Since we are pooling the information on σ^2 obtained from both samples, the degrees of freedom associated with the pooled variance s_p^2 is equal to the sum of the degrees of freedom for the two samples, namely, the denominator of s_p^2; that is, $(n_1 - 1) + (n_2 - 1) = n_1 + n_2 - 2$.

Note that the second formula given for s_p^2 shows that the pooled variance is simply a **weighted average** of the two sample variances, s_1^2 and s_2^2. The weight given each variance is proportional to its degrees of freedom. If the two variances have the same number of degrees of freedom (i.e., if the sample sizes are equal), then the pooled variance is a simple average of the two sample variances. The result is an average or "pooled" variance that is a better estimate of σ^2 than either s_1^2 or s_2^2 alone.

To obtain the **small-sample test statistic** for testing H_0: $(\mu_1 - \mu_2) = D_0$, substitute the pooled estimate of σ^2 into the formula for the two-sample z statistic (Section 9.1) to obtain

$$t = \frac{(\bar{x}_1 - \bar{x}_2) - D_0}{\sqrt{s_p^2 \left(\dfrac{1}{n_1} + \dfrac{1}{n_2} \right)}}$$

It can be shown that this statistic has a t-distribution, but with $(n_1 + n_2 - 2)$ degrees of freedom.

We use the television viewer example to outline the final steps for this t-test: The hypothesized difference in mean number of viewers is $D_0 = 0$. The rejection region will be two-tailed and will be based on a t-distribution with $(n_1 + n_2 - 2)$ or $(13 + 15 - 2) = 26$ df. For $\alpha = .05$, the rejection region for the test would be

$$t < -t_{\alpha/2} \quad \text{or} \quad t > t_{\alpha/2}$$

The value for $t_{.025}$ given in Table VI of Appendix A is 2.056. Thus, the rejection region for the television example is

$$t < -2.056 \quad \text{or} \quad t > 2.056$$

This rejection region is shown in Figure 9.5.

FIGURE 9.5

Rejection region for a two-tailed t-test: $\alpha = .05$, df $= 26$

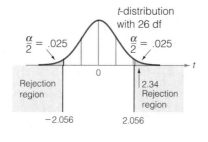

EXAMPLE 9.5

Suppose the television network's samples produce the following results:

SPORTS	MOVIE
$n_1 = 13$	$n_2 = 15$
$\bar{x}_1 = 6.8$ million	$\bar{x}_2 = 5.3$ million
$s_1 = 1.8$ million	$s_2 = 1.6$ million

Do the data provide sufficient evidence to indicate a difference between the mean numbers of viewers for major sports events and first-run movies shown in prime time? Test using $\alpha = .05$.

Solution

We must assume that these independent samples were selected from normal distributions (i.e., the number of viewers per evening is normally distributed for each type of program) with equal variances. [*Note:* Although the sample estimates of variance are not equal, the assumption that the population variances are equal may still be valid. We present a method for checking this assumption statistically in Section 9.6.] We now calculate

$$s_p^2 = \frac{(n_1 - 1)s_1^2 + (n_2 - 1)s_2^2}{n_1 + n_2 - 2} = \frac{(13 - 1)(1.8)^2 + (15 - 1)(1.6)^2}{13 + 15 - 2} = \frac{74.72}{26}$$
$$= 2.87$$

Then

$$t = \frac{(\bar{x}_1 - \bar{x}_2) - D_0}{\sqrt{s_p^2\left(\dfrac{1}{n_1} + \dfrac{1}{n_2}\right)}} = \frac{(6.8 - 5.3) - 0}{\sqrt{2.87\left(\dfrac{1}{13} + \dfrac{1}{15}\right)}} = \frac{1.5}{.64} = 2.34$$

Since the observed value of t ($t = 2.34$) falls in the rejection region (see Figure 9.5), the samples provide sufficient evidence to indicate that the mean numbers of viewers differ for major sports events and first-run movies shown in prime time. Or, we can say that the test results are statistically significant at the $\alpha = .05$ level of significance. Because the rejection was in the positive or upper tail of the *t*-distribution, it appears that the mean number of viewers for sports events exceeds that for movies.

EXAMPLE 9.6

Solution

Find the approximate observed significance level for the test in Example 9.5.

The observed significance level (*p*-value) for the test is the probability of observing a value of the test statistic that is *at least* as contradictory to H_0: $\mu_1 = \mu_2$ as the value observed if in fact H_0 is true. Thus, since the test was two-sided, we have

$$p\text{-value} = P(t < -2.34 \quad \text{or} \quad t > 2.34)$$
$$= 2P(t > 2.34) \qquad \text{(because the } t\text{-distribution is symmetric about its mean, 0)}$$

Turning to Table VI in Appendix A for 26 degrees of freedom, you can see that

$$P(t < -2.479 \quad \text{or} \quad t > 2.479) = 2P(t > 2.479) = 2(.01) = .02$$

and

$$P(t < -2.056 \quad \text{or} \quad t > 2.056) = 2P(t > 2.056) = 2(.025) = .05$$

Since the observed value of *t* lies between 2.479 and 2.056, the *p*-value for the test lies between .02 and .05. We have agreed to choose the larger of these as the approximate observed significance level and would thus give the approximate *p*-value for the test as .05. A reader of this reported *p*-value would reject the null hypothesis (and conclude that a difference exists between the mean numbers of sports and movie viewers) for values of α larger than or equal to .05.

The same *t* statistic can also be used to construct confidence intervals for the difference between population means. Both the confidence interval and the test of hypothesis procedures are summarized in the accompanying boxes.

Small-Sample Confidence Interval for $(\mu_1 - \mu_2)$ (Independent Samples)

$$(\bar{x}_1 - \bar{x}_2) \pm t_{\alpha/2} \sqrt{s_p^2 \left(\frac{1}{n_1} + \frac{1}{n_2} \right)}$$

where

$$s_p^2 = \frac{(n_1 - 1)s_1^2 + (n_2 - 1)s_2^2}{n_1 + n_2 - 2}$$

and $t_{\alpha/2}$ is based on $(n_1 + n_2 - 2)$ degrees of freedom.

Assumptions: 1. Both sampled populations have relative frequency distributions that are approximately normal.
2. The population variances are equal.
3. The samples are randomly and independently selected from the populations.

Small-Sample Test of Hypothesis for $(\mu_1 - \mu_2)$ (Independent Samples)

ONE-TAILED TEST

H_0: $(\mu_1 - \mu_2) = D_0$

H_a: $(\mu_1 - \mu_2) < D_0$
      [or H_a: $(\mu_1 - \mu_2) > D_0$]

Test statistic:

$$t = \frac{(\bar{x}_1 - \bar{x}_2) - D_0}{\sqrt{s_p^2\left(\dfrac{1}{n_1} + \dfrac{1}{n_2}\right)}}$$

Rejection region:

$t < -t_\alpha$

[or $t > t_\alpha$ when
H_a: $(\mu_1 - \mu_2) > D_0$]

TWO-TAILED TEST

H_0: $(\mu_1 - \mu_2) = D_0$

H_a: $(\mu_1 - \mu_2) \neq D_0$

Test statistic:

$$t = \frac{(\bar{x}_1 - \bar{x}_2) - D_0}{\sqrt{s_p^2\left(\dfrac{1}{n_1} + \dfrac{1}{n_2}\right)}}$$

Rejection region:

$t < -t_{\alpha/2}$ or $t > t_{\alpha/2}$

where t_α and $t_{\alpha/2}$ are based on $(n_1 + n_2 - 2)$ degrees of freedom.

Assumptions: Same as for the small-sample confidence interval for $(\mu_1 - \mu_2)$ in the previous box.

EXAMPLE 9.7

Suppose you wish to compare a new method of teaching reading to "slow learners" to the current standard method. You decide to base this comparison on the results of a reading test given at the end of a learning period of 6 months. Of a random sample of 20 slow learners, 8 are taught by the new method and 12 are taught by the standard method. All 20 children are taught by qualified instructors under similar conditions for a 6-month period. The results of the reading test at the end of this period are summarized in the table. Estimate the true mean difference $(\mu_1 - \mu_2)$ between the test scores for the new method and the standard method. Use a 90% confidence interval, and interpret the interval. What assumptions must be made in order that the estimate be valid?

NEW METHOD	STANDARD METHOD
$n_1 = 8$	$n_2 = 12$
$\bar{x}_1 = 76.9$	$\bar{x}_2 = 72.7$
$s_1 = 4.85$	$s_2 = 6.35$

Solution

The objective of this experiment is to obtain a 90% confidence interval for $(\mu_1 - \mu_2)$. To use the small-sample confidence interval for $(\mu_1 - \mu_2)$, the following assumptions must be satisfied:

1. We assume that the populations of test scores are normally distributed for both the new and standard methods of instruction. Since a test score can be viewed as a sum of the results on the various components of the test, the Central Limit Theorem lends credence to this assumption.

2. The variance of the test scores is assumed to be the same for the two populations. Under the circumstances, we might expect the variation in test scores to be approximately the same for both methods.
3. The samples are randomly and independently selected from the two populations. We have randomly chosen 20 different slow learners for the two samples in such a way that the test score for one child is not dependent on the test score for any other child. Therefore, this assumption would probably be valid.

The first step in constructing the confidence interval is to calculate the pooled estimate of variance:

$$s_p^2 = \frac{(n_1 - 1)s_1^2 + (n_2 - 1)s_2^2}{n_1 + n_2 - 2}$$

$$= \frac{(8 - 1)(4.85)^2 + (12 - 1)(6.35)^2}{8 + 12 - 2}$$

$$= 33.7892$$

where s_p^2 is based on $(n_1 + n_2 - 2) = (8 + 12 - 2) = 18$ degrees of freedom. Then the 90% confidence interval for $(\mu_1 - \mu_2)$, the difference between mean test scores for the two methods, is

$$(\bar{x}_1 - \bar{x}_2) \pm t_{\alpha/2}\sqrt{s_p^2\left(\frac{1}{n_1} + \frac{1}{n_2}\right)} = (76.9 - 72.7) \pm t_{.05}\sqrt{33.7892\left(\frac{1}{8} + \frac{1}{12}\right)}$$

$$= 4.20 \pm 1.734(2.653) = 4.20 \pm 4.60$$

or $(-.40, 8.80)$. This means that with a confidence coefficient equal to .90, we estimate the difference in mean test scores between using the new method of teaching and the standard method to fall in the interval $-.40$ to 8.80. In other words, we estimate the mean test score for the new method to be anywhere from .40 less than to 8.80 more than the mean test score for the standard method. Although the sample means seem to suggest that the new method is associated with a higher mean test score, there is insufficient evidence to indicate that $(\mu_1 - \mu_2)$ differs from 0 because the interval includes 0 as a possible value for $(\mu_1 - \mu_2)$. To show a difference in mean test scores (if it exists), you could increase the sample size and thereby narrow the width of the confidence interval for $(\mu_1 - \mu_2)$. An alternative is to design the experiment differently. This possibility is discussed in the next section.

The two-sample t statistic is a powerful tool for comparing population means when the assumptions are satisfied. It has also been shown to retain its usefulness when the sampled populations are only approximately normally distributed. And when the sample sizes are equal, the assumption of equal population variances can be relaxed. That is, if $n_1 = n_2$, then σ_1^2 and σ_2^2 can be quite different and the test statistic will still possess, approximately, a Student t-distribution. When the assumptions are not satisfied, you can select larger samples from the populations or you can use other available statistical tests (nonparametric statistical tests, which are described in Chapter 11).

> **What Should You Do if the Assumptions Are Not Satisfied?**
>
> *Answer:* If you are concerned that the assumptions are not satisfied, use the Wilcoxon rank sum test for independent samples to test for a shift in population distributions. See Chapter 11.

EXERCISES 9.21–9.36

LEARNING THE MECHANICS

9.21 To use the t statistic to test for a difference between the means of two populations, what assumptions must be made about the two populations? About the two samples?

9.22 Two populations are described in each of the following cases. In which cases would it be appropriate to apply the small sample t-test to investigate the difference between the population means?
   a. Population 1: Normal distribution with variance σ_1^2
      Population 2: Skewed to the right with variance $\sigma_2^2 = \sigma_1^2$
   b. Population 1: Normal distribution with variance σ_1^2
      Population 2: Normal distribution with variance $\sigma_2^2 \neq \sigma_1^2$
   c. Population 1: Skewed to the left with variance σ_1^2
      Population 2: Skewed to the left with variance $\sigma_2^2 = \sigma_1^2$
   d. Population 1: Normal distribution with variance σ_1^2
      Population 2: Normal distribution with variance $\sigma_2^2 = \sigma_1^2$
   e. Population 1: Uniform distribution with variance σ_1^2
      Population 2: Uniform distribution with variance $\sigma_2^2 = \sigma_1^2$

9.23 In the t-tests of this section, σ_1^2 and σ_2^2 are assumed to be equal. Thus, we say $\sigma_1^2 = \sigma_2^2 = \sigma^2$. Why is a pooled estimator of σ^2 used instead of either s_1^2 or s_2^2?

9.24 Assume that $\sigma_1^2 = \sigma_2^2 = \sigma^2$. Calculate the pooled estimator of σ^2 for each of the following cases:
   a. $s_1^2 = 100$, $s_2^2 = 90$, $n_1 = n_2 = 25$
   b. $s_1^2 = 12$, $s_2^2 = 16$, $n_1 = 20$, $n_2 = 10$
   c. $s_1^2 = .15$, $s_2^2 = .20$, $n_1 = 8$, $n_2 = 12$
   d. $s_1^2 = 2,500$, $s_2^2 = 2,000$, $n_1 = 16$, $n_2 = 17$
   e. Note that the pooled estimate is a weighted average of the sample variances. Which of the variances does the pooled estimate fall nearer in each of the above cases?

9.25 Independent random samples from normal populations produced the results shown in the table.

SAMPLE 1	SAMPLE 2
1.2	4.2
3.1	2.7
1.7	3.6
2.8	3.9
3.0	

   a. Calculate the pooled estimate of σ^2.
   b. Do the data provide sufficient evidence to indicate that $\mu_2 > \mu_1$? Test using $\alpha = .10$.

c. Find a 90% confidence interval for $(\mu_1 - \mu_2)$.
d. Which of the two inferential procedures, the test of hypothesis in part **b** or the confidence interval in part **c**, provides more information about $(\mu_1 - \mu_2)$?

9.26   Independent random samples selected from two normal populations produced the sample means and standard deviations shown in the table.

SAMPLE 1	SAMPLE 2
$n_1 = 17$	$n_2 = 12$
$\bar{x}_1 = 5.4$	$\bar{x}_2 = 7.9$
$s_1 = 3.4$	$s_2 = 4.8$

a. Test $H_0: (\mu_1 - \mu_2) = 0$ against $H_a: (\mu_1 - \mu_2) \neq 0$. Use $\alpha = .05$.
b. Estimate $(\mu_1 - \mu_2)$ using a 95% confidence interval.

9.27   Independent random samples from approximately normal populations produced the results shown in the table.

SAMPLE 1				SAMPLE 2			
52	33	42	44	52	43	47	56
41	50	44	51	62	53	61	50
45	38	37	40	56	52	53	60
44	50	43		50	48	60	55

a. Do the data provide sufficient evidence to conclude that $(\mu_2 - \mu_1) > 10$? Test using $\alpha = .01$.
b. Construct a 98% confidence interval for $(\mu_2 - \mu_1)$. Interpret your result.

APPLYING THE CONCEPTS

9.28   Amid concerns and protests about its potential dangers, nuclear-generated energy is reportedly on the rise, and the growth rate of nuclear power facilities is expected to continue to increase. One reason is that nuclear power is cheaper to produce than conventional sources. The Atomic Industrial Forum, which represents the nuclear industry, reports that coal-produced energy costs 3.5¢ per kilowatt-hour to produce compared to 3.1¢ for nuclear power (*Time*, February 13, 1984).
       Assume that these costs are averages and that the results were taken over 9 nuclear power plants and 11 coal-powered plants. Assume also that the variances of the cost estimates for nuclear and coal-produced power were .05 and .04, respectively. Does this information provide sufficient evidence to say that nuclear-generated power is less expensive to produce than coal-produced power? Test using $\alpha = .05$. What assumptions are required for the test to be valid?

9.29   An industrial plant wants to determine which of two types of fuel—gas or electric—will produce more useful energy at the lower cost. One measure of economical energy production, called the *plant investment per delivered quad*, is calculated by taking the amount of money (in dollars) invested in the particular utility by the plant and dividing by the delivered amount of energy (in quadrillion British thermal units). The smaller this ratio, the less an industrial plant pays for its delivered energy.
       Random samples of 11 plants using electrical utilities and 16 plants using gas utilities were taken, and the plant investment per quad was calculated for each. The data produced the results shown in the table.

	ELECTRIC	GAS
Sample size	11	16
Mean investment/quad (billions)	$44.5	$34.5
Variance	76.4	63.8

a. Do these data provide sufficient evidence at the $\alpha = .05$ level of significance to indicate a difference in the average investment per quad between the plants using gas and those using electrical utilities?

b. Find a 90% confidence interval for $(\mu_1 - \mu_2)$. Give a practical interpretation of this interval.

c. What assumptions are necessary to assure the validity of the inferential procedures you used in parts **a** and **b**? State them in terms of the specifics of this exercise.

9.30 With the emergence of Japan as an industrial superpower, American businesses have begun taking a close look at Japanese management styles and philosophies. Some of the credit for the high quality of Japanese products has been attributed to the Japanese system of permanent employment for their workers. In the United States, high job turnover rates are common in many industries and are associated with high product defect rates. High turnover rates mean that U.S. plants are more highly populated with inexperienced workers who are unfamiliar with the company's product lines than is the case in Japan. In a study of the room air conditioner industry in Japan and the United States, David Garvin reported that the difference in the average annual turnover rate of workers between American plants and Japanese plants was 3.1%. In a different study, five Japanese and five American plants that manufacture room air conditioners were randomly sampled and their turnover rates were determined as shown in the table.

U.S. PLANTS	JAPANESE PLANTS
7.11%	3.52%
6.06%	2.02%
8.00%	4.91%
6.87%	3.22%
4.77%	1.92%

Source: Garvin, D. "Quality on the line," *Harvard Business Review*, September–October, 1983, pp. 65–75.

a. Do these data provide sufficient evidence to contradict the results of Garvin's study? Test using $\alpha = .10$.

b. Report the observed significance level of the test you conducted in part **a**.

c. List any assumptions necessary to assure the validity of the test you conducted. Use the terminology of this exercise.

9.31 An experiment is conducted to investigate the effect of a drug on the time to complete a task. Twenty people are divided at random into two groups of ten each. One group is given a placebo, while the second experimental group is administered a drug thought to increase the ability to complete the task quickly. For the control group, the times required to complete the task had a mean of 14.8 minutes and a variance of 3.9; for the experimental group, the average was 12.3 minutes and the variance was 4.3.

a. Test the null hypothesis that the drug has no effect in reducing the mean length of time to complete the task against the alternative hypothesis that the mean time is less for those subjects who receive the drug. Conduct the test at the $\alpha = .10$ level of significance.

b. Find the approximate observed significance level for the test and interpret its value.

9.32   Some statistics students complain that pocket calculators give other students an unfair advantage during statistics examinations. To check this contention, 45 students were randomly assigned to two groups, 23 to use calculators and 22 to perform calculations by hand. The students then took a statistics examination that required a modest amount of arithmetic. The means and variances of the test scores for the two groups are shown in the table. Do the data provide sufficient evidence to indicate that students taking *this particular examination* obtain higher scores when using a calculator? Test using $\alpha = .10$.

	n	$\bar{x}$	s^2
Calculators	23	80.7	49.5
No calculators	22	78.9	60.4

9.33   To compare two methods of teaching reading, randomly selected groups of elementary school children were assigned to each of the two teaching methods for a 6-month period. The criterion for measuring achievement was a reading comprehension test. The results are shown in the accompanying table. Do the data provide sufficient evidence to indicate a difference in mean scores on the comprehension test for the two teaching methods? Test using $\alpha = .05$.

	NUMBER OF CHILDREN PER GROUP	$\bar{x}$	s^2
Method 1	11	64	52
Method 2	14	69	71

9.34   Suppose you are the personnel manager for a company and you suspect a difference in the mean length of work time lost due to sickness for two types of employees: those who work at night versus those who work during the day. In particular, you suspect that the mean time lost for the night shift exceeds the mean for the day shift. To check your theory, you randomly sample the records for ten employees for each shift category and record the number of days lost due to sickness within the past year. The data are shown in the table.

NIGHT SHIFT, 1		DAY SHIFT, 2	
21	2	13	18
10	19	5	17
14	6	16	3
33	4	0	24
7	12	7	1
$\bar{x}_1 = 12.8$		$\bar{x}_2 = 10.4$	
$\Sigma x_1^2 = 2{,}436$		$\Sigma x_2^2 = 1{,}698$	

a. Calculate s_1^2 and s_2^2.

b. Show that the pooled estimate of the common population standard deviation, σ, is 8.86. Look at the range of the observations within each of the two samples. Does it appear that the estimate, 8.86, is a reasonable value for σ?

c. If μ_1 and μ_2 represent the mean number of days per year lost due to sickness for the night and day shifts, respectively, test the null hypothesis H_0: $\mu_1 = \mu_2$ against the alternative hypothesis H_a: $\mu_1 > \mu_2$. Use $\alpha = .05$. Do the data provide sufficient evidence to indicate that $\mu_1 > \mu_2$?

d. What assumptions must be satisfied so that the t-test from part c is valid?

9.35 Suppose your plant purifies its liquid waste and discharges the water into a local river. An EPA inspector has collected water specimens of the discharge of your plant and also water specimens in the river upstream from your plant. Each water specimen is divided into five parts, the bacteria count is read on each, and the mean count for each specimen is reported. The average bacteria counts for each of six specimens are reported in the table for the two locations.

PLANT DISCHARGE	UPSTREAM
30.1	29.7
36.2	30.3
33.4	26.4
28.2	27.3
29.8	31.7
34.9	32.3

a. Why might the bacteria counts shown here tend to be approximately normally distributed?
b. Do the data provide sufficient evidence to indicate that the mean of the bacteria count for the discharge exceeds the mean of the count upstream? Use $\alpha = .05$.
c. Find the approximate observed significance level for the test and interpret its value.

9.36 Shorkey, McRoy, and Armendariz examined the relationship between the intensity of parental punishment practices and various demographic characteristics of mothers. The intensity score descriptive statistics are shown in the accompanying table, with higher scores showing a greater intensity of parental punishments.

	SINGLE	SEPARATED
n	8	6
$\bar{x}$	70.75	77.33
s	14.80	13.69

Source: Shorkey, C. T., McRoy, R. G., and Armendariz, J. "Intensity of parental punishments and problem-solving attitudes and behaviors," *Psychological Reports*, 1985, 56.

a. Use a 95% confidence interval to estimate the difference between the mean intensity of parental punishment scores for single and separated mothers. Interpret the interval in terms of this application.
b. Does the confidence interval in part a support an inference that the true means for single and separated mothers differ?
c. Conduct a test of hypothesis comparing the two means using $\alpha = .05$. Does your conclusion agree with that in part b?
d. What assumptions are necessary to assure the validity of the inferences in parts a–c? State them in terms of this application.

9.3 Inferences About the Difference Between Two Population Means: Paired Difference Experiments

In Example 9.7 we compared two methods of teaching reading to slow learners by means of a 90% confidence interval. Suppose it is possible to measure the slow learners' "reading IQ's" *before* they are subjected to a teaching method. Eight pairs of slow learners with similar reading IQ's are found, and one member of each pair is randomly assigned to the standard teaching method while the other

TABLE 9.2 **Reading Test Scores for Eight Pairs of Slow Learners**

PAIR	NEW METHOD	STANDARD METHOD
1	77	72
2	74	68
3	82	76
4	73	68
5	87	84
6	69	68
7	66	61
8	80	76
	$\bar{x}_1 = 76.0$	$\bar{x}_2 = 71.625$
	$s_1^2 = 48.0$	$s_2^2 = 49.1$

is assigned to the new method. Do the data in Table 9.2 support the hypothesis that the population mean reading test score for slow learners taught by the new method is greater than the mean reading test score for those taught by the standard method?

We want to test

$$H_0: \quad (\mu_1 - \mu_2) = 0$$
$$H_a: \quad (\mu_1 - \mu_2) > 0$$

If we use the two-sample t statistic (Section 9.2), we first calculate

$$s_p^2 = \frac{(n_1 - 1)s_1^2 + (n_2 - 1)s_2^2}{n_1 + n_2 - 2}$$

$$= \frac{(8 - 1)(48.0) + (8 - 1)(49.1)}{8 + 8 - 2}$$

$$= 48.55$$

and then the test statistic

$$t = \frac{(\bar{x}_1 - \bar{x}_2) - 0}{\sqrt{s_p^2\left(\dfrac{1}{n_1} + \dfrac{1}{n_2}\right)}} = \frac{76.0 - 71.625}{\sqrt{48.55\left(\dfrac{1}{8} + \dfrac{1}{8}\right)}} = \frac{4.375}{3.485} = 1.26$$

This small t value will not lead to rejection of H_0 when compared to the t-distribution with $n_1 + n_2 - 2 = 14$ df, even if α is chosen as large as .10 ($t_{.10} = 1.345$). Thus, from *this* analysis we might conclude that there is insufficient evidence to infer a difference in the mean test scores for the two methods.

If you carefully examine the data in Table 9.2, however, you will find this result difficult to accept. The test score of the new method is larger than the corresponding test score for the standard method *for every one of the eight pairs of slow learners.* This, in itself, seems to provide strong evidence to indicate that μ_1 exceeds μ_2. Why, then, did the t-test fail to detect this difference? The answer is: *The two-sample t is not a valid procedure to use with this set of data.*

The two-sample t is inappropriate because the assumption of independent samples is invalid. We have randomly chosen *pairs of test scores*, and thus, once we have chosen the sample for the new method, we have *not* independently chosen the sample for the standard method. The dependence between observations within pairs can be seen by examining the pairs of test scores, which tend to rise and fall together as we go from pair to pair. This pattern provides strong visual evidence of a violation of the assumption of independence required for the two-sample t-test of Section 9.2. In this situation, you will note the *large variation within samples* (reflected by the large value of s_p^2) in comparison to the relatively *small difference between the sample means.* Because s_p^2 is so large, the t-test of Section 9.2 is unable to detect a difference between μ_1 and μ_2.

We now consider a valid method of analyzing the data of Table 9.2. In Table 9.3 we add the column of differences between the test scores of the pairs of slow learners. We can regard these differences in test scores as a random sample of differences for all pairs (matched on reading IQ) of slow learners, past and present. Then we can use this sample to make inferences about the mean of the population of differences, μ_D—which is equal to the difference $(\mu_1 - \mu_2)$. That is, the mean

TABLE 9.3

PAIR	NEW METHOD	STANDARD METHOD	DIFFERENCE (NEW METHOD − STANDARD METHOD)
1	77	72	5
2	74	68	6
3	82	76	6
4	73	68	5
5	87	84	3
6	69	68	1
7	66	61	5
8	80	76	4
			$\bar{x}_D = 4.375$
			$s_D = 1.69$

of the population (and sample) of differences equals the difference between the population (and sample) means. Thus, our test becomes

$$H_0: \quad \mu_D = 0 \quad (\mu_1 - \mu_2 = 0)$$
$$H_a: \quad \mu_D > 0 \quad (\mu_1 - \mu_2 > 0)$$

The test statistic is a one-sample t (Section 8.4), since we are now analyzing a single sample of differences:

$$\textit{Test statistic:} \quad t = \frac{\bar{x}_D - 0}{s_D/\sqrt{n_D}}$$

where

$\bar{x}_D$ = Sample mean difference

s_D = Sample standard deviation of differences

n_D = Number of differences = Number of pairs

Assumptions: The population of differences in test scores is approximately normally distributed. The sample differences are randomly selected from the population of differences. [*Note:* We do not need to make the assumption that $\sigma_1^2 = \sigma_2^2$.]

Rejection region: At significance level $\alpha = .05$, we will reject H_0 if $t > t_{.05}$, where $t_{.05}$ is based on $(n_D - 1)$ degrees of freedom.

Referring to Table VI in Appendix A, we find the t value corresponding to $\alpha = .05$ and $n_D - 1 = 8 - 1 = 7$ df to be $t_{.05} = 1.895$. Then we will reject the null hypothesis if $t > 1.895$. Note that the number of degrees of freedom has decreased from $n_1 + n_2 - 2 = 14$ to 7 by using the paired difference experiment rather than the two independent random samples design. Now calculate

$$t = \frac{\bar{x}_D - 0}{s_D/\sqrt{n_D}} = \frac{4.375}{1.69/\sqrt{8}} = 7.32$$

Because this value of t falls in the rejection region, we conclude that the mean test score for slow learners taught by the new method exceeds the mean score for those taught by the standard method. We have confidence in this conclusion because the probability that this test would lead to the rejection of H_0, given that it is true, is only $\alpha = .05$.

This kind of experiment, in which observations are paired and the differences are analyzed, is called a **paired difference experiment**. In many cases, a paired difference experiment can provide more information about the difference between population means than an independent samples experiment. The idea is to compare population means by comparing the differences between pairs of experimental units (objects, people, etc.) that were very similar prior to the experiment. The differencing removes sources of variation that tend to inflate σ^2. For example, when two children are taught to read by two different methods, the observed difference in achievement may be due to a difference in the effectiveness of the two teaching methods *or* it may be due to differences in the initial reading levels and IQ's of the two children (random error). To reduce the effect of differences in the children on the observed differences in reading achievement, the two methods of reading are imposed on two children who are more likely to possess similar intellectual potentials, namely children with nearly equal IQ's. The effect of this pairing is to remove the larger source of variation that would be present if children with different abilities were randomly assigned to the two samples. Making comparisons within groups of similar experimental units is called **blocking**, and the paired difference experiment is a simple example of a **randomized block experiment**. In our example, pairs of children with matching IQ scores represent the blocks.

Some other examples for which the paired difference experiment might be appropriate are the following:

1. Suppose you want to estimate the difference $(\mu_1 - \mu_2)$ in mean price per gallon between two major brands of premium gasoline. If you choose two independent random samples of stations for each brand, the variability in price due to geographic location may be large. To eliminate this source of variability you could choose pairs of stations of similar size, one station for each brand, in close geographic proximity and use the sample of differences between the prices of the brands to make an inference about $(\mu_1 - \mu_2)$.

2. Suppose a college placement center wants to estimate $(\mu_1 - \mu_2)$, the difference in mean starting salaries for men and women graduates who seek jobs through the center. If it independently samples men and women, the starting salaries may vary because of their different college majors and differences in grade-point averages. To eliminate these sources of variability, the placement center could match male and female job-seekers according to their majors and grade-point averages. Then the differences between the starting salaries of each pair in the sample could be used to make an inference about $(\mu_1 - \mu_2)$.

3. Suppose you wish to estimate the difference $(\mu_1 - \mu_2)$ in mean absorption rate into the bloodstream for two drugs that relieve pain. If you independently sample people, the absorption rates might vary because of age, weight, sex, blood pressure, etc. In fact, there are many possible sources of nuisance variability, and pairing individuals who are similar in all the possible sources would be quite difficult. However, it may be possible to obtain two measurements *on the same person*. First, we administer one of the two drugs and record the time until absorption. After a sufficient amount of time, the other drug is administered and a second measurement on absorption time is obtained. The

differences between the measurements for each person in the sample could then be used to estimate $(\mu_1 - \mu_2)$. This procedure would be advisable only if the amount of time allotted between drugs is sufficient to guarantee little or no carryover effect. Otherwise, it would be better to use different people matched as closely as possible on the factors thought to be most important.

The one-tailed and two-tailed hypothesis-testing procedures and the method of forming confidence intervals for the difference between two means using a paired difference experiment are summarized in the boxes.

Paired Difference Confidence Interval

$$\bar{x}_D \pm t_{\alpha/2}\frac{s_D}{\sqrt{n_D}}$$

where $t_{\alpha/2}$ is based on $(n_D - 1)$ degrees of freedom.

Assumptions: 1. The relative frequency distribution of the population of differences is normal.
2. The sample differences are randomly selected from the population of differences.

Paired Difference Test of Hypothesis

ONE-TAILED TEST

H_0: $(\mu_1 - \mu_2) = D_0$;
  i.e., $(\mu_D = D_0)$

H_a: $(\mu_1 - \mu_2) < D_0$;
  i.e., $(\mu_D < D_0)$
  [or H_a: $(\mu_1 - \mu_2) > D_0$;
  i.e., $(\mu_D > D_0)$]

Test statistic: $t = \dfrac{\bar{x}_D - D_0}{s_D/\sqrt{n_D}}$

Rejection region:

  $t < -t_\alpha$
  [or $t > t_\alpha$ when
  H_a: $(\mu_1 - \mu_2) > D_0$]

TWO-TAILED TEST

H_0: $(\mu_1 - \mu_2) = D_0$;
  i.e., $(\mu_D = D_0)$

H_a: $(\mu_1 - \mu_2) \neq D_0$;
  i.e., $(\mu_D \neq D_0)$

Test statistic: $t = \dfrac{\bar{x}_D - D_0}{s_D/\sqrt{n_D}}$

Rejection region:

  $t < -t_{\alpha/2}$ or $t > t_{\alpha/2}$

where t_α and $t_{\alpha/2}$ are based on $(n_D - 1)$ degrees of freedom.

Assumptions: 1. The relative frequency distribution of the population of differences is normal.
2. The differences are randomly selected from the population of differences.

EXAMPLE 9.8

A paired difference experiment is conducted to compare the starting salaries of male and female college graduates who find jobs. Pairs are formed by choosing a male and a female with the same major and similar grade-point averages. Suppose a random sample of ten pairs is formed in this manner and the starting annual salary of each person is recorded. The results are shown in Table 9.4. Test to see whether there is evidence that the mean starting salary, μ_1, for males exceeds the mean starting salary, μ_2, for females. Use $\alpha = .05$.

TABLE 9.4

PAIR	MALE	FEMALE	DIFFERENCE (MALE − FEMALE)
1	$24,300	$23,800	$ 500
2	26,500	26,600	−100
3	25,400	24,800	600
4	23,500	23,500	0
5	28,500	27,600	900
6	22,800	23,000	−200
7	24,500	24,200	300
8	26,200	25,100	1,100
9	23,400	23,200	200
10	24,200	23,500	700

Solution

The elements of the paired difference test are

$$H_0: \quad \mu_D = 0 \qquad (\mu_1 - \mu_2 = 0)$$
$$H_a: \quad \mu_D > 0 \qquad (\mu_1 - \mu_2 > 0)$$

Note that we propose a one-sided research hypothesis, since we are interested in determining whether the data indicate that μ_1 exceeds μ_2—that is, male mean starting salary exceeds female mean starting salary.

Test statistic: $\quad t = \dfrac{\bar{x}_D - 0}{s_D/\sqrt{n_D}}$

Assumptions: 1. The relative frequency distribution for the population of differences is normal.
2. The sample differences are randomly selected from the population.

Since the test is upper-tailed, we will reject H_0 if $t > t_\alpha$, where $t_{.05} = 1.833$ is based upon $(n_D - 1) = 9$ degrees of fredom. The rejection region is shown in Figure 9.6.
We now calculate

$$\sum x_D = 500 + (-100) + \cdots + 700 = 4,000$$

and

$$\sum x_D^2 = 3,300,000$$

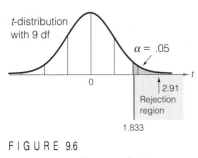

t-distribution
with 9 df

$\alpha = .05$

0

2.91

Rejection
region

1.833

FIGURE 9.6

Rejection region for Example 9.8

Then

$$\bar{x}_D = \frac{\sum x_D}{10} = \frac{4,000}{10} = 400$$

$$s_D^2 = \frac{\sum (x_D - \bar{x}_D)^2}{n_D - 1} = \frac{\sum x_D^2 - \left(\sum x_D\right)^2 / 10}{9}$$

$$= \frac{3,300,000 - (4,000)^2 / 10}{9} = 188,888.89$$

$$s_D = \sqrt{s_D^2} = 434.61$$

Substituting these values into the formula for the test statistic, we find that

$$t = \frac{\bar{x}_D - 0}{s_D / \sqrt{n_D}} = \frac{400}{434.61 / \sqrt{10}} = \frac{400}{137.44} = 2.91$$

As you can see in Figure 9.6, the calculated t falls in the rejection region. Thus, we conclude at the $\alpha = .05$ level of significance that the mean starting salary for males exceeds the mean starting salary for females.

One measure of the amount of information about $(\mu_1 - \mu_2)$ gained by using a paired difference experiment rather than an independent samples experiment in Example 9.8 is the relative widths of the confidence intervals obtained by the two methods. A 95% confidence interval for $(\mu_1 - \mu_2)$ using the paired difference experiment is

$$\bar{x}_D \pm t_{\alpha/2} \frac{s_D}{\sqrt{n_D}} = 400 \pm t_{.025} \frac{434.61}{\sqrt{10}}$$

$$= 400 \pm 2.262 \frac{434.61}{\sqrt{10}}$$

$$= 400 \pm 310.88$$

$$\approx 400 \pm 311 = (\$89, \$711)$$

If we analyzed the same data as though this were an independent samples experiment,* we would first calculate the following quantities:

MALES	FEMALES
$n_1 = 10$	$n_2 = 10$
$\bar{x}_1 = \$24,930$	$\bar{x}_2 = \$24,530$
$s_1^2 = 3,009,000$	$s_2^2 = 2,331,222.22$

*This is done only to provide a measure of the increase in the amount of information obtained by a paired design in comparison to an unpaired design. Actually, if an experiment is designed using pairing, an unpaired analysis would be invalid because the assumption of independent samples would not be satisfied.

Then

$$s_p^2 = \frac{(n_1 - 1)s_1^2 + (n_2 - 1)s_2^2}{n_1 + n_2 - 2} = \frac{9(3,009,000) + 9(2,331,222.22)}{10 + 10 - 2}$$

$$= 2,670,111.11$$

$$s_p = \sqrt{s_p^2} = 1,634.05$$

The 95% confidence interval is

$$(\bar{x}_1 - \bar{x}_2) \pm t_{.025} \sqrt{s_p^2\left(\frac{1}{n_1} + \frac{1}{n_2}\right)} = 400 \pm (2.101) \sqrt{2,670,111.11\left(\frac{1}{10} + \frac{1}{10}\right)}$$

$$= 400 \pm 1,535.35$$

$$\approx 400 \pm 1,535$$

$$= (-\$1,135, \$1,935)$$

The confidence interval for the independent sampling experiment is about five times wider than for the corresponding paired difference confidence interval. Blocking out the variability due to differences in majors and grade-point averages significantly increases the information about the difference in male and female mean starting salaries by providing a much more accurate (smaller confidence interval for the same confidence coefficient) estimate of $(\mu_1 - \mu_2)$.

You may wonder whether conducting a paired difference experiment is always superior to an independent samples experiment. The answer is: Most of the time, but not always. We sacrifice half the degrees of freedom in the t statistic when a paired difference design is used instead of an independent samples design. This is a loss of information, and unless this loss is more than compensated for by the reduction in variability obtained by blocking (pairing), the paired difference experiment will result in a net loss of information about $(\mu_1 - \mu_2)$. Thus, we should be convinced that the pairing will significantly reduce variability before performing the paired difference experiment. Most of the time this will happen.

One final note: The pairing of the observations is determined *before* the experiment is performed (that is, by the *design* of the experiment). A paired difference experiment is *never* obtained by pairing the sample observations after the measurements have been acquired.

What Do You Do When the Assumption of a Normal Distribution for the Population of Differences Is Not Satisfied?

Answer: Use the Wilcoxon signed rank test for the paired difference design (Chapter 11).

CASE STUDY 9.2   MATCHED PAIRING IN STUDYING THE MENTALLY RETARDED

The statistical implications underlying matching procedures have frequently been overlooked in educational research with the mentally retarded population. It is, therefore, the purpose of this paper to point out some of the advantages and disadvantages of different matching procedures.

Stainback and Stainback* describe a number of experimental situations in which blocking (matching) subjects before experimentation might reduce variability and thereby increase the amount of information obtained. One of the matching procedures discussed by the authors can be summarized as follows: Suppose it is desired to form two groups of mentally retarded subjects to compare two methods of educational therapy. Subjects could be randomly selected from an existing (large) group of subjects and ordered (from lowest to highest) on the scores of an appropriate matching variable. Several variables suggested by the authors are "pretest measures of the experimental criterion, measures of learning rate (mental age and Intelligence Quotients), chronological age, personal characteristics (sex, race), environmental conditions (socio-economic level), or combinations of two or more variables." The two highest-ranking subjects on the matching variables would then form pair 1, the next two pair 2, etc., and one member from each pair would be randomly assigned to each therapy group. This experiment is a practical example of a paired difference experiment, and if the matching variables are correctly chosen, the responses of subjects will be more homogeneous within pairs than between pairs. The authors conclude:

> It is important that when comparing two groups on a criterion variable the two groups be as equal as possible on all relevant factors excepting only the independent variables. This consideration deserves particular emphasis in the area of mental retardation since the researchers are constantly dealing with diverse groups. It should be restated, therefore, that researchers in the area of mental retardation should become acutely aware of advantages and disadvantages of matching procedures and their alternatives.

EXERCISES 9.37–9.53

LEARNING THE MECHANICS

9.37 Suppose a paired difference experiment is to be performed in order to test the null hypothesis H_0: $\mu_D = 2$ against H_a: $\mu_D > 2$. Assume that the second observation in each pair will be subtracted from the first. Define μ_D, and explain what is meant by the hypotheses $\mu_D = 2$ and $\mu_D > 2$.

9.38 A paired difference experiment yielded n_D pairs of observations. In each case, what is the rejection region for testing H_0: $\mu_D = 2$ against H_a: $\mu_D > 2$?
   a. $n_D = 10$, $\alpha = .05$ **b.** $n_D = 20$, $\alpha = .10$
   c. $n_D = 5$, $\alpha = .025$ **d.** $n_D = 9$, $\alpha = .01$

PERSON	BEFORE x_1	AFTER x_2
1	83	92
2	60	71
3	55	56
4	99	104
5	77	89

---

*Reprinted from Stainback, S., and Stainback, W. C. "Matched procedures and research in mental retardation," *Training School Bulletin*, May 1973, 70, 33–37.

9.39   A paired difference experiment yielded the data shown in the table.
   a. Compute $\bar{x}_D$ and s_D.
   b. Demonstrate that $\bar{x}_D = \bar{x}_1 - \bar{x}_2$.
   c. Is there sufficient evidence to conclude that $\mu_1 \neq \mu_2$? Test using $\alpha = .05$.
   d. Find and interpret the approximate p-value for the paired difference test.
   e. What assumptions are necessary so that the paired difference test will be valid?

9.40   The data for a random sample of six paired observations are shown in the table.

PAIR	SAMPLE FROM POPULATION 1	SAMPLE FROM POPULATION 2
	Observation 1	Observation 2
1	5	3
2	3	1
3	9	7
4	6	2
5	4	5
6	8	7

   a. Calculate the difference between each pair of observations by subtracting observation 2 from observation 1. Use the differences to calculate $\bar{x}_D$ and s_D^2.
   b. If μ_1 and μ_2 are the means of populations 1 and 2, respectively, express μ_D in terms of μ_1 and μ_2.
   c. Form a 95% confidence interval for μ_D.
   d. Test the null hypothesis H_0: $\mu_D = 0$ against the alternative hypothesis H_a: $\mu_D \neq 0$. Use $\alpha = .05$.

9.41   The data for a random sample of ten paired observations are shown in the table.

PAIR	SAMPLE FROM POPULATION 1	SAMPLE FROM POPULATION 2
1	19	24
2	25	27
3	31	36
4	52	53
5	49	55
6	34	34
7	59	66
8	47	51
9	17	20
10	51	55

   a. Do the data provide sufficient evidence to indicate that μ_2, the mean of population 2, is larger than μ_1? (Test H_0: $\mu_D = 0$ against H_a: $\mu_D < 0$, where $\mu_D = \mu_1 - \mu_2$.) Test using $\alpha = .05$.
   b. Form a 90% confidence interval for μ_D.

9.42   A paired difference experiment produced the following data:

$$n_D = 18 \qquad \bar{x}_1 = 92 \qquad \bar{x}_2 = 95.5 \qquad \bar{x}_D = -3.5 \qquad s_D^2 = 21$$

   a. Determine the values of t for which the null hypothesis, $\mu_1 - \mu_2 = 0$, would be rejected in favor of the alternative hypothesis $\mu_1 - \mu_2 < 0$. Use $\alpha = .10$.
   b. Conduct the paired difference test described in part a. Draw the appropriate conclusions.
   c. What assumptions are necessary so that the paired difference test will be valid?

d. Find a 90% confidence interval for the mean difference μ_D.

e. Which of the two inferential procedures, the confidence interval of part **d** or the test of hypothesis of part **b**, provides more information about the difference between the population means?

9.43 A paired difference experiment yielded the data shown in the table.

PAIR	x	y
1	55	44
2	68	55
3	40	25
4	55	56
5	75	62
6	52	38
7	49	31

a. Test $H_0: \mu_D = 10$ against $H_a: \mu_D \neq 10$, where $\mu_D = (\mu_1 - \mu_2)$. Use $\alpha = .05$.

b. Report the p-value for the test you conducted in part **a**. Interpret the p-value.

APPLYING THE CONCEPTS

9.44 The data shown in the table are part of a study conducted to compare the abilities of men and women to perform the strenuous tasks of a firefighter. They represent the pulling force (in newtons) that a firefighter was able to exert in pulling the starter cord of a P-250 fire pump. Firefighters were matched in pairs according to weight, thus producing data for a matched pairs (or paired difference) experiment.

PAIR	FEMALE	MALE
1	40.03	84.51
2	75.62	80.06
3	53.38	102.30
4	62.27	88.96

Source: Phillips, M. D., and Pepper, R. L. "Shipboard firefighting performance of females and males," *Human Factors*, 1982, 24. Copyright 1982 by The Human Factors Society, Inc. Reproduced by permission.

a. Do the data provide sufficient evidence to indicate a difference in mean pulling force between female and male firefighters? Test using $\alpha = .05$.

b. Find a 95% confidence interval for the difference in mean pulling force between female and male firefighters. Interpret the interval.

9.45 A new weight-reducing technique, consisting of a liquid protein diet, is currently undergoing tests by the Food and Drug Administration (FDA) before its introduction into the market. A typical test performed by the FDA is the following: The weights of a random sample of five people are recorded before they are introduced to the liquid protein diet. The five individuals are then instructed to follow the liquid protein diet for 3 weeks. At the end of this period, their weights (in pounds) are again recorded. The results are listed in the table on page 412. Let μ_1 be the true mean weight of individuals before starting the diet and let μ_2 be the true mean weight of individuals after 3 weeks on the diet. Construct a 95% confidence interval for the difference between the true mean weights before and after the diet is used. What assumptions are necessary to ensure the validity of the procedure you used?

PERSON	WEIGHT BEFORE DIET	WEIGHT AFTER DIET
1	150	143
2	195	190
3	188	185
4	197	191
5	204	200

9.46   Do people generally receive the asking price when they sell their home? The accompanying table lists the asking price and sale price for a random sample of ten 1986 single-family home sales in Eden Prairie, Minnesota.

SINGLE-FAMILY HOME	ASKING PRICE	SALE PRICE
1	$ 98,000	$ 94,300
2	122,500	119,800
3	109,900	105,000
4	98,000	100,000
5	79,900	79,900
6	275,000	265,000
7	99,700	96,000
8	97,900	99,200
9	209,000	209,000
10	173,900	169,000

Source: *MLS Compilation of Listings, Division II*, No. 44, October 31–November 6, 1986, Greater Minneapolis Area Board of Realtors.

a. Explain why the samples are not independent.
b. What is the dimension on which the samples are paired or matched?
c. Given that the mean difference between the asking price and the sale price for the ten homes is $2,660 with standard deviation $3,615, conduct a hypothesis test that will help answer the question of interest for people in Eden Prairie who sold their homes in 1986. Use $\alpha = .05$.
d. Use a 90% confidence interval to estimate the true difference between the asking price and the sale price for homes in Eden Prairie.

9.47   The data shown in the table provide information on the relationship between the mean daily air temperature and the cocoon temperature of wooly-bear caterpillars of the High Arctic. You can see from the table that the data indicate that the caterpillar's body temperature (inside the cocoon) is higher than the outside air temperature. Estimate the mean difference in temperature between the cocoon and the outside air. Use a 95% confidence interval.

DAY	TEMPERATURE (°C)		DAY	TEMPERATURE (°C)		DAY	TEMPERATURE (°C)	
	Air	Cocoon[a]		Air	Cocoon[a]		Air	Cocoon[a]
1	10.4	15.1	5	4.1	8.0	9	3.0	7.0
2	9.2	14.6	6	3.7	8.7	10	3.5	7.1
3	2.2	6.8	7	1.7	3.6	11	4.5	9.6
4	2.6	6.8	8	2.0	5.3	12	4.4	9.5

[a]Each cocoon temperature is the average of the temperatures of two cocoons.
Source: Kevan, P. G., Jensen, T. S., and Shorthouse, J. D. "Body temperatures and behavioral thermoregulation of High Arctic wooly-bear caterpillars and pupae (*Gynaephora rossii*, Lymantridae: Lepidoptera) and the importance of sunshine," *Arctic and Alpine Research*, 1982, 14. Reproduced with permission of the Regents of the University of Colorado.

9.48   In the past, many bodily functions were thought to be beyond conscious control. However, recent experimentation suggests that it may be possible for a person to control certain body functions if that person is trained in a program of *biofeedback* exercises. An experiment is conducted to show that blood pressure levels can be consciously reduced in people trained in this program. The blood pressure measurements (in millimeters of mercury) listed in the table represent readings before and after the biofeedback training of six subjects.

SUBJECT	BEFORE	AFTER
1	136.9	130.2
2	201.4	180.7
3	166.8	149.6
4	150.0	153.2
5	173.2	162.6
6	169.3	160.1

a. Is there sufficient evidence to conclude that the mean blood pressure after people are trained in this program is less than the mean blood pressure before training? Use $\alpha = .05$.

b. Find the approximate observed significance level for the test and interpret its value.

9.49 In the May 21, 1986, edition of *USA Today*, gasoline prices are compared with those of 1 year earlier for the 48 contiguous states. The mean difference between 1985 and 1986 prices reported for the 48 states is 26.4 cents per gallon, with standard deviation 4.6 cents per gallon.

a. Although we have not given the individual states' price data here, within what range would you expect the differences between 1985 and 1986 prices to fall?

b. Use a 95% confidence interval to estimate the true mean difference between the price of gasoline in May 1985 and May 1986 over the 48 contiguous states. Interpret the interval in terms of this application.

c. What assumptions are necessary to assure the validity of the inference you made in part **b**?

9.50 While producing many economic benefits to the state of Florida, gypsum and phosphate mines also produce a harmful byproduct: *radiation*. It has been known for a number of years that the mine tailings (waste) contain radioactive radon 222. In fact, new housing complexes built on top of the leveled piles of residue have shown disturbing radiation levels within the houses. The radiation levels in waste gypsum and phosphate mounds in Polk County, Florida, are regularly monitored by the Eastern Environmental Radiation Facility (EERF) and by the Polk County Health Department (PCHD), Winter Haven, Florida. The accompanying table shows measurements of the exhalation rate (a measure of radiation) for 15 soil samples obtained from waste mounds in Polk County, Florida. The exhalation rate was measured for each soil sample by both the PCHD and the EERF. The objective of selecting the paired measurements was to determine whether a bias exists, a difference in the mean readings, between PCHD and EERF. They represent part of the data contained in a report by Thomas R. Horton of EERF.

CHARCOAL CANISTER NO.	PCHD	EERF	CHARCOAL CANISTER NO.	PCHD	EERF
71	1,709.79	1,479.0	85	393.55	187.7
58	357.17	257.8	46	880.84	630.4
84	1,150.94	1,287.0	4	2,996.49	3,707.0
91	1,572.69	1,395.0	20	2,367.40	2,791.0
44	558.33	416.5	36	599.84	706.8
43	4,132.28	3,993.0	42	538.37	618.5
79	1,489.86	1,351.0	55	2,770.23	2,639.0
61	3,017.48	1,813.0			

Source: Horton, T. R. "Preliminary radiological assessment of radon exhalation from phosphate gypsum piles and inactive uranium mill tailings piles," EPA-520/5-79-004. Washington, D. C.: Environmental Protection Agency, 1979.

a. Considering the relative size of the measurements from canister to canister, explain why a paired difference experiment was conducted rather than an independent samples experiment.

b. Given that the mean difference (PCHD − EERF) for the 15 sampled canisters is 84.17 with standard deviation 408.92, do the data provide sufficient evidence to indicate a difference in the mean exhalation rates between PCHD and EERF? Test using $\alpha = .05$.

c. Find a 95% confidence interval for the difference in mean measurements between PCHD and EERF. Interpret the interval. Does it support the result of the test in part **a**?

9.51 A manufacturer of automobile shock absorbers was interested in comparing the durability of its shocks with that of the biggest competitor. To make the comparison, one of the manufacturer's and one of the competitor's shocks were randomly selected and installed on the rear wheels of six cars. After the cars had been driven 20,000 miles, the strength of each test shock was measured, coded, and recorded. The following are the results of the examination:

CAR NUMBER	MANUFACTURER'S	COMPETITOR'S
1	8.8	8.4
2	10.5	10.1
3	12.5	12.0
4	9.7	9.3
5	9.6	9.0
6	13.2	13.0

a. Do the data present sufficient evidence to conclude there is a difference in the mean strength of the two types of shocks after 20,000 miles of use? Use $\alpha = .05$.

b. What assumptions are necessary in order to apply a paired difference analysis to the data?

c. Construct a 95% confidence interval for $(\mu_1 - \mu_2)$. Interpret the meaning of your confidence interval.

9.52 Suppose the data in Exercise 9.51 are based on independent random samples.

a. Do the data provide sufficient evidence to indicate a difference between the mean strengths for the two types of shocks? Use $\alpha = .05$.

b. Construct a 95% confidence interval for $(\mu_1 - \mu_2)$. Interpret your result.

c. Compare the confidence intervals you obtained in Exercise 9.51 and part **b** of this exercise. Which is larger? To what do you attribute the difference in size? Assuming in each case that the appropriate assumptions are satisfied, which interval provides you with more information about $(\mu_1 - \mu_2)$? Explain.

d. Are the results of an upaired analysis valid when the data have been collected from a paired experiment?

9.53 An experiment was conducted to measure the effects of fructose and glucose on high-endurance performance of athletes. Six trained female runners were used in the experiment. Each was given 300 milliliters of a liquid 45 minutes prior to running for 85 minutes or until they reached a state of exhaustion, whichever occurred first. Various measures of endurance, such as performance time (time to exhaustion), rating of perceived exertion, etc., were recorded at the end of each run. Four liquids (treatments) were used in the experiment. The first contained fructose, the second contained glucose, the third contained water sweetened with a calcium saccharine solution (a placebo designed to suggest the presence of fructose or glucose), and the fourth contained water alone. Each of the six subjects performed the run for each of the four liquids, which were arranged in random order. The table gives the averages of the six runners' times (in minutes) to exhaustion for only two of the mixtures, glucose and the placebo. The table also gives the sample sizes and standard deviations for the two samples.

	GLUCOSE	PLACEBO
n	6	6
$\bar{x}$	63.9	52.2
s	20.3	13.5

Source: McMurry, R. G., Wilson, J. R., and Kitchell, B. S.
"The effects of fructose and glucose on high-endurance per-
formance," *Research Quarterly for Exercise and Sport*, 1983, 54.
Reprinted by permission of the American Alliance for Health,
Physical Education, Recreation, and Dance, 1900 Association
Drive, Reston, VA 20091.

a. Describe the experiment and explain how and why it does or does not satisfy the assumptions of independent random sampling.

b. Suppose that the data were based on independent random samples. Would the data provide sufficient evidence to indicate a difference in the mean time to exhaustion between runners given the glucose mixture and those given the placebo? Test using $\alpha = .05$. Interpret your result.

c. Consider the manner in which the experiment was actually conducted. What do you gain or lose by analyzing the data using the method of part b?

9.4 Inferences About the Difference Between Population Proportions: Independent Binomial Experiments

Suppose a presidential candidate wants to compare the preference of registered voters in the northeastern United States (NE) to those in the southeastern United States (SE). Such a comparison would help determine where to concentrate campaign efforts. The candidate hires a professional pollster to randomly choose 1,000 registered voters in the northeast and 1,000 in the southeast and interview each to learn her or his voting preference. The objective is to use this sample information to make an inference about the difference $(p_1 - p_2)$ between the proportion p_1 of *all* registered voters in the northeast and the proportion p_2 of *all* registered voters in the southeast who plan to vote for the presidential candidate.

The two samples represent independent binomial experiments. (See Section 4.4 for the characteristics of binomial experiments.) The binomial random variables are the numbers x_1 and x_2 of the 1,000 sampled voters in each area who indicate they will vote for the candidate. The results are summarized in the accompanying table. We can now calculate the sample proportions $\hat{p}_1$ and $\hat{p}_2$ of the voters in favor of the candidate in the northeast and southeast, respectively:

NE	SE
$n_1 = 1,000$	$n_2 = 1,000$
$x_1 = 546$	$x_2 = 475$

$$\hat{p}_1 = \frac{x_1}{n_1} = \frac{546}{1,000} = .546$$

$$\hat{p}_2 = \frac{x_2}{n_2} = \frac{475}{1,000} = .475$$

The difference between the sample proportions, $(\hat{p}_1 - \hat{p}_2)$ makes an intuitively appealing point estimator of the difference between the population parameters, $(p_1 - p_2)$. For our example, the estimate is

$$(\hat{p}_1 - \hat{p}_2) = .546 - .475 = .071$$

To judge the reliability of the estimator $(\hat{p}_1 - \hat{p}_2)$, we must observe its performance in repeated sampling from the two populations. That is, we need to know the sampling distribution of $(\hat{p}_1 - \hat{p}_2)$. The properties of the sampling distribution are given in the box. Remember that $\hat{p}_1$ and $\hat{p}_2$ can be viewed as means of the number of successes per trial in the respective samples, so the Central Limit Theorem applies when the sample sizes are large.

Since the distribution of $(\hat{p}_1 - \hat{p}_2)$ in repeated sampling is approximately normal, we can use the z statistic to derive confidence intervals for $(p_1 - p_2)$ or to test a hypothesis about $(p_1 - p_2)$.

| Properties of the Sampling Distribution of $(\hat{p}_1 - \hat{p}_2)$

1. If the sample sizes n_1 and n_2 are large (see Section 7.4 for a guideline), the sampling distribution of $(\hat{p}_1 - \hat{p}_2)$ is approximately normal.
2. The mean of the sampling distribution of $(\hat{p}_1 - \hat{p}_2)$ is $(p_1 - p_2)$, that is,

$$E(\hat{p}_1 - \hat{p}_2) = p_1 - p_2$$

Thus, $(\hat{p}_1 - \hat{p}_2)$ is an unbiased estimator of $(p_1 - p_2)$.
3. The standard deviation of the sampling distribution of $(\hat{p}_1 - \hat{p}_2)$ is

$$\sigma_{(\hat{p}_1 - \hat{p}_2)} = \sqrt{\frac{p_1 q_1}{n_1} + \frac{p_2 q_2}{n_2}}$$

For the voter example, a 95% confidence interval for the difference $(p_1 - p_2)$ is

$$(\hat{p}_1 - \hat{p}_2) \pm 1.96 \sigma_{(\hat{p}_1 - \hat{p}_2)}$$

or

$$(\hat{p}_1 - \hat{p}_2) \pm 1.96 \sqrt{\frac{p_1 q_1}{n_1} + \frac{p_2 q_2}{n_2}}$$

The quantities $p_1 q_1$ and $p_2 q_2$ must be estimated in order to complete the calculation of the standard deviation, $\sigma_{(\hat{p}_1 - \hat{p}_2)}$, and hence the calculation of the confidence interval. In Section 8.3 we showed that the value of pq is relatively insensitive to the value chosen to approximate p. Therefore, $\hat{p}_1 \hat{q}_1$ and $\hat{p}_2 \hat{q}_2$ will provide satisfactory estimates to approximate $p_1 q_1$ and $p_2 q_2$, respectively. Then

$$\sqrt{\frac{p_1 q_1}{n_1} + \frac{p_2 q_2}{n_2}} \approx \sqrt{\frac{\hat{p}_1 \hat{q}_1}{n_1} + \frac{\hat{p}_2 \hat{q}_2}{n_2}}$$

and we will approximate the 95% confidence interval by

$$(\hat{p}_1 - \hat{p}_2) \pm 1.96 \sqrt{\frac{\hat{p}_1 \hat{q}_1}{n_1} + \frac{\hat{p}_2 \hat{q}_2}{n_2}}$$

Substituting the sample quantities yields

$$(.546 - .475) \pm 1.96 \sqrt{\frac{(.546)(.454)}{1,000} + \frac{(.475)(.525)}{1,000}}$$

or $.071 \pm .044$. Thus, we are 95% confident that the interval from .027 to .115 contains $(p_1 - p_2)$.

We infer that there are between 2.7% and 11.5% more registered voters in the northeast than in the southeast who plan to vote for the presidential candidate. It seems that the candidate should direct a greater campaign effort in the southeast compared to the northeast.

The general form of a confidence interval for the difference $(p_1 - p_2)$ between binomial proportions is given in the box.

Large-Sample $100(1 - \alpha)$% Confidence Interval for $(p_1 - p_2)$

$$(\hat{p}_1 - \hat{p}_2) \pm z_{\alpha/2}\sigma_{(\hat{p}_1 - \hat{p}_2)} = (\hat{p}_1 - \hat{p}_2) \pm z_{\alpha/2}\sqrt{\frac{p_1q_1}{n_1} + \frac{p_2q_2}{n_2}}$$

$$\approx (\hat{p}_1 - \hat{p}_2) \pm z_{\alpha/2}\sqrt{\frac{\hat{p}_1\hat{q}_1}{n_1} + \frac{\hat{p}_2\hat{q}_2}{n_2}}$$

Assumptions: The two samples are independent random samples from binomial distributions. Both samples should be large enough that the normal distribution provides an adequate approximation to the sampling distribution of $\hat{p}_1$ and $\hat{p}_2$ (see Section 8.5).

The z statistic,

$$z = \frac{(\hat{p}_1 - \hat{p}_2) - (p_1 - p_2)}{\sigma_{(\hat{p}_1 - \hat{p}_2)}}$$

is used to test the null hypothesis that $(p_1 - p_2)$ equals some specified difference, say D_0. For the special case where $D_0 = 0$, i.e., where we want to test the null hypothesis H_0: $(p_1 - p_2) = 0$ (or, equivalently, H_0: $p_1 = p_2$), the best estimate of $p_1 = p_2 = p$ is obtained by dividing the total number of successes $(x_1 + x_2)$ for the two samples by the total number of observations $(n_1 + n_2)$; that is,

$$\hat{p} = \frac{x_1 + x_2}{n_1 + n_2}$$

or

$$\hat{p} = \frac{n_1\hat{p}_1 + n_2\hat{p}_2}{n_1 + n_2}$$

The second equation shows that $\hat{p}$ is a weighted average of $\hat{p}_1$ and $\hat{p}_2$, with the larger sample receiving more weight. If the sample sizes are equal, then $\hat{p}$ is a simple average of the two sample proportions of successes.

We now substitute the weighted average $\hat{p}$ for both p_1 and p_2 in the formula for the standard deviation of $(\hat{p}_1 - \hat{p}_2)$:

$$\sigma_{(\hat{p}_1 - \hat{p}_2)} = \sqrt{\frac{p_1q_1}{n_1} + \frac{p_2q_2}{n_2}} \approx \sqrt{\frac{\hat{p}\hat{q}}{n_1} + \frac{\hat{p}\hat{q}}{n_2}} = \sqrt{\hat{p}\hat{q}\left(\frac{1}{n_1} + \frac{1}{n_2}\right)}$$

The test is summarized in the box on page 418.

Large-Sample Test of Hypothesis About $(p_1 - p_2)$

ONE-TAILED TEST

H_0: $(p_1 - p_2) = 0$*

H_a: $(p_1 - p_2) < 0$
   [or H_a: $(p_1 - p_2) > 0$]

Test statistic: $z = \dfrac{(\hat{p}_1 - \hat{p}_2)}{\sigma_{(\hat{p}_1 - \hat{p}_2)}}$

Rejection region: $z < -z_\alpha$
   [or $z > z_\alpha$ when
   H_a: $(p_1 - p_2) > 0$]

TWO-TAILED TEST

H_0: $(p_1 - p_2) = 0$

H_a: $(p_1 - p_2) \neq 0$

Test statistic: $z = \dfrac{(\hat{p}_1 - \hat{p}_2)}{\sigma_{(\hat{p}_1 - \hat{p}_2)}}$

Rejection region: $z < -z_{\alpha/2}$
   or $z > z_{\alpha/2}$

Note: $\sigma_{(\hat{p}_1 - \hat{p}_2)} = \sqrt{\dfrac{p_1 q_1}{n_1} + \dfrac{p_2 q_2}{n_2}} \approx \sqrt{\hat{p}\hat{q}\left(\dfrac{1}{n_1} + \dfrac{1}{n_2}\right)}$

where $\hat{p} = \dfrac{x_1 + x_2}{n_1 + n_2}$

Assumption: Same as for large-sample confidence interval for $(p_1 - p_2)$. (See previous box.)

EXAMPLE 9.9

1985	1990
$n_1 = 1{,}500$	$n_2 = 2{,}000$
$x_1 = 576$	$x_2 = 652$

Solution

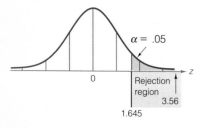

FIGURE 9.7

Rejection region for Example 9.9

In the past decade there have been intensive antismoking campaigns sponsored by both federal and private agencies. Suppose the American Cancer Society randomly sampled 1,500 adults in 1985 and then sampled 2,000 adults in 1990 to determine whether there was evidence that the percentage of smokers had decreased. The results of the two sample surveys are shown in the table, where x_1 and x_2 represent the numbers of smokers in the 1985 and 1990 samples, respectively. Do these data indicate that the fraction of smokers decreased over this 5-year period? Use $\alpha = .05$.

If we define p_1 and p_2 as the true proportions of adult smokers in 1985 and 1990 the elements of our test are

H_0: $(p_1 - p_2) = 0$

H_a: $(p_1 - p_2) > 0$

(The test is one-tailed since we are interested only in determining whether the proportion of smokers *decreased*.)

Test statistic: $z = \dfrac{(\hat{p}_1 - \hat{p}_2) - 0}{\sigma_{(\hat{p}_1 - \hat{p}_2)}}$

Rejection region: $\alpha = .05$
   $z > z_\alpha = z_{.05} = 1.645$ (see Figure 9.7)

---

*The test can be adapted to test for a difference $D_0 \neq 0$. Because most applications call for a comparison of p_1 and p_2, implying $D_0 = 0$, we will confine our attention to this case.

We now calculate the sample proportions of smokers

$$\hat{p}_1 = \frac{576}{1,500} = .384 \qquad \hat{p}_2 = \frac{652}{2,000} = .326$$

Then

$$z = \frac{(\hat{p}_1 - \hat{p}_2) - 0}{\sigma_{(\hat{p}_1 - \hat{p}_2)}} \approx \frac{(\hat{p}_1 - \hat{p}_2)}{\sqrt{\hat{p}\hat{q}\left(\frac{1}{n_1} + \frac{1}{n_2}\right)}}$$

where

$$\hat{p} = \frac{x_1 + x_2}{n_1 + n_2} = \frac{576 + 652}{1,500 + 2,000} = .351$$

Note that $\hat{p}$ is a weighted average of $\hat{p}_1$ and $\hat{p}_2$, with more weight given to the larger (1990) sample.

Thus, the computed value of the test statistic is

$$z = \frac{.384 - .326}{\sqrt{(.351)(.649)\left(\frac{1}{1,500} + \frac{1}{2,000}\right)}} = \frac{.058}{.0164} = 3.56$$

There is sufficient evidence at the $\alpha = .05$ level to conclude that the proportion of adults who smoke has decreased over the 1985–1990 period. We could place a confidence interval on $(p_1 - p_2)$ if we were interested in estimating the extent of the decrease.

EXERCISES 9.54–9.75

LEARNING THE MECHANICS

9.54   What are the characteristics of a binomial experiment?

9.55   Explain why the Central Limit Theorem is important in finding an approximate distribution for $(\hat{p}_1 - \hat{p}_2)$. (See Example 6.9.)

9.56   In each case, determine whether the sample sizes are large enough to conclude that the sampling distribution of $(\hat{p}_1 - \hat{p}_2)$ is approximately normal.
a. $n_1 = 10, \quad n_2 = 12, \quad \hat{p}_1 = .50, \quad \hat{p}_2 = .50$ b. $n_1 = 10, \quad n_2 = 12, \quad \hat{p}_1 = .10, \quad \hat{p}_2 = .08$
c. $n_1 = n_2 = 30, \quad \hat{p}_1 = .20, \quad \hat{p}_2 = .30$ d. $n_1 = 100, \quad n_2 = 200, \quad \hat{p}_1 = .05, \quad \hat{p}_2 = .09$
e. $n_1 = 100, \quad n_2 = 200, \quad \hat{p}_1 = .95, \quad \hat{p}_2 = .91$

9.57   In each case, find the values of z for which $H_0: (p_1 - p_2) = 0$ would be rejected in favor of $H_a:$ $(p_1 - p_2) < 0$.
a. $\alpha = .01$ b. $\alpha = .025$ c. $\alpha = .05$ d. $\alpha = .10$

9.58   Independent random samples, each containing 800 observations, were selected from two binomial populations. The samples from populations 1 and 2 produced 320 and 400 successes, respectively.
a. Test the null hypothesis $H_0: (p_1 - p_2) = 0$ against the alternative hypothesis $H_a:$ $(p_1 - p_2) \neq 0$. Use $\alpha = .05$.
b. Test $H_0: (p_1 - p_2) = 0$ against $H_a: (p_1 - p_2) \neq 0$. Use $\alpha = .01$.
c. Test $H_0: (p_1 - p_2) = 0$ against $H_a: (p_1 - p_2) < 0$. Use $\alpha = .01$.
d. Form a 90% confidence interval for $(p_1 - p_2)$.

9.59   Construct a 95% confidence interval for $(p_1 - p_2)$ in each of the following situations:
        a. $n_1 = 400$, $\hat{p}_1 = .65$; $n_2 = 400$, $\hat{p}_2 = .58$
        b. $n_1 = 180$, $\hat{p}_1 = .31$; $n_2 = 250$, $\hat{p}_2 = .25$
        c. $n_1 = 100$, $\hat{p}_1 = .46$; $n_2 = 120$, $\hat{p}_2 = .61$

9.60   Sketch the sampling distribution of $(\hat{p}_1 - \hat{p}_2)$ based on independent random samples of $n_1 = 100$ and $n_2 = 200$ observations from two binomial populations with success probabilities $p_1 = .1$ and $p_2 = .5$, respectively.

9.61   Random samples of size $n_1 = 50$ and $n_2 = 60$ were drawn from populations 1 and 2, respectively. The samples yielded $\hat{p}_1 = .4$ and $\hat{p}_2 = .2$. Test H_0: $(p_1 - p_2) = .1$ against H_a: $(p_1 - p_2) > .1$ using $\alpha = .05$.

APPLYING THE CONCEPTS

9.62   Two surgical procedures are widely used to treat a certain type of cancer. To compare the sucess rates of the two procedures, random samples of the two types of surgical patients were obtained and the numbers of patients who showed no recurrence of the disease after a 1-year period were recorded. The data are shown in the table. Do the data provide sufficient evidence to indicate a difference in the success rates of the two procedures? Use $\alpha = .05$.

	n	NUMBER OF SUCCESSES
Procedure A	100	78
Procedure B	100	87

9.63   A new insect spray, type A, is to be compared with a spray, type B, that is currently in use. Two rooms of equal size are sprayed with the same amount of spray, one room with A, the other with B. Two hundred insects are released into each room, and after 1 hour the numbers of dead insects are counted. The results are given in the table.

	SPRAY A	SPRAY B
Number of insects	200	200
Number of dead insects	120	80

        a. Do the data provide sufficient evidence to indicate that spray A is more effective than spray B in controlling the insects? Test using $\alpha = .05$.
        b. Find a 90% confidence interval for $(p_1 - p_2)$, the difference in the rates of kill for the two sprays. Interpret this interval.

9.64   What makes entrepreneurs different from chief executive officers (CEOs) of *Fortune 500* companies? The *Wall Street Journal* hired the Gallup Organization to investigate this question. For the study, entrepreneurs were defined as chief executive officers of companies listed by *Inc.* magazine as among the 500 fastest-growing smaller companies in the United States. The Gallup Organization sampled 207 CEOs of *Fortune 500* companies and 153 entrepreneurs. They obtained the results shown in the table.

	FORTUNE 500 CEOs	ENTREPRENEURS
Age: Under 45 years old	19	96
Education: Completed four years of college	195	116
Employment record: Have been fired or dismissed from a job	19	47

Source: Graham, E. "The entrepreneurial mystique," *Wall Street Journal*, May 20, 1985, p. 1c.

    a. In each of the three areas—age, education, and employment record—are the sample sizes large enough to use the inferential methods of this section to investigate the differences between *Fortune 500* CEOs and entrepreneurs? Justify your answer.

    b. Do the data indicate that *Fortune 500* CEOs and entrepreneurs differ in terms of education? Test using $\alpha = .05$.

    c. What assumption(s) must be satisfied in order for your test of part **b** to be valid?

9.65 Refer to Exercise 9.64. Use a 95% confidence interval to estimate the difference in the proportion of *Fortune 500* CEOs and entrepreneurs who are under 45 years of age.

9.66 Refer to Exercise 9.64.

    a. Test to determine whether the data indicate that the fractions of CEOs and entrepreneurs who have been fired or dismissed from a job differ at the $\alpha = .01$ level of significance.

    b. Construct a 99% confidence interval for the difference between the fraction of CEOs and entrepreneurs who have been fired or dismissed from a job.

    c. Which inferential procedure provides more information about the difference between employment records, the test of hypothesis of part **a** or the confidence interval of part **b**? Explain.

9.67 The *Orlando Sentinel* report (cited in Exercise 9.16) on a 1983 survey of the American public provides many other interesting statistics concerning our reading habits. One outcome of the sampling of 1,961 showed that 56% claimed to read a book occasionally—a percentage, the article notes, that has "changed barely in 5 years." In 1978, the percentage was 55%. Despite the apparent small change in the sample percentages, small changes in the actual percentages could be very important for the publishing industry. Assume that the 1978 and 1983 surveys were based on independent random samples of 1,961 persons each.

    a. Do the data provide sufficient evidence to indicate a change from 1978 to 1983 in the proportion of people in the literate American population who claim that they occasionally read a book? Test using $\alpha = .01$.

    b. The survey also found that the percentage of heavy readers, those who read at least one book per week, increased from 18% in 1978 to 35% in 1983. Find a 99% confidence interval for the actual change in population proportions from 1978 to 1983. Interpret your results.

9.68 It is estimated that more than half the votes cast in upcoming elections will be cast by women. Of concern to the prevalent political parties is the possibility of a "gender gap." This hypothesis states that there is a difference between male and female perceptions of which political party best addresses the nation's problems. Particularly, those who favor this hypothesis believe that the Democratic platform may appear to be more in line with philosophies of the majority of women. Suppose a Republican campaign manager wishes to test this theory. Assume that 300 registered voters (150 men, 150 women) were given the platforms of both the Republicans and Democrats for five major issues and then asked to give their preference for one of the parties. Suppose 79 men and 71 women preferred the Republican views.

    a. Does this result provide sufficient evidence to indicate that a gender gap exists? Test at $\alpha = .05$.

    b. If there were no undecided responses, can we say that women prefer the Democrats' responses to issues? Test at $\alpha = .01$.

9.69 A *U.S. News and World Report* article (June 21, 1982) describes a new method of treating a major form of blindness in elderly people. The process, using laser beams to seal abnormal blood vessels in the eye, was tried on 224 patients, and of these only 14% went blind in 1 year. In a control group of similar untreated patients, 42% went blind in 1 year. Assume that the control group contained the same number of patients as the treated group.

    a. Do the data provide sufficient evidence to indicate that the laser beam treatment was effective in reducing the probability that a patient will go blind after 1 year? Test using $\alpha = .05$.

    b. Find the *p*-value for the test and interpret its value.

    c. Find a 95% confidence interval for the reduction in the probability of going blind after 1 year for a patient receiving the laser beam treatment.

9.70   The Reserve Mining Company of Minnesota commissioned a team of physicians to study the breathing patterns of its miners who were exposed to taconite dust. The physicians compared the breathing of 307 miners who had been employed in the Reserve's Babbit, Minnesota, mine for more than 20 years with 35 Duluth area men with no history of exposure to taconite dust. The physicians concluded that "there is no significant difference in respiratory symptoms or breathing ability between the group of men who have worked in the taconite industry for more than 20 years and a group of men of similar smoking habits but without exposure to taconite dust." Using the statistical procedures you have learned in this chapter, design a hypothesis test (give H_0, H_a, test statistic, etc.) that would have been appropriate for use in the physicians' study. [Source: Associated Press, *Minneapolis Tribune*, February 20, 1977.]

9.71   Refer to Exercise 9.70. Suppose the physicians determined that 61 of the 307 miners had breathing irregularities and that five of the 35 Duluth men had breathing irregularities.

    a. Test to determine whether these data indicate that a higher proportion of breathing irregularities exists among those who have been exposed to taconite dust than among those who have not been exposed.

    b. Find the observed significance level for the test and interpret its value.

9.72   As part of their research on the enjoyment of work by full-time workers in the United States, N. D. Glenn and C. N. Weaver give a comparison of the 1955 and the 1980 percentages of full-time workers who claimed to enjoy their work so much that they had a difficult time putting it aside. The sample sizes and sample percentages, based on a national sampling, are shown in the table for workers in three age groups. Test H_0: $p_1 = p_2$ against H_a: $p_1 \neq p_2$ for each of the age groups and calculate the p-values for your tests. State the practical implications of your p-values.

AGE	1955		1980	
	n	%	n	%
29 and under	136	44.1	387	25.6
30–49	413	50.8	577	32.1
50 and over	232	57.8	323	44.9

Source: Glenn, N. D., and Weaver, C. N. "Enjoyment of work by full-time workers in the U.S., 1955 and 1980," *Public Opinion Quarterly*, 1982, 46.

9.73   In 1984 the *Orlando Sentinel* (March 8, 1984) suggested that coffee drinking was dropping in favor of soda. The article noted that the "Winter Coffee Drinking Study" showed that 55.2% of all Americans drank coffee as compared with 74.7% in 1962. No sample sizes are given, but let us assume that both surveys were based on samples of 1,000 adult Americans. Use this information and the results just cited to find a 95% confidence interval for the percentage drop in coffee-drinking adult Americans between 1962 and 1984.

9.74   The number of practicing attorneys more than doubled in the decade 1973–1983, and the percentage of female attorneys increased from 5% to 15% of the legal profession. A random sample of 400 female attorneys and 200 male attorneys, from among the approximately 606,000 total number of attorneys in the United States, revealed that 25% of the women finished in the top 10% of their classes versus 18% of the males. The survey also found that women tended to enter law school at a later age than men. (Nearly 33% of women began practicing after age 30 compared with 14% for men.) For older women and men beginning practice, women's starting salaries tended to be higher than men of the same age. Finally, the median salary for different age groups increased with age, peaking for men at age 51–55 and leveling off thereafter (*American Bar Association Journal*, October 1983). Based on the independent random samples of 400 female and 200 male attorneys selected from all attorneys in the United States, do the data provide sufficient evidence to indicate that the probabilities of finishing in the upper 10% of their law school class differ between male and female lawyers?

9.75 According to an article in the *Bakersfield Californian* (November 18, 1982), beta blockers given to heart patients can reduce or increase the death rate, depending on when the patient receives the drug. A beta blocker is a drug that blocks the body's adrenaline from activating the heart. In a British study, 632 heart attack patients received a beta blocker while 471 others received a placebo (a pill that contains no drugs and is designed to make patients think they are taking a drug). According to the article, the medicine significantly improved the survival of people who took the drug within 4 months of their heart attack. Ninety-five percent were alive after 6 years versus 77% of those who took the placebos. For people who started taking the blocker between 1 and $7\frac{1}{2}$ years after a heart attack, the situation was reversed. Those who received the beta blocker had a 79% survival rate versus 92% who received the placebo. The *Bakersfield Californian* does not give the sample sizes involved in these two comparisons, nor does it answer a very curious question. Why is there an apparent difference in the percentages of survival among patients who received placebos? Theoretically they should be identical. Let us suppose that both samples of patients who received placebos contained 471 people.

a. What is the probability that the percentages of survival for the two experiments, 77% and 92%, could differ by as much as 15%, given that the population survival rates were equal to the average of the two percentages, .845?

b. Suppose you were to test the null hypothesis that the percentages of survival were equal for the two populations of patients who received placebos. Find the *p*-value for the test, and state your conclusion.

9.5 Determining the Sample Size

You can find the appropriate sample size to estimate the difference between a pair of parameters with a specified degree of reliability by using the method described in Sections 7.2 and 7.5. That is, to estimate the difference between a pair of parameters correct to within B units with probability $(1 - \alpha)$, let $z_{\alpha/2}$ standard deviations of the sampling distribution of the estimator equal B. Then solve for the sample size. To do this, you have to solve the problem for a specific ratio between n_1 and n_2. Most often, you will want to have equal sample sizes, that is, $n_1 = n_2 = n$. We will illustrate the procedure with two examples.

EXAMPLE 9.10

New fertilizer compounds are often advertised with the promise of increased yields. Suppose we want to compare the mean yield μ_1 of wheat when a new fertilizer is used to the mean yield μ_2 with a fertilizer in common use. The estimate of the difference in mean yield per acre is to be correct to within .25 bushel with a confidence coefficient of .95. If the sample sizes are to be equal, find $n_1 = n_2 = n$, the number of 1-acre plots of wheat assigned to each fertilizer.

Solution

To solve the problem, you need to know something about the variation in the bushels of yield per acre. Suppose from past records you know the yields of wheat possess a range of approximately 10 bushels per acre. You could then approximate $\sigma_1 = \sigma_2 = \sigma$ by letting the range equal 4σ. Thus,

$$4\sigma \approx 10 \text{ bushels}$$

$$\sigma \approx 2.5 \text{ bushels}$$

The next step is to solve the equation

$$z_{\alpha/2}\sigma_{(\bar{x}_1-\bar{x}_2)} = B \quad \text{or} \quad z_{\alpha/2}\sqrt{\frac{\sigma_1^2}{n_1} + \frac{\sigma_2^2}{n_2}} = B$$

for n, where $n = n_1 = n_2$. Since we want the estimate to lie within $B = .25$ of $(\mu_1 - \mu_2)$ with confidence coefficient equal to .95, we have $z_{\alpha/2} = z_{.025} = 1.96$. Then, letting $\sigma_1 = \sigma_2 = 2.5$ and solving for n, we have

$$1.96\sqrt{\frac{(2.5)^2}{n} + \frac{(2.5)^2}{n}} = .25$$

$$1.96\sqrt{\frac{2(2.5)^2}{n}} = .25$$

$$n = 768.32 \approx 769 \text{ (rounding up)}$$

Consequently, you will have to sample 769 acres of wheat for each fertilizer to estimate the difference in mean yield per acre to within .25 bushel. Since this would necessitate extensive and costly experimentation, you might decide to allow a larger bound (say, $B = .50$ or $B = 1$) in order to reduce the sample size, or you might decrease the confidence coefficient. The point is that we can obtain an idea of the experimental effort necessary to achieve a specified precision in our final estimate by determining the approximate sample size *before* the experiment is begun.

EXAMPLE 9.11

A production supervisor suspects a difference exists between the proportions p_1 and p_2 of defective items produced by two different machines. Experience has shown that the proportion defective for each of the two machines is in the neighborhood of .03. If the supervisor wants to estimate the difference in the proportions using a 95% confidence interval of width .01, how many items must be randomly sampled from the production of each machine? (Assume that the supervisor wants $n_1 = n_2 = n$.)

Solution

For the specified level of reliability, $z_{\alpha/2} = z_{.025} = 1.96$. Then, letting $p_1 = p_2 = .03$ and $n_1 = n_2 = n$, we find the required sample size per machine by solving the following equation for n:

$$z_{\alpha/2}\sigma_{(\hat{p}_1-\hat{p}_2)} = B$$

or

$$z_{\alpha/2}\sqrt{\frac{p_1q_1}{n_1} + \frac{p_2q_2}{n_2}} = B$$

$$1.96\sqrt{\frac{(.03)(.97)}{n} + \frac{(.03)(.97)}{n}} = .005$$

$$1.96\sqrt{\frac{2(.03)(.97)}{n}} = .005$$

$$n = 8,943.2$$

You can see that this may be a tedious sampling procedure. If the supervisor insists on estimating $(p_1 - p_2)$ correct to within .005 with probability equal to .95, approximately 9,000 items will have to be inspected for each machine.

You can see from the calculations in Example 9.11 that $\sigma_{(\hat{p}_1 - \hat{p}_2)}$ (and hence the solution, $n_1 = n_2 = n$) depends on the actual (but unknown) values of p_1 and p_2. In fact, the required sample size $n_1 = n_2 = n$ is largest when $p_1 = p_2 = \frac{1}{2}$. Therefore, if you have no prior information on the approximate values of p_1 and p_2, use $p_1 = p_2 = \frac{1}{2}$ in the formula for $\sigma_{(\hat{p}_1 - \hat{p}_2)}$. If p_1 and p_2 are in fact close to $\frac{1}{2}$, then the values of n_1 and n_2 that you have calculated will be correct. If p_1 and p_2 differ substantially from $\frac{1}{2}$, then your solutions for n_1 and n_2 will be larger than needed. Consequently, using $p_1 = p_2 = \frac{1}{2}$ when solving for n_1 and n_2 is a conservative procedure because the sample sizes n_1 and n_2 will be at least as large as (and probably larger than) needed.

The procedures for determining sample sizes necessary for estimating $(\mu_1 - \mu_2)$ or $(p_1 - p_2)$ for the case $n_1 = n_2$ are given in the box.

Determination of Sample Size for Two-Sample Procedures

1. To estimate $(\mu_1 - \mu_2)$ to within a given bound B with probability $(1 - \alpha)$ or, equivalently, with a $100(1 - \alpha)\%$ confidence interval of width $W = 2B$, use the following formula to solve for equal sample sizes that will achieve the desired reliability:

$$n_1 = n_2 = \frac{(z_{\alpha/2})^2(\sigma_1^2 + \sigma_2^2)}{B^2} = \frac{4(z_{\alpha/2})^2(\sigma_1^2 + \sigma_2^2)}{W^2}$$

You will need to substitute estimates for the values of σ_1^2 and σ_2^2 before solving for the sample size. These estimates might be sample variances s_1^2 and s_2^2 from prior sampling (e.g., a pilot sample), or from an educated (and conservatively large) guess based on the range—i.e., $s \approx R/4$.

2. To estimate $(p_1 - p_2)$ to within a given bound B with probability $(1 - \alpha)$ or, equivalently, with a $100(1 - \alpha)\%$ confidence interval of width $W = 2B$, use the following formula to solve for equal sample sizes that will achieve the desired reliability:

$$n_1 = n_2 = \frac{(z_{\alpha/2})^2(p_1 q_1 + p_2 q_2)}{B^2} = \frac{4(z_{\alpha/2})^2(p_1 q_1 + p_2 q_2)}{W^2}$$

You will need to substitute estimates for the values of p_1 and p_2 before solving for the sample size. These estimates might be based on prior samples, obtained from educated guesses or, most conservatively, specified as $p_1 = p_2 = .5$.

EXERCISES 9.76–9.87

LEARNING THE MECHANICS

9.76   Suppose you want to estimate the difference between two population means correct to within 1.5 with probability .95. If prior information suggests that the population variances are approximately equal to $\sigma_1^2 = \sigma_2^2 = 12$ and you want to select independent random samples of equal size from the populations, how large should the sample sizes, n_1 and n_2, be?

9.77   A pollster wants to estimate the difference between the proportions of men and women who favor a particular national candidate using a 90% confidence interval of width .06. Suppose the pollster has no prior information about the proportions. If equal numbers of men and women are to be polled, how large should the sample sizes be?

9.78   Find the appropriate values of n_1 and n_2 (assume $n_1 = n_2$) needed to estimate $(\mu_1 - \mu_2)$ with:
    a. A bound on the error of estimation equal to 2.8 with 95% confidence. From prior experience it is known that $\sigma_1 \approx 13$ and $\sigma_2 \approx 14$.
    b. A bound on the error of estimation equal to 5 with 99% confidence. The range of each population is 40.
    c. A 90% confidence interval of width 1.0. Assume that $\sigma_1^2 \approx 4.4$ and $\sigma_2^2 \approx 6.7$.

9.79   Assuming that $n_1 = n_2$, find the sample sizes needed to estimate $(p_1 - p_2)$ for each of the following situations:
    a. Bound = .01 with 99% confidence. Assume that $p_1 \approx .2$ and $p_2 \approx .5$.
    b. A 90% confidence interval of width .05. Assume there is no prior information available to obtain approximate values of p_1 and p_2.
    c. Bound = .05 with 90% confidence. Assume that $p_1 \approx .1$ and $p_2 \approx .3$.

9.80   Enough money has been budgeted to collect independent random samples of size $n_1 = n_2 = 100$ from populations 1 and 2 in order to estimate $(\mu_1 - \mu_2)$. Prior information indicates that $\sigma_1 = \sigma_2 = 10$. Have sufficient funds been allocated to construct a 90% confidence interval for $(\mu_1 - \mu_2)$ of width 4 or less? Justify your answer.

APPLYING THE CONCEPTS

9.81   Nationally televised home shopping was introduced in 1985. Overnight it became the hottest craze in television programming. By December 1986 there were 34 home shopping cable services (Covert, 1986). Who uses these home shopping services? Are the shoppers primarily men or women? Suppose you wish to estimate the difference in the proportions of men and women who say they have used or expect to use televised home shopping using an 80% confidence interval of width .06 or less.
    a. Approximately how many people should be included in your samples?
    b. Suppose you want to obtain individual estimates for the two proportions of interest. Will the sample size found in part a be large enough to provide estimates of each proportion correct to within .02 with probability equal to .90? Justify your response.

9.82   One reason high school seniors are encouraged to attend college is that the job opportunities are much better for those with college degrees than for those without. A high school counselor wants to estimate the difference in mean income per day between high school graduates who have a college education and those who have not gone on to college. Suppose it is decided to compare the daily incomes of 30-year-olds, and the range of daily incomes for both groups is approximately $200 per day. How many people from each group should be sampled in order to estimate the true difference between mean daily incomes correct to within $10 per day with probability .9? Assume that $n_1 = n_2$.

9.83   Rat damage creates a large financial loss in the production of sugarcane. One aspect of the problem that has been investigated by the U.S. Department of Agriculture concerns the optimal place to locate rat poison. To be most effective in reducing rat damage, should the poison be located in the middle of the field or on the outer perimeter? One way to answer this question is to determine where the greater amount of damage occurs. If damage is measured by the proportion of cane stalks that have been damaged by rats, how many stalks from each section of the field should be sampled in order to estimate the true difference between proportions of stalks damaged in the two sections to within .02 with probability .95?

9.84 Suppose you are interested in the growth rate of dividends. Consider investing $1,000 in a stock and suppose you want to estimate the dividend rate on your $1,000 investment at the end of 5 years. Particularly, you want to compare two types of stocks, electrical utilities and oil companies. To conduct your study, you plan to select n oil stocks and n electrical utility stocks at random. For each stock, you will check the records, calculate the number of shares of stock you could have purchased 5 years ago for $1,000, and then calculate the dividend rate (in percent) that the stock would be paying today on your $1,000 investment. Suppose you think the dividend rates will vary over a range of roughly 25%. How large should n be if you want to estimate the difference in mean rates of dividend return with a 95% confidence interval of width 6%?

9.85 A television manufacturer wants to compare with a competitor the proportions of its best sets that need repair within 1 year. If it is desired to estimate the difference between proportions to within .05 with 90% confidence, and if the manufacturer plans to sample twice as many buyers (n_1) of its sets as buyers (n_2) of the competitor's sets, how many buyers of each brand must be sampled? Assume that the proportion of sets that need repair will be about .2 for both brands.

9.86 In seeking a good professional football running back, a coach is looking for a player with a large mean yards gained per carry and a small standard deviation. Suppose the coach wishes to compare the mean yards gained per carry for two major prospects based on independent random samples of their yards gained per carry in the early part of the coming pro football season. Suppose data from last year indicate that $\sigma_1 = \sigma_2 \approx 5$ yards. If the coach wants to estimate the difference in means correct to within 1 yard with probability equal to .9, how many runs would have to be observed for each player? (Assume equal sample sizes.)

9.87 In Exercise 8.28 we compared the mean costs of electrical energy produced by nuclear and coal-powered electrical power plants. In that exercise, we were given the information that the variances in costs for the nuclear and coal-powered plants were $\sigma_1^2 = .05$ and $\sigma_2^2 = .04$, respectively. How many plants of each type should be sampled in order to estimate the difference in mean costs per kilowatt-hour with a 95% confidence interval of width .2 cent?

9.6 Comparing Two Population Variances: Independent Random Samples

Many times it is of practical interest to use the techniques developed previously in this chapter to compare the means or proportions of two populations. However, there are also important instances when it is desired to compare two population variances. For example, when two devices are available for producing precision measurements (scales, calipers, thermometers, etc.), we might want to compare the variability of the measurements of the devices before deciding which one to purchase. Or when two standardized tests can be used to rate applicants, the variability of the scores for both tests should be taken into consideration before deciding which test to use.

For problems like these we need to develop a statistical procedure to compare population variances. The common statistical procedure for comparing population variances, σ_1^2 and σ_2^2, makes an inference about the ratio σ_1^2/σ_2^2. In this section, we will show how to test the null hypothesis that the ratio σ_1^2/σ_2^2 equals

1 (the variances are equal) against the alternative hypothesis that the ratio differs from 1 (the variances differ):

$$H_0: \quad \frac{\sigma_1^2}{\sigma_2^2} = 1 \qquad (\sigma_1^2 = \sigma_2^2)$$

$$H_a: \quad \frac{\sigma_1^2}{\sigma_2^2} \neq 1 \qquad (\sigma_1^2 \neq \sigma_2^2)$$

To make an inference about the ratio σ_1^2/σ_2^2 it seems reasonable to collect sample data and use the ratio of the sample variances, s_1^2/s_2^2. We will use the test statistic

$$F = \frac{s_1^2}{s_2^2}$$

To establish a rejection region for the test statistic, we need to know the sampling distribution of s_1^2/s_2^2. As you will subsequently see, the sampling distribution of s_1^2/s_2^2 is based on two of the assumptions already required for the t-test:

1. The two sampled populations are normally distributed.
2. The samples are randomly and independently selected from their respective populations.

When these assumptions are satisfied and when the null hypothesis is true (that is, $\sigma_1^2 = \sigma_2^2$), the sampling distribution of $F = s_1^2/s_2^2$ is the **F-distribution** with $(n_1 - 1)$ numerator degrees of freedom and $(n_2 - 1)$ denominator degrees of freedom, respectively. The shape of the F-distribution depends on the degrees of freedom associated with s_1^2 and s_2^2—that is, on $(n_1 - 1)$ and $(n_2 - 1)$. An F-distribution with 7 and 9 df is shown in Figure 9.8. As you can see, the distribution is skewed to the right, since s_1^2/s_2^2 cannot be less than 0 but can increase without bound.

FIGURE 9.8

An F-distribution with 7 numerator and 9 denominator degrees of freedom

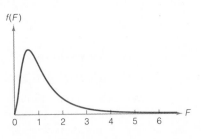

We need to be able to find F values corresponding to the tail areas of this distribution in order to establish the rejection region for our test of hypothesis, because when the population variances are unequal, we expect the ratio F of the sample variances to be either very large or very small. The upper-tail F values for $\alpha = .10, .05, .025,$ and $.01$ can be found in Tables VIII, IX, X, and XI of Appendix A. Table IX is partially reproduced in Figure 9.9. It gives F values that correspond to $\alpha = .05$ upper-tail areas for different degrees of freedom. The columns of Tables VIII, IX, X, and XI correspond to the degrees of freedom ν_1 for the numerator sample variance, s_1^2, whereas the rows correspond to the degrees of freedom ν_2 for the denominator sample variance, s_2^2. Thus, if the numerator

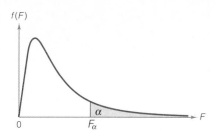

	NUMERATOR DEGREES OF FREEDOM								
ν_2 \ ν_1	1	2	3	4	5	6	7	8	9
1	161.4	199.5	215.7	224.6	230.2	234.0	236.8	238.9	240.5
2	18.51	19.00	19.16	19.25	19.30	19.33	19.35	19.37	19.38
3	10.13	9.55	9.28	9.12	9.01	8.94	8.89	8.85	8.81
4	7.71	6.94	6.59	6.39	6.26	6.16	6.09	6.04	6.00
5	6.61	5.79	5.41	5.19	5.05	4.95	4.88	4.82	4.77
6	5.99	5.14	4.76	4.53	4.39	4.28	4.21	4.15	4.10
7	5.59	4.74	4.35	4.12	3.97	3.87	3.79	3.73	3.68
8	5.32	4.46	4.07	3.84	3.69	3.58	3.50	3.44	3.39
9	5.12	4.26	3.86	3.63	3.48	3.37	3.29	3.23	3.18
10	4.96	4.10	3.71	3.48	3.33	3.22	3.14	3.07	3.02
11	4.84	3.98	3.59	3.36	3.20	3.09	3.01	2.95	2.90
12	4.75	3.89	3.49	3.25	3.11	3.00	2.91	2.85	2.80
13	4.67	3.81	3.41	3.18	3.03	2.92	2.83	2.77	2.71
14	4.60	3.74	3.34	3.11	2.96	2.85	2.76	2.70	2.65

DENOMINATOR DEGREES OF FREEDOM

FIGURE 9.9

Reproduction of part of Table IX in Appendix A: $\alpha = .05$

degrees of freedom is $\nu_1 = 7$ and the denominator degrees of freedom is $\nu_2 = 9$, we look in the seventh column and ninth row to find $F_{.05} = 3.29$. As shown in Figure 9.10, $\alpha = .05$ is the tail area to the right of 3.29 in the F-distribution with 7 and 9 df. That is, if $\sigma_1^2 = \sigma_2^2$ then the probability that the F statistic will exceed 3.29 is $\alpha = .05$.

FIGURE 9.10

An F-distribution for $\nu_1 = 7$ and $\nu_2 = 9$ df: $\alpha = .05$

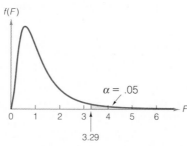

EXAMPLE 9.12

An experimenter wants to compare the metabolic rates of white mice subjected to different drugs. The weights of the mice may affect their metabolic rates, and thus the experimenter wishes to obtain mice that are relatively homogeneous with regard to weight. Five hundred mice will be needed to complete the study. Presently, 18 mice from supplier 1 and another 13 mice from supplier 2 are available for comparison. The experimenter weighs these mice and obtains the following summary information:

SUPPLIER 1	SUPPLIER 2
$n_1 = 18$	$n_2 = 13$
$\bar{x}_1 = 4.21$ ounces	$\bar{x}_2 = 4.18$ ounces
$s_1^2 = .019$	$s_2^2 = .049$

Do these data provide sufficient evidence to indicate a difference in the variability of weights of mice obtained from the two suppliers? (Use $\alpha = .10$.) Using the results of this analysis, what would you suggest to the experimenter?

Solution

Let

$\sigma_1^2 =$ Population variance of weights of white mice from supplier 1

$\sigma_2^2 =$ Population variance of weights of white mice from supplier 2

The hypotheses of interest are then

$$H_0: \quad \frac{\sigma_1^2}{\sigma_2^2} = 1 \qquad (\sigma_1^2 = \sigma_2^2)$$

$$H_a: \quad \frac{\sigma_1^2}{\sigma_2^2} \neq 1 \qquad (\sigma_1^2 \neq \sigma_2^2)$$

The nature of the F-tables given in Appendix A affects the form of the test statistic. To form the rejection region for a two-tailed F-test, we want to make certain that the upper tail is used, because only the upper-tail values of F are shown in Tables VIII, IX, X, and XI. To accomplish this, *we will always place the larger sample variance in the numerator of the F-test statistic.* This has the effect of doubling the tabulated value for α, since we double the probability that the F-ratio will fall in the upper tail by always placing the larger sample variance in the numerator. That is, we establish a one-tailed rejection region by putting the larger variance in the numerator rather than establishing rejection regions in both tails.

Thus, for our example, we have a denominator s_1^2 with df $= \nu_2 = n_1 - 1 = 17$ and a numerator s_2^2 with df $= \nu_1 = n_2 - 1 = 12$. Therefore, the test statistic will be

$$F = \frac{\text{Larger sample variance}}{\text{Smaller sample variance}} = \frac{s_2^2}{s_1^2}$$

and we will reject $H_0: \sigma_1^2 = \sigma_2^2$ for $\alpha = .10$ when the calculated value of F exceeds the tabulated value:

$$F_{\alpha/2} = F_{.05} = 2.38 \qquad \text{(see Figure 9.11)}$$

We can now calculate the value of the test statistic and complete the analysis:

$$F = \frac{s_2^2}{s_1^2} = \frac{.049}{.019} = 2.58$$

When we compare this result to the rejection region shown in Figure 9.11, we see that $F = 2.58$ falls in the rejection region. Therefore, the data provide sufficient evidence to indicate that the population variances differ. It appears that the

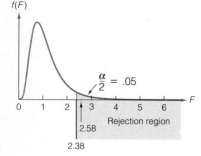

FIGURE 9.11

Rejection region for Example 9.12

weights of mice obtained from supplier 1 tend to be more homogeneous than the weights of mice obtained from supplier 2. On the basis of this evidence, we would advise the experimenter to purchase the mice from supplier 1.

What would you have concluded if the value of F calculated from the samples had not fallen in the rejection region? Would you conclude that the null hypothesis of equal variances is true? No, because then you risk the possibility of a Type II error (accepting H_0 if H_a is true) without knowing the value of β, the probability of accepting H_0: $\sigma_1^2 = \sigma_2^2$ if in fact it is false. Since we will not consider the calculation of β for specific alternatives in this text, when the F statistic does not fall in the rejection region we simply conclude that insufficient sample evidence exists to refute the null hypothesis that $\sigma_1^2 = \sigma_2^2$.

The F-test for equal population variances is summarized in the box.

F-Test for Equal Population Variances

ONE-TAILED TEST

H_0: $\sigma_1^2 = \sigma_2^2$

H_a: $\sigma_1^2 < \sigma_2^2$

(or H_a: $\sigma_1^2 > \sigma_2^2$)

Test statistic:

$$F = \frac{s_2^2}{s_1^2}$$

$$\left(\text{or } F = \frac{s_1^2}{s_2^2} \quad \text{when } H_a: \ \sigma_1^2 > \sigma_2^2\right)$$

Rejection region:

$F > F_\alpha$

(or $F > F_\alpha$ when H_a: $\sigma_1^2 > \sigma_2^2$)

TWO-TAILED TEST

H_0: $\sigma_1^2 = \sigma_2^2$

H_a: $\sigma_1^2 \neq \sigma_2^2$

Test statistic:

$$F = \frac{\text{Larger sample variance}}{\text{Smaller sample variance}}$$

$$= \frac{s_1^2}{s_2^2} \quad \text{when } s_1^2 > s_2^2$$

$$\left(\text{or } \frac{s_2^2}{s_1^2} \quad \text{when } s_2^2 > s_1^2\right)$$

Rejection region:

$F > F_{\alpha/2}$ when $s_1^2 > s_2^2$

(or $F > F_{\alpha/2}$ when $s_2^2 > s_1^2$)

where F_α and $F_{\alpha/2}$ are based on ν_1 = numerator degrees of freedom and ν_2 = denominator degrees of freedom; ν_1 and ν_2 are the degrees of freedom for the numerator and denominator sample variances, respectively.

Assumptions: 1. Both sampled populations are normally distributed.
2. The samples are random and independent.

EXAMPLE 9.13

Find the approximate p-value for the test in Example 9.12.

Solution

Since the observed value of the F statistic in Example 9.12 was 2.58, the observed significance level of the test would equal the probability of observing a value of F at least as contradictory to H_0: $\sigma_1^2 = \sigma_2^2$ as $F = 2.58$, if in fact H_0 is true. Since we give the F-tables in Appendix A only for values of α equal to .10, .05, .025, and .01, we can only approximate the observed significance level. Checking Tables VIII, IX, X, and XI, we find $F_{.05} = 2.38$ and $F_{.025} = 2.82$. Since the observed value of F exceeds $F_{.05}$ but is less than $F_{.025}$, the observed significance level for the test is less than

Approximate p-value $= 2(.05) = .10$

Note that we double the α value shown in Table IX because this is a two-tailed test.

EXAMPLE 9.14

STOCK 1	STOCK 2
$n_1 = 25$	$n_2 = 25$
$\bar{x}_1 = .250$	$\bar{x}_2 = .125$
$s_1 = .76$	$s_2 = .46$

Solution

An investor believes that although the price of stock 1 usually exceeds that of stock 2, stock 1 represents a riskier investment, where the risk of a given stock is measured by the variation in daily price changes. Suppose we obtain a random sample of 25 daily price changes for stock 1 and 25 for stock 2. The sample results are summarized in the table. Compare the risks associated with the two stocks by testing the null hypothesis that the variances of the price changes for the stocks are equal against the alternative that the price variance of stock 1 exceeds that of stock 2. Use $\alpha = .05$.

H_0: $\sigma_1^2 = \sigma_2^2$

H_a: $\sigma_1^2 > \sigma_2^2$

Test statistic: $F = \dfrac{s_1^2}{s_2^2}$

Assumptions: 1. The changes in daily stock prices have relative frequency distributions that are approximately normal.
2. The stock samples are randomly and independently selected from a set of daily stock reports.

Rejection region: $F > F_\alpha = F_{.05} = 1.98$, where $F_{.05}$ is based on $\nu_1 = 24$ df and $\nu_2 = 24$ df.

We calculate

$$F = \frac{s_1^2}{s_2^2} = \frac{(.76)^2}{(.46)^2} = 2.73$$

The calculated F exceeds the rejection value of 1.98. Therefore, we conclude that the variance of the daily price change for stock 1 exceeds that for stock 2. It appears that stock 1 is a riskier investment than stock 2. How much reliability can we place in this inference? Only one time in 20 (since $\alpha = .05$), on the average, would this statistical test lead us to conclude erroneously that σ_1^2 exceeds σ_2^2 if in fact they are equal.

Since this is a one-tailed test, the p-value is equal to the probability that F exceeds the computed value, 2.73; that is,

p-value $= P(F > 2.73)$

Checking Tables X and XI, we find that $F_{.025} = 2.27$ and $F_{.01} = 2.66$. Since the computed value of the F statistic exceeds 2.66, the observed significance level for the test is slightly less:

Approximate p-value $= .01$

As a final example of an application, consider the comparison of population variances as a check of the assumption $\sigma_1^2 = \sigma_2^2$ needed for the two-sample t-test. Rejection of the null hypothesis $\sigma_1^2 = \sigma_2^2$ would indicate that the assumption is invalid. [*Note:* Nonrejection of the null hypothesis does *not* imply that the assumption is valid.] We illustrate with an example.

EXAMPLE 9.15

In Example 9.7 (Section 9.2) we used the two-sample t statistic to compare the mean reading scores of two groups of slow learners who had been taught to read using two different methods. The data are repeated here for convenience. The use of the t statistic was based on the assumption that the population variances of the test scores were equal for the two methods. Check this assumption by using $\alpha = .10$.

NEW METHOD	STANDARD METHOD
$n_1 = 8$	$n_2 = 12$
$\bar{x}_1 = 76.9$	$\bar{x}_2 = 72.7$
$s_1 = 4.85$	$s_2 = 6.35$

Solution

We can test

$$H_0: \quad \sigma_1^2 = \sigma_2^2$$
$$H_a: \quad \sigma_1^2 \neq \sigma_2^2$$

using

Test statistic: $\quad F = \dfrac{\text{Larger sample variance}}{\text{Smaller sample variance}} = \dfrac{s_2^2}{s_1^2}$

Rejection region: $\quad \alpha = .10$

$$F > F_{\alpha/2} = F_{.05} = 3.6$$

where $F_{.05}$ is based on $\nu_1 = 11$ df and $\nu_2 = 7$ df. (We interpolate between 10 and 12 for the numerator degrees of freedom.) We calculate

$$F = \frac{(6.35)^2}{(4.85)^2} = 1.71$$

Since $F < 3.6$, we do not reject the null hypothesis that the population variances of the reading test scores are equal. It is here that the temptation to misuse the F-test is strongest. *We cannot conclude that the data justify the use of the t statistic.* This is equivalent to accepting H_0, and we have repeatedly warned against this conclusion because the probability of a Type II error, β, is unknown. The α level of .10 protects us only against rejecting H_0 if it is true. This use of the F-test may prevent us from abusing the t-procedure when we obtain a value of F that leads to a rejection of the assumption that $\sigma_1^2 = \sigma_2^2$. But when the F statistic does not fall in the rejection region, we know little more about the validity of the assumption than before we conducted the test.

We have presented the F-test as a test of a hypothesis of equality of variances, that is, $\sigma_1^2 = \sigma_2^2$. Although this is the most common application of the test, it can also be used to test a hypothesis that the ratio between the population variances is equal to some specified value, $H_0: \sigma_1^2/\sigma_2^2 = k$. The test would be conducted in exactly the same way as a test of a hypothesis concerning the equality of variances except that we would use the test statistic

$$F = \frac{s_1^2}{s_2^2}\left(\frac{1}{k}\right)$$

What Do You Do if the Assumption of Normal Population Distributions Is Not Satisfied?

Answer: The F-test is much less robust (i.e., much more sensitive) to departures from normality than the t-test for comparing population means (Section 9.2). If you have doubts about the normality of the population frequency distributions, use a nonparametric method for comparing the two population variances. A method can be found in the references listed at the end of this chapter.

Rather than test a hypothesis concerning σ_1^2/σ_2^2, you may wish to gain insight into the relative magnitudes of σ_1^2 and σ_2^2 by estimating their ratio. A confidence interval for the ratio of two population variances can be obtained by using the formula shown in the box.

Confidence Interval for σ_1^2/σ_2^2

$$\left(\frac{s_1^2}{s_2^2}\right)\left(\frac{1}{F_{L,\alpha/2}}\right) < \frac{\sigma_1^2}{\sigma_2^2} < \left(\frac{s_1^2}{s_2^2}\right)F_{U,\alpha/2}$$

where $F_{L,\alpha/2}$ is the tabulated value of F that places an area $\alpha/2$ in the upper tail of the F-distribution and that is based on $\nu_1 = (n_1 - 1)$ numerator and $\nu_2 = (n_2 - 1)$ denominator degrees of freedom, and $F_{U,\alpha/2}$ is the tabulated upper-tail value of F based on $\nu_1 = (n_2 - 1)$ numerator and $\nu_2 = (n_1 - 1)$ denominator degrees of freedom.

Assumptions: The samples were randomly and independently selected from populations that possess approximately normal distributions.

EXAMPLE 9.16

Refer to Example 9.12 and find a 90% confidence interval for the ratio of the variances of the weights of mice obtained from the two suppliers.

Solution

In Example 9.12, we were given the following information:

SUPPLIER 1	SUPPLIER 2
$n_1 = 18$	$n_2 = 13$
$\bar{x}_1 = 4.21$ ounces	$\bar{x}_2 = 4.18$ ounces
$s_1^2 = .019$	$s_2^2 = .049$

Then

$$n_1 - 1 = 18 - 1 = 17 \qquad n_2 - 1 = 13 - 1 = 12$$

and, from Table IX of Appendix A,

$$F_{L,\alpha/2} = F_{L,.05} \approx 2.59 \qquad \text{(where } \nu_1 = 17 \text{ and } \nu_2 = 12\text{)}$$
$$F_{U,\alpha/2} = F_{U,.05} = 2.38 \qquad \text{(where } \nu_1 = 12 \text{ and } \nu_2 = 17\text{)}$$

Substituting these values into the formula for the confidence interval, we obtain

$$\left(\frac{s_1^2}{s_2^2}\right)\left(\frac{1}{F_{L,\alpha/2}}\right) < \frac{\sigma_1^2}{\sigma_2^2} < \left(\frac{s_1^2}{s_2^2}\right)F_{U,\alpha/2}$$

$$\left(\frac{.019}{.049}\right)\left(\frac{1}{2.59}\right) < \frac{\sigma_1^2}{\sigma_2^2} < \left(\frac{.019}{.049}\right)2.38$$

$$.150 < \frac{\sigma_1^2}{\sigma_2^2} < .923$$

According to this confidence interval, we estimate that σ_1^2, the variance in the weights of mice obtained from supplier 1, could be as smaller as .150 or as large as .923 times the size of σ_2^2, the variance in the weights of mice obtained from supplier 2.

EXERCISES 9.88–9.103

LEARNING THE MECHANICS

9.88 Under what conditions is the sampling distribution of s_1^2/s_2^2 an F-distribution?

9.89 Use Tables VIII, IX, X, and XI of Appendix A to find each of the following F values:
a. $F_{.05}$ where $\nu_1 = 9$ and $\nu_2 = 6$ b. $F_{.01}$ where $\nu_1 = 18$ and $\nu_2 = 14$
c. $F_{.025}$ where $\nu_1 = 11$ and $\nu_2 = 4$ d. $F_{.10}$ where $\nu_1 = 20$ and $\nu_2 = 5$

9.90 Given ν_1 and ν_2, find the following probabilities:
a. $\nu_1 = 2$, $\nu_2 = 30$, $P(F \geq 5.39)$ b. $\nu_1 = 24$, $\nu_2 = 10$, $P(F < 2.74)$
c. $\nu_1 = 7$, $\nu_2 = 1$, $P(F \leq 236.8)$ d. $\nu_1 = 40$, $\nu_2 = 40$, $P(F > 2.11)$

9.91 For each of the following cases, identify the rejection region that should be used to test $H_0: \sigma_1^2 = \sigma_2^2$ against $H_a: \sigma_1^2 > \sigma_2^2$. Assume $\nu_1 = 15$ and $\nu_2 = 20$.
a. $\alpha = .10$ b. $\alpha = .05$ c. $\alpha = .025$ d. $\alpha = .01$

9.92 For each of the following cases, identify the rejection region that should be used to test $H_0: \sigma_1^2 = \sigma_2^2$ against $H_a: \sigma_1^2 \neq \sigma_2^2$. Assume $\nu_1 = 24$ and $\nu_2 = 18$.
a. $\alpha = .20$ b. $\alpha = .10$ c. $\alpha = .05$ d. $\alpha = .02$

9.93 Specify the appropriate rejection region for testing $H_0: \sigma_1^2 = \sigma_2^2$ in each of the following situations:
a. $H_a: \sigma_1^2 > \sigma_2^2$; $\alpha = .05$, $n_1 = 21$, $n_2 = 20$
b. $H_a: \sigma_1^2 < \sigma_2^2$; $\alpha = .05$, $n_1 = 10$, $n_2 = 21$
c. $H_a: \sigma_1^2 \neq \sigma_2^2$; $\alpha = .10$, $n_1 = 16$, $n_2 = 25$
d. $H_a: \sigma_1^2 < \sigma_2^2$; $\alpha = .01$, $n_1 = 31$, $n_2 = 41$
e. $H_a: \sigma_1^2 \neq \sigma_2^2$; $\alpha = .05$, $n_1 = 9$, $n_2 = 25$

9.94 Give the appropriate tabled F values to form a 95% confidence interval for σ_1^2/σ_2^2 for each of the following combinations of n_1 and n_2:
a. $n_1 = 10$, $n_2 = 8$ b. $n_1 = 13$, $n_2 = 9$
c. $n_1 = 16$, $n_2 = 41$ d. $n_1 = 16$, $n_2 = 13$

9.95 Independent random samples were selected from each of two normally distributed populations, $n_1 = 16$ from population 1 and $n_2 = 25$ from population 2. The means and variances for the two samples are shown in the table.

SAMPLE 1	SAMPLE 2
$n_1 = 16$	$n_2 = 25$
$\bar{x}_1 = 22.5$	$\bar{x}_2 = 28.2$
$s_1^2 = 3.56$	$s_2^2 = 10.24$

a. Test the null hypothesis H_0: $\sigma_1^2 = \sigma_2^2$ against the alternative hypothesis H_a: $\sigma_1^2 \neq \sigma_2^2$. Use $\alpha = .05$.

b. Form a 95% confidence interval for σ_1^2/σ_2^2.

c. Test H_0: $\sigma_1^2 = \sigma_2^2$ against H_a: $\sigma_1^2 < \sigma_2^2$. Use $\alpha = .05$.

9.96 Independent random samples were selected from each of two normally distributed populations, $n_1 = 6$ from population 1 and $n_2 = 4$ from population 2. The data are shown in the table.

SAMPLE 1	SAMPLE 2
3.1	2.3
4.3	1.4
1.2	3.7
1.7	8.9
.6	
3.4	

a. Form a 90% confidence interval for σ_1^2/σ_2^2.

b. Test H_0: $\sigma_1^2 = \sigma_2^2$ against H_a: $\sigma_1^2 < \sigma_2^2$. Use $\alpha = .01$.

c. Test H_0: $\sigma_1^2 = \sigma_2^2$ against H_a: $\sigma_1^2 \neq \sigma_2^2$. Use $\alpha = .10$.

APPLYING THE CONCEPTS

9.97 Tests of product quality can be completely automated or can be conducted using human inspectors or human inspectors aided by mechanical devices. Although human inspection is frequently the most economical alternative, it can lead to serious inspection error problems (Benson and Ohta, 1986). Numerous studies have demonstrated that inspectors are rarely able to detect as many as 85% of the defective items that they inspect and that performance varies across inspectors (Sinclair, 1978). To evaluate the performance of inspectors in a new company, a quality manager had a sample of 12 novice inspectors evaluate 200 finished products. The same 200 items were evaluated by 12 experienced inspectors. The quality of each item—whether defective or nondefective—was known to the manager. The table lists the number of inspection errors (classifying a defective item as nondefective or vice versa) made by each inspector.

NOVICE INSPECTORS		EXPERIENCED INSPECTORS	
30	45	31	19
35	31	15	18
26	33	25	24
40	29	19	10
36	21	28	20
20	48	17	21

a. Prior to conducting this experiment, the manager believed the variance in inspection errors was lower for experienced inspectors than for novice inspectors. Do the sample data support her belief? Test using $\alpha = .05$.

b. What is the approximate p-value of the test you conducted in part **a**?

9.98 A series of experiments has been conducted to compare the quantity of hemoglobin in the blood of men and women who are between the ages of 20 and 30 years. One phase of the study deals with a comparison of the variability in the hemoglobin measurements between the two groups. In random samples of 25 women and 20 men, all between the ages of 20 and 30 years, the researcher recorded the amount of hemoglobin in the blood of each. (The quantity of the hemoglobin is measured as a percentage of the total volume of blood.) The results are summarized in the table. Form a 90% confidence interval for the ratio of the variance of the amount of hemoglobin in women's blood to the corresponding variance for men.

WOMEN	MEN
$n = 25$	$n = 20$
$\bar{x} = 42.7$	$\bar{x} = 41.8$
$s^2 = 18.3$	$s^2 = 8.5$

9.99 The goalie is generally regarded as the most important player on a hockey team. One measure of a goalie's ability is the "goals against" average (GA average), i.e., the average number of goals the goalie gives up per game. However, most National Hockey League coaches agree that consistency in performance is just as important as the GA average. A consistent goalie is one whose number of goals given up per game varies only slightly from game to game. Two goalies with nearly equal GA averages are competing for the starting position on a hockey team. The coach will choose the starter on the basis of the better GA average only if there is no evidence of a difference in the consistency of the two goalies based on their performances in ten exhibition games (ten games per goalie). Otherwise, the more consistent goalie will win the starting position. The results of the exhibition games are given in the table. What decision does the coach make? Test at the $\alpha = .05$ level of significance.

	GOALIE A	GOALIE B
Number of games	10	10
$\bar{x}$ (GA average)	3.3	3.1
s^2	.68	2.77

9.100 In Exercise 9.86 we planned to compare the mean yards gained per run for two professional football running backs. Suppose a record of early season runs shows that the standard deviations of yards gained per run, based on samples of 50 runs per player, are $s_1 = 3.2$ and $s_2 = 5.7$ yards, respectively. Do these standard deviations indicate that the variation in the distributions of yards gained per carry differs for the two players? Test using $\alpha = .05$.

9.101 Suppose a firm has been experimenting with two different physical arrangements of its assembly line. It has been determined that both arrangements yield approximately the same average number of finished units per day. To obtain an arrangement that produces greater process control, you suggest that the arrangement with the smaller variance in the number of finished units produced per day be permanently adopted. Two independent random samples yield the results shown in the table on page 438. Do the samples provide sufficient evidence at the .10 significance level to conclude that the variances of the two arrangements differ? If so, which arrangement would you choose? If not, what would you suggest the firm do?

ASSEMBLY LINE 1	ASSEMBLY LINE 2
$n_1 = 21$ days	$n_2 = 21$ days
$s_1^2 = 1,432$	$s_2^2 = 3,761$

9.102   In Exercise 9.33 we used a two-sample t-test to compare two methods of teaching reading. Independent random samples of 11 and 14 children were assigned to methods 1 and 2, respectively. The resulting sample variances were 52 and 71, respectively.

  a. One of the required assumptions for the test is that the population variances be equal. Can this assumption possibly be true when the two sample variances, 52 and 71, are unequal?
  b. Test the null hypothesis that the assumption of equal variances is true. Use $\alpha = .10$.

9.103   The quality control department of a paper company measures the brightness (a measure of reflectance) of finished paper on a periodic basis throughout the day. Two instruments that are available to measure the paper specimens are subject to error, but they can be adjusted so that the mean readings for a control paper specimen are the same for both instruments. Suppose you are concerned about the precision of the two instruments and want to compare the variability in the readings of instrument 1 to those of instrument 2. Five brightness measurements were made on a single paper specimen using each of the two instruments. The data are shown in the table.

INSTRUMENT 1	INSTRUMENT 2
29	26
28	34
30	30
28	32
30	28

  a. Form a 95% confidence interval for the ratio of the variance of the measurements obtained by instrument 1 to the variance of the measurements obtained using instrument 2.
  b. Interpret the interval you found in part **a** in terms of the precision of the two instruments.
  c. What assumptions must be satisfied for the confidence interval in part **a** to be valid?

Summary

We have presented various techniques for using the information in two samples to make inferences about the difference between population parameters. As you would expect, we are able to make reliable inferences with fewer assumptions about the sampled populations when the sample sizes are large. When we cannot take large samples from the populations, the **two-sample t statistic** permits us to use the limited sample information to make inferences about the **difference between means** when the assumptions of normality and equal population variances are at least approximately true. The **paired difference experiment** offers the possibility of increasing the information about $(\mu_1 - \mu_2)$ by pairing similar observational units to control variability. In designing a paired difference experiment, we expect that the reduction in variability will more than compensate for the loss in degrees of freedom.

Two other inferential procedures for making comparisons between population parameters were presented in this chapter. A method for comparing two binomial parameters, p_1 and p_2, was presented. Practical examples of such comparisons are numerous and appear frequently in the analysis of surveys. (Applications of this and other techniques discussed in the chapter are suggested by the exercises that follow.)

The chapter concluded with a procedure for comparing two population variances, σ_1^2 and σ_2^2. We use an **F statistic** to test the null hypothesis that two population variances are equal. This test is of practical importance because variances often represent measures of risk, error, or precision.

SUPPLEMENTARY EXERCISES 9.104–9.140

LEARNING THE MECHANICS

[*Note: List the assumptions necessary to ensure the validity of the statistical procedures you use to work these exercises.*]

9.104 Independent random samples were selected from two normally distributed populations with means μ_1 and μ_2, respectively. The sample sizes, means, and variances are shown in the table.

SAMPLE 1	SAMPLE 2
$n_1 = 12$	$n_2 = 14$
$\bar{x}_1 = 17.8$	$\bar{x}_2 = 15.3$
$s_1^2 = 74.2$	$s_2^2 = 60.5$

a. Test the null hypothesis H_0: $(\mu_1 - \mu_2) = 0$ against the alternative hypothesis. H_a: $(\mu_1 - \mu_2) > 0$. Use $\alpha = .05$.
b. Form a 99% confidence interval for $(\mu_1 - \mu_2)$.
c. How large must n_1 and n_2 be if you wish to estimate $(\mu_1 - \mu_2)$ to within 2 units with 99% confidence? Assume that $n_1 = n_2$.

9.105 Two independent random samples were selected from normally distributed populations with means and variances (μ_1, σ_1^2) and (μ_2, σ_2^2), respectively. The sample sizes, means, and variances are shown in the table.

SAMPLE 1	SAMPLE 2
$n_1 = 20$	$n_2 = 15$
$\bar{x}_1 = 123$	$\bar{x}_2 = 116$
$s_1^2 = 31.3$	$s_2^2 = 120.1$

a. Form a 95% confidence interval for σ_1^2/σ_2^2.
b. Test H_0: $\sigma_1^2 = \sigma_2^2$ against H_a: $\sigma_1^2 \neq \sigma_2^2$. Use $\alpha = .05$.
c. Would you be willing to use a *t*-test to test the null hypothesis H_0: $(\mu_1 - \mu_2) = 0$ against the alternative hypothesis H_a: $(\mu_1 - \mu_2) \neq 0$? Why?

9.106 Two independent random samples are taken from two populations. The results of these samples are summarized in the table on the next page.

SAMPLE 1	SAMPLE 2
$n_1 = 135$	$n_2 = 148$
$\bar{x}_1 = 12.2$	$\bar{x}_2 = 8.3$
$s_1^2 = 2.1$	$s_2^2 = 3.0$

a. Form a 90% confidence interval for $(\mu_1 - \mu_2)$.

b. Test H_0: $(\mu_1 - \mu_2) = 0$ against H_a: $(\mu_1 - \mu_2) \neq 0$. Use $\alpha = .01$.

c. What sample sizes would be required if you wish to estimate $(\mu_1 - \mu_2)$ to within .2 with 90% confidence? Assume that $n_1 = n_2$.

9.107 Independent random samples were selected from two binomial populations. The sizes and number of observed successes for each sample are shown in the table.

SAMPLE 1	SAMPLE 2
$n_1 = 200$	$n_2 = 200$
$x_1 = 110$	$x_2 = 130$

a. Test the null hypothesis H_0: $(p_1 - p_2) = 0$ against the alternative hypothesis H_a: $(p_1 - p_2) < 0$. Use $\alpha = .10$.

b. Form a 95% confidence interval for $(p_1 - p_2)$.

c. What sample sizes would be required if we wish to use a 95% confidence interval of width .01 to estimate $(p_1 - p_2)$?

9.108 A random sample of five pairs of observations were selected, one of each pair from a population with mean μ_1, the other from a population with mean μ_2. The data are shown in the accompanying table.

PAIR	VALUE FROM POPULATION 1	VALUE FROM POPULATION 2
1	28	22
2	31	27
3	24	20
4	30	27
5	22	20

a. Test the null hypothesis H_0: $\mu_D = 0$ against H_a: $\mu_D \neq 0$, where $\mu_D = \mu_1 - \mu_2$. Use $\alpha = .05$.

b. Form a 95% confidence interval for μ_D.

c. When are the procedures you used in parts **a** and **b** valid?

9.109 List the assumptions necessary for each of the following inferential techniques:

a. Large-sample inferences about the difference $(\mu_1 - \mu_2)$ between population means using a two-sample z statistic

b. Small-sample inferences about $(\mu_1 - \mu_2)$ using an independent samples design and a two-sample t statistic

c. Small-sample inferences about $(\mu_1 - \mu_2)$ using a paired difference design and a single-sample t statistic to analyze the differences

d. Large-sample inferences about the difference $(p_1 - p_2)$ between binomial proportions using a two-sample z statistic

APPLYING THE CONCEPTS

9.110   A pupillometer is a device used to observe changes in an individual's pupil dilations as he or she is exposed to different visual stimuli. Since there is a direct correlation between the amount an individual's pupil dilates and his or her interest in the stimuli, marketing organizations sometimes use pupillometers to help them evaluate potential consumer interest in new products, alternative package designs, and other factors. The Design and Market Research Laboratories of the Container Corporation of America used a pupillometer to evaluate consumer reaction to different silverware patterns for one of its clients. Suppose 15 consumers were chosen at random, and each was shown two different silverware patterns. The pupillometer readings for each consumer, with the means and standard deviations of each sample of observations and their differences, are shown in the tables (in millimeters).

CONSUMER	PATTERN 1	PATTERN 2	CONSUMER	PATTERN 1	PATTERN 2
1	1.00	.80	9	.98	.91
2	.97	.66	10	1.46	1.10
3	1.45	1.22	11	1.85	1.60
4	1.21	1.00	12	.33	.21
5	.77	.81	13	1.77	1.50
6	1.32	1.11	14	.85	.65
7	1.81	1.30	15	.15	.05
8	.91	.32			

	MEAN	STANDARD DEVIATION
Pattern 1	1.12	.50
Pattern 2	.88	.45
Difference (1 − 2)	.24	.16

a. Which type of experiment does this represent—independent samples or paired difference? Explain.
b. Use a 90% confidence interval to estimate the difference in mean pupil dilation per consumer for silverware patterns 1 and 2. Interpret the confidence interval, assuming that the pupillometer indeed measures consumer interest.
c. Test the hypothesis that the mean dilation differs for the two patterns. Use $\alpha = .10$. Does the test conclusion support your interpretation of the confidence interval in part b?
d. What assumptions are necessary to assure the validity of the inferences in parts b and c?

9.111   To compare the rate of return an investor can expect on tax-free municipal bonds with the rate of return on taxable bonds, an investment advisory firm randomly samples ten bonds of each type and computes the annual rate of return over the past 3 years for each bond. The rate of return is then adjusted for taxes, assuming the investor is in a 30% tax bracket. The means and standard deviations for the adjusted returns are shown in the table.

TAX-FREE BONDS	TAXABLE BONDS
$\bar{x}_1 = 9.8\%$	$\bar{x}_2 = 9.3\%$
$s_1 = 1.1\%$	$s_2 = 1.0\%$

a. Test to see whether there is a difference in the mean rates of return between tax-free and taxable bonds for investors in the 30% tax bracket. Use $\alpha = .05$.
b. What assumptions were necessary for the validity of the testing procedure you used in part a?

9.112 Refer to Exercise 9.111.
a. Test the hypothesis that the two population variances are equal. Use $\alpha = .10$.
b. Form a 90% confidence interval for σ_1^2/σ_2^2.

9.113 Management training programs are often instituted in order to teach supervisory skills and thereby increase productivity. Suppose a company psychologist administers a set of examinations to each of ten supervisors before such a training program begins and then administers similar examinations at the end of the program. The examinations are designed to measure supervisory skills, with higher scores indicating increased skill. The results of the tests are shown in the table. Test to see whether the data indicate that the training program is effective. Use $\alpha = .10$.

SUPERVISOR	BEFORE TRAINING PROGRAM	AFTER TRAINING PROGRAM
1	63	78
2	93	92
3	84	91
4	72	80
5	65	69
6	72	85
7	91	99
8	84	82
9	71	81
10	80	87

9.114 Lack of motivation is a problem of many students in inner-city schools. To cope with this problem, an experiment was conducted to determine whether motivation could be improved by allowing students greater choice in the structures of their curricula. Two schools with similar student populations were chosen, and 50 students were randomly selected from each to participate in the experiment. School A permitted its 50 students to choose only the courses they wanted to take. School B permitted its students to choose their courses and also to choose when and from which instructors to take the courses. The measure of student motivation was the number of times each student was absent from or late for a class during a 20-day period. The means and variances for the two samples are shown in the table. Do the data provide sufficient evidence to indicate that students from school B were late or absent less frequently than those from school A? (Use $\alpha = .1$.)

SCHOOL A	SCHOOL B
$\bar{x}_A = 20.5$	$\bar{x}_B = 19.6$
$s_A^2 = 26.2$	$s_B^2 = 24.1$

9.115 Two banks, bank 1 and bank 2, each independently sampled 40 and 50 of their business accounts, respectively, and determined the number of the bank's services (loans, checking, savings, investment counseling, etc.) each sampled business was using. Both banks offer the same services. A summary of the data supplied by the samples is given in the table. Do the samples yield sufficient evidence to conclude that the average number of services used by bank 1's business customers is significantly greater (at the $\alpha = .10$ level) than the average number of services used by bank 2's business customers?

BANK 1	BANK 2
$\bar{x}_1 = 2.2$	$\bar{x}_2 = 1.8$
$s_1 = 1.15$	$s_2 = 1.10$

9.116 Find a 99% confidence interval for $(\mu_1 - \mu_2)$ in Exercise 9.115. Does the interval include 0? Interpret the confidence interval.

9.117   Was the average amount spent by firms in the electronics industry on company-sponsored research and development (R&D) higher in 1989 than it was in 1988? The table lists R&D expenditures (in millions of dollars) for a sample of firms in the electronics industry.

FIRM	1988	1989	FIRM	1988	1989
Adams-Russell	6.3	5.1	Compudyn	2.3	1.8
Harris	116.9	104.0	Raytheon	271.0	274.7
Aydin	8.5	6.6	Varian Associates	80.2	83.1
Andrew	14.1	17.0	General Instr.	37.5	46.9

Source: *Business Week*, Special Issue on Innovation in America, 1989 and 1990.

   a. A securities analyst who follows the electronics industry believes R&D expenditures have increased. Do the data support the analyst's beliefs? Test using $\alpha = .10$.
   b. In the context of this exercise, what are the Type I and Type II errors associated with the hypothesis test of part **a**?
   c. What assumptions must hold in order for your test of part **a** to be valid?

9.118   Refer to Exercise 9.117. Use a 95% confidence interval to estimate the mean difference between 1985 and 1984 R&D expenditures. Interpret the interval.

9.119   Radio stations sometimes conduct prize giveaways in an attempt to increase their share of the listening audience. Suppose a station manager calls 300 randomly selected households in a city and finds that 65 have members who regularly listen to the station. The station then conducts a 2-month promotional contest and follows it with a survey of 500 randomly chosen households. The survey shows that 154 households have members who regularly listen to the station.
   a. Use a 90% confidence interval to estimate the difference between the proportions of those who regularly listen to the station before and after the promotional contest.
   b. Construct a 95% confidence interval for the proportion of those who listen to the station after the promotion is over.

9.120   A consumer protection agency wants to compare the work of two electrical contractors in order to evaluate their safety records. The agency plans to inspect residences in which each of these contractors has done the wiring in order to estimate the difference in the proportions of residences that are electrically deficient. Suppose the proportions of residences with deficient work are expected to be about .10 for both contractors. How many homes should be sampled in order to estimate the difference in proportions using a 90% confidence interval of width .05?

9.121   The interocular pressure of glaucoma patients is often reduced by treatment with adrenaline. To compare a new synthetic drug with adrenaline, seven glaucoma patients were treated with both drugs, one eye with adrenaline and one with the synthetic drug. The reduction in pressure in each eye was then recorded, as shown in the table. Do the data provide sufficient evidence to indicate a difference in the mean reductions in eye pressure for the two drugs? Test using $\alpha = .10$.

PATIENT	ADRENALINE	SYNTHETIC
1	3.5	3.2
2	2.6	2.8
3	3.0	3.1
4	1.9	2.4
5	2.9	2.9
6	2.4	2.2
7	2.0	2.2

9.122   Some power plants are located near rivers or oceans so that the available water can be used for cooling the condensers. As part of an environmental impact study, suppose a power company wants to estimate the difference in mean water temperature between the discharge of its plant and the offshore waters. How many sample measurements must be taken at each site in order to estimate the true difference between means to within .2°C with 95% confidence? Assume that the range in readings will be about 4°C at each site and the same number of readings will be taken at each site.

9.123   The use of preservatives by food processors has become a controversial issue. Suppose two preservatives are extensively tested and determined safe for use in meats. A processor wants to compare the preservatives for their effects on retarding spoilage. Suppose 15 cuts of fresh meat are treated with preservative A and 15 with B, and the number of hours until spoilage begins is recorded for each of the 30 cuts of meat. The results are summarized in the table.

PRESERVATIVE A	PRESERVATIVE B
$\bar{x}_1 = 106.4$ hours	$\bar{x}_2 = 96.5$ hours
$s_1 = 10.3$ hours	$s_2 = 13.4$ hours

a. Is there evidence of a difference in mean time until spoilage begins between the two preservatives at the $\alpha = .05$ level?

b. Can you recommend an experimental design that the processor could have used to reduce the variability in the data?

9.124   Refer to Exercise 9.123.

a. Construct a 95% confidence interval for the difference between the mean times until spoilage for the two preservatives.

b. Form a 95% confidence interval for σ_1^2/σ_2^2.

9.125   A physiologist wishes to study the effect of birth-control pills on exercise capacity. Five female subjects who have never taken the pill have their maximal oxygen uptake measured (in milliliters per kilogram of their body weight) during a treadmill session. The five subjects then take the pill for a specified length of time and their uptakes are measured again, as given in the table. Do the data provide sufficient evidence to indicate that the mean maximal oxygen uptake after taking birth-control pills is less than the mean uptake before taking the pill? Use $\alpha = .01$.

SUBJECT	MAXIMAL OXYGEN UPTAKE	
	Before	After
1	35.0	29.5
2	36.5	33.5
3	36.0	32.0
4	39.0	36.5
5	37.5	35.0

9.126   An automobile manufacturer wants to estimate the difference in the mean miles per gallon rating for two models it produces. If the range of ratings is expected to be about 6 miles per gallon for each model, how many cars of each model must be tested in order to estimate the difference in mean rating with a 90% confidence interval with a width of 1 mile per gallon?

9.127   Since tourism is one of the largest industries in the state of Florida, the economy of the state depends heavily on the number of tourists who visit Florida annually and on the mean amount of money tourists spend while they are in the state. Suppose a study is conducted during two consecutive years, say 1990 and 1991, to compare the mean expenditure of tourists in Florida. Random samples of 325 and 375 tourists (a family is treated as one tourist) are selected in 1990 and 1991, respectively, and the total expenditure in the state is recorded for each. The results are given in the table.

1990	1991
$\bar{x}_1 = 927$	$\bar{x}_2 = 1{,}053$
$s_1 = 850$	$s_2 = 959$
$n_1 = 325$	$n_2 = 375$

a. Form a 90% confidence interval for the difference between the mean expenditures per tourist in 1990 and 1991. Interpret the interval in terms of this problem.

b. What assumptions are necessary for the validity of the procedure in part a?

9.128 A large shipment of produce contains Valencia and navel oranges. To determine whether there is a difference in the proportions of nonmarketable fruit between the two varieties, random samples of 850 Valencia oranges and 1,500 navel oranges are independently selected and the number of non-marketable oranges of each type is counted. It is found that 30 Valencia and 90 navel oranges from these samples are nonmarketable. Do these data provide sufficient evidence to indicate a difference between the proportions of nonmarketable Valencia and navel oranges? Test at the $\alpha = .05$ level of significance.

9.129 In 1986, the American Society for Quality Control commissioned the Gallup Organization to explore the attitudes, beliefs, and experiences of upper level executives in U.S. businesses with respect to the quality and quality practices related to their companies' products and services. The following is one of the questions asked of the executives: "Poor quality—as measured by repair, rework and scrap costs, lost sales, and so on—is said to cost American business billions of dollars annually. How much does poor quality cost your company, as a percent of gross sales?" The table describes the responses (rounded to the nearest percent) of 387 service-company executives and 311 industrial-company executives.

	SERVICE COMPANIES	INDUSTRIAL COMPANIES
Less than 5%	45%	47%
5–10%	24	23
11–19%	5	12
20–29%	3	6
30–49%	2	2
50% or more	1	0
Don't know	20	10

Source: "Gallup survey: Top executives talk quality," *Quality Progress*, December 1986, pp. 49–54.

a. Use a 90% confidence interval to estimate the proportion of all executives in U.S. service companies who believe poor quality costs their firms 10% or less of gross sales.

b. Use a 90% confidence interval to estimate the difference between the proportion of all executives in service companies and the proportion of all executives in industrial companies who believe poor quality costs their companies 10% or less of gross sales.

c. What assumptions must hold in order for your confidence intervals of parts a and b to be valid?

9.130 Refer to Exercise 9.129. Do the data provide sufficient evidence to indicate that a difference exists between the proportion of service-company executives and the proportion of industrial-company executives who do not know how much poor quality costs their companies? Test using $\alpha = .10$.

9.131   When new instruments are developed to perform chemical analyses of products (food, medicine, etc.), they are usually evaluated with respect to two criteria: accuracy and precision. *Accuracy* refers to the ability of the instrument to identify correctly the nature and amounts of a product's components. *Precision* refers to the consistency with which the instrument will identify the components of the same material. Thus, a large variability in the identification of a single batch of a product indicates a lack of precision. Suppose a pharmaceutical firm is considering two brands of an instrument designed to identify the components of certain drugs. As part of a comparison of precision, ten test-tube samples of a well-mixed batch of a drug are selected and then five are analyzed by instrument A and five by instrument B. The data shown in the table are the percentages of the primary component of the drug given by the instruments. Do these data provide evidence of a difference in the precision of the two machines? Use $\alpha = .10$.

INSTRUMENT A	INSTRUMENT B
43	46
48	49
37	43
52	41
45	48

9.132   Two basketball players engage in a foul-shooting contest in which each player takes 100 shots. It is desired to compare the percentages of shots made by each player.
   a. What is the parameter of interest?
   b. In this contest, player A made 93 shots and player B made 86 shots. Estimate the parameter of interest using a 95% confidence interval.

9.133   A political candidate conducted a sample survey to determine whether a television advertising campaign was worthwhile. Both before and after the advertising campaign, random samples were taken from the candidate's constituency, and each person was asked his or her voter preference. The results of the survey are shown in the table. Using a 95% confidence interval, estimate the difference in the proportion of voters who favor the candidate before and after the campaign.

	SAMPLE SIZE	NUMBER WHO PREFER THE CANDIDATE
Before advertising campaign	200	85
After advertising campaign	300	139

9.134   Refer to Exercise 9.133. What size samples need to be taken in order to estimate the difference between proportions to within .04 with probability .95? Assume that $n_1 = n_2$.

9.135   The federal government is interested in determining whether salary discrimination exists between men and women in the private sector. Suppose random samples of 15 women and 22 men are drawn from the population of first-level managers in the private sector. The information is summarized in the table.

WOMEN	MEN
$\bar{x}_1 = \$28,400$	$\bar{x}_2 = \$30,300$
$s_1 = \$4,600$	$s_2 = \$5,206$
$n_1 = 15$	$n_2 = 22$

    a. Do these data provide sufficient evidence to indicate that the mean salary of male first-level managers exceeds the mean salary of females in that position? Use $\alpha = .10$.

    b. What assumptions are necessary for the validity of the test used in part **a**?

9.136   Refer to Exercise 9.135. Conduct a test to determine whether the assumption of equal salary variances is reasonable. Use $\alpha = .10$.

9.137   The state of Florida now requires all high school students to pass a literacy test before they receive a high school diploma. A student who fails the test can enroll in a refresher course and retake the test at a later date. To evaluate the effectiveness of the refresher course, eight students' test scores were compared, before and after, with the results shown in the table. Do the data provide sufficient evidence to conclude that the mean test score has increased? (Use $\alpha = .05$.)

STUDENT	BEFORE	AFTER
1	45	49
2	52	50
3	63	70
4	68	71
5	57	53
6	55	61
7	60	62
8	59	67

9.138   A number of computer programs are available to conduct two-sample tests of hypotheses to compare the means of two populations, for both independent and paired samples. Most of these report both the test statistic and the observed significance level for the test, but some report only the observed significance level of the test. Suppose you use one of these programs to test the null hypothesis H_0: $(\mu_1 - \mu_2) = 0$ against H_a: $(\mu_1 - \mu_2) \neq 0$ with independent samples of size 12 and 10, respectively. Assuming you are using $\alpha = .05$, what conclusion would you reach in each of the following instances of an observed significance level reported by the program?

    a. P-VALUE = .0429   **b.** P-VALUE = .1984

    c. P-VALUE = .0001   **d.** P-VALUE = .0344

    e. P-VALUE = .0545   **f.** P-VALUE = .9633

    g. You should always be sure that the program is performing the calculations correctly, especially programs with which you do not have much experience. Even programs that perform the calculations correctly usually do not remind you of the assumptions necessary for the validity of the procedure. What assumptions are necessary for this test?

9.139   Some power plants are located near rivers or oceans so the water can be used for cooling their condensers. As part of an environmental impact study, suppose a power company wants to estimate the mean difference in water temperature between the discharge of its plant and the offshore waters. How many sample measurements must be taken at each site to obtain a 95% confidence interval of width .4°C? Assume the range in readings will be about 4°C at each site and the same number of readings will be taken at each site.

9.140   A drug currently being used to reduce the heart rate of patients before surgery works very well in some patients while having little effect in others. Researchers have developed a new drug that they hope will produce more consistent results. To test the new drug, 24 dogs were randomly chosen and divided into two groups of 12 dogs each. One group of dogs was injected with the new drug, while the other group was injected with the old drug. The reduction in heart rate for each dog was recorded, with the results summarized in the table.

OLD DRUG	NEW DRUG
$n_1 = 12$	$n_2 = 12$
$s_1^2 = 14.3$	$s_2^2 = 8.2$

a. Do the data provide sufficient evidence to conclude that the variation in the heart rate reductions is less for the new drug than the old? Use $\alpha = .05$.

b. Find the approximate observed significance level for the test and interpret its value.

ON YOUR OWN...

We have now discussed two methods of collecting data to compare two population means. In many experimental situations a decision must be made either to collect two independent samples or to conduct a paired difference experiment. The importance of this decision cannot be overemphasized, since the amount of information obtained and the cost of the experiment are both directly related to the method of experimentation that is chosen.

Choose two populations (pertinent to your major area) that have unknown means and for which you could both collect two independent samples and collect paired observations. Before conducting the experiment, state which method of sampling you think will provide more information (and why). To compare the two methods, first perform the independent sampling procedure by collecting ten observations from each population (a total of 20 measurements), and then perform the paired difference experiment by collecting ten pairs of observations.

Construct two 95% confidence intervals, one for each experiment you conduct. Which method provides the shorter confidence interval and thus more information on this performance of the experiment? Does this result agree with your preliminary expectations?

USING THE COMPUTER...

Select two of the census regions given in Appendix C, and consider the sports purchasing index. Suppose a marketing firm wants to target one of the two regions for a sports magazine marketing campaign.

a. Treat the sports index measurements for the zip codes in the regions as a random sample of all the zip codes for the regions. Test the null hypothesis that the populations' mean sports purchasing indexes are equal using $\alpha = .01$, and place a 99% confidence interval on the true difference between the mean purchasing index for the two regions.

b. Repeat part a using the following pairs of α and confidence levels for the tests and confidence intervals: (.05, 95%), (.10, 90%), and (.20, 80%). Describe what happens to the tests and confidence intervals as α is increased and the confidence level is decreased. Which do you think is more informative—the tests or the confidence intervals?

References

Benson, P. G., and Ohta, H. "Classifying sensory inspectors with heterogeneous inspection-error probabilities," *Journal of Quality Technology*, Vol. 18, No. 2, April 1986, pp. 79–90.

Covert, C. "Television viewers snapping up Home Shopping Network," *Minneapolis Star and Tribune*, December 16, 1986, p. 1C.

Freedman, D., Pisani, R., and Purves, R. *Statistics*. New York: W. W. Norton and Co., 1978.

Mendenhall, W. *Introduction to Probability and Statistics*, 7th ed. Boston: Duxbury, 1987. Chapters 8 and 9.

Sinclair, M. A. "A collection of modelling approaches for visual inspection in industry," *International Journal of Production Research*, Vol. 16, No. 4, 1978, pp. 275–292.

Snedecor, G. W., and Cochran, W. *Statistical Methods*, 7th ed. Ames: Iowa State University Press, 1980.

Analysis of Variance: Comparing More Than Two Means

CONTENTS

WHERE WE'VE BEEN...

As we have seen in preceding chapters, the solutions of many practical problems are based on inferences about population means. Methods for estimating and testing hypotheses about a single mean and the comparison of two means were presented in Chapters 7–9.

WHERE WE'RE GOING...

In this chapter we extend the methodology of Chapters 7–9 in two important ways. First, we discuss the critical elements in the *design* of a sampling experiment. Then we show how to *analyze* the experiment in order to compare more than two populations. We will present several of the more popular experimental designs, and will show how to use the computer to analyze designed experiments using both analysis of variance and regression programs.

Most of the data we have analyzed in previous chapters were collected in **observational** sampling experiments rather than **designed** sampling experiments. In observational experiments the analyst has little or no control over the variables under study and merely observes their values. In contrast, designed experiments are those in which the analyst attempts to control the levels of one or more variables to determine their effect on a variable of interest. Although many practical situations do not present the opportunity for such control, it is instructive, even for observational experiments, to have a working knowledge of the analysis and interpretation of data that result from designed experiments and to know the basics of how to design experiments when the opportunity arises.

We first present the basic elements of an experimental design in Section 10.1. We then discuss two of the simpler, and more popular, experimental designs in Sections 10.2 and 10.3. Slightly more complex experiments are discussed in Section 10.4.

10.1   Elements of a Designed Experiment

Certain elements are common to almost all designed experiments, regardless of the specfic area of application. For example, the *response* is the variable of interest in the experiment. The response might be the SAT score of a high school senior, the total sales of a firm last year, or the total income of a particular household this year. We will also refer to the response as the *dependent variable.*

> **Definition 10.1**
>
> The **response** is the variable of interest to be measured in the experiment. We also refer to the response as the **dependent variable**.

The intent of most statistical experiments is to determine the effect of one or more variables on the response. These variables are usually referred to as the *factors* in a designed experiment. Factors are either *quantitative* or *qualitative*, depending on whether the variable is measured on a numerical scale or not. For example, we might want to explore the effect of the qualitative factor Gender on the response SAT score. In other words, we want to compare the SAT scores of male and female high-school seniors. Or, we might wish to determine the effect of the quantitative factor Number of salespeople on the response Total sales for retail firms. Often two or more factors are of interest. For example, we might want to determine the effect of the quantitative factor Number of wage earners and the qualitative factor Location on the response Household income.

> **Definition 10.2**
>
> **Factors** are those variables whose effect on the response is of interest to the experimenter. **Quantitative** factors are measured on a numerical scale, whereas **qualitative** factors are those that are not (naturally) measured on a numerical scale.

Levels are the values of the factors that are utilized in the experiment. The levels of qualitative factors are usually nonnumerical. For example, the levels of Gender are Male and Female, and the levels of Location might be North, East, South, and West.* The levels of quantitative factors are the numerical values of the variable utilized in the experiment. The Number of salespeople for each of a set of companies, the Number of wage earners in each of a set of households, and the High-school GPAs for a set of high school seniors all represent levels of the respective quantitative factors.

Definition 10.3

The **levels** of a factor are the values of the factor utilized in the experiment.

When a *single factor* is employed in an experiment, the *treatments* of the experiment are the levels of the factor. For example, if the effect of the factor Gender on the response SAT score is being investigated, the treatments of the experiment are the two levels of Gender—Female and Male. Or, if the effect of the Number of wage earners on Household income is the subject of the experiment, then the numerical values assumed by the quantitative factor Number of wage earners are the treatments. If *two or more factors* are utilized in an experiment, the *treatments* are the factor-level combinations used. For example, if the effects of the factors Gender and GPA on the response SAT score are being investigated, then the treatments are the combinations of the levels of Gender and GPA used; (Female, 2.61), (Male, 3.43), and (Female, 3.82) are all treatments if those particular factor-level combinations are utilized in the experiment.

Definition 10.4

The **treatments** of an experiment are the factor-level combinations utilized.

The objects on which the response variable and factors are observed are the *experimental units*. For example, SAT score, High school GPA, and Gender are all variables that can be observed on the same experimental unit—a high school senior. Or, the Total sales, the Earnings per share, and the Number of salespeople can be measured on a particular firm in a particular year, and the firm–year combination is the experimental unit. The Total income, the Number of female wage-earners, and the Location can be observed for a household at a particular point in time, and the household–time combination is the experimental unit. Every experiment, whether observational or designed, has experimental units on which the variables are observed. However, the identification of the experimental

---

*The levels of a qualitative variable may bear numerical labels. For example, the Locations could be numbered 1, 2, 3, and 4. However, in such cases the numerical labels for a qualitative variable will usually be codes representing nonnumerical levels.

units is more important in designed experiments, when the experimenter must actually sample the experimental units and measure the variables.

Definition 10.5

An **experimental unit** is the object on which the response and factors are observed or measured.*

When the specification of the treatments and the method of assigning the experimental units to each of the treatments is controlled by the analyst, the experiment is said to be *designed*. In contrast, if the analyst is just an observer of the treatments on a sample of experimental units, the experiment is *observational*. For example, if you specify the number of Female and Male high school students within each GPA range to be randomly selected in order to evaluate the effect of Gender and GPA on SAT scores, then you are designing the experiment. If, on the other hand, you simply observe the SAT scores, Gender, and GPA for all students who took the SAT test last month at a particular high school, the experiment is observational.

Definition 10.6

A **designed experiment** is one for which the analyst controls the specification of the treatments and the method of assigning the experimental units to each treatment. An **observational experiment** is one for which the analyst simply observes the treatments and the response on a sample of experimental units.

The diagram in Figure 10.1 provides an overview of the experimental process and a summary of the terminology we have introduced in this section. Note that the experimental unit is at the core of the process. The method by which the sample of experimental units is selected from the population determines the type of experiment. The level of every factor (the treatment) and the response are all variables that are observed or measured on each experimental unit.

| E X A M P L E 10.1

The USGA (United States Golf Association) regularly tests golf equipment to assure that it conforms to USGA standards. Suppose it wishes to compare the mean distance traveled by four different brands of golf balls when struck by a

---

*Recall (Chapter 1) that the set of all experimental units is the population.

FIGURE 10.1

Sampling experiments: process and terminology

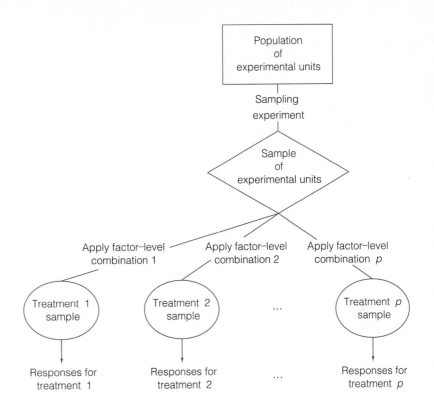

driver (the club with least loft used to maximize distance). The following experiment is conducted: Ten balls of each brand are randomly selected. Each is struck by "Iron Byron" (the USGA's golf robot named for the famous golfer, Byron Nelson) using a driver, and the distance traveled is recorded. Identify each of the following elements in this experiment: response, factors, factor types, levels, treatments, and experimental units.

Solution

The response is the variable of interest, Distance traveled. The only factor being investigated is Brand of golf ball, and it is nonnumerical and therefore qualitative. The four brands (say A, B, C, and D) represent the levels of this factor. Since only one factor is utilized, the treatments are the four levels of this factor—that is, the four brands. The experimental unit is a golf ball; more specifically, it is a golf ball at a particular position in the striking sequence, since the distance traveled can be recorded only when the ball is struck, and we would expect the distance to be different (due to random factors such as wind resistance, landing place, and so forth) if the same ball were struck a second time. Note that ten experimental units are sampled for each treatment, generating a total of 40 observations.

This experiment, like many real applications, is a blend of designed and observational: The analyst cannot control the assignment of the brand to each golf ball (observational), but he or she can control the assignment of each ball to the position in the striking sequence (designed).

EXAMPLE 10.2

Refer to Example 10.1. Suppose the USGA is also interested in comparing the mean distances the four brands of golf balls travel when struck by a five-iron. Ten balls of each brand are randomly selected, five to be struck by the driver, and five by the five-iron. Identify the elements of the experiment, and construct a schematic diagram similar to Figure 10.1 to provide an overview of this experiment.

Solution

The response is unchanged—Distance traveled. The experiment now has two factors, Brand of golf ball and Club utilized. There are four levels of Brand (A, B, C, and D) and two of Club (driver and five-iron, or 1 and 5). Treatments are factor-level combinations, so there are $4 \times 2 = 8$ treatments in this experiment: (A, 1), (A, 5), (B, 1), (B, 5), (C, 1), (C, 5), (D, 1), and (D, 5). The experimental units are still the combinations of golf ball and hitting position. Note that five experimental units are sampled per treatment, generating 40 observations. The experiment is summarized in Figure 10.2.

FIGURE 10.2

Two-factor golf experiment summary: Example 10.2

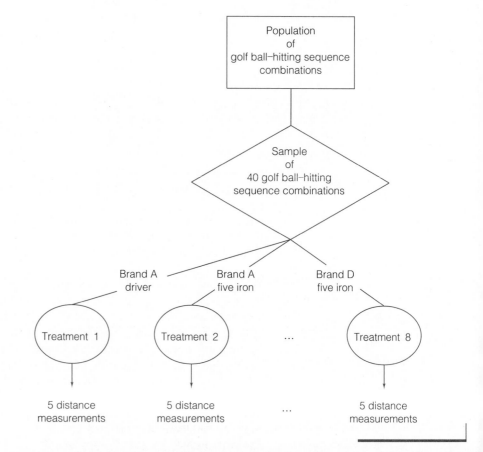

Our objective in designing an experiment is usually to maximize the amount of information obtained about the relationship between the treatments and the response. Of course, we are almost always subject to constraints on budget, time, and even the availability of experimental units. Nevertheless, designed experiments are generally preferred to observational experiments. Not only do we have better control of the amount and quality of the information collected, but also observational experiments are subject to biases in the selection of the experimental units representing each treatment. Inferences based on observational experiments always carry the implicit assumption that the sample has no hidden bias that was not considered in the statistical analysis. Better understanding of the potential problems with observational experiments is a by-product of our study of experimental design in the remainder of this chapter.

EXERCISES 10.1–10.8

LEARNING THE MECHANICS

10.1   What are the treatments for a designed experiment that utilizes one qualitative factor with four levels—A, B, C, and D?

10.2   What are the treatments for a designed experiment with two factors, one qualitative with two levels (A and B) and one quantitative with five levels (50, 60, 70, 80, and 90)?

10.3   What are the experimental units on which each of the following responses are observed:
a. College GPA
b. Statewide unemployment rate in December
c. Gasoline mileage rating for a model of automobile
d. Number of defective sectors on a computer diskette

10.4   What is the difference between an observational and a designed experiment?

APPLYING THE CONCEPTS

10.5   Suppose the mean cholesterol levels are to be compared for adults in four socioeconomic classes—poverty, low income, middle income, and high income. If random samples of each socioeconomic class are selected to make the comparison, identify each of the following elements of the experiment:
a. Response
b. Factor(s) and factor type(s)
c. Treatments
d. Experimental units

10.6   R. H. Brockhaus (1980) compared the risk-taking propensity of three types of managers—entrepreneurs, newly hired managers, and newly promoted managers. Samples of individuals in each group were administered a test that measures one's propensity for taking risks. Identify each of the following elements:
a. Response
b. Factor(s) and factor type(s)
c. Treatments
d. Experimental units

10.7   A quality control supervisor measures the quality of a steel ingot on a scale from 0 to 10. He designs an experiment in which three different temperatures (ranging from 1100°F to 1200°F) and five different pressures (ranging from 500 psi to 600 psi) are utilized, with 20 ingots produced at each Temperature–Pressure combination. Identify the following elements of the experiment:
a. Response
b. Factor(s) and factor type(s)
c. Treatments
d. Experimental units

10.8   Brief descriptions of a number of experiments are given next. Determine whether each is observational or designed, and explain your reasoning.

a. An economist obtains the unemployment rate and gross state product for a sample of states over the past 10 years, with the objective of examining the relationship between the unemployment rate and the gross state product by census region.

b. A psychologist tests the effects of three different feedback programs by randomly assigning five rats to each program and recording their response times at specified intervals during the program.

c. A marketer of microcomputers runs ads in each of four national publications for one quarter, and keeps track of the number of sales that are attributable to each publication's ad.

d. An electric utility engages a consultant to monitor the discharge from its smokestack on a monthly basis over a 1-year period in order to relate the level of sulfur dioxide in the discharge to the load on the facility's generators.

e. Intrastate trucking rates are compared before and after governmental deregulation of prices charged, with the comparison also taking into account distance of haul, goods hauled, and the price of diesel fuel.

f. An agriculture student compares the amount of rainfall in four different states over the past 5 years.

10.2 The Completely Randomized Design

The simplest experimental design, a *completely randomized design*, consists of the *independent random selection* of experimental units representing each treatment. For example, we could independently select random samples of 20 female and 15 male high school seniors to compare their mean SAT scores. Or, we could independently select random samples of 30 households from each of four census districts to compare the mean income per household among the districts. In both examples our objective is to compare treatment means by selecting random, independent samples for each treatment.

Definition 10.7

A **completely randomized design** is a design for which independent random samples of experimental units are selected for each treatment.*

The objective of a completely randomized design is usually to compare the treatment means. If we denote the true, or population, means of the p treatments as $\mu_1, \mu_2, \ldots, \mu_p$, then we will test the null hypothesis that the treatment means are all equal against the alternative that at least two of the treatment means differ:

H_0: $\mu_1 = \mu_2 = \cdots = \mu_p$

H_a: At least two of the p treatment means differ

The μ's might represent the means of *all* female and male high school seniors' SAT scores or the means of *all* households' income in each of four census regions.

---

*We use *completely randomized design* to refer to both designed and observational experiments. Thus, the only requirement is that the experimental units to which treatments are applied (designed) or on which treatments are observed (observational) are independently selected for each treatment.

To conduct a statistical test of these hypotheses, we will use the means of the independent random samples selected from the treatment populations using the completely randomized design. That is, we compare the p sample means $\bar{x}_1$, $\bar{x}_2$, ..., $\bar{x}_p$.

For example, suppose you select independent random samples of five female and five male high school seniors and obtain sample mean SAT scores of 550 and 590, respectively. Can we conclude that males score 40 points higher, on average, than females? To answer this question, we must consider the amount of sampling variability among the experimental units (students). If the scores are as depicted in the dot diagram shown in Figure 10.3, then the difference between the means is small relative to the sampling variability of the scores within the treatments, Female and Male. We would be inclined to not reject the null hypothesis of equal population means in this case.

FIGURE 10.3

Dot diagram of SAT scores: Difference between means dominated by sampling variability

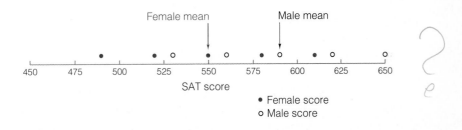

In contrast, if the data are as depicted in the dot diagram of Figure 10.4, then the sampling variability is small relative to the difference between the two means. We would be inclined to favor the alternative hypothesis that the population means differ in this case.

FIGURE 10.4

Dot diagram of SAT scores: Difference between means large relative to sampling variability

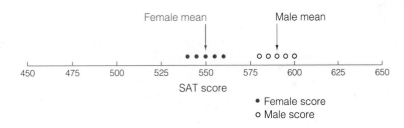

You can see that the key is to compare the difference between the treatment means to the amount of sampling variability. To conduct a formal statistical test of the hypotheses requires numerical measures of the difference between the treatment means and the sampling variability within each treatment. The variation between the treatment means is measured by the **Sum of Squares for Treatments (SST)**, which is calculated by squaring the distance between each treatment mean and the overall mean of *all* sample measurements, multiplying each squared distance by the number of sample measurements for the treatment, and adding the results over all treatments:

$$\text{SST} = \sum_{i=1}^{p} n_i(\bar{x}_i - \bar{x})^2 = 5(550 - 570)^2 + 5(590 - 570)^2$$

$$= 4,000$$

where we use $\bar{x}$ to represent the overall mean response of all sample measurements, i.e., the mean of the combined samples. The symbol n_i is used to denote the sample size for the ith treatment. You can see that the value of SST is 4,000 for the two samples of five female and five male SAT scores depicted in Figures 10.3 and 10.4.

Next, we must measure the sampling variability within the treatments. We call this the **Sum of Squares for Error** (SSE) because it measures the variability around the treatment means that is attributed to sampling error. Suppose the ten measurements in the first dot diagram (Figure 10.3) are 490, 520, 550, 580, and 610 for females and 530, 560, 590, 620, and 650 for males. Then the value of SSE is computed by summing the squared distance between each response measurement and the corresponding treatment mean, and then adding the squared differences over all measurements in the entire sample:

$$SSE = \sum_{j=1}^{n_1} (x_{1j} - \bar{x}_1)^2 + \sum_{j=1}^{n_2} (x_{2j} - \bar{x}_2)^2 + \cdots + \sum_{j=1}^{n_p} (x_{pj} - \bar{x}_p)^2$$

where the symbol x_{1j} is the jth measurement in sample 1, x_{2j} is the jth measurement in sample 2, and so on. This rather complex-looking formula is not difficult to interpret in practice. For our samples of SAT scores,

$$
\begin{aligned}
SSE &= [(490 - 550)^2 + (520 - 550)^2 + (550 - 550)^2 + (580 - 550)^2 + (610 - 550)^2] \\
&\quad + [(530 - 590)^2 + (560 - 590)^2 + (590 - 590)^2 + (620 - 590)^2 + (650 - 590)^2] \\
&= 18,000
\end{aligned}
$$

To make the two measures of variability comparable, we divide each by the degrees of freedom to convert the sums of squares to mean squares. First, the **Mean Square for Treatments** (MST), which measures the variability *among* the treatment means, is equal to

$$MST = \frac{SST}{p - 1} = \frac{4,000}{2 - 1} = 4,000$$

where the number of degrees of freedom for the p treatments is $(p - 1)$. Next, the **Mean Square for Error** (MSE), which measures the sampling variability *within* the treatments, is

$$MSE = \frac{SSE}{n - p} = \frac{18,000}{10 - 2} = 2,250$$

Finally, we calculate the ratio of MST to MSE, an F statistic:

$$F = \frac{MST}{MSE} = \frac{4,000}{2,250} = 1.78$$

Values of the F statistic near 1 indicate that the two sources of variation, between treatment means and within treatments, are approximately equal. In this case, the difference between the treatment means may well be attributable to sampling error, which provides little support for the alternative hypothesis that the population treatment means differ. Values of F well in excess of 1 indicate that the variation among treatment means well exceeds that within means and therefore support the alternative hypothesis that the population treatment means differ.

When does F exceed 1 by enough to reject the null hypothesis that the means are equal? This depends on the degrees of freedom for treatments and for error, and on the value of α selected for the test. We compare the calculated F value to a table F value (Tables VIII–XI of Appendix A) with $\nu_1 = (p - 1)$ degrees of freedom in the numerator and $\nu_2 = (n - p)$ degrees of freedom in the denominator and corresponding to a Type I error probability of α. For the SAT score example, the F statistic has $\nu_1 = (2 - 1) = 1$ numerator degree of freedom and $\nu_2 = (10 - 2) = 8$ denominator degrees of freedom. Thus, for $\alpha = .05$ we find (Table IX of Appendix A):

$$F_{.05} = 5.32$$

The implication is that MST would have to be 5.32 times greater than MSE before we could conclude at the .05 level of significance that the two population treatment means differ. Since the data yielded $F = 1.78$, our initial impressions for the dot diagram in Figure 10.3 are confirmed—there is insufficient information to conclude that the mean SAT scores differ for the populations of female and male high school seniors. The rejection region and the calculated F value are shown in Figure 10.5.

FIGURE 10.5

Rejection region and calculated F values for SAT score samples

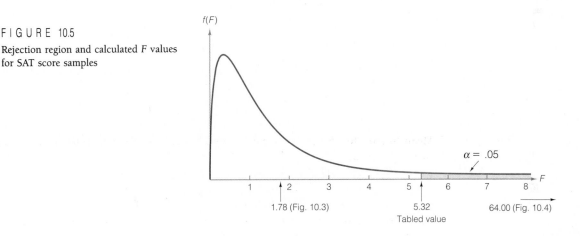

In contrast, consider the dot diagram in Figure 10.4. Since the means are the same as in the first example, 550 and 590, respectively, the variation between the means is the same, MST = 4,000. But the variation within the two treatments appears to be considerably smaller. The observed SAT scores are 540, 545, 550, 555, and 560 for females and 580, 585, 590, 595, and 600 for males. The variation within the treatments is measured by

$$\begin{aligned} \text{SSE} = &[(540 - 550)^2 + (545 - 550)^2 + (550 - 550)^2 + (555 - 550)^2 + (560 - 550)^2] \\ &+ [(580 - 590)^2 + (585 - 590)^2 + (590 - 590)^2 + (595 - 590)^2 + (600 - 590)^2] \\ = &500 \end{aligned}$$

$$\text{MSE} = \frac{\text{SSE}}{n - p} = \frac{500}{8} = 62.5$$

Then the F-ratio is

$$F = \frac{\text{MST}}{\text{MSE}} = \frac{4,000}{62.5} = 64.0$$

Again, our visual analysis of the dot diagram is confirmed statistically: $F = 64.0$ well exceeds the tabled F value, 5.32, corresponding to the .05 level of significance. We would therefore reject the null hypothesis at that level and conclude that the SAT mean score of Males differs from that of Females.

Recall that we performed a hypothesis test for the difference between two means in Section 9.2 using a two-sample t statistic for two independent samples. When two independent samples are being compared, the t- and F-tests are equivalent. To see this, recall the formula

$$t = \frac{\bar{x}_1 - \bar{x}_2}{\sqrt{s_p^2 \left(\frac{1}{n_1} + \frac{1}{n_2} \right)}} = \frac{590 - 550}{\sqrt{(62.5) \left(\frac{1}{5} + \frac{1}{5} \right)}} = \frac{40}{5} = 8$$

where we used the fact that $s_p^2 = \text{MSE}$, which you can verify by comparing the formulas. Note that the calculated F for these samples ($F = 64$) equals the square of the calculated t for the same samples ($t = 8$). Likewise, the tabled F value (5.32) equals the square of the tabled t value at the two-sided .05 level of significance ($t_{.025} = 2.306$ with 8 df). Since both the rejection region and the calculated values are related in the same way, the tests are equivalent. Moreover, the assumptions that must be met to assure the validity of the t- and F-tests are the same:

1. The probability distributions of the populations of responses associated with each treatment must all be normal.
2. The probability distributions of the populations of responses associated with each treatment must have equal variances.
3. The samples of experimental units selected for the treatments must be random and independent.

In fact, the only real difference between the tests is that the F-test can be used to compare *more than two* treatment means, whereas the t-test is applicable to two samples only. The F-test is summarized in the box.

Test to Compare p Treatment Means for a Completely Randomized Design

H_0: $\mu_1 = \mu_2 = \cdots = \mu_p$

H_a: At least two treatment means differ

Test statistic: $F = \dfrac{\text{MST}}{\text{MSE}}$

Assumptions: 1. All p population probability distributions are normal.
2. The p population variances are equal.
3. Samples are selected randomly and independently from the respective populations.

Rejection region: $F > F_\alpha$, where F_α is based on $(p - 1)$ numerator degrees of freedom (associated with MST) and $(n - p)$ denominator degrees of freedom (associated with MSE).

Computational formulas for MST and MSE are given in Appendix B. We will rely on some of the many computer programs available to compute the F statistic, concentrating on the interpretation of the results rather than their calculations.

EXAMPLE 10.3

Suppose the United States Golf Association (USGA) wants to compare the mean distances associated with four different brands of golf balls when struck with a driver. A completely randomized design is employed, with Iron Byron, the USGA's robotic golfer, using a driver to hit a random sample of ten balls of each brand in a random sequence. The distance is recorded for each hit, and the results are shown in Table 10.1, organized by brand.

TABLE 10.1 **Results of completely randomized design: Iron Byron driver**

	BRAND A	BRAND B	BRAND C	BRAND D
	251.2	263.2	269.7	251.6
	245.1	262.9	263.2	248.6
	248.0	265.0	277.5	249.4
	251.1	254.5	267.4	242.0
	265.5	264.3	270.5	246.5
	250.0	257.0	265.5	251.3
	253.9	262.8	270.7	262.8
	244.6	264.4	272.9	249.0
	254.6	260.6	275.6	247.1
	248.8	255.9	266.5	245.9
Means	251.3	261.1	270.0	249.4

a. Set up the test to compare the mean distances for the four brands. Use $\alpha = .10$.

b. Use the SAS System Analysis of Variance program to obtain the test statistic. Interpret the results.

Solution

a. To compare the mean distances of the four brands, we first specify the hypotheses to be tested. Denoting the population mean of the ith brand by μ_i, we test

H_0: $\mu_1 = \mu_2 = \mu_3 = \mu_4$

H_a: The mean distances differ for at least two of the brands

The test statistic compares the variation among the four treatment (Brand) means to the sampling variability within each of the treatments.

Test statistic: $F = \dfrac{\text{MST}}{\text{MSE}}$

Rejection region: $F > F_\alpha = F_{.10}$
with $\nu_1 = (p - 1) = 3$ df and $\nu_2 = (n - p) = 36$ df

From Table VIII of Appendix A, we find $F_{.10} \approx 2.25$ for 3 and 36 df. Thus, we will reject H_0 if $F > 2.25$.

The assumptions necessary to assure the validity of the test are as follows: (1) The probability distributions of the distances for each brand are normal. (2) The variances of the distance probability distributions for each brand are equal. (3) The samples of ten golf balls for each brand are selected randomly and independently.

b. The SAS printout for the data in Table 10.1 resulting from this completely randomized design is given in Figure 10.6. The Total Sum of Squares is designated the "Corrected Total," and it is partitioned into the "Model" and "Error" Sums of Squares. The bottom part of the printout further partitions the "Model" component into the factors that comprise the model. In this single-factor experiment, the Model and Brand sums of squares are the same. The Sum of Squares column is headed "Type I SS," one of four types of sums of squares that SAS will calculate. The distinction becomes important only in multifactor experiments with unequal numbers of observations per treatment; we will need to utilize only the Type I sum of squares. See the references at the end of this chapter for a more complete discussion of using SAS to analyze more complex experiments.

FIGURE 10.6

SAS analysis of variance printout for golf ball distance data: Completely randomized design

Dependent Variable: DISTANCE

Source	DF	Sum of Squares	Mean Square	F Value	Pr > F
Model	3	2709.19875	903.06625	35.81	0.0001
Error	36	907.86100	25.21836		
Corrected Total	39	3617.05975			

R-Square	C.V.	Root MSE	DISTANCE Mean
0.749006	1.9469768	5.02179	257.927500

Source	DF	Type I SS Mean Square		F Value	Pr > F
BRAND	3	2709.20	903.07	35.81	0.0001

The values of the mean squares, MST and MSE (shaded on the printout), are 903.07 and 25.22, respectively. The *F*-ratio, 35.81, also shaded on the printout, exceeds the tabled value of 2.25. We therefore reject the null hypothesis at the .10 level of significance, concluding that at least two of the brands differ with respect to mean distance traveled when struck by the driver.

The observed significance level of the *F*-test is also given on the printout: .0001. This is the area to the right of the calculated *F* value and it implies that we would reject the null hypothesis that the means are equal at any α level greater than .0001. The result of the test is summarized in Figure 10.7.

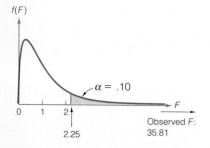

FIGURE 10.7

F-test for completely randomized design: Golf ball experiment

The results of an analysis of variance (ANOVA) can be summarized in a simple tabular format similar to that obtained from the SAS program in Example 10.3. The general form of the table is shown in Table 10.2, where the symbols df, SS,

TABLE 10.2 ANOVA Summary Table for a Completely Randomized Design

SOURCE	df	SS	MS	F
Treatments	$p - 1$	SST	$\text{MST} = \dfrac{\text{SST}}{p - 1}$	$\dfrac{\text{MST}}{\text{MSE}}$
Error	$n - p$	SSE	$\text{MSE} = \dfrac{\text{SSE}}{n - p}$	
Total	$n - 1$	SS(Total)		

and MS stand for degrees of freedom, Sum of Squares, and Mean Square, respectively. Note that the two sources of variation, Treatments and Error, add to the Total Sum of Squares, SS(Total). The ANOVA summary table for Example 10.3 is given in Table 10.3, and the partitioning of the Total Sum of Squares into its two components is illustrated in Figure 10.8.

TABLE 10.3 ANOVA Summary Table for Example 10.3

SOURCE	df	SS	MS	F
Brands	3	2,709.20	903.07	35.81
Error	36	907.86	25.22	
Total	39	3,617.06		

FIGURE 10.8

Partitioning of the total sum of squares for the completely randomized design

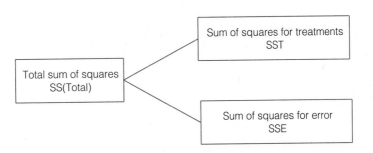

Suppose the F-test results in a rejection of the null hypothesis that the treatment means are equal. Is the analysis complete? Usually, the conclusion that at least two of the treatment means differ leads to another question. Which of the means differ, and by how much? For example, the F-test in Example 10.3 leads to the conclusion that at least two of the brands of golf balls have different mean distances traveled when struck with a driver. Now the question is, which of the brands differ? How are the brands ranked with respect to mean distance?

We can place confidence intervals on the difference between the various pairs of treatment means in the experiment. If there are p treatment means, then there are $c = p(p - 1)/2$ pairs of means that can be compared. If we want to have $100(1 - \alpha)\%$ confidence that each of the c confidence intervals contains the true difference it is intended to estimate, then each individual confidence interval will have to be formed using a smaller value of α than would a single interval. For example, suppose we want to compare the $4(3)/2 = 6$ pairs of golf ball brand means and we want 95% confidence that all six confidence intervals comparing the means contain the true differences between the brand means. Then each individual confidence interval will need to be constructed using a smaller level of significance than .05 in order to have 95% confidence that the six intervals collectively include the true differences.*

There are a number of procedures available for making **multiple comparisons** of a set of treatment means which, under various assumptions, assure that the overall confidence level associated with all the comparisons remains at or above the specified $100(1 - \alpha)\%$ level. Perhaps the simplest of these to apply is the **Bonferroni procedure**, which calls for specifying the level of significance of each comparison at α/c, where α is the overall level of significance desired and c is the number of pairs of means to be compared. The result is a set of confidence intervals in which we can be *at least* $100(1 - \alpha)\%$ confident. The Bonferroni procedure is conservative since the confidence level is generally greater than what we specify. Of course, this means that the intervals are somewhat wider than they need to be in order to have the specified confidence level. Procedures that are more exact are also more complex than the Bonferroni procedure. The Bonferroni procedure is summarized in the box. Other multiple comparison procedures can be found in the references at the end of the chapter.

Bonferroni Procedure for Multiple Comparisons:
Completely Randomized Design

1. Specify the overall level of significance α or, equivalently, the overall confidence level $100(1 - \alpha)\%$.

2. If there are c pairs of treatment means to be compared, specify the level of significance for each comparison at α/c.

3. Calculate the $100(1 - \alpha/c)\%$ confidence interval for each of the c pairs of means using the following formula:

$$(\bar{x}_i - \bar{x}_j) \pm t_{\alpha/(2c)} s \sqrt{\left(\frac{1}{n_i} + \frac{1}{n_j}\right)}$$

where $s = \sqrt{\text{MSE}}$ and $t_{\alpha/(2c)}$ is the table value of t (Table VI of Appendix A) that locates area $\alpha/(2c)$ in the upper tail of the t-distribution with $(n - p)$ degrees of freedom (the number of degrees of freedom associated with error in the ANOVA).

4. Summarize the results of the multiple comparisons by ranking the treatment means, showing which pairs are significantly different.

Assumptions: Same as for the ANOVA.

---

*The reason each interval must be formed at a higher confidence level than that specified for the collection of intervals can be demonstrated as follows:

$P\{$At least one of c intervals fails to contain the true difference$\}$

$= 1 - P\{$All c intervals contain the true differences$\}$

$\geq 1 - (1 - \alpha)^c \geq \alpha$

Thus, to make this probability of at least one failure equal to α, we must specify the individual levels of significance to be less than α.

EXAMPLE 10.4

Refer to Example 10.3, in which we concluded that at least two of the four brands of golf balls are associated with different mean distances traveled when struck with a driver. Use the Bonferroni procedure to compare the $4(3)/2 = 6$ pairs of treatment means. Use an overall confidence level of 90%.

Solution

The level of significance specified for the multiple comparisons is $\alpha = .10$. The Bonferroni procedure specifies the level of significance for each comparison at $\alpha/c = .10/6 = .0167$. The t statistic to be used in the formation of the six confidence intervals splits this area into two tails, with $\alpha/(2c) = .0167/2 = .0083$ in each tail. The degrees of freedom associated with error are $(n - p) = (40 - 4) = 36$, which is more than the maximum (29) contained in the t-table (Table VI of Appendix A). We will use the standard normal z-table (Table IV of Appendix A) to find $z_{.0083} \approx 2.4$. That is, as shown in Figure 10.9, we will form six confidence intervals using plus and minus 2.4 standard deviations in order to have 90% confidence in the overall results of all six intervals. Contrast this with the fact that we would be using plus and minus $z = 1.645$ standard deviations if we were forming a single 90% confidence interval.*

FIGURE 10.9

Confidence level for Bonferroni multiple comparisons: Example 10.4

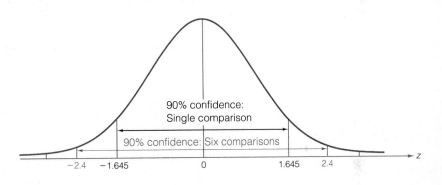

We are now ready to form the confidence intervals. After calculating $s = \sqrt{\text{MSE}} = \sqrt{25.22} = 5.02$ (also given in the SAS printout shown in Figure 10.6 as "Root MSE"), we begin with a comparison of brands A and B:

$$(\bar{x}_A - \bar{x}_B) \pm t_{.0083} s \sqrt{\left(\frac{1}{10} + \frac{1}{10}\right)}$$

$$(251.3 - 261.1) \pm (2.4)(5.02)(.447)$$

$$-9.8 \pm 5.4 \quad \text{or} \quad (-15.2, -4.4)$$

Thus, we conclude that the brand B mean distance exceeds the brand A mean distance by between 4.4 and 15.2 yards. The remainder of the intervals are easier to calculate. The interval half-widths are all the same (5.4), since the sample

---

*If the t-table is used, we may have to round $\alpha/(2c)$ *down* to enable the use of Table VI. In Example 10.4, we would approximate $\alpha/(2c) = .0083 \approx .005$. This would provide slightly wider intervals than necessary, but assures an overall confidence level of at least $100(1 - \alpha)\%$. Alternatively, many computer programs will provide a more exact tabled t value corresponding to a given level of significance and a specified number of degrees of freedom.

sizes are the same for each brand. The rest of the comparisons are summarized as follows:

$$(A - C): \quad -18.7 \pm 5.4 \quad \text{or} \quad (-24.1, -13.3)$$
$$(A - D): \quad 1.9 \pm 5.4 \quad \text{or} \quad (-3.5, 7.3)$$
$$(B - C): \quad -8.9 \pm 5.4 \quad \text{or} \quad (-14.3, -3.5)$$
$$(B - D): \quad 11.7 \pm 5.4 \quad \text{or} \quad (6.3, 17.1)$$
$$(C - D): \quad 20.6 \pm 5.4 \quad \text{or} \quad (15.2, 26.0)$$

We are 90% confident that the intervals collectively contain all the differences between the true brand mean distances. Note that intervals that contain 0, such as the (Brand A − Brand D) interval from −3.5 to 7.3, do not support a conclusion that the true brand mean distances differ. If both endpoints of the interval are positive, as with the (Brand B − Brand D) interval, the implication is that the first brand (B) mean distance exceeds the second (D). Conversely, if both end-points of the interval are negative, as with the (Brand A − Brand C) interval, the implication is that the second brand C mean distance exceeds the first brand A mean distance.

A convenient summary of the results of the Bonferroni multiple comparisons is a listing of the brand means from highest to lowest, with a solid line connecting those that are *not* significantly different. This summary is shown in Figure 10.10. The interpretation is that brand C's mean distance exceeds all others, brand B's mean exceeds that of brands A and D; and the means of brands A and D do not differ significantly. All these inferences are made with 90% confidence, the overall confidence level of the Bonferroni multiple comparisons.

FIGURE 10.10

Summary of Bonferroni multiple comparisons

BRAND	MEAN
C	270.0
B	261.1
A	251.3
D	249.4

Since the samples were selected independently in a completely randomized design, we can also form a confidence interval for an individual treatment mean. We use the one-sample t confidence interval of Section 7.3 to form the interval, again using the standard deviation, $s = \sqrt{\text{MSE}}$, as the measure of sampling variability for the experiment. For example, if we wish to place a 90% confidence interval on the mean distance traveled by brand C (apparently the "longest ball" of those tested), we calculate

$$\bar{x}_C \pm t_{.05} s \sqrt{\frac{1}{10}}$$
$$270.0 \pm (1.645)(5.02)(.32)$$
$$270.0 \pm 2.6 \quad \text{or} \quad (267.4, 272.6)$$

Thus, we are 90% confident that the true mean distance traveled for brand C is between 267.4 and 272.6 yards, when hit with a driver by Iron Byron.

The procedure for conducting an analysis of variance for a completely randomized design is summarized in the accompanying box. Remember that the hallmark of this design is independent random samples of experimental units associated with each treatment. We discuss a design with dependent samples in the next section.

Steps for Conducting an ANOVA for a Completely Randomized Design

1. Be sure the design is truly completely randomized, with independent random samples for each treatment.

2. Create an ANOVA summary table that specifies the variability attributable to treatments and error, and that leads to the calculation of the F statistic for testing the null hypothesis that the treatment means are equal in the population. Use either an ANOVA computer program or the calculation formulas in Appendix B to obtain the necessary numerical ingredients.

3. If the F-test leads to the conclusion that the means differ:
   a. Conduct multiple comparisons of as many of the pairs of means as you wish to compare by using the Bonferroni (or some other) multiple comparisons procedure. Use the results to summarize the statistically significant differences among the treatment means.
   b. If desired, form confidence intervals for one or more individual treatment means.

4. If the F-test leads to the nonrejection of the null hypothesis that the treatment means differ, several possibilities exist:
   a. The treatment means are equal—that is, the null hypothesis is true.
   b. The treatment means really differ, but other important factors affecting the response are not accounted for by the completely randomized design. These factors inflate the sampling variability, as measured by MSE, resulting in smaller values of the F statistic. Either increase the sample size for each treatment, or use a different experimental design (as in Section 10.3) that accounts for the other factors affecting the response.

Be careful not to reach conclusion **a** automatically, since the possibility of a Type II error must be considered if you accept H_0.

CASE STUDY 10.1

COMPARING THE STRENGTHS OF WOMEN ATHLETES

Muscle strength and endurance are two important aspects of neuromuscular performance.

This quote from Vivian Heyward and Leslie McCreary appeared in the introduction to an article in which they discussed the strength and endurance of women athletes.

A completely randomized design was used to compare the maximal grip strength of women athletes who engaged in eight different sports. The means and standard deviations for these data are shown in Table 10.4. The eight sports are the treatments, and the sample sizes for each sport are shown in parentheses.

The ANOVA table given in the article appears in Table 10.5. As can be seen, the response (maximal strength) is referred to as the *dependent variable* and the sources of variation, which we have called *treatment* and *error*, are referred to as *between sports* and *within sports*, respectively. (This terminology is often used in connection with completely randomized designs.) The F statistic is equal to 3.15; it is significant at the .05 level of significance, since the tabulated value of $F_{.05}$ corresponding to $\nu_1 = 7$ numerator and $\nu_2 = 42$ denominator degrees of freedom is $F_{.05} \approx 2.25$. Thus, there is sufficient evidence to indicate that there is a difference in mean maximal grip strength among the eight sports.

According to Table 10.4, the largest (strongest) mean grip strengths were associated with women in tennis and golf, whereas the smallest was associated with women in swimming. Heyward and McCreary attribute this result to the extensive use of forearm and hand muscles in tennis and golf. Statistical comparison of specific pairs of means could be made by using a multiple comparison procedure such as the Bonferroni procedure.

T A B L E 10.4 **Means and Standard Deviations of Neuromuscular Performance Data for Each Sport ($n = 50$)**

SPORT	MAXIMAL STRENGTH (KILOGRAMS)	
	$\bar{x}$	s
Basketball ($n_1 = 7$)	40.68	5.96
Field hockey ($n_2 = 7$)	39.16	4.84
Golf ($n_3 = 6$)	42.74	4.62
Gymnastics ($n_4 = 5$)	39.14	2.80
Swimming ($n_5 = 6$)	33.38	4.63
Tennis ($n_6 = 5$)	45.44	5.68
Track and field ($n_7 = 8$)	37.26	3.69
Volleyball ($n_8 = 6$)	38.37	5.24
Totals ($n = 50$)	39.29	5.51

Source: Heyward, V., and McCreary, L. "Analysis of the static strength and relative endurance of women athletes," *Research Quarterly*, 1977, 48, pp. 703–709.

T A B L E 10.5 **Summary of ANOVA for Neuromuscular Data ($n = 50$)**

SOURCE OF VARIATION	df	SS	MS	F
Dependent variable, maximal strength				
Between sports (treatments)	7	511.50	73.07	3.15
Within sports (error)	42	974.88	23.21	

In concluding a discussion of the analysis of variance for a completely randomized design, we should comment on the assumptions that are required for its validity, i.e., the assumptions of equal population variances and normally distributed populations. Moderate departures from normality do not have much effect on the significance levels of tests or on the confidence coefficients associated with confidence intervals. Departures from the assumption of equal population variances can affect those measures of reliability, but the effect is less when the sample sizes are equal. When in doubt, use a nonparametric statistical method such as the Kruskal–Wallis H-test of Section 11.4.

> **What Do You Do When the Assumptions Are Not Satisfied for the Analysis of Variance for a Completely Randomized Design?**
>
> *Answer:* Use a nonparametric statistical method such as the Kruskal–Wallis H-test of Section 11.4.

EXERCISES 10.9–10.28

LEARNING THE MECHANICS

10.9 Use Tables VIII, IX, X, and XI of Appendix A to find each of the following F values:
   a. $F_{.05}$, $\nu_1 = 2$, $\nu_2 = 2$ **b.** $F_{.01}$, $\nu_1 = 2$, $\nu_2 = 2$
   c. $F_{.10}$, $\nu_1 = 20$, $\nu_2 = 40$ **d.** $F_{.025}$, $\nu_1 = 12$, $\nu_2 = 9$

10.10 Find the following probabilities:
   a. $P(F \leq 2.88)$ for $\nu_1 = 20$, $\nu_2 = 21$
   b. $P(F > 3.52)$ for $\nu_1 = 15$, $\nu_2 = 15$
   c. $P(F > 2.40)$ for $\nu_1 = 15$, $\nu_2 = 15$
   d. $P(F \leq 1.69)$ for $\nu_1 = 40$, $\nu_2 = 40$

10.11 In which dot diagram is the difference between the sample means small relative to the variability within the sample observations? Justify your answer.

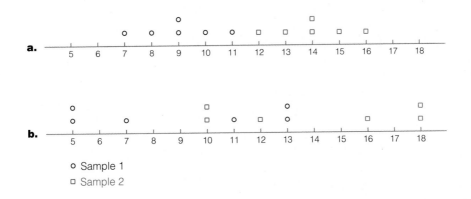

o Sample 1
□ Sample 2

10.12 Refer to Exercise 10.11. Assume that the two samples represent independent, random samples corresponding to two treatments in a completely randomized design.
   a. Calculate the treatment means, i.e., the means of samples 1 and 2, for both dot diagrams.
   b. Use the means to calculate the Sum of Squares for Treatments (SST) for each dot diagram.

c. Calculate the sum of squared differences between each sample measurement and the corresponding sample mean, and sum the squared differences over the two samples to obtain the Sum of Squares for Error (SSE) for each dot diagram.

d. Calculate the Total Sum of Squares [SS(Total)] for the two dot diagrams by adding the Sums of Squares for Treatment and Error. What percentage of SS(Total) is accounted for by the treatments—that is, what percentage of the Total Sum of Squares is the Sum of Squares for Treatment—in each case?

e. Convert the Sums of Squares for Treatment and Error to mean squares by dividing each by the appropriate number of degrees of freedom. Calculate the F-ratio of the Mean Square for Treatment (MST) to the Mean Square for Error (MSE) for each dot diagram.

f. Use the F-ratios to test the null hypothesis that the two samples are drawn from populations with equal means. Use $\alpha = .05$.

g. What assumptions must be made about the probability distributions corresponding to the responses for each treatment in order to assure the validity of the F-tests conducted in part f?

10.13   Refer to Exercises 10.11 and 10.12. Conduct a two-sample t-test (Section 10.2) of the null hypothesis that the two treatment means are equal for each dot diagram. Use $\alpha = .05$ and two-tailed tests. In the course of the test, compare each of the following with the F-tests in Exercise 10.12:

a. The pooled variances and the MSEs

b. The t- and the F-test statistics

c. The tabled values of t and F that determine the rejection regions

d. The conclusions of the t- and F-tests

e. The assumptions that must be made in order to assure the validity of the t- and F-tests

10.14   Refer to Exercises 10.11 and 10.12. Complete the following ANOVA table for each of the two dot diagrams:

SOURCE	df	SS	MS	F
Treatments				
Error				
Total				

10.15   Suppose the Total Sum of Squares for a completely randomized design with $p = 6$ treatments and $n = 36$ total measurements (six per treatment) is equal to 500. In each of the following cases, conduct an F-test of the null hypothesis that the six treatment means are the same. Use $\alpha = .10$.

a. Sum of Squares for Treatment (SST) is 20% of SS(Total)

b. SST is 50% of SS(Total)

c. SST is 80% of SS(Total)

d. What happens to the F-ratio as the percentage of the Total Sum of Squares attributable to treatments is increased?

10.16   A partially completed ANOVA table for a completely randomized design is shown here:

SOURCE	df	SS	MS	F
Treatments	6	16.9		
Error				
Total	41	45.2		

a. Complete the ANOVA table.

b. How many treatments are involved in the experiment?

c. Do the data provide sufficient evidence to indicate a difference among the population means? Test using $\alpha = .10$.

d. Find the approximate observed significance level for the test in part c, and interpret it.

e. Suppose that $\bar{x}_1 = 3.7$ and $\bar{x}_2 = 4.1$. Do the data provide sufficient evidence to indicate a difference between μ_1 and μ_2? Assume that there are seven observations for each treatment. Test using $\alpha = .10$.

f. Refer to part e. Find a 90% confidence interval for $(\mu_1 - \mu_2)$.

g. Refer to part e. Find a 90% confidence interval for μ_1.

10.17 The Minitab printout for an experiment utilizing a completely randomized design is shown here.

Printout for Exercise 10.17

ANALYSIS OF VARIANCE				
SOURCE	DF	SS	MS	F
FACTOR	3	57258	19086	14.80
ERROR	34	43836	1289	
TOTAL	37	101094		

Note: Minitab uses "FACTOR" instead of "Treatments."

a. How many treatments are involved in the experiment? What is the total sample size?

b. Conduct a test of the null hypothesis that the treatment means are equal. Use $\alpha = .01$.

c. What additional information is needed in order to be able to compare specific pairs of treatment means?

10.18 Refer to Exercise 10.17. Suppose the treatment means are 190.8, 260.1, 191.7, and 279.4, with sample sizes 8, 12, 10, and 8, respectively. Use the Bonferroni technique to compare all pairs of treatment means with an overall level of significance of $\alpha = .10$.

10.19 The data in the accompanying table resulted from an experiment that utilized a completely randomized design.

TREATMENT 1	TREATMENT 2	TREATMENT 3
3.8	5.4	1.3
1.2	2.0	.7
4.1	4.8	2.2
5.5	3.8	
2.3		

a. Use the appropriate calculation formulas in Appendix B to complete the following ANOVA table:

SOURCE	df	SS	MS	F
Treatments				
Error				
Total				

b. Test the null hypothesis that $\mu_1 = \mu_2 = \mu_3$, where μ_i represents the true mean for treatment i, against the alternative that at least two of the means differ. Use $\alpha = .01$.

c. Use the Bonferroni technique to compare all pairs of means. Use an $\alpha = .05$ overall level of significance.

APPLYING THE CONCEPTS

10.20   Studies conducted at the University of Melbourne (Australia) indicate that there may be a difference between the pain thresholds of blonds and brunettes (*Family Weekly, Gainesville Sun*, Gainesville, Florida, February 5, 1978). Men and women of various ages were divided into four categories according to hair color: light blond, dark blond, light brunette, and dark brunette. The purpose of the experiment was to determine whether hair color is related to the amount of pain produced by common types of mishaps and assorted types of trauma. Each person in the experiment was given a pain threshold score based on his or her performance in a pain sensitivity test (the higher the score, the higher the person's pain tolerance). The results are given in the table.

HAIR COLOR			
Light blond	Dark blond	Light brunette	Dark brunette
62	63	42	32
60	57	50	39
71	52	41	51
55	41	37	30
48	43		35

a. Based on the given information, what type of experimental design appears to have been employed?

b. SAS was used to conduct the analysis of variance calculations, resulting in the printout shown here. Conduct a test to determine whether the mean pain thresholds differ among people possessing the four types of hair color. Use $\alpha = .05$.

Printout for Exercise 10.20

Dependent Variable: PAIN

Source	DF	Sum of Squares	Mean Square	F Value	Pr > F
Model	3	1360.7263158	453.5754386	6.79	0.0041
Error	15	1001.8000000	66.7866667		
Corrected Total	18	2362.5263158			

R-Square	C.V.	Root MSE	PAIN Mean
0.575962	17.081838	8.1723110	47.84210526

Source	DF	Type I SS	Mean Square	F Value	Pr > F
COLOR	3	1360.726316	453.575439	6.79	0.0041

c. What is the observed significance level for the test in part b? Interpret it.

d. Use the Bonferroni technique to compare all pairs of means. Use $\alpha = .05$ as the overall level of significance.

e. What assumptions must be met in order to assure the validity of the inferences you made in parts b and d?

10.21   A researcher at St. Louis University conducted a study to determine whether entrepreneurs, newly hired managers, and newly promoted managers differ in their risk-taking propensities. For the purpose of this study, entrepreneurs were defined as individuals who, within 3 months prior to the study, had ceased working for their employers in order to manage their own business ventures. Thirty-one individuals of each type were selected to participate in the study. Each was asked to complete a questionnaire which required the respondent to choose between a safe alternative and a more attractive but risky one. Test scores were designed to measure risk-taking propensity. (Lower scores are associated with greater conservatism in risk-taking situations.) Summary statistics for the test scores of the three groups are given in the table.

GROUP	SAMPLE SIZE	SAMPLE MEAN	STANDARD DEVIATION	GROUP TOTALS
Entrepreneurs	31	71.00	11.94	2,201
Newly hired managers	31	72.52	12.19	2,248
Promoted managers	31	66.97	10.84	2,076
	93			6,525

Source: Brockhaus, R. H. "Risk-taking propensity of entrepreneurs," *Academy of Management Journal*, Vol. 23, September 1980, pp. 509–520.

a. The following partial ANOVA table is calculated from the data. Complete the table.

SOURCE	df	SS	MS	F
Treatments		509.87		
Error		12,259.96		
Total				

b. Do the data provide sufficient evidence to indicate differences in the mean risk-taking propensities among the three groups? Test using $\alpha = .05$.

c. What assumptions must be satisfied in order for the test of part **b** to be valid?

d. Would you advise conducting tests to compare the individual pairs of means? Explain.

e. Would you classify this experiment as observational or designed? Explain.

10.22 Some varieties of nematodes (roundworms that live in the soil and are frequently so small they are invisible to the naked eye) feed on the roots of lawn grasses and crops such as strawberries and tomatoes. This pest, which is particularly troublesome in warm climates, can be treated by the application of nematocides. However, because of the size of the worms, it is very difficult to measure the effectiveness of these pesticides directly. To compare four nematocides, the yields of equal-size plots of one variety of tomatoes were collected. The data (yields in pounds per plot) are shown in the table.

NEMATOCIDE			
1	2	3	4
18.6	18.7	19.4	19.0
18.2	19.3	19.9	18.5
17.6	18.9	19.7	18.6
		19.1	

a. Given that SST = 3.45609 and SSE = 1.20083, construct an ANOVA table for this experiment.

b. Do the data provide sufficient evidence to conclude there is a difference in mean yields of tomatoes per plot for the four nematocides? Test using $\alpha = .05$.

c. Use the Bonferroni technique for comparing the pairs of means, using an overall significance level of .05.

10.23   A research psychologist wishes to investigate the difference in maze test scores for a strain of laboratory mice trained under different laboratory conditions. The experiment is conducted using 18 randomly selected mice of this strain, with six receiving no training at all (control group), six trained under condition 1, and six trained under condition 2. Then each of the mice is given a test score between 0 and 100, depending on its performance in a test maze. The experiment produced the results in the table.

CONTROL	CONDITION 1	CONDITION 2
58	73	53
32	70	74
59	68	72
64	71	62
55	60	58
49	62	61

a. Identify the following elements of this experiment: experimental units, response, factor(s) and factor type(s), and treatments.
b. Is this a designed or observational experiment?
c. Is there sufficient evidence to indicate a difference among mean maze test scores for mice trained under the three different laboratory conditions? Use Appendix B or a computer program to perform the calculations, and test at $\alpha = .10$.
d. Use the Bonferroni technique to compare the control mean to each of the condition means. How many comparisons are involved? Use an overall significance level of $\alpha = .10$.

10.24   A company that employs a large number of salespeople is interested in learning which of the salespeople sell the most: those strictly on commission, those with a fixed salary, or those with a reduced fixed salary plus a commission. The previous month's records for a sample of salespeople are inspected and the amount of sales (in dollars) is recorded for each, as shown in the table.

COMMISSIONED	FIXED SALARY	COMMISSION PLUS SALARY
$425	$420	$430
507	448	492
450	437	470
483	432	501
466	444	
492		

a. Construct a dot diagram for these data. Indicate the location of each sample mean.
b. The Minitab ANOVA printout for these data is shown here. Do the data provide sufficient evidence to indicate that the mean sales differ among the three types of compensation?

Printout for Exercise 10.24

```
ANALYSIS OF VARIANCE
SOURCE     DF      SS        MS       F
FACTOR      2     4195      2098     3.17
ERROR      12     7945       662
TOTAL      14    12140
```

c. Use a 90% confidence interval to estimate the mean sales for salespeople who receive a commission plus salary.

10.25   How does flextime, which allows workers to set their individual work schedules, affect worker job satisfaction? Researchers recently conducted a study to compare a measure of job satisfaction for workers using three types of work scheduling: flextime, staggered starting hours, and fixed hours. Workers in each group worked according to their specified work scheduling system for 4 months. Although each worker filled out job satisfaction questionnaires both before and after the 4-month test period, we will examine only the post-test-period scores. The sample sizes, means, and standard deviations of the scores for the three groups are shown in the table.

| | GROUP | | |
	Flextime	Staggered	Fixed
Sample size	27	59	24
Mean	35.22	31.05	28.71
Standard deviation	10.22	7.22	9.28

a. Assume that the data were collected according to a completely randomized design. Identify the response, the factor, the factor type, the treatments, and the experimental units.
b. Use the sample means to calculate the Sum of Squares for Treatments, SST.
c. Use the standard deviations to calculate the Sum of Squares for Error, SSE. Remember that SSE is the sum of squared differences between each measurement in the experiment and the corresponding treatment mean.
d. Construct an ANOVA table for this experiment.
e. Do the data provide sufficient evidence that the three groups differ with respect to their mean job satisfaction? Test using $\alpha = .05$.
f. Use the Bonferroni technique to compare the pairs of means corresponding to the three treatments. Use $\alpha = .05$ as the overall significance level for the comparisons.

10.26   An accounting firm that specializes in auditing the financial records of large corporations is interested in evaluating the appropriateness of the fees it charges for its services. As part of its evaluation it wants to compare the costs it incurs in auditing corporations of different sizes. The accounting firm decided to measure the size of its client corporations in terms of their yearly sales. Accordingly, its population of client corporations was divided into three subpopulations:

A: Those with sales over $250 million

B: Those with sales between $100 million and $250 million

C: Those with sales under $100 million

The firm chose random samples of ten corporations from each of the subpopulations and determined the costs (in thousands of dollars) given in the table from its records.

| COSTS INCURRED IN AUDITS | | |
A	B	C
250	100	80
150	150	125
275	75	20
100	200	186
475	55	52
600	80	92
150	110	88
800	160	141
325	132	76
230	233	200

a. Construct a dot diagram for the sample data using different types of dots for each of the three samples. Indicate the location of each of the sample means. Based on the information reflected in your dot diagram, do you believe that a significant difference exists among the subpopulation means? Explain.

b. SAS was used to conduct the analysis of variance calculations, resulting in the printout shown. Conduct a test to determine whether the three classes of firms have different mean costs incurred in audits. Use $\alpha = .05$.

Printout for Exercise 10.26

General Linear Models Procedure

Dependent Variable: COST

Source	DF	Sum of Squares	Mean Square	F Value	Pr > F
Model	2	318861.667	159430.833	8.44	0.0014
Error	27	510163.000	18894.926		
Corrected Total	29	829024.667			

R-Square	C.V.	Root MSE	COST Mean
0.384623	72.220043	137.459	190.333333

Source	DF	Type I SS	Mean Square	F Value	Pr > F
TREATMNT	2	318861.67	159430.83	8.44	0.0014

c. What is the observed significance level for the test in part b? Interpret it.

d. Use the Bonferroni technique to compare all pairs of means. Use $\alpha = .05$ as the overall level of significance.

e. What assumptions must be met in order to assure the validity of the inferences you made in parts b and d?

10.27 An experiment is conducted to determine whether there is a difference among the mean increases in growth produced by five inoculins of growth hormones for plants. The experimental material consists of 20 cuttings of a shrub (all of equal weight), with four cuttings randomly assigned to each of the five different inoculins. The results of the experiment are given in the table; all measurements represent an increase in weight (in grams).

	INOCULIN			
A	B	C	D	E
15	21	22	10	6
18	13	19	14	11
9	20	24	21	15
16	17	21	13	8

a. Given that SST = 292.80 and SS(Total) = 500.55, complete an ANOVA table for this experiment.

b. Is there evidence of a difference among the mean increases in weight for the five inoculins of growth hormone? Test using $\alpha = .05$.

c. Find the approximate observed significance level for the test in part **a**, and interpret its value.

d. What assumptions are necessary to assure the validity of the inferences in parts **a** and **b**?

10.28 One of the selling points of golf balls is their durability. An independent testing laboratory is commissioned to compare the durability of three different brands of golf balls (brands A, B, and C). Balls of each type will be put into a machine that hits the balls with the same force that a golfer does on the course. The number of hits required until the outer covering is cracked is recorded for each ball. Ten balls from each manufacturer are randomly selected for testing. The results are given in the table.

BRAND		
A	B	C
310	261	233
235	219	289
279	263	301
306	247	264
237	288	273
284	197	208
259	207	245
273	221	271
219	244	298
301	228	276

a. Identify the following elements of this experiment: experimental units, response, factor(s) and factor type(s), and treatments.

b. Is the experiment designed or observational? Explain.

c. Is there evidence that the mean durabilities of the three brands differ? Use the formulas in Appendix B or a computer program to perform the calculations, and test using $\alpha = .05$.

d. If the result of the test in part **c** warrants it, use the Bonferroni technique to compare the mean durabilities for each pair of brands. Use an overall significance level of $\alpha = .10$.

10.3 The Randomized Block Design

If the completely randomized design results in nonrejection of the null hypothesis that the treatment means differ because the sampling variability (as measured by MSE) is large, we may want to consider an experimental design that better controls the variability. In contrast to the selection of independent samples of experimental units specified by the completely randomized design, the *randomized block design* utilizes experimental units that are *matched sets*, assigning one from each set to each treatment. The matched sets of experimental units are called *blocks*. The theory behind the randomized block design is that the sampling variability of the experimental units in each block will be reduced, in turn reducing the measure of error, MSE.

For example, if we wish to compare SAT scores of female and male high school seniors, we could select independent random samples of five females and five

> **Definition 10.8**
>
> The **randomized block design** consists of a two-step procedure:
>
> 1. Matched sets of experimental units, called **blocks**, are formed, each block consisting of p experimental units (where p is the number of treatments). The b blocks should consist of experimental units that are as similar as possible.
> 2. One experimental unit from each block is randomly assigned to each treatment, resulting in a total of $n = bp$ responses.

males, and analyze the results of the completely randomized design as outlined in Section 10.2. Or, we could select matched pairs of females and males according to their scholastic records, and analyze the SAT scores of the pairs. For example, we could select pairs of students with approximately the same GPAs from the same high school. Five such pairs (blocks) are depicted in Table 10.6. Note that this is just a *paired difference experiment*, first discussed in Section 9.3.

T A B L E 10.6   **Randomized Block Design: SAT Score Comparison**

BLOCK	FEMALE SAT SCORE	MALE SAT SCORE	BLOCK MEAN
1 (School A, 2.75 GPA)	540	530	535
2 (School B, 3.00 GPA)	570	550	560
3 (School C, 3.25 GPA)	590	580	585
4 (School D, 3.50 GPA)	640	620	630
5 (School E, 3.75 GPA)	690	690	690
TREATMENT MEAN	606	594	

As before, the variation between the treatment means is measured by squaring the distance between each treatment mean and the overall mean, and multiplying each squared distance by the number of measurements for the treatment, and summing over treatments:

$$\text{SST} = \sum_{i=1}^{p} b(\bar{x}_{T_i} - \bar{x})^2$$
$$= 5(606 - 600)^2 + 5(594 - 600)^2 = 360$$

where $\bar{x}_{T_i}$ represents the sample mean for the ith treatment, b (the number of blocks) is the number of measurements for each treatment, and p is the number of treatments.

The blocks also account for some of the variation among the different responses. That is, just as SST measures the variation between the female and male means, we can calculate a measure of variation among the five block means representing different schools and scholastic abilities. Analogous to the computation of SST, we sum the squares of the differences between each block mean and the overall mean, multiplying each squared difference by the number of measurements for each block, and sum over blocks to calculate the **Sum of Squares for Blocks** (SSB):

$$SSB = \sum_{i=1}^{b} p(\bar{x}_{B_i} - \bar{x})^2$$
$$= 2(535 - 600)^2 + 2(560 - 600)^2 + 2(585 - 600)^2 + 2(630 - 600)^2 + 2(690 - 600)^2$$
$$= 30,100$$

where $\bar{x}_{B_i}$ represents the sample mean for the ith block and p (the number of treatments) is the number of measurements in each block. As we expect, the variation in SAT scores attributable to Schools and Levels of scholastic achievement is apparently large.

As before, we want to compare the variability attributed to treatments with that which is attributed to sampling variability. In a randomized block design, the sampling variability is measured by subtracting that portion attributed to treatments and blocks from the Total Sum of Squares, SS(Total). The total variation is the sum of squared differences of each measurement from the overall mean:

$$SS(Total) = \sum_{i=1}^{n} (x_i - \bar{x})^2$$
$$= (540 - 600)^2 + (530 - 600)^2 + (570 - 600)^2 + (550 - 600)^2 + \cdots + (690 - 600)^2$$
$$= 30,600$$

Then the variation attributable to sampling error is found by subtraction:

$$SSE = SS(Total) - SST - SSB = 30,600 - 360 - 30,100$$
$$= 140$$

In summary, the Total Sum of Squares, 30,600, is divided into three components: 360 attributed to treatments (Gender), 30,100 attributed to blocks (Scholastic ability), and 140 attributed to sampling error.

The mean squares associated with each source of variability are obtained by dividing the sum of squares by the appropriate number of degrees of freedom. The partitioning of the Total Sum of Squares and the total degrees of freedom for a randomized block experiment is summarized in Figure 10.11 (page 482).

To determine whether we can reject the null hypothesis that the treatment means are equal in favor of the alternative that at least two of them differ, we calculate

$$MST = \frac{SST}{p - 1} = \frac{360}{2 - 1} = 360$$
$$MSE = \frac{SSE}{n - b - p + 1} = \frac{140}{10 - 5 - 2 + 1} = 35$$

The *F*-ratio that is used to test the hypothesis is

$$F = \frac{360}{35} = 10.29$$

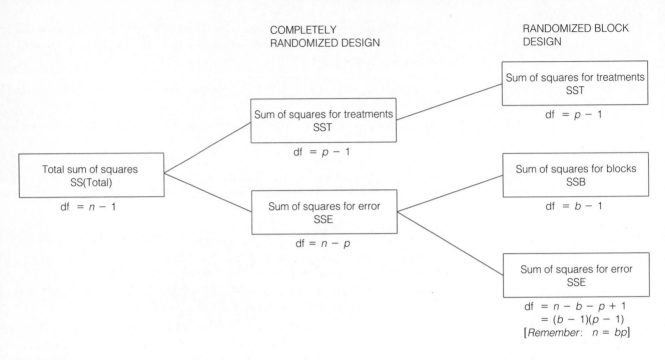

F I G U R E 10.11

Partitioning of the total sum of squares for the randomized block design

Comparing this ratio to the tabled F value corresponding to $\alpha = .05$, with $\nu_1 = (p - 1) = 1$ degree of freedom in the numerator and $\nu_2 = (n - b - p + 1) = 4$ degrees of freedom in the denominator, we find that

$$F = 10.29 > F_{.05} = 7.71$$

which indicates that we should reject the null hypothesis and conclude that the mean SAT scores differ for females and males.

If you review Section 9.3, you will find that the analysis of a paired difference experiment results in a one-sample t-test on the differences between the treatment responses within each block. Applying the procedure to the differences between female and male scores in Table 10.6, we find

$$t = \frac{\bar{x}_D}{s_D/\sqrt{n_D}} = \frac{12}{\sqrt{70}/\sqrt{5}} = 3.207$$

At the .05 level of significance with $(n_D - 1) = 4$ degrees of freedom.

$$t = 3.21 > t_{.025} = 2.776$$

Since $t^2 = (3.207)^2 = 10.29$ and $t_{.025}^2 = (2.776)^2 = 7.71$, we find that the paired difference t-test and the ANOVA F-test are equivalent, with both the calculated test statistics and the rejection region related by the formula $F = t^2$. The difference between the tests is that the paired difference t-test can be used to compare only two treatments in a randomized block design, whereas the F-test can be applied to *two or more* treatments in a randomized block design. The F-test is summarized in the box.

> ### Test to Compare p Treatment Means: Randomized Block Design
>
> H_0: $\mu_1 = \mu_2 = \cdots = \mu_p$
>
> H_a: At least two treatment means differ
>
> Test statistic: $F = \dfrac{\text{MST}}{\text{MSE}}$
>
> Rejection region: $F > F_\alpha$, where F_α is based on $(p - 1)$ numerator degrees of freedom and $(n - b - p + 1)$ denominator degrees of freedom.
>
> Assumptions: 1. The probability distributions of observations corresponding to all the block–treatment combinations are normal.
> 2. The variances of all the probability distributions are equal.

Note that the assumptions concern the probability distributions associated with each block–treatment combination. The experimental unit selected for each combination is assumed to have been randomly selected from all possible experimental units for that combination, and the response is assumed to be normally distributed with the same variance for each of the block–treatment combinations. For example, the F-test comparing female and male SAT score means requires the scores for each combination of gender and scholastic ability (e.g., females with 3.25 GPA) to be normally distributed with the same variance as the other combinations employed in the experiment.

The calculation formulas for randomized block designs are given in Appendix B. We will rely on the computer programs available to analyze randomized block designs and to obtain the necessary ingredients for testing the null hypothesis that the treatment means are equal.

EXAMPLE 10.5

Refer to Examples 10.3 and 10.4. Suppose the USGA wants to compare the mean distances associated with the four brands of golf balls when struck by a driver, but wishes to employ human golfers rather than the robot, Iron Byron. Assume that ten balls of each brand are to be utilized in the experiment.

a. Explain how a completely randomized design could be employed.
b. Explain how a randomized block design could be employed.
c. Which design is likely to provide more information about the differences among the brand mean distances?

Solution

a. Since the completely randomized design calls for independent samples, we can employ such a design by randomly selecting 40 golfers and then randomly assigning ten golfers to each of the four brands. Finally, each golfer will strike the ball of the assigned brand, and the distance will be recorded. The design is illustrated in Figure 10.12(a) on page 484.
b. The randomized block design employs blocks of relatively homogeneous experimental units. For example, we could randomly select ten golfers, and

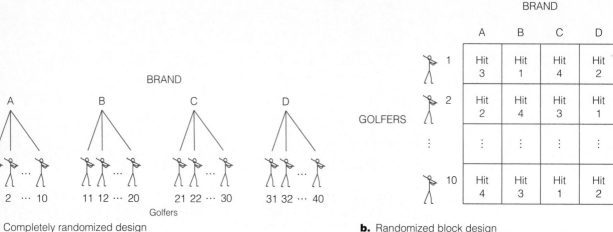

a. Completely randomized design

b. Randomized block design

FIGURE 10.12

Illustration of completely randomized design and randomized block design: Comparison of four golf ball brands

permit each golfer to hit four balls, one of each brand, in a random sequence. Then each golfer is a block, with each treatment (brand) assigned to each block (golfer). The design is summarized in Figure 10.12(b).

c. Because we expect much more variability among distances generated by "real" golfers than by Iron Byron, we would expect the randomized block design to control the variability better than the completely randomized design. That is, with 40 different golfers, we would expect more sampling variability among the measured distances within each brand than we would among the four distances generated by each of ten golfers hitting one ball of each brand.

EXAMPLE 10.6

Refer to Example 10.5. Suppose the randomized block design of part **b** is employed, utilizing a random sample of ten golfers, with each golfer using a driver to hit four balls, one of each brand, in a random sequence.

a. Set up a test of the null hypothesis that the brand mean distances differ. Use $\alpha = .05$.

b. The data for the experiment are given in Table 10.7. Use a computer program to analyze the data, and conduct the test set up in part **a**.

Solution

a. We want to test whether the data in Table 10.7 provide sufficient evidence to conclude that the brand mean distances differ. Denoting the population mean of the ith brand by μ_i, we test

H_0: $\mu_1 = \mu_2 = \mu_3 = \mu_4$

H_a: The mean distances differ for at least two of the brands

The test statistic compares the variation among the four treatment (brand) means to the sampling variability within each of the treatments.

Test statistic: $F = \dfrac{\text{MST}}{\text{MSE}}$

TABLE 10.7 **Distance Data for Randomized Block Design**

GOLFER (Block)	BRAND A	BRAND B	BRAND C	BRAND D
1	202.4	203.2	223.7	203.6
2	242.0	248.7	259.8	240.7
3	220.4	227.3	240.0	207.4
4	230.0	243.1	247.7	226.9
5	191.6	211.4	218.7	200.1
6	247.7	253.0	268.1	244.0
7	214.8	214.8	233.9	195.8
8	245.4	243.6	257.8	227.9
9	224.0	231.5	238.2	215.7
10	252.2	255.2	265.4	245.2
Means	227.0	233.2	245.3	220.7

Rejection region: $F > F_\alpha = F_{.05}$, with $\nu_1 = (p - 1) = 3$ numerator degrees of freedom and $\nu_2 = (n - p - b + 1) = 27$ denominator degrees of freedom. From Table IX of Appendix A, we find $F_{.05} = 2.96$. Thus, we will reject H_0 if $F > 2.96$.

The assumptions necessary to assure the validity of the test are as follows: (1) The probability distributions of the distances for each brand–golfer combination are normal. (2) The variances of the distance probability distributions for each brand–golfer combination are equal.

b. The Minitab program for ANOVA was used to analyze the data in Table 10.7, and the result is shown in Figure 10.13.

FIGURE 10.13

Minitab printout for randomized block design: Golf ball brand comparison

```
ANALYSIS OF VARIANCE ON DISTANCE

SOURCE      DF        SS        MS
GOLFER       9    12073.9    1341.5
BRAND        3     3298.7    1099.6
ERROR       27      546.6      20.2
TOTAL       39    15919.2
```

The values of MST and MSE (shaded on the printout) are 1,099.6 and 20.2, respectively. The F-ratio is not given on the printout, but is $F = \text{MST}/\text{MSE} = 1{,}099.6/20.2 = 54.4$, which exceeds the tabled value of 2.96. We therefore reject the null hypothesis at the .05 level of significance, concluding that at least two of the brands differ with respect to mean distance traveled when struck by the driver.

The results of an ANOVA can be summarized in a simple tabular format, similar to that utilized for the completely randomized design in Section 10.2. The general form of the table is shown in Table 10.8 (page 486), and that for Example 10.6 is given in Table 10.9. Note that the randomized block design is characterized by three sources of variation—Treatments, Blocks, and Error—which sum to the Total Sum of Squares. We hope that employing blocks of experimental units will reduce the error variability and thereby make the test for comparing treatment means more powerful.

T A B L E 10.8   ANOVA Summary Table for a Randomized Block Design

SOURCE	df	SS	MS	F
Treatment	$p - 1$	SST	MST	MST/MSE
Block	$b - 1$	SSB	MSB	
Error	$n - p - b + 1$	SSE	MSE	
Total	$n - 1$	SS(Total)		

T A B L E 10.9   ANOVA Table for Example 10.6

SOURCE	df	SS	MS	F
Treatment	3	3,298.7	1,099.6	54.4
Block	9	12,073.9	1,341.5	
Error	27	546.6	20.2	
Total	39	15,919.2		

When the F-test results in the rejection of the null hypothesis that the treatment means are equal, we will usually want to compare the various pairs of treatment means to determine which specific pairs differ. We can employ the same Bonferroni procedure as in Section 10.2. The differences are that the degrees of freedom for error are $(n - p - b + 1)$, and that the number of measurements per treatment is equal to b, the number of blocks (because each treatment is observed once in each block). The Bonferroni procedure for a randomized block design is summarized in the box.

Bonferroni Procedure for Multiple Comparisons: Randomized Block Design

1. Specify the overall level of significance α or, equivalently, the overall confidence level $100(1 - \alpha)\%$.

2. If there are c pairs of treatment means to be compared, specify the level of significance for each comparison at α/c.

3. Calculate the $100(1 - \alpha/c)\%$ confidence interval for each of the c pairs of means using the following formula:

$$(\bar{x}_i - \bar{x}_j) \pm t_{\alpha/(2c)}s\sqrt{\left(\frac{1}{b} + \frac{1}{b}\right)}$$

where $s = \sqrt{MSE}$ and $t_{\alpha/(2c)}$ is the tabulated value of t (Table VI of Appendix A) that locates $\alpha/(2c)$ in the upper tail of the t-distribution with $(n - p - b + 1)$ degrees of freedom (the number of degrees of freedom associated with error in the ANOVA).

4. Summarize the results of the multiple comparisons by ranking the treatment means, showing which pairs are significantly different.

Assumptions: Same as for the ANOVA.

| EXAMPLE 10.7 | Use the Bonferroni procedure to compare the mean distances of the four golf ball brands in Example 10.6. Use an overall significance level of $\alpha = .05$. |

Solution

The number of confidence intervals necessary to compare all pairs of brand means is $c = 4(3)/2 = 6$. The Bonferroni procedure therefore will use a level of significance equal to $\alpha/c = .05/6 = .00833$. The two-tailed confidence intervals will use a tabled t value with $(n - p - b + 1) = (40 - 4 - 10 + 1) = 27$ degrees of freedom and an area of $\alpha/(2c) = .00417$ in each tail. The closest area in Table VI of Appendix A is .005,* and $t_{.005}$ with 27 degrees of freedom is 2.771.

The standard deviation s used in the comparisons is $s = \sqrt{MSE} = \sqrt{20.2} = 4.5$. Beginning with brands A and B, we calculate

$$(\bar{x}_A - \bar{x}_B) \pm t_{\alpha/(2c)} s \sqrt{\left(\frac{1}{b} + \frac{1}{b}\right)}$$

$$(227.0 - 233.2) \pm (2.771)(4.5)(.447)$$

$$-6.2 \pm 5.6 \quad \text{or} \quad (-11.8, -.6)$$

Thus, we conclude that the mean distance for brand B is between .6 and 11.8 yards greater than for brand A when struck with a driver. All the comparisons have the same confidence interval half-width of 5.6, and they are summarized here:

(A − C):	−18.3 ± 5.6 or	(−23.9, −12.7)
(A − D):	6.3 ± 5.6 or	(.7, 11.9)
(B − C):	−12.1 ± 5.6 or	(−17.7, −6.5)
(B − D):	12.5 ± 5.6 or	(6.9, 18.1)
(C − D):	24.6 ± 5.6 or	(19.0, 30.2)

Note that we are 95% confident that all the brand means differ because none of the intervals contains 0. The listing of the brand means in Figure 10.14 has no vertical lines connecting them because there are no nonsignificant differences at the .05 level.

FIGURE 10.14

Listing of brand means for randomized block design

BRAND	MEAN
C	245.3
B	233.2
A	227.0
D	220.7

Note: All differences are statistically significant.

Unlike the completely randomized design, the randomized block design cannot, in general, be used to estimate individual treatment means. Whereas the completely randomized design employs a random sample for each treatment, the

*Note that we have rounded up slightly, from .00417 to .005. Thus, we are not *guaranteed* an overall significance level of .05, but the conservative nature of the Bonferroni method virtually assures it.

randomized block design does not necessarily employ a random sample of experimental units for each treatment. The experimental units within the blocks are assumed to be randomly selected, but the blocks themselves may not be randomly selected.

We can, however, test the hypothesis that the block means are significantly different. We simply compare the variability attributable to differences among the block means to that associated with sampling variability. The ratio of MSB to MSE is an F-ratio similar to that formed in testing treatment means. The F statistic is compared to a tabled value for a specific value of α, with numerator degrees of freedom $(b - 1)$ and denominator degrees of freedom $(n - p - b + 1)$. The test is usually given on the same printout as the test for treatment means. Refer to the Minitab printout in Figure 10.13, and note that the test statistic for comparing the block means is

$$F = \frac{\text{MSB}}{\text{MSE}} = \frac{\text{MS(Golfers)}}{\text{MS(Error)}}$$

$$= \frac{1,341.5}{20.2} = 66.4$$

Since this calculated value exceeds 2.25, the tabled values of $F_{.05}$ with $\nu_1 = 10 - 1 = 9$ df and $\nu_2 = 27$ df, we conclude that the block means are different at the $\alpha = .05$ level of significance. The results of the test are summarized in Table 10.10.

TABLE 10.10   **ANOVA Table for Randomized Block Design: Test for Blocks Included**

SOURCE	df	SS	MS	F
Treatments	3	3,298.7	1,099.6	54.4
Blocks	9	12,073.9	1,341.5	66.4
Error	27	546.6	20.2	
Total	39	15,919.2		

In the golf example, the test for block means confirms our suspicion that the golfers vary significantly, and therefore that the use of the block design was a good decision. However, be careful not to conclude that the block design was a mistake if the F-test for blocks does not result in rejection of the null hypothesis that the block means are the same. Remember that the possibility of a Type II error exists, and we are not controlling its probability as we are the probability α of a Type I error. If the experimenter believes that the experimental units are more homogeneous within blocks than between blocks, then he or she should use the randomized block design regardless of the results of a single test comparing the block means.

The procedure for conducting an analysis of variance for a randomized block design is summarized in the box. Remember that the hallmark of this design is the utilization of blocks of homogeneous experimental units in which each treatment is represented.

> **Steps for Conducting an ANOVA for a Randomized Block Design**
>
> 1. Be sure the design consists of blocks of homogeneous experimental units, and that each treatment is randomly assigned to one experimental unit in each block.
>
> 2. Create an ANOVA summary table that specifies the variability attributable to Treatments, Blocks, and Error, and that leads to the calculation of the F statistic to test the null hypothesis that the treatment means are equal in the population. Use either an ANOVA computer program or the calculation formulas in Appendix B to obtain the necessary numerical ingredients.
>
> 3. If the F-test leads to the conclusion that the means differ, use the Bonferroni procedure to conduct multiple comparisons of as many of the pairs of means as you wish. Use the results to summarize the statistically significant differences among the treatment means. Remember that, in general, the randomized block design cannot be used to form confidence intervals for individual treatment means.
>
> 4. If the F-test leads to the nonrejection of the null hypothesis that the treatment means are equal, several possibilities exist:
>   a. The treatment means are equal—that is, the null hypothesis is true.
>   b. The treatment means really differ, but other important factors affecting the response are not accounted for by the randomized block design. These factors inflate the sampling variability, as measured by MSE, resulting in smaller values of the F statistic. Either increase the sample size for each treatment, or conduct an experiment that accounts for the other factors affecting the response (as in Section 10.4).
>
>   Be careful not to automatically reach conclusion **a**, since the possibility of a Type II error must be considered if you accept H_0.
>
> 5. If desired, conduct the F-test of the null hypothesis that the block means are equal. Rejection of this hypothesis lends statistical support to the utilization of the randomized block design.

EXERCISES 10.29–10.41

LEARNING THE MECHANICS

10.29 A randomized block design yielded the following ANOVA table:

SOURCE	df	SS	MS	F
Treatment	4	501	125.25	9.109
Block	2	225	112.50	8.182
Error	8	110	13.75	
Total	14	836		

    a. How many blocks and treatments were used in the experiment?

    b. How many observations were collected in the experiment?

    c. Specify the null and alternative hypotheses you would use to compare the treatment means.

    d. What test statistic should be used to conduct the hypothesis test of part c?

    e. Specify the rejection region for the test of parts c and d. Use $\alpha = .01$.

    f. Conduct the test of parts c–e, and state the proper conclusion.

    g. What assumptions are necessary to assure the validity of the test you conducted in part f?

10.30 An experiment was conducted using a randomized block design. The data from the experiment are displayed in the table.

TREATMNT	BLOCK		
	1	2	3
1	2	3	5
2	8	6	7
3	7	6	5

    a. Fill in the missing entries in the ANOVA table.

SOURCE	df	SS	MS	F
Treatments		21.5555		
Blocks				
Error				
Total		30.2222		

    b. Specify the null and alternative hypotheses you would use to investigate whether a difference exists among the treatment means.

    c. What test statistic should be used in conducting the test of part b?

    d. Describe the Type I and Type II errors associated with the hypothesis test of part b.

    e. Conduct the hypothesis test of part b using $\alpha = .05$.

10.31 A randomized block design was used to compare the mean responses for three treatments. Four blocks of three homogeneous experimental units were selected, and each treatment was randomly assigned to one experimental unit within each block. The data are shown in the table.

TREATMENT	BLOCK			
	1	2	3	4
A	3.4	5.5	7.9	1.3
B	4.4	5.8	9.6	2.8
C	2.2	3.4	6.9	.3

The Minitab ANOVA printout for this experiment is given here:

Printout for Exercise 10.31

```
ANALYSIS OF VARIANCE ON RESPONSE

SOURCE        DF         SS         MS
TREATMNT       2     12.032      6.016
BLOCK          3     71.749     23.916
ERROR          6      0.708      0.118
TOTAL         11     84.489
```

a. Use the printout to fill in the entries in the following ANOVA table.

SOURCE	df	SS	MS	F
Treatments				
Blocks				
Error				
Total				

b. Do the data provide sufficient evidence to indicate that the treatment means differ? Use $\alpha = .05$.

c. Do the data provide sufficient evidence to indicate that blocking was effective in reducing the experimental error? Use $\alpha = .05$.

d. Use the Bonferroni technique to compare all pairs of treatment means. Use an overall significance level of $\alpha = .10$. How do the equal sample sizes for each treatment reduce the computations involved?

e. What assumptions are necessary to assure the validity of the inferences made in parts **b**, **c**, and **d**?

10.32 The analysis of variance for a randomized block design produced the ANOVA table entries shown here.

SOURCE	df	SS	MS	F
Treatments	3	28.2		
Blocks	5		13.80	
Error		34.1		
Total				

a. Complete the ANOVA table.

b. Do the data provide sufficient evidence to indicate a difference among the treatment means? Test using $\alpha = .01$.

c. Do the data provide sufficient evidence to indicate that blocking was a useful design strategy for this experiment? Explain.

d. If the sample means for treatments A and B are $\bar{x}_A = 9.7$ and $\bar{x}_B = 12.1$, respectively, find a 90% confidence interval for $(\mu_A - \mu_B)$. Interpret the interval.

10.33 Suppose an experiment utilizing a randomized block design has four treatments and nine blocks, for a total of $4 \times 9 = 36$ observations. Assume that the Total Sum of Squares for the response is SS(Total) = 500. For each of the following partitions of SS(Total), test the null hypothesis that the treatment means are equal, and the null hypothesis that the block means are equal. Use $\alpha = .05$ for each test.

a. The Sum of Squares for Treatments (SST) is 20% of SS(Total), and the Sum of Squares for Blocks (SSB) is 30% of SS(Total).

b. SST is 50% of SS(Total), and SSB is 20% of SS(Total).

c. SST is 20% of SS(Total), and SSB is 50% of SS(Total).

d. SST is 40% of SS(Total), and SSB is 40% of SS(Total).

e. SST is 20% of SS(Total), and SSB is 20% of SS(Total).

APPLYING THE CONCEPTS

10.34 An evaluation of diffusion bonding of zircaloy components is performed. The main objective is to determine which of three elements—nickel, iron, or copper—is the best bonding agent. A series of zircaloy components are bonded with each of the possible bonding agents. Since there is a great deal of variation in components machined from different ingots, a randomized block design is used, blocking on the ingots. A pair of components from each ingot are bonded together using each of the three agents, and the pressure (in units of 1,000 pounds per square inch) required to separate the bonded components is measured. The data in the table are obtained.

INGOT	BONDING AGENT		
	Nickel	Iron	Copper
1	67.0	71.9	72.2
2	67.5	68.8	66.4
3	76.0	82.6	74.5
4	72.7	78.1	67.3
5	73.1	74.2	73.2
6	65.8	70.8	68.7
7	75.6	84.9	69.0

a. Identify the following elements of the experiment: experimental units, blocks, response, factor(s) and factor type(s), and treatments.
b. Is the experiment designed or observational? Explain.
c. The SAS printout for this experiment is shown here. Construct an ANOVA summary table using the printout.

Printout for Exercise 10.34

Dependent Variable: PRESSURE

Source	DF	Sum of Squares	Mean Square	F Value	Pr > F
Model	8	400.19047619	50.02380952	4.82	0.0076
Error	12	124.45904762	10.37158730		
Corrected Total	20	524.64952381			

	R-Square	C.V.	Root MSE	PRESSURE Mean
	0.762777	4.4484899	3.2204949	72.39523810

Source	DF	Type I SS	Mean Square	F Value	Pr > F
BONDING	2	131.90095	65.95048	6.36	0.0131
INGOT	6	268.28952	44.71492	4.31	0.0151

d. Is there sufficient evidence that the mean pressure required to separate the components differs for the three bonding agents? Use $\alpha = .05$.
e. Use the Bonferroni technique to compare all pairs of treatment combinations. Use an overall significance level of $\alpha = .05$.
f. What assumptions must hold to assure the validity of the inferences in parts d and e?

10.35   A large clothing manufacturer conducted an experiment to study the effect on productivity of increases in its employees' hourly wages. Four treatments were used in the experiment:

Treatment 1:   No increase in hourly wage
Treatment 2:   Increase hourly wage by $.50
Treatment 3:   Increase hourly wage by $1.00
Treatment 4:   Increase hourly wage by $1.50

Twelve employees were selected and grouped into three blocks of size four according to the length of time they had been with the company. The four treatments were randomly assigned to the four employees in each block. The employees were observed for 3 weeks, and their productivity was measured as the average number of nondefective garments each produced per hour. The resulting productivity measures appear in the table.

| | TREATMENT | | | |
	1	2	3	4
Group 1 (less than 1 year)	2.4	3.0	3.1	3.2
Group 2 (1–5 years)	4.8	6.1	5.9	5.7
Group 3 (over 5 years)	5.1	7.0	7.2	7.3

a. What type of experimental design was used in this study? Why do you think such a design was employed?
b. The SPSS printout for this experiment is shown here. Is there evidence that the mean productivity levels differ among the four pay programs? Use $\alpha = .05$.

Printout for Exercise 10.35

```
* * *   A N A L Y S I S   O F   V A R I A N C E   * * *

            NUMBER
     BY     TREATMNT
            GROUP

                            Sum of              Mean              Signif
    Source of Variation     Squares    DF       Square      F     of F

    Main Effects            33.362      5        6.672    45.236   .000
       TREATMNT              3.740      3        1.247     8.452   .014
       GROUP                29.622      2       14.811   100.412   .000

    Explained               33.362      5        6.672    45.236   .000

    Residual                  .885      6         .148

    Total                   34.247     11        3.113
```

Note: SPSS shows the combined treatment and block effects in both the "Main Effects" and "Explained" rows of the printout. Also, SPSS uses "Residual" instead of "Error."

c. What is the observed significance level for the test you conducted in part b?
d. Use the Bonferroni technique to compare all the pairs of treatment means. Use an overall significance level of $\alpha = .10$.

10.36 A power plant, which uses water from the surrounding bay for cooling its condensers, is required by the EPA to determine whether discharging its heated water into the bay has a detrimental effect on the flora (plant life) in the water. The EPA requests that the power plant make its investigation at three strategically chosen locations, called *stations*. Stations 1 and 2 are located near the plant's discharge tubes; station 3 is located farther out in the bay. During one randomly selected day in each of 4 months, a diver descends to each of the stations, randomly samples a square meter area of the bottom, and counts the number of blades of the different types of grasses present. The results for one important grass type are given in the table on page 894.
a. Use Appendix B to perform the computations necessary to create an ANOVA summary table for this experiment.
b. Is there sufficient evidence to indicate that the mean number of blades found per square meter per month differs for at least two of the three stations? Use $\alpha = .05$.

MONTH	STATION		
	1	2	3
May	28	31	53
June	25	22	61
July	37	30	56
August	20	26	48

c. Is there sufficient evidence to indicate that the mean number of blades found per square meter differs among the 4 months? Use $\alpha = .05$.

d. Place a 90% confidence interval on the difference in means between stations 1 and 3.

10.37 Two drugs, A and B, used for the treatment of glaucoma (an eye disease) were tested for effectiveness on ten diseased dogs. Drug A was administered to one eye (chosen randomly) of each dog and drug B to the other eye. Pressure measurements were taken 1 hour later on both eyeballs of each dog. The ten diseased dogs act as the blocks for comparing the two treatments, drugs A and B. Pressure measurements are given in the table. (The smaller the measurement, the less serious the eye disease.)

DOG	TREATMENT	
	Drug A	Drug B
1	.17	.15
2	.20	.18
3	.14	.13
4	.18	.18
5	.23	.19
6	.19	.12
7	.12	.07
8	.10	.09
9	.16	.14
10	.13	.08

a. Perform an analysis of variance for these data. Do the data provide sufficient evidence to indicate a difference in mean pressure readings for the two treatments (i.e., is one of the glaucoma drugs better than the other)? Use $\alpha = .05$.

b. What is the purpose of using the dogs as blocks in this experiment?

c. Recall that a randomized block design with $p = 2$ treatments is a paired difference experiment (Chapter 9). Analyze the data as a paired difference experiment using a t-test to compare the treatment means. Use $\alpha = .05$.

d. Compare the computed F and t values from parts **a** and **c**, and verify that $F = t^2$. Also verify that for the rejection region values of F and t, $F_\alpha = t_{\alpha/2}^2$.

e. Find the approximate observed significance level for the test in part **a**, and interpret its value.

10.38 AT&T's long-distance phone charges may appear to be exorbitant when compared with some of its competitors, but this is because a comparison of charges between competing companies is often analogous to comparing apples and eggs. AT&T's charges for individuals are on a per-call basis. In contrast, its competitors often charge a monthly minimum long-distance fee, reduce the charges as the usage rises, or both. Shown in the table is a sampling of long-distance charges from Orlando, Florida, to 12 cities for three other companies offering long-distance service. The data were contained in an advertisement in the *Orlando Sentinel*, March 19, 1984. A note in fine print below the advertisement states that the rates are based on "30 hours of usage" for each of the servicing companies.

The data in the table are pertinent for companies making phone calls to large cities. Therefore, assume that the cities receiving the calls were randomly selected from among all large cities in the United States.

FROM ORLANDO TO:	TIME	LENGTH OF CALL (Minutes)	COMPANY		
			1	2	3
New York	Day	2	$ .77	$ .79	$ .66
Chicago	Evening	3	.69	.71	.59
Los Angeles	Day	2	.87	.88	.66
Atlanta	Evening	1	.22	.23	.20
Boston	Day	3	1.15	1.19	.99
Phoenix	Day	5	1.92	1.98	1.65
West Palm Beach	Evening	2	.49	.42	.40
Miami	Day	3	1.12	1.05	.99
Denver	Day	10	3.85	3.96	3.30
Houston	Evening	1	.22	.23	.20
Tampa	Day	3	1.06	1.00	.99
Jacksonville	Day	3	1.06	1.00	.99
Means			1.12	1.12	.97

a. What type of design was used for the data collection?

b. The Minitab ANOVA printout for this experiment is shown here. Do the data provide sufficient evidence to indicate that the mean charges differ among the three companies? Use $\alpha = .05$.

Printout for Exercise 10.38

```
ANALYSIS OF VARIANCE ON CHARGE

SOURCE      DF        SS         MS
COMPANY      2    0.1820    0.0910
CITY        11   29.2598    2.6600
ERROR       22    0.2208    0.0100
TOTAL       35   29.6626
```

c. Company 3 placed the advertisement. Compare all the pairs of means using the Bonferroni technique with an overall significance level of $\alpha = .05$, and interpret the results.

10.39 A construction firm employs three cost estimators. Usually only one estimator works on each potential job, but it is advantageous to the company if the estimators are consistent enough that it does not matter which of the three estimators is assigned to a particular job. To check on the consistency of the estimators, several jobs are selected and all three estimators are asked to make estimates. The estimates (in thousands of dollars) for each job by each estimator are given in the table.

JOB	ESTIMATOR		
	A	B	C
1	27.3	26.5	28.2
2	66.7	67.3	65.9
3	104.8	102.1	100.8
4	87.6	85.6	86.5
5	54.5	55.6	55.9
6	58.7	59.2	60.1

The values of SST = .941, SSB = 10,239.969, and SSE = 13.919 are computed using the formulas in Appendix B.

a. Do these data provide sufficient evidence that the means for at least two of the estimators differ? Use $\alpha = .05$.

    b. Find the approximate observed significance level for the test, and interpret it.

    c. Present the complete ANOVA summary table for this experiment.

    d. Does the analysis indicate that a comparison of the pairs of estimator means is worthwhile? Explain.

10.40 The table lists the number of strikes that occurred per year in five U.S. manufacturing industries over the period from 1976 to 1981. Only work stoppages that continued for at least 1 day and involved six or more workers were counted.

YEAR	FOOD AND KINDRED PRODUCTS	PRIMARY METAL INDUSTRY	ELECTRICAL EQUIPMENT AND SUPPLIES	FABRICATED METAL PRODUCTS	CHEMICALS AND ALLIED PRODUCTS
1976	227	197	204	309	129
1977	221	239	199	354	111
1978	171	187	190	360	113
1979	178	202	195	352	143
1980	155	175	140	280	89
1981	109	114	106	203	60

Source: *Statistical Abstract of the United States.*

    a. Is this a designed or an observational experiment? Explain.

    b. If the objective is to compare the mean number of strikes for the five industries, what type of ANOVA should be utilized? Explain.

    c. The SAS ANOVA printout for these data is shown here. Is there evidence that the mean number of strikes differs among the five industries? Use $\alpha = .10$.

Printout for Exercise 10.40

```
                    General Linear Models Procedure

Dependent Variable: STRIKES
                                 Sum of          Mean
Source                  DF       Squares         Square      F Value     Pr > F

Model                    9     170505.667      18945.074      45.96      0.0001

Error                   20       8243.533        412.177

Corrected Total         29     178749.200

                R-Square         C.V.        Root MSE           STRIKES Mean

                0.953882      10.662886       20.3021            190.400000


Source                  DF     Type I SS Mean Square      F Value     Pr > F

YEAR                     5       40726.80      8145.36      19.76      0.0001
INDUSTRY                 4      129778.87     32444.72      78.72      0.0001
```

    d. Use the Bonferroni technique to compare the means for all pairs of industries, with an overall significance level of $\alpha = .10$. Interpret the results.

10.41 Refer to Exercise 9.53 (McMurray, Wilson, & Kitchell). The table presents the mean and the estimated standard deviation of the mean (also called standard error of the mean, SEM) for each of the four treatments and for seven response variables (performance time, rating of perceived exertion, etc.) observed in the experiment. Since each mean is based on six observations, the standard error of a sample mean is $s/\sqrt{6}$—that is, the sample standard deviation divided by $\sqrt{6}$. Therefore, the standard deviation for each of the samples can be calculated by multiplying the standard error by $\sqrt{6} = 2.45$. Focus your attention on any one of the response variables in the table, say performance time, and ignore the other response variables.

Metabolic and Performance Responses of Six Female Subjects Running Until Exhaustion After the Ingestion of Fructose, Glucose, Water, and a Placebo (Mean ± SEM)

	FRUCTOSE	GLUCOSE	PLACEBO	CONTROL (H$_2$O)
Performance time (min)	61.9 ± 8.3	63.9 ± 8.5	52.2 ± 5.5[a]	65.6 ± 7.6[a]
Rating of perceived exertion[b]	11.6 ± 1.3	13.1 ± 1.1	14.6 ± 1.3	11.9 ± 1.3
Heart rate (beats/min)	173.4 ± 7.8	172.0 ± 5.9	175.4 ± 6.1	172.1 ± 6.5
Resting oxygen uptake (l/min)	.31 ± .06	.26 ± .08	.30 ± .03	.31 ± .05
Exercise oxygen uptake (l/min)	3.15 ± .14	2.96 ± .07[a]	3.46 ± .11[a]	3.19 ± .14
Resting R value[c]	.94 ± .02	.99 ± .02	.73 ± .02	.71 ± .02
Exercise R value[c]	.93 ± .02	.99 ± .02	.81 ± .03	.85 ± .03

[a]Significant difference ($p < .05$) from each other.

[b]Significant difference ($p < .05$) placebo vs. fructose or control trials.

[c]Significant difference ($p < .05$) fructose and glucose vs. control and placebo.

Source: McMurray, R. G., Wilson, J. R., and Kitchell, B. S. "The effects of fructose and glucose on high-endurance performance," *Research Quarterly for Exercise and Sport*, 1983, 54. Reprinted by permission of the American Alliance for Health, Physical Education, Recreation, and Dance, 1900 Association Drive, Reston, VA 22091.

a. What experimental design was used to collect the data?

b. Exercise 10.25 illustrates that it is possible to perform an analysis of variance for some data sets when given the sample means and standard deviations. Is it possible to perform an analysis of variance for the performance data based on the information in the table?

c. Suppose the information on performance times given in the table is derived from a completely randomized design. Perform the analysis of variance, and present your results in an ANOVA table. Do the data provide sufficient evidence to indicate differences in mean performance times among the four treatments? Test using $\alpha = .05$ and explain the practical consequences of the test results.

d. Refer to part c. What is gained or lost by analyzing data as if they had been collected according to a completely randomized design?

10.4 Factorial Experiments

All the experiments discussed in Sections 10.2 and 10.3 were **single-factor** experiments. The treatments were levels of a single factor, with the sampling of experimental units performed using either a completely randomized or randomized block design. However, most responses are affected by more than one factor, and we will therefore often wish to design experiments involving more than one factor.

Consider an experiment in which the effects of two factors on the response are being investigated. Assume that factor A is to be investigated at a levels, and factor B at b levels. Recalling that treatments are factor-level combinations, you can see that the experiment has, potentially, ab treatments that could be included in the experiment. A *complete factorial experiment* is one in which all possible ab treatments are utilized.

> **Definition 10.9**
>
> **A complete factorial experiment** is one for which every factor-level combination is utilized. That is, the number of treatments in the experiment equals the total number of factor-level combinations.

For example, suppose the USGA wants to determine not only the relationship between distance and brand of golf ball, but also between distance and the club used to hit the ball. If they decide to use four brands and two clubs (say, driver and five-iron) in the experiment, then a complete factorial would call for utilizing all $4 \times 2 = 8$ Brand–Club combinations. This experiment is referred to more specifically as a **complete 4 × 2 factorial**. A layout for a two-factor factorial experiment (we are henceforth referring to a *complete factorial* when we use the term *factorial*) is given in Table 10.11. The factorial experiment is also referred to as a **two-way classification** because it can be arranged in the row–column format exhibited in Table 10.11.

T A B L E 10.11   **Schematic Layout of Two-Factor Factorial Experiment**

	LEVEL	FACTOR *B* AT *b* LEVELS				
		1	2	3	$\cdots$	*b*
FACTOR A AT *a* LEVELS	1	Trt. 1	Trt. 2	Trt. 3	$\cdots$	Trt. *b*
	2	Trt. $b + 1$	Trt. $b + 2$	Trt. $b + 3$	$\cdots$	Trt. $2b$
	3	Trt. $2b + 1$	Trt. $2b + 2$	Trt. $2b + 3$	$\cdots$	Trt. $3b$
	$\vdots$	$\vdots$	$\vdots$	$\vdots$	$\cdots$	$\vdots$
	a	Trt. $(a - 1)b + 1$	Trt. $(a - 1)b + 2$	Trt. $(a - 1)b + 3$	$\cdots$	Trt. ab

In order to complete the specification of the experimental design, the treatments must be assigned to the experimental units. If the assignment of the *ab* treatments in the factorial experiment is random and independent, the design is completely randomized. For example, if the machine Iron Byron is used to hit 80 golf balls, ten for each of the eight Brand–Club combinations, in a random sequence, the design would be completely randomized. On the other hand, if the assignment is made within homogeneous blocks of experimental units, then the design is a randomized block. For example, if ten golfers are employed to hit each of the eight golf balls, and each golfer hits all eight Brand–Club combinations in a random sequence, then the design is randomized block, with the golfers serving as blocks. In the remainder of this section, we confine our attention to factorial experiments utilizing completely randomized designs.

If we utilize a completely randomized design to conduct a factorial experiment with *ab* treatments, we can proceed with the analysis in exactly the same way as we did in Section 10.2. That is, we calculate (or let the computer calculate) the measure of treatment mean variability (MST) and the measure of sampling variability (MSE) and use the F-ratio of these two quantities to test the null hypothesis that the treatment means are equal. However, if this hypothesis is rejected, so that we conclude some differences exist among the treatment means, important questions remain. Are both factors affecting the response, or only one? If both, do they affect the response independently, or do they interact to affect the response?

For example, suppose the distance data indicate that at least two of the eight treatment (Brand–Club combinations) means differ in the golf experiment. Does the brand of ball (factor A) or the club utilized (factor B) affect mean distance, or do both affect it? Several possibilities are shown in Figure 10.15. In Figure 10.15(a), the brand means are equal (only three are shown for the purpose of illustration), but the distances differ for the two levels of factor B (Club). Thus, there is no effect of Brand on distance, but a Club main effect is present. In Figure 10.15(b), the Brand means differ, but the Club means are equal for each Brand. Here a Brand main effect is present, but no effect of Club is present.

FIGURE 10.15
Illustration of possible treatment effects:
Factorial experiment

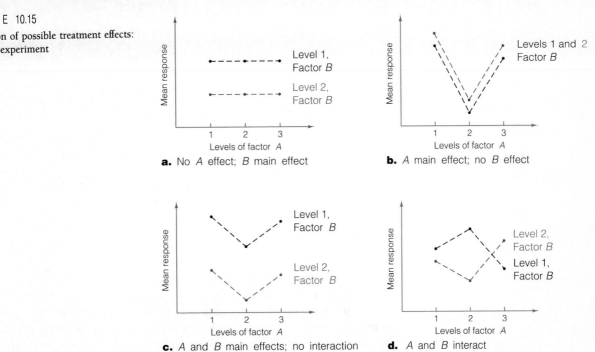

a. No A effect; B main effect

b. A main effect; no B effect

c. A and B main effects; no interaction

d. A and B interact

Figures 10.15(c) and 10.15(d) illustrate cases in which both factors affect the response. In Figure 10.15(c) the mean distances between clubs does not change for the three Brands, so that the effect of Brand on distance is independent of Club. That is, the two factors Brand and Club *do not interact*. In contrast, Figure 10.15(d) shows that the difference between mean distances between Clubs varies with Brand. Thus, the effect of Brand on distance depends on Club, and therefore the two factors *do interact*.

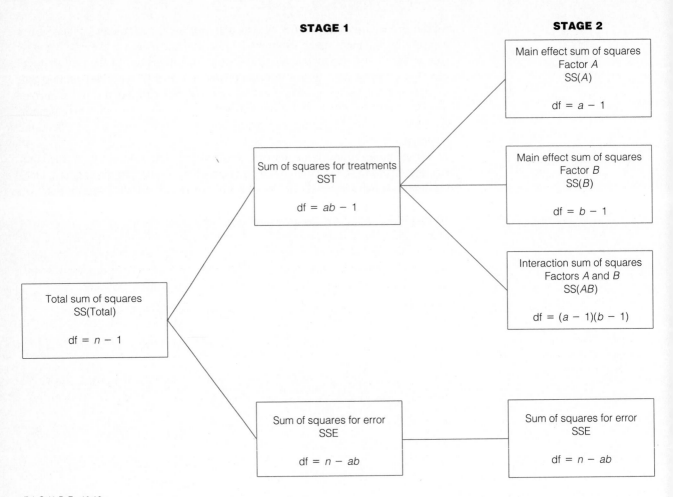

FIGURE 10.16

Partitioning the total sum of squares for a two-factor factorial

In order to determine the nature of the treatment effect, if any, on the response in a factorial experiment, we need to break the treatment variability into three components: Interaction Between Factors A and B, Main Effect of Factor A, and Main Effect of Factor B. The Interaction component is used to test whether the factors combine to affect the response, while the Main Effect components are used to determine whether the factors separately affect the response.

The partitioning of the Total Sum of Squares into its various components is illustrated in Figure 10.16. Notice that at stage 1 the components are identical to those in the one-factor completely randomized designs of Section 10.2; the Sums of Squares for Treatment and Error sum to the Total Sum of Squares. The degrees of freedom for treatments is equal to $(ab - 1)$, one less than the number of treatments. The degrees of freedom for error is equal to $(n - ab)$, the total sample size minus the number of treatments. Only at stage 2 of the partitioning is the factorial experiment differentiated from those previously discussed. Here we divide the Treatment Sum of Squares into its three components: Interaction and the two Main Effects. These components can then be used to test the nature of the differences, if any, among the treatment means.

There are a number of ways to proceed in the testing and estimation of factors in a factorial experiment. We present one approach in the box.

| Procedure for Analysis of Two-Factor Factorial Experiment

1. Partition the Total Sum of Squares into the Treatment and Error components (stage 1 of Figure 10.16). Use either a computer program or the calculation formulas in Appendix B to accomplish the partitioning.

2. Use the F-ratio of Mean Square for Treatments to Mean Square for Error to test the null hypothesis that the treatment means are equal.*
   a. If the test results in nonrejection of the null hypothesis, consider refining the experiment by increasing the number of replications or introducing other factors. Also consider the possibility that the response is unrelated to the two factors.
   b. If the test results in rejection of the null hypothesis, then proceed to step 3.

3. Partition the Treatment Sum of Squares into the Main Effect and Interaction Sum of Squares (stage 2 of Figure 10.16). Use either a computer program or the calculation formulas in Appendix B to accomplish the partitioning.

4. Test the null hypothesis that factors A and B do not interact to affect the response by computing the F-ratio of the Mean Square for Interaction to the Mean Square for Error.
   a. If the test results in nonrejection of the null hypothesis, proceed to step 5.
   b. If the test results in rejection of the null hypothesis, conclude that the two factors interact to affect the mean response. Then proceed to step 6a.

5. Conduct tests of two null hypotheses that the mean response is the same at each level of factor A and factor B. Compute two F-ratios by comparing the Mean Square for each Factor Main Effect to the Mean Square for Error.
   a. If one or both tests result in rejection of the null hypothesis, conclude that the factor affects the mean response. Proceed to step 6b.
   b. If both tests result in nonrejection, an apparent contradiction has occurred. Although the treatment means apparently differ (step 2 test), the interaction (step 4) and main effect (step 5) tests have not supported that result. Further experimentation is advised.

6. a. If the test for interaction (step 4) is significant, use the Bonferroni multiple comparisons procedure to compare any or all pairs of the treatment means.
   b. If the test for one or both main effects (step 5) is significant, use the Bonferroni technique to compare the pairs of means corresponding to the levels of the significant factor(s).

---

*Some analysts prefer to proceed directly to test the interaction and main effect components, skipping the test of treatment means. We begin with this test to be consistent with our approach in the one-factor completely randomized design.

Tests Conducted in Analyses of Factorial Experiments:
Completely Randomized Design, r Replicates Per Treatment

Test for Treatment Means

H_0:   No difference among the ab treatment means

H_a: At least two treatment means differ

Test statistic: $F = \dfrac{\text{MST}}{\text{MSE}}$

Rejection region: $F \geq F_\alpha$, based on $(ab - 1)$ numerator and $(n - ab)$ denominator degrees of freedom [*Note:* $n = abr$]

Test for Factor Interaction

H_0:   Factors A and B do not interact to affect the response mean

H_a: Factors A and B do interact to affect the response mean

Test statistic: $F = \dfrac{\text{MS}(AB)}{\text{MSE}}$

Rejection region: $F \geq F_\alpha$, based on $(a - 1)(b - 1)$ numerator and $(n - ab)$ denominator degrees of freedom

Test for Main Effect of Factor A

H_0:   No difference among the a mean levels of factor A

H_a: At least two factor A mean levels differ

Test statistic: $F = \dfrac{\text{MS}(A)}{\text{MSE}}$

Rejection region: $F \geq F_\alpha$, based on $(a - 1)$ numerator and $(n - ab)$ denominator degrees of freedom

Test for Main Effect of Factor B

H_0:   No difference among the b mean levels of factor B

H_a: At least two factor B mean levels differ

Test statistic: $F = \dfrac{\text{MS}(B)}{\text{MSE}}$

Rejection region: $F \geq F_\alpha$, based on $(b - 1)$ numerator and $(n - ab)$ denominator degrees of freedom

Assumptions for All Tests

1. The response distribution for each factor-level combination (treatment) is normal.

2. The response variance is constant for all treatments.

3. Random and independent samples of experimental units are associated with each treatment.

We assume the completely randomized design is **balanced**, meaning that the same number of observations are made for each treatment. That is, we assume that r experimental units are randomly and independently selected for each treatment. The numerical value of r must exceed 1 in order to have any degrees of freedom with which to measure the sampling variability. [Note that if $r = 1$, then $n = ab$, and the degrees of freedom associated with Error (Figure 10.16) is df $= n - ab = 0$.] The value of r is often referred to as the number of **replicates** of the factorial experiment since we assume that all ab treatments are repeated, or replicated, r times. Whatever approach is adopted in the analysis of a factorial experiment, several tests of hypotheses are usually conducted. The tests are summarized in the preceding box.

EXAMPLE 10.8

Suppose the USGA tests four different brands (A, B, C, D) of golf ball and two different clubs (driver, five-iron) in a completely randomized design. Each of the eight Brand–Club combinations (treatments) is randomly and independently assigned to four experimental units, each experimental unit consisting of a specific position in the sequence of hits by Iron Byron. The distance response is recorded for each of the 32 hits, and the results are shown in Table 10.12.

TABLE 10.12 **Distance Data for 4 × 2 Factorial Golf Experiment**

		BRAND			
		A	B	C	D
CLUB	Driver	226.4	238.3	240.5	219.8
		232.6	231.7	246.9	228.7
		234.0	227.7	240.3	232.9
		220.7	237.2	244.7	237.6
	Five-iron	163.8	184.4	179.0	157.8
		179.4	180.6	168.0	161.8
		168.6	179.5	165.2	162.1
		173.4	186.2	156.5	160.3

a. Use a computer program to partition the Total Sum of Squares into the components necessary to analyze this 4 × 2 factorial experiment.

b. Follow the steps for analyzing a two-factor factorial experiment, and interpret the results of your analysis. Use $\alpha = .10$ for the tests you conduct.

Solution

a. The SAS printout that partitions the Total Sum of Squares for this factorial experiment is given in Figure 10.17 (page 504). The partitioning takes place in two stages: First, the Total Sum of Squares is partitioned into the Model (Treatment) and Error Sums of Squares at the top of the printout. Note that SST is 33,659.8 with 7 degrees of freedom, and SSE is 822.2 with 24 degrees of freedom, adding to 34,482.0 and 31 degrees of freedom. In the second stage of partitioning, the Treatment Sum of Squares is further divided into the Main Effect and Interaction Sums of Squares. At the bottom of the printout we see that SS(Club) is 32,093.1 with 1 degree of freedom, SS(Brand) is 800.7 with 3 degrees of freedom, and SS(Club × Brand) is 766.0 with 3 degrees of freedom, adding to 33,659.8 and 7 degrees of freedom.

b. Once partitioning is accomplished, our first test is:

H_0: The eight treatment means are equal

H_a: At least two of the eight means differ

```
Dependent Variable: DISTANCE
                                   Sum of           Mean
Source                 DF         Squares         Square      F Value      Pr > F

Model                   7     33659.8087      4808.5441       140.35       0.0001

Error                  24       822.2400        34.2600

Corrected Total        31     34482.0487

             R-Square          C.V.       Root MSE          DISTANCE Mean

             0.976155     2.8964608        5.85320             202.081250


Source                 DF    Type I SS Mean Square      F Value      Pr > F

CLUB                    1     32093.11      32093.11       936.75       0.0001
BRAND                   3       800.74        266.91         7.79       0.0008
CLUB*BRAND              3       765.96        255.32         7.45       0.0011
```

FIGURE 10.17

SAS printout for factorial golf experiment

Test statistic: $F = \dfrac{\text{MST}}{\text{MSE}} = 140.35$ (top of printout)

Rejection region: $F > F_{.10} = 1.98$, with $(ab - 1) = 7$ numerator degrees of freedom, and $(n - ab) = (32 - 8) = 24$ denominator degrees of freedom.

We reject this null hypothesis and conclude that at least two of the Brand–Club combinations differ in mean distance.

After accepting the hypothesis that the treatment means differ, and therefore that the factors Brand and/or Club somehow affect the mean distance, we want to determine how the factors affect the mean response. We begin with a test of interaction between Brand and Club:

H_0:   The factors Brand and Club do not interact to affect the mean response

H_a: Brand and Club interact to affect mean response

Test statistic: $F = \dfrac{\text{MS}(AB)}{\text{MSE}}$

$$= \dfrac{\text{MS}(\text{Brand} \times \text{Club})}{\text{MSE}} = \dfrac{255.32}{34.26}$$

$$= 7.45 \quad \text{(bottom of printout)}$$

Rejection region: $F > F_{.10} = 2.33$, with $(a - 1)(b - 1) = (4 - 1)(2 - 1) = 3$ numerator degrees of freedom and 24 denominator degrees of freedom.

Since the test statistic falls in the rejection region, we conclude that the factors Brand and Club interact to affect mean distance.

Because the factors interact, we do not test the main effects for Brand and Club. Instead, we compare the treatment means in an attempt to learn the nature of the interaction. Rather than compare all $8(7)/2 = 28$ pairs of treatment means, we test for differences only between pairs of brands within each club. That differences exist *between* clubs can be assumed. Therefore, only $4(3)/2 = 6$ pairs of means need to be compared for each club, or a total of 12 comparisons for the two clubs. If the Bonferroni methodology is used with an overall $\alpha = .10$, each of the comparisons will use $\alpha = .10/12 = .0083$. Then the appropriate t value for the two-sided confidence intervals is based on $\alpha/2 = .0083/2 = .0042 \approx .005$. From Table VI of Appendix A, with 24 degrees of freedom, we find $t_{.005} = 2.797$.

We note that the half-width of each interval will be the same:

$$\text{Half-width} = t_{.005}s\sqrt{\left(\frac{1}{4} + \frac{1}{4}\right)}$$
$$= (2.797)(5.85)(.707)$$
$$= 11.6$$

where we find $s = 5.85$ labeled "Root MSE" on the printout, and we have $n_i = 4$ observations for each Club–Brand combination. We can declare all pairs of treatment means that differ by more than 11.6 yards statistically significantly different at the $\alpha = .10$ level of significance. The means are listed in descending order in Figure 10.18, and those not significantly different are connected by a vertical line.

FIGURE 10.18

Comparison of treatment means for factorial golf experiment

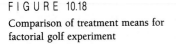

CLUB	BRAND	MEAN	
Driver	C	243.10	
	B	233.72	
	D	229.75	
	A	228.42	
Five-iron	B	182.68	
	A	171.30	
	C	167.18	
	D	160.50	

As shown in Figure 10.18, the picture is unclear with respect to Brand means. For the driver, the brand C mean significantly exceeds all except brand B. However, when hit with a five-iron, brand B's mean distance exceeds all except brand A's. Note the nontransitive nature of the multiple comparisons. For example, for the five-iron the brand B mean can be "the same" as the brand A mean, and the brand A mean can be "the same" as the brand C mean, and yet the brand B mean can significantly exceed the brand C mean. The reason lies in the definition of "the same"—we must be careful not to conclude two means are equal simply because they are connected by a vertical line. The line indicates only that *the connected means are not significantly different*. You should conclude (at the overall α level of significance) only that means *not* connected are different, while withholding judgment on those that are connected. The picture of which means differ and by how much will become clearer as we increase the number of replicates of the factorial experiment.

The Club $\times$ Brand interaction can be seen in the plot of means in Figure 10.19. Note that the brand C mean drops considerably more than the others

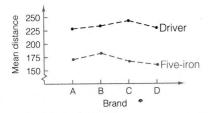

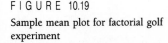

FIGURE 10.19

Sample mean plot for factorial golf experiment

when going from driver to five-iron, whereas brand A gains relative to the others. Only brands B and D seem relatively consistent in their positions relative to the others, with brand B near the top and brand D near the bottom for both clubs.

EXAMPLE 10.9

Refer to Example 10.8. Suppose the same factorial experiment is performed on four other brands (E, F, G, and H), and the results are as shown in Table 10.13. Repeat the factorial analysis and interpret the results.

T A B L E 10.13   **Distance Data for Second Factorial Golf Experiment**

| | | BRAND | | | |
		E	F	G	H
CLUB	Driver	238.6	261.4	264.7	235.4
		241.9	261.3	262.9	239.8
		236.6	254.0	253.5	236.2
		244.9	259.9	255.6	237.5
	Five-iron	165.2	179.2	189.0	171.4
		156.9	171.0	191.2	159.3
		172.2	178.0	191.3	156.6
		163.2	182.7	180.5	157.4

Solution

The printout for the second factorial experiment is shown in Figure 10.20. The F-ratio for Treatments is $F = 290.1$, which exceeds the tabled value of $F_{.10} = 1.98$ for 7 numerator and 24 denominator degrees of freedom. (Note that the same rejection regions will apply in this example as in Example 10.8 since the factors, treatments, and replicates are the same.) We conclude that at least two of the Brand–Club combinations have different mean distances.

We next test for interaction between Brand and Club:

$$F = \frac{MS(\text{Brand} \times \text{Club})}{MSE} = 1.42$$

FIGURE 10.20

SAS printout for second factorial golf experiment

Dependent Variable: DISTANCE

Source	DF	Sum of Squares	Mean Square	F Value	Pr > F
Model	7	49959.3747	7137.0535	290.12	0.0001
Error	24	590.4075	24.6003		
Corrected Total	31	50549.7822			

R-Square	C.V.	Root MSE	DISTANCE Mean
0.988320	2.3515897	4.95987	210.915625

Source	DF	Type I SS	Mean Square	F Value	Pr > F
CLUB	1	46443.90	46443.90	1887.94	0.0001
BRAND	3	3410.32	1136.77	46.21	0.0001
CLUB*BRAND	3	105.16	35.05	1.42	0.2600

Since this F-ratio does *not* exceed the tabled value of $F_{.10} = 2.33$ with 3 and 24 df, we cannot conclude at the .10 level of significance that the factors interact. In fact, note that the observed significance level (on the SAS printout) for the test of interaction is .26. Thus, at any level of significance lower than $\alpha = .26$, we could not conclude that the factors interact. We therefore test the main effects for Brand and Club.

We first test the Brand main effect:

H_0:   No difference exists among the true Brand mean distances

H_a: At least two Brand mean distances differ

Test statistic: $F = \dfrac{\text{MS(Brand)}}{\text{MSE}} = \dfrac{1{,}136.77}{24.60} = 46.21$

Rejection region: $F > F_{.10} = 2.33$, with $(a - 1) = (4 - 1) = 3$ numerator degrees of freedom, and 24 denominator degrees of freedom.

Since 46.21 exceeds 2.33, we conclude that at least two of the brand means differ. We will subsequently determine which brand means differ using the Bonferroni technique. But first, we want to test the Club main effect:

H_0:   No differences exist between the Club mean distances

H_a: The Club mean distances differ

Test statistic: $F = \dfrac{\text{MS(Club)}}{\text{MSE}} = \dfrac{46{,}443.9}{24.60} = 1{,}887.94$

Rejection region: $F > F_{.10} = 2.93$, with $(b - 1) = (2 - 1) = 1$ numerator degree of freedom, and 24 denominator degrees of freedom.

Since $F = 1{,}887.94$ exceeds 2.93, we conclude that the two clubs are associated with different mean distances. Since only two levels of Club were utilized in the experiment, this F-test leads to the inference that the mean distance differs for the two clubs. It is no surprise (to golfers) that the mean distance for balls hit with the driver is significantly greater than the mean distance for those hit with the five-iron.

To determine which of the Brands' mean distances differ, we wish to compare the four Brand means. Since $4(3)/2 = 6$ pairs of means are to be compared, the Bonferroni multiple comparisons procedure requires that we specify $\alpha = .10/6 = .0167$ in order to have an overall significance level of (at least) .10. The two-tailed tabled t statistic has $\alpha/2 = .0167/2 = .0083 \approx .005$, with 24 degrees of freedom—that is, $t_{.005} = 2.797$. Each of the comparison intervals will have the same half-width:

$$\text{Half-width} = t_{.005}s\sqrt{\left(\frac{1}{8} + \frac{1}{8}\right)}$$
$$= (2.797)(4.96)(.5)$$
$$= 6.9$$

BRAND	MEAN	
G	223.59	│
F	218.44	│
E	202.44	│
H	199.20	│

FIGURE 10.21

Comparison of mean distances by brand

where we find $s = 4.96$ labeled "Root MSE" on the printout, and we have $n_i = 8$ observations for each brand. The Brand means are shown in descending order in Figure 10.21, with a vertical line connecting the means that are not significantly different. Brands F and G are apparently associated with significantly greater mean distances than brands E and H, but we cannot distinguish between brands F and G or between brands E and H using these data. Since the interaction between Brand and Club was not significant, we conclude that this difference among brands applies to both clubs. The sample means for all Club–Brand combinations are shown in Figure 10.22 and appear to support the conclusions of the tests and comparisons. Note that the Brand means maintain their relative positions for each Club—brands F and G dominate brands E and H for both the driver and five-iron.

FIGURE 10.22

Sample mean plot for second factorial golf experiment

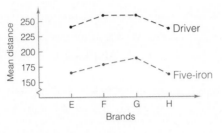

The analysis of factorial experiments can become complex if the number of factors is increased. Even the two-factor experiment becomes more difficult to analyze if some factor combinations have different numbers of observations than others. We have provided an introduction to these important experiments using two-factor factorials with equal numbers of observations for each treatment. Although similar principles apply to most factorial experiments, you should consult the references at the end of the chapter if you need to design and analyze more complex factorials.

CASE STUDY 10.2

ANXIETY LEVELS AND MATHEMATICAL ACHIEVEMENT

If you have a high level of anxiety when you take a mathematics examination, it may come as no surprise that there appears to be a relationship between the level of your anxiety and your achievement. Writing in the *Journal for Research in Mathematics Education*, Pamela S. Clute describes her doctoral research on this subject. Clute's research involved a study of the effect of three factors on a student's mathematics examination score taken after completing a survey course in mathematics. One factor was the college where the course was taken. Students took the course at both the University of California at Riverside and California State College, San Bernardino. Therefore, the factor College was at two levels. The second factor was the anxiety level of the student. Students were tested to determine their anxiety level and placed into one of three anxiety level groups. Therefore, the factor Anxiety level was at three levels. The third factor, at two levels, was the manner in which the instructional material was presented. Two methods were employed; a standard direct instruction method and a direct instruction discovery method. This latter method developed the subject by a sequence of questions, eventually leading the student to the mathematical principle that was the object of the lesson.

Since all the students were taught by the same instructor (P. S. Clute), one would not expect to see differences in research results between the two colleges and none were found in an analysis of the data. Consequently, we will combine the data for the two colleges and view Clute's experiment as a two-factor factorial experiment. The two factors are the mathematics anxiety level (three levels) and the method of instruction (two levels). The sample sizes, examination test score means, and standard deviations (in parentheses) for the six factor-level combinations are shown in Table 10.14. In the table, you can see that for both teaching methods the mean test score drops as the mathematics anxiety level increases. You may also note the unusually high mean score for students with a low anxiety level who were taught by the discovery method.

T A B L E 10.14 **Means and Standard Deviations on the Mathematics Achievement Test by Instructional Method and Level of Anxiety**

METHOD	MATHEMATICS ANXIETY LEVEL						TOTAL
	Low		Medium		High		
	n	$\bar{x}$	n	$\bar{x}$	n	$\bar{x}$	
Discovery	14	352.50 (27.31)	14	299.71 (29.60)	13	225.38 (59.09)	294.17 (65.68)
Expository	14	288.86 (79.94)	13	265.69 (81.24)	13	260.69 (54.17)	272.17 (72.26)
Total		320.68 (66.98)		283.33 (61.52)		243.04 (58.38)	283.31 (69.46)

Note: Maximum score = 400.

Source: Clute, P. S. "Mathematics anxiety: Instructional method and achievement in a survey course in college mathematics," *Journal for Research in Mathematics Education*, 1984, 15.

Clute's analysis of variance of the three-factor factorial experiment confirms our examination of the table of means (Table 10.14). Her analysis of variance is shown in Table 10.15. Although we have not explained how to compute the sums of squares for a three-factor factorial experiment, the interpretation of the ANOVA table is the same as the interpretation for a two-factor experiment. Sources of variation shown in Table 10.15 include the main effects for methods (M), anxiety (A), and college (C). They also include the three two-way interactions, $M \times A$, $M \times C$, and $A \times C$, and the three-way interaction $M \times A \times C$. Examining the computed values of the F statistics, you can see that only one interaction, the $M \times A$ interaction, is statistically significant ($F = 4.96$) at the .01 level of significance. This result tells us that the teaching method to use for a group of students depends on their

T A B L E 10.15 **Analysis of Variance of Mathematics Achievement Scores**

SOURCE	df	MS	F
Method (M)	1	7,945	2.30
Anxiety (A)	2	34,935	10.11[a]
College (C)	1	2,239	.65
$M \times A$	2	17,125	4.96[a]
$M \times C$	1	516	.15
$A \times C$	2	7,914	2.29
$M \times A \times C$	2	1,985	.58
Residual	69	3,455	

[a]$p < .01$.

Source: Clute (1984).

anxiety level. Note that none of the statistical tests involving colleges (interactions or main effects) were statistically significant. While this result does not imply that there is no difference in mean scores between colleges, it supports our initial thought that there is no reason to expect students at one school to score higher or lower than those at the other.

Clute concludes that "the most interesting finding of the study is the significant interaction between instructional method and the level of anxiety. The students with low mathematics anxiety tended to do better with the discovery method, whereas students with high mathematics anxiety tended to do better under the expository treatment."

EXERCISES 10.42–10.53

LEARNING THE MECHANICS

10.42 Suppose you conduct a 4 × 3 factorial experiment.
a. How many factors are used in the experiment?
b. Can you determine the factor type(s)—qualitative or quantitative—from the information given? Explain.
c. Can you determine the number of levels used for each factor? Explain.
d. Describe a treatment for this experiment, and determine the number of treatments used.
e. What problem is caused by using a single replicate of this experiment? How is the problem solved?

10.43 The partially completed ANOVA table for a 3 × 4 factorial experiment with two replications is shown here.

SOURCE	df	SS	MS	F
A		.8		
B		5.3		
AB		9.6		
Error				
Total		17.0		

a. Complete the ANOVA table.
b. Which sums of squares are combined to find the Sum of Squares for Treatment? Do the data provide sufficient evidence to indicate that the treatment means differ? Use $\alpha = .05$.
c. Does the result of the test in part b warrant further testing? Explain.
d. What is meant by factor interaction, and what is the practical implication if it exists?
e. Test to determine whether these factors interact to affect the response mean. Use $\alpha = .05$, and interpret the result.
f. Does the result of the interaction test warrant further testing? Explain.

10.44 The partially complete ANOVA table given here is for a two-factor factorial experiment.

SOURCE	df	SS	MS	F
A	3		.75	
B	1	.95		
AB			.30	
Error				
Total	23	6.5		

a. Give the number of levels for each factor.
b. How many observations were collected for each factor-level combination?
c. Complete the ANOVA table.
d. Test to determine whether the treatment means differ. Use $\alpha = .10$.
e. Conduct the tests of factor interaction and main effects, each at the $\alpha = .10$ level of significance. Which of the tests are warranted as part of the factorial analysis? Explain.

10.45 The two-way table gives data for a 2×3 factorial experiment with two observations for each factor-level combination.

		FACTOR B		
	LEVEL	1	2	3
FACTOR A	1	3.1, 4.0	4.6, 4.2	6.4, 7.1
	2	5.9, 5.3	2.9, 2.2	3.3, 2.5

a. Identify the treatments for this experiment. Calculate and plot the treatment means, using the response variable as the y-axis, and the levels of factor B as the x-axis. Use the levels of factor A as plotting symbols. Do the treatment means appear to differ? Do the factors appear to interact?
b. The Minitab ANOVA printout for this experiment is shown here. Test to determine whether the treatment means differ at the $\alpha = .05$ level of significance. Does the test support your visual interpretation from part a?

Printout for Exercise 10.45

ANALYSIS OF VARIANCE ON RESPONSE

SOURCE	DF	SS	MS
A	1	4.441	4.441
B	2	4.127	2.063
INTERACTION	2	18.007	9.003
ERROR	6	1.475	0.246
TOTAL	11	28.049	

c. Does the result of the test in part b warrant a test for interaction between the two factors? If so, perform it using $\alpha = .05$.
d. Do the results of the previous tests warrant tests of the two factor main effects? If so, perform them using $\alpha = .05$.
e. Interpret the results of the tests. Do they support your visual interpretation from part a?
f. Use the Bonferroni technique to compare all pairs of treatment means. Use an overall significance level of $\alpha = .10$.

10.46 The accompanying two-way table gives data for a 2×2 factorial experiment with two observations per factor-level combination.

		FACTOR B	
	LEVEL	1	2
FACTOR A	1	29.6 35.2	47.3 42.1
	2	12.9 17.6	28.4 22.7

a. Identify the treatments for this experiment. Calculate and plot the treatment means, using the response variable as the y-axis, and the levels of factor B as the x-axis. Use the levels of factor A as plotting symbols. Do the treatment means appear to differ? Do the factors appear to interact?

b. Use the computational formulas in Appendix B to create an ANOVA table for this experiment.

c. Test to determine whether the treatment means differ at the $\alpha = .05$ level of significance. Does the test support your visual interpretation from part **a**?

d. Does the result of the test in part **b** warrant a test for interaction between the two factors? If so, perform it using $\alpha = .05$.

e. Do the results of the previous tests warrant tests of the two factor main effects? If so, perform them using $\alpha = .05$.

f. Interpret the results of the tests. Do they support your visual interpretation from part **a**?

g. Given the results of your tests, which pairs of means, if any, should be compared? Use the Bonferroni technique to compare them, with an overall significance level of $\alpha = .05$. Interpret the results.

10.47 Suppose a 3×3 factorial experiment is conducted with three replications. Assume that SS(Total) = 1,000. For each of the following scenarios, form an ANOVA table, conduct the appropriate tests, and interpret the results.

a. The Sum of Squares of factor A main effect [SS(A)] is 20% of SS(Total), the Sum of Squares for factor B main effect [SS(B)] is 10% of SS(Total), and the Sum of Squares for interaction [SS(AB)] is 10% of SS(Total).

b. SS(A) is 10%, SS(B) is 10%, and SS(AB) is 50% of SS(Total).

c. SS(A) is 40%, SS(B) is 10%, and SS(AB) is 20% of SS(Total).

d. SS(A) is 40%, SS(B) is 40%, and SS(AB) is 10% of SS(Total).

APPLYING THE CONCEPTS

10.48 Most short-run supermarket strategies such as price reductions, media advertising, and in-store promotions and displays are designed to increase unit sales of particular products temporarily. Factorial designs have been employed by Curhan (1974) and Wilkinson, Mason, and Paksoy (1982) to evaluate the effectiveness of such strategies. Two of the factors examined by Wilkinson et al. were Price level (regular, reduced price, cost to supermarket) and Display level (normal display space, normal display space plus end-of-aisle display, twice the normal display space). A complete factorial design based on these two factors involves nine treatments. Suppose each treatment was applied three times to a particular product at a particular supermarket. Each application lasted a full week and the dependent variable of interest was unit sales for the week. To minimize treatment carryover effects, each treatment was preceded and followed by a week in which the product was priced at its regular price and was displayed in its normal manner. The accompanying table reports the data collected.

			PRICE	
		Regular	Reduced	Cost to supermarket
	Normal	989	1,211	1,577
		1,025	1,215	1,559
		1,030	1,182	1,598
DISPLAY	Normal plus	1,191	1,860	2,492
		1,233	1,910	2,527
		1,221	1,926	2,511
	Twice normal	1,226	1,516	1,801
		1,202	1,501	1,833
		1,180	1,498	1,852

a. The SAS ANOVA printout for this experiment is given. Use the printout to complete the following diagram partitioning the Total Sum of Squares:

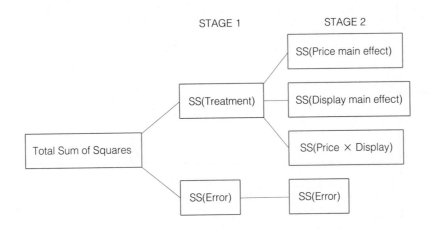

b. Do the data indicate that the mean sales differ among the nine treatments? Test using $\alpha = .10$.

c. Is the test of interaction between the factors Price and Display warranted as a result of the test in part **b**? If so, conduct the test using $\alpha = .10$.

d. Are the tests of the main effects for Price and Display warranted as a result of the previous tests? If so, conduct them using $\alpha = .10$.

e. Which pairs of treatment means should be compared as a result of the tests in parts **b**–**d**? Use the Bonferroni technique with an overall significance level of $\alpha = .10$ to compare them.

f. The treatment means are graphically portrayed in the accompanying SAS printout. Use the graph on page 514 to interpret the results of your tests.

SAS ANOVA printout for Exercise 10.48

Dependent Variable: SALES

Source	DF	Sum of Squares	Mean Square	F Value	Pr > F
Model	8	5291151.19	661393.90	1336.85	0.0001
Error	18	8905.33	494.74		
Corrected Total	26	5300056.52			

	R-Square	C.V.	Root MSE	SALES Mean
	0.998320	1.4344689	22.2428	1550.59259

Source	DF	Type I SS	Mean Square	F Value	Pr > F
DISPLAY	2	1691392.5	845696.3	1709.37	0.0001
PRICE	2	3089053.9	1544526.9	3121.89	0.0001
DISPLAY*PRICE	4	510704.8	127676.2	258.07	0.0001

g. What assumptions must hold in order for the inferential procedures you used to be appropriate?

Graph of treatment means for Exercise 10.48

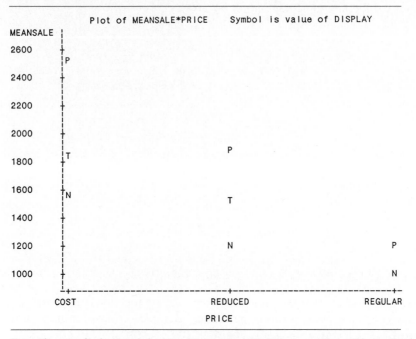

Note: The means for the Normal plus (P) and Twice normal (T) displays are nearly equal for Regular price. Both are represented by the symbol P at this level of price.

10.49   A beverage distributor wanted to determine the combination of advertising agency (two levels) and advertising medium (three levels) that would produce the largest increase in sales per advertising dollar. Each of the advertising agencies prepared copy or film, as required for each of the media— newspaper, radio, and television. Twelve small towns of roughly the same size were selected for the experiment, and two each were assigned to receive an advertisement prepared and transmitted by each of the six Agency–Medium combinations. The dollar increases in sales per advertising dollar, based on a 1-month sales period, are shown in the table.

| | | ADVERTISING MEDIUM | | |
		Newspaper	Radio	Television
AGENCY	1	15.3	20.1	12.7
		12.7	17.4	16.2
	2	18.9	24.3	12.5
		22.4	28.8	9.4

a. Describe the experiment. Include the identification of the response variable, factors, factor types, factor levels, treatments, and experimental units. What is the experiment called? What type of design was employed?

b. The SPSS ANOVA printout for this experiment is given. Use it to perform a complete analysis of the experiment. Be sure to conduct relevant tests, and use the Bonferroni technique to compare relevant pairs of means. Use $\alpha = .10$ throughout.

c. Graph the treatment means, placing advertisement medium on the horizontal axis, and using agency as a plotting symbol. Do the results of your analysis in part **b** appear reasonable? Use the graph to interpret the results of your analysis.

Printout for Exercise 10.49

```
          * * *   A N A L Y S I S   O F   V A R I A N C E   * * *

              SALES
          BY  AGENCY
              MEDIUM


                             Sum of              Mean              Signif
     Source of Variation     Squares    DF       Square      F      of F

     Main Effects            238.299     3       79.433    13.934   .004
        AGENCY                39.967     1       39.967     7.011   .038
        MEDIUM               198.332     2       99.166    17.395   .003

     2-way Interactions       77.345     2       38.672     6.784   .029
        AGENCY    MEDIUM      77.345     2       38.672     6.784   .029

     Explained               315.644     5       63.129    11.074   .005

     Residual                 34.205     6        5.701

     Total                   349.849    11       31.804
```

Note: SPSS uses "Explained" instead of "Treatment" in the factorial analysis. Also, SPSS uses "Residual" instead of "Error."

10.50 How do women compare with men in their ability to perform laborious tasks that require strength? Some information on this question is provided in a study of the firefighting ability of men and women. The researchers conducted a 2×2 factorial experiment to investigate the effect of the factor Sex (male or female) and the factor Weight (light or heavy) on the length of time required for a person to perform a particular firefighting task. Eight persons were selected for each of the $2 \times 2 = 4$ Sex–Weight categories of the 2×2 factorial experiment and the length of time needed to complete the task was recorded for each of the 32 persons. The means and standard deviations of the four samples are shown in the table.

	LIGHT		HEAVY	
	Mean	Standard deviation	Mean	Standard deviation
FEMALE	18.30	6.81	14.50	2.93
MALE	13.00	5.04	12.25	5.70

Source: Phillips, M. D., and Pepper, R. L. "Shipboard firefighting performance of females and males," *Human Factors*, 1982, 24. Copyright 1982 by The Human Factors Society, Inc. Reproduced by permission.

a. Calculate the total of the $n = 8$ time measurements for each of the four categories of the 2×2 factorial experiment.

b. Calculate CM. (See Appendix B for computational formulas.)

c. Use the results of parts **a** and **b** to calculate the sums of squares for Sex, Weight, and for the Sex $\times$ Weight interaction.

d. Calculate each sample variance. Then calculate the sum of squares of deviations *within* each sample for each of the four samples.

e. Calculate SSE. [*Hint:* SSE is the pooled sum of squares of the deviations calculated in part **d**.]

f. Now that you know SS(Sex), SS(Weight), SS(Sex $\times$ Weight), and SSE, find SS(Total).

g. Summarize the calculations in an ANOVA table.

h. Conduct a complete analysis of these data. Use $\alpha = .05$ for any inferential techniques you employ. Interpret your conclusions graphically.

i. What assumptions are necessary to assure the validity of the inferential techniques you utilized? State them in terms of this experiment.

10.51   Refer to Exercise 10.50. The researchers give data on another 2×2 factorial experiment utilizing 20 males and 20 females. The experiment involved the same treatments with ten persons assigned to each Sex–Weight category. The response measured for each person was the pulling force the person was able to exert on the starter cord of a P-250 fire pump. The means and standard deviations of the four samples (corresponding to the $2 \times 2 = 4$ categories of the experiment) are shown in the next table.

| | LIGHT | | HEAVY | |
	Mean	Standard deviation	Mean	Standard deviation
FEMALES	46.26	14.23	62.72	13.97
MALES	88.07	8.32	86.29	12.45

Source: Phillips, M. D., and Pepper, R. L. "Shipboard firefighting performance of females and males," *Human Factors*, 1982, 24. Copyright 1982 by The Human Factors Society, Inc. Reproduced by permission.

a. Use the procedures outlined in Exercise 10.50 to perform an analysis of variance for the experiment. Display your results in an ANOVA table.

b. Conduct a complete analysis of these data. Use $\alpha = .05$ for any inferential techniques you employ. Interpret your conclusions graphically.

c. Note that the observed standard deviations for the four treatments vary from 8.32 to 14.23. Since the validity of the inferences requires that the treatment standard deviations be equal, how do you refute the criticism that the unequal standard deviations invalidate your analysis?

10.52   If you are overweight, how will your friends respond to a request for a favor? Researchers sought an answer to this question by experimentation. Each subject in the experiment contacted a second person (a confederate) who was either male or female, overweight or normal weight. After following a standard procedure, the confederate requested a favor of the subject and a measure of the subject's compliance was recorded. The portion of the data summary that gives the measure of compliance of male subjects to the weight and sex of confederates is shown in the table.

| | | CONFEDERATE'S WEIGHT | |
		Overweight	Normal
CONFEDERATE'S SEX	Male	32.0 (28.65)	34.67 (36.52)
	Female	24.27 (12.94)	36.67 (14.48)

Source: Steinberg, C. L., and Birk, J. M. *Journal of General Psychology*, 1983, 109. A publication of the Helen Dwight Reid Educational Foundation.

As the table suggests, the study was conducted as a 2×2 factorial experiment. The two factors were the confederate's sex (two levels) and the confederate's weight (two levels). Fifteen male subjects were assigned to each of the confederate Sex–Weight groups. The means and standard deviations of each of the four samples are shown in the table (standard deviations are in parentheses).

a. Identify the following elements of this experiment: experimental units, response, factor(s) and factor type(s), and treatments.

b. What type of experimental design was employed?

c. Given that SS(Total) = 36,772.85, SS(Sex) = 123.12, SS(Weight) = 851.64, and SS(Sex × Weight) = 355.02, complete an ANOVA table for this experiment.

d. Perform a statistical analysis of the experiment. Plot the means to assist in your interpretation of the results of your analysis.

10.53   The data summary in Exercise 10.52 based on Steinberg and Birk's weight and compliance experiment, contained only half of the experimental results. The other half of the data was produced by the same experiment described in Exercise 10.52, except that 60 female subjects were used. Therefore, this data summary, shown in the table, enables us to determine the measure of compliance of female subjects to the four Sex × Weight factor-level combinations defining the characteristics of the confederates. (Standard deviations are in parentheses.)

		CONFEDERATE'S WEIGHT	
		Overweight	Normal
CONFEDERATE'S SEX	Male	21.33 (15.86)	38.31 (20.04)
	Female	20.33 (16.42)	27.80 (16.21)

Source: Steinberg, C. L., and Birk, J. M. *Journal of General Psychology*, 1983, 109. A publication of the Helen Dwight Reid Educational Foundation.

Given that SS(Total) = 19,675.01, SS(Sex) = 496.80, SS(Weight) = 2,241.76, and SS(Sex × Weight) = 339.15, complete an ANOVA table and perform an analysis of the experiment. Plot the means to assist in your interpretation of the results of your analysis.

Summary

Designed experiments are those for which the analyst controls the assignment of treatments to experimental units. **Observational experiments** are those in which the analyst simply observes the treatments and the response of the experimental units. The **response, factors, levels, treatments,** and **experimental units** are elements of experiments with which you should now be familiar. A **completely randomized design** is one for which treatment assignments are made to experimental units in a random and independent manner. A **randomized block design** is one for which treatment assignments are made within homogeneous groups, or **blocks**, of experimental units, with the assignments made randomly within the blocks. The analysis and interpretation of completely randomized and randomized block designs are useful even when the assignment of treatments to experimental units is beyond the analyst's control—i.e., for observational as well as designed experiments.

Factorial experiments are those that involve more than one factor, and **complete factorials** utilize all combinations of the factors. *F*-tests can be conducted to determine whether treatment means differ, and, if so, whether the factors interact or independently affect the response mean. The **Bonferroni multiple comparisons procedure** can be used to compare pairs of treatment means.

SUPPLEMENTARY EXERCISES 10.54–10.83

[*Note: List the assumptions necessary to ensure the validity of the procedure you use to solve these problems. Starred exercises (*) require the use of a computer.*]

LEARNING THE MECHANICS

10.54   Explain the difference between an experiment that utilizes a completely randomized design and one that utilizes a randomized block design.

10.55   What are the treatments in a two-factor experiment, with factor A at three levels and factor B at two levels?

10.56   Why does the overall significance level of the Bonferroni multiple comparisons procedure differ from the significance level for each comparison (assuming the experiment has more than two treatments)?

10.57   An experiment utilizing a randomized block design was conducted to compare the mean responses for four treatments, A, B, C, and D. The treatments were randomly assigned to the four experimental units in each of five blocks. The data are shown in the table.

TREATMENT	BLOCK				
	1	2	3	4	5
A	8.6	7.5	8.7	9.8	7.4
B	7.3	6.3	7.3	8.4	6.3
C	9.1	8.3	9.0	9.9	8.2
D	9.3	8.2	9.2	10.0	8.4

a. Given that SS(Total) = 22.308, SS(Block) = 10.688, and SSE = .288, complete an ANOVA table for the experiment.

b. Do the data provide sufficient evidence to indicate a difference among treatment means? Test using $\alpha = .05$.

c. Does the result of the test in part **b** warrant further comparison of the treatment means? If so, use the Bonferroni technique to compare the pairs of means. Use an overall significance level of $\alpha = .10$.

d. Is there evidence that the block means differ? Use $\alpha = .05$.

10.58   A completely randomized design is utilized to compare four treatment means. The data are shown in the table.

TREATMENT 1	TREATMENT 2	TREATMENT 3	TREATMENT 4
8	6	9	12
10	9	10	13
9	8	8	10
10	8	11	11
11	7	12	11

a. Given that SST = 36.95 and SS(Total) = 62.55, complete an ANOVA table for this experiment.

b. Is there evidence that the treatment means differ? Use $\alpha = .10$.

c. Place a 90% confidence interval on the mean response for treatment 4.

10.59   The table shows a partially completed ANOVA table for a two-factor factorial experiment.

a. Complete the ANOVA table.

b. How many levels were used for each factor? How many treatments were used? How many replications were performed?

SOURCE	df	SS	MS	F
A	3	2.6		
B	5	9.2		
A × B			3.1	
Error		18.7		
Total	47			

c. Find the value of the Sum of Squares for Treatments. Test to determine whether the data provide evidence that the treatment means differ. Use $\alpha = .05$.

d. Is further testing of the nature of factor effects warranted? If so, test to determine whether the factors interact. Use $\alpha = .05$. Interpret the result.

APPLYING THE CONCEPTS

10.60   Researchers hypothesized that in evaluating salespeople, sales managers underutilize data on the salesperson's territory and instead rely on data associated with how much effort the salesperson expended. To investigate this hypothesis, they developed four scenarios based on the four factor-level combinations associated with the factors Salesperson's effort (high and low) and Sales territory difficulty (high and low). A random sample of 120 sales managers was selected, and 30 were randomly assigned to each of the four factor-level combinations. Each group was presented with the scenario corresponding to the appropriate factor-level combination, and then each sales manager was asked to evaluate the performance of a salesperson using a 7-point scale (1 = extremely high, 7 = extremely low). The authors summarized their results in the accompanying partial ANOVA table.

SOURCE	df	F
Sales territory difficulty	1	.39
Effort	1	53.27
Interaction	1	1.95

Source: Mowen, J. C., Keith, J. E., Brown, S. W., and Jackson, D. W., Jr., "Utilizing effort and task difficulty information in evaluating salespeople," *Journal of Marketing Research*, Vol. 22, May 1985, pp. 185–191.

a. What type of experimental design was used in this study?

b. How many degrees of freedom are associated with SSE?

c. Do the data suggest that a sales manager's perceptions of a salesperson's performance are subject to an interaction effect between the difficulty of the salesperson's territory and the salesperson's level of effort? Test using $\alpha = .05$.

d. Do the data indicate that sales managers' performance ratings are influenced by sales territory data? Test using $\alpha = .05$.

e. Do the data indicate that the level of effort expended by a salesperson influences perceived performance? Test using $\alpha = .05$.

f. Do the results of parts d and e tend to confirm or to refute the researchers' hypothesis? Explain.

g. What assumptions must hold in order for the tests you conducted in parts c, d, and e to be valid?

10.61   Higher wholesale beef prices over the past few years have resulted in the sale of ground beef with higher fat content in an attempt to keep retail prices down. Four different supermarket chains were chosen, and four 1-pound packages of ground beef were randomly selected from each. The percentage of fat content was measured for each package, with the results shown in the table on the next page.

	SUPER	MARKET	
A	B	C	D
22	25	30	18
20	27	20	20
23	24	23	17
25	24	27	17

a. What type of experimental design does this represent?
b. Do the data provide sufficient evidence at the $\alpha = .05$ level of significance that the mean percentage of fat content differs for at least two of the supermarket chains? Use the calculation formulas in Appendix B.
c. Does the result of the test in part b warrant the comparison of the individual pairs of means? If so, use the Bonferroni technique to compare all pairs of means, using an overall level of significance $\alpha = .10$.

10.62   Advertising agencies are continually faced with the problem of creating attractive gimmicks to use in advertising campaigns. One agency recently decided to run an experiment to compare the preferences of financial analysts for three different brands of financial calculators, including its client's brand (brand A). It hoped to show that (1) financial analysts were not indifferent toward the three brands of calculators and (2) its client's brand was preferred to the industry leader, brand C. Three financial analysts were selected to perform an identical series of calculations on each of the three brands of calculators, A, B, and C. To avoid the possibility of fatigue, a suitable time period separated each set of calculations, and the calculators were used in random order by each analyst. A preference rating, based on a 0–100 scale, was recorded for each machine–analyst combination. These data are shown in the table.

ANALYST	BRAND		
	A	B	C
1	85	90	95
2	70	70	75
3	65	60	80

a. What type of design was employed?
b. Why is Analyst treated as a block rather than a factor in this experiment?
c. The Minitab printout for these data is given here. Do the data provide sufficient evidence to indicate a difference among the mean preferences for the three brands? Use $\alpha = .10$.

Printout for Exercise 10.62

ANALYSIS OF VARIANCE ON RATING

SOURCE	DF	SS	MS
BRAND	2	200.0	100.0
ANALYST	2	816.7	408.3
ERROR	4	83.3	20.8
TOTAL	8	1100.0	

d. Use the Bonferroni technique with an overall 90% confidence level to compare both brands B and C to brand A. Should the advertising agency use the results of this experiment in its advertising campaign for brand A?
e. Why did the experiment call for each analyst to test all three brands of calculators, rather than assigning different analysts to each calculator?

*10.63 It has been hypothesized that treatment, after casting, of a plastic used in optic lenses will improve wear. Four different treatments are to be tested. To determine whether any differences in mean wear exist among treatments, 28 castings from a single formulation of the plastic were made, and seven castings were randomly assigned to each of the treatments. Wear was determined by measuring the increase in "haze" after 200 cycles of abrasion (better wear being indicated by small increases). The results are given in the table.

| | TREATMENT | | |
A	B	C	D
9.16	11.95	11.47	11.35
13.29	15.15	9.54	8.73
12.07	14.75	11.26	10.00
11.97	14.79	13.66	9.75
13.31	15.48	11.18	11.71
12.32	13.47	15.03	12.45
11.78	13.06	14.86	12.38

a. What type of experimental design was utilized? Identify the response, factors(s), factor type(s), treatments, and experimental units.

b. Use either a computer program or the computational formulas in Appendix B to analyze the data. Is there evidence of a difference in mean wear among the treatments? Use $\alpha = .05$.

c. What is the observed significance level of the test? Interpret it.

d. Use the Bonferroni technique to compare all the pairs of treatment means with an overall significance level of $\alpha = .10$.

e. Use a 90% confidence interval to estimate the mean wear for lenses receiving treatment A.

10.64 A large citrus products company is interested in purchasing several new orange juice extractors. To help them with their purchase decision, six different manufacturers of juice extractors have agreed to let the company conduct an experiment to compare the yields of juices for the six different brands of extractors. Because of the possibility of a variation in the amount of juice per orange from one truckload of oranges to another, equal weights of oranges from a single truckload were assigned to each extractor, and this process was repeated for 15 loads. The amount of juice recorded for each extractor for each truckload produced the sums of squares shown in the table.

SOURCE	df	SS	MS	F
Extractors		84.71		
Truckloads		159.29		
Error		95.33		
Total		339.33		

a. Complete the ANOVA table.

b. Do the data provide sufficient evidence to indicate a difference among the mean amounts of juice extracted by the six extractors? Use $\alpha = .05$.

10.65 A professor wishes to determine whether performance in an introductory statistics course depends on the student's year in college. Random samples of the final grade-point averages of students from each of the freshman, sophomore, junior, and senior years produce the following data summary:

	FRESHMAN	SOPHOMORE	JUNIOR	SENIOR
Σx	52.55	168.25	760.50	363.10
Σx^2	108.25	360.19	1,630.75	1,045.33
n	25	75	350	150

a. Calculate and plot the four sample means. Based on the plot, do the sample means appear to indicate that the true mean performance differs by class?

b. Use the appropriate formulas in Appendix B to create the ANOVA table for this experiment.

c. Test to see whether the professor should conclude that the mean performance varies depending on the year of the student. Use $\alpha = .05$.

d. If warranted, use the Bonferroni procedure to compare all pairs of means. Use an overall significance level of $\alpha = .10$, and interpret the results.

10.66   Sixteen workers were randomly selected to participate in an experiment to determine the effects of work scheduling and method of payment on attitude toward the job. Two types of scheduling were employed, the standard 8–5 workday and a modification whereby the worker was permitted to start the day at either 7 or 8 A.M. and to vary the starting time as desired; in addition, the worker was allowed to choose, on a daily basis, either a $\frac{1}{2}$-hour or 1-hour lunch period. The two methods of payment were a standard hourly rate and a reduced hourly rate with an added piece rate based on the worker's production. Four workers were randomly assigned to each of the four Scheduling–Payment combinations, and each completed an attitude test after 1 month on the job. The test scores are shown in the table.

| | | PAYMENT | |
		Hourly rate	Hourly and piece rate
SCHEDULING	8–5	54, 68 55, 63	89, 75 71, 83
	Worker-modified schedule	79, 65 62, 74	83, 94 91, 86

a. What type of experiment was performed? Identify the response, factor(s), factor type(s), treatments, and experimental units.

b. The SAS printout for this experiment is shown here. Is there evidence that the treatment means differ? Use $\alpha = .05$.

Printout for Exercise 10.66

General Linear Models Procedure

Dependent Variable: SCORE

Source	DF	Sum of Squares	Mean Square	F Value	Pr > F
Model	3	1806.00000	602.00000	12.29	0.0006
Error	12	588.00000	49.00000		
Corrected Total	15	2394.00000			

R-Square	C.V.	Root MSE	SCORE Mean
0.754386	9.3959732	7.00000	74.5000000

Source	DF	Type I SS	Mean Square	F Value	Pr > F
SCHEDULE	1	361.0000	361.0000	7.37	0.0188
PAYMENT	1	1444.0000	1444.0000	29.47	0.0002
SCHEDULE*PAYMENT	1	1.0000	1.0000	0.02	0.8888

c. If the test in part **b** warrants further analysis, conduct the appropriate tests of interaction and main effects. Interpret your results.

d. What assumptions are necessary to assure the validity of the inferences? State the assumptions in terms of this experiment.

10.67 In a nutrition experiment, an investigator studied the effects of different rations on the growth of young rats. Forty rats from the same inbred strain were divided at random into four groups of ten and used for the experiment. A different ration was fed to each group and, after a specified length of time, the increase in growth of each rat was measured (in grams). The data are shown in the table.

RATION A		RATION B		RATION C		RATION D	
10	6	13	9	12	10	15	21
8	6	15	10	16	12	13	18
12	9	14	8	13	10	15	20
11	5	13	10	11	9	10	19
9	6	17	8	15	9	12	22

a. Identify the experimental units, the response, the factor(s) and factor type(s), and the treatments for this experiment.

b. Is the experiment designed or observational? Explain.

c. Given that SS(Ration) = 348.675 and SSE = 342.300, complete an ANOVA table for this experiment.

d. Do the data provide sufficient evidence to conclude that a difference exists in the mean growth of rats receiving the different rations? Test using $\alpha = .05$.

e. If warranted, use the Bonferroni procedure to compare the pairs of mean growths corresponding to the four rations. Use an overall significance level of $\alpha = .10$.

f. Suppose that prior to conducting the experiment, the investigator expected the mean increase in weight for rats receiving ration B to exceed 10 grams. Construct a 95% confidence interval for the true mean increase for ration B. Does the interval support the investigator's theory?

10.68 A farmer wants to determine the effect of five different concentrations of lime on the pH (acidity) of the soil on a farm. Fifteen soil samples are to be used in the experiment, five from each of three different locations. The five soil samples from each location are then randomly assigned to the five concentrations of lime, and 1 week after the lime is applied the pH of the soil is measured. The data are shown in the table.

LOCATION	LIME CONCENTRATION				
	0	1	2	3	4
I	3.2	3.6	3.9	4.0	4.1
II	3.6	3.7	4.2	4.3	4.3
III	3.5	3.9	4.0	3.9	4.2

a. What type of experimental design was utilized? Explain.

b. Given that SS(Total) = 1.42933, SS(Lime) = 1.14267, and SSE = .11733, complete an ANOVA table for this experiment.

c. What is the estimated standard deviation for the experiment? Interpret it.

d. Is there evidence that the mean soil pH levels differ over the five lime concentrations? Use $\alpha = .05$.

e. Is there evidence that the mean pH levels vary over the three locations? Use $\alpha = .05$. What is the practical significance of the result of this test?

10.69   Several companies are experimenting with the concept of paying production workers (generally paid by the hour) on a salary basis. It is believed that absenteeism and tardiness will increase under this plan, yet some companies feel that the working environment and overall productivity will improve. Fifty production workers under the salary plan are monitored at company A, and likewise 50 under the hourly plan at company B. The number of work-hours missed due to tardiness or absenteeism over a 1-year period is recorded for each worker. The results are partially summarized in the table.

SOURCE	df	SS	MS	F
Company		3,237.2		
Error		16,167.7		
Total	99			

a. Fill in the missing information.
b. Is there evidence at the $\alpha = .05$ level of significance that the mean number of hours missed differs for employees of the two companies?
c. Is there sufficient information given to form a confidence interval for the difference between the mean number of hours missed at the two companies?

10.70   From time to time, one branch office of a company must make shipments to a certain branch office in another state. There are three package delivery services between the two cities where the branch offices are located. Since the price structures for the three delivery services are quite similar, the company wants to compare the delivery times. The company plans to make several different types of shipments to its branch office. To compare the carriers, each shipment will be sent in triplicate, one with each carrier. The results listed in the table are the delivery times in hours.

SHIPMENT	CARRIER		
	I	II	III
1	15.2	16.9	17.1
2	14.3	16.4	16.1
3	14.7	15.9	15.7
4	15.1	16.7	17.0
5	14.0	15.6	15.5

a. What type of experiment was utilized? Is it a designed or observational experiment?
b. Given that SST = 8.8573, SSB = 3.9773, and SSE = .4227, complete an ANOVA summary table.
c. Is there evidence that the mean delivery times differ among the three carriers? Use $\alpha = .05$.
d. What assumptions are necessary to assure the validity of the test conducted in part c?

*10.71   The data shown in the table are for a 4×3 factorial experiment with two replications.

		LEVEL OF A		
		1	2	3
LEVEL OF A	1	2 4	5 6	1 3
	2	5 4	2 2	10 9
	3	7 10	1 0	5 3
	4	8 7	12 11	7 4

a. Use a computer program or the calculation formulas in Appendix B to perform an analysis of variance for the data. Display the results in an ANOVA table.

b. Do the data indicate that the treatment means differ? Use $\alpha = .05$.

c. Is a test of factor interaction warranted? If so, perform it using $\alpha = .05$.

d. Is it necessary to perform tests of the main effects? If so, perform the tests using $\alpha = .05$.

e. Use the Bonferroni technique to compare all pairs of treatment means using an $\alpha = .10$ level of significance.

10.72 England has experimented with different 40-hour work weeks to maximize production and minimize expenses. A factory tested a 5-day week (8 hours per day), a 4-day week (10 hours per day), and a $3\frac{1}{3}$-day week (12 hours per day), with the weekly production results shown in the table (in thousands of dollars worth of items produced).

8-HOUR DAY	10-HOUR DAY	12-HOUR DAY
87	75	95
96	82	76
75	90	87
90	80	82
72	73	65
86		

a. What type of experimental design was used?

b. Use the accompanying SPSS printout to test the null hypothesis that the mean productivity level is the same for the three lengths of workday. Use $\alpha = .10$.

c. Is further comparison of the pairs of means warranted? If so, use the Bonferroni technique to compare the pairs of mean productivity levels. Use an overall $\alpha = .10$.

Printout for Exercise 10.72

Variable PRDCTION

By Variable DAY

Analysis of Variance

Source	D.F.	Sum of Squares	Mean Squares	F Ratio	F Prob.
Between Groups	2	57.6042	28.8021	.3375	.719
Within Groups	13	1109.3333	85.3333		
Total	15	1166.9375			

10.73 An agricultural research unit wished to study the yields of three strains of wheat, A, B, and C. Because of the direction of drainage, it was suspected that the field used for the experiment varied substantially in fertility along the north–south line. Consequently, to reduce data variation the three strains of wheat were planted in plots laid out in east–west blocks. One plot of each type of wheat was assigned to each block, and the strains were assigned in random order within each block. The layout of the design and the data on wheat yields for each strain within each block (in bushels) are shown at the top of page 526.

A 53	B 62	C 47	A 56	B 58
B 60	C 44	A 49	B 55	C 45
C 49	A 57	B 61	C 43	A 51

a. Identify the following elements of this experiment: experimental units, response, factor(s) and factor type(s), blocks, treatments.
b. Use the appropriate formulas in Appendix B or a computer to construct an ANOVA summary table for this experiment.
c. Is there evidence that the mean yield differs among the three strains of wheat? Use $\alpha = .10$.
d. Can you use the results of this experiment to obtain a valid confidence interval for the mean yield for strain C? Explain.

10.74   To be able to provide its clients with comparative information on two large suburban residential communities, a realtor wants to know the average home value in each community. Eight homes are selected at random within each community and are appraised by the realtor. The appraisals are given in the table (in thousands of dollars). Can you conclude that the average home value is different in the two communities? You have two ways of analyzing this problem.

a. Use the two-sample t statistic (Section 9.2) to test $H_0: \mu_A = \mu_B$.
b. The SAS ANOVA printout for this experiment is shown here. Is there evidence that the treatment means differ? Use $\alpha = .05$.
c. Using the results of the two tests in parts a and b, verify that the tests are equivalent.

Printout for Exercise 10.74

COMMUNITY	
A	B
43.5	73.5
49.5	62.0
38.0	47.5
66.5	36.5
57.5	44.5
32.0	56.0
67.5	68.0
71.5	63.5

General Linear Models Procedure

Dependent Variable: PRICE

Source	DF	Sum of Squares	Mean Square	F Value	Pr > F
Model	1	40.6406250	40.6406250	0.21	0.6501
Error	14	2648.718750	189.1941964		
Corrected Total	15	2689.359375			

R–Square	C.V.	Root MSE	PRICE Mean
0.015112	25.079956	13.7548	54.8437500

Source	DF	Type I SS	Mean Square	F Value	Pr > F
COMMUNTY	1	40.6406	40.6406	0.21	0.6501

10.75   One indicator of employee morale is the length of time employees stay with a company. A large corporation has three factories located in similar areas of the country. While the corporation management attempts to maintain uniformity in management, working conditions, employee relations, etc., at its various factories, it realizes that differences may exist between the various factories. To study this phenomenon, employee records are randomly selected at each of the three factories and the length of employee service with the company is recorded. The following is a summary of the data:

Factory	1	2	3
Number in sample	15	21	17

SST = 421.74   SSE = 3,574.06

Is there evidence of a difference in mean length of service at the three factories? Use $\alpha = .05$.

10.76 Three anticoagulant drugs are studied to compare their effectiveness in dissolving blood clots. Each of five subjects receives the drugs at equally spaced time intervals and in random order. Time periods between drug applications permit a drug to be passed out of a subject's body before the subject receives the next drug. After each drug is in the bloodstream, the length of time (in seconds) required for a cut of specified size to stop bleeding is recorded. The results are shown in the table.

PERSON	DRUG		
	A	B	C
1	127.5	129.0	135.5
2	130.6	129.1	138.0
3	118.3	111.7	110.1
4	155.5	144.3	162.3
5	180.7	174.4	181.8

a. What type of experimental design was used in this study? Identify the response, factor(s), factor type(s), treatments, and experimental units.
b. The SAS printout for this experiment is shown here. Is there evidence of a difference in mean clotting time among the three drugs? Test using $\alpha = .01$.

Printout for Exercise 10.76

Dependent Variable: TIME

Source	DF	Sum of Squares	Mean Square	F Value	Pr > F
Model	6	7802.1746667	1300.3624444	64.97	0.0001
Error	8	160.1093333	20.0136667		
Corrected Total	14	7962.2840000			

	R-Square	C.V.	Root MSE		TIME Mean
	0.979892	3.1522433	4.4736637		141.92000000

Source	DF	Type I SS	Mean Square	F Value	Pr > F
DRUG	2	156.36400	78.18200	3.91	0.0655
PERSON	4	7645.81067	1911.45267	95.51	0.0001

c. What is the observed significance level of the test you conducted in part a? Interpret it.
d. Was blocking effective in reducing the variation among the data? That is, do the data support the contention that the mean clotting time varies from person to person?

e. If warranted, use the Bonferroni technique to determine whether one of the drugs is most effective. Use an overall significance level of $\alpha = .10$.

10.77   In-town mileage tests were performed to compare three different brands of regular gasoline. Four different automobiles were used in the experiment, and each brand of gas was used in each car until the mileage was determined. The results (in miles per gallon) are shown in the table.

BRAND	AUTOMOBILE			
	1	2	3	4
A	20.2	18.7	19.7	17.9
B	19.7	19.0	20.3	19.0
C	18.3	18.5	17.9	21.1

a. What type of experimental design was used in this study?
b. Use a computer program or the computational formulas in Appendix B to construct an ANOVA summary table for this experiment.
c. Is there evidence of a difference in the mean mileage rating among the three brands of gasoline? Use $\alpha = .05$.
d. Is there evidence of a difference in the mean mileage among the four models; i.e., is blocking important in this type of experiment? Use $\alpha = .05$.

10.78   Refer to Exercise 10.77. Use the Bonferroni technique to compare the pairs of Brand means. Use an overall significance level of $\alpha = .10$.

10.79   In hopes of attracting more riders, a city transit company plans to have express bus service from a suburban terminal to the downtown business district. These buses will travel along a major city street where there are numerous traffic lights that affect travel time. The city decides to perform a study of the effect of four different plans (a special bus lane, traffic signal progression, etc.) on the travel times for the buses. Travel times (in minutes) are measured for several weekdays during a morning rush-hour trip while each plan is in effect. The results are recorded in the table.

	PLAN		
1	2	3	4
27	25	34	30
25	28	29	33
29	30	32	31
26	27	31	
	24	36	

a. What experimental design was used?
b. Use the appropriate formulas in Appendix B or a computer program to construct an ANOVA summary table for this experiment.
c. Is there evidence of a difference between the mean travel times for the four plans? Use $\alpha = .01$.
d. If warranted, use the Bonferroni multiple comparisons procedure to compare the pairs of mean travel times corresponding to the four plans.

10.80   Methods of displaying goods can affect their sales. The manager of a large produce market would like to try three different display types for a certain fruit. The manager chooses the locations for the three displays so that each display type will be equally accessible to the customers. The three displays will be set up for five 1-week periods. Between each of these periods will be a 2-week period when a standard display is used. For each of the 5 experimental weeks, the sales (in dollars) of fruit from each display is determined, with the results in the table.

PERIOD	DISPLAY		
	A	B	C
1	$525	$653	$408
2	450	500	380
3	400	510	335
4	602	699	475
5	306	422	315

a. What type of experimental design was used in this study? Identify the response, factor(s), factor type(s), treatments, and experimental units.

b. The SPSS printout for this experiment is shown here. Is there evidence of a difference among mean sales for the three types of displays? Use $\alpha = .05$.

Printout for Exercise 10.80

```
* * *   A N A L Y S I S   O F   V A R I A N C E   * * *
```

Source of Variation	Sum of Squares	DF	Mean Square	F	Signif of F
Main Effects	187191.467	6	31198.578	24.761	.000
DISPLAY	76436.133	2	38218.067	30.332	.000
PERIOD	110755.333	4	27688.833	21.976	.000
Explained	187191.467	6	31198.578	24.761	.000
Residual	10079.867	8	1259.983		
Total	197271.333	14	14090.810		

c. Do the data indicate that the use of weeks as blocks was necessary? Test at $\alpha = .05$.

d. Use the Bonferroni technique to compare the mean sales for all pairs of displays. Use an overall significance level of $\alpha = .05$.

10.81 Refer to Case Study 10.2. Since there was no evidence to indicate any difference in mathematics scores between the two colleges, Clute combined the data for the two colleges and analyzed the portion of the mathematics examination scores that pertained to the low-level items on the test.

Means and Standard Deviations on the Low-Level Items by Instructional Method and Level of Anxiety

METHOD		MATHEMATICS ANXIETY LEVEL					TOTAL
	n	Low	n	Medium	n	High	
Discovery	14	218.79 (20.81)	14	183.43 (27.17)	13	146.85 (36.93)	183.90 (40.77)
Expository	14	183.43 (52.27)	13	182.46 (41.64)	13	185.92 (33.72)	183.92 (42.38)
Total		201.11 (43.99)		182.96 (34.20)		166.38 (39.97)	183.91 (41.31)

Source: Clute, P. S. "Mathematics anxiety: Instructional method and achievement in a survey course in college mathematics," *Journal for Research in Mathematics Education*, 1984, 15.

a. The sample sizes, mean scores, and standard deviations (in parentheses) for the six Method–Anxiety combinations are shown in the table. Plot the mean score on the vertical axis against the anxiety level on the horizontal axis, using the first letter of the instructional method as a plotting symbol. Does it appear that teaching method and anxiety level interact to affect the mean test score?

b. Clute's ANOVA table is shown below. Interpret the results, and discuss whether they support your impressions from part **a**.

Analysis of Variance for Low-Level Items

SOURCE	df	MS	F
Method (M)	1	17	.01
Anxiety (A)	2	8,146	6.02[a]
$M \times A$	2	9,341	6.90[a]
Residual	75	1,354	

[a]$p < .01$.

Source: Clute (1984).

10.82   Refer to Case Study 10.2 and Exercise 10.81. The table shown here gives data for the portion of the mathematics examination scores that pertained to the high-level items on the test (again combining the two colleges). Clute's summary ANOVA table for this portion of the test is also given. Perform the same analysis described in parts **a** and **b** of Exercise 10.81 for the high-level item data. Compare the results to those obtained for the low-level items.

Means and Standard Deviations on the High-Level Items by Instructional Method and Level of Anxiety

METHOD		MATHEMATICS ANXIETY LEVEL					TOTAL
	n	Low	n	Medium	n	High	
Discovery	14	134.43 (23.06)	14	109.14 (12.88)	13	78.08 (28.67)	107.93 (31.77)
Expository	14	104.71 (31.42)	13	91.69 (47.92)	13	74.77 (37.39)	90.75 (40.26)
Total		119.57 (30.99)		100.74 (34.95)		76.42 (32.69)	99.44 (37.01)

Note: Maximum score = 240.
Source: Clute (1984).

Analysis of Variance for High-Level Items

SOURCE	df	MS	F
Method (M)	1	5,724	5.65[a]
Anxiety (A)	2	12,573	12.39[b]
$M \times A$	2	1,176	1.16
Residual	75	1,015	

[a]$p < .05$.
[b]$p < .01$.

Source: Clute (1984).

10.83   Refer to Exercises 10.81 and 10.82. Use the Bonferroni technique to compare the appropriate pairs of means in each analysis (low-level and high-level items). Use an overall significance level of $\alpha = .10$ for each analysis, and interpret the results in terms of the specifics of this application.

ON YOUR OWN...

Due to ever-increasing food costs, consumers are becoming more discerning in their choice of supermarkets. It is usually more convenient to shop at only one market, as opposed to buying different items at different markets. Thus, it would be useful to compare the mean food expenditure for a market basket of food items from store to store. Since there is a great deal of variability in the prices of products sold at any supermarket, we will consider an experiment that blocks on products.

Choose three (or more) supermarkets in your area that you want to compare. Then choose approximately ten (or more) food products you typically purchase. For each food item, record the price each store charges in the following manner:

FOOD ITEM 1	FOOD ITEM 2	. . .	FOOD ITEM 10
Price store 1	Price store 1	. . .	Price store 1
Price store 2	Price store 2	. . .	Price store 2
Price store 3	Price store 3	. . .	Price store 3

Use the data you obtain to test

H_0: Mean expenditures at the stores are the same

H_a: Mean expenditures for at least two of the stores are different

Also, test to determine whether blocking on food items is advisable in this kind of experiment. Fully interpret the results of your analysis.

USING THE COMPUTER...

Do home values differ among the four census regions in the United States? Treat the 1,000 zip codes in the data set described in Appendix C as if they were the population of all U.S. zip codes. Select random samples of 50 zip code areas using a completely randomized design.

a. Identify the following elements of this experiment: experimental units, response, factor(s) and factor levels, and treatments.
b. Construct a dot diagram and stem and leaf display for each of the four samples. What do your graphs suggest about the differences in mean home values between census regions?
c. Conduct an analysis of variance to determine whether differences exist among the four regions' mean home values.
d. Use the Bonferroni multiple comparisons procedure to compare the four region (treatment) means.
e. What assumptions are necessary to make in order to employ the inferential procedures of parts b and c? Which of these assumptions are suspect in this exercise?

References

Brockhaus, R. H. "Risk-taking propensity of entrepreneurs." *Academy of Management Journal*, Vol. 23, Sept. 1980, pp. 509–520.

Curhan, R. C. "The effects of merchandising and temporary promotional activities on the sales of fresh fruit and vegetables in supermarkets," *Journal of Marketing Research*, Vol. 11, Aug. 1974, pp. 286–294.

Mendenhall, W. *Introduction to Linear Models and the Design and Analysis of Experiments.* Belmont, Calif.: Wadsworth, 1968, Chapter 8.

Snedecor, G. W., and Cochran, W. G. *Statistical Methods*, 7th ed. Ames: Iowa State University Press, 1980.

Steele, R. G. D., and Torrie, J. H. *Principles and Procedures of Statistics: A Biometrical Approach*, 2d ed. New York: McGraw-Hill, 1980.

Wilkinson, J. B., Mason, J. B., and Paksoy, C. H. "Assessing the impact of short-term supermarket strategy variables." *Journal of Marketing Research*, Vol. 19, Feb. 1982, pp. 72–86.

Winer, B. J. *Statistical Principles in Experimental Design*, 2d ed. New York: McGraw-Hill, 1971.

Nonparametric Statistics

CONTENTS

WHERE WE'VE BEEN...

Chapters 7–10 presented techniques for making inferences about the mean of a single population and for comparing the means of two or more populations. Most of the techniques discussed in those chapters were based on the assumption that the sampled populations have probability distributions that are approximately normal with equal variances. But how can you analyze data that evolve from populations that do not satisfy these assumptions? How can you make comparisons between populations when you cannot assign specific numerical values to your observations?

WHERE WE'RE GOING...

In this chapter, we present statistical techniques for comparing two or more populations that are based on an ordering of the sample measurements according to their relative magnitudes. These techniques, which require fewer or less stringent assumptions concerning the nature of the probability distributions of the populations, are called *nonparametric statistical methods*.

The t- and F-tests for making inferences about means, presented in Chapters 7–10, are unsuitable for analyzing some types of data. These data fall into two categories. The first are data from populations that do not satisfy the assumptions upon which the t- and F-tests are based. For both tests, we assume that the random variables being measured had normal probability distributions with equal variances. Yet in practice, the observations from one population may exhibit much greater variability than those from another, or the probability distributions may be decidedly nonnormal. For example, the distribution might be very flat, peaked, or strongly skewed to the right or left. When any of the assumptions required for the t- and F-tests are seriously violated, the computed t and F statistics may not follow the standard t- and F-distributions. In this case, the tabulated values of t and F (Tables VI, VIII, IX, X, and XI in Appendix A) are not applicable, the correct value of α for the test is unknown, and the t- and F-tests are of dubious value.

The second type of data for which t- and F-tests are inappropriate are responses that are not susceptible to numerical measurement but can be *ranked in order of magnitude*. That is, in the terminology introduced in Chapter 2, the t- and F-tests are generally not appropriate for *ordinal data*. For example, if we want to compare the teaching ability of two college instructors based on subjective evaluations of students, we cannot assign a number that exactly measures the teaching ability of a single instructor. But we can compare two instructors by ranking them according to a subjective evaluation of their teaching abilities. If instructors A and B are evaluated by each of ten students, we have the standard problem of comparing the probability distributions for two populations of ratings—one for instructor A and one for B—but the t-test of Chapter 9 would be inappropriate, because the only data that can be recorded are preference statements of ten students—i.e., each student decides that either A is better than B or vice versa.

Consider another example of this type of data. Most firms that plan to market a new product nationally first test it in a few cities or regions to determine its acceptability. For a food product this may entail taste tests in which consumers rank the new product in order of preference with respect to one or more currently popular brands. A consumer probably has a preference for each product, but the strength of the preference is difficult, if not impossible, to measure. Consequently, the best we can do is to have each consumer examine the new product, along with a few established products, and rank them according to preference: 1 for the most preferred, 2 for the second, etc. The result is ordinal data for which the t- and F-tests would be inappropriate.

The *nonparametric* counterparts of the t- and F-tests compare the probability distributions of the sampled populations rather than specific parameters of these populations (such as the means or variances). For example, nonparametric tests can be used to compare the probability distribution of the strengths of preferences for a new product to the probability distributions of the strengths of preferences for the currently popular brands. If it can be inferred that the distribution for the new product lies above (to the right of) the others (see Figure 11.1), the implication is that the new product tends to be more preferred than the currently popular products. Such an inference might lead to a decision to market the product nationally.

Most nonparametric methods use the **relative ranks** of the sample observations, rather than their actual numerical values. These tests are particularly valuable when we are unable to obtain numerical measurements of some phenomena but

FIGURE 11.1

Probability distributions of strengths of preference measurements (new product is preferred)

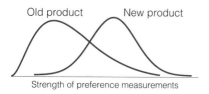

are able to rank them in comparison to each other. Statistics based on ranks of measurements are called **rank statistics**. In Section 11.1, we develop a test to make inferences about the central tendency of a single population. In Sections 11.2 and 11.4, we present rank statistics for comparing two probability distributions using independent samples. In Sections 11.3 and 11.5, the matched-pairs and randomized block designs are used to make nonparametric comparisons of populations. Finally, in Section 11.6, we present a nonparametric measure of correlation between two variables—**Spearman's rank correlation coefficient**.

11.1 Single Population Inferences: The Sign Test

In Chapter 8 we utilized the z and t statistics for testing hypotheses about a population mean. The z statistic is appropriate for large random samples selected from "general" populations—that is, with few limitations on the probability distribution of the underlying population. The t statistic was developed for small-sample tests when the sample is selected at random from a *normal* distribution. The question is: How can we conduct a test of hypothesis when we have a small sample from a *nonnormal* distribution?

The **sign test** is a relatively simple nonparametric procedure for testing hypotheses about the central tendency of a nonnormal probability distribution. Note that we used the phrase *central tendency* rather than *population mean*. This is because the sign test, like many nonparametric procedures, provides inferences about the population *median*, M, rather than the population mean, μ. Remember (Chapter 2) that the median is the 50th percentile of the distribution (Figure 11.2) and as such is less affected by the skewness of the distribution and the presence of outliers (extreme observations). Since the nonparametric test must be suitable for all distributions, not just the normal, it is reasonable for nonparametric tests to focus on the more robust (less sensitive to extreme values) measure of central tendency, the median.

For example, increasing numbers of both private and public agencies are requiring their employees to submit to tests for substance abuse. One laboratory that conducts the testing has developed a system with a normalized measurement scale, in which values less than 1.00 indicate "normal" ranges and values equal to or greater than 1.00 are indicative of potential substance abuse. The lab reports a normal result as long as the median level for an individual is less than 1.00.

FIGURE 11.2

Location of the population median, M

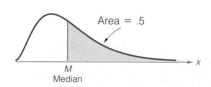

Eight independent measurements of each individual's sample are made; suppose that one individual's results were as follows:

.78   .51   3.79   .23
.77   .98   .96   .89

If the objective is to determine whether the *population* median (that is, the true median level if an indefinitely large number of measurements were made on the same individual sample) is less than 1.00, we establish that as our alternative hypothesis and test

H_0: $M = 1.00$

H_a: $M < 1.00$

The one-tailed sign test is conducted by counting the number of sample measurements that "favor" the alternative hypothesis—in this case, the number that are less than 1.00. If the null hypothesis is true, we expect approximately half of the measurements to fall on each side of the hypothesized median, and if the alternative is true, we expect significantly more than half to favor the alternative—that is, to be less than 1.00. Thus,

Test statistic: S = Number of measurements less than 1.00, the null hypothesized median

If we wish to conduct the test at the $\alpha = .05$ level of significance, then the rejection region can be expressed in terms of the observed significance level, or *p*-value, of the test:

Rejection region: *p*-value $\leq .05$

In this example, $S = 7$ of the 8 measurements are less than 1.00. To determine the observed significance level associated with this outcome, we note that the number of measurements less than 1.00 is a binomial random variable (check the binomial characteristics presented in Chapter 4), and *if H_0 is true*, the binomial probability p that a measurement lies below (or above) the median 1.00 is equal to .5 (Figure 11.2). What is the probability that a result *is as contrary to or more contrary* to H_0 than the one observed? That is, what is the probability that 7 *or more* of 8 binomial measurements will result in Success (be less than 1.00) if the probability of Success is .5? Binomial Table II in Appendix A (using $n = 8$ and $p = .5$) indicates that

$$P(x \geq 7) = 1 - P(x \leq 6) = 1 - .965 = .035$$

Thus, the probability that at least 7 of 8 measurements would be less than 1.00 *if the true median were* 1.00 is only .035. The *p*-value of the test is therefore .035, which is less than $\alpha = .05$. We therefore conclude that this sample provides sufficient evidence to reject the null hypothesis. The implication of this rejection is that the laboratory can conclude at the $\alpha = .05$ level of significance that the true median level for the tested individual is less than 1.00. However, we note that one of the measurements greatly exceeds the others, with a value of 3.79, and deserves special attention. Note that this large measurement is an outlier that would make the use of a *t*-test and its concomitant assumption of normality dubious. The only assumption necessary to ensure the validity of the sign test is that the probability distribution of measurements is continuous.

The use of the sign test for testing hypotheses about population medians is summarized in the box.

Sign Test for a Population Median M

ONE-TAILED TEST

H_0: $M = M_0$

H_a: $M > M_0$
   [or H_a: $M < M_0$]

Test statistic:

S = Number of sample measurements greater than M_0

[or S = Number of sample measurements less than M_0]

Observed significance level:

p-value = $P(x \geq S)$

TWO-TAILED TEST

H_0: $M = M_0$

H_a: $M \neq M_0$

Test statistic:

S = Larger of S_1 and S_2, where S_1 is the number of measurements less than M_0 and S_2 is the number of measurements greater than M_0

Observed significance level:

p-value = $2P(x \geq S)$

where x has a binomial distribution with parameters n and $p = .5$. (Use Table II, Appendix A.)

Rejection region: Reject H_0 if p-value $\leq .05$.

Assumption: The sample is selected randomly from a continuous probability distribution. [*Note:* No assumptions need to be made about the shape of the probability distribution.]

Recall that the normal probability distribution provides a good approximation for the binomial distribution when the sample size is large. For tests about the median of a distribution, the null hypothesis implies that $p = .5$, and the normal distribution provides a good approximation if $n \geq 10$. (Samples with $n \geq 10$ satisfy the condition that $np \pm 3\sqrt{npq}$ is contained in the interval 0 to n.) Thus, we can use the standard normal z-distribution to conduct the sign test for large samples. The large-sample sign test is summarized in the box.

Large-Sample Sign Test for a Population Median M

ONE-TAILED TEST

H_0: $M = M_0$

H_a: $M > M_0$
   [or H_a: $M < M_0$]

TWO-TAILED TEST

H_0: $M = M_0$

H_a: $M \neq M_0$

Test statistic: $z = \dfrac{(S - .5) - .5n}{.5\sqrt{n}}$

(*continued*)

[*Note:* *S* is calculated as shown in the previous box. We subtract .5 from *S* as the "correction for continuity." The null hypothesized mean value is $np = .5n$, and the standard deviation is

$$\sqrt{npq} = \sqrt{n(.5)(.5)} = .5\sqrt{n}$$

See Chapter 6 for details on the normal approximation to the binomial distribution.]

Rejection region: *Rejection region:*

$$z > z_\alpha$$ $$z > z_{\alpha/2}$$

where tabulated *z* values can be found inside the front cover.

EXAMPLE 11.1

A manufacturer of compact disk (CD) players has established that the median time to failure for its players is 5,250 hours of utilization. A sample of 20 CDs from a competitor is obtained, and they are continuously tested until each fails. The 20 failure times range from 5 hours (a "defective" player) to 6,575 hours, and 14 of the 20 exceed 5,250 hours. Is there evidence that the median failure time of the competitor differs from 5,250 hours? Use $\alpha = .10$.

Solution

The null and alternative hypotheses of interest are

H_0: $M = 5,250$ hours

H_a: $M \neq 5,250$ hours

Test statistic: Since $n \geq 10$, we use the standard normal *z* statistic:

$$z = \frac{(S - .5) - .5n}{.5\sqrt{n}}$$

where *S* is the maximum of S_1, the number of measurements greater than 5,250, and S_2, the number of measurements less than 5,250.

Rejection region: $z > 1.645$, where $z_{\alpha/2} = z_{.05} = 1.645$

Assumptions: The distribution of the failure times is continuous (time is a continuous variable), but nothing is assumed about the shape of its probability distribution.

Since the number of measurements exceeding 5,250 is $S_2 = 14$ and thus the number of measurements less than 5,250 is $S_1 = 6$, then $S = 14$, the greater of S_1 and S_2. The calculated *z* statistic is therefore

$$z = \frac{(S - .5) - .5n}{.5\sqrt{n}} = \frac{13.5 - 10}{.5\sqrt{20}} = \frac{3.5}{2.236} = 1.565$$

The value of *z* is not in the rejection region, so we cannot reject the null hypothesis at the $\alpha = .10$ level of significance. Thus, the CD manufacturer should not conclude, on the basis of this sample, that its competitor's CDs have a median failure time that differs from 5,250 hours.

The one-sample nonparametric sign test for a median provides an alternative to the *t*-test for small samples from nonnormal distributions. However, if the distribution is approximately normal, the *t*-test provides a more powerful test about the central tendency of the distribution.

LEARNING THE MECHANICS

11.1   Under what circumstances is the sign test preferred to the *t*-test for making inferences about the central tendency of a population?

11.2   What is the probability that a randomly selected observation exceeds the
a. Mean of a normal distribution?   b. Median of a normal distribution?
c. Mean of a nonnormal distribution?   d. Median of a nonnormal distribution?

11.3   Use Table II of Appendix A to calculate the following binomial probabilities:
a. $P(x \geq 6)$ when $n = 7$ and $p = .5$ b. $P(x \geq 5)$ when $n = 9$ and $p = .5$
c. $P(x \geq 8)$ when $n = 8$ and $p = .5$
d. $P(x \geq 10)$ when $n = 15$ and $p = .5$. Also use the normal approximation to calculate this probability, and compare the approximation with the exact value.
e. $P(x \geq 15)$ when $n = 25$ and $p = .5$. Also use the normal approximation to calculate this probability, and compare the approximation with the exact value.

11.4   Consider the following sample of ten measurements:

   12.1   8.5   15.8   17.7   9.6   4.0   25.2   10.3   6.2   13.9

Use these data to conduct each of the following sign tests using the binomial tables (Table II, Appendix A) and $\alpha = .05$:
a. $H_0: M = 10$ versus $H_a: M > 10$ b. $H_0: M = 10$ versus $H_a: M \neq 10$
c. $H_0: M = 18$ versus $H_a: M < 18$ d. $H_0: M = 18$ versus $H_a: M \neq 18$
e. Repeat each of the preceding tests using the normal approximation to the binomial probabilities. Compare the results.
f. What assumptions are necessary to assure the validity of each of the preceding tests?

11.5   Suppose you wish to conduct a test of the research hypothesis that the median of a population is greater than 75. You randomly sample 25 measurements from the population and determine that 16 of them exceed 75. Set up and conduct the appropriate test of hypothesis at the .10 level of significance. Be sure to specify all necessary assumptions.

APPLYING THE CONCEPTS

11.6   Many states now require that mathematics teachers pass a basic math literacy test in order to qualify to teach math in the state. An important part of the process is the development of a fair test. Suppose that one state board of education has developed a test, and it is submitted to a sample of 20 math teachers. The test will be considered too hard if at least 50% of *all* math teachers in the state score less than 60 (on a scale of 0 to 100).
a. Set up the appropriate null and alternative hypotheses to test whether the test is too hard.
b. Establish the appropriate test statistic and rejection region for the test using $\alpha = .05$.
c. What assumptions are necessary to assure the validity of the test?
d. Suppose that 14 of the 20 sampled teachers score less than 60. What is the appropriate conclusion?
e. Calculate the observed significance level of the test. Interpret it.
f. Considering your answers to parts **d** and **e**, what would your recommendation be concerning the adoption of the test statewide?

11.7   Airline industry analysts expect that the Federal Aviation Administration (FAA) will increase the frequency and thoroughness of its review of aircraft maintenance procedures in response to the admission by Eastern Airlines in the summer of 1990 that it had previously not met some maintenance requirements. Suppose that the FAA samples the records of six aircraft currently utilized by one airline and determines the number of flights between the last two complete engine maintenances for each, with the following results:

24   27   25   94   29   28

The FAA requires that this maintenance be performed at least every 30 flights. Although it is obvious that not all aircraft are meeting the requirement, the FAA wishes to test whether the airline is meeting this particular maintenance requirement "on average."

a. Would you suggest the t-test or sign test to conduct the test? Why?

b. Set up the null and alternative hypotheses such that the "burden of proof" is on the airline to show it is meeting the "on average" requirement.

c. What are the test statistic and rejection region for this test if the level of significance is $\alpha = .01$? Why would the level of significance be set at such a low value?

d. Conduct the test, and state the conclusion in terms of this application.

11.8   A paper company requires that the median height of pine trees exceed 40 feet before they are harvested. A sample of 24 trees in one large plot is selected, and 17 of them are over 40 feet.

a. Test whether the company can conclude at the .05 level of significance that the median height of trees in the plot exceeds 40 feet.

b. Calculate and interpret the p-value of this test.

c. What assumptions must be made about the probability distribution of the tree heights in the plot in order to assure the validity of the test?

11.2   Comparing Two Populations: Wilcoxon Rank Sum Test for Independent Samples

Suppose two independent random samples are to be used to compare two populations and the t-test of Chapter 9 is inappropriate for making the comparison. We may be unwilling to make assumptions about the form of the underlying population probability distributions or we may be unable to obtain exact values of the sample measurements. For either of these situations, if the data can be ranked in order of magnitude then the **Wilcoxon rank sum test** (developed by Frank Wilcoxon) can be used to test the hypothesis that the probability distributions associated with the two populations are equivalent.

Suppose an experimental psychologist wants to compare reaction times for adult males under the influence of drug A to those under the influence of drug B. Experience has shown that populations of reaction time measurements often possess probability distributions that are skewed to the right, as shown in Figure 11.3. Consequently, a t-test should not be used to compare the mean reaction times for the two drugs because the normality assumption that is required for the t-test may not be valid.

Suppose the psychologist randomly assigns seven subjects to each of two groups, one group to receive drug A and the other to receive drug B. The reaction time for each subject is measured at the completion of the experiment. These data (with the exception of the measurement for one subject in group A who was eliminated from the experiment for personal reasons) are shown in Table 11.1.

The population of reaction times for either of the drugs, say drug A, is that which could conceptually be obtained by giving drug A to all adult males. To compare the probability distributions for populations A and B, *we first rank the*

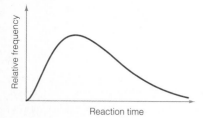

FIGURE 11.3

Typical probability distribution of reaction times

TABLE 11.1 **Reaction Times of Subjects Under the Influence of Drug A or B**

DRUG A		DRUG B	
Reaction time (seconds)	Rank	Reaction time (seconds)	Rank
1.96	4	2.11	6
2.24	7	2.43	9
1.71	2	2.07	5
2.41	8	2.71	11
1.62	1	2.50	10
1.93	3	2.84	12
		2.88	13

sample observations as though they were all drawn from the same population. That is, we pool the measurements from both samples and then rank the measurements from the smallest (a rank of 1) to the largest (a rank of 13). The results of this ranking process are also shown in Table 11.1.

If the two populations were identical, we would expect the ranks to be *randomly mixed* between the two samples. If, on the other hand, one population tends to have longer reaction times than the other, we would expect the larger ranks to be mostly in one sample and the smaller ranks mostly in the other. Thus, the test statistic for the Wilcoxon test is based on the totals of the ranks for each of the two samples—that is, on the **rank sums**. When the sample sizes are equal, for example, the greater the difference in the rank sums, the greater will be the weight of evidence to indicate a difference between the probability distributions for populations A and B. In the reaction times example, we denote the rank sum for drug A by T_A and that for drug B by T_B. Then

$$T_A = 4 + 7 + 2 + 8 + 1 + 3 = 25$$
$$T_B = 6 + 9 + 5 + 11 + 10 + 12 + 13 = 66$$

The sum of T_A and T_B will always equal $n(n + 1)/2$, where $n = n_1 + n_2$. So, for this example, $n_1 = 6$, $n_2 = 7$, and

$$T_A + T_B = \frac{13(13 + 1)}{2} = 91$$

Since $T_A + T_B$ is fixed, a small value for T_A implies a large value for T_B (and vice versa) and a large difference between T_A and T_B. Therefore, the smaller the value of one of the rank sums, the greater the evidence to indicate that the samples were selected from different populations.

The test statistic for this test is the rank sum for the smaller sample or, in the case where $n_1 = n_2$, either rank sum can be used. Values that locate the rejection region for this rank sum are given in Table XI of Appendix A. A partial reproduction of this table is shown in Figure 11.4. The columns of the table represent n_1, the first sample size, and the rows represent n_2, the second sample size. *The T_L and T_U entries in the table are the boundaries of the lower and upper regions,*

FIGURE 11.4

Reproduction of part of Table XII in Appendix A

a. $\alpha = .025$ one-tailed; $\alpha = .05$ two-tailed

n_2 \ n_1	3		4		5		6		7		8		9		10	
	T_L	T_U	T_L	T_U	T_L	T_U	T_L	T_U	T_L	T_U	T_L	T_U	T_L	T_U	T_L	T_U
3	5	16	6	18	6	21	7	23	7	26	8	28	8	31	9	33
4	6	18	11	25	12	28	12	32	13	35	14	38	15	41	16	44
5	6	21	12	28	18	37	19	41	20	45	21	49	22	53	24	56
6	7	23	12	32	19	41	26	52	28	56	29	61	31	65	32	70
7	7	26	13	35	20	45	28	56	37	68	39	73	41	78	43	83
8	8	28	14	38	21	49	29	61	39	73	49	87	51	93	54	98
9	8	31	15	41	22	53	31	65	41	78	51	93	63	108	66	114
10	9	33	16	44	24	56	32	70	43	83	54	98	66	114	79	131

respectively, for the rank sum associated with the sample that has fewer measurements. If the sample sizes n_1 and n_2 are the same, either rank sum may be used as the test statistic. To illustrate, suppose $n_1 = 8$ and $n_2 = 10$. For a two-tailed test with $\alpha = .05$, we consult part **a** of the table and find that the null hypothesis will be rejected if the rank sum of sample 1 (the sample with fewer measurements), T, is less than or equal to $T_L = 54$ *or* greater than or equal to $T_U = 98$. The Wilcoxon rank sum test is summarized in the accompanying box.

Wilcoxon Rank Sum Test: Independent Samples*

ONE-TAILED TEST

H_0: Two sampled populations have identical probability distributions

H_a: The probability distribution for population A is shifted to the right of that for B

Test statistic: The rank sum T associated with the sample with fewer measurements (if sample sizes are equal, either rank sum can be used)

Rejection region: Assuming the smaller sample size is associated with distribution A (if sample sizes are equal, we use the rank sum T_A), we reject the null hypothesis if

$$T_A \geq T_U$$

where T_U is the upper value given by Table XII in Appendix A for the chosen *one-tailed* α value.

TWO-TAILED TEST

H_0: Two sampled populations have identical probability distributions

H_a: The probability distribution for population A is shifted to the left *or* to the right of that for B

Test statistic: The rank sum T associated with the sample with fewer measurements (if the sample sizes are equal, either rank sum can be used)

Rejection region: $T \leq T_L$ or $T \geq T_U$, where T_L is the lower value given by Table XII in Appendix A for the chosen *two-tailed* α value and T_U is the upper value from Table XII.

[*Note:* If the one-sided alternative is that the probability distribution for A is shifted to the *left* of B (and T_A is the test statistic), we reject the null hypothesis if $T_A \leq T_L$.]

Assumptions: 1. The two samples are random and independent.
 2. The two probability distributions from which the samples are drawn are continuous.

Ties: Assign tied measurements the average of the ranks they would receive if they were unequal but occurred in successive order. For example, if the third-ranked and fourth-ranked measurements are tied, assign each a rank of $(3 + 4)/2 = 3.5$.

Note that the assumptions necessary for the validity of the Wilcoxon rank sum test do not specify the shape or type of probability distribution. However, the distributions are assumed to be continuous so that the probability of tied measurements is 0 (see Chapter 5), and each measurement can be assigned a unique rank. In practice, however, rounding of continuous measurements will sometimes produce ties. As long as the number of ties is small relative to the sample sizes, the Wilcoxon test procedure will still have approximate significance

---

*Another statistic used for comparing two populations based on independent random samples is the **Mann–Whitney** U **statistic**. The U statistic is a simple function of the rank sums. It can be shown that the Wilcoxon rank sum test and the Mann–Whitney U-test are equivalent.

level α. The test is not recommended to compare discrete distributions for which many ties are expected.

EXAMPLE 11.2

Do the data given in Table 11.1 provide sufficient evidence to indicate a shift in the probability distributions for drugs A and B, that is, that the probability distribution corresponding to drug A lies either to the right or left of the probability distribution corresponding to drug B? Test at the .05 level of significance.

Solution

H_0: The two populations of reaction times corresponding to drug A and drug B have the same probability distribution

H_a: The probability distribution for drug A is shifted to the right or left of the probability distribution corresponding to drug B*

Test statistic: Since drug A has fewer subjects than drug B, the test statistic is T_A, the rank sum of drug A's reaction times.

Rejection region: Since the test is two-sided, we consult part **a** of Table XII for the rejection region corresponding to $\alpha = .05$. We will reject H_0 for $T_A \leq T_L$ or $T_A \geq T_U$. Thus, we will reject H_0 if $T_A \leq 28$ or $T_A \geq 56$.

Since T_A, the rank sum of drug A's reaction times in Table 11.1, is 25, it is in the rejection region (see Figure 11.5).[†]

FIGURE 11.5

Alternative hypothesis and rejection region for Example 11.2

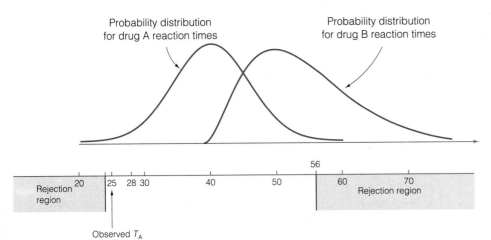

We can conclude that the probability distributions for drugs A and B are not identical. In fact, it appears that drug B tends to be associated with reaction times that are larger than those associated with drug A (because T_A fell in the lower tail of the rejection region).

---

*The alternative hypotheses in this chapter will be stated in terms of a difference in the *location* of the distributions. However, since the shapes of the distributions may also differ under H_a, some of the figures (e.g., Figure 11.5) depicting the alternative hypothesis will show probability distributions with different shapes.

[†]Figure 11.5 depicts only one side of the two-sided alternative hypothesis. The other would show distribution A shifted to the right of distribution B.

Table XII in Appendix A gives values of T_L and T_U for values of n_1 and n_2 less than or equal to 10. When both sample sizes, n_1 and n_2, are 10 or larger, the sampling distribution of T_A can be approximated by a normal distribution with mean and variance

$$E(T_A) = \frac{n_1(n_1 + n_2 + 1)}{2} \quad \text{and} \quad \sigma^2_{T_A} = \frac{n_1 n_2(n_1 + n_2 + 1)}{12}$$

Therefore, for $n_1 \geq 10$ and $n_2 \geq 10$ we can conduct the Wilcoxon rank sum test using the familiar z-test of Chapters 9 and 10. The test is summarized in the box.

Wilcoxon Rank Sum Test: Large Independent Samples

ONE-TAILED TEST

H_0: Two sampled populations have identical probability distributions

H_a: The probability distribution for population A is shifted to the right of that for B

Test statistic: $z = \dfrac{T_A - \dfrac{n_1(n_1 + n_2 + 1)}{2}}{\sqrt{\dfrac{n_1 n_2(n_1 + n_2 + 1)}{12}}}$

Rejection region: $z > z_\alpha$

Assumptions: $n_1 \geq 10$ and $n_2 \geq 10$

TWO-TAILED TEST

H_0: Two sampled populations have identical probability distributions

H_a: The probability distribution for population A is shifted to the left *or* to the right of that for B

Test statistic: $z = \dfrac{T_A - \dfrac{n_1(n_1 + n_2 + 1)}{2}}{\sqrt{\dfrac{n_1 n_2(n_1 + n_2 + 1)}{12}}}$

Rejection region: $z < -z_{\alpha/2}$ or $z > z_{\alpha/2}$

EXERCISES 11.9–11.22

LEARNING THE MECHANICS

11.9 Specify the test statistic and the rejection region for the Wilcoxon rank sum test for independent samples in each of the following situations:

a. H_0: Two probability distributions, A and B, are identical

H_a: Probability distribution for population A is shifted to the right or left of the probability distribution for population B

$n_A = 8$, $n_B = 6$, $\alpha = .10$

b. H_0: Two probability distributions, A and B, are identical

H_a: Probability distribution for population A is shifted to the right of the probability distribution for population B

$n_A = 5$, $n_B = 6$, $\alpha = .05$

c. H_0: Two probability distributions, A and B, are identical

H_a: Probability distribution for population A is shifted to the left of the probability distribution for population B

$n_A = 10$, $n_B = 8$, $\alpha = .025$

   d. H_0: Two probability distributions, A and B, are identical
      H_a: Probability distribution for population A is shifted to the right or left of the probability
          distribution for population B
      $n_A = 20, \quad n_B = 20, \quad \alpha = .05$

11.10   Suppose you want to compare two treatments, A and B. In particular, you wish to determine whether
   the distribution for population B is shifted to the right of the distribution for population A. You plan
   to use the Wilcoxon rank sum test.
   a. Specify the null and alternative hypotheses you would test.
   b. Suppose you obtained the following independent random samples of observations on experimental
      units subjected to the two treatments:

      A: 36, 39, 33, 29, 42, 33, 35, 28, 34

      B: 35, 48, 52, 66

      Conduct a test of the hypotheses described in part a. Test using $\alpha = .05$.

11.11   Explain the difference between the one-tailed and two-tailed versions of the Wilcoxon rank sum test
   for independent random samples.

11.12   Random samples of sizes $n_1 = 20$ and $n_2 = 15$ were drawn from populations 1 and 2, respectively.
   The measurements obtained are listed in the table.

POPULATION 1				POPULATION 2		
9.0	15.6	25.6	31.1	10.1	11.1	13.5
21.1	26.9	24.6	20.0	12.0	18.2	10.3
24.8	16.5	26.0	25.1	9.2	7.0	14.2
17.2	30.1	18.7	26.1	15.8	13.6	13.2
18.9	25.4	22.0	23.3	8.8	12.5	21.5

   a. Conduct a hypothesis test to determine whether the probability distribution for population 2 is
      shifted to the left of the probability distribution for population 1. Use $\alpha = .05$.
   b. What is the approximate p-value of the test of part a?

11.13   Independent random samples are selected from two populations. The data are shown in the table.

SAMPLE FROM POPULATION 1	SAMPLE FROM POPULATION 2
15	6
16	13
13	8
14	9
12	7
17	5
	4
	10

   a. Use the Wilcoxon rank sum test to determine whether the data provide sufficient evidence to
      indicate a shift in the locations of the probability distributions of the sampled populations. Test
      using $\alpha = .05$.
   b. Do the data provide sufficient evidence to indicate that the probability distribution for population
      1 is shifted to the right of the probability distribution for population 2? Use the Wilcoxon rank
      sum test with $\alpha = .05$.

11.14   Suppose you wish to compare two treatments, A and B, based on independent random samples of 15 observations selected from each of the two populations. If $T_A = 173$, do the data provide sufficient evidence to indicate that distribution A is shifted to the left of distribution B? Test using $\alpha = .05$.

APPLYING THE CONCEPTS

11.15   The property taxes levied by local governments are based on a government tax assessor's judgment of the value of the property in question. Frequently, property owners challenge the tax assessor's judgment. As a result, tax assessors must be able to demonstrate the equity or fairness of their judgments across neighborhoods, property classes, and other property groups. One measure that is used in the evaluation of fairness is the *assessment ratio*. The assessment ratio is calculated by dividing a property's assessed value by its market value (or a proxy for market value such as a recent sale price) (Freedman, 1985). Equity issues are evaluated by private real estate appraisers and government agencies using procedures such as the Wilcoxon rank sum test and the Kruskal–Wallis *H*-test (described in Section 11.4) to compare the distributions of the assessment ratios for different groups of properties (*Improving Real Property Assessment: A Reference Manual*, [1978]). The table lists the assessment ratios for random samples of ten properties in neighborhood A and eight properties in neighborhood B.

NEIGHBORHOOD A		NEIGHBORHOOD B	
.850	.880	.911	.835
1.060	.895	.770	.800
.910	.844	.815	.793
.813	.965	.748	.796
.737	.875		

a. Use the Wilcoxon rank sum test to investigate the fairness of the assessments between the two neighborhoods. Use $\alpha = .05$ and interpret your findings in the context of the problem.
b. Under what circumstances could the two-sample *t*-test of Chapter 9 be used to investigate the fairness issue of part a?
c. What assumptions are necessary to assure the validity of the test you conducted in part a?

11.16   Recall that the variance of a binomial sample proportion $\hat{p}$ depends on the value of the population parameter p. As a consequence, the variance of a sample percentage, $(100\hat{p})\%$, also depends on p. Thus, if you conduct an unpaired *t*-test (Section 9.2) to compare the means of two populations of percentages, you may be violating the assumption that $\sigma_1^2 = \sigma_2^2$, upon which the *t*-test is based. If the disparity in the variances is large, you will obtain more reliable test results using the Wilcoxon rank sum test for independent samples. In Exercise 9.30, we used a Student *t*-test to compare the mean annual percentages of labor turnover between U.S. and Japanese manufacturers of air conditioners. The annual percentage turnover rates for five U.S. and five Japanese plants are shown in the table. Do the data provide sufficient evidence to indicate that the mean annual percentage turnover for U.S. plants exceeds the corresponding mean for Japanese plants? Test using the Wilcoxon rank sum test with $\alpha = .05$.

U.S. PLANTS	JAPANESE PLANTS
7.11%	3.52%
6.06%	2.02%
8.00%	4.91%
6.87%	3.22%
4.77%	1.92%

11.17 An educational psychologist claims that the order in which test questions are asked affects a student's ability to answer correctly. To investigate this assertion, a professor randomly divides a class of 13 students into two groups—7 in one group and 6 in the other. The professor prepares one set of test questions but arranges the questions in two different orders. On test A the questions are arranged in order of increasing difficulty (that is, from easiest to most difficult), while on test B the order is reversed. One group of students is given test A, the other test B, and the test score is recorded for each student. The results are as follows:

Test A: 90, 71, 83, 82, 75, 91, 65

Test B: 66, 78, 50, 68, 80, 60

Do the data provide sufficient evidence to indicate a difference (a shift in location) in the probability distributions of student scores on the two tests? Test using $\alpha = .05$.

11.18 A major razor blade manufacturer advertises that its twin-blade disposable razor will "get you a lot more shaves" than any single-blade disposable razor on the market. A rival blade company that has been very successful in selling single-blade razors wishes to test this claim. Independent random samples of eight single-blade shavers and eight twin-blade shavers are taken, and the number of shaves that each gets before indicating a preference to change blades is recorded. The results are shown in the table.

TWIN BLADES		SINGLE BLADES	
8	15	10	13
17	10	6	14
9	6	3	5
11	12	7	7

a. Do the data support the twin-blade manufacturer's claim? Use $\alpha = .05$.
b. Do you think this experiment was designed in the best possible way? If not, what design might have been better?
c. What assumptions are necessary for the validity of the test you performed in part a? Do the assumptions seem reasonable for this application?

11.19 A realtor wants to determine whether a difference exists between home prices in two subdivisions. Six homes from subdivision A and eight homes from subdivision B are sampled, and the prices (in thousands of dollars) are recorded in the table.
a. Use the two-sample t-test to compare the population mean price per house in the two subdivisions. What assumptions are necessary for the validity of this procedure? Do you think they are reasonable in this case?
b. Use the Wilcoxon rank sum test to see whether there is a difference (a shift in location) of the probability distributions of house prices in the two subdivisions.

| SUBDIVISION | |
A	B
43	57
48	39
42	55
60	52
39	88
47	46
	41
	64

11.20   Fourteen rats were used in an experiment aimed at comparing two deprivation schedules, A and B, for their effect on hoarding behavior. An independent sampling design was used, with seven rats randomly assigned to each schedule. At the end of the deprivation period, the rats were permitted free access to food pellets and the number of pellets hoarded (taken but not eaten) during a given time period was recorded. The data are given in the table. Is there sufficient evidence to indicate that rats on one of the deprivation schedules have a greater tendency to hoard than those on the other schedule? Test using $\alpha = .05$.

SCHEDULE A	SCHEDULE B
15	5
10	1
5	2
7	8
4	2
9	6
7	3

11.21   In a comparison of visual acuity of deaf and hearing children, eye movement rates are taken on ten deaf and ten hearing children. (See the table.) A clinical psychologist believes that deaf children have greater visual acuity than hearing children. Test the psychologist's claim by using the data in the table. (The larger a child's eye movement rate, the more visual acuity the child possesses.) Use $\alpha = .05$.

DEAF CHILDREN	HEARING CHILDREN
2.75	1.15
3.14	1.65
3.23	1.43
2.30	1.83
2.64	1.75
1.95	1.23
2.17	2.03
2.45	1.64
1.83	1.96
2.23	1.37

11.22   Conduct the test in Exercise 11.21 by using the large-sample approximation for the Wilcoxon rank sum test. Compare the results with those found in Exercise 11.21.

11.3 Comparing Two Populations: Wilcoxon Signed Rank Test for the Paired Difference Experiment

Nonparametric techniques may also be employed to compare two probability distributions when a paired difference design is used. For example, consumer preferences for two competing products are often compared by having each of a sample of consumers rate both products. Thus, the ratings have been paired on each consumer. Here is an example of this type of experiment.

For some paper products, softness of the paper is an important consideration in determining consumer acceptance. One method of determining softness is to have judges give a sample of the products a softness rating. Suppose each of ten judges is given a sample of two products that a company wants to compare. Each judge rates the softness of each product on a scale from 1 to 10, with higher ratings implying a softer product. The results of the experiment are shown in Table 11.2.

T A B L E 11.2   **Softness Ratings of Paper**

JUDGE	PRODUCT A	PRODUCT B	DIFFERENCE (A − B)	ABSOLUTE VALUE OF DIFFERENCE	RANK OF ABSOLUTE VALUE
1	6	4	2	2	5
2	8	5	3	3	7.5
3	4	5	−1	1	2
4	9	8	1	1	2
5	4	1	3	3	7.5
6	7	9	−2	2	5
7	6	2	4	4	9
8	5	3	2	2	5
9	6	7	−1	1	2
10	8	2	6	6	10

T_+ = Sum of positive ranks = 46
T_- = Sum of negative ranks = 9

Since this is a paired difference experiment, we analyze the differences between the measurements (see Section 9.3). However, the nonparametric approach requires that we calculate the ranks of the absolute values of the differences between the measurements, i.e., the ranks of the differences after removing any minus signs. *Note that tied absolute differences are assigned the average of the ranks they would receive if they were unequal but successive measurements.* After the absolute differences are ranked, the sum of the ranks of the positive differences of the original measurements, T_+, and the sum of the ranks of the negative differences of the original measurements, T_-, are computed.

We are now prepared to test the nonparametric hypothesis:

H_0: The probability distributions of the ratings for products A and B are identical.

H_a: The probability distributions of the ratings differ (in location) for the two products. (Note that this is a two-sided alternative and that it implies a two-tailed test.)

Test statistic: $T = $ Smaller of the positive and negative rank sums T_+ and T_-

The smaller the value of T, the greater the evidence to indicate that the two probability distributions differ in location. The rejection region for T can be determined by consulting Table XIII in Appendix A (part of the table is shown in Figure 11.6). This table gives a value T_0 for both one-tailed and two-tailed tests for each value of n, the number of matched pairs. For a two-tailed test with $\alpha = .05$, we will reject H_0 if $T \le T_0$. You can see in Figure 11.6 (page 550) that the value of T_0 that locates the boundary of the rejection region for the judges'

FIGURE 11.6

Reproduction of part of Table XIII of Appendix A

ONE-TAILED	TWO-TAILED	$n = 5$	$n = 6$	$n = 7$	$n = 8$	$n = 9$	$n = 10$
$\alpha = .05$	$\alpha = .10$	1	2	4	6	8	11
$\alpha = .025$	$\alpha = .05$		1	2	4	6	8
$\alpha = .01$	$\alpha = .02$			0	2	3	5
$\alpha = .005$	$\alpha = .01$				0	2	3
		$n = 11$	$n = 12$	$n = 13$	$n = 14$	$n = 15$	$n = 16$
$\alpha = .05$	$\alpha = .10$	14	17	21	26	30	36
$\alpha = .025$	$\alpha = .05$	11	14	17	21	25	30
$\alpha = .01$	$\alpha = .02$	7	10	13	16	20	24
$\alpha = .005$	$\alpha = .01$	5	7	10	13	16	19
		$n = 17$	$n = 18$	$n = 19$	$n = 20$	$n = 21$	$n = 22$
$\alpha = .05$	$\alpha = .10$	41	47	54	60	68	75
$\alpha = .025$	$\alpha = .05$	35	40	46	52	59	66
$\alpha = .01$	$\alpha = .02$	28	33	38	43	49	56
$\alpha = .005$	$\alpha = .01$	23	28	32	37	43	49
		$n = 23$	$n = 24$	$n = 25$	$n = 26$	$n = 27$	$n = 28$
$\alpha = .05$	$\alpha = .10$	83	92	101	110	120	130
$\alpha = .025$	$\alpha = .05$	73	81	90	98	107	117
$\alpha = .01$	$\alpha = .02$	62	69	77	85	93	102
$\alpha = .005$	$\alpha = .01$	55	61	68	76	84	92

ratings for $\alpha = .05$ and $n = 10$ pairs of observations is 8. Therefore, the rejection region for the test (see Figure 11.7) is

Rejection region: $T \leq 8$ for $\alpha = .05$

Since the smaller rank sum for the paper data, $T_- = 9$, does not fall within the rejection region, the experiment has not provided sufficient evidence to indicate that the two paper products differ with respect to their softness ratings at the $\alpha = .05$ level.

FIGURE 11.7

Rejection region for paired difference experiment

Note that if a significance level of $\alpha = .10$ had been used, the rejection region would have been $T \leq 11$ and we would have rejected H_0. In other words, the samples do provide evidence that the probability distributions of the softness ratings differ at the $\alpha = .10$ significance level.

The Wilcoxon signed rank test is summarized in the box. Note that the difference measurements are assumed to have a continuous probability distribution so that the absolute differences will have unique ranks. Although tied (absolute) differences can be assigned ranks by averaging, the number of ties should be small relative to the number of observations to assure the validity of the test.

EXAMPLE 11.3

Suppose the police commissioner in a small community must choose between two plans for patrolling the town's streets. Plan A, the less expensive plan, uses voluntary citizen groups to patrol certain high-risk neighborhoods. In contrast,

---

Wilcoxon Signed Rank Test for a Paired Difference Experiment

ONE-TAILED TEST

H_0: Two sampled populations have identical probability distributions

H_a: The probability distribution for population A is shifted to the right of that for population B

Test statistic: T_-, the negative rank sum (we assume the differences are computed by subtracting each paired B measurement from the corresponding A measurement)

Rejection region: $T_- \leq T_0$, where T_0 is found in Table XIII (in Appendix A) for the one-tailed significance level α and the number of untied pairs, n.

TWO-TAILED TEST

H_0: Two sampled populations have identical probability distributions

H_a: The probability distribution for population A is shifted to the right *or* to the left of that for population B

Test statistic: T, the smaller of the positive and negative rank sums, T_+ and T_-

Rejection region: $T \leq T_0$, where T_0 is found in Table XIII (in Appendix A) for the two-tailed significance level α and the number of untied pairs, n.

[*Note:* If the alternative hypothesis is that the probability distribution for A is shifted to the left of B, we use T_+ as the test statistic and reject H_0 if $T_+ \leq T_0$.]

Assumptions: 1. The sample of differences is randomly selected from the population of differences.
2. The probability distribution from which the sample of paired differences is drawn is continuous.

Ties: Assign tied absolute differences the average of the ranks they would receive if they were unequal but occurred in successive order. For example, if the third-ranked and fourth-ranked differences are tied, assign both a rank of $(3 + 4)/2 = 3.5$.

---

plan B would utilize police patrols. As an aid in reaching a decision, both plans are examined by ten trained criminologists, each of whom is asked to rate the plans on a scale from 1 to 10. (High ratings imply a more effective crime prevention plan.) The city will adopt plan B (and hire extra police) only if the data provide sufficient evidence that criminologists tend to rate plan B more effective than plan A.

The results of the survey are shown in Table 11.3 on page 552. Do the data provide evidence at the $\alpha = .05$ level that the distribution of ratings for plan B lies above that for plan A?

Solution

The null and alternative hypotheses are

H_0: The two probability distributions of effectiveness ratings are identical

H_a: The effectiveness ratings of the more expensive plan (B) tend to exceed those of plan A

Observe that the alternative hypothesis is one-sided (i.e., we only wish to detect a shift in the distribution of the B ratings to the right of the distribution of A ratings) and therefore it implies a one-tailed test of the null hypothesis (see Figure 11.8, page 552). If the alternative hypothesis is true, the B ratings will tend to be larger than the paired A ratings, more negative differences in pairs will occur, T_- will be large, and T_+ will be small. Because Table XIII is constructed

TABLE 11.3 **Effectiveness Ratings by Ten Qualified Crime Prevention Experts**

CRIME PREVENTION EXPERT	PLAN A	PLAN B	DIFFERENCE (A − B)	RANK OF ABSOLUTE DIFFERENCE
1	7	9	−2	4.5
2	4	5	−1	2
3	8	8	0	(Eliminated)
4	9	8	1	2
5	3	6	−3	6
6	6	10	−4	7.5
7	8	9	−1	2
8	10	8	2	4.5
9	9	4	5	9
10	5	9	−4	7.5

Positive rank sum $= T_+ = 15.5$

FIGURE 11.8

The alternative hypothesis for Example 11.3. We expect T_+ to be small

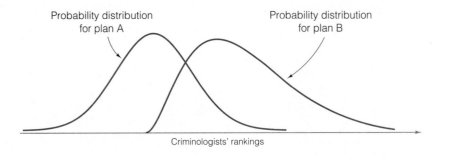

Probability distribution for plan A

Probability distribution for plan B

Criminologists' rankings

to give lower-tail values of T_0, we will use T_+ as the test statistic and reject H_0 for $T_+ \le T_0$.

The differences in ratings for the pairs (A − B) are shown in Table 11.3. Note that one of the differences equals 0. Consequently, we eliminate this pair from the ranking and reduce the number of pairs to $n = 9$. Looking in Table XIII, for a one-tailed test with $\alpha = .05$ and $n = 9$, we have $T_0 = 8$. Therefore, the test statistic and rejection region for the test are:

Test statistic: T_+, the positive rank sum

Rejection region: $T_+ \le 8$

Summing the ranks of the positive differences from Table 11.3, we find $T_+ = 15.5$. Since this value exceeds the critical value, $T_0 = 8$, we conclude that this sample provides insufficient evidence at the $\alpha = .05$ level to support the alternative hypothesis. The commissioner *cannot* conclude that the plan utilizing police patrols tends to be rated higher than the plan using citizen volunteers. That is, on the basis of this study, extra police will not be hired.

As is the case for the rank sum test for independent samples, the sampling distribution of the signed rank statistic can be approximated by a normal distribution when the number n of paired observations is large (say $n \ge 25$). The large-sample z-test is summarized in the box.

| Wilcoxon Signed Rank Test for a Paired Difference Experiment: Large Sample

ONE-TAILED TEST

H_0: Two sampled populations have identical probability distributions

H_a: The probability distribution for population A is shifted to the right of that for population B

Test statistic: $z = \dfrac{T_+ - \dfrac{n(n+1)}{4}}{\sqrt{\dfrac{n(n+1)(2n+1)}{24}}}$

Rejection region: $z > z_\alpha$

Assumptions: $n \geq 25$

TWO-TAILED TEST

H_0: Two sampled populations have identical probability distributions

H_a: The probability distribution for population A is shifted to the right or to the left of that for population B

Test statistic: $z = \dfrac{T_+ - \dfrac{n(n+1)}{4}}{\sqrt{\dfrac{n(n+1)(2n+1)}{24}}}$

Rejection region: $z < -z_{\alpha/2}$ or $z > z_{\alpha/2}$

EXERCISES 11.23–11.38

LEARNING THE MECHANICS

11.23 Specify the test statistic and the rejection region for the Wilcoxon signed rank test for the paired difference design in each of the following situations:

a. H_0: Two probability distributions, A and B, are identical
   H_a: Probability distribution for population A is shifted to the right or left of probability distribution for population B
   $n = 25$, $\alpha = .10$

b. H_0: Two probability distributions, A and B, are identical
   H_a: Probability distribution for population A is shifted to the right of the probability distribution for population B
   $n = 41$, $\alpha = .05$

c. H_0: Two probability distributions, A and B, are identical
   H_a: Probability distribution for population A is shifted to the left of the probability distribution for population B
   $n = 8$, $\alpha = .005$

11.24 Suppose you want to test a hypothesis that two treatments, A and B, are equivalent against the alternative hypothesis that the responses for A tend to be larger than those for B. You plan to use a paired difference experiment and to analyze the resulting data using the Wilcoxon signed rank test.

PAIR OF EXPERIMENTAL UNITS	TREATMENT A	TREATMENT B	PAIR OF EXPERIMENTAL UNITS	TREATMENT A	TREATMENT B
1	56	42	6	76	75
2	62	45	7	74	63
3	98	87	8	29	30
4	45	31	9	63	59
5	82	71	10	80	82

a. Specify the null and alternative hypotheses you would test.

b. Suppose the paired difference experiment yielded the data in the table. Conduct the test of part a. Test using $\alpha = .025$.

11.25   Explain the difference between the one- and two-tailed versions of the Wilcoxon signed rank test for the paired difference experiment.

11.26   In order to conduct the Wilcoxon signed rank test, why do we need to assume the probability distribution of differences is continuous?

11.27   Suppose you wish to test a hypothesis that two treatments, A and B, are equivalent against the alternative that the responses for A tend to be larger than those for B.

a. If $n = 8$ and $\alpha = .01$, give the rejection region for a Wilcoxon signed rank test.

b. Suppose you wish to detect a difference in the locations of the distributions of the responses for A and B if such a difference exists. If $n = 7$ and $\alpha = .10$, give the rejection region for the Wilcoxon signed rank test.

11.28   A random sample of nine pairs of measurements is shown in the table.

PAIR	SAMPLE DATA FROM POPULATION 1	SAMPLE DATA FROM POPULATION 2
1	8	7
2	10	1
3	6	4
4	10	10
5	7	4
6	8	3
7	4	6
8	9	2
9	8	4

a. Use the Wilcoxon signed rank test to determine whether the data provide sufficient evidence to indicate that the probability distribution for population 1 is shifted to the right of the probability distribution for population 2. Test using $\alpha = .05$.

b. Use the Wilcoxon signed rank test to determine whether the data provide sufficient evidence to indicate that the probability distribution for population 1 is shifted either to the right or to the left of the probability distribution for population 2. Test using $\alpha = .05$.

11.29   Suppose you wish to test a hypothesis that two treatments, A and B, are equivalent against the alternative that the responses for A tend to be larger than those for B.

a. If the number of pairs equals 25, give the rejection region for the large-sample Wilcoxon signed rank test for $\alpha = .05$.

b. Suppose that $T_+ = 273$. State your test conclusions.

c. Find the p-value for the test and interpret it.

11.30   A paired difference experiment with $n = 30$ pairs yielded $T_+ = 354$.

a. Specify the null and alternative hypotheses that should be used in conducting a hypothesis test to determine whether the probability distribution for population A is located to the right of that for population B.

b. Conduct the test of part a using $\alpha = .05$.

c. What is the approximate p-value of the test of part b?

d. What assumptions are necessary to assure the validity of the test you performed in part b?

APPLYING THE CONCEPTS

11.31   Which is the more effective means of dealing with complex group problem-solving tasks—face-to-face meetings or video teleconferencing? In an experiment similar to the one described in this exercise, Daniel K. Rosetti and Theodore J. Surynt of Stetson University concluded that video teleconferencing may be the more effective method. Ten groups of four people each were randomly assigned both to a specific communication setting (face-to-face or video teleconferencing) and to one of two specific complex problems. Upon completion of the problem-solving task, the same groups were placed in the alternative communication setting and asked to complete the second problem-solving task. The percentage of each problem task correctly completed was recorded for each group, with the results given in the accompanying table.

GROUP	FACE-TO-FACE	VIDEO TELECONFERENCING	GROUP	FACE-TO-FACE	VIDEO TELECONFERENCING
1	65%	75%	6	85%	90%
2	82	80	7	98	98
3	54	60	8	35	40
4	69	65	9	85	89
5	40	55	10	70	80

a. What type of experimental design was used in this study?
b. Specify the null and alternative hypotheses that should be used in determining whether the data provide sufficient evidence to conclude (as did Rosetti and Surynt) that the problem-solving performance of video teleconferencing groups is superior to that of groups that interact face-to-face.
c. Conduct the hypothesis test of part b. Use $\alpha = .05$. Interpret the results of your test in the context of the problem.
d. What is the p-value of the test in part c?

11.32   According to the American Bar Association, in 1982 there were 612,593 lawyers in the United States. About 70% of these lawyers were in private practice; about 15% worked in government as judges, prosecutors, legislators, etc.; and about 9% worked for businesses. Because of mushrooming government regulation, high outside legal fees, and complex litigation, the number of corporate lawyers has been growing at a rapid pace. The data shown in the table are the average salaries for lawyers with 8 years experience for a sample of ten U.S. cities.

CITY	CORPORATE LAWYERS	LAWYERS WITH LAW FIRMS
Atlanta	$45,500	$45,500
Chicago	43,000	48,000
Cincinnati	43,500	45,000
Dallas/Ft. Worth	49,500	46,500
Los Angeles	47,000	60,000
Milwaukee	37,500	50,000
Minneapolis/St. Paul	47,500	43,500
New York	43,500	54,000
Pittsburgh	42,000	44,000
San Francisco	47,500	59,500

Source: Reprinted by permission of Avon Books from *The American Almanac of Jobs and Salaries* by John W. Wright. Copyright 1982, 1984, 1987 by John W. Wright.

a. Use the Wilcoxon signed rank test to determine whether the data provide sufficient evidence to conclude that the salaries of corporate lawyers differ from those of lawyers working for law firms. Test using $\alpha = .05$.

b. Under what circumstances would it be appropriate to conduct the test in part **a** using the paired difference t-test described in Chapter 9?

11.33   A 1974 Supreme Court decision (*Milliken v. Bradley*) held that desegregation could not extend beyond the boundary of the school systems that were found to be segregated. One dissenting justice feared that the decision would trigger white flight, the migration of white families out of the inner cities to the suburbs. In discussing this decision, Charles T. Clotfelter (1976) presented data showing the percentages of minority students for 1968, 1970, and 1972 in the school systems of 12 large cities. Examine the last two columns of the table, which give the percentage change in white students over two time periods, 1968–1970 and 1970–1972. Was the shift of white families out of the inner cities larger in the time period 1970–1972 than for the comparable period, 1968–1970? Test using a Wilcoxon signed rank test with $\alpha = .05$.

Racial Change and White Flight in Selected City School Systems, 1968–1972

CITY	PERCENTAGE MINORITY			PERCENTAGE CHANGE IN WHITE STUDENTS[a]	
	1968	1970	1972	1968–1970	1970–1972
Boston	31.5	35.9	40.4	−3.9	−7.4
Philadelphia	61.8	63.6	64.8	−5.7	−2.2
Baltimore	65.1	67.1	69.3	−5.6	−9.3
St. Louis	63.8	65.9	69.1	−9.4	−13.8
Chicago	62.3	65.4	69.2	−9.0	−14.7
Detroit	60.7	65.5	69.5	−15.6	−14.0
Atlanta	61.8	68.7	77.4	−22.3	−34.4
Charlotte	29.5	31.1	32.8	−3.1	−5.6
Jacksonville	28.2	29.4	32.6	−1.8	−11.4
Houston	46.7	50.6	56.4	−9.1	−17.5
San Francisco	58.8	63.1	68.2	−13.5	−22.4
San Diego	23.9	24.6	26.3	−1.1	−5.5

[a]Percentages based on beginning years.

Source: U.S. Dept. of Health, Education, and Welfare, Office for Civil Rights, *Directory of Public Elementary and Secondary Schools in Selected Districts, Fall 1968, Fall 1970, and Fall 1972.*

11.34   Hypoglycemia is a condition in which blood sugar is below normal limits. To compare two compounds, X and Y, for treating hypoglycemia, each compound is applied to half the diaphragms of each of seven white mice. Blood glucose uptake in milligrams per gram of tissue is measured for each half, producing the results listed in the table. Do the data provide sufficient evidence to indicate that one of the compounds tends to produce higher blood sugar uptake readings than the other? Test using $\alpha = .10$.

MOUSE	COMPOUND		MOUSE	COMPOUND	
	X	Y		X	Y
1	4.7	5.1	5	7.0	6.1
2	3.3	4.6	6	4.7	4.1
3	8.5	8.7	7	5.2	5.1
4	3.9	3.6			

11.35   Children completing the sixth grade at a school located in a large city have the choice of going to one of two junior high schools, A or B. Members of the school board want to compare the academic effectiveness of the two schools. The parents of six sets of identical twins agree to send one child to school A and the other to school B. Since each set of twins is in the same class at each grade level through the sixth grade, a paired difference design could be employed. Near the end of the ninth grade, an achievement test is given to each child in the experiment. The results are given in the table. Test to determine whether there is evidence of a difference (shift in location) in the probability distributions of achievement test scores at the two schools. Use $\alpha = .10$.

TWIN PAIR	SCHOOL A	B
1	65	69
2	72	72
3	86	74
4	50	52
5	60	47
6	81	72

11.36 Twelve sets of identical twins are given psychological tests to determine whether the firstborn of the twins tends to be more aggressive than the secondborn. The results are shown in the table, where the higher score indicates greater aggressiveness. Do the data provide sufficient evidence to indicate that the firstborn of a pair of twins is more aggressive than the other? Test using $\alpha = .05$.

SET	FIRSTBORN	SECONDBORN	SET	FIRSTBORN	SECONDBORN
1	86	88	7	77	65
2	71	77	8	91	90
3	77	76	9	70	65
4	68	64	10	71	80
5	91	96	11	88	81
6	72	72	12	87	72

11.37 In Exercise 9.50 we compared matched pairs of measurements on the exhalation rate (a measure of radiation) of 15 soil samples from waste gypsum and phosphate mounds in Polk County, Florida. Each soil sample was measured for exhalation rate by the Polk County Health Department (PCHD) and the Eastern Environmental Radiation Facility (EERF). The data are reproduced in the table. Do the data provide sufficient evidence to indicate that one of the measuring facilities, PCHD or EERF, tends to read higher or lower than the other? Test using the Wilcoxon signed rank test with $\alpha = .05$.

CHARCOAL CANISTER NO.	PCHD	EERF	CHARCOAL CANISTER NO.	PCHD	EERF
71	1,709.79	1,479.0	85	393.55	187.7
58	357.17	257.8	46	880.84	630.4
84	1,150.94	1,287.0	4	2,996.49	3,707.0
91	1,572.69	1,395.0	20	2,367.40	2,791.0
44	558.33	416.5	36	599.84	706.8
43	4,132.28	3,993.0	42	538.37	618.5
79	1,489.86	1,351.0	55	2,770.23	2,639.0
61	3,017.48	1,813.0			

Source: Horton, T. R. "Preliminary radiological assessment of radon exhalation from phosphate gypsum piles and inactive uranium mill tailings piles," EPA-520/5-79-004. Washington, D. C.: Environmental Protection Agency, 1979.

11.38   Economic indexes provide measures of economic change. The table lists the producer commodity price indexes for January 1985 and January 1986, for six product categories. By comparing these two sets of indexes, you can obtain information regarding changes in the economy that occurred during 1985.

PRODUCT CATEGORY	JANUARY 1985	JANUARY 1986
Processed poultry	198.8	192.4
Concrete ingredients	331.0	339.0
Lumber	343.0	329.6
Gas fuels	1,073.0	1,034.3
Drugs and pharmaceuticals	247.4	265.9
Synthetic fibers	157.6	151.1

Source: *Standard & Poor's Statistical Service, Current Statistics,* January 1987, pp. 12–13.

a. Conduct a paired difference *t*-test to compare the mean values of these indexes for January 1985 and January 1986. Use $\alpha = .05$. What assumptions are necessary for the validity of this procedure? Why might these assumptions be in doubt?

b. Use the Wilcoxon signed rank test to determine whether the data provide evidence that the probability distribution of the economic indexes has changed. Use $\alpha = .05$. What assumptions are necessary to assure the validity of this test?

11.4   Kruskal–Wallis *H*-Test for a Completely Randomized Design

In Chapter 10 we used an analysis of variance and the *F*-test to compare the means of *p* populations (treatments) based on random sampling from populations that were normally distributed with a common variance σ^2. We now present a nonparametric technique for comparing the populations that requires no assumptions concerning the population probability distributions.

Suppose a health administrator wants to compare the unoccupied bed space for three hospitals located in the same city. She randomly selects ten different days from the records of each hospital and lists the number of unoccupied beds for each day (see Table 11.4). Because the number of unoccupied beds per day may occasionally be quite large, it is conceivable that the population distributions

TABLE 11.4   **Number of Available Beds**

HOSPITAL 1		HOSPITAL 2		HOSPITAL 3	
Beds	Rank	Beds	Rank	Beds	Rank
6	5	34	25	13	9.5
38	27	28	19	35	26
3	2	42	30	19	15
17	13	13	9.5	4	3
11	8	40	29	29	20
30	21	31	22	0	1
15	11	9	7	7	6
16	12	32	23	33	24
25	17	39	28	18	14
5	4	27	18	24	16
$R_1 = 120$		$R_2 = 210.5$		$R_3 = 134.5$	

of data may be skewed to the right and that this type of data may not satisfy the assumptions necessary for a parametric comparison of the population means. We therefore use a nonparametric analysis and base our comparison on the rank sums for the three sets of sample data. Just as with two independent samples (Section 11.2), the ranks are computed for each observation according to the relative magnitude of the measurements *when the data for all the samples are combined* (see Table 11.4). Ties are treated as they were for the Wilcoxon rank sum and signed rank tests by assigning the average value of the ranks to each of the tied observations.

We test

H_0: The probability distributions of the number of unoccupied beds are the same for all three hospitals

H_a: At least two of the three hospitals have probability distributions of number of unoccupied beds that differ in location

If we denote the three sample rank sums by R_1, R_2, and R_3, the test statistic is given by

$$H = \frac{12}{n(n + 1)} \sum \frac{R_j^2}{n_j} - 3(n + 1)$$

where n_j is the number of measurements in the *j*th sample and n is the total sample size ($n = n_1 + n_2 + \cdots + n_p$). For the data in Table 11.4, we have $n_1 = n_2 = n_3 = 10$ and $n = 30$. The rank sums are $R_1 = 120$, $R_2 = 210.5$, and $R_3 = 134.5$. Thus,

$$H = \frac{12}{30(31)} \left[\frac{(120)^2}{10} + \frac{(210.5)^2}{10} + \frac{(134.5)^2}{10} \right] - 3(31)$$
$$= 99.097 - 93 = 6.097$$

The *H* statistic measures the extent to which the *p* samples differ with respect to their relative ranks. This is more easily seen by writing *H* in an alternative but equivalent form:

$$H = \frac{12}{n(n + 1)} \sum n_j(\bar{R}_j - \bar{R})^2$$

where $\bar{R}_j$ is the mean rank corresponding to sample *j* and $\bar{R}$ is the mean of all the ranks [i.e., $\bar{R} = \frac{1}{2}(n + 1)$]. Thus, the *H* statistic is 0 if all samples have the same mean rank and becomes increasingly large as the distance between the sample mean ranks grows.

If the null hypothesis is true, the distribution of *H* in repeated sampling is approximately a χ^2 (chi-square) distribution. This approximation for the sampling distribution of *H* is adequate as long as each of the *p* sample sizes exceeds 5. (See the references for more detail.) The χ^2 probability distribution is characterized by a single parameter called the *degrees of freedom associated with the distribution.* Several χ^2 probability distributions with different degrees of freedom are shown in Figure 11.9. The degrees of freedom corresponding to the approximate sampling distribution of *H* will always be $(p - 1)$, one less than the number of probability distributions being compared. Because large values of *H* support the alternative hypothesis that the populations have different probability distributions, the rejection region for the test is located in the upper tail of the χ^2-distribution, as shown in Figure 11.10 on the next page.

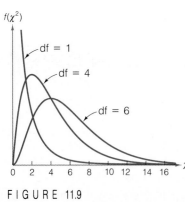

FIGURE 11.9

Several χ^2 probability distributions

FIGURE 11.10

Reproduction of part of Table VII in Appendix A

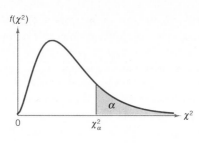

DEGREES OF FREEDOM	$\chi^2_{.100}$	$\chi^2_{.050}$	$\chi^2_{.025}$	$\chi^2_{.010}$	$\chi^2_{.005}$
1	2.70554	3.84146	5.02389	6.63490	7.87944
2	4.60517	5.99147	7.37776	9.21034	10.5966
3	6.25139	7.81473	9.34840	11.3449	12.8381
4	7.77944	9.48773	11.1433	13.2767	14.8602
5	9.23635	11.0705	12.8325	15.0863	16.7496
6	10.6446	12.5916	14.4494	16.8119	18.5476
7	12.0170	14.0671	16.0128	18.4753	20.2777
8	13.3616	15.5073	17.5346	20.0902	21.9550
9	14.6837	16.9190	19.0228	21.6660	23.5893
10	15.9871	18.3070	20.4831	23.2093	25.1882
11	17.2750	19.6751	21.9200	24.7250	26.7569
12	18.5494	21.0261	23.3367	26.2170	28.2995
13	19.8119	22.3621	24.7356	27.6883	29.8194
14	21.0642	23.6848	26.1190	29.1413	31.3193
15	22.3072	24.9958	27.4884	30.5779	32.8013
16	23.5418	26.2962	28.8454	31.9999	34.2672
17	24.7690	27.5871	30.1910	33.4087	35.7185
18	25.9894	28.8693	31.5264	34.8053	37.1564
19	27.2036	30.1435	32.8523	36.1908	38.5822

For the data of Table 11.4, the approximate distribution of the test statistic H is χ^2 with $(p - 1) = 2$ df. To determine how large H must be before we will reject the null hypothesis, we consult Table VII in Appendix A. (Part of this table is shown in Figure 11.10.) Entries in the table give an upper-tail value of χ^2, call it χ^2_α, such that $P(\chi^2 > \chi^2_\alpha) = \alpha$. The columns of the table identify the value of α associated with the tabulated value of χ^2_α, and the rows correspond to the degrees of freedom. Thus, for $\alpha = .05$ and df $= 2$, we can reject the null hypothesis that the three probability distributions are the same if

$$H > \chi^2_{.05} \quad \text{where } \chi^2_{.05} = 5.99147$$

The rejection region is pictured in Figure 11.11. Since the calculated $H = 6.097$ exceeds the critical value of 5.99147, we conclude that at least one of the three hospitals tends to have a larger number of unoccupied beds than the others.

Note that prior to conducting the experiment we might have decided to compare the daily occupancy rates for a specific pair of hospitals. The Wilcoxon rank sum test presented in Section 11.2 could be used for this purpose. The Kruskal–Wallis H-test for comparing more than two probability distributions is summarized in the accompanying box. Note that we can use the Wilcoxon rank sum test to compare the separate pairs of populations if the Kruskal–Wallis H-test

FIGURE 11.11

Rejection region for the comparison
of three probability distributions

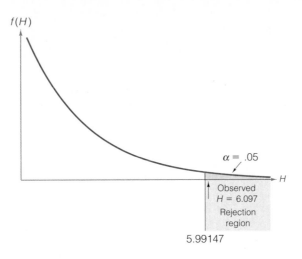

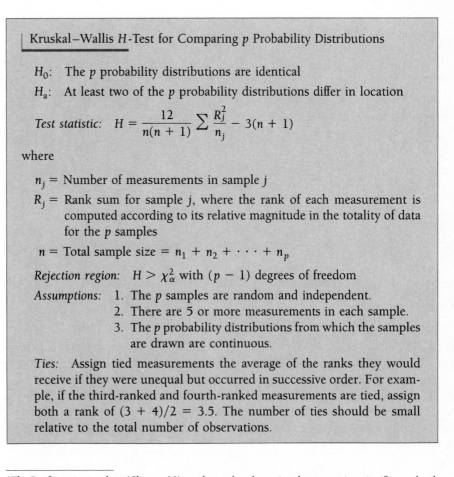

supports the alternative hypothesis that at least two of the probability distributions
differ.*

Kruskal–Wallis *H*-Test for Comparing *p* Probability Distributions

H_0: The *p* probability distributions are identical

H_a: At least two of the *p* probability distributions differ in location

Test statistic: $H = \dfrac{12}{n(n+1)} \sum \dfrac{R_j^2}{n_j} - 3(n+1)$

where

n_j = Number of measurements in sample *j*

R_j = Rank sum for sample *j*, where the rank of each measurement is
computed according to its relative magnitude in the totality of data
for the *p* samples

n = Total sample size = $n_1 + n_2 + \cdots + n_p$

Rejection region: $H > \chi_\alpha^2$ with $(p-1)$ degrees of freedom

Assumptions: 1. The *p* samples are random and independent.
2. There are 5 or more measurements in each sample.
3. The *p* probability distributions from which the samples
are drawn are continuous.

Ties: Assign tied measurements the average of the ranks they would
receive if they were unequal but occurred in successive order. For example, if the third-ranked and fourth-ranked measurements are tied, assign
both a rank of $(3+4)/2 = 3.5$. The number of ties should be small
relative to the total number of observations.

---

*The Bonferroni procedure (Chapter 10) can be used to determine the appropriate significance level
of each test such that the overall level of significance is less than or equal to a prespecified α.

EXAMPLE 11.4

Suppose a dairy farmer wants to compare the amount of milk produced by dairy cattle fed on four different diets. Five cattle are randomly assigned and maintained on each of the diets for a 3-month period. At the end of the 3 months, each cow's milk production is recorded for a 1-week period. The numbers given in Table 11.5 are the ranks of the productivity data. Do the data provide sufficient evidence to indicate that at least one of the diets tends to achieve greater milk production than the others? Test at the .05 level of significance.

TABLE 11.5 **Ranks for Milk Production Data**

DIET 1	DIET 2	DIET 3	DIET 4
1	12	8	14
5	2	9	15
6	17	3	16
7	19	11	4
10	20	13	18
$R_1 = 29$	$R_2 = 70$	$R_3 = 44$	$R_4 = 67$

Solution

The elements of the test are as follows:

H_0:   The population probability distributions of milk production for the four diets are identical

H_a: At least two of the diets have probability distributions with different locations

Test statistic: $H = \dfrac{12}{n(n+1)} \sum \dfrac{R_j^2}{n_j} - 3(n+1)$

$$= \dfrac{12}{20(21)} \left[\dfrac{(29)^2}{5} + \dfrac{(70)^2}{5} + \dfrac{(44)^2}{5} + \dfrac{(67)^2}{5} \right] - 3(21)$$

$$= 69.5 - 63 = 6.5$$

Rejection region: Since we are comparing four probability distributions, there are $p - 1 = (4 - 1) = 3$ df associated with the test statistic. Thus, we will reject H_0 if $H > \chi_{.05}^2 = 7.81473$.

Since 6.5 is less than 7.81473, we have insufficient evidence to indicate that at least one of the diets tends to achieve greater milk productivity than the others.

EXERCISES 11.39–11.52

LEARNING THE MECHANICS

11.39   Use Table VII in Appendix A to find each of the following χ^2 values:
a. $\chi_{.05}^2$,   df = 20   b. $\chi_{.025}^2$,   df = 15   c. $\chi_{.01}^2$,   df = 30
d. $\chi_{.10}^2$,   df = 80   e. $\chi_{.05}^2$,   df = 2   f. $\chi_{.005}^2$,   df = 10

11.40   Use Table VII in Appendix A to find each of the following probabilities:
a. $P(\chi^2 \geq 3.07382)$,   where df = 12   b. $P(\chi^2 \leq 24.4331)$,   where df = 40
c. $P(\chi^2 \geq 14.6837)$,   where df = 9   d. $P(\chi^2 < 34.1696)$,   where df = 20
e. $P(\chi^2 < 6.26214)$,   where df = 15   f. $P(\chi^2 \leq .584375)$,   where df = 3

11.41 Data were collected from three populations, A, B, and C, using a completely randomized design. The following describes the sample data:

$$n_A = n_B = n_C = 15$$
$$R_A = 235 \qquad R_B = 439 \qquad R_C = 361$$

a. Specify the null and alternative hypotheses that should be used in conducting a test of hypothesis to determine whether the probability distributions of populations A, B, and C differ in location.
b. Conduct the test of part **a**. Use $\alpha = .05$.
c. What is the approximate *p*-value of the test of part **b**?
d. Calculate the mean rank for each sample, and compute *H* according to the formula on page 559 that utilizes these means. Verify that this formula yields the same value of *H* that you obtained in part **b**.

11.42 Suppose you want to use the Kruskal–Wallis *H*-test to compare the probability distributions of three populations. The following are independent random samples selected from the three populations:

I: 66   33   55   88   58   62   69   49
II: 22   31   16   25   30   33   40
III: 75   96   102   75   88   78

a. What experimental design was used?
b. Specify the null and alternative hypotheses you would test.
c. Specify the rejection region you would use for your hypothesis test at $\alpha = .01$.
d. Conduct the test at $\alpha = .01$.

11.43 Under what circumstances does the χ^2-distribution provide an appropriate characterization of the sampling distribution of the Kruskal–Wallis *H* statistic?

APPLYING THE CONCEPTS

11.44 Random samples of seven lawyers employed by corporations were selected from each of three major cities. Their salaries were determined and are recorded in the table. You have been hired to determine whether differences exist among the salary distributions for corporate lawyers in the three cities.

ATLANTA	LOS ANGELES	WASHINGTON, D.C.
$45,500	$52,000	$41,500
47,900	72,000	40,100
43,100	41,000	39,000
42,000	54,000	56,500
49,000	33,000	37,000
52,000	42,000	49,000
39,000	50,000	43,500

Source: Based on *The American Almanac of Jobs and Salaries*, 1984, pp. 365–374.

a. Under what circumstances would it be appropriate to use the *F*-test for a completely randomized design to perform the required analysis?
b. Which assumptions required by the *F*-test are likely to be violated in this problem? Explain.
c. Use the Kruskal–Wallis *H*-test to determine whether the salary distributions differ among the three cities. Specify your null and alternative hypotheses, and state your conclusions in the context of the problem. Use $\alpha = .05$. What assumptions are necessary to assure the validity of the nonparametric test?

11.45   Sixth graders in an elementary school are taught reading by three different methods. Students were randomly assigned to three classes. One class used programmed instruction, a second used standard memorization techniques, and the third used an open classroom approach. The increases in reading levels attained by five students randomly selected from each of the three classes are shown in the table. Do the data provide sufficient evidence to indicate that the probability distributions of increases in reading level differ for at least two of the methods? Use $\alpha = .05$.

PROGRAMMED	STANDARD	OPEN
.9	1.0	1.7
1.5	.8	.5
.7	.9	1.6
1.1	1.2	1.4
.5	1.4	1.0

11.46   Random samples of six senior computer systems analysts were selected from each of three industries: banking, federal government, and retail sales. Their salaries are recorded in the table. You have been hired to determine whether differences exist among the salary distributions for senior systems analysts in the three industries.

BANKING	FEDERAL GOVERNMENT	RETAIL SALES
$30,000	$28,000	$20,100
24,500	34,000	19,200
27,100	39,000	20,500
26,000	35,000	20,600
23,800	34,100	21,100
25,800	36,200	19,300

Source: Based on *The American Almanac of Jobs and Salaries*, 1982, p. 420.

a. Under what circumstances would it be appropriate to use the F-test for a completely randomized design to perform the required analysis?

b. Which assumption(s) required by the F-test may be violated in this problem? Explain.

c. Use the Kruskal–Wallis H-test to determine whether the salary distributions differ among the three industries. Specify your null and alternative hypotheses, and state your conclusions in the context of the problem. Use $\alpha = .05$.

11.47   Three lists of words, representing three levels of abstractness, are randomly assigned to 21 experimental subjects so that seven subjects receive each list. The subjects are asked to respond to each word on their list with as many associated words as possible within a given period of time. A subject's score is the total number of associates, summing over all words in the list. Scores for each list are given in the table. Do the data provide sufficient evidence to indicate a difference (shift in location) between at least two of the probability distributions of the numbers of word associates that subjects can name for the three lists? Use $\alpha = .05$.

LIST 1	LIST 2	LIST 3
48	41	18
43	36	42
39	29	28
57	40	38
21	35	15
47	45	33
58	32	31

11.48 An experiment was conducted to compare the length of time it takes a human to recover from each of the three types of influenza—Victoria A, Texas, and Russian. Twenty-one human subjects were selected at random from a group of volunteers and divided into three groups of seven each. Each group was randomly assigned a strain of the virus, and the influenza was induced in the subjects. All the subjects were then cared for under identical conditions, and the recovery time (in days) was recorded. The results are shown in the accompanying table.

VICTORIA A	TEXAS	RUSSIAN
12	9	7
6	10	3
13	5	7
10	4	5
8	9	6
11	8	4
7	11	8

a. Do the data provide sufficient evidence to indicate that the recovery times for one or more types of influenza tend to be longer than for the other types? Test using $\alpha = .05$.

b. Do the data provide sufficient evidence to indicate a difference in locations of the distributions of recovery times for the Victoria A and Russian types? Test using $\alpha = .05$.

11.49 The EPA wants to determine whether temperature changes in the ocean's water caused by a nuclear power plant will have a significant effect on the animal life in the region. Recently hatched specimens of a certain species of fish are randomly divided into four groups. The groups are placed in separate simulated ocean environments that are identical in every way except for water temperature. Six months later, the specimens are weighed. The results (in ounces) are given in the table. Do the data provide sufficient evidence to indicate that one or more of the temperatures tend to produce larger weight increases than the other temperatures? Test using $\alpha = .10$.

38°F	42°F	46°F	50°F
22	15	14	17
24	21	28	18
16	26	21	13
18	16	19	20
19	25	24	21
	17	23	

11.50 An experiment was conducted to determine whether a test designed to identify a certain form of mental illness could be easily interpreted with little psychological training. Thirty judges were selected to review the results of 100 tests, half of which were given to disturbed patients and half to normal people. Of the 30 judges chosen, ten were staff members of a mental hospital, ten were trainees at the hospital, and ten were undergraduate psychology majors. The results in the table give the number of the 100 tests correctly classified by each judge. Do the data provide sufficient evidence of a difference in the probability distributions of the number of correct identifications among the three types of judges? Use $\alpha = .05$.

STAFF		TRAINEES		UNDERGRADUATES	
78	76	80	69	65	74
79	86	75	81	70	80
85	88	72	76	74	73
93	84	68	72	78	75
90	81	75	76	68	73

11.51   Three different brands of magnetron tubes (the key components in microwave ovens) were subjected to stressful testing, and the number of hours each operated without repair was recorded. (See the table.) Although these times do not represent typical life lengths, they do indicate how well the tubes can withstand extreme stress.

| | BRAND | |
A	B	C
36	49	71
48	33	31
5	60	140
67	2	59
53	55	42

a. Use the F-test for a completely randomized design (Chapter 10) to test the hypothesis that the mean length of life under stress is the same for the three brands. Use $\alpha = .05$. What assumptions are necessary for the validity of this procedure? Is there any reason to doubt these assumptions?

b. Use the Kruskal–Wallis H-test to determine whether evidence exists to conclude that the brands of magnetron tubes tend to differ in length of life under stress. Use $\alpha = .05$.

11.52   In choosing a mutual fund, an investor should compare his or her personal investment goals with those of the mutual fund. Among the hundreds of available mutual funds, three of the more prevalent advertised goals are income, growth, and maximum growth. *Income funds* seek to maximize current income; *growth funds* seek long-term capital appreciation, with current income a secondary goal; and *maximum growth funds* seek greater capital appreciation by taking larger risks. The table lists the total rate of return to investors for samples of seven mutual funds in each of these three categories.

INCOME		GROWTH		MAXIMUM GROWTH	
Mutual Fund	Rate of Return	Mutual Fund	Rate of Return	Mutual Fund	Rate of Return
Am. Nat. Income	8.3%	Babson Growth	20.2%	Fairfield	2.2%
Bull & Bear		Oppenheimer	1.9	44 Wall St. Equity	16.9
Equity–Income	19.0	Midamerica Mutual	11.2	IDS Strategy–	
Oppenheimer		Investment		Aggressive Equity	23.3
Equity–Income	15.6	Portfolios–Equity	4.5	Omega	12.1
T. Rowe Price		Franklin Equity	19.3	Pilot	10.4
Equity–Income	26.8	Fidelity Contrafund	13.3	Strong Opportunity	59.9
Seligman Income	17.1	Thomson McKinnon		Lowry Market	
Safeco Income	20.1	Growth	23.1	Timing	−9.3
National Stock	14.9				

Source: "Mutual fund scoreboard," *Business Week*, Feb. 23, 1987, pp. 70–103. Reprinted by special permission. © 1987 by McGraw-Hill, Inc.

a. Do the data provide sufficient evidence to conclude that the rate-of-return distributions differ among the three types of mutual funds? Test using $\alpha = .05$.

b. What assumptions must hold for your test of part a to be valid?

c. Describe in the context of the problem the Type I and Type II errors associated with the test of part a.

d. Under what circumstances could the F-test of Section 10.2 be employed to address the question of part a?

11.5 The Friedman F_r-Test for a Randomized Block Design

In Section 10.3 we employed an analysis of variance to compare p population (treatment) means when the data were collected using a randomized block design. The **Friedman F_r-test** provides another method for testing to detect a shift in location of a set of p populations.* Like other nonparametric tests, it requires no assumptions concerning the nature of the populations other than that you be able to rank the individual observations.

In Section 11.2, we gave an example where a completely randomized design was used to compare the reaction times of subjects under the influence of one of two drugs. When the effect of a drug is short-lived (there is no carryover effect) and when the drug effect varies greatly from person to person, it may be beneficial to employ a *randomized block design*. Using the subjects as blocks, we would hope to eliminate the variability among subjects and thereby increase the amount of information in the experiment. Suppose that three drugs, A, B, and C, are to be compared using a randomized block design. Each of the three drugs is administered to the *same subject* with suitable time lags between the three doses. The order in which the drugs are administered is randomly determined for each subject. Thus, one drug would be administered to a subject, a reaction time noted, and after a sufficient length of time the second drug administered, etc.

Suppose six subjects are chosen and that the reaction times for each drug are as shown in Table 11.6. To compare the three drugs, we rank the observations within each subject (block) and then compute the rank sums for each of the drugs (treatments). Tied observations within blocks are handled in the usual manner by assigning the average value of the ranks to each of the tied observations.

T A B L E 11.6 **Reaction Times for Three Drugs**

SUBJECT	DRUG A	RANK	DRUG B	RANK	DRUG C	RANK
1	1.21	1	1.48	2	1.56	3
2	1.63	1	1.85	2	2.01	3
3	1.42	1	2.06	3	1.70	2
4	2.43	2	1.98	1	2.64	3
5	1.16	1	1.27	2	1.48	3
6	1.94	1	2.44	2	2.81	3
		$R_1 = 7$		$R_2 = 12$		$R_3 = 17$

The null and alternative hypotheses are

H_0: The populations of reaction times are identically distributed for all three drugs

H_a: At least two of the drugs have probability distributions of reaction times that differ in location

The Friedman F_r-test statistic, which is based on the rank sums for each treatment, is

$$F_r = \frac{12}{bp(p + 1)} \sum R_j^2 - 3b(p + 1)$$

---

*The Friedman F_r-test was developed by the Nobel prize winning economist Milton Friedman.

where b is the number of blocks, p is the number of treatments, and R_j is the jth rank sum. For the data in Table 11.6,

$$F_r = \frac{12}{(6)(3)(4)}[(7)^2 + (12)^2 + (17)^2] - 3(6)(4)$$

$$= 80.33 - 72 = 8.33$$

The Friedman F_r statistic measures the extent to which the p samples differ with respect to their relative ranks within the blocks. This is more easily seen by writing F_r in an alternative but equivalent form:

$$F_r = \frac{12}{bp(p + 1)} \sum b(\bar{R}_j - \bar{R})^2$$

where $\bar{R}_j$ is the mean rank corresponding to treatment j and $\bar{R}$ is the mean of all the ranks [i.e., $\bar{R} = \frac{1}{2}(p + 1)$]. Thus, the F_r statistic is 0 if all treatments have the same mean rank and becomes increasingly large as the distance between the sample mean ranks grows.

As for the Kruskal–Wallis H statistic, the Friedman F_r statistic has approximately a χ^2 sampling distribution with $(p - 1)$ degrees of freedom. Empirical results show the approximation to be adequate if either b (the number of blocks) or p (the number of treatments) exceeds 5. The Friedman F_r-test for a randomized block design is summarized in the box.

Friedman F_r-Test for a Randomized Block Design

H_0: The probability distributions for the p treatments are identical

H_a: At least two of the probability distributions differ in location

Test statistic: $F_r = \dfrac{12}{bp(p + 1)} \sum R_j^2 - 3b(p + 1)$

where

b = Number of blocks

p = Number of treatments

R_j = Rank sum of the jth treatment, where the rank of each measurement is computed relative to its position *within its own block*

Rejection region: $F_r > \chi_\alpha^2$ with $(p - 1)$ degrees of freedom

Assumptions: 1. The treatments are randomly assigned to experimental units within the blocks.
2. The measurements can be ranked within blocks.
3. The p probability distributions from which the samples within each block are drawn are continuous.

Ties: Assign tied measurements within a block the average of the ranks they would receive if they were unequal but occurred in successive order. For example, if the third-ranked and fourth-ranked measurements are tied, assign each a rank of $(3 + 4)/2 = 3.5$. The number of ties should be small relative to the total number of observations.

For the drug example, we will use $\alpha = .05$ to form the rejection region:

$$F_r > \chi^2_{.05} = 5.99147$$

where $\chi^2_{.05}$ is based on $(p - 1) = 2$ degrees of freedom. Consequently, because the observed value, $F_r = 8.33$, exceeds 5.99147, we conclude that at least two of the three drugs have probability distributions of reaction times that differ in location.

EXAMPLE 11.5

Suppose a marketing firm wants to compare the relative effectiveness of three different modes of advertising: direct-mail, newspaper ads, and magazine ads. For 15 clients, all three modes are used over a 1-year period and the marketing firm records the year's percentage response to each type of advertising. That is, the firm divides the number of responses to a certain type of advertising by the total number of potential customers reached by the advertisements of that type. The results are shown in Table 11.7. Do these data provide sufficient evidence to indicate a difference in the locations of the probability distributions of response rates? Use the Friedman F_r-test at the $\alpha = .10$ level of significance.

TABLE 11.7 **Percentage Response to Three Types of Advertising for 15 Different Companies**

COMPANY	DIRECT-MAIL	RANK	NEWSPAPER	RANK	MAGAZINE	RANK
1	7.3	1	15.7	3	10.1	2
2	9.4	2	18.3	3	8.2	1
3	4.3	1	11.2	3	5.1	2
4	11.3	2	19.1	3	6.5	1
5	3.3	1	9.2	3	8.7	2
6	4.2	1	10.5	3	6.0	2
7	5.9	1	8.7	2	12.3	3
8	6.2	1	14.3	3	11.1	2
9	4.3	2	3.1	1	6.0	3
10	10.0	1	18.8	3	12.1	2
11	2.2	1	5.7	2	6.3	3
12	6.3	2	20.2	3	4.3	1
13	8.0	1	14.1	3	9.1	2
14	7.4	2	6.2	1	18.1	3
15	3.2	1	8.9	3	5.0	2
		$R_1 = 20$		$R_2 = 39$		$R_3 = 31$

Solution

The 15 companies act as blocks in this experiment; thus, we rank the observations within each company (block) and then compute the rank sums for each of the three types of advertising (treatments). The null and alternative hypotheses are

H_0: The probability distributions for the response rates are the same for all three types of advertising

H_a: At least two of the probability distributions of response rates differ in location

Friedman F_r-test statistic:

$$F_r = \frac{12}{bp(p + 1)} \sum R_j^2 - 3b(p + 1)$$

$$= \frac{12}{(15)(3)(4)} (R_1^2 + R_2^2 + R_3^2) - (3)(15)(4)$$

$$= \frac{12}{(15)(3)(4)} [(20)^2 + (39)^2 + (31)^2] - (3)(15)(4)$$

$$= 192.13 - 180 = 12.13$$

Rejection region: $F_r > \chi_{.10}^2 = 4.60517$ where $\chi_{.10}^2$ is based on $p - 1 = 2$ degrees of freedom

Since the calculated $F_r = 12.13$ exceeds the critical value of 4.60517 (see Figure 11.12), we conclude that the probability distributions of response rates differ in location for at least two of the three types of advertising.

FIGURE 11.12

Rejection region for Example 11.5

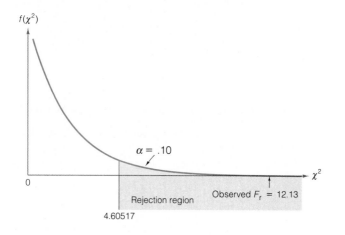

Clearly, the assumptions that the measurements are ranked within blocks and that the number of blocks (companies) is greater than 5 are satisfied. However, the experimenter must be sure that the treatments are randomly assigned to blocks. For the procedure to be valid, we assume that the three modes of advertising are used in a random order by each company. If this were not true, the difference in the response rates for the three advertising modes might be due to the order in which the modes are used.

CASE STUDY 11.1   CONSUMER RANKINGS OF PRODUCTS

Since consumers' images reflect, to some extent, actions taken by marketers in dealing with many marketing variables, it is frequently desirable to determine if these images have patterns.

McClure studied the pattern of consumers' images of three appliances. Five attributes—price, looks, need for repair, ease of use, and familiarity—were examined for each of two brands. For example, a consumer was asked to rank brand A's refrigerators, ranges, and automatic clothes washers in terms of the attribute "ease of use." McClure states:

> Conceptually, it was an investigation of the "halo effect" . . . which refers to the individual's supposed tendency to imbue his evaluations of specific character-istics of an appliance with the same direction of general feeling expressed about the brand. The halo effect would be considered operative to the extent that the individual's general image of Brand X influences his rating of an individual appliance of that brand when asked to evaluate it.*

Responses of 282 female heads of households were obtained in a large midwestern city. For each of the ten responses (five attributes for two brands) a Friedman analysis of variance was conducted, with the three types of appliances representing the treatments, and the 282 consumers acting as blocks. A signifi-cant value of the test statistic F_r would indicate consistent ranking of the three appliances by the 282 subjects for the attribute—i.e., that the probability distributions of ranks given the three appliances differ. A small value of F_r would lend credence to the hypothesis that the rankings are randomly performed—i.e., that they have approximately the same probability distributions. McClure found that the rank probability distributions differ for all attributes except price for brand A (at the $\alpha = .10$ level); brand A refrigerators consistently tend to obtain the highest ranking. However, brand B has no such clear pattern, with only the familiarity rankings showing significant differences. The subjects seemed to be most familiar with the brand B clothes washer, but they could not agree on the ranking of the other attributes. McClure concluded his article:

> Many marketing researchers have not been aware of the Friedman two-way analysis of variance by ranks. However, it has potential for use in situations common to many consumer surveys in which sets of ordinal [rank] data are generated by each respondent.

EXERCISES 11.53–11.65

LEARNING THE MECHANICS

11.53   Use Table VII in Appendix A to find each of the following χ^2 values:
  a. $\chi^2_{.05}$,   df $= 10$ b. $\chi^2_{.10}$,   df $= 15$
  c. $\chi^2_{.005}$,   df $= 50$ d. $\chi^2_{.01}$,   df $= 24$

11.54   Use Table VII in Appendix A to find each of the following probabilities:
  a. $P(\chi^2 \geq 18.3070)$ where df $= 10$
  b. $P(\chi^2 < 25.9894)$ where df $= 18$
  c. $P(\chi^2 \geq 39.9968)$ where df $= 20$

11.55   Data were collected using a randomized block design with four treatments (A, B, C, and D) and $b = 6$. The following rank sums were obtained:

  $R_A = 11$ $R_B = 21$ $R_C = 21$ $R_D = 7$

---

*Reprinted from McClure, P. "Analyzing consumer image data using Friedman two-way analysis of variance by ranks," *Journal of Marketing Research*, 1971, 8, pp. 370–371.

a. How many blocks were used in the experimental design?

b. Specify the null and alternative hypotheses that should be used in conducting a hypothesis test to determine whether the probability distributions for at least two of the treatments differ in location.

c. Conduct the test of part b. Use $\alpha = .10$.

d. What is the approximate p-value of the test of part c?

e. Calculate the mean rank for each of the four treatments, and compute the value of the F_r-test statistic according to the formula (page 568) that utilizes those means. Verify that the test statistic is the same as that you obtained in part c.

11.56 Suppose you have used a randomized block design to help you compare the effectiveness of three different treatments, A, B, and C. You obtained the data given in the table and plan to conduct a Friedman F_r-test.

BLOCK	TREATMENT		
	A	B	C
1	9	11	18
2	13	13	13
3	11	12	12
4	10	15	16
5	9	8	10
6	14	12	16
7	10	12	15

a. Specify the null and alternative hypotheses you would test.

b. Specify the rejection region for the test. Use $\alpha = .10$.

c. Conduct the test and interpret the results.

11.57 An experiment was conducted using a randomized block design with four treatments and six blocks. The ranks of the measurements within each block are shown in the table. Use the Friedman F_r-test for a randomized block design to determine whether the data provide sufficient evidence to indicate that at least two of the treatment probability distributions differ in location. Test using $\alpha = .05$.

TREATMENT	BLOCK					
	1	2	3	4	5	6
1	3	3	2	3	2	3
2	1	1	1	2	1	1
3	4	4	3	4	4	4
4	2	2	4	1	3	2

APPLYING THE CONCEPTS

11.58 An *optical mark reader* (OMR) is a machine that is able to "read" pencil marks that have been entered on a scannable form. When connected to a computer, such systems are able to read and analyze data in one step. As a result, the keypunching of data into a machine-readable code can be eliminated. Eliminating this step reduces the possibility that the data will be contaminated by human error. OMRs are used by schools to grade exams and by survey research organizations to compile data from questionnaires. A manufacturer of OMRs believes its product can operate equally well in a variety of temperature and humidity environments. To determine whether operating data contradict this belief, the manufacturer asks a well-known industrial testing laboratory to test its product. Five recently produced OMRs were randomly selected and each was operated in five different environments. The number of forms each was able to process in an hour was recorded and used as a measure of the OMR's operating efficiency. These data appear in the table. Use the Friedman F_r-test to determine whether evidence exists to indicate that the probability distributions for the number of forms processed per hour differ in location for at least two of the environments. Test using $\alpha = .10$.

MACHINE NUMBER	ENVIRONMENT				
	1	2	3	4	5
1	8,001	8,025	8,100	8,055	7,991
2	7,910	7,932	7,900	7,990	7,892
3	8,111	8,101	8,201	8,175	8,102
4	7,802	7,820	7,904	7,850	7,819
5	7,500	7,601	7,702	7,633	7,600

11.59   A sociologist conducted an experiment to investigate the general public's perception of certain occupations. Each of a random sample of 15 people was asked to rank, in order of prestige, five leading professions: lawyer, politician, physician, corporate president, and college professor. Do the data shown in the table provide sufficient evidence to indicate a difference in the amount of prestige the public attaches to these five professions? Test using $\alpha = .025$.

PERSON	LAWYER	POLITICIAN	PHYSICIAN	CORPORATE PRESIDENT	COLLEGE PROFESSOR
1	3	5	1	4	2
2	4	5	1	2	3
3	1	4	2	5	3
4	3	5	2	4	1
5	4	5	1	3	2
6	5	4	3	2	1
7	1	5	3	4	2
8	4	5	1	3	2
9	3	5	1	4	2
10	4	5	2	3	1
11	5	4	2	3	1
12	3	5	1	4	2
13	4	5	2	3	1
14	3	4	1	5	2
15	4	5	1	2	3

11.60   Suppose we want to compare the mean flood crests on the Suwannee and Santa Fe rivers for the flood level years 1984, 1983, 1973, and 1948. The data (in feet), matched by year at 6 locations on the rivers, are shown in the table. Do the data indicate that the flood crests on the rivers were higher or lower in any one of the flood level years than any other? Test using the Friedman F_r-test with $\alpha = .05$.

STATION	FLOOD STAGE	1984	1983	1973	1948
Suwannee River					
White Springs	77.0	79.9	74.4	88.5	85.2
Ellaville	54.0	59.8	52.8	65.0	68.1
Branford	29.0	29.3	29.2	35.6	38.9
Wilcox	14.0	11.6	13.0	18.6	22.3
Santa Fe River					
Three Rivers	19.0	24.1	24.3	27.2	30.6
US 129 Bridge	21.0	23.8	23.8	30.7	34.2

Source: *Gainesville Sun*, March 15, 1984.

11.61   Corrosion of different metals is a problem in many mechanical devices. Three sealers used to help retard the corrosion of metals were tested to see whether there were any differences among them. Samples of ten different metal compositions were treated with each of the three sealers, and the amount of corrosion was measured after exposure to the same environmental conditions for 1 month. The data are given in the table. Is there any evidence of a difference in the probability distributions of the amounts of corrosion among the three types of sealer? Use $\alpha = .05$.

METAL	SEALER		
	I	II	III
1	4.6	4.2	4.9
2	7.2	6.4	7.0
3	3.4	3.5	3.4
4	6.2	5.3	5.9
5	8.4	6.8	7.8
6	5.6	4.8	5.7
7	3.7	3.7	4.1
8	6.1	6.2	6.4
9	4.9	4.1	4.2
10	5.2	5.0	5.1

11.62   A serious drought-related problem for farmers is the spread of aflatoxin, a highly toxic substance caused by mold, which contaminates field corn. In higher levels of contamination, aflatoxin is potentially hazardous to animal and possibly human health. (Officials of the Food and Drug Administration have set a maximum limit of 20 parts per billion aflatoxin as safe for interstate marketing.) Three sprays, A, B, and C, have been developed to control aflatoxin in field corn. To determine whether differences exist among the sprays, ten ears of corn are randomly chosen from a contaminated corn field and each is divided into three pieces of equal size. The sprays are then randomly assigned to the pieces for each ear of corn, thus setting up a randomized block design. The table gives the amount (in parts per billion) of aflatoxin present in the corn samples after spraying.

EAR	SPRAY		
	A	B	C
1	21	23	15
2	29	30	21
3	16	19	18
4	20	19	18
5	13	10	14
6	5	12	6
7	18	18	12
8	26	32	21
9	17	20	9
10	4	10	2

a. Use the Friedman F_r-test to determine whether there is evidence that the distributions of the levels of aflatoxin in corn differ for at least two of the three sprays. Test at $\alpha = .05$.

b. Do the results of the test warrant further comparisons of the pairs of sprays? If so, conduct the comparison of the pairs of sprays using the Wilcoxon signed rank test and $\alpha = .01$ for each comparison. Can you conclude that one spray appears to best control the level of aflatoxin?

c. What assumptions are necessary to assure the validity of the procedures used in parts **a** and **b**?

11.63   In recent years, domestic car manufacturers have devoted more attention to the small car market. To compare the popularity of four domestic small cars within a city, a local trade organization obtained the information given in the table from four car dealers—one dealer for each of the four car makes. Is there evidence of a difference in location among the probability distributions of the number of cars sold for each type? Use $\alpha = .10$.

MONTH	MAKE OF CAR			
	A	B	C	D
1	9	17	14	8
2	10	20	16	9
3	13	15	19	12
4	11	12	19	11
5	7	18	13	8

11.64 An experiment is conducted to investigate the toxic effect of three chemicals, A, B, and C, on the skin of rats. Three adjacent 1-inch squares are marked on the backs of eight rats, and each of the three chemicals is applied to each rat. The squares of skin are then scored from 0 to 10, depending on the degree of irritation. The data are given in the table. Is there sufficient evidence to support the alternative hypothesis that at least two of the probability distributions of skin irritation scores corresponding to the three chemicals differ in location? Use $\alpha = .01$.

RAT	CHEMICAL		
	A	B	C
1	6	5	3
2	9	8	4
3	6	9	3
4	5	8	6
5	7	8	9
6	5	7	6
7	6	7	5
8	6	5	7

11.65 The table lists the values of five industries' inventories (in millions of dollars) as stated by the manufacturers. These figures represent book values of stocks on hand at the end of the period and include materials and supplies, goods in process, and finished goods. Suppose you wish to compare the locations of the probability distributions of the inventories among the five industries.

YEAR	FOOD AND KINDRED PRODUCTS	TOBACCO PRODUCTS	PAPER & ALLIED PRODUCTS	TEXTILE MILL PRODUCTS	CHEMICALS & ALLIED PRODUCTS
1979	$19,903	$3,434	$6,936	$6,131	$17,265
1980	21,737	3,649	7,812	6,623	19,621
1981	21,481	4,375	8,613	6,871	22,039
1982	20,769	4,411	8,559	6,192	19,907
1983	20,869	3,935	8,728	6,908	19,616
1984	21,500	3,558	9,691	7,017	21,872

Source: *Business Statistics: 1984.* Dept. of Commerce, Bureau of Economic Analysis, p. 16.

a. What can be learned about the five industries by conducting a Friedman F_r-test? Identify the blocks and treatments of the experiment.

b. Conduct a Friedman F_r-test using $\alpha = .05$. Specify the null and alternative hypotheses, and state your conclusion in the context of the problem.

c. Find the approximate *p*-value for the test of part **b**, and interpret its value.

d. What assumptions must be made to assure the validity of the test?

e. Use the Wilcoxon signed rank test to compare all pairs of industries. Use $\alpha = .01$ for each comparison.

11.6   Spearman's Rank Correlation Coefficient

Suppose ten new paintings are shown to two art critics and each critic ranks the paintings from 1 (best) to 10 (worst). We want to determine whether the critics' ranks are related. Does a correspondence exist between their ratings? If a painting is rated high by critic 1, is it likely to be rated high by critic 2? Or do high rankings by one critic correspond to low rankings by the other? That is, we wish to determine whether the rankings of the critics are **correlated**.

If the rankings are as shown in the "Perfect Agreement" columns of Table 11.8, we immediately notice that the critics agree on the rank of every painting. High ranks correspond to high ranks and low ranks to low ranks. This is an example of perfect **positive correlation** between the ranks. In contrast, if the rankings appear as shown in the "Perfect Disagreement" columns of Table 11.8, high ranks for one critic correspond to low ranks for the other. This is an example of perfect **negative correlation**.

TABLE 11.8   **Rankings of Ten Paintings by Two Critics**

PAINTING	PERFECT AGREEMENT		PERFECT DISAGREEMENT	
	Critic 1	Critic 2	Critic 1	Critic 2
1	4	4	9	2
2	1	1	3	8
3	7	7	5	6
4	5	5	1	10
5	2	2	2	9
6	6	6	10	1
7	8	8	6	5
8	3	3	4	7
9	10	10	8	3
10	9	9	7	4

In practice, you will rarely see perfect positive or negative correlation between the ranks. In fact, it is quite possible for the critics' ranks to appear as shown in Table 11.9. You will note that these rankings indicate some agreement between the critics, but not perfect agreement, thus indicating a need for a measure of rank correlation.

Spearman's rank correlation coefficient, r_s, provides a measure of correlation between ranks. The formula for this measure of correlation is given in the accompanying box. We also give a formula that is identical to r_s when there are no ties in rankings; this provides a good approximation to r_s when the number of ties is small relative to the number of pairs.

Note that if the ranks for the two critics are identical, as in the second and third columns of Table 11.8, the differences between the ranks, *d*, will all be 0.

TABLE 11.9 **Rankings of Paintings: Less Than Perfect Agreement**

PAINTINGS	CRITIC		DIFFERENCE BETWEEN RANK 1 AND RANK 2	
	1	2	d	d^2
1	4	5	-1	1
2	1	2	-1	1
3	9	10	-1	1
4	5	6	-1	1
5	2	1	1	1
6	10	9	1	1
7	7	7	0	0
8	3	3	0	0
9	6	4	2	4
10	8	8	0	0
				$\sum d^2 = 10$

Thus,

$$r_s = 1 - \frac{6 \sum d^2}{n(n^2 - 1)} = 1 - \frac{6(0)}{10(99)} = 1$$

That is, *perfect positive correlation* between the pairs of ranks is characterized by a Spearman correlation coefficient of $r_s = 1$. When the ranks indicate perfect disagreement, as in the fourth and fifth columns of Table 11.8, $\Sigma d_i^2 = 330$ and

$$r_s = 1 - \frac{6(330)}{10(99)} = -1$$

Thus, *perfect negative correlation* is indicated by $r_s = -1$.

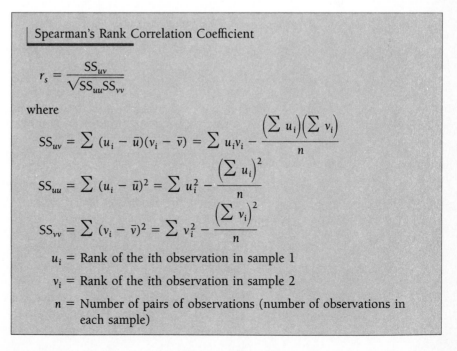

Spearman's Rank Correlation Coefficient

$$r_s = \frac{SS_{uv}}{\sqrt{SS_{uu}SS_{vv}}}$$

where

$$SS_{uv} = \sum (u_i - \bar{u})(v_i - \bar{v}) = \sum u_i v_i - \frac{\left(\sum u_i\right)\left(\sum v_i\right)}{n}$$

$$SS_{uu} = \sum (u_i - \bar{u})^2 = \sum u_i^2 - \frac{\left(\sum u_i\right)^2}{n}$$

$$SS_{vv} = \sum (v_i - \bar{v})^2 = \sum v_i^2 - \frac{\left(\sum v_i\right)^2}{n}$$

u_i = Rank of the *i*th observation in sample 1

v_i = Rank of the *i*th observation in sample 2

n = Number of pairs of observations (number of observations in each sample)

> **Shortcut Formula for r_s***
>
> $$r_s = 1 - \frac{6 \sum d_i^2}{n(n^2 - 1)}$$
>
> where
>
> $d_i = u_i - v_i$ (difference in the ranks of the ith observations for samples 1 and 2)

For the data of Table 11.9,

$$r_s = 1 - \frac{6 \sum d^2}{n(n^2 - 1)} = 1 - \frac{6(10)}{10(99)} = 1 - \frac{6}{99} = .94$$

The fact that r_s is *close* to 1 indicates that the critics tend to agree, but the agreement is not perfect.

The value of r_s always falls between -1 and $+1$, with $+1$ indicating perfect positive correlation and -1 indicating perfect negative correlation. The closer r_s falls to $+1$ or -1, the greater the correlation between the ranks. Conversely, the nearer r_s is to 0, the less the correlation.

Note that the concept of correlation implies that two responses are obtained for each experimental unit. In the art critics' example, each painting received two ranks (one for each critic) and the objective of the study was to determine the degree of positive correlation between the two rankings. Rank correlation methods can be used to measure the correlation between any pair of variables. If two variables are measured on each of n experimental units, we rank the measurements associated with each variable separately. Ties receive the average of the ranks of the tied observations. Then we calculate the value of r_s for the two rankings. This value measures the rank correlation between the two variables. We illustrate the procedure in Example 11.6.

EXAMPLE 11.6

A study is conducted to investigate the relationship between cigarette smoking during pregnancy and the weights of newborn infants. A sample of 15 women smokers kept accurate records of the number of cigarettes smoked during their pregnancies, and the weights of their children were recorded at birth. The data are given in Table 11.10. Calculate and interpret Spearman's rank correlation coefficient for the data.

Solution

We first rank the number of cigarettes smoked per day, assigning a 1 to the smallest number (12) and a 15 to the largest (46). Note that the two ties receive the averages of their respective ranks. Similarly, we assign ranks to the 15 babies'

---

*The shortcut formula is not exact when there are tied measurements, but it is a good approximation when the total number of ties is not large relative to n.

TABLE 11.10   **Data and Calculations for Example 11.6**

WOMAN	CIGARETTES PER DAY	RANK	BABY'S WEIGHT (Pounds)	RANK	d	d^2
1	12	1	7.7	5	−4	16
2	15	2	8.1	9	−7	49
3	35	13	6.9	4	9	81
4	21	7	8.2	10	−3	9
5	20	5.5	8.6	13.5	−8	64
6	17	3	8.3	11.5	−8.5	72.25
7	19	4	9.4	15	−11	121
8	46	15	7.8	6	9	81
9	20	5.5	8.3	11.5	−6	36
10	25	8.5	5.2	1	7.5	56.25
11	39	14	6.4	3	11	121
12	25	8.5	7.9	7	1.5	2.25
13	30	12	8.0	8	4	16
14	27	10	6.1	2	8	64
15	29	11	8.6	13.5	−2.5	6.25
					Total =	795

weights. Since the number of ties is relatively small, we will use the shortcut formula to calculate r_s. The differences between the ranks of the babies' weights and the ranks of the number of cigarettes smoked per day are shown in Table 11.10. The squares of the differences are also given. Thus,

$$r_s = 1 - \frac{6 \sum d_i^2}{n(n^2 - 1)} = 1 - \frac{6(795)}{15(15^2 - 1)} = 1 - 1.42 = -.42$$

This negative correlation coefficient indicates that in this sample an increase in the number of cigarettes smoked per day is associated with (but is not necessarily the *cause* of) a decrease in the weight of the newborn infant. Can this conclusion be generalized from the sample to the population? That is, can we conclude that weights of newborns and the number of cigarettes smoked per day are negatively correlated for the populations of observations for *all* smoking mothers?

If we define ρ_s as the **population Spearman rank correlation coefficient** (i.e., the rank correlation coefficient that could be calculated from all (x, y) values in the population), this question can be answered by conducting the test

H_0: $\rho_s = 0$ (no population correlation between ranks)

H_a: $\rho_s < 0$ (negative population correlation between ranks)

Test statistic: r_s, the *sample* Spearman rank correlation coefficient

To determine a rejection region, we consult Table XIV in Appendix A, which is partially reproduced in Figure 11.13 on page 580. Note that the left-hand column gives values of n, the number of pairs of observations. The entries in the table are values for an upper-tail rejection region, since only positive values are given. Thus, for $n = 15$ and $\alpha = .05$, the value .441 is the boundary of the upper-tailed rejection region, so that $P(r_s > .441) = .05$ if H_0: $\rho_s = 0$ is true. Similarly, for negative values of r_s, we have $P(r_s < -.441) = .05$ if $\rho_s = 0$. That is, we expect to see $r_s < -.441$ only 5% of the time if there is really no relationship between the ranks of the variables. The lower-tailed rejection region is therefore

Rejection region: $\alpha = .05$ $r_s < -.441$

FIGURE 11.13

Reproduction of part of Table XIV in Appendix A

n	$\alpha = .05$	$\alpha = .025$	$\alpha = .01$	$\alpha = .005$
5	.900	—	—	—
6	.829	.886	.943	—
7	.714	.786	.893	—
8	.643	.738	.833	.881
9	.600	.683	.783	.833
10	.564	.648	.745	.794
11	.523	.623	.736	.818
12	.497	.591	.703	.780
13	.475	.566	.673	.745
14	.457	.545	.646	.716
15	.441	.525	.623	.689
16	.425	.507	.601	.666
17	.412	.490	.582	.645
18	.399	.476	.564	.625
19	.388	.462	.549	.608
20	.377	.450	.534	.591

Since the calculated $r_s = -.42$ is not less than $-.441$, we cannot reject H_0 at the $\alpha = .05$ level of significance. That is, this sample of 15 smoking mothers provides insufficient evidence to conclude that a negative correlation exists between number of cigarettes smoked and the weight of newborns for the populations of measurements corresponding to all smoking mothers. This does not, of course, mean that no relationship exists. A study using a larger sample of smokers and taking other factors into account (father's weight, sex of newborn child, etc.) would be more likely to discover whether smoking and the weight of a newborn child are related.

A summary of Spearman's nonparametric test for correlation is given in the box.

Spearman's Nonparametric Test for Rank Correlation

ONE-TAILED TEST

H_0: $\rho_s = 0$

H_a: $\rho_s > 0$
    (or H_a: $\rho_s < 0$)

Test statistic: r_s, the sample rank correlation (see the formulas for calculating r_s)

Rejection region: $r_s > r_{s,\alpha}$
    (or $r_s < -r_{s,\alpha}$ when H_a: $\rho_s < 0$)

TWO-TAILED TEST

H_0: $\rho_s = 0$

H_a: $\rho_s \neq 0$

Test statistic: r_s, the sample rank correlation (see the formulas for calculating r_s)

Rejection region: $r_s < -r_{s,\alpha/2}$ or $r_s > r_{s,\alpha/2}$

where $r_{s,\alpha}$ is the value from Table XIV corresponding to the upper-tail area α and n pairs of observations

where $r_{s,\alpha/2}$ is the value from Table XIV corresponding to the upper-tail area $\alpha/2$ and n pairs of observations.

Assumptions: 1. The sample of experimental units on which the two variables are measured is randomly selected.

2. The probability distributions of the two variables are continuous.

Ties: Assign tied measurements the average of the ranks they would receive if they were unequal but occurred in successive order. For example, if the third-ranked and fourth-ranked measurements are tied, assign each a rank of $(3 + 4)/2 = 3.5$. The number of ties should be small relative to the total number of observations.

EXAMPLE 11.7

Manufacturers of perishable foods often use preservatives to retard spoilage. One concern is that using too much preservative will change the flavor of the food. Suppose an experiment is conducted using samples of a food product with varying amounts of preservative added. The length of time until the food shows signs of spoiling and a taste rating are recorded for each sample. The taste rating is the average rating for three tasters, each of whom rates each sample on a scale from 1 (good) to 5 (bad). Twelve sample measurements are shown in Table 11.11. Use a nonparametric test to find whether the spoilage times and taste ratings are correlated. Use $\alpha = .05$.

TABLE 11.11   **Data for Example 11.7**

SAMPLE	DAYS UNTIL SPOILAGE	RANK	TASTE RATING	RANK
1	30	2	4.3	11
2	47	5	3.6	7.5
3	26	1	4.5	12
4	94	11	2.8	3
5	67	7	3.3	6
6	83	10	2.7	2
7	36	3	4.2	10
8	77	9	3.9	9
9	43	4	3.6	7.5
10	109	12	2.2	1
11	56	6	3.1	5
12	70	8	2.9	4

Note: Tied measurements are assigned the average of the ranks that would be given the measurements if they were different but consecutive.

Solution

The test is two-tailed, with

H_0: $\rho_s = 0$

H_a: $\rho_s \neq 0$

Test statistic: $r_s = 1 - \dfrac{6 \sum d_i^2}{n(n^2 - 1)}$

Rejection region: Since the test is two-tailed, we need to halve the α value before consulting Table XIV. For $\alpha = .05$, we calculate $\alpha/2 = .025$ and look up the value of $r_{s,.025}$ corresponding to $n = 12$ pairs of observations:

$$r_{s,\alpha/2} = r_{s,.025} = .591$$

We will reject H_0 if

$$r_s < -.591 \quad \text{or} \quad r_s > .591$$

The first step in the computation of r_s is to sum the squares of the differences between ranks:

$$\sum d_i^2 = (2 - 11)^2 + (5 - 7.5)^2 + \cdots + (8 - 4)^2 = 536.5$$

Then

$$r_s = 1 - \frac{6(536.5)}{12(144 - 1)} = -.876$$

Since $-.876 < -.591$, we reject H_0 and conclude that the preservative is associated with the taste ratings of the food. The fact that r_s is negative suggests that the preservative has an adverse effect on the taste.

CASE STUDY 11.2

DOES PERCEIVED PRESTIGE OF OCCUPATION DEPEND ON TITLES OF JOBS?

Wayne J. Villemez and Burton B. Silver argue that the prestige of occupations may depend on the title used to describe the occupation. This suggestion has important implications because of "the preoccupation of American sociology with status."

Table 11.12 shows ten occupational categories described in three slightly different ways. Each set of descriptions is accompanied by a ranking of those categories obtained from a random sample of 144 college students.* Villemez and Silver report Spearman's rank correlation coefficients between the different occupational descriptions. For description 1 versus description 2, $r_s = .59$; for description 1 versus description 3, $r_s = .49$; and for description 2 versus description 3, $r_s = .33$. The authors state:

> These moderate (and nonsignificant) correlations resulting from using different words suggest at least that situs (category) ranking is not entirely consistent. . . . That is, when rating broad categories of type of work, raters probably do so with sets of specific familiar occupations in mind, which biases their ratings. The changing titles in this study may have brought to mind different occupational sets.

---

*In this study the lowest rank refers to the occupation rated *highest*. Although this method of assigning ranks is opposite to the method we have been using, the numerical value of r_s, the rank correlation, is the same for both methods.

TABLE 11.12   **Occupational Descriptions and Ranks**

OCCUPATIONAL CATEGORY	DESCRIPTION 1	RANK	DESCRIPTION 2	RANK	DESCRIPTION 3	RANK
1	Building and maintenance	9	Constructing and maintaining	7.5	Construction	6.5
2	Arts and entertainment	3	Culture and information	9	Aesthetics and information	3
3	Transportation	8	Movement of goods and people	7.5	Movement of materials and people	8
4	Extraction	10	Removing materials from the earth	10	Farming and mining	9
5	Health and welfare	6.5	Citizen well-being	5.5	Social, physical, and moral well-being	1
6	Commerce	4	Business	2	Trade	5
7	Legal authority	1	Law and control	5.5	Law enforcement	6.5
8	Finance and records	5	Monetary affairs	3	Records and accounts	4
9	Manufacturing	6.5	Production	4	Production and assembly	10
10	Education and research	2	Knowledge	1	Training and education	2

Source: Villemez, W. J., and Silver, B. B. "Occupational situs as horizontal social position: A reconsideration," *Sociology and Social Research*, 1976, 61, 320–335.

CASE STUDY 11.3

THE PROBLEM OF NONRESPONSE BIAS IN MAIL SURVEYS

Researchers who collect their sample data via mail questionnaires always run the risk that their respondents will not be a representative sample of the entire population. The reason for this, according to Rosenthal and Rosnow (1975), is the tendency for respondents to be (1) better educated, (2) of higher social status, (3) more intelligent, (4) in need of social approval, (5) more social, and (6) more interested in the research topic than nonrespondents.

Researchers sometimes attempt to verify the representativeness of their sample by obtaining information about the demographic characteristics of the nonrespondents and comparing them to those of the sample of respondents. Finding similarities gives researchers confidence in the representativeness of their sample. Another approach is to contact a small sample of the nonrespondents and obtain responses to the questionnaire. These responses can then be compared with those of the original respondents. The more similar the patterns of responses, the more confident the researcher can be that the original sample of returned questionnaires is representative of the population sampled from.

In a recent marketing research study by David W. Finn, Chih-Kang Wang, and Charles W. Lamb, a random sample of 20 nonrespondents to their mail questionnaire were contacted by telephone and asked to respond to one of the questions on the questionnaire. The question was: "In general, what is your willingness to buy products made in each of the following countries?" They were asked to indicate their willingness by responding on a 5-point scale that ranged from extremely willing to extremely unwilling. The mean willingness score was computed for each country, and the countries were ranked accordingly. Similarly, a rank ordering was developed for the 273 respondents to their mail questionnaire. These rankings are displayed in Table 11.13 on the next page.

TABLE 11.13 **Ranking of Consumer Willingness to Buy Products Made in Indicated Countries**

COUNTRY	RANK ORDER	
	Respondent	Nonrespondent
UK	1	1
Japan	2	3
France	3	2
Taiwan	4	4
Brazil	5	5
India	6	6
Iran	7	7
Angola	8	8
USSR	9	9
Cuba	10	10

Note: Data were collected in Spring 1977.

Source: Finn, D. W., Wang, C.-K., and Lamb, C. W. "An examination of the effects of sample composition bias in a mail survey," *Journal of the Marketing Research Society*, October 1983, 25, 331–338.

Finn, Wang, and Lamb compared the rank orderings for the respondents and nonrespondents using Spearman's rank correlation coefficient. They obtained $r_s = .9879$ (p-value < 0.01) and concluded "that respondents and nonrespondents in this study did not differ attitudinally." Further, they noted that even if respondents and nonrespondents to a mail survey differ demographically, as was the case in their study, the Spearman rank correlation result indicates that such differences should not automatically be interpreted as signaling the existence of nonresponse bias. The sample of opinions obtained from the respondents may, in fact, be representative of the population of opinions even though respondents and nonrespondents differ demographically.

EXERCISES 11.66–11.79

LEARNING THE MECHANICS

11.66   Use Table XIV of Appendix A to find each of the following probabilities:
  a. $P(r_s > .496)$ when $n = 23$ **b.** $P(r_s > .496)$ when $n = 28$
  c. $P(r_s \le .600)$ when $n = 9$ **d.** $P(r_s < -.377 \text{ or } r_s > .377)$ when $n = 20$

11.67   Specify the rejection region for Spearman's nonparametric test for rank correlation in each of the following situations:
  a. $H_0: \rho_s = 0$; $H_a: \rho_s \ne 0$, $n = 9$, $\alpha = .05$
  b. $H_0: \rho_s = 0$; $H_a: \rho_s > 0$, $n = 25$, $\alpha = .025$
  c. $H_0: \rho_s = 0$; $H_a: \rho_s < 0$, $n = 30$, $\alpha = .005$

11.68   Compute Spearman's rank correlation coefficient for each of the following pairs of sample observations:

a.

x	30	55	60	19	40
y	26	36	65	25	35

b.

x	90	100	120	137	41
y	81	95	75	52	136

c.

x	1	15	4	10
y	11	26	15	21

d.

x	5	20	15	10	3
y	80	83	91	82	87

11.69 The following sample data were collected on variables x and y:

x	−1	2	−5	−1	3	0	4
y	−1	1	−4	1	3	1	2

a. Specify the null and alternative hypotheses that should be used in conducting a hypothesis test to determine whether the variables x and y are correlated.

b. Conduct the test of part **a** using $\alpha = .05$.

c. What is the approximate p-value of the test of part **b**?

d. What assumptions are necessary to assure the validity of the test of part **b**?

11.70 Compute r_s for the data in Case Study 11.3, and compare the result with the value given in the case study.

APPLYING THE CONCEPTS

11.71 Public safety and millions of tax dollars are at stake in highway design decisions. Accordingly, Martin Helander (1978) argues that designers should consider both the physical and mental skills of drivers. He reports a study in which he investigated the correlation between the magnitude of brake pressure applied by drivers and various physiological variables such as heart rate and electrodermal response (EDR). The table lists 14 different traffic events to which 60 test drivers were exposed. The events

TRAFFIC EVENT	TRAFFIC EVENTS ORDERED IN RANK BY:	
	Brake Pressure	EDR
Cyclist or pedestrian and oncoming car	1	1
Other car merges in front of own car	2	2
Multiple events	3	3
Leading car diverges	4	4
Cyclist or pedestrian	5	6
Own car passes other car with car following	6	5
Cyclist or pedestrian and car following	7	9
Car following and meeting other car	8	11
Meeting other car	9	8
Car following	10	10
Parked car	11	14
No event	12	12
Other car passes own car	13	7
Parked car and car following	14	13

Source: Helander, M. "Applicability of drivers' electrodermal responses to the design of the traffic environment," *Journal of Applied Psychology*, Vol. 63, No. 4, 1978, pp. 481–488.

are listed in rank order according to the average magnitude of brake pressure associated with each (rank 1 is the highest mean brake pressure and rank 14 is the lowest). The ranks of the events according to the drivers' average electrodermal response are also given (rank 1 is the highest response and rank 14 is the lowest). High electrodermal responses are associated with mental stress. Based on a one-sided hypothesis test employing Spearman's rank correlation coefficient, the researcher concluded that events involving the use of the brake are perceived as stressful and, therefore, highway designs that minimize braking should be preferred.

a. Calculate Spearman's rank correlation and test its significance using $\alpha = .05$. Does your analysis confirm Helander's finding? Explain.

b. Describe the Type I and Type II errors associated with the hypothesis test of part a.

c. What is the approximate p-value of the test of part a?

d. What assumptions are necessary to assure the validity of the test?

11.72   Two expert wine tasters were asked to rank six brands of wine. Their rankings are shown in the table. Do the data present sufficient evidence to indicate a positive correlation in the rankings of the two experts?

BRAND	EXPERT 1	EXPERT 2
A	6	5
B	5	6
C	1	2
D	3	1
E	2	4
F	4	3

11.73   An experiment was designed to study whether eye pupil size is related to a person's attempt at deception. Eight students were asked to respond verbally to a series of questions. Before the questioning began, the pupil size of each student was noted and the students were instructed to answer some of the questions dishonestly. (The number of questions answered dishonestly was left to individual choice.) During questioning, the percentage increase in pupil size was recorded. Each student was then given a deception score based on the proportion of questions answered dishonestly. (High scores indicate a large number of deceptive responses.) The results are given in the table. Can you conclude that the percentage increase in eye pupil size is positively correlated with deception score? Use $\alpha = .05$.

STUDENT	DECEPTION SCORE	PERCENTAGE INCREASE IN EYE PUPIL SIZE
1	87	10
2	63	6
3	95	11
4	50	7
5	43	0
6	89	15
7	33	4
8	55	5

11.74   The decision to build a new plant or to move an existing plant to a new location involves long-term commitment of both human and monetary resources. Accordingly, such decisions should be made only after carefully considering the relevant factors associated with numerous alternative plant sites. A researcher recently examined the relationship between the location factors deemed important by businesses that located in Arkansas and those that considered Arkansas but located elsewhere. A questionnaire that asked manufacturers to rate the importance of 13 general location factors on a 9-point scale was completed by 118 firms that had moved a plant to Arkansas in the period 1955–1977 and by 73 firms that had recently considered Arkansas but located elsewhere. Epping averaged the importance ratings and arrived at the rankings shown in the table. Calculate Spearman's rank correlation coefficient for these data. Do the data present significant evidence to indicate a rank correlation between rankings of factor importance for manufacturers locating in Arkansas and similar rankings by manufacturers deciding to locate elsewhere? Test using $\alpha = .05$.

FACTOR	RANK	
	Manufacturers locating in Arkansas	Manufacturers not locating in Arkansas
Labor	1	1
Taxes	2	2
Industrial site	3	4
Information sources and special inducements	4	5
Legislative laws and structure	5	3
Utilities and resources	6	7
Transportation facilities	7	8
Raw material supplies	8	10
Community	9	6
Industrial financing	10	9
Markets	11	12
Business services	12	11
Personal preferences	13	13

Source: Epping, G. M. "Importance factors in plant locations in 1980," *Growth and Change*, April 1982, 13, 47–51.

11.75 It is frequently conjectured that income is one of the primary determinants of social status for an adult male. To investigate this theory, 15 adult males are chosen at random from a community inhabited primarily by professional people, and their annual gross incomes are noted. Each subject is then asked to complete a questionnaire designed to measure social status within the community. The social status scores (higher scores correspond to higher social status) and gross incomes (in thousands of dollars) are given in the table for the 15 adult males.

SUBJECT	SOCIAL STATUS	INCOME	SUBJECT	SOCIAL STATUS	INCOME
1	92	29.9	9	45	16.0
2	51	18.7	10	72	25.0
3	88	32.0	11	53	17.2
4	65	15.0	12	43	9.7
5	80	26.0	13	87	20.1
6	31	9.0	14	30	15.5
7	38	11.3	15	74	16.5
8	75	22.1			

a. Compute Spearman's rank correlation coefficient for these data.
b. Is there evidence that social status and income are positively correlated? Use $\alpha = .05$.

11.76 Many large businesses send representatives to college campuses to conduct job interviews. To aid the interviewer, one company decides to study the correlation between the strength of an applicant's references (the company requires three references) and the performance of the applicant on the job. Eight recently hired employees are sampled, and independent evaluations of both references and job performance are made on a scale from 1 to 20. The scores are given in the table on page 588.

EMPLOYEE	REFERENCES	JOB PERFORMANCE
1	18	20
2	14	13
3	19	16
4	13	9
5	16	14
6	11	18
7	20	15
8	9	12

a. Compute Spearman's rank correlation coefficient for these data.

b. Is there evidence that strength of references and job performance are positively correlated? Use $\alpha = .05$.

11.77 Recreation therapy is currently being used for the treatment of the mentally retarded. Particularly, psychologists theorize that certain types of recreation have a tranquilizing effect on highly excitable patients. In one experiment, nine mentally retarded patients were monitored for 1 week, and the total number of hours each spent in a specially designed recreation room was recorded. The patients were permitted to spend as much time as they wanted in the room. Also recorded was the number of times during the week that some type of tranquilizing medication had to be given to a patient in a highly excited state. Using the data in the table, determine whether there is sufficient evidence to indicate that recreation is negatively correlated with the number of times a tranquilizer has to be given. Test using $\alpha = .05$.

PATIENT	RECREATION TIME (Hours)	TRANQUILIZERS
1	16	3
2	22	1
3	10	4
4	8	9
5	14	5
6	34	2
7	26	0
8	13	10
9	5	9

11.78 A large manufacturing firm wants to determine whether a relationship exists between the number of work-hours an employee misses per year and the employee's annual wages. A sample of 15 employees produced the data in the table. Do these data provide evidence that the number of work-hours missed is related to annual wages? Use $\alpha = .05$.

EMPLOYEE	WORK-HOURS MISSED	ANNUAL WAGES (Thousands of dollars)	EMPLOYEE	WORK-HOURS MISSED	ANNUAL WAGES (Thousands of dollars)
1	49	15.8	9	191	10.8
2	36	17.5	10	6	18.8
3	127	11.3	11	63	13.8
4	91	13.2	12	79	12.7
5	72	13.0	13	43	15.1
6	34	14.5	14	57	24.2
7	155	11.8	15	82	13.9
8	11	20.2			

11.79 A *negotiable certificate of deposit* is a marketable receipt for funds deposited in a bank for a specified period of time at a specified rate of interest. The table lists the end-of-quarter interest rate for 3-month certificates of deposit during the period January 1979 through June 1986. The table also lists end-of-quarter values of Standard & Poor's 500 Stock Composite Average (an indicator of stock market activity) for the same time period.

YEAR	QUARTER	INTEREST RATE	S&P 500	YEAR	QUARTER	INTEREST RATE	S&P 500
1979	I	10.13%	101.59	1983	I	8.69%	152.96
	II	9.95	102.91		II	9.20	168.11
	III	11.89	109.32		III	9.39	166.07
	IV	13.43	107.94		IV	9.69	164.93
1980	I	17.57	104.69	1984	I	10.08	159.18
	II	8.49	114.24		II	11.34	153.18
	III	11.29	125.46		III	11.29	166.10
	IV	18.65	135.76		IV	8.60	167.24
1981	I	14.43	136.00	1985	I	9.02	180.66
	II	16.90	131.21		II	7.44	191.85
	III	16.84	116.18		III	7.93	182.08
	IV	12.49	122.55		IV	7.80	211.28
1982	I	14.21	111.96	1986	I	7.24	238.90
	II	14.46	109.61		II	6.73	250.84
	III	10.66	120.42				
	IV	8.66	135.28				

Source: *Standard & Poor's Statistical Service, Current Statistics, 1986*, Standard & Poor's Corporation.

a. Compute Spearman's rank correlation coefficient to measure the strength of the relationship between the interest rate on certificates of deposit and the S&P 500.
b. Test the null hypothesis that the interest rate on certificates of deposit and the S&P 500 are not correlated against the alternative hypothesis that these variables are correlated. Use $\alpha = .10$.
c. Repeat parts a and b using monthly data instead of quarterly data. These can be obtained at your library in *Standard & Poor's Current Statistics*. Compare the results with those you obtained using the quarterly data.

Summary

We presented several useful **nonparametric techniques** for testing hypotheses about the median of a single population and for comparing two or more populations. Nonparametric techniques are useful when the underlying assumptions for their parametric counterparts are not justified or when it is impossible to assign specific numerical values to the observations. The **sign test** is useful for testing hypotheses about the central tendency of a population, particularly when the assumptions necessary to conduct a t-test are not satisfied.

Rank sums are the primary tools of nonparametric statistics. The **Wilcoxon rank sum statistic** and the **Wilcoxon signed rank statistic** can be used to compare two populations for either an **independent sampling experiment** or a **paired difference experiment**. The **Kruskal–Wallis H-test** is applied when comparing p populations using a **completely randomized design**. The **Friedman F_r-test** is used to compare p populations when a **randomized block design** is conducted.

The strength of nonparametric statistics lies in their general applicability. Few restrictive assumptions are required, and they may be used for observations that can be ranked but cannot be exactly measured. Therefore, nonparametric tests, in conjunction with the parametric tests of Chapters 7–10, provide a useful set of statistical methods for comparing the locations of two or more population frequency distributions.

SUPPLEMENTARY EXERCISES 11.80–11.113

LEARNING THE MECHANICS

11.80   When is it appropriate to use the t- and F-tests of Chapters 9 and 10 for comparing two or more population means?

11.81   The data for three independent random samples are shown in the table. It is known that the sampled populations are not normally distributed. Use an appropriate test to determine whether the data provide sufficient evidence to indicate that at least two of the populations differ in location. Test using $\alpha = .05$.

SAMPLE FROM POPULATION 1	SAMPLE FROM POPULATION 2	SAMPLE FROM POPULATION 3
18	12	87
32	33	53
43	10	65
15	34	50
63	18	64
		77

11.82   A random sample of nine pairs of observations are recorded on two variables, x and y. The data are shown in the table.

PAIR	x	y
1	19	12
2	27	19
3	15	7
4	35	25
5	13	11
6	29	10
7	16	16
8	22	10
9	16	18

a. Do the data provide sufficient evidence to indicate that ρ_s, the Spearman rank correlation between x and y, differs from 0? Test using $\alpha = .05$.

b. Do the data provide sufficient evidence to indicate that the probability distribution for x is shifted to the right of that for y? Test using $\alpha = .05$.

11.83   Two independent random samples produced the measurements listed in the table. Do the data provide sufficient evidence to conclude that there is a difference between the locations of the probability distributions for the sampled populations? Test using $\alpha = .05$.

SAMPLE FROM POPULATION 1	SAMPLE FROM POPULATION 2
1.2	1.5
1.9	1.3
.7	2.9
2.5	1.9
1.0	2.7
1.8	3.5
1.1	

11.84   An experiment was conducted using a randomized block design with five treatments and four blocks. The data are shown in the table. Do the data provide sufficient evidence to conclude that at least two of the treatment probability distributions differ in location? Test using $\alpha = .05$.

TREATMENT	BLOCK			
	1	2	3	4
1	75	77	70	80
2	65	69	63	69
3	74	78	69	80
4	80	80	75	86
5	69	72	63	77

APPLYING THE CONCEPTS

11.85   A national clothing store franchise operates two stores in one city—one urban and one suburban. To stock the stores with clothing suited to the customers' needs, a survey is conducted to determine the incomes of the customers. Ten customers in each store are offered significant discounts if they will reveal the annual income of their household. The results are listed in the table (in thousands of dollars). Is there evidence that customers of one store tend to have higher incomes than customers of the other store? Use $\alpha = .05$.

STORE 1	STORE 2
23.8	16.3
32.9	23.2
17.2	10.3
90.3	28.5
18.1	15.0
34.5	14.3
21.3	19.6
27.1	13.8
20.7	12.6
29.0	23.3

11.86   An experiment was conducted to compare two print types, A and B, to determine whether type A is easier to read. Ten subjects were randomly divided into two groups of five. Each subject was given the same material to read, one group receiving the material printed with type A, the other group receiving print type B. The times necessary for each subject to read the material (in seconds) are shown at the top of the next page.

Type A: 95, 122, 101, 99, 108

Type B: 110, 102, 115, 112, 120

Do the data provide sufficient evidence to indicate that print type A is easier to read? Test using $\alpha = .05$.

11.87   Refer to Exercise 11.86. Test the research hypothesis that the median of the type A probability distribution exceeds 100 seconds. Repeat the test for the type B probability distribution. Use $\alpha = .05$ for both tests.

11.88   David J. Teece used the Wilcoxon signed rank test to examine the differential performance between organizations using an innovative decentralized and divisional organization structure known as the M-form, and their principal competitors. He identified the first firm in each of 20 industries that adopted the M-form structure. The principal competitor of each of these firms was identified. To qualify for inclusion in the study, the competing firm must also have adopted the M-form structure, but at a later date. A total of 14 pairs of firms qualified for the sample. The difference in performance within the pairs of firms was measured over two 3-year time periods—a "before" period in which only the M-form originator in each pair used the M-form, and an "after" period in which both firms had M-form structures. Differential performance in each time period was measured as the average difference in yearly return on stockholders' equity, where the differences were formed by subtracting the competitor's return on equity from the M-form originator's. The performance data appear in the accompanying table.

INDUSTRY	AVERAGE DIFFERENCE IN YEARLY RETURN ON EQUITY	
	Before	After
Grocery	25.70%	10.78%
Chemicals	6.26	−.39
Textiles	5.84	1.94
Aluminum	4.16	.56
Meat packing	−2.43	−5.80
Packaged foods	.77	−1.18
Can manufacturing	3.79	2.49
Grain milling	3.86	4.81
Petroleum	3.29	2.38
Tires	−3.36	−2.46
Autos	13.48	14.38
Electrical equipment	−10.85	−10.40
Tobacco	1.50	1.90
Retail department stores	12.32	12.37

Source: Teece, D. J. "Internal organization and economic performance: An empirical analysis of the profitability of principal firms," *The Journal of Industrial Economics*, Vol. 30, No. 2, December 1981, pp. 173–199.

a. Teece concluded that ". . . the M-form innovation has been shown to display a statistically significant impact on firm performance." Do you agree? Test using $\alpha = .05$.

b. What is the approximate *p*-value of the test of part **a**?

c. What assumptions are necessary to assure the validity of the test procedure you used in part **a**?

11.89   A study was conducted to determine whether the installation of a traffic light was effective in reducing the number of accidents at a busy intersection. Samples of 6 months prior to installation and 5 months after installation of the light yielded the numbers of accidents per month shown in the table.

BEFORE	AFTER
12	4
5	2
10	7
9	3
14	8
6	

a. Is there sufficient evidence to conclude that the probability distribution of the number of monthly accidents before installation of the traffic light is shifted to the right of the probability distribution of the number of accidents after installation of the light? Test using $\alpha = .025$.

b. Explain why this type of data might or might not be suitable for analysis using the t-test of Chapter 9.

11.90 The length of time required for a human to respond to a new painkiller was tested in the following manner. Seven randomly selected subjects were assigned to receive both aspirin and the new drug. The two treatments were spaced in time and assigned in random order. The length of time (in minutes) required for a subject to indicate that he or she could physically feel pain relief was recorded for both the aspirin and the drug. The data are shown in the table. Do the data provide sufficient evidence to indicate that the probability distribution of the times required to obtain relief with aspirin is shifted to the right of the probability distribution of the times required to obtain relief with the drug? Test using $\alpha = .05$.

SUBJECT	ASPIRIN	DRUG
1	15	7
2	20	14
3	12	13
4	20	11
5	17	10
6	14	16
7	17	11

11.91 A new diet that has recently appeared on the market is supposed to be effective in reducing weight. To determine whether the diet will produce a weight loss within a 1-week period, ten people were randomly chosen and placed on the new diet. Their weights (in pounds) were recorded at the beginning and end of 1 week, as listed in the table. Do the data provide sufficient evidence to indicate that the probability distribution of weights before the diet is shifted to the right of the probability distribution of weights after the diet? Test using $\alpha = .05$.

SUBJECT	BEFORE	AFTER
1	115	112
2	123	124
3	155	153
4	220	219
5	215	216
6	166	166
7	185	180
8	172	172
9	245	241
10	184	182

11.92   Suppose a company wants to study how personality relates to leadership. Four supervisors with different types of personalities are selected. Several employees are then selected from the group supervised by each, and these employees are asked to rate the leader of their group on a scale from 1 to 20 (20 signifies highly favorable). The resulting data are shown in the table.

SUPERVISOR			
1	2	3	4
20	17	16	8
19	11	15	12
20	13	13	10
18	15	18	14
17	14	11	9
	16		10

a. What type of experimental design was employed? Identify the key elements of the experiment: response, factor(s), factor type(s), treatments, and experimental units.
b. Test to determine whether evidence exists that the probability distributions of ratings differ for at least two of the four supervisors. Use $\alpha = .05$.
c. What assumptions are necessary to assure the validity of the test?
d. Do the results of the test warrant further comparisons of the pairs of supervisors? If so, compare all pairs of probability distributions using $\alpha = .01$ for each comparison. Does one supervisor appear to be most popular?

11.93   A manufacturer of household appliances is considering one of two chains of department stores to be the sales merchandiser for its product in the southwest. Before choosing one chain, the manufacturer wants to make a comparison of the product exposure that might be expected for the two chains. Eight locations are selected where both chains have stores and, on a specific day, the number of shoppers entering each store is recorded. The data are shown in the table. Do the data provide sufficient evidence to indicate that one of the chains tends to have more customers per day than the other? Test using $\alpha = .05$.

LOCATION	A	B
1	879	1,085
2	445	325
3	692	848
4	1,565	1,421
5	2,326	2,778
6	857	992
7	1,250	1,303
8	773	1,215

11.94   A union wants to determine the preferences of its members before negotiating with management. Ten union members are randomly selected, and an extensive questionnaire is completed by each member. The responses to the various aspects of the questionnaire will enable the union to rank in order of importance the items to be negotiated. The rankings are shown in the table.
a. What type of experimental design was employed? Identify the key elements of the experiment: response, factor(s), factor type(s), treatments, and experimental units. [*Hint:* Be careful not to confuse the ranking of the response with the response itself.]
b. Test to determine whether evidence exists that the probability distributions of ratings differ for at least two of the four negotiable items. Use $\alpha = .05$.

PERSON	MORE PAY	JOB STABILITY	FRINGE BENEFITS	SHORTER HOURS	PERSON	MORE PAY	JOB STABILITY	FRINGE BENEFITS	SHORTER HOURS
1	2	1	3	4	6	1	3	4	2
2	1	2	3	4	7	2.5	1	2.5	4
3	4	3	2	1	8	3	1	4	2
4	1	4	2	3	9	1.5	1.5	3	4
5	1	2	3	4	10	2	3	1	4

c. What assumptions are necessary to assure the validity of the test?

d. Do the results of the test warrant further comparisons of the pairs of negotiable items? If so, compare all pairs of probability distributions using the Wilcoxon signed rank tests. To determine the significance level of each test, use the Bonferroni technique (Chapter 10) and an overall $\alpha =$ .05. Does one item appear to be most important?

11.95 A state highway patrol was interested in knowing whether frequent patrolling of highways substantially reduces the number of speeders. Two similar interstate highways were selected for the study—one heavily patrolled and the other only occasionally patrolled. After 1 month, random samples of 100 cars were chosen on each highway, and the number of cars exceeding the speed limit was recorded. This process was repeated on 5 randomly selected days. The data are shown in the table.

DAY	HIGHWAY 1 Heavily patrolled	HIGHWAY 2 Occasionally patrolled
1	35	60
2	40	36
3	25	48
4	38	54
5	47	63

a. Do the data provide evidence to indicate that the heavily patrolled highway tends to have fewer speeders per 100 cars than the occasionally patrolled highway? Test using $\alpha = .05$.

b. Use the paired t-test with $\alpha = .05$ to compare the population mean number of speeders per 100 cars for the two highways. What assumptions are necessary for the validity of this procedure?

11.96 Weevils cause millions of dollars worth of damage each year to cotton crops. Three chemicals (A, B, and C) designed to control weevil populations were applied, one to each of three fields of cotton. After 3 months, ten plots of equal size were randomly selected within each field and the percentage of cotton plants with weevil damage was recorded for each. Do the data in the table provide sufficient evidence to indicate a difference in location among the distributions of damage rates corresponding to the three treatments? Use $\alpha = .05$.

A	B	C
10.8	22.3	9.8
15.6	19.5	12.3
19.2	18.6	16.2
17.9	24.3	14.1
18.3	19.9	15.3
9.8	20.4	10.8
16.7	23.6	12.2
19.0	21.2	17.3
20.3	19.8	15.1
19.4	22.6	11.3

11.97   A drug company has synthesized two new compounds to be used in sleeping pills. The data in the table represent the additional hours of sleep gained by ten patients through the use of the two drugs. Do the data present sufficient evidence to indicate that the probability distributions of additional hours of sleep differ for the two drugs? Test using $\alpha = .10$.

PATIENT	DRUG A	DRUG B	PATIENT	DRUG A	DRUG B
1	.4	.7	6	2.9	3.4
2	−.7	−1.6	7	4.0	3.7
3	−.4	−.2	8	.1	.8
4	−1.4	−1.4	9	3.1	.0
5	−1.6	−.2	10	1.9	2.0

11.98   Many water treatment facilities add hydrofluosilicic acid to the water to supplement the natural fluoride concentration in order to reach a target concentration of fluoride in the drinking water. Certain levels are thought to enhance dental health, but very high concentrations can be dangerous. Suppose that one such treatment plant targets .75 milligram per liter (mg/L) for their water. The plant tests 25 samples each day to determine whether the median level differs from the target.
   a. Set up the null and alternative hypotheses.
   b. Set up the test statistic and rejection region using $\alpha = .10$.
   c. Explain the implication of a Type I error in the context of this application. A Type II error.
   d. Suppose that one day's samples result in 18 values that exceed .75 mg/L. Conduct the test and state the appropriate conclusion in the context of this application.
   e. When it was suggested to the plant's supervisor that a t-test should be used to conduct the daily test, she replied that the probability distribution of the fluoride concentrations was "heavily skewed to the right." Show graphically what she meant by this, and explain why this is a reason to prefer the sign test to the t-test.

11.99   Three new traps were tested to compare their ability to trap mosquitoes. Each of three traps, A, B, and C, were placed side-by-side at each of five different locations. After a specified length of time, the number of mosquitoes in each trap was recorded.

LOCATION	TRAP A	TRAP B	TRAP C
1	3	5	0
2	23	17	15
3	11	5	7
4	19	11	5
5	8	4	2

   a. What type of design was employed?
   b. Identify the experimental units, response, factor(s) and factor type(s), treatments, and blocks.
   c. Is there evidence that the three probability distributions associated with the number of mosquitoes trapped by the devices differ? Use $\alpha = .05$.
   d. If warranted, use the Wilcoxon signed rank test to compare the pairs of traps. Use a significance level of $\alpha = .10$ for each comparison.
   e. What assumptions are necessary to assure the validity of the inferential procedures used in parts c and d?

11.100   An insurance company wants to determine whether a relationship exists between the number of claims filed by owners of family policies and the annual incomes of the families. A random sample of ten policies was selected for the study. The data are shown in the table. Do the data on claims and annual income provide sufficient evidence to conclude that a correlation exists between the number of claims per policy and the annual income of the policyholder? Use $\alpha = .10$.

FAMILY	CLAIMS 3-year period	ANNUAL INCOME Averaged over 3 years ($ thousand)
1	5	14.5
2	1	9.6
3	9	62.5
4	0	22.5
5	4	10.3
6	7	16.2
7	0	8.1
8	2	21.2
9	6	17.1
10	3	12.3

11.101   An experiment was conducted to compare three calculators, A, B, and C, according to ease of operation. To make the comparison, six randomly selected students were assigned to perform the same sequence of arithmetic operations on each of the three calculators. The order of use of the calculators varied in a random manner from student to student. The times necessary for the completion of the sequence of tasks (in seconds) are recorded in the table. Do the data provide sufficient evidence to indicate that the task completion times tend to be lower for at least one of the calculator types?

STUDENT	CALCULATOR TYPE		
	A	B	C
1	306	330	300
2	260	265	285
3	281	290	277
4	288	301	305
5	301	309	319
6	262	245	240

11.102   The trend among doctors in some areas of the country is to form a group practice. By combining treatment resources, doctors hope to be more efficient. One doctor who recently joined a group family practice wants to compare the distribution of the number of patients he treated during a day when he was an individual practitioner with the distribution after joining the group practice. Records were checked for 8 days before and 8 days after, with the results listed in the table. Use the Wilcoxon rank sum test to determine whether these samples indicate that the doctor tends to treat more patients per day now than before. Use $\alpha = .05$.

BEFORE	AFTER
26	28
25	30
27	27
26	31
24	29
20	23
22	28
21	29

11.103   A clothing manufacturer employs five inspectors who provide quality control of workmanship. Every item of clothing produced carries with it the number of the inspector who checked it. Thus, the company can evaluate an inspector by keeping records of the number of complaints received about products bearing his or her inspection number. The numbers of returns for 6 months are given in the table.

MONTH	INSPECTOR				
	1	2	3	4	5
1	8	10	7	6	9
2	5	7	4	12	12
3	5	8	6	10	6
4	9	6	8	10	13
5	4	13	3	7	15
6	4	8	2	6	9

a. What type of design was utilized for this experiment?

b. Use the appropriate nonparametric test to compare the treatments. Specify the hypotheses and interpret the results in terms of this experiment. Use $\alpha = .10$.

c. Does the result of your test warrant further comparisons of the pairs of treatments? If so, compare the pairs using the appropriate nonparametric technique and $\alpha = .01$ for each comparison. Interpret the results in terms of this experiment.

d. What assumptions are necessary to assure the validity of the nonparametric procedures you employed? How do they differ from the assumptions that would have to be satisfied in order to use the corresponding parametric technique to analyze this experiment?

11.104   Each applicant to a certain university is judged by his or her high school grade-point average and a score on a standard aptitude test. The GPAs and aptitude scores for eight randomly selected applicants are shown in the table. Do these data provide sufficient evidence to indicate a positive correlation between high school GPA and aptitude test score? Use $\alpha = .05$.

APPLICANT	GPA	APTITUDE TEST	APPLICANT	GPA	APTITUDE TEST
1	3.25	1,200	5	3.10	1,510
2	2.85	890	6	2.90	950
3	3.01	980	7	2.75	1,010
4	4.00	1,150	8	3.35	1,080

11.105   Refer to Exercise 11.104. Test whether this sample indicates that the median GPA of all applicants exceeds 3.0. Use $\alpha = .05$. Calculate and interpret the observed significance level of this test.

11.106   The coach of a mediocre basketball team has one all-star player who attempts the vast majority of the team's shots. For each of the last ten games, the star's number of shots and the team's winning margin (negative number implies a loss) are recorded:

GAME	STAR'S SHOTS	WINNING MARGIN	GAME	STAR'S SHOTS	WINNING MARGIN
1	19	−3	6	22	16
2	15	13	7	14	4
3	21	−8	8	11	1
4	17	−5	9	18	−5
5	25	−2	10	18	−12

Compute Spearman's rank correlation coefficient. Is there evidence of a relationship between the team's performance and the star's number of shots?

11.107   For many years, the Girl Scouts of America have sold cookies using various sales techniques. One troop experimented with several techniques and reported the number of boxes of cookies sold per scout, as listed in the table.

DOOR-TO-DOOR	TELEPHONE	GROCERY STORE STAND	DEPARTMENT STORE STAND
47	63	113	25
93	19	50	36
58	29	68	21
37	24	37	27
62	33	39	18
		77	31

a. Is there sufficient evidence to indicate that at least one sales method tends to produce a larger number of sales per scout than the others? Use $\alpha = .10$.

b. Does the door-to-door technique tend to produce a different number of sales per scout than the grocery store stand? Test using $\alpha = .05$.

11.108   A manufacturer wants to determine whether the number of defectives produced by its employees tends to increase as the day progresses. Unknown to the employees, a complete inspection is made of every item that was produced on one day, and the hourly fraction defective is recorded. The resulting data are given in the table. Do they provide evidence that the fraction defective increases as the day progresses? Test at the $\alpha = .05$ level.

HOUR	FRACTION DEFECTIVE
1	.02
2	.05
3	.03
4	.08
5	.06
6	.09
7	.11
8	.10

11.109   Refer to Exercise 11.108. Assume that the eight measurements of the fraction defective constitute a random sample from the probability distribution of all fraction defective measurements. The manufacturer wants to determine whether the median hourly fraction defective exceeds .05.

a. Conduct the appropriate test to make this determination at the .10 level of significance.

b. Why might the manufacturer wish to conduct the test with this relatively high level of significance?

c. What assumptions are necessary to assure the validity of this test? How do they differ from the assumptions that would be necessary to conduct a t-test about the mean fraction defective?

11.110   A taste test conducted to compare three brands of beer utilized ten randomly selected beer drinkers. Each person was given three unmarked glasses of beer—one containing each brand—and was asked to rate each on a scale from 1 to 10 (a higher score indicates a better taste). Do the data given in the table indicate that one (or more) of the brands of beer is preferred to the others? Test using $\alpha = .05$. If the brands differ, use the appropriate nonparametric procedure to try to determine which brand is preferred.

PERSON	A	B	C	PERSON	A	B	C
1	5	7	3	6	10	9	8
2	8	8	5	7	6	8	7
3	6	7	7	8	5	5	4
4	9	6	7	9	6	8	5
5	9	8	5	10	7	6	4

11.111   A psychologist ranked a random sample of ten children on two subjective scales according to the amount of paranoid behavior they exhibit and the amount of aggressiveness they exhibit. The rankings according to these two criteria are given in the table. Compute Spearman's rank correlation coefficient. Is there evidence of a relationship between aggression and paranoia in children as judged by this psychologist?

CHILD	PARANOIA	AGGRESSION
1	7	5
2	3	1
3	6	4
4	1	2
5	2	8
6	4	7
7	10	9
8	8	3
9	5	6
10	9	10

11.112   Two car-rental companies have long waged an advertising war. An independent testing agency is hired to compare the number of rentals at one major airport. After 10 days, the agency has the data listed in the table. At this point, can either car-rental company claim to be number one at this airport? Use $\alpha = .05$.

DAY	RENTAL COMPANY		DAY	RENTAL COMPANY	
	A	B		A	B
1	29	22	6	16	20
2	26	29	7	35	30
3	19	30	8	43	45
4	28	25	9	29	38
5	27	26	10	32	40

11.113   David K. Campbell, James Gaertner, and Robert P. Vecchio (1983) investigated the perceptions of accounting professors with respect to the present and desired importance of various factors considered in promotion and tenure decisions at major universities. One hundred fifteen professors at universities with accredited doctoral programs responded to a mailed questionnaire. The questionnaire asked the professors to rate (1) the *current* importance placed on 20 factors in the promotion and tenure decisions at their universities and (2) how they believe the factors *should* be weighted. Responses were obtained on a 5-point scale ranging from "no importance" to "extreme importance." The resulting ratings were averaged and converted to the rankings shown in the accompanying table. Calculate Spearman's rank correlation coefficient for the data, and carefully interpret its value in the context of the problem.

FACTOR	CURRENT IMPORTANCE	IDEAL IMPORTANCE
I. Teaching (and related items):		
Teaching performance	6	1
Advising and counseling students	19	15
Students' complaints/praise	14	17
II. Research:		
Number of journal articles	1	6.5
Quality of journal articles	4	2
Refereed publications:		
a. Applied studies	5	4
b. Theoretical empirical studies	2	3
c. Educationally oriented	11	8
Papers at professional meetings	10	12
Journal editor or reviewer	9	10
Other (textbooks, etc.)	7.5	11
III. Service and professional interaction:		
Service to profession	15	9
Professional/academic awards	7.5	6.5
Community service	18	19
University service	16	16
Collegiality/cooperativeness	12	13
IV. Other		
Academic degrees attained	3	5
Professional certification	17	14
Consulting activities	20	20
Grantsmanship	13	18

ON YOUR OWN...

In Chapters 10 and 11 we have discussed two methods of analyzing a randomized block design. When the populations have normal probability distributions and their variances are equal, we can employ the analysis of variance described in Chapter 10. Otherwise, we can use the Friedman F_r-test.

In the "On Your Own" section of Chapter 10, we asked you to conduct a randomized block design to compare supermarket prices, and to use an analysis of variance to interpret the data. Now use the Friedman F_r-test to compare the supermarket prices.

How do the results of the two analyses compare? Explain the similarity (or lack of similarity) between the two results.

USING THE COMPUTER...

Select two states to make a comparison of unemployment rates, perhaps the state in which you currently reside and a neighboring state. Treat the zip codes in the data set described in Appendix C as a random sample of all zip codes in each of the two states.

a. Use the appropriate nonparametric procedure to compare the unemployment rates for the two states. What assumptions are necessary to assure the validity of this inferential procedure?

b. Use the appropriate parametric procedure of Chapter 9 to make the comparison. What assumptions are necessary to assure the validity of this inferential procedure?

c. Add two more states to the two you used in making the comparison of unemployment rates in parts a and b. Use the appropriate nonparametric and parametric procedures to compare the unemployment rates of the four states you selected.

References

Agresti, A., and Agresti, B. F. *Statistical Methods for the Social Sciences*, 2d ed. San Francisco: Dellen, 1986.

Campbell, D. K., Gaertner, J., and Vecchio, R. P. "Perceptions of promotion and tenure criteria: A survey of accounting educators." *Journal of Accounting Education*, Spring 1983, 1, pp. 83–92.

Clotfelter, C. T. "Detroit decision and white flight," *Journal of Legal Studies*, 1976, 5.

Freedman, D. A. "The mean versus the median: A case study in 4-R act litigation." *Journal of Business and Economic Statistics*, Vol. 3, No. 1, January 1985, pp. 1–13.

Gibbons, J. D. *Nonparametric Statistical Inference*, 2d ed. New York: McGraw-Hill, 1985.

Hollander, M., and Wolfe, D. A. *Nonparametric Statistical Methods*. New York: Wiley, 1973.

Improving Real Property Assessment: A Reference Manual. International Association of Assessing Officers, 1978.

Noether, G. E. *Elements of Nonparametric Statistics*. New York: Wiley, 1967.

Rosenthal, R., and Rosnow, R. L. *The Volunteer Subject*. New York: Wiley, 1975.

Rosetti, D. K., and Surynt, T. J. "Video teleconferencing and performance," *Journal of Business Communication*, Vol. 22, No. 4, Fall 1985, pp. 25–31.

Siegel, S. *Nonparametric Statistics for the Behavioral Sciences*. New York: McGraw-Hill, 1956.

The Chi-Square Test and the Analysis of Contingency Tables

CONTENTS

WHERE WE'VE BEEN...

The preceding chapters have presented statistical methods for analyzing many types of data. Most of the techniques presented in Chapters 7–10, except those that pertained to binomial proportions, were appropriate for populations of data on quantitative random variables that were independent and, for small samples, had (at least approximately) normal probability distributions with a common variance. Nonparametric statistical procedures were presented in Chapter 11 to make inferences when the assumptions of normality or common variance were likely to be violated or when the responses could only be ranked according to their relative magnitudes.

WHERE WE'RE GOING...

The methods of this chapter are appropriate for a type of data that is common in many disciplines, a type exemplified by the binomial data of Chapters 7–9. We refer to *count*, or *enumerative*, *data*. Recall that the binomial experiment (Chapter 4) consists of *n* trials, each of which results in *two* outcomes. Thus, we have two classifications into which all the data fall. In this chapter, we consider the analysis of data that fall into *two or more* categories.

Many useful experiments consist of *enumerating* the number of occurrences of some event. For example, we may *count* the number of people who recover from leukemia when receiving a newly developed treatment, or the number of consumers who choose each of three brands of coffee, or the number of students who major in each of five different academic disciplines. In some instances, the objective of collecting the **count data** is to analyze the distribution of the counts in the various **classes**, or **cells**. For example, we may want to estimate the proportion of smokers who prefer each of three different brands of cigarettes by counting the number in a sample of smokers who buy each brand. We say that count data classified on a single scale have a **one-dimensional classification**. The analysis of one-dimensional count data is discussed in Section 12.1.

In other instances, the objective of collecting the count data is to study the relationship between two different categorical factors. For example, we may be interested in investigating whether the style of package purchased is related to the sex of the buyer. Or the relationship between socioeconomic status and political party affiliation could be of interest. When count data are classified in a **two-dimensional table**, we call the result a **contingency table**. The analysis of contingency tables is discussed in Section 12.2.

12.1   One-Dimensional Count Data: The Multinomial Distribution

We first consider experiments that result in classification according to a single criterion, i.e., classification on a one-dimensional scale. Suppose we wish to compare the percentage of voters favoring each of three political candidates running for the same elective position. The voting preferences of a random sample of 150 eligible voters are obtained, and the resulting count data are classified according to a single criterion: candidate preference. The data are shown in Table 12.1. Do you think these data indicate a voter preference for any of the candidates?

To answer this question with a valid statistical analysis, we need to know the underlying probability distribution of these count data. This distribution, called the **multinomial probability distribution**, is an extension of the binomial distribution (Section 4.4). The properties of a multinomial experiment are given in the box.

TABLE 12.1   **Voter-Preference Survey**

CANDIDATE		
1	2	3
61	53	36

| Properties of the Multinomial Experiment

1. The experiment consists of n identical trials.
2. There are k possible outcomes to each trial.
3. The probabilities of the k outcomes, denoted by $p_1, p_2, \ldots, p_k$, remain the same from trial to trial, where $p_1 + p_2 + \cdots + p_k = 1$.
4. The trials are independent.
5. The random variables of interest are the counts $n_1, n_2, \ldots, n_k$ in each of the k cells.

Note that our voter-preference survey satisfies the properties of a multinomial experiment. The experiment consists of randomly sampling $n = 150$ voters from a large population of voters containing an unknown proportion p_1 who favor candidate 1, a proportion p_2 who favor candidate 2, and a proportion p_3 who favor candidate 3. Each voter sampled represents a single trial that can result in one of three outcomes: The voter will favor either candidate 1, 2, or 3 with probabilities p_1, p_2, and p_3, respectively. (Assume that all voters will have a preference.) The voting preference of any single voter in the sample does not affect the preference of another; consequently, the trials are independent. And, finally, you can see that the recorded data are the numbers of voters in each of the three voter-preference categories. Thus, the voter-preference survey satisfies the five properties of a multinomial experiment. You can see that the properties of the multinomial experiment closely resemble those of the binomial experiment and that, in fact, a binomial experiment is a multinomial experiment for the special case where $k = 2$.

In the voter-preference survey, and in most practical applications of the multinomial experiment, the k outcome probabilities $p_1, p_2, \ldots, p_k$ are unknown and we want to use the survey data to make inferences about their values. The unknown probabilities in the voter-preference survey are

p_1 = Proportion of all voters who favor candidate 1

p_2 = Proportion of all voters who favor candidate 2

p_3 = Proportion of all voters who favor candidate 3

To decide whether the voters have a preference for any of the candidates, we will want to test the null hypothesis that the candidates are equally preferred $\left(\text{that is, } p_1 = p_2 = p_3 = \frac{1}{3}\right)$ against the alternative hypothesis that one candidate is preferred $\left(\text{that is, at least one of the probabilities } p_1, p_2, \text{ and } p_3 \text{ exceeds } \frac{1}{3}\right)$. Thus, we want to test

H_0: $p_1 = p_2 = p_3 = \frac{1}{3}$ (no preference)

H_a: At least one of the proportions exceeds $\frac{1}{3}$ (a preference exists)

If the null hypothesis is true and $p_1 = p_2 = p_3 = \frac{1}{3}$, the expected value (mean value) of the number of voters who prefer candidate 1 is given by

$$E(n_1) = np_1 = (n)\tfrac{1}{3} = (150)\tfrac{1}{3} = 50$$

Similarly, $E(n_2) = E(n_3) = 50$ if the null hypothesis is true and no preference exists.

The following test statistic measures the degree of disagreement between the data and the null hypothesis:

$$X^2 = \frac{[n_1 - E(n_1)]^2}{E(n_1)} + \frac{[n_2 - E(n_2)]^2}{E(n_2)} + \frac{[n_3 - E(n_3)]^2}{E(n_3)}$$
$$= \frac{(n_1 - 50)^2}{50} + \frac{(n_2 - 50)^2}{50} + \frac{(n_3 - 50)^2}{50}$$

Note that the farther the observed numbers n_1, n_2, and n_3 are from their expected value (50), the larger X^2 will become. That is, large values of X^2 imply that the null hypothesis is false.

We have to know the distribution of X_2 in repeated sampling before we can decide whether the data indicate that a preference exists. When H_0 is true, X^2 can be shown to have approximately a χ^2-distribution with $(k - 1)$ degrees of freedom.* The properties of the χ^2-distribution and the use of Table VII in Appendix A, which gives the critical values of χ^2, were presented in optional Section 8.7 and again in Section 11.4. The rejection region for the voter-preference survey for $\alpha = .05$ and $k - 1 = 3 - 1 = 2$ df is

Rejection region: $X^2 > \chi^2_{.05}$

This value of $\chi^2_{.05}$ (found in Table VII) is 5.99147. (See Figure 12.1.) The computed value of the test statistic is

$$X^2 = \frac{(n_1 - 50)^2}{50} + \frac{(n_2 - 50)^2}{50} + \frac{(n_3 - 50)^2}{50}$$

$$= \frac{(61 - 50)^2}{50} + \frac{(53 - 50)^2}{50} + \frac{(36 - 50)^2}{50} = 6.52$$

FIGURE 12.1

Rejection region for voter-preference survey

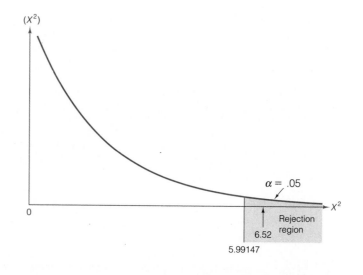

Since the computed $X^2 = 6.51$ exceeds the critical value of 5.99147, we conclude at the $\alpha = .05$ level of significance that there does exist a voter preference for one or more of the candidates.

Now that we have evidence to indicate that the proportions p_1, p_2, and p_3 are unequal, we can make inferences concerning their individual values using the methods of Section 7.4. [*Note:* We cannot use the methods of Section 9.4 to compare two proportions because the cell counts are dependent random variables.] The general form for a test of a hypothesis concerning multinomial probabilities is shown in the box.

---

*The derivation of the degrees of freedom for χ^2 involves the number of linear restrictions imposed on the count data. In the present case, the only constraint is that $\Sigma n_i = n$, where n (the sample size) is fixed in advance. Therefore, df = $k - 1$. For other cases, we will give the degrees of freedom for each usage of χ^2 and refer the interested reader to the references at the end of the chapter for more detail.

> ## A Test of a Hypothesis About Multinomial Probabilities
>
> H_0: $p_1 = p_{1,0}$, $p_2 = p_{2,0}$, $\ldots$, $p_k = p_{k,0}$,
> where $p_{1,0}, p_{2,0}, \ldots, p_{k,0}$ represent the hypothesized values of the multinomial probabilities
>
> H_a: At least one of the multinomial probabilities does not equal its hypothesized value
>
> Test statistic: $X^2 = \sum \dfrac{[n_i - E(n_i)]^2}{E(n_i)}$
>
> where $E(n_i) = np_{i,0}$, the expected number of outcomes of type i assuming that H_0 is true. The total sample size in n.
>
> Rejection region: $X^2 > \chi_\alpha^2$, where χ_α^2 has $(k - 1)$ df
>
> Assumptions: 1. A multinomial experiment has been conducted. This is generally satisfied by taking a random sample from the population of interest.
>
> 2. The sample size n will be large enough so that for every cell, the expected cell count $E(n_i)$ will be equal to 5 or more.*

EXAMPLE 12.1

Suppose an educational television station has broadcast a series of programs on the physiological and psychological effects of smoking marijuana. Now that the series is finished, the station wants to see whether the citizens within the viewing area have changed their minds about how possession of marijuana should be considered legally. Before the series was shown, it was determined that 7% of the citizens favored legalization, 18% favored decriminalization, 65% favored the existing law (a person could be fined or imprisoned), and 10% had no opinion.

A summary of the opinions (after the series was shown) of a random sample of 500 people in the viewing area is given in Table 12.2. Test at the $\alpha = .01$ level to see whether these data indicate that the distribution of opinions differs significantly from the proportions that existed before the educational series was aired.

TABLE 12.2 **Distribution of Opinions About Marijuana Possession**

LEGALIZATION	DECRIMINALIZATION	EXISTING LAWS	NO OPINION
39	99	336	26

---

*The assumption that all expected cell counts are at least 5 is necessary in order to ensure that the χ^2 approximation is appropriate. Exact methods for conducting the test of a hypothesis exist and may be used for small expected cell counts, but these methods are beyond the scope of this text.

Solution

Define the proportions after the airing to be

p_1 = Proportion of citizens favoring legalization

p_2 = Proportion of citizens favoring decriminalization

p_3 = Proportion of citizens favoring existing laws

p_4 = Proportion of citizens with no opinion

Then the null hypothesis representing no change in the distribution of percentages is

H_0: $p_1 = .07$, $p_2 = .18$, $p_3 = .65$, $p_4 = .10$

and the alternative is

H_a: At least one of the proportions differs from its null hypothesized value

Test statistic: $X^2 = \sum \dfrac{[n_i - E(n_i)]^2}{E(n_i)}$

where

$E(n_1) = np_{1,0} = 500(.07) = 35$

$E(n_2) = np_{2,0} = 500(.18) = 90$

$E(n_3) = np_{3,0} = 500(.65) = 325$

$E(n_4) = np_{4,0} = 500(.10) = 50$

Since all these values are larger than 5, the χ^2 approximation is appropriate. Also, if the citizens in the sample were randomly selected, the properties of the multinomial probability distribution are satisfied.

Rejection region: For $\alpha = .01$ and df $= k - 1 = 3$, reject H_0 if $X^2 > \chi^2_{.01}$, where (from Table VII in Appendix A) $\chi^2_{.01} = 11.3449$.

We now calculate the test statistic:

$$X^2 = \frac{(39 - 35)^2}{35} + \frac{(99 - 90)^2}{90} + \frac{(336 - 325)^2}{325} + \frac{(26 - 50)^2}{50} = 13.249$$

Since this value exceeds the table value of χ^2 (11.3449), the data provide sufficient evidence ($\alpha = .01$) that the opinions on legalization of marijuana have changed since the series was aired.

If we focus on one particular outcome of a multinomial experiment, we can use the methods developed in Section 7.4 for a binomial proportion to establish a confidence interval for any one of the multinomial probabilities.* For example, if we want a 95% confidence interval for the proportion of citizens in the viewing area who have no opinion about the issue, we calculate

$\hat{p}_4 \pm 1.96\sigma_{\hat{p}_4}$

---

*Note that focusing on one outcome has the effect of lumping the other $(k - 1)$ outcomes into a single group. Thus, we obtain, in effect, two outcomes—or a binomial experiment.

where

$$\hat{p}_4 = \frac{n_4}{n} = \frac{26}{500} = .052 \quad \text{and} \quad \sigma_{\hat{p}_4} \approx \sqrt{\frac{\hat{p}_4(1 - \hat{p}_4)}{n}}$$

Thus, we get

$$.052 \pm 1.96\sqrt{\frac{(.052)(.948)}{500}} = .052 \pm .019$$

or (.033, .071). Thus, we estimate that between 3.3% and 7.1% of the citizens now have no opinion on the issue of marijuana legalization. The series of programs may have helped citizens who formerly had no opinion on the issue to form an opinion, since it appears that the proportion of "no opinions" is now less than 10%.

EXERCISES 12.1–12.19

12.1 Use Table VII of Appendix A to find each of the following χ^2 values:
a. $\chi^2_{.05}$ for df = 15 b. $\chi^2_{.990}$ for df = 100
c. $\chi^2_{.10}$ for df = 12 d. $\chi^2_{.005}$ for df = 2

12.2 Use Table VII of Appendix A to find the following probabilities:
a. $P(\chi^2 \leq .872085)$ for df = 6 b. $P(\chi^2 > 30.5779)$ for df = 15
c. $P(\chi^2 \geq 82.3581)$ for df = 100 d. $P(\chi^2 < 13.7867)$ for df = 30

12.3 Find the rejection region for a one-dimensional χ^2-test of a null hypothesis concerning p_1, $p_2, \ldots, p_k$ if:
a. $k = 3$; $\alpha = .10$ b. $k = 5$; $\alpha = .01$ c. $k = 4$; $\alpha = .05$

12.4 What are the characteristics of a multinomial experiment? Compare the characteristics to those of a binomial experiment.

12.5 What conditions must n satisfy to make the χ^2-test valid?

12.6 A multinomial experiment with $k = 3$ cells and $n = 300$ produced the data shown in the table. Do these data provide sufficient evidence to contradict the null hypothesis that $p_1 = .25$, $p_2 = .25$, and $p_3 = .50$? Test using $\alpha = .05$.

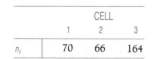

	CELL		
	1	2	3
n_i	70	66	164

12.7 A multinomial experiment with $k = 4$ cells and $n = 206$ produced the data shown in the table.
a. Do these data provide sufficient evidence to conclude that the multinomial probabilities differ? Test using $\alpha = .05$.
b. What are the Type I and Type II errors associated with the test of part a?

	CELL			
	1	2	3	4
n_i	45	54	60	47

12.8   Refer to Exercise 12.7. Construct a 95% confidence interval for the multinomial probability associated with cell 3.

12.9   A multinomial experiment with $k = 4$ cells and $n = 400$ produced the data shown in the table. Do these data provide sufficient evidence to contradict the null hypothesis that $p_1 = .2$, $p_2 = .4$, $p_3 = .1$, and $p_4 = .3$? Test using $\alpha = .05$.

	CELL			
	1	2	3	4
n_i	71	194	46	89

APPLYING THE CONCEPTS

12.10   A 1985 Gallup survey portrays U.S. entrepreneurs as "... the mavericks, dreamers, and loners whose rough edges and uncompromising need to do it their own way set them in sharp contrast to senior executives in major American corporations" (Graham, 1985). One of the many questions put to a sample of $n = 100$ entrepreneurs about their job characteristics, work habits, social activities, etc. concerned the origin of the car they personally drive most frequently. The responses given in the table were obtained.

UNITED STATES	EUROPE	JAPAN
45	46	9

a. Do these data provide evidence of a difference in the preference of entrepreneurs for the cars of the United States, Europe, and Japan? Test using $\alpha = .05$.
b. Do these data provide evidence of a difference in the preference of entrepreneurs for domestic versus foreign cars? Test using $\alpha = .05$.
c. What assumptions must be made in order for your inferences of parts **a** and **b** to be valid?

12.11   Refer to Exercise 12.10. Use a 90% confidence interval to estimate the proportion of U.S. entrepreneurs who drive foreign cars.

12.12   Overweight trucks are responsible for much of the damage sustained by our local, state, and federal highway systems. Though illegal, overweight trucks proliferate. Truckers have learned to avoid weigh stations run by enforcement officers by taking back roads when weigh stations are open and traveling during periods of the week when weigh stations are likely to be closed. A state highway planning agency recently monitored the movements of overweight trucks on an interstate highway using an unmanned, computerized scale that is built into the highway. Unknown to the truckers, the scale weighs their vehicles as they pass over it. For a particular week, each day's proportion of the week's total truck traffic (5-axle tractor truck semitrailers) was as follows:

MONDAY	TUESDAY	WEDNESDAY	THURSDAY	FRIDAY	SATURDAY	SUNDAY
.191	.198	.187	.180	.155	.043	.046

Source: Dahlin, C., and Owen, F. "An analysis of data collected at the I-494 weighing-in-motion site." St. Paul: Minnesota Department of Transportation, 1984, p. 10.

During the same week, the number of overweight trucks per day was as follows:

MONDAY	TUESDAY	WEDNESDAY	THURSDAY	FRIDAY	SATURDAY	SUNDAY
90	82	72	70	51	18	31

Source: Dahlin and Owen (1984).

a. The planning agency would like to know whether the number of overweight trucks per week is distributed over the 7 days of the week in direct proportion to the volume of truck traffic. Test using $\alpha = .05$.

b. Find the approximate *p*-value for the test of part **a**.

12.13   After purchasing a policy from one life insurance company, a person has a certain period of time in which the policy can be canceled without financial obligation. An insurance company is interested in seeing whether those who cancel a policy during this time period are as likely to be in one policy-size category as another. Records for 250 people who canceled policies during this period are selected at random from company files with the results shown in the table. Is there sufficient evidence to conclude that the canceled policies are not distributed equally among the five policy-size categories? Use $\alpha = .05$.

SIZE OF POLICY (THOUSANDS OF DOLLARS)				
10	15	20	25	30
31	39	67	54	59

12.14   In 1960 there were 69,628,000 people in the U.S. civilian labor force, of whom 23,240,000 were women. The table describes the age distribution of those women.

AGE IN YEARS	RELATIVE FREQUENCY IN 1960
16–19	.089
20–24	.111
25–34	.178
35–44	.228
45–54	.227
55–64	.128
65 and over	.039

Source: *Statistical Abstract of the United States: 1986*, p. 392.

In 1984, there were 113,544,000 people in the U.S. civilian labor force, of whom 49,709,000 were women. A random sample of 500 working women in 1984 yielded the age distribution shown in the next table.

AGE IN YEARS	FREQUENCY IN 1984
16–19	38
20–24	75
25–34	143
35–44	110
45–54	72
55–64	50
65 and over	12

Source: Based on the relative frequencies for 1984 cited in *Statistical Abstract of the United States: 1986*, p. 392.

a. Do the sample data provide sufficient evidence to conclude that the 1984 age distribution of the female work force differs from the 1960 age distribution? Test using $\alpha = .05$.

b. In the context of the problem, specify the Type I and Type II errors associated with the hypothesis test of part **a**.

c. Find the approximate p-value for the test in part **a**.

d. Use 95% confidence intervals to estimate the proportion of women between the ages of 25 and 34 (inclusive) in the 1984 labor force and the proportion between 20 and 44 (inclusive) in the 1984 labor force.

12.15   In the game of chess, the first few moves play a very important role in determining the final outcome. Five different opening strategies are highly favored by chess experts. To determine whether one or more of these strategies is most preferred by grand masters in international competition, a random sample of 100 grand masters is taken, and each is asked which of the strategies he or she would prefer to employ. A summary of their responses is shown here:

Strategy:	A	B	C	D	E
Frequency:	17	27	22	15	19

Do these data present sufficient evidence to indicate a preference for one or more of the strategies? Use $\alpha = .05$.

12.16   There are four standard surgical techniques, A, B, C, and D, presently used in abdominal surgery. To find out whether one method is preferred over any other, each of a random sample of 200 surgeons was asked which technique he or she preferred. A summary of the data is shown here:

A	B	C	D
48	68	45	39

Do the data provide sufficient evidence to indicate differences in preferences for the four techniques? Test using $\alpha = .05$.

12.17   According to a geneticist's theory, a crossing of red and white snapdragons should produce offspring that are 25% red, 50% pink, and 25% white. An experiment conducted to test the theory produces 30 red, 78 pink, and 36 white offspring in 144 crossings. Do the data provide sufficient evidence to contradict the geneticist's theory?

12.18   Supermarket chains often carry products with their own brand labels and usually price them lower than the nationally known brands. A supermarket conducted a taste test to determine whether there was a difference in taste among the four brands of ice cream it carried: a local brand (A) and three national brands (B, C, D). A sample of 200 people participated, and they indicated the preferences shown in the table. Is there evidence of a difference in preference for the four brands? Test at $\alpha = .05$.

BRAND			
A	B	C	D
39	57	55	49

12.19   A local manufacturing company utilizes a computerized sales invoice printing system. Each distinct bit of information on the invoices (e.g., sold-to address, ship-to address, sales tax, total sale) is referred to as a *field*. From a handwritten copy of each invoice, a keypunch operator transcribes the field data onto computer cards so that computer-printed invoices can be produced. The manager of data processing believes that the distribution of the number of errors per invoice appearing on printed invoices has changed dramatically since the physical arrangement of the fields on the invoices was changed 6 months ago. The table describes the distribution of the number of errors per invoice when the previous format was used:

ERRORS PER INVOICE	0	1	2	3	4	More than 4
PROPORTION OF FINISHED INVOICES	.90	.04	.03	.02	.005	.005

A random sample of 300 printed invoices was selected from those that were printed during the past week. Each invoice was examined for errors. The following data resulted:

ERRORS PER INVOICE	0	1	2	3	4	More than 4
NUMBER OF INVOICES	150	120	15	7	4	4

a. Do the data provide sufficient evidence to indicate that the proportions of printed invoices in the six error categories differ from the proportions using the previous format?
b. Find the approximate observed significance level for the test in part **a**.

12.2 Contingency Tables

In Section 12.1, we introduced the multinomial probability distribution and considered data classified according to a single criterion. We now consider multinomial experiments in which the data are classified according to two criteria, i.e., *classification with respect to two factors*.

For example, the energy shortage has made many consumers more aware of the size of the automobiles they purchase. Suppose an automobile manufacturer is interested in determining the relationship between the size and manufacturer of newly purchased automobiles. One thousand recent buyers of American-made cars are randomly sampled, and each purchase is classified with respect to the size and manufacturer of the automobile. The data are summarized in the **two-way table** shown in Table 12.3. This table is called a **contingency table**; it presents multinomial count data classified on two scales, or **dimensions**, of classification—namely, automobile size and manufacturer.

TABLE 12.3 **Contingency Table for Automobile Size Example**

	MANUFACTURER				TOTALS
	A	B	C	D	
Small	157	65	181	10	413
Intermediate	126	82	142	46	396
Large	58	45	60	28	191
TOTALS	341	192	383	84	1,000

The symbols representing the cell counts for the multinomial experiment in Table 12.3 are shown in Table 12.4 (page 614), part A, and the corresponding cell, row, and column probabilities are shown in Table 12.4, part B. Thus, n_{11} represents the number of buyers who purchase a small car of manufacturer A and p_{11} represents the corresponding cell probability. Note the symbols for the row and column totals and also the symbols for the probability totals. The latter are called **marginal probabilities** for each row and column. The marginal probability p_{r1} is the probability that a small car is purchased; the marginal probability p_{c1} is the probability that a car by manufacturer A is purchased. Thus, $p_{r1} = p_{11} + p_{12} + p_{13} + p_{14}$ and $p_{c1} = p_{11} + p_{21} + p_{31}$.

T A B L E   12.4   **(A) Observed Counts for Contingency Table 12.3**

	MANUFACTURER				TOTALS
	A	B	C	D	
Small	n_{11}	n_{12}	n_{13}	n_{14}	r_1
Intermediate	n_{21}	n_{22}	n_{23}	n_{24}	r_2
Large	n_{31}	n_{32}	n_{33}	n_{34}	r_3
TOTALS	c_1	c_2	c_3	c_4	n

(B) Probabilities for Contingency Table 12.3

	MANUFACTURER				TOTALS
	A	B	C	D	
Small	p_{11}	p_{12}	p_{13}	p_{14}	p_{r1}
Intermediate	p_{21}	p_{22}	p_{23}	p_{24}	p_{r2}
Large	p_{31}	p_{32}	p_{33}	p_{34}	p_{r3}
TOTALS	p_{c1}	p_{c2}	p_{c3}	p_{c4}	1

Thus, we can see that this really is a multinomial experiment with a total of 1,000 trials, $(3)(4) = 12$ cells or possible outcomes, and probabilities for each cell as shown in Table 12.4, part B. If the 1,000 recent buyers are randomly chosen, the trials are considered independent and the probabilities are viewed as remaining constant from trial to trial.

Suppose we want to know whether the two classifications, manufacturer and size, are dependent. That is, if we know which size car a buyer will choose, does that information give us a clue about the manufacturer of the car the buyer will choose? In a probabilistic sense we know (Chapter 3) that independence of events A and B implies $P(AB) = P(A)P(B)$. Similarly, in the contingency table analysis, if the two classifications are independent, the probability that an item is classified in any particular cell of the table is the product of the corresponding marginal probabilities. Thus, under the hypothesis of independence, in Table 12.4, part B, we must have

$$p_{11} = p_{r1}p_{c1} \qquad p_{12} = p_{r1}p_{c2}$$

and so forth.

To test the hypothesis of independence, we use the same reasoning employed in the one-dimensional tests of Section 12.1. First, we calculate the expected, or mean, count in each cell assuming that the null hypothesis of independence is true. We do this by noting that the expected count in a cell of the table is just the total number of multinomial trials, n, times the cell probability. Recall that n_{ij} represents the observed count in the cell located in the ith row and jth column. Then the expected cell count for the upper-left-hand cell (first row, first column) is

$$E(n_{11}) = np_{11}$$

or, when the null hypothesis (the classifications are independent) is true,

$$E(n_{11}) = np_{r1}p_{c1}$$

Since these true probabilities are not known, we estimate p_{r1} and p_{c1} by the sample proportions $\hat{p}_{r1} = r_1/n$ and $\hat{p}_{c1} = c_1/n$. Thus, the estimate of the expected value $E(n_{11})$ is

$$\hat{E}(n_{11}) = n\left(\frac{r_1}{n}\right)\left(\frac{c_1}{n}\right) = \frac{r_1 c_1}{n}$$

Similarly, for each i, j,

$$\hat{E}(n_{ij}) = \frac{(\text{Row total})(\text{Column total})}{\text{Total sample size}}$$

Thus,

$$\hat{E}(n_{12}) = \frac{r_1 c_2}{n}$$

$$\vdots \qquad \vdots$$

$$\hat{E}(n_{34}) = \frac{r_3 c_4}{n}$$

Using the data in Table 12.3, we find

$$\hat{E}(n_{11}) = \frac{r_1 c_1}{n} = \frac{(413)(341)}{1,000} = 140.833$$

$$\hat{E}(n_{12}) = \frac{r_1 c_2}{n} = \frac{(413)(192)}{1,000} = 79.296$$

$$\vdots \qquad \vdots \qquad \vdots \qquad \vdots$$

$$\hat{E}(n_{34}) = \frac{r_3 c_4}{n} = \frac{(191)(84)}{1,000} = 16.044$$

The observed data and the estimated expected values (in parentheses) are shown in Table 12.5 on page 616.

We now use the X^2 statistic to compare the observed and expected (estimated) counts in each cell of the contingency table:

$$X^2 = \frac{[n_{11} - \hat{E}(n_{11})]^2}{\hat{E}(n_{11})} + \frac{[n_{12} - \hat{E}(n_{12})]^2}{\hat{E}(n_{12})} + \cdots + \frac{[n_{34} - \hat{E}(n_{34})]^2}{\hat{E}(n_{34})}$$

$$= \sum \frac{[n_{ij} - \hat{E}(n_{ij})]^2}{\hat{E}(n_{ij})} {}^*$$

---

*The use of Σ in the context of a contingency table analysis refers to a sum over all cells in the table.

T A B L E 12.5   **Observed and Estimated Expected (in Parentheses) Counts**

| | MANUFACTURER | | | | TOTALS |
	A	B	C	D	
Small	157 (140.833)	65 (79.296)	181 (158.179)	10 (34.692)	413
Intermediate	126 (135.036)	82 (76.032)	142 (151.668)	46 (33.264)	396
Large	58 (65.131)	45 (36.672)	60 (73.153)	28 (16.044)	191
TOTALS	341	192	383	84	1,000

Substituting the data of Table 12.5 into this expression, we get

$$X^2 = \frac{(157 - 140.833)^2}{140.833} + \frac{(65 - 79.296)^2}{79.296} + \cdots + \frac{(28 - 16.044)^2}{16.044} = 45.81$$

Large values of X^2 imply that the observed counts do not closely agree and therefore imply that the hypothesis of independence is false. To determine how large X^2 must be before it is too large to be attributed to chance, we make use of the fact that the sampling distribution of X^2 is approximately a χ^2 probability distribution when the classifications are independent.

When testing the null hypothesis of independence in a two-way contingency table, the appropriate degrees of freedom will be $(r - 1)(c - 1)$, where r is the number of rows and c is the number of columns in the table.

For the size and make of automobiles example, the degrees of freedom for χ^2 is $(r - 1)(c - 1) = (3 - 1)(4 - 1) = 6$. Then, for $\alpha = .05$, we reject the hypothesis of independence when

$$X^2 > \chi^2_{.05} = 12.5916$$

Since the computed $X^2 = 45.81$ exceeds the value 12.5916, we conclude that the size and manufacturer of a car selected by a purchaser are dependent events.

The pattern of dependence can be seen more clearly by expressing the data as percentages. We first select one of the two classifications to be used as the base variable. In the automobile size preference example, suppose we select manufacturer as the classificatory variable to be the base. Next, we represent the responses for each level of the second categorical variable (size of automobile in our example) as a percentage of the subtotal for the base variable. For example, from Table 12.5 we convert the response for small car sales for manufacturer A (157) to a percentage of the total sales for manufacturer A (341) That is,

$$\left(\frac{157}{341}\right)100\% = 46\%$$

The conversions of all Table 12.5 entries are similarly computed, and the values are shown in Table 12.6. The value shown at the right of each row is the row's total expressed as a percentage of the total number of responses in the entire table. Thus, the small car percentage is $\frac{413}{1,000}(100\%) = 41\%$ (rounded to the nearest percent).

T A B L E 12.6 **Percentage of Car Sizes by Manufacturer**

| | MANUFACTURER | | | | ALL |
	A	B	C	D	
Small	46	34	47	12	41
Intermediate	37	43	37	55	40
Large	17	23	16	33	19
TOTALS	100	100	100	100	100

If the size and manufacturer variables are independent, then the percentages in the cells of the table are expected to be approximately equal to the corresponding row percentages. Thus, we would expect the small car percentages for each of the four manufacturers to be approximately 41% if size and manufacturer were independent. The extent to which each manufacturer's percentage departs from this value determines the dependence of the two classifications, with greater variability of the row percentages meaning a greater degree of dependence. A plot of the percentages helps summarize the observed pattern. In Figure 12.2 we show the manufacturer (the base variable) on the horizontal axis, and the size percentages on the vertical axis. The "expected" percentages under the assumption of independence are shown as horizontal lines, and each observed value is represented by a symbol indicating the size category.

F I G U R E 11.2

Size as a percentage of manufacturer subtotals

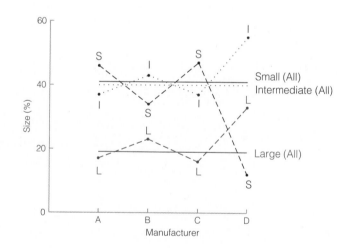

Figure 12.2 clearly indicates the reason that the test resulted in the conclusion that the two classifications in the contingency table are dependent. Note that the sales of manufacturers A, B, and C fall relatively close to the expected percentages under the assumption of independence. However, the sales of manufacturer D deviate significantly from the expected values, with much higher percentages for large and intermediate cars and a much smaller percentage for small cars than expected under independence. Also, manufacturer B deviates slightly from the expected pattern, with a greater percentage of intermediate than small car sales.

Statistical measures of the degree of dependence and procedures for making comparisons of pairs of levels for classifications are available. They are beyond the scope of this text, but can be found in the references at the end of the chapter. We will, however, utilize descriptive summaries such as Figure 12.2 to examine the degree of dependence exhibited by the sample data.

The general form of a two-way contingency table containing r rows and c columns (called an $r \times c$ contingency table) is shown in Table 12.7. Note that the observed count in the (ij) cell is denoted by n_{ij}, the ith row total is r_i, the jth column total is c_j, and the total sample size is n. Using this notation, we give the general form of the contingency table test for independent classifications in the box.

TABLE 12.7 **General $r \times c$ Contingency Table**

		COLUMN				ROW TOTALS
		1	2	$\cdots$	c	
	1	n_{11}	n_{12}	$\cdots$	n_{1c}	r_1
	2	n_{21}	n_{22}	$\cdots$	n_{2c}	r_2
ROW	.	.	.		.	.
	.	.	.		.	.
	r	n_{r1}	n_{r2}	$\cdots$	n_{rc}	r_r
COLUMN TOTALS		c_1	c_2	$\cdots$	c_c	n

General Form of a Contingency Table Analysis: A Test for Independence

H_0: The two classifications are independent

H_a: The two classifications are dependent

Test statistic: $X^2 = \sum \dfrac{[n_{ij} - \hat{E}(n_{ij})]^2}{\hat{E}(n_{ij})}$

where

$$\hat{E}(n_{ij}) = \frac{r_i c_j}{n}$$

Rejection region: $X^2 > \chi_\alpha^2$, where χ_α^2 has $(r-1)(c-1)$ df.

Assumptions: 1. The n observed counts are a random sample from the population of interest. We may then consider this to be a multinomial experiment with $r \times c$ possible outcomes.

2. The sample size, n, will be large enough so that, for every cell, the expected count, $E(n_{ij})$, will be equal to 5 or more.

EXAMPLE 12.2

A social scientist wants to determine whether the marital status (divorced or not divorced) of American men is independent of their religious affiliation (or lack thereof). A sample of 500 American men is surveyed and the results are tabulated as shown in Table 12.8.

TABLE 12.8 **Observed and Estimated (in Parentheses) Expected Counts, Example 12.2**

| | | RELIGIOUS AFFILIATION | | | | | TOTALS |
		A	B	C	D	None	
MARITAL STATUS	Divorced	39 (48.952)	19 (18.560)	12 (12.992)	28 (22.736)	18 (12.760)	116
	Never divorced	172 (162.048)	61 (61.440)	44 (43.008)	70 (75.264)	37 (42.240)	384
TOTALS		211	80	56	98	55	500

a. Test to see whether there is sufficient evidence to indicate that the marital status of men who have been or are currently married is dependent on religious affiliation. Test using $\alpha = .01$.

b. Plot the data and describe the patterns revealed. Is the result of the test supported by the plot?

Solution

a. The first step is to calculate estimated expected cell frequencies under the assumption that the classifications are independent. Thus,

$$\hat{E}(n_{11}) = \frac{r_1 c_1}{n} = \frac{(116)(211)}{500} = 48.952$$

$$\hat{E}(n_{12}) = \frac{r_1 c_2}{n} = \frac{(116)(80)}{500} = 18.560$$

and so forth. All the estimated expected cell counts are shown in Table 12.8. We are now ready to conduct the test for independence:

H_0: The marital status of American men and their religious affiliation are independent

H_a: The marital status of American men and their religious affiliation are dependent

Test statistic: $X^2 = \sum \dfrac{[n_{ij} - \hat{E}(n_{ij})]^2}{\hat{E}(n_{ij})}$

Since all the estimated expected cell frequencies are greater than 5, the χ^2 approximation is appropriate. Assuming the men chosen were randomly selected from all married or previously married American men, the characteristics of the multinomial probability distribution are satisfied.

Rejection region: For $\alpha = .01$ and $(r - 1)(c - 1) = (1)(4) = 4$ df, reject H_0 if $X^2 > \chi^2_{.01}$, where $\chi^2_{.01} = 13.2767$

The calculated value of the test statistic is

$$X^2 = \frac{(39 - 48.952)^2}{48.952} + \frac{(19 - 18.560)^2}{18.560} + \cdots + \frac{(37 - 42.240)^2}{42.240}$$
$$= 7.135$$

Since $X^2 = 7.135$ is less than $\chi^2_{.01} = 13.2767$, we cannot conclude that the marital status of American men depends on their religious affiliation. (Note that we could not reject H_0 even with $\alpha = .10$, since $\chi^2_{.10} = 7.77944$.)

b. The marital status frequencies are expressed as percentages of the number of men in each religious affiliation category in Table 12.9. The expected percentages under the assumption of independence are shown at the right of each row. The plot of the percentage data is shown in Figure 12.3, where horizontal lines represent the expected percentages assuming independence. Note that the response percentages deviate only slightly from those expected under the assumption of independence, supporting the result of the test in part **a**. That is, neither the descriptive plot nor the statistical test provides evidence that the male divorce rate depends on (varies with) religious affiliation.

T A B L E 12.9   **Marital Status as Percentage of Religious Affiliation**

		RELIGIOUS AFFILIATION				ALL
		A	B	C	D	
MARITAL STATUS	Divorced	18	24	21	33	23
	Never divorced	82	76	79	67	77
TOTALS		100	100	100	100	100

F I G U R E 12.3

Plot of marital status–religious affiliation contingency table

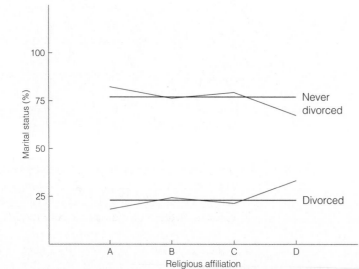

E V A L U A T I N G   A   N E W   M E T H O D   F O R   T R E A T I N G   C A N C E R

The typical method of treating moderately advanced cancer of the larynx (voice box) is removal by surgery. This method achieves initial control of the cancer in approximately 75% of the cases; if the cancer recurs, salvage treatment results in approximately 87% of the cancers ultimately controlled. Recent attempts have been made to treat cancer of the larynx by radiation therapy alone, thereby saving the patient's larynx. W. M. Mendenhall et al. present data on a group of patients treated by this method at the University of Florida's Shands Hospital. Eighteen patients with cancer of the larynx were treated by radiation alone and 23 were treated by surgery alone. Of those treated by radiation, 11 cancers were controlled at the primary site (the larynx). In seven patients the cancer recurred in the larynx; six were treated by surgery and one refused further treatment. Of these, four were controlled so that the number ultimately controlled by radiation therapy alone (11) or by surgical salvage (four) was 15 of 18 patients. Initial control of the cancer was achieved for 18 of the 23 patients treated by surgery alone. Salvage treatment was successful for three of the remaining five. Thus, removal of the larynx by surgery alone achieved ultimate control in 21 of the 23 cases. These data are summarized in Table 12.10.

T A B L E   12.10   **Comparison of Two Methods for Treating Cancer of the Larynx**

	SURGERY	RADIATION THERAPY
NUMBER OF CANCERS ULTIMATELY CONTROLLED	21	15
NUMBER NOT ULTIMATELY CONTROLLED	2	3
TOTALS	23	18

Source: Mendenhall, W. M., Million, R. R., Sharkey, D. E., and Cassisi, N. J. "Stage T3 squamous cell carcinoma of the larynx treated with surgery and/or radiation therapy," *International Journal of Radiation Oncology and Biological Physics*, 1984, 10. Copyright 1984, Pergamon Journals, Ltd. Reprinted with permission.

The data of Table 12.10 cannot be analyzed using the chi-square test to detect differences in rates of control for the two methods of treatment. The expected numbers in the "no ultimate control" cells are too small. However, the same test can be conducted by using a small-sample method known as *Fisher's exact test*. This method calculates the exact probability (*p*-value) of observing sample results at least as contradictory to the hypothesis of independence as those observed for the researchers' data. Mendenhall and coworkers report the *p*-value for this test as .187, a value that indicates little evidence of a difference in the rates of achieving ultimate control for the two methods of treatment.

Mendenhall and colleagues clearly intend their research to be a first step in a study that will require the treatment of many more cancer patients using radiation alone. Small-sample methods are available for constructing a confidence interval for the rate of achieving ultimate control using radiation treatment alone, but the interval width would be quite large. As we learned in Chapter 7, it requires a fairly large sample to estimate a binomial proportion *p* with a small margin of error. This larger sample will be obtained by combining treatment results from other cancer clinics and by collecting data on patients who will be treated in the future.

EXERCISES 12.20–12.36

12.20   Find the rejection region for a test of independence of two classifications where the contingency table contains r rows and c columns and:

a. $r = 5$, $c = 5$, $\alpha = .05$ b. $r = 3$, $c = 6$, $\alpha = .10$ c. $r = 2$, $c = 3$, $\alpha = .01$

12.21   Consider the accompanying 2×3 (i.e., $r = 2$ and $c = 3$) contingency table.

		COLUMN		
		1	2	3
ROW	1	10	32	53
	2	15	30	25

a. Specify the null and alternative hypotheses that should be used in testing the independence of the row and column classifications.

b. Specify the test statistic and the rejection region that should be used in conducting the hypothesis test of part a. Use $\alpha = .01$.

c. Assuming the row classification and the column classification are independent, find estimates for the expected cell counts.

d. Conduct the hypothesis test of part a. Interpret your result.

12.22   Refer to Exercise 12.21.

a. Convert the frequency responses to percentages by calculating the percentage of each column total falling in each row. Also convert the row totals to percentages of the total number of responses. Display the percentages in a table.

b. Create a graph with percentage on the vertical axis and column number on the horizontal axis. Show the row total percentages as horizontal lines on the plot, and plot the cell percentages from part a using the row number as a plotting symbol.

c. What pattern do you expect to see if the rows and columns are independent? Does the plot support the result of the test of independence in Exercise 12.21?

12.23   Test the null hypothesis of independence of the two classifications, A and B, of the 3×3 contingency table shown here. Test using $\alpha = .05$.

		B		
		B_1	B_2	B_3
	A_1	39	75	42
A	A_2	63	51	70
	A_3	30	38	29

12.24   Refer to Exercise 12.23. Convert the responses to percentages by calculating the percentage of each B class total falling into each A classification. Also, calculate the percentage of the total number of responses that constitute each of the A classification totals. Create a graph with percentage on the vertical axis and B classification on the horizontal axis. Show the percentages corresponding to the A classification totals as horizontal lines on the graph, and plot the individual cell percentages using the A class number as a plotting symbol. Does the graph support the result of the test of hypothesis in Exercise 12.23? Explain.

12.25 A chi-square test was applied to each of the following contingency tables and the null hypothesis of independence was rejected. Construct a graph for each table that enables you to interpret the pattern of dependence in each case.

a.

		B	
		B_1	B_2
A	A_1	50	10
	A_2	10	40

b.

		D	
		D_1	D_2
C	C_1	15	70
	C_2	80	15

c.

		F		
		F_1	F_2	F_3
E	E_1	10	30	15
	E_2	30	10	15

APPLYING THE CONCEPTS

12.26 Are all employees equally prone to having accidents? Or, are certain employee groups—say, younger employees—more likely to have particular kinds of accidents? The effective implementation of hiring policies, training programs, and safety programs requires such knowledge. A recent study used contingency table analysis to address these questions for a particular manufacturing company. A portion of the research results is summarized in the accompanying contingency table.

		KIND OF ACCIDENT		
		Sprain	Burn	Cut
AGE	Under 25	9	17	5
	25 and Over	61	13	12

Source: Derived from Parry, A. E. "Changing assumptions about loss frequency," *Professional Safety*, October 1985, pp. 39–43.

a. From this contingency table, the researcher concluded that there is a relationship between an employee's age and the kind of accident that the employee may have. Do you agree? Test using $\alpha = .05$.

b. According to the frequencies of the contingency table, which is the most frequent type of accident? Are younger or older employees more likely to have sprains? Burns? Justify your answers.

c. What assumptions must hold in order for your test of part **a** to be valid?

d. Plot the percentage of employees under 25 who are injured based on the total injuries for each kind of accident. Compare this to the percentage of the total number of employees under 25 who are injured based on the total of all kinds of accidents. What does the plot indicate about the pattern of dependence in these data?

12.27 Researchers conducted a study of 58 children admitted to the child psychiatry ward of the University of Iowa for aggressive conduct. The purpose of the study was to determine the influence of social class, sex, and age on the clinical characteristics of the children. A small portion of their data is shown in the table, which lists the numbers of children exhibiting antisocial behavior for two categories of social class. Classes I to III represent children from middle-class families. Those in Classes IV and V were from poor families.

	CLASSES I–III	CLASSES IV–V
Numbers exhibiting antisocial behavior	24	17
Sample size	28	30

Source: Behar, D., and Stewart, M. A. "Aggressive conduct disorder: The influence of social class, sex, and age on the clinical picture," *Journal of Child Psychology and Psychiatry*, 1984, 25.

a. Do the data provide sufficient evidence to indicate a dependence between antisocial behavior and social class? Test using $\alpha = .05$.

b. The authors give the approximate p-value for the test as .015. Calculate the approximate p-value for the test and compare it with the authors' value. Interpret the p-value.

12.28 Refer to Exercise 12.27 and the study by Behar and Stewart (1984) on aggressive behavior in children. Another comparison given in their paper was the numbers of male and female children for whom physical aggressiveness was a presenting complaint. The data are shown in the table.

	FEMALE	MALE
Number for which physical aggressiveness is a presenting complaint	7	40
Sample size	12	46

Source: Behar and Stewart (1984).

a. Do the data provide sufficient evidence to indicate a dependence between sex of the child and the proportion for which physical aggressiveness is a presenting complaint? Test using $\alpha = .05$.

b. The authors give the p-value for the test as .025. Find the p-value for your test and compare it with the authors' value. Interpret the p-value.

12.29 Many scientists believe that alcoholism is linked to social isolation. One measure of social isolation is marital status, i.e., whether a person is married or not. To test the notion that alcoholics are socially isolated, 280 adults were randomly selected and each was classified as a diagnosed alcoholic, undiagnosed alcoholic, or nonalcoholic and categorized according to his or her marital status. A summary of the responses is shown in the table.

		ALCOHOLIC CLASSIFICATION		
		Diagnosed	Undiagnosed	Nonalcoholic
MARITAL STATUS	Married	21	37	58
	Not married	59	63	42

a. Is there evidence of a relationship between marital status and alcoholic classifications? Test using $\alpha = .05$.

b. Construct a graph showing the percentage of married and not married for each alcoholic classification. Compare these to the percentages based on the total over all alcoholic classifications. Does the plot support the result of the test in part **a**?

12.30 One criterion used to evaluate employees in the assembly section of a large factory is the number of defective pieces per 1,000 parts produced. The quality control department wants to find out whether there is a relationship between years of experience and defect rate. Since the job is repetitious, after the initial training period any improvement due to a learning effect might be offset by a decrease in the motivation of a worker. A defect rate is calculated for each worker for a yearly evaluation. The

		YEARS OF EXPERIENCE (AFTER TRAINING PERIOD)		
		1	2–5	6–10
DEFECT RATE	High	6	9	9
	Average	9	19	23
	Low	7	8	10

results for 100 workers are given in the table. Is there evidence of a relationship between defect rate and years of experience? Use $\alpha = .05$.

12.31 An experimenter wishes to determine whether there is a relationship between hair color and eye color. One hundred people are randomly sampled and the eyes and hair of each person are judged to be light or dark. A summary of the number of people in each of the four categories is shown in the table. Do the data provide sufficient evidence to indicate a relationship between eye and hair color? Test using $\alpha = .10$.

	LIGHT HAIR	DARK HAIR	TOTALS
LIGHT EYES	31	21	52
DARK EYES	14	34	48
TOTALS	45	55	100

12.32 An insurance company that sells hospitalization policies wants to know whether there is a relationship between the amount of hospitalization coverage a person has and the length of stay in the hospital. Records are selected at random at a large hospital by hospital personnel, and the information on length of stay and hospitalization coverage is given to the insurance company. The results are summarized in the table. Can you conclude that there is a relationship between length of stay and hospitalization coverage? Use $\alpha = .01$.

		LENGTH OF STAY IN HOSPITAL (DAYS)			
		5 or under	6–10	11–15	Over 15
	Under 25%	26	30	6	5
HOSPITALIZATION COVERAGE OF COSTS	25 to < 50%[a]	21	30	11	7
	50 to < 75%	25	25	45	9
	75% and over	11	32	17	11

[a]The symbol < is read as "less than."

12.33 Researchers have investigated the impact of marketing on the role of the consumer in influencing the design and development of new products. A total of 107 companies were involved in the study and, based on an analysis, were classified according to the extent to which they were perceived to have adopted the marketing concept (low, medium, or high). New products for these firms were then classified according to whether the new product idea was derived from a consumer-oriented source. The data are shown in the table.

		CONSUMER-ORIENTED SOURCE FOR IDEA	
		Yes	No
	Low	6	17
EXTENT OF ADOPTION OF MARKETING CONCEPT	Medium	10	45
	High	6	23

Source: Lawton, L., and Parasuraman, A. "The impact of the new marketing concept on new product planning," *Journal of Marketing*, 1980, 44, 19–25.

a. Do the data provide sufficient evidence to indicate that the proportion of new product ideas based on consumer sources depends on the extent to which a company is perceived to have adopted the marketing concept? Test using $\alpha = .05$.

b. Plot the percentage of firms for which the new product idea was derived from a consumer-oriented source based on the extent to which the new marketing concept was adopted. Compare these percentages to the percentage of all firms whose idea was derived from a consumer-oriented source. Is the result of the test in part **a** supported by the graph?

12.34   A team of market researchers conducted a study involving 200 inhabitants of the United States to determine what people fear the most. The sex of each person polled was noted, and then each was asked which of the following was his or her greatest fear: speaking before a group, heights, bugs/insects, financial problems, sickness/death, and other. The results of the poll are given in the table. Do the data provide sufficient information to indicate a relationship between sex and greatest fear? Test at the $\alpha = .05$ level.

		GREATEST FEAR					
		Speaking before a group	Heights	Bugs/insects	Financial problems	Sickness/ death	Other
SEX	Male	21	10	7	23	15	21
	Female	16	22	15	9	18	23

12.35   As noted earlier in our discussion, the X^2 statistic possesses approximately a chi-square distribution when the sample size n is sufficiently large. The actual sample size n needed to achieve a satisfactory approximation depends on the application. To be safe, we suggested that n be large enough so that all expected cell counts exceed 5. Case Study 12.1 produced data that did not satisfy this criterion. Calculate the X^2 statistic for the data of Case Study 12.1 and find its p-value. Compare this approximate p-value with the exact p-value given by Mendenhall et al. This comparison will enable you to see how well the results for the two tests agree.

12.36   Over the years pollsters have found that the public's confidence in big business has been closely tied to the economic climate of the country. When businesses are growing and employment is increasing, public confidence is high. When the opposite occurs, public confidence is low. In one study, Harvey Kahalas (1981) explored the relationship between confidence in business and job satisfaction. He hypothesized that there is a relationship between level of confidence and job satisfaction and that this relationship holds true for both union and nonunion workers. To test his hypothesis he used the sample data given in the tables, which were collected by the National Opinion Research Center.

Sample of Union Members

		JOB SATISFACTION			
		Very satisfied	Moderately satisfied	Little dissatisfied	Very dissatisfied
CONFIDENCE IN MAJOR CORPORATIONS	A great deal	26	15	2	1
	Only some	95	73	16	5
	Hardly any	34	28	10	9

Sample of Nonunion Workers

		JOB SATISFACTION			
		Very satisfied	Moderately satisfied	Little dissatisfied	Very dissatisfied
CONFIDENCE IN MAJOR CORPORATIONS	A great deal	111	52	13	4
	Only some	246	142	37	18
	Hardly any	73	51	19	9

a. Kahalas concluded that his hypothesis was not supported by the data. Do you agree? Conduct the appropriate hypothesis tests using $\alpha = .05$. Be sure to specify the null and alternative hypotheses of your tests.

b. Find and interpret the approximate p-values of the tests you conducted in part a.

c. Construct graphs to assist in the interpretation of the test results that you obtained.

12.3 Caution

Because the X^2 statistic for testing hypotheses about multinomial probabilities is one of the most widely applied statistical tools, it is also one of the most abused statistical procedures. The user should always be certain that the experiment satisfies the assumptions given with each procedure. Furthermore, the user should be certain that the sample is drawn from the correct population—that is, from the population about which the inference is to be made.

The use of the χ^2 probability distribution as an approximation to the sampling distribution for X^2 should be avoided when the expected counts are very small. The approximation can become very poor when these expected counts are small, and thus the true α level may be quite different from the tabled value. As a rule of thumb, an expected cell count of at least 5 means that the χ^2 probability distribution can be used to determine an approximate critical value.

If the X^2 value does not exceed the established critic[al value] *accept the hypothesis of independence.* You would be risking a [the error of accept]ing H_0 if it is false), and the probability β of committing such [an error] The usual alternative hypothesis is that the classifications a[re dependent. Since] there is literally an infinite number of ways two classificati[ons can be dependent,] it is difficult to calculate one or even several values of β [to associate with this] broad alternative hypothesis. Therefore, we avoid conclu[ding that two classifi]cations are independent, even when X^2 is small.

Finally, if a contingency table X^2 value does exceed the c[ritical value, we must] be careful to avoid inferring that a causal relationship ex[ists between the clas]sifications. Our alternative hypothesis states that the two [classifications are sta]tistically dependent—and a statistical dependence does [not imply causality.] *Therefore, the existence of a causal relationship cannot be established by a contingency table analysis.*

Summary

The use of **count data** to test hypotheses about **multinomial probabilities** represents a useful statistical technique. In a **one-dimensional table** we can use count data to test the hypothesis that the multinomial probabilities are equal to specified values. In the **two-dimensional contingency table**, we can test the independence of the two classifications. And these by no means exhaust the uses of the X^2 statistic. Many other applications can be found in the references at the end of this chapter.

Caution should be exercised to avoid misuse of the χ^2-procedure. The experiment must be multinomial,* and the expected counts should not be too small if the χ^2 critical value is used. Moreover, the χ^2 statistic should not always be viewed as the final answer. If two classifications are found to be dependent, many measures of association exist for quantifying the nature and strength of their dependence (see the references).

SUPPLEMENTARY EXERCISES 12.37–12.57

LEARNING THE MECHANICS

12.37 A random sample of 250 observations was classified according to the row and column categories shown in the table.

		COLUMN		
		1	2	3
	1	20	20	10
ROW	2	10	20	70
	3	20	50	30

a. Do the data provide sufficient evidence to conclude that the rows and columns are dependent? Test using $\alpha = .05$.
b. Would the analysis change if the row totals were fixed before the data were collected?
c. Do the assumptions required for the analysis to be valid differ according to whether the row (or column) totals are fixed? Explain.
d. Convert the table entries to percentages by using each column total as a base, and calculating each row response as a percentage of the corresponding column total. In addition, calculate the row totals, and convert them to percentages of all 250 observations.
e. Plot the row percentages on the vertical axis against the column number on the horizontal axis. Draw horizontal lines corresponding to the row total percentages. Does the deviation (or lack thereof) of the individual row percentages from the row total percentages support the result of the test conducted in part a?

12.38 A random sample of 150 observations was classified into the categories shown in the table.

	CATEGORY				
	1	2	3	4	5
n_i	28	35	33	25	29

---

*When the row (or column) totals are fixed, each row (or column) represents a separate multinomial experiment.

a. Do the data provide sufficient evidence that the categories are not equally likely? Use $\alpha = .10$.

b. Form a 90% confidence interval for p_2, the probability that an observation will fall in category 2.

APPLYING THE CONCEPTS

12.39 Many investors believe that the stock market's directional change in January signals the market's direction for the remainder of the year. This so-called January indicator is frequently cited in the popular press. But is this indicator valid? If so, the well-known *random walk* and *efficient markets* theories (basically postulating that market movements are unpredictable) of stock-price behavior would be called into question. The accompanying table summarizes the relevant changes in the Dow Jones Industrial Average for the period December 31, 1927, through January 31, 1981. Joseph S. Martinich applied the chi-square test of independence to this data to investigate the January indicator.

		NEXT 11-MONTH CHANGE	
		Up	Down
JANUARY CHANGE	Up	25	10
	Down	9	9

a. Examine the contingency table. Based solely on your visual inspection, do the data appear to confirm the validity of the January indicator? Explain.

b. Construct a plot of the percentage of years for which the 11-month movement is up based on the January change. Compare these two percentages to the percentage of times the market moved up during the last 11 months over all years in the sample. What do you think of the January indicator now?

c. If a chi-square test of independence is to be used to investigate the January indicator, what are the appropriate null and alternative hypotheses?

d. Conduct the test of part c. Use $\alpha = .05$. Interpret your results in the context of the problem.

e. Would you get the same result in part d if $\alpha = .10$ were used? Explain.

12.40 A computer used by a 24-hour banking service is supposed to assign each transaction to one of five memory locations at random. A check at the end of a day's transactions gives the following counts to each of the five memory locations:

1	2	3	4	5
90	78	100	72	85

Is there evidence to indicate a difference among the proportions of transactions assigned to the five memory locations? Test using $\alpha = .025$.

12.41 Despite a good winning percentage, a certain major league baseball team has not drawn as many fans as one would expect. In hopes of finding ways to increase attendance, the management plans to interview fans who come to the games to find out why they come. One thing that the management might want to know is whether there are differences in support for the team among various age groups. Suppose the information in the table was collected during interviews with fans selected at random. Can you conclude that there is a relationship between age and number of games attended per year? Use $\alpha = .05$.

| | | NUMBER OF GAMES ATTENDED PER YEAR | | |
		1 or 2	3–5	Over 5
AGE OF FAN	Under 20	78	107	17
	21–30	147	87	13
	31–40	129	86	19
	41–55	55	103	40
	Over 55	23	74	22

12.42   If a company can identify times of day when accidents are most likely to occur, extra precautions can be instituted during those times. A random sampling of the accident reports over the last year at a plant gives the frequency of occurrence of accidents during the different hours of the workday. Can it be concluded from the data in the table that the proportions of accidents are different for at least two of the four time periods?

HOURS	1–2	3–4	5–6	7–8
NUMBER OF ACCIDENTS	31	28	45	47

12.43   The classification of solder joints as acceptable or rejectable is a particularly difficult inspection task due to its subjective nature. Westinghouse Electric Company has experimented with different means of evaluating the performance of solder inspectors. One approach involves comparing an individual inspector's classifications with those of the group of experts that comprise Westinghouse's Work Standards Committee. In an experiment reported by Joseph J. Meagher and Joseph A. Scazzero (1985) of Westinghouse, 153 solder connections were evaluated by the committee and 111 were classified as acceptable. An inspector evaluated the same 153 connections and classified 124 as acceptable. Of the items rejected by the inspector, the committee agreed with 19.
   a. Construct a contingency table that summarizes the classifications of the committee and the inspectors.
   b. Based on a visual examination of the table you constructed in part a, does it appear that there is a relationship between the inspector's classifications and the committee's? Explain. (A plot of the percentage rejected by committee and inspector will aid your examination.)
   c. Conduct a chi-square test of independence for these data. Use $\alpha = .05$. Carefully interpret the results of your test in the context of the problem.

12.44   A sociologist was interested in knowing whether sons have a tendency to choose the same occupation as their fathers. To investigate this question, 500 males were polled and questioned concerning their occupation and the occupation of their father. A summary of the numbers of father–son pairs falling in each occupational category is shown in the table. Do the data provide sufficient evidence to indicate a dependence between a son's choice of occupation and his father's occupation? Test using $\alpha = .05$.

| | | SON | | | |
		Professional or business	Skilled	Unskilled	Farmer
FATHER	Professional or business	55	38	7	0
	Skilled	79	71	25	0
	Unskilled	22	75	38	10
	Farmer	15	23	10	32

12.45 A study was done on the accuracy of newspaper advertisements by the five types of food stores in a southeastern city. On each of 4 days, items were randomly selected from the advertisements for each type of store and the actual price was compared to the advertised price. Each of the stores in the city was classified as one of the following types: national, regional chain A, regional chain B, regional chain C, or independent. Values in the table represent the number of items that were correctly and incorrectly priced.

TYPE OF STORE	NUMBER CORRECTLY PRICED	NUMBER INCORRECTLY PRICED
National chain	89	10
Regional chain A	53	14
Regional chain B	43	12
Regional chain C	32	13
Independent	41	7

   a. Determine whether these data provide sufficient evidence to conclude that the proportion of correctly priced items differs for at least two types of stores. Use $\alpha = .10$.
   b. Use a 95% confidence interval to estimate the proportion of correctly priced items in the stores in the national chain category.

12.46 Teenage alcoholism is a big problem in the United States. To discover why teenagers are turning to alcohol, a survey was conducted to find out whether a teenager's family status has any relationship to the amount of alcohol he or she consumes. A random sample of 200 teenagers between the ages of 15 and 19 were questioned concerning their use of alcohol. A summary of the responses is shown in the table.

		ALCOHOL		
		None	Occasional	Frequent
	Upper class	4	16	10
FAMILY	Upper middle class	11	40	24
STATUS	Lower middle class	9	47	9
	Lower class	6	17	7

   a. Do the data provide sufficient evidence to indicate a relationship between family status and the use of alcohol? Test using $\alpha = .05$.
   b. What assumptions are necessary to assure the validity of the inferential procedure you applied in part a? Do you think they are satisfied in this application?
   c. Calculate the percentages of the total number of teenagers in each family status classification falling into each alcohol use classification. Graph these percentages on the vertical axis against the family status on the horizontal axis. Show the overall percentages for each alcohol use classification as horizontal lines on the graph. Does the graph support the result of the test in part a? Explain.

12.47 A restaurateur who owns restaurants in four cities is considering the possibility of building separate dining rooms for nonsmokers to accommodate customers who wish to dine in a smokefree environment. Since this change would involve significant expense, the restaurateur plans to survey the customers at each restaurant and ask them the following question: "Would you be more comfortable dining here if there were a separate dining room for nonsmokers only?" Suppose 75 people were randomly selected and surveyed at each restaurant with the results shown in the table. Is there sufficient evidence to indicate a difference among customer preferences for the four restaurants? Use $\alpha = .10$.

		ANSWERS TO QUESTION		
		Yes	No	It makes no difference
RESTAURANT	1	38	32	5
	2	42	26	7
	3	35	34	6
	4	37	30	8

12.48   Five candidates have just entered the race for mayor of a large city. To determine whether any of the candidates has an early lead in popularity, a poll is conducted. Some 2,000 voters are asked to indicate the candidate they prefer. A summary of their responses is shown in the table. Do the data provide sufficient evidence to indicate a preference for at least one of the five candidates? Test using $\alpha = .01$.

	CANDIDATE				
	I	II	III	IV	V
RESPONSES	385	493	628	235	259

12.49   Suppose an industrial security firm wants to conduct a study of criminal cases involving stolen company money in which employees have been found guilty. Among the data they record are the employee's salary (wages) and the amount of money stolen from the company for 400 recent cases. The results are given in the accompanying table.

		AMOUNT STOLEN ($)			
		Under 5,000	5,000–9,999	10,000–19,999	20,000 or more
INCOME OF EMPLOYEE ($ THOUSAND)	Under 15	46	39	17	5
	15–25	78	79	61	19
	Over 25	5	14	25	12

a. Does this information provide evidence of a relationship between employee income and amount stolen? Use $\alpha = .05$.

b. Convert the responses to percentages by calculating the percentage in each income category based on the total number of cases in each amount stolen category. Also calculate the percentage of all 400 cases in each income category. Plot percentage on the vertical axis and amount stolen on the horizontal axis, showing the overall income category percentages as horizontal lines on the plot. Plot the income percentages for each cell using 1, 2, and 3 as the plotting symbols for the three categories. Does the plot support the result of the test in part a?

12.50   A national survey was conducted to determine how the general public views government involvement in domestic projects. Two hundred people from the low and medium income levels and 150 from the high income level were asked if they thought the government was too involved, not involved enough, or involved just enough. A summary of their responses is shown in the table. Do the data provide sufficient information to indicate a relationship between income and view on government involvement in domestic projects? Test using $\alpha = .05$.

		INVOLVEMENT			TOTALS
		Too little	Just enough	Too much	
	Low	125	48	27	200
INCOME	Medium	103	58	39	200
	High	57	54	39	150
TOTALS		285	160	105	550

12.51 Consumers have traditionally viewed products with warranties more favorably than products without warranties. In fact, several studies have demonstrated that, when given the choice between two similar products, one of which is warranted, consumers prefer the warranted product, even at a higher price. Thus, consumers generally perceive warranties as a kind of "value" added to the product. However, a substantial number of firms have been found to perceive their warranties primarily as legal disclaimers of responsibility and nothing more. As a result of the differences in perceptions by consumers and businesses, Congress passed the Magnuson–Moss Warranty Act, which took effect in 1977. Its purpose was to reform consumer product warranty practices. According to this act, all warranties must be designated as "full" or "limited" and must be clearly written in readily understood language. Further, it specified what was to be contained in warranties (McDaniel and Rao, 1982).

Recently, McDaniel and Rao undertook a study to investigate consumer satisfaction with warranty practices since the advent of the Magnuson–Moss Warranty Act. Using a mailed questionnaire, they sampled 237 midwestern consumers who had purchased a major appliance within the past 6 to 18 months. One of the questions they asked the consumers was, "Do most retailers and dealers make a conscientious effort to satisfy their customers' warranty claims?" One hundred fifty-six answered yes, 61 were uncertain, and 20 said no.

The population of consumers from which this sample was drawn had also been investigated 2 years prior to the Magnuson–Moss Warranty Act. At that time, 37.0% of the population answered yes to the same question, 53.3% were uncertain, and 9.7% said no.

a. As reflected in the answers to the above question, have consumer attitudes toward warranties changed since the pre-Magnuson–Moss Act study? Test using $\alpha = .05$.

b. Compare the pre- and post-Magnuson–Moss Warranty Act responses, and describe the changes that have occurred.

12.52 A political scientist wants to determine whether there is a relationship between a person's income and his or her political affiliation. The researcher randomly samples 265 registered voters and determines the income and political affiliation of each. A summary of the data is shown in the table. Do the data provide sufficient evidence to indicate a relationship between political affiliation and annual income? Test using $\alpha = .10$.

	ANNUAL INCOME (THOUSANDS OF DOLLARS)			
	25 or over	$16 < 25$	$8 < 16$	Below 8
REPUBLICAN	50	28	20	12
DEMOCRAT	14	35	35	41
OTHER	6	7	10	7

12.53 A city has three television stations. Each station has its own evening news program from 6:00 to 6:30 P.M. every weekday. An advertising firm wants to know whether there is an unequal breakdown of the evening news audience among the three stations. One hundred people are selected at random from those who watch the evening news on one of these three stations. Each is asked to specify which news program he or she watches. Do the results in the table provide sufficient evidence to indicate that the three stations do not have equal shares of the evening news audience? Use $\alpha = .05$.

STATION	1	2	3
NUMBER OF VIEWERS	35	43	22

12.54   Several life insurance firms have policies geared to college students. To get more information about this group, a major insurance firm interviewed college students to find out the type of life insurance they preferred, if any. The accompanying table was produced after surveying 1,600 students.

	PREFERRED A TERM POLICY	PREFERRED A WHOLE-LIFE POLICY	NO PREFERENCE
FEMALES	116	27	676
MALES	215	33	533

a. Is there evidence that the life insurance preference of students depends on their sex?

b. Find the approximate observed significance level for the test in part **a**.

c. Construct a plot of percentage responses that will help to interpret the result of the test in part **a**.

12.55   In late 1977, many farmers across the United States went on strike, protesting that the prices of farm products, chiefly grains, were less than the cost of production. Although the main strike goal was to receive 100% of parity prices for all farm products, a second controversial goal was to induce farmers to reduce production, thereby reducing surpluses and boosting prices. A sample survey of 100 farmers was conducted to determine whether a relationship exists between a farmer's decision to participate in the strike and the farmer's opinion concerning the necessity for a cutback in production. The results are shown in the table.

		ON STRIKE	
		Yes	No
	Favor	21	7
50% CUTBACK IN PRODUCTION	Undecided	37	2
	Opposed	22	11

a. Is there evidence of a relationship between a farmer's strike position and the stand on a cutback in production? Use $\alpha = .05$.

b. Calculate and graph the percentages on strike based on the number in each cutback opinion classification. Compare these to the percentage of all farmers sampled who are on strike. Use the graph to assist in the interpretation of the result of the test in part **a**.

12.56   A statistical analysis is to be done on a set of data consisting of 1,000 monthly salaries. The analysis requires the assumption that the sample was drawn from a normal distribution. A preliminary test, called the χ^2 *goodness-of-fit test*, can be used to help determine whether it is reasonable to assume that the sample is from a normal distribution. Suppose the mean and standard deviation of the 1,000 salaries are hypothesized to be $1,200 and $200, respectively. Using the standard normal table, we can approximate the probability of a salary being in the intervals listed in the table. The third column represents the expected number of the 1,000 salaries to be found in each interval if the sample was drawn from a normal distribution with $\mu = \$1,200$ and $\sigma = \$200$. Suppose the last column contains the actual observed frequencies in the sample. Large differences between the observed and expected frequencies cast doubt on the normality assumption.

INTERVAL	PROBABILITY	EXPECTED FREQUENCY	OBSERVED FREQUENCY
Less than $800	.023	23	26
$800 < $1,000	.136	136	146
$1,000 < $1,200	.341	341	361
$1,200 < $1,400	.341	341	311
$1,400 < $1,600	.136	136	143
$1,600 or above	.023	23	13

a. Compute the X^2 statistic based on the observed and expected frequencies—just as you did in Section 12.1.
b. Find the tabulated χ^2 value when $\alpha = .05$ and there are 5 df (there are $k - 1 = 5$ df associated with this X^2 statistic).
c. Based on the X^2 statistic and the tabulated χ^2 value, is there evidence that the salary distribution is nonnormal?
d. Find the approximate observed significance level for the test in part c.

12.57   Suppose a random variable is hypothesized to be normally distributed with mean 0 and standard deviation 1. A random sample of 200 observations on the variable yields frequencies in the listed intervals as shown in the table. Do the data provide sufficient evidence to contradict the hypothesis that x is normally distributed with $\mu = 0$ and $\sigma = 1$? Use the technique developed in Exercise 12.56.

INTERVAL	$x < -2$	$-2 \leq x < -1$	$-1 \leq x < 0$	$0 \leq x < 1$	$1 \leq x < 2$	$x \geq 2$
FREQUENCY	7	20	61	77	26	9

ON YOUR OWN...

Many researchers rely on surveys to estimate the proportions of experimental units in populations that possess certain specified characteristics. A political scientist may want to estimate the proportion of an electorate in favor of a certain legislative bill. A social scientist may be interested in the proportions of people in a geographical region who fall in certain socioeconomic classifications. A psychologist might want to compare the proportions of patients who have different psychological disorders.

Choose a specific topic, similar to those described above, that interests you. Clearly define the population of interest, identify data categories of specific interest, and identify the proportions associated with them. Now guesstimate the pro-

portions of the population that you think fall in each of the categories. For example, you might guess that all the proportions are equal, or that the first proportion is twice as large as the second but equal to the third, etc.

You are now ready to collect the data by obtaining a random sample from your population of interest. Select a sample size so that all expected cell counts are at least 5 (preferably larger), and collect the data.

Use the count data you have obtained to test the null hypothesis that the true proportions in the population equal your presampling guesstimates of these actual proportions. Would failure to reject this null hypothesis imply that your guesstimates are correct?

USING THE COMPUTER...

Are monthly homeowner costs dependent on the region of the country in which the homeowner lives? Use the zip code data set described in Appendix C to investigate this question.

a. Form a contingency table using region as one classification, and the following four levels of homeowner costs as the second classification: $0–$200, $200.01–$400, $400.01–$600, and $600.01–$800. The contingency table should contain a count of the total number of zip codes in each region–cost classification combination.

b. Conduct a χ^2 test to determine whether homeowner costs are dependent on region.

c. Calculate the percentage of zip code areas in each cost class using the total number of zip code areas in each region as the base. Then calculate the percentage in each cost class for all 1,000 zip codes. Construct a table like Table 12.9 to summarize the results. Display the percentages in a graph like Figure 12.3. Does the graphical display support the results of the test in part b?

References

Agresti, A., and Agresti, B. F. *Statistical Methods for the Social Sciences*, 2d edition. San Francisco: Dellen, 1986, Chapter 8.

Cochran, W. G. "The χ^2 test of goodness of fit." *Annals of Mathematical Statistics*, 1952, 23, 315–345.

Graham, E. "The entrepreneurial mystique," *Wall Street Journal*, May 20, 1985, p. 1C.

Kahalas, H. "The relationship between confidence in business and job satisfaction for union and nonunion members," *Baylor Business Studies*, February–April 1981, 127, 45–53.

Martinich, J. S. "The January indicator: A nonrandom but unprofitable walk," *Mid-South Business Journal*, Vol. 4, No. 4, 3rd Quarter, 1984.

McDaniel, S. W., and Rao, C. P. "Consumer attitudes toward and satisfaction with warranties and warranty performance—Before and after Magnuson–Moss," *Baylor Business Studies*, November–December 1982, 130, 47–61.

Meagher, J. J., and Scazzero, J. A. "Measuring inspector variability," *1985 ASQC Quality Congress Transaction*, May 1985, pp. 75–81.

Savage, I. R. "Bibliography of nonparametric statistics and related topics," *Journal of the American Statistical Association*, 1953, 48, 844–906.

Siegel, S. *Nonparametric Statistics for the Behavioral Sciences*. New York: McGraw-Hill, 1956, Chapter 9.

Simple Linear Regression

WHERE WE'VE BEEN...

We have learned how to estimate and test hypotheses about population parameters based on a random sample of observations from the population and have extended these methods to allow for a comparison of parameters from two or more populations.

WHERE WE'RE GOING...

For many sampling situations, we have much more information available on a random variable (and the population it generates) than that contained in a single random sample. For example, if we wanted to predict the rainfall at a given location on a given day, we could select a single random sample of n daily rainfalls, use the methods of Chapters 7 and 8 to estimate the mean daily rainfall μ, and then use this quantity to predict the rainfall for any given day. But this method fails to utilize scientific information that is available to any forecaster. We know the daily rainfall is related to barometric pressure, cloud cover, etc. By measuring barometric pressure and cloud cover at the same time we sample the daily rainfall, we hope to establish the relationship between these variables and to utilize them for prediction.

This chapter is devoted to the most elementary situation—relating two variables. The more complex problem of relating more than two variables is the topic of Chapter 14.

In Chapters 7–10 we described methods for making inferences about population means. The mean of a population has been treated as a *constant*, and we have shown how to use sample data to estimate or to test hypotheses about this constant mean. In many business applications, the mean of a population is not viewed as a constant but rather as a variable. For example, the mean sales price of residences in a large city during 1990 can be treated as a constant and might be equal to $150,000. But we might also treat the mean sales price as a variable that depends on the square feet of living space in the residence. For example, the relationship might be

Mean sales price = $30,000 + $60(Square feet)

The formula implies that the mean sales price of 1,000-square-foot homes is $90,000, the mean sales price of 2,000-square-foot homes is $150,000, and the mean sales price of 3,000-square-foot homes is $210,000.

What do we gain by treating the mean as a variable rather than a constant? In many practical applications we will be dealing with highly variable data, data for which the standard deviation is so large that a constant mean is almost "lost" in a sea of variability. For example, if the mean residential sales price is $150,000 but the standard deviation is $75,000, then the actual sales prices will vary considerably, and the mean price is not a very meaningful or useful characterization of the price distribution. On the other hand, if the mean sales price is treated as a variable that depends on the square feet of living space, the standard deviation of sales prices for any given size of home might be only $10,000. In this case, the mean price will provide a much better characterization of sales prices when it is treated as a variable rather than a constant.

In this chapter we discuss situations in which the mean of the population is treated as a variable, dependent on the value of another variable. The preceding example of residential sales price depending on the square feet of living space is one illustration. Other examples are: the mean reaction time depending on the amount of a drug in the bloodstream, the mean starting salary of a college graduate depending on the student's GPA, and the mean number of years to which a criminal is sentenced depending on the number of previous convictions.

In this chapter we discuss the simplest of all models relating a population mean to another variable, the *straight-line model*. We show how to use sample data to estimate the straight-line relationship between the mean value of one variable, *y*, as it relates to a second variable, *x*. The methodology of estimating and using a straight-line relationship is referred to as *simple linear regression analysis*.

13.1 Probabilistic Models

An important consideration when taking a drug is how it may affect one's perception or general awareness. Suppose you want to model the length of time it takes to respond to a stimulus (a measure of awareness) as a function of the percentage of a certain drug in the bloodstream. The first question to be answered is this: "Do you think an exact relationship exists between these two variables?" That is, do you think it is possible to state the exact length of time it takes an individual (subject) to respond if the amount of the drug in the bloodstream is known? We think you will agree with us that this is *not* possible for several reasons. The reaction time depends on many variables other than the percentage of the drug in the bloodstream—for example, the time of day, the amount of sleep the subject had the night before, the subject's visual acuity, the subject's general reaction time without the drug, and the subject's age would all probably affect reaction time. Even if many variables are included in a model (the topic of Chapter 14), it is still unlikely that we would be able to predict *exactly* the subject's reaction time. There will almost certainly be some variation in response times due strictly to *random phenomena* that cannot be modeled or explained.

If we were to construct a model that hypothesized an exact relationship between variables, it would be called a **deterministic model**. For example, if we believe that y, the reaction time (in seconds), will be exactly one and one-half times x, the amount of drug in the blood, we write

$$y = 1.5x$$

This represents a **deterministic relationship** between the variables y and x. It implies that y can always be determined exactly when the value of x is known. *There is no allowance for error in this prediction.*

If, on the other hand, we believe there will be unexplained variation in reaction times—perhaps caused by important but unincluded variables or by random phenomena—we discard the deterministic model and use a model that accounts for this **random error**. This **probabilistic model** includes both a deterministic component and a random error component. For example, if we hypothesize that the response time y is related to the percentage of drug x by

$$y = 1.5x + \text{Random error}$$

we are hypothesizing a **probabilistic relationship** between y and x. Note that the deterministic component of this probabilistic model is $1.5x$.

Figure 13.1(a) shows the possible responses for five different values of x, the percentage of drug in the blood, when the model is deterministic. All the responses must fall exactly on the line because a deterministic model leaves no room for error.

Figure 13.1(b) shows a possible set of responses for the same values of x when we are using a probabilistic model. Note that the deterministic part of the model (the straight line itself) is the same. Now, however, the inclusion of a random error component allows the response times to vary from this line. Since we know that the response time does vary randomly for a given value of x, the probabilistic model provides a more realistic model for y than does the deterministic model.

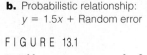

a. Deterministic relationship:
$y = 1.5x$

b. Probabilistic relationship:
$y = 1.5x + \text{Random error}$

FIGURE 13.1

Possible reaction times, y, for five different drug percentages, x

---

General Form of Probabilistic Models

$$y = \text{Deterministic component} + \text{Random error}$$

where y is the variable of interest. We always assume that the mean value of the random error equals 0. This is equivalent to assuming that the mean value of y, $E(y)$, equals the deterministic component of the model, i.e.,

$$E(y) = \text{Deterministic component}$$

We begin with the simplest of probabilistic models—the **straight-line model**—which derives its name from the fact that the deterministic portion of the model graphs as a straight line. Fitting this model to a set of data is an example of **regression analysis**, or **regression modeling**. The elements of the straight-line model are summarized in the box.

A First-Order (Straight-Line) Probabilistic Model

$$y = \beta_0 + \beta_1 x + \varepsilon$$

where

$y =$ *Dependent* or *response* variable (variable to be modeled)

$x =$ *Independent* or *predictor* variable (variable used as a predictor of y)*

ε (epsilon) $=$ Random error component

β_0 (beta zero) $=$ y-intercept of the line—i.e., point at which the line intercepts or cuts through the y-axis (see Figure 13.2)

β_1 (beta one) $=$ Slope of the line—i.e., amount of increase (or decrease) in the deterministic component of y for every 1-unit increase in x. You can see (Figure 13.2) that $E(y)$ increases by the amount β_1 as x increases from 2 to 3.

FIGURE 13.2

The straight-line model

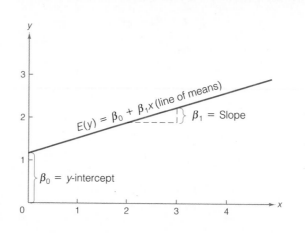

In the probabilistic model, the deterministic component is referred to as the **line of means**, because the mean of y, $E(y)$, is equal to the straight-line component of the model. That is,

$$E(y) = \beta_0 + \beta_1 x$$

Note that the Greek symbols β_0 and β_1 represent the y-intercept and slope of the model. They are population parameters that will be known only if we have

---

*The word *independent* should not be interpreted in a probabilistic sense, as defined in Chapter 3. The phrase *independent variable* is used in regression analysis to refer to a predictor variable for the response y.

access to the entire population of (x, y) measurements. Together with a specific value of the independent variable x, they determine the mean value of y, which is just a specific point on the line of means (Figure 13.2).

The values of β_0 and β_1 will be unknown in almost all practical applications of regression analysis. The process of developing a model, estimating the unknown parameters, and using the model can be viewed as the five-step procedure shown in the box.

STEP 1 Hypothesize the deterministic component of the model that relates the mean, $E(y)$, to the independent variable x (Section 13.1).

STEP 2 Use the sample data to estimate unknown parameters in the model (Section 13.2).

STEP 3 Specify the probability distribution of the random error term, and estimate the standard deviation of this distribution (Sections 13.3 and 13.4).

STEP 4 Statistically evaluate the usefulness of the model (Sections 13.5, 13.6, and 13.7).

STEP 5 When satisfied that the model is useful, use it for prediction, estimation, and other purposes (Section 13.8).

In this chapter only the straight-line model is discussed; more complex models are addressed in Chapters 14 and 15.

EXERCISES 13.1–13.9

LEARNING THE MECHANICS

13.1 In each case, graph the line that passes through the given points.
    a. $(0, 0)$ and $(4, 4)$ **b.** $(0, 2)$ and $(2, 0)$ **c.** $(-2, 2)$ and $(6, 3)$ **d.** $(-4, -1)$ and $(3, 4)$

13.2 Give the slope and y-intercept for each of the lines graphed in Exercise 13.1.

13.3 The equation for a straight line (deterministic) is

$$y = \beta_0 + \beta_1 x$$

If the line passes through the point $(-1, 3)$, then $x = -1$, $y = 3$ must satisfy the equation; i.e.,

$$3 = \beta_0 + \beta_1(-1)$$

Similarly, if the line passes through the point $(3, 4)$, then $x = 3$, $y = 4$ must satisfy the equation; i.e.,

$$4 = \beta_0 + \beta_1(3)$$

Use these two equations to solve for β_0 and β_1, and find the equation of the line that passes through the points $(-1, 3)$ and $(3, 4)$.

13.4 Refer to Exercise 13.3. Find the equations of the lines that pass through the points listed in Exercise 13.1.

13.5   Plot the following lines:
    a. $y = 3 + 2x$ **b.** $y = 3 - 2x$ **c.** $y = -3 + 2x$
    d. $y = -x$ **e.** $y = 2x$ **f.** $y = .50 + 1.25x$

13.6   Give the slope and y-intercept for each of the lines defined in Exercise 13.5.

13.7   Why do we generally prefer a probabilistic model to a deterministic model? Give examples for which the two types of models might be appropriate.

13.8   What is the line of means?

13.9   If a straight-line probabilistic relationship relates the mean $E(y)$ to an independent variable x, does it imply that every value of the variable y will always fall exactly on the line of means? Why or why not?

13.2   Fitting the Model: Least Squares Approach

After the straight-line model has been hypothesized to relate the mean $E(y)$ to the independent variable x, the next step is to collect data and to estimate the (unknown) population parameters, the y-intercept β_0 and the slope β_1.

To begin with a simple example, suppose an experiment involving five subjects is conducted to determine the relationship between the percentage of a certain drug in the bloodstream and the length of time it takes to react to a stimulus. The results are shown in Table 13.1. (The number of measurements and the measurements themselves are unrealistically simple in order to avoid arithmetic confusion in this introductory example.) This set of data will be used to demonstrate the five-step procedure of regression modeling given in Section 13.1. In this section we hypothesize the deterministic component of the model and estimate its unknown parameters (steps 1 and 2). Discussion of the model assumptions and the random error component (step 3) are the subjects of Sections 13.3 and 13.4, whereas Sections 13.5–13.7 assess the utility of the model (step 4). Finally, using the model for prediction and estimation (step 5) is the subject of Section 13.8.

T A B L E 13.1   **Reaction Time Versus Drug Percentage**

SUBJECT	AMOUNT OF DRUG x (%)	REACTION TIME y (seconds)
1	1	1
2	2	1
3	3	2
4	4	2
5	5	4

STEP 1 *Hypothesize the deterministic component of the probabilistic model.* As stated before, we will consider only straight-line models in this chapter, and thus the complete model to relate mean response time $E(y)$ to drug percentage x is given by

$$E(y) = \beta_0 + \beta_1 x$$

STEP 2 *Use sample data to estimate unknown parameters in the model.* This step is the subject of this section—namely, how can we best use the information in the

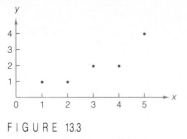

FIGURE 13.3

Scattergram for data in Table 13.1

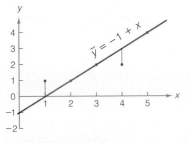

FIGURE 13.4

Visual straight-line fit to the data

sample of five observations in Table 13.1 to estimate the unknown y-intercept β_0 and slope β_1?

To determine whether a linear relationship between y and x is plausible, it is helpful to plot the sample data. Such a plot, called a **scattergram**, locates each of the five data points on a graph, as shown in Figure 13.3. Note that the scattergram suggests a general tendency for y to increase as x increases. If you place a ruler on the scattergram, you will see that a line may be drawn through three of the five points, as shown in Figure 13.4. To obtain the equation of this visually fitted line, note that the line intersects the y-axis at $y = -1$, so the y-intercept is -1. Also, y increases exactly 1 unit for every 1-unit increase in x, indicating that the slope is $+1$. Therefore, the equation is

$$\tilde{y} = -1 + 1(x) = -1 + x$$

where $\tilde{y}$ is used to denote the predicted y from the visual model.

One way to decide quantitatively how well a straight line fits a set of data is to note the extent to which the data points deviate from the line. For example, to evaluate the model in Figure 13.4, we calculate the magnitude of the **deviations**, i.e., the differences between the observed and the predicted values of y. These deviations, or **errors**, are the vertical distances between observed and predicted values (see Figure 13.4). The observed and predicted values of y, their differences, and their squared differences are shown in Table 13.2. Note that the *sum of errors* equals 0 and the *sum of squares of the errors* (SSE), which gives greater emphasis to large deviations of the points from the line, is equal to 2.

TABLE 13.2 **Comparing Observed and Predicted Values for the Visual Model**

x	y	$\tilde{y} = -1 + x$	$(y - \tilde{y})$	$(y - \tilde{y})^2$
1	1	0	$(1 - 0) = 1$	1
2	1	1	$(1 - 1) = 0$	0
3	2	2	$(2 - 2) = 0$	0
4	2	3	$(2 - 3) = -1$	1
5	4	4	$(4 - 4) = 0$	0
			Sum of errors $= 0$	Sum of squared errors (SSE) $= 2$

$\bar{x} = 3 \quad \bar{y} = 2$

You can see by shifting the ruler around the graph that it is possible to find many lines for which the sum of the errors is equal to 0, but it can be shown that there is one (and only one) line for which the SSE is a *minimum*. This line is called the **least squares line**, the **regression line**, or **least squares prediction equation**.

To find the least squares line for a set of data, assume that we have a sample of n data points consisting of pairs of values of x and y, say (x_1, y_1), $(x_2, y_2), \ldots, (x_n, y_n)$. For example, the $n = 5$ data points shown in Table 13.2 are $(1, 1)$, $(2, 1)$, $(3, 2)$, $(4, 2)$, and $(5, 4)$. The fitted line, which we will calculate based on the five data points, is written as

$$\hat{y} = \hat{\beta}_0 + \hat{\beta}_1 x$$

The "hats" can be read as "estimator of." Thus, $\hat{y}$ is an estimator of the mean value of y, $E(y)$, and a predictor of some future value of y; and $\hat{\beta}_0$ and $\hat{\beta}_1$ are estimators of β_0 and β_1, respectively.

For a given data point, say the point (x_i, y_i), the observed value of y is y_i and the predicted value of y would be obtained by substituting x_i into the prediction equation:

$$\hat{y}_i = \hat{\beta}_0 + \hat{\beta}_1 x_i$$

And the deviation of the ith value of y from its predicted value is

$$(y_i - \hat{y}_i) = [y_i - (\hat{\beta}_0 + \hat{\beta}_1 x_i)]$$

Then the sum of squares of the deviations of the y-values about their predicted values for all the n data points is

$$\text{SSE} = \sum [y_i - (\hat{\beta}_0 + \hat{\beta}_1 x_i)]^2$$

The quantities $\hat{\beta}_0$ and $\hat{\beta}_1$ that make the SSE a minimum are called the **least squares estimates** of the population parameters β_0 and β_1, and the prediction equation $\hat{y} = \hat{\beta}_0 + \hat{\beta}_1 x$ is called the **least squares line**.

| Definition 13.1 |

The **least squares line** is one that has a smaller sum of squared errors (SSE) than any other straight-line model.

The values of $\hat{\beta}_0$ and $\hat{\beta}_1$ that minimize the SSE are (proof omitted) given by the formulas in the accompanying box.*

| Formulas for the Least Squares Estimates |

Slope: $\hat{\beta}_1 = \dfrac{\text{SS}_{xy}}{\text{SS}_{xx}}$

y-intercept: $\hat{\beta}_0 = \bar{y} - \hat{\beta}_1 \bar{x}$

where

$$\text{SS}_{xy} = \sum (x_i - \bar{x})(y_i - \bar{y}) = \sum x_i y_i - \frac{\left(\sum x_i\right)\left(\sum y_i\right)}{n}$$

$$\text{SS}_{xx} = \sum (x_i - \bar{x})^2 = \sum x_i^2 - \frac{\left(\sum x_i\right)^2}{n}$$

$n = $ Sample size

---

*Students who are familiar with calculus should note that the values of β_0 and β_1 that minimize SSE $= \sum (y_i - \hat{y}_i)^2$ are obtained by setting the two partial derivatives $\partial\text{SSE}/\partial\beta_0$ and $\partial\text{SSE}/\partial\beta_1$ equal to 0. The solutions to these two equations yield the formulas shown in the box. Furthermore, we denote the *sample* solutions to the equations by $\hat{\beta}_0$ and $\hat{\beta}_1$, where the "hat" denotes that these are sample estimates of the true population intercept β_0 and slope β_1.

Preliminary computations for finding the least squares line for the drug reaction example are presented in Table 13.3. We can now calculate

$$SS_{xy} = \sum x_i y_i - \frac{\left(\sum x_i\right)\left(\sum y_i\right)}{5} = 37 - \frac{(15)(10)}{5}$$

$$= 37 - 30 = 7$$

$$SS_{xx} = \sum x_i^2 - \frac{\left(\sum x_i\right)^2}{5} = 55 - \frac{(15)^2}{5}$$

$$= 55 - 45 = 10$$

TABLE 13.3 **Preliminary Computations for the Drug Reaction Example**

x_i	y_i	x_i^2	$x_i y_i$
1	1	1	1
2	1	4	2
3	2	9	6
4	2	16	8
5	4	25	20
TOTALS $\sum x_i = 15$	$\sum y_i = 10$	$\sum x_i^2 = 55$	$\sum x_i y_i = 37$

Then the slope of the least squares line is

$$\hat{\beta}_1 = \frac{SS_{xy}}{SS_{xx}} = \frac{7}{10} = .7$$

and the y-intercept is

$$\hat{\beta}_0 = \bar{y} - \hat{\beta}_1 \bar{x} = \frac{\sum y_i}{5} - \hat{\beta}_1 \frac{\left(\sum x_i\right)}{5} = \frac{10}{5} - (.7)\frac{(15)}{5}$$

$$= 2 - (.7)(3) = 2 - 2.1 = -.1$$

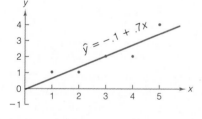

FIGURE 13.5
The line $\hat{y} = -.1 + .7x$ fit to the data

The least squares line is thus

$$\hat{y} = \hat{\beta}_0 + \hat{\beta}_1 x = -.1 + .7x$$

The graph of this line is shown in Figure 13.5.

The predicted value of y for a given value of x can be obtained by substituting into the formula for the least squares line. Thus, when $x = 2$ we predict y to be

$$\hat{y} = -.1 + .7x = -.1 + .7(2) = 1.3$$

We show how to find a prediction interval for y in Section 13.8.

The observed and predicted values of y, the deviations of the y values about their predicted values, and the squares of these deviations are shown in Table 13.4 (page 646). Note that the sum of squares of the deviations, SSE, is 1.10, and (as we would expect) this is less than the SSE = 2.0 obtained in Table 13.2 for the visually fitted line.

TABLE 13.4 **Comparing Observed and Predicted Values for the Least Squares Prediction Equation**

x	y	$\hat{y} = -.1 + .7x$	$(y - \hat{y})$	$(y - \hat{y})^2$
1	1	.6	$(1 - .6) = .4$	.16
2	1	1.3	$(1 - 1.3) = -.3$	.09
3	2	2.0	$(2 - 2.0) = 0$	.00
4	2	2.7	$(2 - 2.7) = -.7$	.49
5	4	3.4	$(4 - 3.4) = .6$	.36
			Sum of errors = 0	SSE = 1.10

It is important that you be able to interpret the intercept and slope in terms of the data being utilized to fit the model. In the drug reaction example, the estimated y-intercept, $\hat{\beta}_0$, is $-.1$. This value would seem to imply that the estimated mean reaction time is equal to $-.1$ second when the amount of drug, x, is equal to 0%. Since negative reaction times are not possible, this seems to make the model nonsensical. However, *the model parameters should be interpreted only within the sampled range of the independent variable*—in this case, for amounts of drug in the bloodstream between 1% and 5%. Thus, the y-intercept, which is, by definition, at $x = 0$ (0% drug), is not within the range of the sampled values of x and is not subject to meaningful interpretation.

The slope, $\hat{\beta}_1$, of the least squares line was calculated to be .7. The implication is that for every unit increase of x, the mean value of y is estimated to increase by .7 unit. In terms of this example, for every 1% increase in the amount of drug in the bloodstream, the mean reaction time is estimated to increase by .7 second *over the sampled range of drug amounts from 1% to 5%*. Thus, the model does not imply that increasing the drug amount from 5% to 10% will result in an increase in mean reaction time of 3.5 seconds, because the range of x in the sample does not extend to 10% ($x = 10$). In fact, 10% might be such a high concentration of the drug that it would kill the subject! Be careful to interpret the estimated parameters only within the sampled range of x.

Even when the interpretations of the estimated parameters are meaningful, it should be remembered that they are only estimates based on the sample. As such, their values will typically change in repeated sampling. How much confidence do we have that the estimated slope, $\hat{\beta}_1$, accurately estimates the true slope, β_1? This requires statistical inference, in the form of confidence intervals and tests of hypotheses, which we address in Section 13.5.

To summarize, we defined the best-fitting straight line to be the one that minimizes the sum of squared errors around the line, and we called it the least squares line. We should interpret the least squares line only within the sampled range of the independent variable. In subsequent sections we show how to make statistical inferences about the model.

EXERCISES 13.10–13.22

LEARNING THE MECHANICS

13.10   The following table is similar to Table 13.3. It is used for making the preliminary computations for finding the least squares line for the given pairs of x and y values.

	x_i	y_i	x_i^2	$x_i y_i$
	7	2		
	4	4		
	6	2		
	2	5		
	1	7		
	1	6		
	3	5		
TOTALS	$\Sigma x_i =$	$\Sigma y_i =$	$\Sigma x_i^2 =$	$\Sigma x_i y_i =$

a. Complete the table. b. Find SS_{xy}. c. Find SS_{xx}.
d. Find $\hat{\beta}_1$. e. Find $\bar{x}$ and $\bar{y}$. f. Find $\hat{\beta}_0$.
g. Find the least squares line.

13.11   Refer to Exercise 13.10. After the least squares line has been obtained, the following table (which is similar to Table 13.4) can be used for (1) comparing the observed and the predicted values of y, and (2) computing SSE.

x	y	$\hat{y} =$	$(y - \hat{y})$	$(y - \hat{y})^2$
7	2			
4	4			
6	2			
2	5			
1	7			
1	6			
3	5			
			$\Sigma (y - \hat{y}) =$	$SSE = \Sigma (y - \hat{y})^2 =$

a. Complete the table.

b. Plot the least squares line on a scattergram of the data. Plot the following line on the same graph:

$$\hat{y} = 14 - 2.5x$$

c. Show that SSE is larger for the line part b than it is for the least squares line.

13.12   Construct a scattergram for the data in the table.

x	.5	1	1.5
y	2	1	3

a. Plot the following two lines on your scattergram:

$$y = 3 - x \quad \text{and} \quad y = 1 + x$$

b. Which of these lines would you choose to characterize the relationship between x and y? Explain.
c. Show that the sum of errors for both of these lines equals 0.
d. Which of these lines has the smaller SSE?
e. Find the least squares line for the data, and compare it to the two lines described in part a.

13.13   Consider the following pairs of measurements:

x	5	3	-1	2	7	6	4
y	4	3	0	1	8	5	3

a. Construct a scattergram for these data.
b. What does the scattergram suggest about the relationship between x and y?
c. Given that SS_{xx} = 43.4286, SS_{xy} = 39.8571, $\bar{y}$ = 3.4286, and $\bar{x}$ = 3.7143, calculate the least squares estimates of β_0 and β_1.
d. Plot the least squares line on your scattergram. Does the line appear to fit the data well? Explain.
e. Interpret the y-intercept and slope of the least squares line. Over what range of x are these interpretations meaningful?

13.14   Suppose that n = 100 recent residential home sales in a city are used to fit a least squares straight-line model relating the sales price, y, to the square feet of living space, x. Homes in the sample range from 1,500 square feet to 4,000 square feet of living space, and the resulting least squares equation is

$$\hat{y} = -30{,}000 + 70x$$

a. What is the underlying hypothesized probabilistic model for this application? What does it imply about the relationship between the mean sales price and living space?
b. Identify the least squares estimates of the y-intercept and slope of the model.
c. Interpret the least squares estimate of the y-intercept. Is it meaningful for this application? Why?
d. Interpret the least squares estimate of the slope of the model. Over what range of x is the interpretation meaningful?
e. Use the least squares model to estimate the sales price of a 3,000-square-foot home. Is the estimate meaningful? Explain.
f. Use the least squares model to estimate the sales price of a 5,000-square-foot home. Is the estimate meaningful? Explain.
[*Note:* We show how to measure the statistical reliability of these least squares estimates in subsequent sections.]

APPLYING THE CONCEPTS

13.15   Individuals who report perceived wrongdoing of a corporation or public agency are known as *whistle blowers*. Janet P. Near and Marcia P. Miceli used regression analysis to study factors that affect the extent of the retaliation by the organization against the whistle blower. Among the factors they studied were the pay rate and education level of the whistle blower, the seriousness of the organization's wrongdoing, and the external whistle-blowing channel used. The researchers developed an index to measure the extent of retaliation. The index was based on the number of forms of reprisal experienced by the whistle blower, the number of forms of reprisal with which the whistle blower was threatened, and the number of different types of people within the organization (i.e., coworkers, immediate supervisor, and so forth) who retaliated against them. The table lists the retaliation index (higher numbers indicate more extensive retaliation) and salary for a sample of 15 whistle blowers from federal agencies.
a. Construct a scattergram for the data. Does it appear that the extent of retaliation increases, decreases, or stays the same with an increase in salary? Explain.
b. Use the method of least squares to fit a straight line to these data.
c. Graph the least squares line on your scattergram. Does the least squares line support your answer to the question in part a? Explain.

RETALIATION INDEX	SALARY	RETALIATION INDEX	SALARY
301	$62,000	535	$15,800
550	36,500	455	44,000
755	17,600	615	46,600
327	20,000	700	12,100
500	30,100	650	62,000
377	35,000	630	21,000
290	47,500	360	11,900
452	54,000		

Source: Data adapted from Near, J. P., and Miceli, M. P. "Retaliation against whistle blowers: Predictors and effects," *Journal of Applied Psychology*, Vol. 71, No. 1, 1986, pp. 137–145.

d. Interpret the y-intercept, $\hat{\beta}_0$, of the least squares line in terms of this application. Is the interpretation meaningful?

e. Interpret the slope, $\hat{\beta}_1$, of the least squares line in terms of this application. Over what range of x is this interpretation meaningful?

13.16 Due primarily to the price controls of the Organization of Petroleum Exporting Countries (OPEC), a cartel of crude oil suppliers, the price of crude oil rose dramatically from the mid-1970s to the early 1980s. As a result, motorists were confronted with a similar upward spiral of gasoline prices. The data in the table are typical prices for a gallon of regular leaded gasoline and a barrel of crude oil (refiner acquisition cost) for the indicated years.

YEAR	GASOLINE y (cents/gallon)	CRUDE OIL x ($/bbl)
1973	39	3.89
1975	57	7.67
1976	59	8.19
1977	62	8.57
1978	63	9.00
1979	86	12.64
1980	119	21.59
1981	133	31.77
1982	122	28.52
1983	116	26.19
1984	113	25.88
1985	112	24.09
1986	86	12.51
1987	90	15.41

Source: *Statistical Abstract of the United States: 1989*, pp. 476, 480.

Given that $\Sigma\, y = 1{,}257$, $\Sigma\, x = 235.92$, $\Sigma\, y^2 = 124{,}459$, $\Sigma\, x^2 = 5{,}074.0898$, and $\Sigma\, xy = 24{,}654.87$:

a. Use the data to calculate the least squares line that describes the relationship between the price of a gallon of gasoline and the price of a barrel of crude oil.

b. Plot your least squares line on a scattergram of the data. Does your least squares line appear to be an appropriate characterization of the relationship between y and x? Explain.

c. If the price of crude oil fell to $8 per barrel, to what level (approximately) would the price of regular gasoline fall? Justify your response.

13.17 In recent years, physicians have used the "dividing reflex" to reduce abnormally rapid heartbeats in humans by briefly submerging the patient's face in cold water. The reflex, triggered by cold water temperatures, is an involuntary neural response that shuts off circulation to the skin, muscles, and internal organs and diverts extra oxygen-carrying blood to the heart, lungs, and brain. A research physician conducted an experiment to investigate the effects of various cold water temperatures on the pulse rate of small children. The data for seven 6-year-old children are shown in the accompanying table.

CHILD	TEMPERATURE OF WATER x (°F)	DECREASE IN PULSE RATE y (beats/minute)
1	68	2
2	65	5
3	70	1
4	62	10
5	60	9
6	55	13
7	58	10

a. Find the least squares line for the data.

b. Construct a scattergram for the data; then graph the least squares line as a check on your calculations.

c. If the water temperature is 60°F, predict the drop in pulse rate for a 6-year-old child. [*Note:* A measure of the reliability of these predictions is discussed in Section 13.8.]

13.18   Is the number of games won by a major league baseball team in a season related to the team's batting average? The accompanying table shows the number of games won and the batting averages for the 14 teams in the American League for the 1986 season.

TEAM	NUMBER OF GAMES WON y	TEAM BATTING AVERAGE x
Cleveland	84	.284
New York	90	.271
Boston	95	.271
Toronto	86	.269
Texas	87	.267
Detroit	87	.263
Minnesota	71	.261
Baltimore	73	.258
California	92	.255
Milwaukee	77	.255
Seattle	67	.253
Kansas City	76	.252
Oakland	76	.252
Chicago	72	.247

Source: Official American League Averages 1986. New York: The American League of Professional Baseball Clubs, pp. 2, 24–27.

a. If you were to model the relationship between the number of games won, y, by a major league team and the team's batting average, x, using a straight line, would you expect the slope of the line to be positive or negative? Explain.

b. Construct a scattergram for the data. Does the pattern revealed by the scattergram agree with your answer to part **a**?

c. Given that $\Sigma y = 1,133$, $\Sigma x = 3.658$, $\Sigma y^2 = 92,703$, $\Sigma x^2 = .957118$, and $\Sigma xy = 296.734$, fit a simple linear regression model to the data.

d. Graph the least squares line on your scattergram. Does your least squares line seem to fit the points on your scattergram?

e. Interpret the least squares intercept and slope in terms of this application.

f. Can you explain why the number of games won does not appear to be strongly related to a team's batting average?

13.19 To investigate the relationship between yield of potatoes, y, and level of fertilizer application, x, an experimenter divides a field into eight plots of equal size and applies differing amounts of fertilizer to each. The yield of potatoes (in pounds) and the fertilizer application (in pounds) are recorded for each plot. The data are as follows:

x	1	1.5	2	2.5	3	3.5	4	4.5
y	25	31	27	28	36	35	32	34

a. Construct a scattergram for the data.

b. Find the least squares estimates for β_0 and β_1.

c. According to your least squares line, approximately how many pounds of potatoes would you expect from a plot to which 3.75 pounds of fertilizer has been applied? [*Note:* A measure of the reliability of these predictions is discussed in Section 13.8.]

13.20 In Exercise 8.47, we gave data on the mean daily air temperature and corresponding cocoon temperature of woolly-bear caterpillars of the High Arctic. The data, collected over 12 days, are reproduced in the table.

DAY	TEMPERATURE (°C)		DAY	TEMPERATURE (°C)	
	Air	Cocoon[a]		Air	Cocoon[a]
1	10.4	15.1	7	1.7	3.6
2	9.2	14.6	8	2.0	5.3
3	2.2	6.8	9	3.0	7.0
4	2.6	6.8	10	3.5	7.1
5	4.1	8.0	11	4.5	9.6
6	3.7	8.7	12	4.4	9.5

[a]Each cocoon temperature is the average of the temperatures of two cocoons.

Source: Kevan, P. G., Jensen, T. S., and Shorthouse, J. D. "Body temperatures and behavioral thermoregulation of High Arctic woolly-bear caterpillars and pupae (*Gynaephora rossii*, Lymantridae: Lepidoptera) and the importance of sunshine," *Arctic and Alpine Research*, 1982, 54. Reprinted with permission of the Regents of the University of Colorado.

Given that $SS_{xx} = 83.3425$, $SS_{xy} = 100.0825$, $\bar{x} = 4.2750$, and $\bar{y} = 8.5083$:

a. Fit a least squares line to relate the cocoon temperature y to the outside air temperature x.

b. Plot the data points and graph your least squares line on the same sheet of graph paper. Does your line provide a good fit to the data?

c. Interpret the least squares intercept and slope in terms of this application.

13.21 The sand lance is a small fish that can be found in the Northwest Atlantic from Cape Hatteras to Greenland. In a study of the biological and demographic characteristics of the sand lance, G. H. Winters includes data on the mean length of specimens collected each year for the period 1969 through 1979 for sand lance ages 2 through 8 years:

MEAN LENGTH (MILLIMETERS)	176	194	212	226	236	244	254
AGE (YEARS)	2	3	4	5	6	7	8

Source: Winters, G. H. "Analysis of the biological and demographic parameters of northern sand lance, *Ammodytes dubins*, from the Newfoundland Grand Bank," *Canadian Journal of Fisheries and Aquatic Sciences*, 1983, 40.

a. Find the least squares prediction equation relating the mean length y of the sand lance to its age x.
b. Plot the data points and graph your least squares lines on the same sheet of graph paper. Does your line appear to provide a good fit to the data?
c. Interpret the least squares intercept and slope in terms of this application. Is the intercept's interpretation meaningful in this case?

13.22   The data shown in the accompanying table are part of a series of experiments conducted to investigate the effect of water temperature on the absorption by rainbow trout of sublethal levels of cyanide. This specific set of data is the result of a preliminary experiment to determine the relationship between mean weight gain (percentage of body weight) over a 20-day period as a function of the ration fed the fish (percentage of body weight). Twenty fish were included in the experiment for each ration level and water temperature combination. The mean percentage weight gain for each sample of 20 fish is shown in the table. (Note that, as expected, a 0 level of rations produced a negative weight gain, i.e., a loss.) Since all the sample sizes are equal, we can fit a simple linear regression model to the data for any water temperature level by fitting the model to the four sample means.

RATION	MEAN WET WEIGHT GAIN (%)		
(% body weight per day)	6°C	12°C	18°C
.0	−8.14	−10.33	−13.21
.8	12.31		
1.2		14.29	
1.5	28.19		15.51
2.5	29.12	38.05	
3.5			51.13
4.0		60.13	
4.5			65.49
Maintenance ration	.32	.48	.69

Source: Kovacs, T. G., and Leduc, G. "Sublethal toxicity to rainbow trout (*Salmo gairdneri*) at different temperatures," *Canadian Journal of Fisheries and Aquatic Sciences*, 1982, 39.

a. Find the least squares line relating mean weight gain to ration level for each of the three water temperature levels.
b. Plot the data points and graph the least squares lines on the same sheet of graph paper. Do the least squares lines appear to provide good fits to their respective data sets? Does the relationship between mean weight gain and ration appear to depend on the water temperature? (In Chapter 15 we test to determine whether differences exist.)

13.3   Model Assumptions

In Section 13.2 we assumed that the probabilistic model relating drug reaction time y to the percentage of drug x in the bloodstream is

$$y = \beta_0 + \beta_1 x + \varepsilon$$

and recall that the least squares estimate of the deterministic component of the model, $\beta_0 + \beta_1 x$, is

$$\hat{y} = \hat{\beta}_0 + \hat{\beta}_1 x = -.1 + .7x$$

Now we turn our attention to the random component ε of the probabilistic model and its relation to the errors in estimating β_0 and β_1. We will use a probability distribution to characterize the behavior of ε. We will see how the probability distribution of ε determines how well the model describes the relationship between the dependent variable y and the independent variable x.

Step 3 in a regression analysis requires us to specify the probability distribution of the random error ε. We will make four basic assumptions about the general form of this probability distribution:

ASSUMPTION 1 The mean of the probability distribution of ε is 0. That is, the average of the values of ε over an infinitely long series of experiments is 0 for each setting of the independent variable x. This assumption implies that the mean value of y, $E(y)$, for a given value of x is $E(y) = \beta_0 + \beta_1 x$.

ASSUMPTION 2 The variance of the probability distribution of ε is constant for all settings of the independent variable x. For our straight-line model, this assumption means that the variance of ε is equal to a constant, say σ^2, for all values of x.

ASSUMPTION 3 The probability distribution of ε is normal.

ASSUMPTION 4 The values of ε associated with any two observed values of y are independent. That is, the value of ε associated with one value of y has no effect on the values of ε associated with other y values.

The implications of the first three assumptions can be seen in Figure 13.6, which shows distributions of errors for three values of x, namely, x_1, x_2, and x_3. Note that the relative frequency distributions of the errors are normal with a mean of 0 and a constant variance σ^2. (All the distributions shown have the same amount of spread or variability.) The straight line shown in Figure 13.6 is the line of means. It indicates the mean value of y for a given value of x. We denote this mean value as $E(y)$. Then, the line of means is given by the equation

$$E(y) = \beta_0 + \beta_1 x$$

FIGURE 13.6

The probability distribution of ε

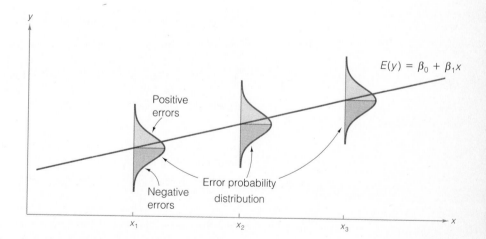

These assumptions make it possible for us to develop measures of reliability for the least squares estimators and to develop hypothesis tests for examining the usefulness of the least squares line. Various techniques exist for checking the validity of these assumptions, and there are remedies to be applied when they appear to be invalid. These topics are beyond the scope of this text, but they are discussed in some of the references listed at the end of the chapter. Fortunately, the assumptions need not hold exactly in order for least squares estimators to be useful. The assumptions will be satisfied adequately for many applications encountered in practice.

13.4   An Estimator of σ^2

It seems reasonable to assume that the greater the variability of the random error ε (which is measured by its variance σ^2), the greater will be the errors in the estimation of the model parameters β_0 and β_1 and in the error of prediction when $\hat{y}$ is used to predict y for some value of x. Consequently, you should not be surprised, as we proceed through this chapter, to find that σ^2 appears in the formulas for all confidence intervals and test statistics that we will be using.

Estimation of σ^2 for a (First-Order) Straight-Line Model

$$s^2 = \frac{\text{SSE}}{\text{Degrees of freedom for error}} = \frac{\text{SSE}}{n - 2}$$

where

$$\text{SSE} = \sum (y_i - \hat{y}_i)^2 = \text{SS}_{yy} - \hat{\beta}_1 \text{SS}_{xy}$$

$$\text{SS}_{yy} = \sum (y_i - \bar{y})^2 = \sum y_i^2 - \frac{\left(\sum y_i\right)^2}{n}$$

To estimate the standard deviation σ of ε, we calculate

$$s = \sqrt{s^2} = \sqrt{\frac{\text{SSE}}{n - 2}}$$

We will refer to s as the **estimated standard error of the regression model**.

Warning: When performing these calculations, you may be tempted to round the calculated values of SS_{yy}, $\hat{\beta}_1$, and SS_{xy}. Be certain to carry at least six significant figures for each of these quantities to avoid substantial errors in calculation of the SSE.

In most practical situations, σ^2 is unknown and we must use our data to estimate its value. The best estimate of σ^2, denoted by s^2, is obtained by dividing the sum of squares of the deviations of the y values from the prediction line,

$$\text{SSE} = \sum (y_i - \hat{y}_i)^2$$

by the number of degrees of freedom associated with this quantity. We use 2 df to estimate the two parameters β_0 and β_1 in the straight-line model, leaving $(n - 2)$ df for the error variance estimation.

In the drug reaction example, we previously calculated SSE = 1.10 for the least squares line $\hat{y} = -.1 + .7x$. Recalling that there were $n = 5$ data points, we have $n - 2 = 5 - 2 = 3$ df for estimating σ^2. Thus,

$$s^2 = \frac{\text{SSE}}{n - 2} = \frac{1.10}{3} = .367$$

is the estimated variance, and

$$s = \sqrt{.367} = .61$$

is the standard error of the regression model.

You may be able to obtain an intuitive feeling for s by recalling the interpretation given to a standard deviation in Chapter 2 and remembering that the least squares line estimates the mean value of y for a given value of x. Since s measures the spread of the distribution of y values about the least squares line, we should not be surprised to find that most of the observations lie within $2s$, or $2(.61) = 1.22$, of the least squares line. For this simple example (only five data points), all five data points fall within $2s$ of the least squares line. In Section 13.8, we use s to evaluate the error of prediction when the least squares line is used to predict a value of y to be observed for a given value of x.

EXERCISES 13.23–13.30

LEARNING THE MECHANICS

13.23 Suppose you fit a least squares line to nine data points and the calculated value of SSE is .429.
  a. Find s^2, the estimator of σ^2 (the variance of the random error term ε).
  b. What is the largest deviation that you might expect between any one of the nine points and the least squares line?

13.24 Calculate SSE and s^2 for each of the following cases:
  a. $n = 18$, $SS_{yy} = 95$, $SS_{xy} = 50$, $\hat{\beta}_1 = .75$
  b. $n = 35$, $\Sigma y^2 = 860$, $\Sigma y = 50$, $SS_{xy} = 2{,}700$, $\hat{\beta}_1 = .2$
  c. $n = 20$, $\Sigma(y_i - \bar{y})^2 = 58$, $SS_{xy} = 91$, $SS_{xx} = 170$

13.25 Refer to Exercises 13.10 and 13.13. Calculate SSE, s^2, and s for the least squares lines obtained in these exercises. Interpret the standard error of the regression model, s, for each.

13.26 Visually compare the following scattergrams. If a least squares line were determined for each data set, which do you think would have the smallest variance, s^2? Explain.

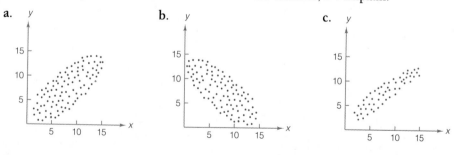

APPLYING THE CONCEPTS

13.27 Although the cable-television industry could provide viewers with 100 or more channels, a study by the A. C. Nielsen Company suggests that that may be many more than viewers want or will ever watch. The Nielsen survey indicates that as the number of television channels increases, the *percentage* of channels viewed for 10 minutes a week or more declines (Landro and Mayer, 1982). In a similar study, 20 households were sampled, and the number of channels available to each household was recorded. In addition, each household was asked to monitor its television viewing for 1 week and report the number of channels watched for 10 minutes or more. The results appear in the table at the top of page 656.

HOUSEHOLD	NUMBER OF CHANNELS AVAILABLE	NUMBER OF CHANNELS WATCHED FOR 10 MINUTES	HOUSEHOLD	NUMBER OF CHANNELS AVAILABLE	NUMBER OF CHANNELS WATCHED FOR 10 MINUTES
1	12	6	11	25	10
2	29	10	12	8	6
3	4	3	13	5	4
4	20	8	14	10	4
5	40	12	15	16	9
6	5	3	16	4	4
7	6	5	17	5	1
8	4	4	18	45	13
9	14	8	19	35	5
10	20	6	20	50	10

a. Do these data tend to support the Nielsen findings? Find the appropriate least squares line, and use it to justify your answer.

b. Plot your least squares line on a scattergram of the data.

c. Calculate SSE, s^2, and s. For the given number of channels available to a particular household, within what approximate bounds would you expect this least squares line to be able to predict the percentage of channels watched for 10 minutes?

13.28   A breeder of thoroughbred horses wishes to model the relationship between the gestation period and the length of life of a horse. The breeder believes that the two variables may follow a linear trend. The information in the table was supplied to the breeder from various thoroughbred stables across the state.

HORSE	GESTATION PERIOD x (days)	LIFE LENGTH y (years)
1	416	24
2	279	25.5
3	298	20
4	307	21.5
5	356	22
6	403	23.5
7	265	21

a. Given that $SS_{xx} = 21,752$, $SS_{xy} = 236.5$, $SS_{yy} = 22$, $\bar{x} = 332$, and $\bar{y} = 22.5$, fit a least squares line to the data. Plot the data points and graph the least squares line as a check on your calculations.

b. According to your least squares line, approximately how long would you expect a horse to live whose gestation period was 400 days?

c. Calculate SSE and s^2.

d. Give an interpretation of the standard deviation s in the context of this problem.

13.29   In order to improve the quality of the output of any production process, it is necessary first to understand the capabilities of the process (Deming, 1982). In a particular manufacturing process, the useful life of a cutting tool is related to the speed at which the tool is operated. It is necessary to understand this relationship in order to predict when the tool should be replaced and how many spare tools should be available. The data in the table were derived from life tests for the two different brands of cutting tools currently used in the production process.

a. Construct a scattergram for each brand of cutting tool.

b. For each brand, use the method of least squares to model the relationship between useful life and cutting speed.

c. Find SSE, s^2, and s for each least squares line.

CUTTING SPEED	USEFUL LIFE (HOURS)		CUTTING SPEED	USEFUL LIFE (HOURS)	
(meters per minute)	Brand A	Brand B	(meters per minute)	Brand A	Brand B
30	4.5	6.0	50	1.0	3.7
30	3.5	6.5	60	4.0	3.8
30	5.2	5.0	60	2.0	3.0
40	5.2	6.0	60	1.1	2.4
40	4.0	4.5	70	1.1	1.5
40	2.5	5.0	70	.5	2.0
50	4.4	4.5	70	3.0	1.0
50	2.8	4.0			

d. For a cutting speed of 70 meters per minute, find $\hat{y} \pm 2s$ for each least squares line.

e. For which brand would you feel more confident in using the least squares line to predict useful life for a given cutting speed? Explain.

13.30   A company keeps extensive records on its new salespeople on the premise that sales should increase with experience. A random sample of seven new salespeople produced the data on experience and sales shown in the table.

MONTHS ON JOB x	MONTHLY SALES y ($ thousands)
2	2.4
4	7.0
8	11.3
12	15.0
1	.8
5	3.7
9	12.0

a. Given that $SS_{xx} = 94.8571$, $SS_{xy} = 124.7571$, $SS_{yy} = 176.5171$, $\bar{x} = 5.8571$, and $\bar{y} = 7.4571$, fit a least squares line to the data.

b. Plot the data and graph the least squares line.

c. Predict the sales that a new salesperson would be expected to generate after 6 months on the job. After 9 months.

d. Calculate SSE, s^2, and s.

e. Within what approximate distance do you expect your predictions in part c to fall from the true number of sales generated by the new salesperson? [*Note:* A more precise measure of reliability for these predictions is discussed in Section 13.8.]

13.5   Assessing the Usefulness of the Model: Making Inferences About the Slope β_1

Now that we have specified the probability distribution of ε and found an estimate of the variance σ^2, we are ready to make statistical inferences about the model's usefulness for predicting the response y. This is step 4 in our regression modeling procedure.

Refer again to the data of Table 13.1 and suppose the reaction times are *completely unrelated* to the percentage of drug in the bloodstream. What could be said about the values of β_0 and β_1 in the hypothesized probabilistic model

$$y = \beta_0 + \beta_1 x + \varepsilon$$

if x contributes no information for the prediction of y? The implication is that the mean of y—that is, the deterministic part of the model $E(y) = \beta_0 + \beta_1 x$— does not change as x changes. In the straight-line model, this means that the true slope, β_1, is equal to 0 (see Figure 13.7). Therefore, to test the null hypothesis that the linear model contributes no information for the prediction of y against the alternative hypothesis that the linear model is useful for predicting y, we test

$$H_0: \quad \beta_1 = 0 \qquad H_a: \quad \beta_1 \neq 0$$

If the data support the alternative hypothesis, we will conclude that x does contribute information for the prediction of y using the straight-line model (although the true relationship between $E(y)$ and x could be more complex than a straight line). Thus, in effect, this is a test of the usefulness of the hypothesized model.

FIGURE 13.7

Graphing the model
$y = \beta_0 + \varepsilon \ (\beta_1 = 0)$

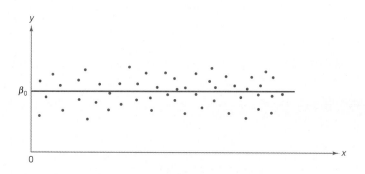

The appropriate test statistic is found by considering the sampling distribution of $\hat{\beta}_1$, the least squares estimator of the slope β_1:

FIGURE 13.8

Sampling distribution of $\hat{\beta}_1$

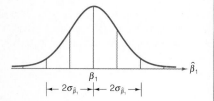

Sampling Distribution of $\hat{\beta}_1$

If we make the four assumptions about ε (see Section 13.3), the sampling distribution of the least squares estimator $\hat{\beta}_1$ of the slope will be normal with mean β_1 (the true slope) and standard deviation

$$\sigma_{\hat{\beta}_1} = \frac{\sigma}{\sqrt{SS_{xx}}} \qquad \text{(see Figure 13.8)}$$

We estimate $\sigma_{\hat{\beta}_1}$ by $s_{\hat{\beta}_1} = \dfrac{s}{\sqrt{SS_{xx}}}$ and refer to this quantity as the estimated standard error of the least squares slope $\hat{\beta}_1$.

Since σ usually is unknown, the appropriate test statistic is a Student t statistic, formed as follows:

$$t = \frac{\hat{\beta}_1 - \text{Hypothesized value of } \beta_1}{s_{\hat{\beta}_1}} \qquad \text{where } s_{\hat{\beta}_1} = \frac{s}{\sqrt{SS_{xx}}}$$

Thus,

$$t = \frac{\hat{\beta}_1 - 0}{s/\sqrt{SS_{xx}}}$$

Note that we have substituted the estimator s for σ and then formed the estimated standard error $s_{\hat{\beta}_1}$ by dividing s by $\sqrt{SS_{xx}}$. The number of degrees of freedom associated with this t statistic is the same as the number of degrees of freedom associated with s. Recall that this number is $(n-2)$ df when the hypothesized model is a straight line (see Section 13.4). The setup of our test of the usefulness of the straight-line model is summarized in the box.

A Test of Model Usefulness

ONE-TAILED TEST

H_0: $\beta_1 = 0$

H_a: $\beta_1 < 0$
(or H_a: $\beta_1 > 0$)

Test statistic: $t = \dfrac{\hat{\beta}_1}{s_{\hat{\beta}_1}} = \dfrac{\hat{\beta}_1}{s/\sqrt{SS_{xx}}}$

Rejection region: $t < -t_\alpha$
(or $t > t_\alpha$ when H_a: $\beta_1 > 0$)

TWO-TAILED TEST

H_0: $\beta_1 = 0$

H_a: $\beta_1 \ne 0$

Test statistic: $t = \dfrac{\hat{\beta}_1}{s_{\hat{\beta}_1}} = \dfrac{\hat{\beta}_1}{s/\sqrt{SS_{xx}}}$

Rejection region: $t < -t_{\alpha/2}$
or $t > t_{\alpha/2}$

where t_α and $t_{\alpha/2}$ are based on $(n-2)$ degrees of freedom

Assumptions: The four assumptions about ε listed in Section 13.3.

For the drug reaction example, we will choose $\alpha = .05$ and, since $n = 5$, t will be based on $n - 2 = 3$ df and the rejection region will be

$$t < -t_{.025} = -3.182 \quad \text{or} \quad t > t_{.025} = 3.182$$

We previously calculated $\hat{\beta}_1 = .7$, $s = .61$, and $SS_{xx} = 10$. Thus,

$$t = \frac{\hat{\beta}_1}{s/\sqrt{SS_{xx}}} = \frac{.7}{.61/\sqrt{10}} = \frac{.7}{.19} = 3.7$$

Since this calculated t value falls in the upper-tail rejection region (see Figure 13.9), we reject the null hypothesis and conclude that the slope β_1 is not 0. The sample evidence indicates that x contributes information for the prediction of y when a linear model is used to characterize the relationship between reaction time and the amount of the drug in the bloodstream.

FIGURE 13.9

Rejection region and calculated t value for testing H_0: $\beta_1 = 0$ versus H_a: $\beta_1 \ne 0$

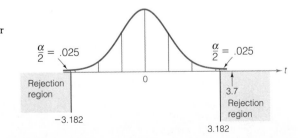

What conclusion can be drawn if the calculated t value does not fall in the rejection region? We know from previous discussions of the philosophy of hypothesis testing that such a t value does *not* lead us to accept the null hypothesis. That is, we do not conclude that $\beta_1 = 0$. Additional data might indicate that β_1 differs from 0, or a more complex relationship may exist between x and y, requiring the fitting of a model other than the straight-line model. We discuss several such models in Chapter 14.

Another way to make inferences about the slope β_1 is to estimate it using a confidence interval. This interval is formed as shown in the box.

A $100(1 - \alpha)\%$ Confidence Interval for the Slope β_1

$$\hat{\beta}_1 \pm t_{\alpha/2} s_{\hat{\beta}_1}$$

where the estimated standard error of $\hat{\beta}_1$ is calculated by

$$s_{\hat{\beta}_1} = \frac{s}{\sqrt{SS_{xx}}}$$

and $t_{\alpha/2}$ is based on $(n - 2)$ degrees of freedom.

Assumptions: The four assumptions about ε listed in Section 13.3.

For the drug reaction example, $t_{\alpha/2}$ is based on $(n - 2) = 3$ degrees of freedom. Therefore, a 95% confidence interval for the slope β_1, the expected change in reaction time for a 1% increase in the amount of drug in the bloodstream, is

$$\hat{\beta}_1 \pm t_{.025} s_{\hat{\beta}_1} = .7 \pm 3.182 \left(\frac{s}{\sqrt{SS_{xx}}} \right)$$

$$= .7 \pm 3.182 \left(\frac{.61}{\sqrt{10}} \right) = .7 \pm .61$$

Thus, the interval estimate of the slope parameter β_1 is .09 to 1.31. In terms of this example, the implication is that we can be 95% confident that the *true* mean increase in reaction time per additional 1% of the drug is between .09 and 1.31 seconds. This inference is meaningful only over the sampled range of x—that is, from 1% to 5% of the drug in the bloodstream. Since all the values in this interval are positive, it appears that β_1 is positive and that the mean of y, $E(y)$, increases as x increases. However, the rather large width of the confidence interval reflects the small number of data points (and, consequently, a lack of information) in the experiment. We would expect a narrower interval if the sample size were increased.

EXERCISES 13.31–13.44

LEARNING THE MECHANICS

13.31   Construct both a 95% and a 90% confidence interval for β_1 for each of the following cases:
  a. $\hat{\beta}_1 = 31$, $s = 3$, $SS_{xx} = 35$, $n = 10$
  b. $\hat{\beta}_1 = 64$, $SSE = 1{,}960$, $SS_{xx} = 30$, $n = 14$
  c. $\hat{\beta}_1 = -8.4$, $SSE = 146$, $SS_{xx} = 64$, $n = 20$

13.32   Consider the following pairs of observations:

x	1	4	3	2	5	6	0
y	1	3	3	1	4	7	2

a. Construct a scattergram for the data.
b. Use the method of least squares to fit a straight line to the seven data points in the table.
c. Plot the least squares line on your scattergram of part a.
d. Specify the null and alternative hypotheses you would use to test whether the data provide sufficient evidence to indicate that x contributes information for the (linear) prediction of y.
e. What is the test statistic that should be used in conducting the hypothesis test of part d? Specify the degrees of freedom associated with the test statistic.
f. Conduct the hypothesis test of part d using $\alpha = .05$.

13.33   Refer to Exercise 13.32. Construct an 80% and a 98% confidence interval for β_1.

13.34   Do the accompanying data provide sufficient evidence to conclude that a straight line is useful for characterizing the relationship between x and y?

y	4	2	4	3	2	4
x	1	6	5	3	2	4

13.35   Suppose that $n = 100$ recent residential home sales in a city are used to fit a least squares straight-line model relating the sales price, y, to the square feet of living space, x. Homes in the sample range from 1,500 square feet to 4,000 square feet of living space, and the resulting least squares equation is

$$\hat{y} = -30,000 + 70x$$

a. When the null hypothesis that the true slope is zero is tested, the result is a t value of 6.572. Give the approximate p-value of this test, and interpret the result in the context of this application.
b. The 95% confidence interval for the slope is calculated to be 49.1 to 90.9. Interpret this interval in the context of this application. What can be done to obtain a narrower confidence interval?

APPLYING THE CONCEPTS

13.36   A breeder of thoroughbred horses wishes to model the relationship between the gestation period and the life span of a horse. The breeder believes that the two variables may follow a linear trend. The information in the table was supplied to the breeder from various thoroughbred stables across the state. (Note that the horse has the greatest variation of gestation period of any species, due to seasonal and nutritional factors.)

HORSE	GESTATION PERIOD x (days)	LIFE SPAN y (years)	HORSE	GESTATION PERIOD x (days)	LIFE SPAN y (years)
1	416	24	5	356	22
2	279	25.5	6	403	23.5
3	298	20	7	265	21
4	307	21.5			

For these data, $SS_{yy} = 22.00$, $SS_{xx} = 21,752.00$, $SS_{xy} = 236.50$, $\bar{y} = 22.50$, and $\bar{x} = 332.00$. The least squares line was obtained in Exercise 13.28.

a. Do the data provide sufficient evidence to support the horse breeder's hypothesis? Test using $\alpha = .05$.

b. Find a 90% confidence interval for β_1. Interpret this interval.

13.37 A group of children, ranging from 10 to 12 years of age, were administered a verbal test in order to study the relationship between the number of words used and the silence interval before response. The tester believes that a linear relationship exists between the two variables. Each subject was asked a series of questions, and the total number of words used in answering was recorded. The time (in seconds) before the subject responded to each question was also recorded. The data for eight children are given in the table.

SUBJECT	TOTAL WORDS y	TOTAL SILENCE TIME x (seconds)
1	61	23
2	70	37
3	42	38
4	52	25
5	91	17
6	63	21
7	71	42
8	55	16

a. Write a simple linear probabilistic model relating total words to total silence time, and use the least squares method to estimate the deterministic part of the model.

b. Does x contribute information for the prediction of y? Test the null hypothesis, slope $\beta_1 = 0$, against the alternative hypothesis, $\beta_1 \neq 0$. Use $\alpha = .05$. Interpret the results of the test.

13.38 In Exercise 13.20, we fit a least squares line to relate the cocoon temperature y of a woolly-bear caterpillar to the outside air temperature x. The data are reproduced in the table.

DAY	TEMPERATURE (°C) Air	TEMPERATURE (°C) Cocoon[a]	DAY	TEMPERATURE (°C) Air	TEMPERATURE (°C) Cocoon[a]
1	10.4	15.1	7	1.7	3.6
2	9.2	14.6	8	2.0	5.3
3	2.2	6.8	9	3.0	7.0
4	2.6	6.8	10	3.5	7.1
5	4.1	8.0	11	4.5	9.6
6	3.7	8.7	12	4.4	9.5

[a]Each cocoon temperature is the average of the temperatures of two cocoons.

Source: Kevan, P. G., Jensen, T. S., and Shorthouse, J. D. "Body temperatures and behavioral thermoregulation of High Arctic woolly-bear caterpillars and pupae (*Gynaephora rossii*, Lymantridae: Lepidoptera) and the importance of sunshine," *Arctic and Alpine Research*, 1982, 54. Reprinted with permission of the Regents of the University of Colorado.

Given that $SS_{xx} = 83.3425$, $SS_{xy} = 100.0825$, $SS_{yy} = 127.5092$, $\bar{x} = 4.2750$, and $\bar{y} = 8.5083$:

a. Calculate SSE, s^2, and s.

b. Interpret the standard error of the regression model.

c. Do the data provide sufficient evidence to indicate that the outside air temperature provides information for predicting the woolly-bear caterpillar cocoon temperature? Test using $\alpha = .05$.

d. Find the observed significance level of the test.

13.39 In Exercise 13.21, we fit a least squares line to relate the length of a sand lance to its age. The data are reproduced in the table.

MEAN LENGTH (MILLIMETERS)	176	194	212	226	236	244	254
AGE (YEARS)	2	3	4	5	6	7	8

Source: Winters, G. H. "Analysis of the biological and demographic parameters of northern sand lance, *Ammodytes dubins*, from the Newfoundland Grand Bank," *Canadian Journal of Fisheries and Aquatic Sciences*, 1983, 40.

a. Do the data provide sufficient evidence to indicate that age x contributes information for the prediction of the length y of a sand lance? Test using $\alpha = .05$.
b. Find the p-value for the test and interpret it.

13.40 In Exercise 13.22, we discussed an experiment conducted to determine the effect of ration level on the growth rate of rainbow trout. Twenty fish were fed at each of four ration levels at 6°C water temperature. The experiment was repeated for water temperatures of 12°C and 18°C. The table gives the means and standard deviations (in parentheses) for each sample of 20 fish. Refer only to the data collected on fish fed in water maintained at 12°C.

RATION	MEAN WET WEIGHT GAIN (%)		
(% body weight per day)	6°C	12°C	18°C
.0	−8.14 (3.27)	−10.33 (2.35)	−13.21 (1.96)
.8	12.31 (3.37)		
1.2		14.29 (5.32)	
1.5	28.19 (5.39)		15.51 (4.48)
2.5	29.12 (5.67)	38.05 (6.99)	
3.5			51.13 (7.14)
4.0		60.13 (12.71)	
4.5			65.49 (13.04)
Maintenance ration	.32	.48	.69

Source: Kovacs, T. G., and Leduc, G. "Sublethal toxicity to rainbow trout (*Salmo gairdneri*) at different temperatures," *Canadian Journal of Fisheries and Aquatic Sciences*, 1982, 39.

a. Find SSE and s^2 for the $n = 4$ data points.
b. Find a 95% confidence interval for the mean gain (percentage of body weight) for a 1% increase in ration level.
c. State the assumptions required for your inference in part b to be valid.
d. Which assumption may not be satisfied? Explain why.

13.41 Based on an observational study of five chief executives, Mintzberg (1973) identified ten managerial roles that can be found in all managerial jobs: figurehead, leader, liaison, monitor, disseminator, spokesperson, entrepreneur, disturbance handler, resource allocator, and negotiator. In a recent observational study of 19 managers from a medium-sized manufacturing plant, Luthans, Rosenkrantz, and Hennessey (1985) extended Mintzberg's work by investigating which activities *successful* managers actually perform. Each manager was observed during eighty 10-minute intervals over a 2-week period and their activities were recorded. The researchers used regression analysis to investigate which of the recorded activities were related to managerial success. To measure success, they devised an index based on the manager's length of time in the organization and his or her level within the firm; the higher the index, the more successful the manager. The table (page 664) presents data (which are representative of the data collected by Luthans, Rosenkrantz, and Hennessey) that can be used to determine whether managerial success can in part be explained by extensiveness of a manager's network-building interactions with people outside the manager's work unit. Such inter-actions include phone and face-to-face meetings with customers and suppliers, attending external meetings, and doing public relations work.

MANAGER	MANAGER SUCCESS INDEX	NUMBER OF INTERACTIONS WITH OUTSIDERS	MANAGER	MANAGER SUCCESS INDEX	NUMBER OF INTERACTIONS WITH OUTSIDERS
	y	x		y	x
1	40	12	11	70	20
2	73	71	12	47	81
3	95	70	13	80	40
4	60	81	14	51	33
5	81	43	15	32	45
6	27	50	16	50	10
7	53	42	17	52	65
8	66	18	18	30	20
9	25	35	19	42	21
10	63	82			

a. Construct a scattergram for the data.
b. Given $SS_{yy} = 7,006.6316$, $SS_{xx} = 10,824.5263$, $SS_{xy} = 2,561.2632$, $\bar{y} = 54.5789$, and $\bar{x} = 44.1579$, use the method of least squares to find a prediction equation for managerial success.
c. Find SSE, s^2, and s for your prediction equation. Interpret the standard deviation s in the context of this problem.
d. Plot the least squares line on your scattergram of part a. Does it appear that the number of interactions with outsiders contributes information for the prediction of managerial success? Explain.
e. Conduct a formal statistical hypothesis test to answer the question posed in part d. Use $\alpha = .05$.
f. Construct a 95% confidence interval for β_1. Interpret the interval in the context of the problem.

13.42 Refer to Exercise 13.15, in which the extent of retaliation against whistle blowers was investigated. Since an individual's salary is a reasonably good indicator of the individual's power within an organization, the data of Exercise 13.15 can be used to investigate whether the extent of retaliation is related to the power of the whistle blower in the organization. Near and Miceli (1986) were unable to reject the hypothesis that the extent of retaliation is unrelated to the whistle blower's power. Do you agree? Test using $\alpha = .05$.

13.43 In Exercise 13.16, the following least squares line was developed to describe the relationship between the price of a gallon of regular leaded gasoline, y, and the price of a barrel of crude oil, x:

$$\hat{y} = 36.514 + 3.1612x$$

Construct a 90% confidence interval to estimate the mean increase in the price of gasoline (in cents) per $1.00 increase per barrel in the price of crude oil.

13.44 Do the data in Exercise 13.27 support the theory that the percentage of channels watched for 10 or more minutes decreases as the number of channels available increases? Test using $\alpha = .10$.

13.6 Correlation: A Measure of the Usefulness of the Model

The claim is often made that the number of cigarettes smoked and the incidence of lung cancer are "highly correlated." Another popular belief is that the crime rate and the unemployment rate are "correlated." Some people even believe that the Dow Jones Industrial Average and the lengths of fashionable skirts are "correlated." In this section we will discuss the concept of **correlation**.* A numerical descriptive measure of the correlation between two variables x and y is provided by the *Pearson product moment coefficient of correlation, r.*

Definition 13.2

The sample **Pearson product moment coefficient of correlation, r**, is defined as

$$r = \frac{SS_{xy}}{\sqrt{SS_{xx}SS_{yy}}}$$

It is a measure of the strength of the linear relationship between two random variables x and y.

Note that the computational formula for the correlation coefficient r given in Definition 13.2 involves the same quantities that were used in computing the least squares prediction equation. In fact, since the numerators of the expressions for $\hat{\beta}_1$ and r are identical, you can see that $r = 0$ when $\hat{\beta}_1 = 0$ (the case where x contributes no information for the prediction of y) and that r is positive when the slope is positive and negative when the slope is negative. Unlike $\hat{\beta}_1$, the correlation coefficient r is *scaleless* and assumes a value between -1 and $+1$, regardless of the units of x and y.

A value of r near or equal to 0 implies little or no linear relationship between y and x. In contrast, the closer r comes to 1 or -1, the stronger the linear relationship between y and x. And if $r = 1$ or $r = -1$, all the sample points fall exactly on the least squares line. Positive values of r imply a positive linear relationship between y and x; that is, y increases as x increases. Negative values of r imply a negative linear relationship between y and x; that is, y decreases as x increases. Each of these situations is portrayed in Figure 13.10, page 666.

We demonstrate how to calculate the coefficient of correlation r using the data in Table 13.1 for the drug reaction example. The quantities needed to calculate r are SS_{xy}, SS_{xx}, and SS_{yy}. The first two quantities have been calculated previously and are repeated here for convenience:

$$SS_{xy} = 7 \qquad SS_{xx} = 10$$

$$SS_{yy} = \sum y^2 - \frac{\left(\sum y\right)^2}{n} = 26 - \frac{(10)^2}{5} = 26 - 20 = 6$$

We now find the coefficient of correlation:

$$r = \frac{SS_{xy}}{\sqrt{SS_{xx}SS_{yy}}} = \frac{7}{\sqrt{(10)(6)}} = \frac{7}{\sqrt{60}}$$
$$= .904$$

---

*Recall that we first discussed the concept of correlation in Section 11.6, where we presented Spearman's rank correlation coefficient.

FIGURE 13.10

Values of r and their implications

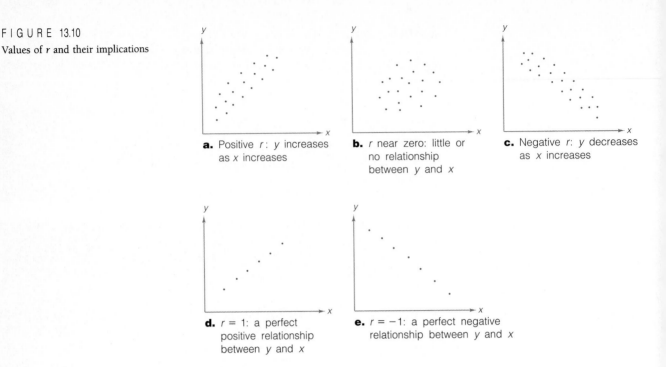

a. Positive r: y increases as x increases

b. r near zero: little or no relationship between y and x

c. Negative r: y decreases as x increases

d. $r = 1$: a perfect positive relationship between y and x

e. $r = -1$: a perfect negative relationship between y and x

The fact that r is positive and near 1 in value indicates that the reaction time tends to increase as the amount of drug in the bloodstream increases—*for this sample of five subjects*. This is the same conclusion we reached when we found the calculated value of the least squares slope to be positive.

EXAMPLE 13.1

A firm wants to know the correlation between the size of its sales force and its yearly sales revenue. The records for the past 10 years are examined, and the results listed in Table 13.5 are obtained. Calculate the coefficient of correlation r for the data.

TABLE 13.5 **Sales Versus Number of Salespeople**

YEAR	NUMBER OF SALESPEOPLE x	SALES y (hundred thousand dollars)
1981	15	1.35
1982	18	1.63
1983	24	2.33
1984	22	2.41
1985	25	2.63
1986	29	2.93
1987	30	3.41
1988	32	3.26
1989	35	3.63
1990	38	4.15

Solution

We need to calculate SS_{xy}, SS_{xx}, and SS_{yy}:

$$SS_{xy} = \sum xy - \frac{\left(\sum x\right)\left(\sum y\right)}{n} = 800.62 - \frac{(268)(27.73)}{10} = 57.456$$

$$SS_{xx} = \sum x^2 - \frac{\left(\sum x\right)^2}{n} = 7{,}668 - \frac{(268)^2}{10} = 485.6$$

$$SS_{yy} = \sum y^2 - \frac{\left(\sum y\right)^2}{n} = 83.8733 - \frac{(27.73)^2}{10} = 6.97801$$

Then, the coefficient of correlation is

$$r = \frac{SS_{xy}}{\sqrt{SS_{xx}SS_{yy}}} = \frac{57.456}{\sqrt{(485.6)(6.97801)}} = \frac{57.456}{58.211} = .99$$

Thus, the size of the sales force and sales revenue are very highly correlated—at least over the past 10 years. The implication is that a strong positive linear relationship exists between these variables (see Figure 13.11). We must be careful, however, not to jump to any unwarranted conclusions. For instance, the firm may be tempted to conclude that the best thing it can do to increase sales is to hire a large number of new salespeople—that is, that there is a *causal relationship* between the two variables. However, high correlation does not imply causality. The fact is, many things have probably contributed both to the increase in the size of the sales force and to the increase in sales revenue. The firm's expertise has undoubtedly grown, the economy has inflated (so that 1990 dollars are not worth as much as 1981 dollars), and perhaps the scope of products and services sold by the firm has widened. *We must be careful not to infer a causal relationship on the basis of high sample correlation. The only safe conclusion when a high correlation is observed in the sample data is that a linear trend may exist between x and y.*

FIGURE 13.11

Scattergram for Example 13.1

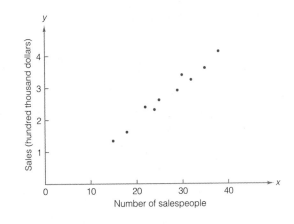

The correlation coefficient r measures the linear correlation between x values and y values in the sample, and a similar linear coefficient of correlation exists for the population from which the data points were selected. The **population correlation coefficient** is denoted by the symbol ρ (rho). As you might expect, ρ is estimated by the corresponding sample statistic, r. Or, rather than estimating ρ, we might want to test the null hypothesis $H_0: \rho = 0$ against $H_a: \rho \neq 0$—i.e., test the hypothesis that x contributes no information for the prediction of y by using the straight-line model against the alternative that the two variables are at least linearly related.

However, we already performed this *identical* test in Section 13.5 when we tested $H_0: \beta_1 = 0$ against $H_a: \beta_1 \neq 0$. That is, the null hypothesis $H_0: \rho = 0$ is equivalent to the hypothesis $H_0: \beta_1 = 0$.* When we tested the null hypothesis $H_0: \beta_1 = 0$ in connection with the drug reaction example, the data led to a rejection of the null hypothesis at the $\alpha = .05$ level. This rejection implies that the null hypothesis of a 0 linear correlation between the two variables (drug and reaction time) can also be rejected at the $\alpha = .05$ level. The only real difference between the least squares slope $\hat{\beta}_1$ and the coefficient of correlation r is the measurement scale. Therefore, the information they provide about the usefulness of the least squares model is to some extent redundant. For this reason, we will use the slope to make inferences about the existence of a positive or negative linear relationship between two variables.

13.7 The Coefficient of Determination

Another way to measure the usefulness of the model is to measure the contribution of x in predicting y. To accomplish this, we calculate how much the errors of prediction of y were reduced by using the information provided by x. To illustrate, consider the sample shown in the scattergram of Figure 13.12(a). If we assume that x contributes no information for the prediction of y, the best prediction for a value of y is the sample mean $\bar{y}$, which is shown as the horizontal line in Figure 13.12(b). The vertical line segments in Figure 13.12(b) are the deviations of the points about the mean $\bar{y}$. Note that the sum of squares of deviations for the model $\hat{y} = \bar{y}$ is

$$SS_{yy} = \sum (y_i - \bar{y})^2$$

Now suppose you fit a least squares line to the same set of data and locate the deviations of the points about the line as shown in Figure 13.12(c). Compare the deviations about the prediction lines in parts (b) and (c) of Figure 13.12. You can see that:

1. If x contributes little or no information for the prediction of y, the sums of squares of deviations for the two lines,

   $$SS_{yy} = \sum (y_i - \bar{y})^2 \quad \text{and} \quad SSE = \sum (y_i - \hat{y}_i)^2$$

   will be nearly equal.
2. If x does contribute information for the prediction of y, the SSE will be smaller than SS_{yy}. In fact, if all the points fall on the least squares line, then SSE $= 0$.

   Then, the reduction in the sum of squares of deviations that can be attributed to x, expressed as a proportion of SS_{yy}, is

   $$\frac{SS_{yy} - SSE}{SS_{yy}}$$

---

*The correlation test statistic that is equivalent to $t = \hat{\beta}_1/s_{\hat{\beta}_1}$ is

$$t = \frac{r}{\sqrt{(1 - r^2)/(n - 2)}}$$

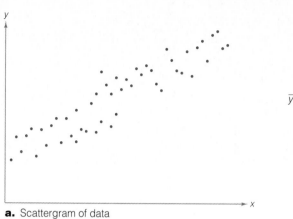

a. Scattergram of data

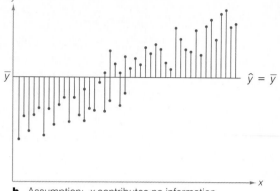

b. Assumption: x contributes no information
for predicting y, $\hat{y} = \bar{y}$

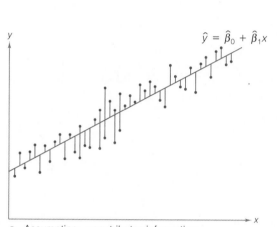

c. Assumption: x contributes information
for predicting y, $\hat{y} = \hat{\beta}_0 + \hat{\beta}_1 x$

FIGURE 13.12

A comparison of the sum of squares of deviations for two models

Note that SS_{yy} is the "total sample variation" of the observations around the mean $\bar{y}$ and that SSE is the remaining "unexplained sample variability" after fitting the line $\hat{y}$. Thus, the difference (SS_{yy} − SSE) is the "explained sample variability" attributable to the linear relationship with x. Then a verbal description of the proportion is

$$\frac{SS_{yy} - SSE}{SS_{yy}} = \frac{\text{Explained sample variability}}{\text{Total sample variability}}$$

$$= \text{Proportion of total sample variability}$$
$$\text{explained by the linear relationship}$$

It can be shown that this proportion is equal to the square of the simple linear coefficient of correlation r (the Pearson product moment coefficient of correlation).

Definition 13.3

The **coefficient of determination** is the square of the coefficient of correlation. It represents the proportion of the total sample variability around $\bar{y}$ that is explained by the linear relationship between y and x. It may be computed as

$$r^2 = \frac{SS_{yy} - SSE}{SS_{yy}} = 1 - \frac{SSE}{SS_{yy}}$$

Note that r^2 is always between 0 and 1, because r is between -1 and $+1$. Thus, an r^2 of .60 means that the sum of squares of deviations of the y values about their predicted values has been reduced 60% by the use of the least squares equation $\hat{y}$, instead of $\bar{y}$, to predict y.

EXAMPLE 13.2

Calculate the coefficient of determination for the drug reaction example using the formula given with Definition 13.3. The data are repeated in Table 13.6 for convenience.

TABLE 13.6

AMOUNT OF DRUG x (%)	REACTION TIME y (seconds)
1	1
2	1
3	2
4	2
5	4

Solution

From previous calculations,

$$SS_{yy} = 6 \quad \text{and} \quad SSE = \sum (y - \hat{y})^2 = 1.10$$

Then the coefficient of determination is given by

$$r^2 = \frac{SS_{yy} - SSE}{SS_{yy}} = \frac{6.0 - 1.1}{6.0} = \frac{4.9}{6.0}$$

$$= .82$$

[In Section 13.6, we calculated $r = .904$. Now we have $r^2 = (.904)^2 = .82$.] So we know that using the amount of drug in the blood, x, to predict y with the least squares line

$$\hat{y} = -.1 + .7x$$

accounts for 82% of the total sum of squares of deviations of the five sample y values about their mean.

CASE STUDY 13.1

PREDICTING THE UNITED STATES CRIME INDEX

Reporting and analyzing crime rates is an important function of many law enforcement agencies. David Heaukulani* comments: "The simple linear regression analysis remains one of the most useful tools for crime prediction." He demonstrates this point by fitting a straight-line model to predict the annual value of the United States crime index as a function of the United States population for each year. The data for the years 1963–1973 are given in Table 13.7, and the least squares line is shown in Figure 13.13.

TABLE 13.7 **United States Population and Crime Index**

YEAR	POPULATION	ACTUAL CRIME INDEX	YEAR	POPULATION	ACTUAL CRIME INDEX
1963	188,531,000	2,259,081	1969	201,921,000	4,989,747
1964	191,334,000	2,604,426	1970	203,184,772	5,568,197
1965	193,818,000	2,780,015	1971	206,256,000	5,995,211
1966	195,857,000	3,243,400	1972	208,232,000	5,891,924
1967	197,864,000	3,802,273	1973	209,851,000	8,638,375
1968	199,861,000	4,466,573			

FIGURE 13.13

Least squares line relating crime index to population size (1963–1973)

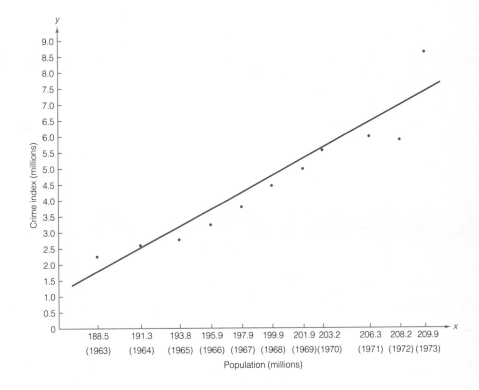

*Reprinted from Heaukulani, D. "The normal distribution of crime," *Journal of Police Science and Administration*, Vol. 3, No. 3, 1975, pp. 312–318.

Heaukulani reports that the correlation is $r = .94$ ($r^2 = .88$), which provides sufficient evidence at $\alpha = .01$ to indicate that the size of the population is useful for predicting the crime index using the straight-line model. Heaukulani concludes that

> most agencies are using too narrow a span of statistical measurement to evaluate the overall crime picture. Year-by-year comparisons and percent analyses do not take average fluctuations into consideration. A lack of knowledge about statistical principles on the part of laymen, especially those in the news media, leads to a distortion or an invalid representation of the crime figures. Any increase is reported either as "bucking a trend" or "in line with the rise in crime."
>
> If we expect to make any progress in solving the crime problem, we must evaluate the crime index objectively. We have to accept the fact that the crime index will probably increase from year to year. We should be willing to accept an average amount of increase and a set maximum limit. If the crime index exceeds the limit, only then should we become concerned and attempt to examine the factors that may have been responsible for the excess deviation.

Note that the 1973 crime rate of 8.6 million greatly exceeded the predicted value of less than 8 million. Perhaps other factors unique to this time period (Vietnam War, Watergate, price controls, and so forth) explain this "excess deviation." In any case, we cannot ascribe causation based only on high correlation between variables.

EXERCISES 13.45–13.57

LEARNING THE MECHANICS

13.45 Explain what each of the following sample correlation coefficients tells you about the relationship between the x and y values in the sample:
a. $r = 1$ b. $r = -1$ c. $r = 0$ d. $r = .90$ e. $r = .10$ f. $r = -.88$

13.46 Describe the slope of the least squares line if:
a. $r = .7$ b. $r = -.7$ c. $r = 0$ d. $r^2 = .64$

13.47 Construct a scattergram for each data set. Then calculate r and r^2 for each data set. Interpret their values.

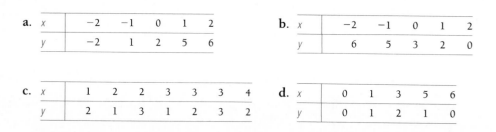

a.

x	−2	−1	0	1	2
y	−2	1	2	5	6

b.

x	−2	−1	0	1	2
y	6	5	3	2	0

c.

x	1	2	2	3	3	3	4
y	2	1	3	1	2	3	2

d.

x	0	1	3	5	6
y	0	1	2	1	0

13.48 Calculate r^2 for the least squares line in each of the following exercises. Interpret their values.
a. Exercise 13.10   b. Exercise 13.13

APPLYING THE CONCEPTS

13.49 Find the correlation coefficient and the coefficient of determination for the sample data listed in the table. Interpret your results, and discuss whether a causal relationship can be inferred.

YEAR	NUMBER OF 18-HOLE AND LARGER GOLF COURSES IN THE UNITED STATES	NUMBER OF DIVORCES IN THE UNITED STATES (in millions)
1960	2,725	2.9
1965	3,769	3.5
1970	4,845	4.3
1975	6,282	6.5
1977	6,551	8.0
1978	6,699	8.6
1979	6,787	8.8
1980	6,856	9.9
1981	6,944	10.8
1982	7,059	11.5
1983	7,125	11.6
1984	7,230	12.3

Source: *U.S. Bureau of the Census, Statistical Abstract of the United States: 1986*, pp. 35, 230.

13.50 In Exercise 13.18, we gave the number of games won, y, and the batting average, x, for the 14 American League baseball teams at the end of the 1986 season. They are repeated here in the table.

TEAM	NUMBER OF GAMES WON y	TEAM BATTING AVERAGE x
Cleveland	84	.284
New York	90	.271
Boston	95	.271
Toronto	86	.269
Texas	87	.267
Detroit	87	.263
Minnesota	71	.261
Baltimore	73	.258
California	92	.255
Milwaukee	77	.255
Seattle	67	.253
Kansas City	76	.252
Oakland	76	.252
Chicago	72	.247

Source: *Official American League Averages: 1986*. New York: The American League of Professional Baseball Clubs, pp. 2, 24–27.

Recall that $\Sigma y = 1,133$, $\Sigma x = 3.658$, $\Sigma y^2 = 92,703$, $\Sigma x^2 = .957118$, and $\Sigma xy = 296.734$.

a. Calculate the correlation coefficient, r, and the coefficient of determination, r^2, for the data. Interpret their values.

b. Do the data provide sufficient evidence to conclude that a correlation exists between a team's number of wins and its batting average? Test using $\alpha = .05$. [*Hint:* See the last paragraph of Section 13.6.]

13.51   Calculate the coefficient of determination and the sample coefficient of correlation for the woolly-bear caterpillar data in Exercise 13.20. Interpret them.

13.52   Is the maximal oxygen uptake, a measure often used by physiologists to indicate an individual's state of cardiovascular fitness, related to the performance of distance runners? Six long-distance runners submitted to treadmill tests for determination of their maximal oxygen uptake. The results, along with each runner's best mile time (in seconds), are shown in the table.

ATHLETE	MAXIMAL OXYGEN UPTAKE (Milliliters/kilogram)	MILE TIME (Seconds)
1	63.3	241.5
2	60.1	249.8
3	53.6	246.1
4	58.8	232.4
5	67.5	237.2
6	62.5	238.4

a. Calculate r and r^2.

b. Do the data provide sufficient evidence to indicate that mile time is negatively correlated with maximal oxygen uptake? Test H_0: $\rho = 0$ against the alternative hypothesis, H_a: $\rho < 0$, using $\alpha = .05$.*

13.53   Find the correlation coefficient and the coefficient of determination for the sample data listed in the table and interpret your results. For these data, $SS_{xx} = 13,143,088.00$, $SS_{yy} = 1,552,011.334$, $SS_{xy} = 49,153.00$, $\bar{x} = 2,023.00$, and $\bar{y} = 1,543.67$.

YEAR	GROSS NATIONAL PRODUCT x (billions of dollars)	NEW HOUSING STARTS y (thousands)	YEAR	GROSS NATIONAL PRODUCT x (billions of dollars)	NEW HOUSING STARTS y (thousands)
1960	507	1,296	1978	2,164	2,036
1965	691	1,510	1979	2,418	1,760
1970	993	1,469	1980	2,632	1,313
1973	1,326	2,057	1981	2,958	1,100
1974	1,434	1,353	1982	3,069	1,072
1975	1,549	1,171	1983	3,305	1,712
1976	1,718	1,548	1984	3,663	1,756
1977	1,918	2,002			

Source: U.S. Bureau of the Census, *Statistical Abstract of the United States: 1986*, pp. 431, 724.

13.54   Is there a correlation between the amount of education received by people living in urban centers and that received by people living in the urban fringes? To determine whether a correlation exists, a sociologist compared the percentages of people with 4 years of high school education or more for the two groups. The data are given in the table.

a. Find the correlation coefficient for the data.

b. Find the coefficient of determination, and explain its meaning in terms of this problem.

c. Is there sufficient evidence to indicate a nonzero correlation between x and y? Test using $\alpha = .05$.†

---

*Recall (Section 13.6) that the test of H_0: $\rho = 0$ is equivalent to the test of H_0: $\beta_1 = 0$.

†Recall (Section 13.6) that the test of H_0: $\rho = 0$ is equivalent to the test of H_0: $\beta_1 = 0$.

CITY	URBAN CENTER y	URBAN FRINGE x	CITY	URBAN CENTER y	URBAN FRINGE x
Baltimore	28.2	42.3	Milwaukee	39.7	54.4
Boston	44.6	55.8	New Orleans	33.3	44.6
Chicago	35.3	53.9	New York	36.4	48.7
Cleveland	30.1	55.5	Philadelphia	30.7	48.0
Dallas	48.9	56.4	St. Louis	26.3	43.3
Detroit	34.4	47.5	San Francisco	49.4	57.9
Houston	45.2	50.1	Washington	47.8	67.5
Los Angeles	53.4	53.4			

Source: Computed from U.S. Bureau of the Census, *U.S. Census of Population: 1960, General Social and Economic Characteristics*, and *U.S. Census of Population and Housing: 1960, Census Tracts* (Washington, D.C.: Government Printing Office, 1961).

13.55 In the summer of 1981, the Minnesota Department of Transportation installed a state-of-the-art weigh-in-motion scale in the concrete surface of the eastbound lanes of Interstate 494 in Bloomington, Minnesota. The system is computerized and monitors traffic continuously. It is capable of distinguishing among 13 different types of vehicles (car, five-axle semi, five-axle twin trailer, etc.). The primary purpose of the system is to provide traffic counts and weights for use in the planning and design of future roadways. After installation, a study was undertaken to determine whether the scale's readings correspond with the static weights of the vehicles being monitored. (Studies of this type are known as *calibration studies*.) After some preliminary comparisons using a two-axle-six-tire truck carrying different loads (see table), calibration adjustments were made in the software of the weigh-in-motion system and the scales were reevaluated.

TRIAL NUMBER	STATIC WEIGHT OF TRUCK x (thousand pounds)	WEIGH-IN-MOTION READING PRIOR TO CALIBRATION ADJUSTMENT y_1 (thousand pounds)	WEIGH-IN-MOTION READING AFTER CALIBRATION ADJUSTMENT y_2 (thousand pounds)
1	27.9	26.0	27.8
2	29.1	29.9	29.1
3	38.0	39.5	37.8
4	27.0	25.1	27.1
5	30.3	31.6	30.6
6	34.5	36.2	34.3
7	27.8	25.1	26.9
8	29.6	31.0	29.6
9	33.1	35.6	33.0
10	35.5	40.2	35.0

Source: Adapted from data in Wright, J. L., Owen, F., and Pena, D. "Status of MN/DOT's weigh-in-motion program," St. Paul: Minnesota Department of Transportation, January 1983.

a. Construct two scattergrams, one of y_1 versus x and the other of y_2 versus x.

b. Use the scattergram of part a to evaluate the performance of the weigh-in-motion scale both before and after the calibration adjustment.

c. Calculate the correlation coefficient for both sets of data, and interpret their values. Explain how these correlation coefficients can be used to evaluate the weigh-in-motion scale.

d. Suppose the sample correlation coefficient for y_2 and x were 1. Could this happen if the static weights and the weigh-in-motion readings disagreed? Explain.

13.56 A problem of economic and social concern in the United States is the importation and sale of illicit drugs. The data shown in the table are part of a larger body of data collected by the Florida attorney general's office in an attempt to relate the incidence of drug seizures and drug arrests to the characteristics of the Florida counties. Given are the number, y, of drug arrests per county in 1982, the density, x_1, of the county (population per square mile), and the number, x_2, of law enforcement employees. In order to simplify the calculations, we show data for only ten counties.

| | COUNTY | | | | | | | | | |
	1	2	3	4	5	6	7	8	9	10
POPULATION DENSITY, x_1	169	68	278	842	18	42	112	529	276	613
NUMBER OF LAW ENFORCEMENT EMPLOYEES, x_2	498	35	772	5,788	18	57	300	1,762	416	520
NUMBER OF ARRESTS IN 1982, y	370	44	716	7,416	25	50	189	1,097	256	432

a. Fit a least squares line to relate the number, y, of drug arrests per county in 1982 to the county population density, x_1.

b. We might expect the mean number of arrests to increase as the population density increases. Do the data support this theory? Test using $\alpha = .05$.

c. Calculate the coefficient of determination for this regression analysis and interpret its value.

13.57 Repeat parts **a**, **b**, and **c** of Exercise 13.56 using the number x_2 of county law enforcement employees as the independent variable. Then answer the following questions:

d. Which least squares line has the lower SSE?

e. Which independent variable explains more of the variation in y? Explain.

13.8 Using the Model for Estimation and Prediction

If we are satisfied that a useful model has been found to describe the relationship between reaction time and amount of drug in the bloodstream, we are ready for step 5 in our regression modeling procedure: using the model for estimation and prediction.

The most common uses of a probabilistic model for making inferences can be divided into two categories. The first is the use of the model for estimating the mean value of y, $E(y)$, for a specific value of x.

For our drug reaction example, we may want to estimate the mean response time for all people whose blood contains 4% of the drug.

The second use of the model entails predicting a new individual y value for a given x.

That is, we may want to predict the reaction time for a specific person who possesses 4% of the drug in the bloodstream.

In the first case, we are attempting to estimate the mean value of y for a very large number of experiments at the given x value. In the second case, we are trying to predict the outcome of a single experiment at the given x value. Which of these model uses—estimating the mean value of y or predicting an individual new value of y (for the same value of x)—can be accomplished with the greater accuracy?

Before answering this question, we first consider the problem of choosing an estimator (or predictor) of the mean (or a new individual) y value. We will use the least squares prediction equation

$$\hat{y} = \hat{\beta}_0 + \hat{\beta}_1 x$$

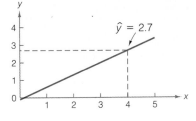

FIGURE 13.14

Estimated mean value and predicted individual value of reaction time y for $x = 4$

both to estimate the mean value of y and to predict a specific new value of y for a given value of x. For our example, we found

$$\hat{y} = -.1 + .7x$$

so that the estimated mean reaction time for all people when $x = 4$ (drug is 4% of blood content) is

$$\hat{y} = -.1 + .7(4) = 2.7 \text{ seconds}$$

The same value is used to predict a new y value when $x = 4$. That is, both the estimated mean and the predicted value of y are $\hat{y} = 2.7$ when $x = 4$, as shown in Figure 13.14.

The difference between these two model uses lies in the relative accuracy of the estimate and the prediction. These accuracies are best measured by using the sampling errors of the least squares line when it is used as an estimator and as a predictor, respectively. These errors are reflected in the standard deviations given in the next box.

Sampling Errors for the Estimator of the Mean of y and the Predictor of an Individual New Value of y

1. The standard deviation of the sampling distribution of the estimator $\hat{y}$ of the mean value of y at a specific value of x, say x_p, is

$$\sigma_{\hat{y}} = \sigma \sqrt{\frac{1}{n} + \frac{(x_p - \bar{x})^2}{SS_{xx}}}$$

where σ is the standard deviation of the random error ε. We refer to $\sigma_{\hat{y}}$ as the standard error of $\hat{y}$.

2. The standard deviation of the prediction error for the predictor $\hat{y}$ of an individual new y value at a specific value of x is

$$\sigma_{(y-\hat{y})} = \sigma \sqrt{1 + \frac{1}{n} + \frac{(x_p - \bar{x})^2}{SS_{xx}}}$$

where σ is the standard deviation of the random error ε. We refer to $\sigma_{(y-\hat{y})}$ as the standard error of prediction.

The true value of σ is rarely known, so we estimate σ by s and calculate the estimation and prediction intervals as shown in the next two boxes.

A $100(1 - \alpha)\%$ Confidence Interval for the Mean Value of y at $x = x_p$

$$\hat{y} \pm t_{\alpha/2} \text{ (Estimated standard error of } \hat{y})$$

or

$$\hat{y} \pm t_{\alpha/2}\, s \sqrt{\frac{1}{n} + \frac{(x_p - \bar{x})^2}{SS_{xx}}}$$

where $t_{\alpha/2}$ is based on $(n - 2)$ degrees of freedom.

> A $100(1 - \alpha)\%$ Prediction Interval* for an Individual New Value of y at $x = x_p$
>
> $\hat{y} \pm t_{\alpha/2}$ (Estimated standard error of prediction)
>
> or
>
> $$\hat{y} \pm t_{\alpha/2}\, s\sqrt{1 + \frac{1}{n} + \frac{(x_p - \bar{x})^2}{SS_{xx}}}$$
>
> where $t_{\alpha/2}$ is based on $(n - 2)$ degrees of freedom.

EXAMPLE 13.3

Find a 95% confidence interval for the mean reaction time when the concentration of the drug in the bloodstream is 4%.

Solution

For a 4% concentration, $x = 4$ and the confidence interval for the mean value of y is

$$\hat{y} \pm t_{\alpha/2}\, s\sqrt{\frac{1}{n} + \frac{(x_p - \bar{x})^2}{SS_{xx}}} = \hat{y} \pm t_{.025}\, s\sqrt{\frac{1}{5} + \frac{(4 - \bar{x})^2}{SS_{xx}}}$$

where $t_{.025}$ is based on $n - 2 = 5 - 2 = 3$ degrees of freedom. Recall that $\hat{y} = 2.7$, $s = .61$, $\bar{x} = 3$, and $SS_{xx} = 10$. From Table VI in Appendix A, $t_{.025} = 3.182$. Thus, we have

$$2.7 \pm (3.182)(.61)\sqrt{\frac{1}{5} + \frac{(4 - 3)^2}{10}} = 2.7 \pm (3.182)(.61)(.55)$$

$$= 2.7 \pm (3.182)(.34)$$

$$= 2.7 \pm 1.1$$

Therefore, when the percentage of drug in the bloodstream is 4%, the standard error of $\hat{y}$ is .34 and the corresponding 95% confidence interval for the mean reaction time for all possible subjects is 1.6 to 3.8 seconds. Note that we used a small amount of data (small sample size) for purposes of illustration in fitting the least squares line. The interval would probably be narrower if more information had been obtained from a larger sample.

EXAMPLE 13.4

Predict the reaction time for the next performance of the experiment for a subject with a drug concentration of 4%. Use a 95% prediction interval.

Solution

To predict the response time for an individual new subject for whom $x = 4$, we calculate the 95% prediction interval as

$$\hat{y} \pm t_{\alpha/2}\, s\sqrt{1 + \frac{1}{n} + \frac{(x_p - \bar{x})^2}{SS_{xx}}} = 2.7 \pm (3.182)(.61)\sqrt{1 + \frac{1}{5} + \frac{(4 - 3)^2}{10}}$$

$$= 2.7 \pm (3.182)(.61)(1.14)$$

$$= 2.7 \pm (3.182)(.70)$$

$$= 2.7 \pm 2.2$$

---

*The term *prediction interval* is used when the interval formed is intended to enclose the value of a random variable. The term *confidence interval* is reserved for estimation of population parameters (such as the mean).

Therefore, the standard error of prediction when the drug concentration is 4% is .70, and we predict with 95% confidence that the reaction time for this new individual will fall in the interval from .5 to 4.9 seconds. Like the confidence interval for the mean value of y, the prediction interval for y is quite large. This is because we have chosen a simple example (only five data points) to fit the least squares line. The width of the prediction interval could be reduced by using a larger number of data points.

A comparison of the confidence interval for the mean value of y and the prediction interval for a new value of y for 4% drug concentration ($x = 4$) is illustrated in Figure 13.15. Note that the prediction interval for an individual new value of y is *always* wider than the corresponding confidence interval for the mean value of y. You can see this by examining the formulas for the two intervals and by studying Figure 13.15.

FIGURE 13.15

A 95% confidence interval for mean sales and a prediction interval for sales when $x = 4$

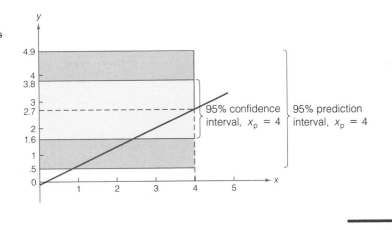

The error in estimating the mean value of y, $E(y)$, for a given value of x, say x_p, is the distance between the least squares line and the true line of means, $E(y) = \beta_0 + \beta_1 x$. This error, $[\hat{y} - E(y)]$, is shown in Figure 13.15. In contrast, *the error* $(y_p - \hat{y})$ *in predicting some future value of y is the sum of two errors—* the error of estimating the mean of y, $E(y)$, shown in Figure 13.16 on page 680, plus the random error that is a component of the value of y to be predicted (see Figure 13.17). Consequently, the error of predicting a particular value of y will be larger than the error of estimating the mean value of y for a particular value of x. Note from their formulas that both the error of estimation and the error of prediction take their smallest values when $x_p = \bar{x}$. The farther x_p lies from $\bar{x}$, the larger will be the errors of estimation and prediction. You can see why this is true by noting the deviations for different values of x_p between the line of means $E(y) = \beta_0 + \beta_1 x$ and the predicted line of means $\hat{y} = \hat{\beta}_0 + \hat{\beta}_1 x$ shown in Figure 13.17. The deviation is larger at the extremities of the interval where the largest and smallest values of x in the data set occur.

Both the confidence intervals for mean values and the prediction intervals for new values are depicted over the entire range of the regression line in Figure 13.18 (page 680). You can see that the confidence interval is always narrower than the prediction interval, and that they are both narrowest at the mean $\bar{x}$, increasing steadily as the distance $|x - \bar{x}|$ increases.

FIGURE 13.16

Error of estimating the mean value of y for a given value of x

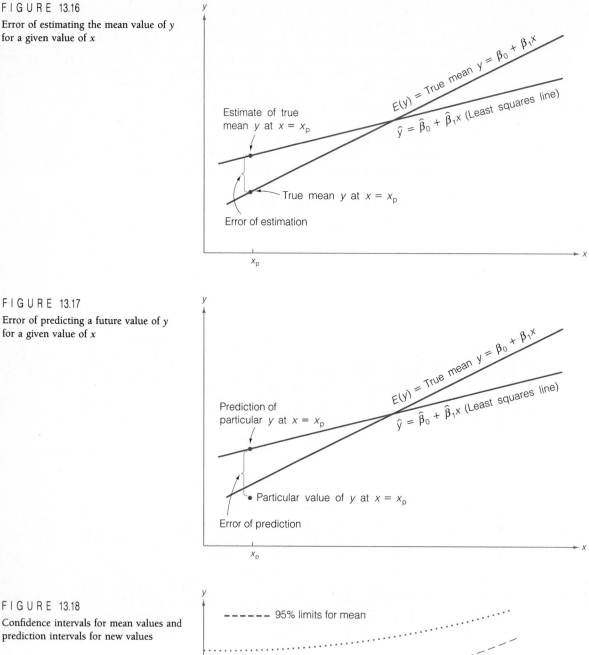

FIGURE 13.17

Error of predicting a future value of y for a given value of x

FIGURE 13.18

Confidence intervals for mean values and prediction intervals for new values

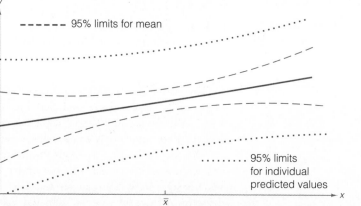

The confidence interval width grows smaller as n is increased; thus, in theory, you can obtain as precise an estimate of the mean value of y as desired (at any given x) by selecting a large enough sample. The prediction interval for a new value of y also grows smaller as n increases, but there is a lower limit on its width. If you examine the formula for the prediction interval (page 678), you will see that the interval can get no smaller than $\hat{y} \pm z_{\alpha/2}\sigma$.* Thus, the only way to obtain more accurate predictions for new values of y is to reduce the standard deviation of the regression model, σ. This can be accomplished only by improving the model, either by using a curvilinear (rather than linear) relationship with x or by adding new independent variables to the model, or both. Methods of improving the model are discussed in Chapters 14 and 15.

EXERCISES 13.58–13.68

LEARNING THE MECHANICS

13.58   Consider the following pairs of measurements:

x	y
1	3
2	5
3	4
4	6
5	7
6	7
7	10

a. Construct a scattergram for these data.
b. Find the least squares line, and plot it on your scattergram.
c. Find s^2.
d. Find a 90% confidence interval for the mean value of y when $x = 4$. Plot the upper and lower bounds of the confidence interval on your scattergram.
e. Find a 90% prediction interval for a new value of y when $x = 4$. Plot the upper and lower bounds of the prediction interval on your scattergram.
f. Compare the widths of the intervals you constructed in parts **d** and **e**. Which is wider and why?

13.59   Consider the following pairs of measurements:

x	y	x	y
1	−1	−2	−6
−1	−5	4	4
2	1	−2	−5
0	−3	5	4
4	6	1	0

---

*The result follows from the facts that, for large n, $t_{\alpha/2} \approx z_{\alpha/2}$, $s \approx \sigma$, and the last two terms under the radical in the standard error of the predictor are approximately 0.

Given that $SS_{xx} = 57.6$, $SS_{yy} = 162.5$, $SS_{xy} = 94.0$, and $\hat{y} = -2.4583 + 1.6319x$:

a. Construct a scattergram for the data.

b. Plot the least squares line on your scattergram.

c. Use a 95% confidence interval to estimate the mean value of y when $x = 5$. Plot the upper and lower bounds of the interval on your scattergram.

d. Repeat part c for $x = 1.2$ and $x = -2$.

e. Compare the widths of the three confidence intervals you constructed in parts c and d and explain why they differ.

13.60   Refer to Exercise 13.59.

a. Using no information about x, estimate and calculate a 95% confidence interval for the mean value of y. [Hint: Use the one-sample t methodology of Section 7.3.]

b. Plot the estimated mean value and the confidence interval as horizontal lines on your scattergram of Exercise 13.59.

c. Compare the confidence intervals you calculated in parts c and d of Exercise 13.59 with the one you calculated in part a of this exercise. Does x appear to contribute information about the mean value of y?

d. Check the answer you gave in part c with a statistical test of the null hypothesis H_0: $\beta_1 = 0$ against H_a: $\beta_1 \neq 0$. Use $\alpha = .05$.

APPLYING THE CONCEPTS

13.61   A large supermarket chain has its own store brand for many grocery items. These tend to be priced lower than other brands. For a particular item, the chain wants to study the effect of varying the price for the major competing brand on the sales of the store brand item, while the prices for the store brand and all other brands are held fixed. The experiment is conducted at one of the chain's stores over a 7-week period, and the results are shown in the table.

WEEK	COMPETITOR'S PRICE x (cents)	STORE BRAND SALES y
1	37	122
2	32	107
3	29	99
4	35	110
5	33	113
6	31	104
7	35	116

a. Find the least squares line relating store brand sales y to major competitor's price x.

b. Plot the data and graph the line as a check on your calculations.

c. Does x contribute information for the prediction of y?

d. Calculate r and r^2 and interpret their values.

e. Find a 90% confidence interval for mean store brand sales when the competitor's price is 33¢.

f. Suppose you were to set the competitor's price at 33¢. Find a 90% prediction interval for next week's sales.

13.62   Certain dosages of a new drug developed to reduce a smoker's reliance on nicotine may reduce one's pulse rate to dangerously low levels. To investigate the drug's effect on pulse rate, different dosages of the drug were administered to six randomly selected patients, and 30 minutes later the decrease in each patient's pulse rate was recorded.

PATIENT	DOSAGE x (cubic centimeters)	DECREASE IN PULSE RATE y (beats/minute)
1	2.0	15
2	1.5	9
3	3.0	18
4	2.5	16
5	4.0	23
6	3.0	20

a. Is there evidence of a linear relationship between drug dosage and change in pulse rate? Test at $\alpha = .10$.

b. Find a 95% confidence interval for the slope β_1.

c. Find a 99% confidence interval for the mean decrease in pulse rate corresponding to a dosage of 3.5 cubic centimeters.

d. Find a 99% prediction interval for the decrease in pulse rate corresponding to a dosage of 3.5 cubic centimeters.

13.63 Refer to Exercises 13.20 and 13.38. The woolly-bear caterpillar data are shown in the table.

DAY	TEMPERATURE (°C) Air	Cocoon[a]	DAY	TEMPERATURE (°C) Air	Cocoon[a]
1	10.4	15.1	7	1.7	3.6
2	9.2	14.6	8	2.0	5.3
3	2.2	6.8	9	3.0	7.0
4	2.6	6.8	10	3.5	7.1
5	4.1	8.0	11	4.5	9.6
6	3.7	8.7	12	4.4	9.5

[a]Each cocoon temperature is the average of the temperatures of two cocoons.

a. Find a 95% confidence interval for the mean cocoon temperature when the air temperature is 7°C. Interpret the interval.

b. Suppose you were to place a single woolly-bear caterpillar cocoon in a controlled environment of 7°C. Find a 95% prediction interval for the cocoon temperature. Interpret the interval.

13.64 The reasons given by workers for quitting their jobs generally fall into one of two categories: (1) worker quits to seek or take a different job; or (2) worker quits to withdraw from the labor force. Economic theory suggests that wages and quit rates are related. Evidence regarding this relationship is important for understanding how labor markets operate [Koch and Ragan (1986)]. The table lists quit rates (quits per 100 employees) and the average hourly wage in a sample of 15 manufacturing industries.

INDUSTRY	QUIT RATE	AVERAGE WAGE	INDUSTRY	QUIT RATE	AVERAGE WAGE
1	1.4	$ 8.20	9	2.0	7.99
2	.7	10.35	10	3.8	5.54
3	2.6	6.18	11	2.3	7.50
4	3.4	5.37	12	1.9	6.43
5	1.7	9.94	13	1.4	8.83
6	1.7	9.11	14	1.8	10.93
7	1.0	10.59	15	2.0	8.80
8	.5	13.29			

a. Construct a scattergram for these data.

b. Use the method of least squares to model the quit rate as a function of average hourly wage.

c. Do the data present sufficient evidence to conclude that average hourly wage rate contributes useful information for the prediction of quit rates? What does your model suggest about the relationship between quit rates and wages?

d. Find a 95% prediction interval for the quit rate in an industry with an average hourly wage of $9.00.

e. Estimate the mean quit rate for industries with an average hourly wage of $6.50. Use a 95% confidence interval.

13.65   In planning for an orientation meeting with new accounting majors, the chairperson of the Accounting Department wants to emphasize the importance of doing well in the major courses in order to get better-paying jobs after graduation. To support this point, the chairperson plans to show that there is a strong positive correlation between starting salaries for recent accounting graduates and their grade-point averages in the major courses. Records for seven of last year's accounting graduates are selected at random and are given in the table.

GRADE-POINT AVERAGE IN MAJOR COURSES x	STARTING SALARY y (thousands of dollars)
2.58	11.5
3.27	13.8
3.85	14.5
3.50	14.2
3.33	13.5
2.89	11.6
2.23	10.6

a. Find the least squares prediction equation.

b. Plot the data and graph the line as a check on your calculations.

c. Find the values of r and r^2 and interpret these values.

d. Find a 95% prediction interval for the starting salary of a graduate whose grade-point average is 3.2.

e. What is the mean starting salary for graduates with grade-point averages equal to 3.0? Use a 95% confidence interval.

13.66   In Exercise 13.21, we found the least squares line relating the age x of a sand lance to its length y. The data are reproduced in the table.

MEAN LENGTH (MILLIMETERS)	176	194	212	226	236	244	254
AGE (YEARS)	2	3	4	5	6	7	8

a. Find a 95% confidence interval for the mean length of sand lances that are 4 years old. Interpret the interval.

b. Find a 95% prediction interval for the length of a sand lance that is 4 years old. Interpret the interval. Explain the difference between this interval and the interval obtained in part a.

13.67   Refer to Exercise 13.40. Find a 95% confidence interval for the mean weight gain of rainbow trout (percentage of body weight) over a 20-day period when the ration level is 2.5% of body weight and the water temperature is 12°C. Interpret the confidence interval.

13.68 Refer to Exercise 13.29. The data are shown in the accompanying table.

CUTTING SPEED	USEFUL LIFE (HOURS)		CUTTING SPEED	USEFUL LIFE (HOURS)	
(meters per minute)	Brand A	Brand B	(meters per minute)	Brand A	Brand B
30	4.5	6.0	50	1.0	3.7
30	3.5	6.5	60	4.0	3.8
30	5.2	5.0	60	2.0	3.0
40	5.2	6.0	60	1.1	2.4
40	4.0	4.5	70	1.1	1.5
40	2.5	5.0	70	.5	2.0
50	4.4	4.5	70	3.0	1.0
50	2.8	4.0			

a. Use a 90% confidence interval to estimate the mean useful life of a brand A cutting tool when the cutting speed is 45 meters per minute. Repeat for brand B. Compare the widths of the two intervals, and comment on the reasons for any difference.

b. Use a 90% prediction interval to predict the useful life of a brand A cutting tool when the cutting speed is 45 meters per minute. Repeat for brand B. Compare the widths of the two intervals to each other, and to the two intervals you calculated in part a. Comment on the reasons for any differences.

c. Note that the estimation and prediction you performed in parts a and b were for a value of x that was not included in the original sample. That is, the value $x = 45$ was not part of the sample. However, the value is within the range of x values in the sample, so that the regression model spans the x value for which the estimation and prediction were made. In such situations, estimation and prediction represent **interpolations**.

Suppose you were asked to predict the useful life of a brand A cutting tool for a cutting speed of $x = 100$ meters per minute. Since the given value of x is outside the range of the sample x values, the prediction is an example of **extrapolation**. Predict the useful life of a brand A cutting tool that is operated at 100 meters per minute, and construct a 95% confidence interval for the actual useful life of the tool. What additional assumption do you have to make in order to assure the validity of an extrapolation?

13.9 Simple Linear Regression: An Example

In the preceding sections we have presented the basic elements necessary to fit and use a straight-line regression model. In this section we will assemble these elements by applying them in an example.

Suppose a fire insurance company wants to relate the amount of fire damage in major residential fires to the distance between the residence and the nearest fire station. The study is to be conducted in a large suburb of a major city; a sample of 15 recent fires in this suburb is selected. The amount of damage, y, and the distance, x, between the fire and the nearest fire station are recorded for each fire. The results are given in Table 13.8 on page 686.

STEP 1 First, we hypothesize a model to relate fire damage, y, to the distance from the nearest fire station, x. We hypothesize a straight-line probabilistic model:

$$y = \beta_0 + \beta_1 x + \varepsilon$$

TABLE 13.8   **Fire Damage Data**

DISTANCE FROM FIRE STATION x (miles)	FIRE DAMAGE y (thousands of dollars)
3.4	26.2
1.8	17.8
4.6	31.3
2.3	23.1
3.1	27.5
5.5	36.0
.7	14.1
3.0	22.3
2.6	19.6
4.3	31.3
2.1	24.0
1.1	17.3
6.1	43.2
4.8	36.4
3.8	26.1

STEP 2 Next, we use the data to estimate the unknown parameters in the deterministic component of the hypothesized model. We make some preliminary calculations:

$$SS_{xx} = \sum x^2 - \frac{\left(\sum x\right)^2}{n} = 196.16 - \frac{(49.2)^2}{15}$$

$$= 196.160 - 161.376 = 34.784$$

$$SS_{yy} = \sum y^2 - \frac{\left(\sum y\right)^2}{n} = 11{,}376.48 - \frac{(396.2)^2}{15}$$

$$= 11{,}376.480 - 10{,}464.96267 = 911.517334$$

$$SS_{xy} = \sum xy - \frac{\left(\sum x\right)\left(\sum y\right)}{n} = 1{,}470.65 - \frac{(49.2)(396.2)}{15}$$

$$= 1{,}470.650 - 1{,}299.536 = 171.114$$

Then the least squares estimate of the slope β_1 and intercept β_0 are

$$\hat{\beta}_1 = \frac{SS_{xy}}{SS_{xx}} = \frac{171.114}{34.784} = 4.919331$$

$$\hat{\beta}_0 = \bar{y} - \hat{\beta}_1\bar{x} = \frac{396.2}{15} - 4.919331\left(\frac{49.2}{15}\right)$$

$$= 26.413333 - (4.919331)(3.28) = 26.413333 - 16.135406$$

$$= 10.277927$$

and the least squares equation is

$$\hat{y} = 10.278 + 4.919x$$

This prediction equation is graphed in Figure 13.19 along with a plot of the data points.

FIGURE 13.19

Least squares model for the fire damage data

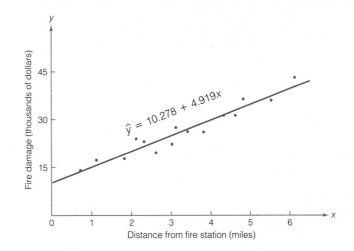

The least squares estimate of the slope β_1 is 4.919, which implies that the estimated mean damage increases by \$4,919 for each additional mile from the fire station. This interpretation is valid over the range of x, or from .7 to 6.1 miles from the station. The estimated y-intercept, $\hat{\beta}_0 = 10.278$, has the interpretation that a fire 0 miles from the fire station has an estimated mean damage of \$10,278. Although this would seem to apply to the fire station itself, remember that the y-intercept is meaningfully interpretable only if $x = 0$ is within the sampled range of the independent variable.

STEP 3 Now we specify the probability distribution of the random error component ε. The assumptions about the distribution are identical to those listed in Section 13.3. Although we know that these assumptions are not completely satisfied (they rarely are for practical problems), we are willing to assume they are approximately satisfied for this example. We have to estimate the variance σ^2 of ε, so we calculate

$$SSE = \sum (y - \hat{y})^2 = SS_{yy} - \hat{\beta}_1 SS_{xy}$$

where the last expression represents a shortcut formula for SSE. Thus,

$$SSE = 911.517334 - (4.919331)(171.114)$$
$$= 911.517334 - 841.766405 = 69.750929*$$

To estimate σ^2, we divide SSE by the degrees of freedom available for error, $n - 2$. Thus,

$$s^2 = \frac{SSE}{n - 2} = \frac{69.750929}{15 - 2} = 5.3655 \qquad s = \sqrt{5.3655} = 2.32$$

STEP 4 We can now check the usefulness of the hypothesized model—that is, whether x really contributes information for the prediction of y using the straight-line model. First test the null hypothesis that the slope β_1 is 0, i.e., that there is no linear relationship between fire damage and the distance from the nearest fire

---

*The values for SS_{yy}, $\hat{\beta}_1$, and SS_{xy} used to calculate SSE are exact for this example. For other problems where rounding is necessary, at least six significant figures should be carried for these quantities. Otherwise, the calculated value of SSE may be substantially in error.

station, against the alternative hypothesis that fire damage increases as the distance increases. We test

H_0: $\beta_1 = 0$

H_a: $\beta_1 > 0$

Test statistic: $t = \dfrac{\hat{\beta}_1 - 0}{s_{\hat{\beta}_1}} = \dfrac{\hat{\beta}_1}{s/\sqrt{SS_{xx}}}$

Assumptions: The same assumptions made about ε in Section 13.3

Rejection region: For $\alpha = .05$, we will reject H_0 if $t > t_\alpha$, where, for $n - 2 = 13$ df, we find $t_{.05} = 1.771$.

We then calculate the t statistic:

$$t = \frac{\hat{\beta}_1}{s_{\hat{\beta}_1}} = \frac{\hat{\beta}_1}{s/\sqrt{SS_{xx}}} = \frac{4.919}{2.32/\sqrt{34.784}} = \frac{4.919}{.393} = 12.5$$

This large t value leaves little doubt that mean fire damage and distance between the fire and the station are at least linearly related, with mean fire damage increasing as the distance increases.

We gain additional information about the relationship by forming a confidence interval for the slope β_1. A 95% confidence interval is

$$\hat{\beta}_1 \pm t_{.025}\, s_{\hat{\beta}_1} = 4.919 \pm (2.160)(.393)$$
$$= 4.919 \pm .849 = (4.070, 5.768)$$

We estimate that the interval from \$4,070 to \$5,768 encloses the mean increase (β_1) in fire damage per additional mile distance from the fire station.

Another measure of the usefulness of the model is the coefficient of correlation r. We have

$$r = \frac{SS_{xy}}{\sqrt{SS_{xx}\, SS_{yy}}} = \frac{171.114}{\sqrt{(34.784)(911.517)}}$$
$$= \frac{171.114}{178.062} = .96$$

The high correlation confirms our conclusion that β_1 is greater than 0; it appears that fire damage and distance from the fire station are positively correlated.

The coefficient of determination is

$$r^2 = (.96)^2 = .92$$

which implies that 92% of the sum of squares of deviations in the sample of y values about $\bar{y}$ is explained by the distance x between the fire and the fire station. All signs point to a strong relationship between x and y.

STEP 5 We are now prepared to use the least squares model. Suppose the insurance company wants to predict the fire damage if a major residential fire were to occur 3.5 miles from the nearest fire station. The predicted value is

$$\hat{y} = \hat{\beta}_0 + \hat{\beta}_1 x = 10.278 + (4.919)(3.5)$$
$$= 10.278 + 17.216 = 27.5$$

(We round to the nearest tenth to be consistent with the units of the original data in Table 13.8.) If we want a 95% prediction interval, we calculate

$$\hat{y} \pm t_{.025}\, s \sqrt{1 + \frac{1}{n} + \frac{(x - \bar{x})^2}{SS_{xx}}} = 27.5 \pm (2.16)(2.32)\sqrt{1 + \frac{1}{15} + \frac{(3.5 - 3.28)^2}{34.784}}$$

$$= 27.5 \pm (2.16)(2.32)\sqrt{1.0681}$$

$$= 27.5 \pm 5.2 = (22.3, 32.7)$$

The model yields a 95% prediction interval of $22,300 to $32,700 for fire damage in a major residential fire 3.5 miles from the nearest station.

One caution before closing: We would not use this prediction model to make predictions for homes less than .7 mile or more than 6.1 miles from the nearest fire station. A look at the data in Table 13.8 reveals that all the x values fall between .7 and 6.1. It is dangerous to use the model to make predictions outside the region in which the sample data fall. A straight line might not provide a good model for the relationship between the mean value of y and the value of x when stretched over a wider range of x values.

13.10   Using the Computer for Simple Linear Regression

All of the examples of simple linear regression that we have presented thus far in Chapter 13 have required rather tedious calculations involving SS_{yy}, SS_{xx}, SS_{xy}, and so forth. Even with the use of a pocket calculator, the process is laborious and susceptible to error. Fortunately, the use of computers can significantly reduce the labor involved in regression calculations. In this final section, we introduce the regression output from one statistical software package, the SAS System. Though this is just one of the many statistical packages available that provide simple linear regression output, most produce essentially the same quantities, differing only in format and labeling. The regression output of other packages is presented in Chapter 14.

The SAS output for the fire damage example of Section 13.9 is presented in Figure 13.20 (page 690). We have shaded the parts of the printout corresponding to most of the key simple linear regression quantities introduced in this chapter.

First, the estimates of the y-intercept and the slope are found about halfway down the printout on the left-hand side, under the column labeled Parameter Estimate and in the rows labeled INTERCEP and X, respectively. The values are $\hat{\beta}_0 = 10.277929$ and $\hat{\beta}_1 = 4.919331$. When rounded to three decimal places, these quantities agree with our calculations in Section 13.9.

Next, we find the measures of variability: SSE, s^2, and s. They are shaded in the upper portion of the printout. SSE is found under the column heading Sum of Squares and in the row labeled Error: SSE = 69.75098. The estimate of the error variance σ^2 is under the column heading Mean Square and in the row labeled Error: $s^2 = 5.36546$. The estimate of the standard deviation σ is next to the heading Root MSE: $s = 2.31635$. Again, all values (after rounding) agree with the corresponding quantities we calculated in Section 13.9.

The coefficient of determination is shown (shaded) next to the heading R-Square in the upper portion of the printout: $r^2 = .9235$. Again, to two decimal places, this agrees with our calculation in Section 13.9. The coefficient of correlation r is not given on the printout.

FIGURE 13.20

SAS printout for fire damage regression
analysis

Dep Variable: Y

Analysis of Variance

Source	DF	Sum of Squares	Mean Square	F Value	Prob>F
Model	1	841.76636	841.76636	156.886	0.0001
Error	13	69.75098	5.36546		
C Total	14	911.51733			

Root MSE	2.31635	R-Square	0.9235	
Dep Mean	26.41333	Adj R-Sq	0.9176	
C.V.	8.76961			

Parameter Estimates

Variable	DF	Parameter Estimate	Standard Error	T for HO: Parameter=0	Prob > |T|
INTERCEP	1	10.277929	1.42027781	7.237	0.0001
X	1	4.919331	0.39274775	12.525	0.0001

Obs	X	Y	Predict Value	Residual	Lower95% Predict	Upper95% Predict
1	3.4	26.2000	27.0037	-0.8037	21.8344	32.1729
2	1.8	17.8000	19.1327	-1.3327	13.8141	24.4514
3	4.6	31.3000	32.9068	-1.6068	27.6186	38.1951
4	2.3	23.1000	21.5924	1.5076	16.3577	26.8271
5	3.1	27.5000	25.5279	1.9721	20.3573	30.6984
6	5.5	36.0000	37.3342	-1.3342	31.8334	42.8351
7	0.7	14.1000	13.7215	0.3785	8.1087	19.3342
8	3	22.3000	25.0359	-2.7359	19.8622	30.2097
9	2.6	19.6000	23.0682	-3.4682	17.8678	28.2686
10	4.3	31.3000	31.4311	-0.1311	26.1908	36.6713
11	2.1	24.0000	20.6085	3.3915	15.3442	25.8729
12	1.1	17.3000	15.6892	1.6108	10.1999	21.1785
13	6.1	43.2000	40.2858	2.9142	34.5906	45.9811
14	4.8	36.4000	33.8907	2.5093	28.5640	39.2175
15	3.8	26.1000	28.9714	-2.8714	23.7843	34.1585
16	3.5	.	27.4956	.	22.3239	32.6672

Sum of Residuals	-3.73035E-14
Sum of Squared Residuals	69.7510
Predicted Resid SS (Press)	93.2117

The t statistic for testing H_0: $\beta_1 = 0$ against H_a: $\beta_1 \neq 0$ is given (shaded) in the center of the page under the column heading T for H0: Parameter = 0 in the row corresponding to X. The value $t = 12.525$ agrees with our computed value when we used the formula

$$t = \frac{\hat{\beta}_1}{s_{\hat{\beta}_1}} = \frac{\hat{\beta}_1}{s/\sqrt{SS_{xx}}}$$

To determine which hypothesis this test statistic supports, we can establish a rejection region using the t table (Table VI of Appendix A), just as we did in Section 13.9. However, the printout makes this unnecessary because the *observed significance level*, or *p-value*, is shown (shaded) immediately to the right of the *t* statistic, under the column heading Prob > |T|. Remember that if the observed significance level is less than the α-value we select, then the test statistic supports the alternative hypothesis at that level. For example, if we select $\alpha = .05$ in this example, the observed significance level of .0001 given on the printout indicates that we should reject H_0. We can conclude that there is sufficient evidence at $\alpha = .05$ to infer that a linear relationship between fire damage and distance from the station is useful for predicting damage. If we wish to conduct a one-tailed test, the observed significance level is half that given on the printout (assuming the sign of the test statistic agrees with the alternative hypothesis). Thus, if we were testing H_0: $\beta_1 = 0$ against H_a: $\beta_1 > 0$ in this example, the *t* statistic is positive, agreeing with H_a, so that the observed significance level would be $.0001/2 = .00005$.

Predicted *y* values and the corresponding prediction intervals are given in the lower portion of the SAS printout. To find the 95% prediction interval for the fire damage *y* when the distance from the fire station is $x = 3.5$ miles, first locate the value 3.5 in the column labeled X (the last value in the column). The prediction is given in the center column labeled Predict Value in the row corresponding to 3.5, $\hat{y} = 27.4956$. The lower and upper confidence bounds are given in the columns headed Lower and Upper 95% Predict, respectively:

Lower = 22.3239   and   Upper = 32.6672

Again, all values agree after rounding with our calculations in Section 13.9.

Although much more information is given on the SAS printout, we have discussed only those aspects that have been presented in the chapter. In the next two chapters we expand our discussion to include other important components of the SAS printout, as well as other statistical packages. The point here is that the computer can alleviate much of the burden of calculation involved in a regression analysis, and enable us to spend more time on the interpretation of the model. The time spent in learning how to read computer regression output is a good investment.

EXERCISES 13.69–13.70

LEARNING THE MECHANICS

13.69   Refer to Exercise 13.15, in which the extent of retaliation (as measured by the retaliation index) is related to the salary of 15 whistle blowers using a least squares model. A computer printout for this regression model is given on the next page.
a. Find the least squares estimates of the intercept β_0 and the slope β_1.
b. Identify the values of SSE, s^2, and s. Based on the standard deviation, how accurately do you expect to be able to predict the retaliation index based on the salary of the whistle blower?
c. Find and interpret the coefficient of determination r^2.
d. Find the *t* statistic and its observed significance level for testing the usefulness of the model. Interpret the values.

SAS simple linear regression printout for Exercise 13.69: retaliation index vs. salary

Dep Variable: R_INDEX

Analysis of Variance

Source	DF	Sum of Squares	Mean Square	F Value	Prob>F
Model	1	20853.93900	20853.93900	0.925	0.3538
Error	13	293208.46100	22554.49700		
C Total	14	314062.40000			

Root MSE	150.18155	R-Square	0.0664	
Dep Mean	499.80000	Adj R-Sq	-0.0054	
C.V.	30.04833			

Parameter Estimates

Variable	DF	Parameter Estimate	Standard Error	T for H0: Parameter=0	Prob > |T|
INTERCEP	1	575.028672	87.31829499	6.585	0.0001
SALARY	1	-0.002186	0.00227386	-0.962	0.3538

Obs	SALARY	R_INDEX	Predict Value	Residual	Lower 95% Mean	Upper 95% Mean
1	62000	301.0	439.5	-138.5	280.1	598.8
2	36500	550.0	495.2	54.7770	410.8	579.6
3	17600	755.0	536.5	218.5	418.9	654.2
4	20000	327.0	531.3	-204.3	421.6	641.0
5	30100	500.0	509.2	-9.2163	422.8	595.6
6	35000	377.0	498.5	-121.5	414.7	582.3
7	47500	290.0	471.2	-181.2	365.6	576.8
8	54000	452.0	457.0	-4.9600	329.4	584.6
9	15800	535.0	540.5	-5.4827	416.5	664.5
10	44000	455.0	478.8	-23.8246	382.7	574.9
11	46600	615.0	473.1	141.9	370.2	576.1
12	12100	700.0	548.6	151.4	410.6	686.5
13	62000	650.0	439.5	210.5	280.1	598.8
14	21000	630.0	529.1	100.9	422.6	635.7
15	11900	360.0	549.0	-189.0	410.3	687.7

Sum of Residuals   2.842171E-13
Sum of Squared Residuals   293208.4610
Predicted Resid SS (Press)   414392.8150

---

e. Find and interpret the 95% confidence interval for the mean retaliation index of whistle blowers with salaries of $35,000.

13.70 Refer to Exercise 13.20, in which the cocoon temperature of a woolly-bear caterpillar was related to the outside air temperature. The simple linear regression printout for these data is given here.
   a. Find the least squares equation, and compare it to the value you calculated in Exercise 13.20.
   b. Find the standard deviation s, and interpret the value in terms of this exercise.
   c. Find and interpret r^2.
   d. Is there sufficient evidence to indicate that the model is useful for predicting y? What is the observed significance level of the test?
   e. Locate and interpret the predicted cocoon temperature and corresponding 95% prediction interval when the air temperature is 7°C.

SAS simple linear regression printout for Exercise 13.70; cocoon temperature vs. air temperature

Dep Variable: COCOON

Analysis of Variance

Source	DF	Sum of Squares	Mean Square	F Value	Prob>F
Model	1	120.18486	120.18486	164.090	0.0001
Error	10	7.32431	0.73243		
C Total	11	127.50917			

Root MSE	0.85582	R-Square	0.9426
Dep Mean	8.50833	Adj R-Sq	0.9368
C.V.	10.05863		

Parameter Estimates

Variable	DF	Parameter Estimate	Standard Error	T for H0: Parameter=0	Prob > |T|
INTERCEP	1	3.374666	0.47079265	7.168	0.0001
AIR	1	1.200858	0.09374540	12.810	0.0001

Obs	AIR	COCOON	Predict Value	Residual	Lower95% Predict	Upper95% Predict
1	10.4	15.1000	15.8636	-0.7636	13.5022	18.2250
2	9.2	14.6000	14.4226	0.1774	12.1870	16.6581
3	2.2	6.8000	6.0166	0.7834	3.9850	8.0481
4	2.6	6.8000	6.4969	0.3031	4.4815	8.5123
5	4.1	8.0000	8.2982	-0.2982	6.3131	10.2833
6	3.7	8.7000	7.8178	0.8822	5.8294	9.8062
7	1.7	3.6000	5.4161	-1.8161	3.3598	7.4725
8	2	5.3000	5.7764	-0.4764	3.7355	7.8172
9	3	7.0000	6.9772	0.0228	4.9747	8.9798
10	3.5	7.1000	7.5777	-0.4777	5.5863	9.5690
11	4.5	9.6000	8.7785	0.8215	6.7932	10.7638
12	4.4	9.5000	8.6584	0.8416	6.6735	10.6434
13	7	.	11.7807	.	9.7159	13.8454

Sum of Residuals	9.325873E-15
Sum of Squared Residuals	7.3243
Predicted Resid SS (Press)	11.6790

Summary

We have introduced an extremely useful tool in this chapter—the **method of least squares** for fitting a prediction equation to a set of data. This procedure, along with the associated statistical tests and estimations, is called a **regression analysis**. In five steps we showed how to use sample data to build a model relating a dependent variable y to a single independent variable x:

1. The first step is to hypothesize a **probabilistic model**. In this chapter, we confined our attention to the straight-line model, $y = \beta_0 + \beta_1 x + \varepsilon$.
2. The second step is to use the method of least squares to estimate the unknown parameters in the **deterministic component**, $\beta_0 + \beta_1 x$. The least squares estimates yield a model $\hat{y} = \hat{\beta}_0 + \hat{\beta}_1 x$ with a sum of squared errors (SSE) that is smaller than that produced by any other straight-line model.

3. The third step is to specify the probability distribution of the **random error component, ε**.
4. The fourth step is to assess the usefulness of the hypothesized model by making inferences about the slope, β_1, calculating the **coefficient of correlation**, r, and calculating the **coefficient of determination**, r^2.
5. Finally, if we are satisfied with the model we are prepared to use it. We used the model to estimate the mean y value, $E(y)$, for a given x value and to predict an individual y value for a new value of x.

The following two chapters develop the concepts introduced in this chapter.

SUPPLEMENTARY EXERCISES 13.71–13.87

LEARNING THE MECHANICS

13.71   In fitting a least squares line to $n = 15$ data points, the following quantities were computed: $SS_{xx} = 50$, $SS_{yy} = 25$, $SS_{xy} = -30$, $\bar{x} = 1.3$, and $\bar{y} = 27$.
   a. Find the least squares line.   b. Graph the least squares line.
   c. Calculate SSE.   d. Calculate s^2.
   e. Find a 90% confidence interval for β_1. Interpret this estimate.
   f. Find a 90% confidence interval for the mean value of y when $x = 1.8$.
   g. Find a 90% prediction interval for y when $x = 1.8$.

13.72   Consider the following sample data:

y	5	1	3
x	5	1	3

   a. Construct a scattergram for the data.
   b. It is possible to find many lines for which $\Sigma\,(y - \hat{y}) = 0$. For this reason, the criterion $\Sigma\,(y - \hat{y}) = 0$ is not used for identifying the "best"-fitting straight line. Find two lines that have $\Sigma\,(y - \hat{y}) = 0$.
   c. Find the least squares line.
   d. Compare the value of SSE for the least squares line to that of the two lines you found in part **b**. What principle of least squares is demonstrated by this comparison?

13.73   Consider the following ten data points:

x	3	5	6	4	3	7	6	5	4	7
y	4	3	2	1	2	3	3	5	4	2

   a. Plot the data on a scattergram.
   b. Calculate the values of r and r^2.
   c. Is there sufficient evidence to indicate that x and y are linearly correlated? Test at the $\alpha = .10$ level of significance.

APPLYING THE CONCEPTS

13.74   A study was conducted to determine whether the final grade in an introductory sociology course was related to a student's performance on the verbal ability test administered before college entrance. The verbal test scores and final grades for a random sample of ten students are shown in the table.

STUDENT	VERBAL ABILITY TEST SCORE x	FINAL SOCIOLOGY GRADE y
1	39	65
2	43	78
3	21	52
4	64	82
5	57	92
6	47	89
7	28	73
8	75	98
9	34	56
10	52	75

Note that $SS_{yy} = 2{,}056$, $SS_{xy} = 1{,}894$, $SS_{xx} = 2{,}474$, $\bar{y} = 76$, and $\bar{x} = 46$.

a. Assuming that a linear relationship exists between verbal scores and final grades, find the least squares line relating y to x.

b. Plot the data points and graph the least squares line.

c. Do the data provide sufficient evidence to indicate that a positive correlation exists between verbal score and final grade? (Use $\alpha = .01$.)

d. Find a 95% confidence interval for the slope β_1.

e. Predict a student's final grade in the introductory course when his or her verbal test score is 50. (Use a 90% prediction interval.)

f. Find a 95% confidence interval for the mean final grade for students scoring 35 on the college entrance verbal exam.

13.75 In placing a weekly order, a concessionaire who provides services at a baseball stadium must know what size crowd is expected during the coming week in order to know how much food, etc., to order. Since advanced ticket sales give an indication of expected attendance, food needs might be predicted on the basis of the advanced sales. Data from seven previous weeks of games are given in the table.

HOT DOGS PURCHASED DURING WEEK y (thousands)	ADVANCED TICKET SALES FOR WEEK x (thousands)
39.1	54.0
35.9	48.1
20.8	28.8
42.4	62.4
46.0	64.4
40.7	59.5
29.9	42.3

a. Use the data in the table to develop a simple linear model for hot dogs purchased as a function of advanced ticket sales.

b. Plot the data and graph the line as a check on your calculations.

c. Do the data provide sufficient information to indicate that advanced ticket sales provide information for the prediction of hot dog demand?

d. Calculate r^2 and interpret its value.

e. Find a 90% confidence interval for the mean number of hot dogs purchased when the advanced ticket sales equal 50,000.

f. If the advanced ticket sales this week equal 55,000, find a 90% prediction interval for the number of hot dogs that will be purchased this week at the game.

13.76   Emotional exhaustion, or *burnout*, is a significant problem for people with careers in the field of human services. It seriously affects productivity and feelings of job satisfaction. Michael P. Leiter and Kimberly Ann Meechan (1986) used regression analysis to investigate the relationship between burnout and different aspects of the human services professional's job and job-related behavior. To measure emotional exhaustion, they used the Maslach Burnout Inventory (a questionnaire from which an index of exhaustion can be developed). One of the independent variables they considered was the proportion of the person's social contacts with individuals who belong to the person's work group. They called this variable *concentration*. The table lists the values of the emotional exhaustion index (higher values indicate greater emotional exhaustion) and concentration for a sample of 25 human services professionals who work in a large public hospital. Given that $SS_{yy} = 1,800,417.44$, $SS_{xx} = 14,026.16$, $SS_{xy} = 124,348.520$, $\bar{y} = 578.32$, and $\bar{x} = 68.560$:

a. Construct a scattergram for the data. Do these variables x and y appear to be related?

b. Calculate the correlation coefficient for the data and interpret its value.

c. Use the method of least squares to estimate the straight-line model relating emotional exhaustion and concentration, and plot it on your scattergram.

d. Calculate the coefficient of determination for your least squares line and interpret it.

e. Test the usefulness of the straight-line relationship with concentration for predicting burnout. Use $\alpha = .05$.

EXHAUSTION INDEX y	CONCENTRATION x (%)	EXHAUSTION INDEX y	CONCENTRATION x (%)
100	20	493	86
525	60	892	83
300	38	527	79
980	88	600	75
310	79	855	81
900	87	709	75
410	68	791	77
296	12	718	77
120	35	684	77
501	70	141	17
920	80	400	85
810	92	970	96
506	77		

f. Is there evidence that the correlation between emotional exhaustion and concentration differs from 0? Does your conclusion mean that concentration causes emotional exhaustion? Explain.

g. Use a 95% prediction interval to forecast the level of emotional exhaustion for a human services professional who has 80% of her social contacts within her work group. Interpret the forecast in the context of the problem.

h. Use a 95% confidence interval to estimate the mean exhaustion level for all professionals who have 80% of their social contacts within their work groups. Interpret the interval, and compare its width with that of the prediction interval in part **g**.

13.77   At temperatures approaching absolute zero (273° below 0°C), helium exhibits traits that defy many laws of conventional physics. An experiment has been conducted with helium in solid form at various temperatures near absolute zero. The solid helium is placed in a dilution refrigerator along with a solid impure substance, and the fraction (in weight) of the impurity passing through the solid helium is recorded. (This phenomenon of solids passing directly through solids is known as *quantum tunneling*.) The data are given in the table.

TEMPERATURE x (°C)	PROPORTION OF IMPURITY PASSING THROUGH HELIUM y	TEMPERATURE x (°C)	PROPORTION OF IMPURITY PASSING THROUGH HELIUM y
−262.0	.315	−272.0	.935
−265.0	.202	−272.4	.957
−256.0	.204	−272.7	.906
−267.0	.620	−272.8	.985
−270.0	.715	−272.9	.987

The least squares printout for the simple linear regression model $E(y) = \beta_0 + \beta_1 x$ is shown.

SAS simple linear regression printout for Exercise 13.77

Dep Variable: IMPURITY

Analysis of Variance

Source	DF	Sum of Squares	Mean Square	F Value	Prob>F
Model	1	0.83089	0.83089	46.728	0.0001
Error	8	0.14225	0.01778		
C Total	9	0.97315			

Root MSE	0.13335	R-Square	0.8538	
Dep Mean	0.68260	Adj R-Sq	0.8356	
C.V.	19.53519			

Parameter Estimates

Variable	DF	Parameter Estimate	Standard Error	T for H0: Parameter=0	Prob > ¦T¦
INTERCEP	1	−13.490347	2.07377160	−6.505	0.0002
TEMP	1	−0.052829	0.00772828	−6.836	0.0001

Obs	TEMP	IMPURITY	Predict Value	Residual	Lower95% Predict	Upper95% Predict
1	−262	0.3150	0.3508	−0.0358	0.00946	0.6922
2	−265	0.2020	0.5093	−0.3073	0.1816	0.8371
3	−256	0.2040	0.0339	0.1701	−0.3559	0.4236
4	−267	0.6200	0.6150	0.00502	0.2917	0.9383
5	−270	0.7150	0.7735	−0.0585	0.4495	1.0974
6	−272	0.9350	0.8791	0.0559	0.5499	1.2084
7	−272.4	0.9570	0.9003	0.0567	0.5695	1.2310
8	−272.7	0.9060	0.9161	−0.0101	0.5841	1.2481
9	−272.8	0.9850	0.9214	0.0636	0.5890	1.2538
10	−272.9	0.9870	0.9267	0.0603	0.5938	1.2595
11	−273	.	0.9320	.	0.5987	1.2653

Sum of Residuals	6.578071E−15
Sum of Squared Residuals	0.1423
Predicted Resid SS (Press)	0.3402

a. Find the least squares estimates of the intercept and slope. Interpret them.

b. Use a 95% confidence interval to estimate the slope β_1. Interpret the interval in terms of this application. Does the interval support the hypothesis that temperature contributes information about the proportion of impurity passing through helium?

c. Interpret the coefficient of determination for this model.

d. Find the 95% prediction interval for the percentage of impurity passing through solid helium at $-273°C$. (Note that this value of x is outside the experimental region, where use of the model for prediction may be unreliable.)

13.78 The data shown in the table give the consumer purchases of oranges, y, and the price per box for an 8-month period.

MONTH	U.S. CONSUMER PURCHASES OF ORANGES y (thousands of boxes)	PRICE PER BOX x (dollars)
October	6,100	6.20
November	6,800	6.05
December	8,400	6.20
January	6,000	8.10
February	5,800	8.70
March	5,500	8.75
April	4,400	9.40
May	4,000	9.75

a. Construct a scattergram of the data.

b. Find the least squares line for the data and plot it on your scattergram. The least squares line may be viewed as an estimate of the short-run demand function for oranges.

c. Define β_1 in the context of this problem.

d. Test the hypothesis that the price per box of oranges contributes no information for the prediction of the number of boxes consumed when a linear model of short-run demand is used (let $\alpha = .05$). Draw the appropriate conclusions.

e. Find a 90% confidence interval for β_1. Interpret your results.

f. Find the coefficient of correlation for the given data.

g. Find the coefficient of determination for the linear model of demand you constructed in part **b**. Interpret your result.

h. Find a 90% prediction interval for the number of boxes of oranges that will be consumed if the price per box is $8.

i. Find a 95% confidence interval for the expected number of boxes that will be consumed at a price of $8 per box.

13.79 Refer to Exercise 13.41, in which managerial success, y, was modeled as a function of the number of contacts a manager makes with people outside his or her work unit, x, during a specific period of time. The data are repeated in the table.

MANAGER	MANAGER SUCCESS INDEX	NUMBER OF INTERACTIONS WITH OUTSIDERS	MANAGER	MANAGER SUCCESS INDEX	NUMBER OF INTERACTIONS WITH OUTSIDERS
1	40	12	11	70	20
2	73	71	12	47	81
3	95	70	13	80	40
4	60	81	14	51	33
5	81	43	15	32	45
6	27	50	16	50	10
7	53	42	17	52	65
8	66	18	18	30	20
9	25	35	19	42	21
10	63	82			

Recall that $SS_{yy} = 7,006.6316$, $SS_{xx} = 10,824.5263$, $SS_{xy} = 2,561.2632$, $\bar{y} = 54.5789$, and $\bar{x} = 44.1579$.

a. A particular manager was observed for 2 weeks as in the Luthans, Rosenkrantz, and Hennessey (1985) study. She made 55 contacts with people outside her work unit. Predict the value of the manager's success index. Use a 90% prediction interval.

b. A second manager was observed for 2 weeks. This manager made 110 contacts with people outside his work unit. Give two reasons why caution should be exercised in using the least squares model developed from the given data set to construct a prediction interval for this manager's success index.

c. In the context of this problem, determine the value of x whose associated prediction interval for y is the narrowest.

13.80　Spiraling energy costs have generated interest in energy conservation in businesses of all sizes. Consequently, firms planning to build new plants or make additions to existing facilities have become very conscious of the energy efficiency of proposed new structures. As a result, such firms are interested in knowing the relationship between a building's yearly energy consumption and the factors that influence heat loss. Some of these factors are the number of building stories above and below ground, the materials used in the construction of the building shell, the number of square feet of building shell, and climatic conditions. The table lists the energy consumption in British thermal units (a BTU is the amount of heat required to raise 1 pound of water 1°F) for 1990 for 22 buildings that were all subjected to the same climatic conditions. The SAS printout that fits the straight-line model relating BTU consumption, y, to building shell area, x, is given on page 700.

BTU/YEAR (Thousands)	SHELL AREA (Square feet)	BTU/YEAR (Thousands)	SHELL AREA (Square feet)
3,870,000	30,001	2,680,000	23,680
1,371,000	13,530	337,500	5,650
2,422,000	26,060	567,500	8,001
672,200	6,355	555,300	6,147
233,100	4,576	239,400	2,660
218,900	24,680	2,629,000	19,240
354,000	2,621	1,102,000	10,700
3,135,000	23,350	423,500	9,125
1,470,000	18,770	423,500	6,510
1,408,000	12,220	1,691,000	13,530
2,201,000	25,490	1,870,000	18,860

a. Find the least squares estimates of the intercept β_0 and the slope β_1.

b. Investigate the usefulness of the model you developed in part a. Is yearly energy consumption positively linearly related to the shell area of the building? Test using $\alpha = .10$.

c. Calculate the observed significance level of the test of part b using the printout. Interpret its value.

d. Find the coefficient of determination r^2 and interpret its value.

e. A company wishes to build a new warehouse that will contain 8,000 square feet of shell area. Find the predicted value of energy consumption and a 95% prediction interval on the printout. Comment on the usefulness of this interval.

f. The application of the model you developed in part a to the warehouse problem of part e is appropriate only if certain assumptions can be made about the new warehouse. What are these assumptions?

SAS simple linear regression printout for Exercise 13.80

Dep Variable: BTU

Analysis of Variance

Source	DF	Sum of Squares	Mean Square	F Value	Prob>F
Model	1	1.658498E+13	1.658498E+13	42.028	0.0001
Error	20	7.89232E+12	394616010047		
C Total	21	2.44773E+13			

Root MSE	628184.69422	R-Square	0.6776	
Dep Mean	1357904.54545	Adj R-Sq	0.6614	
C.V.	46.26133			

Parameter Estimates

Variable	DF	Parameter Estimate	Standard Error	T for H0: Parameter=0	Prob > |T|
INTERCEP	1	-99045	261617.65980	-0.379	0.7090
AREA	1	102.814048	15.85924082	6.483	0.0001

Obs	AREA	BTU	Predict Value	Residual	Lower95% Predict	Upper95% Predict
1	30001	3870000	2985479	884521	1546958	4424000
2	13530	1371000	1292029	78971.2	-47949.3	2632007
3	26060	2422000	2580289	-158289	1183940	3976637
4	6355	672200	554338	117862	-810192	1918868
5	4576	233100	371432	-138332	-1005463	1748327
6	24680	218900	2438405	-2219505	1054223	3822588
7	2621	354000	170430	183570	-1222796	1563657
8	23350	3135000	2301663	833337	927871	3675455
9	18770	1470000	1830774	-360774	482352	3179196
10	12220	1408000	1157342	250658	-184021	2498706
11	25490	2201000	2521685	-320685	1130530	3912840
12	23680	2680000	2335591	344409	959345	3711838
13	5650	337500	481854	-144354	-887287	1850995
14	8001	567500	723570	-156070	-631698	2078838
15	6147	555300	532953	22347.3	-832898	1898804
16	2660	239400	174440	64959.9	-1218433	1567313
17	19240	2629000	1879097	749903	528832	3229362
18	10700	1102000	1001065	100935	-343656	2345786
19	9125	423500	839133	-415633	-511035	2189301
20	6510	423500	570274	-146774	-793294	1933842
21	13530	1691000	1292029	398971	-47949.3	2632007
22	18860	1870000	1840028	29972.3	491266	3188789
23	8000	.	723467	.	-631806	2078740

Sum of Residuals		1.6298145E-9
Sum of Squared Residuals		7.89232E+12
Predicted Resid SS (Press)		1.012747E+13

13.81 Sometimes it is known from theoretical considerations that the straight-line relationship between two variables, x and y, passes through the origin of the xy-plane. Consider the relationship between the total weight of a shipment of 50-pound bags of flour, y, and the number of bags in the shipment, x. Since a shipment containing $x = 0$ bags (i.e., no shipment at all) has a total weight of $y = 0$, a straight-line model of the relationship between x and y should pass through the point $x = 0$, $y = 0$. In such a case you could assume $\beta_0 = 0$ and characterize the relationship between x and y with the following model:

$$y = \beta_1 x + \varepsilon$$

The least squares estimate of β_1 for this model is

$$\hat{\beta}_1 = \frac{\sum x_i y_i}{\sum x_i^2}$$

From the records of past flour shipments, 15 shipments were randomly chosen and the data shown in the table were recorded.

WEIGHT OF SHIPMENT	NUMBER OF 50-POUND BAGS IN SHIPMENT	WEIGHT OF SHIPMENT	NUMBER OF 50-POUND BAGS IN SHIPMENT
5,050	100	7,162	150
10,249	205	24,000	500
20,000	450	4,900	100
7,420	150	14,501	300
24,685	500	28,000	600
10,206	200	17,002	400
7,325	150	16,100	400
4,958	100		

a. Find the least squares line for the given data under the assumption that $\beta_0 = 0$. Plot the least squares line on a scattergram of the data.

b. Find the least squares line for the given data using the model

$$y = \beta_0 + \beta_1 x + \varepsilon$$

(i.e., do not restrict β_0 to equal 0). Plot this line on the same scatterplot you constructed in part a.

c. Refer to part b. Why might $\hat{\beta}_0$ be different from 0 even though the true value of β_0 is known to be 0?

d. The estimated standard error of $\hat{\beta}_0$ is equal to

$$s\sqrt{\frac{1}{n} + \frac{\bar{x}^2}{SS_{xx}}}$$

Use the t statistic

$$t = \frac{\hat{\beta}_0 - 0}{s\sqrt{(1/n) + (\bar{x}^2/SS_{xx})}}$$

to test the null hypothesis H_0: $\beta_0 = 0$ against the alternative H_a: $\beta_0 \neq 0$. Use $\alpha = .10$. Should you include β_0 in your model?

13.82 A study was conducted to determine whether there is a linear relationship between the breaking strength, y, of wooden beams and the specific gravity, x, of the wood. Ten randomly selected beams of the same cross-sectional dimensions were stressed until they broke. The breaking strengths and the density of the wood are shown in the table (page 702) for each of the ten beams.

BEAM	SPECIFIC GRAVITY x	STRENGTH y	BEAM	SPECIFIC GRAVITY x	STRENGTH y
1	.499	11.14	6	.528	12.60
2	.558	12.74	7	.418	11.13
3	.604	13.13	8	.480	11.70
4	.441	11.51	9	.406	11.02
5	.550	12.38	10	.467	11.41

Given that $\bar{x} = .4951$, $\bar{y} = 11.8760$, $SS_{xx} = .03776$, $SS_{xy} = .4089$, and $SS_{yy} = 5.3102$:

a. Fit the model $y = \beta_0 + \beta_1 x + \varepsilon$, and interpret the estimates.

b. Calculate and interpret the standard error of regression for the model in part **a**.

c. Test $H_0: \beta_1 = 0$ against the alternative hypothesis $H_a: \beta_1 \neq 0$.

d. Calculate and interpret the coefficient of determination.

e. Estimate the mean strength for beams with specific gravity .590. Use a 90% confidence interval.

13.83 Although the income tax system is structured so that people with higher incomes should pay a higher percentage of their incomes in taxes, there are many loopholes and tax shelters available for people with higher incomes. A sample of seven individual 1991 tax returns gave the data listed in the table on income and taxes paid.

INDIVIDUAL	GROSS INCOME x (thousands of dollars)	TAXES PAID y (percentage of total income)
1	35.8	16.7
2	80.2	21.4
3	14.9	15.2
4	7.3	10.1
5	9.1	12.2
6	150.7	19.6
7	25.9	17.3

a. Fit a least squares line to the data.

b. Plot the data and graph the line as a check on your calculations.

c. Calculate r and r^2 and interpret each.

d. Find a 90% confidence interval for the mean percent taxes paid for individuals with gross incomes of $70,000.

13.84 The data in the accompanying table were collected to calibrate a new instrument for measuring interocular pressure. The interocular pressure for each of ten glaucoma patients was measured by the new instrument and by a standard, reliable, but more time-consuming method.

PATIENT	RELIABLE METHOD x	NEW INSTRUMENT y	PATIENT	RELIABLE METHOD x	NEW INSTRUMENT y
1	20.2	20.0	6	21.8	22.1
2	16.7	17.1	7	19.1	18.9
3	17.1	17.2	8	22.9	22.2
4	26.3	25.1	9	23.5	24.0
5	22.2	22.0	10	17.0	18.1

a. Fit a least squares line to the data.

b. Calculate r and r^2. Interpret each of these quantities.

c. Predict the pressure measured by the new instrument when the reliable method gives a reading of 20.0. Use a 90% prediction interval.

13.85   As a result of the increase in the number of suburban shopping centers, many center-city stores are suffering financially. A downtown department store thinks that increased advertising might help lure more shoppers into the area. To study the effect of advertising on sales, records were obtained for several mid-year months during which the store varied advertising expenditures. Those records are shown in the table.

ADVERTISING EXPENSE x ($ thousands)	SALES y ($ thousands)
.9	30
1.1	34
.8	32
1.2	37
.7	31

a. Estimate the coefficient of correlation between sales and advertising expenditures.
b. Do the data provide sufficient evidence to indicate a nonzero correlation between sales, y, and advertising expense, x?
c. Can you use the results of parts a and b to conclude that additional advertising expense will *cause* sales to increase? Why or why not?

13.86   In the late 1970s and early 1980s the prices of single-family homes in the United States rose faster than the rate of inflation. As a result, many investors directed their funds to the housing market as a hedge against inflation. One way an investor can assess the value of a specific house is to compare it to the sale prices of similar homes that have recently been sold. Another popular approach, according to Cho and Reichert (1980), involves the use of a regression analysis to model the relationship between price and the variables that influence price. Independent variables that could be utilized are total living area, number of rooms, number of baths, age of property, and so on. Of these factors, Cho and Reichert indicate that total living area provides the most information for determining the worth of a house. The table lists the final selling price and total living area for a sample of 24 homes in the same geographic area that were sold during the last 3 months of 1990. The regression printout using a straight-line model to relate price to area for these data is given on page 704.

AREA (sq. ft.)	PRICE ($)	AREA (sq. ft.)	PRICE ($)
2,100	150,000	3,000	205,000
1,455	114,900	1,400	79,400
1,630	106,500	2,750	200,000
2,600	195,000	2,900	215,100
1,210	75,500	2,500	180,800
1,857	126,600	1,535	120,900
2,000	135,400	1,333	70,000
2,400	178,650	2,455	165,200
2,256	145,100	3,010	185,000
1,290	62,600	2,180	160,000
2,332	168,200	1,870	119,900
1,725	138,100	1,582	99,900

a. Construct a scattergram for the data.
b. Find the least squares line and plot it on your scattergram.

Linear regression printout for Exercise 13.86

SELLING PRICE VS. TOTAL LIVING AREA

Model: MODEL1
Dep Variable: PRICE

Analysis of Variance

Source	DF	Sum of Squares	Mean Square	F Value	Prob>F
Model	1	42825909688	42825909688	247.857	0.0001
Error	22	3801265207.6	172784782.16		
C Total	23	46627174896			

Root MSE	13144.76254	R-Square	0.9185	
Dep Mean	141572.91667	Adj R-Sq	0.9148	
C.V.	9.28480			

Parameter Estimates

Variable	DF	Parameter Estimate	Standard Error	T for H0: Parameter=0	Prob > ¦T¦
INTERCEP	1	−15124	10308.485043	−1.467	0.1565
AREA	1	76.174547	4.83848421	15.743	0.0001

Obs	AREA	PRICE	Predict Value	Residual	Lower95% Mean	Upper95% Mean
1	2100	150000	144842	5157.9	139261	150423
2	1455	114900	95709	19190.5	87495.9	103923
3	1630	106500	109040	−2540.0	102017	116064
4	2600	195000	182929	12070.7	175142	190717
5	1210	75500.0	77046.7	−1546.7	66887.4	87206.1
6	1857	126600	126332	268.3	120416	132247
7	2000	135400	137225	−1824.6	131631	142819
8	2400	178650	167694	10955.6	161152	174237
9	2256	145100	156725	−11625.3	150814	162637
10	1290	62600.0	83140.7	−20540.7	73642.8	92638.6
11	2332	168200	162515	5685.4	156304	168725
12	1725	138100	116277	21823.4	109791	122763
13	3000	205000	213399	−8399.2	202423	224376
14	1400	79400.0	91519.9	−12119.9	82892.2	100148
15	2750	200000	194356	5644.5	185450	203261
16	2900	215100	205782	9318.3	195657	215906
17	2500	180800	175312	5488.1	168190	182433
18	1535	120900	101803	19096.5	94160.9	109446
19	1333	70000.0	86416.2	−16416.2	77264.5	95568
20	2455	165200	171884	−6684.0	165035	178733
21	3010	185000	214161	−29160.9	203098	225224
22	2180	160000	150936	9064.0	145236	156636
23	1870	119900	127322	−7421.9	121449	133195
24	1582	99900	105384	−5483.7	98056	112711
25	2200	.	152460	.	146713	158206

c. Find r^2 and interpret its value in the context of the problem.

d. Do the data provide evidence that living area contributes information for predicting the price of a home? Use $\alpha = .05$.

e. Find a 95% confidence interval for β_1. Does your confidence interval support the conclusion you reached in part **d**? Explain. (Note that the standard error of $\hat{\beta}_1$ is given on the printout in the column headed Standard Error in the row corresponding to AREA.)

f. Find the observed significance level for the test in part **d**, and interpret its value.
g. Estimate the mean selling price for homes with a total living area of 2,200 square feet. Use a 95% confidence interval.

13.87   In Exercise 13.22, we discussed an experiment conducted to determine the effect of ration level on the growth rate of rainbow trout. Twenty fish were fed at each of four ration levels at 6°C water temperature. The experiment was repeated for water temperatures of 12°C and 18°C. The table gives the means and standard deviations (in parentheses) for each sample of 20 fish. Refer only to data collected on fish fed in water maintained at 18°C.

RATION	MEAN WET WEIGHT GAIN (%)		
(% body weight per day)	6°C	12°C	18°C
.0	−8.14 (3.27)	−10.33 (2.35)	−13.21 (1.96)
.8	12.31 (3.37)		
1.2		14.29 (5.32)	
1.5	28.19 (5.39)		15.51 (4.48)
2.5	29.12 (5.67)	38.05 (6.99)	
3.5			51.13 (7.14)
4.0		60.13 (12.71)	
4.5			65.49 (13.04)
Maintenance ration	.32	.48	.69

Source: Kovacs, T. G., and Leduc, G. "Sublethal toxicity to rainbow trout (*Salmo gairdneri*) at different temperatures," *Canadian Journal of Fisheries and Aquatic Sciences*, 1982, 39.

a. Find SSE and s^2 for the $n = 4$ data points.
b. Find a 95% confidence interval for the mean gain (percentage of body weight) for a 1% increase in ration level.
c. State the assumptions required for your inference in part **b** to be valid.
d. Find a 95% confidence interval for the mean weight gain of rainbow trout (percentage of body weight) over a 20-day period when the ration level is 4% of body weight.

ON YOUR OWN...

There are many dependent variables in all areas of research that are the subject of regression modeling efforts. We list five such variables here:

1. Crime rate in various communities
2. Daily maximum temperature in your town
3. Grade-point average of students who have completed one academic year at your college
4. Gross National Product of the United States
5. Points scored by your favorite football team in a single game

Choose one of these dependent variables that is of particular interest to you or choose some other dependent variable for which you want to construct a prediction model. There may be a large number of independent variables that should be included in a prediction equation for the depen-

dent variable you choose. List three potentially important independent variables, x_1, x_2, and x_3, that you think might be (individually) strongly related to your dependent variable. Next, obtain ten data values, each of which consists of a measure of your dependent variable y and the corresponding values of x_1, x_2, and x_3.

a. Use the least squares formulas given in this chapter to fit three straight-line models—one for each independent variable—for predicting y.
b. Interpret the sign of the estimated slope coefficient $\hat{\beta}_1$ in each case, and test the utility of each model by testing H_0: $\beta_1 = 0$ against H_a: $\beta_1 \neq 0$. What assumptions must be satisfied to assure the validity of these tests?
c. Calculate the coefficient of determination r^2 for each model. Which of the independent variables

predicts y best for the ten sampled sets of data? Is this variable necessarily best in general (i.e., for the entire population)? Explain.

Be sure to keep the data and the results of your calculations, since you will need them for the **On Your Own** sections in Chapters 14 and 15.

USING THE COMPUTER...

Suppose we wish to estimate the relationship between the median household income y and the percentage of college graduates x in a zip code using a straight-line model.

a. Draw a random sample of 100 zip codes from the 1,000 described in Appendix C, and extract y and x for each zip code sampled. Use a software package that includes a regression program to obtain the least squares fit of the straight-line model of interest.
   1. Graph the fitted model.
   2. Interpret the estimated intercept and slope.

3. Interpret the estimated standard deviation of the error term.
4. Evaluate the usefulness of the model.
5. Estimate the mean median income for all zip codes having 15% college graduates using a 95% confidence interval.
6. Predict the mean median income for a particular zip code having 15% college graduates using a 95% prediction interval.

b. Repeat part a using the entire set of 1,000 zip codes. Compare the results to those you obtained in part a.

References

Cho, C. C., and Reichert, A. "An application of multiple regression analysis for appraising single-family housing values," *Business Economics*, January 1980, 15, pp. 47–52.

Deming, W. E. *Out of the Crisis*. Cambridge, Mass.: MIT Center for Advanced Engineering Study, 1986.

Graybill, F. *Theory and Application of the Linear Model*. North Scituate, Mass.: Duxbury, 1976, Chapter 5.

Koch, P. D., and Ragan, J. F., Jr. "Investigating the causal relationship between quits and wages: An exercise in comparative dynamics," *Economic Inquiry*, Vol. 24, January 1986, pp. 61–83.

Landro, L., and Mayer, J. "Cable-TV viewing study dims prospect of large increase in number of channels." *Wall Street Journal*, Nov. 16, 1982, p. 10.

Leiter, M. P., and Meechan, K. A. "Role structure and burnout in the field of human services," *Journal of Applied Behavioral Science*, Vol. 22, No. 1, 1986, pp. 47–52.

Luthans, F., Rosenkrantz, S. A., and Hennessey, H. W. "What do successful managers really do? An observational study of managerial activities," *Journal of Applied Behavioral Science*, Vol. 21, No. 3, August 1985, pp. 255–270.

Mendenhall, W. *Introduction to Linear Models and the Design and Analysis of Experiments*. Belmont, Ca.: Wadsworth, 1968.

Mendenhall, W., and Sincich, T. *A Second Course in Business Statistics: Regression Analysis*, 2d ed. San Francisco: Dellen, 1986.

Mintzberg, H. *The Nature of Managerial Work*. New York: Harper & Row, 1973.

Near, J. P., and Miceli, M. P. "Retaliation against whistle blowers: Predictors and effects." *Journal of Applied Psychology*, Vol. 71, No. 1, 1986, pp. 137–145.

Neter, J., and Wasserman, W. *Applied Linear Statistical Models*. Homewood, Ill.: Richard Irwin, 1974, Chapters 2–6.

Younger, M. S. *A First Course in Linear Regression*, 2d ed. Boston, Mass.: Duxbury, 1985.

Multiple Regression

CONTENTS

WHERE WE'VE BEEN...

In Chapter 13 we demonstrated how to model the relationship between a dependent variable y and an independent variable x using a straight line. We fit the straight line to the data points, used r and r^2 to measure the strength of the relationship between y and x, and used the resulting prediction equation to estimate the mean value of y or to predict some future value of y for a given value of x.

WHERE WE'RE GOING...

This chapter converts the basic concept of Chapter 13 into a powerful estimation and prediction device by modeling the mean value of y as a function of two or more independent variables. The techniques developed will enable you to model a response, y, as a function of both quantitative and qualitative variables. As in the case of a simple linear regression, a multiple regression analysis involves fitting the model to a data set, testing the utility of the model, and using it for estimation and prediction.

14.1 Multiple Regression: The Model and the Procedure

Most practical applications of regression anlaysis utilize models that are more complex than the simple straight-line model. For example, a realistic probabilistic model for reaction time to a stimulus would include more than just the amount of a particular drug in the bloodstream. Factors such as age, a measure of visual perception, and sex of the subject are a few of the many variables that might be related to reaction time. Thus, we would want to incorporate these and other potentially important independent variables into the model in order to make accurate predictions.

Probabilistic models that include terms involving x^2, x^3 (or higher-order terms), or more than one independent variable are called **multiple regression models**. The general form of these models is

$$y = \beta_0 + \beta_1 x_1 + \beta_2 x_2 + \cdots + \beta_k x_k + \varepsilon$$

The dependent variable y is now written as a function of k independent variables, $x_1, x_2, \ldots, x_k$. The random error term is added to make the model probabilistic rather than deterministic. The value of the coefficient β_i determines the contribution of the independent variable x_i, and β_0 is the y-intercept. The coefficients $\beta_0, \beta_1, \ldots, \beta_k$ are usually unknown, because they represent population parameters.

At first glance it might appear that the regression model shown above would not allow for anything other than straight-line relationships between y and the independent variables, but this is not true. Actually, $x_1, x_2, \ldots, x_k$ can be functions of variables as long as the functions do not contain unknown parameters. For example, the dollar sales, y, in new housing in a region could be a function of the independent variables

$x_1 =$ Mortgage interest rate

$x_2 =$ (Mortgage interest rate)$^2 = x_1^2$

$x_3 =$ Unemployment rate in the region

and so on. You could even insert a cyclical term (if it would be useful) of the form $x_4 = \sin t$, where t is a time variable. The multiple regression model is quite versatile and can be made to model many different types of response variables.

As shown in the box, we use the same steps to develop the multiple regression model as we used for the simple regression model.

STEP 1 Hypothesize the deterministic component of the model. This component relates the mean, $E(y)$, to the independent variables $x_1, x_2, \ldots, x_k$. This involves the choice of the independent variables to be included in the model (Chapter 15).

STEP 2 Use the sample data to estimate the unknown model parameters $\beta_0, \beta_1, \beta_2, \ldots, \beta_k$ in the model (Section 14.2).

STEP 3 Specify the probability distribution of the random error term, ε, and estimate the standard deviation of this distribution, σ (Sections 14.3 and 14.9).

> STEP 4 Statistically evaluate the usefulness of the model (Sections 14.4 and 14.5).
>
> STEP 5 When satisfied that the model is useful, use it for prediction, estimation, and other purposes (Section 14.6).

Although we introduce several different types of models in this chapter, we defer formal discussion of model building (step 1) until Chapter 15.

CASE STUDY 14.1

PREDICTING CORPORATE EXECUTIVE COMPENSATION

Towers, Perrin, Forster & Crosby (TPF&C), an international management consulting firm, has developed a unique application of multiple regression analysis. Many firms are interested in evaluating their management salary structure, and TPF&C uses multiple regression models to accomplish this evaluation. The Compensation Management Service, as TPF&C calls it, measures both the internal and external consistency of a company's pay policies to determine whether they reflect the management's intent.

The dependent variable y used to measure executive compensation is annual salary. The independent variables used to explain salary structure include the executive's age, education, rank, and bonus eligibility; number of employees under the executive's direct supervision; as well as variables that describe the company for which the executive works, such as annual sales, profit, and total assets.

The initial step in developing models for executive compensation is to obtain a sample of executives from various client firms, which TPF&C calls the Compensation Data Bank. The data for these executives are used to estimate the model coefficients (the β parameters), and these estimates are then substituted into the linear model to form a prediction equation. To predict a particular executive's compensation, TPF&C substitutes into the prediction equation the values of the independent variables that pertain to the executive (age, rank, etc.). This application of multiple regression analysis is developed more fully in Section 14.7.

14.2 Fitting the Model: Least Squares Approach

The method of fitting multiple regression models is identical to that of fitting the simple straight-line model: the method of least squares. That is, we choose the estimated model

$$\hat{y} = \hat{\beta}_0 + \hat{\beta}_1 x_1 + \cdots + \hat{\beta}_k x_k$$

that minimizes

$$\text{SSE} = \sum (y - \hat{y})^2$$

As in the case of the simple linear model, the sample estimates $\hat{\beta}_0, \hat{\beta}_1, \ldots, \hat{\beta}_k$ are obtained as a solution to a set of simultaneous linear equations.*

The primary difference between fitting the simple and multiple regression models is computational difficulty. The $(k + 1)$ simultaneous linear equations that must be solved to find the $(k + 1)$ estimated coefficients $\hat{\beta}_0, \hat{\beta}_1, \ldots, \hat{\beta}_k$ are difficult (sometimes nearly impossible) to solve with a calculator. Consequently, we resort to the use of computers. Many computer packages have been developed to fit a multiple regression model using the method of least squares. We will present output from the SAS System† computer package instead of presenting the tedious hand calculations required to fit the models. The regression output of the SAS System is similar to that of most other package regression programs. We will compare the SAS output with two other regression program outputs in Section 14.8. We demonstrate the SAS regression procedure with the following example.

In all-electric homes the amount of electricity expended is of interest to consumers, builders, and groups involved with energy conservation. Suppose we wish to investigate the monthly electrical usage, y, in all-electric homes and its relationship to the size, x, of the home. Moreover, suppose we think that monthly electrical usage in all-electric homes is related to the size of the home by the model

$$y = \beta_0 + \beta_1 x + \beta_2 x^2 + \varepsilon$$

To estimate the unknown parameters β_0, β_1, and β_2, values of y and x are collected for ten homes during a particular month. The data are shown in Table 14.1.

TABLE 14.1   **Home Size–Electrical Usage Data**

SIZE OF HOME x (sq. ft.)	MONTHLY USAGE y (kilowatt-hours)	SIZE OF HOME x (sq. ft.)	MONTHLY USAGE y (kilowatt-hours)
1,290	1,182	1,840	1,711
1,350	1,172	1,980	1,804
1,470	1,264	2,230	1,840
1,600	1,493	2,400	1,956
1,710	1,571	2,930	1,954

Notice that we include a term involving x^2 in this model because we expect curvature in the graph of the response model relating y to x. The term involving x^2 is called a **quadratic term**. Figure 14.1 illustrates that the electrical usage appears to increase in a curvilinear manner with the size of the home. This provides some support for the inclusion of the quadratic term x^2 in the model.

Part of the output from the SAS multiple regression routine for the data in Table 14.1 is reproduced in Figure 14.2. The least squares estimates of the β parameters appear in the column labeled Parameter Estimate. You can see that

---

*Students who are familiar with calculus should note that $\hat{\beta}_0, \hat{\beta}_1, \ldots, \hat{\beta}_k$ are the solutions to the set of equations $\partial \text{SSE}/\partial \hat{\beta}_0 = 0, \partial \text{SSE}/\partial \hat{\beta}_1 = 0, \ldots, \partial \text{SSE}/\partial \hat{\beta}_k = 0$. The solution is usually given in matrix form, but we do not present the details here. See the references for details.

†SAS is the registered trademark of SAS Institute Inc., Cary, N.C., U.S.A.

FIGURE 14.1

Scattergram of the home size–electrical usage data

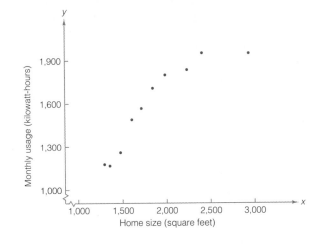

FIGURE 14.2

SAS output for the home size–electrical usage data

Dep Variable: Y

Analysis of Variance

Source	DF	Sum of Squares	Mean Square	F Value	Prob>F
Model	2	831069.54637	415534.77319	189.710	0.0001
Error	7	15332.55363	2190.36480		
C Total	9	846402.10000			

Root MSE	46.80133	R-Square	0.9819
Dep Mean	1594.70000	Adj R-Sq	0.9767
C.V.	2.93480		

Parameter Estimates

Variable	DF	Parameter Estimate	Standard Error	T for H0: Parameter=0	Prob > |T|
INTERCEP	1	-1216.143887	242.80636850	-5.009	0.0016
X	1	2.398930	0.24583560	9.758	0.0001
XSQ	1	-0.000450	0.00005908	-7.618	0.0001

$\hat{\beta}_0 = -1,216.1$, $\hat{\beta}_1 = 2.3989$, and $\hat{\beta}_2 = -.00045$. Therefore, the equation that minimizes the SSE for the data is

$$\hat{y} = -1,216.1 + 2.3989x - .00045x^2$$

The minimum value of the SSE, 15,332.6, also appears in the printout. [*Note:* Throughout this chapter we shade the aspects of the printout that are under discussion.]

Note that the graph of the multiple regression model (Figure 14.3, page 712, a response curve) provides a good fit to the data of Table 14.1. Furthermore, the small value of $\hat{\beta}_2$ does *not* imply that the curvature is insignificant, since the numerical value of $\hat{\beta}_2$ depends on the scale of the measurements. We will test the contribution of the quadratic coefficient $\hat{\beta}_2$ in Section 14.4.

The ultimate goal of this multiple regression analysis is to use the fitted model to predict electrical usage *y* for a home of a specific size (area) *x*. And, of course,

FIGURE 14.3

Least squares model for the home
size–electrical usage data

FIGURE 14.3

Least squares model for the home
size–electrical usage data

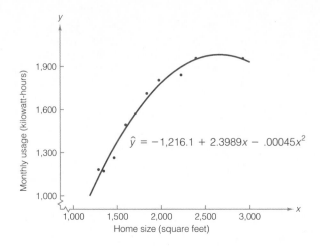

$$\hat{y} = -1{,}216.1 + 2.3989x - .00045x^2$$

we will want to give a prediction interval for y so that we will know how much faith we can place in the prediction. That is, if the prediction model is used to predict electrical usage y for a given size of home, x, what will be the error of prediction? To answer this question, we need to estimate σ^2, the variance of ε.

The interpretation of the estimated coefficients in a quadratic model must be undertaken cautiously. First, the estimated y-intercept, $\hat{\beta}_0$, can be meaningfully interpreted only if the range of the independent variable includes zero—i.e., if $x = 0$ is included in the sampled range of x. In the electrical usage example, $\hat{\beta}_0 = -1{,}216.1$, which would seem to imply that the estimated electrical usage is negative when $x = 0$, but this zero point represents a home with 0 square feet. The zero point is, of course, not in the range of the sample (the lowest value of x is 1,290 square feet), and thus the interpretation of $\hat{\beta}_0$ is not meaningful.

The estimated coefficient of x is $\hat{\beta}_1$, but it no longer represents a slope in the presence of the quadratic term x^2.* The estimated coefficient of the linear term x will not, in general, have a meaningful interpretation in the quadratic model.

The sign of the coefficient, $\hat{\beta}_2$, of the quadratic term, x^2, is the indicator of whether the curve is concave downward (mound-shaped) or concave upward (bowl-shaped). A negative $\hat{\beta}_2$ implies downward concavity, as in the electrical usage example (Figure 14.3), and a positive $\hat{\beta}_2$ implies upward concavity. Rather than interpreting the numerical value of $\hat{\beta}_2$ itself, we utilize a graphical representation of the model, as in Figure 14.3, to describe the model.

Note that Figure 14.3 implies that the estimated electrical usage is leveling off as the home sizes increase beyond 2,500 square feet. In fact, the convexity of the model would lead to decreasing usage estimates if we were to display the model out to 4,000 square feet and beyond. However, model interpretations are not meaningful outside the range of the independent variable, which has a maximum value of 2,930 square feet in this example. Thus, although the model appears to support the hypothesis that the **rate of increase** per square foot **decreases** as the home sizes near the high end of the sampled values, the conclusion that usage will actually begin to decrease for very large homes would be

---

*For students with knowledge of calculus, note that the slope of the quadratic model is the first derivative $\partial y / \partial x = \beta_1 + 2\beta_2 x$. Thus, the slope varies as a function of x, rather than the constant slope associated with the straight-line model.

a **misuse** of the model, since no homes of 3,000 square feet or more were included in the sample.

All the interpretations of the electrical usage model were based on the sample estimates of unknown model parameters. As has been our theme throughout, we want to measure the reliability of these interpretations with regard to the real relationship between usage and home size. Additionally, we will want to use the model to estimate usage for homes not included in the sample, and to measure the reliability of these estimates. All this requires an assessment of the error in the model—i.e., of the probability distribution of the random error component, ε. This is the subject of the next section.

14.3 Model Assumptions

We noted in Section 14.1 that the multiple regression model is of the form

$$y = \beta_0 + \beta_1 x_1 + \beta_2 x_2 + \cdots + \beta_k x_k + \varepsilon$$

where y is the response variable that you wish to predict; $\beta_0, \beta_1, \ldots, \beta_k$ are parameters with unknown values; $x_1, x_2, \ldots, x_k$ are information-contributing variables that are measured without error; and ε is a random error component. Since $\beta_0, \beta_1, \ldots, \beta_k$ and $x_1, x_2, \ldots, x_k$ are nonrandom, the quantity

$$\beta_0 + \beta_1 x_1 + \beta_2 x_2 + \cdots + \beta_k x_k$$

represents the deterministic portion of the model. Therefore, y is composed of two components—one fixed and one random—and, consequently, y is a random variable.

$$y = \overbrace{\beta_0 + \beta_1 x_1 + \cdots + \beta_k x_k}^{\substack{\text{Deterministic} \\ \text{portion of model}}} + \overbrace{\varepsilon}^{\substack{\text{Random} \\ \text{error}}}$$

We will assume (as in Chapter 13) that the random error can be positive or negative and that for any setting of the x values, $x_1, x_2, \ldots, x_k$, the random error ε has a normal probability distribution with mean equal to 0 and variance equal to σ^2. Further, we assume that the random errors associated with any (and every) pair of y values are probabilistically independent. That is, the error, ε, associated with any one y value is independent of the error associated with any other y value. These assumptions are summarized as follows:

Assumptions for Random Error ε

1. For any given set of values of $x_1, x_2, \ldots, x_k$, the random error ε has a normal probability distribution with mean equal to 0 and variance equal to σ^2.
2. The random errors are independent (in a probabilistic sense).

Note that σ^2 represents the variance of the random error, ε. As such, σ^2 is an important measure of the usefulness of the model for the estimation of the mean and the prediction of actual values of y. If $\sigma^2 = 0$, all the random errors will equal 0 and the predicted values, $\hat{y}$, will be identical to $E(y)$; that is $E(y)$ will be estimated without error. In contrast, a large value of σ^2 implies large (absolute) values of ε and larger deviations between the predicted values, $\hat{y}$, and the mean value, $E(y)$. Consequently, the larger the value of σ^2, the greater will be the error in estimating the model parameters $\beta_0, \beta_1, \ldots, \beta_k$ and the error in predicting a value of y for a specific set of values of $x_1, x_2, \ldots, x_k$. Thus, σ^2 plays a major role in making inferences about $\beta_0, \beta_1, \ldots, \beta_k$, in estimating $E(y)$, and in predicting y for specific values of $x_1, x_2, \ldots, x_k$.

Since the variance, σ^2, of the random error, ε, will rarely be known, we must use the results of the regression analysis to estimate its value. You will recall that σ^2 is the variance of the probability distribution of the random error, ε, for a given set of values for $x_1, x_2, \ldots, x_k$, and hence that it is the mean value of the squares of the deviations of the y values (for given values of $x_1, x_2, \ldots, x_k$) about the mean value $E(y)$.* Since the predicted value, $\hat{y}$, estimates $E(y)$ for each of the data points, it seems natural to use

$$\text{SSE} = \sum (y_i - \hat{y}_i)^2$$

to construct an estimator of σ^2.

For example, in the second-order (quadratic) model describing electrical usage as a function of home size, we found that $\text{SSE} = 15,332.6$. We now want to use this quantity to estimate the variance of ε. Recall that the estimator for the straight-line model was $s^2 = \text{SSE}/(n - 2)$ and note that the denominator is $(n - \text{Number of estimated } \beta \text{ parameters})$, which is $(n - 2)$ in the first-order (straight-line) model. Since we must estimate one more parameter, β_2, for the second-order model, the estimator of σ^2 is

$$s^2 = \frac{\text{SSE}}{n - 3}$$

That is, the denominator becomes $(n - 3)$ because there are now three β parameters in the model. The numerical estimate for this example is

$$s^2 = \frac{\text{SSE}}{10 - 3} = \frac{15,332.6}{7} = 2,190.36$$

In many computer printouts and textbooks, s^2 is called the **mean square for error (MSE)**. This estimate of σ^2 is shown in the column titled MEAN SQUARE in the SAS printout in Figure 14.2.

The units of the estimated variance are squared units of the dependent variable, y. In the electrical usage example, the units of s^2 are (kilowatt-hours)^2. This makes meaningful interpretation of s^2 difficult, and we use the standard deviation s to provide a more meaningful measure of variability. In the electrical usage example,

$$s = \sqrt{2,190.36} = 46.8,$$

which is given on the computer printout in Figure 14.2 under ROOT MSE. One useful interpretation of the estimated standard deviation s is that the interval $\pm 2s$ will provide a rough approximation to the accuracy with which the model

---

*Since $y = E(y) + \varepsilon$, ε is equal to the deviation $y - E(y)$. Also, by definition, the variance of a random variable is the expected value of the square of the deviation of the random variable from its mean. According to our model, $E(\varepsilon) = 0$. Therefore, $\sigma^2 = E(\varepsilon^2)$.

will predict future values of y for given values of x. Thus, in the electrical usage example, we expect the model to provide predictions of electrical usage to within about $\pm 2s = \pm 93.6$ kilowatt-hours.*

For the general multiple regression model

$$y = \beta_0 + \beta_1 x_1 + \beta_2 x_2 + \cdots + \beta_k x_k + \varepsilon$$

we must estimate the $(k + 1)$ parameters $\beta_0, \beta_1, \beta_2, \ldots, \beta_k$. Thus, the estimator of σ^2 is SSE divided by the quantity $(n -$ Number of estimated β parameters$)$.

We will use the estimator of σ^2 both to check the utility of the model (Sections 14.4 and 14.5) and to provide a measure of reliability of predictions and estimates when the model is used for those purposes (Section 14.6). Thus, you can see that the estimation of σ^2 plays an important part in the development of a regression model.

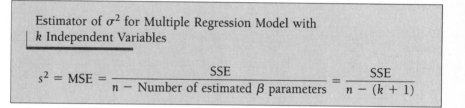

Estimator of σ^2 for Multiple Regression Model with k Independent Variables

$$s^2 = \text{MSE} = \frac{\text{SSE}}{n - \text{Number of estimated } \beta \text{ parameters}} = \frac{\text{SSE}}{n - (k + 1)}$$

14.4 Estimating and Testing Hypotheses About the β Parameters

Sometimes the individual β parameters in a model have practical significance and we want to estimate their values or test hypotheses about them. For example, if electrical usage y is related to home size x by the straight-line relationship

$$y = \beta_0 + \beta_1 x + \varepsilon$$

then β_1 has a very practical interpretation. That is, you saw in Chapter 13 that β_1 is the mean increase in electrical usage, y, for a 1-unit increase in home size x.

As proposed in the preceding sections, suppose that electrical usage y is related to home size x by the quadratic model

$$y = \beta_0 + \beta_1 x + \beta_2 x^2 + \varepsilon$$

Then the mean value of y for a given value of x is

$$E(y) = \beta_0 + \beta_1 x + \beta_2 x^2$$

What is the practical interpretation of β_2? As noted earlier, the parameter β_2 measures the curvature of the response curve shown in Figure 14.3. If $\beta_2 > 0$, the slope of the curve will increase as x increases (upward concavity), as shown in Figure 14.4(a), page 716. If $\beta_2 < 0$, the slope of the curve will decrease as x increases (downward concavity), as shown in Figure 14.4(b).

---

*The $\pm 2s$ approximation will improve as the sample size is increased. We will provide more precise methodology for the construction of prediction intervals in Section 14.6.

FIGURE 14.4

The interpretation of β_2 for a second-order model

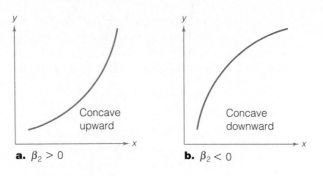

a. $\beta_2 > 0$ **b.** $\beta_2 < 0$

Intuitively, we would expect the electrical usage, y, to rise almost proportionally to home size, x. Then, eventually, as the size of the home increases, the increase in electrical usage for a 1-unit increase in home size might begin to decrease. Thus, a forecaster of electrical usage would want to determine whether this type of curvature actually was present in the response curve, or, equivalently, the forecaster would want to test the null hypothesis

H_0: $\beta_2 = 0$ (No curvature in the response curve)

against the alternative hypothesis

H_a: $\beta_2 < 0$ (Downward concavity exists in the response curve)

A test of this hypothesis can be performed using a Student t-test.

The t-test utilizes a test statistic analogous to that used to make inferences about the slope of the straight-line model (Section 13.5). The t statistic is formed by dividing the sample estimate, $\hat{\beta}_2$, of the parameter, β_2, by the estimated standard deviation of the sampling distribution of $\hat{\beta}_2$:

Test statistic: $t = \dfrac{\hat{\beta}_2}{s_{\hat{\beta}_2}}$

We use the symbol $s_{\hat{\beta}_2}$ to represent the estimated standard deviation of $\hat{\beta}_2$. The formula for computing $s_{\hat{\beta}_2}$ is very complex and its presentation is beyond the scope of this text,* but this omission will not cause difficulty. Most computer packages list the estimated standard deviation $s_{\hat{\beta}_i}$ for each of the estimated model coefficients $\hat{\beta}_i$. Moreover, they usually give the calculated t values for each coefficient in the model.

The rejection region for the test is found in exactly the same way as the rejection regions for the t-tests in previous chapters. That is, we consult Table VI in Appendix A to obtain an upper-tail value of t. This is a value t_α such that $P(t > t_\alpha) = \alpha$. We can then use this value to construct rejection regions for either one-tailed or two-tailed tests. To illustrate, in the electrical usage example the error degrees of freedom is $(n - 3) = 7$, the denominator of the estimate of

---

*Because most of the formulas in a multiple regression analysis are so complex, the only reasonable way to present them is by using matrix algebra. We do not assume a prerequisite of matrix algebra for this text and, in any case, we think the formulas can be omitted in an introductory course without serious loss. They are programmed into almost all standard multiple regression computer packages and are presented in some of the texts listed in the references.

σ^2. Then the rejection region (shown in Figure 14.5) for a one-tailed test with $\alpha = .05$ is

$$\textit{Rejection region:} \quad t < -t_\alpha$$
$$t < -1.895$$

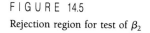

FIGURE 14.5

Rejection region for test of β_2

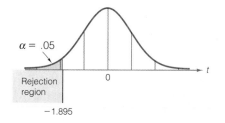

$\alpha = .05$

Rejection region

0

t

-1.895

In Figure 14.6, we again show a portion of the SAS printout for the electrical usage example. The following quantities are shaded:

1. The estimated coefficients, $\hat{\beta}_0$, $\hat{\beta}_1$, and $\hat{\beta}_2$.
2. The SSE and the MSE (estimate of σ^2, variance of ε)
3. The t statistic, observed significance level, and standard error of $\hat{\beta}_2$ for testing $H_0\colon \beta_2 = 0$

FIGURE 14.6

Output from the SAS system

Dep Variable: Y

Analysis of Variance

Source	DF	Sum of Squares	Mean Square	F Value	Prob>F
Model	2	831069.54637	415534.77319	189.710	0.0001
Error	7	15332.55363	2190.36480		
C Total	9	846402.10000			

Root MSE	46.80133	R-Square	0.9819	
Dep Mean	1594.70000	Adj R-Sq	0.9767	
C.V.	2.93480			

Parameter Estimates

Variable	DF	Parameter Estimate	Standard Error	T for H0: Parameter=0	Prob > |T|
INTERCEP	1	-1216.143887	242.80636850	-5.009	0.0016
X	1	2.398930	0.24583560	9.758	0.0001
XSQ	1	-0.000450	0.00005908	-7.618	0.0001

The estimated standard deviations for the model coefficients appear under the column labeled Standard Error. The t statistics for testing the null hypothesis that the true coefficients are equal to 0 appear under the column headed T for H0: Parameter = 0. The t value corresponding to the test of the null hypothesis $H_0\colon \beta_2 = 0$ is the last one in the column, that is, $t = -7.618$. Since this value is less than -1.895, we conclude that the quadratic term $\beta_2 x^2$ makes a contribution to the prediction model of electrical usage.

The SAS printout shown in Figure 14.6 also lists the two-tailed significance levels for each t value. These values appear under the column headed Prob > |T|. The significance level .0001 corresponds to the quadratic term, and this implies that we would reject H_0: $\beta_2 = 0$ in favor of H_a: $\beta_2 \neq 0$ at any α level larger than .0001. Since our alternative was one-sided, H_a: $\beta_2 < 0$, the significance level is half that given in the printout, that is, $\frac{1}{2}(.0001) = .00005$. Thus, there is very strong evidence that the mean electrical usage increases more slowly per square foot for large houses than for small houses.

We can also form a confidence interval for the parameter β_2 as follows:

$$\hat{\beta}_2 \pm t_{\alpha/2}s_{\hat{\beta}_2} = -.000450 \pm (2.365)(.0000591)$$

or $(-.000590, -.000310)$. Note that the t value 2.365 corresponds to $\alpha/2 = .025$ and $(n - 3) = 7$ df. This interval constitutes a 95% confidence interval for β_2 and can be used to estimate the rate of curvature in mean electrical usage as home size is increased. Note that all values in the interval are negative, reconfirming the conclusion of our test.

Note that the computer printout in Figure 14.6 also provides the t-test statistic and corresponding two-tailed p-values for the tests of H_0: $\beta_0 = 0$ and H_0: $\beta_1 = 0$. Since the interpretation of these parameters is not meaningful for this model, the tests are not of interest.

Testing a hypothesis about a single β parameter that appears in any multiple regression model is accomplished in exactly the same manner as described for the quadratic electrical usage model. The t-test and a confidence interval for a β parameter are shown in the next two boxes.

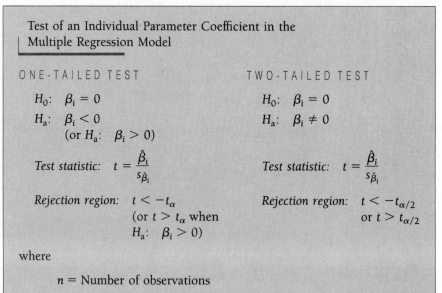

Test of an Individual Parameter Coefficient in the Multiple Regression Model

ONE-TAILED TEST

H_0: $\beta_i = 0$

H_a: $\beta_i < 0$
(or H_a: $\beta_i > 0$)

Test statistic: $t = \dfrac{\hat{\beta}_i}{s_{\hat{\beta}_i}}$

Rejection region: $t < -t_{\alpha}$
(or $t > t_{\alpha}$ when
H_a: $\beta_i > 0$)

TWO-TAILED TEST

H_0: $\beta_i = 0$

H_a: $\beta_i \neq 0$

Test statistic: $t = \dfrac{\hat{\beta}_i}{s_{\hat{\beta}_i}}$

Rejection region: $t < -t_{\alpha/2}$
or $t > t_{\alpha/2}$

where

n = Number of observations

$k + 1$ = Number of β parameters in the model

and t_{α} and $t_{\alpha/2}$ are based on $n - (k + 1)$ degrees of freedom.

Assumptions: See Section 14.3 for assumptions about the probability distribution for the random error component ε.

> A 100($1 - \alpha$)% Confidence Interval for a β Parameter
>
> $$\hat{\beta}_i \pm t_{\alpha/2} s_{\hat{\beta}_i}$$
>
> where $t_{\alpha/2}$ is based on $n - (k + 1)$ degrees of freedom and
>
> $\qquad$ n = Number of observations
>
> $\qquad$ $k + 1$ = Number of β parameters in the model

EXAMPLE 14.1

A collector of antique grandfather clocks knows that the price received for the clocks increases linearly with the age of the clocks. Moreover, the collector hypothesizes that the auction price of the clocks will increase linearly as the number of bidders increases. Thus, the following model is hypothesized:

$$y = \beta_0 + \beta_1 x_1 + \beta_2 x_2 + \varepsilon$$

where

$\qquad$ y = Auction price

$\qquad$ x_1 = Age of clock (years)

$\qquad$ x_2 = Number of bidders

TABLE 14.2 **Auction Price Data**

AGE x_1	NUMBER OF BIDDERS x_2	AUCTION PRICE y	AGE x_1	NUMBER OF BIDDERS x_2	AUCTION PRICE y
127	13	$1,235	170	14	$2,131
115	12	1,080	182	8	1,550
127	7	845	162	11	1,884
150	9	1,522	184	10	2,041
156	6	1,047	143	6	854
182	11	1,979	159	9	1,483
156	12	1,822	108	14	1,055
132	10	1,253	175	8	1,545
137	9	1,297	108	6	729
113	9	946	179	9	1,792
137	15	1,713	111	15	1,175
117	11	1,024	187	8	1,593
137	8	1,147	111	7	785
153	6	1,092	115	7	744
117	13	1,152	194	5	1,356
126	10	1,336	168	7	1,262

A sample of 32 auction prices of grandfather clocks, along with their age and the number of bidders, is given in Table 14.2. The model $y = \beta_0 + \beta_1 x_1 + \beta_2 x_2 + \varepsilon$ is fit to the data, and a portion of the SAS printout is shown in Figure 14.7. Test the hypothesis that the mean auction price of a clock increases as the number of bidders increases when age is held constant, that is, $\beta_2 > 0$. Use $\alpha = .05$.

Solution

The hypotheses of interest concern the parameter β_2. Specifically,

H_0: $\beta_2 = 0$

H_a: $\beta_2 > 0$

Test statistic: $t = \dfrac{\hat{\beta}_2}{s_{\hat{\beta}_2}} = \dfrac{85.8151}{8.7058} = 9.857$

Rejection region: For $\alpha = .05$ and $n - (k + 1) = 32 - (2 + 1) = 29$ df, reject H_0 if $t > 1.699$.

FIGURE 14.7

SAS printout for Example 14.1

Dep Variable: Y

Analysis of Variance

Source	DF	Sum of Squares	Mean Square	F Value	Prob>F
Model	2	4277159.7034	2138579.8517	120.651	0.0001
Error	29	514034.51534	17725.32812		
C Total	31	4791194.2187			

Root MSE	133.13650	R-Square	0.8927		
Dep Mean	1327.15625	Adj R-Sq	0.8853		
C.V.	10.03171				

Parameter Estimates

Variable	DF	Parameter Estimate	Standard Error	T for H0: Parameter=0	Prob > ¦T¦
INTERCEP	1	-1336.722052	173.35612607	-7.711	0.0001
X1	1	12.736199	0.90238049	14.114	0.0001
X2	1	85.815133	8.70575681	9.857	0.0001

This rejection region is shown in Figure 14.8. The calculated t value, $t = 9.857$, exceeds 1.699 and therefore falls in the rejection region. Thus, the collector can conclude that the mean auction price of a clock increases as the number of bidders increases, when age is held constant.

FIGURE 14.8

Rejection region for H_0: $\beta_2 = 0$

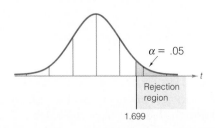

Note that the values $\hat{\beta}_1 = 12.74$ and $\hat{\beta}_2 = 85.82$ (shaded in Figure 14.7) are easily interpreted. We estimate that the mean auction price increases \$12.74 per

year of age of the clock (when the number of bidders is held constant), and that the mean price increases by \$85.82 per additional bidder (when the age of the clock is held constant).

Be careful not to try to interpret the estimated intercept $\hat{\beta}_0 = -1,336.72$ in the same way as we interpreted $\hat{\beta}_1$ and $\hat{\beta}_2$. You might think that this implies a negative price for clocks 0 years of age with 0 bidders. However, these zeros are meaningless numbers in this example, since the ages range from 108 to 194 and the number of bidders ranges from 5 to 15. Interpretations of predicted y values for values of the independent variables outside their sampled range can be very misleading.

Some computer programs use an F statistic (SAS does not) to conduct two-tailed tests concerning the individual β parameters, because the square of a Student t with ν degrees of freedom is equal to an F statistic with 1 degree of freedom in the numerator and ν degrees of freedom in the denominator. If you conduct a two-tailed t-test and reject the null hypothesis if $t > t_{\alpha/2}$ or $t < -t_{\alpha/2}$, the corresponding F-test implies rejection if the computed value of F (which is equal to the square of the computed t statistic) is larger than F_α. Thus,

$$t_{\alpha/2}^2 = F_\alpha$$

As an example, when we tested a hypothesis about the curvature parameter β_2 in the quadratic model relating electrical usage to home size, the computed t value was -7.62 (see Figure 14.6). The equivalent F-test yields a value $F = t^2 = (-7.62)^2 = 58.06$. And the upper-tail rejection region for a two-tailed test with $\alpha = .05$, 1 df in the numerator, and 7 df in the denominator is

$$F > F_{.05} = 5.59$$

Note that the F value, 5.59, is equal to the square of 2.365, the value of t that corresponds to $t_{.025}$ based on 7 df. You can see that the conclusion is the same no matter which test statistic is used: There is very strong evidence that curvature is present in the model.

EXERCISES 14.1–14.11

[*Note: Starred (*) exercises require the use of a computer.*]

LEARNING THE MECHANICS

14.1 SAS was used to fit the model $y = \beta_0 + \beta_1 x_1 + \beta_2 x_2 + \varepsilon$ to $n = 20$ data points and the printout on page 722 was obtained.
a. What are the sample estimates of β_0, β_1, and β_2?
b. What is the least squares prediction equation?
c. Find SSE, MSE, and s. Interpret the standard deviation in the context of the problem.
d. Test $H_0: \beta_1 = 0$ against $H_a: \beta_1 \neq 0$. Use $\alpha = .05$.
e. Use a 95% confidence interval to estimate β_2.

SAS printout for Exercise 14.1

Dep Variable: Y

Analysis of Variance

Source	DF	Sum of Squares	Mean Square	F Value	Prob>F
Model	2	128329.27624	64164.63812	7.223	0.0054
Error	17	151015.72376	8883.27787		
C Total	19	279345.00000			

Root MSE	94.25114	R-Square	0.4594	
Dep Mean	360.50000	Adj R-Sq	0.3958	
C.V.	26.14456			

Parameter Estimates

Variable	DF	Parameter Estimate	Standard Error	T for H0: Parameter=0	Prob > ¦T¦
INTERCEP	1	506.346067	45.16942487	11.210	0.0001
X1	1	-941.900226	275.08555975	-3.424	0.0032
X2	1	-429.060418	379.82566485	-1.130	0.2743

14.2   Suppose you fit the multiple regression model

$$y = \beta_0 + \beta_1 x_1 + \beta_2 x_2 + \beta_3 x_3 + \varepsilon$$

to $n = 30$ data points and obtain the following result:

$$\hat{y} = 3.4 - 4.6x_1 + 2.7x_2 + .93x_3$$

The estimated standard errors of $\hat{\beta}_2$ and $\hat{\beta}_3$ are 1.86 and .29, respectively.
a. Test the null hypothesis H_0: $\beta_2 = 0$ against the alternative hypothesis H_a: $\beta_2 \neq 0$. Use $\alpha = .05$.
b. Test the null hypothesis H_0: $\beta_3 = 0$ against the alternative hypothesis H_a: $\beta_3 \neq 0$. Use $\alpha = .05$.
c. The null hypothesis H_0: $\beta_2 = 0$ is not rejected. In contrast, the null hypothesis H_0: $\beta_3 = 0$ is rejected. Explain how this can happen even though $\hat{\beta}_2 > \hat{\beta}_3$.

14.3   Suppose you fit the second-order model,

$$y = \beta_0 + \beta_1 x + \beta_2 x^2 + \varepsilon$$

to $n = 25$ data points. Your estimate of β_2 is $\hat{\beta}_2 = .47$, and the estimated standard error of the estimate is $s_{\hat{\beta}_2} = .15$.
a. Test the null hypothesis that the mean value of y is related to x by the (*first-order*) linear model

$$E(y) = \beta_0 + \beta_1 x$$

(H_0: $\beta_2 = 0$) against the alternative hypothesis that the true relationship is given by the quadratic model (a *second-order* linear model),

$$E(y) = \beta_0 + \beta_1 x + \beta_2 x^2$$

(H_a: $\beta_2 \neq 0$). Use $\alpha = .05$.
b. Suppose you wanted to determine only whether the quadratic curve opens upward; i.e., as x increases, the slope of the curve increases. Give the test statistic and the rejection region for the test for $\alpha = .05$. Do the data support the theory that the slope of the curve increases as x increases? Explain.

c. What is the value of the F statistic for testing the null hypothesis H_0: $\beta_2 = 0$ against H_a: $\beta_2 \neq 0$?

d. Could the F statistic in part **c** be used to conduct the test in part **b**? Explain.

14.4 How is the number of degrees of freedom available for estimating σ^2 (the variance of ε) related to the number of independent variables in a regression model?

*14.5 Use a computer to fit a second-order model to the following data:

x	0	1	2	3	4	5	6
y	1	2.7	3.8	4.5	5.0	5.3	5.2

a. Find SSE and s^2.

b. Do the data provide sufficient evidence to indicate that the second-order term provides information for the prediction of y? [*Hint:* Test H_0: $\beta_2 = 0$.]

c. Find the least squares prediction equation.

d. Plot the data points and graph $\hat{y}$. Does your prediction equation provide a good fit to the data?

APPLYING THE CONCEPTS

14.6 A researcher wishes to investigate the effects of two independent variables on a teacher's attitude toward handicapped students. A study is conducted involving 40 randomly selected teachers. The response y, a teacher's attitude toward handicapped students, is measured with a standardized attitude scale. Independent variables in the study are

x_1 = Average number of handicapped children taught per year

x_2 = Number of years of teaching experience

The researcher fits the model

$$y = \beta_0 + \beta_1 x_1 + \beta_2 x_2 + \beta_3 x_2^2 + \varepsilon$$

to the data with the following results:

$$\hat{y} = 50 + 1.5x_1 + 5x_2 - .1x_2^2$$
$$s_{\hat{\beta}_3} = .03$$

a. Do these data provide sufficient evidence to indicate that the quadratic term in years of experience, x_2^2, is useful for predicting attitude score? Use $\alpha = .05$.

b. Sketch the predicted attitude score $\hat{y}$ as a function of the number of years of experience x_2 for $x_1 = 5$. Repeat this for $x_1 = 10$. [*Note:* For each value of x_1 ($x_1 = 5$ and $x_1 = 10$), plot $\hat{y}$ for $x_2 = 0, 2, 4, 6, 8$, and 10. Observe that (for this model), the vertical distance between the two prediction curves is the same for all values of x_2.]

14.7 To run a manufacturing operation efficiently, it is necessary to know how long it takes employees to manufacture the product. Without such information, the cost of making the product cannot be determined. Furthermore, management would not be able to establish an effective incentive plan for its employees because it would not know how to set work standards. Estimates of production time are frequently obtained using time studies. The data in the table on page 724 were obtained from a recent time study of a sample of 15 employees on an automobile assembly line.

TIME TO COMPLETE TASK y (minutes)	MONTHS OF EXPERIENCE x	TIME TO COMPLETE TASK y (minutes)	MONTHS OF EXPERIENCE x
10	24	17	3
20	1	18	1
15	10	16	7
11	15	16	9
11	17	17	7
19	3	18	5
11	20	10	20
13	9		

a. The SAS computer printout for fitting the model $y = \beta_0 + \beta_1 x + \beta_2 x^2 + \varepsilon$ is shown here. Find the least squares prediction equation.

SAS Printout for Exercise 14.7

```
DEPENDENT VARIABLE: Y

SOURCE                      DF    SUM OF SQUARES    MEAN SQUARE     F VALUE

MODEL                        2      156.11947722    78.05973861      65.59
ERROR                       12       14.28052278     1.19004356      PR > F
CORRECTED TOTAL             14      170.40000000                     0.0001

R-SQUARE          C.V.           ROOT MSE         Y MEAN

0.916194         7.3709         1.09089118       14.80000000

                                 T FOR H0:      PR > !T!     STD ERROR OF
PARAMETER        ESTIMATE     PARAMETER=0                      ESTIMATE

INTERCEPT      20.09110757        27.72         0.0001        0.72470507
X              -0.67052219        -4.33         0.0010        0.15470634
X*X             0.00953474         1.51         0.1576        0.00632580
```

b. Plot the fitted equation on a scattergram of the data. Is there sufficient evidence to support the inclusion of the quadratric term in the model? Explain.

c. Test the null hypothesis that $\beta_2 = 0$ against the alternative that $\beta_2 \neq 0$. Use $\alpha = .01$. Does the quadratic term make an important contribution to the model?

d. Your conclusion in part **c** should have been to drop the quadratic term from the model. Do so and fit the "reduced model," $y = \beta_0 + \beta_1 x + \varepsilon$, to the data.

e. Define β_1 in the context of this exercise. Find a 90% confidence interval for β_1 in the reduced model of part **d**.

14.8   To project personnel needs for the winter holiday shopping season, a department store wants to project sales for the season. The sales for the previous holiday season are an indication of what to expect for the current season. However, the projection should also reflect the current economic environment by taking into consideration sales for a more recent period. The following model might be appropriate:

$$y = \beta_0 + \beta_1 x_1 + \beta_2 x_2 + \varepsilon$$

where

   x_1 = Previous winter holiday sales

   x_2 = Sales for August of current year

   y = Sales for current winter holiday

Data for 10 previous years were used to fit the prediction equation, and the following were calculated (all units are in thousands of dollars):

$$\hat{\beta}_1 = .62 \qquad s_{\hat{\beta}_1} = .273$$
$$\hat{\beta}_2 = .55 \qquad s_{\hat{\beta}_2} = .181$$

and the regression model standard deviation is $s = 2.5$.

a. Test to determine whether evidence exists to indicate that August sales contribute information for predicting winter holiday sales.

b. If this model were used to predict winter holiday sales, what would the approximate precision of the prediction be? [*Note:* We show how to determine the precision of a multiple regression prediction in Section 14.7.]

14.9 Research was conducted by Tanner (1983) to discover the factors in a person's education that determine future wages. A first-order model was fit to a set of $n = 60$ data points and the following prediction equation and t-test values were obtained:

$$y = 0 - .0945x_1 - .032x_2 + .009x_3 - .0028x_4 + .007x_5 + .105x_6 + .469x_7$$
$$\quad\; (-2.61) \quad (-.96) \quad (2.74) \quad (-2.23) \quad (5.71) \quad (4.02) \quad (2.21)$$

where

y = Future wages

x_1 = Amount of business course work in high school

x_2 = Amount of college prep work in high school

x_3 = Math aptitude

x_4 = High school grade-point average

x_5 = Measure of socioeconomic status

x_6 = 1 if the individual is married and 0 if not

x_7 = Amount of on-the-job training

The t values used to test the individual model parameters are shown in parentheses below their respective estimates.

a. What are the interpretations of the coefficients?

b. Are they statistically significant at the $\alpha = .01$ level?

c. What happens to the intercept term when x_6 is 1?

d. What happens to the intercept term when x_6 is 0?

14.10 Economists have two major types of data available to them: *time series data* and *cross-sectional data*. For example, an economist estimating a consumption function, say, household food consumption as a function of household income and household size, might measure the variables of interest for a particular sample of households at a particular point in time. In this case, the economist is using *cross-sectional data*. If instead, the economist is interested in how total consumption in the United States is related to national income, the economist probably would track these variables over time. In this case, the economist is using *time series data*. The cross-sectional data in the table on the next page have been collected for a random sample of 25 households in Washington, D.C.

a. It has been hypothesized that household food consumption, y, is related to household income, x_1, and to the size of the household, x_2, as follows:

$$y = \beta_0 + \beta_1 x_1 + \beta_2 x_2 + \varepsilon$$

The SAS computer printout for fitting the model to the data is also shown on page 726. Give the least squares prediction equation.

HOUSEHOLD	FOOD CONSUMPTION DURING 1990 ($ thousands)	1990 HOUSEHOLD INCOME ($ thousands)	NUMBER OF PERSONS IN HOUSEHOLD AT END OF 1990	HOUSEHOLD	FOOD CONSUMPTION DURING 1990 ($ thousands)	1990 HOUSEHOLD INCOME ($ thousands)	NUMBER OF PERSONS IN HOUSEHOLD AT END OF 1990
1	3.2	31.1	4	14	3.1	85.2	2
2	2.4	20.5	2	15	4.5	35.6	9
3	3.8	42.3	4	16	3.5	68.5	3
4	1.9	18.9	1	17	4.0	10.5	5
5	2.5	26.5	2	18	3.5	21.6	4
6	3.0	29.8	4	19	1.8	29.9	1
7	2.6	24.3	3	20	2.9	28.6	3
8	3.2	38.1	4	21	2.6	20.2	2
9	3.9	52.0	5	22	3.6	38.7	5
10	1.7	16.0	1	23	2.8	11.2	3
11	2.9	41.9	3	24	4.5	14.3	7
12	1.7	9.9	1	25	3.5	16.9	5
13	4.5	33.1	7				

SAS Printout for Exercise 14.10

```
DEPENDENT VARIABLE: FOOD

SOURCE                      DF    SUM OF SQUARES   MEAN SQUARE   F VALUE

MODEL                        2      15.46228509    7.73114255   100.80
ERROR                       22       1.68731491    0.07669613
CORRECTED TOTAL             24      17.14960000                 PR > F
                                                                0.0001

R-SQUARE                  C.V.            ROOT MSE      FOOD MEAN

0.901612                 8.9221          0.27694067    3.10400000

                              T FOR H0:    PR > !T!   STD ERROR OF
PARAMETER     ESTIMATE     PARAMETER=0              ESTIMATE

INTERCEPT    1.43260377       9.76        0.0001     0.14673751
INCOME       0.00999062       3.15        0.0046     0.00316806
SIZE         0.37928986      13.68        0.0001     0.02772472
```

b. Do the data provide sufficient evidence to conclude that food consumption increases with household income? Test using $\alpha = .01$.

c. As a check on your conclusion of part b, construct a scattergram of household food consumption versus household income. Does the plot support your conclusion in part b? Explain.

d. In Chapter 13, we used the method of least squares to fit a straight line to a set of data points that were plotted in two dimensions. In this exercise, we are fitting a plane to a set of points plotted in three dimensions. We are attempting to determine the plane, $\hat{y} = \hat{\beta}_0 + \hat{\beta}_1 x_1 + \hat{\beta}_2 x_2$, that, according to the principle of least squares, best fits the data points. Sketch the least squares plane you developed in part a. Be sure to label all three axes of your graph.

e. Find the standard error of the regression model on the printout, and interpret it in terms of these data.

*14.11 The owner of an apartment building in Minneapolis believed that her 1990 property tax bill was too high due to an overassessment of the property's value by the city tax assessor. The owner hired an independent real estate appraiser to investigate the appropriateness of the city's assessment. The appraiser used regression analysis to explore the relationship between the sale prices of apartments sold in Minneapolis during 1990 and various characteristics of the properties. Twenty-five apartment buildings were randomly sampled from all apartment buildings that were sold during 1990. The table lists the data collected by the appraiser. The real estate appraiser hypothesized that the sale price (i.e., market value) of an apartment building is related to the other variables in the table according to the following model:

$$y = \beta_0 + \beta_1 x_1 + \beta_2 x_2 + \beta_3 x_3 + \beta_4 x_4 + \beta_5 x_5 + \varepsilon$$

a. Fit the real estate appraiser's model to the data in the table. Report the least squares prediction equation.
b. Find the standard deviation of the regression model, and interpret its value in the context of this problem.
c. Do the data provide sufficient evidence to conclude that value increases with the number of units in an apartment building? Report the observed significance level, and reach a conclusion using $\alpha = .05$.
d. Interpret the value of $\hat{\beta}_1$ in terms of these data. Remember that your interpretation must recognize the presence of the other variables in the model.
e. Construct a scattergram of sale price versus age. What does your scattergram suggest about the relationship between these variables?
f. Test $H_0: \beta_2 = 0$ against $H_a: \beta_2 < 0$ using $\alpha = .01$. Interpret the result in the context of the problem. Does the result agree with your observation in part e? Why is it reasonable to conduct a one-tailed rather than a two-tailed test of this null hypothesis?
g. What is the observed significance level of the hypothesis test of part f?

CODE NO.	SALE PRICE y ($)	NUMBER OF APARTMENT UNITS x_1	AGE OF STRUCTURE x_2 (years)	LOT SIZE x_3 (sq. ft.)	NUMBER OF ON-SITE PARKING SPACES x_4	GROSS BUILDING AREA x_5
0229	90,300	4	82	4,635	0	4,266
0094	384,000	20	13	17,798	0	14,391
0043	157,500	5	66	5,913	0	6,615
0079	676,200	26	64	7,750	6	34,144
0134	165,000	5	55	5,150	0	6,120
0179	300,000	10	65	12,506	0	14,552
0087	108,750	4	82	7,160	0	3,040
0120	276,538	11	23	5,120	0	7,881
0246	420,000	20	18	11,745	20	12,600
0025	950,000	62	71	21,000	3	39,448
0015	560,000	26	74	11,221	0	30,000
0131	268,000	13	56	7,818	13	8,088
0172	290,000	9	76	4,900	0	11,315
0095	173,200	6	21	5,424	6	4,461
0121	323,650	11	24	11,834	8	9,000
0077	162,500	5	19	5,246	5	3,828
0060	353,500	20	62	11,223	2	13,680
0174	134,400	4	70	5,834	0	4,680
0084	187,000	8	19	9,075	0	7,392
0031	155,700	4	57	5,280	0	6,030
0019	93,600	4	82	6,864	0	3,840
0074	110,000	4	50	4,510	0	3,092
0057	573,200	14	10	11,192	0	23,704
0104	79,300	4	82	7,425	0	3,876
0024	272,000	5	82	7,500	0	9,542

Source: Robinson Appraisal Co., Inc., Mankato, Minnesota.

14.5 Checking the Usefulness of a Model: R^2 and the Analysis of Variance F-Test

Conducting t-tests on each β parameter in a model is *not* a good way to determine whether a model is contributing information for the prediction of y. If we were to conduct a series of t-tests to determine whether the independent variables are contributing to the predictive relationship, we would be very likely to make one or more errors in deciding which terms to retain in the model and which to exclude. For example, even if all the β parameters (except β_0) are equal to 0, $100(\alpha)\%$ of the time you will reject the null hypothesis and conclude that some β parameter differs from 0. Thus, in multiple regression models for which a large number of independent variables is being considered, conducting a series of t-tests may include a large number of insignificant variables and exclude some useful ones. If we want to test the utility of a multiple regression model, we will need a global test (one that encompasses all the β parameters). We would also like to find some statistical quantity that measures how well the model fits the data.

We commence with the easier problem—finding a measure of how well a linear model fits a set of data. For this we use the multiple regression equivalent of r^2, the coefficient of determination for the straight-line model (Chapter 13). Thus, we define the **multiple coefficient of determination, R^2**, as

$$R^2 = 1 - \frac{\sum (y - \hat{y})^2}{\sum (y - \bar{y})^2} = 1 - \frac{\text{SSE}}{\text{SS}_{yy}} = \frac{\text{SS}_{yy} - \text{SSE}}{\text{SS}_{yy}} = \frac{\text{Explained variability}}{\text{Total variability}}$$

where $\hat{y}$ is the predicted value of y for the model. Just as for the simple linear model, R^2 represents the fraction of the sample variation of the y values (measured by SS_{yy}) that is explained by the least squares prediction equation. Thus, $R^2 = 0$ implies a complete lack of fit of the model to the data and $R^2 = 1$ implies a perfect fit with the model passing through every data point. In general, the larger the value of R^2, the better the model fits the data.

To illustrate, the value $R^2 = .982$ for the electrical usage example is indicated in Figure 14.9. This very high value of R^2 implies that using the independent variable home size in a quadratic model explains 98.2% of the total **sample variation** (measured by SS_{yy}) of electrical usage y. Thus, R^2 is a sample statistic that tells how well the model fits the data and thereby represents a measure of the usefulness of the entire model.

The fact that R^2 is a sample statistic implies that it can be used to make inferences about the usefulness of the entire model for predicting the population of y values at each setting of the independent variables. In particular, for the electrical usage data the test

H_0: $\beta_1 = \beta_2 = 0$

H_a: At least one of the coefficients is nonzero

would formally test the global usefulness of the model.

The test statistic used to test this hypothesis is an F statistic, and several equivalent versions of the formula can be used (although we will usually rely on the computer to calculate the F statistic):

Test statistic: $F = \dfrac{(\text{SS}_{yy} - \text{SSE})/k}{\text{SSE}/[n - (k + 1)]} = \dfrac{R^2/k}{(1 - R^2)/[n - (k + 1)]}$

FIGURE 14.9

SAS printout for electrical usage example

Dep Variable: Y

Analysis of Variance

Source	DF	Sum of Squares	Mean Square	F Value	Prob>F
Model	2	831069.54637	415534.77319	189.710	0.0001
Error	7	15332.55363	2190.36480		
C Total	9	846402.10000			

Root MSE	46.80133	R-Square	0.9819
Dep Mean	1594.70000	Adj R-Sq	0.9767
C.V.	2.93480		

Parameter Estimates

Variable	DF	Parameter Estimate	Standard Error	T for HO: Parameter=0	Prob > |T|
INTERCEP	1	-1216.143887	242.80636850	-5.009	0.0016
X	1	2.398930	0.24583560	9.758	0.0001
XSQ	1	-0.000450	0.00005908	-7.618	0.0001

Both these formulas indicate that the F statistic is the ratio of the *explained* variability divided by the model degrees of freedom to the *unexplained* variability divided by the error degrees of freedom. Thus, the larger the proportion of the total variability accounted for by the model, the larger the F statistic.

To determine when the ratio becomes large enough that we can confidently reject the null hypothesis and conclude that the model is more useful than no model at all for predicting y, we compare the calculated F statistic to a tabled F value with k df in the numerator and $[n - (k + 1)]$ df in the denominator. Tables of the F-distribution for various values of α are given in Tables VIII, IX, X, and XI of Appendix A.

Rejection region: $F > F_\alpha$, where F is based on k numerator and $n - (k + 1)$ denominator degrees of freedom.

For the electrical usage example $[n = 10, k = 2, n - (k + 1) = 7$, and $\alpha = .05]$, we will reject H_0: $\beta_1 = \beta_2 = 0$ if

$$F > F_{.05} = 4.74$$

From the computer printout (Figure 14.9), we find that the computed F is 189.71. Since this value greatly exceeds the tabulated value of 4.74, we conclude that at least one of the model coefficients β_1 and β_2 is nonzero. Therefore, this global F-test indicates that the quadratic model $y = \beta_0 + \beta_1 x + \beta_2 x^2 + \varepsilon$ is useful for predicting electrical usage.

The F statistic is also given as a part of most regression printouts, usually in a portion of the printout called the "Analysis of Variance." This is an appropriate descriptive term, since the F statistic relates the explained and unexplained portions of the total variance of y. For example, the elements of the SAS computer printout in Figure 14.9 that lead to the calculation of the F value are:

$$F \text{ Value} = \frac{\text{Sum of Squares(Model)}/\text{df(Model)}}{\text{Sum of Squares(Error)}/\text{df(Error)}}$$

$$= \frac{\text{Mean Square(Model)}}{\text{Mean Square(Error)}}$$

From Figure 14.9 we see that F Value = 189.71. Note, too, that the observed significance level for the F statistic is given under the heading Prob > F as .0001, which means that we would reject the null hypothesis H_0: $\beta_1 = \beta_2 = 0$ at any α value greater than .0001.

The analysis of variance F-test for testing the usefulness of the model is summarized in the next box.

Testing Global Usefulness of the Model: The Analysis of Variance F-Test

H_0: $\beta_1 = \beta_2 = \cdots = \beta_k = 0$ (All model terms are unimportant for predicting y)

H_a: At least one $\beta_i \neq 0$ (At least one model term is useful for predicting y)

Test statistic:

$$F = \frac{(SS_{yy} - SSE)/k}{SSE/[n - (k + 1)]} = \frac{R^2/k}{(1 - R^2)/[n - (k + 1)]}$$

$$= \frac{\text{Mean Square(Model)}}{\text{Mean Square(Error)}}$$

where n is the sample size and k is the number of terms in the model.

Rejection region: $F > F_\alpha$, with k numerator degrees of freedom and $[n - (k + 1)]$ denominator degrees of freedom.

Assumptions: The standard regression assumptions about the random error component (Section 14.1).

Caution: A rejection of the null hypothesis leads to the conclusion [with $100(1 - \alpha)\%$ confidence] that the model is useful. However, "useful" does not necessarily mean "best." Another model may prove even more useful in terms of providing more reliable estimates and predictions. This global F-test is usually regarded as a test that the model *must* pass to merit further consideration.

EXAMPLE 14.2

Refer to Example 14.1, in which an antique collector modeled the auction price y of grandfather clocks as a function of the age of the clock, x_1, and the number of bidders, x_2. The hypothesized model is

$$y = \beta_0 + \beta_1 x_1 + \beta_2 x_2 + \varepsilon$$

A sample of 32 observations is obtained, with the results summarized in the SAS printout repeated in Figure 14.10. Discuss the coefficient of determination R^2 for this example, and then conduct the global F-test of model usefulness at the $\alpha = .05$ level of significance.

Solution

The R^2 value is .89 (see Figure 14.10). This implies that the least squares model has explained 89% of the total variation, SS_{yy}. We now test

H_0: $\beta_1 = \beta_2 = 0$ $(k = 2)$

H_a: At least one of the two model coefficients is nonzero

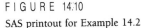

FIGURE 14.10

SAS printout for Example 14.2

Dep Variable: Y

Analysis of Variance

Source	DF	Sum of Squares	Mean Square	F Value	Prob>F
Model	2	4277159.7034	2138579.8517	120.651	0.0001
Error	29	514034.51534	17725.32812		
C Total	31	4791194.2187			

Root MSE	133.13650	R-Square	0.8927	
Dep Mean	1327.15625	Adj R-Sq	0.8853	
C.V.	10.03171			

Parameter Estimates

Variable	DF	Parameter Estimate	Standard Error	T for H0: Parameter=0	Prob > |T|
INTERCEP	1	-1336.722052	173.35612607	-7.711	0.0001
X1	1	12.736199	0.90238049	14.114	0.0001
X2	1	85.815133	8.70575681	9.857	0.0001

Test statistic: $F = \dfrac{R^2/k}{(1 - R^2)/[n - (k + 1)]}$

Rejection region: $F > F_\alpha$

For this example, $n = 32$, $k = 2$, and $n - (k + 1) = 32 - 3 = 29$. Then, for $\alpha = .05$, we will reject H_0: $\beta_1 = \beta_2 = 0$ if $F > F_{.05}$, where F is based on $k = 2$ numerator and $n - (k + 1) = 29$ denominator degrees of freedom—i.e., if $F > 3.33$. The computed value of the F-test statistic is 120.65 (see Figure 14.10). Since this value of F falls in the rejection region ($F = 120.65$ greatly exceeds $F_{.05} = 3.33$), the data provide strong evidence that at least one of the model coefficients is nonzero. The model appears to be useful for predicting auction prices.

Can we be sure that the best prediction model has been found if the global F-test indicates that a model is useful? Unfortunately, we cannot. The addition of other independent variables may improve the usefulness of the model, as Example 14.3 indicates.

EXAMPLE 14.3

Refer to Examples 14.1 and 14.2. Suppose the collector, having observed many auctions, believes that the *rate of increase* of the auction price with age will be driven upward by a large number of bidders. Thus, instead of a relationship like that shown in Figure 14.11(a) (page 732), in which the rate of increase in price with age is the same for any number of bidders, the collector believes the relationship is like that shown in Figure 14.11(b). Note that as the number of bidders increases from 5 to 15, the slope of the price versus age line increases. When the slope of the relationship between y and one independent variable (x_1) depends

FIGURE 14.11

Examples of no interaction and
interaction models

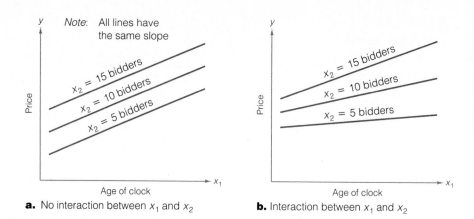

a. No interaction between x_1 and x_2 **b.** Interaction between x_1 and x_2

on the value of a second independent variable (x_2), as is the case here, we say
that x_1 and x_2 **interact**.* The model that includes interaction is written

$$y = \beta_0 + \beta_1 x_1 + \beta_2 x_2 + \beta_3 x_1 x_2 + \varepsilon$$

Note that the increase in the mean price, $E(y)$, for each 1-year increase in age,
x_1, is no longer given by the constant β_1 but is now $\beta_1 + \beta_3 x_2$. *That is, the
amount $E(y)$ increases for each 1-unit increase in x_1 is dependent on the number of
bidders, x_2. Thus, the two variables x_1 and x_2 interact to affect y.*

The 32 data points listed in Table 14.2 were used to fit the model with
interaction. A portion of the SAS printout is shown in Figure 14.12. Test the
hypothesis that the price–age slope increases as the number of bidders increases—
i.e., that age and number of bidders, x_2, interact positively.

FIGURE 14.12

Portion of the SAS printout for the model
with interaction

Dep Variable: Y

Analysis of Variance

Source	DF	Sum of Squares	Mean Square	F Value	Prob>F
Model	3	4572547.9872	1524182.6624	195.188	0.0001
Error	28	218646.23157	7808.79398		
C Total	31	4791194.2187			

Root MSE	88.36738	R-Square	0.9544	
Dep Mean	1327.15625	Adj R-Sq	0.9495	
C.V.	6.65840			

Parameter Estimates

Variable	DF	Parameter Estimate	Standard Error	T for H0: Parameter=0	Prob > |T|
INTERCEP	1	322.754353	293.32514660	1.100	0.2806
X1	1	0.873288	2.01965115	0.432	0.6688
X2	1	-93.409920	29.70767946	-3.144	0.0039
X1X2	1	1.297898	0.21102602	6.150	0.0001

*A detailed discussion of interaction is given in Chapter 15.

Solution

The model is

$$y = \beta_0 + \beta_1 x_1 + \beta_2 x_2 + \beta_3 x_1 x_2 + \varepsilon$$

and the hypotheses of interest to the collector concern the parameter β_3. Specifically,

H_0: $\beta_3 = 0$

H_a: $\beta_3 > 0$

Test statistic: $t = \dfrac{\hat{\beta}_3}{s_{\hat{\beta}_3}}$

Rejection region: For $\alpha = .05$, $t > t_{.05}$

where $n = 32$, $k = 3$, and $t_{.05} = 1.701$, based on $n - (k + 1) = 28$ df. [Remember: $(k + 1) = 4$ is the number of parameters in the regression model.]

The t value corresponding to $\hat{\beta}_3$ is indicated in Figure 14.12. The value $t = 6.15$ exceeds 1.701 and therefore falls in the rejection region. Thus, the collector can conclude that the rate of change of the mean price of the clocks with age increases as the number of bidders increases; that is, x_1 and x_2 interact. Thus, it appears that the interaction term should be included in the model.

One note of caution: Although the coefficient of x_2 is negative ($\hat{\beta}_2 = -93.41$), this does *not* imply that auction price decreases as the number of bidders increases. Since interaction is present, the rate of change (slope) of mean auction price with the number of bidders *depends on* x_1, the age of the clock. Thus, for example, the estimated rate of change of y for a unit increase in x_2 (one new bidder) for a 150-year-old clock is

$$\text{Estimated } x_2 \text{ slope} = \hat{\beta}_2 + \hat{\beta}_3 x_1 = -93.41 + 1.30(150)$$
$$= 101.59$$

In other words, we estimate that the auction price of a 150-year-old clock will *increase* by about \$101.59 for every additional bidder. Although the rate of increase will vary as x_1 is changed, it will remain positive for the range of values of x_1 included in the sample. Extreme care is needed in interpreting the signs and sizes of coefficients in a multiple regression model.

To summarize the discussion in this section, the value of R^2 is an indicator of how well the prediction equation fits the data. More importantly, it can be used in the F-statistic to determine whether the data provide sufficient evidence to indicate that the model contributes information for the prediction of y. Intuitive evaluations of the contribution of the model based on the computed value of R^2 must be examined with care. The value of R^2 increases as more and more variables are added to the model. Consequently, you could force R^2 to take a value very close to 1 even though the model contributes no information for the prediction of y. In fact, R^2 equals 1 when the number of terms in the model equals the number of data points. Therefore, you should not rely solely on the value of R^2 to tell you whether the model is useful for predicting y. Use the F-test.

EXERCISES 14.12–14.27

[*Note: Starred (*) exercises require the use of a computer.*]

14.12   The model $y = \beta_0 + \beta_1 x + \beta_2 x^2 + \varepsilon$ was fit to $n = 19$ data points with the results shown in the accompanying printout.

SAS printout for Exercise 14.12

Dep Variable: Y

Analysis of Variance

Source	DF	Sum of Squares	Mean Square	F Value	Prob>F
Model	2	24.22335	12.11167	65.478	0.0001
Error	16	2.95955	0.18497		
C Total	18	27.18289			

Root MSE	0.43008	R-Square	0.8911	
Dep Mean	3.56053	Adj R-Sq	0.8775	
C.V.	12.07921			

Parameter Estimates

Variable	DF	Parameter Estimate	Standard Error	T for H0: Parameter=0	Prob > |T|
INTERCEP	1	0.734606	0.29313351	2.506	0.0234
X	1	0.765179	0.08754136	8.741	0.0001
XSQ	1	-0.030810	0.00452890	-6.803	0.0001

a. Find R^2 and interpret its value.

b. Test the null hypothesis that $\beta_1 = \beta_2 = 0$ against the alternative hypothesis that at least one of β_1 and β_2 is nonzero. Calculate the test statistic using the two formulas given in this section, and compare your results to each other and to that given on the printout. Use $\alpha = .05$ and interpret the result of your test.

c. Find the observed significance level for this test on the printout, and interpret it.

d. Test H_0: $\beta_2 = 0$ against H_a: $\beta_2 \neq 0$. Use $\alpha = .05$ and interpret the result of your test. Report and interpret the observed significance level of the test.

14.13   Suppose you fit the model

$$y = \beta_0 + \beta_1 x_1 + \beta_2 x_2 + \beta_3 x_1 x_2 + \beta_4 x_1^2 + \beta_5 x_2^2 + \varepsilon$$

to $n = 30$ data points and obtain

$$SSE = .33 \qquad R^2 = .92$$

a. Do the values of SSE and R^2 suggest that the model provides a good fit to the data? Explain.

b. Is the model of any use in predicting y? Test the null hypothesis that $E(y) = \beta_0$, that is,

$$H_0: \quad \beta_1 = \beta_2 = \cdots = \beta_5 = 0$$

against the alternative hypothesis

H_a: At least one of the parameters $\beta_1, \beta_2, \ldots, \beta_5$ is nonzero

Use $\alpha = .05$.

14.14   Suppose you fit the model

$$y = \beta_0 + \beta_1 x_1 + \beta_2 x_2 + \varepsilon$$

to $n = 20$ data points and obtain

$$\sum (y_i - \hat{y}_i)^2 = 1.47 \qquad \sum (y_i - \bar{y})^2 = 2.73$$

a. Construct an analysis of variance table for this regression analysis, using the printout in Exercise 14.12 as a model. Be sure to include the sources of variability, the degrees of freedom, the sums of squares, the mean squares, and the F statistic. Calculate R^2 for the regression analysis.

b. Test the null hypothesis that $\beta_1 = \beta_2 = 0$ against the alternative hypothesis that at least one of the parameters differs from 0. Calculate the test statistic in two different ways and compare the results. Use $\alpha = .05$ to reach a conclusion about whether the model contributes information for the prediction of y.

14.15   If the analysis of variance F-test leads to the conclusion that at least one of the model parameters is nonzero, can you conclude that the model is the best predictor for the dependent variable y? Can you conclude that all of the terms in the model are important for predicting y? What is the appropriate conclusion?

APPLYING THE CONCEPTS

14.16   K. D. Fox and S. Y. Nickols report on a study of the impact of a wife's employment on the number of hours available for household tasks. Some 206 families were employed in the study. One independent variable measured for each family was the total time y (in minutes per day) that the wife spent on household chores. Two independent variables were also recorded:

WEMP: $x_1 = $ Wife's hours of employment per week

YAGE: $x_2 = $ Age of the youngest child in the family

The authors explain that they first attempted to fit the model

$$E(y) = \beta_0 + \beta_1 x_1 + \beta_2 x_2 + \beta_3 x_1 x_2$$

If the regression analysis did not indicate statistical significance for the interaction term, it was eliminated from the model. Looking at the computer printout on page 736, you can see that the interaction term was eliminated from their final regression analysis.

a. Based on Fox and Nickols's computer printout, does the model contribute information for the prediction of y?

b. Find R^2 and give its practical implications.

c. Do the data provide sufficient information to indicate that the model contributes information for the prediction of y? Test using $\alpha = .05$.

d. In concluding, the researchers state that "for each additional hour of employment, wives decreased household work time 4 minutes per day." Do you agree? [*Note*: We will test H_0: $\beta_1 = 0$ in Exercise 15.37.]

Regression analyses of wives' and husbands' time spent in household work

DEPENDENT VARIABLE: WIFE'S HOUSEHOLD WORK		(MEAN=401.3[a])			
SOURCE OF VARIATION	DF	SUMS OF SQUARES	MEAN SQUARE	F	R²
TOTAL	205	5,432,296		62.10***	0.38
MODEL	2	2,061,918	1,030,959		
ERROR	203	3,370,378	16,603		
SOURCE WITHIN MODEL	DF	SEQUENTIAL S.S.	PARTIAL S.S.	b-ESTIMATES	
WEMP	1	1,810,154***	1,110,288***	-4.08	
YAGE	1	251,765***	251,765***	-7.02	

DEPENDENT VARIABLE: HUSBAND'S WORK   (MEAN=105.2[a])
REGRESSION ANALYSIS NOT SIGNIFICANT

Note: In this particular printout "b-estimate" is our β estimate, or parameter estimate. Also, the asterisks indicate the level of significance of test statistics (or sum of squares used to compute test statistics). Here, *** means $p < .001$.

[a]Minutes per day.

Source: Fox, K. D., and Nickols, S. Y. "The time crunch," *Journal of Family Issues.* March 1983, Vol. 4, No. 1, pp. 61–82. Copyright 1983 by Sage Publications. Reprinted by permission of Sage Publications, Inc.

14.17 In hopes of increasing the company's share of the fine food market, researchers for a meat-processing firm are working to improve the quality of its hickory-smoked hams. One of their studies concerns the effect on the flavor of the ham of time spent in the smokehouse. Hams that were in the smokehouse for varying amounts of time were each subjected to a taste test by a panel of ten food experts. The following model was thought to be appropriate by the researchers:

$$y = \beta_0 + \beta_1 t + \beta_2 t^2 + \varepsilon$$

where

y = Mean of the taste scores for the ten experts

t = Time in the smokehouse (hours)

A sample of 20 hams yielded the following least squares model:

$$\hat{y} = 20.3 + 5.2t - .0025t^2$$

where $s_{\hat{\beta}_2} = .0011$. The coefficient of determination is $R^2 = .79$.

a. Is there evidence to indicate that the overall model is useful? Test at $\alpha = .05$.
b. Is there evidence to indicate that the quadratic term is important in this model? Test at $\alpha = .05$.

14.18 Refer to Exercise 14.6. Recall that the dependent variable y is a teacher's attitude toward handicapped students as measured by a standardized attitude scale. The independent variables are the average number of handicapped children taught per year (x_1) and the teacher's years of experience (x_2). Suppose the interaction terms ($x_1 x_2$ and $x_1 x_2^2$) are added to the model proposed in Exercise 14.6 to produce

$$y = \beta_0 + \beta_1 x_1 + \beta_2 x_2 + \beta_3 x_2^2 + \beta_4 x_1 x_2 + \beta_5 x_1 x_2^2 + \varepsilon$$

This model is fit to the same 40 observations used in Exercise 14.6 with the result

$$\hat{y} = 50 + x_1 + 6x_2 - .2x_2^2 + x_1 x_2 - .05 x_1 x_2^2 \quad \text{and} \quad R^2 = .87$$

a. Interpret the value of R^2.

b. Is there sufficient evidence to indicate that this model is useful for predicting attitude score? Test $H_0: \beta_1 = \beta_2 = \beta_3 = \beta_4 = \beta_5 = 0$ using $\alpha = .05$.

c. Sketch the predicted attitude score $\hat{y}$ as a function of the number of years of experience, x_2, for teachers who have taught an average of ten handicapped children per year ($x_1 = 10$). Repeat this for $x_1 = 5$. Compare these sketches with those obtained for the noninteraction model of Exercise 14.6. [*Note:* In Chapter 15 we test to determine whether this model provides more information for the prediction of y than the noninteraction model.]

14.19   Carson Bays used regression analysis to investigate the determinants of survival size of nonprofit hospitals. For a given sample of hospitals, survival size, y, is defined as the largest size hospital (in terms of number of beds) exhibiting growth in market share over a specific time interval. Suppose ten states are randomly selected and the survival size for all nonprofit hospitals in each state is determined for the time period 1981–1982 and for 1984–1985, yielding two observations per state. The 20 survival sizes are listed in the accompanying table, along with the following data for each state, for the second year in each time interval:

x_1 = Percentage of beds that are in for-profit hospitals

x_2 = Ratio of the number of persons enrolled in health maintenance organizations (HMOs) to the number of persons covered by hospital insurance

x_3 = State population (in thousands)

x_4 = Percent of state that is urban

STATE	TIME PERIOD	SURVIVAL SIZE y	x_1	x_2	x_3	x_4
1	1	370	.13	.09	5,800	89
1	2	390	.15	.09	5,955	87
2	1	455	.08	.11	17,648	87
2	2	450	.10	.16	17,895	85
3	1	500	.03	.04	7,332	79
3	2	480	.07	.05	7,610	78
4	1	550	.06	.005	11,731	80
4	2	600	.10	.005	11,790	81
5	1	205	.30	.12	2,932	44
5	2	230	.25	.13	3,100	45
6	1	425	.04	.01	4,148	36
6	2	445	.07	.02	4,205	38
7	1	245	.20	.01	1,574	25
7	2	200	.30	.01	1,560	28
8	1	250	.07	.08	2,471	38
8	2	275	.08	.10	2,511	38
9	1	300	.09	.12	4,060	52
9	2	290	.12	.20	4,175	54
10	1	280	.10	.02	2,902	37
10	2	270	.11	.05	2,925	38

Source: Adapted from Bays, C. W. "The determinants of hospital size: A survivor analysis." *Applied Economics*, Vol. 18, 1986, pp. 359–377.

Bays hypothesized that the following model characterizes the relationship between survival size and the four variables just listed:

$$y = \beta_0 + \beta_1 x_1 + \beta_2 x_2 + \beta_3 x_3 + \beta_4 x_4 + \varepsilon$$

a. The model was fit to the data in the table using the SAS System with the results given in the printout. Report the least squares prediction equation.

SAS printout for Exercise 14.19

Dep Variable: Y

Analysis of Variance

Source	DF	Sum of Squares	Mean Square	F Value	Prob>F
Model	4	246537.05939	61634.26485	28.180	0.0001
Error	15	32807.94061	2187.19604		
C Total	19	279345.00000			

Root MSE	46.76747	R-Square	0.8826
Dep Mean	360.50000	Adj R-Sq	0.8512
C.V.	12.97295		

Parameter Estimates

Variable	DF	Parameter Estimate	Standard Error	T for H0: Parameter=0	Prob > |T|
INTERCEP	1	295.327091	40.17888737	7.350	0.0001
X1	1	-480.837576	150.39050364	-3.197	0.0060
X2	1	-829.464955	196.47303539	-4.222	0.0007
X3	1	0.007934	0.00355335	2.233	0.0412
X4	1	2.360769	0.76150774	3.100	0.0073

b. Find the regression standard deviation s, and interpret its value in the context of the problem.
c. Use an F-test to investigate the usefulness of the hypothesized model. Report the observed significance level, and use $\alpha = .025$ to reach your conclusion.
d. Prior to collecting the data it was hypothesized that increases in the number of for-profit hospital beds would decrease the survival size of nonprofit hospitals. Do the data support this hypothesis? Test using $\alpha = .05$.

14.20   Because the coefficient of determination R^2 always increases when a new independent variable is added to the model, it is tempting to include many variables in a model to force R^2 to be near 1. However, doing so reduces the degrees of freedom available for estimating σ^2, which adversely affects our ability to make reliable inferences. Suppose you want to use 18 economic indicators to predict next year's GNP. You fit the model

$$y = \beta_0 + \beta_1 x_1 + \beta_2 x_2 + \cdots + \beta_{17} x_{17} + \beta_{18} x_{18} + \varepsilon$$

where y = GNP and $x_1, x_2, \ldots, x_{18}$ are indicators. Only 20 years of data ($n = 20$) are used to fit the model, and you obtain $R^2 = .95$. Test to see whether this impressive-looking R^2 is large enough for you to infer that the model is useful, i.e., that at least one term in the model is important for predicting GNP. Use $\alpha = .05$.

14.21   The length of a mosquito's proboscis plays a large role in determining its feeding habits. An entomologist has proposed the following model for predicting the length of a mosquito's proboscis:

$$E(y) = \beta_0 + \beta_1 x_1 + \beta_2 x_2 + \beta_3 x_3$$

where

    y = Length of proboscis (millimeters)
    x_1 = Dry weight (milligrams)

$x_2 = $ Length of wing (millimeters)

$x_3 = $ Width of wing (millimeters)

An analysis of data obtained from a sample of 44 mosquitos of a certain species produces the least squares model

$$\hat{y} = .968 + .292x_1 + .614x_2 - .201x_3$$

Also,

$$s_{\hat{\beta}_1} = .248 \qquad s_{\hat{\beta}_2} = .131 \qquad s_{\hat{\beta}_3} = .267 \qquad R^2 = .536$$

a. Do these statistics indicate that the model is useful in predicting proboscis length? Use $\alpha = .10$.

b. Given the first-order model specified above, do the data provide sufficient evidence to indicate that x_3 (width of wing) is an important variable for predicting y? Test using $\alpha = .05$.

*14.22 In Exercise 13.21, we found the least squares line relating the age x of a sand lance to its length y. The data are shown in the table.

MEAN LENGTH (MILLIMETERS)	176	194	212	226	236	244	254
AGE (YEARS)	2	3	4	5	6	7	8

a. Use a computer to fit a second-order model to the data.

b. Do the data provide sufficient evidence to indicate that the second-order term contributes information for the prediction of sand lance length? Test using $\alpha = .05$.

c. Find the p-value for the test in part b.

d. Find the coefficient of determination for the second-order model and interpret it.

14.23 In Exercise 13.75, advanced ticket sales were used to predict the number of hot dogs purchased at a baseball stadium during the coming week. Another variable that might indicate attendance and help predict food needs is the visiting team's standing in their division. The data in the table were collected for eight games at the stadium.

HOT DOGS PURCHASED DURING WEEK y (thousands)	ADVANCED TICKET SALES FOR WEEK x_1 (thousands)	VISITING TEAM'S STANDING x_2
54.2	55.6	1
48.9	63.5	3
50.3	60.1	2
56.1	58.9	1
20.0	45.4	6
53.8	66.6	2
46.4	59.3	4
52.7	64.8	2

a. Fit the following model to the data:

$$y = \beta_0 + \beta_1 x_1 + \beta_2 x_2 + \beta_3 x_1 x_2 + \varepsilon$$

b. Find R^2 for the least squares equation of part a. Interpret your result.

c. Is there sufficient evidence to indicate that the overall model is useful for predicting y? Test using $\alpha = .05$.

d. Is there sufficient evidence to indicate that the interaction term should be included in the model? Test using $\alpha = .05$.

e. Interpret the values of $\hat{\beta}_0$, $\hat{\beta}_1$, $\hat{\beta}_2$, and $\hat{\beta}_3$ found in part **a**.

14.24   If producers (providers) of goods (services) are able to reduce the unit cost of their goods by increasing the scale of their operation, they are the beneficiaries of an economic force known as *economies of scale*. Economies of scale cause a firm's long-run average costs to decline. The question of whether economies of scale, diseconomies of scale, or neither (i.e., constant economies of scale) exist in the U.S. motor freight, common carrier industry has been debated for years. In an effort to settle the debate within a specific subsection of the trucking industry, Sugrue, Ledford, and Glaskowsky (1982) used regression analysis to model the relationship between each of a number of profitability/cost measures and the size of the operation. In one case, they modeled expense per vehicle-mile, y_1, as a function of the firm's total revenue, x. In another case, they modeled expense per ton-mile, y_2, as a function of x. Data were collected from 264 firms and the least squares results obtained are shown in the accompanying table.

DEPENDENT VARIABLE	$\hat{\beta}_0$	$\hat{\beta}_1$	r	F
Expense per vehicle-mile	2.279	$-.00000069$	$-.0783$	1.616
Expense per ton-mile	.1680	$-.000000066$	$-.0902$	2.148

a. Investigate the usefulness of the two models estimated by Sugrue, Ledford, and Glaskowsky. Use $\alpha = .05$. Draw the appropriate conclusions in the context of the problem.

b. Are the observed significance levels of the hypothesis tests you conducted in part **a** greater than .10 or less than .10? Explain.

c. What do your hypothesis tests of part **a** suggest about economies of scale in the subsection of the trucking industry investigated by Sugrue, Ledford, and Glaskowsky—the long-haul, heavy-load, intercity-general-freight common carrier sector? Explain.

*14.25   Refer to Exercise 14.11, in which a regression model was used to explore the valuation of apartment buildings in Minneapolis.

a. Find R^2 for the least squares prediction equation found in part **a** of Exercise 14.11. Interpret its value in the context of the problem.

b. Use an F-test to investigate the usefulness of the model hypothesized by the real estate appraiser. Use $\alpha = .01$, and carefully state your conclusion in the context of this application.

c. What is the observed significance level of the F-test?

14.26   Writing in *Accounting Review*, Benston (1966) describes how multiple regression can be used by accountants in cost analysis. He points out that multiple regression models can be used to shed light on "the factors that cause costs to be incurred and the magnitudes of their effects" (p. 658). The independent variables of such a regression model are the factors believed to be related to cost, the dependent variable. The estimates of the coefficients of the regression model provide measures of the magnitude of the factors' effects on cost. In some instances, however, Benston notes that it may be desirable to use physical units instead of cost as the dependent variable in a cost analysis. This would be the case if most of the cost associated with the activity of interest is a function of some physical unit, such as hours of labor. The advantage of this approach is that the regression model will provide estimates of the number of labor hours required under different circumstances, and these hours can then be costed at the current labor rate.

The sample data shown in the table have been collected from a firm's accounting and production records to provide cost information about the firm's shipping department. The variables for which data were collected were suggested by Benston.

The SAS printout for fitting the model $y = \beta_0 + \beta_1 x_1 + \beta_2 x_2 + \beta_3 x_3 + \varepsilon$ to the data is also shown.

WEEK	HOURS OF LABOR y	THOUSANDS OF POUNDS SHIPPED x_1	PERCENTAGE OF UNITS SHIPPED BY TRUCK x_2	AVERAGE NUMBER OF POUNDS PER SHIPMENT x_3
1	100	5.1	90	20
2	85	3.8	99	22
3	108	5.3	58	19
4	116	7.5	16	15
5	92	4.5	54	20
6	63	3.3	42	26
7	79	5.3	12	25
8	101	5.9	32	21
9	88	4.0	56	24
10	71	4.2	64	29
11	122	6.8	78	10
12	85	3.9	90	30
13	50	3.8	74	28
14	114	7.5	89	14
15	104	4.5	90	21
16	111	6.0	40	20
17	110	8.1	55	16
18	100	2.9	64	19
19	82	4.0	35	23
20	85	4.8	58	25

SAS printout for Exercise 14.26

DEPENDENT VARIABLE: LABOR

SOURCE		DF	SUM OF SQUARES	MEAN SQUARE	F VALUE
MODEL		3	5158.31382780	1719.43794260	17.87
ERROR		16	1539.88617220	96.24288576	PR > F
CORRECTED TOTAL		19	6698.20000000		0.0001

R-SQUARE	C.V.	ROOT MSE	LABOR MEAN
0.770104	10.5148	9.81034585	93.30000000

PARAMETER	ESTIMATE	T FOR H0: PARAMETER=0	PR > !T!	STD ERROR OF ESTIMATE
INTERCEPT	131.92425208	5.13	0.0001	25.69321439
WEIGHT	2.72608977	1.20	0.2483	2.27500488
TRUCK	0.04721841	0.51	0.6199	0.09334856
AVGSHIP	-2.58744391	-4.03	0.0010	0.64281819

a. Find the least squares prediction equation.
b. Use an F-test to investigate the usefulness of the model specified in part a. Use $\alpha = .01$, and state your conclusion in the context of the problem.
c. Test $H_0: \beta_2 = 0$ versus $H_a: \beta_2 \neq 0$ using $\alpha = .05$. What do the results of your test suggest about the magnitude of the effects of x_2 on labor costs?
d. Find R^2, and interpret its value in the context of the problem.
e. If shipping department employees are paid \$7.50 per hour, how much less, on average, will it cost the company per week if the average number of pounds per shipment increases from a level of 20 to 21? Assume that x_1 and x_2 remain unchanged. Your answer to this question is an estimate of what is known in economics as the *expected marginal cost* associated with a 1-pound increase in x_3.

f. With what approximate precision can this model be used to predict the hours of labor? [*Note:* The precision of multiple regression predictions is discussed in Section 14.7.]

*14.27   Regression analysis can be used to model the relationship between the selling price of a house and its total living area (for more details, see Exercise 13.86). The table contains the final selling prices and total living areas for a sample of 30 houses that were sold during 1990 in the same geographic area.

AREA (sq. ft.)	PRICE	AREA (sq. ft.)	PRICE	AREA (sq. ft.)	PRICE
1,600	$101,900	1,910	$104,000	2,390	$123,300
1,980	108,700	1,810	104,500	2,560	133,000
2,130	114,500	2,035	112,000	2,370	125,000
1,990	110,300	3,000	210,000	2,420	123,300
1,710	100,900	2,500	160,000	2,300	119,200
2,357	121,100	2,120	113,600	2,320	120,000
2,335	118,700	1,880	106,500	2,470	126,800
2,650	137,200	1,775	105,000	2,460	127,100
2,070	109,300	1,650	99,000	3,650	225,100
1,930	108,400	2,250	116,100	2,750	205,600

a. Construct a scattergram for the data. Is there evidence to suggest that it would be inappropriate to represent the relationship between price and area with a straight line? Explain.

b. Fit the following model to the data:

$$y = \beta_0 + \beta_1 x + \beta_2 x^2 + \varepsilon$$

where y = price and x = area. Plot the fitted equation on the scattergram of part **a**.

c. Find the standard deviation for the least squares equation of part **b**. Interpret the value in the context of this application.

d. Use an F-test to investigate whether the model is useful for predicting y. Test using $\alpha = .05$.

e. Do the data provide sufficient evidence to indicate that the second-order term (x^2) contributes information for the prediction of y? Test using $\alpha = .05$.

14.6   Using the Model for Estimation and Prediction

In Section 13.8 we discussed the use of the least squares line for estimating the mean value of y, $E(y)$, for some particular value of x, say $x = x_p$. We also showed how to use the same fitted model to predict, when $x = x_p$, some new value of y to be observed in the future. Recall that the least squares line yielded the same value for both the estimate of $E(y)$ and the prediction of some future value of y. That is, both are the result of substituting x_p into the prediction equation, $\hat{y} = \hat{\beta}_0 + \hat{\beta}_1 x$ and calculating $\hat{y}_p$. There the equivalence ends. The confidence interval for the mean $E(y)$ is narrower than the prediction interval for y because of the additional uncertainty attributable to the random error ε when predicting some future value of y.

These same concepts carry over to the multiple regression model. Suppose we want to estimate the mean electrical usage for a given home size, say $x_p = 1,500$ square feet. Assuming that the quadratic model represents the true relationship between electrical usage and home size, we want to estimate

$$E(y) = \beta_0 + \beta_1 x_p + \beta_2 x_p^2$$
$$= \beta_0 + \beta_1 (1,500) + \beta_2 (1,500)^2$$

Substituting into the least squares prediction equation, we find the estimate of $E(y)$ to be

$$\hat{y} = \hat{\beta}_0 + \hat{\beta}_1(1,500) + \hat{\beta}_2(1,500)^2$$
$$= -1,216.144 + 2.3989(1,500) - .00045004(1,500)^2$$
$$= 1,369.7$$

To form a confidence interval for the mean, we need to know the standard deviation of the sampling distribution for the estimator $\hat{y}$. For multiple regression models, the form of this standard deviation is rather complex. However, the SAS regression package allows us to obtain the confidence intervals for mean values of y for any given combination of values of the independent variables. This portion of the SAS output for the electrical usage example is shown in Figure 14.13. The mean value and corresponding 95% confidence interval for $x_p = 1,500$ are shown in the columns labeled ESTIMATED MEAN VALUE, LOWER 95% CL FOR MEAN, and UPPER 95% CL FOR MEAN. Note that

$$\hat{y} = 1,369.7$$

which agrees with our earlier calculation. The 95% confidence interval for the true mean of y is shown to be 1,325.0 to 1,414.3 (see Figure 14.14).

FIGURE 14.13

SAS printout for estimated mean value and corresponding confidence interval for $x_p = 1,500$

X	ESTIMATED MEAN VALUE	LOWER 95% CL FOR MEAN	UPPER 95% CL FOR MEAN
1500	1369.66088739	1324.98831001	1414.33346477

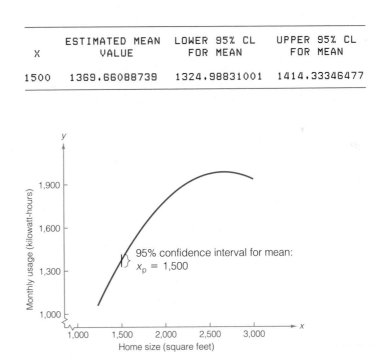

FIGURE 14.14

Confidence interval for mean electrical usage

If we were interested in predicting the electrical usage for a particular 1,500-square-foot home, $\hat{y} = 1,369.7$ would be used as the predicted value. However, the prediction interval for a new value of y is wider than the confidence interval for the mean value. This is reflected by the printout shown in Figure 14.15 (page 744), which gives the predicted value of y and corresponding 95% prediction interval when $x_p = 1,500$ square feet. The prediction interval for $x_p = 1,500$ is 1,250.3 to 1,489.0 (see Figure 14.16, next page).

FIGURE 14.15

SAS printout for predicted value and corresponding prediction interval for $x_p = 1,500$

X	PREDICTED VALUE	LOWER 95% CL INDIVIDUAL	UPPER 95% CL INDIVIDUAL
1500	1369.66088739	1250.31627944	1489.00549533

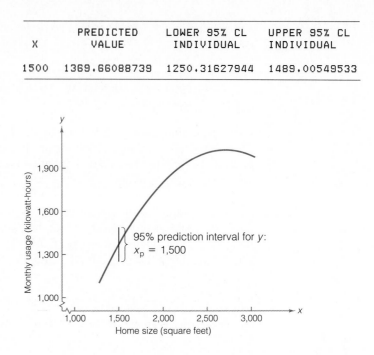

FIGURE 14.16

Prediction interval for electrical usage

Unfortunately, not all computer packages have the capability to produce confidence intervals for means and prediction intervals for specific y values. This is a rather serious oversight, since the estimation of mean values and the prediction of specific values represent the culmination of our model-building efforts: using the model to make inferences about the dependent variable y.

14.7 Multiple Regression: An Example

In Case Study 14.1, we described an application of regression analysis—its use by Towers, Perrin, Forster & Crosby (TPF&C) to develop a prediction equation for corporate executive compensation. We will now use this interesting application of multiple regression to demonstrate all the ideas we have introduced in this chapter. Suppose the list of independent variables given in Table 14.3 is to be used to build a model for the salaries of corporate executives.

TABLE 14.3　**List of Independent Variables for Executive Compensation Example**

INDEPENDENT VARIABLE	DESCRIPTION
x_1	Years of experience
x_2	Years of education
x_3	1 if male; 0 if female
x_4	Number of employees supervised
x_5	Corporate assets (millions of dollars)
x_6	x_1^2
x_7	$x_3 x_4$

STEP 1　The first step is to hypothesize a model relating executive salary to the independent variables listed in Table 14.3. TPF&C have found that executive

compensation models that use the logarithm of salary as the dependent variable provide better predictive power than those using the salary as the dependent variable. This is probably because salaries tend to be incremented in *percentages* rather than dollar values. When a response variable undergoes percentage changes as the independent variables are varied, the logarithm of the response variable will be more suitable as a dependent variable. The model we propose is

$$y = \beta_0 + \beta_1 x_1 + \beta_2 x_2 + \beta_3 x_3 + \beta_4 x_4 + \beta_5 x_5 + \beta_6 x_6 + \beta_7 x_7 + \varepsilon$$

where $y = \log(\text{executive salary})$, $x_6 = x_1^2$ (quadratic term in years of experience), and $x_7 = x_3 x_4$ (cross product or interaction term between sex and number of employees supervised). The variable x_3 is a **dummy variable**; it is used to describe an independent variable that is not measured on a numerical scale but instead is qualitative (categorical) in nature. Sex is such a variable, since its values, male and female, are categories rather than numbers. Thus, we assign the value $x_3 = 1$ if the executive is male and $x_3 = 0$ if the executive is female. For more detail on the use and interpretation of dummy variables, see Chapter 15. The interaction term $x_3 x_4$ accounts for the fact that the relationship between corporate salary and the number of employees supervised, x_4, is dependent on sex, x_3. That is, as the number of supervised employees increases, a woman's salary (with all other factors being equal) might rise more (or less) rapidly than a man's. This concept (interaction), too, is explained in more detail in Chapter 15.

STEP 2 Now we estimate the model coefficients $\beta_0, \beta_1, \ldots, \beta_7$. Suppose a sample of 100 executives is selected, and the variables y and $x_1, x_2, \ldots, x_7$ are recorded (or, in the case of x_6 and x_7, calculated). The sample is then used as input for the SAS regression routine; the output is shown in Figure 14.17. The least squares model is

$$\hat{y} = 9.88 + .045 x_1 + .033 x_2 + .119 x_3 + .00033 x_4$$
$$+ .0020 x_5 - .00072 x_6 + .00031 x_7$$

FIGURE 14.17 SAS printout for executive compensation example

SOURCE	DF	SUM OF SQUARES	MEAN SQUARE	F VALUE	PR > F
MODEL	7	27.06425564	3.85632223	1819.30	0.0001
ERROR	92	0.19551523	0.00212517		STD DEV
CORRECTED TOTAL	99	27.25977087		R-SQUARE	0.0460995
				0.992828	

PARAMETER	ESTIMATE	T FOR H0: PARAMETER = 0	PR > |T|	STD ERROR OF ESTIMATE
INTERCEPT	9.87878688	192.49	0.0001	0.04612667
X1 (EXPERIENCE)	0.04460301	26.83	0.0001	0.00166257
X2 (EDUCATION)	0.03326230	12.31	0.0001	0.00270306
X3 (SEX)	0.11892473	6.89	0.0001	0.01724977
X4 (EMPLOYEES SUPERVISED)	0.00033216	19.97	0.0001	0.00001664
X5 (ASSETS)	0.00201021	73.25	0.0001	0.00002744
X6 (= X1*X1)	-0.00071702	-15.11	0.0001	0.00004746
X7 (= X3*X4)	0.00031244	16.16	0.0001	0.00001933

STEP 3 The next step is to specify the probability distribution of ε, the random error component. We assume that ε is normally distributed with a mean of 0 and a constant variance σ^2. Furthermore, we assume that the errors are independent. The estimate of the variance σ^2 is given in the SAS printout as

$$s^2 = MSE = \frac{SSE}{n - (k + 1)} = \frac{SSE}{100 - (7 + 1)} = .0021$$

STEP 4 We now want to see how well the model predicts salaries. First, note that $R^2 = .993$. This implies that 99.3% of the variation in y (the logarithm of salaries) for these 100 sampled executives is accounted for by the model. The significance of this can be tested:

H_0: $\beta_1 = \beta_2 = \cdots = \beta_7 = 0$

H_a: At least one of the model coefficients is nonzero

Test statistic: $F = \dfrac{R^2/k}{(1 - R^2)/[n - (k + 1)]}$

Rejection region: For $\alpha = .05$, $F > F_{.05}$,

where from Table IX of Appendix A the tabulated value of F for $\alpha = .05$ and based on $k = 7$ and $n - (k + 1) = 92$ df is $F_{.05} \approx 2.1$. The test statistic is given on the SAS printout. Since $F = 1{,}819.3$ exceeds the tabulated value of F and, in fact, has an observed significance level of .0001, we conclude that the model does contribute information for predicting executive salaries. It appears that at least one of the β parameters in the model differs from 0.

We may be particularly interested in whether the data provide evidence that the mean salary of executives increases as the asset value of the company increases when all other variables (experience, education, etc.) are held constant. In other words, we may want to know whether the data provide sufficient evidence to show that $\beta_5 > 0$. We use the following test:

H_0: $\beta_5 = 0$

H_a: $\beta_5 > 0$

Test statistic: $t = \dfrac{\hat{\beta}_5}{s_{\hat{\beta}_5}}$

For $\alpha = .05$, $n = 100$, $k = 7$, and $n - (k + 1) = 92$, we will reject H_0 if $t > t_{.05}$, where (because the degrees of freedom, 92, of t is so large) $t_{.05} \approx z_{.05} = 1.645$. Thus, we reject H_0 if

$t > 1.645$

The t value is indicated in Figure 14.17. The value corresponding to the independent variable x_5 is 73.25. Since this value exceeds 1.645 (having an observed significance level of $.0001/2 = .00005$, we find evidence that the mean salaries of executives do increase as the firm's assets increase.

STEP 5 The culmination of the modeling effort is the use of the model for estimation and/or prediction. Suppose a firm is trying to determine fair compensation for an executive with the characteristics shown in Table 14.4. The least squares

TABLE 14.4 Values of Independent Variables for an Exectutive

$x_1 = 12$ years of experience
$x_2 = 16$ years of education
$x_3 = 0$ (female)
$x_4 = 400$ employees supervised
$x_5 = \$160.1$ million (the firm's asset value)
$x_6 = x_1^2 = 144$
$x_7 = x_3 x_4 = 0$

model can be used to obtain a predicted value for the logarithm of salary. That is,

$$\hat{y} = \hat{\beta}_0 + \hat{\beta}_1(12) + \hat{\beta}_2(16) + \hat{\beta}_3(0) + \hat{\beta}_4(400)$$
$$+ \hat{\beta}_5(160.1) + \hat{\beta}_6(144) + \hat{\beta}_7(0)$$

This predicted value, $\hat{y} = 11.298$, is given in Figure 14.18, a partial reproduction of the SAS regression printout for this problem. The 95% prediction interval is also given: from 11.203 to 11.392. To predict the salary of an executive with these characteristics we take the antilog of these values. That is, the predicted salary is $e^{11.298} = \$80,700$ (rounded to the nearest hundred) and the 95% prediction interval is from $e^{11.203}$ to $e^{11.392}$ (from \$73,400 to \$88,600). Thus, an executive with the characteristics in Table 14.4 should be paid between \$73,400 and \$88,600 to be consistent with the sample data.

FIGURE 14.18

SAS printout for executive compensation problem

X1	X2	X3	X4	X5	X6	X7	PREDICTED VALUE	LOWER 95% CL INDIVIDUAL	UPPER 95% CL INDIVIDUAL
12	16	0	400	160.1	144	0	11.29766682	11.20298295	11.39235070

14.8 Statistical Computer Programs

There are a number of different statistical program packages; some of the most popular are Biomed, Minitab, SAS, and SPSS. (See the references at the end of the chapter.) Some can be used on all large computers produced by a specific manufacturer. Consequently, you may have access to one or more of these packages at your computer center.

The multiple regression computer programs for these packages may differ in what they can do, how they do it, and the appearance of their computer printouts, but all of them print the basic outputs needed for a regression analysis. Some will compute confidence intervals for $E(y)$ and prediction intervals for y; others will not. Some test the null hypotheses that the individual β parameters equal 0 using Student t-tests; others use F-tests. But all give the least squares estimates, the values of SSE, s^2, etc.

To illustrate, the Minitab, SAS, and SPSS regression analysis printouts for Example 14.3 are shown in Figure 14.19 (page 748). For that example, we fit the model

$$y = \beta_0 + \beta_1 x_1 + \beta_2 x_2 + \beta_3 x_1 x_2 + \varepsilon$$

to $n = 32$ data points. The variables in the model are

y = Auction price

x_1 = Age of clock (years)

x_2 = Number of bidders

Notice that the Minitab printout gives the prediction equation at the top of the printout. The independent variables, shown in the prediction equation and listed at the left side of the printout, are AGE, BIDDERS, and AGE-BID. Each program treats the product AGE-BID = AGE × BIDDERS as a third independent variable, which must be computed before the fitting begins. For this reason, the prediction equation will always appear on the printout as first-order even though some of the independent variables shown in the prediction equation may actually be the squares or cross products of other independent variables. The SPSS printout in Figure 14.19 lists the independent variables in a different order: AGE-BID, AGE, and BIDDERS. The order is determined statistically rather than by the user, and the user must therefore pay attention to the name of each variable when

FIGURE 14.19

Computer printouts for Example 14.3

(a) Minitab regression printout

THE REGRESSION EQUATION IS
PRICE = 323 + 0.87 AGE − 93.4 BIDDERS + 1.30 AGE-BID

COLUMN	COEFFICIENT	ST. DEV. OF COEF.	T-RATIO = COEF/S.D.
	322.8	293.3	1.10
AGE	0.873	2.020	0.43
BIDDERS	−93.41	29.71	−3.14
AGE-BID	1.2979	0.2110	6.15

S = 88.37

R−SQUARED = 95.4 PERCENT
R−SQUARED = 94.9 PERCENT, ADJUSTED FOR D.F.

ANALYSIS OF VARIANCE

DUE TO	DF	SS	MS=SS/DF
REGRESSION	3	4572548	1524183
RESIDUAL	28	218646	7809
TOTAL	31	4791194	

(b) SAS regression printout

Dep Variable: Y

Analysis of Variance

Source	DF	Sum of Squares	Mean Square	F Value	Prob>F
Model	3	4572547.9872	1524182.6624	195.188	0.0001
Error	28	218646.23157	7808.79398		
C Total	31	4791194.2187			

Root MSE	88.36738	R-Square	0.9544	
Dep Mean	1327.15625	Adj R-Sq	0.9495	
C.V.	6.65840			

Parameter Estimates

Variable	DF	Parameter Estimate	Standard Error	T for H0: Parameter=0	Prob > |T|
INTERCEP	1	322.754353	293.32514660	1.100	0.2806
X1	1	0.873288	2.01965115	0.432	0.6688
X2	1	−93.409920	29.70767946	−3.144	0.0039
X1X2	1	1.297898	0.21102602	6.150	0.0001

(c) SPSS regression printout

```
Equation Number 1     Dependent Variable..    PRICE

Beginning Block Number  1.  Method:  Enter
    AGE        BIDDERS  AGE_BID


Variable(s) Entered on Step Number
    1..    AGE_BID
    2..    AGE
    3..    BIDDERS


Multiple R              .97692
R Square                .95436
Adjusted R Square       .94948
Standard Error        88.36738

Analysis of Variance
                      DF      Sum of Squares      Mean Square
Regression             3       4572547.98718    1524182.66239
Residual              28        218646.23157       7808.79398

F =      195.18797        Signif F =   .0000


------------------ Variables in the Equation ------------------

Variable               B         SE B        Beta        T   Sig T

AGE_BID          1.29790      .21103     1.37032     6.150   .0000
AGE               .87329     2.01965      .06085      .432   .6688
BIDDERS        -93.40992    29.70768     -.67471    -3.144   .0039
(Constant)     322.75435   293.32515                 1.100   .2806
```

interpreting the SPSS printout. Note that SPSS also prints the intercept last rather than first, using the label of (Constant). Minitab prints the intercept first, but with no label.

The estimates of the regression coefficients appear opposite the identifying variable in the Minitab column titled COEFFICIENT, in the SAS column titled Parameter Estimate, and in the SPSS column titled B. Compare the estimates given in these three columns. Note that the Minitab printout gives the estimates with a lesser degree of accuracy (fewer decimal places) than the SAS and SPSS printouts. (Ignore the column titled Beta in the SPSS printout. These are standardized estimates and will not be discussed in this text.)

The estimated standard errors of the estimates are given in the Minitab column titled ST. DEV. OF COEF., in the SAS column titled Standard Error, and in the SPSS column titled SE B.

The values of the test statistics for testing H_0: $\beta_i = 0$, where $i = 1, 2, 3$, are shown in the Minitab column titled T-RATIO = COEF/S.D., in the SAS column titled T for H0: Parameter = 0, and in the SPSS column titled T. Note that the computed t values shown are identical (except for the number of decimal places), but Minitab does not give the observed significance level of the test. Consequently, to draw conclusions from the Minitab printout, you must compare the computed values of t with the critical values given in a t-table (Table VI in Appendix A).

In contrast, the SAS printout gives the observed significance level for each t-test in the column titled Prob $> |T|$, and the SPSS printout in the column titled Sig T. Note that these observed significance levels have been computed assuming that the tests are two-tailed. The observed significance levels for one-tailed tests would equal half of these values.

The Minitab printout gives the value SSE $= 218646$ under the ANALYSIS OF VARIANCE column headed SS and in the row identified as RESIDUAL. The value of $s^2 = 7809$ is shown in the same row under the column headed MS $=$ SS/DF, and the degrees of freedom DF, appears in the same row as 28. The corresponding values are shown at the top of the SAS printout in the row labeled Error and in the columns designated as Sum of Squares, Mean Square, and DF, respectively. These quantities appear with similar headings in the SPSS printout, but Error is called Residual in SPSS. The standard deviation s for the regression equation is given on each printout, but with different labels. Minitab shows $s = 88.37$ in the center of the printout; SAS labels it Root MSE in the left center of the printout. SPSS calls the standard deviation the Standard Error and prints it in the center of the printout. All the estimates are identical, except for the number of decimal places.

The value of R^2, as defined in Section 14.5, is given in the Minitab printout as 95.4 PERCENT (we defined this quantity as a ratio where $0 \le R^2 \le 1$). It is given in the SAS printout as 0.9544, and shown in the left column of the SPSS printout as 0.95436. (Ignore the quantities shown in the Minitab printout as R^2 ADJUSTED FOR D.F., in the SAS printout as Adj R Sq, and in the SPSS printout as Adjusted R Square. These quantities are adjusted for the degrees of freedom associated with the total SS and SSE and are not used or discussed in this text.)

The F statistic for testing the usefulness of the model (Section 14.5)—i.e., testing the null hypothesis that all model parameters (except β_0) equal zero— is shown under the title F VALUE as 195.188 at the top of the SAS printout. In addition, the SAS printout gives the observed significance level of this F-test under Prob $<$ F as 0.0001. This F value, 195.18797, is also printed at the center of the SPSS printout with the observed significance level given as Signif F. The F statistic for testing the usefulness of the model is not given in the Minitab printout. If you are using Minitab and wish to obtain the value of this statistic, you must compute it using the analysis of variance formula given in Section 14.5:

$$F = \frac{\text{MEAN SQUARE(MODEL)}}{\text{MEAN SQUARE(ERROR)}}$$

These quantities are given in the Minitab printout under the column marked MS $=$ SS/DF. The rows are labeled REGRESSION and RESIDUAL, which are synonymous with MODEL and ERROR, respectively. Thus,

$$F = \frac{1,524,183}{7,809} = 195.18$$

a value that agrees with the values given in the SAS and SPSS printouts. The logic behind this test and other tests of hypotheses concerning sets of the β parameters are further discussed in Section 15.4.

Although we will henceforth use the three popular statistical packages in this chapter (SAS, Minitab, and SPSS), there are many others available. The rapid expansion of the microcomputer market has been accompanied by significant growth in the availability of statistical software. It is important to evaluate any new package carefully to assure that the results it produces are complete and correct. Fortunately, there is usually much similarity in the computer printouts produced by the packages, so you will find it relatively easy to learn how to read a new regression printout after you have become familiar with those used in this text. We will intermix the different packages in the examples and exercises to help familiarize you with different formats.

14.9 Residual Analysis: Checking the Regression Assumptions

When we apply regression analysis to a set of data, we never know for certain whether the assumptions of Section 14.3 are satisfied. How far can we deviate from the assumptions and still expect regression analysis to yield results that will have the reliability stated in this chapter? How can we detect departures (if they exist) from the assumptions of Section 14.3 and what can we do about them? We provide some partial answers to these questions in this section and direct you to further discussion in the next chapter.

Remember from Section 14.3 that

$$y = E(y) + \varepsilon$$

where the expected value $E(y)$ of y for a given set of values of $x_1, x_2, \ldots, x_k$ is

$$E(y) = \beta_0 + \beta_1 x_1 + \beta_2 x_2 + \cdots + \beta_k x_k$$

and ε is a random error. The first assumption we made was that the mean value of the random error for *any* given set of values of $x_1, x_2, \ldots, x_k$ is $E(\varepsilon) = 0$. One consequence of this assumption is that the mean $E(y)$ for a specific set of values of $x_1, x_2, \ldots, x_k$ is

$$E(y) = \beta_0 + \beta_1 x_1 + \beta_2 x_2 + \cdots + \beta_k x_k$$

That is,

$$
y \quad = \quad \underbrace{E(y)}_{\substack{\text{Mean value of } y \\ \text{for specific values} \\ \text{of } x_1, x_2, \ldots, x_k}} \quad + \quad \underbrace{\varepsilon}_{\substack{\text{Random} \\ \text{error}}}
$$

The second consequence of the assumption is that the least squares estimators of the model parameters, $\beta_0, \beta_1, \beta_2, \ldots, \beta_k$, will be unbiased regardless of the remaining assumptions that we attribute to the random errors and their probability distributions.

The properties of the sampling distributions of the parameter estimators $\hat{\beta}_0$, $\hat{\beta}_1, \ldots, \hat{\beta}_k$ will depend on the remaining assumptions that we specify concerning the probability distributions of the random errors. Recall that we assumed that for any given set of values of $x_1, x_2, \ldots, x_k$, ε has a normal probability distribution with mean equal to 0 and variance equal to σ^2. Also, we assumed that the random errors are probabilistically independent.

It is unlikely that these assumptions are ever satisfied exactly in a practical application of regression analysis. Fortunately, experience has shown that least squares regression analysis produces reliable statistical tests, confidence intervals, and prediction intervals as long as the departures from the assumptions are not too great. In this section we present some methods for determining whether the data indicate significant departures from the assumptions.

Because the assumptions all concern the random error component, ε, of the model, the first step is to estimate the random error. Since the actual random error associated with a particular value of y is the difference between the actual y value and its unknown mean, we estimate the error by the difference between the actual y value and the *estimated* mean. This estimated error is called the **regression residual**, or simply the **residual**, and is denoted by $\hat{\varepsilon}$. The actual error ε and residual $\hat{\varepsilon}$ are shown in Figure 14.20.

$$\text{Actual random error} = \varepsilon$$
$$= (\text{Actual } y \text{ value}) - (\text{Mean of } y)$$
$$= y - E(y) = y - (\beta_0 + \beta_1 x_1 + \beta_2 x_2 + \cdots + \beta_k x_k)$$

$$\text{Estimated random error (residual)} = \hat{\varepsilon}$$
$$= (\text{Actual } y \text{ value}) - (\text{Estimated mean of } y)$$
$$= y - \hat{y} = y - (\hat{\beta}_0 + \hat{\beta}_1 x_1 + \hat{\beta}_2 x_2 + \cdots + \hat{\beta}_k x_k)$$

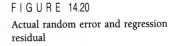

FIGURE 14.20

Actual random error and regression residual

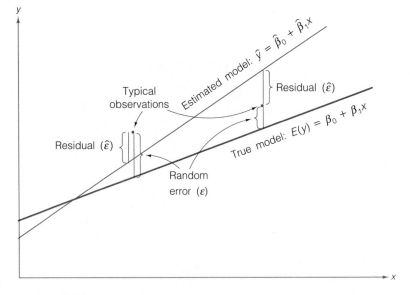

Since the true mean of y (i.e., the true regression model) is not known, the actual random error cannot be calculated. However, because the residual is based on the estimated mean (the least squares regression model), it can be calculated and used to estimate the random error and to check the regression assumptions. Such checks are generally referred to as **residual analyses**. Some useful properties of residuals are given in the next box.

The following examples show how the analysis of regression residuals can be used to verify the assumptions associated with the model and to improve the model when the assumptions do not appear to be satisfied. Although the residuals can be calculated and plotted by hand, we rely on the computer for these tasks in the examples and exercises. Most statistical computer packages now include residual analyses as a standard component of their regression modeling programs.

Properties of Regression Residuals

1. A residual is equal to the difference between the observed y value and its estimated (regression) mean:

   $$\text{Residual} = y - \hat{y}$$

2. The mean of the residuals is equal to 0. This property follows from the fact that the sum of the differences between the observed y values and the least squares regression model is equal to 0.

   $$\sum (\text{Residuals}) = \sum (y - \hat{y}) = 0$$

3. The standard deviation of the residuals is equal to the standard deviation of the fitted regression model, s. This property follows from the fact that the sum of the squared residuals is equal to SSE, which when divided by the error degrees of freedom is equal to the variance of the fitted regression model, s^2. The square root of the variance is both the standard deviation of the residuals and the standard deviation of the regression model.

   $$\sum (\text{Residuals})^2 = \sum (y - \hat{y})^2 = \text{SSE}$$

   $$s = \sqrt{\frac{\sum (\text{Residuals})^2}{n - (k + 1)}} = \sqrt{\frac{\text{SSE}}{n - (k + 1)}}$$

EXAMPLE 14.4

The data for the home size–electrical usage example used throughout this chapter are repeated in Table 14.5. SAS printouts for a straight-line model and a quadratic model fitted to the data are shown in Figure 14.21(a) and 14.21(b) (pages 754 and 755), respectively. The residuals from these models are shaded in the printouts. The residuals are then plotted on the vertical axis against the variable x, size of home, on the horizontal axis in Figure 14.22(a) and 14.22(b) (page 756), respectively.

a. Verify that each residual is equal to the difference between the observed y value and the estimated mean value, y.

b. Analyze the residual plots.

TABLE 14.5 **Home Size–Electrical Usage Data**

SIZE OF HOME x (sq. ft.)	MONTHLY USAGE y (kilowatt-hours)	SIZE OF HOME x (sq. ft.)	MONTHLY USAGE y (kilowatt-hours)
1,290	1,182	1,840	1,711
1,350	1,172	1,980	1,804
1,470	1,264	2,230	1,840
1,600	1,493	2,400	1,956
1,710	1,571	2,930	1,954

FIGURE 14.21

SAS printouts for electrical usage example

(a) Straight-line model

Dep Variable: Y

Analysis of Variance

Source	DF	Sum of Squares	Mean Square	F Value	Prob>F
Model	1	703957.18342	703957.18342	39.536	0.0002
Error	8	142444.91658	17805.61457		
C Total	9	846402.10000			

Root MSE	133.43768	R-Square	0.8317	
Dep Mean	1594.70000	Adj R-Sq	0.8107	
C.V.	8.36757			

Parameter Estimates

Variable	DF	Parameter Estimate	Standard Error	T for H0: Parameter=0	Prob > |T|
INTERCEP	1	578.927752	166.96805715	3.467	0.0085
X	1	0.540304	0.08592981	6.288	0.0002

Obs	Y	Predict Value	Residual
1	1182.0	1275.9	-93.9204
2	1172.0	1308.3	-136.3
3	1264.0	1373.2	-109.2
4	1493.0	1443.4	49.5852
5	1571.0	1502.8	68.1517
6	1711.0	1573.1	137.9
7	1804.0	1648.7	155.3
8	1840.0	1783.8	56.1935
9	1956.0	1875.7	80.3417
10	1954.0	2162.0	-208.0

Sum of Residuals	0
Sum of Squared Residuals	142444.9166

Solution

a. For the straight-line model the residual is calculated for the first y value as follows:

$$\text{Residual} = (\text{Observed } y \text{ value}) - (\text{Estimated mean})$$
$$= y - \hat{y} = 1,182 - 1,275.9 = -93.9$$

where the estimated mean is the first number in the column labeled Predict Value (shaded) on the SAS printout in Figure 14.21(a). Similarly, the residual for the first y value using the quadratic model is

$$\text{Residual} = 1,182 - 1,129.6 = 52.4$$

Both residuals agree (after rounding) with the first values given in the column labeled Residual in Figure 14.21(a) and 14.21(b), respectively. Although the residuals both correspond to the same observed y value, 1,182, they differ because the estimated mean value changes depending on whether the straight-line model or quadratic model is used. Similar calculations produce the remaining residuals.

FIGURE 14.21

SAS printouts for electrical usage example

(b) Quadratic model

Dep Variable: Y

Analysis of Variance

Source	DF	Sum of Squares	Mean Square	F Value	Prob>F
Model	2	831069.54637	415534.77319	189.710	0.0001
Error	7	15332.55363	2190.36480		
C Total	9	846402.10000			

Root MSE	46.80133	R-Square	0.9819	
Dep Mean	1594.70000	Adj R-Sq	0.9767	
C.V.	2.93480			

Parameter Estimates

Variable	DF	Parameter Estimate	Standard Error	T for H0: Parameter=0	Prob > |T|
INTERCEP	1	−1216.143887	242.80636850	−5.009	0.0016
X	1	2.398930	0.24583560	9.758	0.0001
XSQ	1	−0.000450	0.00005908	−7.618	0.0001

Obs	Y	Predict Value	Residual
1	1182.0	1129.6	52.4359
2	1172.0	1202.2	−30.2136
3	1264.0	1337.8	−73.7916
4	1493.0	1470.0	22.9586
5	1571.0	1570.1	0.9359
6	1711.0	1674.2	36.7685
7	1804.0	1769.4	34.5998
8	1840.0	1895.5	−55.4654
9	1956.0	1949.1	6.9431
10	1954.0	1949.2	4.8287

Sum of Residuals	−2.27374E−12
Sum of Squared Residuals	15332.5536

b. The plot of the residuals for the straight-line model [Figure 14.22(a)] reveals a nonrandom pattern. The residuals exhibit a mound shape, with the residuals for the small values of x below the horizontal 0 (mean of the residuals) line, the residuals corresponding to the middle values of x above the 0 line, and the residual for the largest value of x again below the 0 line. The indication is that the mean value of the random error ε *within* each of these ranges of x (small, medium, large) may not be equal to 0. Such a pattern usually indicates that curvature needs to be added to the model.

When the second-order term is added to the model, the nonrandom pattern disappears. In Figure 14.22(b), the residuals appear to be randomly distributed around the 0 line, as expected. Note, too, that the ± 2 standard deviation lines are at about ± 95 on the quadratic residual plot, compared to (about) ± 275 on the straight-line plot. The implication is that the quadratic model provides a considerably better model for predicting electrical usage, verifying our conclusions from previous analyses in this chapter.

FIGURE 14.22

Residual plots for electrical usage example

(a) Straight-line model

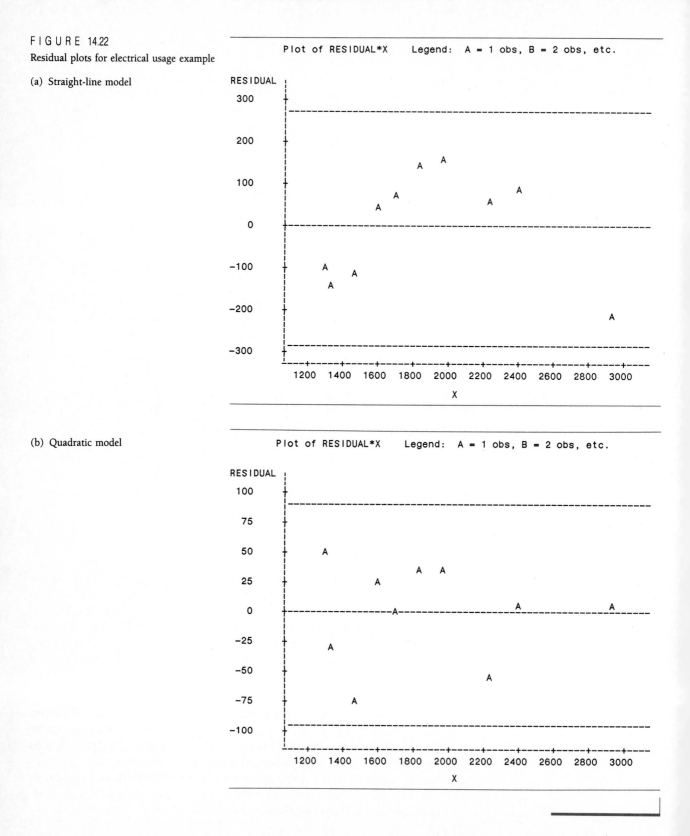

(b) Quadratic model

Residual analyses are also useful for detecting one or more observations that deviate significantly from the regression model. We expect approximately 95% of the residuals to fall within 2 standard deviations of the 0 line, and all or almost all of them to lie within 3 standard deviations of their mean of 0. Residuals that are extremely far from the 0 line, and disconnected from the bulk of the other residuals, are called **outliers**, and should receive special attention from the régression analyst.

EXAMPLE 14.5

The data for the grandfather clock example used throughout this chapter are repeated in Table 14.6, with one important difference: the auction price of the clock at the top of the second column has been changed from $2,131 to $1,131 (shaded in Table 14.6). The interaction model

$$E(y) = \beta_0 + \beta_1 x_1 + \beta_2 x_2 + \beta_3 x_1 x_2$$

is again fit to these (modified) data, with the printout shown in Figure 14.23 on page 758. The residuals are shown shaded in the printout and then plotted against the number of bidders, x_2, in Figure 14.24 (page 759). Analyze the residual plot.

TABLE 14.6 **Auction Price Data**

AGE x_1	NUMBER OF BIDDERS x_2	AUCTION PRICE y	AGE x_1	NUMBER OF BIDDERS x_2	AUCTION PRICE y
127	13	$1,235	170	14	$1,131
115	12	1,080	182	8	1,550
127	7	845	162	11	1,884
150	9	1,522	184	10	2,041
156	6	1,047	143	6	854
182	11	1,979	159	9	1,483
156	12	1,822	108	14	1,055
132	10	1,253	175	8	1,545
137	9	1,297	108	6	729
113	9	946	179	9	1,792
137	15	1,713	111	15	1,175
117	11	1,024	187	8	1,593
137	8	1,147	111	7	785
153	6	1,092	115	7	744
117	13	1,152	194	5	1,356
126	10	1,336	168	7	1,262

Solution

The residual plot dramatically reveals the one altered measurement. Note that one of the two residuals at $x_2 = 14$ bidders falls more than 3 standard deviations below 0. Note that no other residual falls more than 2 standard deviations from 0.

What do we do with outliers once we identify them? First, we try to determine the cause. Were the data entered into the computer incorrectly? Was the observation recorded incorrectly when the data were collected? If so, correct the

FIGURE 14.23

Printout for grandfather clock example with altered data

THE REGRESSION EQUATION IS
PRICE = - 511 + 8.16 AGE + 19.7 BIDDERS + 0.320 AGE-BID

COLUMN	COEFFICIENT	ST. DEV. OF COEF.	T-RATIO = COEF/S.D.
	-510.5	664.8	-0.77
AGE	8.160	4.577	1.78
BIDDERS	19.74	67.33	0.29
AGE-BID	0.3197	0.4783	0.67

S = 200.3

R-SQUARED = 73.0 PERCENT
R-SQUARED = 70.1 PERCENT, ADJUSTED FOR D.F.

ANALYSIS OF VARIANCE

DUE TO	DF	SS	MS=SS/DF
REGRESSION	3	3029141	1009714
RESIDUAL	28	1123116	40111
TOTAL	31	4152256	

ROW	AGE	Y PRICE	PRED. Y VALUE	ST.DEV. PRED. Y	RESIDUAL
1	127	1235.0	1310.3	59.2	-75.3
2	115	1080.0	1106.0	62.0	-26.0
3	127	845.0	948.2	61.0	-103.2
4	150	1522.0	1322.8	37.1	199.2
5	156	1047.0	1180.2	60.2	-133.2
6	182	1979.0	1831.8	82.8	147.2
7	156	1822.0	1597.9	61.8	224.1
8	132	1253.0	1186.1	39.7	66.9
9	137	1297.0	1179.3	39.0	117.7
10	113	946.0	914.4	58.5	31.6
11	137	1713.0	1560.6	78.3	152.4
12	117	1024.0	1072.8	53.0	-48.8
13	137	1147.0	1115.8	44.2	31.2
14	153	1092.0	1149.9	58.9	-57.9
15	117	1152.0	1187.1	69.6	-35.1
16	126	1336.0	1117.9	43.4	218.1
17	170	1131.0	1914.0	116.5	-783.0
18	182	1550.0	1598.1	62.7	-48.1
19	162	1884.0	1598.3	56.9	285.7
20	184	2041.0	1776.6	70.6	264.4
21	143	854.0	1049.2	58.8	-195.2
22	159	1483.0	1422.1	40.5	60.9
23	108	1055.0	1130.6	97.8	-75.6
24	175	1545.0	1523.0	55.3	22.0
25	108	729.0	696.4	99.5	32.6
26	179	1792.0	1642.9	57.5	149.1
27	111	1175.0	1223.7	107.1	-48.7
28	187	1593.0	1651.7	68.5	-58.7
29	111	785.0	781.9	80.7	3.1
30	115	744.0	823.5	75.4	-79.5
31	194	1356.0	1481.4	133.3	-125.4
32	168	1262.0	1374.6	57.6	-112.6

FIGURE 14.24

Residual plot against number of bidders

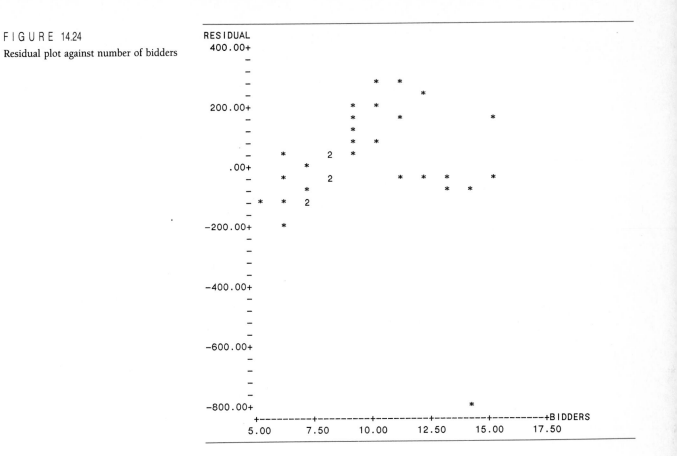

observation and rerun the program. Another possibility is that the observation is not representative of the conditions you are trying to model. For example, in this case the low price may be attributable to extreme damage to the clock, or to a clock of inferior quality compared to the others. In these cases we probably would exclude the observation from the analysis. In many cases you may not be able to determine the cause of the outlier. Even so, you may want to rerun the regression analysis excluding the outlier in order to assess the effect of that observation on the results of the analysis.

Figure 14.25 (page 760) shows the printout when the outlier observation is excluded from the grandfather clock analysis, and Figure 14.26 shows the new plot of the residuals against the number of bidders. Now only one of the residuals lies beyond 2 standard deviations from 0, and none of them lies beyond 3 standard deviations. Also, the model statistics indicate a much better model without the outlier. Most notably, the standard deviation (s) has decreased from 200.3 to 85.28, indicating a model that will provide more precise estimates and predictions (narrower confidence and prediction intervals) for clocks that are similar to those in the reduced sample. Remember, though, that if you decide to remove the outlier from the analysis, and in fact it belongs to the same population as the rest of the sample, the resulting model may provide misleading estimates and predictions.

FIGURE 14.25

Minitab regression printout for
Example 14.5

```
THE REGRESSION EQUATION IS
PRICE-2 = 476 - 0.46 AGE-2 - 114 BIDDERS2 + 1.48 AGE-BID2

                              ST. DEV.    T-RATIO =
COLUMN        COEFFICIENT     OF COEF.    COEF/S.D.
                 475.8          296.2        1.61
AGE-2           -0.465          2.093       -0.22
BIDDERS2      -114.19          31.03       -3.68
AGE-BID2         1.4775         0.2280       6.48

S = 85.28

R-SQUARED = 95.2 PERCENT
R-SQUARED = 94.7 PERCENT, ADJUSTED FOR D.F.


ANALYSIS OF VARIANCE

  DUE TO        DF           SS        MS=SS/DF
REGRESSION      3        3927844       1309282
RESIDUAL       27         196341          7272
TOTAL          30        4124186
```

FIGURE 14.26

Minitab residual plot for Example 14.5

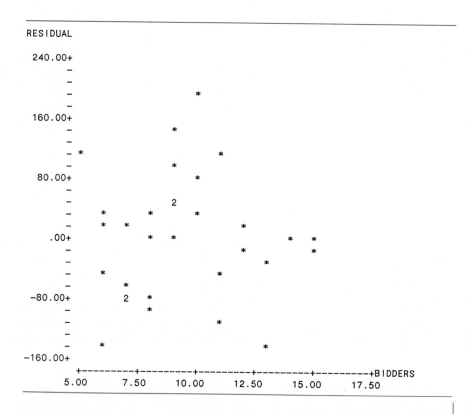

Outlier analysis is another example of testing the assumption that the expected (mean) value of the random error component is 0, since this assumption is in doubt for the error terms corresponding to the outliers. The last example in this section checks the assumption of the normality of the random error component.

EXAMPLE 14.6

Solution

Refer to Example 14.5. Use a stem and leaf display (Section 2.2) to plot the frequency distribution of the residuals in the grandfather clock example, both before and after the outlier residual is removed. Analyze the plots and determine whether the assumption of normality of the error distribution is reasonable.

The stem and leaf displays for the two sets of residuals are constructed using Minitab and are shown in Figure 14.27.* Note that the outlier appears to skew the frequency distribution in Figure 14.27(a), whereas the stem and leaf display in Figure 14.27(b) appears to be more mound-shaped. Although the displays do not provide formal statistical tests of normality, they do provide a descriptive display. Relative frequency histograms can also be used to check the normality assumption. In this example the normality assumption appears to be more plausible after the outlier is removed. Consult the references for methods to conduct statistical tests of normality using the residuals.

FIGURE 14.27

Stem and leaf displays for grandfather clock example

(a) Outlier included

```
STEM-AND-LEAF DISPLAY OF RESIDUAL
LEAF DIGIT UNIT =  10.0000
1 2 REPRESENTS 120.

        STEM   LEAF

     1   -7   8
     1   -6
     1   -5
     1   -4
     1   -3
     1   -2
     6   -1   93210
    16   -0   7775544432
    16    0   0233366
     9    1   14459
     4    2   1268
```

(b) Outlier excluded

```
STEM-AND-LEAF DISPLAY OF RESIDUAL
LEAF DIGIT UNIT =  10.0000
1 2 REPRESENTS 120.

     3   -1*  331
     9   -0.  987765
    (7)   -0*  4321000
    15   +0*  011223344
     6   +0.  79
     4    1*  004
     1    1.  9
```

Residual analysis is a useful tool for the regression analyst, not only to check the assumptions, but also to provide information about how the model can be improved. A summary of the residual analyses presented in this section to check the assumption that the random error ε is normally distributed with mean 0 is presented in the next box.

---

*Recall that the left column of the Minitab printout shows the number of measurements at least as extreme as the stem. In Figure 14.27(a), for example, the 6 corresponding to the STEM = −1 means that six measurements are less than or equal to −100. If one of the numbers in the leftmost column is enclosed in parentheses, the number in parentheses is the number of measurements in that row, and the median is contained in that row.

> ### Steps in a Residual Analysis
>
> 1. Calculate and plot the residuals against each of the independent variables, preferably with the assistance of a computer program.
> 2. Analyze each plot, looking for curvature—either a mound or bowl shape. Both shapes are distinguished by groups of residuals at the low and high values of the x variable on one side of the 0 line, and the residuals for the medium x values on the opposite side of the 0 line. This shape signals the need for a curvature term in the model. Try a second-order term in the variable against which the residuals are plotted.
> 3. Examine the residual plots for outliers. Draw lines on the residual plots at 2 and 3 standard deviation distances below and above the 0 line. Examine residuals outside the 3 standard deviation lines as potential outliers, and check to see that approximately 5% of the residuals exceed the 2 standard deviation lines. Determine whether each outlier can be explained as an error in data collection or transcription, or corresponds to a member of a population different from that of the remainder of the sample, or simply represents an unusual observation. If the observation is determined to be an error, fix it or remove it. Even if cause cannot be determined, you may want to rerun the regression analysis without the observation to determine its effect on the analysis.
> 4. Plot a frequency distribution of the residuals, using a stem and leaf display or a histogram. Check to see if obvious departures from normality exist. Extreme skewness of the frequency distribution may indicate the need for a transformation of the dependent variable, a topic which is beyond the scope of this book but which can be found in the references.

EXERCISES 14.28–14.36

[*Note: Starred (*) exercises require the use of a computer.*]

LEARNING THE MECHANICS

14.28   When a multiple regression model is used for estimating the mean of the dependent variable and for predicting a new value of y, which will be narrower—the confidence interval for the mean or the prediction interval for the new y value?

14.29   Refer to Exercise 14.1, in which the model

$$y = \beta_0 + \beta_1 x_1 + \beta_2 x_2 + \varepsilon$$

was fit to $n = 20$ data points. The Minitab regression printout for these data is shown at the top of page 763.

a. Find the least squares prediction equation.

b. Find and interpret the standard deviation of the regression model.

c. Calculate the F statistic. Does the model contribute information for the prediction of y? Test using $\alpha = .05$.

d. Place a 90% confidence interval on the coefficient β_2.

e. Find and interpret the coefficient of determination R^2.

Minitab printout for Exercise 14.29

COLUMN	COEFFICIENT	ST. DEV. OF COEF.	T-RATIO = COEF/S.D.
	506.35	45.17	11.21
X1	-941.9	275.1	-3.42
X2	-429.1	379.8	-1.13

S = 94.25

R-SQUARED = 45.9 PERCENT
R-SQUARED = 39.6 PERCENT, ADJUSTED FOR D.F.

ANALYSIS OF VARIANCE

DUE TO	DF	SS	MS=SS/DF
REGRESSION	2	128329	64165
RESIDUAL	17	151016	8883
TOTAL	19	279345	

14.30   Refer to Exercise 14.29. The SPSS printout for the same data set is given here. Repeat parts **a–e** of Exercise 14.29 using the SPSS printout.

SPSS printout for Exercise 14.30

Multiple R	.67779
R Square	.45939
Adjusted R Square	.39579
Standard Error	94.25114

Analysis of Variance

	DF	Sum of Squares	Mean Square
Regression	2	128329.27624	64164.63812
Residual	17	151015.72376	8883.27787

F =   7.22308   Signif F =   .0054

------------------ Variables in the Equation ------------------

Variable	B	SE B	Beta	T	Sig T
X1	-941.90023	275.08556	-.61728	-3.424	.0032
X2	-429.06042	379.82566	-.20365	-1.130	.2743
(Constant)	506.34607	45.16942		11.210	.0000

14.31   Refer to Exercise 14.12, in which a quadratic model was fit to $n = 19$ data points. Two residual plots are shown on page 764—one corresponding to the fitting of a straight-line model to the data and the other to the quadratic model fit in Exercise 14.12. Analyze the two plots. Is the need for a quadratic term evident from the residual plot for the straight-line model? Does your conclusion agree with your test of the quadratic term in Exercise 14.12?

Residual plot for straight-line model for
Exercise 14.31

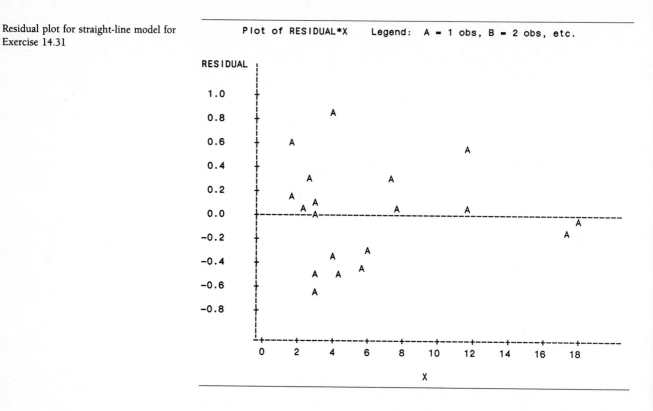

Residual plot for quadratic model for
Exercise 14.31

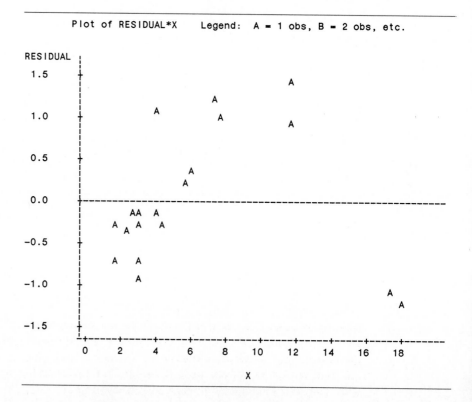

APPLYING THE CONCEPTS

14.32 Refer to Exercise 14.10, in which 1990 food consumption expenditure, y, of a sample of 25 households in Washington, D.C., was related to the household income, x_1, and size of household, x_2, by the model

$$y = \beta_0 + \beta_1 x_1 + \beta_2 x_2 + \varepsilon$$

Plots of the residuals from this model are shown—one against x_1 (below) and one against x_2 (top of page 766). Analyze the plots. Is there visual evidence of a need for a quadratic term in either x_1 or x_2? Explain.

Exercise 14.32: Printout of residuals against household income

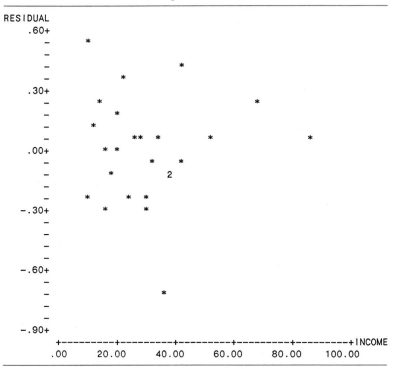

14.33 Refer to Exercise 14.32. Suppose a 26th household is added to the sample, with the following characteristics:

Food consumption: 6.5

Income: 62.3

Persons in household: 5

Two Minitab printouts are shown on page 766—one for the same model fit to the first 25 observations, the second fit to all 26 observations.
a. Record the least squares estimates of the model parameters for each model, and note the differences in the estimates. Interpret each estimate.
b. Find and interpret the standard deviation for each model.
c. Conduct the analysis of variance F-test for each model using $\alpha = .05$.
d. Place a 95% confidence interval on the mean rate of change in food consumption per additional person in the household for each model.
e. According to the results of parts a–d, how much influence does the additional observation have on the model?

Exercise 14.32: Printout of residuals against household size

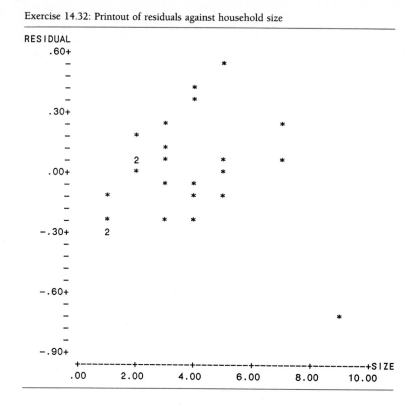

Exercise 14.33: Printout using 25 observations

THE REGRESSION EQUATION IS
FOOD = 1.43 + 0.00999 INCOME + 0.379 SIZE

COLUMN	COEFFICIENT	ST. DEV. OF COEF.	T-RATIO = COEF/S.D.
	1.4326	0.1467	9.76
INCOME	0.009991	0.003168	3.15
SIZE	0.37929	0.02772	13.68

S = 0.2769

R-SQUARED = 90.2 PERCENT
R-SQUARED = 89.3 PERCENT, ADJUSTED FOR D.F.

ANALYSIS OF VARIANCE

DUE TO	DF	SS	MS=SS/DF
REGRESSION	2	15.4623	7.7311
RESIDUAL	22	1.6873	0.0767
TOTAL	24	17.1496	

Exercise 14.33: Printout using 26 observations

THE REGRESSION EQUATION IS
FOOD = 1.16 + 0.0187 INCOME + 0.406 SIZE

COLUMN	COEFFICIENT	ST. DEV. OF COEF.	T-RATIO = COEF/S.D.
	1.1554	0.2882	4.01
INCOME	0.018735	0.006034	3.11
SIZE	0.40577	0.05551	7.31

S = 0.5581

R-SQUARED = 74.6 PERCENT
R-SQUARED = 72.4 PERCENT, ADJUSTED FOR D.F.

ANALYSIS OF VARIANCE

DUE TO	DF	SS	MS=SS/DF
REGRESSION	2	21.076	10.538
RESIDUAL	23	7.163	0.311
TOTAL	25	28.239	

14.34   Refer to Exercises 14.32 and 14.33. Residual plots against household income and size of household for the models corresponding to 26 observations are shown.
   a. Analyze the plots. Are there any outliers? If so, identify them.
   b. What are possible explanations for any outliers you identified in part **a**?

Exercise 14.34: Printout of residuals against household size

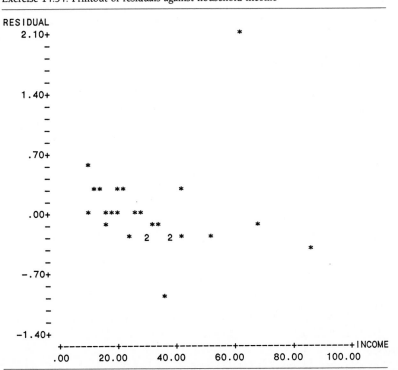

Exercise 14.34: Printout of residuals against household income

14.35 Refer to Exercises 14.32–14.34. The accompanying stem and leaf displays represent the frequency distributions of the residuals for the two data sets, one with $n = 25$ and one with $n = 26$. Analyze the displays, especially with regard to the normality assumption.

Stem and leaf display for 25 residuals

```
STEM-AND-LEAF DISPLAY OF RESIDUAL
LEAF DIGIT UNIT =     .1000
1 2 REPRESENTS 1.2


   1   -0S   7
   1   -0F
   6   -0T   32222
  11   -0*   11100
  (8)  +0*   00000001
   6   +0T   2223
   2   +0F   45
```

Stem and leaf display for 26 residuals

```
STEM-AND-LEAF DISPLAY OF RESIDUAL
LEAF DIGIT UNIT =     .1000
1 2 REPRESENTS 1.2


   1   -0.  9
 (16)  -0*  4333222211110000
   9   +0*  0022223
   2   +0.  6
   1    1*
   1    1.
   1    2*  1
```

14.36 Refer to Exercise 14.7, in which a quadratic model was used to relate the time to complete a task, y, to the months of experience, x, for a sample of 15 employees on an automobile assembly line. The SPSS printouts for both straight-line and quadratic models are given.

  a. Find and interpret the standard deviation of each regression model in the context of this application.
  b. Find and interpret R^2 for each model.
  c. Conduct the analysis of variance F-test for each model. Use $\alpha = .05$ in each case. Interpret the results of the tests.
  d. Test the quadratic term in the second-order model. Use $\alpha = .05$. Does the quadratic term for experience appear to provide information for the prediction of time to complete the task?

Exercise 14.36: Straight-line model

```
Multiple R          .94886
R Square            .90033
Adjusted R Square   .89266
Standard Error      1.14301
```

Analysis of Variance

	DF	Sum of Squares	Mean Square
Regression	1	153.41583	153.41583
Residual	13	16.98417	1.30647

F =   117.42733   Signif F = .0000

------------------ Variables in the Equation ------------------

Variable	B	SE B	Beta	T	Sig T
X	-.44494	.04106	-.94886	-10.836	.0000
(Constant)	19.27908	.50788		37.960	.0000

Exercise 14.36: Quadratic model

```
Multiple R             .95718
R Square               .91619
Adjusted R Square      .90223
Standard Error        1.09089

Analysis of Variance
                    DF    Sum of Squares    Mean Square
Regression           2         156.11948       78.05974
Residual            12          14.28052        1.19004

F  =       65.59402      Signif F  =  .0000


------------------ Variables in the Equation ------------------

Variable           B           SE B        Beta        T    Sig T

X                 -.67052        .15471   -1.42992    -4.334  .0010
XSQ      9.534744E-03  6.32580E-03     .49728     1.507  .1576
(Constant)    20.09111        .72471               27.723  .0000
```

14.10   Some Pitfalls: Estimability, Multicollinearity, and Extrapolation

There are several problems you should be aware of when constructing a prediction model for some response y. A few of the most important are discussed in this section.

| PROBLEM 1

Parameter Estimability

Suppose you want to fit a model relating annual yield y to the total expenditure for fertilizer, x. We propose the first-order model

$$E(y) = \beta_0 + \beta_1 x$$

Now suppose we have 3 years of data and $1,000 is spent on fertilizer each year. The data are shown in Figure 14.28. You can see the problem: The parameters of the model cannot be estimated when all the data are concentrated at a single x value. Recall that it takes two points (x values) to fit a straight line. Thus, the parameters are not estimable when only one x value is observed.

FIGURE 14.28

Yield and fertilizer expenditure data: 3 years

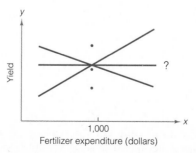

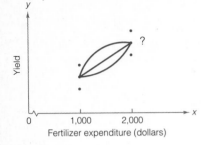

FIGURE 14.29

Only two x values observed—quadratic model is not estimable

| PROBLEM 2

Multicollinearity

A similar problem would occur if we attempted to fit the quadratic model

$$E(y) = \beta_0 + \beta_1 x + \beta_2 x^2$$

to a set of data for which only one or two different x values were observed (see Figure 14.29). At least three different x values must be observed before a quadratic model can be fit to a set of data (that is, before all three parameters are estimable).

In general, the number of levels of observed x values must be one more than the order of the polynomial in x that you want to fit.

For controlled experiments, the researcher can select experimental designs that will permit estimation of the model parameters. Even when the values of the independent variables cannot be controlled by the researcher, the independent variables are almost always observed at a sufficient number of levels to permit estimation of the model parameters. When the computer program you use suddenly refuses to fit a model, however, the problem is probably inestimable parameters.

Often, two or more of the independent variables used in the model for $E(y)$ contribute redundant information. That is, the independent variables are correlated with each other. Suppose we want to construct a model to predict the gasoline mileage rating of a truck as a function of its load, x_1, and the horsepower, x_2, of its engine. In general, you would expect heavy loads to require greater horsepower and to result in lower mileage ratings. Thus, although both x_1 and x_2 contribute information for the prediction of mileage rating, some of the information is overlapping because x_1 and x_2 are correlated.

If the model

$$E(y) = \beta_0 + \beta_1 x_1 + \beta_2 x_2$$

were fit to a set of data, we might find that the t values for both $\hat{\beta}_1$ and $\hat{\beta}_2$ (the least squares estimates) are nonsignificant. However, the F-test for H_0: $\beta_1 = \beta_2 = 0$ would probably be highly significant. The tests may seem to produce contradictory conclusions, but really they do not. The t-tests indicate that the contribution of one variable, say $x_1 = $ Load, is not significant after the effect of $x_2 = $ Horsepower has been taken into account (because x_2 is also in the model). The significant F-test, on the other hand, tells us that at least one of the two variables is making a contribution to the prediction of y (i.e., either β_1, β_2, or both differ from 0). In fact, both are probably contributing, but the contribution of one overlaps with that of the other.

When highly correlated independent variables are present in a regression model, the results are confusing. The researcher may want to include only one of the variables in the final model. One way of deciding which one to include is by using stepwise regression. Generally, only one of a set of multicollinear independent variables is included in a stepwise regression model, since at each step every variable is tested in the presence of all the variables already in the model. For example, if at one step the variable truck load is included as a significant variable in the prediction of the mileage rating, the variable horsepower will probably never be added in a future step. Thus, if a set of independent variables is thought to be multicollinear, some screening by stepwise regression may be helpful.

Note that it would be fallacious to conclude that an independent variable x_1 is unimportant for predicting y *only* because it is not chosen by a stepwise regression procedure. The independent variable x_1 may be correlated with another one, x_2, that the stepwise procedure did select. The implication is that x_2 contributes *more* for predicting y (in the sample being analyzed), but it may still be true that x_1 alone contributes information for the prediction of y.

PROBLEM 3

Prediction Outside the Experimental Region

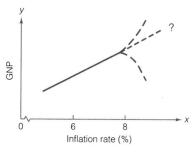

FIGURE 14.30
Using a regression model outside the experimental region

By the late 1960s many research economists had developed highly technical models to relate the state of the economy to various economic indices and other independent variables. Many of these models were multiple regression models, where, for example, the dependent variable y might be next year's growth in GNP and the independent variables might include this year's rate of inflation, this year's CPI, etc. In other words, the model might be constructed to predict next year's economy using this year's knowledge.

Unfortunately, these models were almost all unsuccessful in predicting the recession in the early 1970s. What went wrong? One of the problems was that many of the regression models were used to predict y for values of the independent variables that were outside the region in which the model was developed. For example, the inflation rate in the late 1960s, when the models were developed, ranged from 6% to 8%. When the double-digit inflation of the early 1970s became a reality, some researchers attempted to use the same models to predict future growth in GNP. As you can see in Figure 14.30, the model may be very accurate for predicting y when x is in the range of experimentation, but the use of the model outside that range is a dangerous practice.

PROBLEM 4

Correlated Errors

Another problem associated with using a regression model to predict a variable y based on independent variables x_1, x_2, ... , x_k arises from the fact that the data are frequently **time series**. That is, the values of both the dependent and independent variables are observed sequentially over a period of time. The observations tend to be correlated over time, which in turn often causes the prediction errors of the regression model to be correlated. Thus, the assumption of independent errors is violated, and the model tests and prediction intervals are no longer valid. The solution to this problem is to construct a **time series model**; the interested reader should consult the references for details of time series analysis.

CASE STUDY 14.2

AN APPARENT INCONSISTENCY: THE CULPRIT, MULTICOLLINEARITY

Studying the relationship between worker strike activity and union membership, Neil A. Palomba* encountered a common problem in modeling. Palomba attempted to relate the strike activity in a state (the percentage y of total working hours lost due to strikes) to three independent variables:

---

*Reprinted from Palomba, N. A. "Strike activity and union membership: An empirical approach," *University of Washington Business Review*, Winter, 1969.

x_1 = Percentage of union members in nonagricultural establishments

x_2 = Percentage of all nonagricultural employment that is manufacturing

x_3 = Hourly earnings of workers on manufacturing payrolls

Fitting a series of first-order models involving x_1, x_2, and x_3 to y, he got conflicting signals concerning the importance of these independent variables in predicting strike activity.

The first step in Palomba's modeling process, relating x_1 to y by using a simple linear regression model, produced the prediction equation

$$\hat{y} = .018201 + .005927x_1 \qquad \text{with } R^2 = .137 \tag{1}$$
$$(.002167)$$
$$2.735$$

The standard error of $\hat{\beta}_1$, .002167, is shown in parentheses below $\hat{\beta}_1$ and the t value for testing H_0: $\beta_1 = 0$, 2.735, is shown beneath the standard error. Since this t value is highly significant, Palomba says it "indicates that union membership has a statistically significant positive effect on the level of strike activity."

Does Palomba's statement imply a cause and effect relationship between x_1 and y? If this is the suggestion, then a problem quickly develops when he fits a first-order model to the data using all three independent variables. The prediction equation, with the standard errors (in parentheses) and computed t values for the parameter estimates, is

$$\hat{y} = -.040205 + .000308x_1 + .003905x_2 + .184774x_3 \qquad \text{with } R^2 = .232$$
$$\qquad\qquad\quad (.003179) \quad (.002260) \quad (.081428)$$
$$\qquad\qquad\quad .097 \qquad\quad 1.728 \qquad\quad 2.269$$

As Palomba observes, this equation "reveals some surprising results." He notes that R^2 increases dramatically (although it is still very small). But above all he notes that union membership, so important in prediction equation (1), is no longer statistically significant. Neither is x_2, the measure of the labor force involved in manufacturing. The statistically significant variable in the prediction equation is hourly wage x_3, "meaning that the level of wages has a positive effect upon strike activity, while the level of union membership has no effect." Since this conclusion is the reverse of the conclusion reached in analyzing model (1), we wonder how many conflicting conclusions we could create if we fit many models and, more important, we might wonder why the conflicting conclusions occur.

Based on Palomba's analysis, which variable—union membership x_1 or hourly earnings x_3—causes strike activity y to increase? A regression analysis cannot answer this question. All that a regression analysis can do is indicate whether an independent variable, as it is entered into the model, contributes information for the prediction of y. It does not indicate a cause and effect relationship. In Palomba's modeling, it seems reasonable to suspect a relationship between union membership x_1 and hourly earnings x_3. Perhaps workers who are union members receive higher hourly earnings. In any case, it appears that x_1 and x_3 are correlated and therefore contribute overlapping information for the prediction of y. This common situation is called **multicollinearity**. When multicollinearity is present, a regression analysis introduces the variables into the model in a way that ensures the best prediction equation. If changes in one of the variables actually cause the strike activity y to change, it may or may not appear to be statistically significant. Some variable highly correlated to the causative variable may replace it in the prediction equation.

EXERCISES 14.37—14.38

APPLYING THE CONCEPTS

14.37 A physiologist wishes to investigate the relationship between the physical characteristics of pre-adolescent boys and their maximal oxygen uptake (measured in milliliters of oxygen per kilogram of body weight). The data shown in the table are collected on a random sample of ten preadolescent boys.

MAXIMAL OXYGEN UPTAKE y	AGE x_1 (years)	HEIGHT x_2 (centimeters)	WEIGHT x_3 (kilograms)	CHEST DEPTH x_4 (centimeters)
1.54	8.4	132.0	29.1	14.4
1.74	8.7	135.5	29.7	14.5
1.32	8.9	127.7	28.4	14.0
1.50	9.9	131.1	28.8	14.2
1.46	9.0	130.0	25.9	13.6
1.35	7.7	127.6	27.6	13.9
1.53	7.3	129.9	29.0	14.0
1.71	9.9	138.1	33.6	14.6
1.27	9.3	126.6	27.7	13.9
1.50	8.1	131.8	30.8	14.5

As a first step in the data analysis, the researcher fits the regression model

$$y = \beta_0 + \beta_1 x_1 + \beta_2 x_2 + \beta_3 x_3 + \beta_4 x_4 + \varepsilon$$

to the data. The output for a SAS regression analysis is shown.

SAS printout for Exercise 14.37

Source	DF	Sum of Squares	Mean Square	F Value	Prob>F
Model	4	0.20604	0.05151	37.204	0.0007
Error	5	0.00692	0.00138		
C Total	9	0.21296			

Root MSE	0.03721	R-Square	0.9675	
Dep Mean	1.49200	Adj R-Sq	0.9415	
C.V.	2.49391			

Parameter Estimates

Variable	DF	Parameter Estimate	Standard Error	T for H0: Parameter=0	Prob > |T|
INTERCEP	1	−4.774739	0.86281773	−5.534	0.0026
AGE	1	−0.035214	0.01538630	−2.289	0.0708
HEIGHT	1	0.051637	0.00621522	8.308	0.0004
WEIGHT	1	−0.023417	0.01342835	−1.744	0.1416
CHEST	1	0.034489	0.08523877	0.405	0.7025

a. Give the value of R^2 and interpret its value.
b. It seems reasonable to assume that the greater a child's chest depth, the greater should be the maximal oxygen uptake. But note that $\hat{\beta}_4$, the estimated coefficient of chest depth, x_4, is negative. Give an explanation for this result.
c. It would seem that the weight of a child should be positively correlated to lung volume and hence to maximal oxygen uptake. Can you explain the small t value associated with $\hat{\beta}_3$?
d. If you have access to the appropriate computer package, find a 95% prediction interval for maximal oxygen uptake for a boy with age = 8.8, height = 128.1, weight = 28.0, and chest depth = 13.7.

14.38   Refer to Exercise 14.37. Calculate the Pearson product moment correlation coefficients between all pairs of the independent variables x_1–x_4. Do these correlations provide an explanation for the confusing signs and small t values associated with the estimated regression coefficients of the model fit in Exercise 14.37?

Summary

We have discussed some of the methodology of multiple regression analysis, a technique for modeling a dependent variable y as a function of several independent variables $x_1, x_2, \ldots, x_k$. The steps we follow in constructing and using multiple regression models are much the same as those for the simple straight-line models:

1. The form of the probabilistic model is hypothesized.
2. The model coefficients are estimated using least squares.
3. The probability distribution of ε is specified and σ^2 is estimated.
4. The utility of the model is checked.
5. If the model is deemed useful, it may be used to make estimates of $E(y)$ and to predict values of y to be observed in the future.

We have covered steps 2 to 5 in Chapter 14, assuming that the model was specified. Model construction—step 1—is discussed in Chapter 15.

SUPPLEMENTARY EXERCISES 14.39–14.65

[Note: Starred (*) exercises require the use of a computer.]

LEARNING THE MECHANICS

14.39   Suppose you fit the model

$$y = \beta_0 + \beta_1 x_1 + \beta_2 x_1^2 + \beta_3 x_2 + \beta_4 x_1 x_2 + \varepsilon$$

to $n = 25$ data points and find that

$$\hat{\beta}_0 = 1.26 \qquad \hat{\beta}_1 = -2.43 \qquad \hat{\beta}_2 = .05 \qquad \hat{\beta}_3 = .62 \qquad \hat{\beta}_4 = 1.81$$

$$\text{SSE} = .41 \qquad R^2 = .83$$

$$s_{\hat{\beta}_1} = 1.21 \qquad s_{\hat{\beta}_2} = .16 \qquad s_{\hat{\beta}_3} = .26 \qquad s_{\hat{\beta}_4} = 1.49$$

a. Is there sufficient evidence to conclude that at least one of the parameters β_1, β_2, β_3, or β_4 is nonzero? Test using $\alpha = .05$.
b. Test H_0: $\beta_1 = 0$ against H_a: $\beta_1 < 0$. Use $\alpha = .05$.
c. Test H_0: $\beta_2 = 0$ against H_a: $\beta_2 > 0$. Use $\alpha = .05$.
d. Test H_0: $\beta_3 = 0$ against H_a: $\beta_3 \neq 0$. Use $\alpha = .05$.

14.40   Suppose you used Minitab to fit the model

$$y = \beta_0 + \beta_1 x_1 + \beta_2 x_2 + \varepsilon$$

to $n = 15$ data points and you obtained the printout shown.
a. What is the least squares prediction equation?
b. Find R^2 and interpret its value.
c. Is there sufficient evidence to indicate that the model is useful for predicting y? Conduct an F-test using $\alpha = .05$.

Minitab printout for Exercise 14.40

```
THE REGRESSION EQUATION IS
Y =     90.1 -  1.84 X1 +  .285 X2

                              ST. DEV.    T-RATIO =
          COLUMN   COEFFICIENT   OF COEF.    COEF/S.D.

          --              90.1        23.1         3.90
X1   C2            -1.836        .367        -5.01
X2   C3             .285         .231         1.24

THE ST. DEV. OF Y ABOUT REGRESSION LINE IS
S =      10.7
WITH (  15- 3) =  12 DEGREES OF FREEDOM

R-SQUARED = 91.6 PERCENT
R-SQUARED = 90.2 PERCENT, ADJUSTED FOR D.F.

ANALYSIS OF VARIANCE

   DUE TO      DF      SS     MS=SS/DF

REGRESSION     2    14801.     7400.
RESIDUAL      12     1364.      114.
TOTAL         14    16165.
```

d. Test the null hypothesis H_0: $\beta_1 = 0$ against the alternative hypothesis H_a: $\beta_1 \neq 0$. Test using $\alpha = .05$. Draw the appropriate conclusions.

e. Find the standard deviation of the regression model, and interpret it.

14.41 After a regression model is fit to a set of data, a confidence interval for the mean value of y at a given setting of the independent variables is *always* narrower than the corresponding prediction interval for a particular value of y at the same setting of the independent variables. Why?

*14.42 A response variable y was observed for $n = 15$ settings of two independent variables, x_1 and x_2. The data are shown in the table.

y	x_1	x_2
−12	3	9
28	8	7
−15	3	1
−25	2	6
2	9	1
−11	5	1
25	7	7
−47	0	8
−28	1	6
12	5	9
−29	1	4
−3	5	5
5	7	3
7	6	4
6	5	9

a. Fit the model $y = \beta_0 + \beta_1 x_1 + \beta_2 x_2 + \varepsilon$ to the data.

b. Fit the model $y = \beta_0 + \beta_1 x_1 + \beta_2 x_2 + \beta_3 x_1 x_2 + \varepsilon$ to the data.

c. Which of the models in parts **a** and **b** best describes the relationship between y and x_1 and x_2? Explain.

d. Repeat part **a** using a different statistical program package. For example, if you originally used Minitab, now use SAS, SPSS, or another available package. Compare these results with those found in part **a**.

14.43 Suppose you have developed a regression model to explain the relationship between y and x_1, x_2, and x_3. A set of $n = 15$ data points is used to find the least squares prediction equation. The ranges of the variables you observed were as follows: $10 \le y \le 100$, $5 \le x_1 \le 55$, $.5 \le x_2 \le 1$, and $1,000 \le x_3 \le 2,000$. Will the error of prediction be smaller when you use the least squares equation to predict y when $x_1 = 30$, $x_2 = .6$, and $x_3 = 1,300$, or when $x_1 = 60$, $x_2 = .4$, and $x_3 = 900$? Why?

APPLYING THE CONCEPTS

14.44 Several states now require all high school seniors to pass an achievement test before they can graduate. On the test, the seniors must demonstrate their familiarity with basic verbal and mathematical skills. Suppose the educational testing company that creates and administers these exams wants to model the score, y, on one of its exams as a function of the student's IQ, x_1, and socioeconomic status (SES). The SES is a *categorical* (or *qualitative*) variable with three levels: low, medium, and high. As we will demonstrate in Chapter 15, two *dummy* (*indicator*) variables are needed to describe a qualitative independent variable with three levels. Thus, we define

$$x_2 = \begin{cases} 1 & \text{if SES is medium} \\ 0 & \text{if SES is low or high} \end{cases} \qquad x_3 = \begin{cases} 1 & \text{if SES is high} \\ 0 & \text{if SES is low or medium} \end{cases}$$

Data were collected for a random sample of sixty seniors who have taken the test, and the model

$$E(y) = \beta_0 + \beta_1 x_1 + \beta_2 x_2 + \beta_3 x_3$$

was fit to the data, with the results shown in the SAS printout.

Printout for Exercise 14.44

SOURCE	DF	SUM OF SQUARES	MEAN SQUARE	F VALUE	PR > F
MODEL	3	12268.56439492	4089.52146497	188.33	0.0001
ERROR	56	1216.01893841	21.71462390		
CORRECTED TOTAL	59	13484.58333333		R-SQUARE	ROOT MSE
				0.909822	4.65989527

PARAMETER	ESTIMATE	T FOR H0: PARAMETER=0	PR > ¦T¦	STD ERROR OF ESTIMATE
INTERCEPT	-13.06166081	-3.21	0.0022	4.07101383
X1	0.74193946	17.56	0.0001	0.04224805
X2	18.60320572	12.49	0.0001	1.48895324
X3	13.40965415	8.97	0.0001	1.49417069

a. Identify the least squares equation.

b. Interpret the value of R^2 and test to determine whether the data provide sufficient evidence to indicate that this model is useful for predicting achievement test scores.

c. Sketch the relationship between predicted achievement test score and IQ for the three levels of SES. [*Note:* Three graphs of $\hat{y}$ versus x_1 must be drawn: the first for the low SES model ($x_2 = x_3 = 0$), the second for the medium SES model ($x_2 = 1$, $x_3 = 0$), and the third for the high SES model ($x_2 = 0$, $x_3 = 1$). The increase in predicted achievement test score per unit increase in IQ is the same for all three levels of SES; i.e., all three lines are parallel.]

14.45   Refer to Exercise 14.44. We now use the same data to fit the model

$$E(y) = \beta_0 + \beta_1 x_1 + \beta_2 x_2 + \beta_3 x_3 + \beta_4 x_1 x_2 + \beta_5 x_1 x_3$$

Thus, we now add the interaction between IQ and SES to the model. The SAS printout for this model is shown here.

Printout for Exercise 14.45

SOURCE	DF	SUM OF SQUARES	MEAN SQUARE	F VALUE	PR > F
MODEL	5	12515.10021009	2503.02004202	139.42	0.0001
ERROR	54	969.48312324	17.95339117		
CORRECTED TOTAL	59	13484.58333333		R-SQUARE	ROOT MSE
				0.928104	4.23714422

PARAMETER	ESTIMATE	T FOR H0: PARAMETER = 0	PR > !T!	STD ERROR OF ESTIMATE
INTERCEPT	0.60129643	0.11	0.9096	5.26818519
X1	0.59526252	10.70	0.0001	0.05563379
X2	-3.72536406	-0.37	0.7115	10.01967496
X3	-16.23196444	-1.90	0.0631	8.55429931
X1*X2	0.23492147	2.29	0.0260	0.10263908
X1*X3	0.30807756	3.53	0.0009	0.08739554

a. Identify the least squares prediction equation.
b. Interpret the value of R^2 and test to determine whether the data provide sufficient evidence to indicate that this model is useful for predicting achievement test scores.
c. Sketch the relationship between predicted achievement test score and IQ for the three levels of SES. [*Note:* The interaction terms in the model allow nonparallelism among the three SES models; i.e., the mean increase in achievement test score per unit increase in IQ differs for the three levels of SES. To determine whether the interaction between IQ and SES is contributing to the prediction of achievement test score, we must test the null hypothesis $H_0: \beta_4 = \beta_5 = 0$ against the alternative that at least one of the coefficients of the interaction terms is nonzero. The method for testing portions of a regression model involving more than one β parameter (but less than all of them) is discussed in Chapter 15.]

14.46   A large government agency would like to predict the number of people it will hire within the next year to fill the 30 positions that are currently open. Historically, the agency has been unable to fill all its job openings. It has been decided to model the number of positions filled in a year, y, as a function of the number of positions open, x_1, and the recruiting budget (in dollars) for the year, x_2 (e.g., for advertising the positions, paying travel expenses, etc.). A random sample of 10 years of recruiting records was drawn from the agency's 30 years of records. The model

$$E(y) = \beta_0 + \beta_1 x_1 + \beta_2 x_2$$

was fit to the data using the Minitab regression computer program package. The results shown in the computer printout on page 778 were obtained.
a. Identify the least squares prediction equation.
b. Is there sufficient evidence to indicate that the model contributes information for predicting the number y of positions that will be filled? Conduct an F-test using $\alpha = .05$.
c. Test the null hypothesis $H_0: \beta_2 = 0$ against the alternative hypothesis $H_a: \beta_2 \neq 0$ using $\alpha = .05$. Interpret the results of your test in the context of the problem.
d. Use the least squares prediction equation to predict how many of the 30 positions the agency will fill next year if the recruiting budget is $10,000.

Printout for Exercise 14.46

```
THE REGRESSION EQUATION IS
Y=    .0562+   .273 X1+   .0006 X2
                                        ST. DEV.        T-RATIO=
            COLUMN      COEFFICIENT     OF COEF.        COEF/S.D.

            --              .056          .902            .06
X1          C2             .2733         .0971           2.81
X2          C3             .000560       .000129         4.34

THE ST. DEV. OF Y ABOUT REGRESSION LINE IS
S=      1.33
WITH (  10-3)=    7 DEGREES OF FREEDOM

R-SQUARED = 97.9 PERCENT
R-SQUARED = 97.3 PERCENT, ADJUSTED FOR D.F.

ANALYSIS OF VARIANCE

DUE TO       DF      SS    MS=SS/DF

REGRESSION   2    583.18    291.59
RESIDUAL     7     12.42      1.77
TOTAL        9    595.60
```

e. Can you think of a situation for which the prediction in part **d** might possibly be inaccurate? Explain.

f. Which (if any) of the assumptions we make about ε in a regression analysis are likely to be violated in this problem? Explain.

*14.47   The data set shown here gives the number y of births per year (in thousands), the number x_1 of marriages per year (in thousands), the number x_2 of women in the work force (in thousands), and the availability (yes) or nonavailability (no) of the birth control pill over the years 1949–1982.

YEAR	BIRTHS	MARRIAGES	WOMEN WORKING	TAKING THE PILL	YEAR	BIRTHS	MARRIAGES	WOMEN WORKING	TAKING THE PILL
1949	3,560	1,580	18,030	No	1966	3,606	1,857	26,820	Yes
1950	3,632	1,667	18,680	No	1967	3,521	1,927	27,545	Yes
1951	3,750	1,595	19,309	No	1968	3,502	2,069	28,778	Yes
1952	3,847	1,539	19,559	No	1969	3,606	2,145	29,898	Yes
1953	3,902	1,546	19,668	No	1970	3,731	2,159	31,233	Yes
1954	4,017	1,490	19,970	No	1971	3,556	2,190	31,778	Yes
1955	4,097	1,518	20,842	No	1972	3,258	2,282	33,152	Yes
1956	4,168	1,569	21,808	No	1973	3,137	2,284	34,195	Yes
1957	4,255	1,518	22,097	No	1974	3,160	2,230	35,708	Yes
1958	4,204	1,451	22,482	No	1975	3,144	2,153	36,981	Yes
1959	4,245	1,494	22,865	No	1976	3,168	2,155	38,399	Yes
1960	4,258	1,523	23,619	No	1977	3,327	2,178	40,053	Yes
1961	4,268	1,548	24,199	No	1978	3,333	2,282	41,747	Yes
1962	4,167	1,580	23,978	No	1979	3,494	2,331	43,844	Yes
1963	4,098	1,654	24,675	Yes	1980	3,598	2,413	44,934	Yes
1964	4,027	1,725	25,399	Yes	1981	3,646	2,438	46,415	Yes
1965	3,760	1,800	25,952	Yes	1982	3,704	2,495	47,095	Yes

Source: *Statistical Abstract of the U.S.*

a. Use a computer to fit the model

$$E(y) = \beta_0 + \beta_1 x_1 + \beta_2 x_2 + \beta_3 x_3$$

to the data. The independent variable x_3 is a dummy (or indicator) variable defined as follows:

$$x_3 = \begin{cases} 1 & \text{if birth control pills were available in a given year} \\ 0 & \text{if not available} \end{cases}$$

Dummy variables enable us to enter qualitative independent variables into a model. Their use and interpretation are discussed in Chapter 15.

b. Interpret the results of your regression analysis.

14.48 Most companies institute rigorous programs to assure employee safety. Suppose 60 accident reports over the last year at a company are sampled, and the number of hours the employee had worked before the accident occurred, x, and the amount of time the employee lost from work, y, are recorded. A quadratic model is proposed to investigate a fatigue hypothesis that more serious accidents occur near the end of workdays than near the beginning. Thus, the proposed model is

$$E(y) = \beta_0 + \beta_1 x + \beta_2 x^2$$

A portion of the computer printout appears as shown.

Printout for Exercise 14.48

SOURCE	DF	SUM OF SQUARES	MEAN SQUARE	F VALUE
MODEL	2	112.110	56.055	1.28
ERROR	57	2496.201	43.793	R-SQUARE
TOTAL	59	2608.311		0.0430

a. Do these data support the fatigue hypothesis? Use $\alpha = .05$ to test whether the proposed model is useful in predicting the lost work time y.

b. Does the result of the test in part **a** necessarily mean that no fatigue factor exists? Explain.

14.49 Refer to Exercise 14.48. Suppose the company persists in using the quadratic model despite its apparent lack of utility. The fitted model is

$$\hat{y} = 12.3 + .25x - .0033x^2$$

where $\hat{y}$ is the predicted time lost (days) and x is the number of hours worked prior to an accident.

a. Use the model to predict the number of days missed by an employee who has an accident after 6 hours of work.

b. Suppose the 95% prediction interval for the predicted value in part **a** (1.35, 26.01). What is the interpretation of this interval? Does this interval support your conclusion about this model in Exercise 14.48?

*14.50 Refer to Exercises 14.11 and 14.25, in which regression analysis was used to explore the valuation of apartment buildings in Minneapolis.

a. Verify that number of apartment units, x_1, is highly correlated with gross building area, x_5.

b. Eliminate x_1 from the appraiser's model and refit the model to the data. Compare the standard deviations of the two models.

c. Use an F-test to investigate the usefulness of this model. Use $\alpha = .05$.

d. Use the least squares equation of part **b** to estimate the value of an apartment building that is 50 years old with gross area 20,000 square feet, lot size 15,000 square feet, and 10 parking spaces. Use a computer program to compute a 95% confidence interval for the model's estimate.

14.51 To increase the motivation and productivity of workers, an electronics manufacturer decides to experiment with a new pay incentive structure at one of two plants. The experimental plan will be tried at plant A for 6 months, while workers at plant B will remain on the original pay plan. To evaluate the effectiveness of the new plan, the average assembly time for part of an electronic system is measured for employees at both plants at the beginning and end of the 6-month period. Suppose the following model is proposed:

$$y = \beta_0 + \beta_1 x_1 + \beta_2 x_2 + \varepsilon$$

where

y = Assembly time (hours) at end of 6-month period

x_1 = Assembly time (hours) at beginning of 6-month period

$x_2 = \begin{cases} 1 & \text{if plant A} \\ 0 & \text{if plant B} \end{cases}$ (dummy variable)

A sample of $n = 42$ observations yields

$$\hat{y} = .11 + .98x_1 - .53x_2$$

where

$$s_{\hat{\beta}_1} = .231 \qquad s_{\hat{\beta}_2} = .48$$

Test to determine whether, after allowing for the effect of initial assembly time, plant A had a lower mean assembly time than plant B. Use $\alpha = .01$. [*Note:* When the (0, 1) coding is used to define a dummy variable, the coefficient of the variable represents the difference between the mean response at the two levels represented by the variable. Thus, the coefficient β_2 is the difference in mean assembly time between plant A and plant B at the end of the 6-month period and $\hat{\beta}_2$ is the sample estimator of that difference.]

14.52   The EPA wants to model the gas mileage ratings, y, of automobiles as a function of their engine size, x. A quadratic (second-order) model

$$y = \beta_0 + \beta_1 x + \beta_2 x^2 + \varepsilon$$

is proposed. A sample of 50 engines of varying sizes is selected and the miles per gallon rating of each is determined. The least squares prediction equation is

$$\hat{y} = 51.3 - 10.1x + .15x^2$$

The size x of the engine is measured in hundreds of cubic inches. Moreover, $s_{\hat{\beta}_2} = .0037$ and $R^2 = .93$.

a. Sketch the prediction equation for values of x between $x = 1$ and $x = 4$.
b. Is there evidence that the quadratic term in the model is contributing to the prediction of the miles per gallon rating, y? Use $\alpha = .05$.
c. Use the model to estimate the mean miles per gallon rating for all cars with 350 cubic inch engines ($x = 3.5$).
d. Suppose a 95% confidence interval for the quantity estimated in part c is (17.2, 18.4). Interpret this interval.
e. Suppose you purchase an automobile with a 350 cubic inch engine and determine that the miles per gallon rating is 14.7. Is the fact that this value lies outside the confidence interval given in part d surprising? Explain.

14.53   To determine whether extra personnel are needed for the day, the owners of a water adventure park would like to find a model that would allow them to predict the day's attendance each morning before opening based on the day of the week and weather conditions. The model is of the form

$$E(y) = \beta_0 + \beta_1 x_1 + \beta_2 x_2 + \beta_3 x_3$$

where

$$y = \text{Daily admissions}$$

$$x_1 = \begin{cases} 1 & \text{if weekend} \\ 0 & \text{otherwise} \end{cases} \quad \text{(dummy variable)}$$

$$x_2 = \begin{cases} 1 & \text{if sunny} \\ 0 & \text{if overcast} \end{cases} \quad \text{(dummy variable)}$$

$$x_3 = \text{Predicted daily high temperature (°F)}$$

These data were recorded for a random sample of 30 days and a regression model was fit to the data. The least squares analysis produced the following results:

$$\hat{y} = -105 + 25x_1 + 100x_2 + 10x_3$$

with

$$s_{\hat{\beta}_1} = 10 \qquad s_{\hat{\beta}_2} = 30 \qquad s_{\hat{\beta}_3} = 4 \qquad R^2 = .65$$

a. Interpret the estimated model coefficients.
b. Is there sufficient evidence to conclude that this model is useful for the prediction of daily attendance? Use $\alpha = .05$.
c. Is there sufficient evidence to conclude that mean attendance increases on weekends? Use $\alpha = .10$.
d. Use the model to predict the attendance on a sunny weekday with a predicted high temperature of 95°F.
e. Suppose the 90% prediction interval for part **d** is (645, 1,245). Interpret this interval.

14.54 Refer to Exercise 14.53. The owners of the water adventure park are advised that the prediction model could probably be improved if interaction terms were added. In particular, it is thought that the *rate* that mean attendance increases as predicted high temperature increases will be greater on weekends than on weekdays. The following model is therefore proposed:

$$E(y) = \beta_0 + \beta_1 x_1 + \beta_2 x_2 + \beta_3 x_3 + \beta_4 x_1 x_3$$

The same 30 days of data used in Exercise 14.53 are again used to obtain the least squares model

$$\hat{y} = 250 - 700x_1 + 100x_2 + 5x_3 + 15x_1 x_3$$
$$s_{\hat{\beta}_4} = 3.0 \qquad R^2 = .96$$

a. Graph the predicted day's attendance, y, against the day's predicted high temperature, x_3, for a sunny weekday and for a sunny weekend day. Plot both on the same paper for x_3 between 70°F and 100°F. Note the increase in slope for the weekend day. Interpret this.
b. Do the data indicate that the interaction term is a useful addition to the model? Use $\alpha = .05$.
c. Use this model to predict the attendance for a sunny weekday with a predicted high temperature of 95°F.
d. Suppose the 90% prediction interval for part **c** is (800, 850). Compare this result with the prediction interval for the model without interaction in Exercise 14.53, part **e**. Do the relative widths of the confidence intervals support or refute your conclusion about the utility of the interaction term (part **b**)?

14.55 Refer to Exercise 14.54. The owners, noting that the coefficient $\hat{\beta}_1 = -700$, conclude the model is ridiculous because it seems to imply that the mean attendance will be 700 less on weekends than on weekdays. Explain why this is *not* the case.

*14.56 Refer to Exercise 13.28. The breeder of thoroughbred horses has been advised that the prediction model could probably be improved if a quadratic term were added. The following model is therefore proposed:

$$y = \beta_0 + \beta_1 x + \beta_2 x^2 + \varepsilon$$

where, as before

y = Lifetime of horse (years)

x = Gestation period of horse (days)

a. Find the least squares prediction equation and test its adequacy.
b. Has the addition of the quadratic term contributed significant information for the prediction of a thoroughbred horse's lifetime? Test H_0: $\beta_2 = 0$ against the alternative H_a: $\beta_2 \neq 0$ using $\alpha = .05$.
c. Construct residual plots for the straight-line and quadratic models, displaying gestation period on the horizontal axis. Does the result of the residual analysis agree with the test you conducted in part b? Which is more reliable, and why?

14.57 Recent increases in gasoline prices have increased interest in modes of transportation other than the automobile. A metropolitan bus company wants to know if changes in numbers of bus riders are related to changes in gasoline prices. By using information from company files and gasoline price information obtained from fuel distributors, the company plans to fit the following model:

$$y = \beta_0 + \beta_1 x_1 + \beta_2 x_2 + \beta_3 x_1 x_2 + \varepsilon$$

where

x_1 = Average wholesale price for regular gas in a given month

$x_2 = \begin{cases} 1 & \text{if the bus travels a city route only} \\ 0 & \text{if the bus travels a suburb–city route} \end{cases}$

y = Total number of riders in a bus over the month

a. For this model, how would you test to determine whether the relationship between the mean number of riders and gasoline price is different for the two different bus routes?
b. Suppose 12 months of data are kept, and the least squares model is

$$\hat{y} = 500 + 50x_1 + 5x_2 - 10x_1 x_2$$

Graph the predicted relationship between number of riders and gas price for city buses and for suburb–city buses. Compare the slopes.
c. If $s_{\hat{\beta}_3} = 3.0$, do the data indicate that gas price affects the number of riders differently for city and suburb–city buses? Use $\alpha = .05$.

14.58 During the winter months a sample of 100 homes is taken to obtain information concerning the relationship between kilowatt usage, y, and total window and glass area, x (measured as a percentage of total wall area). The correlation coefficient for the data is equal to .24. Test to determine whether the data indicate that a linear relationship exists between y and x. Use $\alpha = .05$. [Hint: Use the methods of Section 14.5.]

14.59 Plastics made under different environmental conditions are known to have differing strengths. A scientist would like to know which combination of temperature and pressure yields a plastic with a high breaking strength. A small preliminary experiment is run at two pressure levels and two temperature levels. The following model is proposed:

$$E(y) = \beta_0 + \beta_1 x_1 + \beta_2 x_2 + \beta_3 x_1 x_2$$

where

$$y = \text{Breaking strength (pounds)}$$
$$x_1 = \text{Temperature (°F)}$$
$$x_2 = \text{Pressure (pounds per square inch)}$$

A sample of $n = 16$ observations yields

$$\hat{y} = 226.8 + 4.9x_1 + 1.2x_2 - .7x_1x_2$$

with

$$s_{\hat{\beta}_1} = 1.11 \qquad s_{\hat{\beta}_2} = .27 \qquad s_{\hat{\beta}_3} = .34$$

Do the data indicate an interaction between temperature and pressure? Test using $\alpha = .05$.

*14.60 A florist is interested in how sunlight and misting affect the thickness of the leaves on a new variety of ivy. Sixteen ivy plants of the same age and appearance are randomly divided into four groups of four plants each. One group receives no misting and indirect sunlight; one, misting and indirect sunlight; one, no misting and direct sunlight; and the last, misting and direct sunlight. After 3 months, the thickness of a center leaf from each plant is measured with the results (in millimeters) listed in the table.

		SUNLIGHT	
		Direct	Indirect
MISTING	No mist	.64, .63, .61, .62	.60, .61, .58, .59
	Mist	.65, .63, .64, .64	.62, .61, .60, .60

a. Fit the model $y = \beta_0 + \beta_1x_1 + \beta_2x_2 + \beta_3x_1x_2 + \varepsilon$ to the data, where

$$x_1 = \begin{cases} 1 & \text{if mist was present} \\ 0 & \text{if no mist} \end{cases} \qquad x_2 = \begin{cases} 1 & \text{if direct sunlight} \\ 0 & \text{if indirect sunlight} \end{cases}$$

b. Is there sufficient evidence to conclude the model is useful in predicting thickness of leaves on this variety of ivy? Use $\alpha = .05$.

c. Is there evidence of an interaction between sunlight conditions and misting conditions? Test using $\alpha = .05$.

14.61 Many students must work part-time to help finance their college education. A survey of 100 students was completed at a university to determine whether the number of hours worked per week, x, was affecting their grade-point average, y. A quadratic model was proposed:

$$y = \beta_0 + \beta_1x + \beta_2x^2 + \varepsilon$$

The 100 observations yielded the least squares model

$$\hat{y} = 2.8 - .005x - .0002x^2$$

with $R^2 = .12$ and $s = .5$.

a. Do these statistics indicate that the model is useful in explaining grade-point averages? Use $\alpha = .05$.

b. Interpret the values of R^2 and s in terms of this application of regression analysis. Do you think the quadratic relationship between y and x is strong? Approximately how precisely would you expect to be able to predict a particular student's grade-point average if you knew the number of hours he or she worked per week?

c. If you changed the model to a straight-line model relating grade-point average to number of hours worked, would the value of R^2 increase, thereby providing a better prediction model for grade-point average? Explain.

14.62   Many colleges and universities develop regression models for predicting the grade-point average (GPA) of incoming freshmen. This predicted GPA can then be used to make admission decisions. Although most models use many independent variables to predict GPA, we will illustrate by choosing two variables:

x_1 = Verbal score on college entrance examination (percentile)

x_2 = Mathematics score on college entrance examination (percentile)

The data in the table are obtained for a random sample of 40 freshmen at one college.

VERBAL x_1	MATHEMATICS x_2	GPA y	VERBAL x_1	MATHEMATICS x_2	GPA y
81	87	3.49	79	75	3.45
68	99	2.89	81	62	2.76
57	86	2.73	50	69	1.90
100	49	1.54	72	70	3.01
54	83	2.56	54	52	1.48
82	86	3.43	65	79	2.98
75	74	3.59	56	78	2.58
58	98	2.86	98	67	2.73
55	54	1.46	97	80	3.27
49	81	2.11	77	90	3.47
64	76	2.69	49	54	1.30
66	59	2.16	39	81	1.22
80	61	2.60	87	69	3.23
100	85	3.30	70	95	3.82
83	76	3.75	57	89	2.93
64	66	2.70	74	67	2.83
83	72	3.15	87	93	3.84
93	54	2.28	90	65	3.01
74	59	2.92	81	76	3.33
51	75	2.48	84	69	3.06

The SPSS printout corresponding to the model

$$y = \beta_0 + \beta_1 x_1 + \beta_2 x_2 + \varepsilon$$

is shown at the top of page 785.

a. Interpret the least squares estimates $\hat{\beta}_1$ and $\hat{\beta}_2$ in the context of this application.

b. Interpret the standard deviation and the coefficient of determination of the regression model in the context of this application.

c. Is this model useful for predicting GPA? Conduct a statistical test to justify your answer.

d. Sketch the relationship between predicted GPA, $\hat{y}$, and verbal score, x_1, for the following mathematics scores: x_2 = 60, 75, and 90.

14.63   Refer to Exercise 14.62. The residuals from the first-order model are plotted against x_1 at the bottom of page 785 and against x_2 at the top of page 786. Analyze the two plots, and determine whether visual evidence exists that curvature (a quadratic term) for either x_1 or x_2 should be added to the model.

Printout for Exercise 14.62

```
Multiple R              .82527
R Square                .68106
Adjusted R Square       .66382
Standard Error          .40228

Analysis of Variance
                   DF       Sum of Squares      Mean Square
Regression          2          12.78595          6.39297
Residual           37           5.98755           .16183

F =      39.50530      Signif F =   .0000


----------------- Variables in the Equation -----------------
                                                      .
Variable          B        SE B        Beta        T    Sig T

X1           .02573 4.02357E-03      .59719      6.395   .0000
X2           .03361 4.92751E-03      .63702      6.822   .0000
(Constant)  -1.57054     .49375                 -3.181   .0030
```

Plot of residuals against x_1 for Exercise 14.63

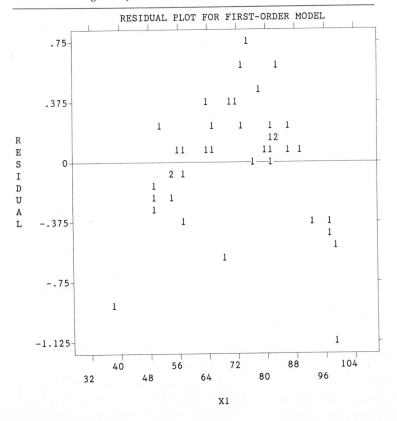

40 cases plotted.

Plot of residuals against x_2 for Exercise 14.63

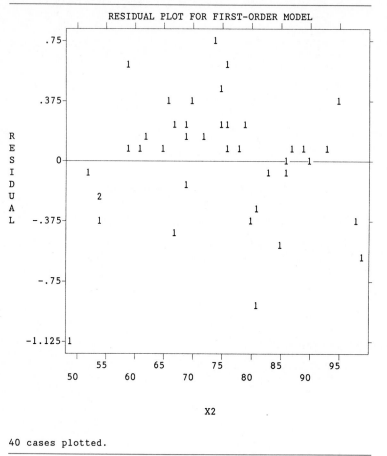

40 cases plotted.

14.64   Refer to Exercises 14.62 and 14.63. The complete second-order model

$$y = \beta_0 + \beta_1 x_1 + \beta_2 x_2 + \beta_3 x_1^2 + \beta_4 x_2^2 + \beta_5 x_1 x_2 + \varepsilon$$

is fit to the data given in Exercise 14.62. The resulting SPSS printout is given at the top of page 787.

a. Compare the standard deviations of the first- and second-order regression models. With what relative precision will these two models predict GPA?

b. Test whether this model is useful for predicting GPA. Use $\alpha = .05$.

c. Test whether the interaction term, $\beta_5 x_1 x_2$, is important for the prediction of GPA. Use $\alpha = .10$. Note that this term permits the distance between three mathematics score curves for GPA versus verbal score to change as the verbal score changes.

14.65   Refer to Exercises 14.62–14.64. The residuals of the second-order model are plotted against x_1 at the bottom of page 787 and against x_2 at the top of page 788. Compare the residual plots to those of the first-order model in Exercise 14.63. Given the analyses you have performed in Exercises 14.62–14.65, which of the two models do you think is preferable as a predictor of GPA: the first- or second-order model? [*Note:* In Chapter 15 we show how to conduct a statistical test to compare two models.]

Printout for Exercise 14.64

```
Multiple R          .96777
R Square            .93657
Adjusted R Square   .92724
Standard Error      .18714

Analysis of Variance
                    DF      Sum of Squares      Mean Square
Regression          5           17.58274          3.51655
Residual            34           1.19076           .03502

F =     100.40901      Signif F =  .0000


------------------- Variables in the Equation -------------------

Variable             B          SE B        Beta         T    Sig T

X1                .16681       .02124     3.87132      7.852   .0000
X2                .13760       .02673     2.60754      5.147   .0000
X1SQ       -1.10825E-03  1.17288E-04    -3.71359     -9.449   .0000
X2SQ       -8.43267E-04  1.59423E-04    -2.37284     -5.290   .0000
X1X2        2.410891E-04 1.43974E-04      .49600      1.675   .1032
(Constant)     -9.91676       1.35441                -7.322   .0000
```

Plot of residuals against x_1 for Exercise 14.65

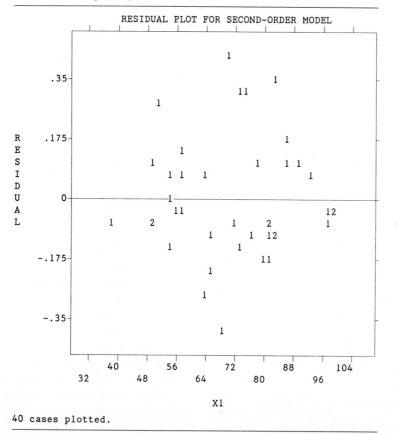

40 cases plotted.

Plot of residuals against x_2 for Exercise 14.65

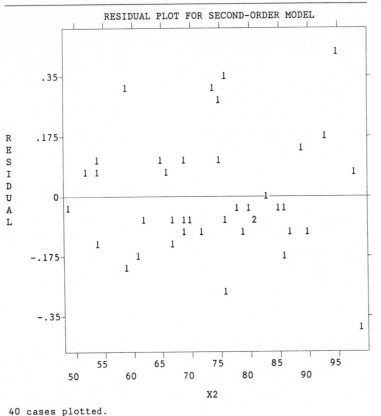

40 cases plotted.

ON YOUR OWN...

[*Note*: The use of a computer is required for this study.]

This is a continuation of the **On Your Own** section in Chapter 13, in which you selected three independent variables as predictors of a dependent variable of your choosing and obtained at least ten data values. Now fit the multiple regression model (use an available computer package, if possible):

$$y = \beta_0 + \beta_1 x_1 + \beta_2 x_2 + \beta_3 x_3 + \varepsilon$$

where

y = Dependent variable you chose

x_1 = First independent variable you chose

x_2 = Second independent variable you chose

x_3 = Third independent variable you chose

a. Compare the coefficients $\hat{\beta}_1$, $\hat{\beta}_2$, and $\hat{\beta}_3$ to their corresponding slope coefficients in the Chapter

13 **On Your Own**, where you fit three separate straight-line models. How do you account for the differences?

b. Calculate the coefficient of determination R^2, and conduct the F-test of the null hypothesis H_0: $\beta_1 = \beta_2 = \beta_3 = 0$. What is your conclusion?

If the independent variables you chose are themselves highly correlated, you may encounter some results that are difficult to explain. For example, the coefficients $\hat{\beta}_1$, $\hat{\beta}_2$, and $\hat{\beta}_3$ may assume signs that run counter to what you expected. Or you may get a highly significant F value in part **b**, but the individual t statistics for x_1, x_2, and x_3 may all be nonsignificant. This phenomenon—a high correlation between the independent variables in a regression model—is known as **multicollinearity**. We discuss it in more detail in Chapter 15.

USING THE COMPUTER...

Suppose we wish to model the relationship of the median income y to the percentage of college graduates x_1 and the percentage of women in the work force x_2 in a zip code.

a. Draw a random sample of 100 zip codes from the 1,000 described in Appendix C, and extract y, x_1, and x_2 for each zip code sampled. Use a software package that includes a regression program to obtain the least squares fit of the following models:

$$E(y) = \beta_0 + \beta_1 x_1 + \beta_2 x_2$$
$$E(y) = \beta_0 + \beta_1 x_1 + \beta_2 x_2 + \beta_3 x_1 x_2$$

1. Interpret the estimated β's of the models.
2. Interpret the estimated standard deviations of the models.
3. Evaluate the usefulness of the models. Is there evidence of interaction between percentage of college graduates and percentage of women in the work force? Which is the preferred model? Why?

4. Use the preferred model to estimate the mean median income for all zip codes having 20% college graduates and 40% women in the work force using a 95% confidence interval.
5. Use the same model to predict the median income for a particular zip code having 20% college graduates and 40% women in the work force using a 95% prediction interval.
6. Plot the residuals against x_1 and x_2 for each model. Check for evidence that curvature is present for either x_1 or x_2.
7. Plot a histogram or stem and leaf display of the residuals for each model. Comment on the assumption of normality for the random error component.
8. Use the residual plots to determine whether outliers exist.

b. Repeat part **a** using the entire set of 1,000 zip codes. Compare the results to those you obtained in part **a**.

References

Benston, G. J. "Multiple regression analysis of cost behavior," *Accounting Review*, October 1966, 41, pp. 657–672.

Brown, M. B., ed. *Biomedical Computer Programs*. Berkeley: University of California Press, 1977.

Draper, N., and Smith, H. *Applied Regression Analysis*. New York: Wiley, 1966.

Graybill, F. *Theory and Application of the Linear Model*. North Scituate, Mass.: Duxbury, 1976, Chapters 8 and 10–13.

Kleinbaum, D., and Kupper, L. *Applied Regression Analysis and Other Multivariable Methods*. North Scituate, Mass.: Duxbury, 1978, Chapters 9 and 10.

Mendenhall, W. *Introduction to Linear Models and the Design and Analysis of Experiments*. Belmont, Ca.: Wadsworth, 1968.

Mendenhall, W., and Sincich, T. *A Second Course in Business Statistics: Regression Analysis*, 2d ed. San Francisco: Dellen, 1986.

Neter, J., and Wasserman, W. *Applied Linear Statistical Models*, 2d ed. Homewood, Ill.: Richard Irwin, 1989.

Nie, N., Hull, C. H., Jenkins, J. G., Steinbrenner, K., and Bent, D. H. *Statistical Package for the Social Sciences*, 2d ed. New York: McGraw-Hill, 1975.

Ray, A. A., ed. *SAS User's Guide: Statistics*. Cary, N.C.: SAS Institute, Inc. 1982.

Ryan, T. A., Joiner, B. L., and Ryan, B. F. *Minitab Handbook*. Boston: PWS-Kent, 1985.

Sugrue, P. K., Ledford, M. H., and Glaskowsky, N. A. "Operating economies of scale in the U.S. long-haul common carrier, motor freight industry," *Transportation Journal*, Fall 1982, 22, pp. 27–41. Published by the American Society of Transportation and Logistics, Inc.

Tanner, M. B. "Vocational education and earnings for white males," *Southern Economic Journal*, 1983, 50.

Younger, M. S. *A First Course in Linear Regression*, 2d ed. Boston: Duxbury, 1985.

Model Building

WHERE WE'VE BEEN...

One of the most important topics in applied statistics, regression analysis, was presented in Chapters 13 and 14. Simple linear regression, using a straight line to model the relationship between the mean of a population and a single independent variable, was the topic of Chapter 13. Multiple regression, relating a population mean to any number of qualitative or quantitative independent variables, was the topic of Chapter 14. In both chapters, we learned how to fit regression models to a set of data and how to use the model to estimate the mean value of y or to predict a future value of y for a given value of x.

WHERE WE'RE GOING...

Chapters 13 and 14 discussed, in general, how to use regression analysis to solve many practical problems. But an important problem was circumvented—the selection of a model that is appropriate for the given data. No matter how much you know about regression analysis, or how well you can fit a model to a set of data and interpret the results, the information will be of little value if you choose an inappropriate model to relate the mean value of y to the independent variables. How to choose a reasonable model and how to use the data to modify and improve it is called model building. We introduce you to this topic in Chapter 15.

We indicated in Chapters 13 and 14 that the first step in the construction of a regression model is to hypothesize the form of the deterministic portion of the probabilistic model. This **model building**, or model construction, is the key to the success (or failure) of the regression analysis. If the regression model does not reflect, at least approximately, the true nature of the relationship between the mean response $E(y)$ and the independent variables $x_1, x_2, \ldots, x_k$, the modeling effort will usually be unrewarded.

By model building, we mean developing a model that provides a good fit to a set of data and gives good estimates of the mean value of y and good predictions of future values of y for given values of the independent variables. To illustrate, suppose you wish to relate a student's score y on an introductory sociology examination to the amount of time x the student has spent studying for the examination and that (unknown to you) the second-order model

$$E(y) = \beta_0 + \beta_1 x + \beta_2 x^2$$

would permit you to predict y with a very small error of prediction; see Figure 15.1(a). However, suppose you erroneously choose the first-order model

$$E(y) = \beta_0 + \beta_1 x$$

to explain the relationship between y and x; see Figure 15.1(b).

FIGURE 15.1

Two models for relating examination score y to amount of study time x

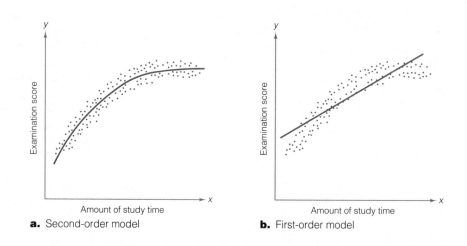

a. Second-order model   **b.** First-order model

The consequence of choosing the wrong model is clearly demonstrated by comparing Figures 15.1(a) and (b). The errors of prediction for the second-order model in Figure 15.1(a) are relatively small in comparison to those for the first-order model shown in Figure 15.1(b). The lesson to be learned from this simple example is very clear. Choosing a good set of independent (predictor) variables, $x_1, x_2, \ldots, x_k$, does not guarantee a good multivariable prediction equation. In addition to selecting independent variables that contain information about y, you must also develop an equation relating y to $x_1, x_2, \ldots, x_k$ that will provide a good fit to your data.

In the following sections, we present several useful models for relating a response y to one or more predictor variables. In addition, we provide tests and procedures for determining which model is appropriate.

15.1 The Two Types of Independent Variables: Quantitative and Qualitative

In Chapter 2 we defined two types of data and, correspondingly, two types of variables that may be observed: quantitative and qualitative. In a regression analysis the dependent variable is always quantitative, but the independent variables may be either quantitative or qualitative. As you will see, the way an independent variable enters the model depends on its type.

Definition 15.1

A **quantitative** independent variable is one that assumes numerical values. An independent variable that is not quantitative is called **qualitative**.

The age of a person, the number of cigarettes smoked per day, the dissolved oxygen content of a river, and the amount of fertilizer applied to an experimental agricultural plot are all examples of quantitative independent variables. On the other hand, suppose three different types of fertilizer, A, B, and C, are used in an agricultural experiment. This independent variable is qualitative, since it is not measured on a numerical scale. Certainly, the type of fertilizer applied to a crop is an independent variable that may affect its yield, and we would want to include it in a model describing the crop yield y.

Definition 15.2

The **levels** of an independent variable are its different intensity settings.

For a quantitative independent variable, the levels correspond to the numerical values it assumes. For example, the number x of cigarettes smoked per day by a person may assume values 0, 1, 2, 3, 4, ...* Therefore, the independent variable x assumes levels corresponding to the numbers, 0, 1, 2, 3, 4, ...

The levels of a qualitative variable are not numerical. Type of fertilizer, the independent variable, was observed at three levels, one corresponding to each of the types, A, B, and C.

EXAMPLE 15.1

In Chapter 14 we considered the problem of predicting executive salaries as a function of several independent variables. Consider the following four independent variables that may affect executive salaries:

a. Number of years of experience
b. Gender of the employee
c. Firm's net asset value
d. Rank of the employee

---

*There undoubtedly is some positive integer that represents the maximum number of cigarettes a person might smoke per day. Since this number is unknown, however, we give no limit to the number of values that x may assume.

For each of these independent variables, give its type and describe the nature of the levels you would expect to observe.

Solution

a. The independent variable for the number of years of experience is quantitative since its values are numerical. We would expect to observe levels ranging from 0 to 40 (approximately) years.

b. The independent variable for gender is qualitative, since its levels can be described only by the nonnumerical labels "female" and "male."

c. The independent variable for the firm's net asset value is quantitative, with a very large number of possible levels corresponding to the range of dollar values representing various firms' net asset values.

d. Suppose the independent variable for the rank of the employee is observed at three levels: supervisor, assistant vice president, and vice president. Since we cannot assign a realistic numerical measure of relative importance to each position, rank is a qualitative independent variable.

Quantitative independent variables are treated differently than qualitative variables in regression modeling. In the next section, we begin our discussion of how quantitative variables are used.

EXERCISES 15.1–15.5

APPLYING THE CONCEPTS

15.1 An experiment was conducted to investigate the effect of the following independent variables on the heart rate of humans:
a. Pounds over (or under) optimal weight
b. A subjective rating of physical condition on a scale of 1 to 10
c. Sex   **d.** Age   **e.** Race
Classify each of the variables as quantitative or qualitative, and describe the levels the variables may assume.

15.2 Companies keep personnel files that contain important information on each employee's background. These data could be used to predict an employee's score on a psychological test designed to measure the person's level of aggression. Identify the independent variables listed below as qualitative or quantitative. For qualitative variables, suggest several levels that might be observed. For quantitative variables, give a range of values (levels) for which the variable might be observed.
a. Age   **b.** Years of experience with the company
c. Highest educational degree   **d.** Job classification
e. Marital status   **f.** Religious preference
g. Salary   **h.** Sex

15.3 Which of the assumptions about ε (Section 13.3) prohibits the use of a qualitative variable as a dependent variable?

15.4 The marketing department of a large food products company conducted a study to investigate the effect of the following independent variables on the total monthly sales of one of the company's new products:
a. Advertising expenditures   **b.** Type of container
c. Color of container   **d.** Media used for advertising
e. Net weight of the product

Classify each of the variables as quantitative or qualitative, and describe the levels that the variables may assume.

15.5 The process of quality planning and control requires a continuous interaction between the customer and the manufacturer. In order to be able to design, monitor, and control the quality of a product, the manufacturer needs to know which attributes of the product play an important role in the customers' decision to buy and use the product (Schroeder, 1985). The following quality attributes were identified for a particular product:

a. Weight   **b.** Color   **c.** Age   **d.** Hardness   **e.** Package design

Regression analysis can be used to explore the relationship between these variables and sales of the product. Classify each of the variables as quantitative or qualitative, and describe the levels that the variables may assume.

15.2 Models with a Single Quantitative Independent Variable

The most common linear models relating y to a single quantitative independent variable x are those derived from a polynomial expression of the type shown in the box. Specific models, obtained by assigning values to p, are listed next.

Formula for a pth-Order Polynomial with One Independent Variable

$$E(y) = \beta_0 + \beta_1 x + \beta_2 x^2 + \beta_3 x^3 + \cdots + \beta_p x^p$$

where p is an integer and $\beta_0, \beta_1, \ldots, \beta_p$ are unknown parameters that must be estimated.

---

1. FIRST-ORDER MODEL

$$E(y) = \beta_0 + \beta_1 x \qquad \text{(Figure 15.2)}$$

Comments on model parameters

β_0: y-intercept

β_1:   Slope of the line

FIGURE 15.2

Graph of a first-order model

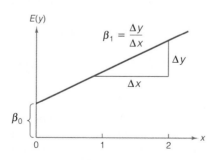

General comments   The first-order model is used when you expect the rate of change in y per unit change in x to remain fairly stable over the range of values of x over which you wish to predict y.* Most relationships between $E(y)$ and x are curvilinear, but the curvature over the range of values of x for which you wish to predict y may be very slight. When this occurs, a first-order (straight-line) model should provide a good fit to your data.

2. SECOND-ORDER MODEL

$$E(y) = \beta_0 + \beta_1 x + \beta_2 x^2 \qquad \text{(Figure 15.3)}$$

Comments on the parameters

β_0: y-intercept

β_1: Changing the value of β_1 shifts the parabola to the right or left. Increasing the value of β_1 causes the parabola to shift to the left.

β_2: Rate of curvature

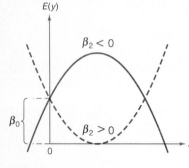

FIGURE 15.3

Graphs for two second-order models

General comments   A second-order model traces a parabola, one that opens either downward ($\beta_2 < 0$) or upward ($\beta_2 > 0$). Since most relationships possess some curvature, a second-order model is often a good choice to relate y to x.

3. THIRD-ORDER MODEL

$$E(y) = \beta_0 + \beta_1 x + \beta_2 x^2 + \beta_3 x^3 \qquad \text{(Figure 15.4)}$$

Comments on model parameters

β_0: y-intercept

β_3: The magnitude of β_3 controls the rate of reversal of curvature for the curve.

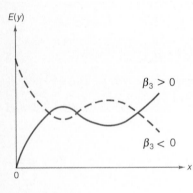

FIGURE 15.4

Graphs for two third-order models

General comments   Reversals in curvature are not common, but such relationships can be modeled by third and higher-order polynomials. As can be seen in Figure 15.3, a second-order model contains no reversal in curvature. The slope either continues to increase or continues to decrease as x increases and produces either a trough or a peak. A third-order model (see Figure 15.4) contains one reversal in curvature and produces one peak and one trough; in general, the graph of a kth-order polynomial contains a total of $(k - 1)$ peaks and troughs.

Most functional relationships in nature seem to be smooth (except for random error), i.e., they are not subject to rapid and irregular reversals in direction. Consequently, the second-order polynomial model is perhaps the most useful of those described here. To develop a better understanding of how this model is employed, consider the following example.

EXAMPLE 15.2

Power companies have to be able to predict the peak power load at their various stations in order to operate effectively. The peak power load is the maximum amount of power that must be generated each day in order to meet demand.

---

*The symbol Δy represents the change in y corresponding to a change in x equal to Δx.

Suppose a power company located in the southern part of the United States decides to model daily peak power load, y, as a function of the daily high temperature, x, and the model is to be constructed for the summer months when demand is greatest. Although we would expect the peak power load to increase as the high temperature increases, the *rate* of increase in $E(y)$ might also increase as x increases. That is, a 1-unit increase in high temperature from 100 to 101°F might result in a larger increase in power demand than would a 1-unit increase from 80 to 81°F. Therefore, we postulate the second-order model

$$E(y) = \beta_0 + \beta_1 x + \beta_2 x^2$$

and we expect β_2 to be positive.

A random sample of 25 summer days is selected, and the data are shown in Table 15.1. Fit a second-order model using these data, and test the hypothesis that the power load increases at an increasing *rate* with temperature—i.e., that $\beta_2 > 0$.

T A B L E 15.1 **Power Load Data**

TEMPERATURE (°F)	PEAK LOAD (Megawatts)	TEMPERATURE (°F)	PEAK LOAD (Megawatts)	TEMPERATURE (°F)	PEAK LOAD (Megawatts)
94	136.0	106	178.2	76	100.9
96	131.7	67	101.6	68	96.3
95	140.7	71	92.5	92	135.1
108	189.3	100	151.9	100	143.6
67	96.5	79	106.2	85	111.4
88	116.4	97	153.2	89	116.5
89	118.5	98	150.1	74	103.9
84	113.4	87	114.7	86	105.1
90	132.0				

Solution

The SAS printout shown in Figure 15.5 gives the least squares fit of the second-order model using the data in Table 15.1. The prediction equation is

$$\hat{y} = 385.048 - 8.293x + .05982x^2$$

A plot of this equation and the observed values is given in Figure 15.6 on page 798.

F I G U R E 15.5   Portion of the SAS printout for the second-order model of Example 15.2

SOURCE	DF	SUM OF SQUARES	MEAN SQUARE	F VALUE	PR > F
MODEL	2	15011.77199776	7505.88599888	259.69	0.0001
ERROR	22	635.87840224	28.90356374		
CORRECTED TOTAL	24	15647.65040000			

R-SQUARE   ROOT MSE
0.959363   5.37620347

PARAMETER	ESTIMATE	STD ERROR OF ESTIMATE	T FOR H0: PARAMETER = 0	PR > !T!
INTERCEPT	385.04809323	55.17243578	6.98	0.0001
TEMP	-8.29252680	1.29904502	-6.38	0.0001
TEMP*TEMP	0.05982337	0.00754855	7.93	0.0001

Plot of the observations and the second-order least squares fit, Example 15.2

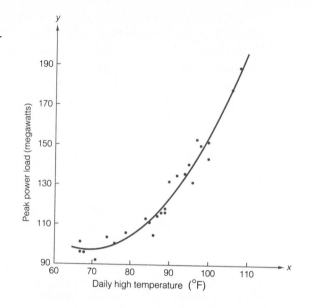

We now test to determine whether the sample value, $\hat{\beta}_2 = .05982$, is large enough to conclude *in general* that the power load increases at an increasing rate with temperature:

$H_0: \quad \beta_2 = 0$

$H_a: \quad \beta_2 > 0$

Test statistic: $\quad t = \dfrac{\hat{\beta}_2}{s_{\hat{\beta}_2}}$

For $\alpha = .05$, $n = 25$, and $k = 2$, we will reject H_0 if

$t > t_{.05}$

where $t_{.05} = 1.717$ (from Table VI in Appendix A) possesses $n - (k + 1) = 22$ degrees of freedom. From Figure 15.5 the calculated value of t is 7.93. Since this value exceeds $t_{.05} = 1.717$, we reject H_0 and conclude that the mean power load increases at an increasing rate with temperature.

EXERCISES 15.6–15.20

[*Note: Starred (*) exercises require the use of a computer.*]

LEARNING THE MECHANICS

15.6   The graphs depict *p*th-order polynomials with one independent variable. For each graph, identify the order of the polynomial. Find the value of β_0 and β_1 for each.

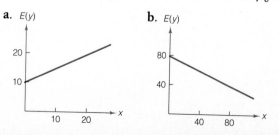

15.7   The graphs depict pth-order polynomials for one independent variable. For each graph, identify the order of the polynomial, the value of β_0, and the sign of β_2.

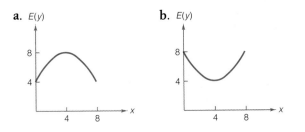

a. $E(y)$ **b.** $E(y)$

15.8   Graph the following polynomials and identify the order of each on your graph:
   a. $E(y) = 2 + 3x$ **b.** $E(y) = 2 + 3x^2$
   c. $E(y) = 1 + 2x + 2x^2 + x^3$ **d.** $E(y) = 2x + 2x^2 + x^3$
   e. $E(y) = 2 - 3x^2$ **f.** $E(y) = -2 + 3x$

15.9   The following graphs depict pth-order polynomials for one independent variable:

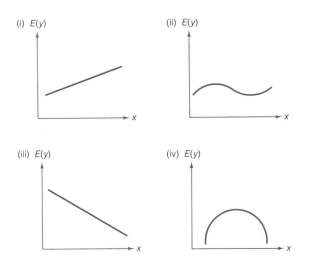

(i) $E(y)$ (ii) $E(y)$

(iii) $E(y)$ (iv) $E(y)$

   a. For each graph, identify the order of the polynomial.
   b. Using the parameters β_0, β_1, β_2, etc., write an appropriate model relating $E(y)$ to x for each graph.
   c. By examining the graphs, you can determine the signs ($+$ or $-$) of many of the parameters in the models of part **b**. Give the signs of those parameters that can be determined.

15.10   Minitab was used to fit the model

$$y = \beta_0 + \beta_1 x + \beta_2 x^2 + \varepsilon$$

to $n = 17$ data points. The printout on page 800 was obtained. Note that on the printout, C21 is y, C22 is x, and C23 is x^2.
   a. Report the least squares prediction equation.
   b. Plot the prediction equation.
   c. Specify the null and alternative hypotheses you would use to determine whether y decreases at an increasing rate as x increases.
   d. Conduct the hypothesis test of part **c** using $\alpha = .05$. Interpret your result.

Printout for Exercise 15.10

```
THE REGRESSION EQUATION IS
C21 = 63.1 + 1.17 C22 - 0.0374 C23

                                ST. DEV.      T-RATIO =
COLUMN        COEFFICIENT       OF COEF.      COEF/S.D.
                  63.11           11.92          5.30
C22              1.1683          0.8276          1.41
C23             -0.03742         0.01234        -3.03

S = 9.289

R-SQUARED = 86.0 PERCENT
R-SQUARED = 84.0 PERCENT, ADJUSTED FOR D.F.

ANALYSIS OF VARIANCE

  DUE TO      DF          SS        MS=SS/DF
REGRESSION     2        7434.4       3717.2
RESIDUAL      14        1208.1         86.3
TOTAL         16        8642.5
```

15.11   Suppose $E(y)$ can best be modeled by a second-order polynomial in x, where x is a quantitative variable. Write the probabilistic model for y.

15.12   Consider the following polynomial model:

$$E(y) = 5 - 3x + x^2$$

   a. Give the order of this polynomial.
   b. Sketch the curve corresponding to the equation for $E(y)$.
   c. How would the graph change if the coefficient of x^2 were negative rather than positive?

15.13   Consider the following polynomial model:

$$E(y) = 2 - 4x$$

   a. Give the order of this polynomial.
   b. Sketch the curve corresponding to the equation for $E(y)$.
   c. How would the graph change if the coefficient of x were positive instead of negative?

APPLYING THE CONCEPTS

15.14   The amount of pressure used to produce a certain plastic is thought to be related to the strength of the plastic. Researchers believe that as pressure is increased, the strength of the plastic increases until, at some point, increases in pressure will have a detrimental effect on strength. Write a model to relate the strength y of the plastic to pressure x that would reflect these beliefs. Sketch the model.

15.15   A company is considering having the employees on its assembly line work 4 days for 10 hours each instead of 5 days for 8 hours. The management is concerned that the effect of fatigue due to longer afternoons of work might increase assembly times to an unsatisfactory level. An experiment with the 4-day week is planned in which time studies will be conducted on some of the workers during the afternoons. It is believed that an adequate model of the relationship between assembly time y and time since lunch, x, should allow for the average assembly time to decrease for a while after lunch (as workers get back in the groove) before it starts to increase as the workers become tired.

  a. Give a model relating $E(y)$ to x that would reflect management's belief.
  b. Define all terms in the model given in part **a.**
  c. Sketch the curve described by the hypothesized model.

15.16   Underinflated or overinflated tires can increase tire wear. A new tire was tested for wear at different pressures with the results shown in the table.

PRESSURE	MILEAGE
x	y
(pounds per square inch)	(thousands)
30	29
31	32
32	36
33	38
34	37
35	33
36	26

  a. Plot the data on a scattergram.
  b. If you were given only the information for $x = 30, 31, 32, 33$, what kind of model would you suggest? For $x = 33, 34, 35, 36$? For all the data?

15.17   An experiment was conducted to relate the number of visits per patient per year to a doctor's office to the age x of the patient. It is suspected that young and old patients tend to seek medical assistance more frequently than those of middle age. Write an appropriate model relating mean number of visits per patient per year to the patient's age.

*15.18   A veterinarian who works for a large midwestern pig cooperative has developed a daily vitamin pellet that she believes will substantially increase the weight of mature pigs within 1 month. However, she is uncertain of the relationship between the daily dosage (amount of the vitamin) and the percentage gain in weight after 1 month. To understand this relationship better, she randomly selects 16 pigs of the same age and weight and feeds them different dosage levels for a 1-month trial period. The data are given in the table.

WEIGHT GAIN (% of original weight)	DAILY PELLET DOSE	WEIGHT GAIN (% of original weight)	DAILY PELLET DOSE
82	0	90	4
78	0	95	4
80	1	89	5
87	1	93	5
87	2	90	6
95	2	85	6
97	3	84	7
90	3	90	7

  a. Plot the data on a scattergram.
  b. Fit the model $E(y) = \beta_0 + \beta_1 x + \beta_2 x^2$ to the data.
  c. Is there evidence to support the inclusion of the quadratic (second-order) term in the model of part **b**? Test using $\alpha = .05$.
  d. Plot the fitted model of part **b** on the scattergram of part **a**.

e. Use the fitted model to estimate the optimum daily dosage (i.e., the dosage that encourages the greatest weight gain). [*Note:* We would want to express this estimate as a confidence interval, but its computation is beyond the scope of this text. The procedure is described in the references. You may also find that it can be obtained by using your computer program package.]

15.19 An economist has proposed the following model to describe the relationship between the number of items produced per day (output) and the number of work-hours expended per day (input) in a particular production process:

$$y = \beta_0 + \beta_1 x + \beta_2 x^2 + \varepsilon$$

where

y = Number of items produced per day

x = Number of work-hours per day

A portion of the Minitab computer printout that results from fitting this model to a sample of 25 weeks of production data is shown here. Test the hypothesis that as the amount of input increases, the amount of output also increases but at a decreasing rate. Do the data provide sufficient evidence to indicate that the *rate* of increase in output per unit increase of input decreases as the input increases? Test using $\alpha = .05$.

Printout for Exercise 15.19

```
THE REGRESSION EQUATION IS
Y =   -6.17  +  2.04 X1  - .0323 X2

                              ST. DEV.   T-RATIO =
           COLUMN   COEFFICIENT   OF COEF.   COEF/S.D.

             --        -6.173       1.666     -3.71
X1   C2                2.036         .185     11.02
X2   C3               -.03231       .00489    -6.60

THE ST. DEV. OF Y ABOUT REGRESSION LINE IS
S =      1.243
WITH (25-3) =  22 DEGREES OF FREEDOM

R-SQUARED = 95.5 PERCENT
R-SQUARED = 95.1 PERCENT, ADJUSTED FOR D.F.

ANALYSIS OF VARIANCE

DUE TO       DF        SS    MS=SS/DF

REGRESSION    2    718.168    359.084
RESIDUAL     22     33.992      1.545
TOTAL        24    752.160
```

15.20 Automobile accidents result in a tragic loss of life and, in addition, they represent a serious dollar loss to the nation's economy. Shown in the accompanying table are the number of highway deaths (to the nearest hundred) and the number of licensed vehicles (in hundreds of thousands) for the years 1950–1985. (The years are coded 1–36 for convenience.) During the years 1974–1985 (years 25–36 in the table), the nationwide 55-mile-per-hour speed limit was in effect.

a. Write a second-order model relating the number, y, of highway deaths for a year to the number, x_1, of licensed vehicles.

YEAR	DEATHS y	NUMBER OF VEHICLES x_1	YEAR	DEATHS y	NUMBER OF VEHICLES x_1
1	34.8	49.2	19	54.9	103.1
2	37.0	51.9	20	55.8	107.4
3	37.8	53.3	21	54.6	111.2
4	38.0	56.3	22	54.4	116.3
5	35.6	58.6	23	56.3	122.3
6	38.4	62.8	24	55.5	129.8
7	39.6	65.2	25	46.4	134.9
8	38.7	67.6	26	45.9	137.9
9	37.0	68.8	27	47.0	143.5
10	37.9	72.1	28	49.5	148.8
11	38.1	74.5	29	52.4	153.6
12	38.1	76.4	30	53.5	159.6
13	40.8	79.7	31	53.1	161.6
14	43.6	83.5	32	51.4	164.1
15	47.7	87.3	33	45.8	165.2
16	49.2	91.8	34	44.5	169.4
17	53.0	95.9	35	46.2	172.0
18	52.9	98.9	36	45.6	175.7

Source: *Accident Facts*, National Safety Council.

Printout for Exercise 15.20

DEP VARIABLE: Y

ANALYSIS OF VARIANCE

SOURCE	DF	SUM OF SQUARES	MEAN SQUARE	F VALUE	PROB>F
MODEL	2	1296.53234	648.26617	50.527	0.0001
ERROR	33	423.39322	12.83009748		
C TOTAL	35	1719.92556			

ROOT MSE	3.581913	R-SQUARE	0.7538
DEP MEAN	45.86111	ADJ R-SQ	0.7389
C.V.	7.810348		

PARAMETER ESTIMATES

VARIABLE	DF	PARAMETER ESTIMATE	STANDARD ERROR	T FOR H0: PARAMETER=0	DF	PROB > |T|
INTERCEP	1	-0.85749696	5.38997832	-0.159	1	0.8746
X1	1	0.82992304	0.10629013	7.808	1	0.0001
X1SQ	1	-0.003218869	0.000469453	-6.857	1	0.0001

b. The SAS computer printout for fitting the model to the data is shown above. Is there sufficient evidence to indicate that the model provides information for the prediction of the number of annual highway deaths? Test using $\alpha = .05$.

c. Give the observed significance level for the test of part b, and interpret it.

d. Does the second-order term contribute information for the prediction of y? Test using $\alpha = .05$.

e. Give the observed significance level for the test of part d, and interpret it.

15.3 Models with Two or More Quantitative Independent Variables

The best way to understand how to write models relating a mean response $E(y)$ to a set of quantitative independent variables is to examine the different types of models when only two independent variables are involved. You can see geometrically (using three-dimensional figures) the relationships implied by the different models and will more easily understand the relationships you will encounter when the model contains three or more quantitative independent variables.

1. FIRST-ORDER MODEL WITH TWO INDEPENDENT VARIABLES

$$E(y) = \beta_0 + \beta_1 x_1 + \beta_2 x_2$$

Comments on the parameters

β_0: y-intercept, the value of $E(y)$ when $x_1 = x_2 = 0$

β_1:   Change in $E(y)$ for a 1-unit increase in x_1 when x_2 is held fixed

β_2:   Change in $E(y)$ for a 1-unit increase in x_2 when x_1 is held fixed

General comments   The graph in Figure 15.7 traces a **response surface** (in contrast to the response curve that is used to relate $E(y)$ to a *single* quantitative variable). Particularly, a first-order model relating $E(y)$ to two independent quantitative variables, x_1 and x_2, graphs as a plane in a three-dimensional space. The plane traces the value of $E(y)$ for every combination of values (x_1, x_2) that correspond to points in the x_1, x_2-plane. Many response surfaces in the real world are well behaved (smooth), and they possess curvature. Consequently, a first-order model is appropriate only if the response surface is fairly flat over the (x_1, x_2) region that is of interest.

FIGURE 15.7

Graph of a first-order model

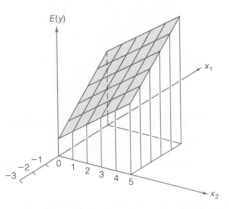

The assumption that a first-order model will adequately characterize the relationship between $E(y)$ and the variables x_1 and x_2 is equivalent to assuming that x_1 and x_2 do not interact; i.e., you assume that the effect on $E(y)$ of changes in x_1 are the same, regardless of the value of x_2 (and vice versa). Thus, no interaction means that the effect of changes in one variable (say x_1) on $E(y)$ is *independent* of the value of the second variable (say x_2). For example, if we assign values to x_2 in a first-order model, the graph of $E(y)$ as a function of x_1 would produce parallel lines as shown in Figure 15.8. These lines, called **contour lines**, show the contours of the surface when it is sliced by three planes, each of which is parallel to the $E(y)$, x_1-plane, at distances $x_2 = 1$, 2, and 3 from the origin.

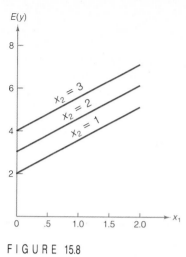

$E(y)$

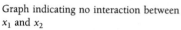

FIGURE 15.8

Graph indicating no interaction between x_1 and x_2

> ### Definition 15.3
>
> Two variables x_1 and x_2 are said to **interact** if the change in $E(y)$ for a 1-unit increase in x_1 (when x_2 is held fixed) is dependent on the value of x_2.

---

2. INTERACTION MODEL (SECOND-ORDER) WITH TWO INDEPENDENT VARIABLES

$$E(y) = \beta_0 + \beta_1 x_1 + \beta_2 x_2 + \beta_3 x_1 x_2$$

Comments on the parameters

β_0: y-intercept, the value of $E(y)$ when x_1 and x_2 equal 0

β_1, β_2: Changing β_1 and β_2 causes the surface to shift along the x_1- and x_2-axes

β_3: Controls the rate of twist in the surface (Figure 15.9, page 806)

General comments   The model described here is said to be second-order because the highest-order term $(x_1 x_2)$ in x_1 and x_2 is of order 2; i.e., the sum of the exponents of x_1 and x_2 equals 2. This interaction model traces a twisted surface in a three-dimensional space (Figure 15.9). A graph of $E(y)$ as a function of x_1 for given values of x_2 (say $x_2 = 1$, 2, and 3) produces nonparallel contour lines (see Figure 15.10), thus indicating that the change in $E(y)$ for a given change in x_1 is dependent on the value of x_2 and, therefore, that x_1 and x_2 interact. Interaction is an extremely important concept because it is easy to get in the habit of fitting first-order models and individually examining the relationships between $E(y)$ and a set of independent variables, $x_1, x_2, \ldots, x_k$. Such a procedure is meaningless when interaction exists (which is, at least to some extent, almost always the case), and it can lead to gross errors in interpretation. For example, suppose the relationship between $E(y)$ and x_1 and x_2 is as shown in Figure 15.10 and you have observed y for each of the $n = 9$ combinations of values of x_1 and x_2, $x_1 = 1$, 2, 3 and $x_2 = 1$, 2, 3. If you fit a first-order model in x_1 and x_2 to the data, the fitted plane would be (except for random error) approximately parallel to the x_1, x_2-plane, thus suggesting that x_1 and x_2 contribute very little information about $E(y)$. Figure 15.10 clearly indicates that this is not the case. Fitting a first-order model to the data would not allow for the twist in the true surface and would therefore give a false impression of the relationship between $E(y)$ and x_1 and x_2. The procedure for detecting interaction between two independent variables can be seen by examining the model. The interaction model differs from the noninteraction first-order model only in the inclusion of the $\beta_3 x_1 x_2$ term.*

---

*The *order of a term* is equal to the sum of the exponents of the variables included in the term. Thus, the $x_1 x_2$ term is second-order, because the exponents of x_1 and x_2 are both equal to 1 and their sum is 2. The *order of a model* is determined by its highest-order term. Therefore, the interaction model is second-order.

FIGURE 15.9

FIGURE 15.9
Graph for an interaction model
(second-order)

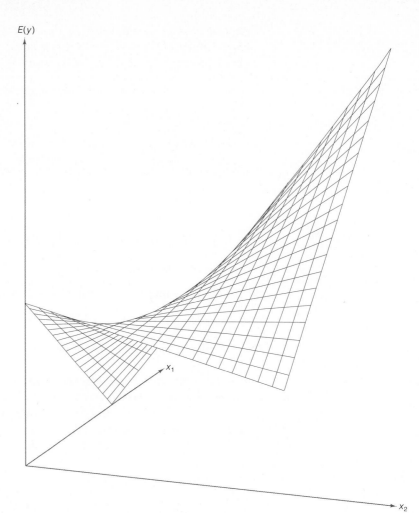

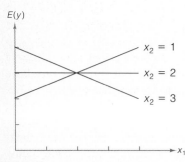

FIGURE 15.10
Graph indicating interaction between x_1
and x_2

Interaction model: $E(y) = \beta_0 + \beta_1 x_1 + \beta_2 x_2 + \beta_3 x_1 x_2$
First-order model: $E(y) = \beta_0 + \beta_1 x_1 + \beta_2 x_2$

Therefore, to test for the presence of interaction we test

H_0: $\beta_3 = 0$ (No interaction)

against the alternative hypothesis

H_a: $\beta_3 \neq 0$ (Interaction)

using the familiar Student t-test of Section 14.4.

3. COMPLETE SECOND-ORDER MODEL WITH TWO INDEPENDENT VARIABLES

$$E(y) = \beta_0 + \beta_1 x_1 + \beta_2 x_2 + \beta_3 x_1 x_2 + \beta_4 x_1^2 + \beta_5 x_2^2$$

Comments on the parameters

β_0: y-intercept, the value of $E(y)$ when $x_1 = x_2 = 0$

β_1, β_2:   Changing β_1 and β_2 causes the surface to shift along the x_1-axis and the x_2-axis

β_3:   Value of β_3 controls the rotation of the surface

β_4, β_5:   Signs and values of these parameters control the type of surface and the rates of curvature

The parameters β_4 and β_5 produce the following three types of surfaces:

β_4 and β_5 positive: Paraboloid that opens upward [Figure 15.11(a)]

β_4 and β_5 negative: Paraboloid that opens downward [Figure 15.11(b)]

β_4 and β_5 differ in sign: Saddle-shaped surface [Figure 15.11(c)]

FIGURE 15.11

Graphs for three second-order surfaces

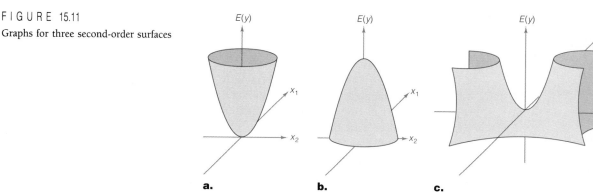

a.   b.   c.

General comments   A complete second-order model is the three-dimensional equivalent of a second-order model in a single quantitative variable. Instead of tracing parabolas, it traces paraboloids and saddle surfaces. Since only a portion of the complete surface is used to fit the data, it provides a very large variety of gently curving surfaces that can be used to fit data. It is a good choice for a model if you expect curvature in the response surface relating $E(y)$ to x_1 and x_2.

EXAMPLE 15.3

A social scientist would like to relate the number of hours worked per week (outside the home) by a married woman to the number of years of formal education completed by the woman and the number of children in the family.

a. Identify the dependent variable and the independent variables.
b. Write the first-order model for this example.
c. Modify the model in part **b** to include an interaction term.
d. Write a complete second-order model for $E(y)$.

Solution

a. The dependent variable is

y = Number of hours worked per week by a married woman

The two independent variables, both quantitative in nature, are

x_1 = Number of years of formal education completed by the woman

x_2 = Number of children in the family

b. The first-order model is

$$E(y) = \beta_0 + \beta_1 x_1 + \beta_2 x_2$$

This model would probably not be appropriate in this situation because it is quite possible that x_1 and x_2 may interact and that second-order terms corresponding to x_1^2 and x_2^2 may be needed to obtain a good model for $E(y)$.

c. Adding the interaction term, we obtain

$$E(y) = \beta_0 + \beta_1 x_1 + \beta_2 x_2 + \beta_3 x_1 x_2$$

This model should be better than the model in part **b**, since we have now allowed for interaction between x_1 and x_2.

d. The complete second-order model is

$$E(y) = \beta_0 + \beta_1 x_1 + \beta_2 x_2 + \beta_3 x_1 x_2 + \beta_4 x_1^2 + \beta_5 x_2^2$$

Since it would not be surprising to find curvature in the response surface, the complete second-order model would be preferred to the models in parts **b** and **c**. How can we tell whether the complete second-order model really does provide better predictions of hours worked than the models in parts **b** and **c**? The answers to these and similar questions are examined in Section 15.4.

Models relating a mean response $E(y)$ to more than two quantitative variables are extensions of the corresponding models in two quantitative variables. For example, a first-order model for three quantitative independent variables, x_1, x_2, and x_3, is

First-order model: $E(y) = \beta_0 + \beta_1 x_1 + \beta_2 x_2 + \beta_3 x_3$

As in the case of two independent variables, this model implies that the relationship between $E(y)$ and any one variable, say x_1, is linear when x_2 and x_3 are held constant. Further, it implies that there is no interaction among x_1, x_2, and x_3; i.e., the slope of the linear relation is exactly the same, regardless of the values of x_2 and x_3. If one of the independent variables, say x_3, is held constant, the graph of $E(y)$ as a function of x_1 and x_2 is a plane. (See Figure 15.7.) Thus, for given values of β_0, β_1, β_2, and β_3, $E(y)$ will graph as a set of parallel planes, one for each different value of x_3.

Similarly, as in the case of two quantitative independent variables, we can add second-order cross-product terms, $x_1 x_2$, $x_1 x_3$, and $x_2 x_3$, to the first-order model and obtain a second-order model that allows interaction among x_1, x_2, and x_3:

Interaction model:

$$E(y) = \beta_0 + \beta_1 x_1 + \beta_2 x_2 + \beta_3 x_3 + \beta_4 x_1 x_2 + \beta_5 x_1 x_3 + \beta_6 x_2 x_3$$

If one of the independent variables, say x_3, is held constant, the graph of $E(y)$ as a function of x_1 and x_2 is a twisted plane of the type shown in Figure 15.9.

Finally, we could add second-order terms involving x_1^2, x_2^2, and x_3^2 to the interaction model and obtain the complete second-order model:

Complete second-order model:

$$E(y) = \beta_0 + \beta_1 x_1 + \beta_2 x_2 + \beta_3 x_3 + \beta_4 x_1 x_2 + \beta_5 x_1 x_3 + \beta_6 x_2 x_3 + \beta_7 x_1^2 + \beta_8 x_2^2 + \beta_9 x_3^2$$

If one of the independent variables, say x_3, is held constant, the graph of $E(y)$ as a function of x_1 and x_2 will trace a conic surface. (See Figure 15.11.)

Most relationships between $E(y)$ and two or more quantitative independent variables are second-order and require the use of either the interactive or the complete second-order model to obtain a good fit to a data set. As in the case of a single quantitative independent variable, however, the curvature in the response surface may be very slight over the range of values of the variables in the data set. When this happens, a first-order model may provide a good fit to the data.

How can you tell which model to use in a particular situation? We answer this question in Section 15.4.

CASE STUDY 15.1

MODELING STRIKE ACTIVITY

Case Study 14.2 describes Neil A. Palomba's efforts to model strike activity (the percentage y of total working hours lost due to strikes) as a function of three independent variables:

x_1 = Percentage of union members in nonagricultural establishments

x_2 = Percentage of all nonagricultural employment that is manufacturing

x_3 = Hourly earnings of workers on manufacturing payrolls

The regression analysis using Palomba's most complete model, a first-order model involving all three independent variables, produced an R^2 that was equal to only .232. In other words, Palomba's best model accounted for only 23% of the variability of the y values about their mean. How can we improve the model so that we can account for a substantial portion of the remaining 77%?

There are two possibilities. Palomba's model may not contain the key information-contributing variables. Perhaps there are other independent variables that are highly related to strike activity but have not been included in the model. Locating these variables, if they exist, is a task for a sociologist or a specialist in industrial relations. The other possibility is that Palomba's three independent variables are the most important variables in explaining strike activity and the form of the model is wrong. Instead of being first-order, perhaps the model should be second-order or a more complex mathematical function of x_1, x_2, and x_3.

Palomba's data giving the values of y, x_1, x_2, and x_3 for 49 states (excluding Maryland) in 1964 are shown in Table 15.2 on page 810. You will have an opportunity to improve on Palomba's model in Exercise 15.39.

TABLE 15.2   **Strike Activity Data, Case Study 15.1**

STATE	1964				STATE	1964			
	y	x_1	x_2	x_3		y	x_1	x_2	x_3
Alabama	.14	18.0	30.5	2.17	Nebraska	.05	19.3	16.6	2.36
Alaska	.11	32.2	8.9	3.54	Nevada	.36	32.8	4.6	3.16
Arizona	.09	20.08	15.3	2.72	New Hampshire	.03	20.9	40.9	2.00
Arkansas	.10	26.2	29.2	1.78	New Jersey	.27	37.7	37.1	2.67
California	.16	33.8	24.9	2.96	New Mexico	.09	13.4	6.8	2.29
Colorado	.04	21.6	15.8	2.74	New York	.11	39.4	28.2	2.60
Connecticut	.08	24.6	42.5	2.62	North Carolina	.01	6.7	41.6	1.75
Delaware	.41	21.5	36.1	2.65	North Dakota	.03	14.8	5.8	2.28
Florida	.20	13.1	15.5	2.11	Ohio	.38	35.7	39.0	2.91
Georgia	.13	12.7	31.8	1.92	Oklahoma	.01	13.7	15.5	2.35
Hawaii	.02	24.2	12.1	2.14	Oregon	.12	34.8	26.5	2.85
Idaho	.11	19.2	18.8	2.50	Pennsylvania	.14	38.4	37.8	2.55
Illinois	.18	37.9	33.5	2.76	Rhode Island	.09	29.6	38.2	2.11
Indiana	.16	34.1	40.8	2.81	South Carolina	.01	7.9	42.7	1.80
Iowa	.16	20.8	25.4	2.71	South Dakota	.16	9.5	8.8	2.34
Kansas	.11	18.8	20.6	2.65	Tennessee	.23	17.6	34.6	2.03
Kentucky	.17	25.7	26.6	2.43	Texas	.06	13.3	19.4	2.42
Louisiana	.10	17.1	17.8	2.49	Utah	.66	19.7	17.6	2.77
Maine	.15	20.3	36.6	2.00	Vermont	.26	19.3	30.9	2.08
Massachusetts	.07	29.1	33.0	2.37	Virginia	.04	15.5	26.5	2.04
Michigan	.83	38.9	40.7	3.11	Washington	.16	43.1	25.6	2.98
Minnesota	.02	33.0	24.0	2.64	West Virginia	.45	42.0	27.4	2.67
Mississippi	.14	11.6	30.5	1.76	Wisconsin	.21	31.5	37.0	2.66
Missouri	.14	39.8	28.5	2.53	Wyoming	.01	19.2	7.7	2.82
Montana	.28	36.2	12.2	2.71	Maryland				

EXERCISES 15.21–15.32

[*Note: Starred (*) exercises require the use of a computer.*]

LEARNING THE MECHANICS

15.21   Consider an application in which you are trying to relate a response y to two independent variables, x_1 and x_2.
  a. Write a first-order model relating $E(y)$ to x_1 and x_2.
  b. Modify the model you constructed in part **a** to include an interaction term.
  c. Modify the model you constructed in part **b** to make it a complete second-order model.

15.22   Suppose the true relationship between $E(y)$ and the quantitative independent variables x_1 and x_2 is described by the first-order model:

$$E(y) = 3 + x_1 - 2x_2$$

  a. Describe the corresponding response surface.
  b. Plot the contour lines of the response surface for $x_1 = 2, 3, 4$, where $0 \le x_2 \le 5$.
  c. Plot the contour lines of the response surface for $x_2 = 2, 3, 4$, where $0 \le x_1 \le 5$.
  d. Use the contour lines you plotted in parts **b** and **c** to explain how changes in the settings of x_1 and x_2 affect $E(y)$.
  e. Use your graph from part **b** to determine how much $E(y)$ changes when x_1 is changed from 4 to 2 and x_2 is simultaneously changed from 1 to 2.

15.23   Suppose the true relationship between $E(y)$ and the quantitative independent variables x_1 and x_2 is

$$E(y) = 3 + x_1 + 2x_2 - x_1x_2$$

   a. Describe the corresponding response surface.
   b. Plot the contour lines of the response surface for $x_1 = 0, 1, 2$, where $0 \leq x_2 \leq 5$.
   c. Explain why the contour lines you plotted in part b are not parallel.
   d. Use the contour lines you plotted in part b to explain how changes in the settings of x_1 and x_2 affect $E(y)$.
   e. Use your graph from part b to determine how much $E(y)$ changes when x_1 is changed from 2 to 0 and x_2 is simultaneously changed from 4 to 5.

15.24   If two variables, x_1 and x_2, do not interact, how would you describe their effect on the mean response $E(y)$?

15.25   Minitab was used to fit the model

$$y = \beta_0 + \beta_1x_1 + \beta_2x_2 + \beta_3x_1x_2 + \varepsilon$$

   to $n = 15$ data points. In the accompanying printout C1 is y, C2 is x_1, C3 is x_2, and C4 is x_1x_2.

Printout for Exercise 15.25

```
THE REGRESSION EQUATION IS
C1 = - 2.55 + 3.82 C2 + 2.63 C3 - 1.29 C4

                        ST. DEV.    T-RATIO =
COLUMN      COEFFICIENT  OF COEF.   COEF/S.D.
            -2.550       1.142      -2.23
C2          3.8150       0.5286      7.22
C3          2.6300       0.3443      7.64
C4          -1.2850      0.1594     -8.06

S = 0.7127

R-SQUARED = 85.6 PERCENT
R-SQUARED = 81.6 PERCENT, ADJUSTED FOR D.F.

ANALYSIS OF VARIANCE

DUE TO      DF          SS        MS=SS/DF
REGRESSION  3           33.149     11.050
RESIDUAL    11          5.587       0.508
TOTAL       14          38.736
```

   a. What is the prediction equation for the response surface?
   b. Describe the geometric form of the response surface of part a.
   c. Plot the prediction equation for the case when $x_2 = 1$. Do this twice more on the same graph for the cases when $x_2 = 3$ and $x_2 = 5$.
   d. Explain what it means to say that x_1 and x_2 interact. Explain why your graph of part c suggests that x_1 and x_2 interact.
   e. Specify the null and alternative hypotheses you would use to test whether x_1 and x_2 interact.
   f. Conduct the hypothesis test of part e using $\alpha = .01$.

APPLYING THE CONCEPTS

15.26   Baseball fans are involved in a never-ending search for variables (and a model) that will enable them to predict a professional baseball team's success for a given baseball season. Adam Clymer in an article titled "New Statistics for Baseball" (*New York Times*, May 20, 1984) notes that John Thorn and Pete Palmer in their book (*The Hidden Game of Baseball*, Doubleday, 1984) "measure players' offensive contributions by what they call a linear weights measure, which gives separate values to at bats, singles, doubles, . . . , home runs, . . . , stolen bases, and times caught stealing."

a. Would a regression analysis, using a first-order model, produce Clymer's description of a "linear weights measure"? Explain.

b. List some of the independent variables that you think might be related to a professional baseball team's percentage of games won. Construct a first-order model to relate percentage of games won to these variables.

c. Can you propose a better model than the model in part **b**? Explain.

15.27   The Department of Energy wants to develop a regression model to help forecast annual gasoline consumption in the United States, y. It has been decided to model $E(y)$ as a function of two independent variables:

x_1 = Number of cars (millions) in use during year

x_2 = Number of trucks (millions) in use during year

a. Identify the independent variables as quantitative or qualitative.

b. Write the first-order model for $E(y)$.

c. Write the complete second-order model for $E(y)$.

d. With respect to the model in part **c**, specify the null and alternative hypotheses you would use to test whether the second-order terms contribute information for the prediction of annual gasoline consumption.

15.28   The dissolved oxygen content y in rivers and streams is related to the amount x_1 of nitrogen compounds per liter of water and the temperature x_2 of the water. Write the complete second-order model relating $E(y)$ to x_1 and x_2.

15.29   Some corporations, instead of owning a fleet of cars, rent cars from a rental agency. A corporation may do this because it is sometimes more economical to rent new cars for a year than to buy new cars each year. A major rental agency wants to develop a model that will allow it to estimate the average annual cost to the prospective customer of renting cars, y, as a function of two independent variables:

x_1 = Number of cars rented

x_2 = Average number of miles driven per car during year (in thousands)

a. Identify the independent variables as quantitative or qualitative.

b. Write the first-order model for $E(y)$.

c. Write a model for $E(y)$ that contains all first-order and interaction terms. Sketch typical response curves showing $E(y)$, the mean cost, versus x_2, the average mileage driven, for different values of x_1. (Assume that x_1 and x_2 interact.)

d. Write the complete second-order model for $E(y)$.

15.30   Refer to Exercise 15.29. Suppose the model from part **c** is fit with the following result:

$$\hat{y} = 1 + .05x_1 + x_2 + .05x_1x_2$$

(The units of $\hat{y}$ are thousands of dollars.) Graph the estimated cost $\hat{y}$ as a function of the average number of miles driven, x_2, over the range $x_2 = 10$ to $x_2 = 50$ (10,000 to 50,000 miles) for $x_1 = 1, 5$, and 10. Do these functions agree (approximately) with the graphs you drew for Exercise 15.29, part **c**?

15.31   An economist is interested in modeling the relationship between quarterly sales of central air-conditioning systems (in thousands) for single-family homes in the United States and two quantitative independent variables, housing starts in the previous quarter (in thousands) and the gross national product (in billions of 1972 dollars). Data were collected, and a model was fit. The displayed portion of the resulting Minitab printout describes the least squares prediction equation. In the prediction equation, $x_3 = x_1^2$ and $x_4 = x_2^2$.

Printout for Exercise 15.31

```
THE REGRESSION EQUATION IS
Y =    149. + .472 X1 - .0993 X2
       - .0005 X3 + .0000 X4

                                ST. DEV.   T-RATIO =
        COLUMN   COEFFICIENT   OF COEF.   COEF/S.D.

        --             148.5      224.5        .66
X1  C2                 .472       .126        3.74
X2  C3                -.099       .162        -.61
X3  C12           -.000535    .000359       -1.49
X4  C13           .0000153   .0000294         .52

THE ST. DEV. OF Y ABOUT REGRESSION LINE IS
S =        6.884
WITH (  16-5) = 11 DEGREES OF FREEDOM

R-SQUARED = 92.0 PERCENT
R-SQUARED = 89.1 PERCENT, ADJUSTED FOR D.F.

ANALYSIS OF VARIANCE

   DUE TO        DF        SS   MS=SS/DF

REGRESSION     4    6002.08    1500.52
RESIDUAL      11     521.36      47.40
TOTAL         15    6523.44
```

a. Write the prediction equation for the response surface.
b. Describe the geometric form of the response surface of part a.
c. Do the data provide sufficient evidence to conclude that the model hypothesized by the economist is useful for predicting quarterly sales of central air-conditioning systems? Test using $\alpha = .01$.
d. Does it appear that the variation in air conditioner sales could be adequately explained by a less complex regression model? Explain. [Note: In the next section we discuss a formal procedure for making this inference.]

*15.32 A supermarket chain is interested in exploring the relationship between the sales of its store-brand vegetables (y), the amount spent on promotion of the vegetables in local newspapers (x_1), and the amount of shelf space allocated to the brand (x_2). One of the chain's supermarkets was randomly selected, and over a 20-week period x_1 and x_2 were varied as reported in the table on page 814.
a. Fit the following model to the data:

$$y = \beta_0 + \beta_1 x_1 + \beta_2 x_2 + \beta_3 x_1 x_2 + \varepsilon$$

b. Conduct an F-test to investigate the overall usefulness of this model. Use $\alpha = .05$.
c. Test for the presence of interaction between advertising expenditure and shelf space. Use $\alpha = .05$.
d. Explain what it means to say that advertising expenditures and shelf space interact.
e. Explain how you could be misled by using a first-order model instead of an interaction model to explain how advertising expenditure and shelf space influence sales.

WEEK	SALES ($)	ADVERTISING EXPENDITURES ($)	SHELF SPACE (sq. ft.)	WEEK	SALES ($)	ADVERTISING EXPENDITURES ($)	SHELF SPACE (sq. ft.)
1	2,010	201	75	11	5,005	996	75
2	1,850	205	50	12	2,500	625	50
3	2,400	355	75	13	3,005	860	50
4	1,575	208	30	14	3,480	1,012	50
5	3,550	590	75	15	5,500	1,135	75
6	2,015	397	50	16	1,995	635	30
7	3,908	820	75	17	2,390	837	30
8	1,870	400	30	18	4,390	1,200	50
9	4,877	997	75	19	2,785	990	30
10	2,190	515	30	20	2,989	1,205	30

15.4 Model Building: Testing Portions of a Model

The presentation of models with one and two quantitative independent variables raises a very general question. Do certain terms in the model contribute information for the prediction of y?

To illustrate, suppose you have collected data on a response y and two independent quantitative variables, x_1 and x_2, and you are considering the use of either a first-order or a second-order model to relate $E(y)$ to x_1 and x_2. Will the second-order model provide better predictions of y than the first-order model? To answer this question, examine the two models. Note that the second-order model contains all terms present in the first-order model plus three additional terms, those involving β_3, β_4, and β_5:

$$\text{First-order model:} \quad E(y) = \beta_0 + \beta_1 x_1 + \beta_2 x_2$$

$$\text{Second-order model:} \quad E(y) = \beta_0 + \beta_1 x_1 + \beta_2 x_2 + \overbrace{\beta_3 x_1 x_2 + \beta_4 x_1^2 + \beta_5 x_2^2}^{\text{Second-order terms}}$$

Therefore, asking whether the second-order model contributes more information for the prediction of y than the first-order model is equivalent to asking whether $\beta_3 = \beta_4 = \beta_5 = 0$—i.e., whether the terms involving β_3, β_4, and β_5 should be retained in the model. To test whether the second-order terms should be included in the model, we test the null hypothesis

$$H_0: \quad \beta_3 = \beta_4 = \beta_5 = 0$$

(i.e., the second-order terms do not contribute information for the prediction of y) against the alternative hypothesis

H_a: At least one of the parameters, β_3, β_4, or β_5, differs from 0

(i.e., at least one of the second-order terms contributes information for the prediction of y).

In Section 14.4 we presented the t-test for a single coefficient, and in Section 14.5 we gave the F-test for *all* the β parameters (except β_0) in the model. We now need a test for *some* of the β parameters in the model. The test procedure is intuitive: First, we use the method of least squares to fit the first-order model and calculate the corresponding sum of squares for error, SSE_1 (the sum of squares of the deviations between observed and predicted y values). Next, we fit the second-order model and calculate its sum of squares for error, SSE_2. Then, we compare SSE_1 to SSE_2 by calculating $SSE_1 - SSE_2$. If the second-order terms contribute to the model, then SSE_2 should be much smaller than SSE_1, and the difference $SSE_1 - SSE_2$ will be large. That is, the larger the difference, the greater the weight of evidence that the second-order model provides better predictions of y than does the first-order model.

The sum of squares for error always decreases when new terms are added to the model. The question is whether this decrease is large enough to conclude that it is due to more than just an increase in the number of model terms and to chance. To test the null hypothesis that the parameters of the second-order terms, β_3, β_4, and β_5, simultaneously equal 0, we use an F statistic calculated as follows:

$$F = \frac{(SSE_1 - SSE_2)/3}{SSE_2/[n - (5 + 1)]}$$

$$= \frac{\text{Drop in SSE/Number of } \beta \text{ parameters being tested}}{s^2 \text{ for the complete second-order model}}$$

When the assumptions listed in Section 14.1 about the error term ε are satisfied and the β parameters for the second-order terms are all 0 (H_0 is true), this F statistic has an F-distribution with $\nu_1 = 3$ numerator df and $\nu_2 = n - 6$ denominator df. Note that ν_1 is the number of β parameters being tested and ν_2 is the number of degrees of freedom associated with s^2 in the complete second-order model.

If the second-order terms do contribute to the model (H_a is true), we expect the F statistic to be large. Thus, we use a one-tailed test and reject H_0 when F exceeds some critical value, F_α, as shown in Figure 15.12. The steps employed in testing the null hypothesis that a set of model parameters are all equal to 0 are summarized in the box.

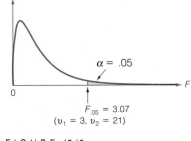

$f(F)$

$\alpha = .05$

0

$F_{.05} = 3.07$
$(\nu_1 = 3, \nu_2 = 21)$

FIGURE 15.12

Rejection region for the F-test of
H_0: $\beta_3 = \beta_4 = \beta_5 = 0$

F-Test for Testing the Null Hypothesis: Set of β Parameters Equal 0

Reduced model: $E(y) = \beta_0 + \beta_1 x_1 + \cdots + \beta_g x_g$

Complete model:

$$E(y) = \beta_0 + \beta_1 x_1 + \cdots + \beta_g x_g + \beta_{g+1} x_{g+1} + \cdots + \beta_k x_k$$

H_0: $\beta_{g+1} = \beta_{g+2} = \cdots = \beta_k = 0$

H_a: At least one of the β parameters being tested is nonzero

Test statistic: $F = \dfrac{(SSE_1 - SSE_2)/(k - g)}{SSE_2/[n - (k + 1)]}$

where

SSE_1 = Sum of squared errors for the reduced model

SSE_2 = Sum of squared errors for the complete model

$k - g$ = Number of β parameters specified in H_0

$k + 1$ = Number of β parameters in the complete model

n = Total sample size

Rejection region: $F > F_\alpha$

where F_α is based on $\nu_1 = (k - g)$ numerator df and $\nu_2 = [n - (k + 1)]$ denominator df.

EXAMPLE 15.4

Suppose you wish to study the growth of carnations as a function of the temperature x_1 in a greenhouse and the amount of fertilizer x_2 applied to the soil. A total of 27 plots of equal size are treated with fertilizer in amounts varying between 50 and 60 kilograms per plot, and these plots are mechanically kept at constant temperatures between 80 and 100°F. Small carnation plants (approximately 15 centimeters in height) are planted in each plot, and their height y is measured after a 6-week growing period. The resulting data are shown in Table 15.3.

TABLE 15.3   **Temperature, Amount of Fertilizer, and Height of Carnations**

x_1 (°F)	x_2 (kilograms per plot)	y (centimeters)	x_1 (°F)	x_2 (kilograms per plot)	y (centimeters)	x_1 (°F)	x_2 (kilograms per plot)	y (centimeters)
80	50	50.8	90	50	63.4	100	50	46.6
80	50	50.7	90	50	61.6	100	50	49.1
80	50	49.4	90	50	63.4	100	50	46.4
80	55	93.7	90	55	93.8	100	55	69.8
80	55	90.9	90	55	92.1	100	55	72.5
80	55	90.9	90	55	97.4	100	55	73.2
80	60	74.5	90	60	70.9	100	60	38.7
80	60	73.0	90	60	68.8	100	60	42.5
80	60	71.2	90	60	71.3	100	60	41.4

a. Fit a complete second-order model to the data.

b. Sketch the response surface.

c. Do the data provide sufficient evidence to indicate that the second-order terms contribute information for the prediction of y?

Solution

a. The complete second-order model is

$$E(y) = \beta_0 + \beta_1 x_1 + \beta_2 x_2 + \beta_3 x_1 x_2 + \beta_4 x_1^2 + \beta_5 x_2^2$$

The data in Table 15.3 were used to fit this model, and a portion of the SAS output is shown in Figure 15.13.

FIGURE 15.13

Portion of the SAS printout for Example 15.4

SOURCE	DF	SUM OF SQUARES	MEAN SQUARE	F VALUE	PR > F
MODEL	5	8402.26453714	1680.45290743	596.32	0.0001
ERROR	21	59.17842582	2.81802028		ROOT MSE
				R-SQUARE	1.67869601
CORRECTED TOTAL	26	8461.44296296		0.993006	

PARAMETER	ESTIMATE	STD ERROR OF ESTIMATE	T FOR H0: PARAMETER = 0	PR > |T|
INTERCEPT	-5127.89907417	110.29601483	-46.49	0.0001
X1	31.09638889	1.34441322	23.13	0.0001
X2	139.74722222	3.14005411	44.50	0.0001
X1*X2	-0.14550000	0.00969196	-15.01	0.0001
X1*X1	-0.13338889	0.00685325	-19.46	0.0001
X2*X2	-1.14422222	0.02741299	-41.74	0.0001

The least squares prediction equation is

$$\hat{y} = -5{,}127.90 + 31.10x_1 + 139.75x_2 - .146x_1x_2 - .133x_1^2 - 1.14x_2^2$$

b. A three-dimensional graph of this prediction model is shown in Figure 15.14. Note that the height seems to be greatest for temperatures of about 85–90°F and for applications of about 55–57 kilograms of fertilizer per plot.* Further experimentation in these ranges might lead to a more precise determination of the optimal temperature–fertilizer combination.

FIGURE 15.14

Plot of second-order least squares model for Example 15.4

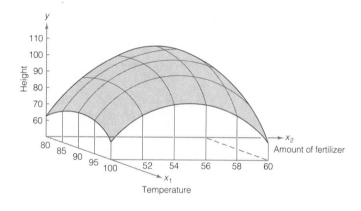

c. To determine whether the data provide sufficient information to indicate that the second-order terms contribute information for the prediction of y, we wish to test

$$H_0: \quad \beta_3 = \beta_4 = \beta_5 = 0$$

against the alternative hypothesis

$$H_a: \quad \text{At least one of the parameters, } \beta_3, \beta_4, \beta_5, \text{ differs from } 0$$

The first step in conducting the test is to drop the second-order terms out of the complete (second-order) model and fit the reduced model

$$E(y) = \beta_0 + \beta_1x_1 + \beta_2x_2$$

to the data. The SAS computer printout for this procedure is shown in Figure 15.15, page 818.

You can see that the sums of squares for error, given in Figures 15.13 and 15.15 for the complete and reduced models, are

$$SSE_2 = 59.17842582$$
$$SSE_1 = 6{,}671.50851852$$

and that s^2 for the complete model is

$$s_2^2 = 2.81802028$$

---

*Students with knowledge of calculus should note that we can solve for the exact temperature and amount of fertilizer that maximize height in the least squares model by solving $\partial \hat{y} / \partial x_1 = 0$ and $\partial \hat{y} / \partial x_2 = 0$ for x_1 and x_2. Sample estimates of these estimated optimal values are $x_1 = 86.25$°F and $x_2 = 55.58$ kilograms per plot.

FIGURE 15.15

SAS computer printout for the reduced (first-order) model, Example 15.4

DEPENDENT VARIABLE: Y

SOURCE	DF	SUM OF SQUARES	MEAN SQUARE	F VALUE	PR > F
MODEL	2	1789.93444444	894.96722222	3.22	0.0577
ERROR	24	6671.50851852	277.97952160		
CORRECTED TOTAL	26	8461.44296296			

R-SQUARE	C.V.	ROOT MSE	Y MEAN
0.211540	24.8984	16.67271788	66.96296296

PARAMETER	ESTIMATE	STD ERROR OF ESTIMATE	T FOR H0: PARAMETER=0	PR > !T!
INTERCEPT	106.08518519	55.94500427	1.90	0.0700
X1	-0.91611111	0.39297973	-2.33	0.0285
X2	0.78777778	0.78595946	1.00	0.3262

Recall that $n = 27$, $k = 5$, and $g = 2$. Therefore, the calculated value of the F statistic, based on $\nu_1 = (k - g) = 3$ numerator df and $\nu_2 = [n - (k + 1)] = 21$ denominator df, is

$$F = \frac{(SSE_1 - SSE_2)/(k - g)}{SSE_2/[n - (k + 1)]} = \frac{(SSE_1 - SSE_2)/(k - g)}{s_2^2}$$

where $\nu_1 = (k - g)$ is equal to the number of parameters involved in H_0 and s_2^2 is the value of s^2 for the complete model. Therefore,

$$F = \frac{(6{,}671.50851852 - 59.17842582)/3}{2.81802028}$$

$$= 782.1$$

The final step in the test is to compare this computed value of F with the tabulated value based on $\nu_1 = 3$ and $\nu_2 = 21$ df. If we choose $\alpha = .05$, then $F_{.05} = 3.07$. Since the computed value of F falls in the rejection region (see Figure 15.12), i.e., it exceeds $F_{.05} = 3.07$, we reject H_0 and conclude that at least one of the second-order terms contributes information for the prediction of y. In other words, the data support the contention that the curvature we see in the response surface is not due simply to random variation in the data. The second-order model appears to provide better predictions of y than a first-order model.

Example 15.4 demonstrates the motivation for testing a hypothesis that each one of a set of β parameters equals 0; it also demonstrates the procedure. Other applications of this test appear in the following sections.

EXERCISES 15.33–15.45

[*Note: Starred (*) exercises require the use of a computer.*]

LEARNING THE MECHANICS

15.33 Suppose you fit the regression model

$$y = \beta_0 + \beta_1 x_1 + \beta_2 x_2 + \beta_3 x_1 x_2 + \beta_4 x_1^2 + \beta_5 x_2^2 + \varepsilon$$

to $n = 30$ data points and you wish to test

$$H_0: \beta_3 = \beta_4 = \beta_5 = 0$$

a. State the alternative hypothesis H_a.
b. Explain in detail how you would find the quantities necessary to compute the F statistic for this test of hypothesis.
c. What are the numerator and denominator degrees of freedom associated with the F statistic?

15.34 Suppose you fit the complete and reduced models for the hypothesis test described in Exercise 15.33 and obtain $SSE_1 = 246.1$ and $SSE_2 = 215.2$. Conduct the hypothesis test and interpret the results of your test. Test using $\alpha = .05$.

15.35 Explain why the F-test used to compare complete and reduced models is a one-tailed, upper-tailed test.

15.36 Minitab was used to fit the complete model

$$y = \beta_0 + \beta_1 x_1 + \beta_2 x_2 + \beta_3 x_3 + \beta_4 x_4 + \varepsilon$$

to $n = 20$ data points. Note that the dependent variable y is represented by C1, and the independent variables x_1–x_4 are represented by C2–C5, respectively.

Printout for Exercise 15.36, complete model

```
THE REGRESSION EQUATION IS
C1 = 14.6 - 0.611 C2 + 0.439 C3 - 0.080 C4 - 0.064 C5
```

COLUMN	COEFFICIENT	ST. DEV. OF COEF.	T-RATIO = COEF/S.D.
	14.575	4.887	2.98
C2	-0.6113	0.1775	-3.44
C3	0.4388	0.2199	2.00
C4	-0.0796	0.1083	-0.74
C5	-0.0636	0.1247	-0.51

```
S = 3.190

R-SQUARED = 84.5 PERCENT
R-SQUARED = 80.3 PERCENT, ADJUSTED FOR D.F.
```

ANALYSIS OF VARIANCE

DUE TO	DF	SS	MS=SS/DF
REGRESSION	4	831.09	207.77
RESIDUAL	15	152.66	10.18
TOTAL	19	983.75	

The independent variables x_3 and x_4 were dropped from the preceding model, and Minitab was used to fit the resulting reduced model (see printout at top of page 820).

Printout for Exercise 15.36, reduced model
_____

THE REGRESSION EQUATION IS
C1 = 14.0 - 0.642 C2 + 0.396 C3

                              ST. DEV.   T-RATIO =
COLUMN   COEFFICIENT   OF COEF.   COEF/S.D.
               13.968   4.626   3.02
C2   -0.6422   0.1675   -3.84
C3   0.3959   0.2061   1.92

S = 3.072

R-SQUARED = 83.7 PERCENT
R-SQUARED = 81.8 PERCENT, ADJUSTED FOR D.F.

ANALYSIS OF VARIANCE

  DUE TO   DF   SS   MS=SS/DF
REGRESSION   2   823.31   411.66
RESIDUAL   17   160.44   9.44
TOTAL   19   983.75
_____

a. Report the least squares prediction equations for the complete and reduced models.

b. Find SSE_1 and SSE_2. Interpret each of these quantities.

c. How many β parameters are in the complete model? The reduced model?

d. Specify the null and alternative hypotheses you would use to investigate whether the complete model contributes more information for the prediction of y than the reduced model.

e. Conduct the hypothesis test of part **d**. Use $\alpha = .05$.

f. What is the approximate p-value of the test of part **e**?

APPLYING THE CONCEPTS

15.37   Exercise 14.16 discussed one aspect of the analysis of working wives' "time crunch" data. In their article "The Time Crunch," K. D. Fox and S. Y. Nickols (1983) report on a study of the impact of a wife's employment on the number of hours available for household tasks. Some 206 families were employed in the study. One independent variable measured for each family was the total time y (in minutes per day) that the wife spent on household chores. Two independent variables were also recorded:

WEMP: x_1 = Wife's hours of employment per week

YAGE: x_2 = Age of the youngest child in the family

The computer printout for the regression analysis is repeated at the top of page 821. The sums of squares shown under SEQUENTIAL SS are those you would obtain by sequentially adding variables to the model. For example, the sum of squares opposite WEMP is for the model

$$E(y) = \beta_0 + \beta_1 x_1$$

The sum of squares opposite YAGE is the drop in SSE for the model obtained by adding x_2, that is, for

$$E(y) = \beta_0 + \beta_1 x_1 + \beta_2 x_2$$

Therefore, this sum of squares can be used in the F statistic to test H_0: $\beta_2 = 0$.

In contrast, the sums of squares shown under PARTIAL SS are the differences in the sums of squares between reduced and complete models. These are the quantities needed to conduct individual F-tests on both of the parameters, β_1 and β_2.

```
DEPENDENT VARIABLE: WIFE'S HOUSEHOLD WORK        (MEAN=401.3ᵃ)

SOURCE OF VARIATION   DF      SUMS OF SQUARES        MEAN SQUARE          F           R²

TOTAL                 205       5,432,296                             62.10***      0.38

   MODEL                2       2,061,918            1,030,959

   ERROR              203       3,370,378              16,603

SOURCE WITHIN MODEL   DF      SEQUENTIAL S.S.        PARTIAL S.S.        b-ESTIMATES

WEMP                   1       1,810,154***          1,110,288***         -4.08

YAGE                   1         251,765***            251,765***         -7.02
---------------------------------------------------------------------------
DEPENDENT VARIABLE: HUSBAND'S WORK      (MEAN=105.2ᵃ)
                    REGRESSION ANALYSIS NOT SIGNIFICANT
```

Note: In this particular printout, b-ESTIMATE is our β-estimate, or parameter estimate. Also, the asterisks indicate the level of significance of test statistics (or sum of squares used to compute test statistics). Here, *** means $p < .001$.

For example, 1,110,288 is the difference in the SSE values for the models

$$E(y) = \beta_0 + \beta_2 x_2 \quad \text{and} \quad E(y) = \beta_0 + \beta_1 x_1 + \beta_2 x_2$$

Therefore, it can be used to compute the F statistic for testing H_0: $\beta_1 = 0$. [*Note:* This sum of squares differs substantially from the corresponding sum of squares, 1,810,154, shown under the SEQUENTIAL SS column.]

  a. Use the information in the printout to test H_0: $\beta_1 = 0$. Test using $\alpha = .05$ and interpret your results.

  b. State the alternative hypothesis implied in the test in part **a**.

  c. Use the information in the printout to test H_0: $\beta_2 = 0$. Test using $\alpha = .05$ and interpret your results.

  d. In concluding, Fox and Nickols state that "the wife's employment explained substantially more variation in the wife's housework time than age of the younger child." Do you agree? Explain.

15.38   A large hospital rates the performance of each member of its technical staff once a year. Each person is rated on a scale of 0 to 100 by his or her immediate supervisor, and this merit rating is used to determine the size of the person's pay raise for the coming year. The hospital's personnel department is interested in developing a regression model to help them forecast the merit rating that an applicant for a technical position will receive after being employed 3 years. The hospital proposes to use the following model to forecast the merit ratings of applicants who have just completed their graduate studies and have no prior related job experience:

$$E(y) = \beta_0 + \beta_1 x_1 + \beta_2 x_2 + \beta_3 x_1 x_2 + \beta_4 x_1^2 + \beta_5 x_2^2$$

where

  y = Applicant's merit rating after 3 years

  x_1 = Applicant's grade-point average (GPA) in graduate school

  x_2 = Applicant's total score (verbal plus quantitative) on the
  Graduate Record Examination (GRE)

A random sample of $n = 40$ employees who have been on the technical staff of the hospital more than 3 years is selected. Each employee's merit rating after 3 years, graduate school GPA, and total score on the GRE are recorded. The model is fit to these data with the aid of a computer. A portion of the resulting computer printout is shown.

Printout for Exercise 15.38, complete model

SOURCE	DF	SUM OF SQUARES	MEAN SQUARE
MODEL	5	4911.56	982.31
ERROR	34	1830.44	53.84
TOTAL	39	6742.00	R-SQUARE
			0.73

The reduced model $E(y) = \beta_0 + \beta_1 x_1 + \beta_2 x_2$ is fit to the same data. The resulting computer printout is partially reproduced here:

Printout for Exercise 15.38, reduced model

SOURCE	DF	SUM OF SQUARES	MEAN SQUARE
MODEL	2	3544.84	1772.42
ERROR	37	3197.16	86.41
TOTAL	39	6742.00	R-SQUARE
			0.53

a. Identify the appropriate null and alternative hypotheses to test whether the complete (second-order) model contributes information for the prediction of y.

b. Conduct the test of hypothesis given in part **a**. Test using $\alpha = .05$. Interpret the results in the context of this problem.

c. Identify the appropriate null and alternative hypotheses to test whether the complete model contributes more information than the reduced (first-order) model for the prediction of y.

d. Conduct the test of hypothesis given in part **c**. Test using $\alpha = .05$. Interpret the results in the context of this problem.

e. Which model, if either, would you use to predict y? Explain.

*15.39 Case Study 15.1 presents Neil A. Palomba's data giving a measure of strike activity (the percentage y of total working hours lost due to strikes) and three quantitative independent variables:

x_1 = Percentage of union members in nonagricultural establishments

x_2 = Percentage of all nonagricultural employment that is manufacturing

x_3 = Hourly earnings of workers on manufacturing payrolls

The SAS computer printout for fitting a first-order model to Palomba's data is shown. Note that $R^2 = .231906$. Can you improve on Palomba's model?

a. Use a computer to fit an interaction model to the data, and test to see whether this model provides more information for the prediction of y than does the first-order model. Test using $\alpha = .05$.

b. Fit a complete second-order model to the data. Test to see whether this model provides more information for the prediction of y than the interaction model. Test using $\alpha = .05$.

c. Comment on the results of parts **a** and **b**.

Printout for Exercise 15.39

DEPENDENT VARIABLE: Y SOURCE	DF	SUM OF SQUARES	MEAN SQUARE	F VALUE
MODEL	3	0.28839448	0.09613149	4.53
ERROR	45	0.95518919	0.02122643	PR > F
CORRECTED TOTAL	48	1.24358367		0.0074
R-SQUARE	C.V.	ROOT MSE	Y MEAN	
0.231906	88.7929	0.14569292	0.16408163	

*15.40   The data in the table were obtained from an experiment designed to investigate the relationship between the yield y of potatoes and the levels of three minerals in the soil, x_1, x_2, and x_3. [*Note:*   The mineral levels have been coded by subtracting an appropriate constant from each of the x values.]

y	x_1	x_2	x_3	y_1	x_1	x_2	x_3
16.40	−1	−1	−1	2.75	1	1	1
13.51	1	−1	−1	14.33	−1.682	0	0
14.41	−1	1	−1	5.44	1.682	0	0
9.38	1	1	−1	19.80	0	−1.682	0
10.77	−1	−1	1	20.00	0	1.682	0
11.78	1	−1	1	9.37	0	0	−1.682
4.11	−1	1	1	10.03	0	0	1.682

a. Fit a second-order polynomial,

$$y = \beta_0 + \beta_1 x_1 + \beta_2 x_2 + \beta_3 x_3 + \beta_4 x_1 x_2 + \beta_5 x_1 x_3 + \beta_6 x_2 x_3 + \beta_7 x_1^2 + \beta_8 x_2^2 + \beta_9 x_3^2 + \varepsilon$$

by least squares.

b. Test the hypothesis H_0: $\beta_4 = \beta_5 = \cdots = \beta_9 = 0$. That is, test whether the data provide sufficient evidence to indicate that a second-order model contributes more information for the prediction of yield than a first-order model. Test using $\alpha = .05$.

15.41   An energy conservationist wants to develop a model that will estimate the mean annual gasoline consumption y in the United States (in millions of barrels) as a function of two independent variables:

x_1 = Number of cars (millions) in use during year

x_2 = Number of trucks (millions) in use during year

a. Identify the independent variables as quantitative or qualitative.
b. Write the first-order model for $E(y)$.
c. Write the complete second-order model for $E(y)$.
d. With respect to the model of part c, specify the null and alternative hypotheses you would employ in testing for the presence of interaction between x_1 and x_2.
e. Based on a sample of $n = 15$ years, the resulting values of SSE_1 and SSE_2 were 1,065.9 and 400.6, respectively. Conduct the test to determine whether the data present sufficient evidence to indicate interaction between x_1 and x_2. Test using $\alpha = .05$.

*15.42   A firm would like to be able to forecast its yearly sales in each of its sales regions. The firm has decided to base its forecasts on regional population size and its yearly regional advertising expenditures. The population data in the table (page 824) were obtained from the U.S. Bureau of the Census, and the advertising and sales data were obtained from the firm's internal records.

SALES REGION	SALES (Thousands of units)	POPULATION OF REGION (Thousands)	ADVERTISING EXPENDITURES (Thousands of dollars)
1	65	200	8
2	80	210	10
3	85	205	9
4	100	300	8.5
5	108	320	12
6	114	290	10
7	40	90	6
8	45	85	8
9	150	450	9
10	42	87	9
11	220	480	13
12	200	500	15

a. Fit a complete second-order model to these data.

b. Is the complete second-order model useful for forecasting sales? Test using $\alpha = .05$.

c. The firm is planning to market its product in a new sales region next year. The region has a population of 400,000, and the firm plans to spend \$12,000 on advertising. Use the fitted model you obtained in part **a** to forecast next year's sales in this new region. [*Note:* We would want to express this estimate as a prediction interval, but its computation is beyond the scope of this text. The procedure is described in the references. You may also find that it can be obtained using your computer program package.]

15.43   Refer to Exercise 15.42, in which a firm would like to develop a regression model to forecast its yearly sales in each of its sales regions. The model under consideration is a complete second-order model:

$$E(y) = \beta_0 + \beta_1 x_1 + \beta_2 x_2 + \beta_3 x_1 x_2 + \beta_4 x_1^2 + \beta_5 x_2^2$$

where

y = Yearly regional sales

x_1 = Population of sales region

x_2 = Yearly regional advertising expenditures

A portion of the computer printout that results from fitting this model to the $n = 12$ data points given in Exercise 15.42 is shown here.

Printout for Exercise 15.43, complete model

SOURCE	DF	SUM OF SQUARES	MEAN SQUARE
MODEL	5	38638.97	7727.79
ERROR	6	159.94	26.66
TOTAL	11	38798.91	R-SQUARE
			0.996

The reduced first-order model

$$E(y) = \beta_0 + \beta_1 x_1 + \beta_2 x_2$$

was fit to the same data, and the resulting computer printout is partially reproduced here.

Printout for Exercise 15.43, reduced model

SOURCE	DF	SUM OF SQUARES	MEAN SQUARE
MODEL	2	36704.5	18352.2
ERROR	9	2094.4	232.7
TOTAL	11	38798.9	R-SQUARE
			0.946

Do these data present sufficient evidence to conclude that a second-order model contributes more information for the prediction of y than does a first-order model? Test using $\alpha = .05$.

*15.44 In Exercise 14.10 we found that a first-order model was useful for characterizing the relationship between household food consumption and the two independent variables, household income and size of household. The sample data are repeated in the accompanying table and the SAS printout summarizes the results of fitting a first-order model to the data.

HOUSEHOLD	FOOD CONSUMPTION DURING 1990 y ($ thousands)	1990 HOUSEHOLD INCOME x_1 ($ thousands)	NUMBER OF PERSONS IN HOUSEHOLD AT END OF 1990 x_2	HOUSEHOLD	FOOD CONSUMPTION DURING 1990 y ($ thousands)	1990 HOUSEHOLD INCOME x_1 ($ thousands)	NUMBER OF PERSONS IN HOUSEHOLD AT END OF 1990 x_2
1	3.2	31.1	4	14	3.1	85.2	2
2	2.4	20.5	2	15	4.5	35.6	9
3	3.8	42.3	4	16	3.5	68.5	3
4	1.9	18.9	1	17	4.0	10.5	5
5	2.5	26.5	2	18	3.5	21.6	4
6	3.0	29.8	4	19	1.8	29.9	1
7	2.6	24.3	3	20	2.9	28.6	3
8	3.2	38.1	4	21	2.6	20.2	2
9	3.9	52.0	5	22	3.6	38.7	5
10	1.7	16.0	1	23	2.8	11.2	3
11	2.9	41.9	3	24	4.5	14.3	7
12	1.7	9.9	1	25	3.5	16.9	5
13	4.5	33.1	7				

SAS printout for Exercise 15.44

DEPENDENT VARIABLE: FOOD

SOURCE	DF	SUM OF SQUARES	MEAN SQUARE	F VALUE
MODEL	2	15.46228509	7.73114255	100.80
ERROR	22	1.68731491	0.07669613	PR > F
CORRECTED TOTAL	24	17.14960000		0.0001

R-SQUARE	C.V.	ROOT MSE	FOOD MEAN
0.901612	8.9221	0.27694067	3.10400000

PARAMETER	ESTIMATE	T FOR H0: PARAMETER=0	PR > ¦T¦	STD ERROR OF ESTIMATE
INTERCEPT	1.43260377	9.76	0.0001	0.14673751
INCOME	0.00999062	3.15	0.0046	0.00316806
SIZE	0.37928986	13.68	0.0001	0.02772472

a. Fit a complete second-order model to the data.

b. Is the complete second-order model useful for explaining the variation in household food consumption? Test using $\alpha = .05$.

c. Is there sufficient evidence to conclude that at least one of the β parameters associated with the second-order terms differs from 0? Test using $\alpha = .05$.

d. Do the data provide sufficient evidence to conclude that household income and household size interact? Test using $\alpha = .05$.

e. Describe the danger involved in conducting a series of hypothesis tests (such as in parts **b**, **c**, and **d**) for determining which terms to retain in a regression model and which to exclude.

*15.45 In Exercise 14.11, a real estate appraiser used regression analysis to explore the relationship between the sale prices of apartments and various characteristics of the apartments. The data are repeated in the table.

CODE NO.	SALE PRICE y ($)	NUMBER OF APARTMENT UNITS x_1	AGE OF STRUCTURE x_2 (years)	LOT SIZE x_3 (sq. ft.)	NUMBER OF ON-SITE PARKING SPACES x_4	GROSS BUILDING AREA x_5
0229	90,300	4	82	4,635	0	4,266
0094	384,000	20	13	17,798	0	14,391
0043	157,500	5	66	5,913	0	6,615
0079	676,200	26	64	7,750	6	34,144
0134	165,000	5	55	5,150	0	6,120
0179	300,000	10	65	12,506	0	14,552
0087	108,750	4	82	7,160	0	3,040
0120	276,538	11	23	5,120	0	7,881
0246	420,000	20	18	11,745	20	12,600
0025	950,000	62	71	21,000	3	39,448
0015	560,000	26	74	11,221	0	30,000
0131	268,000	13	56	7,818	13	8,088
0172	290,000	9	76	4,900	0	11,315
0095	173,200	6	21	5,424	6	4,461
0121	323,650	11	24	11,834	8	9,000
0077	162,500	5	19	5,246	5	3,828
0060	353,500	20	62	11,223	2	13,680
0174	134,400	4	70	5,834	0	4,680
0084	187,000	8	19	9,075	0	7,392
0031	155,700	4	57	5,280	0	6,030
0019	93,600	4	82	6,864	0	3,840
0074	110,000	4	50	4,510	0	3,092
0057	573,200	14	10	11,192	0	23,704
0104	79,300	4	82	7,425	0	3,876
0024	272,000	5	82	7,500	0	9,542

Source: Robinson Appraisal Co., Inc., Mankato, Minnesota.

a. Fit a first-order model to the data. (You may already have done this in Exercise 14.11.)

b. Do the data provide sufficient evidence to conclude that the model of part **a** is useful for predicting sale price? Test using $\alpha = .05$.

c. Drop x_3 and x_4 from the model of part **a** and refit the model to the data.

d. Do the data provide sufficient evidence to conclude that the model of part **c** is useful for predicting sale price? Test using $\alpha = .05$.

15.5 Models with One Qualitative Independent Variable

Suppose we want to develop a model for the mean operating cost per mile, $E(y)$, of compact cars as a function of the car's manufacturer. (For the purpose of explanation, we ignore other independent variables that might affect the response.) Further suppose there are three manufacturers of interest, which we identify as A, B, and C. Then the automobile manufacturer is a single qualitative variable with three levels, A, B, and C. Note that with a qualitative independent variable, we cannot attach a quantitative measure to a given level. Even if we were to call the manufacturers 1, 2, and 3, the numbers would simply be identifiers of the manufacturers and would have no meaningful quantitative interpretation.

To simplify our notation, let μ_A be the mean cost per mile for compact cars produced by manufacturer A, and let μ_B and μ_C be the corresponding mean costs per mile for those produced by B and C. Our objective is to write a single equation that will give the mean value of y (cost per mile) for the three manufacturers. This can be done as follows:

$$E(y) = \beta_0 + \beta_1 x_1 + \beta_2 x_2$$

where

$$x_1 = \begin{cases} 1 & \text{if the car is manufactured by B} \\ 0 & \text{if the car is not manufactured by B} \end{cases}$$

$$x_2 = \begin{cases} 1 & \text{if the car is manufactured by C} \\ 0 & \text{if the car is not manufactured by C} \end{cases}$$

The variables x_1 and x_2 are not meaningful independent variables as for the case of the models with quantitative independent variables. Instead, they are **dummy** (or **indicator**) **variables** that make the model function. To see how they work, let $x_1 = 0$ and $x_2 = 0$. This condition will apply when we are seeking the mean response A. (Neither B nor C will be the manufacturer; hence, it must be A.) Then the mean cost per mile, $E(y)$, when A is the manufacturer is

$$\mu_A = E(y) = \beta_0 + \beta_1(0) + \beta_2(0) = \beta_0$$

This tells us that the mean cost per mile for A is β_0. Or, using our notation, it means that $\mu_A = \beta_0$.

Now suppose we want to represent the mean cost per mile, $E(y)$, for manufacturer B. Checking the dummy variable definitions, we see that we should let $x_1 = 1$ and $x_2 = 0$:

$$\mu_B = E(y) = \beta_0 + \beta_1 x_1 + \beta_2 x_2 = \beta_0 + \beta_1(1) + \beta_2(0) = \beta_0 + \beta_1$$

or, since $\beta_0 = \mu_A$,

$$\mu_B = \mu_A + \beta_1$$

Then it follows that the interpretation of β_1 is

$$\beta_1 = \mu_B - \mu_A$$

which is the difference between the mean costs per mile for manufacturers B and A.

Finally, if we want the mean value of y when C is the manufacturer, we let $x_1 = 0$ and $x_2 = 1$:

$$\mu_C = E(y) = \beta_0 + \beta_1(0) + \beta_2(1) = \beta_0 + \beta_2$$

or, since $\beta_0 = \mu_A$,

$$\mu_C = \mu_A + \beta_2$$

Then it follows that the interpretation of β_2 is

$$\beta_2 = \mu_C - \mu_A$$

Note that we were able to describe *three levels* of the qualitative variable with only *two dummy variables*. This is because the mean of the base level (manufacturer A, in this case) is accounted for by the intercept β_0.

Since the automobile manufacturer is a qualitative variable, we will use a bar graph to show the value of mean operating cost per mile $E(y)$ for the three levels of automobile manufacturer (see Figure 15.16). Particularly, note that the height of the bar, $E(y)$, for each level of automobile manufacturer is equal to the sum of the model parameters shown in the preceding equations. You can see that the height of the bar corresponding to manufacturer A is β_0, that is, $E(y) = \beta_0$. Similarly, the heights of the bars corresponding to manufacturers B and C are $E(y) = \beta_0 + \beta_1$ and $E(y) = \beta_0 + \beta_2$, respectively.*

FIGURE 15.16

Bar chart comparing $E(y)$ for three automobile manufacturers

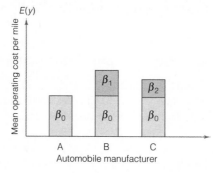

Now carefully examine the model with a single qualitative independent variable at three levels, because we will use exactly the same pattern for any number of levels. Moreover, the interpretation of the parameters will always be the same.

One level is selected as the base level. (We used manufacturer A as the base level.) Then for the 1–0 system of coding[†] for the dummy variables,

$$\mu_A = \beta_0$$

The coding for all dummy variables is as follows: To represent the mean value of y for a particular level, let that dummy variable equal 1; otherwise, the dummy variable is set equal to 0. Using this system of coding,

$$\mu_B = \beta_0 + \beta_1 \qquad \mu_C = \beta_0 + \beta_2$$

---

*Note that either β_1, β_2, or both could be negative. If, for example, β_1 were negative, the height of the bar corresponding to manufacturer B would be *reduced* (rather than increased) from the height of the bar for manufacturer A by the amount β_1. Figure 15.16 is constructed assuming that β_1 and β_2 are positive quantities.

[†]You do not have to use a 1–0 system of coding for the dummy variables. Any two-value system will work, but the interpretation given to the model parameters will depend on the code. Using the 1–0 system makes the model parameters easy to interpret.

and so on. Because $\mu_A = \beta_0$, any other model parameter will represent the difference between means for that level and the base level:

$$\beta_1 = \mu_B - \mu_A \qquad \beta_2 = \mu_C - \mu_A$$

and so on.

> **Procedure for Writing a Model With One Qualitative Independent Variable at k Levels**
>
> Always use one less dummy variable than the number of levels of the qualitative variable. Thus, for a qualitative variable with k levels, use $k - 1$ dummy variables:
>
> $$y = \beta_0 + \beta_1 x_1 + \beta_2 x_2 + \cdots + \beta_{k-1} x_{k-1} + \varepsilon$$
>
> where x_i is the dummy variable for level $i + 1$ and
>
> $$x_i = \begin{cases} 1 & \text{if } y \text{ is observed at level } i + 1 \\ 0 & \text{otherwise} \end{cases}$$
>
> Then, for this system of coding
>
> $$\mu_A = \beta_0 \qquad \text{and} \quad \beta_1 = \mu_B - \mu_A$$
> $$\mu_B = \beta_0 + \beta_1 \qquad\qquad \beta_2 = \mu_C - \mu_A$$
> $$\mu_C = \beta_0 + \beta_2 \qquad\qquad \beta_3 = \mu_D - \mu_A$$
> $$\mu_D = \beta_0 + \beta_3 \qquad\qquad\qquad \vdots$$
> $$\vdots$$

EXAMPLE 15.5

Suppose a sociologist wants to compare the mean dollar amounts owed by delinquent credit card customers in three different annual income groups: under \$12,000, \$12,000–\$25,000, and over \$25,000. A sample of ten customers with delinquent accounts is selected from each group, and the amount owed by each is recorded, as shown in Table 15.4. Do the data provide sufficient evidence to indicate that the mean dollar amounts owed by customers differ for the three annual income groups?

TABLE 15.4   **Dollars Owed**

GROUP 1 Under \$12,000	GROUP 2 \$12,000–\$25,000	GROUP 3 Over \$25,000
\$148	\$513	\$335
76	264	643
393	433	216
520	94	536
236	535	128
134	327	723
55	214	258
166	135	380
415	280	594
153	304	465

Solution

Note that income is ordinarily a quantitative variable, but in this example only the group (low, medium, high) in which the customer's income falls is known. We therefore treat income as a qualitative variable. The model relating $E(y)$ to the single qualitative variable, annual income group, is

$$E(y) = \beta_0 + \beta_1 x_1 + \beta_2 x_2$$

where

$$x_1 = \begin{cases} 1 & \text{if group 2} \\ 0 & \text{if not} \end{cases} \qquad x_2 = \begin{cases} 1 & \text{if group 3} \\ 0 & \text{if not} \end{cases}$$

and

$$\beta_1 = \mu_2 - \mu_1 \qquad \beta_2 = \mu_3 - \mu_1$$

where μ_1, μ_2, and μ_3 are the mean responses for income groups 1, 2, and 3, respectively. Testing the null hypothesis that the means for the three groups are equal, that is, $\mu_1 = \mu_2 = \mu_3$, is equivalent to testing

$$H_0: \quad \beta_1 = \beta_2 = 0$$

because if $\beta_1 = \mu_2 - \mu_1 = 0$ and $\beta_2 = \mu_3 - \mu_1 = 0$, then μ_1, μ_2, and μ_3 must be equal. The alternative hypothesis is

$$H_a: \quad \text{At least one of the parameters, } \beta_1 \text{ or } \beta_2, \text{ differs from 0}$$

which implies that at least two of the three means (μ_1, μ_2, and μ_3) differ.

There are two ways to conduct this test. We can fit the complete model shown above and the reduced model (deleting the terms involving β_1 and β_2)

$$E(y) = \beta_0$$

and conduct the F-test described in the preceding section. (We leave this as an exercise for you.) Or we can use the F-test of the complete model (Section 14.5), which tests the null hypothesis that all parameters in the model, with the exception of β_0, equal 0. Either way you conduct the test, you will obtain the same computed value of F, the value shown on the SAS printout. The SAS printout for fitting the complete model,

$$E(y) = \beta_0 + \beta_1 x_1 + \beta_2 x_2$$

FIGURE 15.17

SAS computer printout for Example 15.5

DEPENDENT VARIABLE: Y

SOURCE	DF	SUM OF SQUARES	MEAN SQUARE	F VALUE	PR > F	R-SQUARE
MODEL	2	198772.46666667	99386.23333333	3.48	0.0452	0.205038
ERROR	27	770670.90000000	28543.36666667		ROOT MSE	
CORRECTED TOTAL	29	969443.36666667			168.94782232	

PARAMETER	ESTIMATE	STD ERROR OF ESTIMATE	T FOR H0: PARAMETER=0	PR > :T:
INTERCEPT	229.60000000	53.42599243	4.30	0.0002
X1	80.30000000	75.55576307	1.06	0.2973
X2	198.20000000	75.55576307	2.62	0.0141

is shown in Figure 15.17; the value of the F statistic for testing the complete model, $F = 3.48$, is shaded. We wish to compare this value with the tabulated value of F based on $\nu_1 = 2$ numerator df and $\nu_2 = 27$ denominator df. If we choose $\alpha = .05$, we will reject $H_0: \beta_1 = \beta_2 = 0$ if the computed value of F exceeds $F_{.05} = 3.35$. Since the computed value of F ($F = 3.48$) exceeds $F_{.05} = 3.35$, we reject H_0 and conclude that at least one of the parameters, β_1 or β_2, differs from 0. Or, equivalently, we conclude that the data provide sufficient evidence to indicate that the mean indebtedness does vary from one income group to another.

We must make two additional comments about Example 15.5. First, regression analysis is not the easiest way to analyze these data (unless you have ready access to a computer and a good regression program). A simpler procedure for calculating the value of the F statistic is the ANOVA (analysis of variance) procedure described in Chapter 10. Second, if you choose to analyze the data by fitting complete and reduced models (Section 15.4), you will find that the least squares estimate of β_0 in the reduced model

$$E(y) = \beta_0$$

is $\bar{y}$, the mean of all $n = 30$ observations, and the sum of squares for error for the reduced model is

$$SSE_1 = \sum(y_i - \hat{y}_i)^2 = \sum(y_i - \bar{y})^2 = 969{,}443.367$$

This value is shown in the SAS printout (Figure 15.17) as the sum of squares corresponding to CORRECTED TOTAL. We leave the remaining steps—calculating the drop in SSE and the resulting F statistic—to you. You will find that the value you obtain is exactly the same as the value of F shown in the SAS printout.

EXERCISES 15.46–15.58

LEARNING THE MECHANICS

15.46   Write a regression model relating the mean value of y to a qualitative independent variable that can assume two levels. Interpret all the terms in the model.

15.47   Write a regression model relating $E(y)$ to a qualitative independent variable that can assume three levels. Interpret all the terms in the model.

15.48   The following model was used to relate $E(y)$ to a single qualitative variable with four levels:

$$E(y) = \beta_0 + \beta_1 x_1 + \beta_2 x_2 + \beta_3 x_3$$

where

$$x_1 = \begin{cases} 1 & \text{if level 2} \\ 0 & \text{if not} \end{cases} \qquad x_2 = \begin{cases} 1 & \text{if level 3} \\ 0 & \text{if not} \end{cases} \qquad x_3 = \begin{cases} 1 & \text{if level 4} \\ 0 & \text{if not} \end{cases}$$

This model was fit to $n = 30$ data points and the following result was obtained:

$$\hat{y} = 10.2 - 4x_1 + 12x_2 + 2x_3$$

a. Use the least squares prediction equation to find the estimate of $E(y)$ for each level of the qualitative independent variable.

b. Specify the null and alternative hypotheses you would use to test whether $E(y)$ is the same for all four levels of the independent variable.

15.49 Minitab was used to fit the following model to $n = 15$ data points:

$$y = \beta_0 + \beta_1 x_1 + \beta_2 x_2 + \varepsilon$$

where

$$x_1 = \begin{cases} 1 & \text{if level 2} \\ 0 & \text{if not} \end{cases} \qquad x_2 = \begin{cases} 1 & \text{if level 3} \\ 0 & \text{if not} \end{cases}$$

Note in the accompanying printout that C1 represents y and C2 and C3 represent x_1 and x_2, respectively.

Printout for Exercise 15.49

```
THE REGRESSION EQUATION IS
C1 = 80.0 + 16.8 C2 + 40.4 C3


                                 ST. DEV.     T-RATIO =
   COLUMN       COEFFICIENT      OF COEF.     COEF/S.D.
                  80.000           4.082         19.60
   C2             16.800           5.774          2.91
   C3             40.400           5.774          7.00


   S = 9.129

   R-SQUARED = 80.5 PERCENT
   R-SQUARED = 77.2 PERCENT, ADJUSTED FOR D.F.

   ANALYSIS OF VARIANCE

   DUE TO        DF          SS        MS=SS/DF
   REGRESSION     2        4118.9        2059.5
   RESIDUAL      12        1000.0          83.3
   TOTAL         14        5118.9
```

a. Report the least squares prediction equation.
b. Interpret the values of $\hat{\beta}_1$ and $\hat{\beta}_2$.
c. Interpret the following hypotheses in terms of μ_1, μ_2, and μ_3:

$$H_0: \quad \beta_1 = \beta_2 = 0$$

H_a: At least one of the parameters β_1 and β_2 differs from 0

d. Conduct the hypothesis test of part c.

APPLYING THE CONCEPTS

15.50 In 1989–1990, 4-year private colleges charged an average of $8,446 for tuition and fees for the year; 4-year public colleges charged $1,781 (*Chronicle of Higher Education*, Sept. 1990). In order to estimate the difference in the mean amounts charged for the 1991–1992 academic year, random samples of 40 private colleges and 40 public colleges were contacted and questioned about their tuition structures.

a. Which of the procedures described in Chapter 9 could be used to estimate the difference in mean charges between private and public colleges?

b. Propose a regression model involving the qualitative independent variable—type of college—that could be used to investigate the difference between the means. Be sure to specify the coding scheme for the dummy variable in the model.

c. Explain how the regression model you developed in part **b** could be used to estimate the difference between the population means.

15.51 The manager of a radio and television retail store would like to compare the life span in months of four different brands of color television picture tubes. Data are gathered on ten television sets selected at random from each of four brands—Sony, Zenith, Magnavox, and RCA. Write a model that will give the mean life span for the four brands, and interpret all parameters used in the model.

15.52 Five varieties of peas are currently being tested by a large agribusiness cooperative in Ohio to determine which is best suited for production. A field was divided into 20 plots, with each variety of peas planted in four plots. The yields (in bushels of peas) produced from each plot are shown in the table.

VARIETY OF PEAS				
A	B	C	D	E
26.2	29.2	29.1	21.3	20.1
24.3	28.1	30.8	22.4	19.3
21.8	27.3	33.9	24.3	19.9
28.1	31.2	32.8	21.8	22.1

SAS was used to fit the model

$$y = \beta_0 + \beta_1 x_1 + \beta_2 x_2 + \beta_3 x_3 + \beta_4 x_4 + \varepsilon$$

to these data, using the coding $x_1 = 1$ for variety A, $x_2 = 1$ for variety B, $x_3 = 1$ for variety C, and $x_4 = 1$ for variety D. The resulting printout is shown here.

Printout for Exercise 15.52

Dep Variable: Y

Analysis of Variance

Source	DF	Sum of Squares	Mean Square	F Value	Prob>F
Model	4	342.04000	85.51000	23.966	0.0001
Error	15	53.52000	3.56800		
C Total	19	395.56000			

Root MSE	1.88892	R-Square	0.8647	
Dep Mean	25.70000	Adj R-Sq	0.8286	
C.V.	7.34986			

Parameter Estimates

Variable	DF	Parameter Estimate	Standard Error	T for H0: Parameter=0	Prob > |T|
INTERCEP	1	20.350000	0.94445752	21.547	0.0001
X1	1	4.750000	1.33566463	3.556	0.0029
X2	1	8.600000	1.33566463	6.439	0.0001
X3	1	11.300000	1.33566463	8.460	0.0001
X4	1	2.100000	1.33566463	1.572	0.1367

a. Write the model relating mean yield to pea variety, and interpret all the parameters in the model.
b. Report the least squares model from the SAS printout.
c. What null and alternative hypotheses are tested by the global F-test for this model? Interpret the hypotheses both in terms of the β parameters and the mean yields for the five varieties of peas.
d. Test the hypotheses of part c using $\alpha = .05$.
e. Place a 95% confidence interval on the difference between the mean yields of varieties D and E.

15.53   Refer to Exercise 15.52. Note that this represents a completely randomized design for comparing the four varieties (treatments), as discussed in Section 10.2. The SAS analysis of variance (ANOVA) printout for this experiment is shown here.

Printout for Exercise 15.53

General Linear Models Procedure

Dependent Variable: Y

Source	DF	Sum of Squares	Mean Square	F Value	Pr > F
Model	4	342.0400000	85.5100000	23.97	0.0001
Error	15	53.5200000	3.5680000		
Corrected Total	19	395.5600000			

R-Square	C.V.	Root MSE	Y Mean
0.864698	7.3498639	1.888915	25.70000000

Source	DF	Type I SS	Mean Square	F Value	Pr > F
VARIETY	4	342.04000	85.51000	23.97	0.0001

a. What are the null and alternative hypotheses tested by the ANOVA? How do they compare with those tested in parts c and d of Exercise 15.52?
b. Find and interpret the test statistic and significance level on the ANOVA printout. Compare the results to those obtained in Exercise 15.52.

15.54   The manager of a supermarket wants to model the total weekly sales of beer, y, as a function of brand. (This model will enable the manager to plan the store's inventory.) The market carries three brands, B_1, B_2, and B_3.
a. What type of independent variable is brand of beer?
b. Write the model relating mean weekly beer sales, $E(y)$, as a function of brand of beer. Be sure to explain any dummy variables you use.
c. Interpret the parameters (β's) of your model in part b.
d. In terms of the model parameters, what is the mean weekly sales for brand B_3?

15.55   Refer to Exercise 15.54. Suppose the manager uses brand B_1 as the base level and obtains the model

$$\hat{y} = 450 + 60x_1 - 30x_2$$

where

$$x_1 = \begin{cases} 1 & \text{if brand } B_2 \\ 0 & \text{otherwise} \end{cases}$$

$$x_2 = \begin{cases} 1 & \text{if brand } B_3 \\ 0 & \text{otherwise} \end{cases}$$

a. What is the estimated difference between the mean* weekly sales for brands B_2 and B_1?

b. What is the estimated mean weekly sales for brand B_2?

15.56 The amount of fructose present in an athlete's bloodstream is critical to performance. A researcher ran an experiment to determine whether diet had any effect on the level of fructose in the blood after 10 minutes of running on a treadmill. Twenty-one subjects were selected, and each subject's fructose level was measured after 10 minutes on a treadmill. Each subject was randomly assigned to one of three diets. One diet was high in protein, one high in carbohydrates, and one high in fruits. After 1 month on the diet, each subject was again run on the treadmill and the fructose level in the bloodstream was measured. The dependent variable was the difference in the level of fructose in the blood between the second and the first runs on the treadmill.

a. Identify the independent variables in the experiment.

b. Write an appropriate regression model relating mean difference in fructose in the blood, $E(y)$, to the independent variables. Identify and code all dummy variables.

15.57 A fisheries researcher believes that the growth rate of fish is related to the age of the fish, its feeding location, and other factors. Suppose the researcher wishes to collect fish of a certain species from four different locations.

a. Write a model relating the mean growth rate $E(y)$ to location, a qualitative variable.

b. Use your model to find the mean growth rate of the fish at location 1.

c. Use your model to find the mean growth rate of the fish at location 4.

d. In terms of your model, what is the difference in mean growth rates between locations 2 and 3?

15.58 The director of marketing of a company that sells business machines is interested in modeling mean monthly sales (in thousands of dollars) per salesperson, $E(y)$, as a function of the type of sales incentive plan that is in effect: commission only, straight salary, or salary plus commission on each sale. The director has proposed the following model:

$$E(y) = \beta_0 + \beta_1 x_1 + \beta_2 x_2$$

where

$$x_1 = \begin{cases} 1 & \text{if salesperson is paid a straight salary} \\ 0 & \text{otherwise} \end{cases}$$

$$x_2 = \begin{cases} 1 & \text{if salesperson is paid a salary plus commission} \\ 0 & \text{otherwise} \end{cases}$$

A portion of the computer printout that results from using Minitab to fit this model to the sales data collected from a sample of 15 salespersons (five from each incentive plan) is shown on page 836.

a. Do the data provide sufficient evidence to conclude that there is a difference in mean monthly sales among the three incentive plans? Test using $\alpha = .05$.

b. Use the least squares prediction equation to estimate the mean sales for salespersons working on a straight salary basis.

c. Use the least squares prediction equation to estimate the mean sales for salespersons working on a commission only basis.

[*Note:* We would prefer to use confidence intervals for the estimates in parts **b** and **c**, but their calculation is beyond the scope of this text. This procedure can be found in the references at the end of the chapter.]

---

*We would generally form confidence intervals to assess the reliability of these estimates. Our objective in these exercises is to develop the ability to use the models to obtain the estimates. The corresponding confidence intervals can be obtained by using the methods of Chapter 14.

Printout for Exercise 15.58

```
THE REGRESSION EQUATION IS
Y =    20.0 -  8.60 X1 + 3.80 X2

                                 ST. DEV.    T-RATIO =
          COLUMN   COEFFICIENT   OF COEF.    COEF/S.D.

            --       20.000       2.898        6.90
    X1  C21          -8.60        4.10        -2.10
    X2  C22           3.80        4.10          .93

THE ST. DEV. OF Y ABOUT REGRESSION LINE IS
S =      6.481
WITH (.15- 3) = 12 DEGREES OF FREEDOM

R-SQUARED = 44.5 PERCENT
R-SQUARED = 35.2 PERCENT, ADJUSTED FOR D.F.

ANALYSIS OF VARIANCE

    DUE TO      DF      SS     MS=SS/DF

    REGRESSION   2    403.60    201.80
    RESIDUAL    12    504.00     42.00
    TOTAL       14    907.60
```

15.6 Comparing the Slopes of Two or More Lines

Suppose you wish to relate the mean monthly sales $E(y)$ of a company to monthly advertising expenditure x for three different advertising media (say newspaper, radio, and television) and you wish to use first-order (straight-line) models to model the responses for all three media. Graphs of these three relationships might appear as shown in Figure 15.18.

Since the lines in Figure 15.18 are hypothetical, a number of practical business questions arise. Is one advertising medium as effective as any other? That is, do the three mean sales lines differ for the three advertising media? Do the increases in mean sales per dollar input in advertising differ for the three advertising media? That is, do the slopes of the three lines differ? Note that the two practical business questions have been rephrased into questions about the parameters that define the three lines of Figure 15.18. To answer them, we must write a single linear statistical model that will characterize the three lines of Figure 15.18 and that, by testing hypotheses about the lines, will answer the practical business questions.

The response described previously, monthly sales, is a function of *two* independent variables, one quantitative (advertising expenditure x_1) and one qualitative (type of medium). We will proceed, in stages, to build a model relating $E(y)$ to these variables and will show graphically the interpretation we would give to the model at each stage. This will help you to see the contributions of the various terms in the model.

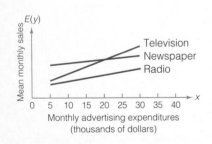

FIGURE 15.18

Graphs of the relationship between mean sales $E(y)$ and advertising expenditure x

1. The straight-line relationship between mean sales $E(y)$ and advertising expenditure is the same for all three media—i.e., a single line will describe the relationship between $E(y)$ and advertising expenditure x_1 for all the media (see Figure 15.19).

$$E(y) = \beta_0 + \beta_1 x_1 \quad \text{where } x_1 = \text{Advertising expenditure}$$

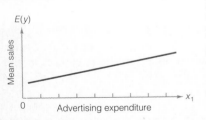

FIGURE 15.19

The relationship between $E(y)$ and x_1 is the same for all media

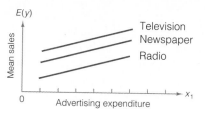

E(y)

Mean sales

0   Advertising expenditure x_1

Television
Newspaper
Radio

FIGURE 15.20

Parallel response lines for the three media

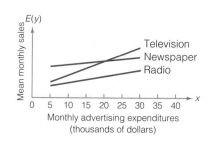

E(y)

Mean monthly sales

0   5   10   15   20   25   30   35   40   x

Monthly advertising expenditures
(thousands of dollars)

Television
Newspaper
Radio

FIGURE 15.21

Different response lines for the three media

2. The straight lines relating mean sales $E(y)$ to advertising expenditure x_1 differ from one medium to another, but the rate of increase in mean sales per increase in dollar advertising expenditure x_1 is the same for all media—i.e., the lines are parallel but possess different y-intercepts (see Figure 15.20).

$$E(y) = \beta_0 + \beta_1 x_1 + \beta_2 x_2 + \beta_3 x_3$$

where

$x_1 =$ Advertising expenditure

$$x_2 = \begin{cases} 1 & \text{if radio medium} \\ 0 & \text{if not} \end{cases}$$

$$x_3 = \begin{cases} 1 & \text{if television medium} \\ 0 & \text{if not} \end{cases}$$

Notice that this model is essentially a combination of a first-order model with a single quantitative variable and the model with a single qualitative variable:

First-order model with a single
quantitative variable: $E(y) = \beta_0 +$ $\boxed{\beta_1 x_1}$

Model with a single
qualitative variable
at three levels: $E(y) = \beta_0 +$ $\boxed{\beta_2 x_2 + \beta_3 x_3}$

where x_1, x_2, and x_3 are as just defined. The model described here implies no interaction between the two independent variables, advertising expenditure x_1, and the qualitative variable, type of advertising medium. The change in $E(y)$ for a 1-unit increase in x_1 is identical (the slopes of the lines are equal) for all three advertising media. The terms corresponding to each of the independent variables are called **main effect** terms because they imply no interaction.

3. The straight lines relating mean sales $E(y)$ to advertising expenditure x_1 differ for the three advertising media—i.e., both line intercepts and slopes differ (see Figure 15.21). As you will see, this interaction model is obtained by adding terms involving the cross-product terms, one each from each of the two independent variables:

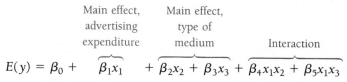

Main effect,
advertising
expenditure

Main effect,
type of
medium

Interaction

$$E(y) = \beta_0 + \overbrace{\beta_1 x_1} + \overbrace{\beta_2 x_2 + \beta_3 x_3} + \overbrace{\beta_4 x_1 x_2 + \beta_5 x_1 x_3}$$

Note that each of the preceding models is obtained by adding terms to model 1, the single first-order model used to model the responses for all three media. Model 2 is obtained by adding the main effect terms for type of medium, the qualitative variable. Model 3 is obtained by adding the interaction terms to model 2.

Will a single line (Figure 15.19) characterize the responses for all three media? Or do all three response lines differ as shown in Figure 15.21? A test of the null hypothesis that a single first-order model adequately describes the relationship between $E(y)$ and advertising expenditure x_1 for all three media is a test of the null hypothesis that the parameters of model 3, β_2, β_3, β_4, and β_5, equal 0; i.e.,

$$H_0: \quad \beta_2 = \beta_3 = \beta_4 = \beta_5 = 0$$

This hypothesis can be tested by fitting the complete model (model 3) and the reduced model (model 1) and conducting an F-test.

Suppose we assume that the response lines for the three media will differ but wonder whether the data present sufficient evidence to indicate a difference among the slopes of the lines. To test the null hypothesis that model 2 adequately describes the relationship between $E(y)$ and advertising expenditure x_1, we wish to test

$$H_0: \quad \beta_4 = \beta_5 = 0$$

i.e., that the two independent variables, advertising expenditure x_1 and type of medium, do not interact. This test can be conducted by fitting the complete model (model 3) and the reduced model (model 2), calculating the drop in the sum of squares for error, and conducting an F-test as described in Section 15.4.

EXAMPLE 15.6

Substitute the appropriate values of the dummy variables in model 3 to obtain the equations of the three response lines in Figure 15.21.

Solution

The complete model that characterizes the three lines in Figure 15.21 is

$$E(y) = \beta_0 + \beta_1 x_1 + \beta_2 x_2 + \beta_3 x_3 + \beta_4 x_1 x_2 + \beta_5 x_1 x_3$$

where

$$x_1 = \text{Advertising expenditure}$$
$$x_2 = \begin{cases} 1 & \text{if radio medium} \\ 0 & \text{if not} \end{cases}$$
$$x_3 = \begin{cases} 1 & \text{if television medium} \\ 0 & \text{if not} \end{cases}$$

Examining the coding, you can see that $x_2 = x_3 = 0$ when the advertising medium is newspaper. Substituting these values into the expression for $E(y)$, we obtain the newspaper medium line:

$$E(y) = \beta_0 + \beta_1 x_1 + \beta_2(0) + \beta_3(0) + \beta_4 x_1(0) + \beta_5 x_1(0)$$
$$= \beta_0 + \beta_1 x_1$$

Similarly, we substitute the appropriate values of x_2 and x_3 into the expression for $E(y)$ to obtain the radio medium line:

$$E(y) = \beta_0 + \beta_1 x_1 + \beta_2(1) + \beta_3(0) + \beta_4 x_1(1) + \beta_5 x_1(0)$$

$$= \overbrace{(\beta_0 + \beta_2)}^{y\text{-intercept}} + \overbrace{(\beta_1 + \beta_4)x_1}^{\text{Slope}}$$

and the television medium line:

$$E(y) = \beta_0 + \beta_1 x_1 + \beta_2(0) + \beta_3(1) + \beta_4 x_1(0) + \beta_5 x_1(1)$$

$$= \overbrace{(\beta_0 + \beta_3)}^{y\text{-intercept}} + \overbrace{(\beta_1 + \beta_5)x_1}^{\text{Slope}}$$

If you were to fit the general equation (model 3), obtain estimates of $\beta_0, \beta_1, \beta_2,$ $\ldots, \beta_5$, and substitute them into the equations for the three media lines shown, you would obtain exactly the same prediction equations as you would obtain if you were to fit three separate straight lines, one to each of the three sets of media data. You may ask why we would not fit the three lines separately. Why bother fitting a model that combines all three lines (model 3) into the same equation? The answer is that you need to use this procedure if you wish to use statistical tests to compare the three media lines. We need to be able to express a practical question about the lines in terms of a hypothesis that a set of parameters in the model equal 0. You could not do this if you performed three separate regression analyses and fit a line to each set of media data.

EXAMPLE 15.7

An industrial psychologist conducted an experiment to investigate the relationship between worker productivity and a measure of salary incentive for two manufacturing plants; one, plant A, had union representation and the other, plant B, with nonunion representation. The productivity y per worker was measured by recording the number of machined castings that a worker could produce in a 4-week period of 40 hours per week. The incentive was the amount x_1 of bonus (in cents per casting) paid for all castings produced in excess of 1,000 per worker for the 4-week period. Nine workers were selected from each plant, and three from each group of nine were assigned to receive a 20¢ bonus per casting, three a 30¢ bonus, and three a 40¢ bonus. The productivity data for the nine workers, three for each plant type and incentive combination, are shown in the table.

| TYPE OF PLANT | INCENTIVE | | |
	20¢/casting	30¢/casting	40¢/casting
Union	1,435, 1,512, 1,491	1,583, 1,529, 1,610	1,601, 1,574, 1,636
Nonunion	1,575, 1,512, 1,488	1,635, 1,589, 1,661	1,645, 1,616, 1,689

a. Assume that the relationship between mean productivity and incentive is first-order. Plot the data points and graph the prediction equations for the two productivity lines.
b. Do the data provide sufficient evidence to indicate a difference in mean worker responses to incentives between the two plants?

Solution

If we assume that a first-order model* is adequate to detect a change in mean productivity as a function of incentive x_1, then the model that produces two productivity lines, one for each plant, is

$$E(y) = \beta_0 + \beta_1 x_1 + \beta_2 x_2 + \beta_3 x_1 x_2$$

where

$$x_1 = \text{Incentive} \qquad x_2 = \begin{cases} 1 & \text{if nonunion plant} \\ 0 & \text{if union plant} \end{cases}$$

a. The SAS printout for the regression analysis is shown in Figure 15.22 (page 840). Reading the parameter estimates from the printout, you can see that

$$\hat{y} = 1{,}365.833 + 6.217x_1 + 47.778x_2 + .033x_1 x_2$$

---

*Although the model contains a term involving $x_1 x_2$, it is first-order (graphs as a straight line) in the quantitative variable x_1. The variable x_2 is a dummy variable that introduces or deletes terms in the model. The order of a model is determined only by the quantitative variables that appear in the model.

FIGURE 15.22

SAS computer printout for the complete model, Example 15.7

DEPENDENT VARIABLE: Y

SOURCE	DF	SUM OF SQUARES	MEAN SQUARE	F VALUE	PR > F	R-SQUARE
MODEL	3	57332.38888889	19110.79629630	11.46	0.0005	0.710600
ERROR	14	23349.22222223	1667.80158730		ROOT MSE	
CORRECTED TOTAL	17	80681.61111112			40.83872656	

PARAMETER	ESTIMATE	STD ERROR OF ESTIMATE	T FOR H0: PARAMETER=0	PR > |T|
INTERCEPT	1365.8333333	51.83641257	26.35	0.0001
X1	6.21666667	1.66723403	3.73	0.0022
X2	47.77777778	73.30775769	0.65	0.5251
X1X2	0.03333333	2.35782498	0.01	0.9889

The prediction equation for the union plant can be obtained (see the coding) by substituting $x_2 = 0$ into the general prediction equation. Then

$$\hat{y} = \hat{\beta}_0 + \hat{\beta}_1 x_1 + \hat{\beta}_2(0) + \hat{\beta}_3 x_1(0)$$
$$= \hat{\beta}_0 + \hat{\beta}_1 x_1$$
$$= 1,365.833 + 6.217 x_1$$

Similarly, the prediction equation for the nonunion plant is obtained by substituting $x_2 = 1$ into the general prediction equation. Then

$$\hat{y} = \hat{\beta}_0 + \hat{\beta}_1 x_1 + \hat{\beta}_2 x_2 + \hat{\beta}_3 x_1 x_2$$
$$\hat{y} = \hat{\beta}_0 + \hat{\beta}_1 x_1 + \hat{\beta}_2(1) + \hat{\beta}_3 x_1(1)$$

$$= \overbrace{(\hat{\beta}_0 + \hat{\beta}_2)}^{y\text{-intercept}} + \overbrace{(\hat{\beta}_1 + \hat{\beta}_3)x_1}^{\text{Slope}}$$
$$= (1,365.833 + 47.778) + (6.217 + .033)x_1$$
$$= 1,413.611 + 6.250 x_1$$

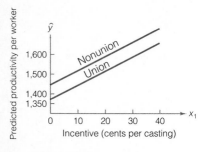

FIGURE 15.23

Graphs of the prediction equations for the two productivity lines, Example 15.7

A graph of these prediction equations is shown in Figure 15.23.

b. To determine whether the data provide sufficient evidence to indicate a difference in mean worker responses to incentives for the two plants, we test the null hypothesis that a *single* line characterizes the relationship between productivity per worker y and the amount of incentive x_1 against the alternative hypothesis that we need two separate lines to characterize the relationship—one for each plant. If there is no difference in mean response $E(y)$ to x_1 between the two plants, then we do not need the variable Type of plant in the model, i.e., we do not need the terms involving x_2. Therefore, we wish to test

$$H_0: \quad \beta_2 = \beta_3 = 0$$

against the alternative hypothesis

$$H_a: \quad \text{At least one of the two parameters, } \beta_2 \text{ or } \beta_3, \text{ is not equal to } 0$$

The SAS computer printout for fitting the reduced model

$$E(y) = \beta_0 + \beta_1 x_1$$

to the data is shown in Figure 15.24. Reading SSE_2 and SSE_1 from Figures 15.22 and 15.24, respectively, we obtain

Complete model:	$SSE_2 = 23{,}349.22$
Reduced model:	$SSE_1 = 34{,}056.28$
Drop in SSE: $SSE_1 - SSE_2 = 10{,}707.06$	

The value of s^2 for the complete model (Figure 15.22) is 1,667.80. Substituting these values, along with $k = 3$ and $g = 1$, into the formula for the F statistic yields

$$F = \frac{(SSE_1 - SSE_2)/(k - g)}{s^2} = \frac{(10{,}707.06)/2}{1{,}667.80}$$

$$= 3.21$$

FIGURE 15.24

SAS computer printout for the reduced model, Example 15.7

```
DEPENDENT VARIABLE: Y

SOURCE              DF   SUM OF SQUARES     MEAN SQUARE    F VALUE       PR > F    R-SQUARE

MODEL                1   46625.33333333    46625.33333333   21.91        0.0003   0.577893
ERROR               16   34056.27777778     2128.51736111                ROOT MSE
CORRECTED TOTAL     17   80681.61111112
                                                                       46.13585765


                                          STD ERROR OF     T FOR H0:      PR > !T!
PARAMETER              ESTIMATE             ESTIMATE     PARAMETER=0

INTERCEPT          1389.72222222          41.40819949       33.56         0.0001
X1                    6.23333333           1.33182749        4.68         0.0003
```

The numerator degrees of freedom (the number of parameters involved in H_0) is $\nu_1 = 2$ and the denominator degrees of freedom (the degrees of freedom associated with s^2 in the complete model) is $\nu_2 = 14$. If we choose $\alpha = .05$, the tabulated value of $F_{.05}$ given in Table IX of Appendix A is 3.74. Since the computed value, $F = 3.21$, is less than the tabulated value, $F_{.05} = 3.74$, there is insufficient evidence (at the $\alpha = .05$ significance level) to indicate a difference in mean worker responses to incentives between the two plants. Therefore, there is no evidence to indicate that two different lines, one for each plant, are needed to describe the relationship between mean productivity per worker $E(y)$ and the amount of incentive x_1.

EXAMPLE 15.8

Refer to Example 15.7 and explain how you would determine whether the data provide sufficient evidence to indicate that the incentive x_1 affects mean productivity.

Solution

If the variable incentive did not affect mean productivity, we would not need terms involving x_1 in the model. Therefore, we would test the null hypothesis

$$H_0: \quad \beta_1 = \beta_3 = 0$$

against the alternative hypothesis

H_a: At least one of the parameters, β_1 or β_3, differs from 0

We would fit the reduced model

$$E(y) = \beta_0 + \beta_2 x_2$$

to the data and find SSE_1. The values of SSE_2 and s^2 for the complete model would be the same as those used in Example 15.7. Finally, you would calculate the value of the F statistic and compare it with a tabulated value of F based on $\nu_1 = 2$ numerator df and $\nu_2 = 14$ denominator df. If the test leads to rejection of H_0, you have evidence to indicate that the increase in mean productivity that appears to be present in the graphs in Figure 15.23 is due to random variation in the data.

EXERCISES 15.59–15.69

LEARNING THE MECHANICS

[*Note:* *Starred (*) exercises require the use of a computer.*]

15.59 Suppose you are interested in an application with a response y, one quantitative independent variable x_1, and one qualitative variable at three levels.
a. Write a first-order model that relates the mean response $E(y)$ to the quantitative independent variable.
b. Add the main effect terms for the qualitative independent variable to the model of part **a**. Specify the coding scheme you use.
c. Add terms to the model of part **b** to allow for interaction between the quantitative and qualitative independent variables.
d. Under what circumstances will the response lines of the model in part **c** be parallel?
e. Under what circumstances will the model in part **c** have only one response line?

15.60 SAS was used to fit the following model to $n = 15$ data points:

$$y = \beta_0 + \beta_1 x_1 + \beta_2 x_2 + \beta_3 x_3 + \varepsilon$$

where x_1 is a quantitative variable and x_2 and x_3 are dummy variables describing a qualitative variable at three levels using the coding scheme

$$x_2 = \begin{cases} 1 & \text{if level 2} \\ 0 & \text{otherwise} \end{cases} \qquad x_3 = \begin{cases} 1 & \text{if level 3} \\ 0 & \text{otherwise} \end{cases}$$

The resulting printout is shown at the top of page 843.
a. What is the response line (equation) for $E(y)$ when $x_2 = x_3 = 0$? When $x_2 = 1$ and $x_3 = 0$? When $x_2 = 0$ and $x_3 = 1$?
b. What is the general least squares prediction equation for the preceding model?
c. What is the least squares prediction equation associated with level 1? Level 2? Level 3? Plot these on the same graph.

Printout for Exercise 15.60, complete model

Dep Variable: Y

Analysis of Variance

Source	DF	Sum of Squares	Mean Square	F Value	Prob>F
Model	3	4747.70480	1582.56827	46.894	0.0001
Error	11	371.22853	33.74805		
C Total	14	5118.93333			

Root MSE	5.80931	R-Square	0.9275	
Dep Mean	99.06667	Adj R-Sq	0.9077	
C.V.	5.86404			

Parameter Estimates

Variable	DF	Parameter Estimate	Standard Error	T for H0: Parameter=0	Prob > |T|
INTERCEP	1	44.802703	8.55817652	5.235	0.0003
X1	1	2.172673	0.50335248	4.316	0.0012
X2	1	9.412913	4.05315974	2.322	0.0404
X3	1	15.631532	6.81368981	2.294	0.0425

d. Specify the null and alternative hypotheses that should be used to investigate whether a difference exists between the response lines of the qualitative variable levels 1, 2, and 3.

e. Use the next printout with only the quantitative variable x_1 in the model to conduct the hypothesis test of part **d.** Use $\alpha = .05$.

Printout for Exercise 15.60, reduced model

Dep Variable: Y

Analysis of Variance

Source	DF	Sum of Squares	Mean Square	F Value	Prob>F
Model	1	4522.90741	4522.90741	98.650	0.0001
Error	13	596.02592	45.84815		
C Total	14	5118.93333			

Root MSE	6.77113	R-Square	0.8836	
Dep Mean	99.06667	Adj R-Sq	0.8746	
C.V.	6.83492			

Parameter Estimates

Variable	DF	Parameter Estimate	Standard Error	T for H0: Parameter=0	Prob > |T|
INTERCEP	1	33.904568	6.78960378	4.994	0.0002
X1	1	3.083380	0.31044102	9.932	0.0001

APPLYING THE CONCEPTS

15.61   Researchers for a dog food company have developed a new puppy food they hope will compete with the major brands. One premarketing test involved the comparison of the new food with that of two competitors in terms of weight gain. Fifteen 8-week-old German shepherd puppies, each from a different litter, were divided into three groups of five puppies each. Each group was fed one of the three brands of food.

a. Set up a model that assumes the final weight y is linearly related to initial weight x but does not allow for differences among the three brands; i.e., assume the response curve is the same for the three brands of dog food. Sketch the response curve as it might appear.

b. Set up a model that assumes the final weight is linearly related to initial weight and allows the intercepts of the lines to differ for the three brands. In other words, assume the initial weight and brand both affect final weight, but the two variables do not interact. Sketch typical response curves.

c. Now write the main effects with interaction model. For this model we assume the final weight is linearly related to initial weight, but both the slopes and the intercepts of the lines depend on the brand. Sketch typical response curves.

15.62   A company is studying three different safety programs, A, B, and C, in an attempt to reduce the number of work-hours lost due to accidents. Each program is to be tried at three of the company's nine factories, and the plan is to monitor the lost work-hours, y, for a 1-year period beginning 6 months after the new safety program is instituted.

a. Write a main effects model relating $E(y)$ to the lost work-hours, x_1, the year before the plan is instituted and to the type of program that is instituted.

b. In terms of the model parameters from part a, what hypothesis would you test to determine whether the mean work-hours lost differ for the three safety programs?

15.63   Refer to Exercise 15.62. After the three safety programs have been in effect for 18 months, the complete main effects model is fit to the $n = 9$ data points. Using safety program A as the base level, the following results were obtained:

$$\hat{y} = -2.1 + .88x_1 - 150x_2 + 35x_3 \qquad SSE = 1{,}527.27$$

Then the reduced model $E(y) = \beta_0 + \beta_1 x_1$ is fit, with the result

$$\hat{y} = 15.3 + .84x_1 \qquad SSE = 3{,}113.14$$

Test to determine whether the mean work-hours lost differ for the three programs. Use $\alpha = .05$.

15.64   An insurance company is experimenting with three different training programs, A, B, and C, for its salespeople. The following main effects model is proposed:

$$E(y) = \beta_0 + \beta_1 x_1 + \beta_2 x_2 + \beta_3 x_3$$

where

y = Monthly sales (in thousands of dollars)

x_1 = Number of months experience

$x_2 = \begin{cases} 1 & \text{if training program B was used} \\ 0 & \text{otherwise} \end{cases}$

$x_3 = \begin{cases} 1 & \text{if training program C was used} \\ 0 & \text{otherwise} \end{cases}$

Training program A is the base level.

a. What hypothesis would you test to determine whether the mean monthly sales differ for salespeople trained by the three programs?

b. After experimenting with 50 salespeople over a 5-year period, the complete model is fit with the following result:

$$\hat{y} = 10 + .5x_1 + 1.2x_2 - .4x_3 \qquad SSE = 140.5$$

Then the reduced model $E(y) = \beta_0 + \beta_1 x_1$ is fit to the same data with the following result:

$$\hat{y} = 11.4 + .4x_1 \qquad SSE = 183.2$$

Test the hypothesis you formulated in part **a**. Use $\alpha = .05$.

*15.65   In Exercises 14.11 and 15.45, a real estate appraiser used regression analysis to explore the relationship between the sale prices of apartments and various characteristics of the apartments. Some of the data from those exercises are reproduced in the table, which also contains data on the physical condition of each apartment building (E: excellent; G: good; and F: fair).

CODE NO.	SALE PRICE	NUMBER OF APARTMENT UNITS	CONDITION OF APARTMENT BLDG.
0229	$ 90,300	4	F
0094	384,000	20	G
0043	157,500	5	G
0079	676,200	26	E
0134	165,000	5	G
0179	300,000	10	G
0087	108,750	4	G
0120	276,538	11	G
0246	420,000	20	G
0025	950,000	62	G
0015	560,000	26	G
0131	268,000	13	F
0172	290,000	9	E
0095	173,200	6	G
0121	323,650	11	G
0077	162,500	5	G
0060	353,500	20	F
0174	134,400	4	E
0084	187,000	8	G
0031	155,700	4	E
0019	93,600	4	F
0074	110,000	4	G
0057	573,200	14	E
0104	79,300	4	F
0024	272,000	5	E

Source: Robinson Appraisal Co., Inc., Mankato, Minnesota.

a. Write a model that describes the relationship between sale price and number of apartment units as three parallel lines, one for each level of physical condition. Be sure to specify the dummy variable coding scheme you use.

b. Plot y against x_1 (number of apartment units) for all buildings in excellent condition. On the same graph, plot y against x_1 for all buildings in good condition. Do this again for all buildings in fair condition. Does it appear that the model you specified in part **a** is appropriate? Explain.

c. Fit the model from part **a** to the data. Report the least squares prediction equation for each of the three building condition levels.

d. Plot the three prediction equations of part **c** on a scattergram of the data.

e. Do the data provide sufficient evidence to conclude that the relationship between sale price and number of units differs depending on the physical condition of the apartments? Test using $\alpha = .05$.

15.66   Many women suffer from anemia. A female physician, who is also an avid jogger, wanted to know if women who exercise regularly have a different mean red blood cell count than women who do not. She also wanted to know if the amount of a particular iron supplement a woman takes has any effect and whether the effect is the same for both groups. Write a model that will reflect the relationship between red blood cell count and the two independent variables described previously.

a. Assume that the effect of the iron supplement on mean blood cell count is the same regardless of whether a woman exercises regularly.

b. Assume that the effect of the iron supplement on mean blood cell count depends on whether a woman exercises regularly.

15.67   A research physician wishes to find a model that will predict the mean time to relief after the administration of a certain drug. Two important independent variables are considered to be good predictors of relief time. They are age of the patient and method of administration. Physicians administer a standard dose of a drug to a patient in one of three different ways: orally in liquid form (method 1), orally in pill form (method 2), and intravenously (method 3). Consequently, we might use the model

$$E(y) = \beta_0 + \beta_1 x_1 + \beta_2 x_2 + \beta_3 x_3 + \beta_4 x_1 x_2 + \beta_5 x_1 x_3$$

where

y = Time to relief (in minutes)

x_1 = Age of patient (in years)

$x_2 = \begin{cases} 1 & \text{if drug is administered orally in pill form} \\ 0 & \text{if not} \end{cases}$

$x_3 = \begin{cases} 1 & \text{if drug is administered intravenously} \\ 0 & \text{if not} \end{cases}$

The data in the table, obtained on 12 patients, were used to fit this model.

METHOD 1		METHOD 2		METHOD 3	
Age	Time to relief	Age	Time to relief	Age	Time to relief
51	22	46	28	37	19
36	25	40	24	60	17
31	20	26	23	25	21
20	25	32	25	38	20

a. A multiple regression analysis produced the results shown here. Test whether the model is useful in predicting mean time to relief. Use $\alpha = .05$.

Printout for Exercise 15.67

SOURCE	DF	SUM OF SQUARES	MEAN SQUARE	F	PR > F	R-SQUARE
MODEL	5	87.473	17.495	4.90	0.0394	0.803120
ERROR	6	21.443	3.574			
CORRECTED TOTAL	11	108.916				

b. What hypothesis would you use to test whether the age–drug method interaction terms contribute to the prediction of mean time to relief?

c. The reduced model

$$E(y) = \beta_0 + \beta_1 x_1 + \beta_2 x_2 + \beta_3 x_3$$

was fit to the data and produced an SSE of 38.289. Does this result provide sufficient evidence at the $\alpha = .05$ level of significance to indicate that the interaction terms should be kept in the model?

d. Use an available computer program to check the results given in parts **a** and **c**.

15.68 An economist is interested in modeling the mean monthly demand $E(y)$ (in thousands of units) for a certain product as a function of its price (in dollars) and the season of the year. The following model has been proposed:

$$E(y) = \beta_0 + \beta_1 x_1 + \beta_2 x_2 + \beta_3 x_3 + \beta_4 x_4$$

where

x_1 = Price

$$x_2 = \begin{cases} 1 & \text{if spring} \\ 0 & \text{otherwise} \end{cases} \qquad x_3 = \begin{cases} 1 & \text{if summer} \\ 0 & \text{otherwise} \end{cases} \qquad x_4 = \begin{cases} 1 & \text{if fall} \\ 0 & \text{otherwise} \end{cases}$$

A portion of the Minitab computer printout that results from fitting this model to a sample of 16 months of sales data selected from among the last 2 years is shown here.

Printout for Exercise 15.68, complete model

```
THE REGRESSION EQUATION IS
Y =    12.1 -   1.11 X1 +   3.94 X2
    +   7.17 X3 +   3.72 X4

                            ST. DEV.    T-RATIO =
        COLUMN COEFFICIENT  OF COEF.    COEF/S.D.

        --          12.067    1.473       8.19
X1  C2              -1.113     .187      -5.93
X2  C3               3.942     .858       4.59
X3  C4               7.166     .815       8.80
X4  C5               3.724     .819       4.55

THE ST. DEV. OF Y ABOUT REGRESSION LINE IS
S =       1.135
WITH ( 16- 5) =   11 DEGREES OF FREEDOM

R-SQUARED = 93.1 PERCENT
R-SQUARED = 90.6 PERCENT, ADJUSTED FOR D.F.

ANALYSIS OF VARIANCE

    DUE TO      DF      SS    MS=SS/DF

REGRESSION     4  191.590    47.697
RESIDUAL      11   14.160     1.287
TOTAL         15  205.750
```

The reduced model $E(y) = \beta_0 + \beta_1 x_1$ was also fit to the same data and the resulting computer printout is partially reproduced on page 848.

a. Do these data present sufficient evidence to conclude that mean monthly demand depends on the season of the year? Test using $\alpha = .05$.

b. Find an estimate for $E(y)$ when the price is $8 and it is summer. Interpret your result.

[*Note:* We prefer to use a confidence interval for this estimate, but its calculation is beyond the scope of this text. This procedure can be found in the references.]

Printout for Exercise 15.68, reduced model

```
THE REGRESSION EQUATION IS
Y =   17.4 -   1.35 X1

                                   ST. DEV.   T-RATIO =
           COLUMN   COEFFICIENT   OF COEF.   COEF/S.D.

             --         17.394     2.833       6.08
X1         C2           -1.348      .403      -3.34

THE ST. DEV. OF Y ABOUT REGRESSION LINE IS
S =    2.859
WITH ( 16- 2) =   14 DEGREES OF FREEDOM

R-SQUARED = 44.4 PERCENT
R-SQUARED = 40.4 PERCENT, ADJUSTED FOR D.F.

ANALYSIS OF VARIANCE

  DUE TO      DF      SS     MS=SS/DF

REGRESSION    1    91.345    91.345
RESIDUAL     14   114.405     8.172
TOTAL        15   205.750
```

*15.69   In Exercise 13.22, we fit a regression line to relate the mean gain in weight of a rainbow trout to the trout's ration level. Twenty fish were fed at each of four ration levels while living in water maintained at 6°C. We also fit regression lines to similar data for fish fed in water at 12°C and at 18°C. The table gives the means and standard deviations (in parentheses) for each sample of 20 fish.

RATION	MEAN WET WEIGHT GAIN (%)		
(% body weight per day)	6°C	12°C	18°C
.0	−8.14 (3.27)	−10.33 (2.35)	−13.21 (1.96)
.8	12.31 (3.37)		
1.2		14.29 (5.32)	
1.5	28.19 (5.39)		15.51 (4.48)
2.5	29.12 (5.67)	38.05 (6.99)	
3.5			51.13 (7.14)
4.0		60.13 (12.71)	
4.5			65.49 (13.04)
Maintenance ration	.32	.48	.69

Source: Kovacs, T. G., and Leduc, G. "Sublethal toxicity to rainbow trout (*Salmo gairdneri*) at different temperatures," *Canadian Journal of Fisheries and Aquatic Sciences*, 1982, 39.

a. Write a single model to relate the mean weight gain to ration level and water temperature (°C). A graphic representation of your model should depict three straight lines with differing intercepts and equal slopes.

b. Use a computer to fit the model in part a to the sample means.

c. Write a single model to relate the mean weight gain to ration level and water temperature (°C). A graphic representation of your model should depict three straight lines with differing intercepts and slopes.

d. Use a computer to fit the model in part c to the sample means.

e. Do the data provide sufficient evidence to indicate differences in mean gain in weight for a 1% change in the ration level (i.e., the slopes of the lines) for the three levels of water temperatures? Test using $\alpha = .05$.

15.7 Comparing Two or More Response Curves

Suppose we think that the relationship between mean monthly sales $E(y)$ and advertising expenditure x_1 (Section 15.6) is second-order. We will construct, stage by stage, a model relating $E(y)$ to the quantitative variable x_1 and the qualitative variable, Type of medium. This will enable you to compare the procedure with the stage-by-stage construction of the first-order model (Section 15.6). The graphic interpretations will help you understand the contributions of the model terms.

1. The mean sales curves are identical for all three advertising media, i.e., a single second-order curve will suffice to describe the relationship between $E(y)$ and x_1 for all the media (see Figure 15.25):

$$E(y) = \beta_0 + \beta_1 x_1 + \beta_2 x_1^2 \qquad \text{where } x_1 = \text{Advertising expenditure}$$

FIGURE 15.25

The relationship between $E(y)$ and x_1 is the same for all media

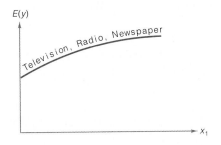

2. The response curves possess the same shapes but different y-intercepts (see Figure 15.26):

$$E(y) = \beta_0 + \beta_1 x_1 + \beta_2 x_1^2 + \beta_3 x_2 + \beta_4 x_3$$

where

$$x_1 = \text{Advertising expenditure}$$

$$x_2 = \begin{cases} 1 & \text{if radio medium} \\ 0 & \text{if not} \end{cases} \qquad x_3 = \begin{cases} 1 & \text{if television medium} \\ 0 & \text{if not} \end{cases}$$

FIGURE 15.26

The response curves have the same shapes but different y-intercepts

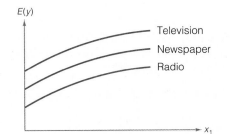

3. The response curves for the three advertising media are different (i.e., Advertising expenditure and Type of medium interact), as shown in Figure 15.27 on page 850:

$$E(y) = \beta_0 + \beta_1 x_1 + \beta_2 x_1^2 + \beta_3 x_2 + \beta_4 x_3 \\ + \beta_5 x_1 x_2 + \beta_6 x_1 x_3 + \beta_7 x_1^2 x_2 + \beta_8 x_1^2 x_3$$

F I G U R E 15.27

The response curves for the three media differ

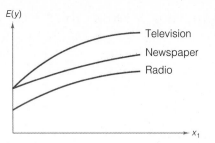

Now that you know how to write a model with two independent variables—one qualitative and one quantitative—we ask a question. Why do it? Why not write a separate second-order model for each type of medium where $E(y)$ is a function of only advertising expenditure? One reason we wrote the single model representing all three response curves is so that we can test to determine whether the curves are different. We illustrate this procedure in Examples 15.10 and 15.11. A second reason for writing a single model is that we obtain a pooled estimate of σ^2, the variance of the random error component ε. If the variance of ε is truly the same for each type of medium, the pooled estimate is superior to calculating three separate estimates by fitting a separate model for each type of medium.

EXAMPLE 15.9

Give the equation of the second-order model for the radio advertising medium.

Solution

Model 3 characterizes the relationship between $E(y)$ and x_1 for the radio advertising medium (see the coding) when $x_2 = 1$ and $x_3 = 0$. Substituting these values into model 3, we obtain

$$E(y) = \beta_0 + \beta_1 x_1 + \beta_2 x_1^2 + \beta_3 x_2 + \beta_4 x_3 + \beta_5 x_1 x_2 + \beta_6 x_1 x_3 + \beta_7 x_1^2 x_2 + \beta_8 x_1^2 x_3$$

$$= \beta_0 + \beta_1 x_1 + \beta_2 x_1^2 + \beta_3(1) + \beta_4(0) + \beta_5 x_1(1) + \beta_6 x_1(0) + \beta_7 x_1^2(1) + \beta_8 x_1^2(0)$$

$$= (\beta_0 + \beta_3) + (\beta_1 + \beta_5)x_1 + (\beta_2 + \beta_7)x_1^2$$

EXAMPLE 15.10

What null hypothesis about the parameters of model 3 would you test if you wished to determine whether the second-order curves for the three media are identical?

Solution

If the curves were identical, we would not need the independent variable Type of medium in the model; i.e., we would delete all terms involving x_2 and x_3. This would produce model 1

$$E(y) = \beta_0 + \beta_1 x_1 + \beta_2 x_1^2$$

and the null hypothesis would be

$$H_0: \quad \beta_3 = \beta_4 = \beta_5 = \beta_6 = \beta_7 = \beta_8 = 0$$

EXAMPLE 15.11

Suppose we assume that the response curves for the three media differ, but we want to know whether the second-order terms contribute information for the prediction of y. Or, equivalently, will a second-order model give better predictions than a first-order model?

Solution

The only difference between model 3 and a first-order model are those terms involving x_1^2. Therefore, the null hypothesis, "the second-order terms contribute no information for the prediction of y," is equivalent to

$$H_0: \quad \beta_2 = \beta_7 = \beta_8 = 0$$

Examples 15.10 and 15.11 identify two tests that answer practical questions concerning a collection of second-order models. Other comparisons between the curves can be made by testing appropriate sets of model parameters (see the exercises).

The models described in the preceding sections provide only an introduction to statistical modeling. Models can be constructed to relate $E(y)$ to any number of quantitative and qualitative independent variables. You can compare response curves and surfaces for different levels of a qualitative variable or for different combinations of levels of two or more qualitative independent variables. A general explanation of how to write linear statistical models can be found in Mendenhall and Sinich (1986, Chapter 6).

EXERCISES 15.70–15.80

[Note: Starred (*) exercises require the use of a computer.]

LEARNING THE MECHANICS

15.70   Consider an application for which you want to relate the response variable y to one quantitative variable and one qualitative variable at three levels.
a. Write a complete second-order model that relates $E(y)$ to the quantitative variable.
b. Add the main effect terms for the qualitative variable (at three levels) to the model of part a.
c. Add terms to the model of part b to allow for interaction between the quantitative and qualitative independent variables.

15.71   Refer to Exercise 15.70, part c.
a. Under what circumstances will the response curves of the model have the same shape but different y-intercepts?
b. Under what circumstances will the response curves of the model be parallel lines?
c. Under what circumstances will the response curves of the model be identical?

15.72   Write a model that relates $E(y)$ to two independent variables—one quantitative and one qualitative at four levels. Construct a model that allows the associated response curves to be second-order but does not allow for interaction between the two independent variables.

15.73   Minitab was used to fit the following model to $n = 25$ data points:

$$y = \beta_0 + \beta_1 x_1 + \beta_2 x_1^2 + \beta_3 x_2 + \beta_4 x_3 + \beta_5 x_1 x_2 + \beta_6 x_1 x_3 + \beta_7 x_1^2 x_2 + \beta_8 x_1^2 x_3 + \varepsilon$$

where x_1 is a quantitative variable and

$$x_2 = \begin{cases} 1 & \text{if level 2} \\ 0 & \text{otherwise} \end{cases} \qquad x_3 = \begin{cases} 1 & \text{if level 3} \\ 0 & \text{otherwise} \end{cases}$$

Note in the printout at the top of page 852 that C1 represents y, and C2–C9 represent the eight independent variable terms in the model, in the same order as specified previously.
Minitab was also used to fit the model

$$y = \beta_0 + \beta_1 x_1 + \beta_2 x_1^2 + \varepsilon$$

Printout for Exercise 15.73, complete model
_____

THE REGRESSION EQUATION IS
C1 = 48.8 - 3.36 C2 + 0.0749 C3 - 2.36 C4 - 7.60 C5 + 3.71 C6
        + 2.66 C7 - 0.0183 C8 - 0.0372 C9

| | | ST. DEV. | T-RATIO = |
COLUMN	COEFFICIENT	OF COEF.	COEF/S.D.
	48.780	4.186	11.65
C2	-3.3618	0.4708	-7.14
C3	0.07492	0.01088	6.89
C4	-2.364	6.922	-0.34
C5	-7.595	6.102	-1.24
C6	3.7144	0.8295	4.48
C7	2.6590	0.6469	4.11
C8	-0.01826	0.02175	-0.84
C9	-0.03724	0.01453	-2.56

S = 3.190

R-SQUARED = 98.9 PERCENT
R-SQUARED = 98.4 PERCENT, ADJUSTED FOR D.F.
CONTINUE?

ANALYSIS OF VARIANCE

DUE TO	DF	SS	MS=SS/DF
REGRESSION	8	15256.9	1907.1
RESIDUAL	16	162.9	10.2
TOTAL	24	15419.8	

_____

Printout for Exercise 15.73, reduced model
_____

THE REGRESSION EQUATION
C1 = 33.9 + 0.86 C2 - 0.0110 C3

| | | ST. DEV. | T-RATIO = |
COLUMN	COEFFICIENT	OF COEF.	COEF/S.D.
	33.93	20.74	1.64
C2	0.862	2.218	0.39
C3	-0.01102	0.05103	-0.22

S = 26.10

R-SQUARED = 2.8 PERCENT
R-SQUARED =   .0 PERCENT, ADJUSTED FOR D.F.

ANALYSIS OF VARIANCE

DUE TO	DF	SS	MS=SS/DF
REGRESSION	2	433.7	216.9
RESIDUAL	22	14986.0	681.2
TOTAL	24	15419.8	

_____

to the same data set. The printout is shown above.

a. What is the general least squares prediction equation for the complete model?

b. What is the equation of the response curve for $E(y)$ when $x_2 = 0$ and $x_3 = 0$? When $x_2 = 1$ and $x_3 = 0$? When $x_2 = 0$ and $x_3 = 1$?

c. On the same graph, plot the least squares prediction equation associated with level 1, with level 2, and with level 3.

d. Specify the null and alternative hypotheses that should be used to investigate whether the second-order response curves for the three levels differ.

e. Conduct the hypothesis test of part d. Use $\alpha = .05$.

APPLYING THE CONCEPTS

15.74 An equal rights group has charged that women are being discriminated against in terms of the salary structure in a state university system. It is thought that a complete second-order model will be adequate to describe the relationship between salary and years of experience for both men and women. A sample is to be taken from the records for faculty members (all of equal rank) within the system and the following model is to be fit:*

$$E(y) = \beta_0 + \beta_1 x_1 + \beta_2 x_1^2 + \beta_3 x_2 + \beta_4 x_1 x_2 + \beta_5 x_1^2 x_2$$

where

y = Annual salary (in thousands of dollars)

x_1 = Experience (years) $x_2 = \begin{cases} 1 & \text{if female} \\ 0 & \text{if male} \end{cases}$

a. What hypothesis would you test to determine whether the *rate* of increase of mean salary with experience is different for males and females?

b. What hypothesis would you test to determine whether there are differences in mean salaries that are attributable to sex?

15.75 Refer to Exercise 15.74 and the model that was proposed to describe the relationship between salary and years of experience for both men and women. A portion of the computer printout that results from fitting this model to a sample of 200 faculty members in the university system is shown below (left). The reduced model $E(y) = \beta_0 + \beta_1 x_1 + \beta_2 x_1^2$ is fit to the same data, and the resulting computer printout is partially reproduced below (right). Do the data provide sufficient evidence to support the claim that the mean salary of faculty members is dependent on sex? Use $\alpha = .05$.

Printout for Exercise 15.74, complete model

SOURCE	DF	SUM OF SQUARES	MEAN SQUARE
MODEL	5	2351.70	470.34
ERROR	194	783.90	.04
TOTAL	199	3135.60	R-SQUARE
			0.750

Printout for Exercise 15.74, reduced model

SOURCE	DF	SUM OF SQUARES	MEAN SQUARE
MODEL	2	2340.37	1170.185
ERROR	197	795.23	4.04
TOTAL	199	3135.60	R-SQUARE
			0.746

15.76 A medical researcher wants to model the extent of lung damage in emphysema patients as a function of two variables: the number of years the patient has smoked (x_1) and the sex of the patient (x_2). Fifty emphysema patients are used in the study, and the response y for each patient is a subjective score ranging from 0 to 50. (High scores indicate extensive lung damage.)

---

*In practice, we would include other variables in the model. We include only two here to simplify the exercise.

a. Identify the independent variables as qualitative or quantitative.

b. Write the main effects (first-order) model for predicting the mean lung damage score, $E(y)$, from the two variables identified in part **a**.

c. Using a female patient as the base level, the researcher obtained the prediction equation

$$\hat{y} = -.85 + .65x_1 + .09x_2$$

What is the estimated difference between the mean lung damage scores for a male and female who have both smoked for 20 years?

d. Find the estimated mean lung damage score for a female patient who has smoked for 15 years.

[*Note:*   With the original data and an appropriate computer program package, you would be able to produce confidence intervals for the estimates in parts **c** and **d**.]

e. It would not be surprising if the functional relation between lung damage and years of smoking were second-order. Further, it is conceivable that years of smoking might affect men differently than women. Write a model that allows for these conditions.

15.77   An operations manager is interested in modeling $E(y)$, the expected length of time per month (in hours) that a machine will be shut down for repairs, as a function of the type of machine (001 or 002) and the age of the machine (in years). The manager has proposed the following model:

$$E(y) = \beta_0 + \beta_1 x_1 + \beta_2 x_1^2 + \beta_3 x_2$$

where

$$x_1 = \text{Age of machine} \qquad x_2 = \begin{cases} 1 & \text{if machine type 001} \\ 0 & \text{if machine type 002} \end{cases}$$

Data obtained on $n = 20$ machine breakdowns were used to estimate the parameters of this model. A portion of the regression analysis computer printout is shown here.

Printout for Exercise 15.77, complete model

SOURCE	DF	SUM OF SQUARES	MEAN SQUARE
MODEL	3	2396.364	798.788
ERROR	16	128.586	8.037
TOTAL	19	2524.950	R-SQUARE
			0.949

The reduced model $E(y) = \beta_0 + \beta_1 x_1 + \beta_3 x_2$ was fit to the same data. The regression analysis computer printout is partially reproduced. Do these data provide sufficient evidence to conclude that the second-order (x_1^2) term in the model proposed by the operations manager is necessary? Test using $\alpha = .05$.

Printout for Exercise 15.77, reduced model

SOURCE	DF	SUM OF SQUARES	MEAN SQUARE
MODEL	2	2342.42	1171.21
ERROR	17	182.53	10.74
TOTAL	19	2524.95	R-SQUARE
			0.928

*15.78   Refer to Exercise 15.77. The data used to fit the operations manager's complete and reduced models are displayed in the table.

DOWNTIME (Hours per month)	MACHINE AGE x_1 (years)	MACHINE TYPE	x_2	DOWNTIME (Hours per month)	MACHINE AGE x_1 (years)	MACHINE TYPE	x_2
10	1.0	001	1	10	2.0	002	0
20	2.0	001	1	20	4.0	002	0
30	2.7	001	1	30	5.0	002	0
40	4.1	001	1	44	8.0	002	0
9	1.2	001	1	9	2.4	002	0
25	2.5	001	1	25	5.1	002	0
19	1.9	001	1	20	3.5	002	0
41	5.0	001	1	42	7.0	002	0
22	2.1	001	1	20	4.0	002	0
12	1.1	001	1	13	2.1	002	0

a. Use these data to test the null hypothesis that $\beta_1 = \beta_2 = 0$. Test using $\alpha = .10$.
b. Interpret the results of the test in the context of the problem.

15.79   Refer to Exercise 15.20, where we presented data on the number of highway deaths and the number of licensed vehicles on the road for the years 1950–1985. We mentioned that the number of deaths, y, may also have been affected by the existence of the national 55-mile-per-hour speed limit during the years 1974–1985 (i.e., years 25–36). Define the dummy variable

$$x_2 = \begin{cases} 1 & \text{if 55-mile-per-hour speed limit was in effect} \\ 0 & \text{if not} \end{cases}$$

a. Introduce the variable x_2 into the second-order model of Exercise 15.20 to account for the presence or absence of the 55-mile-per-hour speed limit in a given year. Include terms involving the interaction between x_2 and x_1.
b. Refer to your model for part a. Sketch on a single piece of graph paper your visualization of the two response curves, the second-order curves relating y to x_1 before and after the imposition of the 55-mile-per-hour speed limit.
c. Suppose that x_1 and x_2 do not interact. How would that affect the graphs of the two response curves of part b?
d. Refer to part c. Suppose that x_1 and x_2 do interact. How would this affect the graphs of the two response curves?

15.80   In Exercise 15.20 we fit a second-order model to data relating the number, y, of U.S. highway deaths per year to the number, x_1, of licensed vehicles on the road. In Exercise 15.79 we added a qualitative variable, x_2, to account for the presence or absence of the 55-mile-per-hour national speed limit. The SAS computer printout at the top of the next page gives the results of fitting the model

$$E(y) = \beta_0 + \beta_1 x_1 + \beta_2 x_1^2 + \beta_3 x_2 + \beta_4 x_1 x_2 + \beta_5 x_1^2 x_2$$

to the data. Use this printout and the printout for Exercise 15.20 to determine whether the data provide sufficient evidence to indicate that the qualitative variable (speed limit) contributes information for the prediction of the annual number of highway deaths. Test using $\alpha = .05$. Discuss the practical implications of your test results.

Printout for Exercise 15.80

DEP VARIABLE: Y

ANALYSIS OF VARIANCE

SOURCE	DF	SUM OF SQUARES	MEAN SQUARE	F VALUE	PROB>F
MODEL	5	1504.25099	300.85020	41.848	0.0001
ERROR	30	215.67457	7.18915219		
C TOTAL	35	1719.92556			

ROOT MSE	2.681259	R-SQUARE	0.8746
DEP MEAN	45.86111	ADJ R-SQ	0.8537
C.V.	5.846477		

PARAMETER ESTIMATES

VARIABLE	DF	PARAMETER ESTIMATE	STANDARD ERROR	T FOR H0: PARAMETER=0	DF	PROB > |T|
INTERCEP	1	16.89448080	7.85600429	2.151	1	0.0397
X1	1	0.35042735	0.18952193	1.849	1	0.0743
X1SQ	1	-0.000175175	0.001079489	-0.162	1	0.8722
X2	1	-361.16301	128.57354	-2.809	1	0.0087
X1X2	1	4.76208073	1.67614061	2.841	1	0.0080
X1SQX2	1	-0.01634485	0.005479575	-2.983	1	0.0056

15.8 Model Building: Stepwise Regression

The problem of predicting executive salaries was discussed in Chapter 14. Perhaps the biggest problem in building a model to describe executive salaries is choosing the important independent variables to be included in the model. The list of potentially important independent variables is extremely long, and we need some objective method of screening out those that are not important.

The problem of deciding which of a large set of independent variables to include in a model is a common one. Trying to determine which variables influence the profit of a firm, affect blood pressure of humans, or are related to a student's performance in college are only a few examples.

A systematic approach to building a model with a large number of independent variables is difficult because the interpretation of multivariable interactions and higher-order polynomials is tedious. We therefore turn to a screening procedure known as **stepwise regression**.

The most commonly used stepwise regression procedure, available in most popular computer packages, works as follows: The user first identifies the response, y, and the set of potentially important independent variables, x_1, x_2, ..., x_k, where k is generally large. [*Note:* This set of variables could include both first-order and higher-order terms. However, we may often include only the main effects of both quantitative variables (first-order terms) and qualitative variables (dummy variables), since the inclusion of second-order terms greatly increases the number of independent variables.] The response and independent variables are then entered into the computer, and the stepwise procedure begins.

STEP 1 The computer fits all possible one-variable models of the form

$$E(y) = \beta_0 + \beta_1 x_i$$

to the data, where x_i is the ith independent variable, $i = 1, 2, \ldots, k$. For each model, the test of the null hypothesis

$$H_0: \quad \beta_1 = 0$$

against the alternative hypothesis

$$H_a: \quad \beta_1 \neq 0$$

is conducted using the t- (or the equivalent F-) test for a single β parameter. The independent variable that produces the largest (absolute) t value is declared the best one-variable predictor of y.* Call this independent variable x_1.

STEP 2 The stepwise program now begins to search through the remaining $(k - 1)$ independent variables for the best two-variable model of the form

$$E(y) = \beta_0 + \beta_1 x_1 + \beta_2 x_i$$

This is done by fitting all two-variable models containing x_1 and each of the other $(k - 1)$ options for the second variable x_i. The t values for the test H_0: $\beta_2 = 0$ are computed for each of the $(k - 1)$ models (corresponding to the remaining independent variables x_i, $i = 2, 3, \ldots, k$), and the variable having the largest t is retained. Call this variable x_2.

At this point, some computer packages diverge in methodology. The better packages now go back and check the t value of $\hat{\beta}_1$ after $\hat{\beta}_2 x_2$ has been added to the model. If the t value has become nonsignificant at some specified α level (say $\alpha = .10$), the variable x_1 is removed and a search is made for the independent variable with a β parameter that will yield the most significant t value in the presence of $\hat{\beta}_2 x_2$. Other packages do not recheck the significance of $\hat{\beta}_1$ but proceed directly to step 3.

The reason the t value for x_1 may change from step 1 to step 2 is that the meaning of the coefficient β_1 changes. In step 2, we are approximating a complex response surface in two variables with a plane. The best-fitting plane may yield a different value for $\hat{\beta}_1$ than that obtained in step 1. Thus, both the value of $\hat{\beta}_1$ and, therefore, its significance usually change from step 1 to step 2. For this reason, the computer packages that recheck the t values at each step are preferred.

STEP 3 The stepwise procedure now checks for a third independent variable to include in the model with x_1 and x_2. That is, we seek the best model of the form

$$E(y) = \beta_0 + \beta_1 x_1 + \beta_2 x_2 + \beta_3 x_i$$

To do this, we fit all the $(k - 2)$ models using x_1, x_2, and each of the $(k - 2)$ remaining variables, x_i, as a possible x_3. The criterion is again to include the independent variable with the largest t value. Call this best third variable x_3.

The better programs now recheck the t values corresponding to the x_1 and x_2 coefficients, replacing the variables with t values that have become nonsignificant. This procedure is continued until no further independent variables can be found that yield significant t values (at the specified α level) in the presence of the variables already in the model.

---

*Note that the variable with the largest t value is also the one with the largest (absolute) Pearson product moment correlation, r (Section 13.6), with y.

The result of the stepwise procedure is a model containing only those terms with t values that are significant at the specified α level. Thus, in most practical situations only several of the large number of independent variables remain. However, it is very important *not* to jump to the conclusion that all the independent variables important for predicting y have been identified or that the unimportant independent variables have been eliminated. Remember, the stepwise procedure is using only *sample estimates* of the true model coefficients (β's) to select the important variables. An extremely large number of single β parameter t-tests have been conducted, and the probability is very high that one or more errors have been made in including or excluding variables. That is, we have very probably included some unimportant independent variables in the model (Type I errors) and eliminated some important ones (Type II errors).

There is a second reason why we might not have arrived at a good model. When we choose the variables to be included in the stepwise regression, we may often omit high-order terms (to keep the number of variables manageable). Consequently, we may have initially omitted several important terms from the model. Thus, we should recognize stepwise regression for what it is: an objective screening procedure.

Now, we will consider second-order terms (for quantitative variables) and other interactions among variables screened by the stepwise procedure. It would be best to develop this response surface model with a second set of data independent of that used for the screening, so the results of the stepwise procedure can be partially verified with new data. This is not always possible, however, because in many modeling situations only a small amount of data is available.

Do not be deceived by the impressive-looking t values that result from the stepwise procedure—it has retained only the independent variables with the largest t values. Also, be certain to consider second-order terms in systematically developing the prediction model. Finally, if you have used a first-order model for your stepwise procedure, remember that it may be greatly improved by the addition of higher-order terms.

EXAMPLE 15.12

In Section 14.7 we fit a multiple regression model for executive salaries as a function of experience, education, sex, etc. A preliminary step in the construction of this model was the determination of the most important independent variables. Ten independent variables were considered, as shown in Table 15.5. It would be very difficult to construct a second-order model with ten independent variables.

TABLE 15.5   **Independent Variables in the Executive Salary Example**

INDEPENDENT VARIABLE	DESCRIPTION
x_1	Experience (years)—quantitative
x_2	Education (years)—quantitative
x_3	Sex (1 if male, 0 if female)—qualitative
x_4	Number of employees supervised—quantitative
x_5	Corporate assets (millions of dollars)—quantitative
x_6	Board member (1 if yes, 0 if no)—qualitative
x_7	Age (years)—quantitative
x_8	Company profits (past 12 months, millions of dollars)—quantitative
x_9	Has international responsibility (1 if yes, 0 if no)—qualitative
x_{10}	Company's total sales (past 12 months, millions of dollars)—quantitative

Therefore, use the sample of 100 executives from Section 14.7 to decide which of the ten variables should be included in the construction of the final model for executive salaries.

Solution

We will use stepwise regression with the main effects of the ten independent variables to identify the most important variables. The dependent variable y is the natural logarithm of the executive salaries. The SAS stepwise regression printout is shown in Figure 15.28 on pages 860 and 861.* Note that the first variable included in the model is x_4, the number of employees supervised by the executive. At the second step, x_5, corporate assets, enters the model. At the sixth step, x_6, a dummy variable for the qualitative variable Board member or not, is brought into the model. However, because the significance (.2295) of the F statistic (SAS uses the $F = t^2$ statistic rather than the t statistic in the stepwise procedure) for x_6 is above the preassigned $\alpha = .10$, x_6 is then removed from the model. Thus, at step 7 the procedure indicates that the five-variable model including x_1, x_2, x_3, x_4, and x_5 is best. That is, none of the other independent variables can meet the $\alpha = .10$ criterion for admission to the model.

Thus, in our final modeling effort (Section 14.7) we concentrated on these five independent variables and determined that several second-order terms were important in the prediction of executive salaries.

| CASE STUDY 15.2 | A STATISTICAL METHOD FOR LAND APPRAISAL |

TABLE 15.6 Stepwise Regression Analysis of Price per Acre

VARIABLE NAME	t VALUES
Residential land (yes–no)	10.466
Seedlings and saplings (number)	6.692
Percent ponds (%)	4.141
Distance to state park (miles)	3.985
Branches or springs (yes–no)	3.855
Site index (ratio)	3.160
Size (acres)	1.142
Farmland (yes–no)	2.288

Source: Wise, J. O., and Dover, H. J. "An evaluation of a statistical method of appraising rural property," *Appraisal Journal*, 1974, 42, pp. 103–113.

New factors and a lack of knowledge about the importance of factors that affect value continue to complicate the job of the rural appraiser. In order to provide knowledge on the subject, this article reports on and evaluates a study in which multiple linear regression equations were used to evaluate and quantify factors affecting value. . . . It is believed that the findings obtained with these equations, and the relationships they indicate, will be of value to the appraiser.

The authors of this statement, James O. Wise and H. Jackson Dover,[†] use stepwise regression to identify a number of important factors (variables) that can be used to predict rural property values. They obtained their results by analyzing a sample of 105 cases from seven counties in the state of Georgia. Some of their findings are duplicated in Table 15.6. The variable names are listed in the order in which the stepwise regression procedure identified their importance, and the t values found at each step are given for each variable. Note that both qualitative and quantitative variables have been included. Since each qualitative variable is at two levels, only one main effect term could be included in the model for each factor.

---

*Note that there are several slight changes in the SAS printout labels for the stepwise procedure. For example, the $\hat{\beta}$ values, labeled ESTIMATE in the multiple regression procedure, are labeled B VALUE in the stepwise procedure.

†The opinions and statements set forth herein do not necessarily reflect the viewpoint of the American Institute of Real Estate Appraisers or its individual members, and neither the Institute nor its editors and staff assume responsibility for such expressions of opinion or statements.

FIGURE 15.28

Stepwise regression printout for Example 15.12

STEP 1
VARIABLE X4 ENTERED   R-SQUARE = 0.42071677

	DF	SUM OF SQUARES	MEAN SQUARE	F	PROB>F
REGRESSION	1	11.46854285	11.46864285	71.17	0.0001
ERROR	98	15.79112802	0.16113396		
TOTAL	99	27.25977087			

	B VALUE	STD ERROR	F	PROB>F
INTERCEPT	10.20077500			
X4 (EMPLOYEES SUPERVISED)	0.00057284	0.00006790	71.17	0.0001

STEP 2
VARIABLE X5 ENTERED   R-SQUARE = 0.78299675

	DF	SUM OF SQUARES	MEAN SQUARE	F	PROB>F
REGRESSION	2	21.34431198	10.67215599	175.00	0.0001
ERROR	97	5.91545889	0.06098411		
TOTAL	99	27.25977087			

	B VALUE	STD ERROR	F	PROB>F
INTERCEPT	9.87702903			
X4 (EMPLOYEES SUPERVISED)	0.00058353	0.00004178	195.06	0.0001
X5 (ASSETS)	0.00183730	0.00014438	161.94	0.0001

STEP 3
VARIABLE X1 ENTERED   R-SQUARE = 0.89667614

	DF	SUM OF SQUARES	MEAN SQUARE	F	PROB>F
REGRESSION	3	24.44318616	8.14772872	277.71	0.0001
ERROR	96	2.81658471	0.02933942		
TOTAL	99	27.25977087			

	B VALUE	STD ERROR	F	PROB>F
INTERCEPT	9.66449288			
X1 (EXPERIENCE)	0.01870784	0.00182032	105.62	0.0001
X4 (EMPLOYEES SUPERVISED)	0.00055251	0.00002914	359.59	0.0001
X5 (ASSETS)	0.00191195	0.00010041	362.60	0.0001

FIGURE 15.28
(continued)

STEP 4
VARIABLE X3 ENTERED   R-SQUARE = 0.94815717

	DF	SUM OF SQUARES	MEAN SQUARE	F	PROB>F
REGRESSION	4	25.84654710	6.46163678	434.37	0.0001
ERROR	95	1.41322377	0.01487604		
TOTAL	99	27.25977087			

	B VALUE	STD ERROR	F	PROB>F
INTERCEPT	9.400077349			
X1 (EXPERIENCE)	0.02074868	0.00131310	249.68	0.0001
X3 (SEX)	0.30011726	0.03089939	94.34	0.0001
X4 (EMPLOYEES SUPERVISED)	0.00055288	0.00002075	710.15	0.0001
X5 (ASSETS)	0.00190876	0.00007150	712.74	0.0001

STEP 5
VARIABLE X2 ENTERED   R-SQUARE = 0.96039323

	DF	SUM OF SQUARES	MEAN SQUARE	F	PROB>F
REGRESSION	5	26.18009940	5.23601988	455.87	0.0001
ERROR	94	1.07967147	0.01148587		
TOTAL	99	27.25977087			

	B VALUE	STD ERROR	F	PROB>F
INTERCEPT	8.85387930			
X1 (EXPERIENCE)	0.02141724	0.00116047	340.61	0.0001
X2 (EDUCATION)	0.03315807	0.00615303	29.04	0.0001
X3 (SEX)	0.31927842	0.02738298	135.95	0.0001
X4 (EMPLOYEES SUPERVISED)	0.00056061	0.00001829	939.84	0.0001
X5 (ASSETS)	0.00193684	0.00006304	943.98	0.0001

(continued)

FIGURE 15.28
(continued)

STEP 6
VARIABLE X6 ENTERED   R-SQUARE = 0.96100666

	DF	SUM OF SQUARES	MEAN SQUARE	F	PROB>F
REGRESSION	6	26.19682148	4.36613691	382.00	0.0001
ERROR	93	1.06294939	0.01142956		
TOTAL	99	27.25977087			

	B VALUE	STD ERROR	F	PROB>F
INTERCEPT	8.87509152			
X1 (EXPERIENCE)	0.02133460	0.00115963	338.48	0.0001
X2 (EDUCATION)	0.03272195	0.00614851	28.32	0.0001
X3 (SEX)	0.31093801	0.02817264	121.81	0.0001
X4 (EMPLOYEES SUPERVISED)	0.00055820	0.00001835	925.32	0.0001
X5 (ASSETS)	0.00193764	0.00006289	949.31	0.0001
X6 (BOARD)	0.03866226	0.03196369	1.46	0.2295

STEP 7
VARIABLE X6 REMOVED   R-SQUARE = 0.96039323

	DF	SUM OF SQUARES	MEAN SQUARE	F	PROB>F
REGRESSION	5	26.18009940	5.23601988	455.87	0.0001
ERROR	94	1.07967147	0.01148587		
TOTAL	99	27.25977087			

	B VALUE	STD ERROR	F	PROB>F
INTERCEPT	8.85387930			
X1 (EXPERIENCE)	0.02141724	0.00116047	340.61	0.0001
X2 (EDUCATION)	0.03315807	0.00615303	29.04	0.0001
X3 (SEX)	0.31927842	0.02738298	135.95	0.0001
X4 (EMPLOYEES SUPERVISED)	0.00056061	0.00001829	939.84	0.0001
X5 (ASSETS)	0.00193684	0.00006304	943.98	0.0001

Since there were 105 cases used in the study, a large number of degrees of freedom are associated with each t statistic (first 103, then 102, etc.). Thus, we should compare the value of the test statistic to a corresponding z value (1.645 for $\alpha = .10$ and the two-sided alternative hypothesis H_a: $\beta_i \neq 0$) when we judge the importance of each variable. Although Wise and Dover imply that the variable Size is important, we might not include it, since the t value is only 1.142.

Finally, we have now only isolated a set of important variables. We still must decide exactly how each should be entered into our prediction equation, remembering that second-order terms often improve the model.

EXERCISES 15.81–15.84

[*Note: Starred (*) exercises require the use of a computer.*]

LEARNING THE MECHANICS

15.81 There are six independent variables, x_1, x_2, x_3, x_4, x_5, and x_6, that might be useful in predicting a response y. A total of $n = 50$ observations are available, and it is decided to employ stepwise regression to help in selecting the independent variables that appear to be useful. The computer fits all possible one-variable models of the form

$$E(y) = \beta_0 + \beta_1 x_i$$

where x_i is the ith independent variable, $i = 1, 2, \ldots, 6$. The information in the table is provided from the computer printout.

INDEPENDENT VARIABLE	$\hat{\beta}_i$	$s_{\hat{\beta}_i}$
x_1	1.6	.42
x_2	−.9	.01
x_3	3.4	1.14
x_4	2.5	2.06
x_5	−4.4	.73
x_6	.3	.35

a. Which independent variable is declared the best one-variable predictor of y? Explain.
b. Would this variable be included in the model at this stage? Explain.
c. Describe the next phase that a stepwise procedure would execute.

APPLYING THE CONCEPTS

15.82 Debra A. G. Robinson and Stewart E. Cooper provide an analysis of data collected on 230 male and 72 female freshmen attending a midwestern university specializing in engineering and technology. Data collected on each student included a score on a self-concept of ability test (SCPT), scores on the SAT, ACT, and various other achievement tests, a standardized measure (called a T-score) of the student's rank in high school class, and the student's first-semester grade-point average (GPA).

One aspect of the study involved fitting by stepwise regression involving first-order terms for the test scores shown in the first column of the table. The variables are listed in the order in which they entered the model. Columns 2 and 3 give the values of R and R^2 at each stage of the stepwise regression. Column 4 gives the simple coefficient of correlation between each variable and GPA.

Column 5 gives the F value at each stage of the stepwise regression for testing the contribution of the last variable included in the stepwise model for the prediction of GPA.

VARIABLE	R	R^2	r	F	VARIABLE	R	R^2	r	F
T-score	.37	.14	.37	18.44	Verbal test	.70	.50	.30	3.93
Mathematics test	.42	.18	.34	10.93	SCPT	.72	.52	.29	1.89
SAT	.49	.24	.18	29.57	Trigonometry test	.72	.52	.28	.16
ACT	.66	.44	.27	14.15					

Source: Robinson, D. A. G., and Cooper, S. E. "The influence of self-concept on academic success in technological careers," *Journal of College Student Personnel*, 1984. Copyright 1984 AACD. Reproduced with permission. No further reproduction authorized without further permission of AACD.

a. Write the models used in the first, second, third, and fourth stages of the fitting process.

b. Does the self-concept of ability (SCPT) score come into the model?

c. Compare the values of r between GPA and the individual variables. How do you explain your answer to part **b** given that the correlation coefficient for the SCPT score is larger than that for the SAT score?

d. Use the value of R^2 given in stage 4 of the stepwise regression to calculate the value of the F statistic for testing the utility of the model.

e. Calculate the approximate p-value for the F statistic in part **d** and interpret it.

15.83 Many power plants dump hot waste water into surrounding rivers, streams, and oceans, an action that may have an adverse effect on the marine life in the dumping areas. A marine biologist was hired by the EPA to determine whether the hot-water runoff from a particular power plant located near a large gulf is having an adverse effect on the marine life in the area. In the initial phase of the study, the biologist's goal is to acquire a prediction equation for the number of marine animals located at certain predesignated areas, or stations, in the gulf. Based on past experience, the biologist considered the following environmental factors as predictors for the number of animals at a particular station:

x_1 = Temperature of water (TEMP)

x_2 = Salinity of water (SAL)

x_3 = Dissolved oxygen content of water (DO)

x_4 = Turbidity index, a measure of the turbidity of the water (TI)

x_5 = Depth of the water at the station (ST_DEPTH)

x_6 = Total weight of sea grasses in sampled area (TGRSWT)

As a preliminary step in the construction of this model, the biologist used a stepwise regression procedure to identify the most important of these six variables. A total of 716 samples were taken at different stations in the gulf, producing the SAS printout shown. (The response measured was y, the logarithm of the number of marine animals found in the sampled area.)

a. According to the SAS printout, which of the six independent variables should be used in the model? (Use $\alpha = .10$.)

b. Are we able to assume that the marine biologist has identified all the important independent variables for the prediction of y? Why?

c. Using the variables identified in part **a**, write the first-order model with interaction that may be used to predict y.

d. How would the marine biologist determine whether the model specified in part **c** is better than the first-order model?

e. Note the small value of R^2. What action might the biologist take to improve the model?

SAS printout for Exercise 15.83

STEP 1
VARIABLE ST_DEPTH ENTERED   R-SQUARE = 0.12227337

	DF	SUM OF SQUARES	MEAN SQUARE	F	PROB>F
REGRESSION	1	57.44041114	57.44041114	99.47	0.0001
ERROR	714	412.32998120	0.57749297		
TOTAL	715	469.77039233			

	B VALUE	STD ERROR	F	PROB>F
INTERCEPT	8.38559344			
ST_DEPTH	-0.43678519	0.04379580	99.47	0.0001

STEP 2
VARIABLE TGRSWT ENTERED   R-SQUARE = 0.18211026

	DF	SUM OF SQUARES	MEAN SQUARE	F	PROB>F
REGRESSION	2	85.55000871	42.77500435	79.38	0.0001
ERROR	713	384.22038363	0.53887852		
TOTAL	715	469.77039233			

	B VALUE	STD ERROR	F	PROB>F
INTERCEPT	8.07681529			
ST_DEPTH	-0.35355301	0.04384775	65.02	0.0001
TGRSWT	0.00271332	0.00037568	52.16	0.0001

STEP 3
VARIABLE TI ENTERED   R-SQUARE = 0.18700047

	DF	SUM OF SQUARES	MEAN SQUARE	F	PROB>F
REGRESSION	3	87.84728446	29.28242815	54.59	0.0001
ERROR	712	381.92310787	0.53640886		
TOTAL	715	469.77039233			

	B VALUE	STD ERROR	F	PROB>F
INTERCEPT	7.38863937			
TI	0.65773503	0.31782817	4.28	0.0389
ST_DEPTH	-0.31451227	0.04764144	43.58	0.0001
TGRSWT	0.00261166	0.00037802	47.73	0.0001

(continued)

SAS printout for Exercise 15.83, continued

STEP 4
VARIABLE DO_B ENTERED   R-SQUARE = 0.18892367

	DF	SUM OF SQUARES	MEAN SQUARE	F	PROB>F
REGRESSION	4	88.75074870	22.18768717	41.40	0.0001
ERROR	711	381.01964364	0.53589261		
TOTAL	715	469.77039233			

	B VALUE	STD ERROR	F	PROB>F
INTERCEPT	7.22576380			
DO	0.01769145	0.01362532	1.69	0.1946
TI	0.67347023	0.31790625	4.49	0.0345
ST_DEPTH	-0.30417372	0.04827962	39.69	0.0001
TGRSWT	0.00266958	0.00038047	49.23	0.0001

STEP 5
VARIABLE DO_B REMOVED   R-SQUARE = 0.18700047

	DF	SUM OF SQUARES	MEAN SQUARE	F	PROB>F
REGRESSION	3	87.84728446	29.28242815	54.59	0.0001
ERROR	712	381.92310787	0.53640886		
TOTAL	715	469.77039233			

	B VALUE	STD ERROR	F	PROB>F
INTERCEPT	7.38863937			
TI	0.65773503	0.31782817	4.28	0.0389
ST_DEPTH	-0.31451227	0.04764144	43.58	0.0001
TGRSWT	0.00261166	0.00037802	47.73	0.0001

15.84   [*Note:*   This exercise is for students who have access to a stepwise regression computer program.] Literacy rate is a reflection of the educational facilities and quality of education available in a country, and mass communication plays a large part in the educational process. In an effort to relate the literacy rate of a country to various mass communication outlets, a demographer has proposed to relate the response

y = Literacy rate

to the independent variables

x_1 = Number of daily newspaper copies (per 1,000 population)

x_2 = Number of radios (per 1,000 population)

x_3 = Number of television sets (per 1,000 population)

Use stepwise regression to find a suitable model relating y to x_1, x_2, and x_3.

COUNTRY	NEWSPAPER COPIES Per 1,000	RADIOS Per 1,000	TELEVISION SETS Per 1,000	LITERACY RATE
Czechoslovakia	280	266	228	.98
Italy	142	230	201	.93
Kenya	10	114	2	.25
Norway	391	313	227	.99
Panama	86	329	82	.79
Philippines	17	42	11	.72
Tunisia	21	49	16	.32
USA	314	1,695	472	.99
USSR	333	430	185	.99
Venezuela	91	182	89	.82

Summary

Although this chapter provides only an introduction to the very important topic of **model building**, it will enable you to construct many interesting and useful models. You can build on this foundation and, with experience, develop competence in this fascinating area of statistics. Successful model building requires a delicate blend of knowledge of the phenomenon being modeled, geometry, and formal statistical testing.

The first step in model building is to **identify the response variable y and a set of independent variables**. Each independent variable is then classified as either **quantitative** or **qualitative**, and **dummy variables** are defined to represent the qualitative independent variables. If the total number of independent variables is large, you may want to use **stepwise regression** to screen out those that do not seem important for the prediction of y.

When the number of independent variables is manageable, the model builder is ready to begin a systematic effort. At least **second-order models**, those containing **two-way interactions and quadratic terms** in the quantitative variables, should be considered. Remember that a model with no interaction terms implies that each of the independent variables affects the response independently of the other independent variables. **Quadratic terms add curvature** to the contour lines when $E(y)$ is plotted as a function of the independent variable. The F-test for testing a set of β parameters aids in deciding the final form of the prediction model.

Many problems can arise in building a regression model, and the process is often tedious and frustrating. However, the end result of a careful and determined modeling effort is very rewarding—you will have a better understanding of the process and a predictive model for the dependent variable y.

SUPPLEMENTARY EXERCISES 15.85–15.104

[*Note:* *Starred (*) exercises require the use of a computer.*]

LEARNING THE MECHANICS

15.85   Why is the model-building step the key to the success or failure of a regression analysis?

15.86   Suppose you fit the regression model

$$E(y) = \beta_0 + \beta_1 x_1 + \beta_2 x_2 + \beta_3 x_2^2 + \beta_4 x_1 x_2 + \beta_5 x_1 x_2^2$$

to $n = 35$ data points and wish to test the null hypothesis

$$H_0:\quad \beta_4 = \beta_5 = 0$$

a. State the alternative hypothesis.
b. Explain in detail how to compute the F statistic needed to test the null hypothesis.
c. What are the numerator and denominator degrees of freedom associated with the F statistic in part **b**?
d. Give the rejection region for the test if $\alpha = .05$.

15.87   Graph each of the following polynomials, and give the order of each:
a. $E(y) = 6 + 3x - 2x^2$ b. $E(y) = 4x$
c. $E(y) = x^2$ d. $E(y) = 2 - 4x$
e. $E(y) = 1 + x - x^2 + x^3$ f. $E(y) = 1 - 4x$

15.88   Write a model relating $E(y)$ to one qualitative independent variable that is at four levels. Define all the terms in your model.

15.89   Explain the difference between qualitative and quantitative variables.

15.90   It is desired to relate $E(y)$ to a quantitative variable x_1 and a qualitative variable at three levels.
a. Write a first-order model.
b. Write a model that will graph as three different second-order curves—one for each level of the qualitative variable.

15.91   Explain why stepwise regression is used. What is its value in the model-building process?

15.92   a. Write a first-order model relating $E(y)$ to two quantitative independent variables, x_1 and x_2.
b. Write a complete second-order model.

15.93   To model the relationship between y, a dependent variable, and x, an independent variable, a researcher has taken one measurement on y at each of three different x values. Drawing on his mathematical expertise, the researcher realizes that he can fit the second-order polynomial model

$$E(y) = \beta_0 + \beta_1 x + \beta_2 x^2$$

and it will pass exactly through all three points, yielding SSE = 0. The researcher, delighted with the "excellent" fit of the model, eagerly sets out to use it to make inferences. What problems will he encounter in attempting to make inferences?

APPLYING THE CONCEPTS

15.94   The number of practicing attorneys more than doubled in the decade 1973–1983, with the percentage of female attorneys increasing from 5% to 15% of the legal profession. A random sample of 400 female attorneys and 200 male attorneys, from among the approximately 606,000 total number of attorneys in the United States, revealed that 25% of the women finished in the top 10% of their classes versus 18% of the males. The survey also found that women tended to enter law school at a later age than men. (Nearly 33% of women began practicing after age 30, compared with 14% for men.) For older women and men beginning practice, women's starting salaries tended to be higher than those of men of the same age. Finally, the median salary for different age groups increased with age, peaking for men at ages 51–55 and leveling off thereafter (*American Bar Association Journal*, October 1983).

Construct a model that estimates attorneys' salaries as a function of age, sex, years of experience, and rank in class. Then write another model and include other variables you would consider useful in estimating attorneys' salaries. Explain why they would be useful.

15.95 Due to the increase in gasoline prices, many service stations are offering self-service gasoline at reduced prices. Suppose an oil company wants to model the mean monthly gasoline sales, $E(y)$, of its affiliated stations as a function of the type of service they offer: self-service, full service, or both.

a. How many dummy variables will be needed to describe the qualitative independent variable Type of service?

b. Write the main effects model relating $E(y)$ to the type of service. Describe the coding of the dummy variables.

15.96 An experiment was conducted to compare three weight-reducing programs. Ten people were assigned to each of the three diets and their weights, at the beginning and end of a 1-month period, are recorded (in pounds) in the table.

DIET A		DIET B		DIET C	
x_1 (weight before)	y (weight loss)	x_1 (weight before)	y (weight loss)	x_1 (weight before)	y (weight loss)
227	14	255	19	206	7
286	16	193	8	222	9
180	−2	186	4	168	2
176	8	145	15	132	0
204	15	219	16	173	−3
155	5	273	19	210	8
303	17	289	25	269	10
146	7	168	6	275	15
215	15	194	12	241	8
187	6	248	21	219	5

a. Construct a regression model to relate the weight loss y at the end of the 1-month period to

$$x_1 = \text{Weight before program} \qquad x_2 = \begin{cases} 1 & \text{if diet B} \\ 0 & \text{otherwise} \end{cases} \qquad x_3 = \begin{cases} 1 & \text{if diet C} \\ 0 & \text{otherwise} \end{cases}$$

Assume that the effect of initial weight on weight loss is identical for each of the three diets. Sketch the type of response curve you would expect to observe, showing y as a function of x_1 for each of the three diets.

b. Suppose the effect of initial weight on weight loss varies from diet to diet. Write the appropriate regression model for this case. Sketch typical response curves depicting this situation.

*c. Fit the models in parts a and b to the data given in the table.

*d. Do the data provide sufficient evidence to indicate an interaction between initial weight and type of diet? That is, is the model of part b preferable to the model of part a?

e. If you wish to test the null hypothesis of no difference among diets, which model parameters would be included in the hypothesis? [*Note:* Assume you are using the model from part b.]

*f. Conduct the test in part e using $\alpha = .05$.

*15.97 The data set shown here (also presented in Exercise 14.47) gives the number y of births per year (in thousands), the number x_1 of marriages per year (in thousands), the number x_2 of women in the work force (in thousands), and the availability (yes) or nonavailability (no) of the birth control pill over the years 1949–1982.

YEAR	BIRTHS	MARRIAGES	WOMEN WORKING	TAKING THE PILL	YEAR	BIRTHS	MARRIAGES	WOMEN WORKING	TAKING THE PILL
1949	3,560	1,580	18,030	No	1966	3,606	1,857	26,820	Yes
1950	3,632	1,667	18,680	No	1967	3,521	1,927	27,545	Yes
1951	3,750	1,595	19,309	No	1968	3,502	2,069	28,778	Yes
1952	3,847	1,539	19,559	No	1969	3,606	2,145	29,898	Yes
1953	3,902	1,546	19,668	No	1970	3,731	2,159	31,233	Yes
1954	4,017	1,490	19,970	No	1971	3,556	2,190	31,778	Yes
1955	4,097	1,518	20,842	No	1972	3,258	2,282	33,152	Yes
1956	4,168	1,569	21,808	No	1973	3,137	2,284	34,195	Yes
1957	4,255	1,518	22,097	No	1974	3,160	2,230	35,708	Yes
1958	4,204	1,451	22,482	No	1975	3,144	2,153	36,981	Yes
1959	4,245	1,494	22,865	No	1976	3,168	2,155	38,399	Yes
1960	4,258	1,523	23,619	No	1977	3,327	2,178	40,053	Yes
1961	4,268	1,548	24,199	No	1978	3,333	2,282	41,747	Yes
1962	4,167	1,580	23,978	No	1979	3,494	2,331	43,844	Yes
1963	4,098	1,654	24,675	Yes	1980	3,598	2,413	44,934	Yes
1964	4,027	1,725	25,399	Yes	1981	3,646	2,438	46,415	Yes
1965	3,760	1,800	25,952	Yes	1982	3,704	2,495	47,095	Yes

Source: *Statistical Abstract of the U.S.*

a. Specify a model relating the number y of annual births to one or more of the independent variables in the data set.
b. Use a computer to fit your model of part **a** to the data set.
c. Do the data provide sufficient evidence to indicate that your model provides information for the prediction of y? Test using $\alpha = .05$.
d. Give the p-value for the test in part **c**.
e. Find R^2 and interpet its value.
f. How might you improve your model of part **a**? Explain.

15.98 To make a product more appealing to the consumer, an automobile manufacturer is experimenting with a new type of paint that is supposed to help the car maintain its new car look. The durability of this paint depends on the length of time the car body is in the oven after its has been painted. In the initial experiment, three groups of ten car bodies each are baked for three different lengths of time—12, 24, and 36 hours—at the standard temperature setting. Then the paint finish of each of the 30 cars is analyzed to determine a durability rating, y.
a. Write a second-order model relating the mean durability $E(y)$ to the length of baking.
b. Could a third-order model be fit to the data? Explain.

15.99 Many companies must accurately estimate their costs before a job is begun in order to acquire a contract and make a profit. For example, a heating and plumbing contractor may base cost estimates for new homes on the total area of the house and whether central air conditioning is to be installed.
a. Write a main effects model relating the mean cost of material and labor, $E(y)$, to the area and central air conditioning variables.
b. Write a complete second-order model for the mean cost as a function of the same two variables.

c. What hypothesis would you test to determine whether the second-order terms are useful for predicting mean cost?

d. Explain how you would compute the F statistic needed to test the hypothesis of part c.

15.100   Refer to Exercise 15.99. The contractor samples 25 recent jobs and fits both the complete second-order model (part b) and the reduced main effects model (part a), so that a test can be conducted to determine whether the additional complexity of the second-order model is necessary. The resulting SSE and R^2 values are shown in the table.

	SSE	R^2
Main effects	8.548	.950
Second-order	6.133	.964

a. Is there sufficient evidence to conclude that the second-order terms are important for predicting the mean cost? Use $\alpha = .05$.

b. Suppose the contractor decides to use the main effects model to predict costs. Use the global F-test (Section 14.5) to determine whether the main effects model is useful for predicting costs.

15.101   One factor that must be considered in developing a shipping system that is beneficial to both the customer and the seller is time of delivery. A manufacturer of farm equipment can ship its products by either rail or truck. Quadratic models are thought to be adequate in relating time of delivery to distance traveled for both modes of transportation. Consequently, it has been suggested that the following model be fit:

$$E(y) = \beta_0 + \beta_1 x_1 + \beta_2 x_1^2 + \beta_3 x_2 + \beta_4 x_1 x_2 + \beta_5 x_1^2 x_2$$

where

y = Shipping time

x_1 = Distance to be shipped $x_2 = \begin{cases} 1 & \text{if rail} \\ 0 & \text{if truck} \end{cases}$

a. What hypothesis would you test to determine whether the data indicate that the quadratic distance terms are useful in the model, i.e., whether curvature is present in the relationship between mean delivery time and distance?

b. What hypothesis would you test to determine whether there is a difference between mean delivery times by rail and by truck?

15.102   Refer to Exercise 15.101. Suppose the model is fit to a total of 50 observations on delivery time. The sum of squared errors is SSE = 226.12. Then, the reduced model

$$E(y) = \beta_0 + \beta_1 x_1 + \beta_2 x_1^2$$

is fit to the same data, and SSE = 259.34. Test to determine whether the data indicate that the mean delivery time differs for rail and truck deliveries. Use $\alpha = .05$.

*15.103   A firm has developed a new type of light bulb and is interested in evaluating its performance in order to decide whether to market it. It is known that the light output of the bulb depends on the cleanliness of its surface area and the length of time the bulb has been in operation. Use the data in the table on page 872 and the procedures you learned in this chapter to build a regression model that relates drop in light output to bulb surface cleanliness and length of operation.

DROP IN LIGHT OUTPUT (% original output)	BULB SURFACE (C = Clean, D = Dirty)	LENGTH OF OPERATION (Hours)	DROP IN LIGHT OUTPUT (% original output)	BULB SURFACE (C = Clean, D = Dirty)	LENGTH OF OPERATION (Hours)
0	C	0	0	D	0
16	C	400	4	D	400
22	C	800	6	D	800
27	C	1,200	8	D	1,200
32	C	1,600	9	D	1,600
36	C	2,000	11	D	2,000
38	C	2,400	12	D	2,400

15.104   A fast-food restaurant chain is interested in modeling the mean weekly sales of a restaurant, $E(y)$, as a function of the weekly traffic flow on the street where the restaurant is located and the city in which the restaurant is located. The table contains data that were collected on 24 restaurants in four cities. The model that has been proposed is

$$E(y) = \beta_0 + \beta_1 x_1 + \beta_2 x_2 + \beta_3 x_3 + \beta_4 x_4$$

where

$$x_1 = \text{Traffic flow}$$

$$x_2 = \begin{cases} 1 & \text{if city 1} \\ 0 & \text{otherwise} \end{cases} \quad x_3 = \begin{cases} 1 & \text{if city 2} \\ 0 & \text{otherwise} \end{cases} \quad x_4 = \begin{cases} 1 & \text{if city 3} \\ 0 & \text{otherwise} \end{cases}$$

CITY	TRAFFIC FLOW (Thousands of cars)	WEEKLY SALES y ($ thousands)	CITY	TRAFFIC FLOW (Thousands of cars)	WEEKLY SALES y ($ thousands)
1	59.3	6.3	3	75.8	8.2
1	60.3	6.6	3	48.3	5.0
1	82.1	7.6	3	41.4	3.9
1	32.3	3.0	3	52.5	5.4
1	98.0	9.5	3	41.0	4.1
1	54.1	5.9	3	29.6	3.1
1	54.4	6.1	3	49.5	5.4
1	51.3	5.0	4	73.1	8.4
1	36.7	3.6	4	81.3	9.5
2	23.6	2.8	4	72.4	8.7
2	57.6	6.7	4	88.4	10.6
2	44.6	5.2	4	23.2	3.3

A portion of the computer printout that results from fitting the model to the data in the table is shown at the top of page 873. The reduced model, $E(y) = \beta_0 + \beta_1 x_1$, was then fit to the same data and the resulting computer printout is partially reproduced also.

a. Test the null hypothesis that $\beta_1 = \beta_2 = \beta_3 = \beta_4 = 0$ using $\alpha = .05$. Interpret the results of your test.

b. Is mean weekly sales, $E(y)$, dependent on the city where a restaurant is located? Test using $\alpha = .05$. Interpret the results of your test.

c. Describe the nature of the response lines that the complete model, $E(y) = \beta_0 + \beta_1 x_1 + \beta_2 x_2 + \beta_3 x_3 + \beta_4 x_4$, would generate. Does the model imply interaction between city and traffic flow?

Printout for Exercise 15.104, complete model

Dep Variable: Y

Analysis of Variance

Source	DF	Sum of Squares	Mean Square	F Value	Prob>F
Model	4	116.65552	29.16388	222.173	0.0001
Error	19	2.49407	0.13127		
C Total	23	119.14958			

Root MSE	0.36231	R-Square	0.9791	
Dep Mean	5.99583	Adj R-Sq	0.9747	
C.V.	6.04265			

Parameter Estimates

Variable	DF	Parameter Estimate	Standard Error	T for H0: Parameter=0	Prob > |T|
INTERCEP	1	1.083388	0.32100795	3.375	0.0032
X1	1	0.103673	0.00409449	25.320	0.0001
X2	1	-1.215762	0.20538681	-5.919	0.0001
X3	1	-0.530757	0.28481946	-1.863	0.0779
X4	1	-1.076525	0.22650014	-4.753	0.0001

Printout for Exercise 15.104, reduced model

SOURCE	DF	SUM OF SQUARES	MEAN SQUARE
MODEL	1	111.3423	111.3423
ERROR	22	7.8073	0.3549
CORRECTED TOTAL	23	119.1496	R-SQUARE
			0.934

d. Use the prediction equation based on the complete model to graph the response lines that relate predicted weekly sales, $\hat{y}$, to traffic flow, x_1 (for each of the four cities). Do the graphed response lines suggest an interaction between city and traffic flow?

e. Write a model that includes interaction between city and traffic flow.

*****f.** Fit the model of part **e** to the data.

*****g.** Do the data provide sufficient evidence to indicate that the slopes of the lines differ for at least two of the four cities? Test using $\alpha = .05$.

ON YOUR OWN...

We continue our **On Your Own** theme from Chapters 13 and 14. Remember that you selected three independent variables related to the annual GNP. Now increase your list of three variables to include approximately ten that you think would be useful in predicting the GNP. Obtain data for as many years as possible for the new list of variables and the GNP. With the aid of a computer analysis package, employ a stepwise regression program to choose the important variables among those you have listed. To test your intuition, list the variables in the order you think they will be selected before you conduct the analysis. How does your list compare with the stepwise regression results?

After the group of ten variables has been narrowed to a smaller group of variables by the stepwise analysis, try to improve the model by including interactions and quadratic terms. Be sure to consider the meaning of each interaction or quadratic term before adding it to the model—a quick sketch can be very helpful. See if you can systematically construct a useful model for predicting the GNP. You might want to hold out the last several years of data to test the predictive ability of your model after it is constructed. (As noted in Section 15.8, using the same data to construct *and* to evaluate predictive ability can lead to invalid statistical tests and a false sense of security.)

USING THE COMPUTER...

Consider the relationship between the mean median household income y of a zip code, the percentage of college graduates x_1 in the zip code, and the census region of the country in which the zip code is located. Randomly select 50 zip codes from each region.

a. Fit each of the following models to your sample:
   1. Complete second-order model
   2. Complete second-order model *without* the qualitative variable census region
   3. Main effects plus interaction (complete first-order) model
   4. Main effects model including both percentage college graduates and census region
   5. Main effects model including only percentage college graduates

b. Test the complete second-order model (1) against each of the reduced models, (2) and (3). Be sure to write the hypotheses you are testing in terms of this exercise. Which of the models is preferred as a result of your testing? If model (1) is preferred, proceed to part **c**. If model (2) is preferred, test it against model (5). If model (3) is preferred, test it against model (4). Select the preferred model among the five as a result of the testing.

c. Interpret the preferred model. Discuss the value of R^2 and the standard deviation of the model. Plot the predicted value of mean median income against the percentage of college graduates, using a different plotting symbol for each census region (assuming the variable census region is in your preferred model). Interpret your plot.

d. Use a prediction interval to predict the mean median income for a zip code in the South region with 20% college graduates.

e. What assumptions are necessary to assure the validity of the inferences in parts **b**–**d**? Perform a residual analysis to determine whether outliers exist. If any are found, remove them and refit the preferred model. Compare the models with and without the outliers to determine its sensitivity to the extreme observations.

f. Repeat this exercise using all the measurements in the data base, rather than your sample. Compare the results to those based on your sample.

References

Draper, N., and Smith, H. *Applied Regression Analysis.* New York: Wiley, 1966.

Fox, K. D., and Nickols, S. Y. "The time crunch," *Journal of Family Issues*, 1983, 4.

Graybill, F. A. *Theory and Application of the Linear Model.* North Scituate, Mass.: Duxbury, 1976.

Mendenhall, W. *Introduction to Linear Models and the Design and Analysis of Experiments.* Belmont, Ca.: Wadsworth, 1968.

Mendenhall, W., and Sincich, T. *A Second Course in Business Statistics: Regression Analysis*, 2d ed. San Francisco: Dellen, 1986.

Schroeder, R. G. *Operations Management: Decision Making in the Operations Function*, 3d ed. New York: McGraw-Hill, 1989.

Tables

CONTENTS

TABLE I Random Numbers

ROW	1	2	3	4	5	6	7	8	9	10	11	12	13	14
1	10480	15011	01536	02011	81647	91646	69179	14194	62590	36207	20969	99570	91291	90700
2	22368	46573	25595	85393	30995	89198	27982	53402	93965	34095	52666	19174	39615	99505
3	24130	48360	22527	97265	76393	64809	15179	24830	49340	32081	30680	19655	63348	58629
4	42167	93093	06243	61680	07856	16376	39440	53537	71341	57004	00849	74917	97758	16379
5	37570	39975	81837	16656	06121	91782	60468	81305	49684	60672	14110	06927	01263	54613
6	77921	06907	11008	42751	27756	53498	18602	70659	90655	15053	21916	81825	44394	42880
7	99562	72905	56420	69994	98872	31016	71194	18738	44013	48840	63213	21069	10634	12952
8	96301	91977	05463	07972	18876	20922	94595	56869	69014	60045	18425	84903	42508	32307
9	89579	14342	63661	10281	17453	18103	57740	84378	25331	12566	58678	44947	05585	56941
10	85475	36857	53342	53988	53060	59533	38867	62300	08158	17983	16439	11458	18593	64952
11	28918	69578	88231	33276	70997	79936	56865	05859	90106	31595	01547	85590	91610	78188
12	63553	40961	48235	03427	49626	69445	18663	72695	52180	20847	12234	90511	33703	90322
13	09429	93969	52636	92737	88974	33488	36320	17617	30015	08272	84115	27156	30613	74952
14	10365	61129	87529	85689	48237	52267	67689	93394	01511	26358	85104	20285	29975	89868
15	07119	97336	71048	08178	77233	13916	47564	81056	97735	85977	29372	74461	28551	90707
16	51085	12765	51821	51259	77452	16308	60756	92144	49442	53900	70960	63990	75601	40719
17	02368	21382	52404	60268	89368	19885	55322	44819	01188	65255	64835	44919	05944	55157
18	01011	54092	33362	94904	31273	04146	18594	29852	71585	85030	51132	01915	92747	64951
19	52162	53916	46369	58586	23216	14513	83149	98736	23495	64350	94738	17752	35156	35749
20	07056	97628	33787	09998	42698	06691	76988	13602	51851	46104	88916	19509	25625	58104
21	48663	91245	85828	14346	09172	30168	90229	04734	59193	22178	30421	61666	99904	32812
22	54164	58492	22421	74103	47070	25306	76468	26384	58151	06646	21524	15227	96909	44592
23	32639	32363	05597	24200	13363	38005	94342	28728	35806	06912	17012	64161	18296	22851
24	29334	27001	87637	87308	58731	00256	45834	15398	46557	41135	10367	07684	36188	18510
25	02488	33062	28834	07351	19731	92420	60952	61280	50001	67658	32586	86679	50720	94953

ROW/COLUMN	1	2	3	4	5	6	7	8	9	10	11	12	13	14
26	81525	72295	04839	96423	24878	82651	66566	14778	76797	14780	13300	87074	79666	95725
27	29676	20591	68086	26432	46901	20849	89768	81536	86645	12659	92259	57102	80428	25280
28	00742	57392	39064	66432	84673	40027	32832	61362	98947	96067	64760	64584	96096	98253
29	05366	04213	25669	26422	44407	44048	37937	63904	45766	66134	75470	66520	34693	90449
30	91921	26418	64117	94305	26766	25940	39972	22209	71500	64568	91402	42416	07844	69618
31	00582	04711	87917	77341	42206	35126	74087	99547	81817	42607	43808	76655	62028	76630
32	00725	69884	62797	56170	86324	88072	76222	36086	84637	93161	76038	65855	77919	88006
33	69011	65795	95876	55293	18988	27354	26575	08625	40801	59920	29841	80150	12777	48501
34	25976	57948	29888	88604	67917	48708	18912	82271	65424	69774	33611	54262	85963	03547
35	09763	83473	73577	12908	30883	18317	28290	35797	05998	41688	34952	37888	38917	88050
36	91576	42595	27958	30134	04024	86385	29880	99730	55536	84855	29080	09250	79656	73211
37	17955	56349	90999	49127	20044	59931	06115	20542	18059	02008	73708	83517	36103	42791
38	46503	18584	18845	49618	02304	51038	20655	58727	28168	15475	56942	53389	20562	87338
39	92157	89634	94824	78171	84610	82834	09922	25417	44137	48413	25555	21246	35509	20468
40	14577	62765	35605	81263	39667	47358	56873	56307	61607	49518	89656	20103	77490	18062
41	98427	07523	33362	64270	01638	92477	66969	98420	04880	45585	46565	04102	46880	45709
42	34914	63976	88720	82765	34476	17032	87589	40836	32427	70002	70663	88863	77775	69348
43	70060	28277	39475	46473	23219	53416	94970	25832	69975	94884	19661	72828	00102	66794
44	53976	54914	06990	67245	68350	82948	11398	42878	80287	88267	47363	46634	06541	97809
45	76072	29515	40980	07391	58745	25774	22987	80059	39911	96189	41151	14222	60697	59583
46	90725	52210	83974	29992	65831	38857	50490	83765	55657	14361	31720	57375	56228	41546
47	64364	67412	33339	31926	14883	24413	59744	92351	97473	89286	35931	04110	23726	51900
48	08962	00358	31662	25388	61642	34072	81249	35648	56891	69352	48373	45578	78547	81788
49	95012	68379	93526	70765	10592	04542	76463	54328	02349	17247	28865	14777	62730	92277
50	15664	10493	20492	38391	91132	21999	59516	81652	27195	48223	46751	22923	32261	85653
51	16408	81899	04153	53381	79401	21438	83035	92350	36693	31238	59649	91754	72772	02338
52	18629	81953	05520	91962	04739	13092	97662	24822	94730	06496	35090	04822	86774	98289
53	73115	35101	47498	87637	99016	71060	88824	71013	18735	20286	23153	72924	35165	43040
54	57491	16703	23167	49323	45021	33132	12544	41035	80780	45393	44812	12515	98931	91202
55	30405	83946	23792	14422	15059	45799	22716	19792	09983	74353	68668	30429	70735	25499
56	16631	35006	85900	98275	32388	52390	16815	69298	82732	38480	73817	32523	41961	44437
57	96773	20206	42559	78985	05300	22164	24369	54224	35083	19687	11052	91491	60383	19746
58	38935	64202	14349	82674	66523	44133	00697	35552	35970	19124	63318	29686	03387	59846
59	31624	76384	17403	53363	44167	64486	64758	76554	76554	31601	12614	33072	60332	92325
60	78919	19474	23632	27889	47914	02584	37680	20801	72152	39339	34806	08930	85001	87820
61	03931	33309	57047	74211	63445	17361	62825	39908	05607	91284	68833	25570	38818	46920
62	74426	33278	43972	10119	89917	15665	52872	73823	73144	88662	88970	74492	51805	99378

(continued)

TABLE I **Continued**

ROW \ COLUMN	1	2	3	4	5	6	7	8	9	10	11	12	13	14
63	09066	00903	20795	95452	92648	45454	09552	88815	16553	51125	79375	97596	16296	66092
64	42238	12426	87025	14267	20979	04508	64535	31355	86064	29472	47689	05974	52468	16834
65	16153	08002	26504	41744	81959	65642	74240	56302	00033	67107	77510	70625	28725	34191
66	21457	40742	29820	96783	29400	21840	15035	34537	33310	06116	95240	15957	16572	06004
67	21581	57802	02050	89728	17937	37621	47075	42080	97403	48626	68995	43805	33386	21597
68	55612	78095	83197	33732	05810	24813	86902	60397	16489	03264	88525	42786	05269	92532
69	44657	66999	99324	51281	84463	60563	79312	93454	68876	25471	93911	25650	12682	73572
70	91340	84979	46949	81973	37949	61023	43997	15263	80644	43942	89203	71795	99533	50501
71	91227	21199	31935	27022	84067	05462	35216	14486	29891	68607	41867	14951	91696	85065
72	50001	38140	66321	19924	72163	09538	12151	06878	91903	18749	34405	56087	82790	70925
73	65390	05224	72958	28609	81406	39147	25549	48542	42627	45233	57202	94617	23772	07896
74	27504	96131	83944	41575	10573	08619	64482	73923	36152	05184	94142	25299	84387	34925
75	37169	94851	39117	89632	00959	16487	65536	49071	39782	17095	02330	74301	00275	48280
76	11508	70225	51111	38351	19444	66499	71945	05422	13442	78675	84081	66938	93654	59894
77	37449	30362	06694	54690	04052	53115	62757	95348	78662	11163	81651	50245	34971	52924
78	46515	70331	85922	38329	57015	15765	97161	17869	45349	61796	66345	81073	49106	79860
79	30986	81223	42416	58353	21532	30502	32305	86482	05174	07901	54339	58861	74818	46942
80	63798	64995	46583	09785	44160	78128	83991	42865	92520	83531	80377	35909	81250	54238
81	82486	84846	99254	67632	43218	50076	21361	64816	51202	88124	41870	52689	51275	83556
82	21885	32906	92431	09060	64297	51674	64126	62570	26123	05155	59194	52799	28225	85762
83	60336	98782	07408	53458	13564	59089	26445	29789	85205	41001	12535	12133	14645	23541
84	43937	46891	24010	25560	86355	33941	25786	54990	71899	15475	95434	98227	21824	19585
85	97656	63175	89303	16275	07100	92063	21942	18611	47348	20203	18534	03862	78095	50136
86	03299	01221	05418	38982	55758	92237	26759	86367	21216	98442	08303	56613	91511	75928
87	79626	06486	03574	17668	07785	76020	79924	25651	83325	88428	85076	72811	22717	50585
88	85636	68335	47539	03129	65651	11977	02510	26113	99447	68645	34327	15152	55230	93448
89	18039	14367	61337	06177	12143	46609	32989	74014	64708	00533	35398	58408	13261	47908
90	08362	15656	60627	36478	65648	16764	53412	09013	07832	41574	17639	82163	60859	75567
91	79556	29068	04142	16268	15387	12856	66207	38358	22478	73373	88732	09443	82558	05250
92	92608	82674	27072	32534	17075	27698	98204	63863	11951	34648	88022	56148	34925	57031
93	23982	25835	40055	67006	12293	02753	14827	23235	35071	99704	37543	11601	35503	85171
94	09915	96306	05908	97901	28395	14186	00821	80703	70426	75647	76310	88717	37890	40129
95	59037	33300	26695	62247	69927	76123	50842	43834	86654	70959	79725	93872	28117	19233
96	42488	78077	69882	61657	34136	79180	97526	43092	04098	73571	80799	76536	71255	64239
97	46764	86273	63003	93017	31204	36692	40202	35275	57306	55543	53203	18098	47625	88684
98	03237	45430	55417	63282	90816	17349	88298	90183	36600	78406	06216	95787	42579	90730
99	86591	81482	52667	61582	14972	90053	89534	76036	49199	43716	97548	04379	46370	28672
100	38534	01715	94964	87288	65680	43772	39560	12918	86537	62738	19636	51132	25739	56947

Source: Abridged from W. H. Beyer (ed.). *CRC Standard Mathematical Tables*, 24th edition. (Cleveland: The Chemical Rubber Company), 1976. Reproduced by permission of the publisher.

TABLE II **Binomial Probabilities**

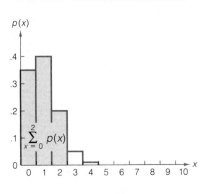

Tabulated values are $\sum_{x=0}^{k} p(x)$. (*Computations are rounded at the third decimal place.*)

a. $n = 5$

k \ p	.01	.05	.10	.20	.30	.40	.50	.60	.70	.80	.90	.95	.99
0	.951	.774	.590	.328	.168	.078	.031	.010	.002	.000	.000	.000	.000
1	.999	.977	.919	.737	.528	.337	.188	.087	.031	.007	.000	.000	.000
2	1.000	.999	.991	.942	.837	.683	.500	.317	.163	.058	.009	.001	.000
3	1.000	1.000	1.000	.993	.969	.913	.812	.663	.472	.263	.081	.023	.001
4	1.000	1.000	1.000	1.000	.998	.990	.969	.922	.832	.672	.410	.226	.049

b. $n = 6$

k \ p	.01	.05	.10	.20	.30	.40	.50	.60	.70	.80	.90	.95	.99
0	.941	.735	.531	.262	.118	.047	.016	.004	.001	.000	.000	.000	.000
1	.999	.967	.886	.655	.420	.233	.109	.041	.011	.002	.000	.000	.000
2	1.000	.998	.984	.901	.744	.544	.344	.179	.070	.017	.001	.000	.000
3	1.000	1.000	.999	.983	.930	.821	.656	.456	.256	.099	.016	.002	.000
4	1.000	1.000	1.000	.998	.989	.959	.891	.767	.580	.345	.114	.033	.001
5	1.000	1.000	1.000	1.000	.999	.996	.984	.953	.882	.738	.469	.265	.059

c. $n = 7$

k \ p	.01	.05	.10	.20	.30	.40	.50	.60	.70	.80	.90	.95	.99
0	.932	.698	.478	.210	.082	.028	.008	.002	.000	.000	.000	.000	.000
1	.998	.956	.850	.577	.329	.159	.063	.019	.004	.000	.000	.000	.000
2	1.000	.996	.974	.852	.647	.420	.227	.096	.029	.005	.000	.000	.000
3	1.000	1.000	.997	.967	.874	.710	.500	.290	.126	.033	.003	.000	.000
4	1.000	1.000	1.000	.995	.971	.904	.773	.580	.353	.148	.026	.004	.000
5	1.000	1.000	1.000	1.000	.996	.981	.937	.841	.671	.423	.150	.044	.002
6	1.000	1.000	1.000	1.000	1.000	.998	.992	.972	.918	.790	.522	.302	.068

TABLE II **Continued**

d. $n = 8$

k \ p	.01	.05	.10	.20	.30	.40	.50	.60	.70	.80	.90	.95	.99
0	.923	.663	.430	.168	.058	.017	.004	.001	.000	.000	.000	.000	.000
1	.997	.943	.813	.503	.255	.106	.035	.009	.001	.000	.000	.000	.000
2	1.000	.994	.962	.797	.552	.315	.145	.050	.011	.001	.000	.000	.000
3	1.000	1.000	.995	.944	.806	.594	.363	.174	.058	.010	.000	.000	.000
4	1.000	1.000	1.000	.990	.942	.826	.637	.406	.194	.056	.005	.000	.000
5	1.000	1.000	1.000	.999	.989	.950	.855	.685	.448	.203	.038	.006	.000
6	1.000	1.000	1.000	1.000	.999	.991	.965	.894	.745	.497	.187	.057	.003
7	1.000	1.000	1.000	1.000	1.000	.999	.996	.983	.942	.832	.570	.337	.077

e. $n = 9$

k \ p	.01	.05	.10	.20	.30	.40	.50	.60	.70	.80	.90	.95	.99
0	.914	.630	.387	.134	.040	.010	.002	.000	.000	.000	.000	.000	.000
1	.997	.929	.775	.436	.196	.071	.020	.004	.000	.000	.000	.000	.000
2	1.000	.992	.947	.738	.463	.232	.090	.025	.004	.000	.000	.000	.000
3	1.000	.999	.992	.914	.730	.483	.254	.099	.025	.003	.000	.000	.000
4	1.000	1.000	.999	.980	.901	.733	.500	.267	.099	.020	.001	.000	.000
5	1.000	1.000	1.000	.997	.975	.901	.746	.517	.270	.086	.008	.001	.000
6	1.000	1.000	1.000	1.000	.996	.975	.910	.768	.537	.262	.053	.008	.000
7	1.000	1.000	1.000	1.000	1.000	.996	.980	.929	.804	.564	.225	.071	.003
8	1.000	1.000	1.000	1.000	1.000	1.000	.998	.990	.960	.866	.613	.370	.086

f. $n = 10$

k \ p	.01	.05	.10	.20	.30	.40	.50	.60	.70	.80	.90	.95	.99
0	.904	.599	.349	.107	.028	.006	.001	.000	.000	.000	.000	.000	.000
1	.996	.914	.736	.376	.149	.046	.011	.002	.000	.000	.000	.000	.000
2	1.000	.988	.930	.678	.383	.167	.055	.012	.002	.000	.000	.000	.000
3	1.000	.999	.987	.879	.650	.382	.172	.055	.011	.001	.000	.000	.000
4	1.000	1.000	.998	.967	.850	.633	.377	.166	.047	.006	.000	.000	.000
5	1.000	1.000	1.000	.994	.953	.834	.623	.367	.150	.033	.002	.000	.000
6	1.000	1.000	1.000	.999	.989	.945	.828	.618	.350	.121	.013	.001	.000
7	1.000	1.000	1.000	1.000	.998	.988	.945	.833	.617	.322	.070	.012	.000
8	1.000	1.000	1.000	1.000	1.000	.998	.989	.954	.851	.624	.264	.086	.004
9	1.000	1.000	1.000	1.000	1.000	1.000	.999	.994	.972	.893	.651	.401	.096

T A B L E II **Continued**
g. *n* = 15

p k	.01	.05	.10	.20	.30	.40	.50	.60	.70	.80	.90	.95	.99
0	.860	.463	.206	.035	.005	.000	.000	.000	.000	.000	.000	.000	.000
1	.990	.829	.549	.167	.035	.005	.000	.000	.000	.000	.000	.000	.000
2	1.000	.964	.816	.398	.127	.027	.004	.000	.000	.000	.000	.000	.000
3	1.000	.995	.944	.648	.297	.091	.018	.002	.000	.000	.000	.000	.000
4	1.000	.999	.987	.836	.515	.217	.059	.009	.001	.000	.000	.000	.000
5	1.000	1.000	.998	.939	.722	.403	.151	.034	.004	.000	.000	.000	.000
6	1.000	1.000	1.000	.982	.869	.610	.304	.095	.015	.001	.000	.000	.000
7	1.000	1.000	1.000	.996	.950	.787	.500	.213	.050	.004	.000	.000	.000
8	1.000	1.000	1.000	.999	.985	.905	.696	.390	.131	.018	.000	.000	.000
9	1.000	1.000	1.000	1.000	.996	.966	.849	.597	.278	.061	.002	.000	.000
10	1.000	1.000	1.000	1.000	.999	.991	.941	.783	.485	.164	.013	.001	.000
11	1.000	1.000	1.000	1.000	1.000	.998	.982	.909	.703	.352	.056	.005	.000
12	1.000	1.000	1.000	1.000	1.000	1.000	.996	.973	.873	.602	.184	.036	.000
13	1.000	1.000	1.000	1.000	1.000	1.000	1.000	.995	.965	.833	.451	.171	.010
14	1.000	1.000	1.000	1.000	1.000	1.000	1.000	1.000	.995	.965	.794	.537	.140

h. *n* = 20

p k	.01	.05	.10	.20	.30	.40	.50	.60	.70	.80	.90	.95	.99
0	.818	.358	.122	.012	.001	.000	.000	.000	.000	.000	.000	.000	.000
1	.983	.736	.392	.069	.008	.001	.000	.000	.000	.000	.000	.000	.000
2	.999	.925	.677	.206	.035	.004	.000	.000	.000	.000	.000	.000	.000
3	1.000	.984	.867	.411	.107	.016	.001	.000	.000	.000	.000	.000	.000
4	1.000	.997	.957	.630	.238	.051	.006	.000	.000	.000	.000	.000	.000
5	1.000	1.000	.989	.804	.416	.126	.021	.002	.000	.000	.000	.000	.000
6	1.000	1.000	.998	.913	.608	.250	.058	.006	.000	.000	.000	.000	.000
7	1.000	1.000	1.000	.968	.772	.416	.132	.021	.001	.000	.000	.000	.000
8	1.000	1.000	1.000	.990	.887	.596	.252	.057	.005	.000	.000	.000	.000
9	1.000	1.000	1.000	.997	.952	.755	.412	.128	.017	.001	.000	.000	.000
10	1.000	1.000	1.000	.999	.983	.872	.588	.245	.048	.003	.000	.000	.000
11	1.000	1.000	1.000	1.000	.995	.943	.748	.404	.113	.010	.000	.000	.000
12	1.000	1.000	1.000	1.000	.999	.979	.868	.584	.228	.032	.000	.000	.000
13	1.000	1.000	1.000	1.000	1.000	.994	.942	.750	.392	.087	.002	.000	.000
14	1.000	1.000	1.000	1.000	1.000	.998	.979	.874	.584	.196	.011	.000	.000
15	1.000	1.000	1.000	1.000	1.000	1.000	.994	.949	.762	.370	.043	.003	.000
16	1.000	1.000	1.000	1.000	1.000	1.000	.999	.984	.893	.589	.133	.016	.000
17	1.000	1.000	1.000	1.000	1.000	1.000	1.000	.996	.965	.794	.323	.075	.001
18	1.000	1.000	1.000	1.000	1.000	1.000	1.000	.999	.992	.931	.608	.264	.017
19	1.000	1.000	1.000	1.000	1.000	1.000	1.000	1.000	.999	.988	.878	.642	.182

(*continued*)

TABLE II **Continued**
i. $n = 25$

k \ p	.01	.05	.10	.20	.30	.40	.50	.60	.70	.80	.90	.95	.99
0	.778	.277	.072	.004	.000	.000	.000	.000	.000	.000	.000	.000	.000
1	.974	.642	.271	.027	.002	.000	.000	.000	.000	.000	.000	.000	.000
2	.998	.873	.537	.098	.009	.000	.000	.000	.000	.000	.000	.000	.000
3	1.000	.966	.764	.234	.033	.002	.000	.000	.000	.000	.000	.000	.000
4	1.000	.993	.902	.421	.090	.009	.000	.000	.000	.000	.000	.000	.000
5	1.000	.999	.967	.617	.193	.029	.002	.000	.000	.000	.000	.000	.000
6	1.000	1.000	.991	.780	.341	.074	.007	.000	.000	.000	.000	.000	.000
7	1.000	1.000	.998	.891	.512	.154	.022	.001	.000	.000	.000	.000	.000
8	1.000	1.000	1.000	.953	.677	.274	.054	.004	.000	.000	.000	.000	.000
9	1.000	1.000	1.000	.983	.811	.425	.115	.013	.000	.000	.000	.000	.000
10	1.000	1.000	1.000	.994	.902	.586	.212	.034	.002	.000	.000	.000	.000
11	1.000	1.000	1.000	.998	.956	.732	.345	.078	.006	.000	.000	.000	.000
12	1.000	1.000	1.000	1.000	.983	.846	.500	.154	.017	.000	.000	.000	.000
13	1.000	1.000	1.000	1.000	.994	.922	.655	.268	.044	.002	.000	.000	.000
14	1.000	1.000	1.000	1.000	.998	.966	.788	.414	.098	.006	.000	.000	.000
15	1.000	1.000	1.000	1.000	1.000	.987	.885	.575	.189	.017	.000	.000	.000
16	1.000	1.000	1.000	1.000	1.000	.996	.946	.726	.323	.047	.000	.000	.000
17	1.000	1.000	1.000	1.000	1.000	.999	.978	.846	.488	.109	.002	.000	.000
18	1.000	1.000	1.000	1.000	1.000	1.000	.993	.926	.659	.220	.009	.000	.000
19	1.000	1.000	1.000	1.000	1.000	1.000	.998	.971	.807	.383	.033	.001	.000
20	1.000	1.000	1.000	1.000	1.000	1.000	1.000	.991	.910	.579	.098	.007	.000
21	1.000	1.000	1.000	1.000	1.000	1.000	1.000	.998	.967	.766	.236	.034	.000
22	1.000	1.000	1.000	1.000	1.000	1.000	1.000	1.000	.991	.902	.463	.127	.002
23	1.000	1.000	1.000	1.000	1.000	1.000	1.000	1.000	.998	.973	.729	.358	.026
24	1.000	1.000	1.000	1.000	1.000	1.000	1.000	1.000	1.000	.996	.928	.723	.222

TABLE III **Poisson Probabilities**

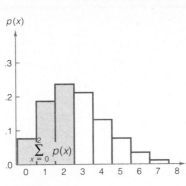

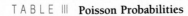

Tabulated values are $\sum\limits_{x=0}^{k} p(x)$. (*Computations are rounded at the third decimal place.*)

λ \ x	0	1	2	3	4	5	6	7	8	9
.02	.980	1.000								
.04	.961	.999	1.000							
.06	.942	.998	1.000							
.08	.923	.997	1.000							
.10	.905	.995	1.000							
.15	.861	.990	.999	1.000						
.20	.819	.982	.999	1.000						
.25	.779	.974	.998	1.000						
.30	.741	.963	.996	1.000						
.35	.705	.951	.994	1.000						
.40	.670	.938	.992	.999	1.000					
.45	.638	.925	.989	.999	1.000					
.50	.607	.910	.986	.998	1.000					
.55	.577	.894	.982	.998	1.000					
.60	.549	.878	.977	.997	1.000					
.65	.522	.861	.972	.996	.999	1.000				
.70	.497	.844	.966	.994	.999	1.000				
.75	.472	.827	.959	.993	.999	1.000				
.80	.449	.809	.953	.991	.999	1.000				
.85	.427	.791	.945	.989	.998	1.000				
.90	.407	.772	.937	.987	.998	1.000				
.95	.387	.754	.929	.981	.997	1.000				
1.00	.368	.736	.920	.981	.996	.999	1.000			
1.1	.333	.699	.900	.974	.995	.999	1.000			
1.2	.301	.663	.879	.966	.992	.998	1.000			
1.3	.273	.627	.857	.957	.989	.998	1.000			
1.4	.247	.592	.833	.946	.986	.997	.999	1.000		
1.5	.223	.558	.809	.934	.981	.996	.999	1.000		

(*continued*)

T A B L E III **Continued**

λ \ x	0	1	2	3	4	5	6	7	8	9
1.6	.202	.525	.783	.921	.976	.994	.999	1.000		
1.7	.183	.493	.757	.907	.970	.992	.998	1.000		
1.8	.165	.463	.731	.891	.964	.990	.997	.999	1.000	
1.9	.150	.434	.704	.875	.956	.987	.997	.999	1.000	
2.0	.135	.406	.677	.857	.947	.983	.995	.999	1.000	
2.2	.111	.355	.623	.819	.928	.975	.993	.998	1.000	
2.4	.091	.308	.570	.779	.904	.964	.988	.997	.999	1.000
2.6	.074	.267	.518	.736	.877	.951	.983	.995	.999	1.000
2.8	.061	.231	.469	.692	.848	.935	.976	.992	.998	.999
3.0	.050	.199	.423	.647	.815	.916	.966	.988	.996	.999
3.2	.041	.171	.380	.603	.781	.895	.955	.983	.994	.998
3.4	.033	.147	.340	.558	.744	.871	.942	.977	.992	.997
3.6	.027	.126	.303	.515	.706	.844	.927	.969	.988	.996
3.8	.022	.107	.269	.473	.668	.816	.909	.960	.984	.994
4.0	.018	.092	.238	.433	.629	.785	.889	.949	.979	.992
4.2	.015	.078	.210	.395	.590	.753	.867	.936	.972	.989
4.4	.012	.066	.185	.359	.551	.720	.844	.921	.964	.985
4.6	.010	.056	.163	.326	.513	.686	.818	.905	.955	.980
4.8	.008	.048	.143	.294	.476	.651	.791	.887	.944	.975
5.0	.007	.040	.125	.265	.440	.616	.762	.867	.932	.968
5.2	.006	.034	.109	.238	.406	.581	.732	.845	.918	.960
5.4	.005	.029	.095	.213	.373	.546	.702	.822	.903	.951
5.6	.004	.024	.082	.191	.342	.512	.670	.797	.886	.941
5.8	.003	.021	.072	.170	.313	.478	.638	.771	.867	.929
6.0	.002	.017	.062	.151	.285	.446	.606	.744	.847	.916

λ \ x	10	11	12	13	14	15	16
2.8	1.000						
3.0	1.000						
3.2	1.000						
3.4	.999	1.000					
3.6	.999	1.000					
3.8	.998	.999	1.000				
4.0	.997	.999	1.000				
4.2	.996	.999	1.000				
4.4	.994	.998	.999	1.000			
4.6	.992	.997	.999	1.000			
4.8	.990	.996	.999	1.000			
5.0	.986	.995	.998	.999	1.000		
5.2	.982	.993	.997	.999	1.000		
5.4	.977	.990	.996	.999	1.000		
5.6	.972	.988	.995	.998	.999	1.000	
5.8	.965	.984	.993	.997	.999	1.000	
6.0	.957	.980	.991	.996	.999	.999	1.000

TABLE III **Continued**

λ \ x	0	1	2	3	4	5	6	7	8	9
6.2	.002	.015	.054	.134	.259	.414	.574	.716	.826	.902
6.4	.002	.012	.046	.119	.235	.384	.542	.687	.803	.886
6.6	.001	.010	.040	.105	.213	.355	.511	.658	.780	.869
6.8	.001	.009	.034	.093	.192	.327	.480	.628	.755	.850
7.0	.001	.007	.030	.082	.173	.301	.450	.599	.729	.830
7.2	.001	.006	.025	.072	.156	.276	.420	.569	.703	.810
7.4	.001	.005	.022	.063	.140	.253	.392	.539	.676	.788
7.6	.001	.004	.019	.055	.125	.231	.365	.510	.648	.765
7.8	.000	.004	.016	.048	.112	.210	.338	.481	.620	.741
8.0	.000	.003	.014	.042	.100	.191	.313	.453	.593	.717
8.5	.000	.002	.009	.030	.074	.150	.256	.386	.523	.653
9.0	.000	.001	.006	.021	.055	.116	.207	.324	.456	.587
9.5	.000	.001	.004	.015	.040	.089	.165	.269	.392	.522
10.0	.000	.000	.003	.010	.029	.067	.130	.220	.333	.458

λ \ x	10	11	12	13	14	15	16	17	18	19
6.2	.949	.975	.989	.995	.998	.999	1.000			
6.4	.939	.969	.986	.994	.997	.999	1.000			
6.6	.927	.963	.982	.992	.997	.999	.999	1.000		
6.8	.915	.955	.978	.990	.996	.998	.999	1.000		
7.0	.901	.947	.973	.987	.994	.998	.999	1.000		
7.2	.887	.937	.967	.984	.993	.997	.999	.999	1.000	
7.4	.871	.926	.961	.980	.991	.996	.998	.999	1.000	
7.6	.854	.915	.954	.976	.989	.995	.998	.999	1.000	
7.8	.835	.902	.945	.971	.986	.993	.997	.999	1.000	
8.0	.816	.888	.936	.966	.983	.992	.996	.998	.999	1.000
8.5	.763	.849	.909	.949	.973	.986	.993	.997	.999	.999
9.0	.706	.803	.876	.926	.959	.978	.989	.995	.998	.999
9.5	.645	.752	.836	.898	.940	.967	.982	.991	.996	.998
10.0	.583	.697	.792	.864	.917	.951	.973	.986	.993	.997

λ \ x	20	21	22
8.5	1.000		
9.0	1.000		
9.5	.999	1.000	
10.0	.998	.999	1.000

(continued)

TABLE III   **Continued**

λ \ x	0	1	2	3	4	5	6	7	8	9
10.5	.000	.000	.002	.007	.021	.050	.102	.179	.279	.397
11.0	.000	.000	.001	.005	.015	.038	.079	.143	.232	.341
11.5	.000	.000	.001	.003	.011	.028	.060	.114	.191	.289
12.0	.000	.000	.001	.002	.008	.020	.046	.090	.155	.242
12.5	.000	.000	.000	.002	.005	.015	.035	.070	.125	.201
13.0	.000	.000	.000	.001	.004	.011	.026	.054	.100	.166
13.5	.000	.000	.000	.001	.003	.008	.019	.041	.079	.135
14.0	.000	.000	.000	.000	.002	.006	.014	.032	.062	.109
14.5	.000	.000	.000	.000	.001	.004	.010	.024	.048	.088
15.0	.000	.000	.000	.000	.001	.003	.008	.018	.037	.070

λ	10	11	12	13	14	15	16	17	18	19
10.5	.521	.639	.742	.825	.888	.932	.960	.978	.988	.994
11.0	.460	.579	.689	.781	.854	.907	.944	.968	.982	.991
11.5	.402	.520	.633	.733	.815	.878	.924	.954	.974	.986
12.0	.347	.462	.576	.682	.772	.844	.899	.937	.963	.979
12.5	.297	.406	.519	.628	.725	.806	.869	.916	.948	.969
13.0	.252	.353	.463	.573	.675	.764	.835	.890	.930	.957
13.5	.211	.304	.409	.518	.623	.718	.798	.861	.908	.942
14.0	.176	.260	.358	.464	.570	.669	.756	.827	.883	.923
14.5	.145	.220	.311	.413	.518	.619	.711	.790	.853	.901
15.0	.118	.185	.268	.363	.466	.568	.664	.749	.819	.875

λ	20	21	22	23	24	25	26	27	28	29
10.5	.997	.999	.999	1.000						
11.0	.995	.998	.999	1.000						
11.5	.992	.996	.998	.999	1.000					
12.0	.988	.994	.997	.999	.999	1.000				
12.5	.983	.991	.995	.998	.999	.999	1.000			
13.0	.975	.986	.992	.996	.998	.999	1.000			
13.5	.965	.980	.989	.994	.997	.998	.999	1.000		
14.0	.952	.971	.983	.991	.995	.997	.999	.999	1.000	
14.5	.936	.960	.976	.986	.992	.996	.998	.999	.999	1.000
15.0	.917	.947	.967	.981	.989	.994	.997	.998	.999	1.000

TABLE III **Continued**

λ \ x	4	5	6	7	8	9	10	11	12	13
16	.000	.001	.004	.010	.022	.043	.077	.127	.193	.275
17	.000	.001	.002	.005	.013	.026	.049	.085	.135	.201
18	.000	.000	.001	.003	.007	.015	.030	.055	.092	.143
19	.000	.000	.001	.002	.004	.009	.018	.035	.061	.098
20	.000	.000	.000	.001	.002	.005	.011	.021	.039	.066
21	.000	.000	.000	.000	.001	.003	.006	.013	.025	.043
22	.000	.000	.000	.000	.001	.002	.004	.008	.015	.028
23	.000	.000	.000	.000	.000	.001	.002	.004	.009	.017
24	.000	.000	.000	.000	.000	.000	.001	.003	.005	.011
25	.000	.000	.000	.000	.000	.000	.001	.001	.003	.006

	14	15	16	17	18	19	20	21	22	23
16	.368	.467	.566	.659	.742	.812	.868	.911	.942	.963
17	.281	.371	.468	.564	.655	.736	.805	.861	.905	.937
18	.208	.287	.375	.469	.562	.651	.731	.799	.855	.899
19	.150	.215	.292	.378	.469	.561	.647	.725	.793	.849
20	.105	.157	.221	.297	.381	.470	.559	.644	.721	.787
21	.072	.111	.163	.227	.302	.384	.471	.558	.640	.716
22	.048	.077	.117	.169	.232	.306	.387	.472	.556	.637
23	.031	.052	.082	.123	.175	.238	.310	.389	.472	.555
24	.020	.034	.056	.087	.128	.180	.243	.314	.392	.473
25	.012	.022	.038	.060	.092	.134	.185	.247	.318	.394

	24	25	26	27	28	29	30	31	32	33
16	.978	.987	.993	.996	.998	.999	.999	1.000		
17	.959	.975	.985	.991	.995	.997	.999	.999	1.000	
18	.932	.955	.972	.983	.990	.994	.997	.998	.999	1.000
19	.893	.927	.951	.969	.980	.988	.993	.996	.998	.999
20	.843	.888	.922	.948	.966	.978	.987	.992	.995	.997
21	.782	.838	.883	.917	.944	.963	.976	.985	.991	.994
22	.712	.777	.832	.877	.913	.940	.959	.973	.983	.989
23	.635	.708	.772	.827	.873	.908	.936	.956	.971	.981
24	.554	.632	.704	.768	.823	.868	.904	.932	.953	.969
25	.473	.553	.629	.700	.763	.818	.863	.900	.929	.950

	34	35	36	37	38	39	40	41	42	43
19	.999	1.000								
20	.999	.999	1.000							
21	.997	.998	.999	.999	1.000					
22	.994	.996	.998	.999	.999	1.000				
23	.988	.993	.996	.997	.999	.999	1.000			
24	.979	.987	.992	.995	.997	.998	.999	.999	1.000	
25	.966	.978	.985	.991	.991	.997	.998	.999	.999	1.000

TABLE IV   **Normal Curve Areas**

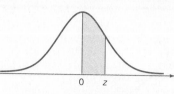

z	.00	.01	.02	.03	.04	.05	.06	.07	.08	.09
.0	.0000	.0040	.0080	.0120	.0160	.0199	.0239	.0279	.0319	.0359
.1	.0398	.0438	.0478	.0517	.0557	.0596	.0636	.0675	.0714	.0753
.2	.0793	.0832	.0871	.0910	.0948	.0987	.1026	.1064	.1103	.1141
.3	.1179	.1217	.1255	.1293	.1331	.1368	.1406	.1443	.1480	.1517
.4	.1554	.1591	.1628	.1664	.1700	.1736	.1772	.1808	.1844	.1879
.5	.1915	.1950	.1985	.2019	.2054	.2088	.2123	.2157	.2190	.2224
.6	.2257	.2291	.2324	.2357	.2389	.2422	.2454	.2486	.2517	.2549
.7	.2580	.2611	.2642	.2673	.2704	.2734	.2764	.2794	.2823	.2852
.8	.2881	.2910	.2939	.2967	.2995	.3023	.3051	.3078	.3106	.3133
.9	.3159	.3186	.3212	.3238	.3264	.3289	.3315	.3340	.3365	.3389
1.0	.3413	.3438	.3461	.3485	.3508	.3531	.3554	.3577	.3599	.3621
1.1	.3643	.3665	.3686	.3708	.3729	.3749	.3770	.3790	.3810	.3830
1.2	.3849	.3869	.3888	.3907	.3925	.3944	.3962	.3980	.3997	.4015
1.3	.4032	.4049	.4066	.4082	.4099	.4115	.4131	.4147	.4162	.4177
1.4	.4192	.4207	.4222	.4236	.4251	.4265	.4279	.4292	.4306	.4319
1.5	.4332	.4345	.4357	.4370	.4382	.4394	.4406	.4418	.4429	.4441
1.6	.4452	.4463	.4474	.4484	.4495	.4505	.4515	.4525	.4535	.4545
1.7	.4554	.4564	.4573	.4582	.4591	.4599	.4608	.4616	.4625	.4633
1.8	.4641	.4649	.4656	.4664	.4671	.4678	.4686	.4693	.4699	.4706
1.9	.4713	.4719	.4726	.4732	.4738	.4744	.4750	.4756	.4761	.4767
2.0	.4772	.4778	.4783	.4788	.4793	.4798	.4803	.4808	.4812	.4817
2.1	.4821	.4826	.4830	.4834	.4838	.4842	.4846	.4850	.4854	.4857
2.2	.4861	.4864	.4868	.4871	.4875	.4878	.4881	.4884	.4887	.4890
2.3	.4893	.4896	.4898	.4901	.4904	.4906	.4909	.4911	.4913	.4916
2.4	.4918	.4920	.4922	.4925	.4927	.4929	.4931	.4932	.4934	.4936
2.5	.4938	.4940	.4941	.4943	.4945	.4946	.4948	.4949	.4951	.4952
2.6	.4953	.4955	.4956	.4957	.4959	.4960	.4961	.4962	.4963	.4964
2.7	.4965	.4966	.4967	.4968	.4969	.4970	.4971	.4972	.4973	.4974
2.8	.4974	.4975	.4976	.4977	.4977	.4978	.4979	.4979	.4980	.4981
2.9	.4981	.4982	.4982	.4983	.4984	.4984	.4985	.4985	.4986	.4986
3.0	.4987	.4987	.4987	.4988	.4988	.4989	.4989	.4989	.4990	.4990

Source: Abridged from Table I of A. Hald, *Statistical Tables and Formulas* (New York: Wiley), 1952. Reproduced by permission of A. Hald and the publisher, John Wiley & Sons, Inc.

$z_{.025} = 1.96$

TABLE V **Exponentials**

λ	$e^{-\lambda}$	λ	$e^{-\lambda}$	λ	$e^{-\lambda}$
.00	1.000000	2.35	.095369	4.70	.009095
.05	.951229	2.40	.090718	4.75	.008652
.10	.904837	2.45	.086294	4.80	.008230
.15	.860708	2.50	.082085	4.85	.007828
.20	.818731	2.55	.078082	4.90	.007447
.25	.778801	2.60	.074274	4.95	.007083
.30	.740818	2.65	.070651	5.00	.006738
.35	.704688	2.70	.067206	5.05	.006409
.40	.670320	2.75	.063928	5.10	.006097
.45	.637628	2.80	.060810	5.15	.005799
.50	.606531	2.85	.057844	5.20	.005517
.55	.576950	2.90	.055023	5.25	.005248
.60	.548812	2.95	.052340	5.30	.004992
.65	.522046	3.00	.049787	5.35	.004748
.70	.496585	3.05	.047359	5.40	.004517
.75	.472367	3.10	.045049	5.45	.004296
.80	.449329	3.15	.042852	5.50	.004087
.85	.427415	3.20	.040762	5.55	.003887
.90	.406570	3.25	.038774	5.60	.003698
.95	.386741	3.30	.036883	5.65	.003518
1.00	.367879	3.35	.035084	5.70	.003346
1.05	.349938	3.40	.033373	5.75	.003183
1.10	.332871	3.45	.031746	5.80	.003028
1.15	.316637	3.50	.030197	5.85	.002880
1.20	.301194	3.55	.028725	5.90	.002739
1.25	.286505	3.60	.027324	5.95	.002606
1.30	.272532	3.65	.025991	6.00	.002479
1.35	.259240	3.70	.024724	6.05	.002358
1.40	.246597	3.75	.023518	6.10	.002243
1.45	.234570	3.80	.022371	6.15	.002133
1.50	.223130	3.85	.021280	6.20	.002029
1.55	.212248	3.90	.020242	6.25	.001930
1.60	.201897	3.95	.019255	6.30	.001836
1.65	.192050	4.00	.018316	6.35	.001747
1.70	.182684	4.05	.017422	6.40	.001661
1.75	.173774	4.10	.016573	6.45	.001581
1.80	.165299	4.15	.015764	6.50	.001503
1.85	.157237	4.20	.014996	6.55	.001430
1.90	.149569	4.25	.014264	6.60	.001360
1.95	.142274	4.30	.013569	6.65	.001294
2.00	.135335	4.35	.012907	6.70	.001231
2.05	.128735	4.40	.012277	6.75	.001171
2.10	.122456	4.45	.011679	6.80	.001114
2.15	.116484	4.50	.011109	6.85	.001059
2.20	.110803	4.55	.010567	6.90	.001008
2.25	.105399	4.60	.010052	6.95	.000959
2.30	.100259	4.65	.009562	7.00	.000912

(*continued*)

T A B L E V   **Continued**

λ	$e^{-\lambda}$	λ	$e^{-\lambda}$	λ	$e^{-\lambda}$
7.05	.000867	8.05	.000319	9.05	.000117
7.10	.000825	8.10	.000304	9.10	.000112
7.15	.000785	8.15	.000289	9.15	.000106
7.20	.000747	8.20	.000275	9.20	.000101
7.25	.000710	8.25	.000261	9.25	.000096
7.30	.000676	8.30	.000249	9.30	.000091
7.35	.000643	8.35	.000236	9.35	.000087
7.40	.000611	8.40	.000225	9.40	.000083
7.45	.000581	8.45	.000214	9.45	.000079
7.50	.000553	8.50	.000204	9.50	.000075
7.55	.000526	8.55	.000194	9.55	.000071
7.60	.000501	8.60	.000184	9.60	.000068
7.65	.000476	8.65	.000175	9.65	.000064
7.70	.000453	8.70	.000167	9.70	.000061
7.75	.000431	8.75	.000158	9.75	.000058
7.80	.000410	8.80	.000151	9.80	.000056
7.85	.000390	8.85	.000143	9.85	.000053
7.90	.000371	8.90	.000136	9.90	.000050
7.95	.000353	8.95	.000130	9.95	.000048
8.00	.000336	9.00	.000123	10.00	.000045

TABLE VI Critical Values of *t*

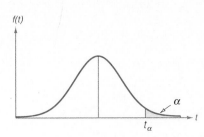

ν	$t_{.100}$	$t_{.050}$	$t_{.025}$	$t_{.010}$	$t_{.005}$	$t_{.001}$	$t_{.005}$
1	3.078	6.314	12.706	31.821	63.657	318.31	636.62
2	1.886	2.920	4.303	6.965	9.925	22.326	31.598
3	1.638	2.353	3.182	4.541	5.841	10.213	12.924
4	1.533	2.132	2.776	3.747	4.604	7.173	8.610
5	1.476	2.015	2.571	3.365	4.032	5.893	6.869
6	1.440	1.943	2.447	3.143	3.707	5.208	5.959
7	1.415	1.895	2.365	2.998	3.499	4.785	5.408
8	1.397	1.860	2.306	2.896	3.355	4.501	5.041
9	1.383	1.833	2.262	2.821	3.250	4.297	4.781
10	1.372	1.812	2.228	2.764	3.169	4.144	4.587
11	1.363	1.796	2.201	2.718	3.106	4.025	4.437
12	1.356	1.782	2.179	2.681	3.055	3.930	4.318
13	1.350	1.771	2.160	2.650	3.012	3.852	4.221
14	1.345	1.761	2.145	2.624	2.977	3.787	4.140
15	1.341	1.753	2.131	2.602	2.947	3.733	4.073
16	1.337	1.746	2.120	2.583	2.921	3.686	4.015
17	1.333	1.740	2.110	2.567	2.898	3.646	3.965
18	1.330	1.734	2.101	2.552	2.878	3.610	3.922
19	1.328	1.729	2.093	2.539	2.861	3.579	3.883
20	1.325	1.725	2.086	2.528	2.845	3.552	3.850
21	1.323	1.721	2.080	2.518	2.831	3.527	3.819
22	1.321	1.717	2.074	2.508	2.819	3.505	3.792
23	1.319	1.714	2.069	2.500	2.807	3.485	3.767
24	1.318	1.711	2.064	2.492	2.797	3.467	3.745
25	1.316	1.708	2.060	2.485	2.787	3.450	3.725
26	1.315	1.706	2.056	2.479	2.779	3.435	3.707
27	1.314	1.703	2.052	2.473	2.771	3.421	3.690
28	1.313	1.701	2.048	2.467	2.763	3.408	3.674
29	1.311	1.699	2.045	2.462	2.756	3.396	3.659
30	1.310	1.697	2.042	2.457	2.750	3.385	3.646
40	1.303	1.684	2.021	2.423	2.704	3.307	3.551
60	1.296	1.671	2.000	2.390	2.660	3.232	3.460
120	1.289	1.658	1.980	2.358	2.617	3.160	3.373
∞	1.282	1.645	1.960	2.326	2.576	3.090	3.291

TABLE VII **Critical Values of χ^2**

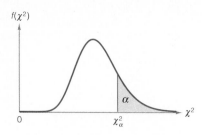

DEGREES OF FREEDOM	$\chi^2_{.995}$	$\chi^2_{.990}$	$\chi^2_{.975}$	$\chi^2_{.950}$	$\chi^2_{.900}$
1	.0000393	.0001571	.0009821	.0039321	.0157908
2	.0100251	.0201007	.0506356	.102587	.210720
3	.0717212	.114832	.215795	.351846	.584375
4	.206990	.297110	.484419	.710721	1.063623
5	.411740	.554300	.831211	1.145476	1.61031
6	.675727	.872085	1.237347	1.63539	2.20413
7	.989265	1.239043	1.68987	2.16735	2.83311
8	1.344419	1.646482	2.17973	2.73264	3.48954
9	1.734926	2.087912	2.70039	3.32511	4.16816
10	2.15585	2.55821	3.24697	3.94030	4.86518
11	2.60321	3.05347	3.81575	4.57481	5.57779
12	3.07382	3.57056	4.40379	5.22603	6.30380
13	3.56503	4.10691	5.00874	5.89186	7.04150
14	4.07468	4.66043	5.62872	6.57063	7.78953
15	4.60094	5.22935	6.26214	7.26094	8.54675
16	5.14224	5.81221	6.90766	7.96164	9.31223
17	5.69724	6.40776	7.56418	8.67176	10.0852
18	6.26481	7.01491	8.23075	9.39046	10.8649
19	6.84398	7.63273	8.90655	10.1170	11.6509
20	7.43386	8.26040	9.59083	10.8508	12.4426
21	8.03366	8.89720	10.28293	11.5913	13.2396
22	8.64272	9.54249	10.9823	12.3380	14.0415
23	9.26042	10.19567	11.6885	13.0905	14.8479
24	9.88623	10.8564	12.4011	13.8484	15.6587
25	10.5197	11.5240	13.1197	14.6114	16.4734
26	11.1603	12.1981	13.8439	15.3791	17.2919
27	11.8076	12.8786	14.5733	16.1513	18.1138
28	12.4613	13.5648	15.3079	16.9279	18.9392
29	13.1211	14.2565	16.0471	17.7083	19.7677
30	13.7867	14.9535	16.7908	18.4926	20.5992
40	20.7065	22.1643	24.4331	26.5093	29.0505
50	27.9907	29.7067	32.3574	34.7642	37.6886
60	35.5346	37.4848	40.4817	43.1879	46.4589
70	43.2752	45.4418	48.7576	51.7393	55.3290
80	51.1720	53.5400	57.1532	60.3915	64.2778
90	59.1963	61.7541	65.6466	69.1260	73.2912
100	67.3276	70.0648	74.2219	77.9295	82.3581

Source: From C. M. Thompson, "Tables of the Percentage Points of the χ^2-Distribution," *Biometrika*, 1941, 32, 188–189. Reproduced by permission of the *Biometrika* Trustees.

DEGREES OF FREEDOM	$\chi^2_{.100}$	$\chi^2_{.050}$	$\chi^2_{.025}$	$\chi^2_{.010}$	$\chi^2_{.005}$
1	2.70554	3.84146	5.02389	6.63490	7.87944
2	4.60517	5.99147	7.37776	9.21034	10.5966
3	6.25139	7.81473	9.34840	11.3449	12.8381
4	7.77944	9.48773	11.1433	13.2767	14.8602
5	9.23635	11.0705	12.8325	15.0863	16.7496
6	10.6446	12.5916	14.4494	16.8119	18.5476
7	12.0170	14.0671	16.0128	18.4753	20.2777
8	13.3616	15.5073	17.5346	20.0902	21.9550
9	14.6837	16.9190	19.0228	21.6660	23.5893
10	15.9871	18.3070	20.4831	23.2093	25.1882
11	17.2750	19.6751	21.9200	24.7250	26.7569
12	18.5494	21.0261	23.3367	26.2170	28.2995
13	19.8119	22.3621	24.7356	27.6883	29.8194
14	21.0642	23.6848	26.1190	29.1413	31.3193
15	22.3072	24.9958	27.4884	30.5779	32.8013
16	23.5418	26.2962	28.8454	31.9999	34.2672
17	24.7690	27.5871	30.1910	33.4087	35.7185
18	25.9894	28.8693	31.5264	34.8053	37.1564
19	27.2036	30.1435	32.8523	36.1908	38.5822
20	28.4120	31.4104	34.1696	37.5662	39.9968
21	29.6151	32.6705	35.4789	38.9321	41.4010
22	30.8133	33.9244	36.7807	40.2894	42.7956
23	32.0069	35.1725	38.0757	41.6384	44.1813
24	33.1963	36.4151	39.3641	42.9798	45.5585
25	34.3816	37.6525	40.6465	44.3141	46.9278
26	35.5631	38.8852	41.9232	45.6417	48.2899
27	36.7412	40.1133	43.1944	46.9630	49.6449
28	37.9159	41.3372	44.4607	48.2782	50.9933
29	39.0875	42.5569	45.7222	49.5879	52.3356
30	40.2560	43.7729	46.9792	50.8922	53.6720
40	51.8050	55.7585	59.3417	63.6907	66.7659
50	63.1671	67.5048	71.4202	76.1539	79.4900
60	74.3970	79.0819	83.2976	88.3794	91.9517
70	85.5271	90.5312	95.0231	100.425	104.215
80	96.5782	101.879	106.629	112.329	116.321
90	107.565	113.145	118.136	124.116	128.299
100	118.498	124.342	129.561	135.807	140.169

TABLE VIII **Percentage Points of the F Distribution, $\alpha = .10$**

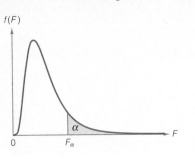

ν_1	NUMERATOR DEGREES OF FREEDOM								
ν_2	1	2	3	4	5	6	7	8	9
1	39.86	49.50	53.59	55.83	57.24	58.20	58.91	59.44	59.86
2	8.53	9.00	9.16	9.24	9.29	9.33	9.35	9.37	9.38
3	5.54	5.46	5.39	5.34	5.31	5.28	5.27	5.25	5.24
4	4.54	4.32	4.19	4.11	4.05	4.01	3.98	3.95	3.94
5	4.06	3.78	3.62	3.52	3.45	3.40	3.37	3.34	3.32
6	3.78	3.46	3.29	3.18	3.11	3.05	3.01	2.98	2.96
7	3.59	3.26	3.07	2.96	2.88	2.83	2.78	2.75	2.72
8	3.46	3.11	2.92	2.81	2.73	2.67	2.62	2.59	2.56
9	3.36	3.01	2.81	2.69	2.61	2.55	2.51	2.47	2.44
10	3.29	2.92	2.73	2.61	2.52	2.46	2.41	2.38	2.35
11	3.23	2.86	2.66	2.54	2.45	2.39	2.34	2.30	2.27
12	3.18	2.81	2.61	2.48	2.39	2.33	2.28	2.24	2.21
13	3.14	2.76	2.56	2.43	2.35	2.28	2.23	2.20	2.16
14	3.10	2.73	2.52	2.39	2.31	2.24	2.19	2.15	2.12
15	3.07	2.70	2.49	2.36	2.27	2.21	2.16	2.12	2.09
16	3.05	2.67	2.46	2.33	2.24	2.18	2.13	2.09	2.06
17	3.03	2.64	2.44	2.31	2.22	2.15	2.10	2.06	2.03
18	3.01	2.62	2.42	2.29	2.20	2.13	2.08	2.04	2.00
19	2.99	2.61	2.40	2.27	2.18	2.11	2.06	2.02	1.98
20	2.97	2.59	2.38	2.25	2.16	2.09	2.04	2.00	1.96
21	2.96	2.57	2.36	2.23	2.14	2.08	2.02	1.98	1.95
22	2.95	2.56	2.35	2.22	2.13	2.06	2.01	1.97	1.93
23	2.94	2.55	2.34	2.21	2.11	2.05	1.99	1.95	1.92
24	2.93	2.54	2.33	2.19	2.10	2.04	1.98	1.94	1.91
25	2.92	2.53	2.32	2.18	2.09	2.02	1.97	1.93	1.89
26	2.91	2.52	2.31	2.17	2.08	2.01	1.96	1.92	1.88
27	2.90	2.51	2.30	2.17	2.07	2.00	1.95	1.91	1.87
28	2.89	2.50	2.29	2.16	2.06	2.00	1.94	1.90	1.87
29	2.89	2.50	2.28	2.15	2.06	1.99	1.93	1.89	1.86
30	2.88	2.49	2.28	2.14	2.05	1.98	1.93	1.88	1.85
40	2.84	2.44	2.23	2.09	2.00	1.93	1.87	1.83	1.79
60	2.79	2.39	2.18	2.04	1.95	1.87	1.82	1.77	1.74
120	2.75	2.35	2.13	1.99	1.90	1.82	1.77	1.72	1.68
∞	2.71	2.30	2.08	1.94	1.85	1.77	1.72	1.67	1.63

DENOMINATOR DEGREES OF FREEDOM

Source: From M. Merrington and C. M. Thompson, "Tables of Percentage Points of the Inverted Beta (F)-Distribution," *Biometrika*, 1943, 33, 73–88. Reproduced by permission of the *Biometrika* Trustees.

ν_1	NUMERATOR DEGREES OF FREEDOM									
ν_2	10	12	15	20	24	30	40	60	120	∞
1	60.19	60.71	61.22	61.74	62.00	62.26	62.53	62.79	63.06	63.33
2	9.39	9.41	9.42	9.44	9.45	9.46	9.47	9.47	9.48	9.49
3	5.23	5.22	5.20	5.18	5.18	5.17	5.16	5.15	5.14	5.13
4	3.92	3.90	3.87	3.84	3.83	3.82	3.80	3.79	3.78	3.76
5	3.30	3.27	3.24	3.21	3.19	3.17	3.16	3.14	3.12	3.10
6	2.94	2.90	2.87	2.84	2.82	2.80	2.78	2.76	2.74	2.72
7	2.70	2.67	2.63	2.59	2.58	2.56	2.54	2.51	2.49	2.47
8	2.54	2.50	2.46	2.42	2.40	2.38	2.36	2.34	2.32	2.29
9	2.42	2.38	2.34	2.30	2.28	2.25	2.23	2.21	2.18	2.16
10	2.32	2.28	2.24	2.20	2.18	2.16	2.13	2.11	2.08	2.06
11	2.25	2.21	2.17	2.12	2.10	2.08	2.05	2.03	2.00	1.97
12	2.19	2.15	2.10	2.06	2.04	2.01	1.99	1.96	1.93	1.90
13	2.14	2.10	2.05	2.01	1.98	1.96	1.93	1.90	1.88	1.85
14	2.10	2.05	2.01	1.96	1.94	1.91	1.89	1.86	1.83	1.80
15	2.06	2.02	1.97	1.92	1.90	1.87	1.85	1.82	1.79	1.76
16	2.03	1.99	1.94	1.89	1.87	1.84	1.81	1.78	1.75	1.72
17	2.00	1.96	1.91	1.86	1.84	1.81	1.78	1.75	1.72	1.69
18	1.98	1.93	1.89	1.84	1.81	1.78	1.75	1.72	1.69	1.66
19	1.96	1.91	1.86	1.81	1.79	1.76	1.73	1.70	1.67	1.63
20	1.94	1.89	1.84	1.79	1.77	1.74	1.71	1.68	1.64	1.61
21	1.92	1.87	1.83	1.78	1.75	1.72	1.69	1.66	1.62	1.59
22	1.90	1.86	1.81	1.76	1.73	1.70	1.67	1.64	1.60	1.57
23	1.89	1.84	1.80	1.74	1.72	1.69	1.66	1.62	1.59	1.55
24	1.88	1.83	1.78	1.73	1.70	1.67	1.64	1.61	1.57	1.53
25	1.87	1.82	1.77	1.72	1.69	1.66	1.63	1.59	1.56	1.52
26	1.86	1.81	1.76	1.71	1.68	1.65	1.61	1.58	1.54	1.50
27	1.85	1.80	1.75	1.70	1.67	1.64	1.60	1.57	1.53	1.49
28	1.84	1.79	1.74	1.69	1.66	1.63	1.59	1.56	1.52	1.48
29	1.83	1.78	1.73	1.68	1.65	1.62	1.58	1.55	1.51	1.47
30	1.82	1.77	1.72	1.67	1.64	1.61	1.57	1.54	1.50	1.46
40	1.76	1.71	1.66	1.61	1.57	1.54	1.51	1.47	1.42	1.38
60	1.71	1.66	1.60	1.54	1.51	1.48	1.44	1.40	1.35	1.29
120	1.65	1.60	1.55	1.48	1.45	1.41	1.37	1.32	1.26	1.19
∞	1.60	1.55	1.49	1.42	1.38	1.34	1.30	1.24	1.17	1.00

DENOMINATOR DEGREES OF FREEDOM

APPENDIX A

TABLE IX Percentage Points of the F Distribution, $\alpha = .05$

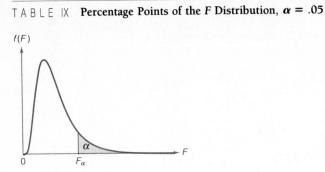

ν_1 ν_2	NUMERATOR DEGREES OF FREEDOM								
	1	2	3	4	5	6	7	8	9
1	161.4	199.5	215.7	224.6	230.2	234.0	236.8	238.9	240.5
2	18.51	19.00	19.16	19.25	19.30	19.33	19.35	19.37	19.38
3	10.13	9.55	9.28	9.12	9.01	8.94	8.89	8.85	8.81
4	7.71	6.94	6.59	6.39	6.26	6.16	6.09	6.04	6.00
5	6.61	5.79	5.41	5.19	5.05	4.95	4.88	4.82	4.77
6	5.99	5.14	4.76	4.53	4.39	4.28	4.21	4.15	4.10
7	5.59	4.74	4.35	4.12	3.97	3.87	3.79	3.73	3.68
8	5.32	4.46	4.07	3.84	3.69	3.58	3.50	3.44	3.39
9	5.12	4.26	3.86	3.63	3.48	3.37	3.29	3.23	3.18
10	4.96	4.10	3.71	3.48	3.33	3.22	3.14	3.07	3.02
11	4.84	3.98	3.59	3.36	3.20	3.09	3.01	2.95	2.90
12	4.75	3.89	3.49	3.26	3.11	3.00	2.91	2.85	2.80
13	4.67	3.81	3.41	3.18	3.03	2.92	2.83	2.77	2.71
14	4.60	3.74	3.34	3.11	2.96	2.85	2.76	2.70	2.65
15	4.54	3.68	3.29	3.06	2.90	2.79	2.71	2.64	2.59
16	4.49	3.63	3.24	3.01	2.85	2.74	2.66	2.59	2.54
17	4.45	3.59	3.20	2.96	2.81	2.70	2.61	2.55	2.49
18	4.41	3.55	3.16	2.93	2.77	2.66	2.58	2.51	2.46
19	4.38	3.52	3.13	2.90	2.74	2.63	2.54	2.48	2.42
20	4.35	3.49	3.10	2.87	2.71	2.60	2.51	2.45	2.39
21	4.32	3.47	3.07	2.84	2.68	2.57	2.49	2.42	2.37
22	4.30	3.44	3.05	2.82	2.66	2.55	2.46	2.40	2.34
23	4.28	3.42	3.03	2.80	2.64	2.53	2.44	2.37	2.32
24	4.26	3.40	3.01	2.78	2.62	2.51	2.42	2.36	2.30
25	4.24	3.39	2.99	2.76	2.60	2.49	2.40	2.34	2.28
26	4.23	3.37	2.98	2.74	2.59	2.47	2.39	2.32	2.27
27	4.21	3.35	2.96	2.73	2.57	2.46	2.37	2.31	2.25
28	4.20	3.34	2.95	2.71	2.56	2.45	2.36	2.29	2.24
29	4.18	3.33	2.93	2.70	2.55	2.43	2.35	2.28	2.22
30	4.17	3.32	2.92	2.69	2.53	2.42	2.33	2.27	2.21
40	4.08	3.23	2.84	2.61	2.45	2.34	2.25	2.18	2.12
60	4.00	3.15	2.76	2.53	2.37	2.25	2.17	2.10	2.04
120	3.92	3.07	2.68	2.45	2.29	2.17	2.09	2.02	1.96
∞	3.84	3.00	2.60	2.37	2.21	2.10	2.01	1.94	1.88

DENOMINATOR DEGREES OF FREEDOM

ν_2 \ ν_1	NUMERATOR DEGREES OF FREEDOM									
	10	12	15	20	24	30	40	60	120	∞
1	241.9	243.9	245.9	248.0	249.1	250.1	251.1	252.2	253.3	254.3
2	19.40	19.41	19.43	19.45	19.45	19.46	19.47	19.48	19.49	19.50
3	8.79	8.74	8.70	8.66	8.64	8.62	8.59	8.57	8.55	8.53
4	5.96	5.91	5.86	5.80	5.77	5.75	5.72	5.69	5.66	5.63
5	4.74	4.68	4.62	4.56	4.53	4.50	4.46	4.43	4.40	4.36
6	4.06	4.00	3.94	3.87	3.84	3.81	3.77	3.74	3.70	3.67
7	3.64	3.57	3.51	3.44	3.41	3.38	3.34	3.30	3.27	3.23
8	3.35	3.28	3.22	3.15	3.12	3.08	3.04	3.01	2.97	2.93
9	3.14	3.07	3.01	2.94	2.90	2.86	2.83	2.79	2.75	2.71
10	2.98	2.91	2.85	2.77	2.74	2.70	2.66	2.62	2.58	2.54
11	2.85	2.79	2.72	2.65	2.61	2.57	2.53	2.49	2.45	2.40
12	2.75	2.69	2.62	2.54	2.51	2.47	2.43	2.38	2.34	2.30
13	2.67	2.60	2.53	2.46	2.42	2.38	2.34	2.30	2.25	2.21
14	2.60	2.53	2.46	2.39	2.35	2.31	2.27	2.22	2.18	2.13
15	2.54	2.48	2.40	2.33	2.29	2.25	2.20	2.16	2.11	2.07
16	2.49	2.42	2.35	2.28	2.24	2.19	2.15	2.11	2.06	2.01
17	2.45	2.38	2.31	2.23	2.19	2.15	2.10	2.06	2.01	1.96
18	2.41	2.34	2.27	2.19	2.15	2.11	2.06	2.02	1.97	1.92
19	2.38	2.31	2.23	2.16	2.11	2.07	2.03	1.98	1.93	1.88
20	2.35	2.28	2.20	2.12	2.08	2.04	1.99	1.95	1.90	1.84
21	2.32	2.25	2.18	2.10	2.05	2.01	1.96	1.92	1.87	1.81
22	2.30	2.23	2.15	2.07	2.03	1.98	1.94	1.89	1.84	1.78
23	2.27	2.20	2.13	2.05	2.01	1.96	1.91	1.86	1.81	1.76
24	2.25	2.18	2.11	2.03	1.98	1.94	1.89	1.84	1.79	1.73
25	2.24	2.16	2.09	2.01	1.96	1.92	1.87	1.82	1.77	1.71
26	2.22	2.15	2.07	1.99	1.95	1.90	1.85	1.80	1.75	1.69
27	2.20	2.13	2.06	1.97	1.93	1.88	1.84	1.79	1.73	1.67
28	2.19	2.12	2.04	1.96	1.91	1.87	1.82	1.77	1.71	1.65
29	2.18	2.10	2.03	1.94	1.90	1.85	1.81	1.75	1.70	1.64
30	2.16	2.09	2.01	1.93	1.89	1.84	1.79	1.74	1.68	1.62
40	2.08	2.00	1.92	1.84	1.79	1.74	1.69	1.64	1.58	1.51
60	1.99	1.92	1.84	1.75	1.70	1.65	1.59	1.53	1.47	1.39
120	1.91	1.83	1.75	1.66	1.61	1.55	1.50	1.43	1.35	1.25
∞	1.83	1.75	1.67	1.57	1.52	1.46	1.39	1.32	1.22	1.00

DENOMINATOR DEGREES OF FREEDOM

T A B L E X   **Percentage Points of the F Distribution, $\alpha = .025$**

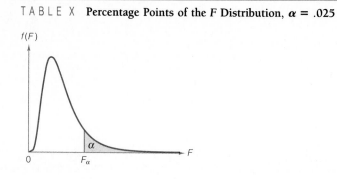

ν_1 ν_2	NUMERATOR DEGREES OF FREEDOM								
	1	2	3	4	5	6	7	8	9
1	647.8	799.5	864.2	899.6	921.8	937.1	948.2	956.7	963.3
2	38.51	39.00	39.17	39.25	39.30	39.33	39.36	39.37	39.39
3	17.44	16.04	15.44	15.10	14.88	14.73	14.62	14.54	14.47
4	12.22	10.65	9.98	9.60	9.36	9.20	9.07	8.98	8.90
5	10.01	8.43	7.76	7.39	7.15	6.98	6.85	6.76	6.68
6	8.81	7.26	6.60	6.23	5.99	5.82	5.70	5.60	5.52
7	8.07	6.54	5.89	5.52	5.29	5.12	4.99	4.90	4.82
8	7.57	6.06	5.42	5.05	4.82	4.65	4.53	4.43	4.36
9	7.21	5.71	5.08	4.72	4.48	4.32	4.20	4.10	4.03
10	6.94	5.46	4.83	4.47	4.24	4.07	3.95	3.85	3.78
11	6.72	5.26	4.63	4.28	4.04	3.88	3.76	3.66	3.59
12	6.55	5.10	4.47	4.12	3.89	3.73	3.61	3.51	3.44
13	6.41	4.97	4.35	4.00	3.77	3.60	3.48	3.39	3.31
14	6.30	4.86	4.24	3.89	3.66	3.50	3.38	3.29	3.21
15	6.20	4.77	4.15	3.80	3.58	3.41	3.29	3.20	3.12
16	6.12	4.69	4.08	3.73	3.50	3.34	3.22	3.12	3.05
17	6.04	4.62	4.01	3.66	3.44	3.28	3.16	3.06	2.98
18	5.98	4.56	3.95	3.61	3.38	3.22	3.10	3.01	2.93
19	5.92	4.51	3.90	3.56	3.33	3.17	3.05	2.96	2.88
20	5.87	4.46	3.86	3.51	3.29	3.13	3.01	2.91	2.84
21	5.83	4.42	3.82	3.48	3.25	3.09	2.97	2.87	2.80
22	5.79	4.38	3.78	3.44	3.22	3.05	2.93	2.84	2.76
23	5.75	4.35	3.75	3.41	3.18	3.02	2.90	2.81	2.73
24	5.72	4.32	3.72	3.38	3.15	2.99	2.87	2.78	2.70
25	5.69	4.29	3.69	3.35	3.13	2.97	2.85	2.75	2.68
26	5.66	4.27	3.67	3.33	3.10	2.94	2.82	2.73	2.65
27	5.63	4.24	3.65	3.31	3.08	2.92	2.80	2.71	2.63
28	5.61	4.22	3.63	3.29	3.06	2.90	2.78	2.69	2.61
29	5.59	4.20	3.61	3.27	3.04	2.88	2.76	2.67	2.59
30	5.57	4.18	3.59	3.25	3.03	2.87	2.75	2.65	2.57
40	5.42	4.05	3.46	3.13	2.90	2.74	2.62	2.53	2.45
60	5.29	3.93	3.34	3.01	2.79	2.63	2.51	2.41	2.33
120	5.15	3.80	3.23	2.89	2.67	2.52	2.39	2.30	2.22
∞	5.02	3.69	3.12	2.79	2.57	2.41	2.29	2.19	2.11

DENOMINATOR DEGREES OF FREEDOM

Source: From M. Merrington and C. M. Thompson, "Tables of Percentage Points of the Inverted Beta (F)-Distribution," *Biometrika*, 1943, 33, 73–88. Reproduced by permission of the *Biometrika* Trustees.

ν_2 \ ν_1	NUMERATOR DEGREES OF FREEDOM									
	10	12	15	20	24	30	40	60	120	∞
1	968.6	976.7	984.9	993.1	997.2	1,001	1,006	1,010	1,014	1,018
2	39.40	39.41	39.43	39.45	39.46	39.46	39.47	39.48	39.49	39.50
3	14.42	14.34	14.25	14.17	14.12	14.08	14.04	13.99	13.95	13.90
4	8.84	8.75	8.66	8.56	8.51	8.46	8.41	8.36	8.31	8.26
5	6.62	6.52	6.43	6.33	6.28	6.23	6.18	6.12	6.07	6.02
6	5.46	5.37	5.27	5.17	5.12	5.07	5.01	4.96	4.90	4.85
7	4.76	4.67	4.57	4.47	4.42	4.36	4.31	4.25	4.20	4.14
8	4.30	4.20	4.10	4.00	3.95	3.89	3.84	3.78	3.73	3.67
9	3.96	3.87	3.77	3.67	3.61	3.56	3.51	3.45	3.39	3.33
10	3.72	3.62	3.52	3.42	3.37	3.31	3.26	3.20	3.14	3.08
11	3.53	3.43	3.33	3.23	3.17	3.12	3.06	3.00	2.94	2.88
12	3.37	3.28	3.18	3.07	3.02	2.96	2.91	2.85	2.79	2.72
13	3.25	3.15	3.05	2.95	2.89	2.84	2.78	2.72	2.66	2.60
14	3.15	3.05	2.95	2.84	2.79	2.73	2.67	2.61	2.55	2.49
15	3.06	2.96	2.86	2.76	2.70	2.64	2.59	2.52	2.46	2.40
16	2.99	2.89	2.79	2.68	2.63	2.57	2.51	2.45	2.38	2.32
17	2.92	2.82	2.72	2.62	2.56	2.50	2.44	2.38	2.32	2.25
18	2.87	2.77	2.67	2.56	2.50	2.44	2.38	2.32	2.26	2.19
19	2.82	2.72	2.62	2.51	2.45	2.39	2.33	2.27	2.20	2.13
20	2.77	2.68	2.57	2.46	2.41	2.35	2.29	2.22	2.16	2.09
21	2.73	2.64	2.53	2.42	2.37	2.31	2.25	2.18	2.11	2.04
22	2.70	2.60	2.50	2.39	2.33	2.27	2.21	2.14	2.08	2.00
23	2.67	2.57	2.47	2.36	2.30	2.24	2.18	2.11	2.04	1.97
24	2.64	2.54	2.44	2.33	2.27	2.21	2.15	2.08	2.01	1.94
25	2.61	2.51	2.41	2.30	2.24	2.18	2.12	2.05	1.98	1.91
26	2.59	2.49	2.39	2.28	2.22	2.16	2.09	2.03	1.95	1.88
27	2.57	2.47	2.36	2.25	2.19	2.13	2.07	2.00	1.93	1.85
28	2.55	2.45	2.34	2.23	2.17	2.11	2.05	1.98	1.91	1.83
29	2.53	2.43	2.32	2.21	2.15	2.09	2.03	1.96	1.89	1.81
30	2.51	2.41	2.31	2.20	2.14	2.07	2.01	1.94	1.87	1.79
40	2.39	2.29	2.18	2.07	2.01	1.94	1.88	1.80	1.72	1.64
60	2.27	2.17	2.06	1.94	1.88	1.82	1.74	1.67	1.58	1.48
120	2.16	2.05	1.94	1.82	1.76	1.69	1.61	1.53	1.43	1.31
∞	2.05	1.94	1.83	1.71	1.64	1.57	1.48	1.39	1.27	1.00

DENOMINATOR DEGREES OF FREEDOM

TABLE XI   **Percentage Points of the F Distribution, $\alpha = .01$**

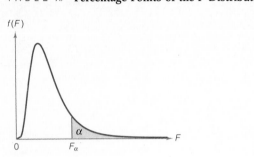

ν_1	NUMERATOR DEGREES OF FREEDOM								
ν_2	1	2	3	4	5	6	7	8	9
1	4,052	4,999.5	5,403	5,625	5,764	5,859	5,928	5,982	6,022
2	98.50	99.00	99.17	99.25	99.30	99.33	99.36	99.37	99.39
3	34.12	30.82	29.46	28.71	28.24	27.91	27.67	27.49	27.35
4	21.20	18.00	16.69	15.98	15.52	15.21	14.98	14.80	14.66
5	16.26	13.27	12.06	11.39	10.97	10.67	10.46	10.29	10.16
6	13.75	10.92	9.78	9.15	8.75	8.47	8.26	8.10	7.98
7	12.25	9.55	8.45	7.85	7.46	7.19	6.99	6.84	6.72
8	11.26	8.65	7.59	7.01	6.63	6.37	6.18	6.03	5.91
9	10.56	8.02	6.99	6.42	6.06	5.80	5.61	5.47	5.35
10	10.04	7.56	6.55	5.99	5.64	5.39	5.20	5.06	4.94
11	9.65	7.21	6.22	5.67	5.32	5.07	4.89	4.74	4.63
12	9.33	6.93	5.95	5.41	5.06	4.82	4.64	4.50	4.39
13	9.07	6.70	5.74	5.21	4.86	4.62	4.44	4.30	4.19
14	8.86	6.51	5.56	5.04	4.69	4.46	4.28	4.14	4.03
15	8.68	6.36	5.42	4.89	4.56	4.32	4.14	4.00	3.89
16	8.53	6.23	5.29	4.77	4.44	4.20	4.03	3.89	3.78
17	8.40	6.11	5.18	4.67	4.34	4.10	3.93	3.79	3.68
18	8.29	6.01	5.09	4.58	4.25	4.01	3.84	3.71	3.60
19	8.18	5.93	5.01	4.50	4.17	3.94	3.77	3.63	3.52
20	8.10	5.85	4.94	4.43	4.10	3.87	3.70	3.56	3.46
21	8.02	5.78	4.87	4.37	4.04	3.81	3.64	3.51	3.40
22	7.95	5.72	4.82	4.31	3.99	3.76	3.59	3.45	3.35
23	7.88	5.66	4.76	4.26	3.94	3.71	3.54	3.41	3.30
24	7.82	5.61	4.72	4.22	3.90	3.67	3.50	3.36	3.26
25	7.77	5.57	4.68	4.18	3.85	3.63	3.46	3.32	3.22
26	7.72	5.53	4.64	4.14	3.82	3.59	3.42	3.29	3.18
27	7.68	5.49	4.60	4.11	3.78	3.56	3.39	3.26	3.15
28	7.64	5.45	4.57	4.07	3.75	3.53	3.36	3.23	3.12
29	7.60	5.42	4.54	4.04	3.73	3.50	3.33	3.20	3.09
30	7.56	5.39	4.51	4.02	3.70	3.47	3.30	3.17	3.07
40	7.31	5.18	4.31	3.83	3.51	3.29	3.12	2.99	2.89
60	7.08	4.98	4.13	3.65	3.34	3.12	2.95	2.82	2.72
120	6.85	4.79	3.95	3.48	3.17	2.96	2.79	2.66	2.56
∞	6.63	4.61	3.78	3.32	3.02	2.80	2.64	2.51	2.41

DENOMINATOR DEGREES OF FREEDOM

Source: From M. Merrington and C. M. Thompson, "Tables of Percentage Points of the Inverted Beta (F)-Distribution," *Biometrika*, 1943, 33, 73–88. Reproduced by permission of the *Biometrika* Trustees.

ν_1 ν_2	NUMERATOR DEGREES OF FREEDOM									
	10	12	15	20	24	30	40	60	120	∞
1	6,056	6,106	6,157	6,209	6,235	6,261	6,287	6,313	6,339	6,366
2	99.40	99.42	99.43	99.45	99.46	99.47	99.47	99.48	99.49	99.50
3	27.23	27.05	26.87	26.69	26.60	26.50	26.41	26.32	26.22	26.13
4	14.55	14.37	14.20	14.02	13.93	13.84	13.75	13.65	13.56	13.46
5	10.05	9.89	9.72	9.55	9.47	9.38	9.29	9.20	9.11	9.02
6	7.87	7.72	7.56	7.40	7.31	7.23	7.14	7.06	6.97	6.88
7	6.62	6.47	6.31	6.16	6.07	5.99	5.91	5.82	5.74	5.65
8	5.81	5.67	5.52	5.36	5.28	5.20	5.12	5.03	4.95	4.86
9	5.26	5.11	4.96	4.81	4.73	4.65	4.57	4.48	4.40	4.31
10	4.85	4.71	4.56	4.41	4.33	4.25	4.17	4.08	4.00	3.91
11	4.54	4.40	4.25	4.10	4.02	3.94	3.86	3.78	3.69	3.60
12	4.30	4.16	4.01	3.86	3.78	3.70	3.62	3.54	3.45	3.36
13	4.10	3.96	3.82	3.66	3.59	3.51	3.43	3.34	3.25	3.17
14	3.94	3.80	3.66	3.51	3.43	3.35	3.27	3.18	3.09	3.00
15	3.80	3.67	3.52	3.37	3.29	3.21	3.13	3.05	2.96	2.87
16	3.69	3.55	3.41	3.26	3.18	3.10	3.02	2.93	2.84	2.75
17	3.59	3.46	3.31	3.16	3.08	3.00	2.92	2.83	2.75	2.65
18	3.51	3.37	3.23	3.08	3.00	2.92	2.84	2.75	2.66	2.57
19	3.43	3.30	3.15	3.00	2.92	2.84	2.76	2.67	2.58	2.49
20	3.37	3.23	3.09	2.94	2.86	2.78	2.69	2.61	2.52	2.42
21	3.31	3.17	3.03	2.88	2.80	2.72	2.64	2.55	2.46	2.36
22	3.26	3.12	2.98	2.83	2.75	2.67	2.58	2.50	2.40	2.31
23	3.21	3.07	2.93	2.78	2.70	2.62	2.54	2.45	2.35	2.26
24	3.17	3.03	2.89	2.74	2.66	2.58	2.49	2.40	2.31	2.21
25	3.13	2.99	2.85	2.70	2.62	2.54	2.45	2.36	2.27	2.17
26	3.09	2.96	2.81	2.66	2.58	2.50	2.42	2.33	2.23	2.13
27	3.06	2.93	2.78	2.63	2.55	2.47	2.38	2.29	2.20	2.10
28	3.03	2.90	2.75	2.60	2.52	2.44	2.35	2.26	2.17	2.06
29	3.00	2.87	2.73	2.57	2.49	2.41	2.33	2.23	2.14	2.03
30	2.98	2.84	2.70	2.55	2.47	2.39	2.30	2.21	2.11	2.01
40	2.80	2.66	2.52	2.37	2.29	2.20	2.11	2.02	1.92	1.80
60	2.63	2.50	2.35	2.20	2.12	2.03	1.94	1.84	1.73	1.60
120	2.47	2.34	2.19	2.03	1.95	1.86	1.76	1.66	1.53	1.38
∞	2.32	2.18	2.04	1.88	1.79	1.70	1.59	1.47	1.32	1.00

DENOMINATOR DEGREES OF FREEDOM

TABLE XII **Critical Values of T_L and T_U for the Wilcoxon Rank Sum Test: Independent Samples**

Test statistic is the rank sum associated with the smaller sample (if equal sample sizes, either rank sum can be used).

a. $\alpha = .025$ one-tailed; $\alpha = .05$ two-tailed

n_2 \ n_1	3 T_L	3 T_U	4 T_L	4 T_U	5 T_L	5 T_U	6 T_L	6 T_U	7 T_L	7 T_U	8 T_L	8 T_U	9 T_L	9 T_U	10 T_L	10 T_U
3	5	16	6	18	6	21	7	23	7	26	8	28	8	31	9	33
4	6	18	11	25	12	28	12	32	13	35	14	38	15	41	16	44
5	6	21	12	28	18	37	19	41	20	45	21	49	22	53	24	56
6	7	23	12	32	19	41	26	52	28	56	29	61	31	65	32	70
7	7	26	13	35	20	45	28	56	37	68	39	73	41	78	43	83
8	8	28	14	38	21	49	29	61	39	73	49	87	51	93	54	98
9	8	31	15	41	22	53	31	65	41	78	51	93	63	108	66	114
10	9	33	16	44	24	56	32	70	43	83	54	98	66	114	79	131

b. $\alpha = .05$ one-tailed; $\alpha = .10$ two-tailed

n_2 \ n_1	3 T_L	3 T_U	4 T_L	4 T_U	5 T_L	5 T_U	6 T_L	6 T_U	7 T_L	7 T_U	8 T_L	8 T_U	9 T_L	9 T_U	10 T_L	10 T_U
3	6	15	7	17	7	20	8	22	9	24	9	27	10	29	11	31
4	7	17	12	24	13	27	14	30	15	33	16	36	17	39	18	42
5	7	20	13	27	19	36	20	40	22	43	24	46	25	50	26	54
6	8	22	14	30	20	40	28	50	30	54	32	58	33	63	35	67
7	9	24	15	33	22	43	30	54	39	66	41	71	43	76	46	80
8	9	27	16	36	24	46	32	58	41	71	52	84	54	90	57	95
9	10	29	17	39	25	50	33	63	43	76	54	90	66	105	69	111
10	11	31	18	42	26	54	35	67	46	80	57	95	69	111	83	127

Source: From F. Wilcoxon and R. A. Wilcox, "Some Rapid Approximate Statistical Procedures," 1964, 20–23. Reproduced with the permission of American Cyanamid Company.

T A B L E XIII **Critical Values of T_0 in the Wilcoxon Paired Difference Signed Rank Test**

ONE-TAILED	TWO-TAILED	$n = 5$	$n = 6$	$n = 7$	$n = 8$	$n = 9$	$n = 10$
$\alpha = .05$	$\alpha = .10$	1	2	4	6	8	11
$\alpha = .025$	$\alpha = .05$		1	2	4	6	8
$\alpha = .01$	$\alpha = .02$			0	2	3	5
$\alpha = .005$	$\alpha = .01$				0	2	3
		$n = 11$	$n = 12$	$n = 13$	$n = 14$	$n = 15$	$n = 16$
$\alpha = .05$	$\alpha = .10$	14	17	21	26	30	36
$\alpha = .025$	$\alpha = .05$	11	14	17	21	25	30
$\alpha = .01$	$\alpha = .02$	7	10	13	16	20	24
$\alpha = .005$	$\alpha = .01$	5	7	10	13	16	19
		$n = 17$	$n = 18$	$n = 19$	$n = 20$	$n = 21$	$n = 22$
$\alpha = .05$	$\alpha = .10$	41	47	54	60	68	75
$\alpha = .025$	$\alpha = .05$	35	40	46	52	59	66
$\alpha = .01$	$\alpha = .02$	28	33	38	43	49	56
$\alpha = .005$	$\alpha = .01$	23	28	32	37	43	49
		$n = 23$	$n = 24$	$n = 25$	$n = 26$	$n = 27$	$n = 28$
$\alpha = .05$	$\alpha = .10$	83	92	101	110	120	130
$\alpha = .025$	$\alpha = .05$	73	81	90	98	107	117
$\alpha = .01$	$\alpha = .02$	62	69	77	85	93	102
$\alpha = .005$	$\alpha = .01$	55	61	68	76	84	92
		$n = 29$	$n = 30$	$n = 31$	$n = 32$	$n = 33$	$n = 34$
$\alpha = .05$	$\alpha = .10$	141	152	163	175	188	201
$\alpha = .025$	$\alpha = .05$	127	137	148	159	171	183
$\alpha = .01$	$\alpha = .02$	111	120	130	141	151	162
$\alpha = .005$	$\alpha = .01$	100	109	118	128	138	149
		$n = 35$	$n = 36$	$n = 37$	$n = 38$	$n = 39$	
$\alpha = .05$	$\alpha = .10$	214	228	242	256	271	
$\alpha = .025$	$\alpha = .05$	195	208	222	235	250	
$\alpha = .01$	$\alpha = .02$	174	186	198	211	224	
$\alpha = .005$	$\alpha = .01$	160	171	183	195	208	
		$n = 40$	$n = 41$	$n = 42$	$n = 43$	$n = 44$	$n = 45$
$\alpha = .05$	$\alpha = .10$	287	303	319	336	353	371
$\alpha = .025$	$\alpha = .05$	264	279	295	311	327	344
$\alpha = .01$	$\alpha = .02$	238	252	267	281	297	313
$\alpha = .005$	$\alpha = .01$	221	234	248	262	277	292
		$n = 46$	$n = 47$	$n = 48$	$n = 49$	$n = 50$	
$\alpha = .05$	$\alpha = .10$	389	408	427	446	466	
$\alpha = .025$	$\alpha = .05$	361	379	397	415	434	
$\alpha = .01$	$\alpha = .02$	329	345	362	380	398	
$\alpha = .005$	$\alpha = .01$	307	323	339	356	373	

Source: From F. Wilcoxon and R. A. Wilcox, "Some Rapid Approximate Statistical Procedures," 1964, p. 28. Reproduced with the permission of American Cyanamid Company.

TABLE XIV **Critical Values of Spearman's Rank Correlation Coefficient**

The α values correspond to a one-tailed test of H_0: $\rho_S = 0$. The value should be doubled for two-tailed tests.

n	$\alpha = .05$	$\alpha = .025$	$\alpha = .01$	$\alpha = .005$	n	$\alpha = .05$	$\alpha = .025$	$\alpha = .01$	$\alpha = .005$
5	.900	—	—	—	18	.399	.476	.564	.625
6	.829	.886	.943	—	19	.388	.462	.549	.608
7	.714	.786	.893	—	20	.377	.450	.534	.591
8	.643	.738	.833	.881	21	.368	.438	.521	.576
9	.600	.683	.783	.833	22	.359	.428	.508	.562
10	.564	.648	.745	.794	23	.351	.418	.496	.549
11	.523	.623	.736	.818	24	.343	.409	.485	.537
12	.497	.591	.703	.780	25	.336	.400	.475	.526
13	.475	.566	.673	.745	26	.329	.392	.465	.515
14	.457	.545	.646	.716	27	.323	.385	.456	.505
15	.441	.525	.623	.689	28	.317	.377	.448	.496
16	.425	.507	.601	.666	29	.311	.370	.440	.487
17	.412	.490	.582	.645	30	.305	.364	.432	.478

Source: From E. G. Olds, "Distribution of Sums of Squares of Rank Differences for Small Samples," *Annals of Mathematical Statistics*, 1938, 9. Reproduced with the permission of the Editor, *Annals of Mathematical Statistics*.

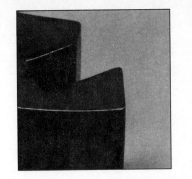

Calculation Formulas for Analysis of Variance

B.1 Formulas for the Calculations in the Completely Randomized Design

$$CM = \text{Correction for mean}$$

$$= \frac{(\text{Total of all observations})^2}{\text{Total number of observations}} = \frac{\left(\sum y_i\right)^2}{n}$$

$$SS(\text{Total}) = \text{Total sum of squares}$$

$$= (\text{Sum of squares of all observations}) - CM = \sum y_i^2 - CM$$

$$SST = \text{Sum of squares for treatments}$$

$$= \left(\begin{array}{c}\text{Sum of squares of treatment totals with} \\ \text{each square divided by the number of} \\ \text{observations for that treatment}\end{array}\right) - CM$$

$$= \frac{T_1^2}{n_1} + \frac{T_2^2}{n_2} + \cdots + \frac{T_p^2}{n_p} - CM$$

$$SSE = \text{Sum of squares for error} = SS(\text{Total}) - SST$$

$$MST = \text{Mean square for treatments} = \frac{SST}{p - 1}$$

$$MSE = \text{Mean square for error} = \frac{SSE}{n - p}$$

$$F = \text{Test statistic} = \frac{MST}{MSE}$$

where

$$n = \text{Total number of observations}$$
$$p = \text{Number of treatments}$$
$$T_i = \text{Total for treatment } i \ (i = 1, 2, \ldots, p)$$

B.2 Formulas for the Calculations in the Randomized Block Design

$$CM = \text{Correction for mean}$$

$$= \frac{(\text{Total of all observations})^2}{\text{Total number of observations}} = \frac{\left(\sum y_i\right)^2}{n}$$

$$SS(\text{Total}) = \text{Total sum of squares}$$

$$= (\text{Sum of squares of all observations}) - CM = \sum y_i^2 - CM$$

$$SST = \text{Sum of squares for treatments}$$

$$= \left(\begin{array}{c}\text{Sum of squares of treatment totals with} \\ \text{each square divided by } b, \text{ the number of} \\ \text{observations for that treatment}\end{array}\right) - CM$$

$$= \frac{T_1^2}{b} + \frac{T_2^2}{b} + \cdots + \frac{T_p^2}{b} - CM$$

$$SSB = \text{Sum of squares for blocks}$$

$$= \left(\begin{array}{c} \text{Sum of squares of block totals with} \\ \text{each square divided by } k, \text{ the number} \\ \text{of observations in that block} \end{array} \right) - CM$$

$$= \frac{B_1^2}{p} + \frac{B_2^2}{p} + \cdots + \frac{B_p^2}{p} - CM$$

$$SSE = \text{Sum of squares for error} = SS(\text{Total}) - SST - SSB$$

$$MST = \text{Mean square for treatments} = \frac{SST}{p - 1}$$

$$MSB = \text{Mean square for blocks} = \frac{SSB}{b - 1}$$

$$MSE = \text{Mean square for error} = \frac{SSE}{n - p - b + 1}$$

$$F = \text{Test statistic} = \frac{MST}{MSE}$$

where

$$n = \text{Total number of observations}$$
$$b = \text{Number of blocks}$$
$$p = \text{Number of treatments}$$
$$T_i = \text{Total for treatment } i \ (i = 1, 2, \ldots, p)$$
$$B_i = \text{Total for block } i \ (i = 1, 2, \ldots, b)$$

B.3 Formulas for the Calculations for a Two-Factor Factorial Experiment

$$CM = \text{Correction for the mean}$$

$$= \frac{(\text{Total of all } n \text{ measurements})^2}{n} = \frac{\left(\sum\limits_{i=1}^{n} y_i \right)^2}{n}$$

$$SS(\text{Total}) = \text{Total sum of squares}$$

$$= \text{Sum of squares of all } n \text{ measurements} - CM = \sum_{i=1}^{n} y_i^2 - CM$$

$$SS(A) = \text{Sum of squares for main effects, factor } A$$

$$= \left(\begin{array}{c} \text{Sum of squares of the totals } A_1, A_2, \ldots, A_a \\ \text{divided by the number of measurements} \\ \text{in a single total, namely } br \end{array} \right) - CM$$

$$= \frac{\sum\limits_{i=1}^{a} A_i^2}{br} - CM$$

$$SS(B) = \text{Sum of squares for main effects, factor } B$$

$$= \left(\begin{array}{c} \text{Sum of squares of the totals } B_1, B_2, \ldots, B_b \\ \text{divided by the number of measurements} \\ \text{in a single total, namely } ar \end{array} \right) - CM$$

$$= \frac{\displaystyle\sum_{i=1}^{b} B_i^2}{ar} - CM$$

$$SS(AB) = \text{Sum of squares for } AB \text{ interaction}$$

$$= \left(\begin{array}{c} \text{Sum of squares of the cell} \\ \text{totals } AB_{11}, AB_{12}, \ldots, AB_{ab} \\ \text{divided by the number of} \\ \text{measurements in a single} \\ \text{total, namely } r \end{array} \right) - SS(A) - SS(B) - CM$$

$$= \frac{\displaystyle\sum_{j=1}^{b} \sum_{i=1}^{a} AB_{ij}^2}{r} - SS(A) - SS(B) - CM$$

where

$a = $ Number of levels of factor A

$b = $ Number of levels of factor B

$r = $ Number of replicates (observations per treatment)

$A_i = $ Total for level i of factor A $(i = 1, 2, \ldots, a)$

$B_i = $ Total of level i of factor B $(i = 1, 2, \ldots, b)$

$AB_{ij} = $ Total for Treatment (i, j), i.e., for ith level of factor A and jth level of factor B

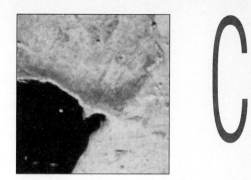

Demographic Data Set

A demographic data set was assembled based on a systematic random sample of 1,000 United States zip codes. To obtain the sample, the more than 30,000 zip codes were sorted, and approximately every 30th was selected. The map in Figure C.1 shows the number of zip codes selected in each state. Note that each state is classified according to its census region: North Central, Northeast, South, and West.

Demographic data for each zip code area selected were supplied by CACI, an international demographic and market information firm, and are reproduced with its permission. CACI produces interim estimates of many of the demographic variables measured decennially by the Bureau of the Census. CACI also produces market information based on the Bureau of the Census Consumer Expenditure

FIGURE C.1

Number of zip code areas selected by state and census region

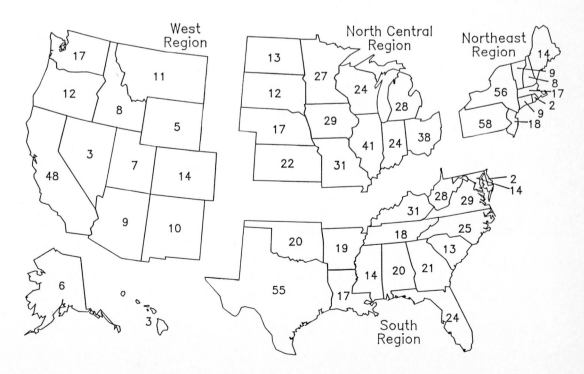

Survey (such as the four purchasing indexes given below for each zip code), which measure the relative propensity of a given market (zip code) to purchase the goods, a higher index value indicating a higher propensity to buy such goods as compared to other similar zip codes in the same census region. In general, the 1980 measurements are official U.S. Bureau of the Census estimates, while the 1986 measurements are CACI estimates.

Fifteen demographic measurements are presented for each zip code area. Portions of the data are referenced at the end of each chapter in "Using the Computer." The objectives are to enable the student to analyze real data in a relatively large sample using the computer, and to gain experience using statistical techniques and concepts on real data. Of course, neither the student nor the instructor need be bound by the suggestions in "Using the Computer"; the data are rich enough to support many more analyses than could be listed (or imagined) by the authors.

The following 15 measurements are reported for each zip code in the sample:

1. Population (1986): Total population for the zip code, 1986.
2. Number of Households (1986): Number of households for the zip code in 1986.
3. Age (Median, 1986): Median age for the zip code in 1986.
4. Household Income (Median, 1986): Median household income for the zip code in 1986.
5. Home Value (Median, 1980): Median residential owner-occupied home value for the zip code in 1980.
6. Monthly Cost of Housing (Median, 1980): Median monthly homeowner cost, including mortgage payments, real estate taxes, property insurance, utilities, and fuels, for the zip code in 1980.
7. Household Size (Average, 1980): Average number of persons per household in the zip code, 1980.
8. Years of Education (Median, 1980): Median years of education for adults (persons 25 years of age and over) in the zip code, 1980.
9. College Education (Percentage, 1980): Percentage of adults in the zip code having a college education.
10. Women in Labor Force (Percentage, 1980): Percentage of women age 16 and over in the zip code who are in the labor force (either having jobs or actively seeking one).
11. Unemployment (Percentage, 1980): Percentage of the labor force in the zip code that is unemployed and actively seeking work.
12. Average Purchasing Potential (1986): Purchasing potential index for all consumer items over the zip code area, based on the Bureau of Labor Statistics Consumer Expenditure Survey. The index compares a zip code to similar urban or rural zip codes in the same census region, with 100 set as the average value.
13. Sporting Goods Purchasing Potential (1986): Purchasing potential index for sporting goods over the zip code area.
14. Groceries Purchasing Potential (1986): Purchasing potential index for groceries over the zip code area.

15. Home Improvement Purchasing Potential (1986): Purchasing potential index for home improvements over the zip code area.

Numerical codes have been assigned to each state and census region to facilitate computer utilization of these data. The codes were assigned alphabetically as shown in Tables C.1 and C.2.

T A B L E C.1 **Census Region Codes**

CODE	CENSUS REGION
1	North Central region
2	Northeast region
3	South region
4	West region

T A B L E C.2 **State Codes**

CODE	STATE		CODE	STATE	
01	AK:	Alaska	26	MT:	Montana
02	AL:	Alabama	27	NC:	North Carolina
03	AR:	Arkansas	28	ND:	North Dakota
04	AZ:	Arizona	29	NE:	Nebraska
05	CA:	California	30	NH:	New Hampshire
06	CO:	Colorado	31	NJ:	New Jersey
07	CT:	Connecticut	32	NM:	New Mexico
08	DE:	Delaware	33	NV:	Nevada
09	FL:	Florida	34	NY:	New York
10	GA:	Georgia	35	OH:	Ohio
11	HI:	Hawaii	36	OK:	Oklahoma
12	IA:	Iowa	37	OR:	Oregon
13	ID:	Idaho	38	PA:	Pennsylvania
14	IL:	Illinois	39	RI:	Rhode Island
15	IN:	Indiana	40	SC:	South Carolina
16	KS:	Kansas	41	SD:	South Dakota
17	KY:	Kentucky	42	TN:	Tennessee
18	LA:	Louisiana	43	TX:	Texas
19	MA:	Massachusetts	44	UT:	Utah
20	MD:	Maryland	45	VA:	Virginia
21	ME:	Maine	46	VT:	Vermont
22	MI:	Michigan	47	WA:	Washington
23	MN:	Minnesota	48	WI:	Wisconsin
24	MO:	Missouri	49	WV:	West Virginia
25	MS:	Mississippi	50	WY:	Wyoming

The demographic database is contained in two ASCII files, which are both sorted by census region and state within census region. The file names and corresponding layouts are given in Tables C.3 and C.4 (page 912). The data files are available on magnetic tape or diskette from the publisher.

T A B L E C.3 **ZIPCOD01.DAT**

COLUMNS	DESCRIPTION
1–4	Observation number
7	Census region number
9–10	State number
12–16	Zip code
18–22	Population in '86
24–28	Number of households in '86
30–33	Median age in '86
35–39	Median household income in '86
41–46	Median home value in '80
48–50	Median monthly homeowner cost in '80
52–54	Median household size in '80
56–59	Median years of education in '80

T A B L E C.4 **ZIPCOD02.DAT**

COLUMNS	DESCRIPTION
1–4	Observation number
7	Census region number
9–10	State number
12–16	Zip code
18–21	Percent college education in '80
23–26	Percent women in work force in '80
28–31	Percent unemployed in '80
33–37	Purchasing potential index '86: Overall average
39–43	Purchasing potential index '86: Sporting goods
45–49	Purchasing potential index '86: Grocery
51–55	Purchasing potential index '86: Home improvement

Answers to Selected Exercises

Chapter 2

2.3 Qualitative, qualitative **2.5a.** Nominal **b.** Ratio **c.** Ratio **d.** Ratio **e.** Interval **2.7a.** Nominal **b.** Ratio **c.** Nominal **d.** Ordinal
2.21b. .15 **2.23b.** .65 **2.27** Mode = 15; mean = 14.545; median = 15 **2.31a.** 8.5 **b.** 25 **c.** .78 **d.** 13.44 **2.33a.** Mean is smaller.
b. Mean is larger. **c.** They are equal. **2.35c.** 81.15; 83; 83 **2.37a.** 180; 162 **b.** Skewed to the right **c.** No **2.39a.** No **b.** 110
2.41 $.35 **2.43b.** Skewed to the right; yes **c.** Shifted downward **2.47a.** 4; 2.30; 1.52 **b.** 6; 3.62; 1.90 **c.** 10; 7.11; 2.67
d. 5; 1.62; 1.27 **2.49a.** 5; 3.70; 1.92 **b.** 99; 1,949.25; 44.15 **c.** 98; 1,307.84; 36.16 **2.51a.** 5.6; 17.3; 4.1593
b. 13.75 feet; 152.25 square feet; 12.339 feet **c.** −2.5; 4.3; 2.0736 **d.** .33 ounce; .0587 square ounces; .2422 ounce
2.57a. Germany: 1.4; Italy: 4.7; France: 4.1; UK: 8.8; Belgium: 1.5 **b.** Germany: .7024; Italy: 2.3544; France: 2.0599; UK: 4.4; Belgium: .7506
c. Ranges: Germany, Belgium, France, Italy, UK; standard deviations: Germany, Belgium, France, Italy, UK **d.** Yes; no
2.59a. Dollars; ratio **b.** At least $\frac{3}{4}$; at least $\frac{8}{9}$; nothing; nothing **2.61a.** ≈68% **b.** ≈95% **c.** ≈All **2.63** Between 104.17 and 156.25; no
2.65a. Chebyshev's **c.** At least .75 (actual 34 measurements); at most .111 (actual 1 measurement) **d.** 4.26
2.67 ≈68% in (131.16, 235.70); ≈95% in (78.89, 287.97); ≈all in (26.62, 340.24) **2.69** Do not buy **2.71a.** At least $\frac{8}{9}$ **b.** ≈.84 **c.** Yes
2.73 At least $\frac{3}{4}$ had performance times in (21.30, 102.50); at least $\frac{8}{9}$ had performance times in (1.00, 122.80) **2.75a.** 25% above; 75% below
b. 50% above; 50% below **c.** 80% above; 20% below **d.** 16% above; 84% below
2.77a. 2 (most above) **b.** −3 (most below) **c.** −2 **d.** 1.67 **2.79a.** 3.7; 2.2; 2.95; 1.45 **b.** 1.9
c. Cum laude: $z > 1$, GPA > 3.2; summa cum laude: $z > 2$; GPA > 3.7; mound-shaped distribution
2.81b. 90% of test scores are below 660. **c.** 94% of test scores are below your score. **2.83a.** 8.04; 3.68 **b.** .42; .53; −1.45
c. + means score is above mean; − means score is below mean **2.85** Yes; $z = 35$ **2.93a.** ≈770; ≈1,540; ≈3,078 **b.** Skewed to the right
c. Yes **2.105a.** −1; 1; 2 **b.** −2; 2; 4 **c.** 1; 3; 4 **d.** .1; .3; .4 **2.107a.** 3.1234 **b.** 9.0233 **c.** 9.7857 **2.109a.** 5.67; 1.067; 1.03
b. −$1.5; 11.5 dollars squared; $3.39 **c.** .4125%; .088% squared; .30% **d.** 3; $10; .7375% **2.113** Decrease; smaller; smaller
2.115 $z = 1.90$; no **2.117c.** 58.24; 65.44; 8.0895 **d.** ≈68%; ≈95%; ≈all **e.** 64%; 96%; 100%
2.119a. Stem is hundreds and tens digits; leaf is ones digit **b.** 925 New Hampshire, 790 South Carolina; 939 New Hampshire, 815 South
Carolina **c.** 888; 896.5 **2.121a.** ≈68% **b.** ≈47.5% **c.** ≈16% **2.125a.** ≈0 **b.** No **c.** Yes ($z = −1.33$) **2.127a.** Magazines
b. Magazines; newspapers **c.** Magazines; newspapers **e.** No

Chapter 3

3.1a. .4 **b.** .25 **c.** .6 **3.3** $P(A) = .60$; $P(B) = .50$; $P(C) = .75$ **3.7a.** $\frac{4}{9}$ **b.** $\frac{1}{3}$ **c.** 0 **3.9** $P(A) = \frac{1}{12}$; $P(B) = \frac{1}{4}$; $P(C) = \frac{1}{2}$; $P(D) = \frac{1}{2}$
3.13a. $\frac{1}{4}$ **b.** $\frac{1}{2}$ **c.** $\frac{1}{4}$ **3.15a.** Visa or Mastercard; Diners Club; Carte Blanche; Choice **b.** .632; .232; .032; .105 **c.** .369
3.17a. $\frac{1}{6}$ **b.** $\frac{1}{6}$ **c.** $\frac{2}{3}$ **3.19** $\frac{3}{4}$; no; probability of being odd man out remains $\frac{1}{4}$ **3.21a.** $\frac{1}{10}$ **b.** $\frac{3}{10}$ **c.** $\frac{7}{10}$ **3.23a.** 1 to 2 **b.** .5 **c.** .4
3.25a. A: {(1, 6), (2, 5), (3, 4), (4, 3), (5, 2), (6, 1)}; B: {(1, 4), (2, 4), (3, 4), (4, 4), (5, 4), (6, 4), (4, 1), (4, 2), (4, 3), (4, 5), (4, 6)};
$A \cap B$: {(3, 4), (4, 3)}; $A \cup B$: {(1, 4), (2, 4), (3, 4), (4, 4), (5, 4), (6, 4), (4, 1), (4, 2), (4, 3), (4, 5), (4, 6), (1, 6), (2, 5), (5, 2), (6, 1)}
A': 30 simple events not in A. **b.** $\frac{1}{6}$; $\frac{11}{36}$; $\frac{1}{18}$; $\frac{15}{36}$; $\frac{5}{6}$ **3.27a.** .5 **b.** .19 **c.** .5 **d.** 1.00 **e.** .31 **f.** .69 **3.29a.** $A \cap F$ **b.** $B \cup C$; A'
3.31a. $B \cap C$ **b.** A' **c.** $C \cup B$ **d.** $A \cap C'$ **3.33a.** $A \cap B$: {11, 13, 15, 17, 29, 31, 33, 35}
b. $A \cup B$: {1, 2, 3, 4, 5, 6, 7, 8, 9, 10, 11, 13, 15, 17, 19, 20, 21, 22, 23, 24, 25, 26, 27, 28, 29, 31, 33, 35} **c.** $\frac{9}{19}$; $\frac{9}{19}$; $\frac{4}{19}$; $\frac{14}{19}$; $\frac{9}{19}$
d. $A \cap B \cap C$: {11, 13, 15, 17} **e.** $\frac{2}{19}$ **f.** $A \cup B \cup C$: {all simple events except 00, 0, 30, 32, 34, 36} **g.** $\frac{16}{19}$ **3.35a.** .03 **b.** .62 **c.** .28
d. .32 **e.** .72 **3.37a.** .5; .5; .4 **b.** .2; .2; .6 **c.** .8 **3.39a.** .55; .40; .35 **b.** .30; 0; 0 **c.** $\frac{4}{11}$; $\frac{1}{11}$; $\frac{6}{11}$ **d.** $\frac{6}{11}$ **e.** 0; $\frac{7}{9}$ **3.41** $\frac{1}{3}$; 0; $\frac{1}{14}$; $\frac{1}{7}$; 1
3.43a. .0105 **b.** .0028; .9920 **c.** .2394; .1397 **d.** .0224 **3.45** .60 **3.47a.** .105 **b.** .638 **c.** .517 **3.49a.** .37 **b.** .68 **c.** .15 **d.** .221
e. 0 **f.** 0 **3.51a.** .625 **b.** .5 **c.** .8 **d.** .625 **e.** Dependent **f.** No **3.53a.** A and C, $P(A \cap C) = 0$; B and C, $P(B \cap C) = 0$
b. None are independent **c.** .65; .90 **3.55a.** .48 **b.** .32 **c.** .16 **d.** .64 **e.** .5 **f.** .47

3.57a. $A \cap B$; $A \cap B'$; A' **b.** $P(A \cap B) = .891$; $P(A \cap B') = .009$; $P(A') = .10$ **c.** .891   **3.59a.** .729   **b.** .001   **c.** .027
d. .9   **3.61a.** $\frac{2}{9}$ **b.** $\frac{1}{36}$ **c.** $\frac{11}{36}$ **3.63a.** $\frac{27}{64}$ **b.** $\frac{1}{64}$ **c.** $\frac{27}{64}$ **3.65b.** $(.001)^3$ **c.** .997
3.67a. ABC, ABD, ABE, ABF, ACD, ACE, ACF, ADE, ADF, AEF, BCD, BCE, BCF, BDE, BDF, BEF, CDE, CDF, CEF, DEF; 20   **b.** $\frac{1}{20}$
3.73a. 20   **b.** 20   **c.** 12   **d.** 4,950   **e.** 1   **f.** 1   **g.** 60   **h.** 1   **3.75a.** 6   **b.** 36   **c.** 1,296   **d.** 6^n **3.77** 6   **3.79** 2,598,960   **3.81a.** 3,003
b. .0186   **c.** .5734   **3.83a.** 24   **b.** 12   **3.85a.** 17,576,000   **b.** 35,152,000   **3.87a.** 56   **b.** 1,680   **c.** 6,720   **3.89a.** $\frac{21}{252}$ **b.** $\frac{21}{252}$ **c.** $\frac{105}{252}$
3.91a. (1, H), (1, T), (2, 1), (2, 2), (2, 3), (2, 4), (2, 5), (2, 6), (3, H), (3, T), (4, 1), (4, 2), (4, 3), (4, 4), (4, 5), (4, 6), (5, H), (5, T), (6, 1),
(6, 2), (6, 3), (6, 4), (6, 5), (6, 6)   **b.** Probability of a simple event with odd number first is $\frac{1}{12}$; otherwise $\frac{1}{36}$.   **c.** $\frac{1}{4}$; $\frac{1}{2}$
d. A' contains all simple events except (1, H), (3, H), (5, H);
B': {(2, 1), (2, 2), (2, 3), (2, 4), (2, 5), (2, 6), (4, 1), (4, 2), (4, 3), (4, 4), (4, 5), (4, 6), (6, 1), (6, 2), (6, 3), (6, 4), (6, 5), (6, 6)};
$A \cap B$: {(1, H), (3, H), (5, H)}; $A \cup B$: {(1, H), (1, T), (3, H), (3, T), (5, H), (5, T)}   **e.** $\frac{3}{4}$; $\frac{1}{2}$; $\frac{1}{4}$; $\frac{1}{2}$; $\frac{1}{2}$; 1   **f.** No; no
3.93a. Not mutually exclusive   **b.** Mutually exclusive   **c.** Not mutually exclusive   **3.95a.** No; $P(A \cap B) = .03$ **b.** .3; .1   **c.** .37
3.97a. 720   **b.** 10   **c.** 10   **d.** 30   **e.** 20   **f.** 1   **g.** 5,040   **h.** 2,450   **3.99a.** .64; .32; .04   **b.** .72; .22; .06   **c.** Dependent
3.103b. .95   **c.** .25   **d.** .5   **3.105a.** $\frac{1}{32}$ **b.** $\frac{5}{32}$ **3.107a.** .810; .725   **b.** .660; .124   **c.** .216; .784   **3.109** .79   **3.111a.** $\frac{48}{1,326} \approx .036$
b. $\left(\frac{48}{1,326}\right)\left(\frac{1,192}{1,225}\right) \approx .035$ **3.113a.** .9639   **b.** .3078   **c.** .0361   **d.** Three subsystems guarantee the system will work 99.31% of the time.
3.115a. $\frac{1}{5}$ **b.** $\frac{5}{29}$ **c.** .634   **3.117a.** 184,756   **b.** $\frac{1}{184,756}$ **c.** $\frac{1}{184,756}$ **d.** $\frac{101}{184,756}$ **3.119a.** 142,506   **b.** $\frac{33,649}{142,506} \approx .236$ **c.** $\frac{95,634}{142,506} \approx .671$
3.121 4.4739×10^{-28}

Chapter 4

4.3a. Discrete   **b.** Continuous   **c.** Continuous   **d.** Discrete   **e.** Continuous   **f.** Continuous   **4.5a.** Continuous   **b.** Discrete   **c.** Discrete
d. Discrete   **e.** Continuous   **f.** Continuous   **4.7a.** .30   **b.** .35   **c.** .85   **4.9a.** .8   **b.** .2   **c.** 1   **d.** .1   **e.** .5

4.11a.

Simple event	HHH	HHT	HTH	THH	HTT	THT	TTH	TTT
x	3	2	2	2	1	1	1	0

b.   **d.** $\frac{1}{2}$

x	0	1	2	3
$p(x)$	$\frac{1}{8}$	$\frac{3}{8}$	$\frac{3}{8}$	$\frac{1}{8}$

4.13a.   **c.** $\frac{37}{64}$

x	0	1	2	3
$p(x)$	$\frac{27}{64}$	$\frac{27}{64}$	$\frac{9}{64}$	$\frac{1}{64}$

4.15a. .118   **b.** .302   **c.** .580   **4.17a.**   **c.** .0272

x	0	1	2	3	4
$p(x)$	.0016	.0256	.1536	.4096	.4096

4.19a. 31; 169; 13   **c.** .95   **4.21a.** 2.7   **b.** 6.61   **c.** 2.571   **e.** No   **f.** Yes   **4.23a.** 0; 2.94; 1.715   **c.** .96   **4.25** 1.29

4.27a.   **b.** $90,000

x	$0	$300,000
$p(x)$	.7	.3

4.29 $11,500; yes   **4.31** $.25   **4.33** $882   **4.35a.** Discrete   **b.** Binomial   **d.** 2.4; 1.2

4.37b.

x	0	1	2	3	4
$p(x)$	.4096	.4096	.1536	.0256	.0016

4.39a. .234   **b.** .151   **c.** .263   **d.** 0   **e.** .722   **f.** .285   **4.41a.** .005   **b.** .973   **c.** .403   **d.** .966   **e.** .009   **f.** .207   **4.43a.** No   **b.** Yes
c. No   **d.** For part b: $\mu = 4$; $\sigma^2 = 3.2$; $\sigma = 1.789$ **4.45a.** .001   **b.** $\mu = 3,000$; $\sigma = 45.826$; $\approx .5$ **4.47** .748   **4.49a.** .009   **c.** Yes
4.51 .098   **4.53a.** 0   **b.** .151   **4.55** 15 correct   **4.57a.** 0; .000; .000; .098; .873; 1   **c.** 0; .000; .000; .234; .966; 1   **4.59a.** .920   **b.** .677
c. .423   **d.** Decreases; yes   **4.61b.** $\mu = 2$; $\sigma = 1.414$ **c.** .947   **4.63** Binomial: $p(0) = .277$; $p(1) = .365$; $p(2) = .231$;
Poisson: $p(0) = .287$; $p(1) = .358$; $p(2) = .224$ **4.65a.** 2   **b.** No; probability = .003   **4.67** .040   **4.69a.** .00026   **b.** 8.25; 2.872
c. $z = 4.09$; very small probability   **4.71a.** .080   **b.** No   **4.73a.** Poisson   **b.** Binomial   **c.** Binomial   **4.75a.** .243   **b.** .131   **c.** .36
d. .157   **e.** .128   **f.** .121   **4.77a.** 16.2; 18.76; 4.33   **b.** .4   **c.** (7.54, 24.86)   **d.** 1   **4.79a.** .180   **b.** .015   **c.** .076   **4.81b.** 4; 1.789
c. .956   **d.** Yes; $P(x = 0) = .012$ **4.83a.** .027   **b.** .376   **c.** No   **4.85a.** 1.25; 1.09; no   **b.** .007   **4.87a.** .4019   **b.** .1608   **d.** $50,000
4.89a. A: 4.60; B: 3.70   **b.** A: $46,000; B: $55,500   **c.** A: $\sigma^2 = 1.34$, $\sigma = 1.16$; B: $\sigma^2 = 1.21$, $\sigma = 1.1$ **d.** A: .95; B: .95
4.91a. 5; 4   **b.** .617   **c.** .006   **4.93a.** .006   **4.95** 4.664%   **4.97a.** .230   **b.** .143   **c.** .100   **4.99a.** .0314   **b.** .1066

Chapter 5

5.1a. $f(x) = \frac{1}{25}$ for $20 \leq x \leq 45$ **b.** 32.5; 7.217 **c.** (18.066, 46.934) **5.3a.** $f(x) = \frac{1}{2}$ for $2 \leq x \leq 4$ **b.** 3; .577 **c.** .577 **d.** .61 **e.** .65
f. 0 **5.5a.** 0 **b.** 1 **c.** 1 **5.7** $\mu = 2$; $\sigma = 1.155$; .3125 **5.9b.** .5; .2; .2; 0 **5.11a.** .5 **b.** 1 **c.** .25 **d.** .25 **e.** 0 **f.** .75 **5.13a.** .667
b. .333 **c.** 82.5°F **5.15a.** .4772 **b.** .3413 **c.** .4987 **d.** .2190 **e.** .4772 **f.** .3413 **g.** .4545 **h.** .2190 **5.19a.** .1151 **b.** .0808
c. .2092 **d.** .2459 **e.** .5000 **f.** .8544 **5.21a.** .6826 **b.** .9500 **c.** .9000 **d.** .9544 **5.23a.** 1.645 **b.** 1.96 **c.** −1.96 **d.** 1.28
e. 1.28 **5.25a.** 0 **b.** 2.00 **c.** 2.06 **d.** −1.75 **5.27a.** −2.03 **b.** 1.35 **c.** .74 **d.** 1.41 **e.** .51 **f.** 1.80 **5.29a.** 0 **b.** 1 **c.** 2.5
d. −3 **e.** .5 **f.** 1.4 **5.31a.** .0456; .0026 **b.** .6826; .9544 **c.** 1,008.4; 987.2 **5.33a.** .6915 **b.** .1736 **c.** .6476 **d.** .1054 **e.** .9756
f. .0244 **5.35a.** .9544 **b.** .0228 **c.** .1587 **d.** .8185 **e.** .1498 **f.** .9974 **5.37a.** .2514 **b.** 90.375 **5.39a.** 2.28%
b. A few extremely high numbers **5.41a.** .0150 **b.** (5.1, 7.5) **5.43a.** .1151 **b.** .6554 **c.** \$10.50 **5.45** 5.068 **5.47** 473 **5.49a.** 7,667
b. 45.62% **5.53** b, d, e, and f **5.55a.** .500; .5 **b.** .212; .2119 **c.** .831; .8301 **5.57a.** ≈.4880 **b.** ≈.2334 **c.** ≈0 **5.59a.** ≈.0559
c. No; $\mu \pm 3\sigma$ lies outside 0 to n **5.61a.** ≈.0011 **b.** Yes; $z = 27.74$ **c.** No **5.63a.** 6,750; 3,712.5 **b.** ≈.0068 **c.** No; $P(x \geq 7,000) \approx 0$
5.65a. .0125 **b.** Yes **5.67a.** ≈.9808 **b.** ≈0 **5.69a.** ≈.4681 **b.** ≈.0436 **c.** ≈.9822 **5.73a.** .367879 **b.** .997521 **c.** .049787
d. .999955 **5.75** .5; .5; .950213 **5.77a.** 20 **b.** .223130 **c.** .049787; .950213 **5.79a.** 13.80 **b.** .82 **5.81a.** .135335 **b.** .393469
5.83 .223130 **5.87a.** .4750 **b.** .9500 **c.** .8064 **d.** .1261 **e.** .2074 **f.** .8980 **5.89a.** .85 **b.** 1.29 **c.** 0 **d.** .88 **e.** −1.04 **f.** 2.54
5.91a. 60 **b.** 78.96 **c.** 37.84 **d.** 76 **e.** 49.76 **f.** 53.28 **5.93a.** .393469 **b.** .606531 **c.** 0 **d.** .950213 **e.** .361141 **5.95a.** .8790
b. .7967 **c.** About 163 **5.97a.** .0456 **b.** .9082 **c** .0023 **5.99** ≈.982 **5.101a.** .8861 **b.** .0301 **5.103a.** .1469 **b.** .0216
5.105a. ≈.0901 **b.** ≈.0384 **c.** ≈.9573 **5.107a.** .0548 **b.** .6006 **c.** .3446 **d.** \$6,503.80 **5.109a.** ≈1.0 **b.** ≈.8830 **5.111a.** .606531
b. .632121 **5.113** $P(x \geq 400$ when $p = .2) \approx 0$ **5.115a.** ≈.4880 **b.** ≈.1093 **c.** ≈.8926 **5.117a.** ≈.1922 **b.** ≈.4681 **c.** ≈.1026

Chapter 6

6.1a.–b.

Samples	(0, 0)	(0, 2) (2, 0)	(0, 4) (4, 0)	(0, 6) (6, 0)	(2, 2)	(2, 4) (4, 2)	(2, 6) (6, 2)	(4, 4)	(4, 6) (6, 4)	(6, 6)
$\bar{x}$	0	1	2	3	2	3	4	4	5	6

c. $\frac{1}{16}$

d.

x	0	1	2	3	4	5	6
$p(\bar{x})$	$\frac{1}{16}$	$\frac{2}{16}$	$\frac{3}{16}$	$\frac{4}{16}$	$\frac{3}{16}$	$\frac{2}{16}$	$\frac{1}{16}$

6.3a

$\bar{x}$	1	1.5	2	2.5	3	3.5	4	4.5	5
$p(\bar{x})$	.04	.12	.17	.20	.20	.14	.08	.04	.01

c. .05 **d.** No

6.5a. Identical to sampling distribution of $\bar{x}$. **6.7a.**

m	1	2	3
$p(m)$	.648	.324	.028

6.9

m	1	1.5	2	2.5	3
$p(m)$	.4752	.3240	.1701	.0270	.0037

6.13a. $\mu = 5$ **b.**

$\bar{x}$	2	$\frac{8}{3}$	$\frac{10}{3}$	4	$\frac{13}{3}$	5	$\frac{17}{3}$	$\frac{20}{3}$	$\frac{22}{3}$	9
$p(\bar{x})$	$\frac{1}{27}$	$\frac{3}{27}$	$\frac{3}{27}$	$\frac{1}{27}$	$\frac{3}{27}$	$\frac{6}{27}$	$\frac{3}{27}$	$\frac{3}{27}$	$\frac{3}{27}$	$\frac{1}{27}$

$E(\bar{x}) = \frac{135}{28} = 5 = \mu$

c.

m	2	4	9
$p(m)$	$\frac{7}{27}$	$\frac{13}{27}$	$\frac{7}{27}$

$E(m) = \frac{129}{27} = 4.778 < \mu = 5$ **d.** Sample mean, $\bar{x}$; it is unbiased

6.17a.

s^2	0	.333	1	1.333
$p(s^2)$	.244	.522	.108	.126

b. .45 **c.** $E(s^2) = .45$

d.

s	0	.577	1	1.155
$p(s)$	.244	.522	.108	.126

e. $E(s) = .555$

6.19a. 50; 3.536   **b.** 50; 1.414   **c.** 50; .707   **d.** 50; 1   **e.** 50; .316   **f.** 50; .224

6.21a. 3.1; 5.69; 2.385   **b.**

$\bar{x}$	1	1.5	2	2.5	3	5.5	6	6.5	10
$p(\bar{x})$	.01	.08	.24	.32	.16	.02	.08	.08	.01

c. $\mu_{\bar{x}} = 3.1 = \mu$; $\sigma_{\bar{x}} = 1.687 = \dfrac{\sigma}{\sqrt{2}}$ **6.23a.** $\mu_{\bar{x}} = 15$; $\sigma_{\bar{x}} = \dfrac{\sigma}{\sqrt{n}} = .5$ **b.** Approximately normal; yes   **c.** 1   **d.** -2

6.25a. .8944   **b.** .0228   **c.** .1292   **d.** .9699   **6.29a.** Approximately normal; no   **b.** .0336   **c.** No   **6.31a.** $\mu_{\bar{x}} = 1.3$; $\sigma_{\bar{x}} = .24$
b. Yes; large random sample   **c.** .1056   **d.** .0062   **6.33a.** .0031; claim is invalid   **b.** More likely; less likely   **c.** Less likely; more likely
6.35a. Decreases   **c.** $\bar{x}$ **d.** 1.25; 2.5   **6.37a.** $\mu_{\bar{x}} = 2.5$; $\sigma_{\bar{x}} = .1904$ **b.** Approximately normal; yes   **6.39a.** $\mu_{\bar{x}} = 120$; $\sigma_{\bar{x}} = 2.338$
b. Approximately normal; no   **c.** .1949   **d.** .8835   **e.** .9756   **f.** .1251

6.41a.

$\bar{x}$	1	1.5	2	2.5	3	3.5	4	4.5	5	5.5	6
$p(\bar{x})$	$\frac{1}{36}$	$\frac{2}{36}$	$\frac{3}{36}$	$\frac{4}{36}$	$\frac{5}{36}$	$\frac{6}{36}$	$\frac{5}{36}$	$\frac{4}{36}$	$\frac{3}{36}$	$\frac{2}{36}$	$\frac{1}{36}$

b. $\frac{33}{36}$

6.43c.

$\bar{x}$	0	$\frac{1}{2}$	1
$p(\bar{x})$	$\frac{1}{4}$	$\frac{1}{2}$	$\frac{1}{4}$

6.47

n	1	5	10	20	30	40	50
$\dfrac{\sigma}{\sqrt{n}}$	10	4.472	3.162	2.236	1.826	1.581	1.414

6.49a. .9887   **b.** Normal distribution for $\bar{x}$ **6.51a.** .2327   **b.** .2215   **6.53a.** All single-family homes in 1987 in a particular county   **b.** .500
c. .8904   **6.55a.** .3830   **b.** It would decrease.   **6.57a.** 45 samples: $\{(1, 2), (1, 3), \ldots, (9, 10)\}$
b. 21 samples: $\{(1, 4), (1, 5), \ldots, (8, 10)\}$; $\frac{21}{45}$ **c.** $p(0) = \frac{3}{45}$; $p(1) = \frac{21}{45}$; $p(2) = \frac{21}{45}$ **6.59a.** Approximately normal   **b.** .0069   **c.** No
6.61a. .0082   **b.** None   **6.63** .9332   **6.65a.** Approximately normal   **b.** $\mu_{\bar{x}} = \theta$; $\sigma_{\bar{x}} = \dfrac{\theta}{\sqrt{n}}$

Chapter 7

7.1a. 1.645   **b.** 2.575   **c.** 1.96   **d.** 1.28   **7.3** (5.969, 9.656)   **7.5a.** (75.96, 77.84)   **c.** (75.66, 78.14)   **d.** Increases
e. Yes; central limit theorem guarantees it   **7.9** Yes, for a large random sample   **7.11** (109.23, 116.77)   **7.13** (8.9, 10.3)   **7.15** (.67, .77)
7.17 (18.575, 21.025)   **7.19a.** (21.266, 25.594)   **c.** The population of $\bar{x}$'s is approximately normal.   **7.21** 2,344   **7.23a.** $n = 68$ **b.** $n = 31$
7.25a. $n = 1,083$ **b.** Wider   **c.** 38.3%   **7.27** $n = 531$ **7.29** $n = 139$ **7.33a.** 2.306   **b.** 2.764   **c.** -2.898 **d.** -1.761
7.35a. (3.075, 6.591)   **b.** (2.590, 7.076)   **c.** (1.315, 8.351)   **d.** (4.102, 5.564); (3.951, 5.715); (3.638, 6.028); decreases
7.37a. (51.13, 54.07)   **7.39a.** (16.575, 31.025)   **7.41a.** (20.16, 56.70)   **7.43a.** Must be approximately normal
b. ($57,709.03, $77,588.13)   **7.47a.** Yes; $\hat{p} \pm 3\sigma_{\hat{p}}$ lies in (0, 1)   **b.** (.372, .468)   **7.49a.** Yes   **b.** No   **c.** Yes   **d.** No
7.51a. (.537, .709)   **7.53a.** Small business owners and managers   **b.** Yes, $\hat{p} \pm 3\sigma_{\hat{p}}$ lies in (0, 1)   **c.** (.34, .46)
d. Narrower, $\sigma_{\hat{p}}$ is smaller for $p = .30$ than for $p = .40$
7.55a. All sample sizes are large enough   **b.** KOCO: (.698, .848); WCBS: (.438, .762); WXIA: (.373, .519)   **c.** (.535, .625)

7.57 $1.96\sqrt{\dfrac{\hat{p}\hat{q}}{n}} = .04$ **7.59a.** $n = 1,083$ **b.** $n = 1,692$ **7.61** $n = 34$ **7.63** No; n needed to have been 984

7.65 $n = 32,270$ **7.67** $n = 322$ **7.69a.** $t = 2.074$ **b.** $z = 1.96$ **c.** $z = 1.96$ **d.** $z = 1.96$ **e.** Neither z nor t
7.73a. (10.555, 13.845)   **b.** $n = 166$ **7.75a.** (.038, .112)   **b.** No   **7.77a.** (47.865, 60.135)   **b.** Maximum depth of snowfall
7.79 (.030, .220)   **7.81** (24.86, 27.94)   **7.83** (.562, .664)   **7.85a.** 42,250   **b.** (41,743.43, 42,756.57)   **7.87** $n = 818$
7.89a. $n = 853$

Chapter 8

8.1a. Null hypothesis; alternative hypothesis   **8.7** No   **8.9a.** H_0: Drug unsafe vs H_a: Drug safe   **c.** α **8.11a.** Reject H_0 if $z > 1.50$
b. $\alpha = .0668$ **8.13a.** $z = -1.57$; reject H_0 **b.** Do not reject H_0 **8.15** $z = 1.80$; reject H_0 **8.17a.** .1660   **b.** .0853
c. $z = 2.36$; reject H_0; yes   **8.19** $z = 2.00$; reject H_0; yes   **8.21** $z = 2.44$; reject H_0; yes   **8.23a.** No   **b.** Yes   **c.** Yes   **d.** No   **e.** No
8.25 α just greater than .06   **8.27** .0376   **8.29** .0643   **8.31a.** $z = -1.29$; do not reject H_0; no   **b.** .0985   **c.** Small   **8.33** p-value = .0359
8.35 p-value = .1922   **8.37** p-value = .0228   **8.43a.** $t = -1.63$; do not reject H_0 **b.** $t = -1.63$; do not reject H_0
8.45a. $t = -2.01$; reject H_0; yes   **b.** Type II error   **c.** Type I error   **d.** (19.73, 19.99)   **8.47a.** $t = .87$; do not reject H_0; no
b. p-value $> .10$ **8.49a.** $t = -3.63$; reject H_0; yes   **b.** $.001 < p$-value $< .002$ **8.51b.** $t = -1.39$; do not reject H_0 **c.** p-value = .0823
8.53a. $z = 1.89$; do not reject H_0 **b.** $z = 1.89$; reject H_0 **c.** $z = -5.33$; reject H_0 **d.** (.65, .83)   **e.** (.63, .85)
8.55a. Yes; $p_0 \pm 3\sigma_{\hat{p}}$ is in (0, 1)   **b.** $z = 2.31$; reject H_0; yes   **c.** p-value = .0104   **8.57b.** $z = 1.72$; reject H_0; yes   **c.** (.1144, .1274)

8.59a. $z = 1.25$; do not reject H_0; no **b.** p-value $= .1056$ **8.63a.** $.3594$ **b.** $.6406$ **8.65** $.3156$; smaller
8.67a. $.8106$; $.5319$; $.2358$; $.0643$; $.0102$ **d.** $.1894$; $.4681$; $.7642$; $.9357$; $.9898$ **8.69a.** $.1251$; $.2578$; $.4404$; $.6368$; $.8023$ **c.** $.5438$
d. Power ≈ 1; ≈ 0 **8.71a.** $.1963$; Type II error **b.** $.05$; Type I error **c.** $.8037$ **8.73a.** Reject H_0 if $\chi^2 > 30.1910$ or $\chi^2 < 7.56418$
b. Reject H_0 if $\chi^2 > 36.1908$ **c.** Reject H_0 if $\chi^2 > 21.0642$ **d.** Reject H_0 if $\chi^2 < 3.05347$ **e.** Reject H_0 if $\chi^2 > 14.0671$ or $\chi^2 < 2.16735$
f. Reject H_0 if $\chi^2 < 13.8484$ **8.75a.** Sampled population normal **b.** $\chi^2 = 19.36$; reject H_0 **c.** $\chi^2 = 19.36$; reject H_0 **d.** $(2.04, 27.24)$
8.77a. $\chi^2 = 1.64$; do not reject H_0 **b.** $(3.46, 46.15)$ **8.79** $\chi^2 = 22.54$; do not reject H_0; no **8.81** $\chi^2 = 4.5$; do not reject H_0; no
8.83 $\chi^2 = 133.90$; reject H_0 for $\alpha = .05$; yes **8.89** Small **8.91a.** $(70.90, 74.30)$ **b.** $t = -7.51$; reject H_0 **c.** $t = -7.51$; reject H_0
d. $(69.78, 75.42)$ **e.** 75 **8.93a.** $(8.08, 8.32)$ **b.** $z = -1.67$; do not reject H_0 **c.** $z = -3.35$; reject H_0
8.95a. $z = -2.99$; reject H_0 **b.** p-value $= .0028$ **8.97a.** $t = 1.56$; reject H_0; yes **b.** $.05 < p$-value $< .10$ **8.99** $z = 3.79$; reject H_0; yes
8.101a. $z = 1.41$; do not reject H_0; no **8.103a.** No **b.** $\beta = .5910$; power $= .4090$ **c.** Increase **8.105** p-value $= .0793$
8.107 $z = 1.07$; do not reject H_0; no **8.109** $n = 48$ **8.111** $n = 119$ **8.113b.** $.0152$ **8.115a.** $z = -3.03$; reject H_0; yes
8.117 $(.0028, .0105)$ **8.119** $\chi^2 = 23.66$; do not reject H_0; no

Chapter 9

9.1a. $(144, 156)$ **b.** $(142, 158)$ **c.** Mean $= 0$; standard deviation $= 5$ **d.** $(-10, 10)$ **9.3a.** $z = -.8$ **b.** $.7881$ **c.** $.2119$ **d.** $.9452$
9.5 $(17.49, 22.51)$ **9.7** Yes, for large, random, and independent samples **9.9b.** $z = -3.76$; reject H_0; yes **c.** $(-4.89, -.91)$ **d.** Narrower
9.11 $(-5.36, 12.16)$ **9.13a.** $z = -4.22$; reject H_0; yes **b.** p-value ≈ 0 **c.** $(-1.53, -.67)$ **9.15a.** H_0: $\mu_1 - \mu_2 \leq 5$ vs H_a: $\mu_1 - \mu_2 > 5$
b. $z = 1.70$; reject H_0; yes **d.** p-value $= .0446$ **9.17a.** $z = 5.68$; reject H_0 **b.** p-value ≈ 0 **9.19a.** $z = 10.63$; reject H_0 **b.** p-value ≈ 0
9.25a. $.5989$ **b.** $t = -2.39$; reject H_0; yes **c.** $(-2.22, -.26)$ **d.** Confidence interval **9.27a.** $t = .013$; do not reject H_0; no
b. $(5.208, 14.842)$ **9.29a.** $t = 3.08$; reject H_0; yes **b.** $(4.45, 15.55)$ **9.31a.** $t = 2.76$; reject H_0 **b.** $.005 < p$-value $< .01$
9.33 $t = -1.57$; do not reject H_0; no **9.35b.** $t = 1.53$; do not reject H_0; no **c.** $.05 < p$-value $< .10$ **9.39a.** -7.6; 4.56
c. $t = -3.73$; reject H_0; yes **d.** $.02 < p$-value $< .05$ **9.41a.** $t = -5.29$; reject H_0; yes **b.** $(-4.98, -2.42)$ **9.43a.** $t = .81$; do not reject H_0
b. p-value $> .20$ **9.45** $(3.04, 6.96)$ **9.47** $(3.60, 4.86)$ **9.49a.** $(17.2, 35.6)$ **b.** $(25.1, 27.7)$ **9.51a.** $t = 7.68$; reject H_0; yes **c.** $(.28, .56)$
9.53a. Not independent random samples **b.** $t = 1.18$; do not reject H_0; no **9.57a.** $z < -2.33$ **b.** $z < -1.96$ **c.** $z < -1.645$
d. $z < -1.28$ **9.59a.** $(.003, .137)$ **b.** $(-.026, .146)$ **c.** $(-.281, -.019)$ **9.61** $z = 1.16$; do not reject H_0 **9.63a.** $z = 4$; reject H_0; yes
b. $(.12, .28)$ **9.65** $(-.621, -.449)$ **9.67a.** $z = -.63$; do not reject H_0; no **b.** $(-.206, -.134)$ **9.69a.** $z = -6.60$; reject H_0; yes
b. p-value ≈ 0 **c.** $(-.359, -.201)$ **9.71a.** $z = .79$; do not reject H_0 **b.** p-value $= .2148$ **9.73** $(-.236, -.154)$ **9.75a.** ≈ 0
b. $z = -6.36$; p-value ≈ 0; reject H_0 **9.77** $n_1 = n_2 = 1,504$ **9.79a.** $n_1 = n_2 = 27,186$ **b.** $n_1 = n_2 = 2,165$ **c.** $n_1 = n_2 = 325$
9.81a. $n_1 = n_2 = 911$ **b.** $n_1 = n_2 = 1,692$; no **9.83** $n_1 = n_2 = 4,802$ **9.85** $n_1 = 520$; $n_2 = 260$ **9.87** $n_1 = 35$
9.89a. 4.10 **b.** 3.57 **c.** 8.795 **d.** 3.21 **9.91a.** $F > 1.84$ **b.** $F > 2.20$ **c.** $F > 2.57$ **d.** $F > 3.09$
9.93a. $F > 2.16$ (if $s_1^2 > s_2^2$) **b.** $F > 2.94$ (if $s_2^2 > s_1^2$) **c.** $F > 2.11$ (if $s_2^2 > s_1^2$); $F > 2.29$ (if $s_1^2 > s_2^2$) **d.** $F > 2.30$ (if $s_2^2 > s_1^2$)
e. $F > 2.78$ (if $s_1^2 > s_2^2$); $F > 3.95$ ($s_2^2 > s_1^2$) **9.95a.** $F = 2.88$; reject H_0 **b.** $(.142, .939)$ **c.** $F = 2.88$; reject H_0
9.97a. $F = 2.26$; do not reject H_0 **b.** p-value $\approx .10$ **9.99** $F = 4.07$; reject H_0 (choose A) **9.101** $F = 2.63$; reject H_0 (choose line 1)
9.103a. $(.01, .96)$ **b.** Instrument 1 is more precise **9.105a.** $(.09, .69)$ **b.** $F = 3.84$; reject H_0 **c.** No; conclude $\sigma_1^2 \neq \sigma_2^2$
9.107a. $z = -2.04$; reject H_0 **b.** $(-.196, -.004)$ **c.** $n_1 = n_2 = 72,991$ **9.111a.** $t = 1.064$; do not reject H_0 **9.113** $t = -4.02$; reject H_0
9.115 $z = 1.67$; reject H_0; yes **9.117a.** $t = .13$; do not reject H_0; no **9.119a.** $(-.14, -.04)$ **b.** $(.27, .35)$
9.121 $t = -.70$; do not reject H_0; no **9.123a.** $t = 2.27$; reject H_0; yes **9.125** $t = 6.14$; reject H_0; yes **9.127a.** $(-238.48, -13.52)$
9.129a. $(.651, .729)$ **b.** $(-.068, .048)$ **9.131** $F = 2.79$; do not reject H_0; no **9.133** $(-.127, .051)$
9.135a. $t = -1.14$; do not reject H_0; no **9.137** $t = -2.0$; reject H_0; yes **9.139** $n_1 = n_2 = 193$

Chapter 10

10.1 The treatments are the 4 levels (A, B, C, D) of the qualitative factor. **10.3a.** Students **b.** States **c.** Automobiles
d. Computer diskettes **10.5a.** Cholesterol levels **b.** Socioeconomic class; qualitative **c.** The 4 levels of socioeconomic class **d.** People
10.7a. Quality **b.** Temperature, pressure; quantitative **c.** The 15 levels of temperature and pressure **d.** Ingots **10.9a.** 19.00 **b.** 99.00
c. 1.61 **d.** 3.87 **10.11** The second diagram
10.13

	t-Test	F-Test
a.	2; 14.4	2; 14.4
b.	-6.12; -2.28	37.5; 5.21
c.	$t < -2.228$ or $t > 2.228$	$F > 4.96$
d.	Reject; reject	Reject; reject

10.15a. $F = 1.5$; do not reject H_0 **b.** $F = 6.0$; reject H_0 **c.** $F = 24$; reject H_0 **d.** Increases **10.17a.** 4; 38 **b.** $F = 14.80$; reject H_0
c. Individual sample means

10.19a.

Source	df	SS	MS	F
Treatments	2	12.301	6.1505	2.931
Error	9	18.888	2.0987	
Total	11	31.189		

b. $F = 2.931$; do not reject H_0 **c.** No means differ at $\alpha = .05$

10.21a.

Source	df	SS	MS	F
Treatments	2	509.87	254.94	1.87
Error	90	12,259.96	136.22	
Total	92	12,769.83		

b. $F = 1.87$; do not reject H_0; no   **d.** No   **e.** Designed

10.23a. Mice; test scores; training; qualitative; the 3 levels of training   **b.** Designed   **c.** $F = 4.55$; reject H_0
d. 2; condition 1 > control; condition 2 > control   **10.25b.** 571.9661   **c.** 7,719.8288

d.

Source	df	SS	MS	F
Treatments	2	571.97	285.985	3.96
Error	107	7,719.83	72.15	
Total	109	8,291.80		

e. $F = 3.96$; reject H_0; yes
f. The population means of flextime hours and fixed hours differ.

10.27a.

Source	df	SS	MS	F
Treatments	4	292.80	73.2	5.29
Error	15	207.75	13.85	
Total	19	500.55		

b. $F = 5.29$; reject H_0; yes   **c.** p-value < .01

10.29a. 3; 5   **b.** 15   **c.** $H_0: \mu_1 = \mu_2 = \mu_3 = \mu_4 = \mu_5$ vs H_a: At least 2 treatment means differ   **d.** $F = 9.109$ **e.** Reject H_0 if $F > 7.01$
f. Reject H_0 at $\alpha = .01$

10.31a.

Source	df	SS	MS	F
Treatments	2	12.032	6.016	50.983
Blocks	3	71.749	23.916	202.68
Error	6	.708	.118	
Total	11	84.489		

b. $F = 50.983$; reject H_0; yes   **c.** $F = 202.68$; reject H_0; yes
d. All 3 treatment means differ

10.33a. $F = 3.20$, reject H_0; $F = 1.80$, do not reject H_0 **b.** $F = 13.33$, reject H_0; $F = 2.00$, do not reject H_0
c. $F = 5.33$, reject H_0; $F = 5.00$, reject H_0 **d.** $F = 16.0$, reject H_0; $F = 6.00$, reject H_0
e. $F = 2.67$, do not reject H_0; $F = 1.00$, do not reject H_0 **10.35a.** Randomized block design   **b.** $F = 8.452$; reject H_0; yes
c. p-value $= .014$ **d.** Treatment 1 differs from all others   **10.37a.** $F = 16.86$; reject H_0; yes   **c.** $t = 4.105$; reject H_0 **e.** p-value < .01

10.39a. $F = .338$; do not reject H_0; no   **b.** p-value > .10   **c.**

Source	df	SS	MS	F
Treatments	2	.941	.4705	.338
Blocks	5	10,239.969	2,047.9938	1,471.37
Error	10	13.919	1.3919	
Total	17	10,254.829		

d. No

10.41a. Randomized block design   **b.** No   **c.**

Source	df	SS	MS	F
Treatments	3	646.68	215.56	.627
Error	20	6,874.50	343.725	
Total	23	7,521.18		

$F = .627$; do not reject H_0; no

10.43a.

Source	df	SS	MS	F
A	2	.8	.4000	3.69
B	3	5.3	1.7667	16.31
AB	6	9.6	1.6000	14.77
Error	12	1.3	.1083	
Total	23	17.0		

b. $F = 13.18$; reject H_0; yes   **c.** Yes   **e.** $F = 14.77$; reject H_0 **f.** No

10.45b. $F = 21.61$; reject H_0 **c.** $F = 36.60$; reject H_0 **d.** No
f. Half-width: 2.583; pairs of treatment means that differ: A_1B_1 and A_1B_3, A_1B_3 and A_2B_2, A_1B_3 and A_2B_3, A_2B_1 and A_2B_2, A_2B_1 and A_2B_3

10.47a.

Source	df	SS	MS	F
A	2	200	100	3.00
B	2	100	50	1.50
AB	4	100	25	.75
Error	18.	600	33.333	
Total	26	1,000		

$F(\text{treatment}) = 1.50$; do not reject H_0 ($\alpha = .05$)

b.

Source	df	SS	MS	F
A	2	100	50	3.00
B	2	100	50	3.00
AB	4	500	125	7.50
Error	18	300	16.667	
Total	26	1,000		

$F(\text{treatment}) = 5.25$, reject H_0; $F(AB) = 7.50$, reject H_0

c.

Source	df	SS	MS	F
A	2	400	200	12.00
B	2	100	50	3.00
AB	4	200	50	3.00
Error	18	300	16.667	
Total	26	1,000		

$F(\text{treatment}) = 5.25$, reject H_0; $F(AB) = 3.00$, reject H_0

d.

Source	df	SS	MS	F
A	2	400	200	36.00
B	2	400	200	36.00
AB	4	100	25	4.50
Error	18	100	5.556	
Total	26	1,000		

$F(\text{treatment}) = 20.25$, reject H_0; $F(AB) = 4.50$, reject H_0

10.49b. $F(\text{treatment}) = 11.073$, reject H_0; $F(\text{interaction}) = 6.789$, reject H_0; half-width: 12.44; pairs of treatment means that differ: A_1N and A_2R, A_2R and A_2T

10.51a.

Source	df	SS	MS	F
Sex	1	10,686.361	10,686.361	68.74
Weight	1	538.756	538.756	3.47
Sex × Weight	1	831.744	831.744	5.35
Error	36	5,596.9083	155.4697	
Total	39	17,653.7693		

b. $F(\text{treatment}) = 25.85$, reject H_0; $F(S \times W) = 5.35$, reject H_0; half-width: 18.624

10.53

Source	df	SS	MS	F
Sex	1	496.80	496.80	1.68
Weight	1	2,241.76	2,241.76	7.56
Sex × Weight	1	339.15	339.15	1.14
Error	56	16,597.30	296.38	
Total	59	19,675.01		

$F(\text{treatment}) = 3.46$, reject H_0 at $\alpha = .05$; $F(S \times W) = 1.14$, do not reject H_0; $F(W) = 7.56$, reject H_0; $F(S) = 1.68$, do not reject H_0; normal > overweight

10.55 A_1B_1, A_1B_2, A_2B_1, A_2B_2, A_3B_1, A_3B_2

10.57a.

Source	df	SS	MS	F
Treatments	3	11.332	3.777	157.38
Blocks	4	10.688	2.672	111.33
Error	12	.288	.024	
Total	19	22.308		

b. $F = 157.38$; reject H_0 **c.** Half-width: .299; C > A, B; D > A, B; A > B

d. $F = 111.33$; reject H_0

10.59a.

Source	df	SS	MS	F
A	3	2.6	.8667	1.11
B	5	9.2	1.84	2.36
AB	15	46.5	3.1	3.98
Error	24	18.7	.7792	
Total	47	77.0		

b. 4 for A; 6 for B; 24; 2   **c.** SST = 58.3; F = 3.25; reject H_0 **d.** Yes; F = 3.98; reject H_0

10.61a. Completely randomized design   **b.** F = 6.30; reject H_0 **c.** Yes; half-width: 5.683; pairs that differ: B and D, C and D

10.63a. Completely randomized design   **b.** F = 4.88; reject H_0 **c.** p-value < .01   **d.** Half-width: 2.369; pairs that differ: B and D
e. (10.965, 13.015)

10.65b.

Source	df	SS	MS	F
Treatments	3	6.951	2.317	11.03
Error	596	125.217	.2101	
Total	599	132.168		

c. F = 11.03; reject H_0; yes   **d.** Se > F, So, J

10.67b. Designed   **c.**

Source	df	SS	MS	F
Treatments	3	348.675	116.225	12.22
Error	36	342.300	9.508	
Total	39	690.975		

d. F = 12.22; reject H_0 **e.** Half-width: 3.3096; pairs of treatment means that differ: A and B, A and C, A and D, B and D, C and D
f. (9.72, 13.68); no

10.69a.

Source	df	SS	MS	F
Companies	1	3,237.2	3,237.2	19.62
Error	98	16,167.7	164.9765	
Total	99	19,404.9		

b. F = 19.62; reject H_0; yes   **c.** No; need sample means

10.71a.

Source	df	SS	MS	F
A	3	74.333	24.778	16.52
B	2	4.083	2.041	1.36
AB	6	168.917	28.153	18.77
Error	12	18.000	1.500	
Total	23	265.333		

b. F = 14.99; reject H_0 **c.** F = 18.77; reject H_0; yes   **d.** No   **e.** Half-width: 5.288

10.73b.

Source	df	SS	MS	F
Treatments	2	464.533	232.267	25.06
Blocks	4	24.667	6.167	.67
Error	8	74.133	9.267	
Total	14	563.333		

c. F = 25.06; reject H_0 **d.** No

10.75 F = 2.95; do not reject H_0; no   **10.77a.** Randomized block design

b.

Source	df	SS	MS	F
Brands	2	.6317	.31585	.193
Automobiles	3	.8558	.2853	.174
Error	6	9.8417	1.6403	
Total	11	11.3292		

c. F = .193; do not reject H_0; no   **d.** F = .174; do not reject H_0; no

10.79a. Completely randomized design

b.

Source	df	SS	MS	F
Treatments	3	117.642	39.214	7.79
Error	13	65.417	5.032	
Total	16	183.059		

c. $F = 7.79$; reject H_0; yes   **d.** At overall $\alpha = .10$, $3 > 1, 2$

10.81b. $F(\text{treatment}) = 5.17$, reject H_0; $F(\text{interaction}) = 6.90$, reject H_0 ($\alpha = .05$)
10.83 Low level: DL > DH (overall $\alpha = .10$); high Level: Disc. > exp.; L > M, H; M > H

Chapter 11

11.3a. .063   **b.** .500   **c.** .004   **d.** .151; .1515   **e.** .212; .2119   **11.5** $S = 16$; p-value $= .115$; do not reject H_0 **11.7a.** Sign test
b. H_0: $M = 30$ vs H_a: $M < 30$ **c.** $S = 5$; reject H_0 if p-value $< .01$ **d.** $S = 5$; p-value $= .109$; do not reject H_0
11.9a. Reject H_0 if $T_B \leq 32$ or $T_B \geq 58$ **b.** Reject H_0 if $T_A \geq 40$ **c.** Reject H_0 if $T_B \geq 98$ **d.** Reject H_0 if $z < -1.96$ or $z > 1.96$
11.13a. $T_A = 67.5$; reject H_0 **b.** $T_A = 67.5$; reject H_0 **11.15a.** $T_B = 53$; reject H_0 **11.17** $T_B = 29$; do not reject H_0
11.19a. $t = -1.258$; do not reject H_0 ($\alpha = .05$); independent samples from normally distributed populations with equal variances
b. $T_A = 37.5$; do not reject H_0 ($\alpha = .05$)   **11.21** $T_A = 150.5$; reject H_0 **11.23a.** Reject H_0 if $T \leq 101$ ($T =$ smaller of T_+ and T_-)
b. Reject H_0 if $T_- \leq 303$ **c.** Reject H_0 if $T_+ \leq 0$ **11.27a.** Reject H_0 if $T_- \leq 2$ **b.** Reject H_0 if $T \leq 4$ ($T =$ smaller of T_+ and T_-)
11.29a. Reject H_0 if $z > 1.645$ **b.** $z = 2.97$; reject H_0 **c.** p-value $= .0015$ **11.31a.** Randomized block design   **c.** $T_+ = 3.5$; reject H_0
d. $.01 < p$-value $< .025$ **11.33** $T_- = 4.5$; reject H_0; yes   **11.35** $T_- = 3$; do not reject H_0 **11.37** $T_- = 36$; do not reject H_0; no
11.39a. 31.4104   **b.** 27.4884   **c.** 50.8922   **d.** 96.5782   **e.** 5.99147   **f.** 25.1882   **11.41b.** $H = 8.190$; reject H_0 **c.** $.01 < p$-value $< .025$
11.45 $H = 1.565$; do not reject H_0; no   **11.47** $H = 7.154$; reject H_0; yes   **11.49** $H = 2.030$; do not reject H_0; no
11.51a. $F = 1.326$; do not reject H_0; lifetimes for the 3 brands are normally distributed with equal variances   **b.** $H = 1.22$; do not reject H_0
11.53a. 18.3070   **b.** 22.3072   **c.** 79.4900   **d.** 42.9798   **11.55a.** 6   **c.** $F_r = 15.2$; reject H_0 **d.** p-value $< .005$ **11.57** $F_r = 13$; reject H_0
11.59 $F_r = 39.413$; reject H_0; yes   **11.61** $F_r = 6.35$; reject H_0; yes   **11.63** $F_r = 12.3$; reject H_0; yes
11.65b. $F_r = 22.93$; reject H_0 **c.** p-value $< .005$

e.

Comparison	Test Statistics	Conclusion
Food, Tobacco	$T_- = 0$	Reject H_0
Food, Paper	$T_- = 0$	Reject H_0
Food, Textiles	$T_- = 0$	Reject H_0
Food, Chemicals	$T_- = 3$	Do not reject H_0
Tobacco, Paper	$T_+ = 0$	Reject H_0
Tobacco, Textiles	$T_+ = 0$	Reject H_0
Tobacco, Chemicals	$T_+ = 0$	Reject H_0
Paper, Textiles	$T_- = 0$	Reject H_0
Paper, Chemicals	$T_+ = 0$	Reject H_0
Textiles, Chemicals	$T_+ = 0$	Reject H_0

11.67a. Reject H_0 if $r_s < -.683$ or $r_s > .683$ **b.** Reject H_0 if $r_s > .400$ **c.** Reject H_0 if $r_s < -.478$ **11.69a.** H_0: $\rho_s = 0$ vs H_a: $\rho_s \neq 0$
b. $r_s = .898$; reject H_0 **c.** p-value $< .02$ **11.71a.** $r_s = .864$; reject H_0 **c.** p-value $< .005$ **11.73** $r_s = .881$; reject H_0; yes
11.75a. $r_s = .861$ **b.** $r_s = .861$; reject H_0; yes   **11.77** $r_s = -.804$; reject H_0; yes   **11.79a.** $r_s = -.668$ **b.** Reject H_0
11.81 $H = 9.859$; reject H_0 **11.83** $T_B = 55.5$; do not reject H_0; no   **11.85** $T_A = 139$; reject H_0; yes
11.87 $S = 3$; p-value $= .500$; do not reject H_0; $S = 5$; p-value $= .031$; reject H_0 **11.89a.** $T_{\text{after}} = 19$; reject H_0; yes
11.91 $T_- = 4$; reject H_0; yes   **11.93** $T = 6$; do not reject H_0 **11.95a.** $T_+ = 1$; reject H_0; yes   **b.** $t = -2.96$; reject H_0
11.97 $T_+ = 19.5$; do not reject H_0; no   **11.99a.** Randomized block design   **c.** $F_r = 6.4$; reject H_0; yes

d.

Comparison	Test Statistic	Conclusion
A, B	$T_- = 1$	Reject H_0
A, C	$T_- = 0$	Reject H_0
B, C	$T_- = 2$	Do not reject H_0

11.101 $F_r = 1.333$; do not reject H_0; no   **11.103a.** Randomized block design   **b.** $F_r = 11.77$; reject H_0 **c.** Yes

Comparison	Test Statistic	Conclusion
1, 2	$T_+ = 3.5$	Do not reject H_0
1, 3	$T_- = 3$	Do not reject H_0
1, 4	$T_+ = 2.5$	Do not reject H_0
1, 5	$T_+ = 0$	Reject H_0
2, 3	$T_- = 1.5$	Do not reject H_0
2, 4	$T_- = 10$	Do not reject H_0
2, 5	$T_+ = 5$	Do not reject H_0
3, 4	$T_+ = 1$	Reject H_0
3, 5	$T_+ = 0$	Reject H_0
4, 5	$T_+ = 4$	Do not reject H_0

11.105 $S = 5$; p-value $= .363$; do not reject H_0 **11.107** $H = 12.7927$; reject H_0; yes   **b.** $T_A = 28.5$; do not reject H_0
11.109a. $S = 5$; p-value $= .363$; do not reject H_0 **11.111** $r_s = .479$; do not reject H_0 ($\alpha = .05$)   **11.113** $r_s = .857$

Chapter 12

12.1a. 24.9958   **b.** 70.0648   **c.** 18.5494   **d.** 10.5966   **12.3a.** $\chi^2 > 4.60517$ **b.** $\chi^2 > 13.2767$ **c.** $\chi^2 > 7.81473$
12.7a. $\chi^2 = 2.738$; do not reject H_0; no   **12.9** $\chi^2 = 17.15$; reject H_0; yes   **12.11** $(.468, .632)$ **12.13** $\chi^2 = 17.36$; reject H_0; yes
12.15 $\chi^2 = 4.4$; do not reject H_0; no   **12.17** $\chi^2 = 1.5$; do not reject H_0; no   **12.19a.** $\chi^2 = 1,037.833$; reject H_0 **b.** p-value $< .005$
12.21a. H_0: The two classifications are independent vs H_a: The two classifications are dependent   **b.** Reject H_0 if $\chi^2 > 9.21034$

c. Expected cell counts:

		Columns		
		1	2	3
Rows	1	14.4	35.7	44.9
	2	10.6	26.3	33.1

d. $\chi^2 = 7.50$; do not reject H_0

12.23 $\chi^2 = 15.27$; reject H_0 **12.27a.** $\chi^2 = 5.91$; reject H_0; yes   **b.** $.01 < p$-value $< .025$ **12.29a.** $\chi^2 = 19.72$; reject H_0; yes
12.31 $\chi^2 = 9.35$; reject H_0; yes   **12.33a.** $\chi^2 = .621$; do not reject H_0; no   **12.35** p-value $> .10$ **12.37a.** $\chi^2 = 54.14$; reject H_0; yes   **b.** No
c. Yes   **12.39a.** Yes   **d.** $\chi^2 = 2.373$; do not reject H_0 **e.** Yes   **12.41** $\chi^2 = 103.08$; reject H_0; yes

12.43a.

		Committee		
		Accept	Not accept	Totals
Inspector	Accept	101	23	124
	Not accept	10	19	29
	Totals	111	42	153

b. Yes   **c.** $\chi^2 = 26.034$; reject H_0

12.45a. $\chi^2 = 9.16$; reject H_0 **b.** $(.84, .96)$ **12.47** $\chi^2 = 2.60$; do not reject H_0; no   **12.49a.** $\chi^2 = 36.86$; reject H_0; yes

b.

	Under 5,000	5,000– 9,999	10,000– 19,999	20,000 or more
Under 15	35.7%	29.5%	16.5%	13.9%
15–25	60.5%	59.8%	59.2%	52.8%
Over 25	3.9%	10.6%	24.3%	33.3%

12.51a. $\chi^2 = 87.38$; reject H_0; yes   **12.53** $\chi^2 = 6.74$; reject H_0; yes   **12.55a.** $\chi^2 = 9.50$; reject H_0; yes
12.57 $\chi^2 = 9.47$; do not reject H_0 ($\alpha = .05$); no

Chapter 13

13.3 $y = \frac{13}{4} + \frac{1}{4}x$ **13.9** No

13.11a.

x	y	$\hat{y} = 7.10 - .78x$	$(y - \hat{y})$	$(y - \hat{y})^2$
7	2	1.64	.36	.1296
4	4	3.98	.02	.0004
6	2	2.42	−.42	.1764
2	5	5.54	−.54	.2916
1	7	6.32	.68	.4624
1	6	6.32	−.32	.1024
3	5	4.76	.24	.0576
			$\Sigma (y - \hat{y}) = 0$	$\Sigma (y - \hat{y})^2 = 1.2204$

13.13c. $.020$; $.918$ **13.15b.** $\hat{y} = 575.03 - .00219x$ **13.17a.** $\hat{y} = 57.912 - .811x$ **c.** 9.252   **13.19a.** 24.45; 2.38   **c.** 33.38
13.21a. $\hat{y} = 156.357 + 12.786x$ **13.23a.** $.0613$ **b.** $2s \approx .495$

13.25

	SSE	s^2	s
13.10	1.2204	.2441	.4940
13.13	5.1349	1.0270	1.0134

13.27a. Yes; $\hat{y} = 74.7129 - 1.2883x$ **c.** 6,341.0883; 352.2827; 18.769; ± 37.538 **13.29b.** A: $\hat{y} = 6.62 - .0727x$; B: $\hat{y} = 9.31 - .1077x$
c. A: 19.056; 1.4658; 1.211; B: 4.833; .3717; .610 **d.** A: 1.531 $\pm$ 2.422; B: 1.771 $\pm$ 1.220 **e.** Brand B
13.31a. (29.83, 32.17); (30.06, 31.94) **b.** (58.92, 69.08); (59.84, 68.16) **c.** (−9.15, −7.65); (−9.02, −7.78)
13.33 (.4889, 1.1539); (.0633, 1.5795) **13.35a.** p-value < .001 **13.37a.** $\hat{y} = 70.8 - .28x$ **b.** $t = -.48$; do not reject H_0; no
13.39a. $t = 13.69$; reject H_0; yes **b.** p-value < .001 **13.41b.** $\hat{y} = 44.1304 + .2366x$ **c.** 6,400.5941; 376.5055; 19.404
e. $t = 1.269$; do not reject H_0 **f.** (−.1569, .6301) **13.43** (2.7746, 3.5478) **13.47a.** .9853; .9709 **b.** −.9934; .9868 **c.** 0, 0 **d.** 0, 0
13.49 .9175; .8418 **13.51** .9426; .9709 **13.53** .0109; .00012 **13.55c.** .9653; .9960 **d.** Yes **13.57a.** $\hat{y} = -235.1 + 1.273x_2$
b. $t = 18.29$; reject H_0; yes **c.** .9767 **d.** Second **e.** x_2 **13.59c.** (4.2449, 7.1575) **d.** (−1.2776, .2776); (−7.0182, −4.4260)
13.61a. $\hat{y} = 22.3248 + 2.6497x$ **c.** $t = 6.280$; reject H_0; yes **d.** .9421; .8875 **e.** (107.6094; 111.9204) **f.** (103.6768, 115.8530)
13.63a. (10.990, 12.574) **b.** (9.7174, 13.8466) **13.65a.** $\hat{y} = 4.5538 + 2.6708x$ **c.** .9669; .9349 **d.** (11.911, 14.289)
e. (12.1402, 12.9922) **13.67** (29.205, 42.119) **13.69a.** 575.029; −.0022 **b.** 293,208.461; 22,554.497; 150.182 **c.** .0664
d. $t = -.962$; p-value = .3538; do not reject H_0 **e.** (414.7, 582.3) **13.71a.** $\hat{y} = 27.78 - .6x$ **c.** 7.0 **d.** .5385 **e.** (−.7838, −.4162)
f. (26.3521, 27.0479) **g.** (25.3546, 28.0454) **13.73b.** −.1245; .0155 **c.** $t = -.35$; do not reject H_0; no **13.75a.** $\hat{y} = 2.0841 + .6682x$
c. $t = 16.28$; reject H_0; yes **d.** .9815 **e.** (34.5155, 36.4727) **f.** (36.0689, 41.6013) **13.77a.** −13.490347; −.052829
b. (−.0706, −.0350); yes **c.** .8538 **d.** (.5987, 1.2653) **13.79a.** (22.33, 91.96) **c.** $x = 44$ **13.81a.** $\hat{y} = 46.3992x$
b. $\hat{y} = 478.4433 + 45.1525x$ **d.** $t = .906$; do not reject H_0; no **13.83a.** $\hat{y} = 13.4815 + .0560x$ **c.** .7402; .5479 **d.** (14.9276, 19.8754)
13.85a. .8429 **b.** $t = 2.71$; reject H_0; yes **c.** No **13.87a.** 7.9416; 3.9708 **b.** (15.1429, 20.0551) **d.** (52.4704, 64.1854)

Chapter 14

14.1a. 506.346; −941.900; −429.060 **b.** $\hat{y} = 506.346 - 941.900x_1 - 429.060x_2$ **c.** 151,015.724; 8,883.278; 94.251
d. $t = -3.424$; reject H_0 **e.** (−1,230.492, 372.372) **14.3a.** $t = 3.13$; reject H_0 **b.** $t = 3.13$; reject H_0; yes **c.** $F = 9.816$ **d.** No
14.5a. SSE = .0438; $s^2 = .01095$ **b.** $t = -13.97$; reject H_0; yes **c.** $\hat{y} = 1.095 + 1.636x - .1595x^2$
14.7a. $\hat{y} = 20.0911 - .6705x + .0095x^2$ **c.** $t = 1.51$; do not reject H_0 **d.** $\hat{y} = 19.279 - .445x$ **e.** (−.5177, −.3723)
14.9b. β_1, β_3, β_5, and β_6 are significantly different from 0. **c.** Intercept is .105 **d.** Intercept is 0
14.11a. $\hat{y} = 93{,}074 + 4{,}152x_1 - 855x_2 + .924x_3 + 2{,}692x_4 + 15.5x_5$ **b.** 33,225.9 **c.** $t = 2.78$; p-value = .0059; reject H_0; yes
f. $t = -2.86$; reject H_0 **g.** p-value = .00495 **14.13a.** Yes **b.** $F = 55.2$; reject H_0; yes **14.15** No; no **14.17a.** $F = 31.98$; reject H_0; yes
b. $t = -2.27$; reject H_0; yes **14.19a.** $\hat{y} = 295.33 - 480.84x_1 - 829.46x_2 + .0079x_3 + 2.3608x_4$ **b.** $s = 46.76747$
c. $F = 28.18$; p-value $\le$ .0001; reject H_0 **d.** $t = -3.197$; reject H_0; yes **14.21a.** $F = 15.40$; reject H_0; yes
b. $t = -.75$; do not reject H_0; no **14.23a.** $\hat{y} = 75.079 - .290x_1 - 19.148x_2 + .268x_1x_2$ **b.** .9909 **c.** $F = 145.55$; reject H_0; yes
d. $t = 3.92$; reject H_0; yes **14.25a.** $R^2 = .9805$ **b.** $F = 190.75$; reject H_0 **c.** p-value $\le$ .0001 **14.27b.** $\hat{y} = 95.521 - 32.99x + .0201x^2$
c. $s = \$13{,}976.41$ **d.** $F = 63.64$; reject H_0 **e.** $t = 2.38$; reject H_0; yes **14.29a.** $\hat{y} = 506.35 - 941.9x_1 - 429.1x_2$ **b.** $s = 94.25$
c. $F = 7.223$; reject H_0; yes **d.** (−1,089.952, 231.752) **e.** $R^2 = .459$ **14.31** Yes; yes
14.33a. $\hat{y}_1 = 1.433 + .00999x_1 + 3793x_2$; $\hat{y}_2 = 1.155 + .0187x_1 + .4058x_2$

	Model 1	Model 2
b.	.2769	.5581
c.	$F = 100.80$; reject H_0	$F = 33.88$; reject H_0
d.	(.3218, .4368)	(.2909, .5207)

14.37a. .9488 **14.39a.** $F = 24.41$; reject H_0; yes **b.** $t = -2.01$; reject H_0 **c.** $t = .31$; do not reject H_0 **d.** $t = 2.38$; reject H_0
14.43 $x_1 = 30$, $x_2 = .6$, and $x_3 = 1{,}300$ are in the experimental region
14.45a. $\hat{y} = .601 + .595x_1 - 3.725x_2 - 16.232x_3 + .235x_1x_2 + .308x_1x_3$ **b.** $R^2 = .928$; $F = 139.42$; reject H_0; model is useful
14.47a. $\hat{y} = 5{,}716.911 - 1.546x_1 + .033x_2 - 32.557x_3$ **14.49a.** 13.6812 **14.51** $t = -1.104$; do not reject H_0
14.53b. $F = 16.10$; reject H_0; yes **c.** $t = 2.5$; reject H_0; yes **d.** 945
14.55 The estimate of the difference in mean attendance between weekends and weekdays is $15x_3 - 700$.
14.57a. H_0: $\beta_2 = \beta_3 = 0$ vs H_a: At least one $\beta_i \ne 0$ **c.** $t = -3.333$; reject H_0; yes **14.59** $t = -2.06$; do not reject H_0; no
14.61a. $F = 6.61$; reject H_0; yes **b.** To within 1.0 **c.** No **14.65** Second-order

Chapter 15

15.1a. Quantitative **b.** Quantitative **c.** Qualitative **d.** Quantitative **e.** Qualitative **15.3** Normality **15.5a.** Quantitative **b.** Qualitative
c. Quantitative **d.** Quantitative **e.** Qualitative **15.7a.** Second-order; $\beta_0 = 4$; $\beta_2 < 0$ **b.** Second-order; $\beta_0 = 8$; $\beta_2 > 0$
15.9a. (i) First (ii) Third (iii) First (iv) Second
b. (i) $E(y) = \beta_0 + \beta_1 x$ **c.** (i) $\beta_0 > 0$; $\beta_1 > 0$
   (ii) $E(y) = \beta_0 + \beta_1 x + \beta_2 x^2 + \beta_3 x^3$ (ii) $\beta_0 > 0$; $\beta_3 > 0$
   (iii) $E(y) = \beta_0 + \beta_1 x$ (iii) $\beta_0 > 0$; $\beta_1 < 0$
   (iv) $E(y) = \beta_0 + \beta_1 x + \beta_2 x^2$ (iv) $\beta_0 < 0$; $\beta_1 > 0$; $\beta_2 < 0$
15.11 $y = \beta_0 + \beta_1 x + \beta_2 x^2 + \varepsilon$ **15.13a.** First **c.** Line would slope upward from left to right. **15.15a.** $E(y) = \beta_0 + \beta_1 x + \beta_2 x^2$; $\beta_2 > 0$
15.17 $E(y) = \beta_0 + \beta_1 x + \beta_2 x^2$; $\beta_2 > 0$ **15.19** $t = -6.60$; reject H_0; yes **15.21a.** $E(y) = \beta_0 + \beta_1 x_1 + \beta_2 x_2$
b. $E(y) = \beta_0 + \beta_1 x_1 + \beta_2 x_2 + \beta_3 x_1 x_2$ **c.** $E(y) = \beta_0 + \beta_1 x_1 + \beta_2 x_2 + \beta_3 x_1 x_2 + \beta_4 x_1^2 + \beta_5 x_2^2$ **15.23e.** Increases from 5 to 13

15.25a. $\hat{y} = -2.55 + 3.82x_1 + 2.63x_2 - 1.29x_1x_2$ **e.** $H_0: \beta_3 = 0$ vs $H_a: \beta_3 \neq 0$ **f.** $t = -8.06$; reject H_0 **15.27a.** Both quantitative
b. $E(y) = \beta_0 + \beta_1x_1 + \beta_2x_2$ **c.** $E(y) = \beta_0 + \beta_1x_1 + \beta_2x_2 + \beta_3x_1x_2 + \beta_4x_1^2 + \beta_5x_2^2$ **d.** $H_0: \beta_3 = \beta_4 = \beta_5 = 0$ vs H_a: At least one $\beta_i \neq 0$
15.29a. Quantitative **b.** $E(y) = \beta_0 + \beta_1x_1 + \beta_2x_2$ **c.** Include $\beta_3x_1x_2$ **d.** Include $\beta_4x_1^2 + \beta_5x_2^2$
15.31a. $\hat{y} = 148.5 + .472x_1 - .099x_2 - .00054x_1^2 + .000015x_2^2$ **c.** $F = 31.66$; reject H_0; yes **d.** Yes
15.33a. H_a: At least one $\beta_i \neq 0$; $i = 3, 4, 5$ **c.** 3; 24 **15.37a.** $F = 66.87$; reject H_0 **b.** $H_a: \beta_1 \neq 0$ **c.** $F = 15.16$; reject H_0
15.39a. $F = 1.32$; do not reject H_0 **b.** $F = .484$; do not reject H_0 **c.** No improvement in model **15.41a.** Both quantitative
b. $E(y) = \beta_0 + \beta_1x_1 + \beta_2x_2$ **c.** $E(y) = \beta_0 + \beta_1x_1 + \beta_2x_2 + \beta_3x_1x_2 + \beta_4x_1^2 + \beta_5x_2^2$ **d.** $H_0: \beta_3 = 0$ vs $H_a: \beta_3 \neq 0$
e. $F = 14.95$; reject H_0 **15.43** $F = 24.19$; reject H_0; yes **15.45a.** $\hat{y} = 93,074 + 4,152x_1 - 855x_2 + .92x_3 + 2,692x_4 + 15.5x_5$
b. $F = 190.749$; reject H_0; yes **c.** $\hat{y} = 114,369 + 5,036x_1 - 1,057x_2 + 15x_5$ **d.** $F = 303.71$; reject H_0; yes
15.47 $E(y) = \beta_0 + \beta_1x_1 + \beta_2x_2$ where $x_1 = 1$ if variable at level 2, $x_1 = 0$ if not; $x_2 = 1$ if variable at level 3, $x_2 = 0$ if not
15.49a. $\hat{y} = 80 + 16.8x_1 + 40.4x_2$ **d.** $F = 24.72$; reject H_0
15.51 $E(y) = \beta_0 + \beta_1x_1 + \beta_2x_2 + \beta_3x_3$ where $x_1 = 1$ if Zenith, $x_1 = 0$ if not; $x_2 = 1$ if Magnavox, $x_2 = 0$ if not; $x_3 = 1$ if RCA,
$x_3 = 0$ if not **15.53a.** $H_0: \mu_1 = \mu_2 = \mu_3 = \mu_4 = \mu_5$ vs H_a: At least two means are unequal **b.** $F = 23.97$; p-value $\leq .0001$; reject H_0
15.55a. 60 **b.** 510
15.57a. $E(y) = \beta_0 + \beta_1x_1 + \beta_2x_2 + \beta_3x_3$ where $x_1 = 1$ if location 2, $x_1 = 0$ if not; $x_2 = 1$ if location 3, $x_2 = 0$ if not; $x_3 = 1$ if location 4,
$x_3 = 0$ if not **b.** β_0 **c.** $\beta_0 + \beta_3$ **d.** $\beta_1 - \beta_2$
15.59a. $E(y) = \beta_0 + \beta_1x_1$ **b.** $E(y) = \beta_0 + \beta_1x_1 + \beta_2x_2 + \beta_3x_3$ where $x_2 = 1$ if level 2, $x_2 = 0$ if not; $x_3 = 1$ if level 3, $x_3 = 0$ if not
c. $E(y) = \beta_0 + \beta_1x_1 + \beta_2x_2 + \beta_3x_3 + \beta_4x_1x_2 + \beta_5x_1x_3$ **d.** $\beta_4 = \beta_5 = 0$ **e.** $\beta_2 = \beta_3 = \beta_4 = \beta_5 = 0$
15.61a. $E(y) = \beta_0 + \beta_1x_1$ **b.** Include $\beta_2x_2 + \beta_3x_3$ where $x_2 = 1$ if brand 2, $x_2 = 0$ if not; $x_3 = 1$ if brand 3, $x_3 = 0$ if not
c. Include $\beta_4x_1x_2 + \beta_5x_1x_3$ **15.63** $F = 2.60$; do not reject H_0
15.65a. $E(y) = \beta_0 + \beta_1x_1 + \beta_2x_2 + \beta_3x_3$ where $x_2 = 1$ if excellent condition, $x_2 = 0$ if not; $x_3 = 1$ if good condition, $x_3 = 0$ if not
c. $\hat{y} = 36,388 + 15,617x_1 + 152,487x_2 + 49,441x_3$; E: $\hat{y} = 188,875 + 15,617x_1$; G: $\hat{y} = 85,829 + 15,617x_1$; F: $\hat{y} = 36,388 + 15,617x_1$
e. $F = 8.43$; reject H_0; yes **15.67a.** $F = 4.90$; reject H_0 **b.** $H_0: \beta_4 = \beta_5 = 0$ vs H_a: At least one $\beta_i \neq 0$ **c.** $F = 2.36$; do not reject H_0; no
15.69a. $E(y) = \beta_0 + \beta_1x_1 + \beta_2x_2 + \beta_3x_3$ where $x_1 = $ ration; $x_2 = 1$ if 12°C, $x_2 = 0$ if not; $x_3 = 1$ if 18°C, $x_3 = 0$ if not
c. $E(y) = \beta_0 + \beta_1x_1 + \beta_2x_2 + \beta_3x_3 + \beta_4x_1x_2 + \beta_5x_1x_3$ **e.** $F = .32$; do not reject H_0; no **15.71a.** $\beta_5 = \beta_6 = \beta_7 = \beta_8 = 0$
b. $\beta_2 = \beta_5 = \beta_6 = \beta_7 = \beta_8 = 0$ **c.** $\beta_3 = \beta_4 = \beta_5 = \beta_6 = \beta_7 = \beta_8 = 0$
15.73a. $\hat{y} = 48.8 - 3.36x_1 + .075x_1^2 - 2.36x_2 - 7.60x_3 + 3.71x_1x_2 + 2.66x_1x_3 - .018x_1^2x_2 - .037x_1^2x_3$
b. $\hat{y} = 48.8 - 3.36x_1 + .075x_1^2$; $\hat{y} = 46.44 + .35x_1 + .057x_1^2$; $\hat{y} = 41.2 - .70x_1 + .038x_1^2$
d. $H_0: \beta_3 = \beta_4 = \beta_5 = \beta_6 = \beta_7 = \beta_8 = 0$ vs H_a: At least one $\beta_i \neq 0$ **e.** $F = 242.65$; reject H_0 **15.75** $F = .93$; do not reject H_0; no
15.77 $F = 6.71$; reject H_0; yes **15.79a.** $E(y) = \beta_0 + \beta_1x_1 + \beta_2x_1^2 + \beta_3x_2 + \beta_4x_1x_2 + \beta_5x_1^2x_2$
15.81a. x_2 ($t = -90$ has the largest absolute value) **b.** Yes **15.83a.** x_4, x_5, x_6
b. No. There may be other important variables as yet unspecified. **c.** $E(y) = \beta_0 + \beta_1x_4 + \beta_2x_5 + \beta_3x_6 + \beta_4x_4x_5 + \beta_5x_4x_6 + \beta_6x_5x_6$
15.87a. Second **b.** First **c.** Second **d.** First **e.** Third **f.** First **15.93** No estimate of σ^2 **15.95a.** Two
b. $E(y) = \beta_0 + \beta_1x_1 + \beta_2x_2$ where $x_1 = 1$ if full service only, $x_1 = 0$ if not; $x_2 = 1$ if both types of service, $x_2 = 0$ if not
15.97a. $E(y) = \beta_0 + \beta_1x_1 + \beta_2x_2 + \beta_3x_3$ where $x_3 = 1$ if not taking pill, $x_3 = 0$ otherwise
b. $\hat{y} = 5,684.986 - 1.5465x_1 + .0328x_2 + 32.8129x_3$ **c.** $F = 22.196$; reject H_0 **d.** p-value $\leq .0001$ **e.** $R_2 = .6894$
15.99a. $E(y) = \beta_0 + \beta_1x_1 + \beta_2x_2$ where $x_1 = $ total area; and $x_2 = 1$ if central air conditioning is installed, $x_2 = 0$ if not
b. $E(y) = \beta_0 + \beta_1x_1 + \beta_2x_1^2 + \beta_3x_2 + \beta_4x_1x_2 + \beta_5x_1^2x_2$ **c.** $H_0: \beta_2 = \beta_4 = \beta_5 = 0$ vs H_a: At least one $\beta_i \neq 0$
15.101a. $H_0: \beta_2 = \beta_5 = 0$ vs H_a: At least one $\beta_i \neq 0$ **b.** $H_0: \beta_3 = \beta_4 = \beta_5 = 0$ vs H_a: At least one $\beta_i \neq 0$

Index